Chapter 6: Topics in Algebra

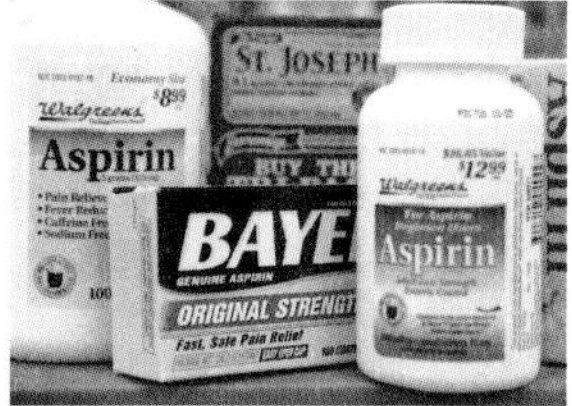

MATH IN Drug Administration

Have you ever looked at the dosage information on a bottle of aspirin and thought "It just doesn't seem reasonable to recommend the same dosage for all adults?" People come in all shapes and sizes, and the effect of a certain dosage is in large part dependent on the size of the individual. If a 105 pound woman and her 230 pound husband both take two aspirin the morning after their wedding reception, she is in effect getting more than twice as much medicine as he is. **Continue reading about Math in Drug Administration on page 277.**

Chapter 7: Additional Topics in Algebra

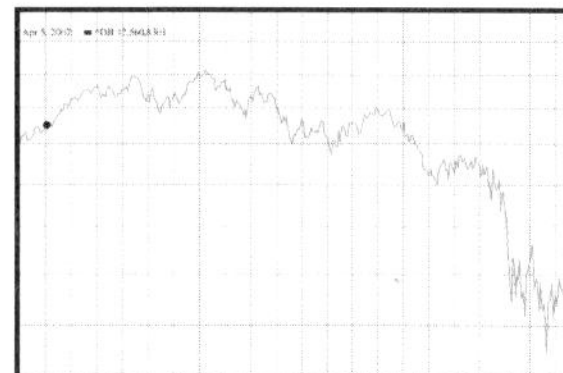

MATH IN The Stock Market

There was a time when the fluctuations of the stock market were exclusively of concern to the rich. Most people weren't directly invested in the market, and felt like it didn't impact their lives very much. But those days are long gone. Most people's retirement savings, at the very least, are invested in stocks and mutual funds. This means that value changes in the market are of more interest to more Americans than ever before, and the ability to monitor those changes is certainly a relevant skill. **Continue reading about Math in the Stock Market on page 347.**

Chapter 8: Consumer Mathematics

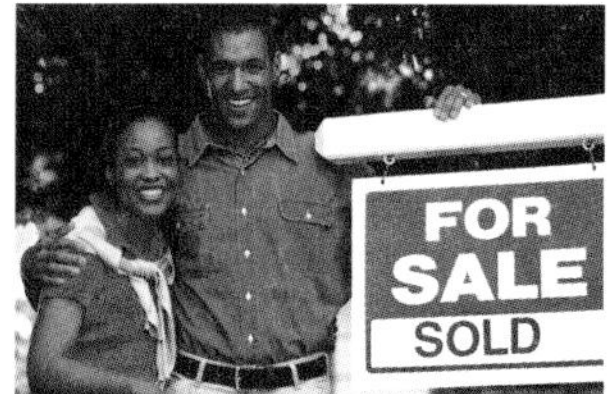

MATH IN Home Buying

Home buying shows have become very popular in the last couple of years. It's interesting to watch people looking for just the right home, and maybe more interesting to find out what homes are going for in other parts of the country. The down side as a viewer is that it can be pretty intimidating to find out just how much houses cost. Almost everyone wonders at some point "Will I ever be able to afford a home of my own?" **Continue reading about Math in Home Buying on page 419.**

Chapter 9: Measurement

MATH IN Travel

To measure something means to assign a number that represents its size. In fact, measurement might be the most common use of numbers in everyday life. Numbers are used to measure height, weight, distance, grades, weather, size of a home, capacity of bottles and cans, and much more. Even in our monetary system, we use numbers to measure sizes: little ones, like the cost of a candy bar, and big ones, like an annual salary. One obvious application of this skill comes into play when traveling outside the United States. Most countries use the metric system almost exclusively, meaning that to understand various bits of information, you need to be able to interpret measurements in an unfamiliar system. **Continue reading about Math in Travel on page 481.**

Chapter 10: Geometry

MATH IN Home Improvement

It probably won't surprise you to learn that one of the most common questions that math students ask is "How can I actually use this stuff?" Of course, how math is used in our world is the main theme of this book, and this chapter fits that framework especially well. The ideas presented are commonly used in everyday things like working around the home, so we'll present some actual projects that geometry was used for in the home of one of the authors. **Continue reading about Math in Home Improvement on page 511.**

Chapter 11: Probability and Counting Techniques

MATH IN Gambling

The fact that you're reading this sentence means that you're probably taking a math class right now. But maybe not...you could be an instructor evaluating the book, or maybe an editor looking for mistakes (unsuccessfully, we hope). Still, I would be willing to bet that you're taking a math class. The word "probably" indicates a certain likelihood of something happening, and that basic idea is the topic of this chapter. We call the study of the likelihood of events occurring probability. **Continue reading about Math in Gambling on page 573.**

Chapter 12: Statistics

MATH IN Sociology

Broadly defined, sociology is the study of human behavior within society. One important branch is criminology. This is not about investigating crimes, but rather studying patterns of criminal behavior and their effect on society. The main tool used by sociologists is statistics. This important area of math allows researchers to study patterns of behavior objectively by analyzing information gathered from a mathematical perspective, not a subjective one. **Continue reading about Math in Sociology on page 647.**

Chapter 13: Other Mathematical Systems

MATH IN Encryption

Cryptography is the study of hiding information using some sort of code. There was a time when codes were of interest mostly to military officers, spies, and grade school kids passing notes they didn't want the teacher to read. But that time passed very quickly with the advent of the computer age. According to some reports, more than half of all Americans made at least one purchase online in 2008. When sensitive financial information and passwords are being transmitted, encryption becomes of supreme importance. Without proper encryption, you could find your identity lost in the endless depths of cyberspace. **Continue reading about Math in Encryption on page 717.**

Chapter 14: Voting Methods

MATH IN College Football

Football and other sports have been used as examples dozens of times throughout this book because sports present an excellent example of a variety of ways that math gets used in our world. There are the obvious ways – keeping score, and adding up statistics like yards gained or home runs hit. But there are also many behind-the-scenes examples of the importance of math in sports. Allocation of salaries, ticket prices, devising a schedule that meets the needs of every team in a league, assigning officials to work games...these are just four of many such examples. **Continue reading about Math in College Football on page 743.**

Chapter 15: Graph Theory

MATH IN Road Trips

The road trip is a great American tradition. There's just something magical about the freedom of heading out to the open road and driving wherever you feel like. It usually takes about an hour for the magic to wear off, though. Then you just want to get to where you're headed as quickly as possible. The branch of mathematics known as graph theory was created to solve problems involving the most efficient way to travel between different locations. **Continue reading about Math in Road Trips on page 791.**

MATH in Our World

SECOND EDITION

Dave Sobecki
Associate Professor
Miami University Hamilton

Allan G. Bluman
Professor Emeritus
Community College of Allegheny County

Angela Schirck-Matthews
Professor
Broward College

MATH IN OUR WORLD, SECOND EDITION

Published by McGraw-Hill, a business unit of The McGraw-Hill Companies, Inc., 1221 Avenue of the Americas, New York, NY 10020.

This book is printed on acid-free paper.

1 2 3 4 5 6 7 8 9 0 DOW/DOW 1 0 9 8 7 6 5 4 3 2 1 0

ISBN 978–0–07–298253–4
MHID 0–07–298253–5

ISBN 978–0–07–310459–1 (Instructor's Edition)
MHID 0–07–310459–0

Vice President, Editor-in-Chief: *Marty Lange*
Vice President, EDP: *Kimberly Meriwether David*
Director of Development: *Kristine Tibbetts*
Editorial Director: *Stewart K. Mattson*
Sponsoring Editor: *John R. Osgood*
Developmental Editor: *Christina A. Lane*
Marketing Manager: *Kevin M. Ernzen*
Senior Project Manager: *April R. Southwood*
Lead Production Supervisor: *Sandy Ludovissy*
Senior Media Project Manager: *Sandra M. Schnee*
Designer: *Tara McDermott*
Cover/Interior Designer: *Greg Nettles/Squarecrow Design*
(USE) Cover Image: *Vasko Miokovic/istock photo (pizza); istock photo (menu); © Greg Nettles (tablecloth)*
Senior Photo Research Coordinator: *Lori Hancock*
Photo Research: *Danny Meldung/Photo Affairs, Inc*
Compositor: *MPS Limited*
Typeface: *10/12 Times*
Printer: *R. R. Donnelley*

All credits appearing on page or at the end of the book are considered to be an extension of the copyright page.

Library of Congress Cataloging-in-Publication Data

Sobecki, Dave.
Math in our world / Dave Sobecki, Allan G. Bluman, Angie Matthews. — 2nd ed.
p. cm.
Rev. ed. of: Mathematics in our world / Allan G. Bluman. c2005.
Includes index.
ISBN 978–0–07–298253–4 — ISBN 0–07–298253–5 (hard copy : alk. paper) 1. Mathematics. I. Bluman, Allan G. II. Matthews, Angie. III. Bluman, Allan G. Mathematics in our world. IV. Title.

QA39.3.B597 2011
510—dc22 2009033824

www.mhhe.com

Brief Contents

CHAPTER 1
Problem Solving 2

CHAPTER 2
Sets 40

CHAPTER 3
Logic 90

CHAPTER 4
Numeration Systems 146

CHAPTER 5
The Real Number System 196

CHAPTER 6
Topics in Algebra 276

CHAPTER 7
Additional Topics in Algebra 346

CHAPTER 8
Consumer Mathematics 418

CHAPTER 9
Measurement 480

CHAPTER 10
Geometry 510

CHAPTER 11
Probability and Counting Techniques 572

CHAPTER 12
Statistics 646

CHAPTER 13
Other Mathematical Systems 716

CHAPTER 14
Voting Methods 742

CHAPTER 15
Graph Theory 790

About the Authors

Dave Sobecki

Dave Sobecki is an associate professor in the Department of Mathematics at Miami University in Hamilton, Ohio. He earned a B.A. in math education from Bowling Green State University before continuing on to earn an M.A. and a Ph.D. in mathematics from Bowling Green State University. He has written or coauthored five journal articles, eleven books, and five interactive CD-ROMs. Dave lives in Fairfield, Ohio, with his wife (Cat) and dogs (Macleod and Tessa).

His passions include Ohio State football, Cleveland Indians baseball, heavy metal music, travel, and home improvement projects.

Dedication: *To two of my biggest supporters: my wife Cat, and my friend Dawn.*

Allan G. Bluman

Allan G. Bluman is a professor emeritus at the Community College of Allegheny County, South Campus, near Pittsburgh, Pennsylvania. He has taught mathematics and statistics for over 35 years. He received an Apple for the Teacher award in recognition of his bringing excellence to the learning environment at South Campus. He has also taught statistics for Penn State University at the Greater Allegheny (McKeesport) Campus and at the Monroeville Center. He received his master's and doctor's degrees from the University of Pittsburgh.

He is the author of two other textbooks, *Elementary Statistics: A Step By Step Approach* and *Elementary Statistics: A Brief Version.* In addition, he is the author of four mathematics books in the McGraw-Hill DeMystified Series. They are *Pre-Algebra, Math Word Problems, Business Math,* and *Probability.*

He is married and has two sons and a granddaughter.

Dedication: *To Betty Bluman, Earl McPeek, and Dr. G. Bradley Seager, Jr.*

Angela Schirck-Matthews

Angela Schirck-Matthews is a Professor of Mathematics at Broward College in Davie, Florida, where she has been teaching mathematics since 1991. Before her employment at Broward College she taught undergraduate mathematics at the University of Miami in Coral Gables, Florida. She is also an alumna of both institutions; Angela earned an Associate of Arts degree in liberal arts at Broward College and went on to earn a Bachelor of Arts degree in mathematics at Florida Atlantic University in Boca Raton, Florida. After graduating from Florida Atlantic University, she continued her education at the University of Miami where she earned a Master of Science degree in mathematics and completed coursework toward a Ph.D. in mathematics.

Angela lives in Hollywood, Florida with her husband, three of her four children, and two dogs.

Dedication: *To my parents Joe and Karen, for their love and encouragement throughout my life.*

Letter from the Authors

Liberal arts math is a challenging course to teach because most of the time, the audience is very different from other college math classes. For most students the course is terminal, and in many cases represents the only math credits the student will take in college. Given this audience, it can often be a great challenge to motivate students and make them see the relevance of math, how math plays a role in our world, and how the thinking and problem-solving strategies that students will learn and practice can be used beyond the classroom.

The first step to gaining interest is to find a way to **engage** students. This is typically done using applications, but too often we rely on the old classics, the applications that we remember from our own education. This is where the two new coauthors come in (Dave and Angie): each is known for making math engaging to students, and together, they bring a fresh, modern approach to the second edition. To fully engage students in the 21st century, we need to think modern, emphasizing applications that show the relevance of math to today's students. We've made every attempt to make that the focus of this edition, from financing a car to figuring out a healthy daily calorie intake to understanding how overwhelming our national debt really is, and what that means to us as citizens. The pizza theme that is carried throughout the book is a constant reminder to students that math isn't a spectator sport, but something that plays an important role in their everyday lives.

No one has ever become stronger by watching someone else lift weights, and our students aren't going to be any better at thinking and problem-solving unless we encourage them to **practice** it on their own. *Math in Our World* includes a plethora of applications and real-world problems for students to hone their skills and give their brains a workout. The critical thinking skills they practice in using this text will serve them well in any future courses whether there is math involved or not. But more important, it will help them to become problem solvers and thinkers beyond the halls of academia.

Once we've drawn your students' attention and offered them plenty of practice, we turn to our ultimate goal in writing this text: helping your students to **succeed.** We're confident that this book offers a fantastic vehicle to drive your classes to higher pass rates because of the pedagogical elements, writing style, and interesting problem sets. Additionally, while no book can prevent the underpreparedness of students in a course like this, we've got you covered there as well with ALEKS Prep. You can read more about it on the back cover and on preface pages x and xi.

Finally, to further emphasize clarity and consistency, we as an author team took it upon ourselves to produce the lecture videos, exercise videos, and solution manuals to accompany *Math in Our World,* allowing our voices to carry throughout these supplements, perfectly complementing the text. We hope that you and your students enjoy using *Math in Our World* as much as we enjoyed writing it and putting all the pieces together. Good luck, and please don't hesitate to contact your local McGraw-Hill representative to let us know what you think!

—Dave, Al, and Angie

Engage

Joint experiences in the classroom have helped each of the authors bring a unique perspective to the text, influencing the **writing style** significantly. Both practical and conversational, the writing functions as a tool to guide learning for students outside of the classroom. The goal is to draw in even the most hesitant students by relating to them through the casual, spoken language they regularly use with friends. By doing so, students gain both a firm initial understanding of the basic concepts fundamental to the curriculum and also a greater degree of retention of the long-term implications of those concepts.

"The writing is in simple, spoken language and seems as if a friend is just talking about the topics to them in a way that would be easy for them to relate to."

—John Ward, Jefferson Community and Technical College

"They paint a broad picture of what the topic is about and provide a possible application. This gives the student a reason for studying the topic and provides the instructor a starting point for the class."

—John Weglarz, Kirkwood Community College

"I like the fact the questions posed in the opener are answered at the end of the chapter with the revisiting of the chapter opener, thus completing the circle!"

—Mark Brenneman, Mesa Community College

CHAPTER 1

Problem Solving

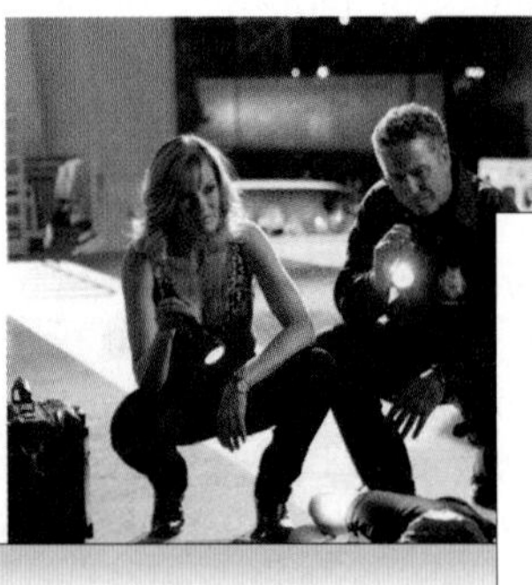

MATH IN Criminal Investigation

In traditional cops-and-robbers movies, crime fighters use guns and fists to catch criminals, but in real life, often it's brain power that brings the bad guys to justice. That's why the TV show *CSI* marked a revolution of sorts when it debuted in October 2000: it featured scientists fighting crime, not tough guys. Solving a case is intimately tied to the process of problem solving: investigators are presented with a mystery, and they use whatever evidence they can find and their reasoning ability to reconstruct the crime. Hopefully, this will lead them to a suspect.

Of course, this isn't limited to just crime fighting. Students in math classes often ask, "When am I going to use what I learn?" The best answer to that question is, "Every day!" Math classes are not only about facts and formulas: they're also about exercising your mind, training your brain to think logically, and learning effective strategies for solv- ... to be useful tools that you can apply in the rest of your education. But more importantly, they can be applied just as well to situations outside the classroom.

And this brings us back to our friends from *CSI*. The logic and reasoning that they use to identify suspects and prove their guilt are largely based on problem-solving skills we'll study in this chapter. By the time you've finished the chapter, you should be able to evaluate the situations below, all based on episodes of *CSI*. In each case, you should identify the type of reasoning that was used and decide whether the conclusion would stand up as proof in a court of law.

- After a violent crime, the investigators identify a recently paroled suspect living in the area who had previously committed three very similar crimes.
- A homeless man is found dead from exposure after ...

MATH IN Criminal Investigation REVISITED

1. The suspect was identified by specific incidents in the past, which makes this inductive reasoning that would not hold up in court without further evidence.
2. Fingerprints positively identify the officer as having had contact with the victim. This is deductive reasoning and would be useful in court.
3. Like fingerprints, DNA can positively show that the suspect had physical contact with the victim. This evidence, based on deductive reasoning, would hurt the suspect badly in court.
4. While this is compelling evidence, it's based on assuming that those five drawings indicate the artist is the killer. While unlikely, it could be a coincidence based on five drawings, so this is inductive reasoning. It might impress a jury to some extent, but wouldn't be sufficient for a conviction.

Review Exercises

Section 1-1

For Exercises 1–4, make reasonable conjectures for the next three numbers or letters in the sequence.

1. 3 4 6 7 9 10 12 13 15 16 ___ ___ ___
2. 2 7 4 9 6 11 8 13 ___ ___ ___

For Exercises 5 and 6, make a reasonable conjecture and draw the next figure.

5. ___

Chapter Openers directly engage student interest by immediately tying mathematical concepts to each of their everyday lives. These vignettes introduce concepts by referencing popular topics familiar to a wide variety of students—travel, demographics, the economy, television, and even college football. By taking slower steps to reach the critical topics of a chapter, students are able to more fully grasp how math really does relate to their own world, further solidifying the connections imperative to retention of information.

Used to clarify concepts and emphasize particularly important points, **Math Notes** provide suggestions for students to keep in mind as they progress through the chapter.

> "Gives helpful hints that need to be reinforced."
> —*Greg Wisloski, Indiana University of Pennsylvania*
>
> "Very useful...looks like they often anticipate students' questions."
> —*Robert Koca, Community College of Baltimore County, Essex*
>
> "I like how they get important details to the student without getting too wordy. I think students would pay attention to them because they are so brief."
> —*Mark Ellis, Central Piedmont Community College*

Math Note

We often use the word *nearest* to describe the place value to round to. Instead of saying, "Round to the hundreds place," we might say, "Round to the nearest hundred."

Sidelights contain carefully chosen material highlighting relevant interdisciplinary connections within math to encourage both curiosity and motivation in students who have a wide variety of interests. These include biographic vignettes about famous mathematicians as well as other interesting facts that emphasize the importance of math in areas like weather, photography, music, and health.

Sidelight **AN ARBITRARY DISCUSSION**

In common usage, the word *arbitrary* is often misinterpreted as a synonym for *random*. When reaching into a bag of potato chips, you make a random selection, and some people would also call this an arbitrary selection. But in math, the word *arbitrary* means something very different. When randomly selecting that chip, you have still chosen a specific chip—it is probably not representative of every chip in the bag. Some chips will be bigger, and others smaller. Still others may have more salt, some less.

When we use *arbitrary* in math, we're referring to a nonspecific item that is able to represent *all* such items. In the series of calculations we looked at above, we could never be sure that the result will always be 3 b numbers. Why? Because we'd have to try which is of course impossible. You have than spend the rest of your life testing nu The value of performing the calculation c ber *x* is that this one calculation proves v *every* number you choose. It is absolutely of mathematics to understand that cho bers can almost never *prove* a result, be every number. Instead, we'll rely on usin and deductive reasoning.

> "They are valuable to explain certain concepts from a practical or a historical view. Faculty could use these as "jumping off" points for projects and/or group work."
> —*Judith Wood, Central Florida Community College*
>
> "They are well written, beautifully illustrated, and relate historical figures and their mathematical genius and discoveries to the topic covered in the text. They are most informative on so many events and disciplines."
> —*Corinna M. Goehring, Jackson State Community College*

> "The variety of real-world applications makes the discrete much more concrete for the students."
> —*Kristin Chatas, Washtenaw Community College*
>
> "Excellent, interesting, current."
> —*Vesna Kilibarda, Indiana University Northwest*

34. Estimate the total cost of the following items for your dorm room:

Loft bed	\$159.95
Beanbag chair	\$49.95
Storage cubes	\$29.95
Lava lamp	\$19.95

35. Estimate the time it would take you in a charity bike tour to ride 86 miles at a rate of 11 miles per hour.
36. If a person earns \$48,300.00 per year, estimate how

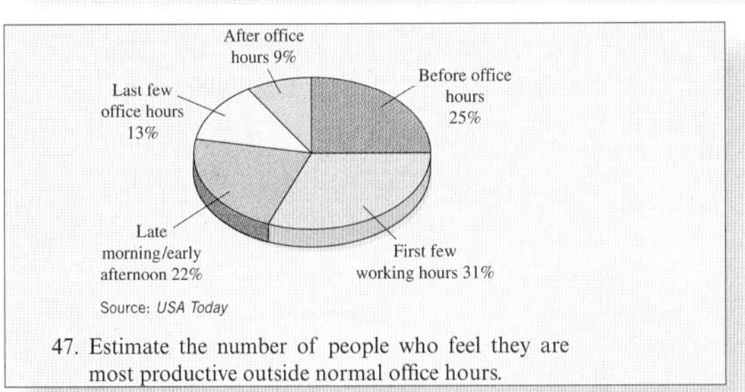

Source: *USA Today*

47. Estimate the number of people who feel they are most productive outside normal office hours.

Highly relevant real-world **Application Exercises and Examples** drawn from the experiences and research of the author team further emphasize the importance that *Math in Our World* places upon students' ability to form a distinct connection with the mathematical content. This helps students relate mathematical concepts presented in a section and the real life relevance of something familiar to them, increasing both student interest and motivation—key factors affecting their comprehension. The new edition brings many brand new and updated application exercises to students in each chapter, ranging in topics from MP3 player usage, college degree majors, elections, relevant business decisions, and scenarios involving popular statistics.

An **Index of Applications** is located immediately after the Detailed Table of Contents and is organized by discipline to help identify applications relevant to specific fields.

Practice

Implementing the **Step-by-Step Approach** used for years in Allan Bluman's *Elementary Statistics: A Step by Step Approach,* this text seeks to guide even the most hesitant student through the basic concepts fundamental to a liberal arts math curriculum. The authors realize that, in order to do so, students need to be introduced to those concepts through a highly relevant example in the opening vignette of each chapter. Carefully chosen questions immediately follow the opening vignette to help students form a connection between the relevant example and the mathematical concepts of the chapter, easing the transition for them. Using the engaging writing style characteristic of the text, the authors support concepts through abundant examples, helpful practice problems, and rich exercise sets.

- **Worked Example Problems** with detailed solutions help students master key concepts. Where solutions consist of more than one step, a numbered, step-by-step procedure is used to guide students through the problem-solving process. Additionally, each example is titled to highlight relevant learning objectives.

- One of our hallmark features, **Try This One** practice exercises provide immediate reinforcement for students. Designed to naturally follow each example, these practice exercises ask students to solve a similar problem, actively involving them in the learning process. All answers to Try This One exercises can be found just prior to the end-of-section exercise sets for students to complete the problem-solving process by confirming their solutions.

EXAMPLE 4 Using Inductive Reasoning to Test a Conjecture

Use inductive reasoning to decide if the following conjecture is true: Any four-digit number is divisible by 11 if the difference between the sum of the first and third digits and the sum of the second and fourth digits is divisible by 11.

SOLUTION

Let's make up a few examples. For 1,738, the sum of the first and third digits is $1 + 3 = 4$, and the sum of the second and fourth digits is $7 + 8 = 15$. The difference is $15 - 4 = 11$, so if the conjecture is true, 1,738 should be divisible by 11. To check: $1{,}738 \div 11 = 158$ (with no remainder).

For 9,273, $9 + 7 = 16$, $2 + 3 = 5$, and $16 - 5 = 11$. So if the conjecture is true, 9,273 should be divisible by 11. To check: $9{,}273 \div 11 = 843$ (with no remainder).

Let's look at one more example. For 7,161, $7 + 6 = 13$, $1 + 1 = 2$, and $13 - 2 + 11$. Also $7{,}161 \div 11 = 651$ (with no remainder), so the conjecture is true for this example as well. While we can't be positive based on three examples, inductive reasoning indicates that the conjecture is true.

▼ Try This One 4

Use inductive reasoning to decide if the following conjecture is true: If the sum

"I think Try This One makes a great addition to the worked examples, and I would definitely integrate them into the classroom lecture process."

—*Marcia Lambert, Pitt Community College*

"Examples are true to life and ones students will relate to."

—*John Ward, Jefferson Community and Technical College*

"The examples and 'Try This One' encourage individual learning."

—*Sam Buckner, North Greenville University*

Calculator Guide

If you use a calculator to try the repeated operations in Example 8, you'll need to press [=] (scientific calculator) or [ENTER] (graphing calculator) after every operation, or you'll get the wrong result. Suppose you simply enter the whole string:

Standard Scientific Calculator

12 [+] 50 [×] 2 [−] 12 [=]

Standard Graphing Calculator

12 [+] 50 [×] 2 [−] 12 [ENTER]

The result is 100, which is incorrect. In Chapter 5, we'll find out why when we study the order of operations.

- Optional **Calculator Guides** are set in the margins to show students how to use a calculator to solve certain examples, enhancing their skill set for solving problems. Keystrokes for both standard scientific calculators as well as standard TI-84 Plus graphing calculators are shown in order to accommodate students with a variety of tools available to them.

The rich variety of problem material found in the **End-of-Section Exercise Sets** helps check student knowledge in a variety of different ways to cater to the varying interests and educational backgrounds of liberal arts math students. In each end-of-section exercise set, there are:

- **Writing Exercises** that nurture the increased emphasis being placed on writing across the curriculum for many schools. These exercises provide an opportunity for students to summarize with words the key mathematical concepts from the section before attempting other exercises, a skill that may come more naturally for your students than computation.

- Traditional **Computational Exercises** provide additional practice for students, reinforcing relevant concepts just learned with increasing levels of difficulty.

- **Critical Thinking Exercises** drive students to take a concept that they have learned in a given section and apply it in a new direction or setting, revealing a deeper level of understanding.

"The writing exercises present a great opportunity for students to summarize the sections, list techniques needed for solving, highlight important rules, and recognize some problems that many times occur when solving problems."

—Lynn Craig, Baton Rouge Community College

"Excellent for developing fluency in the language of mathematics."

—Janet Teeguarden, Ivy Tech Community College

"These are well designed to try to help them connect sections and get the bigger picture."

—Andrew Beiderman, Community College of Baltimore County, Essex

"Chapter Projects are great!!! I use these to get students used to researching topics that require both Mathematics and writing out their solutions."

—Mark Brenneman, Mesa Community College

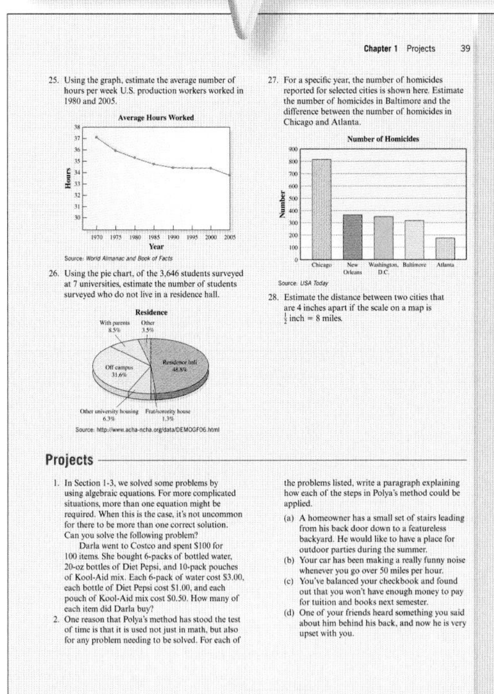

Chapter 1 Projects 39

25. Using the graph, estimate the average number of hours per week U.S. production workers worked in 1980 and 2005.

Average Hours Worked

Source: *World Almanac and Book of Facts*

26. Using the pie chart, of the 3,646 students surveyed at 7 universities, estimate the number of students surveyed who do not live in a residence hall.

Residence

Source: http://www.acha-ncha.org/data/DEMOGF06.html

27. For a specific year, the number of homicides reported for selected cities is shown here. Estimate the number of homicides in Baltimore and the difference between the number of homicides in Chicago and Atlanta.

Number of Homicides

Source: *USA Today*

28. Estimate the distance between two cities that are 4 inches apart if the scale on a map is $\frac{1}{2}$ inch = 8 miles.

Projects

1. In Section 1-3, we solved some problems by using algebraic equations. For more complicated situations, more than one equation might be required. When this is the case, it's not uncommon for there to be more than one correct solution. Can you solve the following problem?
   Darla went to Costco and spent $100 for 100 items. She bought 6-packs of bottled water, 20-oz bottles of Diet Pepsi, and 10-pack pouches of Kool-Aid mix. Each 6-pack of water cost $3.00, each bottle of Diet Pepsi cost $1.00, and each pouch of Kool-Aid mix cost $0.50. How many of each item did Darla buy?
2. One reason that Polya's method has stood the test of time is that it is used not just in math, but also for any problem needing to be solved. For each of the problems listed, write a paragraph explaining how each of the steps in Polya's method could be applied.
   (a) A homeowner has a small set of stairs leading from his back door down to a featureless backyard. He would like to have a place for outdoor parties during the summer.
   (b) Your car has been making a really funny noise whenever you go over 50 miles per hour.
   (c) You've balanced your checkbook and found out that you won't have enough money to pay for tuition and books next semester.
   (d) One of your friends heard something you said about him behind his back, and now he is very upset with you.

Exercises and activities located in the **End-of-Chapter Material** provide opportunities for students to prepare for success on quizzes or tests. At the conclusion of each chapter, students find critical summary information that helps them pull together each step learned while moving through the chapter. In each end-of-chapter segment, students and instructors will find:

- An **End-of-Chapter Summary,** separated by section, reviews the important terms and ideas students should have learned from the material. Answers to the questions posed in the opening vignette are also provided, ushering students back to the central theme that mathematics is truly ingrained in many aspects of their lives. This brief summary helps students to gather details that may have at first seemed disparate and assemble them in a logical progression.

- **Review Exercises** are also presented by section with a combination of both computational and real-world application exercises similar to those found in the end-of-section exercise sets.

- A **Chapter Test** gives students the opportunity to check their knowledge of the chapter as a whole in order to prepare for classroom quizzes, tests, and other larger assessments.

- **Chapter Projects** encourage more in-depth investigation for students working to summarize key concepts from the entire chapter. These projects are more extensive than any of the review exercises or test questions and are valuable assets for instructors looking for ways that students can work collaboratively.

Succeed

Experience Student Success!

ALEKS® ALEKS is a unique online math tool that uses adaptive questioning and artificial intelligence to correctly place, prepare, and remediate students . . . all in one product! Institutional case studies have shown that **ALEKS has improved pass rates by over 20% versus traditional online homework and by over 30% compared to using a text alone.**

By offering each student an individualized learning path, ALEKS directs students to work on the math topics that they are ready to learn. Also, to help students keep pace in their course, instructors can correlate ALEKS to their textbook or syllabus in seconds.

To learn more about how ALEKS can be used to boost student performance, please visit **www.aleks.com/highered/math** or contact your McGraw-Hill representative.

ALEKS Pie

Each student is given his/her own individualized learning path.

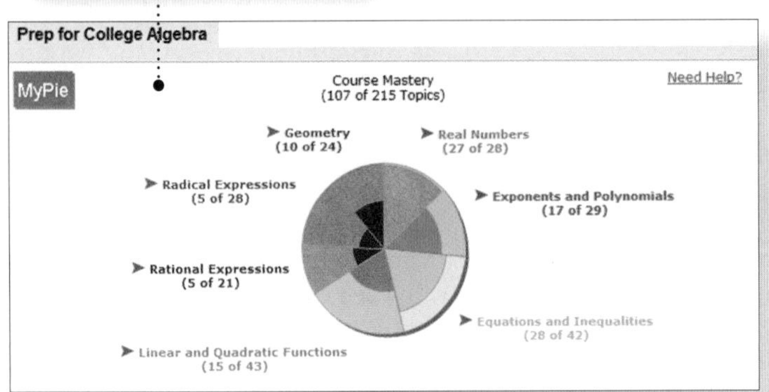

Easy Graphing Utility!

Students can answer graphing problems with ease!

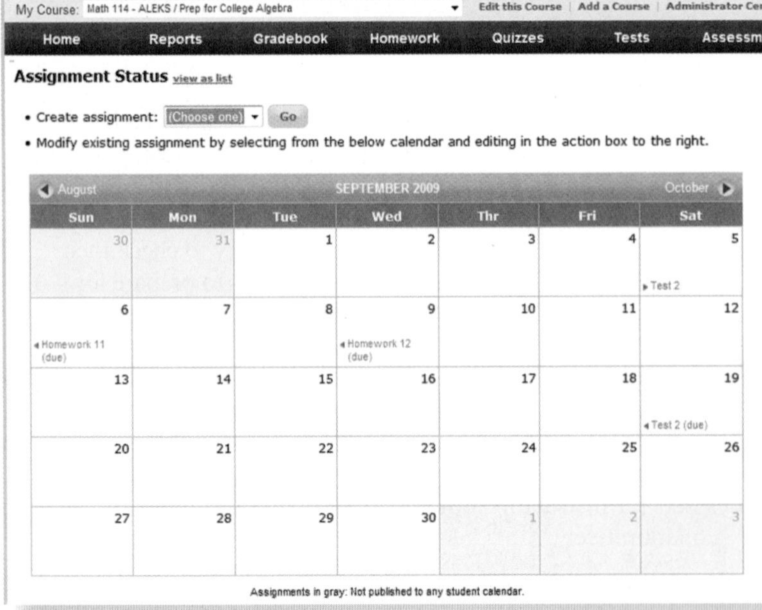

Course Calendar

Instructors can schedule assignments and reminders for students.

New ALEKS Instructor Module

Enhanced Functionality and Streamlined Interface Help to Save Instructor Time

ALEKS® The new ALEKS Instructor Module features enhanced functionality and a streamlined interface based on research with ALEKS instructors and homework management instructors. Paired with powerful assignment driven features, textbook integration, and extensive content flexibility, the new ALEKS Instructor Module simplifies administrative tasks and makes ALEKS more powerful than ever.

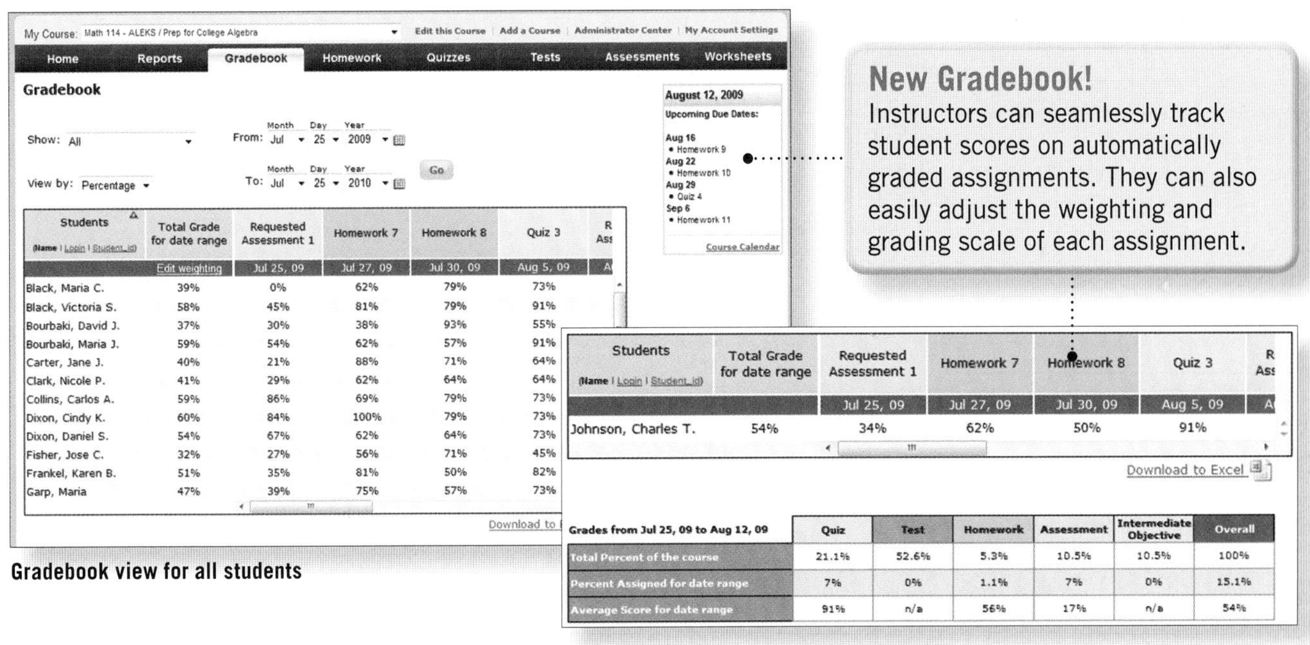

Students (Name | Login | Student_Id)	Total Grade for date range	Requested Assessment 1	Homework 7	Homework 8	Quiz 3
	Edit weighting	Jul 25, 09	Jul 27, 09	Jul 30, 09	Aug 5, 09
Black, Maria C.	39%	0%	62%	79%	73%
Black, Victoria S.	58%	45%	81%	79%	91%
Bourbaki, David J.	37%	30%	38%	93%	55%
Bourbaki, Maria J.	59%	54%	62%	57%	91%
Carter, Jane J.	40%	21%	88%	71%	64%
Clark, Nicole P.	41%	29%	62%	64%	64%
Collins, Carlos A.	59%	86%	69%	79%	73%
Dixon, Cindy K.	60%	84%	100%	79%	73%
Dixon, Daniel S.	54%	67%	62%	64%	73%
Fisher, Jose C.	32%	27%	56%	71%	45%
Frankel, Karen B.	51%	35%	81%	50%	82%
Garp, Maria	47%	39%	75%	57%	73%

Gradebook view for all students

New Gradebook!

Instructors can seamlessly track student scores on automatically graded assignments. They can also easily adjust the weighting and grading scale of each assignment.

Students (Name | Login | Student_Id)	Total Grade for date range	Requested Assessment 1	Homework 7	Homework 8	Quiz 3
		Jul 25, 09	Jul 27, 09	Jul 30, 09	Aug 5, 09
Johnson, Charles T.	54%	34%	62%	50%	91%

Grades from Jul 25, 09 to Aug 12, 09	Quiz	Test	Homework	Assessment	Intermediate Objective	Overall
Total Percent of the course	21.1%	52.6%	5.3%	10.5%	10.5%	100%
Percent Assigned for date range	7%	0%	1.1%	7%	0%	15.1%
Average Score for date range	91%	n/a	56%	17%	n/a	54%

Gradebook view for an individual student

Track Student Progress Through Detailed Reporting

Instructors can track student progress through automated reports and robust reporting features.

Login (Name | Student Id)	Total time in ALEKS	Last login	Last assessment	Performance goal
MBLACK2	15.0	07/09/2009	07/09/2009	5 +8 %
VBLACK	83.6	07/11/2009	07/11/2009	45 +11 %
DBOURBAKI2	54.8	07/11/2009	07/11/2009	30 +9 %
MBOURBAKI2	65.7	07/11/2009	07/11/2009	54 +9 %
JCARTER4	33.9	07/11/2009	07/11/2009	21 +8 %
CCOLLINS2	155.6	07/11/2009	07/11/2009	86 +8 %

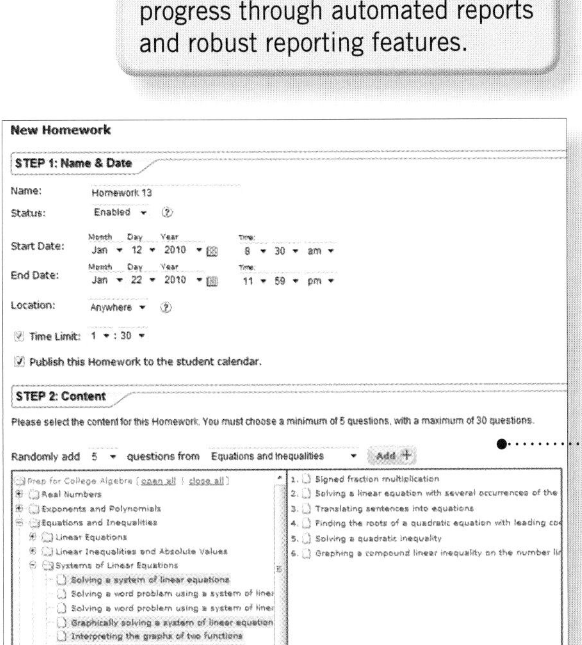

Automatically Graded Assignments

Instructors can easily assign homework, quizzes, tests, and assessments to all or select students. Deadline extensions can also be created for select students.

Learn more about ALEKS by visiting **www.aleks.com/highered/math** or contact your McGraw-Hill representative.

Supplements

MathZone www.mhhe.com/sobecki. McGraw-Hill's MathZone is a complete online homework system for mathematics and statistics. Instructors can assign textbook-specific content from over 40 McGraw-Hill titles as well as customize the level of feedback students receive, including the ability to have students show their work for any given exercise. Assignable content includes an array of videos and other multimedia along with algorithmic exercises, providing study tools for students with many different learning styles.

MathZone also helps ensure consistent assignment delivery across several sections through a course administration function and makes sharing courses with other instructors easy. In addition, instructors can also take advantage of a virtual whiteboard by setting up a Live Classroom for online office hours or a review session with students.

For more information, visit the book's website **www.mhhe.com/sobecki** or contact your local McGraw-Hill sales representative **www.mhhe.com/rep**.

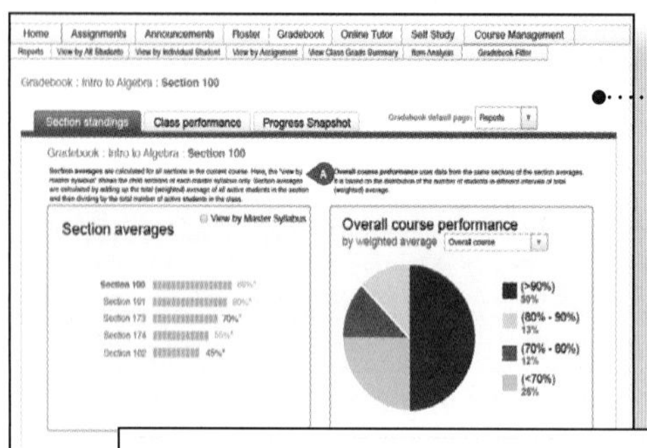

Visual Reporting – The new dashboard-like reports will provide the progress snapshot instructors are looking for to help them make informed decisions about their students.

Item Analysis – Instructors can view detailed statistics on student performance at a learning objective level to understand what students have mastered and where they need additional help.

Managing Assignments for Individual Students – Instructors have greater control over creating individualized assignment parameters for individual students, special populations and groups of students, and for managing specific or ad hoc course events.

New User Interface – Designed by You! Instructors and students will experience a modern, more intuitive layout. Items used most commonly are easily accessible through the menu bar such as assignments, visual reports, and course management options.

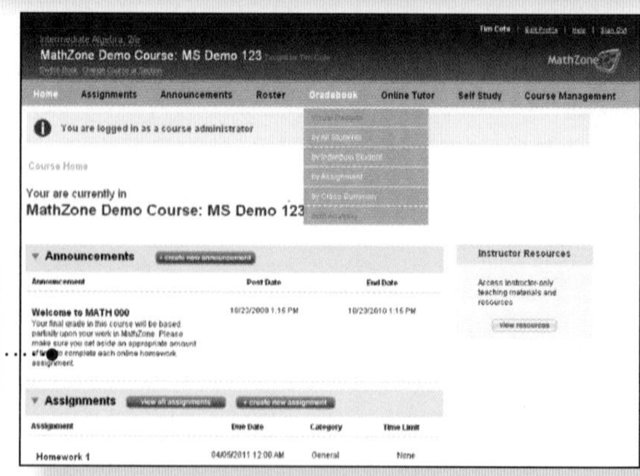

Instructor Supplements

- **Computerized Test Bank Online:** Utilizing Brownstone Diploma® algorithm-based testing software, this supplement enables users to create customized exams quickly.
- **Instructor Solutions Manual:** Written by author Angela Schirck-Matthews, the *Instructor Solutions Manual* provides comprehensive, worked-out solutions to all exercises in the text. The methods used to solve the problems in the manual are the same as those used to solve the examples in the textbook. The manual can be found on the Instructor Center at the book's website: www.mhhe.com/sobecki.
- **PowerPoint Slides:** These slides closely follow the textbook, and are completely editable.

Student Supplements

- **Student Solutions Manual:** Written by author Angela Schirck-Matthews, the *Student Solutions Manual* provides comprehensive, worked-out solutions to all of the odd-numbered exercises. The steps shown in the solutions match the style of the worked examples found in the text.
- **Lecture Videos:** Engaging lectures by author Dave Sobecki introduce concepts, definitions, theorems, formulas, and problem solving procedures to help students better comprehend key topics. The videos can be found online at www.mhhe.com/sobecki.
- **Exercise Videos:** Presented by author Angela Schirck-Matthews, these videos work through selected exercises, following the solution methodology found in the text. The videos can be found online at www.mhhe.com/sobecki.

Mc Graw Hill Tegrity campus

http://tegritycampus.mhhe.com

- Tegrity Campus is a service that makes class time available all the time by automatically capturing every lecture in a searchable format for students to review when they study and complete assignments. With a simple one-click start and stop process, you capture all computer screens and corresponding audio. Students replay any part of any class with easy-to-use browser-based viewing on a PC or Mac.
- Educators know that the more students can see, hear, and experience class resources, the better they learn. With Tegrity Campus, students quickly recall key moments by using Tegrity Campus's unique search feature. This search helps students efficiently find what they need, when they need it across an entire semester of class recordings. Help turn all your students' study time into learning moments immediately supported by your lecture.
- To learn more about Tegrity watch a 2 minute Flash demo at http://tegritycampus.mhhe.com.

ALEKS®

ALEKS (**A**ssessment and **LE**arning in **K**nowledge **S**paces) is a dynamic online learning system for mathematics education, available over the Web 24/7. ALEKS assesses students, accurately determines their knowledge, and then guides them to the material that they are most ready to learn. With a variety of reports, Textbook Integration Plus, quizzes, and homework assignment capabilities, ALEKS offers flexibility and ease of use for instructors.

- ALEKS uses artificial intelligence to determine exactly what each student knows and is ready to learn. ALEKS remediates student gaps and provides highly efficient learning and improved learning outcomes
- ALEKS is a comprehensive curriculum that aligns with syllabi or specified textbooks. Used in conjunction with McGraw-Hill texts, students also receive links to text-specific videos, multimedia tutorials, and textbook pages.
- ALEKS offers a dynamic classroom management system that enables instructors to monitor and direct student progress toward mastery of course objectives.

ALEKS Prep/Remediation:

- Helps instructors meet the challenge of remediating unequally prepared or improperly placed students.
- Assesses students on their prerequisite knowledge needed for the course they are entering (for example, calculus students are tested on precalculus knowledge) and prescribes unique and efficient learning paths specific to their strengths and weaknesses.
- Students can address prerequisite knowledge gaps outside of class, freeing the instructor to use class time pursuing course outcomes.

Electronic Textbook: CourseSmart is a new way for faculty to find and review e-textbooks. It's also a great option for students who are interested in accessing their course materials digitally and saving money. CourseSmart offers thousands of the most commonly adopted textbooks across hundreds of courses from a wide variety of higher education publishers. It is the only place for faculty to review and compare the full text of a textbook online, providing immediate access without the environmental impact of requesting a print exam copy. At CourseSmart, students can save up to 50 percent off the cost of a print book, reduce their impact on the environment, and gain access to powerful Web tools for learning including full text search, notes and highlighting, and email tools for sharing notes between classmates. To learn more, visit **www.CourseSmart.com.**

Create: Craft your teaching resources to match the way you teach! With McGraw-Hill Create, **www.mcgrawhillcreate.com,** you can easily rearrange chapters, combine material from other content sources, and quickly upload content you have written like your course syllabus or teaching notes. Find the content you need in Create by searching through thousands of leading McGraw-Hill textbooks. Arrange your book to fit your teaching style. Create even allows you to personalize your book's appearance by selecting the cover and adding your name, school, and course information. Order a Create book and you'll receive a complimentary print review copy in 3–5 business days or a complimentary electronic review copy (eComp) via email in about one hour. Go to **www.mcgrawhillcreate.com** today and register. Experience how McGraw-Hill Create empowers you to teach *your* students *your* way.

Design

Based on Student Feedback

The development team at McGraw-Hill Higher Education realizes that there are many elements beyond content that contribute to creation of a student-friendly textbook. After hosting several student focus groups, we have learned a great deal about how students use their books, what pedagogical elements are useful, which elements are distracting and not useful, and general feedback on page layout. As a result of this feedback, we have integrated many of the student suggestions in a manner that is helpful for students, and simultaneously does not detract from instructor use of the text.

The following specific choices in *Math in Our World* have been made to increase student engagement based directly upon this feedback.

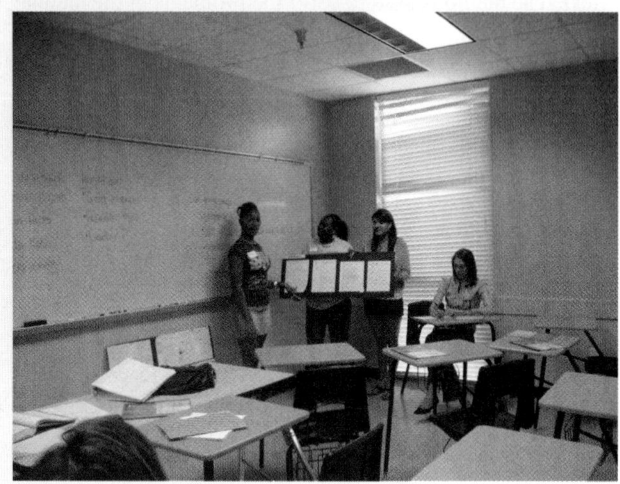

Section 1-1 The Nature of Mathematical Reasoning

Learning Objectives clearly define and set goals for students at the beginning of each section to focus and direct their learning.

LEARNING OBJECTIVES

❑ 1. Identify two types of reasoning.

❑ 2. Use inductive

Part of being an adult is making decisions—every day is full of them, from the simple, like what to eat for breakfast, to the critical, like a choice of major. If you make every decision based on a coin flip, chances are you won't get very far in life. Instead, it's important to be able to analyze a situation and make a decision based on logical thinking. We'll call the process of logical thinking **reasoning.** It doesn't take a lot of imagination to understand how important reasoning is in everyone's life.

EXAMPLE 5 Finding a Counterexample

Find a counterexample that proves the conjecture below is false.
Conjecture: A number is divisible by 3 if the last two digits are divisible by 3.

SOLUTION

We'll pick a few numbers at random whose last two digits are divisible by 3, then divide them by 3, and see if there's a remainder.

1,527: Last two digits, 27, divisible by 3; $1{,}527 \div 3 = 509$
11,745: Last two digits, 45, divisible by 3; $11{,}427 \div 3 = 3{,}809$

At this point, you might start to suspect that the conjecture is true, but you shouldn't! We've only checked two cases.

1,136: Last two digits, 36, divisible by 3; $1{,}136 \div 3 = 378\frac{2}{3}$

This counterexample shows that the conjecture is false.

Check boxes alert students to the completion of each goal once a learning objective has been completed.

☑ 3. Find a counterexample to disprove a conjecture.

▼ Try This One 5

Find a counterexample to disprove the conjecture that the name of every month in English contains either the letter y or the letter r.

CAUTION

Remember: One counterexample is enough to show that a conjecture is false. But one positive example is *never* enough to show that a conjecture is true.

Making caution boxes red and placing them in the body of the text emphasizes their importance for students and makes them a more useful tool.

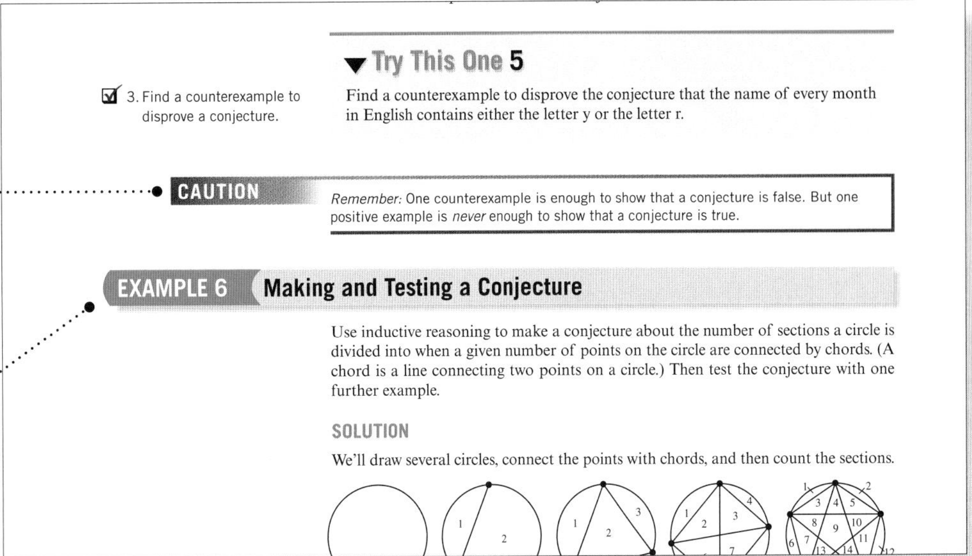

Example headings are shaded and pulled out in to the margins to make them easy for students to identify.

Students said that when they spend a lot of time on the exercise sets, a white background can be hard on their eyes. To help them, we have chosen to use soft colors for both end-of-section and end-of-chapter exercise sets.

EXERCISE SET 1-1

Writing Exercises

1. Explain the difference between inductive and deductive reasoning.
2. What is meant by the term *conjecture?*
3. Give a real-life example of how to use inductive reasoning.
4. What is a counterexample? What are counterexam- ples used for?
5. Explain why you can never be sure that a conclusior you arrived at using inductive reasoning is true.
6. Explain the difference between an arbitrary numbei and a number selected at random.

Computational Exercises

For Exercises 7–14, use inductive reasoning to find a pattern, and then make a reasonable conjecture for the next number or item in the sequence.

7. 1 2 4 7 11 16 22 29 __
8. 5 15 45 135 405 1215
11. 100 99 97 94 90 85 79 __
12. 9 12 11 14 13 16 15 18 __
13. 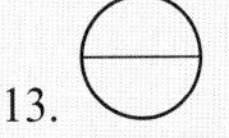    ______

A darker shaded bar runs along all end-of-chapter material to make it easily identifiable for students.

Review Exercises

Section 1-1

For Exercises 1–4, make reasonable conjectures for the next three numbers or letters in the sequence.

1. 3 4 6 7 9 10 12 13 15 16 __ __ __
2. 2 7 4 9 6 11 8 13 __ __ __

For Exercises 5 and 6, make a reasonable conjecture and draw the next figure.

5. 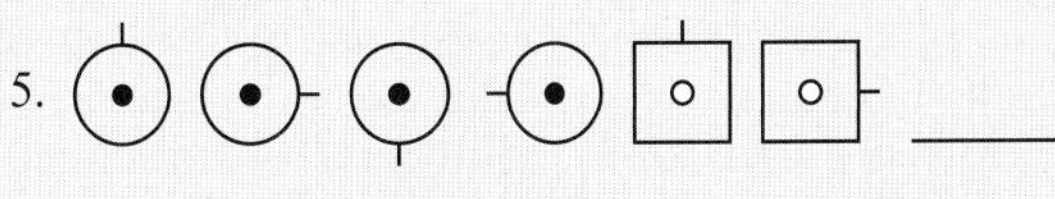 ______

Updated Content

CHAPTER 1 Problem Solving

- Section 1-1 on mathematical reasoning has been expanded from three examples to ten examples.
- New real-world application exercises have been added throughout the chapter to reinforce the real-world relevance of key concepts, particularly in Section 1-1 where the total number of exercises has increased from 23 to 70.
- Sections 1-2 and 1-3 have been rearranged so that estimation is learned first. This allows it to then be used later in the chapter as a tool in problem solving.
- Estimation in particular has many new problems based on real data that will interest modern college students, including more examples than the previous edition.

CHAPTER 2 Sets

- Coverage is greatly increased throughout the chapter, including new sub-sections on cardinality of sets (Section 2-1), set subtraction and Cartesian product (Section 2-2), DeMorgan's laws and cardinality of unions (Section 2-3), and countable and uncountable sets (Section 2-5).
- Eight new applications have been added to Section 2-1, establishing a total of thirty-four questions based on current events and relevant data. Additionally, thirty-six new application exercises have been added to Sections 2-3 and 2-4.
- Two new sidelights on infinite sets in Section 2-5 have been added to increase the interdisciplinary connections students can make to enhance their understanding of key concepts.

CHAPTER 3 Logic

- The coverage of truth tables has been standardized by using the method of building new columns for parts of compound statements as the main presentation. To accommodate a variety of teaching styles, the method of writing truth values underneath connectives is still presented as an alternative method.
- An application of truth tables has been added as a worked example in Section 3-2.
- Content updates include the addition of DeMorgan's laws to Section 3-3 and circular reasoning among common fallacies in Section 3-4.
- A total of thirty-eight new application exercises have been added to Sections 3-2 and 3-3 and the applications in Section 3-4 have been updated to reflect current student interests.

CHAPTER 4 Numeration Systems

- Coverage of tally systems and multiplicative grouping systems has been added and the Chinese system has been expanded from a Sidelight to full coverage in Section 4-1.
- Tools and algorithms in arithmetic, located in Section 4-2, is completely new to the second edition.

CHAPTER 5 The Real Number System

- More worked examples and real-world applications have been created to facilitate student learning, including applications of LCM and GCF (Section 5-1) and arithmetic with integers and order of operations (Section 5-2). Additionally, brand new application exercises have been created in every section, including sixteen new applications in Section 5-5.
- Common "canceling" mistakes are now more prominently featured through a new Sidelight found in Section 5-2.
- Coverage of adding and subtracting of fractions has been increased extensively and the chapter has been rearranged to cover this topic after multiplication and division of fractions to ease the learning progression for students.
- The new presentation of calculator guides for arithmetic with scientific notation are especially helpful in this chapter.

CHAPTER 6 Topics in Algebra

- Formerly Chapter 7, the order of Sections 4 and 5 have been reversed from the previous edition so that ratio, proportion, and variation are now presented directly after applications of solving equations.
- Organization and coverage of topics through much of the chapter has been reevaluated to focus on understanding what it really means to solve an equation. This includes increased coverage of solving linear equations with fractions and three-part inequalities and an introduction to the terms "identity" and "contradiction" in Section 6-2.
- Real-world applications have been added to both worked examples and exercise sets throughout the chapter, including twenty new application exercises in Section 6-2 and a doubling of the number of applications in worked examples in Sections 6-1 to 6-4 from the first edition.

CHAPTER 7 Additional Topics in Algebra

- Formerly Chapter 8, it has been expanded to include a brand new section on solving systems of equations with matrices (Section 7-3). Additionally, coverage of functions has grown to encompass two sections, including a total of thirty new application exercises.
- To provide further reinforcement for critical concepts like basic linear equations, linear systems, and linear inequalities, a total of five new application examples and thirty-one new exercises have been added.
- The second edition presents a more thorough coverage of graphing linear inequalities in two variables before moving on to systems of linear inequalities.

CHAPTER 8 Consumer Mathematics

- Formerly Chapter 9, it has undergone significant changes to content coverage, including new sub-sections on the banker's rule and discounted loans (Section 8-2), a new subsection on fixed installment loans and APR (Section 8-4), and an entirely new section on stocks and bonds (Section 8-6).
- Organizationally, the discussion of interest has been split into two sections, one on simple interest and one on compound interest.
- The coverage of annuities in Section 8-3 has been expanded, and a section formerly focused on markup and markdown has been replaced with a more thorough coverage of percent increase and decrease, including an example on deceptive use of percents in advertising.
- Updates to the discussion of home ownership now reflect current interest rates and trends.

CHAPTER 9 Measurement

- Formerly found only in Appendix A, the discussion of measurement has been expanded into an entirely new chapter.

CHAPTER 10 Geometry

- A new section on non-Euclidean and transformational geometries has been added to the text.
- Expanded coverage is found throughout the chapter, including sections focused on area and volume as well as finding angles in the trigonometry section.
- New applications of parallel lines and transversals using city streets have been added.

CHAPTER 11 Probability and Counting Techniques

- The entire chapter has been completely reorganized and rewritten to include a new section on the binomial theorem. This includes shifting counting techniques to be presented prior to probability and the addition of permutations when some objects are alike.
- Dozens of new application exercises are found throughout the chapter, including comparisons of expected value for common games of chance and real world applications of probability and counting.

CHAPTER 12 Statistics

- Nearly all worked examples and exercises featuring statistics have been replaced with problems featuring newer, more relevant data.
- A table is now included in Section 12-3 to compare the strengths and weaknesses of different measures of average.
- Coverage of the empirical rule can now be found in the discussion of the normal distribution and the discussion of finding area under the normal curve has been streamlined.
- The presentation of correlation is now based on an interesting real-world example of a correlation between MySpace usage and illiteracy and helpful information regarding the use of technology to find correlation coefficients and regression lines has been added.

CHAPTER 13 Other Mathematical Systems

- Formerly Chapter 6, coverage of mathematical systems has been moved to Section 13-1 so that clock and modular arithmetic can be taught in the context of mathematical systems.
- Mathematical systems are now introduced in terms of the familiar example of making turns at intersections and new application exercises in Section 13-1 include the pull chain on a ceiling fan and mixing paint colors.
- Presentation of clock arithmetic now uses the standard clock with numbers from 1 through 12, and also includes clocks with other numbers of hours.
- Congruences in modular arithmetic have been revised so that now all use the $\equiv$ symbol.

CHAPTER 14 Voting Methods

- Formerly Chapter 13, two completely new sections on apportionment and apportionment flaws have been added to the new edition.

CHAPTER 15 Graph Theory

- This is a completely new chapter to the second edition and features an expanded form of the coverage of networks, formerly found in Chapter 10.

Acknowledgements

McGraw-Hill's 360° Development Process is an ongoing, market-oriented approach to building accurate and innovative print and digital products. It is dedicated to continual large-scale and incremental improvement driven by multiple customer feedback loops and checkpoints. This process is initiated during the early planning stages of our new products, intensifies during the development and production stages, and then begins again upon publication, in anticipation of the next edition.

A key principle in the development of any mathematics text is its ability to adapt to teaching specifications in a universal way. The only way to do so is by contacting those universal voices—and learning from their suggestions. We are confident that our book has the most current content the industry has to offer, thus pushing our desire for accuracy to the highest standard possible. In order to accomplish this, we have moved through an arduous road to production. Extensive and open-minded advice is critical in the production of a superior text.

We engaged dozens of instructors and students to provide us with guidance during the development of the second edition. By investing in this extensive endeavor, McGraw-Hill delivers to you a product suite that has been created, refined, tested, and validated to be a successful tool in your course.

The McGraw-Hill mathematics team and the authors wish to thank the following instructors who participated in postpublication reviews of the first edition, the proposed second edition table of contents, and both first and second drafts of the second edition manuscript to give feedback on reworked narrative, design changes, pedagogical enhancements, and organizational changes. This feedback was summarized by the book team and used to guide the direction of the final text.

Gerald Angelichio, *Herkimer County Community College*
Anne Antonippillai, *University of Wisconsin—Stout*
Hamid Attarzadeh, *Jefferson Community and Technical College*
Jon Becker, *Indiana Univ-Northwest-Gary*
Andrew Beiderman, *Community College of Baltimore County Essex*
Laverne Blagmon-Earl, *Univ of Dist of Columbia*
Mike Bosch, *Iowa Lakes Community College*
Caroline Boulis, *Lee University*
Mark Brenneman, *Mesa Community College*
Sam Buckner, *North Greenville University*
Brian Burrell, *University of Massachusetts—Amherst*
Gerald Burton, *Virginia State University*
Fred Butler, *York College of Pennsylvania*
David Capaldi, *Johnson & Wales University*
Blayne Carroll, *Lee University*
Gail Carter, *St. Petersburg College—Tarpon Springs*
Florence Chambers, *Southern Maine Community College*
Kristin Chatas, *Washtenaw Community College*
Sandy Cohen, *Saint Petersburg College*
Lynn Craig, *Baton Rouge Community College*
Karen Crossin, *George Mason University*
Cheryl Davids, *Central Carolina Tech College*
Mark Ellis, *Central Piedmont Comm Coll*
Dan Endres, *University of Central Oklahoma*
Scott Fallstrom, *University of Oregon*
Shurron Farmer, *University of Dist of Columbia*
Nicki Feldman, *Pulaski Technical College*
Scott Garten, *Northwest Missouri State University*

Mahmood Ghamsary, *California State University, Fullerton*
Carrie Goehring, *Jackson State Community College*
John Grant, *Towson University*
Bo Green, *Abilene Christian University*
Sheryl Griffith, *Iowa Central Community College*
Renu Gupta, *Louisiana State University—Alexandria*
Kim Hagens, *Louisiana State University*
Quin Hearn, *Brevard Community College—Cocoa*
Thomas Hoffman, *Coastal Carolina University*
Lori Holdren, *Lake City Community College*
Robert Jajcay, *Indiana State University, Terre Haute*
Maryann Justinger, *Erie Community College South Campus—Orchard Park*
Joseph Kazimir, *East Los Angeles College*
David Keller, *Kirkwood Community College*
Betsy Kiedaisch, *College of Dupage*
Vesna Kilibarda, *Indiana University Northwest*
Bob Koca, *Community College Of Baltimore County, Essex*
Bhaskara Kopparty, *Indiana State University—Terre Haute*
Steve Kristoff, *Ivy Tech Community College of Indiana*
Pamela Krompak, *Owens Community College*
Valerie LaFrance, *Saint Petersburg College*
Marcia Lambert, *Pitt Community College*
Eveline Lapierre, *Johnson & Wales University*
John Lattanzio, *Indiana University of Pennsylvania*
Edith Lester, *Volunteer State Community College*
Antonio Magliaro, *Quinnipiac University*
Barbara Manley, *Jackson State Community College*
Carrie McCammon, *Ivy Tech Community College of Indiana*
Cornelius Nelan, *Quinnipiac University*
Bernard Omolo, *University of South Carolina Upstate*
Eugenia Peterson, *Richard J Daley College*
Becky Pohle, *Ivy Tech Community College of Indiana*
Elaine Previte, *Bristol Community College*
Carolyn Reed, *Austin Community College*
Natalie Rivera, *Estrella Mountain Community College*
Lisa Rombes, *Washtenaw Community College*
Fary Sami, *Harford Community College*
Nancy Schendel, *Iowa Lakes Community College, Estherville*
Mike Skowronski, *University of Wisconsin, Oshkosh*
Zeph Smith, *Salt Lake Community College*
Laura Stapleton, *Marshall University*
Janet Teeguarden, *Ivy Tech Community College of Indiana*
William Thralls, *Johnson & Wales University, North Miami*
David Troidl, *Erie Community College*
John Ward, *Jefferson Community and Technical College*
Susan Warner, *Friends University*
John Weglarz, *Kirkwood Community College*
Ronald White, *Norfolk State University*
Greg Wisloski, *Indiana University of Pennsylvania*
Judith Wood, *Central Florida Community College*

The authors would like to thank the many people who have helped bring the second edition of Math in Our World to life. First and foremost we acknowledge the comments and suggestions we received from all reviewers of the manuscript. Their thoughtful insights allowed us to understand how we can better help their students learn and be engaged by math, and guided us through this substantial revision. We think the result is a truly student-friendly text. Additionally, we thank Cindy Trimble and George Watson whose invaluable contributions have helped us to ensure that we present the most accurate information possible, and Jennifer Siegel of Broward College for providing many application exercises with style and flair.

Finally, at McGraw-Hill Higher Education, we thank John Osgood, Sponsoring Editor; Christina Lane, Developmental Editor; Kevin Ernzen, Marketing Manager; April Southwood, Senior Project Manager; Tara McDermott, Designer; Amber Bettcher, Digital Product Manager; and Sandy Schnee, Senior Media Project Manager. Their expertise helped transform this work from raw manuscript to a finished product we're very proud of.

Detailed Table of Contents

CHAPTER 1

Problem Solving 2

1-1 The Nature of Mathematical Reasoning 4
1-2 Estimation and Interpreting Graphs 16
1-3 Problem Solving 26
Chapter 1 Summary 35

CHAPTER 2

Sets 40

2-1 The Nature of Sets 42
2-2 Subsets and Set Operations 54
2-3 Using Venn Diagrams to Study Set Operations 65
2-4 Using Sets to Solve Problems 74
2-5 Infinite Sets 82
Chapter 2 Summary 86

CHAPTER 3

Logic 90

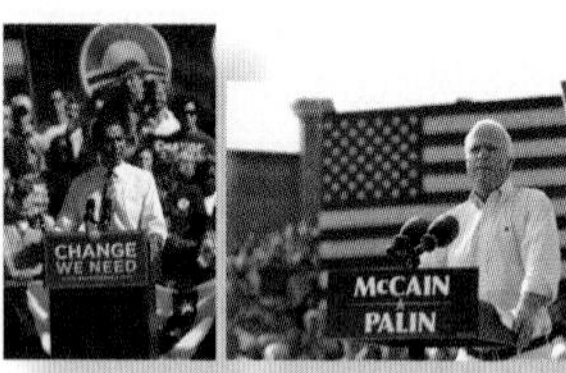

3-1 Statements and Quantifiers 92
3-2 Truth Tables 100
3-3 Types of Statements 115
3-4 Logical Arguments 123
3-5 Euler Circles 134
Chapter 3 Summary 141

CHAPTER 4
Numeration Systems 146

4-1 Early and Modern Numeration Systems 148
4-2 Tools and Algorithms in Arithmetic 163
4-3 Base Number Systems 170
4-4 Operations in Base Number Systems 183
Chapter 4 Summary 193

CHAPTER 5
The Real Number System 196

5-1 The Natural Numbers 198
5-2 The Integers 210
5-3 The Rational Numbers 221
5-4 The Irrational Numbers 235
5-5 The Real Numbers 244
5-6 Exponents and Scientific Notation 251
5-7 Arithmetic and Geometric Sequences 261
Chapter 5 Summary 271

CHAPTER 6
Topics in Algebra 276

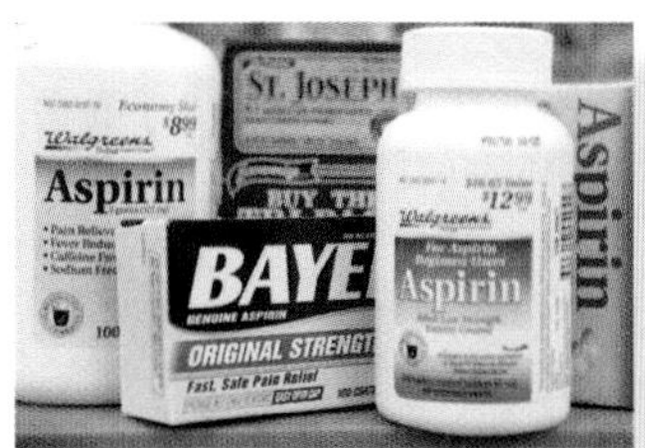

6-1 The Fundamentals of Algebra 278
6-2 Solving Linear Equations 288
6-3 Applications of Linear Equations 301
6-4 Ratio, Proportion, and Variation 308
6-5 Solving Linear Inequalities 319
6-6 Solving Quadratic Equations 328
Chapter 6 Summary 341

CHAPTER 7

Additional Topics in Algebra 346

7-1 The Rectangular Coordinate System and Linear Equations in Two Variables 348

7-2 Systems of Linear Equations 359

7-3 Solving Systems of Linear Equations Using Matrices 374

7-4 Linear Inequalities 382

7-5 Linear Programming 389

7-6 Functions 394

7-7 Linear, Quadratic, and Exponential Functions 401

Chapter 7 Summary 413

Available Online at www.mhhe.com/sobecki: Chapter 7 Supplement: An Application of Functions—Sound

CHAPTER 8

Consumer Mathematics 418

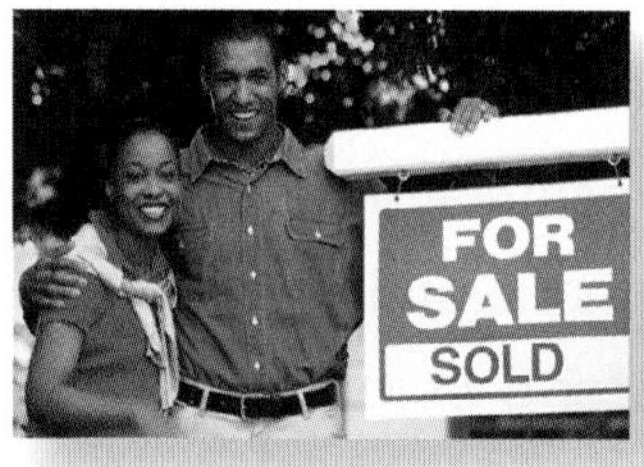

8-1 Percents 420

8-2 Simple Interest 429

8-3 Compound Interest 437

8-4 Installment Buying 446

8-5 Home Ownership 457

8-6 Stocks and Bonds 465

Chapter 8 Summary 474

CHAPTER 9

Measurement 480

9-1 Measures of Length: Converting Units and the Metric System 482

9-2 Measures of Area, Volume, and Capacity 490

9-3 Measures of Weight and Temperature 499

Chapter 9 Summary 506

CHAPTER 10

Geometry 510

10-1 Points, Lines, Planes, and Angles 512

10-2 Triangles 521

10-3 Polygons and Perimeter 531
10-4 Areas of Polygons and Circles 536
10-5 Volume and Surface Area 544
10-6 Right Triangle Trigonometry 552
10-7 A Brief Survey of Non-Euclidean and Transformational Geometries 559
Chapter 10 Summary 564

CHAPTER 11

Probability and Counting Techniques 572

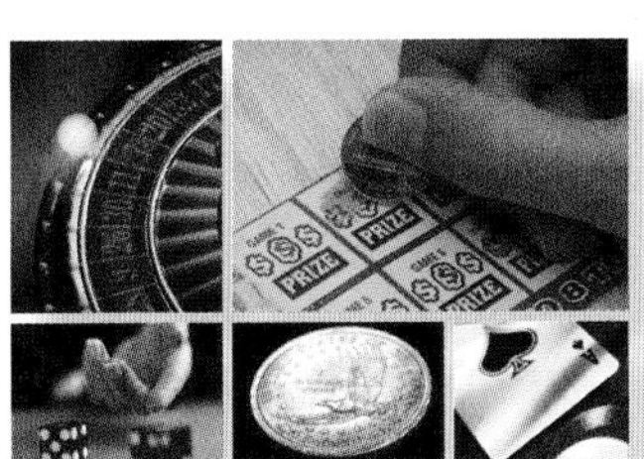

11-1 The Fundamental Counting Principle and Permutations 574
11-2 Combinations 582
11-3 Basic Concepts of Probability 587
11-4 Tree Diagrams, Tables, and Sample Spaces 597
11-5 Probability Using Permutations and Combinations 604
11-6 Odds and Expectation 609
11-7 The Addition Rules for Probability 616
11-8 The Multiplication Rules and Conditional Probability 623
11-9 The Binomial Distribution 632
Chapter 11 Summary 639

CHAPTER 12

Statistics 646

12-1 Gathering and Organizing Data 648
12-2 Picturing Data 657
12-3 Measures of Average 665
12-4 Measures of Variation 675
12-5 Measures of Position 681
12-6 The Normal Distribution 685
12-7 Applications of the Normal Distribution 693
12-8 Correlation and Regression Analysis 699
Chapter 12 Summary 710
Available Online at www.mhhe.com/sobecki:
Chapter 12 Supplement: Misuses of Statistics

CHAPTER 13

Other Mathematical Systems 716

13-1 Mathematical Systems and Groups 718
13-2 Clock Arithmetic 725
13-3 Modular Systems 732
Chapter 13 Summary 738

CHAPTER 14

Voting Methods 742

14-1 Preference Tables and the Plurality Method 744
14-2 The Borda Count Method and the Plurality-with-Elimination Method 751
14-3 The Pairwise Comparison Method and Approval Voting 759
14-4 Apportionment 767
14-5 Apportionment Flaws 778
Chapter 14 Summary 784

CHAPTER 15

Graph Theory 790

15-1 Basic Concepts of Graph Theory 792
15-2 Euler's Theorem 800
15-3 Hamilton Paths and Circuits 806
15-4 Trees 814
Chapter 15 Summary 822

Appendix A Area Under the Standard Normal Distribution A-1

Available Online at www.mhhe.com/sobecki

Appendix B Using the TI-84 Plus Graphing Calculator CA-1

Selected Answers (Student edition only) SA-1

Selected Answers (Instructor edition only) IA-1

Photo Credits PC-1

Index I-1

Index of Applications

Automotive

Car accidents, 656
Dealership capacity, 391
Depreciation, 269
Fuel efficiency, 674
Gas mileage, 30, 34
Gas prices, 680
Loans, 447
Purchase price, 307
Stopping distance, 340
Tire life, 698
Used car purchase, 327

Consumer Information

Advertising statements, 115
Airline tickets, 373
ATM banking, 637
Auction sales, 388
Bar codes, 178
Book prices, 373
Cab fare, 355
Cable TV, 208
Candy prices, 373
Car purchase, 307
Cell phone charges, 34
Cell phone equipment, 621
Cell phone minutes, 300
Cell phone plans, 327
Chicken prices, 373
Coffee prices, 372
Computer purchase, 630
Cost of fast food meal, 23
Cost of milk, 25
Credit card balance, 453, 456
Credit card finance charge, 453
Credit card penalty fees, 25
Credit card statement, 457
Credit cards, 424
Discount cards, 426
Discount price, 284
Down payment, 327, 446, 455, 456
Electric bill, 307
Estimating cost of cell phone, 18
Estimating remodeling costs, 19
Estimating total cost, 17, 24, 36
Exchange rate, 490
Existing home sales, 673
Fast food costs, 373
Food costs, 325
Gas prices, 680
Health club membership, 209
Home buying, 419
Home ownership, 457
Housing prices, 53
Identity theft, 52, 657
Installment buying, 446
Internet service, 208
Inventory levels, 393
Laundry detergent, 674
Lightbulb ratings, 674
Loan payoff, 451, 456
Misleading advertising, 426
Monthly payments, 431, 447, 455, 462
Movie tickets, 381
Newspaper advertisements, 358
Nut prices, 381
Online transactions, 638
Paint cost, 317
Party costs, 342
Pizza cost, 356
Pizza toppings, 81
Price reduction percent, 428
Retail sales, 147, 193
Sale prices, 23, 342
Sales tax, 284, 425, 428
School supplies, 381
Television screen size, 529
Thrift store prices, 373
Train fares, 342

Education

Bachelor's degrees, 52
Class scheduling, 80
Class size, 220
Costs, 428
Course selection, 64, 71, 77
Debt after graduation, 638
Degree completion, 388
Enrollment numbers, 318, 359
Extra credit, 36
Financial aid, 80
Grade point average, 36, 648
Grading structure, 392
Graduate school applications, 684
Groups for a project, 208
IQ testing, 685
Levels, 359
Major selection, 663
Majors, 80
Missed tests, 343
Political science class, 209
Quiz grades, 327
Reasons for failure, 80
Saving for college, 445
Scholarship application, 655
Student loans, 436
Study habits, 622
Teacher salaries, 428, 697
Test grades, 327, 328
Test percentiles, 684
Test times, 387
Textbook costs, 400
Textbook inventory, 381
Textbook resales, 81
Textbook sales, 697
True/false questions, 399
Tuition, 307

Finance

Account balances, 214
Actuarial method, 450
Advertising budget, 233
Amortization schedule, 462
Annual percentage rate, 448
Annuities, 441, 445
Average daily balance, 452, 454
Bank balance, 220
Banker's rule, 433
Bonds, 471
Business expansion, 445
Car loan, 447
Certificate of deposit, 272
Checking account balance, 218, 273
Coins, 29
Combined revenue, 307
Commission, 24, 425, 428
Comparing mortgages, 460
Comparison of effective rates, 440
Compound interest, 437
Computing principal, 432
Credit card balance, 453, 456
Credit card finance charge, 453
Credit card penalty fees, 25
Credit card statement, 457
Credit cards, 424
Debt, 269
Depreciation, 269
Discount, 284
Discounted loan, 434, 436
Down payment, 446, 455, 456
Earnings per share, 469, 472
Effective interest, 287, 439
Estate division, 234
Finance charge, 446, 453

Fixed installment loan, 446
Future value, 287, 430, 441
Grocery shopping, 206
History of money, 424
Home sale, 307, 425
Hourly income, 287
Inheritance, 308
Installment price, 446
Interest in investing, 307
Interest rate, 432
Investing, 270, 381
Investment value, 408
Loan payoff, 451, 456
Market caps, 685
Materials costs, 389
Monthly annuity payment, 444
Monthly interest, 437
Monthly payments, 431, 447, 455, 458, 462
Mortgage payments, 461, 464
Mortgages, 458
Mutual fund, 471
National debt, 197
Net profit, 274
Number of coins, 305
P/E ratio, 468
Packaging goods for sale, 204
Profit equation, 287
Profits, 260
Property value, 287
Quarterly interest, 437
Real estate commission, 399
Rent payments, 24, 307
Retirement, 445, 446
Rule of 72, 443
Rule of 78, 451
Salary increase, 268
Sales commission, 283
Sales tax, 284, 425, 428
Saving for a home, 267
Savings account, 269
Selling stock, 470, 473
Semiannual interest, 437
Simple interest, 287, 318, 429, 436
Simple interest in months, 430
Small business investment, 370
Starting costs, 399, 436
Stock dividends, 473
Stock listing, 467
Stock prices, 347, 472
Stock purchase, 344
Stock yield, 469
Term of loan, 433
Tips, 306, 399
Total cost with tax, 284
Total interest, 460
Unearned interest, 449
Unit cost, 371
Unpaid balance, 452
Yearly earnings, 24
Yearly interest, 437

General Interest

Animal shelter, 394
Balloon inflation, 359
Book price, 698
Border relationships, 793, 805
Breakfast, 318
Building heights, 709
Bungee jumping, 269
Cabinetmaking, 388
Calendar design, 584
Camera use, 89
Car dealership capacity, 391
Car rental, 327, 400
Carpeting room, 493
Ceiling fans, 724
Cell phone plans, 327
Cell phone tower, 269
CEO ages, 698
Chat room use, 88
Check digits, 736
City streets, 794
Civilian military staff, 358
Community garden, 393
Computer use at work, 359
Computer viruses, 359
Concert tickets, 381
Condominium conversion, 436
Converting between English and metric system, 488, 489
Cookie recipe, 234
Cookies weights, 694
Courier service, 818
Crime statistics, 673
Cryptography, 717
Deck building, 343
Domain name reservation, 359
Drive-through waiting time, 698
Drunk driving, 358
E-mails sent, 358
Egg packages, 737
Electrical wiring, 820
Electricity generation sources, 20
Elevator capacity, 505
Emergency response, 696
Encryption, 717
Fast food preference, 637
Females in Congress, 307
Fencing, 405
Fitness training, 232
Floor plans, 793
Food bank supplies, 220
Forensics, 124
Fundraising, 381, 393
Game show contestants, 373
Garbage produced, 695
Garden sculpture, 339
Global priorities, 680
Government spending, 197
Hair care, 394
Halloween candy purchase, 307
Hate crimes, 629
Home appraisal, 411
Home improvement, 304, 511
Home size, 698
Ice cream preference, 317
Inaugural addresses, 682
IQ testing, 685
Irrigation, 817
Jogging speed, 362
Ladder distance from wall, 555
Law practice, 388
Life insurance, 615
Lighthouses, 209
Literacy, 707
Lottery winnings, 359
Major scales, 7–S6
Manufacturing, 287
Measurement conversions, 182
Metric conversions, 486
Milk preference, 317
Modular homes, 388
Mountain height, 483
Movie credits, 162
Movie ticket cost, 697
MP3 players, 53
Municipal waste, 234
Murder statistics, 674
Music preference, 79
Musical notes, 7–S5
National Park acreage, 656
Paint mixing, 724
Paint needed, 343
Painter labor cost, 358
Painting bedroom, 498
Park construction, 821
Parking fines, 269
Patio construction, 498
Perimeter of triangle, 307
Persuasion, 91, 141
Pet food, 314
Pizza boxes, 339
Play tickets, 342
Population growth, 408
Postal codes, 182
Productivity and time of day, 24
Property taxes, 318
Pyramids, 524
Radio tower height, 558
Real estate value, 287
Reality TV, 655
Rug size, 340
Sand at Coney Island, 260
Shelving, 328
Snow melt, 317
Social networking, 698
Sociology, 647
Soda consumption, 273
Soda packages, 737
Soft drink preferences, 76
Storage, 388
Taxicabs, 663

Television preferences, 81
Television set life cycles, 697
Television watching, 81
Textiles, 388
Theater capacity, 394
Tornados, 327
Toymaking, 393
Train tickets, 372
Trombone playing, 263
Twelve tone scale, 7–S5
Vacation planning, 324
Website use, 80
Website visitors, 25
Wills, 343
Woodworking, 337

Geometry

Angle measure, 520
Angle of depression, 557
Area enclosed by track, 541
Base and height of triangle, 339
Baseball diamond size, 529
Baseboard size, 535
Carpet area, 537, 543
Cell phone tower height, 530
Coordinate system, 287
Distance around track, 540
Fence size, 535
Fight cage size, 536
Hedge size, 535
Height from shadow, 527
Height of tower, 530
Length of beam, 530
Object height with trigonometry, 556
Perimeter, 28
Picture frame molding length, 535
Pythagorean theorem, 340, 524
Rectangle dimensions, 339
Rectangle perimeter, 300
Volume of cylinder, 300
Volume of rectangular solid, 300
Volume of sphere, 287, 548
Volume of swimming pool, 545

Health and Nutrition

Alcohol consumption, 52
Alcohol-related traffic deaths, 358
Bacteria, 669
Birth weight, 504
Blood pressure, 694
Blood types, 575
Body composition, 247
Body mass index, 247, 251, 318
Body piercings, 76
Body temperature, 504
Calories burned in exercise, 30, 343, 400
Calories burned in walking, 698
Cells in body, 259
Diet, 328, 359
Disease elimination, 359
Disease spread, 622
Drug administration, 277
Fat in food, 490
Fitness training, 232
Flu breakouts, 680
Gender and exercise, 318
Hearing range, 7–S3
Insurance coverage, 233, 274
Obesity, 247
Occupational injuries, 20
Prescription drug prices, 53
Protein in diet, 388
Red blood cells, 260
Soda consumption, 273
Tattoos, 76
Transplants, 663
Weight loss, 501, 504

Labor

Average hours worked, 39
Career change, 25
Computer use at work, 359
Hourly income, 287, 697
Layoffs, 269, 424
Litigation costs, 446
Manufacturing, 287
Multiple jobs, 372
Overtime, 34
Productivity and time of day, 24
Profit, 340
Raises, 267
Salary, 29
Salary plans, 328
Starting salaries, 270
Teacher salaries, 428
Tips, 306
Training costs, 436
Unemployment, 234, 657
Weekly salary, 37
Workers by age group, 51

Numbers

Binary coding, 182
Data modeling, 404
Morse code, 182

Probability

Animal adoptions, 621
Band competition, 603
Betting, 612
Birthday problem, 607
Blackjack, 595
Book selection, 605
Card games, 595
Cards dealt, 586
Code words, 581
Coin toss, 588
Combination lock, 606
Committee selection, 608
Craps, 604
Dating, 595
Deck of cards, 604
Die roll, 589, 594
Draft lottery, 599
Exam questions, 586
Finalist selection, 587
Fortune teller, 602
Game shows, 594
Gender of children, 594
ID card digits, 581
Insurance, 615
Job applicants, 608
Letters to form word, 581
Lottery tickets, 595, 615
Names at party, 596
Names on list, 595
Opinion survey, 596
Pieces of pie, 617
Poker, 595
Poll results, 596
Presents under tree, 586
Raffle, 581
Scratch-off tickets, 615
Slot machines, 613
Songs played, 581
Struck by lightning, 580
Success of TV shows, 581
Teams in playoffs, 581
Test design, 586
Video game play, 603

Science and Nature

Acceleration, 300
Amperage, 343
Astronomy, 172, 258
Atomic size, 260
Average temperature, 504
Bacteria, 669
Calculators, 163
Carbon dating, 409
Cells in body, 259
Centripetal force, 300
Cicada cycles, 209
Circuits, 102
Comet orbit, 541
Computer design, 102
Computer simulation, 731
Computer storage, 180
Criminal investigation, 3, 35
Current in circuit, 296
Distance from Sun, 266, 489
Distances of stars, 260
Dropped object speed, 243
Dropped object time, 243, 339
Electrical resistance, 300
Frequency, 7–S3
Harmonics, 7–S3

Heat energy, 287
Height of thrown object, 287
High and low temperatures, 220
Highest and lowest points, 220
Irrigation, 104
Kinetic energy, 287
Light illumination, 300
Mass-energy equivalence, 287
Momentum, 300
Periodic cicadas, 209
Planetary distances, 260
Plants in greenhouse, 657
Pollutant releases, 36
Pressure of gas, 318
Psychology experiment, 625
Rat breeding, 220
Reaction times in dogs, 664
Recycling, 80
Search for intelligent life, 172
Sound wave analysis, 7–S2
Speed of light, 259
Speed of sound, 273
Subatomic particles, 259
Supercomputers, 380
Support beam, 316
Temperature conversion, 296
Tornados, 327
Vibrations per second, 344
Volcano height, 656
Voltage and power, 243
Weather prediction, 8
Weight of liquid, 494
Weight on moon, 485
Wildlife population, 313

Sports, Leisure, and Hobbies

Backgammon, 631
Baseball teams in playoffs, 309
Board games, 344
Boat race, 684
Bowling league, 209
Bungee jumping, 269
College football championship, 260
Concert tickets, 373
Contract negotiations, 303
Distance of ball in baseball, 529
Distance ran in race, 489
Earned run average, 300
Football player weights, 680
Franchise values, 656
Free agent players, 260
Gambling by state, 74
Gambling debts, 269
Gambling odds, 573
Game board spinner, 737
Gardening, 27
Golf course renovation, 437
Golf Masters tournament, 671
Golf scores, 148
Home run records, 656
Hot dog eating contest, 428
Jet ski rentals, 411
Karate, 318
Letter to fans, 114
Lottery tickets, 307
Oddsmaking, 4
Olympics attendance, 381
Photography, 543
Playing cards, 64
Playoff statistics, 273
Poker tournament, 400
Quiz shows, 269
Recreational sports, 309
Runners completing race, 423
Ski resorts, 614
Slugging percentage, 300
Sprinting in race, 489
Steel cage matches, 536
Tae Kwon Do, 621
Track distance, 540
Triathlon, 307, 529
Turf for football field, 543
Walkathon, 536
Win rate of baseball team, 209

Statistics and Demographics

Age and population, 358
Data modeling, 404
Educational achievement, 359
Health insurance coverage, 233
Number of immigrants, 51
Population distribution, 234
Population growth, 220
Poverty, 233
Racial demographics, 312
Self-reporting of race, 41
Surveys, 77
Unemployment, 234
Violent crime data by state, 88

Travel

Air, 317, 558
Air fares between cities, 813
Airline routes, 388
Airport passenger totals, 685
Bicycle speed, 208
Bus, 339
Cities visited, 584
Construction on roads, 373
Distance based on speed, 23, 24
Distance formula, 285
Distances between cities, 813
Driving time, 315
Estimating distance using map, 22
Foreign, 481
Fuel consumption, 312
Gas costs, 428
Map scale, 234
Miles driven, 233
Shuttle buses, 209
Speed units, 484
Subway, 339
Tour boats, 410
Traffic, 342, 489
Vacation package, 428

Voting

Age groups, 234
Alabama paradox in apportionment, 783
Campaign survey, 622
Conference location, 757
Department chair, 765
Exam start time, 758
Faculty promotions, 776
Field trip destination, 757
Gaming club, 757
Hamilton's method of apportionment, 783
Homecoming queen, 752
Income brackets, 234
New state paradox in apportionment, 783
Number of voters, 749
Party affiliation, 234
Plurality method, 746
Preference tables, 745
Presidential election, 234
Racial identity, 234
Rankings, 750
Representative elections, 776
Senior class trip, 765
Speaker selection, 758
Votes received, 308

CHAPTER 1

Problem Solving

Outline

1-1 The Nature of Mathematical Reasoning
1-2 Estimation and Interpreting Graphs
1-3 Problem Solving
Summary

MATH IN Criminal Investigation

In traditional cops-and-robbers movies, crime fighters use guns and fists to catch criminals, but in real life, often it's brain power that brings the bad guys to justice. That's why the TV show *CSI* marked a revolution of sorts when it debuted in October 2000: it featured scientists fighting crime, not tough guys. Solving a case is intimately tied to the process of problem solving: investigators are presented with a mystery, and they use whatever evidence they can find and their reasoning ability to reconstruct the crime. Hopefully, this will lead them to a suspect.

Of course, this isn't limited to just crime fighting. Students in math classes often ask, "When am I going to use what I learn?" The best answer to that question is, "Every day!" Math classes are not only about facts and formulas: they're also about exercising your mind, training your brain to think logically, and learning effective strategies for solving problems. And not just math problems. Every day of our lives, we face a wide variety of problems: they pop up in our jobs, in school, and in our personal lives. Which computer should you buy? What should you do when your car starts making an awful noise? What would be a good topic for a research paper? How can you get all your work done in time to go to that party Friday night?

Chapter 1 of this book is dedicated to the most important topic we'll cover: an introduction to some of the classic techniques of problem solving. These techniques will prove to be useful tools that you can apply in the rest of your education. But more importantly, they can be applied just as well to situations outside the classroom.

And this brings us back to our friends from *CSI*. The logic and reasoning that they use to identify suspects and prove their guilt are largely based on problem-solving skills we'll study in this chapter. By the time you've finished the chapter, you should be able to evaluate the situations below, all based on episodes of *CSI*. In each case, you should identify the type of reasoning that was used and decide whether the conclusion would stand up as proof in a court of law.

- After a violent crime, the investigators identify a recently paroled suspect living in the area who had previously committed three very similar crimes.
- A homeless man is found dead from exposure after being roughed up. His wrists look like he had been handcuffed, and fingerprints on his ID lead them to a local police officer.
- A murder victim grabbed a pager from the killer while being attacked and threw it under the couch. With the suspect identified, the investigators found that his DNA matched DNA found under the victim's fingernails.
- A series of five bodies are found posed like mannequins in public places. The lead suspect is an artist that is found to have sketches matching the poses of all five victims.

For answers, see Math in Criminal Investigation Revisited on page 35

Section 1-1 The Nature of Mathematical Reasoning

LEARNING OBJECTIVES

- ☐ 1. Identify two types of reasoning.
- ☐ 2. Use inductive reasoning to form conjectures.
- ☐ 3. Find a counterexample to disprove a conjecture.
- ☐ 4. Explain the difference between inductive and deductive reasoning.
- ☐ 5. Use deductive reasoning to prove a conjecture.

Part of being an adult is making decisions—every day is full of them, from the simple, like what to eat for breakfast, to the critical, like a choice of major. If you make every decision based on a coin flip, chances are you won't get very far in life. Instead, it's important to be able to analyze a situation and make a decision based on logical thinking. We'll call the process of logical thinking **reasoning.** It doesn't take a lot of imagination to understand how important reasoning is in everyone's life.

You may not realize it, but every day in your life, you use two types of reasoning to make decisions and solve problems: *inductive reasoning* or *induction,* and *deductive reasoning* or *deduction.*

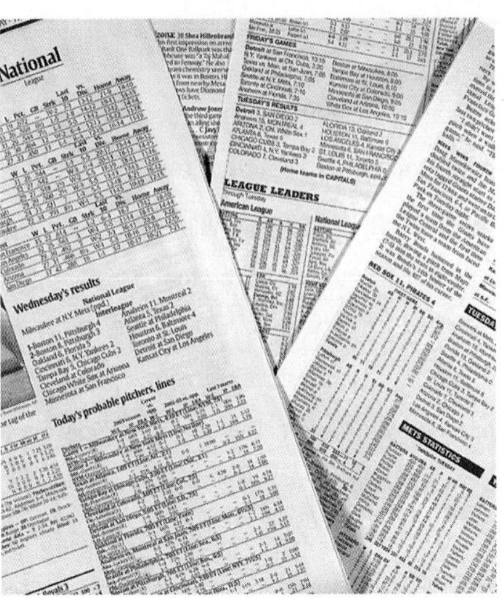

Oddsmakers, who set point spreads for sporting events, predict how teams and individual players will do based on their past performance. This is an example of inductive reasoning.

Inductive reasoning is the process of reasoning that arrives at a general conclusion based on the observation of specific examples.

For example, suppose that your instructor gives a surprise quiz every Friday for the first four weeks of your math class. At this point, you might make a **conjecture,** or educated guess, that you'll have a surprise quiz the next Friday as well. As a result, you'd probably study before that class.

☑ 1. Identify two types of reasoning.

This is an example of inductive reasoning. By observing certain events for four *specific* Fridays, you arrive at a general conclusion. Inductive reasoning is useful in everyday life, and it is also useful as a problem-solving tool in math, as shown in Example 1.

EXAMPLE 1 Using Inductive Reasoning to Find a Pattern

Use inductive reasoning to find a pattern, and then find the next three numbers by using that pattern.

1, 2, 4, 5, 7, 8, 10, 11, 13, __, __, __

SOLUTION

To find the pattern, look at the first number and see how to obtain the second number. Then look at the second number and see how to obtain the third number, etc.

1 2 4 5 7 8 10 11 13 __ __ __

+1 +2 +1 +2 +1 +2 +1 +2 +1 +2 +1

The pattern seems to be to add 1, then add 2, then add 1, then add 2, etc. So a reasonable conjecture for the next three numbers is 14, 16, and 17.

▼ Try This One 1

Use inductive reasoning to find a pattern and make a reasonable conjecture for the next three numbers by using that pattern.

1, 4, 2, 5, 3, 6, 4, 7, 5, __, __, __

EXAMPLE 2 Using Inductive Reasoning to Find a Pattern

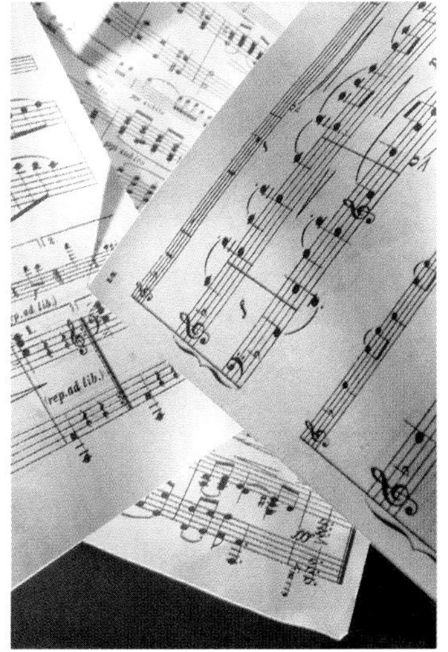

Recognizing, describing, and creating patterns are important in many fields. Many types of patterns are used in music such as following an established pattern, altering an established pattern, and producing variations on a familiar pattern.

Make a reasonable conjecture for the next figure in the sequence.

 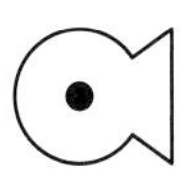  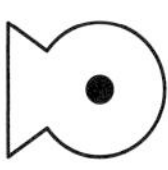  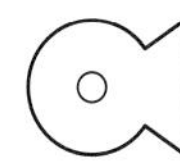  ______

SOLUTION

The flat part of the figure is up, right, down, and then left. There is a solid circle ● in each figure. The sequence then repeats with an open circle ○ in each figure. So we could reasonably expect the next figure to be .

▼ Try This One 2

Make a reasonable conjecture for the next figure in the sequence.

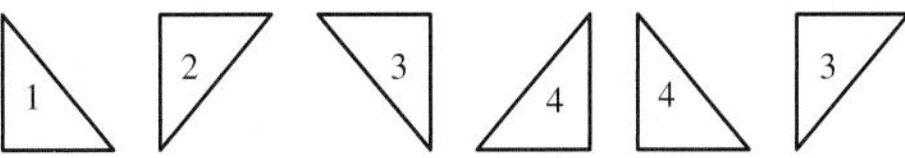

EXAMPLE 3 Using Inductive Reasoning to Make a Conjecture

When two odd numbers are added, will the result always be an even number? Use inductive reasoning to determine your answer.

SOLUTION

We will try several specific examples:

$$3 + 7 = 10 \qquad 25 + 5 = 30$$
$$5 + 9 = 14 \qquad 1 + 27 = 28$$
$$19 + 9 = 28 \qquad 21 + 33 = 54$$

Since all the answers are even, it seems reasonable to conclude that the sum of two odd numbers will be an even number.

Note: Since the sum of *every* pair of odd numbers hasn't been tried, we can't be 100% sure that the answer will always be an even number by using inductive reasoning.

2. Use inductive reasoning to form conjectures.

▼ Try This One 3

If two odd numbers are multiplied, is the result always odd, always even, or sometimes odd and sometimes even? Use inductive reasoning to answer.

One number is *divisible* by another number if the remainder is zero after dividing. For example, 16 is divisible by 8 because $16 \div 8$ has remainder zero. And 17 is not divisible by 8 because $17 \div 8$ has remainder 1.

EXAMPLE 4 Using Inductive Reasoning to Test a Conjecture

Use inductive reasoning to decide if the following conjecture is true: Any four-digit number is divisible by 11 if the difference between the sum of the first and third digits and the sum of the second and fourth digits is divisible by 11.

SOLUTION

Let's make up a few examples. For 1,738, the sum of the first and third digits is $1 + 3 = 4$, and the sum of the second and fourth digits is $7 + 8 = 15$. The difference is $15 - 4 = 11$, so if the conjecture is true, 1,738 should be divisible by 11. To check: $1{,}738 \div 11 = 158$ (with no remainder).

For 9,273, $9 + 7 = 16$, $2 + 3 = 5$, and $16 - 5 = 11$. So if the conjecture is true, 9,273 should be divisible by 11. To check: $9{,}273 \div 11 = 843$ (with no remainder).

Let's look at one more example. For 7,161, $7 + 6 = 13$, $1 + 1 = 2$, and $13 - 2 + 11$. Also $7{,}161 \div 11 = 651$ (with no remainder), so the conjecture is true for this example as well. While we can't be positive based on three examples, inductive reasoning indicates that the conjecture is true.

▼ Try This One 4

Use inductive reasoning to decide if the following conjecture is true: If the sum of the digits of a number is divisible by 3, then the number itself is divisible by 3.

Inductive reasoning can definitely be a useful tool in decision making, and we use it very often in our lives. But it has one very obvious drawback: because you can very seldom verify conclusions for every possible case, you can't be positive that the conclusions you're drawing are correct. In the example of the class in which a quiz is given on four consecutive Fridays, even if that continues for 10 more weeks, there's still a chance that there won't be a quiz the following Friday. And if there's even one Friday on which a quiz is not given, then the conjecture that there will be a quiz every Friday proves to be false.

This is a useful observation: while it's not often easy to prove that a conjecture is true, it's much simpler to prove that one is false. All you need is to find one specific example that contradicts the conjecture. This is known as a **counterexample.** In the quiz example, one Friday without a quiz serves as a counterexample: it proves that your conjecture that there would be a quiz every Friday is false. In Example 5, we'll use this idea to prove that a conjecture is false.

EXAMPLE 5 Finding a Counterexample

Find a counterexample that proves the conjecture below is false.

Conjecture: A number is divisible by 3 if the last two digits are divisible by 3.

SOLUTION

We'll pick a few numbers at random whose last two digits are divisible by 3, then divide them by 3, and see if there's a remainder.

1,527: Last two digits, 27, divisible by 3; $1{,}527 \div 3 = 509$
11,745: Last two digits, 45, divisible by 3; $11{,}427 \div 3 = 3{,}809$

At this point, you might start to suspect that the conjecture is true, but you shouldn't! We've only checked two cases.

1,136: Last two digits, 36, divisible by 3; $1{,}136 \div 3 = 378\frac{2}{3}$

This counterexample shows that the conjecture is false.

▼ Try This One 5

☑ 3. Find a counterexample to disprove a conjecture.

Find a counterexample to disprove the conjecture that the name of every month in English contains either the letter y or the letter r.

CAUTION

Remember: One counterexample is enough to show that a conjecture is false. But one positive example is *never* enough to show that a conjecture is true.

EXAMPLE 6 Making and Testing a Conjecture

Use inductive reasoning to make a conjecture about the number of sections a circle is divided into when a given number of points on the circle are connected by chords. (A chord is a line connecting two points on a circle.) Then test the conjecture with one further example.

SOLUTION

We'll draw several circles, connect the points with chords, and then count the sections.

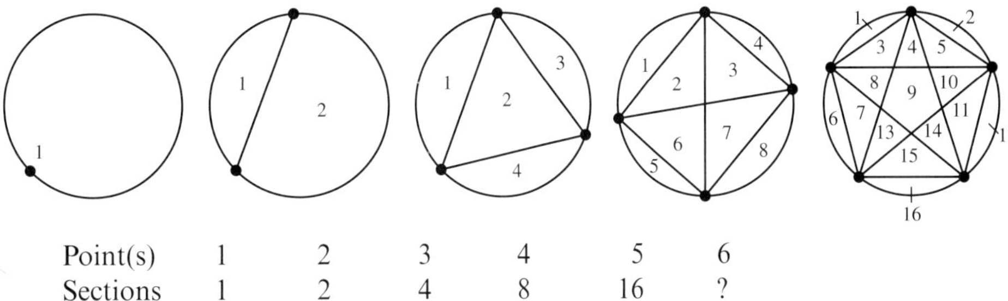

Point(s)	1	2	3	4	5	6
Sections	1	2	4	8	16	?

Looking at the pattern in the number of sections, we see that a logical guess for the next number is 32. In fact, the number appears to be 2 raised to the power of 1 less than the number of points. This will be our conjecture. Let's see how we did by checking with six points.

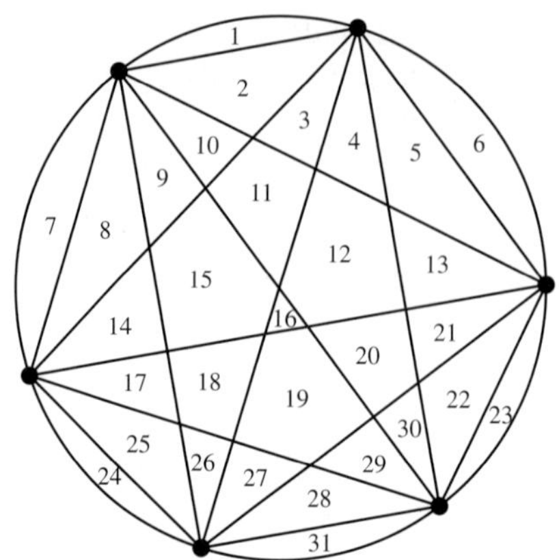

Uh oh . . . there are 31 sections! It looks like our conjecture is not true.

▼ Try This One 6

Given that there are 31 sections with 6 points, guess how many there will be with 7, and then check your answer.

The other method of reasoning that we will study is called *deductive reasoning,* or *deduction.*

Deductive reasoning is the process of reasoning that arrives at a conclusion based on previously accepted general statements. It does not rely on specific examples.

Here's an example of deductive reasoning. At many colleges, a student has to be registered for at least 12 hours to be considered full-time. So we accept the statement "A student is full-time if he or she is registered for at least 12 hours" as true. If we're then told that a particular student has full-time status, we can conclude that she or he is registered for at least 12 hours. The key is that we can be *positive* that

Sidelight MATH AND WEATHER PREDICTION

Weather involves numerous variables, such as temperatures, wind velocity and direction, type and amount of precipitation, cloud cover, and atmospheric pressures, to name just a few. All these variables use numbers.

To create a weather forecast, the National Weather Service collects information from various sources such as satellites, weather balloons, weather stations, and human observers around the world. The information is then put into a computer and a grid with over 200,000 points is drawn up by the computer showing the weather conditions at each point at the time the data were collected.

From this grid, the computer is able to predict the movement of weather fronts at 10-minute intervals into the future, and a forecast of tomorrow's weather can be generated.

Short-term forecasts are generally accurate, but due to variations in atmospheric conditions, long-range forecasts are at best guesses based on current conditions. Sometimes, however, atmospheric conditions change so rapidly that even short-term forecasts are inaccurate.

☑ 4. Explain the difference between inductive and deductive reasoning.

this is true. This is what separates deductive reasoning from inductive reasoning. Example 7 illustrates the difference between using inductive and deductive reasoning in math.

EXAMPLE 7 Using Deductive Reasoning to Prove a Conjecture

Consider the following problem: Think of any number. Multiply that number by 2, then add 6, and divide the result by 2. Next subtract the original number. What is the result?

(a) Use inductive reasoning to make a conjecture for the answer.
(b) Use deductive reasoning to prove your conjecture.

SOLUTION

(a) Inductive reasoning will be helpful in forming a conjecture. We'll choose a few specific numbers at random and perform the given operations to see what the result is.

Number:	12	5	43
Multiply by 2:	$2 \times 12 = 24$	$2 \times 5 = 10$	$2 \times 43 = 86$
Add 6:	$24 + 6 = 30$	$10 + 6 = 16$	$86 + 6 = 92$
Divide by 2:	$30 \div 2 = 15$	$16 \div 2 = 8$	$92 \div 2 = 46$
Subtract the original number:	$15 - 12 = 3$	$8 - 5 = 3$	$46 - 43 = 3$
Result:	3	3	3

At this point, you may be tempted to conclude that the result is always 3. But this is just a conjecture: we've tried only three of infinitely many possible numbers! As usual when using inductive reasoning, we can't be completely sure that our conjecture is always true.

(b) The problem with the inductive approach is that it requires using specific numbers, and we know that we can't check every possible number. Instead, we'll choose an *arbitrary* number and call it x. If we can show that the result is 3 in this case, that will tell us that this is the result for *every* number. Remember, we'll be doing the exact same operations, just on an arbitrary number x.

Number:	x
Multiply by 2:	$2x$
Add 6:	$2x + 6$
Divide by 2:	$\frac{2x + 6}{2} = x + 3$
Subtract the original number:	$x + 3 - x = 3$

Now we know for sure that the result will always be 3, and our conjecture is proved.

▼ Try This One 7

☑ 5. Use deductive reasoning to prove a conjecture.

Consider the following problem: Think of any number. Multiply that number by 3, then add 30, and divide the result by 3. Next subtract the original number. What is the result?

(a) Use inductive reasoning to make a conjecture for the answer.
(b) Use deductive reasoning to prove your conjecture.

Sidelight AN ARBITRARY DISCUSSION

In common usage, the word *arbitrary* is often misinterpreted as a synonym for *random*. When reaching into a bag of potato chips, you make a random selection, and some people would also call this an arbitrary selection. But in math, the word *arbitrary* means something very different. When randomly selecting that chip, you have still chosen a specific chip—it is probably not representative of every chip in the bag. Some chips will be bigger, and others smaller. Still others may have more salt, some less.

When we use *arbitrary* in math, we're referring to a nonspecific item that is able to represent *all* such items. In the series of calculations we looked at above, we could never be sure that the result will always be 3 by choosing specific numbers. Why? Because we'd have to try it for *every* number, which is of course impossible. You have better things to do than spend the rest of your life testing number after number. The value of performing the calculation on an *arbitrary* number x is that this one calculation proves what will happen for *every* number you choose. It is absolutely crucial in the study of mathematics to understand that choosing specific numbers can almost never *prove* a result, because you can't try every number. Instead, we'll rely on using arbitrary numbers and deductive reasoning.

Let's try another example. Try to focus on the difference between inductive and deductive reasoning, and the fact that inductive reasoning is great for giving you an idea about what the truth might be for a given situation, but deductive reasoning is needed for proof.

EXAMPLE 8 Using Deductive Reasoning to Prove a Conjecture

Use inductive reasoning to arrive at a general conclusion, and then prove your conclusion is true by using deductive reasoning.

Select a number:
Add 50:
Multiply by 2:
Subtract the original number:
Result:

SOLUTION

Approach: Induction

Try a couple different numbers and make a conjecture.

Original number:	12	50
Add 50:	$12 + 50 = 62$	$50 + 50 = 100$
Multiply by 2:	$62 \times 2 = 124$	$100 \times 2 = 200$
Subtract the original number:	$124 - 12 = 112$	$200 - 50 = 150$
Result:	112	150

The conjecture is that the final answer is 100 more than the original number.

Approach:	Deduction
Select a number:	x
Add 50:	$x + 50$
Multiply by 2:	$2(x + 50) = 2x + 100$
Subtract the original number:	$2x + 100 - x$
Result:	$x + 100$

Our conjecture was right: the final answer is always 100 more than the original number.

Calculator Guide

If you use a calculator to try the repeated operations in Example 8, you'll need to press [=] (scientific calculator) or [ENTER] (graphing calculator) after every operation, or you'll get the wrong result. Suppose you simply enter the whole string:

Standard Scientific Calculator

12 [+] 50 [×] 2 [−] 12 [=]

Standard Graphing Calculator

12 [+] 50 [×] 2 [−] 12 [ENTER]

The result is 100, which is incorrect. In Chapter 5, we'll find out why when we study the order of operations.

▼ Try This One 8

Arrive at a conclusion by using inductive reasoning, and then try to prove your conclusion by using deductive reasoning.

Select a number:

Add 16:

Multiply by 3:

Add 2:

Subtract twice the original number:

Subtract 50:

Result:

Now that we've seen how inductive and deductive reasoning can be used, let's review by distinguishing between the two types of reasoning in some real-world situations.

EXAMPLE 9 Comparing Inductive and Deductive Reasoning

Determine whether the type of reasoning used is inductive or deductive.

The last six times we played our archrival in football, we won, so I know we're going to win on Saturday.

SOLUTION

This conclusion is based on six specific occurrences, not a general rule that we know to be true. Therefore, inductive reasoning was used.

▼ Try This One 9

Determine whether the type of reasoning used is inductive or deductive.

There is no mail delivery on holidays. Tomorrow is Thanksgiving, so I know my student loan check won't be delivered.

EXAMPLE 10 Comparing Inductive and Deductive Reasoning

Determine whether the type of reasoning used is inductive or deductive.

The syllabus states that any final average between 80 and 90% will result in a B. If I get 78% on my final, my overall average will be 80.1%, so I'll get a B.

SOLUTION

Although we're talking about a specific person's grade, the conclusion that I'll get a B is based on a general rule: all scores in the 80s earn a B. So this is deductive reasoning.

▼ Try This One 10

Determine whether the type of reasoning used is inductive or deductive.

Everyone I know in my sorority got a least a 2.5 GPA last semester, so I'm sure I'll get at least a 2.5 this semester.

Remember that both inductive reasoning and deductive reasoning are useful tools for problem solving. But the biggest difference between them is that conclusions drawn from inductive reasoning, no matter how reasonable, are still at least somewhat uncertain. But conclusions drawn by using deductive reasoning can be considered definitely true, as long as the general rules used to draw the conclusion are known to be true.

In addition, it's worth reviewing the fact that to disprove a conjecture, you only need to find one specific example for which it's not true. But to prove a conjecture, you have to show that it's true in *every* possible case.

Answers to Try This One

1 Pattern: Every entry is 1 more than the one that comes two spots before it. The next three numbers are 8, 6, 9.

2 2

3 Always odd

4 True

5 June has neither a y nor an r.

6 There are 57.

7 10

8 The result is the original number.

9 Deductive

10 Inductive

EXERCISE SET 1-1

Writing Exercises

1. Explain the difference between inductive and deductive reasoning.
2. What is meant by the term *conjecture?*
3. Give a real-life example of how to use inductive reasoning.
4. What is a counterexample? What are counterexamples used for?
5. Explain why you can never be sure that a conclusion you arrived at using inductive reasoning is true.
6. Explain the difference between an arbitrary number and a number selected at random.

Computational Exercises

For Exercises 7–14, use inductive reasoning to find a pattern, and then make a reasonable conjecture for the next number or item in the sequence.

7. 1 2 4 7 11 16 22 29 __
8. 5 15 45 135 405 1215 __
9. 10 20 11 18 12 16 13 14 14 12 15 __
10. 1 4 9 16 19 24 31 34 39 46 __
11. 100 99 97 94 90 85 79 __
12. 9 12 11 14 13 16 15 18 __
13.  ____
14. ____

For Exercises 15–18, find a counterexample to show that each statement is false.

15. The sum of any three odd numbers is even.
16. When an even number is added to the product of two odd numbers, the result will be even.
17. When an odd number is squared and divided by 2, the result will be a whole number.
18. When any number is multiplied by 6 and the digits of the answer are added, the sum will be divisible by 6.

For Exercises 19–22, use inductive reasoning to make a conjecture about a rule that relates the number you selected to the final answer. Try to prove your conjecture by using deductive reasoning.

19. Select a number:
    Double it:
    Subtract 20 from the answer:
    Divide by 2:
    Subtract the original number:
    Result:
20. Select a number:
    Multiply it by 9:
    Add 21:
    Divide by 3:
    Subtract three times the original number:
    Result:
21. Select a number:
    Add 50:
    Multiply by 2:
    Subtract 60:
    Divide by 2:
    Subtract the original number:
    Result:
22. Select a number:
    Multiply by 10:
    Subtract 25:
    Divide by 5:
    Subtract the original number:
    Result:

For Exercises 23–32, use inductive reasoning to find a pattern for the answers. Then use the pattern to guess the result of the final calculation, and perform the operation to see if your answer is correct.

23. $12{,}345{,}679 \times 9 = 111{,}111{,}111$
    $12{,}345{,}679 \times 18 = 222{,}222{,}222$
    $12{,}345{,}679 \times 27 = 333{,}333{,}333$
    $\vdots$
    $12{,}345{,}679 \times 72 = ?$
24. $0^2 + 1 = 1$
    $1^2 + 3 = 2^2$
    $2^2 + 5 = 3^2$
    $3^2 + 7 = 4^2$
    $4^2 + 9 = 5^2$
    $5^2 + 11 = ?$
25. $999{,}999 \times 1 = 0{,}999{,}999$
    $999{,}999 \times 2 = 1{,}999{,}998$
    $999{,}999 \times 3 = 2{,}999{,}997$
    $\vdots$
    $999{,}999 \times 9 = ?$
26. $1 = 1^2$
    $1 + 2 + 1 = 2^2$
    $1 + 2 + 3 + 2 + 1 = 3^2$
    $\vdots$
    $1 + 2 + 3 + 4 + 5 + 6 + 7 + 6 + 5 + 4 + 3 + 2 + 1 = ?$
27. $9 \times 9 = 81$
    $99 \times 99 = 9{,}801$
    $999 \times 999 = 998{,}001$
    $9{,}999 \times 9{,}999 = 99{,}980{,}001$
    $99{,}999 \times 99{,}999 = ?$
28. $1 \times 8 + 1 = 9$
    $12 \times 8 + 2 = 98$
    $123 \times 8 + 3 = 987$
    $1{,}234 \times 8 + 4 = 9{,}876$
    $12{,}345 \times 8 + 5 = ?$
29. $1 \cdot 1 = 1$
    $11 \cdot 11 = 121$
    $111 \cdot 111 = 12{,}321$
    $1{,}111 \cdot 1{,}111 = 1{,}234{,}321$
    $11{,}111 \cdot 11{,}111 = ?$
30. $9 \cdot 91 = 819$
    $8 \cdot 91 = 728$
    $7 \cdot 91 = 637$
    $6 \cdot 91 = 546$
    $5 \cdot 91 = ?$
31. Explain what happens when the number 142,857 is multiplied by the numbers 2 through 8.
32. A Greek mathematician named Pythagoras is said to have discovered the following number pattern. Find the next three sums by using inductive reasoning. Don't just add!
    $1 = 1$
    $1 + 3 = 4$
    $1 + 3 + 5 = 9$
    $1 + 3 + 5 + 7 = 16$
    $1 + 3 + 5 + 7 + 9 = ?$
    $1 + 3 + 5 + 7 + 9 + 11 = ?$
    $1 + 3 + 5 + 7 + 9 + 11 + 13 = ?$
33. Select any two-digit number, such as 62. Multiply it by 9: $62 \times 9 = 558$. Then add the digits: $5 + 5 + 8 = 18$. Keep adding the digits in the answer until you get a single answer. In this case, the answer is $1 + 8 = 9$. Try several other numbers. Using inductive reasoning, what can you conjecture about any whole number multiplied by 9?
34. Select any three-digit number in which all digits are different, such as 571. Reverse the digits and subtract the smaller number from the larger number: $571 - 175 = 396$. Reverse the digits in that number and add the new number to the old number:

396 + 693 = 1,089. (*Note:* If an answer has two digits, place a 0 in the hundreds place.) Now select several other three-digit numbers and try the procedure again. By using inductive reasoning, what can you conjecture?

35. Use inductive reasoning to make a conjecture about the next three sums, and then perform the calculations to verify that your conjecture is true.

$1 + \frac{1}{2} = \frac{3}{2}$

$1 + \frac{1}{2} + \frac{1}{2 \cdot 3} = \frac{5}{3}$

$1 + \frac{1}{2} + \frac{1}{2 \cdot 3} + \frac{1}{3 \cdot 4} = \frac{7}{4}$

$1 + \frac{1}{2} + \frac{1}{2 \cdot 3} + \frac{1}{3 \cdot 4} + \frac{1}{4 \cdot 5} = ?$

$1 + \frac{1}{2} + \frac{1}{2 \cdot 3} + \frac{1}{3 \cdot 4} + \frac{1}{4 \cdot 5} + \frac{1}{5 \cdot 6} = ?$

$1 + \frac{1}{2} + \frac{1}{2 \cdot 3} + \frac{1}{3 \cdot 4} + \frac{1}{4 \cdot 5} + \frac{1}{5 \cdot 6} + \frac{1}{6 \cdot 7} = ?$

36. Use inductive reasoning to determine the unknown sum, then perform the calculation to verify your answer.

$2 = 1(2)$

$2 + 4 = 2(3)$

$2 + 4 + 6 = 3(4)$

$2 + 4 + 6 + 8 = 4(5)$

$2 + 4 + 6 + 8 + 10 + 12 + 14 = ?$

In Exercises 37–40, use inductive reasoning to predict the next equation in the pattern.

37. $5 = \frac{5(2)}{2}$

$5 + 10 = \frac{10(3)}{2}$

$5 + 10 + 15 = \frac{15(4)}{2}$

$5 + 10 + 15 + 20 = \frac{20(5)}{2}$

38. $50{,}505 \times 2 = 101{,}010$
$50{,}505 \times 4 = 202{,}020$
$50{,}505 \times 6 = 303{,}030$
$50{,}505 \times 8 = 404{,}040$

39. $37{,}037 \times 3 = 111{,}111$
$37{,}037 \times 6 = 222{,}222$
$37{,}037 \times 9 = 333{,}333$
$37{,}037 \times 12 = 444{,}444$

40. $4 \cdot 1^3 = 1^2 \cdot (2)^2$
$4 \cdot (1^3 + 2^3) = 2^2 \cdot (3)^2$
$4 \cdot (1^3 + 2^3 + 3^3) = 3^2 \cdot (4)^2$

Real-World Applications

In Exercises 41–64, determine whether the type of reasoning used is inductive or deductive reasoning.

41. I am going to be rich some day. I know this because everyone in my family who graduated from college is rich, and I just graduated from college.
42. If a computer tech says it will cost \$100 to clear the viruses from my PC, it will actually cost \$150. A computer tech told me it would cost \$100 to clean up my PC; therefore I can expect it to cost me \$150.
43. I know I will have to work a double shift today because I have a migraine and every time I have a migraine I get stuck pulling a double.
44. If class is canceled, I go to the beach with my friends. I did not go to the beach with my friends yesterday; therefore, class was not canceled.
45. On Christmas Day, movie theaters and Chinese restaurants are always open, so this Christmas Day we can go to a movie and get some Chinese takeout.
46. For the first three games this year, the parking lot was packed with tailgaters, so we'll have to leave extra early to find a spot this week.
47. Every time Beth sold back her textbooks, she got a mere fraction of what she paid for them; so this semester she realized it would not be worth the effort to sell back her books at all.
48. Experts say that opening email attachments that come from unknown senders is one way you'll get a virus on your computer. Shauna constantly opens attachments from people she doesn't know, so she'll probably end up with a virus on her system.
49. Whenever Marcie let friends set her up on a blind date, the guy turned out to be a total loser. This time, when a friend offered to fix her up, she decided the guy would be a loser, so she declined.
50. Each time I have moved, something has gone wrong. So now that I am moving again next week, I know something has to go wrong.
51. The professor's policy is that any student whose cell phone goes off during class will be asked to leave. So when Ericka forgot to turn hers off and it rang during a quiz, the professor asked her to leave class.
52. Both times Melanie took her state test for teacher certification, she'd been so nervous she blanked out and failed the test. So when she walked into the testing room this time, her heart started beating faster and her palms were sweaty.
53. Since Josie ate a diet of mostly foods high in saturated fat, she was not surprised when her doctor said her cholesterol levels were too high.
54. In the past, even when Chris followed a recipe, her meal was either burned or underdone. Now her party guests know to eat before they attend her dinners so they won't starve all evening.
55. Tim has already lost three cell phones. He didn't spend a lot of money on his newest cell phone because he knows he'll eventually lose that one too.
56. Marathon runners should eat extra carbs before a big race, and since Mark did not eat enough carbs before the race, he felt sluggish the entire time.

57. Organizing chapter contents in your own words before the test will decrease the amount of study you have to do before a test. When Scott tried this method, he was pleasantly surprised at how fast he was able to study.
58. When the printer runs out of paper, a little light on the LCD display flashes. So when the light was flashing, Frank knew he needed to put more paper into the printer.
59. Few people showed up at the last several fund-raisers, which were not publicized very well. So this time, the organizers of the fund-raisers went all out with advertising the event well before the date to ensure more people would attend.
60. Since it gets dark around 6:00 P.M. during the winter months, Connie made her children come in from playing at 5:30 P.M.
61. The last several network dramas I've followed have been canceled just when I started getting into them. So I'm not going to bother watching the new one they're advertising even though it looks good, because I don't want to be disappointed when it gets canceled.
62. Rollerblading without knee pads and a helmet is said to be dangerous. So when I got my first pair of Rollerblades, I made sure to get a helmet and knee pads.
63. Every time Ralph trusted his girlfriend to return the DVDs to Blockbuster on time, she returned them late and he was stuck with a late charge. So this time, he made sure to drop off the DVDs on his way to work.
64. Whenever Sarah drove over the speed bumps near the dorm too fast, her CD player would skip around like crazy. So this time before she entered the dorm parking lot, she paused the CD.

Critical Thinking

65. When a map is drawn, countries that share a common border are always shaded with different colors. Can you draw a map on a flat surface consisting of five different countries so that five different colors are required to meet this criterion?
66. Find the next three numbers in the sequence.
    (a) 1, 1, 2, 3, 5, 8, 13, 21, . . .
    (b) 1, 4, 9, 16, 25, 36, . . .
    (c) 1, 2, 4, 8, 16, 32, . . .
67. The following figure is called *Pascal's triangle*. It has many uses in mathematics and other disciplines. Some uses will be shown in later chapters. Use inductive reasoning to find the numbers in the last row. (*Hint:* The entries that aren't 1 are obtained by using addition.)

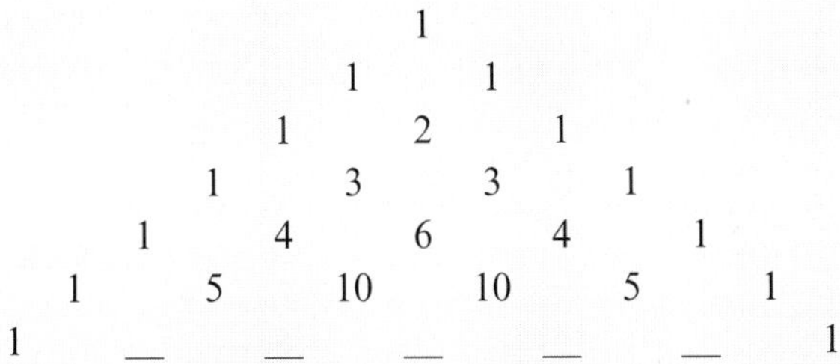

68. The difference of two squares $a^2 - b^2$ can be found by using the pattern illustrated in the following:

$$16^2 - 13^2 = 3(16 + 13) = 3 \cdot 29 = 87$$
$$25^2 - 21^2 = 4(25 + 21) = 4 \cdot 46 = 184$$

Using algebra and deductive reasoning, show that this will always be true.

69. The numbers 1, 3, 6, 10, 15, . . . are called *triangular numbers* since they can be displayed as shown.

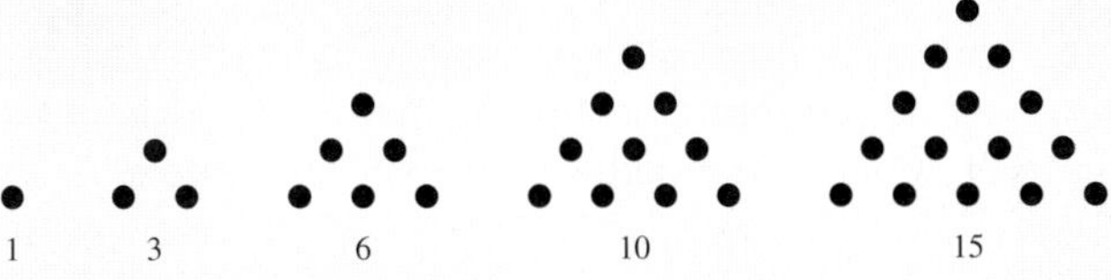

The numbers 1, 4, 9, 16, 25, . . . are called *square numbers* since they can be displayed as shown.

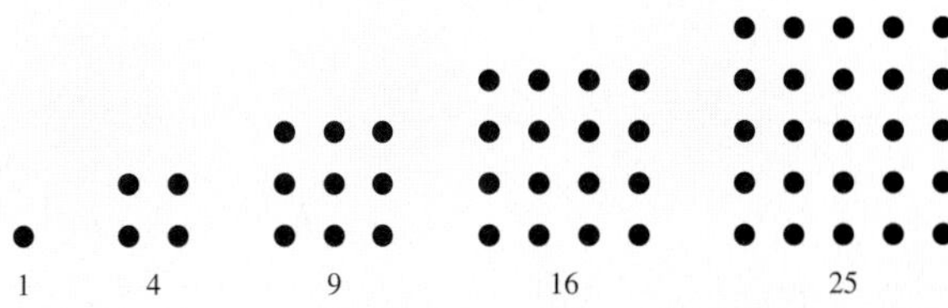

The numbers 1, 5, 12, 22, 35, . . . are called *pentagonal numbers* since they can be displayed as shown.

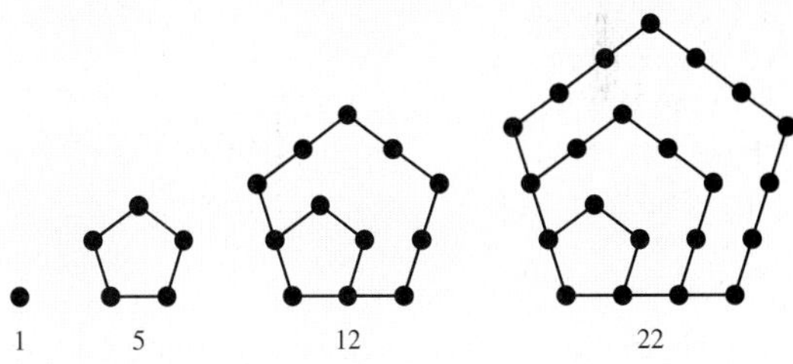

(a) Using inductive reasoning, find the next three triangular numbers.
(b) Using inductive reasoning, find the next three square numbers.
(c) Using inductive reasoning, find the next three pentagonal numbers.
(d) Using inductive reasoning, find the first four hexagonal numbers.

70. Refer to Exercise 69. The formula for finding triangular numbers is $\frac{n[1n-(-1)]}{2}$ or $\frac{n(n+1)}{2}$. The formula for finding square numbers is $\frac{n(2n-0)}{2}$ or n^2. The formula for pentagonal numbers is $\frac{n(3n-1)}{2}$. Find the formula for hexagonal numbers, using inductive reasoning.

Section 1-2 Estimation and Interpreting Graphs

LEARNING OBJECTIVES

- ❑ 1. Identify some uses for estimation.
- ❑ 2. Round numbers to a given level of accuracy.
- ❑ 3. Estimate the answers to real-world problems.
- ❑ 4. Use estimation to obtain information from graphs.

Everyone likes buying items on sale, so we should all be familiar with the idea of finding a rough approximation for a sale price. If you're looking at a pair of shoes that normally sells for \$70 and the store has a 40% off sale, you might figure that the shoes are a little more than half price, which would be \$35, so they're probably around \$40. We will call the process of finding an approximate answer to a math problem **estimation.** Chances are that you use estimation a lot more than you realize, unless you carry a calculator with you everywhere you go. (Okay, so your cell phone has a calculator, but how often do you use it?)

Estimation comes in handy in a wide variety of settings. When the auto repair shop technicians look over your car to see what's wrong, they can't know for sure what the exact cost will be until they've made the repairs, so they will give you an estimate. When you go to the grocery store and have only \$20 to spend, you'll probably keep a rough estimate of the total as you add items to the cart. If you plan on buying carpet for a room, you'd most likely measure the square footage and then estimate the total cost as you looked at different styles of carpet. You could find the exact cost if you really needed to, but often an estimate is good enough for you to make a sound buying decision.

Estimation is also a really useful tool in checking answers to math problems, particularly word problems. Let's say you're planning an outing for a student group you belong to, and lunch is included at \$3.95 per person. If 24 people signed up and you were billed \$94.80, you could quickly figure that this is a reasonable bill. Since 24 is close to 25, and \$3.95 is close to \$4.00, the bill should be close to $25 \times \$4.00 = \$100$.

You also use estimation when rounding numbers for simplicity. If someone asks your height and age, you might say 5′11″ and 20, even if you're actually $5'10\frac{1}{2}''$ (everyone fudges a little) and 20 years, 4 months, and 6 days old.

Since the process of estimating uses rounding, we'll start with a brief review of rounding. Rounding uses the concept of place value. The *place value* of a digit in a number tells the value of the digit in terms of ones, tens, hundreds, etc. For example, in the number 325, the 3 means 3 hundreds or 300 since its place value is hundreds. The 2 means 2 tens, or 20, and the 5 means 5 ones. A place value chart is shown here entitled "Rounding Numbers."

☑ 1. Identify some uses for estimation.

Rounding Numbers

To round numbers, use these rules:

1. Locate the place-value digit of the number that is being rounded. Here is the place-value chart for whole numbers and decimals:

8,	9	8	5,	7	3	0,	2	6	1	.	2	3	5	6	7	5
billions	hundred-millions	ten-millions	millions	hundred-thousands	ten-thousands	thousands	hundreds	tens	ones		tenths	hundredths	thousandths	ten-thousandths	hundred-thousandths	millionths

2a. If the digit to the right of the place-value digit is 0 through 4, then do not change the place-value digit.

2b. If the digit to the right of the place-value digit is 5 through 9, add 1 to the place-value digit.

Note: When you round whole numbers, replace all digits to the right of the digit being rounded with zeros. When you round decimal numbers, drop all digits to the right of the digit that is being rounded.

Math Note

We often use the word *nearest* to describe the place value to round to. Instead of saying, "Round to the hundreds place," we might say, "Round to the nearest hundred."

EXAMPLE 1 Rounding Numbers

Round each number to the place value given.

(a) 7,328 (hundreds)
(b) 15,683 (thousands)
(c) 32.4817 (tenths)
(d) 0.047812 (ten-thousandths)

SOLUTION

(a) In the number 7,328, 3 is in the hundreds place, so it's the digit being rounded. Since the digit to the right is 2, the digit 3 remains the same, and the 2 and 8 are replaced by zeros. The rounded number is 7,300.
(b) In the number 15,683, the 5 is the digit to be rounded. Since the digit to the right is 6, 1 is added to the 5 and the digits 6, 8, and 3 are replaced by zeros. The rounded number is 16,000.
(c) In the number 32.4817, the 4 is the digit to be rounded. Since the digit to the right of the 4 is 8, 1 is added to the 4 to get 5 and all digits to the right of the 4 are dropped. The rounded number is 32.5.
(d) In the number 0.047812, the 8 is the digit to be rounded. Since the digit to the right of 8 is 1, the 8 remains the same. The digits 1 and 2 are dropped. The rounded number is 0.0478.

▼ Try This One 1

☑ 2. Round numbers to a given level of accuracy.

Round each number to the place value given.

(a) 372,651 (hundreds)
(b) 32.971 (ones)
(c) 0.37056 (thousandths)
(d) 1,465.983 (hundredths)

Estimation

When you estimate answers for numerical calculations, two steps are required.

1. Round the numbers being used to numbers that make the calculation simple.
2. Perform the operation or operations involved.

EXAMPLE 2 Estimating Total Cost When Shopping

The owner of an apartment complex needs to buy six refrigerators for a new building. She chooses a model that costs \$579.99 per refrigerator. Estimate the total cost of all six.

SOLUTION

Step 1 Round the cost of the refrigerators. In this case, rounding up to \$600 will make the calculation easy.

Step 2 Perform the calculation: \$600 × 6 = \$3,600. Our estimated cost is \$3,600. (The actual cost will be a little less than \$3,600—note that we rounded the price up, so our estimate will be high.)

Sidelight THE BIGGEST NUMBER IN THE WORLD?

It seems that large numbers are a part of our advancing civilization. Cave people probably relied on their fingers and toes for counting, while large numbers today tend to overwhelm us.

Do you know what the largest number is? A million? A billion? A trillion? Do you know how large a million is? A billion? A trillion?

A million is no longer considered a large number. Many times, we hear the word *billion* being used. Government budgets are in the billions of dollars, the world population is approaching 7 billion, etc. Just how large are these numbers?

If you counted to a million, one number per second with no time off to eat or sleep, it would take you approximately $11\frac{1}{2}$ days. A million pennies would make a stack almost a mile high and a million $1 bills would weigh about a ton.

A million is written as a 1 followed by six zeros (1,000,000). Millions are used for measuring the distance from the earth to the sun, about 93 million miles.

The number 1 followed by nine zeros is called a billion (1,000,000,000). It is equal to 1,000 million.

It would take you 32 years to count to a billion with no rest. A billion dollars in $1 bills would weigh over a thousand tons.The age of the earth is estimated to be almost 4.5 billion years. There are more than 100 billion stars in our galaxy. A billion pennies would make a stack almost 1,000 miles high.

The number 1 followed by 12 zeros is a trillion. It is written as 1,000,000,000,000. Counting to a trillion would take 32 thousand years. The nearest star is 27 trillion miles away.

A quadrillion, a quintillion, and a sextillion are written with 15, 18, and 21 zeros, respectively. The weight of the earth is 6 sextillion, 570 quintillion tons (6.6 sextillion tons). Adding three zeros brings another -illion until a vigintillion is reached. That number has a total of 63 zeros.

Is this the largest number? No. A nine-year-old child invented a number with 100 zeros. It is called a googol. It looks like this:

10,000,000,000,000,000,000,000,000,000,000,000,
000,000,000,000,000,000,000,000,000,000,000,
000,000,000,000,000,000,000,000,000,000,000,000

This number is said to be more than the total number of protons in the universe. Yet an even larger number is the googolplex. It is a 1 followed by a googol of zeros. If you tried to write it, it would take a piece of paper larger than the distance from here to the moon.

Ultimately though, any number you can think of can be made larger by simply adding 1.

▼ Try This One 2

At one ballpark, large frosty beverages cost $6.75 each. Estimate the cost of buying one for each member of a group consisting of four couples.

Students often wonder, "How do I know what digit to round to?" There is no exact answer to that question—it depends on the individual numbers. In Example 2, the cost of the refrigerators could have been rounded from $579.99 to $580. Then the cost estimate would be $580 × 6 = $3,480. This is a much closer estimate because we rounded the cost to the nearest dollar, rather than the nearest $100.

Deciding on how much to round is really a tradeoff: ease of calculation versus accuracy. In most cases you'll get a more accurate result if you round less, but the calculation will be a little harder. Since there's no exact rule, it's important to evaluate the situation and use good old-fashioned common sense. And remember, when you are estimating, there is no one correct answer.

EXAMPLE 3 Estimating the Cost of a Cell Phone

You're considering a new cell phone plan where you have to pay $179 up front for the latest phone, but the monthly charge of $39.99 includes unlimited minutes and messaging. Estimate the cost of the phone for 1 year if there are no additional charges.

SOLUTION

Step 1 We can round the cost of the phone to \$180 and the monthly charge to \$40.

Step 2 The monthly cost estimate for 1 year will be $\$40 \times 12 = \$480$; add the estimated cost of the phone to get an estimated total cost of $\$480 + \$180 = \$660$.

▼ Try This One 3

A rental car company charges a rate of \$78 per week to rent an economy car. For an up-front fee of \$52, you can upgrade to a midsize car. Estimate the cost of renting a midsize car for 3 weeks.

EXAMPLE 4 Estimating Remodeling Costs

The Fabeets family plans on remodeling their living room. They will be replacing 21 square yards of carpet at a cost of \$23 per yard and having 16 linear feet of wall papered at a cost of \$32 per foot. Estimate the total cost of the remodel.

SOLUTION

Step 1 We can round the size and cost of the carpet to 20 square yards and \$25 per square yard. We can also round the length and cost of the wall to 15 feet and \$30 per foot.

Step 2 The estimated carpet cost is $20 \times \$25 = \$500$. The estimated wallpaper cost is $15 \times \$30 = \$450$. So our estimate for the total cost is $\$500 + \$450 = \$950$.

▼ Try This One 4

☑ 3. Estimate the answers to real-world problems.

Estimate the total cost if the Fabeets family decides to remodel a bedroom by installing 16 square yards of carpet at \$18 per square yard and having 21 feet of wall painted at a cost of \$12 per foot.

Sometimes you will need to overestimate or underestimate an answer. For example, if you were buying groceries and had only \$20.00, you would want to overestimate the cost of your groceries so that you would have enough money to pay for them. If you needed to earn money to pay the cost of your college tuition for the next semester, you would want to underestimate how much you could earn each week in order to determine the number of weeks you need to work to pay your tuition.

At times you will need to make estimates from information displayed in various graphs, such as bar graphs, pie graphs, or time series graphs. Examples 5, 6, and 7 illustrate how this can be done.

A **bar graph** is used to compare amounts or percents using either vertical or horizontal bars of various lengths which correspond to the amounts or percents.

EXAMPLE 5 Obtaining Information from a Bar Graph

The graph shown represents the percent of electricity generated by nuclear energy in five countries. Find the approximate percent of electricity generated by nuclear energy in France.

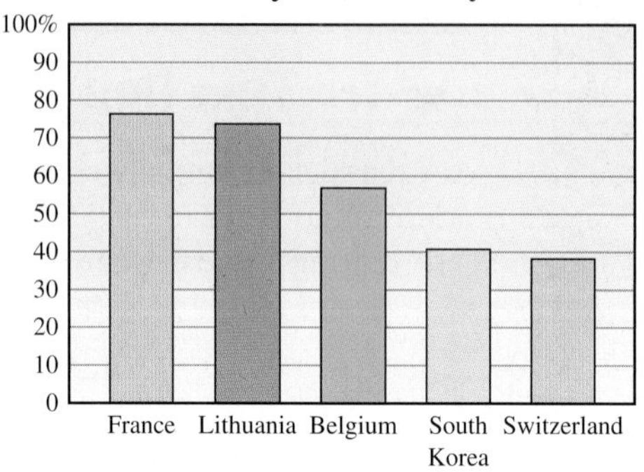

Source: *The World Almanac and Book of Facts*

SOLUTION

The first vertical bar represents the percentage for France. Reading across from the height of the bar to the vertical axis, we see that the height is a little bit more than halfway between 70 and 80%, so we estimate that the answer is about 76%.

▼ Try This One 5

Using the graph shown in Example 5, find the approximate percent of electricity generated by nuclear energy for Switzerland.

A **pie chart,** also called a **circle graph,** is constructed by drawing a circle and dividing it into parts called sectors, according to the size of the percentage of each portion in relation to the whole.

EXAMPLE 6 Obtaining Information from a Pie Chart

The pie chart shown represents the number of fatal occupational injuries in the United States for a selected year. If the total number of fatal injuries was 5,915 for the year, estimate how many resulted from assaults and violent acts.

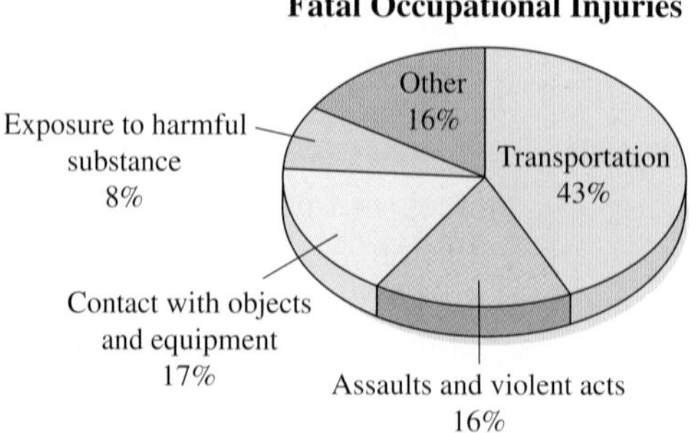

Source: *The World Almanac and Book of Facts*

SOLUTION

The sector labeled "Assaults and violent acts" indicates that 16% of the total fatal occupational injuries resulted from assaults and violent acts. So we need to find 16% of 5,915. To find a percentage, we multiply the decimal equivalent of the percentage by the total amount. In this case, we get $0.16 \times 5{,}915 = 946.4$. Since 0.4 fatal injury makes no sense, we round to 946. (*Note:* To change a percent to decimal form, move the decimal point two places to the left and drop the percent sign: 16 percent means "16 per hundred," which is 16/100, or 0.16. We will study percents in detail in Chapter 8.)

▼ Try This One 6

Using the chart shown in Example 6, find the approximate number of fatal occupational injuries that resulted from transportation accidents.

A **time series graph** or **line graph** represents how something varies or changes over a specific time period.

EXAMPLE 7 Obtaining Information from a Line Graph

The graph shown indicates the number of cable systems in the United States from 1960 to 2005. Find the approximate number of cable systems in 1970.

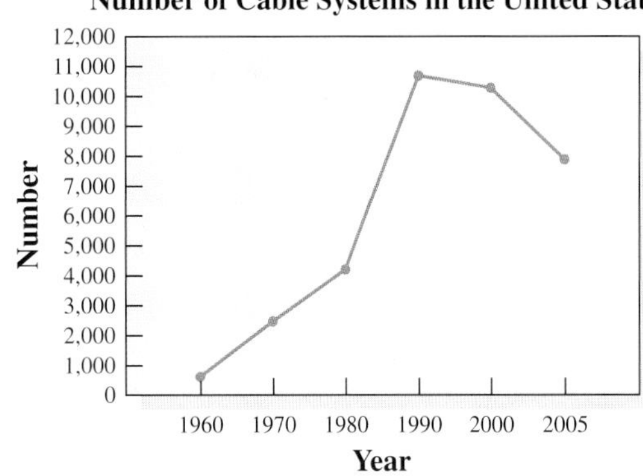

Source: National Cable and Telecommunications Association

SOLUTION

Locate the year 1970 on the horizontal axis and move up to the line on the graph. At this point, move horizontally to the point on the vertical axis as shown.

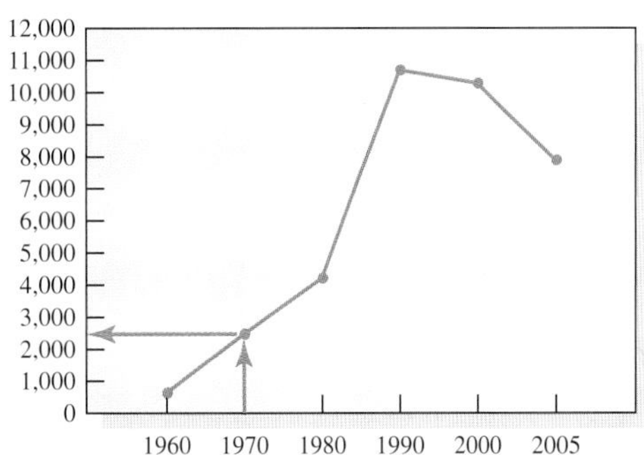

The height is about halfway between 2,000 and 3,000; so there were about 2,500 cable systems in the United States in 1970.

▼ Try This One 7

Using the graph shown in Example 7, find the approximate number of cable systems in the United States for the year 2005.

Distances between two cities can be estimated by using a map and its scale. The procedure is shown in Example 8.

EXAMPLE 8 Estimating Distance by Using a Map

A portion of a map is reproduced below. Fairbanks and Cordova are $2\frac{7}{8}$ inches apart on the map. If the scale is $\frac{1}{4}$ inch = 25 miles, estimate the distance between the cities.

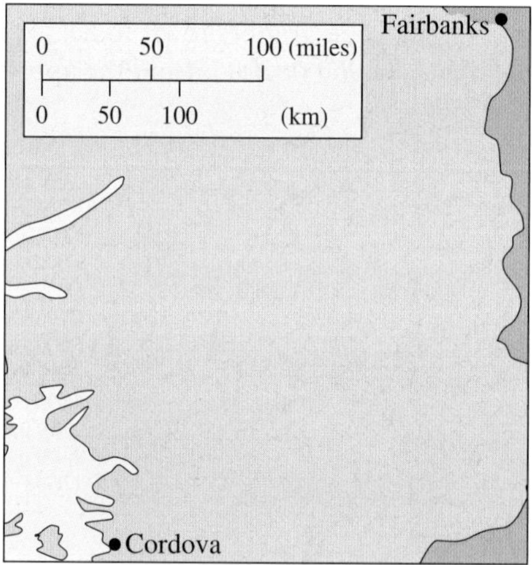

SOLUTION

First, we need to know how many $\frac{1}{4}$-inch units there are in $2\frac{7}{8}$ inches. Three full inches would be 12 quarter-inch units; $2\frac{7}{8}$ is $\frac{1}{8}$ less than 3, which is one-half of a $\frac{1}{4}$-inch unit less than 3. This means that there are $11\frac{1}{2}$ quarter-inch units. (You could also divide $2\frac{7}{8}$ by $\frac{1}{4}$ to get the same result.) Now we multiply: $25 \times 11\frac{1}{2} = 287.5$. So the cities are about 287.5 miles apart.

▼ Try This One 8

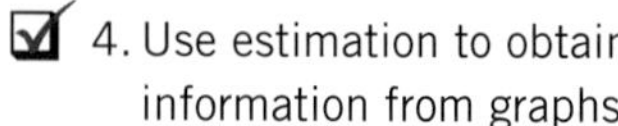

Estimate the distance between two cities that are $2\frac{1}{8}$ inches apart on a map if the scale is $\frac{1}{8}$ inch = 50 miles.

Answers to Try This One

1 (a) 372,700
(b) 33
(c) 0.371
(d) 1465.98

2 About $56

3 About $290

4 About $500

5 About 39%

6 About 2,543

7 About 8,000

8 About 850 miles

EXERCISE SET 1-2

Writing Exercises

1. What is estimation? Describe it in your own words.
2. Think of three real-world situations where you could use estimation.
3. Explain why an exact answer to a math problem isn't always necessary.
4. How can estimation be used as a quick check to see if the answer to a math problem is reasonable?
5. Describe the rules for rounding numbers to a given place.
6. Explain why there is never a single, correct answer to a question that asks you to estimate some quantity.

Computational Exercises

For Exercises 7–26, round the number to the place value given.

7. 2,861 (hundreds)
8. 732.6498 (thousandths)
9. 3,261,437 (ten-thousands)
10. 9,347 (tens)
11. 62.67 (ones)
12. 45,371,999 (millions)
13. 218,763 (hundred-thousands)
14. 923 (hundreds)
15. 3.671 (hundredths)
16. 56.3 (ones)
17. 327.146 (tenths)
18. 83,261,000 (millions)
19. 5,462,371 (ten-thousands)
20. 7.123 (hundredths)
21. 272,341 (hundred-thousands)
22. 63.715 (tenths)
23. 264.97348 (ten-thousandths)
24. 1,655,432 (millions)
25. 563.271 (hundredths)
26. 426.861356 (hundred-thousandths)

Real-World Applications

27. Estimate the total cost of eight energy-saving lightbulbs on sale for $16.99 each.
28. Estimate the cost of five espresso machines that cost $39.95 each.
29. Estimate the time it would take you to drive 237 miles at 37 miles per hour.
30. Estimate the distance you can travel in 3 hours 25 minutes if you drive on average 42 miles per hour.
31. Estimate the sale price of a futon you saw on eBay that costs $178.99 and is now on sale for 60% off.
32. Estimate the sale price of a DVD player that costs $42.99, on sale for 15% off.
33. Estimate the total cost of the following meal at McDonald's:

Quarter pounder with cheese	$2.89
Supersized fries	$1.89
Small shamrock shake	$1.29

34. Estimate the total cost of the following items for your dorm room:

Loft bed	\$159.95
Beanbag chair	\$49.95
Storage cubes	\$29.95
Lava lamp	\$19.95

35. Estimate the time it would take you in a charity bike tour to ride 86 miles at a rate of 11 miles per hour.
36. If a person earns \$48,300.00 per year, estimate how much the person earns per hour. Assume a person works 40 hours per week and 50 weeks per year.
37. If a person earns \$8.75 per hour, estimate how much the person would earn per year. Assume a person works 40 hours per week and 50 weeks per year.
38. If a salesperson earns a commission of 12%, estimate how much money the salesperson would earn on an item that sold for \$529.85.
39. Estimate the cost of putting up a decorative border in your dorm room if your room is 24 feet long and 18 feet wide and the border costs \$5.95 every 10 feet.
40. Estimate the cost of painting a homecoming float if the area to be painted is 12 feet by 16 feet and a quart of paint that covers 53 square feet costs \$11.99.
41. Posters for a rock concert cost \$4.95 each. Estimate the cost of hanging the maximum number of posters along a 30-foot-long hallway if the posters are each 2 feet wide and there is to be 5 feet between posters.
42. Estimate your cost to live in an apartment for 1 year if the rent is \$365.00 per month and utilities are \$62.00 per month.

Use the information shown in the graph for Exercises 43–46. The graph gives the areas of various college campuses in acres.

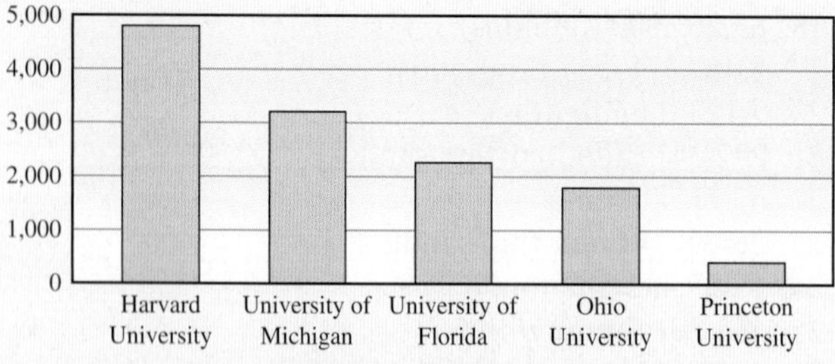

Source: http://colleges.usnews.rankingsandreviews.com

43. Estimate the area of the campus of the University of Michigan.
44. Estimate the area of the campus of Princeton University.
45. Estimate the difference between the largest campus shown and the smallest campus shown.
46. Estimate the areas of the two universities that are approximately the same size.

Use the information shown in the graph for Exercises 47 and 48. The graph represents a survey of 1,385 office workers and shows the percent of people who indicated what time of day is most productive for them.

Most Productive Time of Day

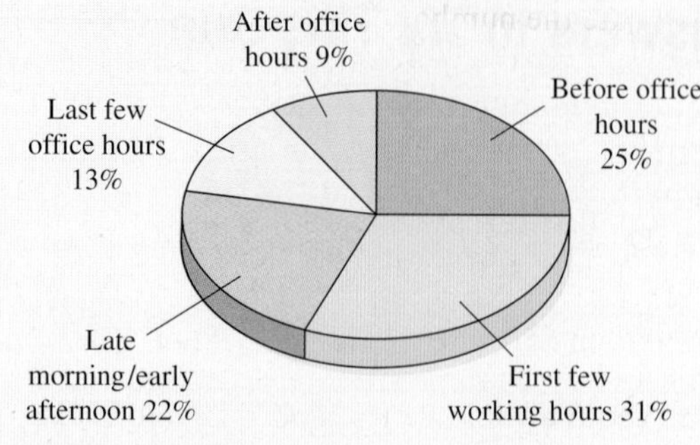

Source: *USA Today*

47. Estimate the number of people who feel they are most productive outside normal office hours.
48. Estimate the number of people who feel they are most productive before late morning.

Use the information shown in the graph for Exercises 49 and 50. The graph represents a survey of undergraduates enrolled in college in the 2003–2004 school year.

Hours per Week Worked by College Students Ages 22 or Younger

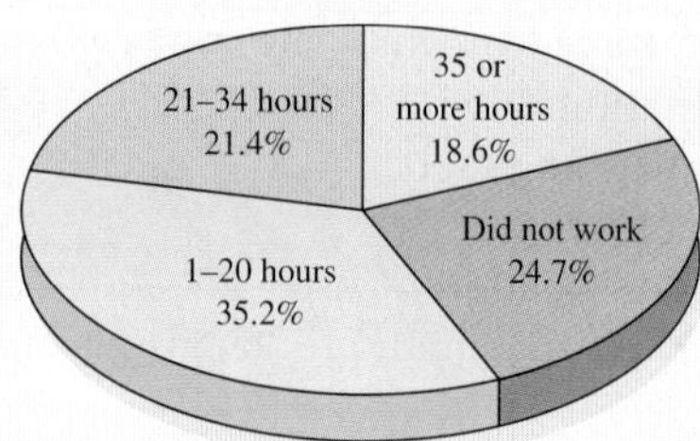

Source: www.acenet.edu

49. Approximately what percentage of students worked more than 21 hours per week?
50. Approximately what percentage of students worked less than 35 hours per week?

Use the information shown in the graph for Exercises 51–54. The graph shows the cigarette consumption (in billions) in the United States for the years 1900 to 2007.

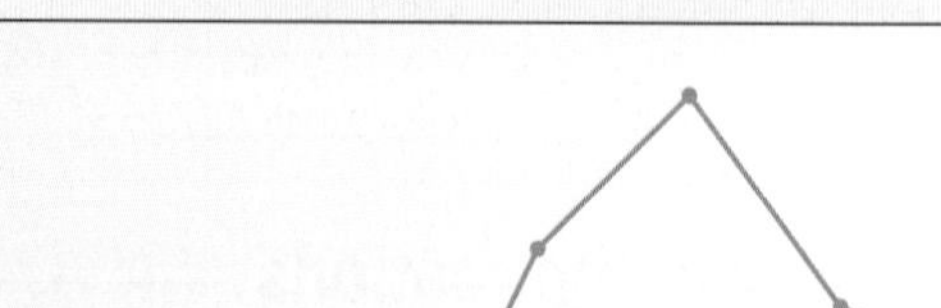

Cigarette Consumption in the United States

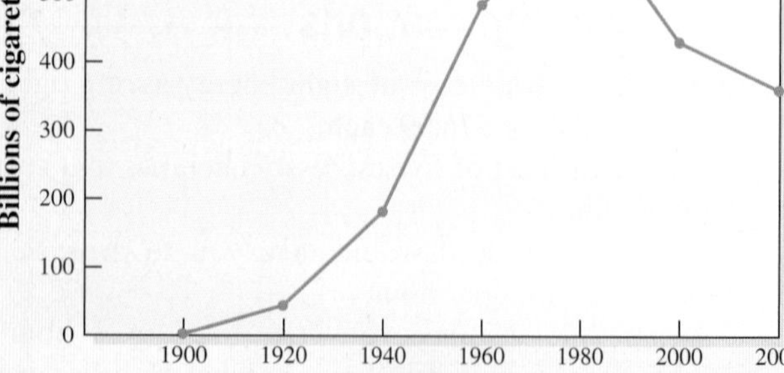

Source: www.infoplease.com

51. Estimate the number of cigarettes smoked in 1950.
52. Estimate the number of cigarettes smoked in 1985.
53. Estimate the year or years in which 200 billion cigarettes were smoked.
54. Estimate the year or years in which 400 billion cigarettes were smoked.
55. In the 2007 atlas distributed by State Farm Insurance, Cincinnati and Columbus are $8\frac{3}{4}$ inches apart. Estimate the distance between the cities if the scale is 1 inch = 11 miles.
56. In the 2007 atlas distributed by State Farm Insurance, Pittsburgh and Johnstown, PA, are $4\frac{7}{8}$ inches apart. Estimate the distance between the cities if the scale is $\frac{1}{2}$ inch = 6 miles.

*Use the information in the graph for Exercises 57–60. The graph shows the average daily reach for the five most visited sites on the Internet for the three-month period ending July 21, 2008. (**Reach** is the percentage of global Internet users who visit a site.)*

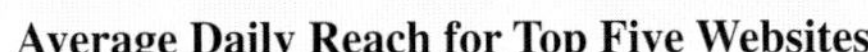

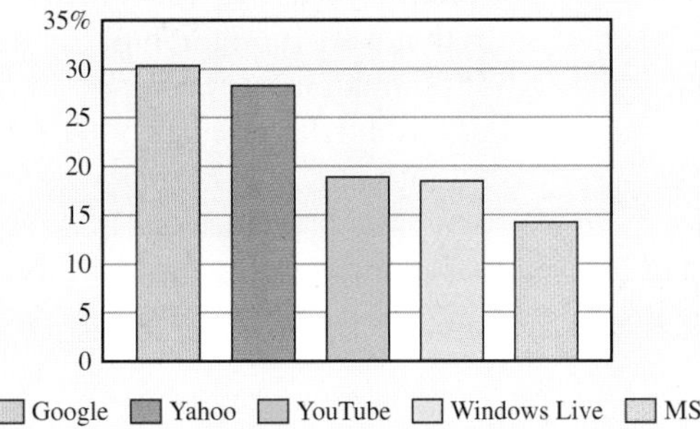

57. Estimate the average daily reach for Yahoo.
58. Estimate the average daily reach for YouTube.
59. Estimate the combined average daily reach for Google and Windows Live.
60. Estimate the difference in average daily reach between the first and fifth most visited sites.

Use the information in the graph for Exercises 61 and 62. The graph shows the number of people who would start over in a completely different career.

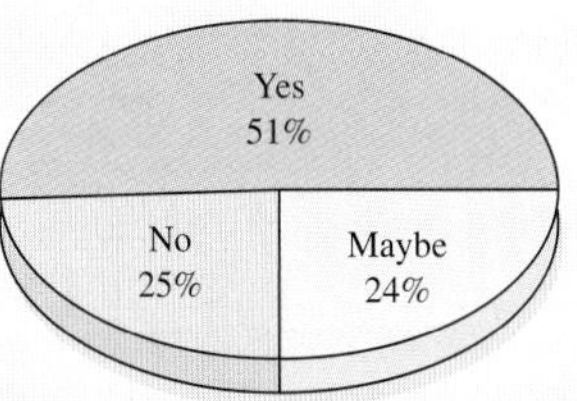

Source: The Konn/Ferry International Survey

61. What percent of people would definitely not choose a different career?
62. The total number of people surveyed was 1,733. Approximately how many would choose a different career?

Use the information in the graph for Exercises 63 and 64. The graph shows the average penalty fees for credit cards for 2002 to 2007.

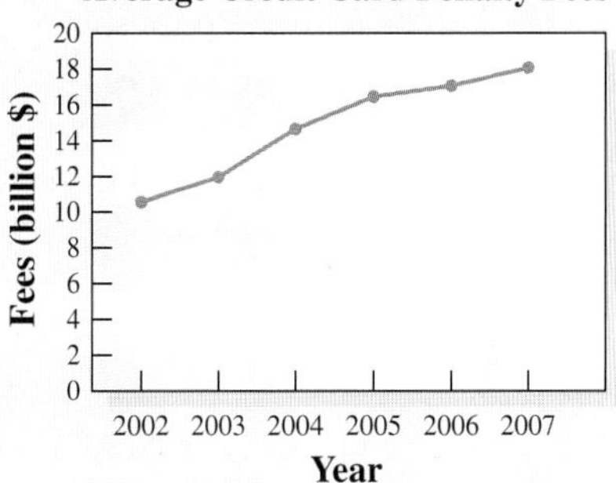

Source: *USA Today*

63. Estimate the credit card penalty fees for 2004.
64. What was the penalty fee increase from 2002 to 2007?

Critical Thinking

65. Sometimes graphs are drawn in such a way as to support a conclusion that may or may not be true. Look at the graph and see if you can find anything misleading.

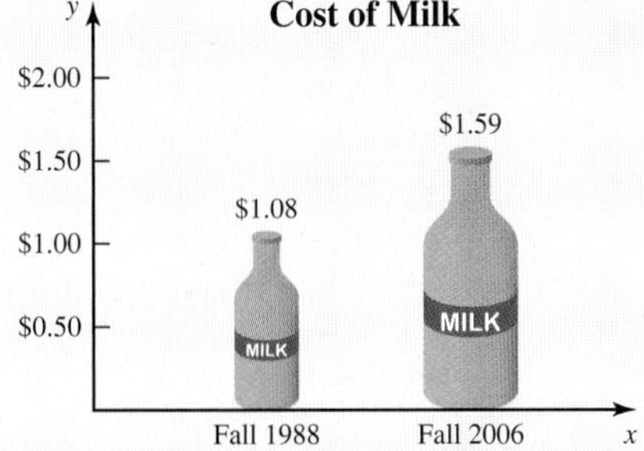

66. The choice of labeling on a graph can have a profound effect on how the information is perceived.

Compare the graph below to the one from Exercises 63 and 64. It contains the same information as the earlier graph. Why does it appear to show a much sharper increase in credit card penalty fees?

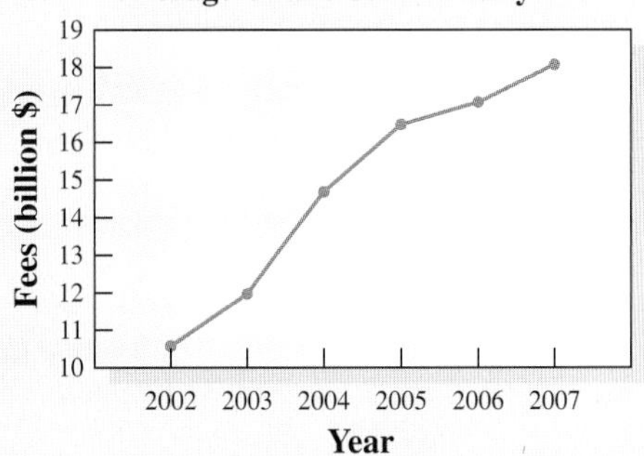

67. Find an advertisement for a store that sells consumer electronics, either from a newspaper or on the Internet. Then use it to estimate the cost of a home entertainment center consisting of a 26-inch flat screen TV, a DVD player, a surround sound system, and a cabinet. Does the total amount sound like a good deal to you?

68. Find an advertisement for a store that sells consumer electronics, either from a newspaper or on the Internet. Then use it to estimate the cost of a home computer center consisting of a computer desk, computer, printer, and 20-inch monitor. Does the total amount sound like a good deal to you?

Section 1-3 Problem Solving

LEARNING OBJECTIVES

- ☐ 1. State the four steps in the basic problem-solving procedure.
- ☐ 2. Solve problems by using a diagram.
- ☐ 3. Solve problems by using trial and error.
- ☐ 4. Solve problems involving money.
- ☐ 5. Solve problems by using calculation.

There's a good reason that questions in math classes are usually called *problems:* the study of mathematics is all about learning and applying problem-solving skills and strategies. For most of recorded history, math and problem solving have been intimately connected, and a tremendous amount of effort has gone into using math as a problem-solving tool. In this section, we'll learn about a simple framework for problem solving and practice a number of useful strategies.

A Hungarian mathematician named George Polya did a lot of research on the nature of problem solving in the first half of the 20th century. His biggest contribution to the field was an attempt to identify a series of steps that were fundamental to problem-solving strategies used by great thinkers throughout human history. One of his books, published in 1945 (and still a big seller on Amazon!), set forth these basic steps. *How to Solve It* is so widely read that it has been translated into at least 17 languages.

Polya's strategy is not necessarily earth-shattering: its brilliance lies in its simplicity. It provides four basic steps that can be used as a framework for problem solving in any area, from mathematics to home improvements.

☑ 1. State the four steps in the basic problem-solving procedure.

Polya's Four-Step Problem-Solving Procedure

Step 1 *Understand the problem.* An excellent way to start is to write down information given in a problem as you come to it. Especially with lengthy problems, if you read the whole thing all at once, you may get a bit overwhelmed and intimidated. If you read the problem slowly and carefully, and you write down information as it's provided, you will always have a head start. It is equally important to carefully note exactly what it is that you are being asked to find, since this will guide you to the next step. You should probably read a problem more than once to make sure you understand what information you are given and what you are asked to find.

Step 2 *Devise a plan to solve the problem.* There are many ways to solve problems. It may be helpful to draw a picture. You could make an organized list. You may be able to solve it by trial and error. You might be able to find a similar problem that has already been solved and apply the technique to the problem that you are trying to solve. You may be able to solve the problem by using the arithmetic operations of addition, subtraction, multiplication, or division. You may be able to solve the problem by using algebraic equations or geometric formulas.

Step 3 *Carry out the plan to solve the problem.* After you have devised a plan, try it out. If you can't get the answer, try a different strategy. Also be advised that sometimes there are several ways to solve a problem, so different students can arrive at the correct answer by different methods.

Step 4 *Check the answer.* Some problems can be checked by using mathematical methods. Sometimes you may have to ask yourself if the answer is reasonable, and sometimes the problem can be checked by using a different method. Finally, you might be able to use estimation to approximate the answer. (See Section 1–2.)

In each of the examples in this section, we'll illustrate the use of Polya's four-step procedure.

EXAMPLE 1 Solving a Problem by Using a Diagram

A gardener is asked to plant eight tomato plants that are 18 inches tall in a straight line with 2 feet between each plant. How much space is needed between the first plant and the last one?

SOLUTION

Be careful—what seems like an obvious solution is not always correct! You might be tempted to just multiply 8 by 2, but instead we will use Polya's method.

Step 1 *Understand the problem.* In this case, the key information given is that there will be eight plants in a line, with 2 feet between each. We are asked to find the total distance from the first to the last.

Step 2 *Devise a plan to solve the problem.* When a situation is described that you can draw a picture of, it's often helpful to do so.

Step 3 *Carry out the plan to solve the problem.* The figure would look like this:

Now we can use the picture to add up the distances:

$$2 + 2 + 2 + 2 + 2 + 2 + 2 = 14 \text{ feet}$$

Step 4 *Check the answer.* There are eight plants, but only seven spaces of 2 feet between them. So $7 \times 2 = 14$ feet is right.

Math Note

Sometimes problems will contain extraneous information. In Example 1, the height of the plants is immaterial to the distance between them, so you can (and should) ignore that information. Students sometimes get into trouble by trying to incorporate *all* the information provided without considering relevance.

▼ Try This One 1

Suppose you want to cut a 4-foot party sub into 10 pieces. How many cuts do you need to make?

Sidelight GEORGE POLYA (1887–1985)

Born in Hungary, George Polya received his Ph.D. from the University of Budapest in 1912. He studied law and literature before turning his interests to mathematics. He came to the United States in 1940 and taught at Brown University. In 1942, he moved to Stanford University. His most famous book entitled *How to Solve It* was published in 1945 by Princeton University Press.

This book explained his method of problem solving, which consisted of four major steps.

1. Understand the problem.
2. Devise a plan.
3. Carry out the plan.
4. Look back.

Over 1 million copies of the book have been sold, and it is still being published today.

His four-step process is used as a blueprint for solving problems not only in mathematics but in other areas of study as well. The book has been translated into at least 17 languages and is sold worldwide.

In addition to his famous book, Polya has written over 200 articles, nine other books, and numerous other papers.

EXAMPLE 2 Solving a Perimeter Problem

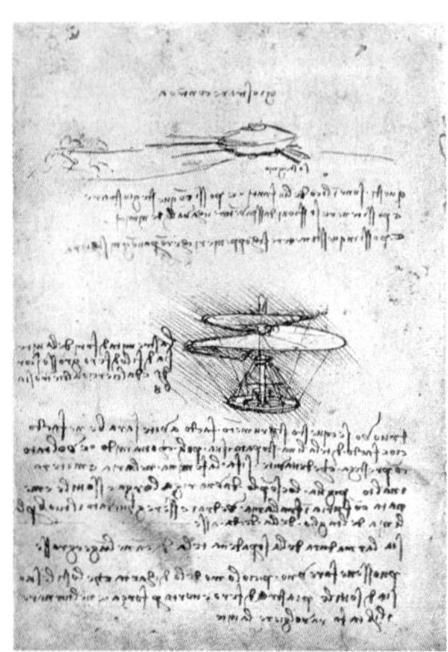

Scientists and inventors often use sketches to organize their thoughts as Leonardo da Vinci did. He wrote backward in Latin to protect his work.

A campus group is setting up a rectangular area for a tailgate bash. They have 100 feet between two roads to use as width and 440 feet of fence to use. What length will use up the total amount of fence and enclose the biggest space?

SOLUTION

Step 1 *Understand the problem.* We're asked to consider a rectangular area, so there will be four sides. We're told that the width is 100 feet and that the four sides add up to 440 feet. (That is, the perimeter is 440 feet.) We're asked to find the length.

Step 2 *Devise a plan to solve the problem.* This is a classic example of a problem where a diagram will be useful. This should help us to figure out all the dimensions.

Step 3 *Carry out the plan to solve the problem.* Our diagram looks like this:

(Since the area is rectangular, the opposite sides have the same length.) Of the 440 feet of fence, 200 feet is accounted for in our diagram. That leaves $440 - 200 = 240$ feet to be divided among the remaining two sides. Each has length 120 feet.

Step 4 *Check the answer.* If there are two sides with width 100 feet and two others with length 120 feet, the perimeter is $100 + 100 + 120 + 120 = 440$ feet.

▼ Try This One 2

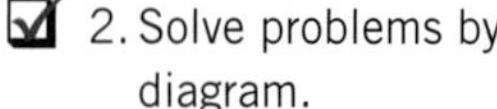

2. Solve problems by using a diagram.

A rectangular poster promoting the tailgate bash in Example 2 is restricted to 10 inches in width to fit bulletin board restrictions. The printer suggests a perimeter of 56 inches for maximum savings in terms of setup. How tall will the posters be?

EXAMPLE 3 Solving a Problem by Using Trial and Error

Math Note

The trial-and-error method works nicely in Example 3, but for many problems, there are just too many possible answers to try. It's not a strategy that you should rely on for solving every problem.

Suppose that you have 10 coins consisting of quarters and dimes. If you have a total of \$1.90, find the number of each type of coin.

SOLUTION

Step 1 *Understand the problem.* We're told that we have a total of 10 coins and that some are dimes (worth \$0.10 each) and the rest are quarters (worth \$0.25 each). The total value is \$1.90. The problem is to find how many quarters and dimes together are worth \$1.90.

Step 2 *Devise a plan to solve the problem.* One strategy that can be used is to make an organized list of possible combinations of 10 total quarters and dimes and see if the sum is \$1.90. For example, you may try one quarter and nine dimes. This gives $1 \times \$0.25 + 9 \times \$0.10 = \$1.15$.

Quarters	Dimes	Amount
1	9	\$1.15

Step 3 *Carry out the plan.* Since one quarter and nine dimes is wrong, try two quarters and eight dimes.

Quarters	Dimes	Amount
2	8	\$1.30

This doesn't work either. We'll continue until we get the correct answer.

Quarters	Dimes	Amount	
1	9	\$1.15	
2	8	\$1.30	
3	7	\$1.45	
4	6	\$1.60	
5	5	\$1.75	
6	4	\$1.90	← Correct

Answer: six quarters and four dimes.

Step 4 *Check the answer.* In this case, our answer can be checked by working out the amounts: $6 \times \$0.25 + 4 \times \$0.10 = \$1.90$.

▼ Try This One 3

☑ 3. Solve problems by using trial and error.

Michelle bought some 42-cent stamps and some 59-cent stamps. If she bought a total of seven stamps and spent a total of \$3.45, how many of each kind did she buy?

EXAMPLE 4 Solving a Problem Involving Salary

So you've graduated from college and you're ready for that first real job. In fact, you have two offers! One pays an hourly wage of \$19.20 per hour, with a 40-hour work week. You work for 50 weeks and get 2 weeks' paid vacation. The second offer is a salaried position, offering \$41,000 per year. Which job will pay more?

SOLUTION

Step 1 *Understand the problem.* The important information is that the hourly job pays $19.20 per hour for 40 hours each week, and that you will be paid for 52 weeks per year. We are asked to decide if that will work out to be more or less than $41,000 per year.

Step 2 *Devise a plan to solve the problem.* We can use multiplication to figure out how much you would be paid each week and then multiply again to get the yearly amount. Then we can compare to the salaried position.

Step 3 *Carry out the plan to solve the problem.* Multiply the hourly wage by 40 hours; this shows that the weekly earnings will be $\$19.20 \times 40 = \$768$. Now we multiply by 52 weeks: $\$768 \times 52 = \$39{,}936$. The salaried position, at $41,000 per year, pays more.

Step 4 *Check the answer.* We can figure out the hourly wage of the job that pays $41,000 per year. We divide by 52 to get a weekly salary of $788.46. Then we divide by 40 to get an hourly wage of $19.71. Again, this job pays more.

▼ Try This One 4

☑ 4. Solve problems involving money.

A condo on the water in Myrtle Beach can be rented for $280 per day, with a nonrefundable application fee of $50. Another one down the beach can be rented for $2,100 per week. Which condo costs less for a week's stay?

EXAMPLE 5 Solving a Problem by Using Calculation

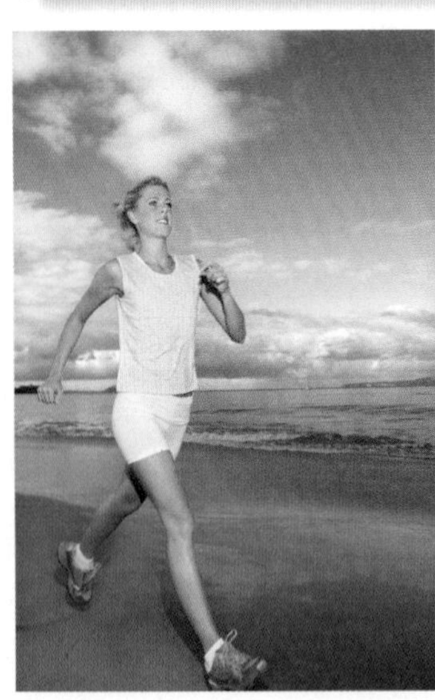

A 150-pound person walking briskly for 1 mile can burn about 100 calories. How many miles per day would the person have to walk to lose 1 pound in 1 week? It is necessary to burn 3,500 calories to lose 1 pound.

SOLUTION

Step 1 *Understand the problem.* A 150-pound person burns 100 calories per 1 mile and needs to burn 3,500 calories in 7 days to lose 1 pound. The problem asks how many miles per day the person has to walk to lose 1 pound in 1 week.

Step 2 *Devise a plan to solve the problem.* We will calculate how many calories need to be burned per day and then divide by 100 to see how many miles need to be walked.

Step 3 *Carry out the plan.* Since 3,500 calories need to be expended in 7 days, divide 3,500 by 7 to get 500 calories per day. Then divide 500 by 100 to get 5 miles. It is necessary to walk briskly 5 miles per day to lose 1 pound in a week.

Step 4 *Check the answer.* Multiply 5 miles per day by 100 calories per mile by 7 days to get 3,500 calories.

▼ Try This One 5

Megan's car gets 28 miles per gallon on long trips. She leaves home, bound for a friend's house 370 miles away. If she makes the drive in 6 hours, how many gallons of gas per hour is the car burning?

Sidelight A LEGENDARY PROBLEM SOLVER

Archimedes is considered by some to be the greatest mathematician of the ancient world. But he is best known for a famous problem he solved in the bathtub.

King Hieron II of Syracuse ordered a solid gold crown from a goldsmith sometime around 250 BCE. He suspected that the crown was not really solid gold, but he didn't want to have it melted down to check, because of course that would ruin it. He asked his friend Archimedes if he could figure out a way to determine whether the crown really was pure gold without destroying it.

After some serious thought, Archimedes decided to relax by taking a bath. While sitting in the tub, he saw that the water level rose in proportion to his weight. Suddenly, he realized that he had discovered the solution to his problem. By placing the crown in a full tub of water, he could tell how much of it was gold by weighing the amount of water that overflowed. Legend has it that Archimedes was so excited by this burst of inspiration that he forgot to clothe himself and ran naked through the streets of Syracuse, shouting "Eureka" ("I have found it!"), a performance that would most likely get him arrested, or maybe institutionalized, today.

It turns out that the crown was not solid gold, and the goldsmith either was placed in jail or had his head removed from his shoulders, depending on which account you believe. The moral of this story is that you never know where inspiration in solving a problem might come from.

EXAMPLE 6 Solving a Problem by Using Calculation

For a date party at a certain restaurant there is a flat fee of \$125 to rent the banquet room plus a charge of \$18 per person attending. How much will the date party cost if 104 people attend?

SOLUTION

Step 1 *Understand the problem.* We are asked to find the cost of the party. We are given that there is a flat rate of \$125 and a per-person charge of \$18.

Step 2 *Devise a plan to solve the problem.* To find the total cost, we should add all the costs involved—the flat rate plus the per-person charge. The flat rate is \$125, and the per-person charge is $\$18 \times 104$.

Step 3 *Carry out the plan.* Add: $\$125 + \$18(104) = \$125 + \$1{,}872 = \$1{,}997$. So it will cost \$1,997 for the date party.

Step 4 *Check the answer.* Here we will use estimation to see if the answer makes sense. The cost of the party is about $\$125 + \$20(100) = \$2{,}125$. So our answer appears to be reasonable.

▼ Try This One 6

☑ 5. Solve problems by using calculation.

If it costs \$35.00 per day and \$0.25 per mile to rent a car, how much will it cost a person to rent a car for a 2-day, 300-mile round trip?

Strategies for Understanding a Problem

One reason that problem solving is challenging is that every problem can be a little different. But here are some suggestions for helping to understand a problem and devise a strategy for solving.

- If the problem describes something that can be diagrammed, a drawing almost always helps.
- Write down all the numeric information in the problem. This often helps to organize your thoughts.
- Sometimes making a chart to organize information is helpful, especially when you're trying to use trial and error.
- Don't expect that you'll know exactly how to solve a problem after reading it once! Most often you'll need to read through a problem several times, and even then you may need to just try some approaches before finding one that works.

Answers to Try This One

1 Nine

2 18 inches

3 Four \$0.42 stamps and three \$0.59 stamps

4 The condo for \$280 per day costs less for a week.

5 About 2.2 gallons per hour

6 \$145.00

EXERCISE SET 1-3

Writing Exercises

1. List and describe the four steps in problem solving.
2. Discuss what you should do first when given a word problem to solve.
3. Discuss different ways you might be able to check your answer to a problem.
4. Why is trial and error not always a good problem-solving strategy?

Computational Exercises

5. One number is 6 more than another number, and their sum is 22. Find the numbers.
6. One number is 7 more than another number. Their sum is 23. Find the numbers.
7. The sum of the digits of a two-digit number is 7. If 9 is subtracted from the number, the answer will be the number with the digits reversed. Find the number.
8. A mother is 4 times as old as her daughter. In 18 years, the mother will be twice as old as her daughter. Find their present ages.
9. In 28 years, Mark will be 5 times as old as he is now. Find his present age.
10. Pete is twice as old as Lashanna. In 5 years, the sum of their ages will be 37. Find their present ages.
11. A piece of rope is 48 inches long and is cut so that one piece is twice as long as the other. Find the length of each piece.
12. At the dog park, there are several dogs with their owners. Counting heads, there are 12; counting legs, there are 38. How many dogs and owners are there?

Real-World Applications

13. Fred has \$5.05 worth of coins floating around in his backpack, all of which are quarters and dimes. If there are 5 more dimes than quarters, how many coins of each are there?
14. A tip jar contains twice as many quarters as dollar bills. If it has a total of \$12 in it, how many quarters and how many dollar bills does it contain?
15. A fraternity charged \$2.00 admission for guys and \$1.00 admission for gals to their finals week bash. The fraternity made \$75 and sold 55 tickets. How many gals attended the party?
16. While sitting in the quad on central campus, I counted 9 students riding 2-wheeled mopeds or 3-wheeled scooters, which made a total of 21 wheels. How many mopeds and how many scooters passed by me?
17. Jane has 10 phone cards worth either \$20 or \$15. How many of each does she have if the phone cards total \$180?
18. May receives \$87 for working one 8-hour day. One day she had to stop after working 5 hours because of a doctor's appointment. How much did she make that day?
19. The manager of the new campus Internet café wants to put six PCs on each table with 3 feet between the PCs and 2 feet on each end. The PCs measure $1\frac{1}{2}$ feet wide and 21 inches high. What length of tables should the Internet café order?
20. Bob is building a deck that is 32 feet long, and he has posts that are 6 inches square to use to build the railing. If he puts the posts 4 feet apart with one on each end, how many posts will he use?
21. Suzie hangs 10 pictures that each measure $8\frac{1}{2}$ inches wide on a wall. She spaces them 6 inches apart with 2 inches on each end, so that they fit perfectly on the wall. How wide is the wall?
22. You want to cut a width of mat board to frame pictures so you can fit six pictures that have a width of 6 inches each with 2 inches between them and a 1-inch border. How wide should you cut the mat board?
23. The campus athletic department is building a new pool that is 60 feet long, and there is enough tile for a 220-foot border. How wide should the pool be to ensure that all tiles are used?
24. If the length of the campus quad is double its width and there's a short retaining wall around the perimeter that measures 300 feet, what are the length and width of the quad?
25. If you have enough caulk to seal 380 feet of window casings, and each window measures 2 feet by 3 feet, how many windows can you caulk?
26. For the Spring Fling gala, Felicia has 500 feet of lights to hang. If the width of the room is 70 feet, what is the length of the room if she can hang all the lights so that they go around the room exactly once?
27. Two students are paid a total of \$60 for leading a campus tour at freshman orientation. The tour lasts approximately 2 hours. If Sam works for $1\frac{1}{2}$ hours, Pete works $\frac{1}{2}$ hour, and each makes the same hourly wage, how much does each receive?
28. You have one-half of an energy drink left. If you drink one-half of that, how much of your energy drink do you have left?
29. At the campus bookstore, a stocker has to move three laptops in bags that weigh 6 pounds each and six backpacks that weigh 8 pounds each. If she lifts them all at once to move them out of the way, how much weight does she lift?
30. A small beverage company has 832 bottles of water to ship. If there are 6 bottles per case, how many cases are needed and how many bottles will be left over?
31. A person's monthly budget includes \$256 for food, \$125 for gasoline, and \$150 for utilities. If this person earns \$1,624 per month after taxes, how much money is left for other expenses?
32. One protein bar has 15 grams of carbs. If a runner wanted to eat 300 grams of carbs the week before a big marathon, how many protein bars would she have to eat?
33. A bag of pretzels contains 1,650 calories. If a person purchased 26 bags of pretzels for a picnic, how many calories would be consumed if all the bags were eaten?
34. An automobile travels 527 miles on 12.8 gallons of gasoline. How many miles per gallon did the car get?
35. If a family borrows \$12,381 for an addition to their home, and the loan is to be paid off in monthly payments over a period of 5 years, how much should each payment be? (Interest has been included in the total amount borrowed.)
36. Assuming that you could average driving 55 miles per hour, how long would it take you to drive from one city to another if the distance was 327 miles?
37. Ticketmaster is selling tickets for an upcoming concert. They are charging \$32 for floor seats, \$25 for pavilion seats, and \$20 for lawn seats (not including the huge handling fee). A group of students purchases 5 floor tickets, 8 pavilion tickets, and 3 lawn tickets. What is the total cost of all the concert tickets purchased by the group of students?
38. Four friends decide to rent an apartment. Because each will be using it for different lengths of time, Mary will pay $\frac{1}{2}$ of the monthly rent, Jean will pay $\frac{1}{4}$ of the monthly rent, Claire will pay $\frac{1}{8}$ of the monthly rent, and Margie will pay the rest. If the monthly rent is \$2,375, how much will each person pay?
39. The Campus Pizza Shack pays its employees 46.5 cents per mile if they use their own cars for

deliveries. In one month, Jenny racked up 106 miles delivering pizzas on campus. How much will she be paid for her mileage?

40. Harry fills up his Jeep with gasoline and notes that the odometer reading is 23,568.7 miles. The next time he fills up his Jeep, he pays for 12.6 gallons of gasoline. He notes his odometer reading is 23,706.3 miles. How many miles per gallon did he get?
41. A clerk earns $9.50 per hour and is paid time and a half for any hours worked over 40. Find the clerk's pay if he worked 46 hours during a specific week.
42. A cell phone company charges 35 cents per minute during the daytime and 10 cents per minute during the evening for those who go over their allotted monthly minutes. If Sally went 32 minutes over her monthly minutes, find out how much she'd save if she went 32 minutes over her monthly minutes during the evening compared to during the daytime.

Critical Thinking

43. With a power mower, Phil can cut a large lawn in 3 hours. His younger brother can cut the same lawn in 6 hours. If both worked together, using two power mowers, how long would it take them to cut the lawn?
44. A king decided to pay a knight one piece of gold for each day's protection on a 6-day trip. The king took a gold bar 6 inches long and paid the knight at the end of each day; however, he made only two cuts. How did he do this?
45. At the finals of the campuswide Brainbowl League Tournament, your team was asked to measure out exactly 1 gallon of water using only a 5-gallon and a 3-gallon container, without wasting any water. Your team completed the task. How did they do it?
46. What is the smallest number of cars that can tailgate in a straight line at a football game so that one car is in front of two cars, one car is behind two cars, and one car is between two cars?
47. A college student with a part-time job budgets $\frac{1}{5}$ of her income for clothes, $\frac{2}{5}$ for living expenses (including food and rent), $\frac{1}{10}$ for entertainment, and $\frac{1}{2}$ of the remainder for savings. If she saves $1,200 per year, what is her yearly income?
48. To purchase a computer for the Student Activities office, the freshman class decides to raise $\frac{1}{3}$ of the money, and the sophomore class decides to raise $\frac{1}{2}$ of the money. The Student Government Association agrees to contribute the rest, which amounted to $400. What was the cost of the computer?

CHAPTER 1 Summary

Section	Important Terms	Important Ideas
1-1	Inductive reasoning Deductive reasoning Conjecture Counterexample	**In mathematics,** two types of reasoning can be used. They are inductive and deductive reasoning. Inductive reasoning is the process of arriving at a general conclusion based on the observation of specific examples. Deductive reasoning is the process of reasoning that arrives at a conclusion based on previously accepted statements.
1-2	Estimation Bar graph Pie chart Time series graph	**In many** cases, it is not necessary to find the exact answer to a problem. When only an approximate answer is needed, you can use estimation. This is often accomplished by rounding the numbers used in the problem and then performing the necessary operation or operations.
1-3	Polya's four-step problem-solving procedure	**A mathematician** named George Polya devised a procedure to solve mathematical problems. The steps of his procedure are (1) understand the problem, (2) devise a plan to solve the problem, (3) carry out the plan to solve the problem, and (4) check the answer.

MATH IN Criminal Investigation REVISITED

1. The suspect was identified by specific incidents in the past, which makes this inductive reasoning that would not hold up in court without further evidence.
2. Fingerprints positively identify the officer as having had contact with the victim. This is deductive reasoning and would be useful in court.
3. Like fingerprints, DNA can positively show that the suspect had physical contact with the victim. This evidence, based on deductive reasoning, would hurt the suspect badly in court.
4. While this is compelling evidence, it's based on assuming that those five drawings indicate the artist is the killer. While unlikely, it could be a coincidence based on five drawings, so this is inductive reasoning. It might impress a jury to some extent, but wouldn't be sufficient for a conviction.

Review Exercises

Section 1-1

For Exercises 1–4, make reasonable conjectures for the next three numbers or letters in the sequence.

1. 3 4 6 7 9 10 12 13 15 16 __ __ __
2. 2 7 4 9 6 11 8 13 __ __ __
3. 4 z 16 w 64 t 256 __ __ __
4. 20 A 18 B 16 C __ __ __

For Exercises 5 and 6, make a reasonable conjecture and draw the next figure.

5. 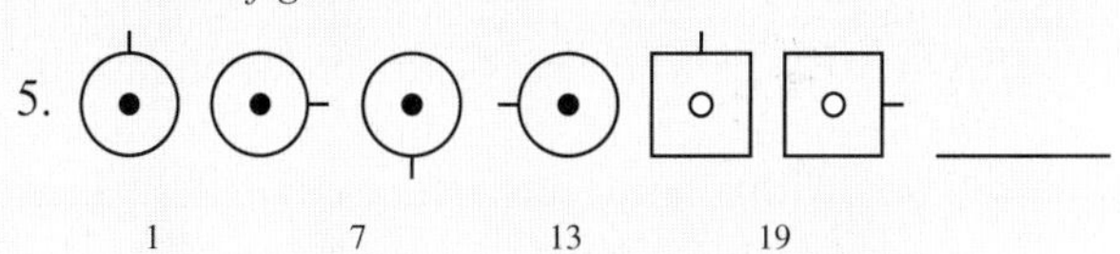

6. 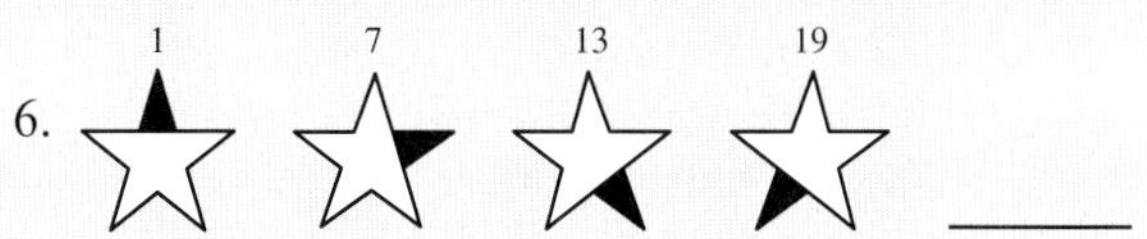

For Exercises 7 and 8, find a counterexample to show that each statement is false.

7. The product of three odd numbers will always be even.
8. The sum of three multiples of 5 will always end in a 5.

For Exercises 9 and 10, use inductive reasoning to find a rule that relates the number selected to the final answer, and then try to prove your conjecture, using deductive reasoning.

9. Select an even number.

   Add 6:
   Divide the answer by 2:
   Add 10:
   Result:

10. Select a number.

    Multiply it by 9:
    Add 18 to the number:
    Divide by 3:
    Subtract 6:
    Result:

In Exercises 11 and 12, use inductive reasoning to find the next two equations in the pattern:

11. $337 \times 3 = 1{,}011$
    $337 \times 6 = 2{,}022$
    $337 \times 9 = 3{,}033$

12. $33 \times 33 = 1{,}089$
    $333 \times 333 = 110{,}889$
    $3333 \times 3333 = 11{,}108{,}889$

13. Use inductive reasoning to solve the last equation.

    $\sqrt{1} = 1$
    $\sqrt{1 + 3} = 2$
    $\sqrt{1 + 3 + 5} = 3$
    $\sqrt{1 + 3 + 5 + 7} = 4$
    $\sqrt{1 + 3 + 5 + 7 + 9 + 11 + 13 + 15 + 17} = ?$

14. Take the number 153, cube each digit, and add the cubes. What is the result? Now cube the digits of the result and add the cubes. What is the result? Use inductive reasoning to determine the result if the procedure is repeated 10 times.

In Exercises 15–18, decide whether inductive or deductive reasoning was used.

15. My professor has given extra credit to his students for contributing canned goods to a food pantry during finals week for the last 5 years, so I know I'll get a chance for some easy extra credit on my final.
16. A GPA of 3.5 is required to make the dean's list. I checked with all my teachers to see what my final grades will be, and my GPA works out to be 3.72, so I'll be on the dean's list this semester.
17. To qualify for bowl games, college football teams have to win at least six games. Our team finished 5–7, so they won't be playing in a bowl game this year.
18. The fastest time that I've ever made it to class from my apartment is 8 minutes, and class starts in 7 minutes, so I'll be late today.

Section 1-2

For Exercises 19–23, round each number to the place value given.

19. 132,356 (thousands)
20. 186.75 (ones)
21. 14.63157 (ten-thousandths)
22. 0.6314 (tenths)
23. 3,725.63 (tens)
24. Estimate the cost of four lawn mowers if each one costs $329.95.
25. Estimate the cost of five textbooks if they cost $115.60, $89.95, $29.95, $62.50, and $43.10.
26. A car averages 14 miles per gallon of gas. Estimate how far the car can travel on 19 gallons of gas.
27. A family of six consists of four people older than 12 and two people 12 or under. Tickets into an amusement park are $57.95 for those over 12 and $53.95 for those 12 and under. Estimate how much it would cost the family to go to the amusement park.
28. At M.T. Wallatts University, it costs a student $689 per credit-hour to attend.
    (a) Estimate the cost for a student to attend one semester if he registers for 9 credit-hours.
    (b) If a student makes $11 an hour at her part-time job (after taxes) and works 30 hours a week, approximately how many weeks will she have to work to afford one semester with 9 credit-hours?
29. A video store sells used DVDs for $8.95, tax included. About how many could you buy with the $130 you got for selling your textbooks back?

Use the information shown in the pie chart for Exercises 30 and 31. The pie chart shows the percent of the pollutants released into the environment for a specific year.

Pollutant Releases in the United States

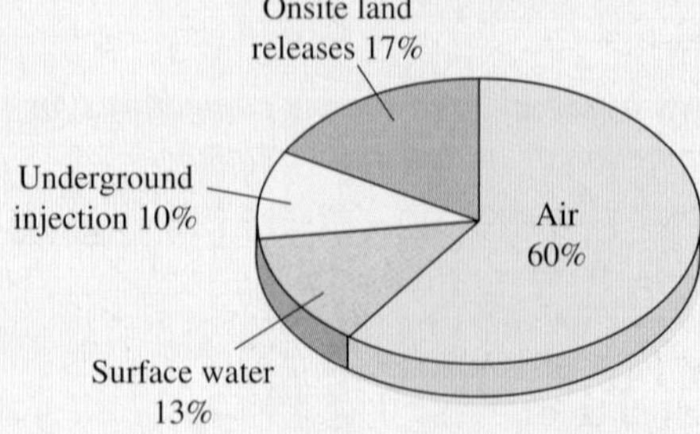

Source: *The World Almanac and Book of Facts*

30. If the total amount of pollutants released is 1,953 million pounds, find the amount of pollutant released in the air.
31. If the total amount of pollutants released is 1,953 million pounds, find the amount of pollutants released in surface water.

Use the information shown in the graph for Exercises 32 and 33. The graph shows the average weekly salary (in dollars) for U.S. production workers from 1970 to 2005.

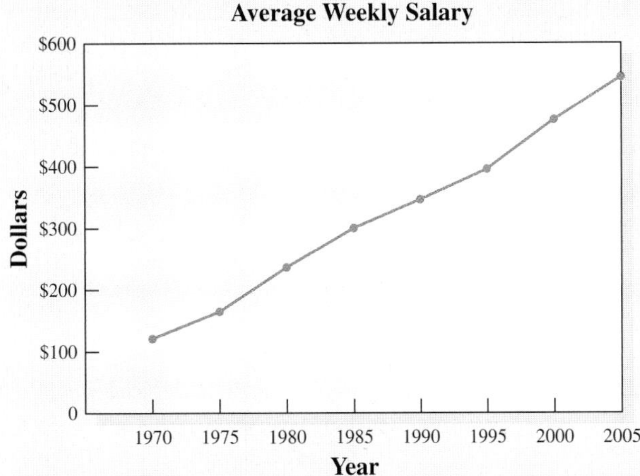

Source: *World Almanac and Book of Facts*

32. Estimate the weekly salary in 1988.
33. Estimate the weekly salary in 2003.
34. Estimate the distance between two cities that are 8 inches apart on a map if the scale is 2 inches = 15 miles.

Section 1-3

35. Cindy has 32 flyers about a campus symposium on environmental issues that she is organizing. She gave away all but 9. How many did she have left?
36. A tennis team played 40 matches. The team won 20 more matches than they lost. How many matches did the team lose?
37. If a person weighs 110 pounds when standing on one foot on a scale, how much will that person weigh when he stands on a scale with both feet?
38. A small mocha latte and biscotti together cost $3.40. If the mocha latte cost $0.40 more than the biscotti, how much did each cost?
39. Harry is 10 years old, and his brother Bill is twice as old. When Harry is 20, how old will Bill be?
40. Mary got $80 in tips last night for her waitressing job at the campus café. She spent $8.00 downloading songs online to her MP3 player and then spent $\frac{1}{3}$ of the remainder on tickets for the spring dance on Saturday. How much did she have left?
41. Your science textbook and its accompanying lab packet cost $120. If the textbook costs twice as much as the lab packet, how much did the lab packet cost?
42. Fill in the squares with digits to complete the problem. There are several correct answers.

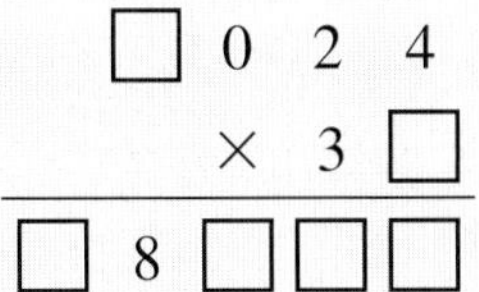

43. In 10 years, Harriet will be 3 times as old as she was 10 years ago. How old is she now?
44. At times during the summer in Alaska, the day is 18 hours longer than the night. How long is each?
45. Babs is 9 years older than Debbie, and Jack is 7 years younger than Babs. The sum of their ages is 20. How old is Debbie?
46. Using +, −, and ×, make a true equation. Do not change the order of the digits.

$$2 \quad 9 \quad 6 \quad 7 \quad = \quad 17$$

47. Tina and Joe are doing homework problems together for their math class. Joe says to Tina, "If I do one more problem, then we'll have both done the same number of problems." Tina says to Joe, "If I do one more problem, than I will have done twice the number you have!" How many problems has each one done so far?
48. Can you divide a pie into 11 pieces with four straight cuts? The cuts must go from rim to rim but not necessarily through the center. The pieces need not be identical.
49. How many triangles are in the figure shown here?

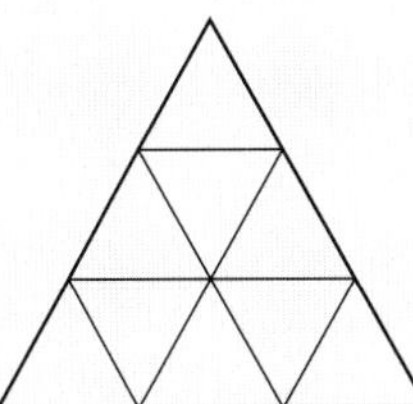

50. In 20 years, Charles will be 3 times as old as he is now. How old is he now?
51. The sum of two numbers is 20 and the difference is 5. Find the numbers.
52. A health food store charges $2.00 per pound for high-protein nature mix and $2.75 for low-carb soy medley. If 10 pounds of the two items mixed together costs $24.50, find the amount of each type in the mixture.
53. Taylor had $1,000 to invest. He invested part of it at 8% and part of it at 6%. If his total simple interest was $76.00, find how much he invested for 1 year at each rate.

Chapter Test

For Exercises 1 and 2, make reasonable conjectures for the next three numbers in the sequence.

1. 2 4 3 6 5 9 8 __ __ __
2. 5 10 20 40 80 __ __ __
3. Use inductive reasoning to find the solution to the problem; then check it by performing the calculation.
   $0 \cdot 9 + 8 = 8$
   $9 \cdot 9 + 7 = 88$
   $98 \cdot 9 + 6 = 888$
   $987 \cdot 9 + 5 = 8{,}888$
   $\vdots$
   $9{,}876{,}543 \cdot 9 + 1 = ?$
4. Use inductive reasoning to find the solution to the problem; then check it by performing the calculation.
   $6 \times 7 = 42$
   $66 \times 67 = 4{,}422$
   $666 \times 667 = 444{,}222$
   $6{,}666 \times 6{,}667 = ?$
5. Use inductive reasoning to find a rule that relates the number you selected to the final answer, and try to prove your conjecture.

   Select a number:
   Add 10 to the number:
   Multiply the answer by 5:
   Add 15 to the answer:
   Divide the answer by 5:
   Result:
6. There were 12 students in line to register for the Underwater Basketweaving 101 course. All but 2 changed their minds. How many remained in line?
7. For a job, a worker receives a salary that doubles each day. After 10 days of work, the job is finished and the person receives \$100.00 for the final day's work. On which day did the worker receive \$25.00?
8. What are the next two letters in the sequence O, T, T, F, F, S, S, . . . ? (*Hint:* It has something to do with numbers.)
9. By moving just one coin, make two lines, each three coins long.

10. A ship is docked in harbor with a rope ladder hanging over the edge, and 9 feet of the ladder is above the waterline. The tide is rising at 8 inches per hour. After 6 hours, how much of the ladder remains above the waterline?
11. The problem here was written by the famous mathematician Diophantus. Can you find the solution?

    The boyhood of a man lasted $\frac{1}{6}$ of his life; his beard grew after $\frac{1}{12}$ more; after $\frac{1}{7}$ more he married; 5 years later his son was born; the son lived to one-half the father's age; and the father died 4 years after the son. How old was the father when he died?
12. A number divided by 3 less than itself gives a quotient of $\frac{8}{5}$. Find the number.
13. The sum of $\frac{1}{2}$ of a number and $\frac{1}{3}$ of the same number is 10. Find the number.
14. Add five lines to the square to make three squares and two triangles.

15. One person works for 3 hours and another person works for 2 hours. They are given a total of \$60.00. How should it be divided up so that each person receives a fair share?
16. The sum of the reciprocals of two numbers is $\frac{5}{6}$ and the difference is $\frac{1}{6}$. Find the numbers. (*Hint:* The reciprocal of a number n is $\frac{1}{n}$.)
17. If Sam scored 87% on his first exam, what score would he need on his second exam to bring his average up to 90%?
18. Mt. McKinley is about 20,300 feet above sea level, and Death Valley is 280 feet below sea level. Find the vertical distance from the top of Mt. McKinley to the bottom of Death Valley.
19. The depth of the ocean near the island of Mindanao is about 36,400 feet. The height of Mt. Everest is about 29,000 feet. Find the vertical distance from the top of Mt. Everest to the floor of the ocean near the island of Mindanao.
20. A person is drawing a map and using a scale of 2 inches = 500 miles. How far apart on the map will two cities be located if they are 1,800 miles apart?
21. Mark's mother is 32 years older than Mark. The sum of their ages is 66 years. How old is each?
22. Round 1,674,253 to the nearest hundred-thousand.
23. Round 1.3752 to the nearest hundredth.
24. Estimate the cost of Stuart's new wardrobe for his campus interview if a blazer costs \$69.95, a new tie costs \$32.54, and new pants cost \$42.99.

25. Using the graph, estimate the average number of hours per week U.S. production workers worked in 1980 and 2005.

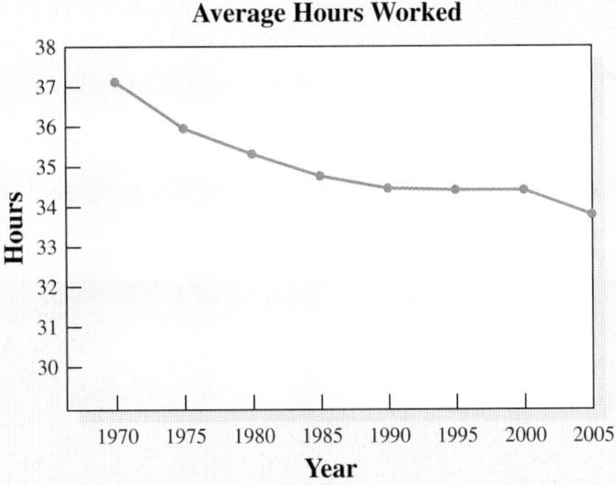

Source: *World Almanac and Book of Facts*

26. Using the pie chart, of the 3,646 students surveyed at 7 universities, estimate the number of students surveyed who do not live in a residence hall.

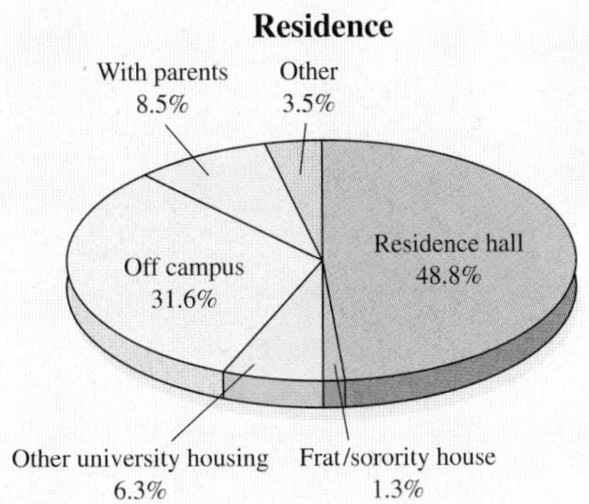

Source: http://www.acha-ncha.org/data/DEMOGF06.html

27. For a specific year, the number of homicides reported for selected cities is shown here. Estimate the number of homicides in Baltimore and the difference between the number of homicides in Chicago and Atlanta.

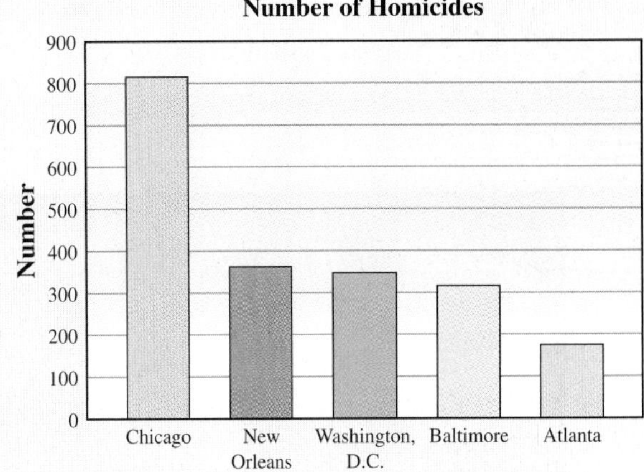

Source: *USA Today*

28. Estimate the distance between two cities that are 4 inches apart if the scale on a map is $\frac{1}{2}$ inch = 8 miles.

Projects

1. In Section 1-3, we solved some problems by using algebraic equations. For more complicated situations, more than one equation might be required. When this is the case, it's not uncommon for there to be more than one correct solution. Can you solve the following problem?

   Darla went to Costco and spent \$100 for 100 items. She bought 6-packs of bottled water, 20-oz bottles of Diet Pepsi, and 10-pack pouches of Kool-Aid mix. Each 6-pack of water cost \$3.00, each bottle of Diet Pepsi cost \$1.00, and each pouch of Kool-Aid mix cost \$0.50. How many of each item did Darla buy?
2. One reason that Polya's method has stood the test of time is that it is used not just in math, but also for any problem needing to be solved. For each of the problems listed, write a paragraph explaining how each of the steps in Polya's method could be applied.
   (a) A homeowner has a small set of stairs leading from his back door down to a featureless backyard. He would like to have a place for outdoor parties during the summer.
   (b) Your car has been making a really funny noise whenever you go over 50 miles per hour.
   (c) You've balanced your checkbook and found out that you won't have enough money to pay for tuition and books next semester.
   (d) One of your friends heard something you said about him behind his back, and now he is very upset with you.

CHAPTER 2

Sets

Outline

2-1 The Nature of Sets
2-2 Subsets and Set Operations
2-3 Using Venn Diagrams to Study Set Operations
2-4 Using Sets to Solve Problems
2-5 Infinite Sets
Summary

MATH IN Demographics

The world certainly has changed a lot in the last 50 years. If you could travel back to the middle of the 20th century, you'd have an awful hard time finding anyone who would believe that one of the most famous, and richest, people in the world in the early 21st century would be a golfer who is part white, part African-American, *and* part Asian. Our society is becoming more and more diverse every year, to the point where even defining what we mean by *race* isn't so easy anymore. The most common racial groups referred to in population statistics are white, black, Asian, and Hispanic, but many people fall into more than one category (including the President of the United States!). But it's even more confusing than that: *Hispanic* isn't really a race, but rather an ethnic group, and many Hispanics also report themselves as either white or black.

Sorting it all out to get any sort of meaningful picture of what we as Americans look like isn't easy, and the techniques of set theory are very useful tools in trying to do so. In this chapter, we'll define what we mean by sets, and we'll study sets and how they can be used to organize information in an increasingly complex world. The concept of sets has been used extensively since people began studying mathematics, but it wasn't until the late 1800s that the theory of sets was studied as a specific branch of math. One of the major tools that we will use to study sets—the Venn diagram—was introduced in an 1880 paper by a man named John Venn. These diagrams allow us to picture complicated relationships between sets of objects, like people of certain races.

Because race and ethnicity are self-reported in a variety of different ways, it's very difficult to find detailed data on the breakdown of races, but there are some reasonable estimates out there. The following estimates were cobbled together from a number of different sources. In a group of 1,000 randomly selected Americans, 777 would self-report as white, 139 as black, and 149 as Hispanic. In addition, 19 would self-report as black and white, 18 as black and Hispanic, and 85 as white and Hispanic. Finally, 7 would self-report as all three. Based on these estimates, after completing this chapter, you should be able to answer the following questions:

1. How many of the original 1,000 report as white only, black only, and Hispanic only?
2. How many report as Hispanic and black, but not white?
3. How many report as either Hispanic or black?
4. How many report as none of white, black, or Hispanic?

For answers, see Math in Demographics Revisited on page 87

Section 2-1 The Nature of Sets

LEARNING OBJECTIVES

- ☐ 1. Define set.
- ☐ 2. Write sets in three different ways.
- ☐ 3. Classify sets as finite or infinite.
- ☐ 4. Define the empty set.
- ☐ 5. Find the cardinality of a set.
- ☐ 6. Decide if two sets are equal or equivalent.

Groups of people, categories of items in stores, laws that apply to traffic—our world is divided into groups of things, or sets. So studying sets from a mathematical standpoint is a good opportunity to study how math is used in our world.

Basic Concepts

We will begin with a basic definition of sets.

A **set** is a well-defined collection of objects.

☑ 1. Define set.

By *well-defined* we mean that given any object, we can definitely decide whether it is or is not in the set. For example, the set "letters of the English alphabet" is well-defined since it consists of the 26 symbols we use to make up our alphabet, and no other objects. The set "tall people in your class" is not well-defined because who exactly belongs to that set is open to interpretation.

Each object in a set is called an **element** or a **member** of the set. One method of designating a set is called the **roster method,** in which elements are listed between braces, with commas between the elements. The order in which we list elements isn't important: $\{2, 5, 7\}$ and $\{5, 2, 7\}$ are the same set. Often, we will name sets by using a capital letter.

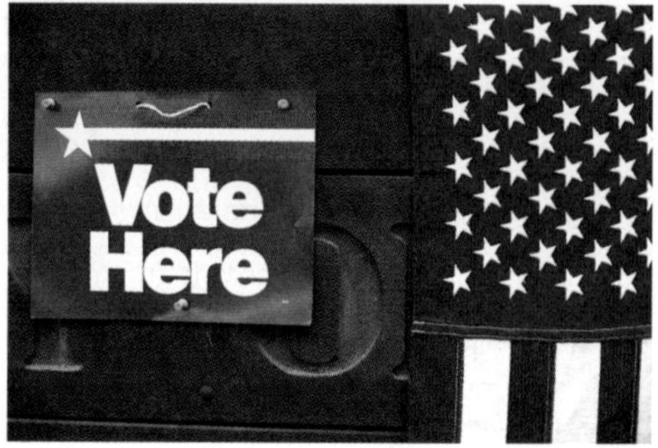

Laws that specify who can vote in a specific election determine a well-defined set of people. There are many other laws that might affect certain sets, like the set of businesses in a certain industry. These laws need to clearly define the set the law applies to—if that set were not well-defined, it would be almost impossible to enforce the law.

EXAMPLE 1 Listing the Elements in a Set

Write the set of months of the year that begin with the letter M.

Math Note

The commas make it clear that it is the words, not the letters, that are the elements of the set.

SOLUTION

The months that begin with M are March and May. So, the answer can be written in set notation as

$$M = \{\text{March, May}\}$$

Each element in the set is separated by a comma.

▼ Try This One 1

Write the set of the Great Lakes.

In math, the set of *counting numbers* or **natural numbers** is defined as $N = \{1, 2, 3, 4, \ldots\}$. (When we are designating sets, the three dots, or *ellipsis,* mean that the list of elements continues indefinitely in the same pattern.)

EXAMPLE 2 Writing Sets Using the Roster Method

Math Note

You can list an element of a set more than once if it means a lot to you, but it's common to choose not to list repeats. For example, the set of letters in the word *letters* is written as {l, e, t, r, s}.

Use the roster method to do the following:

(a) Write the set of natural numbers less than 6.
(b) Write the set of natural numbers greater than 4.

SOLUTION

(a) $\{1, 2, 3, 4, 5\}$
(b) $\{5, 6, 7, 8, \ldots\}$

▼ Try This One 2

Write each set, using the roster method.

(a) The set of even natural numbers from 80 to 90.
(b) The set of odd natural numbers greater than 10.

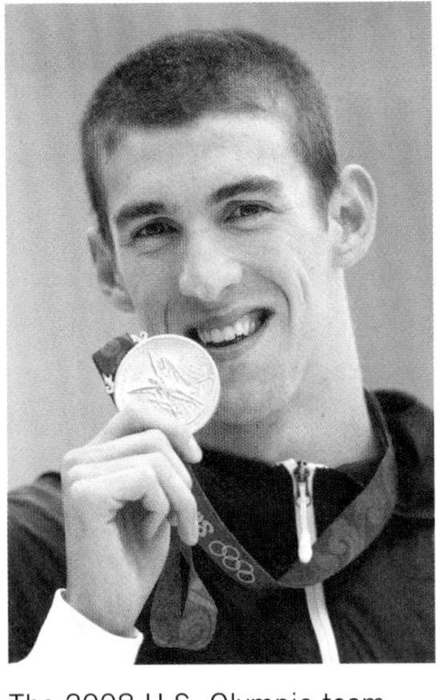

The 2008 U.S. Olympic team is a set of athletes. Swimmer Michael Phelps is an element of that set.

The symbol ∈ is used to show that an object is a member or element of a set. For example, let set $A = \{2, 3, 5, 7, 11\}$. Since 2 is a member of set A, it can be written as

$$2 \in \{2, 3, 5, 7, 11\} \quad \text{or} \quad 2 \in A$$

Likewise,

$$5 \in \{2, 3, 5, 7, 11\} \quad \text{or} \quad 5 \in A$$

When an object is not a member of a set, the symbol ∉ is used. Because 4 is not an element of set A, this fact is written as

$$4 \notin \{2, 3, 5, 7, 11\} \quad \text{or} \quad 4 \notin A$$

EXAMPLE 3 Understanding Set Notation

Math Note

Be sure to use correct symbols when you show membership in a set. For example, the notation $\{6\} \in \{2, 4, 6\}$ is incorrect since the set {6} is not a member of this set: the number 6 is.

Decide whether each statement is true or false.

(a) Oregon $\in A$, where A is the set of states west of the Mississippi River.
(b) $27 \in \{1, 5, 9, 13, 17, \ldots\}$
(c) $z \notin \{v, w, x, y, z\}$

SOLUTION

(a) Oregon is west of the Mississippi, so Oregon is an element of A. The statement is true.
(b) The pattern shows that each element is 4 more than the previous element. So the next three elements are 21, 25, and 29; this shows that 27 is not in the set. The statement is false.
(c) The letter z is an element of the set, so the statement is false.

▼ Try This One 3

Decide whether each statement is true or false.

(a) July $\in A$, where A is the set of months between Memorial Day and Labor Day.
(b) $21 \in \{2, 5, 8, 11, \ldots\}$
(c) map $\notin$ {m, a, p}

There are three common ways to designate sets:

1. The *list* or *roster* method.
2. The *descriptive* method.
3. *Set-builder* notation.

We already know a lot about using the list or roster method; the elements of the set are listed in braces and are separated by commas, as in Examples 1 through 3.

The **descriptive method** uses a short statement to describe the set.

EXAMPLE 4 Describing a Set in Words

Use the descriptive method to describe the set E containing $2, 4, 6, 8, \ldots$.

SOLUTION

The elements in the set are called the even natural numbers. The set E is the set of even natural numbers.

▼ Try This One 4

Use the descriptive method to describe the set A containing $-3, -2, -1, 0, 1, 2, 3$.

> **Math Note**
>
> When you hear *variable*, you might automatically think *letter*, like x or y. But you should think about what the word *variable* really means: something that can change, or vary. A variable is just a symbol that represents some number or object that can change.

The third method of designating a set is **set-builder notation,** and this method uses *variables*.

A **variable** is a symbol (usually a letter) that can represent different elements of a set.

Set-builder notation uses a variable, braces, and a vertical bar | that is read as "such that." For example, the set {1, 2, 3, 4, 5, 6} can be written in set-builder notation as $\{x \mid x \in N \text{ and } x < 7\}$. It is read as "the set of elements x such that x is a natural number and x is less than 7." We can use any letter or symbol for the variable, but it's common to use x.

EXAMPLE 5 Writing a Set Using Set-Builder Notation

Use set-builder notation to designate each set, then write how your answer would be read aloud.

(a) The set R contains the elements 2, 4, and 6.
(b) The set W contains the elements red, yellow, and blue.

SOLUTION

(a) $R = \{x | x \in E \text{ and } x < 7\}$, the set of all x such that x is an even natural number and x is less than 7.
(b) $W = \{x | x \text{ is a primary color}\}$, the set of all x such that x is a primary color.

Math Note

Recall from Example 4 that E represents the set of even natural numbers. Also, you should note that there could be more than one way to write a set in set-builder notation. In Example 5, we could have written $W = \{x | x \text{ is a color in the flag of Colombia}\}$.

▼ Try This One 5

Use set-builder notation to designate each set; then write how your answer would be read aloud.

(a) The set K contains the elements 10, 12, 14, 16, 18.
(b) The set W contains the elements Democratic and Republican.

EXAMPLE 6 Using Different Set Notations

Designate the set S with elements 32, 33, 34, 35, . . . using

(a) The roster method.
(b) The descriptive method.
(c) Set-builder notation.

SOLUTION

(a) $\{32, 33, 34, 35, \ldots\}$
(b) The set S is the set of natural numbers greater than 31.
(c) $\{x | x \in N \text{ and } x > 31\}$

▼ Try This One 6

Designate the set with elements 11, 13, 15, 17, . . . using

(a) The roster method.
(b) The descriptive method.
(c) Set-builder notation.

If a set contains many elements, we can again use an ellipsis to represent the missing elements. For example, the set $\{1, 2, 3, \ldots, 99, 100\}$ includes all the natural numbers from 1 to 100. Likewise, the set $\{a, b, c, \ldots, x, y, z\}$ includes all the letters of the alphabet.

EXAMPLE 7 Writing a Set Using an Ellipsis

Using the roster method, write the set containing all even natural numbers between 99 and 201.

SOLUTION

$\{100, 102, 104, \ldots, 198, 200\}$

▼ Try This One 7

☑ 2. Write sets in three different ways.

Using the roster method, write the set of odd natural numbers between 50 and 500.

Finite and Infinite Sets

Sets can be classified as *finite* or *infinite*.

If a set has no elements or a specific natural number of elements, then it is called a **finite set.** A set that is not a finite set is called an **infinite set.**

The set {p, q, r, s} is called an finite set since it has four members: p, q, r, and s. The set $\{10, 20, 30, \ldots\}$ is called an infinite set since it has an unlimited number of elements: the natural numbers that are multiples of 10.

EXAMPLE 8 Classifying Sets as Finite or Infinite

Classify each set as finite or infinite.

(a) $\{x \mid x \in N \text{ and } x < 100\}$
(b) Set R is the set of letters used to make Roman numerals.
(c) {100, 102, 104, 106, . . .}
(d) Set M is the set of people in your immediate family.
(e) Set S is the set of songs that can be written.

SOLUTION

(a) The set is finite since there are 99 natural numbers that are less than 100.
(b) The set is finite since the letters used are C, D, I, L, M, V, and X.
(c) The set is infinite since it consists of an unlimited number of elements.
(d) The set is finite since there is a specific number of people in your immediate family.
(e) The set is infinite because an unlimited number of songs can be written.

▼ Try This One 8

☑ 3. Classify sets as finite or infinite.

Classify each set as finite or infinite.

(a) Set P is the set of numbers that are multiples of 6.
(b) $\{x \mid x \text{ is a member of the U.S. Senate}\}$.
(c) $\{3, 6, 9, \ldots, 24\}$.

There are some situations in which it is necessary to define a set with no elements. For example, the set of female Presidents of the United States would contain no people, so it would have no elements.

A set with no elements is called an *empty set* or **null set.** The symbols used to represent the null set are { } or ∅.

EXAMPLE 9 Identifying Empty Sets

Which of the following sets are empty?

(a) The set of woolly mammoth fossils in museums.
(b) $\{x \mid x \text{ is a living woolly mammoth}\}$
(c) $\{\varnothing\}$
(d) $\{x \mid x \text{ is a natural number between 1 and 2}\}$

SOLUTION

(a) There is certainly at least one woolly mammoth fossil in a museum somewhere, so the set is not empty.
(b) Woolly mammoths have been extinct for almost 8,000 years, so this set is most definitely empty.
(c) Be careful! Each instance of { } and $\varnothing$ represents the empty set, but $\{\varnothing\}$ is a set with one element: $\varnothing$.
(d) This set is empty because there are no natural numbers between 1 and 2.

▼ Try This One 9

Which of the following sets are empty sets?

(a) $\{x \mid x \text{ is a natural number divisible by 7}\}$
(b) $\{x \mid x \text{ is a human being living on Mars}\}$
(c) $\{+, -, \times, \div\}$
(d) The set Z consists of the living people on earth who are over 200 years old.

CAUTION

Don't write the empty set as $\{\varnothing\}$. This is the set *containing* the empty set.

☑ 4. Define the empty set.

Cardinal Number of a Set

The number of elements in a set is called the *cardinal number* of a set. For example, the set $R = \{2, 4, 6, 8, 10\}$ has a cardinal number of 5 since it has 5 elements. This could also be stated by saying the **cardinality** of set R is 5. Formally defined,

The **cardinal number** of a finite set is the number of elements in the set. For a set A the symbol for the cardinality is $n(A)$, which is read as "n of A."

EXAMPLE 10 Finding the Cardinality of a Set

Find the cardinal number of each set.

(a) $A = \{5, 10, 15, 20, 25, 30\}$
(b) $B = \{10, 12, 14, \ldots, 28, 30\}$
(c) $C = \{16\}$
(d) $\varnothing$

SOLUTION

(a) $n(A) = 6$ since set A has 6 elements
(b) $n(B) = 11$ since set B has 11 elements
(c) $n(C) = 1$ since set C has 1 element
(d) $n(\varnothing) = 0$ since there are no elements in an empty set

5. Find the cardinality of a set.

▼ Try This One 10

Find the cardinal number of each set.

(a) $A = \{z, y, x, w, v\}$
(b) $B = \{1, 3, 5, 7, \ldots, 27, 29\}$
(c) $C = \{\text{Chevrolet}\}$

Math Note

All equal sets are equivalent since both sets will have the same number of members, but not all equivalent sets are equal. For example, the sets {x, y, z} and {10, 20, 30} have three members, but in this case, the members of the sets are not identical. The two sets are equivalent but not equal.

Equal and Equivalent Sets

In set theory, it is important to understand the concepts of *equal* sets and *equivalent* sets.

Two sets A and B are **equal** (written $A = B$) if they have exactly the same members or elements. Two finite sets A and B are said to be **equivalent** (written $A \cong B$) if they have the same number of elements: that is, $n(A) = n(B)$.

For example, the two sets {a, b, c} and {c, b, a} are equal since they have exactly the same members, a, b, and c. Also the set {4, 5, 6} is equal to the set {4, 4, 5, 6} since 4 need not be written twice in the second set. The set $C = \{x, y, z\}$ is equivalent to the set $D = \{10, 20, 30\}$ (i.e., $C \cong D$) since both sets have three elements, but the sets are not equal.

EXAMPLE 11 Deciding If Sets Are Equal or Equivalent

State whether each pair of sets is equal, equivalent, or neither.

(a) {p, q, r, s}; {a, b, c, d}
(b) {8, 10, 12}; {12, 8, 10}
(c) {213}; {2, 1, 3}
(d) {1, 2, 10, 20}; {2, 1, 20, 11}
(e) {even natural numbers less than 10}; {2, 4, 6, 8}

SOLUTION

(a) Equivalent
(b) Equal and equivalent
(c) Neither
(d) Equivalent
(e) Equal and equivalent

Two sets of basketball teams on the court have a one-to-one correspondence. (Assuming each has five healthy players!)

▼ Try This One 11

State whether each pair of sets is equal, equivalent, or neither.

(a) {d, o, g}; {c, a, t}
(b) {run}; {r, u, n}
(c) {t, o, p}; {p, o, t}
(d) {10, 20, 30}; {1, 3, 5}

The elements of two equivalent sets can be paired in such a way that they are said to have a *one-to-one correspondence* between them.

Two sets have a **one-to-one correspondence** of elements if each element in the first set can be paired with exactly one element of the second set and each element of the second set can be paired with exactly one element of the first set.

EXAMPLE 12 Putting Sets in One-to-One Correspondence

Show that (a) the sets {8,16, 24, 32} and {s, t, u, v} have a one-to-one correspondence and (b) the sets {x, y, z} and {5, 10} do not have a one-to-one correspondence.

SOLUTION

(a) We need to demonstrate that each element of one set can be paired with one and only one element of the second set. One possible way to show a one-to-one correspondence is this:

$$\begin{array}{cccc} \{8, & 16, & 24, & 32\} \\ \updownarrow & \updownarrow & \updownarrow & \updownarrow \\ \{s, & t, & u, & v\} \end{array}$$

(b) The elements of the sets {x, y, z} and {5, 10} can't be put in one-to-one correspondence. No matter how we try, there will be an element in the first set that doesn't correspond to any element in the second set.

▼ Try This One 12

Show that the sets {North, South, East, West} and {sun, rain, snow, sleet} have a one-to-one correspondence.

Using one-to-one correspondence, we can decide if two sets are equivalent without actually counting the elements. This can come in handy for large sets, and *really* handy for infinite sets!

Correspondence and Equivalent Sets

Two sets are

- Equivalent if you can put their elements in one-to-one correspondence.
- Not equivalent if you cannot put their elements in one-to-one correspondence.

☑ 6. Decide if two sets are equal or equivalent.

In this section, we defined sets and some other basic terms. In Section 2-2, subsets and set operations will be explained.

Answers to Try This One

1 {Ontario, Erie, Huron, Michigan, Superior}

2 (a) {80, 82, 84, 86, 88, 90}
(b) {11, 13, 15, 17, . . .}

3 (a) True (b) False (c) True

4 The set of integers from −3 to 3

5 (a) $K = \{x | x \in E, x > 9, \text{ and } x < 19\}$, the set of all x such that x is an even natural number, x is greater than 9, and x is less than 19.
(b) $W = \{x | x$ is a major American political party$\}$, the set of all x such that x is a major American political party.

6 (a) $\{11, 13, 15, 17, \ldots\}$
(b) The set of odd natural numbers greater than 10
(c) $\{x | x \in N, x \text{ is odd, and } x > 10\}$

7 $\{51, 53, 55, \ldots, 497, 499\}$

8 (a) Infinite (b) Finite (c) Finite

9 (b) and (d)

10 (a) 5 (b) 15 (c) 1

11 (a) Equivalent (b) Neither (c) Equal and equivalent (d) Equivalent

12

North	South	East	West
↕	↕	↕	↕
Sun	Rain	Snow	Sleet

EXERCISE SET 2-1

Writing Exercises

1. Explain what a set is.
2. List three ways to write sets.
3. What is the difference between equal and equivalent sets?
4. Explain the difference between a finite and an infinite set.
5. What is meant by "one-to-one correspondence between two sets"?
6. Define the empty set and give two examples of an empty set.

Computational Exercises

For Exercises 7–20, write each set, using roster notation. Do not include repeats.

7. S is the set of letters in the word *stress*.
8. A is the set of letters in the word *Alabama*.
9. P is the set of natural numbers between 50 and 60.
10. R is the set of even natural numbers between 10 and 40.
11. Q is the set of odd natural numbers less than 15.
12. M is the set of even natural numbers less than 8.
13. $G = \{x | x \in N \text{ and } x > 10\}$
14. B is the set of natural numbers greater than 100.
15. Y is the set of natural numbers between 2,000 and 3,000.
16. $Z = \{x | x \in N \text{ and } 500 < x < 6{,}000\}$
17. W is the set of days in the week.
18. C is the set of colors in a U.S. flag.
19. D is the set of suits in a deck of cards.
20. F is the set of face cards in a deck of cards.

For Exercises 21–28, write each set, using the descriptive method.

21. $\{2, 4, 6, 8, \ldots\}$
22. $\{1, 3, 5, 7, \ldots\}$
23. $\{9, 18, 27, 36\}$
24. $\{5, 10, 15, 20\}$
25. $\{m, a, r, y\}$
26. $\{t, h, o, m, a, s\}$
27. $\{100, 101, 102, \ldots, 199\}$
28. $\{21, 22, 23, \ldots, 29, 30\}$

For Exercises 29–34, write each set, using set-builder notation.

29. $\{10, 20, 30, 40, \ldots\}$
30. $\{55, 65, 75, 85\}$
31. X is the set of natural numbers greater than 20.
32. Z is the set of even natural numbers less than 12.
33. $\{1, 3, 5, 7, 9\}$
34. $\{18, 21, 24, 27, 30\}$

For Exercises 35–40, list the elements in each set.

35. H is the set of natural numbers less than 0.
36. $\{x | x \in N \text{ and } 70 < x < 80\}$
37. $\{7, 14, 21, \ldots, 63\}$
38. $\{5, 12, 19, \ldots, 40\}$
39. $\{x | x$ is an even natural number between 100 and 120$\}$
40. $\{x | x$ is an odd natural number between 90 and 100$\}$

For Exercises 41–48, state whether each collection is well-defined or not well-defined.

41. J is the set of seasons in the year.
42. $\{x \mid x \in N\}$
43. $\{x \mid x$ is an excellent instructor$\}$
44. $\{10, 15, 20\}$
45. $\{1, 2, 3, \ldots, 100\}$
46. $\{x \mid x$ is a good professional golfer$\}$
47. $\{100, \ldots\}$
48. W is the set of days of the week.

For Exercises 49–54, state whether each is true or false.

Let $A = \{\text{Saturday, Sunday}\}$
$B = \{1, 2, 3, 4, 5\}$
$C = \{\text{p, q, r, s, t}\}$

49. $3 \in B$
50. $\text{a} \in C$
51. Wednesday $\notin A$
52. $7 \notin B$
53. $\text{r} \in C$
54. $\text{q} \in B$

For Exercises 55–62, state whether each set is infinite or finite.

55. $\{x \mid x \in N$ and x is even$\}$
56. $\{1, 2, 3, \ldots, 999, 1{,}000\}$
57. K is the set of letters of the English alphabet.
58. $\{x \mid x \in$ years in which the past Presidents of the United States were born$\}$
59. $\{3, 6, 9, 12, \ldots\}$
60. $\varnothing$
61. $\{x \mid x$ is a current television program$\}$
62. $\{x \mid x$ is a fraction$\}$

For Exercises 63–70, state whether each pair of sets is equal, equivalent, or neither.

63. $\{6, 12, 18, 20\}$ and $\{20, 12, 6, 18\}$
64. $\{\text{p, q, r, s, t}\}$ and $\{5, 3, 4, 2, 1\}$
65. $\{2, 3, 7, 8\}$ and $\{1, 4, 5\}$
66. $\{2, 4, 6, 8\}$ and $\{2, 4, 6, 8, \ldots\}$
67. $\{1, 2, 3, \ldots, 99, 100\}$ and $\{1{,}001, 1{,}002, 1{,}003, \ldots, 1{,}100\}$
68. $\{\text{s, t, o, p}\}$ and $\{\text{p, o, t, s}\}$
69. $\{x \mid x$ is a three-digit number$\}$ and $\{1, 2, 3, \ldots, 100\}$
70. $\{$January, June, July$\}$ and $\{x \mid x$ is a month that begins with J$\}$

For Exercises 71–74, show that each pair of sets is equivalent by using a one-to-one correspondence.

71. $\{10, 20, 30, 40\}$ and $\{40, 10, 20, 30\}$
72. $\{\text{w, x, y, z}\}$ and $\{1, 2, 3, 4\}$
73. $\{1, 2, 3, \ldots, 25, 26\}$ and $\{\text{a, b, c}, \ldots, \text{x, y, z}\}$
74. $\{x \mid x$ is an odd natural number less than $11\}$ and $\{x \mid x$ is an even natural number less than $12\}$

For Exercises 75–82, find the cardinal number for each set.

75. $A = \{18, 24, 32, 63, 48\}$
76. $B = \{1, 2, 5, 7, \ldots, 37\}$
77. $C = \{x \mid x$ is a day of the week$\}$
78. $D = \{x \mid x$ is a month of the year$\}$
79. $E = \{\text{three}\}$
80. $F = \{\text{t, h, r, e, e}\}$
81. $G = \{x \mid x \in N$ and x is negative$\}$
82. $H = \varnothing$

For Exercises 83–90, determine whether each statement is true or false.

83. $\{1, 3, 5\} = \{3, 1, 5\}$
84. $\{2, 4, 6\} \neq \{1, 3, 5\}$
85. All equal sets are equivalent.
86. No equivalent sets are equal.
87. $\varnothing = \{\varnothing\}$
88. $\{2, 6, 10, 12\} = \{2, 6, 10\}$
89. $n\{\varnothing\} = 0$
90. $E = \{2, 4, 6, 8, \ldots\}$ is a finite set

Real-World Applications

91. The table below shows the top 10 states in number of immigrants in 2006.

State	Number of Immigrants	% of Total Immigrants to United States
California	264,677	20.9
New York	180,165	14.2
Florida	155,996	12.3
Texas	89,037	7.0
New Jersey	65,934	5.2
Illinois	52,459	4.1
Virginia	38,488	3.0
Massachusetts	35,560	2.8
Georgia	32,202	2.5
Maryland	30,204	2.4

Source: *The World Almanac and Book of Facts,* 2008.

(a) List the set of states with more than 100,000 immigrants.
(b) List the set of states in the top 10 with fewer than 50,000 immigrants.
(c) List $\{x \mid x$ is a state with at least 5% of the immigrant total$\}$.
(d) List $\{x \mid x$ is a state with between 3% and 9% of the immigrant total$\}$.

92. There is a relatively young workforce in the U.S. information technology industry. The percentage of people by age group working at Internet service providers (ISPs), Web search portals, and data processing companies in 2006 is shown in the table on the next page:

Age Group	Percentage Working at ISPs, Web Search Portals, and Data Processing Companies
16–19	1.1
20–24	10.1
25–34	26.8
35–44	31.8
45–54	19.6
55–64	7.8
65 and older	2.2

Source: http://www.bls.gov/oco/cg/cgs055.htm

(a) List the set of age groups of those whose percentages are over 18%. What can you conclude?
(b) List the set of age groups of those whose percentages are less than 10%. What can you conclude?
(c) List the set of percentages of those who are between 20 and 44 years old.
(d) Find $\{x \mid x$ is the age group with the largest percentage in the industry$\}$.
(e) Find $\{x \mid x$ is the percentage in the industry for those between ages 45 and 64$\}$.
(f) Find $\{x \mid x$ is the percentage in the industry for those under age 16$\}$.

93. Excessive alcohol consumption by those aged 18–24 affects nearly all U.S. college students, whether they choose to drink or not. The consequences of excessive drinking are listed below.

Consequence	Average Number of College Students Aged 18–24 Affected per Year
Death	1,700
Injury	599,000
Assault	696,000
Sexual abuse	97,000
Unsafe sex	500,000
Health problems	150,000
Drunk driving	2,100,000

Source: http://www.collegedrinkingprevention.gov/StatsSummaries/snapshot.aspx

(a) List the set of the three consequences with the most students affected by excessive alcohol consumption.
(b) List the set of consequences that affect between 100,000 and 600,000 college students each year.
(c) Find the set $\{x \mid x$ is a consequence of which over a half million students are affected$\}$.
(d) Find the set $\{x \mid x$ is the average number of college students affected by sexual abuse, death, or health problems$\}$.
(e) Find a set of three elements that do not belong in the set $\{x \mid x$ is a consequence of excessive drinking to college students ages 18–24$\}$.

94. The number of bachelor's degrees awarded by degree-granting institutions in the United States in different years is shown below.

Major	1991	1998	2005
Business	249,200	232,100	311,600
Social sciences	125,100	125,000	156,900
Education	110,800	105,800	105,500
Psychology	58,700	74,100	85,600
Health professions	59,900	86,800	80,700
Engineering	79,800	78,700	78,600
Communications	51,700	49,400	72,700
Computers	25,200	27,800	54,100
Physical sciences	16,300	19,400	18,900
Mathematics	14,400	11,800	14,400
Philosophy	7,400	8,400	11,600

Source: http://nces.ed.gov/programs/coe/2007/section5/table.asp?tableID=739

(a) List the set of majors that increased in popularity every year listed.
(b) List the set of majors that did not increase in popularity from 1998 to 2005.
(c) List the set of majors that had between 50,000 and 110,000 degrees awarded in 1998.
(d) Find the set $\{x \mid x$ increased in popularity between 1991 and 1998$\}$.
(e) Find the set $\{x \mid x$ is a major that starts with the letter M, P, or E$\}$.
(f) To find the percent increase P between an original amount O and a new amount N, use the following formula: $P = (N - O)/O$. Calculate the percent increase for any major that saw an increase in degrees awarded between 1998 and 2005. List the set of majors that increased at least 30%.

95. Identity theft is the fastest-growing crime in the United States, and college students aged 18–24 are the most affected. Of the 675,000 identity theft complaints in 2005 to the Federal Trade Commission, 29% were made by college students in this age group. The following charts show types of identity theft fraud reported in 2005 and the percentage of victims by age.

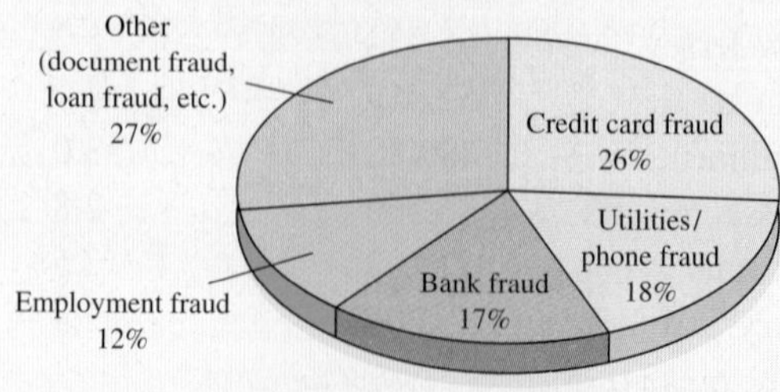

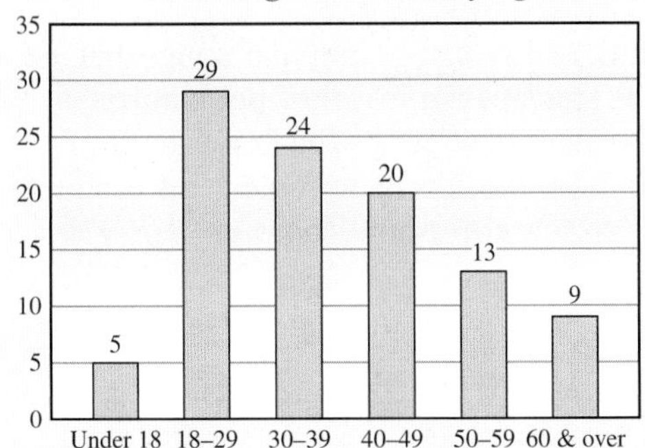

(a) List the set of the two types of identity fraud with the lowest percentage of reported crimes.
(b) List the set of age groups that are above 18%.
(c) List the set of identity fraud types that have more than 17% of reported crimes.
(d) Find the set $\{x | x$ is a percentage of those 40 and over who are victims of identity fraud$\}$.
(e) Find the set $\{x | x$ is a type of fraud that has between 10% and 20% of reported crimes$\}$.

96. Many people are now using MP3 players on their cell phones rather than a separate MP3 player to listen to music. The chart below shows age groups for those who have phones with MP3 players, those who listened to music using an MP3 player on a phone in the last 12 months, and those who downloaded music to an MP3 on their phone in the last 12 months.

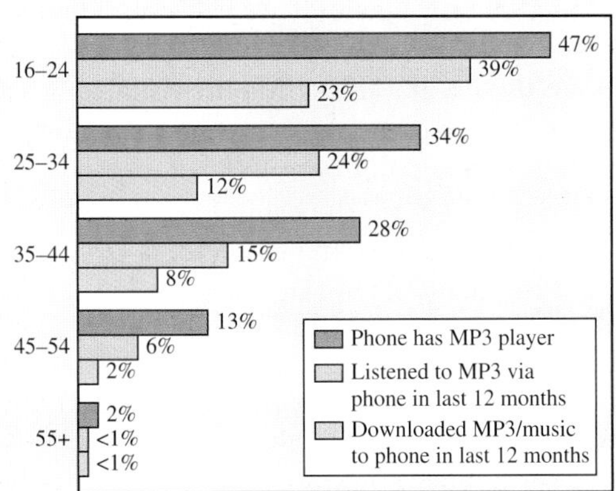

Source: http://www.cellular-news.com/story/20124.php

(a) List the set of age groups of those in which at least 20% listened to an MP3 via cell phone in the last year.
(b) List the set of age groups of those in which less than 10% downloaded music to their phones in the last year.
(c) Find the set $\{x | x$ is an age group in which at least 30% had a phone with an MP3 player$\}$.
(d) Find the set $\{x | x$ is a percentage for those aged 35–44 who have a phone with an MP3 player and listened to MP3 via phone in the last 12 months$\}$.

97. The rising cost of medical care has become one of the biggest burdens on working families in the last 10 years. The graph below represents the average prescription drug price in dollars for the years 2000 to 2006.

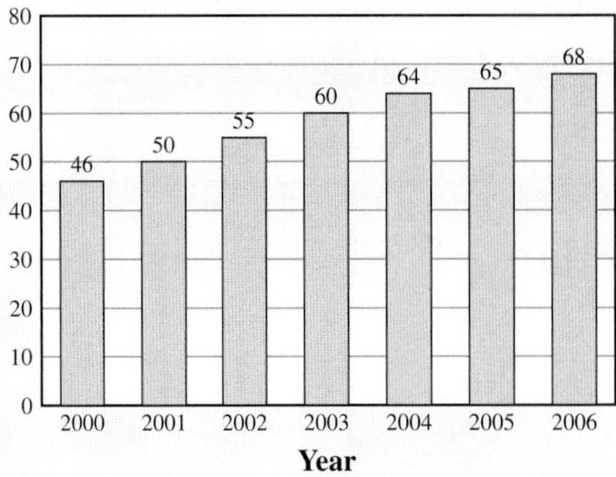

(a) Find the set of years when the average prescription drug price was less than \$60.
(b) Find the set of years when the average prescription drug price was between \$50 and \$68.
(c) Find $\{x | x$ is a year in which the average prescription drug price was greater than \$50$\}$.
(d) Find $\{x | x$ is a year in which the average prescription drug price was less than \$60$\}$.

98. Housing prices have been in the news a lot in recent years, as the boom that began in 2004 has given way to the bust just a few years later. The graph below displays the median housing prices for all houses sold in the United States between 2003 and 2008.

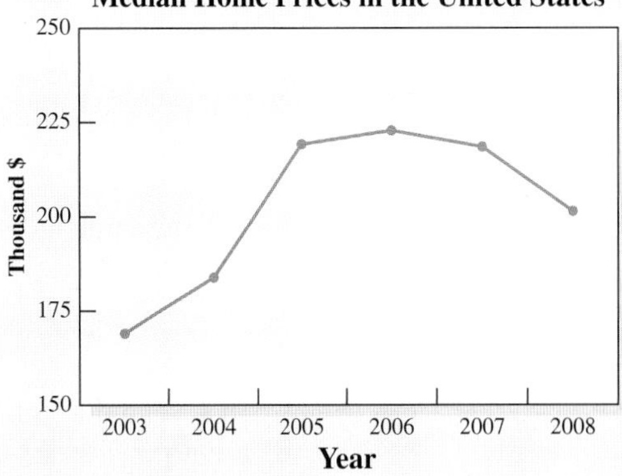

(a) List the set of years in which the median price was above \$200,000.
(b) List the set of years in which the median price was between \$175,000 and \$200,000.
(c) Find $\{x | x$ is a year in which the median price increased from the year before$\}$.
(d) Find $\{x | x$ is a year in which the median price decreased from the year before$\}$.

Critical Thinking

99. If $A \cong B$ and $A \cong C$, is $B \cong C$? Explain your answer.
100. Is $\{0\}$ equivalent to $\varnothing$? Explain your answer.
101. The set of people who attend college is not well-defined. How could you alter the definition to make this set well-defined?
102. The set of students in your math class has many subsets. Think of three subsets of this set.
103. Think of two sets of popular songs that are equivalent. Think of a third set of popular songs that is not equivalent to either of the first two sets.
104. Think of two sets of four elements each that have a one-to-one correspondence.

Section 2-2 Subsets and Set Operations

LEARNING OBJECTIVES

- ☐ 1. Define the complement of a set.
- ☐ 2. Find all subsets of a set.
- ☐ 3. Use subset notation.
- ☐ 4. Find the number of subsets for a set.
- ☐ 5. Find intersections, unions, and differences of sets.
- ☐ 6. Find the Cartesian product of two sets.

When we classify things in the real world, sets often have relationships with one another. For example, you are a member of both the set of college students and the set of students taking a college math course. You could be in the set of sophomores or the set of juniors, but not in both. You might be in the set of students living off campus and the set of students who walk to class. In this section, we'll be studying relationships between sets.

To begin, we need to consider a new concept called a *universal set.*

A **universal set,** symbolized by U, is the set of all potential elements under consideration for a specific situation.

In the examples above, a reasonable choice of U would be the set of all college students, since all the elements under consideration are college students. Once we define a universal set in a given setting, we are restricted to considering only elements from that set. If $U = \{1, 2, 3, 4, 5, 6, 7, 8\}$, then the only elements we can use to define other sets in this setting are the integers from 1 to 8.

In the remainder of this chapter, we'll use a clever method for visualizing sets and their relationships called a *Venn diagram* (so named because it was developed by a man named John Venn in the 1800s). Figure 2-1 shows an example.

You can get a lot of information from this simple diagram. A set called A is being defined. The universal set from which elements of A can be chosen is $U = \{1, 2, 3, 4, 5, 6, 7, 8\}$. The set A is $\{2, 4, 6, 8\}$, and the elements not in A are $\{1, 3, 5, 7\}$. We will call the elements in U that are not in A the *complement* of A, and denote it A'.

Math Note

The complement of the universal set is the empty set: $U' = \varnothing$. The complement of the empty set is the universal set: $\varnothing' = U$.

The **complement** of a set A, symbolized A', is the set of elements contained in the universal set that are *not* in A. Using set-builder notation, the complement of A is $A' = \{x \mid x \in U \text{ and } x \notin A\}$.

In a Venn diagram, the complement of a set A is all the things inside the rectangle that are not inside the circle representing set A. This is shown in Figure 2-2.

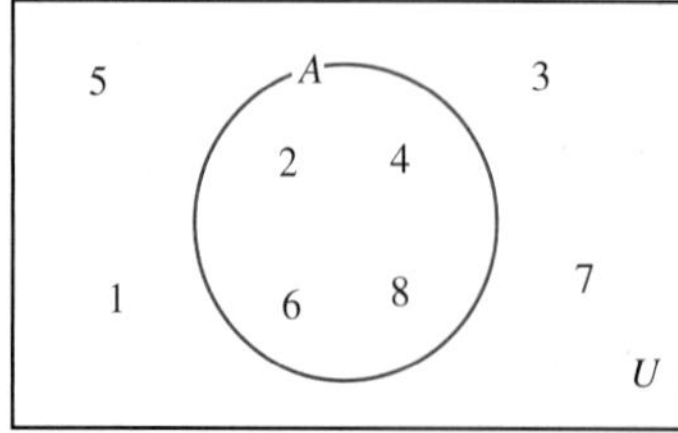

Figure 2-1

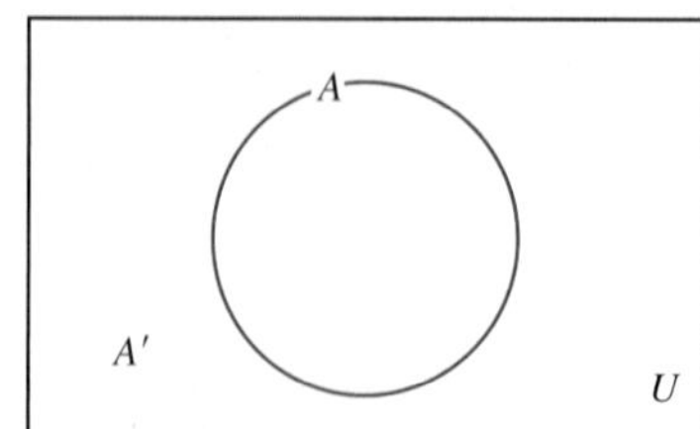

Figure 2-2

EXAMPLE 1 Finding the Complement of a Set

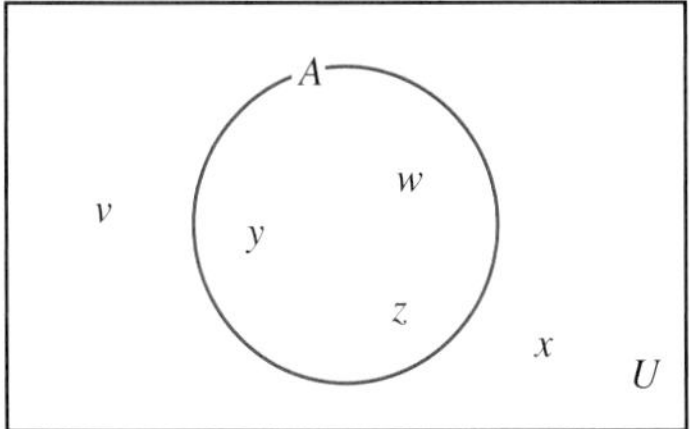

Figure 2-3

Let U = {v, w, x, y, z} and A = {w, y, z}. Find A' and draw a Venn diagram that illustrates these sets.

SOLUTION

Using the list of elements in U, we just have to cross out the ones that are also in A. The elements left over are in A'.

U = {v, ~~w~~, x, ~~y~~, ~~z~~} A' = {v, x}

The Venn diagram is shown in Figure 2-3.

▼ Try This One 1

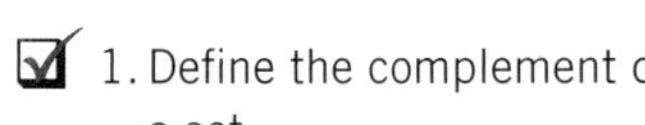

1. Define the complement of a set.

Let U = {10, 20, 30, 40, 50, 60, 70, 80, 90} and A = {10, 30, 50}. Find A' and draw a Venn diagram that illustrates these sets.

Subsets

At the beginning of the section, we pointed out that you're in both the set of college students and the set of students taking a college math course. Notice that everyone in the second set is automatically in the first one. We could say that the set of students taking a college math course is contained in the set of all college students. When one set is contained in a second set, we call the smaller set a *subset* of the larger one.

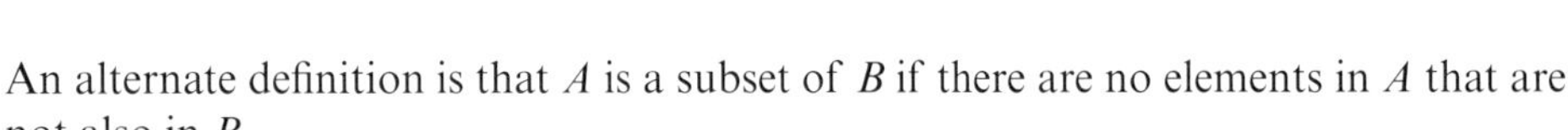

If every element of a set A is also an element of a set B, then A is called a **subset** of B. The symbol $\subseteq$ is used to designate a subset; in this case, we write $A \subseteq B$.

There are many subsets of this set of spring breakers: the subset of female students, the subset of sophomores, the subset of students who had their fake I.D. confiscated by the police, and so on.

An alternate definition is that A is a subset of B if there are no elements in A that are not also in B.

Here are a couple of observations about subsets.

- Every set is a subset of itself. Every element of a set A is of course an element of set A, so $A \subseteq A$.
- The empty set is a subset of every set. The empty set has no elements, so for any set A, you can't find an element of $\varnothing$ that is not also in A.

If we start with the set {x, y, z}, let's look at how many subsets we can form:

Number of Elements in Subset	Subsets with That Number of Elements	
3	{x, y, z}	(One subset)
2	{x, y}, {x, z}, {y, z}	(Three subsets)
1	{x}, {y}, {z}	(Three subsets)
0	$\varnothing$	(One subset)

So for a set with three elements, we can form eight subsets.

EXAMPLE 2 Finding All Subsets of a Set

Find all subsets of A = {*American Idol, Survivor*}.

SOLUTION

The subsets are

{*American Idol, Survivor*}
{*American Idol*}
{*Survivor*}
∅

Note that a set with 2 elements has 4 subsets.

▼ Try This One 2

Find all subsets of B = {Verizon, Nextel, AT&T}.

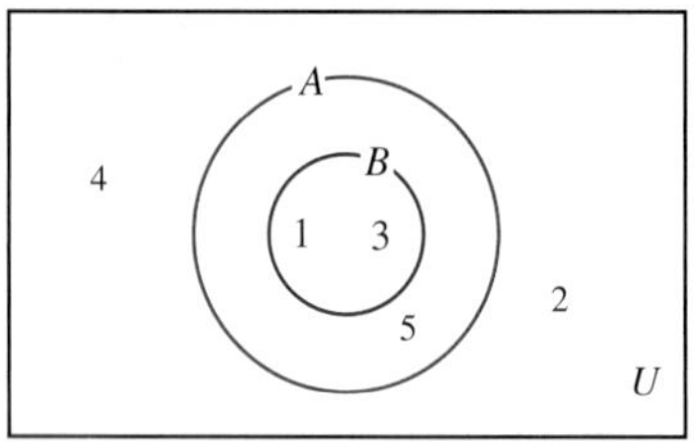

Figure 2-4 $A \subset B$

To indicate that a set is not a subset of another set, the symbol $\not\subseteq$ is used. For example, $\{1, 3\} \not\subseteq \{0, 3, 5, 7\}$ since $1 \notin \{0, 3, 5, 7\}$.

Of the four subsets in Example 2, only one is equal to the original set. We will call the remaining three *proper subsets* of A. The Venn diagram for a proper subset is shown in Figure 2-4. In this case, $U = \{1, 2, 3, 4, 5\}$, $A = \{1, 3, 5\}$, and $B = \{1, 3\}$.

If a set A is a subset of a set B and is not equal to B, then we call A a **proper subset** of B, and write $A \subset B$.

EXAMPLE 3 Finding Proper Subsets of a Set

Find all proper subsets of {x, y, z}.

SOLUTION

{x, y} {x, z} {y, z}
{x} {y} {z}
∅

▼ Try This One 3

Find all proper subsets of {♦, ♥, ♠, ♣}.

☑ 2. Find all subsets of a set.

The symbol $\not\subset$ is used to indicate that the set is not a proper subset. For example, $\{1, 3\} \subset \{1, 3, 5\}$, but $\{1, 3, 5\} \not\subset \{1, 3, 5\}$.

EXAMPLE 4 Understanding Subset Notation

State whether each statement is true or false.

(a) $\{1, 3, 5\} \subseteq \{1, 3, 5, 7\}$

(b) $\{a, b\} \subset \{a, b\}$

(c) $\{x \mid x \in N \text{ and } x > 10\} \subset N$
(d) $\{2, 10\} \not\subseteq \{2, 4, 6, 8, 10\}$
(e) $\{\text{r, s, t}\} \not\subset \{\text{t, s, r}\}$
(f) {Lake Erie, Lake Huron} $\subset$ The set of Great Lakes

Math Note

It is important not to confuse the concept of subsets with the concept of elements. For example, the statement $6 \in \{2, 4, 6\}$ is true since 6 is an element of the set {2, 4, 6}, but the statement $\{6\} \in \{2, 4, 6\}$ is false since it states that the set containing the element 6 is an element of the set containing 2, 4, and 6. However, it *is* correct to say that $\{6\} \subseteq \{2, 4, 6\}$, or that $\{6\} \subset \{2, 4, 6\}$.

SOLUTION

(a) All of 1, 3, and 5 are in the second set, so {1, 3, 5} is a subset of {1, 3, 5, 7}. The statement is true.
(b) Even though {a, b} is a subset of {a, b}, it is not a proper subset, so the statement is false.
(c) Every element in the first set is a natural number, but not all natural numbers are in the set, so that set is a proper subset of the natural numbers. The statement is true.
(d) Both 2 and 10 are elements of the second set, so {2, 10} is a subset, and the statement is false.
(e) The two sets are identical, so {r, s, t} is not a proper subset of {t, s, r}. The statement is true.
(f) Lake Erie and Lake Huron are both Great Lakes, so the statement is true.

▼ Try This One 4

State whether each statement is true or false.

(a) $\{8\} \subseteq \{x \mid x \text{ is an even natural number}\}$
(b) $\{6\} \subseteq \{1, 3, 5, 7, \ldots\}$
(c) $\{2, 3\} \subseteq \{x \mid x \in N\}$
(d) {a, b, c} $\subset$ {letters of the alphabet}
(e) $\varnothing \in \{\text{x, y, z}\}$

EXAMPLE 5 Understanding Subset Notation

State whether each statement is true or false.

(a) $\varnothing \subset \{5, 10, 15\}$
(b) $\{\text{u, v, w, x}\} \subseteq \{\text{x, w, u}\}$
(c) $\{0\} \subseteq \varnothing$
(d) $\varnothing \subset \varnothing$

SOLUTION

(a) True: the empty set is a proper subset of every set.
(b) False: v is an element of {u, v, w, x} but not {x, w, u}.
(c) The set on the left has one element, 0. The empty set has no elements, so the statement is false.
(d) The empty set is a subset of itself (as well as every other set), but not a proper subset of itself. The statement is false.

▼ Try This One 5

State whether each statement is true or false.

(a) $\varnothing \subseteq$ {red, yellow, blue}
(b) $\varnothing \subseteq \varnothing$
(c) $\{100, 200, 300, 400\} \subset \{200, 300, 400\}$
(d) $\{\varnothing\} \subseteq \varnothing$

☑ 3. Use subset notation.

A set with one element has two subsets—itself and the empty set. We have seen that if a set has two elements, there are four subsets, and if a set has three elements, there are eight subsets. This is an excellent opportunity to use the inductive reasoning that we practiced in Chapter 1!

Number of elements	0	1	2	3
Number of subsets	1	2	4	8

Based on this pattern, it's reasonable to conjecture that a set with 4 elements will have 16 subsets, a set with 5 elements will have 32 subsets, and so forth. (Notice that the number of subsets in each case is 2 raised to the number of elements.) It turns out that this is always the case. We also know that the number of proper subsets of a set is always 1 less than the total number of subsets, since only the set itself is excluded when forming proper subsets. We conclude:

The Number of Subsets for a Finite Set

If a finite set has n elements, then the set has 2^n subsets and $2^n - 1$ proper subsets.

EXAMPLE 6 Finding the Number of Subsets of a Set

Find the number of subsets and proper subsets of the set $\{1, 3, 5, 7, 9, 11\}$.

SOLUTION

The set has $n = 6$ elements, so there are 2^n, or $2^6 = 64$, subsets. Of these, $2^n - 1$, or 63, are proper.

▼ Try This One 6

Find the number of subsets and proper subsets of the set {OSU, USC, KSU, MSU, UND, PSU, UT, FSU}.

☑ 4. Find the number of subsets for a set.

The advertisement is looking for people in the *intersection* of three sets of potential employees.

Intersection and Union of Sets

At the beginning of the section, we pointed out that you might be in both the set of students living off campus and the set of students who walk to class. We will identify objects that are common to two or more sets by using the term *intersection*.

The **intersection** of two sets A and B, symbolized by $A \cap B$, is the set of all elements that are in both sets. In set-builder notation, $A \cap B = \{x \mid x \in A \text{ and } x \in B\}$.

For example, if $A = \{10, 12, 14, 15\}$ and $B = \{13, 14, 15, 16, 17\}$, then the intersection $A \cap B = \{14, 15\}$, since 14 and 15 are the elements that are common to both sets. The Venn diagram for $A \cap B$ is shown in Figure 2-5. Notice that the elements of A are placed inside the circle for set A, and the elements of B are inside the circle for set B. The elements in the intersection are placed into the portion where the circles overlap: $A \cap B$ is the shaded portion.

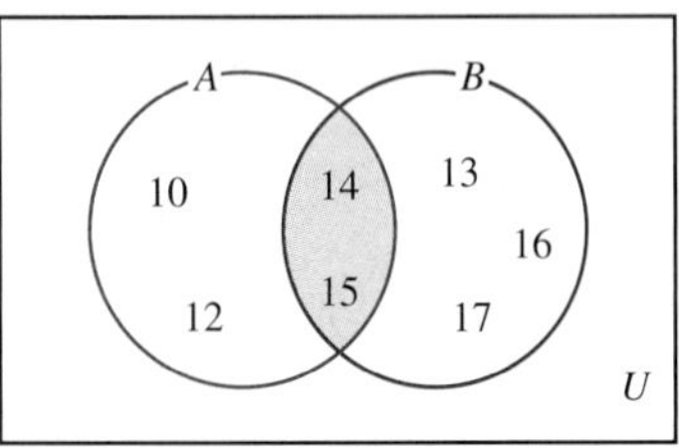

Figure 2-5 $A \cap B = \{14, 15\}$

Intersection is an example of a **set operation**—a rule for combining two or more sets to form a new set. The intersection of three or more sets consists of the set of elements that are in every single set. Note that the word *and* is sometimes used to indicate intersection; $A \cap B$ is the set of elements in A and B.

EXAMPLE 7 Finding Intersections

If $A = \{5, 10, 15, 20, 25\}$, $B = \{0, 10, 20, 30, 40\}$, and $C = \{30, 50, 70, 90\}$, find

(a) $A \cap B$ (b) $B \cap C$ (c) $A \cap C$

SOLUTION

(a) The elements 10 and 20 are in both sets A and B, so $A \cap B = \{10, 20\}$.
(b) The only member of both sets B and C is 30, so $B \cap C = \{30\}$.
(c) There are no elements common to sets A and C, so $A \cap C = \varnothing$.

▼ Try This One 7

If A = {Cleveland, Indianapolis, Chicago, Des Moines, Detroit}, B = {New York, Los Angeles, Chicago, Detroit}, and C = {Seattle, Los Angeles, San Diego}, find $A \cap B$, $B \cap C$, and $A \cap C$.

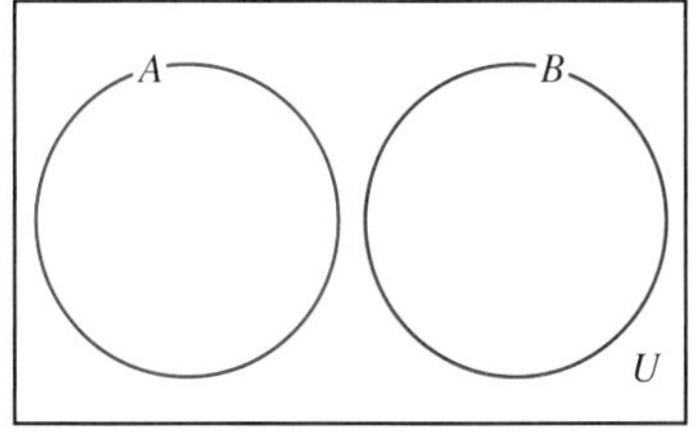

Figure 2-6 $A \cap B = \varnothing$

When the intersection of two sets is the empty set, the sets are said to be *disjoint*. For example, the set of students who stop attending class midway through a term and the set of students earning A's are disjoint, because you can't be a member of both sets. The Venn diagram for a pair of disjoint sets A and B is shown in Figure 2-6. If the sets have no elements in common, the circles representing them don't overlap at all.

Another way of combining sets to form a new set is called *union*.

The **union** of two sets A and B, symbolized by $A \cup B$, is the set of all elements that are in either set A or set B (or both). In set-builder notation,

$$A \cup B = \{x \mid x \in A \quad \text{or} \quad x \in B\}$$

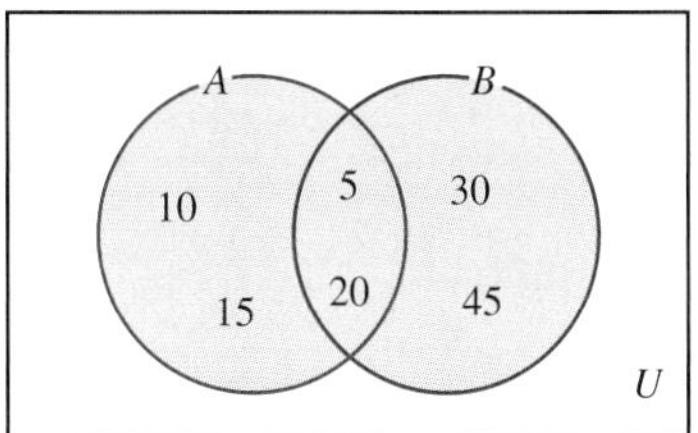

Figure 2-7 $A \cup B$

For example, if $A = \{5, 10, 15, 20\}$ and $B = \{5, 20, 30, 45\}$, then the union $A \cup B = \{5, 10, 15, 20, 30, 45\}$. Even though 5 and 20 are in both sets, we list them only once in the union. The Venn diagram for $A \cup B$ is shown in Figure 2-7. The set $A \cup B$ is the shaded area consisting of all elements in either set.

EXAMPLE 8 Finding Unions

If $A = \{0, 1, 2, 3, 4, 5\}$, $B = \{2, 4, 6, 8, 10\}$, and $C = \{1, 3, 5, 7\}$, find each.

(a) $A \cup B$ (b) $A \cup C$ (c) $B \cup C$

SOLUTION

To find a union, just make a list of all the elements in either set without writing repeats.

(a) $A \cup B = \{0, 1, 2, 3, 4, 5, 6, 8, 10\}$
(b) $A \cup C = \{0, 1, 2, 3, 4, 5, 7\}$
(c) $B \cup C = \{1, 2, 3, 4, 5, 6, 7, 8, 10\}$

▼ Try This One 8

If $A = \{a, b, c, d, e\}$, $B = \{a, c, e, g, i\}$, and $C = \{b, d, f, h, j\}$, find $A \cup B$, $A \cup C$, and $B \cup C$.

What about operations involving more than two sets? Just like with operations involving numbers, we use parentheses to indicate an order of operations. This is illustrated in Example 9.

EXAMPLE 9 Performing Set Operations

Let $A = \{l, m, n, o, p\}$, $B = \{o, p, q, r\}$, and $C = \{r, s, t, u\}$. Find each.

(a) $(A \cup B) \cap C$ (b) $A \cap (B \cup C)$ (c) $(A \cap B) \cup C$

SOLUTION

(a) First find $A \cup B$: $A \cup B = \{l, m, n, o, p, q, r\}$. Then intersect this set with set C; the only common element is r, so $(A \cup B) \cap C = \{r\}$.
(b) First find $B \cup C$: $B \cup C = \{o, p, q, r, s, t, u\}$. Then intersect this set with set A to get $\{o, p\}$.
(c) First find $A \cap B$: $A \cap B = \{o, p\}$. Then find the union of this set with set C to get $\{o, p, r, s, t, u\}$.

▼ Try This One 9

If $A = \{2, 3, 4, 5, 6, 7\}$, $B = \{7, 8, 9, 10\}$, and $C = \{0, 5, 10, 15, 20\}$, find $A \cup (B \cap C)$, $(A \cap B) \cup C$, and $A \cap (B \cup C)$.

CAUTION

When combining union and intersection with complements, we'll have to be extra careful. Pay particular attention to the parentheses and to whether the complement symbol is inside or outside the parentheses.

EXAMPLE 10 Performing Set Operations

If $U = \{10, 20, 30, 40, 50, 60, 70, 80\}$, $A = \{10, 30, 50, 70\}$, $B = \{40, 50, 60, 70\}$, and $C = \{20, 40, 60\}$, find each.

(a) $A' \cap C'$ (b) $(A \cap B)' \cap C$ (c) $B' \cup (A \cap C')$

SOLUTION

(a) First, write A' and C': $A' = \{20, 40, 60, 80\}$ and $C' = \{10, 30, 50, 70, 80\}$. Now note that 80 is the only element common to both: $A' \cap C' = \{80\}$.
(b) The parentheses tell us that we should find $A \cap B$ first: $A \cap B = \{50, 70\}$. Next we find the complement: $(A \cap B)' = \{10, 20, 30, 40, 60, 80\}$. Finally, we find the intersection of this set and C: $(A \cap B)' \cap C = \{20, 40, 60\}$.
(c) First, find $A \cap C'$: $C' = \{10, 30, 50, 70, 80\}$, and all but 80 are also in A, so $A \cap C' = \{10, 30, 50, 70\}$. Next note that $B' = \{10, 20, 30, 80\}$. Now form the union: $B' \cup (A \cap C') = \{10, 20, 30, 50, 70, 80\}$.

▼ Try This One 10

Let $U = \{1, 2, 3, 4, 5, 6, 7, 8\}$, $A = \{1, 3, 5, 7\}$, $B = \{2, 4, 6, 8\}$, and $C = \{2, 3, 5, 7\}$. Find each.

(a) C' (b) $(A \cup B)'$ (c) $A' \cap C'$ (d) $(A \cup B) \cap C'$

The union and intersection of sets are commonly used in real life—it's just that you might not have thought of it in those terms. For example, the intersection of the set of U.S. citizens older than 17 and the set of U.S. citizens who are not convicted felons makes up the set of those eligible to vote. The union of the set of your mom's parents and your dad's parents forms the set of your grandparents.

Set Subtraction

The third set operation we'll study is called the *difference* of sets. We also call it *set subtraction* and use a minus sign to represent it.

> The **difference** of set A and set B is the set of elements in set A that are *not* in set B. In set-builder notation, $A - B = \{x \mid x \in A \text{ and } x \notin B\}$.

EXAMPLE 11 Finding the Difference of Two Sets

Math Note

Sometimes operations can be written in terms of other operations. For example, $3 - 5$ is also $3 + (-5)$. Can you think of a way to write $A - B$ using intersection and complement? Drawing a Venn diagram might help.

Let $U = \{2, 4, 6, 8, 10, 12\}$, $A = \{4, 6, 8, 10\}$, $B = \{2, 6, 12\}$, $C = \{8, 10\}$

Find each.

(a) $A - B$ (b) $A - C$ (c) $B - C$

SOLUTION

(a) Start with the elements in set A and take out the elements in set B that are also in set A. In this case, only 6 is removed, and $A - B = \{4, 8, 10\}$.

(b) Start with the elements in set A and remove the elements in set C that are also in set A. In this case, 8 and 10 are removed, and $A - C = \{4, 6\}$.

(c) Start with the elements in set B and take out the elements in set C that are also in set B. In this case, none of the elements in B are also in C. So $B - C = \{2, 6, 12\}$.

▼ Try This One 11

Let $L = \{p, q, r, s, t, u, v, w\}$, $M = \{p, s, t, u, y, z\}$, and $N = \{u, v, w, z\}$. Find

(a) $L - M$ (b) $L - N$ (c) $M - N$

☑ 5. Find intersections, unions, and differences of sets.

Cartesian Products

The fourth set operation we'll study is called the *Cartesian product* or *cross product*. To define it, we need to first define an *ordered pair*. An ordered pair is a pair of numbers or objects that are associated by writing them together in a set of parentheses, like (3, 5). In this ordered pair, 3 is called the *first component* and 5 is called the *second component*.

As indicated by the term *ordered pair,* the order in which numbers are written is important: (3, 5) is not the same ordered pair as (5, 3).

Math Note

The Cartesian product $A \times B$ is usually read aloud as "A cross B."

The **Cartesian product** (denoted $A \times B$) of two sets A and B is formed by writing all possible ordered pairs in which the first component is an element of A and the second component is an element of B. Using set-builder notation, $A \times B = \{(x, y) \mid x \in A \text{ and } y \in B\}$.

EXAMPLE 12 Finding Cartesian Products

Math Note

It's important to note that the Cartesian product of two sets is also a set, so it should be enclosed within braces.

If $A = \{1, 3, 5\}$ and $B = \{2, 4\}$, find $A \times B$ and $B \times A$.

SOLUTION

To form $A \times B$, first form ordered pairs with first component 1: (1, 2) and (1, 4). Then form pairs with first component 3: (3, 2) and (3, 4). Finally, use 5 as the first component: (5, 2) and (5, 4). $A \times B = \{(1, 2), (1, 4), (3, 2), (3, 4), (5, 2), (5, 4)\}$. For $B \times A$, form all possible ordered pairs with first components from B and second components from A: $B \times A = \{(2, 1), (2, 3), (2, 5), (4, 1), (4, 3), (4, 5)\}$.

▼ Try This One 12

☑ 6. Find the Cartesian product of two sets.

If $R = \{c, d, e\}$ and $S = \{g, h\}$, find $R \times S$, $S \times R$, and $S \times S$.

CAUTION

Even though we use a multiplication sign for Cartesian product, it has nothing to do with multiplying elements!

In this section, we defined subsets, Venn diagrams, and four basic operations for combining sets: intersection, union, subtraction, and Cartesian product. In the next section, we'll examine how Venn diagrams can be used to study set operations in greater depth.

Answers to Try This One

1 $A' = \{20, 40, 60, 70, 80, 90\}$

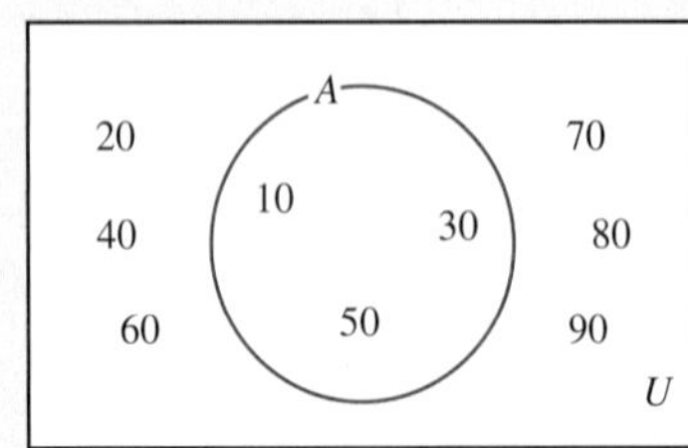

2 {Verizon, Nextel, AT&T}, {Verizon, Nextel}, {Verizon, AT&T}, {Nextel, AT&T}, {Verizon}, {Nextel}, {AT&T}, ∅

3 {♦,♥,♠}, {♦,♥,♣}, {♦,♠,♣}, {♥,♠,♣}, {♦,♥}, {♦,♠}, {♦,♣}, {♥,♠}, {♥,♣}, {♠,♣}, {♦}, {♥}, {♠}, {♣}, ∅

4 (a) True (b) False (c) True (d) True (e) True

5 (a) True (b) True (c) False (d) False

6 Subsets: $2^8 = 256$; proper subsets: 255

7 $A \cap B = \{\text{Chicago, Detroit}\}$; $B \cap C = \{\text{Los Angeles}\}$; $A \cap C = \varnothing$

8 $A \cup B = \{a, b, c, d, e, g, i\}$;
$A \cup C = \{a, b, c, d, e, f, h, j\}$;
$B \cup C = \{a, b, c, d, e, f, g, h, i, j\}$

9 $A \cup (B \cap C) = \{2, 3, 4, 5, 6, 7, 10\}$
$(A \cap B) \cup C = \{0, 5, 7, 10, 15, 20\}$
$A \cap (B \cup C) = \{5, 7\}$

10 (a) $\{1, 4, 6, 8\}$ (b) $\varnothing$ (c) $\{4, 6, 8\}$ (d) $\{1, 4, 6, 8\}$

11 (a) $\{q, r, v, w\}$ (b) $\{p, q, r, s, t\}$ (c) $\{p, s, t, y\}$

12 $R \times S = \{(c, g), (c, h), (d, g), (d, h), (e, g), (e, h)\}$
$S \times R = \{(g, c), (g, d), (g, e), (h, c), (h, d), (h, e)\}$
$S \times S = \{(g, g), (g, h), (h, g), (h, h)\}$

EXERCISE SET 2-2

Writing Exercises

1. What is a subset?
2. Explain the difference between a subset and a proper subset.
3. Explain the difference between a subset and an element of a set.
4. Explain why the empty set is a subset, but not a proper subset, of itself.
5. Explain the difference between the union and intersection of two sets.
6. When are two sets said to be disjoint?
7. What is a universal set?
8. What is the complement of a set?
9. Write an example from real life that represents the union of sets and explain why it represents union.
10. Write an example from real life that represents the difference of sets and explain why it represents difference.

Computational Exercises

For Exercises 11–14, let $U = \{2, 3, 5, 7, 11, 13, 17, 19\}$, $A = \{5, 7, 11, 13\}$, $B = \{2\}$, $C = \{13, 17, 19\}$, and $D = \{2, 3, 5\}$. *Find each set.*

11. A'
12. B'
13. C'
14. D'

For Exercises 15–22, find all subsets of each set.

15. $\{r, s, t\}$
16. $\{2, 5, 7\}$
17. $\{1, 3\}$
18. $\{p, q\}$
19. $\{\ \}$
20. $\varnothing$
21. $\{5, 12, 13, 14\}$
22. $\{m, o, r, e\}$

For Exercises 23–28, find all proper subsets of each set.

23. $\{1, 10, 20\}$
24. {March, April, May}
25. $\{6\}$
26. $\{t\}$
27. $\varnothing$
28. $\{\ \}$

For Exercises 29–38, state whether each is true or false.

29. $\{3\} \subseteq \{1, 3, 5\}$
30. $\{a, b, c\} \subset \{c, b, a\}$
31. $\{1, 2, 3\} \subseteq \{123\}$
32. $\varnothing \subset \varnothing$
33. $\varnothing \in \{\ \}$
34. {Mars, Venus, Sun} $\subset$ {planets in our solar system}
35. $\{3\} \in \{1, 3, 5, 7, \ldots\}$
36. $\{x \mid x \in N \text{ and } x > 10\} \subseteq \{x \mid x \in N \text{ and } x \geq 10\}$
37. $\varnothing \subset \{a, b, c\}$
38. $\varnothing \in \{r, s, t, u\}$

For Exercises 39–44, find the number of subsets and proper subsets each set has. Do not list the subsets.

39. $\{25, 50, 75\}$
40. $\{1, 2, 3, 4, 5, 6, 7, 8, 9, 10\}$
41. $\varnothing$
42. $\{0\}$
43. $\{x, y\}$
44. $\{a, b, c, d, e\}$

For Exercises 45–54, use the Venn diagram and find the elements in each set.

45. U
46. A
47. B
48. $A \cap B$
49. $A \cup B$
50. A'
51. B'
52. $(A \cup B)'$
53. $(A \cap B)'$
54. $A \cap B'$

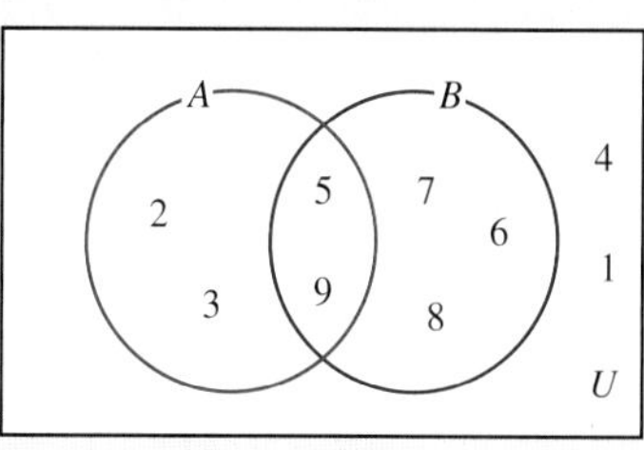

For Exercises 55–64, let

$U = \{10, 20, 30, 40, 50, 60, 70, 80, 90, 100\}$
$A = \{10, 30, 50, 70, 90\}$
$B = \{20, 40, 60, 80, 100\}$
$C = \{30, 40, 50, 60\}$

Find each set.

55. $A \cup C$
56. $A \cap B$
57. A'
58. $(A \cap B) \cup C$
59. $A' \cap (B \cup C)$
60. $(A \cap B) \cap C$
61. $(A \cup B)' \cap C$
62. $A \cap B'$
63. $(B \cup C) \cap A'$
64. $(A' \cup B)' \cup C'$

For Exercises 65–74, let

$U = \{a, b, c, d, e, f, g, h\}$
$P = \{b, d, f, g\}$
$Q = \{a, b, c, d\}$
$R = \{e, f, g\}$

Find each set.

65. $P \cap Q$
66. $Q \cup R$
67. P'
68. Q'
69. $R' \cap P'$
70. $P \cup (Q \cap R)$
71. $(Q \cup P)' \cap R$
72. $P \cap (Q \cap R)$
73. $(P \cup Q) \cap (P \cup R)$
74. $Q' \cup R'$

For Exercises 75–84, let

$U = \{1, 2, 3, 4, 5, 6, 7, 8, 9, 10, 11, 12\}$
$W = \{2, 4, 6, 8, 10, 12\}$
$X = \{1, 3, 5, 7, 9, 11\}$
$Y = \{1, 2, 3, 4, 5, 6\}$
$Z = \{2, 5, 6, 8, 10, 11, 12\}$

Find each set.

75. $W \cap Y$
76. $X \cup Z$
77. $W \cup X$
78. $(X \cap Y) \cap Z$
79. $W \cap X$
80. $(Y \cup Z)'$
81. $(X \cup Y) \cap Z$
82. $(Z \cap Y) \cup W$
83. $W' \cap X'$
84. $(Z \cup X)' \cap Y$

For Exercises 85–88, let

$U = \{1, 2, 3, \ldots\}$
$A = \{3, 6, 9, 12, \ldots\}$
$B = \{9, 18, 27, 36, \ldots\}$
$C = \{2, 4, 6, 8, \ldots\}$

Find each set.

85. $A \cap B$
86. $A' \cap C$
87. $A \cap (B \cup C)'$
88. $A \cup B$

For Exercises 89–94, let

$U = \{20, 40, 60, 80, 100, 110\}$
$A = \{20, 60, 100, 110\}$
$B = \{60, 80, 100\}$
$C = \{80, 100, 110\}$

Find each set.

89. $A - B$
90. $A - C$
91. $B - C$
92. $B - A$
93. $C \cap B'$
94. $A \cap C'$

For Exercises 95–100, let

$U = \{p, q, r, s, t, u, v, w\}$
$A = \{p, q, r, s, t\}$
$B = \{r, s, t, u, v\}$
$C = \{p, r, t, v\}$

Find each set.

95. $C - B$
96. $A - C$
97. $B - C$
98. $B - A$
99. $B \cap C'$
100. $C \cap A'$

For Exercises 101–104, let

$A = \{9, 12, 18\}$
$B = \{1, 2, 3\}$

Find each set.

101. $A \times B$
102. $B \times A$
103. $A \times A$
104. $B \times B$

For Exercises 105–108, let

$A = \{1, 2, 4, 8\}$
$B = \{1, 3\}$

Find each set.

105. $A \times B$
106. $B \times A$
107. $B \times B$
108. $A \times A$

Real-World Applications

109. A student can have a cell phone, a laptop, and an iPod while hanging out on campus between classes. List all the sets of different communication options a student can select, considering all, some, or none of these technologies.
110. If a person is dealt five cards and has a chance of discarding any number including 0, how many choices will the person have?
111. A college freshman can choose one, some, or all of the following classes for her first semester: an English class, a math class, a foreign language class, a science class, a philosophy class, a physical education class, and a history class. How many different possibilities does she have for her new schedule?
112. Since the student union is being remodeled, there is a limited choice of foods and drinks a student can buy for a snack between classes. Students can choose none, some, or all of these items: pizza, fries, big soft pretzels, Coke, Diet Coke, and Hawaiian Punch. How many different selections can be made?
113. Suzie is buying a new laptop for school and can select none, some, or all of the following choices

of peripherals: a laser mouse, a DVD burner, a Web cam, or a jump drive. How many different selections of peripherals are possible for her laptop?

114. To integrate aerobics into her exercise program, Claire can select one, some, or all of these machines: treadmill, cycle, and stair stepper. List all possibilities for her aerobics selection.

Critical Thinking

115. Can you find two sets whose union and intersection are the same set?
116. Write a short paragraph listing five things you admire about your role model and five things you like about yourself. Find the union of the two sets. Find the intersection of the two sets.
117. Select two medications and find a resource on the Internet that lists the possible side effects of each. Find the intersection of the sets.
118. Select two possible careers you may pursue and write the job responsibilities you would have with each one. What responsibilities do the jobs have in common? If you could combine these careers somehow, what would your new job responsibilities be?

Section 2-3 Using Venn Diagrams to Study Set Operations

LEARNING OBJECTIVES

- ☐ 1. Illustrate set statements involving two sets with Venn diagrams.
- ☐ 2. Illustrate set statements involving three sets with Venn diagrams.
- ☐ 3. Use DeMorgan's laws.
- ☐ 4. Use Venn diagrams to decide if two sets are equal.
- ☐ 5. Use the formula for the cardinality of a union.

The world we live in is a pretty complicated place. Everywhere you look, there are interactions between sets of people, businesses, products, objects—we could go on, but you probably get the picture. The more complicated these interactions are, the more challenging it can be to sort them out. One good way to get a handle on a complicated situation is to diagram it. When we are dealing with sets, Venn diagrams will be our tool of choice.

In this section, we'll develop a method for drawing Venn diagrams that will help us to illustrate set operations. We'll start with diagrams involving interactions between two sets, as in Figure 2-8. Notice that there are four distinct regions in a Venn diagram illustrating two sets A and B. Let's label these regions with Roman numerals to designate that they represent regions and not elements in the sets.

The procedure that we will use to illustrate set statements, found in the box below, is demonstrated in Examples 1 and 2.

Illustrating a Set Statement with a Venn Diagram

Step 1 Draw a diagram for the sets, with Roman numerals in each region.

Step 2 Using the Roman numerals, list the regions for each set.

Step 3 Find the set of numerals that correspond to the set given in the set statement.

Step 4 Shade the area corresponding to the set of numerals found in step 3.

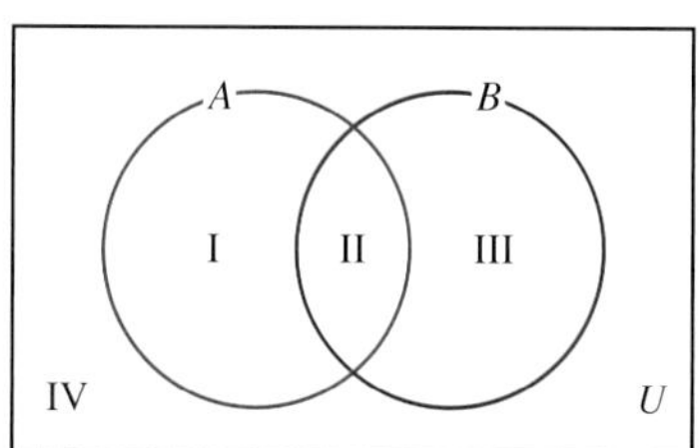

Figure 2-8

Region I represents the elements in set A that are not in set B.
Region II represents the elements in both sets A and B.
Region III represents the elements in set B that are not in set A.
Region IV represents the elements in the universal set that are in neither set A nor set B.

EXAMPLE 1 Drawing a Venn Diagram

Draw a Venn diagram to illustrate the set $(A \cup B)'$.

SOLUTION

Step 1 Draw the diagram and label each area.

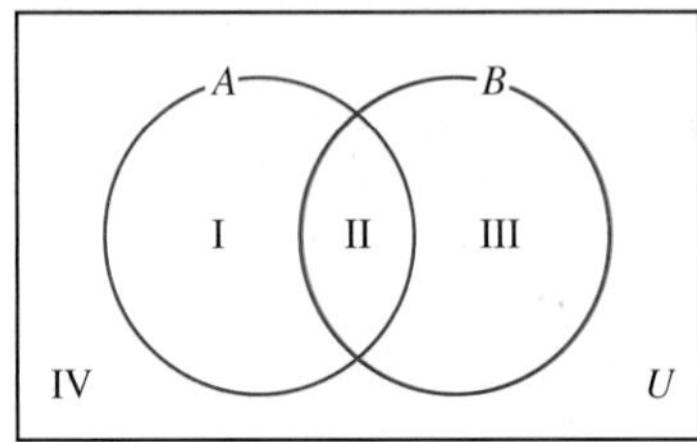

Step 2 From the diagram, list the regions that make up each set.

$U = \{I, II, II, IV\}$
$A = \{I, II\}$
$B = \{II, III\}$

Step 3 Using the sets in step 2, find $(A \cup B)'$.

First, all of I, II, and III are in either A or B, so $A \cup B = \{I, II, II\}$. The complement is $(A \cup B)' = \{IV\}$.

Step 4 Shade region IV to illustrate $(A \cup B)'$.

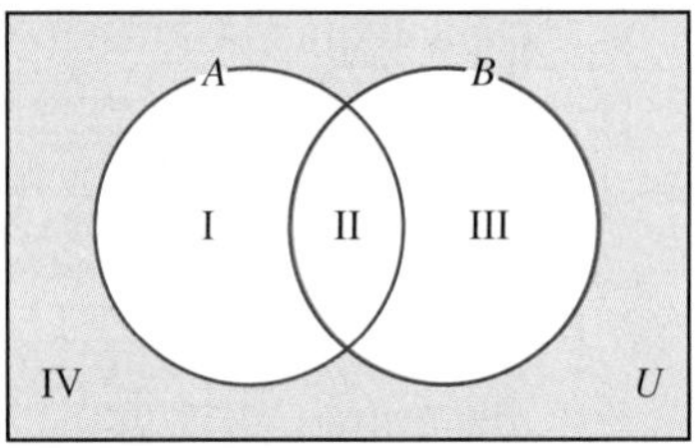

▼ Try This One 1

Draw a Venn diagram to illustrate the set $A' \cap B$.

EXAMPLE 2 Drawing a Venn Diagram

Draw a Venn diagram to illustrate the set $A \cup B'$.

SOLUTION

Step 1 Draw the diagram and label each area. This will be the same diagram as in Example 1.

Step 2 From the diagram, list the regions that make up each set.

$U = \{I, II, II, IV\}$
$A = \{I, II\}$
$B = \{II, III\}$

Step 3 Using the sets in step 2, find $A \cap B'$.

First, regions I and IV are outside of set B. Of these two regions, I is also in set A, so $A \cap B' = \{I\}$.

Step 4 Shade region I to illustrate $A \cap B'$.

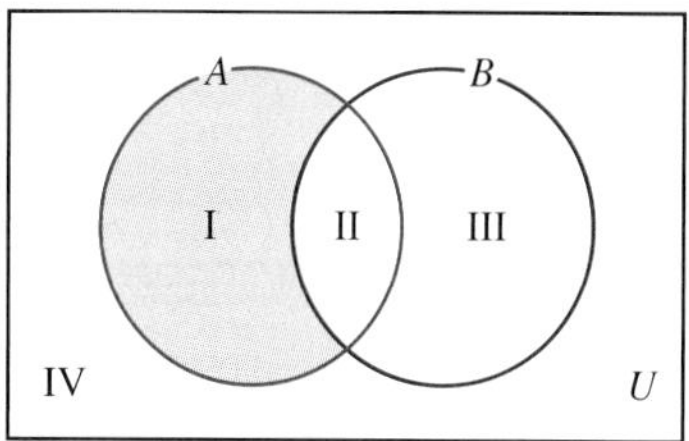

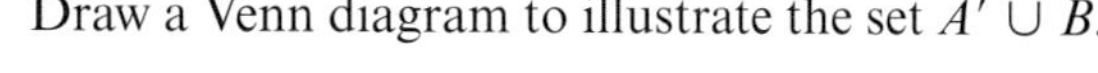

▼ Try This One 2

Draw a Venn diagram to illustrate the set $A' \cup B$.

☑ 1. Illustrate set statements involving two sets with Venn diagrams.

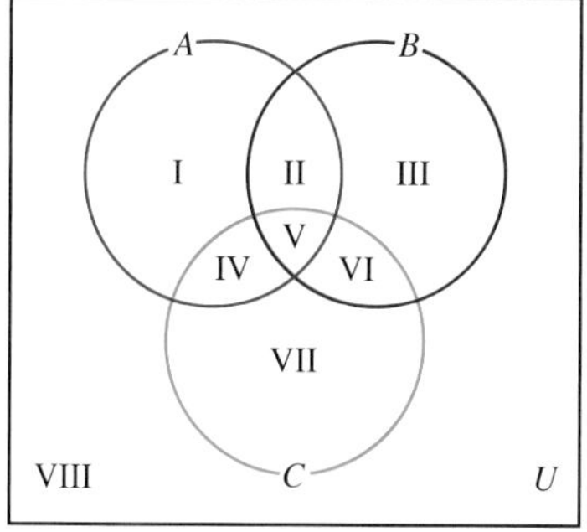

Figure 2-9

A similar procedure can be used to illustrate set statements involving three or more sets. The Venn diagram for three sets is shown in Figure 2-9.

Region I represents the elements in set *A* but not in set *B* or set *C*.
Region II represents the elements in set *A* and set *B* but not in set *C*.
Region III represents the elements in set *B* but not in set *A* or set *C*.
Region IV represents the elements in sets *A* and *C* but not in set *B*.
Region V represents the elements in sets *A*, *B*, and *C*.
Region VI represents the elements in sets *B* and *C* but not in set *A*.
Region VII represents the elements in set *C* but not in set *A* or set *B*.
Region VIII represents the elements in the universal set *U*, but not in set *A*, *B*, or *C*.

We can use the same four-step procedure (see page 65) to illustrate operations involving three sets. The process is demonstrated in Example 3.

EXAMPLE 3 Drawing a Venn Diagram with Three Sets

Draw a Venn diagram to illustrate the set $A \cap (B \cap C)'$.

SOLUTION

Step 1 Draw and label the diagram as in Figure 2-9.

Step 2 From the diagram, list the regions that make up each set.

$U = \{I, II, III, IV, V, VI, VII, VIII\}$
$A = \{I, II, IV, V\}$
$B = \{II, III, V, VI\}$
$C = \{IV, V, VI, VII\}$

Step 3 Using the sets in step 2, find $A \cap (B \cap C)'$.

First, find $B \cap C$: $B \cap C = \{V, VI\}$. The complement is $(B \cap C)' = \{I, II, III, IV, VII, VIII\}$. Regions I, II, and IV are also part of A, so $A \cap (B \cap C)' = \{I, II, IV\}$.

Step 4 Shade regions I, II, and IV to illustrate $A \cap (B \cap C)'$.

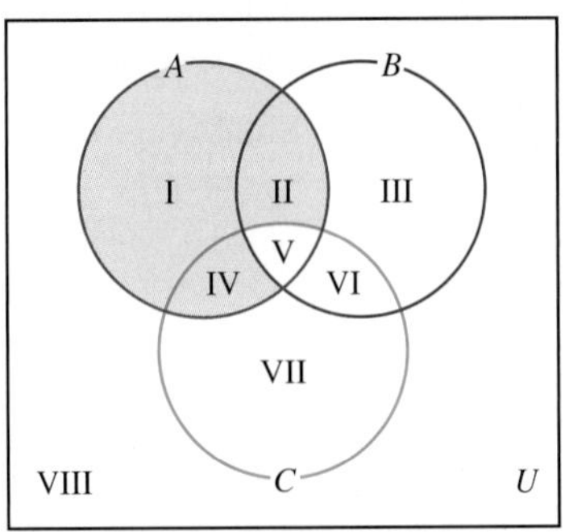

▼ Try This One 3

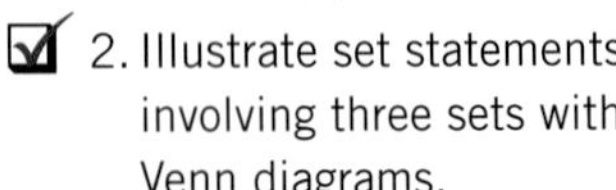
2. Illustrate set statements involving three sets with Venn diagrams.

Draw a Venn diagram to illustrate the set $(A \cap B') \cup C$.

De Morgan's Laws

There are two very well-known formulas that are useful in simplifying some set operations. They're named in honor of a 19th-century mathematician named Augustus De Morgan.

First, we'll write the formulas and illustrate each with an example. Then we'll see how Venn diagrams can be used to prove the formulas.

De Morgan's Laws

For any two sets A and B,

$(A \cup B)' = A' \cap B'$

$(A \cap B)' = A' \cup B'$

The first law states that the complement of the union of two sets will always be equal to the intersection of the complements of each set.

EXAMPLE 4 Using De Morgan's Laws

If $U = \{a, b, c, d, e, f, g, h\}$, $A = \{a, c, e, g\}$, and $B = \{b, c, d, e\}$, find $(A \cup B)'$ and $A' \cap B'$.

SOLUTION

$A \cup B = \{a, b, c, d, e, g\}$ and $(A \cup B)' = \{f, h\}$

$A' = \{b, d, f, h\}$ $B' = \{a, f, g, h\}$ and $A' \cap B' = \{f, h\}$

Notice that $(A \cup B)'$ and $A' \cap B'$ are equal, illustrating the first of De Morgan's laws.

▼ Try This One 4

If $U = \{15, 30, 45, 60, 75, 90, 105\}$, $A = \{30, 60, 90\}$, and $B = \{15, 45, 75, 90\}$, find $(A \cup B)'$ and $A' \cap B'$.

Sidelight NOW AND VENN

Venn diagrams are generally credited to the British mathematician John Venn, who introduced them in an 1880 paper as they are used today. That makes it sound like a pretty old concept, but the general idea can be traced back much further. The great mathematician Leonhard Euler used similar diagrams in the 1700s, and other figures like them can be traced as far back as the 1200s!

The second law states that the complement of the intersection of two sets will equal the union of the complements of the sets.

EXAMPLE 5 Using De Morgan's Laws

If $U = \{10, 11, 12, 13, 14, 15, 16\}$, $A = \{10, 11, 12, 13\}$, and $B = \{12, 13, 14, 15\}$, find $(A \cap B)'$ and $A' \cup B'$.

SOLUTION

$$A \cap B = \{12, 13\} \quad \text{and} \quad (A \cap B)' = \{10, 11, 14, 15, 16\}$$

$$A' = \{14, 15, 16\} \quad B' = \{10, 11, 16\}; \quad A' \cup B' = \{10, 11, 14, 15, 16\}$$

Notice that $(A \cap B)'$ and $A' \cup B'$ are equal, illustrating the second of De Morgan's laws.

▼ Try This One 5

☑ 3. Use De Morgan's laws.

If $U = \{\text{ABC, NBC, CBS, Fox, USA, TBS, TNT, MTV}\}$, $A = \{\text{NBC, Fox, USA, TBS}\}$, and $B = \{\text{ABC, NBC, CBS, Fox}\}$, find $(A \cap B)'$ and $A' \cup B'$.

Venn diagrams can be used to show the equality of two set statements. To do this, we draw a Venn diagram for the set statement on each side of the equation. If the same regions are shaded, then the equation is true. The next example shows how to verify the first of De Morgan's laws by using Venn diagrams. We'll leave the second for you to try.

EXAMPLE 6 Using a Venn Diagram to Show Equality of Sets

Use Venn diagrams to show that $(A \cup B)' = A' \cap B'$.

SOLUTION

Start by drawing the Venn diagram for $(A \cup B)'$.

Step 1 Draw the figure (as shown in Step 4).

Step 2 Set U contains regions I, II, III, and IV. Set A contains regions I and II, and B contains regions II and III.

Step 3 $A \cup B = \{\text{I, II, II}\}$, so $(A \cup B)' = \{\text{IV}\}$.

Step 4 Shade region IV to illustrate $(A \cup B)'$.

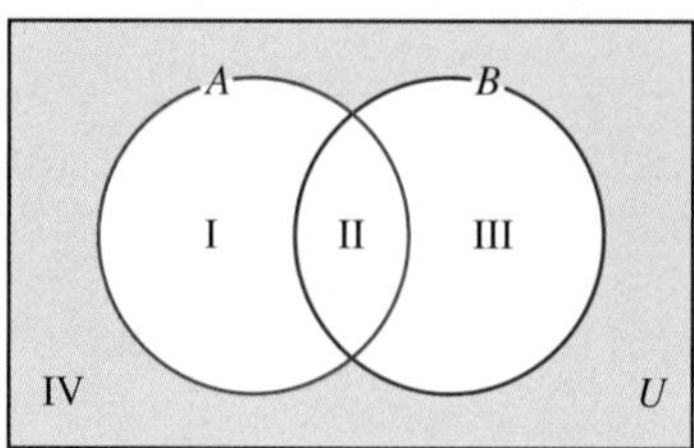

Next draw the Venn diagram for $A' \cap B'$. Steps 1 and 2 are the same as above.

Step 3 $A' = \{\text{III, IV}\}$ and $B' = \{\text{I, IV}\}$, so $A' \cap B' = \{\text{IV}\}$.

Step 4 Shade region IV to illustrate $A' \cap B'$.

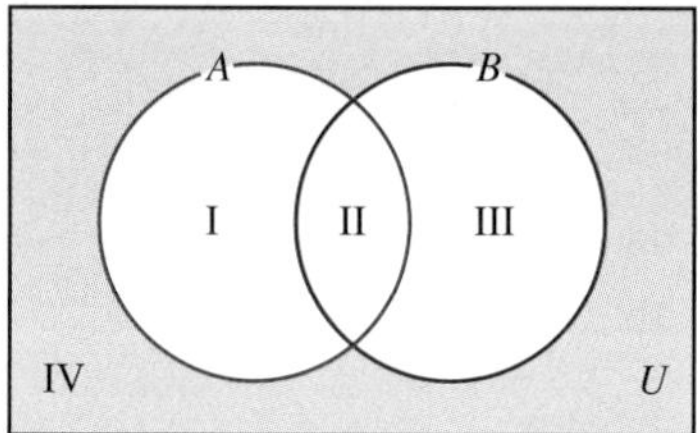

Since the diagrams for each side are identical, we use deductive reasoning to conclude that $(A \cup B)' = A' \cap B'$.

▼ Try This One 6

Use Venn diagrams to show that $(A \cap B)' = A' \cup B'$.

Here's an example using three sets.

EXAMPLE 7 Using Venn Diagrams to Decide If Two Sets Are Equal

Determine if the two sets are equal by using Venn diagrams: $(A \cup B) \cap C$ and $(A \cap C) \cup (B \cap C)$.

SOLUTION

The set $A \cup B$ consists of regions I through VI. Of these, IV, V, and VI are also in C, so $(A \cup B) \cap C$ consists of regions IV, V, and VI.

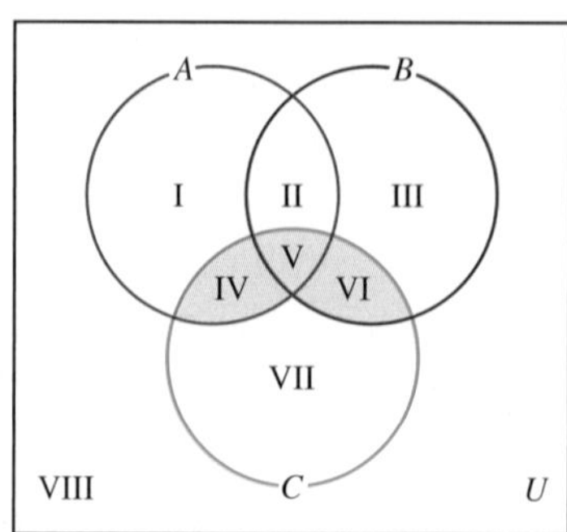

The set $A \cap C$ consists of regions IV and V, and the set $B \cap C$ consists of regions V and VI. Their union is regions IV, V, and VI.

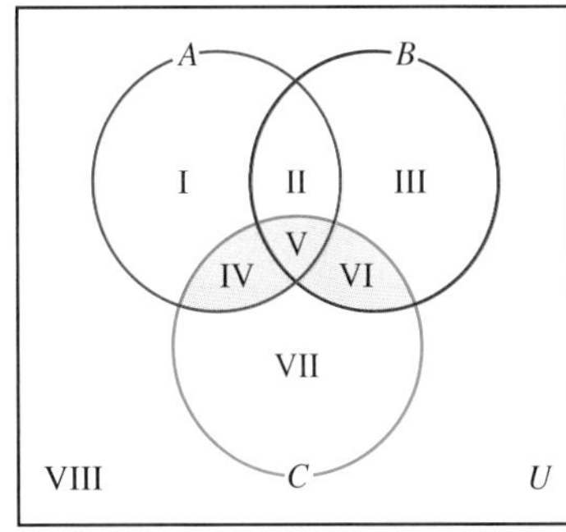

Since the shaded areas are the same, the two sets are equal.

4. Use Venn diagrams to decide if two sets are equal.

Determine if the two sets are equal by using Venn diagrams: $B \cup (A \cap C)$ and $(A \cup B) \cap (B \cup C)$.

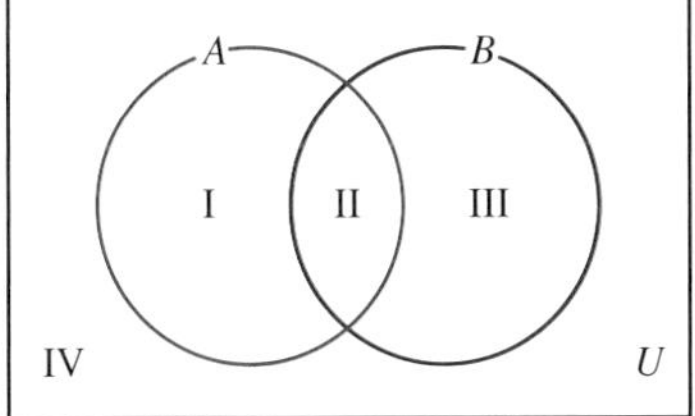

Figure 2-10

The Cardinal Number of a Union

For two sets A and B, how many elements would you guess are in $A \cup B$? Many students guess the number of elements in A plus the number of elements in B, and those students are incorrect when they do. This is an area that Venn diagrams can help us with.

As we can see in Figure 2-10, if we start with the number of elements in A, we're counting all the members in regions I and II. When we add the number of elements in B, we're counting all the members in regions II and III. Do you see the issue? The elements in region II get counted twice. We can fix that by subtracting the number of elements in region II. Since region II represents $A \cap B$, we get the following useful formula:

Math Note

In words, the formula to the right says that to find the number of elements in the union of A and B, you add the number of elements in A and B and then subtract the number of elements in the intersection of A and B.

The Cardinality of a Union

If $n(A)$ represents the cardinal number of set A, then for any two finite sets A and B, $n(A \cup B) = n(A) + n(B) - n(A \cap B)$.

Next, we'll see how this formula can be used in an applied situation.

EXAMPLE 8 Using the Formula for Cardinality of a Union

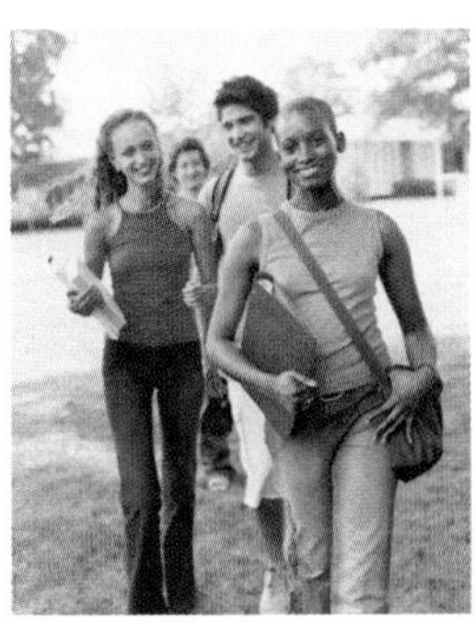

In a survey of 100 randomly selected freshmen walking across campus, it turns out that 42 are taking a math class, 51 are taking an English class, and 12 are taking both. How many students are taking either a math class or an English class?

SOLUTION

If we call the set of students taking a math class A and the set of students taking an English class B, we're asked to find $n(A \cup B)$. We're told that $n(A) = 42$, $n(B) = 51$, and $n(A \cap B) = 12$. So

$$n(A \cup B) = n(A) + n(B) - n(A \cap B) = 42 + 51 - 12 = 81$$

☑ 5. Use the formula for the cardinality of a union.

▼ Try This One 8

A telephone poll of 200 registered Democrats during the 2008 primary season found that 94 supported Barack Obama, 85 supported Hillary Clinton, and 20 supported both. How many supported either Obama or Clinton?

In this section, we saw how Venn diagrams can be used to illustrate sets, prove equality of two sets, and solve problems. We'll explore the problem-solving aspect of Venn diagrams further in Section 2-4.

Answers to Try This One

1

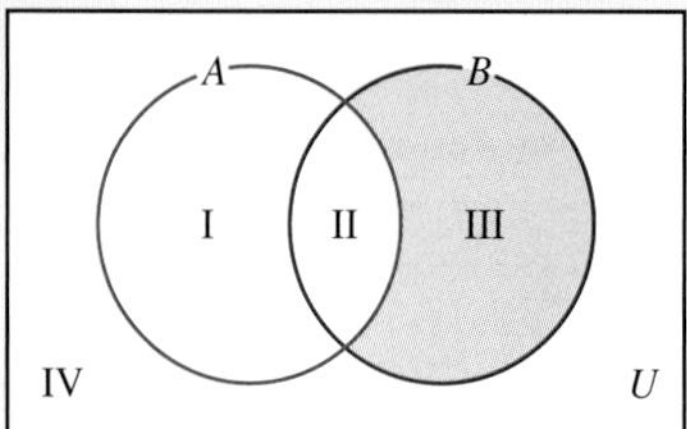

2

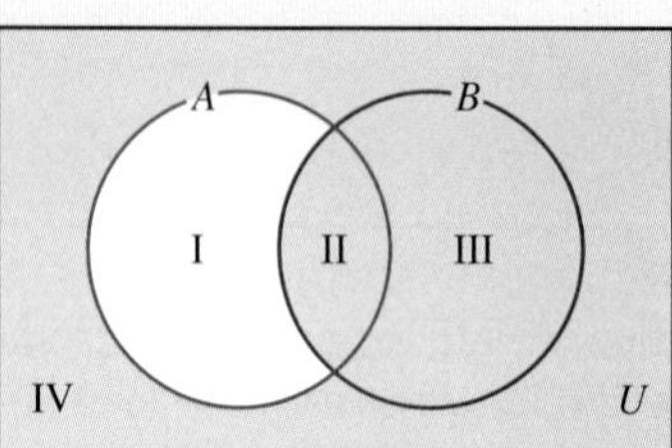

3

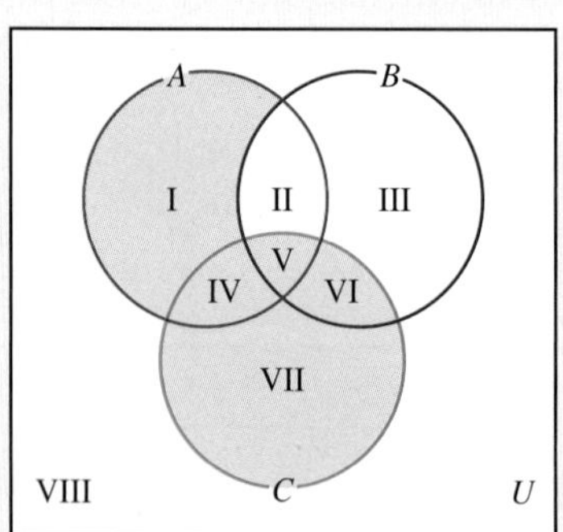

4 Both are {105}.

5 Both are {ABC, CBS, USA, TBS, TNT, MTV}.

6 Both diagrams are

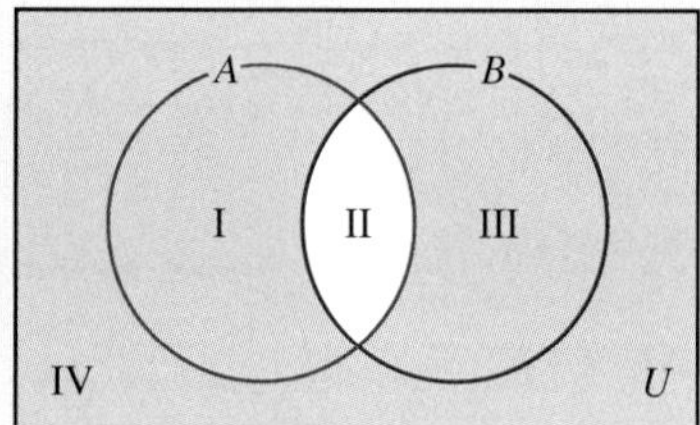

7 Both diagrams are

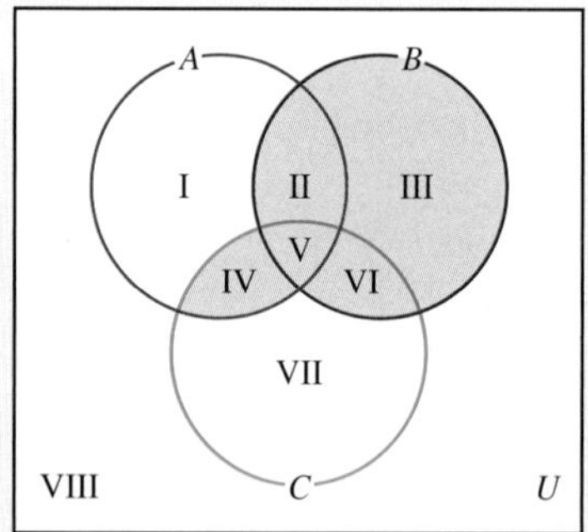

8 159

EXERCISE SET 2-3

Writing Exercises

1. Explain in your own words how to draw a Venn diagram representing the set $A \cup B$.
2. Explain in your own words how to draw a Venn diagram representing the set $A \cap B$.
3. Describe in your own words what De Morgan's laws say.
4. Describe in your own words how to find the cardinal number of the union of two sets.

Computational Exercises

For Exercises 5–28, draw a Venn diagram and shade the sections representing each set.

5. $A \cup B'$
6. $(A \cup B)'$
7. $A' \cup B'$
8. $A' \cup B$
9. $A' \cap B'$
10. $A \cap B'$
11. $A \cup (B \cap C)$
12. $A \cap (B \cup C)$
13. $(A \cup B) \cup (A \cap C)$
14. $(A \cup B) \cap C$
15. $(A \cup B) \cap (A \cup C)$
16. $(A \cap B) \cup C$
17. $(A \cap B)' \cup C$
18. $(A \cup B) \cup C'$
19. $A \cap (B \cup C)'$
20. $A' \cap (B' \cup C')$
21. $(A' \cup B') \cap C$
22. $A \cap (B \cap C)'$
23. $(A \cup B)' \cap (A \cup C)$
24. $(B \cup C) \cup C'$
25. $A' \cap (B' \cap C')$
26. $(A \cup B)' \cap C'$
27. $A' \cap (B \cup C)'$
28. $(A \cup B) \cap (A \cap C)$

For Exercises 29–36, determine whether the two sets are equal by using Venn diagrams.

29. $(A \cap B)'$ and $A' \cup B'$
30. $(A \cup B)'$ and $A' \cup B'$
31. $(A \cup B) \cup C$ and $A \cup (B \cup C)$
32. $A \cap (B \cup C)$ and $(A \cap B) \cup (A \cap C)$
33. $A' \cup (B \cap C')$ and $(A' \cup B) \cap C'$
34. $(A \cap B) \cup C'$ and $(A \cap B) \cup (B \cap C')$
35. $(A \cap B)' \cup C$ and $(A' \cup B') \cap C$
36. $(A' \cup B') \cup C$ and $(A \cap B)' \cap C'$

For Exercises 37–46, use the following Venn diagram to find each set.

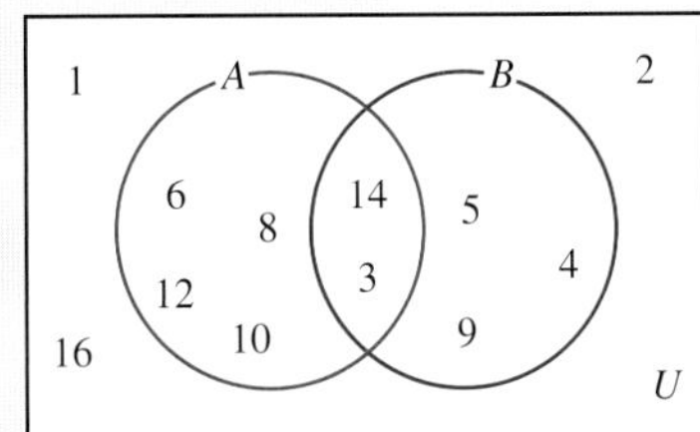

37. $n(A)$
38. $n(B)$
39. $n(A \cap B)$
40. $n(A \cup B)$
41. $n(A')$
42. $n(B')$
43. $n(A' \cap B')$
44. $n(A' \cup B')$
45. $n(A - B)$
46. $n(B - A)$

For Exercises 47–56, use the following information:

$$U = \{x \mid x \text{ is a natural number} < 13\}$$
$$A = \{x \mid x \text{ is an odd natural number}\}$$
$$B = \{x \mid x \text{ is a prime number}\}$$

(*Hint:* The prime numbers less than 13 are 2, 3, 5, 7, and 11.) *Find each.*

47. $n(A)$
48. $n(B)$
49. $n(A \cap B)$
50. $n(A \cup B)$
51. $n(A \cap B')$
52. $n(A' \cup B)$
53. $n(A')$
54. $n(B')$
55. $n(A - B)$
56. $n(B' - A)$

In Exercises 57–60, A = {people who drive an SUV} and B = {people who drive a hybrid vehicle}. Draw a Venn diagram of the following, and write a sentence describing what the set represents.

57. $A \cup B$
58. $A \cap B$
59. A'
60. $(A \cap B)'$

In Exercises 61–64, O = {students in online courses}, B = {students in blended courses}, and T = {students in traditional courses}. Draw a Venn diagram of the following, and write a sentence describing what the set represents.

61. $O \cap (T \cup B)$
62. $B \cup (O \cap T)$
63. $B \cap O \cap T$
64. $(B \cup O) \cap (T \cup O)$

In Exercises 65–68, D = {students voting Democrat}, R = {students voting Republican}, and I = {students voting Independent}. Draw a Venn diagram of the following, and write a sentence describing what the set represents.

65. $D' \cup R$
66. $D' \cap I'$
67. $(D \cup R) \cap I'$
68. $I - (D \cup R)$

In Exercises 69–72, G = {people who regularly use Google}, Y = {people who regularly use Yahoo!}, and M = {people who regularly use MSN Live}. Draw a Venn diagram of the following, and write a sentence describing what the set represents.

69. $G - Y$
70. $G - (Y \cap M)$
71. $G' \cap Y' \cap M'$
72. $(Y \cap M) \cup (Y \cap G)$

The table and Venn diagram below are to be used for Exercises 73–78. The table shows the four baseball teams that made the playoffs in the American League from 2005 to 2007. For each exercise, write the region(s) of the Venn diagram that would include the team listed.

Year	2005	2006	2007
Teams	Boston Red Sox	Detroit Tigers	Boston Red Sox
	Chicago White Sox	Minnesota Twins	Cleveland Indians
	Los Angeles Angels	New York Yankees	Los Angeles Angels
	New York Yankees	Oakland A's	New York Yankees

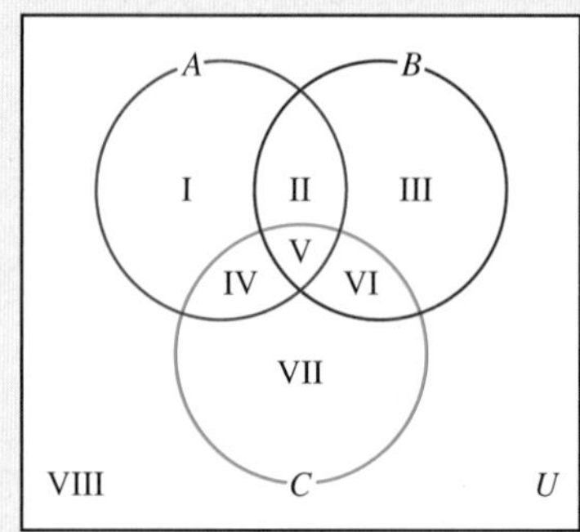

Note: set A represents 2005 playoff teams, set B represents 2006 playoff teams, and set C represents 2007 playoff teams.

73. Boston Red Sox
74. Los Angeles Angels
75. Cleveland Indians
76. Minnesota Twins
77. New York Yankees
78. Oakland A's

Critical Thinking

79. For two finite sets A and B, is $n(A - B)$ equal to $n(A) - n(B)$? If not, can you find a formula for $n(A - B)$?
80. Can you find a formula for $n(A \cap B)$ in terms of only $n(A)$ and $n(B)$? Why or why not? See if you can find a formula for $n(A \cap B)$ using any sets you like.
81. Make a conjecture about another form for the set $(A \cup B \cup C)'$ based on the first of De Morgan's laws. Check out your conjecture by using a Venn diagram.
82. Make a conjecture about another form for the set $(A \cap B \cap C)'$ based on the second of De Morgan's laws. Check out your conjecture by using a Venn diagram.

Section 2-4 Using Sets to Solve Problems

LEARNING OBJECTIVE

☐ 1. Solve problems by using Venn diagrams.

We live in the information age—every time you turn around, somebody somewhere is trying to gather information about you, your opinions, and (most commonly) your spending habits. Surveys are conducted by the thousands every day, and every person, pet, pastime, and product are classified. Some of the things we've learned about sets can be very helpful in interpreting information from surveys and classifications, and that will be the focus of this section.

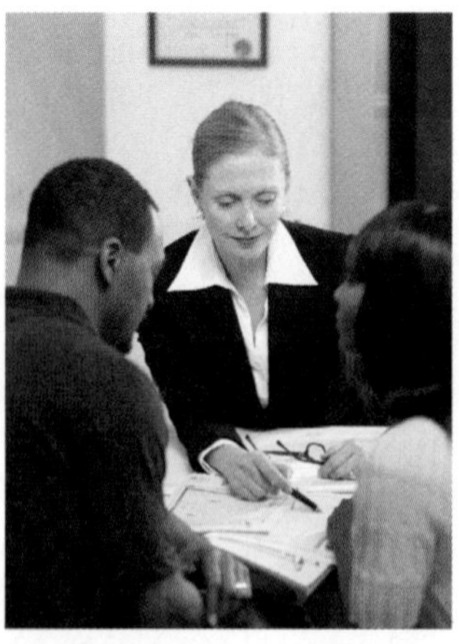

When things are classified into two distinct sets, we can use a two-set Venn diagram to interpret the information. This is illustrated in Example 1.

EXAMPLE 1 Solving a Problem by Using a Venn Diagram

In 2008, there were 36 states that had some form of casino gambling in the state, 42 states that sold lottery tickets of some kind, and 34 states that had both casinos and lotteries. Draw a Venn diagram to represent the survey results, and find how many states have only casino gambling, how many states have only lotteries, and how many states have neither.

SOLUTION

Step 1 Draw a Venn diagram with circles for casino gambling and lotteries, labeling the regions as usual.

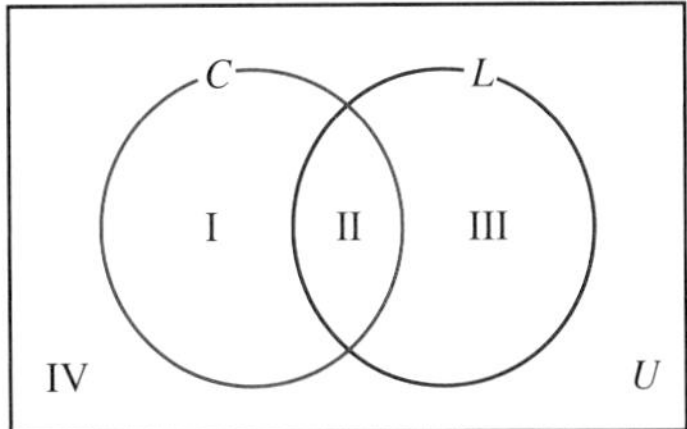

Step 2 Thirty-four states have both, so put 34 in the intersection of C and L, which is region II.

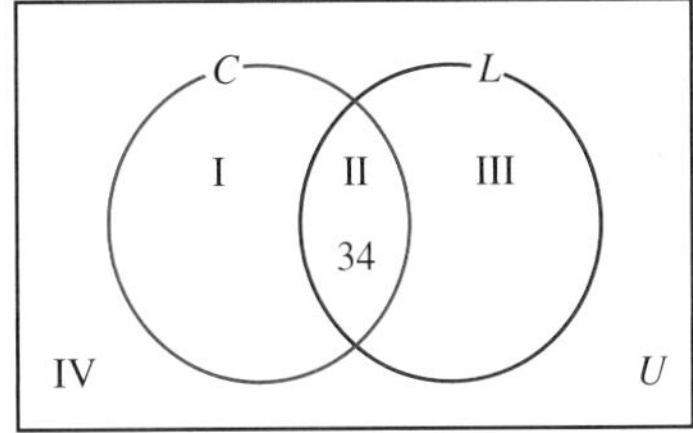

Step 3 Since 36 states have casino gambling and 34 have both, there must be 2 that have only casino gambling. Put 2 in region I. Since 42 states have lotteries and 34 have both, there are 8 that have only lotteries. Put 8 in region III.

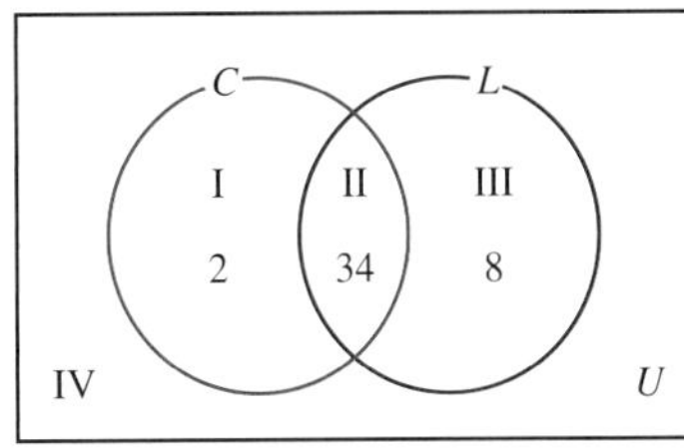

Step 4 Now 44 states are accounted for, so there must be 6 left to put in region IV.

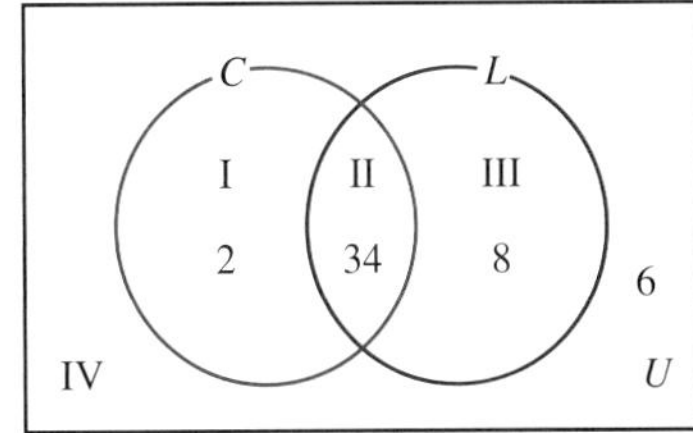

Now we can answer the questions easily. There are only two states that have casino gambling but no lottery (Nevada and Wyoming, in case you're wondering). There are eight states that have lotteries but no casino gambling (region III), and just six states that have neither (region IV).

▼ Try This One 1

In an average year, Columbus, Ohio, has 163 days with some rain, 63 days with some snow, and 24 days with both. Draw a Venn diagram to represent these averages, and find how many days have only rain, only snow, and neither.

We can use the results of Example 1 to write a general procedure for using a Venn diagram to interpret information that can be divided into two sets.

Using Venn Diagrams with Two Sets

Step 1 Find the number of elements that are common to both sets and write that number in region II.

Step 2 Find the number of elements that are in set *A* and not set *B* by subtracting the number in region II from the total number of elements in *A*. Then write that number in region I. Repeat for the elements in *B* but not in region II, and write in region III.

Step 3 Find the number of elements in *U* that are not in either *A* or *B*, and write it in region IV.

Step 4 Use the diagram to answer specific questions about the situation.

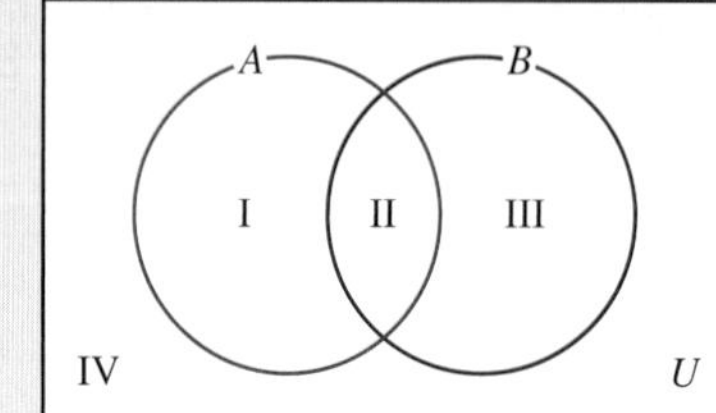

Quick Vote

Are you worse off financially than you were in recent years?

- Yes
- No
- About the same

VOTE or see results

Many news websites include daily surveys, like this one from cnn.com.

One of the most useful applications of Venn diagrams is for studying the results of surveys. Whether they are for business-related research or just to satisfy curiosity, surveys seem to be everywhere these days, especially online. Example 2 analyzes the results of a survey on tattoos and body piercings.

EXAMPLE 2 Solving a Problem by Using a Venn Diagram

In a survey published in the *Journal of the American Academy of Dermatologists,* 500 people were polled by random telephone dialing. Of these, 120 reported having a tattoo, 72 reported having a body piercing, and 41 had both. Draw a Venn diagram to represent these results, and find out how many respondents have only tattoos, only body piercings, and neither.

SOLUTION

In this example, we'll insert just one diagram at the end, and we will refer to it as we go.

Step 1 Place the number of respondents with both tattoos and body piercings (41) in region II.

Step 2 There are 120 people with tattoos and 41 with both, so there are $120 - 41$, or 79, people with only tattoos. This goes in region I. By the same logic, there are $72 - 41$, or 31, people with only piercings. This goes in region III.

Step 3 We now have $41 + 79 + 31 = 151$ of the 500 people accounted for, so $500 - 151 = 349$ goes in region IV.

Step 4 There are 79 people with only tattoos, 31 with only piercings, and 349 with neither.

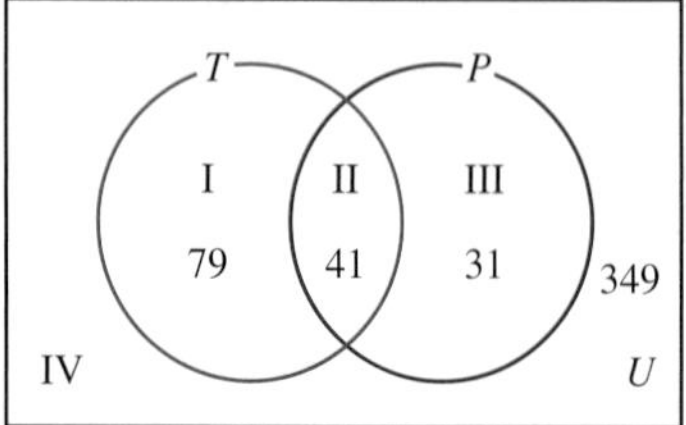

▼ Try This One 2

According to an online survey on Howstuffworks.com, 12,595 people gave their thoughts on Coke versus Pepsi. Of these, 7,642 drink Coke, 5,619 drink Pepsi, and 1,856 drink both. Draw a Venn diagram to represent these results, and find out how many respondents drink only Coke, only Pepsi, and neither.

Sidelight HOW MUCH IS YOUR OPINION WORTH?

Communication in our society is becoming cheaper, easier, and more effective all the time. In the age of cell phones and Internet communication, businesses are finding it simpler than ever to contact people for their opinions, and more and more people are finding out that companies are willing to pay to hear what they have to say. There are literally hundreds of companies in the United States today whose main function is to gather opinions on everything from political candidates to potato chips. In fact, over $6 *billion* is spent on market research in the United States each year. Maybe you'll think twice the next time somebody asks you for your opinion for free.

When a classification problem or a survey consists of three sets, a similar procedure is followed, using a Venn diagram with three sets. We just have more work to do since there are now eight regions instead of four.

EXAMPLE 3 Solving a Problem by Using a Venn Diagram

A survey of 300 first-year students at a large midwestern university was conducted to aid in scheduling for the following year. Responses indicated that 194 were taking a math class, 210 were taking an English class, and 170 were taking a speech course. In addition, 142 were taking both math and English, 111 were taking both English and speech, 91 were taking both math and speech, and 45 were taking all three. Draw a Venn diagram to represent these survey results, and find the number of students taking

(a) Only English.
(b) Math and speech but not English.
(c) Math or English.
(d) None of these three subjects.

SOLUTION

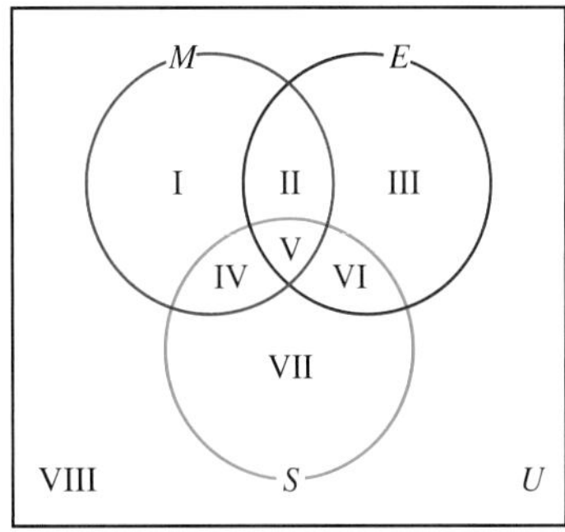

Step 1 The only region we know for sure from the given information is region V—the number of students taking all three classes. So we begin by putting 45 in region V.

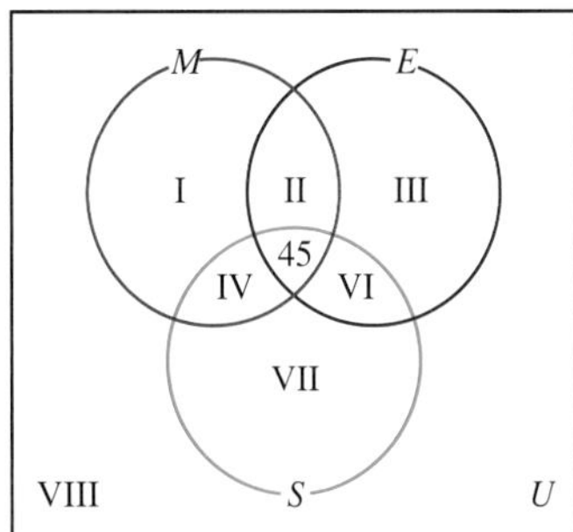

Step 2 There are 142 students taking both math and English, but we must subtract the number in all three classes to find the number in region II: $142 - 45 = 97$. In the same way, we get $91 - 45 = 46$ in region IV (both math and speech) and $111 - 45 = 66$ in region VI (both English and speech).

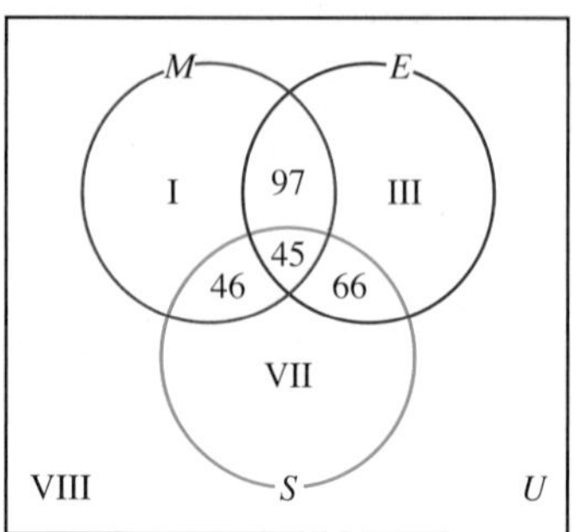

Step 3 Now we can find the number of elements in regions I, III, and VII. There are 194 students in math classes, but $97 + 45 + 46 = 188$ are already accounted for in the diagram, so that leaves 6 in region I. Of the 210 students in English classes, $97 + 45 + 66 = 208$ are already accounted for, leaving just 2 in region III. There are 170 students in speech classes, with $46 + 45 + 66 = 157$ already accounted for. This leaves 13 in region VII.

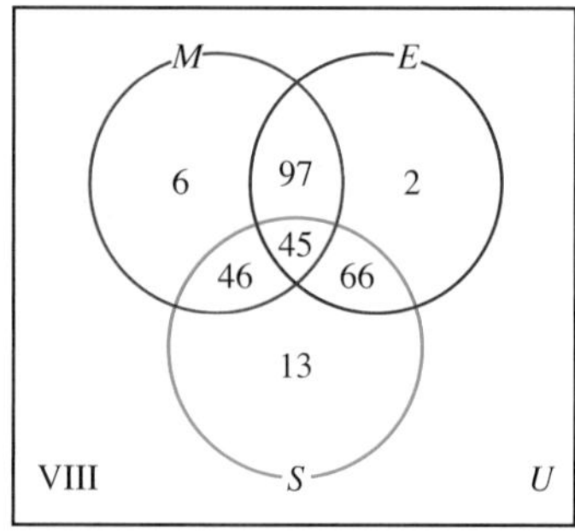

Step 4 Adding up all the numbers in the diagram so far, we get 275. That leaves 25 in region VIII.

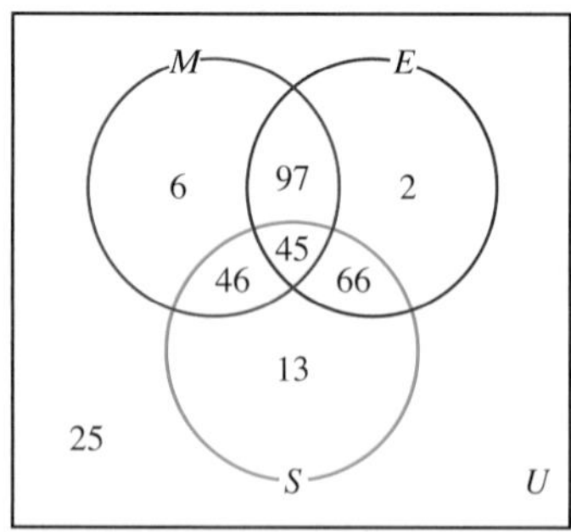

Step 5 Now that we have the diagram completed, we turn our attention to the questions.

(a) Students taking only English are represented by region III—there are only 2.
(b) Math and speech but not English is region IV, so there are 46 students.
(c) Students taking math or English are represented by all but regions VII and VIII. So there are only 38 students not taking either math or English, and $300 - 38 = 262$ who are.
(d) There are 25 students outside of the regions for all of math, English, and speech.

Math Note

Notice that in filling in the Venn diagram in Example 3, we started with the number of elements in the innermost region and worked our way outward.

☑ 1. Solve problems by using Venn diagrams.

▼ Try This One 3

An online music service surveyed 500 customers and found that 270 listen to hip-hop music, 320 listen to rock, and 160 listen to country. In addition, 140 listen to both rock and hip-hop, 120 listen to rock and country, and 80 listen to hip-hop and country. Finally, 50 listen to all three. Draw a Venn diagram to represent the results of the survey and find the number of customers who

(a) Listen to only hip-hop.
(b) Listen to rock and country but not hip-hop.
(c) Don't listen to any of these three types of music.
(d) Don't listen to country music.

We can use the results of Example 3 to write a general procedure for using a Venn diagram to interpret information that can be divided into three sets.

Using Venn Diagrams with Three Sets

Step 1 Put the number of elements common to all three sets in region V.

Step 2 Find the number of elements in $A \cap B$, $A \cap C$, and $B \cap C$. Subtract the number of elements in all three, and place these numbers in regions II, IV, and VI.

Step 3 Subtract the number of elements you now have inside each large set from the total number you were given for each set: these will be the numbers you put in regions I, III, and VII.

Step 4 Add all the numbers inside the circles and subtract that from the total number of elements in U to get the number in region VIII.

Step 5 Use the completed diagram to answer specific questions.

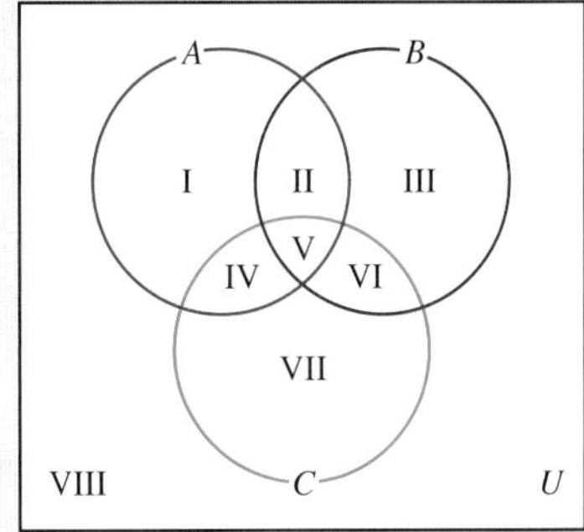

We've come a pretty long way from simply defining what sets and elements are! In this section, we've seen that Venn diagrams can be effectively used to sort out some pretty complicated real-world situations. And the better you are at interpreting information, the better equipped you'll be to survive and thrive in the information age.

Answers to Try This One

1 Only rain: 139; only snow: 39; neither: 163 (assuming it's not leap year!)

2 Only Coke: 5,786; only Pepsi: 3,763; neither: 1,190

3 (a) 100 (b) 70
(c) 40 (d) 340

EXERCISE SET 2-4

Real-World Applications

1. In a survey of 40 students, 20 use myspace.com, 25 use facebook.com, and 7 use both myspace.com and facebook.com.

   (a) How many use myspace.com only?
   (b) How many use facebook.com only?
   (c) How many use neither myspace.com nor facebook.com?

2. In a class of 25 students, 18 were mathematics majors, 12 were computer science majors, and 7 were dual majors in mathematics and computer science.

   (a) How many students were majoring in mathematics only?
   (b) How many students were not majoring in computer science?
   (c) How many students were not mathematics or computer science majors?

3. There were 16 recycling bins in the student union. Eight of them contained only paper, five contained only plastic, and three contained paper and plastic.

   (a) How many contained only paper?
   (b) How many contained only plastic?
   (c) How many contained neither paper nor plastic?

4. In the cafeteria, 25 students were seated at tables. Fifteen were enrolled in psychology, nine were enrolled in physics, and four were enrolled in both psychology and physics.

   (a) How many students were enrolled in psychology only?
   (b) How many students were not enrolled in physics?
   (c) How many students were enrolled in at least one of these courses?

5. The financial aid department at a college surveyed 70 students, asking if they receive any type of financial aid. The results of the survey are summarized in the table below.

Financial Aid	Number of Students
Scholarships	16
Student loans	24
Private grants	20
Scholarships and loans	9
Loans and grants	11
Scholarships and grants	7
Scholarships, loans, and grants	2

   (a) How many students got only scholarships?
   (b) How many got loans and private grants but not scholarships?
   (c) How many didn't get any of these types of financial aid?

6. The manager of a campus gym is planning the schedule of fitness classes for a new school year, and will decide how often to hold certain classes based on the interests of students. She polls 47 students at various times of day, asking what type of classes they'd be interested in attending. The results are summarized in the table below.

Type of Class	Students Interested
Yoga	17
Pilates	13
Spinning	12
Yoga and pilates	9
Pilates and spinning	3
Yoga and spinning	5
All three	2

   (a) How many students are interested in yoga or spinning, but not pilates?
   (b) How many are interested in exactly two of the three classes?
   (c) How many are interested in yoga but not pilates?

7. One semester in a chemistry class, 14 students failed due to a poor attendance, 23 failed due to not studying, 15 failed because they did not turn in assignments, 9 failed because of poor attendance and not studying, 8 failed because of not studying and not turning in assignments, 5 failed because of poor attendance and not turning in assignments, and 2 failed because of all three of these reasons.

   (a) How many failed for exactly two of the three reasons?
   (b) How many failed because of poor attendance and not studying but not because of not turning in assignments?
   (c) How many failed because of exactly one of the three reasons?
   (d) How many failed because of poor attendance and not turning in assignments but not because of not studying?

8. Last Super Bowl weekend, there were 109 pizzas ordered for the sophomore dorm. That weekend

32 customers ordered their pizza with just pepperoni, 40 ordered their pizza with just sausage, 18 ordered theirs with just onions, 13 ordered theirs with pepperoni and sausage, 10 ordered theirs with sausage and onions, 9 ordered theirs with pepperoni and onions, and 7 ordered theirs with all three items.

(a) How many customers ordered their pizza with pepperoni or sausage (or both)?
(b) How many customers ordered their pizza with sausage or onions (or both)?
(c) How many customers ordered a pizza without pepperoni, sausage, or onions?

9. There were 70 students in line at the campus bookstore to sell back their textbooks after finals: 19 had history books to return, 21 had business books to return, and 19 had math books to return. Nine were selling back both history and business books, eight were selling back both business and math books, five were selling back history and math books, and three were selling back history, business, and math books.

(a) How many students were selling back at most two of these three types of books?
(b) How many students were selling back history and math books but not business books?
(c) How many were selling back neither history nor math books?

10. A survey of 96 students on campus showed that 29 read the *Campus Observer* student newspaper that morning, 24 read the news via the Internet that morning, and 20 read the local city paper that morning. Eight read the *Campus Observer* and the Internet news that morning, while four read the Internet news and the local paper, seven read the *Campus Observer* and the local city paper, and one person read the *Campus Observer,* the Internet news, and the local paper.

(a) How many read the Internet news or local paper but not both?
(b) How many read the Internet news and the local paper but not the *Campus Observer?*
(c) How many read the Internet news or the *Campus Observer?* (*Note: Or* means one or the other or both.)

11. Sixty-one students in an art class were surveyed about their musical tastes. Of those asked, 20 liked alternative, 26 liked hip-hop, 20 liked techno, 9 liked alternative and hip-hop, 15 liked hip-hop and techno, 6 liked alternative and techno, and 3 liked all three types of music.

(a) How many liked none of these three types?
(b) How many liked exactly three of these types?
(c) How many liked exactly one of these types?

12. In a survey of 200 college students, 128 watched at least one reality show on television each week, 131 watched at least one video on MTV each week, 114 watched a national news show on CNN each week, 75 watched a reality show and a video, 59 watched a video and a news show, 81 watched a reality show and a news show, and 33 watched a reality show, a video, and a news show each week.

(a) How many watched exactly two of these?
(b) How many watched only one of these?
(c) How many did not watch a reality show?

13. In a survey of 121 students, 39 belonged to a club on campus, 51 regularly attended a campus sporting event, and 26 belonged to a professional organization. Furthermore, 25 regularly attended campus sporting events and belonged to a campus club, 14 belonged to a professional organization and attended campus sporting events, 18 belonged to a professional organization and a campus club, and 10 were involved in all three.

(a) How many students did exactly one of these?
(b) How many students regularly attended a sporting event, but did not belong to a professional organization?
(c) How many students did not regularly attend a campus sporting event or belong to a professional organization?

14. Out of 20 students taking a midterm exam, 15 answered the first of two bonus questions, 13 answered the second bonus question, and 2 answered neither bonus question. How many students answered both bonus questions?

15. In a group of 34 people waiting in line at the local java house, 18 wanted to order a cappuccino, 20 wanted to order a latte, and 2 wanted to order neither of these drinks. How many wanted to order both a cappuccino and a latte?

16. An entertainment magazine conducted an online poll of 230 college students who watch late-night TV, asking what shows they watch regularly. There were 119 who watch David Letterman, 101 who watch Conan O'Brien, and 75 who watch Jimmy Fallon. In addition, 69 watch Letterman and O'Brien, 48 watch Letterman and Fallon, 38 watch Fallon and O'Brien, and 28 watch all three.

(a) How many don't watch any of these three shows?
(b) How many watch only one of the three shows?
(c) How many watch exactly two of the three shows?

Critical Thinking

17. A researcher was hired to examine the drinking habits of energy drink consumers. Explain why he was fired when he published the results below, from a survey of 40 such consumers:

    23 said they drink Red Bull.
    18 said they drink Monster.
    19 said they drink G2.
    12 said they drink Red Bull and Monster.
    6 said they drink Monster and G2.
    7 said they drink Red Bull and G2.
    2 said they drink all three (not at the same time, hopefully).
    2 said they don't drink any of the three brands.

18. The marketing research firm of OUWant12 designed and sent three spam advertisements to 40 e-mail accounts. The first one was an ad for hair removal cream, the second was an ad for Botox treatments, and the third was an ad for a new all lima bean diet. Explain why, when the following results occurred, the sponsors discontinued their services.

    23 recipients deleted the ad for hair removal cream without looking at it.
    18 recipients deleted the ad for Botox treatments.
    19 recipients deleted the ad for the all lima bean diet.
    12 recipients deleted the ads for hair removal cream and Botox treatments.
    6 recipients deleted the ads for Botox treatments and the all lima bean diet.
    7 recipients deleted the ads for the hair removal cream and the all lima bean diet.
    2 recipients deleted all three ads.

Section 2-5 Infinite Sets

LEARNING OBJECTIVES

- ❑ 1. Formally define infinite sets.
- ❑ 2. Show that a set is infinite.
- ❑ 3. Find a general term of an infinite set.
- ❑ 4. Define countable and uncountable sets.

The night sky may look infinite, but have you ever thought about what that really means?

Infinity is a concept that's tremendously difficult for us human beings to wrap our minds around. Because our thoughts are shaped by experiences in a physical world with finite dimensions, things that are infinitely large always seem just out of our grasp. Whether you're a believer in the Big Bang theory or creationism, you probably think that one or the other must be true because the alternative is too far beyond our experience: that time didn't have a beginning at all, but extends infinitely far in each direction. Some philosophers feel that human beings are fundamentally incapable of grasping the concept of something infinitely large at all!

The study of infinity and infinite sets from a mathematical standpoint is a relatively young one compared to the history of math in general. For at least a couple of thousand years, the nature of infinity so confounded the greatest human minds that they chose to not deal with it at all. And yet in working with a set as simple as the natural numbers, we deal with infinite sets in math all the time. It's an interesting paradox.

A Definition of Infinite Sets

Recall from Section 2-1 that a set is considered to be finite if the number of elements is either zero or a natural number. Otherwise, it is considered to be an infinite set. For example, the set $\{10, 20, 30, 40\}$ is finite because the number of elements (four) is a natural number. But the set $\{10, 20, 30, 40, \ldots\}$ is infinite because the number of elements is unlimited, and therefore not a natural number.

You might recognize an infinite set when you see one, but it's not necessarily easy to make a precise definition of what it means for a set to be infinite (other than the obvious definition, "not finite"). The German mathematician Georg Cantor, widely regarded as the father of set theory, is famous for his 19th-century study of infinite sets. (Sadly his work was not highly regarded in his lifetime, and he died in a mental institution.) Cantor's simple and elegant definition of an infinite set is as follows:

> A set is **infinite** if it can be placed into a one-to-one correspondence with a proper subset of itself.

☑ 1. Formally define infinite sets.

First, notice that a finite set definitely does not meet the condition in this definition: if a set has some finite number of elements, let's say 10, then any proper subset has at most 9 elements, and an attempt at one-to-one correspondence will always leave out at least one member.

The trickier thing is to understand how an infinite set can meet this definition. We'll illustrate with an infinite set we know well, the set of natural numbers $\{1, 2, 3, 4, \ldots\}$. The set of even natural numbers $\{2, 4, 6, 8, \ldots\}$ is of course a proper subset: every even number is also a natural number, but there are natural numbers that are not even numbers. Now we demonstrate the clever way to put these two sets in one-to-one correspondence: match each natural number with its double.

$$1 \leftrightarrow 2, \qquad 2 \leftrightarrow 4, \qquad 3 \leftrightarrow 6, \qquad 4 \leftrightarrow 8, \ldots$$

In general, we can define our correspondence as matching any n from the set of natural numbers with a corresponding even number $2n$. This is a one-to-one correspondence because every natural number has a match (its double), and every even number has a match (its half). So we've put the natural numbers into one-to-one correspondence with a proper subset, and they are an infinite set.

Let's try another example.

EXAMPLE 1 Showing That a Set Is Infinite

Show that the set $\{5, 10, 15, 20, 25, \ldots\}$ is an infinite set.

SOLUTION

A simple way to put this set in correspondence with a proper subset of itself is to match every element n with its double $2n$:

$$\begin{matrix} \{5, & 10, & 15, & 20, & 25, \ldots\} \\ \updownarrow & \updownarrow & \updownarrow & \updownarrow & \updownarrow \\ \{10, & 20, & 30, & 40, & 50, \ldots\} \end{matrix}$$

The second set, $\{10, 20, 30, 40, 50, \ldots\}$ is a proper subset of the first, and the two are in one-to-one correspondence, so $\{5, 10, 15, 20, 25, \ldots\}$ is an infinite set.

▼ Try This One 1

☑ 2. Show that a set is infinite.

Show that the set $\{-1, -2, -3, -4, -5, \ldots\}$ is an infinite set.

A General Term of an Infinite Set

One consequence of the way we showed that the set of natural numbers is infinite is that we can find a generic formula for the set of even numbers: $2n$, where n is the set $\{1, 2, 3, 4, \ldots\}$. We will call $2n$ in this case a **general term** of the set of even numbers. Notice that we said "a general term," not "the general term." There are other general terms we could write for this set: $2n - 6$, where n is the set $\{4, 5, 6, 7, \ldots\}$, is another possibility. But in most cases the simplest general term is the one where the first listed number is obtained from substituting in 1 for n, and that's the one we'll typically find.

Sidelight THE INFINITE HOTEL

Suppose a hotel in some far-off galaxy was so immense that it actually had infinitely many rooms, numbered 1, 2, 3, 4, ... There's a big convention of creepy alien creatures in town, so every room is filled. A weary traveler drags into the lobby and asks for a room, and when he's informed that the hotel is full, he protests that the hotel can most definitely accommodate him. Do you agree? Can they find a room for him without kicking someone out?

People tend to be split on this question about half and half: half think they can't accommodate him because all the rooms are full, and half think they can because there are infinitely many rooms. In fact, the traveler is correct—it just takes inconveniencing every other guest! If the manager asks every guest to move into the room whose number is 1 higher than his or her current room, everyone that was originally in a room still has one, and our traveler can rest his weary body in room 1.

This clever little mind exercise is a consequence of the fact that the natural numbers form an infinite set—they can be put in one-to-one correspondence with a proper subset of themselves by corresponding any n with $n + 1$.

EXAMPLE 2 Finding a General Term for an Infinite Set

Math Note

Finding a general term for a set is not always easy. In some cases, it can be very difficult or even impossible. You may need to do some trial and error before finding a formula that works.

Find a general term for the set $\{4, 7, 10, 13, 16, \ldots\}$.

SOLUTION

We should always begin by trying to recognize a pattern in the numbers of the set. In this case, the pattern is that the numbers increase by 3. When this is the case, $3n$ is a good choice, because as n increases by 1, $3n$ increases by 3. But simply using $3n$ will give us the set $\{3, 6, 9, 12, \ldots\}$, which is not quite what we want. We remedy that by adding 1 to our general term, to get $3n + 1$. (We encourage you to check that answer by substituting in $1, 2, 3, \ldots$ for n to see that it generates the set $\{4, 7, 10, 13, 16, \ldots\}$.)

▼ Try This One 2

☑ 3. Find a general term of an infinite set.

Find a general term for the set $\{2, 8, 14, 20, 26, \ldots\}$.

Different Kinds of Infinity?

Quick, which set is bigger, the set of natural numbers or the set of real numbers? You probably answered the set of real numbers. But both sets are infinitely large, so aren't they the same size? Cantor attacked this problem in the late 1800s. He defined a set to be **countable** if it is finite or can be placed into one-to-one correspondence with the natural numbers and an infinite set to be **uncountable** if it cannot. He used the symbol $\aleph_0$, pronounced aleph-null or aleph-naught (aleph is the first letter of the Hebrew alphabet), to represent the cardinality of a countable set.

☑ 4. Define countable and uncountable sets.

One of Cantor's greatest achievements was to show that the set of real numbers is not countable. So in this case, your intuition was correct: there *are* more real numbers than natural numbers. But the study of infinite sets is a strange and interesting one, with unexpected results at nearly every turn. For example, it can be shown that the cardinality of the set of numbers between 0 and 1 is the same as the cardinality of the entire set of real numbers! If you're interested in learning about a *really* unusual infinite set, do an Internet search for "Cantor set."

Answers to Try This One

1 Can be done in many ways: one choice is to correspond -1 with -2, -2 with -4, and in general $-n$ with $-2n$.

2 $6n - 4$

EXERCISE SET 2-5

Writing Exercises

1. Define an infinite set, both in your own words and by using Cantor's definition.
2. What is meant by a general term for an infinite set?
3. What does it mean for a set to be countable?
4. Explain how you can tell that the set of natural numbers and the set of even numbers have the same cardinality.

Computational Exercises

For Exercises 5–20, find a general term for the set.

5. $\{7, 14, 21, 28, 35, \ldots\}$
6. $\{1, 8, 27, 64, 125, \ldots\}$
7. $\{4, 16, 64, 256, 1{,}024, \ldots\}$
8. $\{1, 4, 9, 16, 25, \ldots\}$
9. $\{-3, -6, -9, -12, -15, \ldots\}$
10. $\{22, 44, 66, 88, 110, \ldots\}$
11. $\left\{\frac{1}{2}, \frac{1}{3}, \frac{1}{4}, \frac{1}{5}, \frac{1}{6}, \ldots\right\}$
12. $\left\{\frac{1}{3}, \frac{2}{3}, \frac{3}{3}, \frac{4}{3}, \frac{5}{3}, \ldots\right\}$
13. $\{2, 6, 10, 14, 18, \ldots\}$
14. $\{1, 4, 7, 10, 13, \ldots\}$
15. $\left\{\frac{2}{3}, \frac{3}{4}, \frac{4}{5}, \frac{5}{6}, \frac{6}{7}, \ldots\right\}$
16. $\left\{\frac{1}{1}, \frac{1}{4}, \frac{1}{9}, \frac{1}{16}, \frac{1}{25}, \ldots\right\}$
17. $\{100, 200, 300, 400, 500, \ldots\}$
18. $\{50, 100, 150, 200, 250, \ldots\}$
19. $\{-4, -7, -10, -13, -16, \ldots\}$
20. $\{-3, -5, -7, -9, -11, \ldots\}$

For Exercises 21–30, show each set is an infinite set.

21. $\{3, 6, 9, 12, 15, \ldots\}$
22. $\{10, 15, 20, 25, 30, \ldots\}$
23. $\{9, 18, 27, 36, 45, \ldots\}$
24. $\{4, 10, 16, 22, 28, \ldots\}$
25. $\{2, 5, 8, 11, 14, \ldots\}$
26. $\{20, 24, 28, 32, 36, \ldots\}$
27. $\{10, 100, 1{,}000, 10{,}000, \ldots\}$
28. $\{100, 200, 300, 400, 500, \ldots\}$
29. $\left\{\frac{5}{1}, \frac{5}{2}, \frac{5}{3}, \frac{5}{4}, \frac{5}{5}, \ldots\right\}$
30. $\left\{\frac{1}{2}, \frac{1}{4}, \frac{1}{8}, \frac{1}{16}, \ldots\right\}$

Critical Thinking

31. Recall that $\aleph_0$ is the cardinality of the natural numbers (and any other countable set). What do you think $\aleph_0 + 1$ should be? What about $\aleph_0 + \aleph_0$?
32. Can you think of a one-to-one correspondence that will show that the set of integers is countable? (The set of integers consists of the natural numbers, their negatives, and zero.)
33. The set of rational numbers is the set of all possible fractions that have integer numerators and denominators. Intuitively, do you think there are more rational numbers than natural numbers? Why? Do you think that the set of rational numbers is countable?
34. Can you think of any set of tangible objects that is infinite? Why or why not?

CHAPTER 2 Summary

Section	Important Terms	Important Ideas
2-1	Set Roster method Element Well-defined Natural numbers Descriptive method Set-builder notation Variable Finite set Infinite set Cardinal number Null set Equal sets Equivalent sets One-to-one correspondence	**A set** is a well-defined collection of objects. Each object is called an element or member of the set. We use three ways to identify sets. They are the roster method, the descriptive method, and set-builder notation. Sets can be finite or infinite. A finite set contains a specific number of elements, while an infinite set contains an unlimited number of elements. If a set has no elements, it is called an empty set or a null set. Two sets are equal if they have the same elements, and two finite sets are equivalent if they have the same number of elements. Two sets are said to have a one-to-one correspondence if it is possible to pair the elements of one set with the elements of the other set in such a way that for each element in the first set there exists one and only one element in the second set, and for each element in the second set there exists one and only one element of the first set.
2-2	Universal set Complement Subset Proper subset Intersection Union Subtraction Cartesian product	**The universal** set is the set of all elements used for a specific problem or situation. The complement of a specific set is a set that consists of all elements in the universal set that are not in the specific set. If every element of one set is also an element of another set, then the first set is said to be a subset of the second set. A subset A of a set B is a proper subset if A is not equal to B. The union of two sets is a set containing all elements of one set or the other set, while the intersection of two sets is a set that contains the elements that both sets have in common. The difference of set A and set B, denoted $A - B$, is the set of elements in set A but not in set B. The Cartesian product of two sets A and B is $A \times B = \{(x, y) \mid x \in A \text{ and } y \in B\}$.
2-3	Venn diagram	**A mathematician** named John Venn devised a way to represent sets pictorially. His method uses overlapping circles to represent the sets. De Morgan's laws for two sets A and B are $(A \cup B)' = A' \cap B'$ and $(A \cap B)' = A' \cup B'$. For any two finite sets A and B, $n(A \cup B) = n(A) + n(B) - n(A \cap B)$.
2-4		**Venn diagrams** can be used to solve real-world problems involving surveys and classifications.
2-5	Infinite set Countable set Uncountable set General term	**An infinite** set can be placed in a one-to-one correspondence with a proper subset of itself. A set is called countable if it is finite or if there is a one-to-one correspondence between the set and the set of natural numbers. A set is called uncountable if it is not countable.

MATH IN Demographics REVISITED

The Venn diagram shown to the right is based on the given demographic estimates:

1. White only: 680; black only: 109; Hispanic only: 53
2. Hispanic and black, but not white: 11
3. Hispanic or black: 270
4. None of white, black, or Hispanic: 50

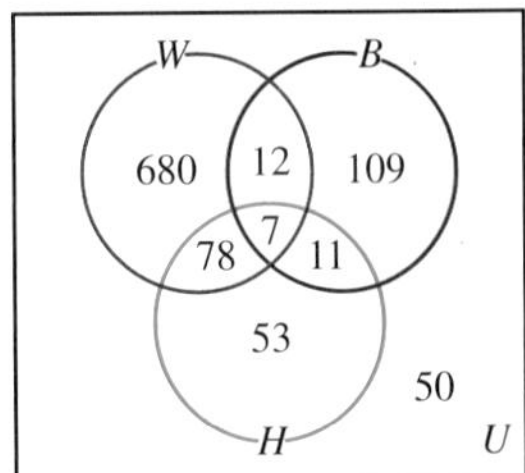

Review Exercises

Section 2-1

For Exercises 1–8, write each set in roster notation.

1. The set D is the set of even numbers between 50 and 60.
2. The set F is the set of odd numbers between 3 and 40.
3. The set L is the set of letters in the word *letter*.
4. The set A is the set of letters in the word *Arkansas*.
5. The set B is $\{x \mid x \in N \text{ and } x > 500\}$.
6. The set C is the set of natural numbers between 5 and 12.
7. The set M is the set of months in the year that begin with the letter P.
8. The set G is the set of days in the week that end with the letter e.

For Exercises 9–12, write each set by using set-builder notation.

9. $\{18, 20, 22, 24\}$
10. $\{5, 10, 15, 20\}$
11. $\{101, 103, 105, 107, \ldots\}$
12. $\{8, 16, 24, \ldots 72\}$

For Exercises 13–18, state whether the set is finite or infinite.

13. $\{x \mid x \in N \text{ and } x \geq 9\}$
14. $\{4, 8, 12, 16, \ldots\}$
15. $\{x, y, z\}$
16. $\{3, 7, 9, 12\}$
17. $\varnothing$
18. {people who have naturally red hair}

Section 2-2

For Exercises 19–22, decide if the statement is true or false.

19. $\{80, 100, 120, \ldots\} \subseteq \{40, 80, 120, \ldots\}$
20. $\{6\} \subset \{6, 12, 18\}$
21. $\{5, 6, 7\} \subseteq \{5, 7\}$
22. $\{a, b, c\} \subset \{a, b, c\}$
23. Find all subsets of $\{r, s, t\}$.
24. Find all subsets of $\{m, n, o\}$.
25. How many subsets and proper subsets does the set $\{p, q, r, s, t\}$ have?
26. How many subsets and proper subsets does the set $\{a, e, i, o, u, y\}$ have?

For Exercises 27–38, let $U = \{p, q, r, s, t, u, v, w, x, y, z\}$, $A = \{p, r, t, u, v\}$, $B = \{t, u, v, x, y\}$, *and* $C = \{s, w, z\}$. *Find each.*

27. $A \cap B$
28. $B \cup C$
29. $(A \cap B) \cap C$
30. B'
31. $A - B$
32. $B - A$
33. $(A \cup B)' \cap C$
34. $B' \cap C'$
35. $(B \cup C) \cap A'$
36. $(A \cup B) \cap C'$
37. $(B' \cap C') \cup A'$
38. $(A' \cap B) \cup C$

For Exercises 39–42, let $M = \{s, t, u\}$ *and* $N = \{v, w, x\}$. *Find each set.*

39. $M \times N$
40. $N \times M$
41. $M \times M$
42. $N \times N$

Section 2-3

For Exercises 43–48, use the Venn diagram below. Describe the region or regions provided in each problem, using set operations on A and B. There may be more than one right answer.

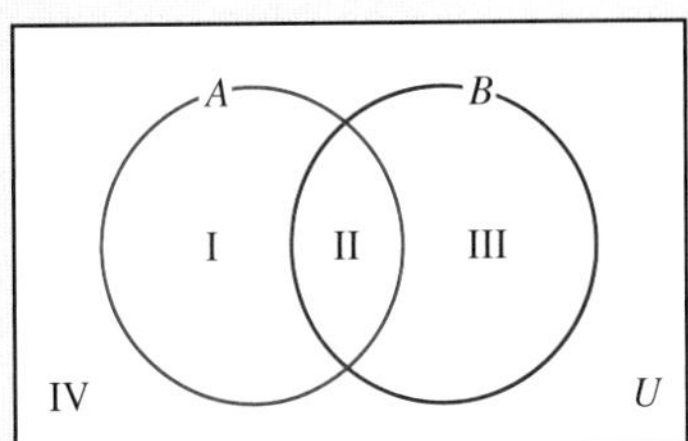

43. Region I
44. Region II
45. Region III
46. Region IV
47. Regions I and III
48. Regions I and IV

For Exercises 49–52, draw a Venn diagram and shade the appropriate area for each.

49. $A' \cap B$
50. $(A \cup B)'$
51. $(A' \cap B') \cup C$
52. $A \cap (B \cup C)'$
53. If $n(A) = 15$, $n(B) = 9$, and $n(A \cap B) = 4$, find $n(A \cup B)$.
54. If $n(A) = 24$, $n(B) = 20$, and $n(A \cap B) = 14$, find $n(A \cup B)$.

The table and Venn diagram below are to be used for Exercises 55–58. The table shows the top five states in terms of number of violent crimes per 100,000 citizens in each year from 2004 to 2006. For each exercise, write the region of the Venn diagram that would include the state listed.

Year	2004	2005	2006
State	1. South Carolina	1. South Carolina	1. South Carolina
	2. Florida	2. Tennessee	2. Tennessee
	3. Maryland	3. Florida	3. Nevada
	4. Tennessee	4. Maryland	4. Florida
	5. New Mexico	5. New Mexico	5. Louisiana

Source: swivel.com

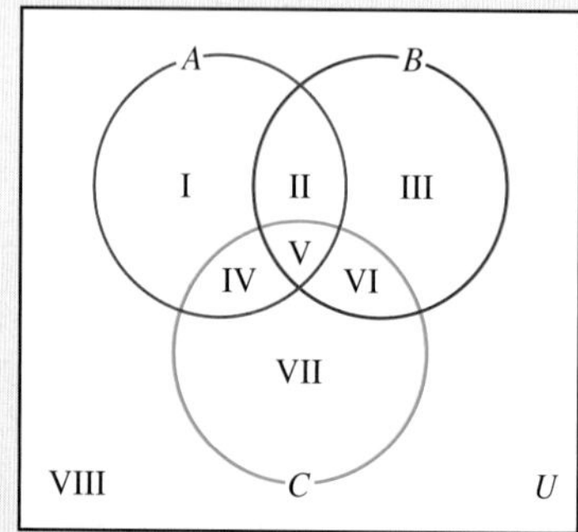

Note: Set A represents 2004, set B represents 2005, and set C represents 2006.

55. Florida
56. Louisiana
57. New Mexico
58. Nevada

Section 2-4

59. In a recent survey of 25 students, 10 had used a chat room that day, 5 posted a new blog that day, and 2 students used a chat room and posted a new blog that day.
    (a) How many did not use a chat room or post a blog that day?
    (b) How many only posted a new blog that day?
60. In a group of 24 students asked what they did last Saturday night, 9 tailgated after the big game, 18 went to a rave, and 3 tailgated and went to a rave.
    (a) How many only went to a rave?
    (b) How many did not go to a rave?
61. Fifty-three callers to the campus radio station were asked what they usually listened to while driving to school. Of those asked, 22 listened to a local radio station, 18 listened to satellite radio, 33 listened to MP3 players, 8 listened to a local radio station and satellite radio, 13 listened to satellite radio and MP3 players, 11 listened to a local radio station and an MP3 player, and 6 listened to all three.
    (a) How many listened to only satellite radio?
    (b) How many listened to local radio stations and MP3 players but not satellite radio?
    (c) How many listened to none of these?
62. Of the last 41 customers at the student bookstore, 15 paid in cash, 16 used a debit card, 20 used a financial aid voucher, 4 used cash and a debit card, 8 used a debit card and a financial aid voucher, 8 used cash and a financial aid voucher, and 1 used all three.
    (a) How many used none of these payment types?
    (b) How many only used a debit card?
    (c) How many used cash and a financial aid voucher but not a debit card?

Section 2-5

63. Find a general term for the set $\{-5, -7, -9, -11, -13, \ldots\}$
64. Show that the set $\{12, 24, 36, 48, 60, \ldots\}$ is an infinite set.

Chapter Test

For Exercises 1–8, write each set in roster notation.

1. The set P is the set of even natural numbers between 90 and 100.
2. The set J is the set of odd natural numbers between 40 and 50.
3. The set K is the set of letters in the word *envelope*.
4. The set W is the set of letters in the word *Washington*.
5. $X = \{x \mid x \in N \text{ and } x < 80\}$
6. $Y = \{x \mid x \in N \text{ and } 16 < x < 25\}$
7. The set J is the set of months of the year that begin with the letter J.
8. The set L is the set of days of the week that end in the letter a.

For Exercises 9–12, write each set using set-builder notation.

9. {12, 14, 16, 18}
10. {30, 35, 40, 45}
11. {201, 203, 205, 207,...}
12. {4, 8, 16,...128}

For Exercises 13–17, state whether the set is finite or infinite.

13. {15, 16, 17, 18, ...}
14. $\{x \mid x \in N$ and x is a multiple of $6\}$
15. {a, b, c}
16. {4, 7, 8, 10}
17. The set V is the set of people with three eyes.
18. Find all subsets of {d, e, f}.
19. Find all subsets of {p, q, r}.
20. How many subsets and proper subsets does the set {a, b, c, d, e} have?

For Exercises 21–30, let U = {a, b, c, d, e, f, g, h, i, j, k}, A = {a, b, d, e, f}, B = {a, g, i, j, k}, *and* C = {e, h, j}. *Find each.*

21. $A \cap B$
22. $B \cup C$
23. B'
24. $(A \cup B)'$
25. $(A \cap B)' \cup C'$
26. $A - B$
27. $B - C$
28. $(A - B) - C$
29. $A - C$
30. Draw and shade a Venn diagram for the area representing $A - B$.

For Exercises 31–34, let $A = \{@, !, \alpha\}$ *and* $B = \{\pi, \#\}$. *Find each set.*

31. $A \times B$
32. $B \times A$
33. $B \times B$
34. $A \times A$

For Exercises 35–37, draw a Venn diagram for each set.

35. $A' \cap B$
36. $(A \cap B)'$
37. $(A' \cup B') \cap C'$
38. If $n(A) = 1{,}500$, $n(B) = 1{,}150$, and $n(A \cap B) = 350$, find $n(A \cup B)$.
39. At a special graduation ceremony of 24 honors students, 8 used a digital camera, 9 used a camera in their cell phones, and 4 used both digital cameras and cell phone cameras.
    (a) How many did not use either of these?
    (b) How many used a digital camera only?
40. Find a general term for the set {15, 30, 45, 60, 75,...}.
41. Show that the set {1, −1, 2, −2, 3, −3,...} is an infinite set.

For Exercises 42–53, state whether each is true or false.

42. {a, b, c} is equal to {x, y, z}.
43. {1, 2, 3, 4} is equivalent to {p, q, r, s}.
44. $\{4, 8, 12, 16, \ldots\} \subseteq \{2, 4, 6, 8, \ldots\}$
45. $\{15\} \subset \{3, 6, 9, 12, \ldots\}$
46. $4 \in$ {even natural numbers}
47. $9 \notin \{2, 4, 5, 6, 10\}$
48. {a, e, i, o, u, y} ⊆ {a, e, i, o, u}
49. $\{12\} \in \{12, 24, 36, \ldots\}$
50. For any set, $\varnothing' = U$
51. $0 \in \varnothing$
52. For any set, $A \cap \varnothing = U$.
53. For any set, $A \cup B = B \cup A$.

Projects

1. Have the students in your class fill out this questionnaire:

   A. Gender: Male _____ Female _____
   B. Age: Under 21 _____ 21 or older _____
   C. Work: Yes _____ No _____

   Draw a Venn diagram, and from the information answer these questions:

   (a) How many students are female?
   (b) How many students are under 21?
   (c) How many students work?
   (d) How many students are under 21 and work?
   (e) How many students are males and do not work?
   (f) How many students are 21 or older and work?
   (g) How many students are female, work, and are under 21?

2. Select five of these set properties. Explain why they are true for all sets. Then illustrate, using Venn diagrams.

   (a) $(A')' = A$
   (b) $\varnothing' = U$ and $U' = \varnothing$
   (c) $A \cup \varnothing = A$ and $A \cap \varnothing = \varnothing$
   (d) $A \cup U = U$ and $A \cap U = A$
   (e) $A \cup A' = U$ and $A \cap A' = \varnothing$
   (f) $A \cup A = A$ and $A \cap A = A$
   (g) $A \cup B = B \cup A$ and $A \cap B = B \cap A$
   (h) $(A \cup B) \cup C = A \cup (B \cup C)$ and $(A \cap B) \cap C = A \cap (B \cap C)$
   (i) $A \cup (B \cap C) = (A \cup B) \cap (A \cup C)$ and $A \cap (B \cup C) = (A \cap B) \cup (A \cap C)$
   (j) $(A \cup B)' = A' \cap B'$ and $(A \cap B)' = A' \cup B'$

CHAPTER 3

Logic

Outline

3-1 Statements and Quantifiers
3-2 Truth Tables
3-3 Types of Statements
3-4 Logical Arguments
3-5 Euler Circles
Summary

MATH IN The Art of Persuasion

Everywhere you turn in modern society, somebody is trying to convince you of something. "Vote for me!" "Buy my product!" "Lease a car from our dealership!" "Bailing out the auto industry is a bad idea!" "You should go out with me this weekend!" "Join our fraternity!" Logic is sometimes defined as correct thinking or correct reasoning. Some people refer to logic by a more casual name: *common sense*. Regardless of what you call it, the ability to think logically is crucial for all of us because our lives are inundated daily with advertisements, contracts, product and service warranties, political debates, and news commentaries, to name just a few. People often have problems processing these things because of misinterpretation, misunderstanding, and faulty logic.

You can look up the truth or falseness of a fact on the Internet, but that won't help you in analyzing whether a certain claim is logically valid. The term *common sense* is misleading, because evaluating logical arguments can be a challenging and involved process. This chapter introduces the basic concepts of formal symbolic logic and shows how to determine whether arguments are valid or invalid by using truth tables. One of our biggest goals will be to examine the form of an argument and determine if its conclusion follows logically from the statements in the argument.

To that end, we've written some claims below. Your job is to determine which of the arguments is valid, meaning that a conclusion can be logically drawn from a set of statements. The skills you learn in this chapter will help you to do so.

- Where there's smoke, there's fire.
- Having a lot of money makes people happy. My neighbor is a really happy guy, so he must have a lot of money.
- Every team in the SEC is good enough to play in a bowl game. Florida State is not in the SEC, so they're not good enough to play in a bowl game.
- Scripture is the word of God. I know this because it says so in the Bible.
- If Iraq has weapons of mass destruction, we should go to war. It turns out that they don't have them, so we should not go to war.
- It will be a snowy day in Hawaii before Tampa Bay makes it to the World Series. Tampa Bay played in the 2008 World Series, so it must have snowed in Hawaii.

For answers, see Math in the Art of Persuasion Revisited on Page 141

Section 3-1 Statements and Quantifiers

LEARNING OBJECTIVES

- ☐ 1. Define and identify statements.
- ☐ 2. Define the logical connectives.
- ☐ 3. Write the negation of a statement.
- ☐ 4. Write statements symbolically.

As the world gets more complex and we are bombarded with more and more information, it becomes more important than ever to be able to make sensible, objective evaluations of that information. One of the most effective tricks that advertisers, politicians, and con artists use is to encourage emotions to enter into these evaluations. They use carefully selected words and images that are designed to keep you from making decisions objectively. The field of **symbolic logic** was designed exactly for this reason. Symbolic logic uses letters to represent statements and special symbols to represent words like *and, or,* and *not*. Use of this symbolic notation in place of the statements themselves allows us to analytically evaluate the validity of the logic behind an argument without letting bias and emotion cloud our judgment. And these unbiased evaluations are the main goal of this chapter.

Statements

In the English language there are many types of sentences, including factual statements, commands, opinions, questions, and exclamations. In the objective study of logic, we will use only factual statements.

> A **statement** is a declarative sentence that can be objectively determined to be true or false, but not both.

For example, sentences like

It is raining.
The United States has sent a space probe to Mars.
$2 + 2 = 4$
$10 - 5 = 4$

are statements because they are either true or false and you don't use an opinion to determine this.

Whether a sentence is true or false doesn't matter in determining if it is a statement. Notice in the above example the last statement "$10 - 5 = 4$" is false, but it's still a statement.

The following sentences, however, are not statements:

Give me onion rings with my order.
What operating system are you running?
Sweet!
The guy sitting next to me is kind of goofy.

The first is not a statement because it is a command. The second is not a statement because it is a question. The third is not a statement because it is an exclamation, and the fourth is not a statement because the word *goofy* is subjective; that is, it requires an opinion.

EXAMPLE 1 Recognizing Statements

Decide which of the following are statements and which are not.

(a) Most scientists agree that global warming is a threat to the environment.
(b) Is that your laptop?
(c) Man, that hurts!
(d) $8 - 2 = 6$
(e) This book is about database management.
(f) Everybody should watch reality shows.

SOLUTION

Parts (a), (d), and (e) are statements because they can be judged as true or false in a nonsubjective manner.
Part (b) is not a statement because it is a question.
Part (c) is not a statement because it is an exclamation.
Part (f) is not a statement because it requires an opinion.

▼ Try This One 1

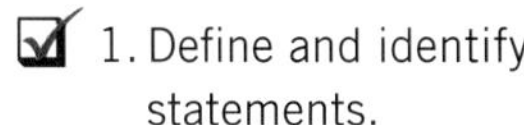

1. Define and identify statements.

Decide which of the following are statements and which are not.

(a) Cool!
(b) $12 - 8 = 5$
(c) Ryan Seacrest is the host of *American Idol*.
(d) Cat can send text messages with her cell phone.
(e) When does the party start?
(f) History is interesting.

Simple and Compound Statements

Statements can be classified as *simple* or *compound*. A **simple statement** contains only one idea. Each of these statements is an example of a simple statement.

These cargo pants are khaki.

My dorm room has three beds in it.

Daytona Beach is in Florida.

A statement such as "I will take chemistry this semester, and I will get an A" is called a **compound statement** since it consists of two simple statements.

Compound statements are formed by joining two simple statements with what is called a *connective*.

The basic **connectives** are *and, or, if . . . then,* and *if and only if.*

Each of the connectives has a formal name: *and* is called a **conjunction,** *or* is called a **disjunction,** *if . . . then* is called a **conditional,** and *if and only if* is called a **biconditional.** Here are some examples of compound statements using connectives.

John studied for 5 hours, and he got an A. (conjunction)

Luisa will run in a mini triathlon or she will play in the campus tennis tournament. (disjunction)

If I get 80% of the questions on the LSAT right, then I will get into law school. (conditional)

We will win the game if and only if we score more points than the other team. (biconditional)

Math Note

In standard usage, the word *then* is often omitted from a conditional statement; instead of "If it snows, then I will go skiing," you'd probably just say, "If it snows, I'll go skiing."

EXAMPLE 2 Classifying Statements as Simple or Compound

Math Note

Technically, we've given the names *conjunction*, *disjunction*, *conditional*, and *biconditional* to the connectives, but from now on, we'll refer to whole statements using these connectives by those names. For example, we would call the compound statement in Example 2d a disjunction.

Classify each statement as simple or compound. If it is compound, state the name of the connective used.

(a) Our school colors are red and white.
(b) If you register for WiFi service, you will get 3 days of free access.
(c) Tomorrow is the last day to register for classes.
(d) I will buy a hybrid or I will buy a motorcycle.

SOLUTION

(a) Don't let use of the word *and* fool you! This is a simple statement.
(b) This if . . . then statement is compound and uses a conditional connective.
(c) This is a simple statement.
(d) This is a compound statement, using a disjunction.

☑ 2. Define the logical connectives.

▼ Try This One 2

Classify each statement as simple or compound. If it is compound, state the name of the connective used.

(a) The jacket is trendy and it is practical.
(b) This is an informative website on STDs.
(c) If it does not rain, then I will go windsurfing.
(d) I will buy a flash drive or I will buy a zip drive.
(e) Yesterday was the deadline to withdraw from a class.

Sidelight A BRIEF HISTORY OF LOGIC

The basic concepts of logic can be attributed to Aristotle, who lived in the fourth century BCE. He used words, sentences, and deduction to prove arguments using techniques we will study in this chapter. In addition, in 300 BCE. Euclid formalized geometry using deductive proofs. Both subjects were considered to be the "inevitable truths" of the universe revealed to rational people.

In the 19th century, people began to reject the idea of inevitable truths and realized that a deductive system like Euclidean geometry is only true based on the original assumptions. When the original assumptions are changed, a new deductive system can be created. This is why there are different types of geometry. (See the Sidelight entitled "Non-Euclidean Geometry" in Chapter 10.)

Eventually, several people developed the use of symbols rather than words and sentences in logic. One such person was George Boole (1815–1864). Boole created the symbols used in this chapter and developed the theory of symbolic logic. He also used symbolic logic in mathematics. His manuscript, entitled "An Investigation into the Laws of Thought, on Which Are Founded the Mathematical Theories of Logic and Probabilities," was published when he was 39 in 1854.

Boole was a friend of Augustus De Morgan, who formulated De Morgan's laws, which we studied in Chapter 2. Much earlier, Leonhard Euler (1707–1783) used circles to represent logical statements and proofs. The idea was refined into Venn diagrams by John Venn (1834–1923).

Quantified Statements

Math Note

We'll worry later about determining whether statements involving quantifiers and connectives are true or false. For now, focus on learning and understanding the terms.

Quantified statements involve terms such as *all, each, every, no, none, some, there exists,* and *at least one*. The first five (*all, each, every, no, none*) are called *universal quantifiers* because they either include or exclude every element of the universal set. The latter three (*some, there exists, at least one*) are called *existential quantifiers* because they show the existence of something, but do not include the entire universal set. Here are some examples of quantified statements:

Every student taking Math for Liberal Arts this semester will pass.

Some people who are Miami Hurricane fans are also Miami Dolphin fans.

There is at least one professor in this school who does not have brown eyes.

No Marlin fan is also a Yankee fan.

The first and the fourth statements use universal quantifiers, and the second and third use existential quantifiers. Note that the statements using existential quantifiers are not "all inclusive" (or all exclusive) as the other two are.

Negation

The *negation* of a statement is a corresponding statement with the opposite truth value. For example, for the statement "My dorm room is blue," the negation is "My dorm room is not blue." It's important to note that the truth values of these two are completely opposite: one is true, and the other is false—period. You can't negate "My dorm room is blue" by saying "My dorm room is yellow," because it's completely possible that *both* statements are false. To make sure that you have a correct negation, check that if one of the statements is true, the other must be false, and vice versa. The typical way to negate a simple statement is by adding the word *not,* as in these examples:

Statement	Negation
Auburn will win Saturday.	Auburn will not win Saturday.
I took a shower today.	I did not take a shower today.
My car is clean.	My car is not clean.

Math Note

The words *each, every,* and *all* mean the same thing, so what we say about *all* in this section applies to the others as well. Likewise, *some, there exists,* and *at least one* are considered to be the same and are treated that way as well.

You have to be especially careful when negating quantified statements. Consider the example statement "All dogs are fuzzy." It's not quite right to say that the negation is "All dogs are not fuzzy," because if some dogs are fuzzy and others aren't, then both statements are false. All we need for the statement "All dogs are fuzzy" to be false is to find at least one dog that is not fuzzy, so the negation of the statement "All dogs are fuzzy" is "Some dogs are not fuzzy." (In this setting, we define the word *some* to mean *at least one*.)

We can summarize the negation of quantified statements as follows:

Statement Contains...	Negation
All do	Some do not, or not all do
Some do	None do, or all do not
Some do not	All do
None do	Some do

EXAMPLE 3 Writing Negations

Write the negation of each of the following quantified statements.

(a) Every student taking Math for Liberal Arts this semester will pass.
(b) Some people who are Miami Hurricane fans are also Miami Dolphin fans.
(c) There is at least one professor in this school who does not have brown eyes.
(d) No Marlin fan is also a Yankee fan.

SOLUTION

(a) Some student taking Math for Liberal Arts this semester will not pass (or, not every student taking Math for Liberal Arts this semester will pass).
(b) No people who are Miami Hurricane fans are also Miami Dolphin fans.
(c) All professors in this school have brown eyes.
(d) Some Marlin fan is also a Yankee fan.

CAUTION

Be especially careful when negating statements. Remember that the negation of "Every student will pass" is *not* "Every student will fail".

3. Write the negation of a statement.

▼ Try This One 3

Write the negation of each of the following quantified statements.

(a) All cell phones have cameras.
(b) No woman can win the lottery.
(c) Some professors have PhDs.
(d) Some students in this class will not pass.

Math Note

For three of the four connectives in Table 3-1, the order of the simple statements doesn't matter: for example, $p \wedge q$ and $q \wedge p$ represent the same compound statement. The same is true for the connectives $\vee$ (disjunction) and $\leftrightarrow$ (biconditional). The one exception is the conditional ($\rightarrow$), where order is crucial.

Symbolic Notation

Recall that one of our goals in this section is to write statements in symbolic form to help us evaluate logical arguments objectively. Now we'll introduce the symbols and methods that will be used. The symbols for the connectives *and, or, if . . . then,* and *if and only if* are shown in Table 3-1.

Simple statements in logic are usually denoted with lowercase letters like p, q, and r. For example, we could use p to represent the statement "I get paid Friday" and q to represent the statement "I will go out this weekend." Then the conditional statement "If I get paid Friday, then I will go out this weekend" can be written in symbols as $p \rightarrow q$.

The symbol ~ (tilde) represents a negation. If p still represents "I get paid Friday," then $\sim p$ represents "I do not get paid Friday."

We often use parentheses in logical statements when more than one connective is involved in order to specify an order. (We'll deal with this in greater detail in the next section.) For example, there is a difference between the compound statements $\sim p \wedge q$ and $\sim(p \wedge q)$. The statement $\sim p \wedge q$ means to negate the statement p first, then use the negation of p in conjunction with the statement q. For example, if p is the statement "Fido is a dog" and q is the statement "Pumpkin is a cat," then $\sim p \wedge q$ reads, "Fido is not a dog and Pumpkin is a cat." The statement $\sim p \wedge q$ could also be written as $(\sim p) \wedge q$.

TABLE 3-1 Symbols for the Connectives

Connective	Symbol	Name
and	$\wedge$	Conjunction
or	$\vee$	Disjunction
if . . . then	$\rightarrow$	Conditional
if and only if	$\leftrightarrow$	Biconditional

The statement $\sim(p \wedge q)$ means to negate the conjunction of the statement p and the statement q. Using the same statements for p and q as before, the statement $\sim(p \wedge q)$ is written, "It is not the case that Fido is a dog and Pumpkin is a cat."

The same reasoning applies when the negation is used with other connectives. For example, $\sim p \rightarrow q$ means $(\sim p) \rightarrow q$.

Example 4 illustrates in greater detail how to write statements symbolically.

EXAMPLE 4 Writing Statements Symbolically

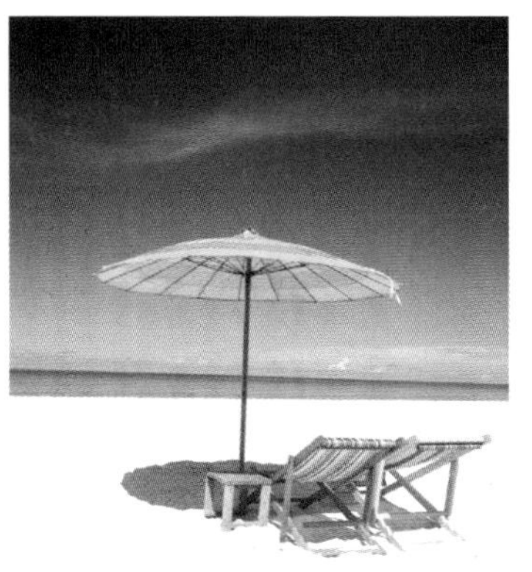

Let p represent the statement "It is cloudy" and q represent the statement "I will go to the beach." Write each statement in symbols.

(a) I will not go to the beach.
(b) It is cloudy, and I will go to the beach.
(c) If it is cloudy, then I will not go to the beach.
(d) I will go to the beach if and only if it is not cloudy.

SOLUTION

(a) This is the negation of statement q, which we write as $\sim q$.
(b) This is the conjunction of p and q, written as $p \wedge q$.
(c) This is the conditional of p and the negation of q: $p \rightarrow \sim q$.
(d) This is the biconditional of p and not q: $p \leftrightarrow \sim q$.

▼ Try This One 4

Let p represent the statement "I will buy a Coke" and q represent the statement "I will buy some popcorn." Write each statement in symbols.

(a) I will buy a Coke, and I will buy some popcorn.
(b) I will not buy a Coke.
(c) If I buy some popcorn, then I will buy a Coke.
(d) I will not buy a Coke, and I will buy some popcorn.

You probably noticed that some of the compound statements we've written sound a little awkward. It isn't always necessary to repeat the subject and verb in a compound statement using *and* or *or*. For example, the statement "It is cold, and it is snowing" can be written "It is cold and snowing." The statement "I will go to a movie, or I will go to a play" can be written "I will go to a movie or a play." Also the words *but* and *although* can be used in place of *and*. For example, the statement "I will not buy a television set, and I will buy a CD player" can also be written as "I will not buy a television set, but I will buy a CD player."

Statements written in symbols can also be written in words, as shown in Example 5.

EXAMPLE 5 Translating Statements from Symbols to Words

Write each statement in words. Let p = "My dog is a golden retriever" and q = "My dog is fuzzy."

(a) $\sim p$ (b) $p \vee q$ (c) $\sim p \rightarrow q$ (d) $q \leftrightarrow p$ (e) $q \wedge p$

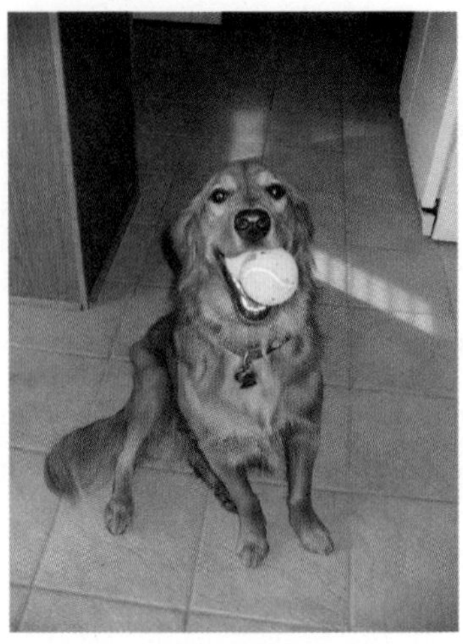

If this is your dog (which it's not, because it's mine), statement (e) describes it pretty well.

SOLUTION

(a) My dog is not a golden retriever.
(b) My dog is a golden retriever or my dog is fuzzy.
(c) If my dog is not a golden retriever, then my dog is fuzzy.
(d) My dog is fuzzy if and only if my dog is a golden retriever.
(e) My dog is fuzzy, and my dog is a golden retriever.

▼ Try This One 5

Write each statement in words. Let p = "My friend is a football player" and q = "My friend is smart."

(a) $\sim p$ (b) $p \vee q$ (c) $\sim p \rightarrow q$ (d) $p \leftrightarrow q$ (e) $p \wedge q$

In this section, we defined the basic terms of symbolic logic and practiced writing statements using symbols. These skills will be crucial in our objective study of logical arguments, so we're off to a good start.

☑ 4. Write statements symbolically.

Answers to Try This One

1 (b), (c), and (d) are statements.

2 (a) (conjunction), (c) (conditional), and (d) (disjunction) are compound; (b) and (e) are simple.

3 (a) Some cell phones don't have cameras.
(b) Some women can win the lottery.
(c) No professors have Ph.Ds.
(d) All students in this class will pass.

4 (a) $p \wedge q$ (b) $\sim p$ (c) $q \rightarrow p$ (d) $\sim p \wedge q$

5 (a) My friend is not a football player.
(b) My friend is a football player or my friend is smart.
(c) If my friend is not a football player, then my friend is smart.
(d) My friend is smart if and only if my friend is a football player.
(e) My friend is smart and my friend is a football player.

EXERCISE SET 3-1

Writing Exercises

1. Define the term *statement* in your own words.
2. Explain the difference between a simple and a compound statement.
3. Describe the terms and symbols used for the four connectives.
4. Explain why the negation of "All spring breaks are fun" is not "All spring breaks are not fun."

Real-World Applications

For Exercises 5–14, state whether the sentence is a statement or not.

5. Please do not use your cell phone in class.
6. $5 + 9 = 14$
7. $9 - 3 = 2$
8. Nicki is a student in vet school.
9. Who will win the student government presidency?
10. Neither Sam nor Mary arrives to the exam on time.
11. You can carry a cell phone with you.
12. Bill Gates is the creator of Microsoft.
13. Go with the flow.
14. Math is not hard.

For Exercises 15–24, decide if each statement is simple or compound.

15. He goes to parties and hangs out at the coffee shop.
16. Sara got her hair highlighted.
17. Raj will buy an iMac or a Dell computer.
18. Euchre is fun if and only if you win.
19. February is when Valentine's Day occurs.
20. Diane is a chemistry major.
21. If you win the Megabucks multistate lottery, then you will be rich.
22. He listened to his iPod and he typed a paper.
23. $8 + 9 = 12$
24. Malcolm and Alisha will both miss the spring break trip.

For Exercises 25–32, identify each statement as a conjunction, disjunction, conditional, or biconditional.

25. Bob and Tom like stand-up comedians.
26. Either he passes the test, or he fails the course.
27. A number is even if and only if it is divisible by 2.
28. Her nails are long, and they have rhinestones on them.
29. I will go to the big game, or I will go to the library.
30. If a number is divisible by 3, then it is an odd number.
31. A triangle is equiangular if and only if three angles are congruent.
32. If your battery is dead, then you need to charge your phone overnight.

For Exercises 33–38, write the negation of the statement.

33. The sky is blue.
34. It is not true that your computer has a virus.
35. The dorm room is not large.
36. The class is not full.
37. It is not true that you will fail this class.
38. He has large biceps.

For Exercises 39–50, identify the quantifier in the statement as either universal or existential.

39. All fish swim in water.
40. Everyone who passes algebra has studied.
41. Some people who live in glass houses throw stones.
42. There is at least one person in this class who won't pass.
43. Every happy dog wags its tail.
44. No men can join a sorority.
45. There exists a four-leaf clover.
46. Each person who participated in the study will get $100.
47. No one with green eyes wears glasses.
48. Everyone in the class was bored by the professor's lecture.
49. At least one of my friends has an iPhone.
50. No one here gets out alive.

For Exercises 51–62, write the negation of the statements in Exercises 39–50.

For Exercises 63–72, write each statement in symbols. Let p = "Sara is a political science major" and let q = "Jane is a quantum physics major."

63. Sara is a political science major, and Jane is a quantum physics major.
64. Sara is not a political science major.
65. If Jane is not a quantum physics major, then Sara is a political science major.
66. It is not true that Jane is a quantum physics major or Sara is a political science major.
67. It is false that Jane is a quantum physics major.
68. It is not true that Sara is a political science major.
69. Jane is a quantum physics major, or Sara is not a political science major.
70. Jane is not a quantum physics major, or Sara is a political science major.
71. Jane is a quantum physics major if and only if Sara is a political science major.
72. If Sara is a political science major, then Jane is a quantum physics major.

For Exercises 73–82, write each statement in symbols. Let p = "My dad is cool" and q = "My mom is cool." Let nerdy *mean* not cool.

73. My mom is not cool.
74. Both my dad and my mom are nerdy.
75. If my mom is cool, then my dad is cool.
76. It is not true that my dad is cool.
77. Either my mom is nerdy, or my dad is cool.
78. It is not true that my mom is nerdy and my dad is cool.
79. My mom is cool if and only if my dad is cool.
80. Neither my mom nor my dad is cool.
81. If my mom is nerdy, then my dad is cool.
82. My dad is nerdy if and only if my mom is not cool.

For Exercises 83–92, write each statement in words. Let p = "The plane is on time." Let q = "The sky is clear."

83. $p \wedge q$
84. $\sim p \vee q$
85. $q \rightarrow p$
86. $q \rightarrow \sim p$
87. $\sim p \wedge \sim q$
88. $q \leftrightarrow p$
89. $p \vee \sim q$
90. $\sim p \leftrightarrow \sim q$
91. $q \rightarrow (p \vee \sim p)$
92. $(p \rightarrow q) \vee \sim p$

For Exercises 93–102, write each statement in words. Let p = "Mark lives on campus." Let q = "Trudy lives off campus."

93. $\sim q$
94. $p \rightarrow q$
95. $p \vee \sim q$
96. $q \leftrightarrow p$
97. $\sim p \rightarrow \sim q$
98. $\sim p$
99. $p \vee q$
100. $(\sim p \vee q) \vee \sim q$
101. $q \vee p$
102. $(p \vee q) \rightarrow \sim(\sim q)$

Critical Thinking

103. Explain why the sentence "All rules have exceptions" is not a statement.

104. Explain why the sentence "This statement is false" is not a statement.

Section 3-2 Truth Tables

LEARNING OBJECTIVES

- ❑ 1. Construct truth tables for negation, disjunction, and conjunction.
- ❑ 2. Construct truth tables for the conditional and biconditional.
- ❑ 3. Construct truth tables for compound statements.
- ❑ 4. Identify the hierarchy of logical connectives.
- ❑ 5. Construct truth tables by using an alternative method.

"You can't believe everything you hear." Chances are you were taught this when you were younger, and it's pretty good advice. In an ideal world, everyone would tell the truth all the time, but in the real world, it is extremely important to be able to separate fact from fiction. When someone is trying to convince you of some point of view, the ability to logically evaluate the validity of an argument can be the difference between being informed and being deceived—and maybe between you keeping and you being separated from your hard-earned money!

This section is all about deciding when a compound statement is or is not true, based not on the situation itself, but simply on the structure of the statement and the truth of the underlying components. We learned about logical connectives in Section 3-1. In this section, we'll analyze these connectives using *truth tables*. A **truth table** is a diagram in table form that is used to show when a compound statement is true or false based on the truth values of the simple statements that make up the compound statement. This will allow us to analyze arguments objectively.

Negation

According to our definition of *statement,* a statement is either true or false, but never both. Consider the simple statement p = "Today is Tuesday." If it is in fact Tuesday, then p is true, and its negation ($\sim p$) "Today is not Tuesday" is false. If it's not Tuesday, then p is false and $\sim p$ is true. The truth table for the negation of p looks like this.

p	$\sim p$
T	F
F	T

There are two possible conditions for the statement p—true or false—and the table tells us that in each case, the negation $\sim p$ has the opposite truth value.

Conjunction

If we have a compound statement with two component statements p and q, there are four possible combinations of truth values for these two statements:

Possibilities	**Symbolic value of each** p	q
1. p and q are both true.	T	T
2. p is true and q is false.	T	F
3. p is false and q is true.	F	T
4. p and q are both false.	F	F

So when setting up a truth table for a compound statement with two component statements, we'll need a row for each of the four possibilities.

Now we're ready to analyze conjunctions. Recall that a conjunction is a compound statement involving the word *and*. Suppose a friend who's prone to exaggeration tells you, "I bought a new computer and a new iPod." This compound statement can be symbolically represented by $p \wedge q$, where p = "I bought a new computer" and q = "I bought a new iPod." When would this conjunctive statement be true? If your friend actually had made both purchases, then of course the statement "I bought a new computer and a new iPod" would be true. In terms of a truth table, that tells us that if p and q are both true, then the conjunction $p \wedge q$ is true as well, as shown below.

p	q	$p \wedge q$
T	T	T

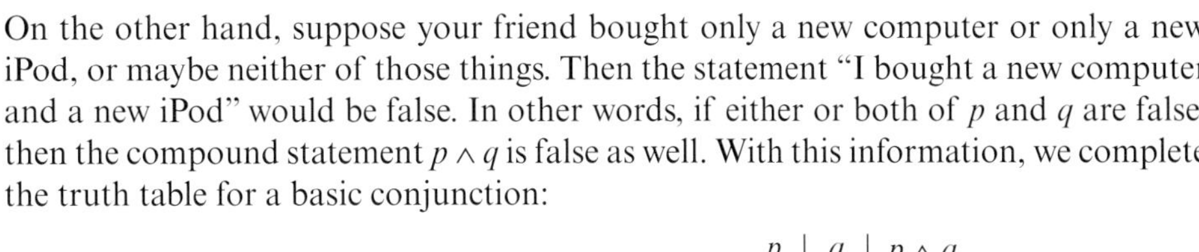

On the other hand, suppose your friend bought only a new computer or only a new iPod, or maybe neither of those things. Then the statement "I bought a new computer and a new iPod" would be false. In other words, if either or both of p and q are false, then the compound statement $p \wedge q$ is false as well. With this information, we complete the truth table for a basic conjunction:

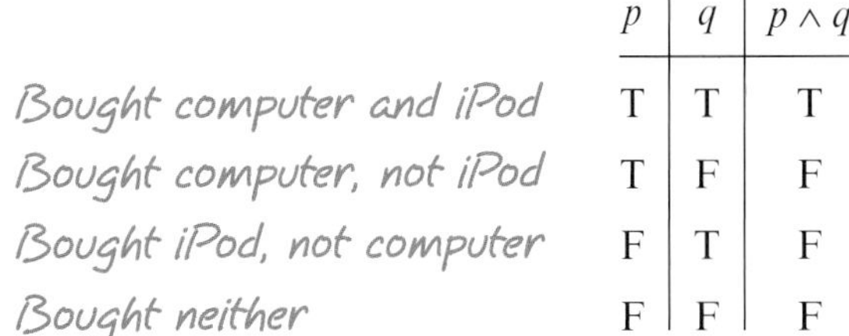

	p	q	$p \wedge q$
Bought computer and iPod	T	T	T
Bought computer, not iPod	T	F	F
Bought iPod, not computer	F	T	F
Bought neither	F	F	F

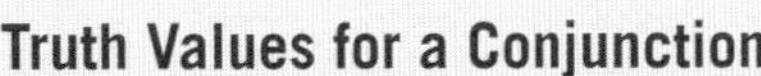

Truth Values for a Conjunction

The conjunction $p \wedge q$ is true only when both p and q are true.

Disjunction

Next, we'll look at truth tables for *or* statements. Suppose your friend from the previous example made the statement, "I bought a new computer *or* a new iPod" (as opposed to *and*). If your friend actually did buy one or the other, then this statement would be true. And if he or she bought neither, then the statement would be false. So a partial truth table looks like this:

	p	q	$p \vee q$
Bought computer and iPod	T	T	
Bought computer, not iPod	T	F	T
Bought iPod, not computer	F	T	T
Bought neither	F	F	F

Sidelight Logical Gates and Computer Design

Logic is used in electrical engineering in designing circuits, which are the heart of computers. The truth tables for *and, or,* and *not* are used for computer gates. These gates determine whether electricity flows through a circuit. When a switch is closed, the current has an uninterrupted path and will flow through the circuit. This is designated by a 1. When a switch is open, the path at the current is broken, and it will not flow. This is designated by a 0. The logical gates are illustrated here—notice that they correspond exactly with our truth tables.

This simple little structure is responsible for the operation of almost every computer in the world—at least until quantum computers become a reality. If you're interested, do a Google search for *quantum computer* to read about the future of computing.

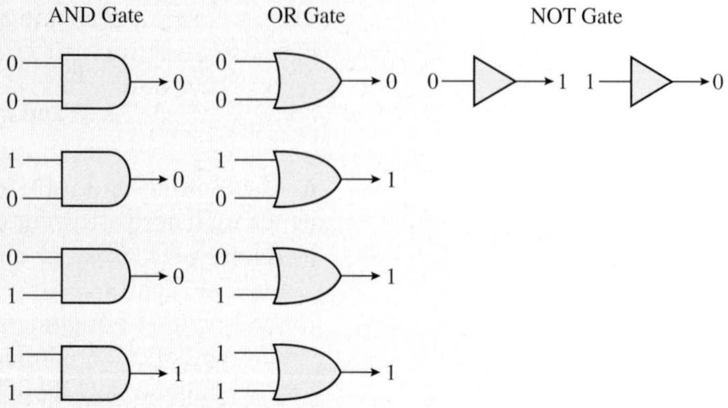

But what if the person actually bought both items? You might lean toward the statement "I bought a new computer or a new iPod" being false. Believe it or not, it depends on what we mean by the word *or*. There are two interpretations of that word, known as the *inclusive or* and the *exclusive or*. The inclusive or has the possibility of both statements being true; but the exclusive or does not allow for this, that is, exactly one of the two simple statements must be true. In English when we use the word *or,* we typically think of the exclusive or. If I were to say, "I will go to work or I will go to the beach," you would assume I am doing one or the other, but not both. In logic we generally use the inclusive or. When the inclusive or is used, the statement "I will go to work or I will go to the beach" would be true if I went to both work and the beach. For the remainder of this chapter we will assume the symbol $\vee$ represents the inclusive or and will drop *inclusive* and just say *or.*

The completed truth table for the disjunction is

	p	q	$p \vee q$
Bought computer and iPod	T	T	T
Bought computer, not iPod	T	F	T
Bought iPod, not computer	F	T	T
Bought neither	F	F	F

1. Construct truth tables for negation, disjunction, and conjunction.

Truth Values for a Disjunction

The disjunction $p \vee q$ is true when either p or q or both are true. It is false only when both p and q are false.

Conditional Statement

A conditional statement, which is sometimes called an *implication,* consists of two simple statements using the connective if ... then. For example, the statement "If I bought a ticket, then I can go to the concert" is a conditional statement. The first component, in this case "I bought a ticket," is called the *antecedent*. The second component, in this case "I can go to the concert," is called the *consequent*.

Conditional statements are used commonly in every area of math, including logic. You might remember statements from high school algebra like "If two lines are parallel, then they have the same slope." Remember that we represent the conditional statement "If p, then q" by using the symbol $p \rightarrow q$.

To illustrate the truth table for the conditional statement, think about the following simple example: "If it is raining, then I will take an umbrella." We'll use p to represent "It is raining" and q to represent "I will take an umbrella," then the conditional statement is $p \rightarrow q$. We'll break the truth table down into four cases.

Case 1: It is raining and I do take an umbrella (p and q are both true). Since I am doing what I said I would do in case of rain, the conditional statement is true. So the first line in the truth table is

	p	q	$p \rightarrow q$
Raining, take umbrella	T	T	T

Case 2: It is raining and I do not take an umbrella (p is true and q is false). Since I am not doing what I said I would do in case of rain, I'm a liar and the conditional statement is false. So the second line in the truth table is

	p	q	$p \rightarrow q$
	T	T	T
Raining, do not take umbrella	T	F	F

Case 3: It is not raining and I do take an umbrella (p is false and q is true). This requires some thought. I never said in the original statement what I would do if it were not raining, so there's no reason to regard my original statement as false. Based on the information given, we consider the original statement to be true, and the third line of the truth table is

	p	q	$p \rightarrow q$
	T	T	T
	T	F	F
Not raining, take umbrella	F	T	T

Case 4: It is not raining, and I do not take my umbrella (p and q are both false). This is essentially the same as case 3—I never said what I would do if it did not rain, so there's no reason to regard my statement as false based on what we know. So we consider the original statement to be true, and the last line of the truth table is

	p	q	$p \rightarrow q$
	T	T	T
	T	F	F
	F	T	T
Not raining, do not take umbrella	F	F	T

For cases 3 and 4, it might help to think of it this way: unless we have definite proof that a statement is false, we will consider it to be true.

Truth Values for a Conditional Statement

The conditional statement $p \rightarrow q$ is false only when the antecedent p is true and the consequent q is false.

Biconditional Statement

A biconditional statement is really two statements in a way; it's the conjunction of two conditional statements. For example, the statement "I will stay in and study Friday if and only if I don't have any money" is the same as "If I don't have any money, then I will stay in and study Friday *and* if I stay in and study Friday, then I don't have any money." In symbols, we can write either $p \leftrightarrow q$ or $(p \rightarrow q) \wedge (q \rightarrow p)$. Since the biconditional is a conjunction, for it to be true, both of the statements $p \rightarrow q$ and $q \rightarrow p$ must be true. We will once again look at cases to build the truth table.

Case 1: Both p and q are true. Then both $p \rightarrow q$ and $q \rightarrow p$ are true, and the conjunction $(p \rightarrow q) \wedge (q \rightarrow p)$, which is also $p \leftrightarrow q$, is true as well.

p	q	$p \leftrightarrow q$
T	T	T

Case 2: p is true and q is false. In this case, the implication $p \rightarrow q$ is false, so it doesn't even matter whether $q \rightarrow p$ is true or false—the conjunction has to be false.

p	q	$p \leftrightarrow q$
T	T	T
T	F	F

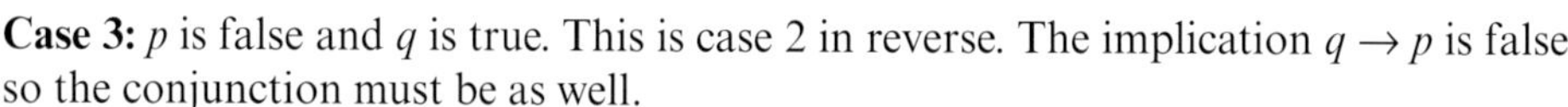

Case 3: p is false and q is true. This is case 2 in reverse. The implication $q \rightarrow p$ is false, so the conjunction must be as well.

p	q	$p \leftrightarrow q$
T	T	T
T	F	F
F	T	F

A technician who designs an automated irrigation system needs to decide whether the system should turn on *if* the water in the soil falls below a certain level or *if and only if* the water in the soil falls below a certain level. In the first instance, other inputs could also turn on the system.

Case 4: p is false and q is false. According to the truth table for a conditional statement, both $p \rightarrow q$ and $q \rightarrow p$ are true in this case, so the conjunction is as well. This completes the truth table.

p	q	$p \leftrightarrow q$
T	T	T
T	F	F
F	T	F
F	F	T

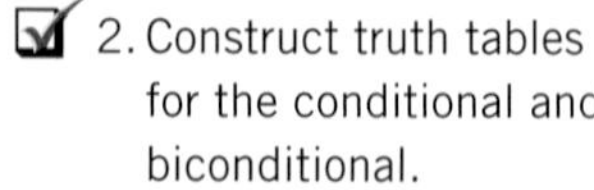

2. Construct truth tables for the conditional and biconditional.

Truth Values for a Biconditional Statement

The biconditional statement $p \leftrightarrow q$ is true when p and q have the same truth value and is false when they have opposite truth values.

Table 3-2 on the next page provides a summary of the truth tables for the basic compound statements and the negation. The last thing you should do is to try and memorize these tables! If you understand how we built them, you can rebuild them on your own when you need them.

Truth Tables for Compound Statements

Once we know truth values for the basic connectives, we can use truth tables to find the truth values for any logical statement. The key to the procedure is to take it step

TABLE 3-2 Truth Tables for the Connectives and Negation

Conjunction "and"

p	q	$p \wedge q$
T	T	T
T	F	F
F	T	F
F	F	F

Disjunction "or"

p	q	$p \vee q$
T	T	T
T	F	T
F	T	T
F	F	F

Conditional "if . . . then"

p	q	$p \rightarrow q$
T	T	T
T	F	F
F	T	T
F	F	T

Biconditional "if and only if"

p	q	$p \leftrightarrow q$
T	T	T
T	F	F
F	T	F
F	F	T

Negation "not"

p	$\sim p$
T	F
F	T

by step, so that in every case, you're deciding on truth values based on one of the truth tables in Table 3-2.

EXAMPLE 1 Constructing a Truth Table

"My leg isn't better, or I'm taking a break" is an example of a statement that can be written as $\sim p \vee q$.

Construct a truth table for the statement $\sim p \vee q$.

SOLUTION

Step 1 Set up a table as shown.

p	q
T	T
T	F
F	T
F	F

The order in which you list the Ts and Fs doesn't matter as long as you cover all the possible combinations. For consistency in this book, we'll always use the order TTFF for p and TFTF for q when these are the only two letters in the logical statement.

Step 2 Find the truth values for $\sim p$ by negating the values for p, and put them into a new column, column 3, marked $\sim p$.

p	q	$\sim p$
T	T	F
T	F	F
F	T	T
F	F	T
	②	③

Truth values for $\sim p$ are opposite those for p.

Step 3 Find the truth values for the disjunction $\sim p \vee q$. Use the T and F values for $\sim p$ and q in columns 2 and 3, and use the disjunction truth table from earlier in the section.

p	q	$\sim p$	$\sim p \vee q$
T	T	F	T
T	F	F	F
F	T	T	T
F	F	T	T
			④

The disjunction is true unless $\sim p$ and q are both false.

The truth values for the statement $\sim p \vee q$ are found in column 4. The statement is true unless p is true and q is false.

▼ Try This One 1

Construct a truth table for the statement $p \vee \sim q$.

When a statement contains parentheses, we find the truth values for the statements in parentheses first, as shown in Example 2. This is similar to the order of operations used in arithmetic and algebra.

EXAMPLE 2 Constructing a Truth Table

"It is not true that if it rains, then we can't go out" is an example of a statement that can be written as $\sim(p \rightarrow \sim q)$.

Construct a truth table for the statement $\sim(p \rightarrow \sim q)$.

SOLUTION

Step 1 Set up the table as shown.

p	q
T	T
T	F
F	T
F	F

Step 2 Find the truth values for $\sim q$ by negating the values for q, and put them into a new column.

p	q	$\sim q$
T	T	F
T	F	T
F	T	F
F	F	T
①		③

Truth values for $\sim q$ are opposite those for q.

Step 3 Find the truth values for the implication $p \rightarrow \sim q$, using the values in columns 1 and 3 and the implication truth table from earlier in the section.

p	q	$\sim q$	$p \rightarrow \sim q$
T	T	F	F
T	F	T	T
F	T	F	T
F	F	T	T
			④

The conditional is true unless p is true and ~q is false.

Step 4 Find the truth values for the negation $\sim(p \rightarrow \sim q)$ by negating the values for $p \rightarrow \sim q$ in column 4.

p	q	$\sim q$	$p \rightarrow \sim q$	$\sim(p \rightarrow \sim q)$
T	T	F	F	T
T	F	T	T	F
F	T	F	T	F
F	F	T	T	F
			④	⑤

The negation has opposite values from column 4.

The truth values for $\sim(p \rightarrow \sim q)$ are in column 5. The statement is true only when p and q are both true.

▼ Try This One 2

Construct a truth table for the statement $p \leftrightarrow (\sim p \wedge q)$.

We can also construct truth tables for compound statements that involve three or more components. For a compound statement with three simple statements p, q, and r, there are eight possible combinations of Ts and Fs to consider. The truth table is set up as shown in step 1 of Example 3.

EXAMPLE 3 Constructing a Truth Table with Three Components

"I'll do my math assignment, or if I think of a good topic, then I'll start my English essay" is an example of a statement that can be written as $p \vee (q \rightarrow r)$.

Construct a truth table for the statement $p \vee (q \rightarrow r)$.

SOLUTION

Step 1 Set up the table as shown.

p	q	r
T	T	T
T	T	F
T	F	T
T	F	F
F	T	T
F	T	F
F	F	T
F	F	F
	②	③

Again, the order of the Ts and Fs doesn't matter as long as all the possible combinations are listed. Whenever there are three letters in the statement, we'll use the order shown above for consistency.

Step 2 Find the truth value for the statement in parentheses, $q \to r$. Use the values in columns 2 and 3 and the conditional truth table from earlier in the section. Put those values in a new column.

p	q	r	$q \to r$
T	T	T	T
T	T	F	F
T	F	T	T
T	F	F	T
F	T	T	T
F	T	F	F
F	F	T	T
F	F	F	T
①			④

The conditional is true unless q is true and r is false.

Step 3 Find the truth values for the disjunction $p \vee (q \to r)$, using the values for p from column 1 and those for $q \to r$ from column 4. Use the truth table for disjunction from earlier in the section, and put the results in a new column.

p	q	r	$q \to r$	$p \vee (q \to r)$
T	T	T	T	T
T	T	F	F	T
T	F	T	T	T
T	F	F	T	T
F	T	T	T	T
F	T	F	F	F
F	F	T	T	T
F	F	F	T	T
				⑤

The disjunction is true unless both p and q → r are false.

The truth values for the statement $p \vee (q \to r)$ are found in column 5. The statement is true unless p and r are false while q is true.

☑ 3. Construct truth tables for compound statements.

▼ Try This One 3

Construct a truth table for the statement $(p \wedge q) \vee \sim r$.

In the method we've demonstrated for constructing truth tables, we begin by setting up a table with all possible combinations of Ts and Fs for the component letters from the statement. Then we build new columns, one at a time, by writing truth values for parts of the compound statement, using the basic truth tables we built earlier in the section.

We have seen that when we construct truth tables, we find truth values for statements inside parentheses first. To avoid having to always use parentheses, a hierarchy of connectives has been agreed upon by those who study logic.

1. Biconditional $\leftrightarrow$
2. Conditional $\rightarrow$
3. Conjunction $\wedge$, disjunction $\vee$
4. Negation $\sim$

Math Note

When parentheses are used to emphasize order, the statement $p \vee q \rightarrow r$ is written as $(p \vee q) \rightarrow r$. The statement $p \leftrightarrow q \wedge r$ is written as $p \leftrightarrow (q \vee r)$.

When we find the truth value for a compound statement without parentheses, we find *the truth value of a lower-order connective first*. For example, $p \vee q \rightarrow r$ is a conditional statement since the conditional ($\rightarrow$) is of a higher order than the disjunction ($\vee$). If you were constructing a truth table for the statement, you would find the truth value for $\vee$ first. The statement $p \leftrightarrow q \wedge r$ is a biconditional statement since the biconditional ($\leftrightarrow$) is of a higher order than the order of the conjunction ($\wedge$). When constructing a truth table for the statement, the truth value for the conjunction ($\wedge$) would be found first. The conjunction and disjunction are of the same order; the statement $p \wedge q \vee r$ cannot be identified unless parentheses are used. In this case, $(p \wedge q) \vee r$ is a disjunction and $p \wedge (q \vee r)$ is a conjunction.

EXAMPLE 4 Using the Hierarchy of Connectives

For each, identify the type of statement using the hierarchy of connectives, and rewrite by using parentheses to indicate order.

(a) $\sim p \vee \sim q$ (b) $p \rightarrow \sim q \wedge r$ (c) $p \vee q \leftrightarrow q \vee r$ (d) $p \rightarrow q \leftrightarrow r$

SOLUTION

(a) The $\vee$ is higher than the $\sim$; the statement is a disjunction and looks like $(\sim p) \vee (\sim q)$ with parentheses.
(b) The $\rightarrow$ is higher than the $\wedge$ or $\sim$; the statement is a conditional and looks like $p \rightarrow (\sim q \wedge r)$ with parentheses.
(c) The $\leftrightarrow$ is higher than $\vee$; the statement is a biconditional and looks like $(p \vee q) \leftrightarrow (q \vee r)$ with parentheses.
(d) The $\leftrightarrow$ is higher than the $\rightarrow$; the statement is a biconditional and looks like $(p \rightarrow q) \leftrightarrow r$ with parentheses.

▼ Try This One 4

☑ 4. Identify the hierarchy of logical connectives.

For each, identify the type of statement using the hierarchy of connectives, and rewrite by using parentheses to indicate order.

(a) $\sim p \vee q$ (c) $p \vee q \leftrightarrow \sim p \vee \sim q$ (e) $p \leftrightarrow q \rightarrow r$
(b) $p \vee \sim q \rightarrow r$ (d) $p \wedge \sim q$

EXAMPLE 5 An Application of Truth Tables

Use the truth value of each simple statement to determine the truth value of the compound statement.

p: O. J. Simpson was convicted in California in 1995.
q: O. J. Simpson was convicted in Nevada in 2008.
r: O. J. Simpson gets sent to prison.

Statement: $(p \vee q) \rightarrow r$

SOLUTION

In probably the most publicized trial of recent times, Simpson was acquitted of murder in California in 1995, so statement p is false. In 2008, however, Simpson was convicted of robbery and kidnapping in Nevada, so statement q is true. Statement r is also true, as Simpson was sentenced in December 2008.

Now we'll analyze the compound statement. First, the disjunction $p \vee q$ is true when either p or q is true, so in this case, $p \vee q$ is true. Finally, the implication $(p \vee q) \rightarrow r$ is true when both r and $p \vee q$ are true, which is the case here. So the compound statement $(p \vee q) \rightarrow r$ is true.

▼ Try This One 5

Using the simple statements in Example 5, find the truth value of the compound statement $(\sim p \wedge \sim q) \rightarrow r$.

An Alternative Method for Constructing Truth Tables

In the next two examples, we will illustrate a second method for constructing truth tables so that you can make a comparison. The problems are the same as Examples 2 and 3.

EXAMPLE 6 Constructing a Truth Table by Using an Alternative Method

Construct a truth table for $\sim(p \rightarrow \sim q)$.

SOLUTION

Step 1 Set up the table as shown.

p	q	$\sim(p \rightarrow \sim q)$
T	T	
T	F	
F	T	
F	F	

Step 2 Write the truth values for p and q underneath the respective letters in the statement as shown, and label the columns as 1 and 2.

p	q	$\sim(p$	$\rightarrow$	$\sim$	$q)$
T	T	T			T
T	F	T			F
F	T	F			T
F	F	F			F
		①			②

Step 3 Find the negation of q since it is inside the parentheses, and place the truth values in column 3. Draw a line through the truth values in column 2 since they will not be used again.

p	q	$\sim(p$	$\rightarrow$	$\sim$	$q)$
T	T	T		F	T
T	F	T		T	F
F	T	F		F	T
F	F	F		T	F
		①		③	②

"It is not true that if it rains, then we can't go out," is an example of a statement that can be written as $\sim(p \rightarrow \sim q)$.

Step 4 Find the truth values for the conditional ($\rightarrow$) by using the T and F values in columns 1 and 3 and the conditional truth table from earlier in the section.

Place these values in column 4 and draw a line through the T and F values in columns 1 and 3, as shown.

p	q	$\sim(p$	$\rightarrow$	$\sim$	$q)$
T	T	T	F	F	T
T	F	T	T	T	F
F	T	F	T	F	T
F	F	F	T	T	F
		(1)	(4)	(3)	(2)

The conditional is true unless p is true and ~q is false.

Step 5 Find the negations of the truth values in column 4 (since the negation sign is outside the parentheses).

p	q	$\sim$	$(p$	$\rightarrow$	$\sim$	$q)$
T	T	T	T	F	F	T
T	F	F	T	T	T	F
F	T	F	F	T	F	T
F	F	F	F	T	T	F
		(5)	(1)	(4)	(3)	(2)

The negation has values opposite those in column 4.

The truth value of $\sim(p \rightarrow \sim q)$ is found in column 5. Fortunately, these are the same values we found in Example 2.

Math Note

It isn't necessary to label the columns with numbers or to draw a line through the truth values in the columns when they are no longer needed; however, these two strategies can help reduce errors.

▼ Try This One 6

Construct a truth table for the statement $p \leftrightarrow (\sim p \wedge q)$, using the alternative method.

EXAMPLE 7 Constructing a Truth Table by Using an Alternative Method

Construct a truth table for the statement $p \vee (q \rightarrow r)$.

SOLUTION

Step 1 Set up the table as shown.

p	q	r	$p \vee (q \rightarrow r)$
T	T	T	
T	T	F	
T	F	T	
T	F	F	
F	T	T	
F	T	F	
F	F	T	
F	F	F	

Step 2 Recopy the values of p, q, and r under their respective letters in the statement as shown.

p	q	r	$p \vee$	$(p \rightarrow$	$r)$
T	T	T	T	T	T
T	T	F	T	T	F
T	F	T	T	F	T
T	F	F	T	F	F
F	T	T	F	T	T
F	T	F	F	T	F
F	F	T	F	F	T
F	F	F	F	F	F
			①	②	③

Step 3 Using the truth values in columns 2 and 3 and the truth table for the conditional ($\rightarrow$), find the values inside the parentheses for the conditional and place them in column 4.

p	q	r	p	$\vee$	$(q$	$\rightarrow$	$r)$
T	T	T	T		T	T	T
T	T	F	T		T	F	F
T	F	T	T		F	T	T
T	F	F	T		F	T	F
F	T	T	F		T	T	T
F	T	F	F		T	F	F
F	F	T	F		F	T	T
F	F	F	F		F	T	F
			①		②	④	③

The conditional is true unless q is true and r is false.

Step 4 Complete the truth table, using the truth values in columns 1 and 4 and the table for the disjunction ($\vee$), as shown.

p	q	r	p	$\vee$	$(q$	$\rightarrow$	$r)$
T	T	T	T	T	T	T	T
T	T	F	T	T	T	F	F
T	F	T	T	T	F	T	T
T	F	F	T	T	F	T	F
F	T	T	F	T	T	T	T
F	T	F	F	F	T	F	F
F	F	T	F	T	F	T	T
F	F	F	F	T	F	T	F
			①	⑤	②	④	③

The disjunction is true unless p and $q \rightarrow r$ are both false.

The truth value for $p \vee (q \rightarrow r)$ is found in column 5. These are the same values we found in Example 3.

▼ Try This One 7

5. Construct truth tables by using an alternative method.

Construct a truth table for the statement $(p \wedge q) \vee \sim r$ using the alternative method.

The best approach to learning truth tables is to try each of the two methods and see which one is more comfortable for you. In any case, we have seen that truth tables are an effective way to organize truth values for statements, allowing us to determine the truth values of some very complicated statements in a systematic way.

Answers to Try This One

1

p	q	$\sim q$	$p \vee \sim q$
T	T	F	T
T	F	T	T
F	T	F	F
F	F	T	T

2

p	q	$\sim p$	$\sim p \wedge q$	$p \leftrightarrow (\sim p \wedge q)$
T	T	F	F	F
T	F	F	F	F
F	T	T	T	F
F	F	T	F	T

3

p	q	r	$p \wedge q$	$\sim r$	$(p \wedge q) \vee \sim r$
T	T	T	T	F	T
T	T	F	T	T	T
T	F	T	F	F	F
T	F	F	F	T	T
F	T	T	F	F	F
F	T	F	F	T	T
F	F	T	F	F	F
F	F	F	F	T	T

4 (a) Disjunction; $(\sim p) \vee q$
(b) Conditional; $(p \vee \sim q) \rightarrow r$
(c) Biconditional; $(p \vee q) \leftrightarrow (\sim p \vee \sim q)$
(d) Conjunction; $p \wedge (\sim q)$
(e) Biconditional; $p \leftrightarrow (q \rightarrow r)$

5 True

6

p	q	p	$\leftrightarrow$	$(\sim$	p	$\wedge$	$q)$
T	T	T	F	F	T	F	T
T	F	T	F	F	T	F	F
F	T	F	F	T	F	T	T
F	F	F	T	T	F	F	F
		①	⑤	③	①	④	②

7

p	q	r	$(p$	$\wedge$	$q)$	$\vee$	$\sim$	r
T	T	T	T	T	T	T	F	T
T	T	F	T	T	T	T	T	F
T	F	T	T	F	F	F	F	T
T	F	F	T	F	F	T	T	F
F	T	T	F	F	T	F	F	T
F	T	F	F	F	T	T	T	F
F	F	T	F	F	F	F	F	T
F	F	F	F	F	F	T	T	F
			①	④	②	⑥	⑤	③

EXERCISE SET 3-2

Writing Exercises

1. Explain the purpose of a truth table.
2. Explain the difference between the inclusive and exclusive disjunctions.
3. Explain the difference between the conditional and biconditional statements.
4. Describe the hierarchy for the basic connectives.

Computational Exercises

For Exercises 5–34, construct a truth table for each.

5. $\sim(p \vee q)$
6. $q \rightarrow p$
7. $\sim p \wedge q$
8. $\sim q \rightarrow \sim p$
9. $\sim p \leftrightarrow q$
10. $(p \vee q) \rightarrow \sim p$
11. $\sim(p \wedge q) \rightarrow p$
12. $(p \vee q) \wedge (q \wedge p)$
13. $(\sim q \wedge p) \rightarrow \sim p$
14. $q \wedge \sim p$
15. $(p \wedge q) \leftrightarrow (q \vee \sim p)$
16. $p \rightarrow (q \vee \sim p)$
17. $(p \wedge q) \vee p$
18. $(q \rightarrow p) \vee \sim r$
19. $(r \wedge q) \vee (p \wedge q)$
20. $(r \rightarrow q) \vee (p \rightarrow r)$
21. $\sim(p \vee q) \rightarrow \sim(p \wedge r)$
22. $(\sim p \vee \sim q) \rightarrow \sim r$
23. $(\sim p \vee q) \wedge r$
24. $p \wedge (q \vee \sim r)$
25. $(p \wedge q) \leftrightarrow (\sim r \vee q)$
26. $\sim(p \wedge r) \rightarrow (q \wedge r)$
27. $r \rightarrow \sim(p \vee q)$
28. $(p \vee q) \vee (\sim p \vee \sim r)$
29. $p \rightarrow (\sim q \wedge \sim r)$
30. $(q \vee \sim r) \leftrightarrow (p \wedge \sim q)$
31. $\sim(q \rightarrow p) \wedge r$
32. $q \rightarrow (p \wedge r)$
33. $(r \vee q) \wedge (r \wedge p)$
34. $(p \wedge q) \leftrightarrow \sim r$

Real-World Applications

For Exercises 35–40, use the truth value of each simple statement to determine the truth value of the compound statement. Use the Internet if you need help determining the truth value of a simple statement.

35. p: Japan bombs Pearl Harbor.
    q: The United States stays out of World War II.
    Statement: $p \rightarrow q$
36. p: Barack Obama wins the Democratic nomination in 2008.
    q: Mitt Romney wins the Republican nomination in 2008.
    Statement: $p \wedge q$
37. p: NASA sends a manned spacecraft to the Moon.
    q: NASA sends a manned spacecraft to Mars.
    Statement: $p \vee q$
38. p: Michael Phelps wins eight gold medals.
    q: Michael Phelps gets a large endorsement deal.
    Statement: $p \rightarrow q$
39. p: Apple builds a portable MP3 player.
    q: Apple stops making computers.
    r: Microsoft releases the Vista operating system.
    Statement: $(p \vee q) \wedge r$
40. p: Hurricane Katrina hits New Orleans.
    q: New Orleans Superdome is damaged.
    r: New Orleans Saints play home games in 2006 in Baton Rouge.
    Statement: $(p \wedge q) \rightarrow r$

Exercises 41–46 are based on the compound statement below.

A new weight loss supplement claims that if you take the product daily and cut your calorie intake by 10%, you will lose at least 10 pounds in the next 4 months.

41. This compound statement is made up of three simple statements. Identify them and assign a letter to each.
42. Write the compound statement in symbolic form, using conjunctions and the conditional.
43. Construct a truth table for the compound statement you wrote in Exercise 42.
44. If you take this product daily and don't cut your calorie intake by 10%, and then don't lose 10 pounds, is the claim made by the advertiser true or false?
45. If you take the product daily, don't cut your calorie intake by 10%, and do lose 10 pounds, is the claim true or false?
46. If you don't take the product daily, cut your calorie intake by 10%, and do lose 10 pounds, is the claim true or false?

Exercises 47–52 are based on the compound statement below.

The owner of a professional baseball team publishes an open letter to fans after another losing season. He claims that if attendance for the following season is over 2 million, then he will add $20 million to the payroll and the team will make the playoffs the following year.

47. This compound statement is made up of three simple statements. Identify them and assign a letter to each.
48. Write the compound statement in symbolic form, using conjunction and the conditional.
49. Construct a truth table for the compound statement you wrote in Exercise 48.
50. If attendance goes over 2 million the next year and the owner raises payroll by $20 million, but the team fails to make the playoffs, is the owner's claim true or false?
51. If attendance is less than 2 million but the owner still raises the payroll by $20 million and the team makes the playoffs, is the owner's claim true or false?
52. If attendance is over 2 million, the owner doesn't raise the payroll, but the team still makes the playoffs, is the owner's claim true or false?

Critical Thinking

53. Using the hierarchy for connectives, write the statement $p \rightarrow q \vee r$ by using parentheses to indicate the proper order. Then construct truth tables for $(p \rightarrow q) \vee r$ and $p \rightarrow (q \vee r)$. Are the resulting truth values the same? Are you surprised? Why or why not?

54. The hierarchy of connectives doesn't distinguish between conjunctions and disjunctions. Does that matter? Construct truth tables for $(p \vee q) \wedge r$ and $p \vee (q \wedge r)$ to help you decide.
55. In 2003, New York City Council was considering banning indoor smoking in bars and restaurants. Opponents of the ban claimed that it would have a negligible effect on indoor pollution, but a huge negative effect on the economic success of these businesses. Eventually, the ban was enacted, and a 2004 study by the city department of health found that there was a sixfold decrease in indoor air pollution in bars and restaurants, but jobs, liquor licenses, and tax revenues all increased. Assign truth values to all the premises of the opponents' claim; then write the claim as a compound statement and determine its validity.

Section 3-3 Types of Statements

LEARNING OBJECTIVES

- ❑ 1. Classify a statement as a tautology, a self-contradiction, or neither.
- ❑ 2. Identify logically equivalent statements.
- ❑ 3. Write negations of compound statements.
- ❑ 4. Write the converse, inverse, and contrapositive of a statement.

It's no secret that weight loss has become big business in the United States. It seems like almost every week, a new company pops into existence with the latest miracle pill to turn you into a supermodel.

A typical advertisement will say something like "Use of our product may result in significant weight loss." That sounds great, but think about what that statement really means. If use of the product "may" result in significant weight loss, then it also may not result in weight loss at all! The statement could be translated as "You will lose weight or you will not lose weight." Of course, this statement is always true. In this section, we will study statements of this type.

Tautologies and Self-Contradictions

In our study of truth tables in Section 3-2, we saw that most compound statements are true in some cases and false in others. What we have not done is think about whether that's true for *every* compound statement. Some simple examples should be enough to convince you that this is most definitely not the case.

Consider the simple statement "I'm going to Cancun for spring break this year." Its negation is "I'm not going to Cancun for spring break this year." Now think about these two compound statements:

> "I'm going to Cancun for spring break this year, or I'm not going to Cancun for spring break this year."
>
> "I'm going to Cancun for spring break this year, and I'm not going to Cancun for spring break this year."

Hopefully, it's pretty clear to you that the first statement is always true, while the second statement is always false (whether you go to Cancun or not). The first is an example of a *tautology,* while the second is an example of a *self-contradiction.*

When a compound statement is always true, it is called a **tautology.**
When a compound statement is always false, it is called a **self-contradiction.**

CAUTION

Don't make the mistake of thinking that every statement is either a tautology or a self-contradiction. We've seen many examples of statements that are sometimes true and other times false.

The sample statements above are simple enough that it's easy to tell that they are always true or always false based on common sense. But for more complicated statements, we'll need to construct a truth table to decide if a statement is a tautology, a self-contradiction, or neither.

EXAMPLE 1 Using a Truth Table to Classify a Statement

Let p = "I am going to a concert" and q = "I will wear black." Translate each statement in Example 1 into a word statement using this choice of p and q. Can you predict which statements are tautologies, self-contradictions, or neither?

Determine if each statement is a tautology, a self-contradiction, or neither.

(a) $(p \wedge q) \to p$ (b) $(p \wedge q) \wedge (\sim p \wedge \sim q)$ (c) $(p \vee q) \to q$

SOLUTION

(a) The truth table for statement (a) is

p	q	$p \wedge q$	$(p \wedge q) \to p$
T	T	T	T
T	F	F	T
F	T	F	T
F	F	F	T

Since the truth table value consists of all Ts, the statement is always true, that is, a tautology.

(b) The truth table for statement (b) is

p	q	$\sim p$	$\sim q$	$p \wedge q$	$\sim p \wedge \sim q$	$(p \wedge q) \wedge (\sim p \wedge \sim q)$
T	T	F	F	T	F	F
T	F	F	T	F	F	F
F	T	T	F	F	F	F
F	F	T	T	F	T	F

Since the truth value consists of all Fs, the statement is always false, that is, a self-contradiction.

(c) The truth table for statement (c) is

p	q	$p \vee q$	$(p \vee q) \to q$
T	T	T	T
T	F	T	F
F	T	T	T
F	F	F	T

Since the statement can be true in some cases and false in others, it is neither a tautology nor a self-contradiction.

▼ Try This One 1

☑ 1. Classify a statement as a tautology, a self-contradiction, or neither.

Determine if each statement is a tautology, a self-contradiction, or neither.

(a) $(p \vee q) \wedge (\sim p \to q)$ (b) $(p \wedge \sim q) \wedge \sim p$ (c) $(p \to q) \vee \sim q$

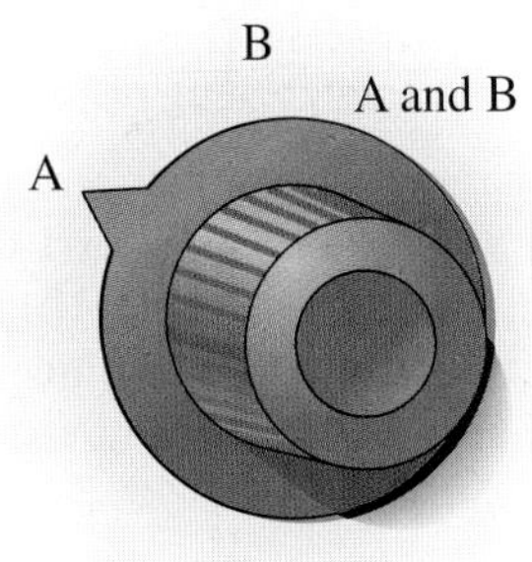

The statements "If the red dial is set to A, then use only speaker A" and "The red dial is not set to speaker A, or only speaker A is used" can be modeled with the logic statements $p \rightarrow q$ and $\sim p \vee q$.

Logically Equivalent Statements

Next, consider the two logical statements $p \rightarrow q$ and $\sim p \vee q$. The truth table for the two statements is

p	q	$\sim p$	$p \rightarrow q$	$\sim p \vee q$
T	T	F	T	T
T	F	F	F	F
F	T	T	T	T
F	F	T	T	T

Notice that the truth values for both statements are *identical,* that is, TFTT. When this occurs, the statements are said to be *logically equivalent;* that is, both compositions of the same simple statements have the same meaning. For example, the statement "If it snows, I will go skiing" is logically equivalent to saying "It is not snowing or I will go skiing." Formally defined,

Two compound statements are **logically equivalent** if and only if they have the same truth table values. The symbol for logically equivalent statements is $\equiv$.

EXAMPLE 2 Identifying Logically Equivalent Statements

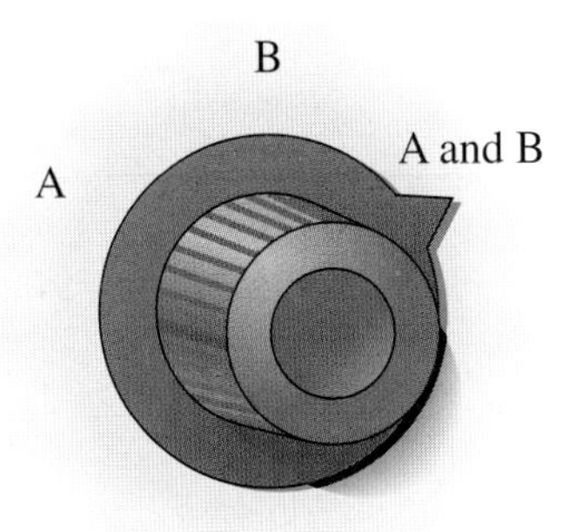

In the red dial example, $\sim q \rightarrow \sim p$ would be "If speaker A is not the only one used, then the red dial is not set to A."

Determine if the two statements $p \rightarrow q$ and $\sim q \rightarrow \sim p$ are logically equivalent.

SOLUTION

The truth table for the statements is

p	q	$\sim p$	$\sim q$	$p \rightarrow q$	$\sim q \rightarrow \sim p$
T	T	F	F	T	T
T	F	F	T	F	F
F	T	T	F	T	T
F	F	T	T	T	T

Since both statements have the same truth values, they are logically equivalent.

☑ 2. Identify logically equivalent statements.

▼ Try This One 2

Determine which two statements are logically equivalent.

(a) $\sim(p \wedge \sim q)$ (b) $\sim p \wedge q$ (c) $\sim p \vee q$

De Morgan's laws for logic give us an example of equivalent statements.

De Morgan's Laws for Logic

For any statements p and q,

$\sim(p \vee q) \equiv \sim p \wedge \sim q$ and $\sim(p \wedge q) \equiv \sim p \vee \sim q$

Notice the similarities between De Morgan's laws for sets and De Morgan's laws for logic. De Morgan's laws can be proved by constructing truth tables; the proofs will be left to you in the exercises.

De Morgan's laws are most often used to write the negation of conjunctions and disjunctions. For example, the negation of the statement "I will go to work or I will go to the beach" is "I will not go to work and I will not go to the beach." Notice that when you negate a conjunction, it becomes a disjunction; and when you negate a disjunction, it becomes a conjunction—that is, the *and* becomes an *or,* and the *or* becomes an *and.*

EXAMPLE 3 Using De Morgan's Laws to Write Negations

Write the negations of the following statements, using De Morgan's laws.

(a) Studying is necessary and I am a hard worker.
(b) It is not easy or I am lazy.
(c) I will pass this test or I will drop this class.
(d) She is angry or she's my friend, and she is cool.

SOLUTION

(a) Studying is not necessary or I am not a hard worker.
(b) It is easy and I am not lazy.
(c) I will not pass this test and I will not drop this class.
(d) She is not angry and she is not my friend, or she is not cool.

▼ Try This One 3

Write the negations of the following statements, using De Morgan's laws.

(a) I will study for this class or I will fail.
(b) I will go to the dance club and the restaurant.
(c) It is not silly or I have no sense of humor.
(d) The movie is a comedy or a thriller, and it is awesome.

Earlier in this section, we saw that the two statements $p \rightarrow q$ and $\sim p \vee q$ are logically equivalent. Now that we know De Morgan's laws, we can use this fact to find the negation of the conditional statement $p \rightarrow q$.

$$\begin{aligned} \sim(p \rightarrow q) &\equiv \sim(\sim p \vee q) \\ &\equiv \sim(\sim p) \wedge \sim q \qquad \text{Note: } \sim(\sim p) \equiv p \\ &\equiv p \wedge \sim q \end{aligned}$$

This can be checked by using a truth table as shown.

p	q	$\sim q$	$p \rightarrow q$	$\sim(p \rightarrow q)$	$p \wedge \sim q$
T	T	F	T	F	F
T	F	T	F	T	T
F	T	F	T	F	F
F	F	T	T	F	F

So the negation of $p \rightarrow q$ is $p \wedge \sim q$.

For example, if you say, "It is false that if it is sunny, then I will go swimming," this is equivalent to the statement "It's sunny and I will not go swimming."

EXAMPLE 4 Writing the Negation of a Conditional Statement

Write the negation of the statement "If I have a computer, then I will use the Internet."

SOLUTION

Let p = "I have a computer" and q = "I will use the Internet." The statement $p \rightarrow q$ can be negated as $p \wedge \sim q$. This translates to "I have a computer and I will not use the Internet."

▼ Try This One 4

☑ 3. Write negations of compound statements.

Write the negation of the statement "If the video is popular, then it can be found on YouTube."

Table 3-3 summarizes the negations of the basic compound statements.

TABLE 3-3 Negation of Compound Statements

Statement	Negation	Equivalent Negation
$p \wedge q$	$\sim(p \wedge q)$	$\sim p \vee \sim q$
$p \vee q$	$\sim(p \vee q)$	$\sim p \wedge \sim q$
$p \rightarrow q$	$\sim(p \rightarrow q)$	$p \wedge \sim q$

Variations of the Conditional Statement

From the conditional statement $p \rightarrow q$, three other related statements can be formed: the **converse,** the **inverse,** and the **contrapositive.** They are shown here symbolically.

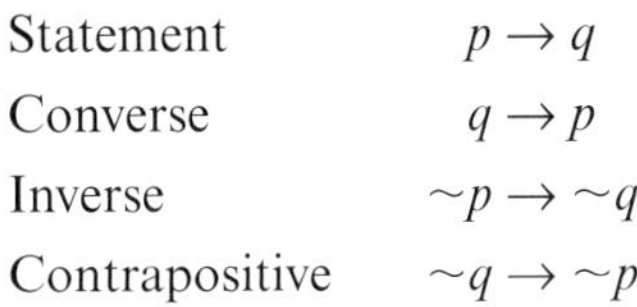

Statement	$p \rightarrow q$
Converse	$q \rightarrow p$
Inverse	$\sim p \rightarrow \sim q$
Contrapositive	$\sim q \rightarrow \sim p$

Using the statement "If Tessa is a chocolate Lab, then Tessa is brown" as our original conditional statement, we find the related statements are as follows:

Converse: If Tessa is brown, then Tessa is a chocolate Lab.

Inverse: If Tessa is not a chocolate Lab, then Tessa is not brown.

Contrapositive: If Tessa is not brown, then Tessa is not a chocolate Lab.

Notice that the original statement is true, but of the three related statements, only the contrapositive is also true. This is an important observation—one that we'll elaborate on shortly.

EXAMPLE 5 Writing the Converse, Inverse, and Contrapositive

Write the converse, the inverse, and the contrapositive for the statement "If you earned a bachelor's degree, then you got a high-paying job."

SOLUTION

It's helpful to write the original implication in symbols: $p \to q$, where p = "You earned a bachelor's degree" and q = "You got a high-paying job."

Converse: $q \to p$. "If you got a high-paying job, then you earned a bachelor's degree."

Inverse: $\sim p \to \sim q$. "If you did not earn a bachelor's degree, then you did not get a high-paying job."

Contrapositive: $\sim q \to \sim p$. "If you did not get a high-paying job, then you did not earn a bachelor's degree."

▼ Try This One 5

Write the converse, the inverse, and the contrapositive for the statement "If you do well in math classes, then you are intelligent."

The relationships between the variations of the conditional statements can be determined by looking at the truth tables for each of the statements.

p	q	$p \to q$	$q \to p$	$\sim p \to \sim q$	$\sim q \to \sim p$
T	T	T	T	T	T
T	F	F	T	T	F
F	T	T	F	F	T
F	F	T	T	T	T

Since the original statement ($p \to q$) and the contrapositive statement ($\sim q \to \sim p$) have the same truth values, they are equivalent. Also note that the converse ($q \to p$) and the inverse ($\sim p \to \sim q$) have the same truth values, so they are equivalent as well. Finally, notice that the original statement is not equivalent to the converse or the inverse since the truth values of the converse and inverse differ from those of the original statement.

Since the conditional statement $p \to q$ is used so often in logic as well as mathematics, a more detailed analysis is helpful. Recall that the conditional statement $p \to q$ is also called an *implication* and consists of two simple statements; the first is the *antecedent p* and the second is the *consequent q*. For example, the statement "If I jump into the pool, then I will get wet" consists of the antecedent p, "I jump into the pool," and the consequent q, "I will get wet," connected by the if . . . then connective.

The conditional can also be stated in these other ways:

p implies q

q if p

p only if q

p is sufficient for q

q is necessary for p

All p are q

In four of these six forms, the antecedent comes first, but for "q if p" and "q is necessary for p," the consequent comes first. So identifying the antecedent and consequent is important.

Math Note

Many people using logic in real life assume that if a statement is true, the converse is automatically true. Consider the following statement: "If a person earns more than $200,000 per year, that person can buy a Corvette." The converse is stated as "If a person can buy a Corvette, then that person earns more than $200,000 per year." This may be far from the truth since a person may have to make very large payments, live in a tent, or work three jobs in order to afford the expensive car.

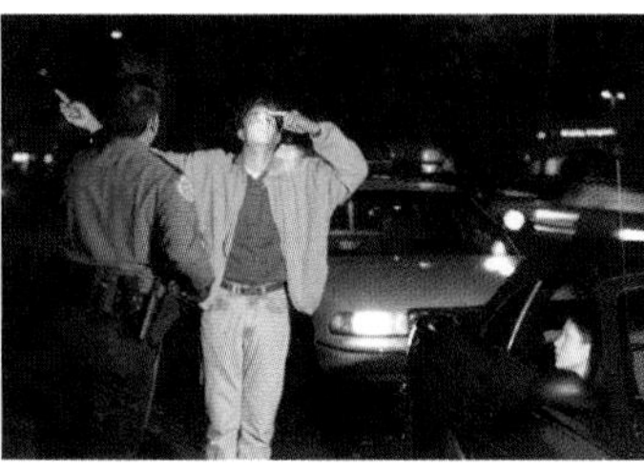

For example, think about the statement "If you drink and drive, you get arrested." Writing it in the different possible forms, we get:

Drinking and driving implies you get arrested.
You get arrested if you drink and drive.
You drink and drive only if you get arrested.
Drinking and driving is sufficient for getting arrested.
Getting arrested is necessary for drinking and driving.
All those who drink and drive get arrested.

Of course, these all say the same thing. To illustrate the importance of getting the antecedent and consequent in the correct order, consider the "q if p" form, in this case "You get arrested if you drink and drive." If we don't start with the consequent, we get "You drink and drive if you get arrested." This is completely false—there are any number of things you could get arrested for other than drinking and driving.

EXAMPLE 6 Writing Variations of a Conditional Statement

Write each statement in symbols. Let p = "A person is over 6′ 6″" and q = "A person is tall."

(a) If a person is over 6′6″, then the person is tall.
(b) Being tall is necessary for being over 6′6″.
(c) A person is over 6′6″ only if the person is tall.
(d) Being 6′6″ is sufficient for being tall.
(e) A person is tall if the person is over 6′6″.

SOLUTION

(a) If p, then q; $p \rightarrow q$
(b) q is necessary for p; $p \rightarrow q$
(c) p only if q; $p \rightarrow q$
(d) p is sufficient for q; $p \rightarrow q$
(e) q if p; $p \rightarrow q$

Actually, these statements all say the same thing!

▼ Try This One 6

☑ 4. Write the converse, inverse, and contrapositive of a statement.

Write each statement in symbols. Let p = "A student comes to class every day" and q = "A student gets a good grade."

(a) A student gets a good grade if a student comes to class every day.
(b) Coming to class every day is necessary for getting a good grade.
(c) A student gets a good grade only if a student comes to class every day.
(d) Coming to class every day is sufficient for getting a good grade.

In this section, we saw that some statements are always true (tautologies) and others are always false (self-contradictions). We also defined what it means for two statements to be logically equivalent—they have the same truth values. Now we're ready to analyze logical arguments to determine if they're legitimate or not.

Answers to Try This One

1 (a) Neither (b) Self-contradiction
(c) Tautology

2 (a) and (c)

3 (a) I will not study for this class and I will not fail.
(b) I will not go to the dance club or the restaurant.
(c) It is silly and I have a sense of humor.
(d) The movie is not a comedy and it is not a thriller, or it is not awesome.

4 The video is popular and it cannot be found on YouTube.

5 Converse: If you are intelligent, then you do well in math classes.
Inverse: If you do not do well in math classes, then you are not intelligent.
Contrapositive: If you are not intelligent, then you do not do well in math classes.

6 (a) $p \to q$ (b) $q \to p$ (c) $q \to p$ (d) $p \to q$

EXERCISE SET 3-3

Writing Exercises

1. Explain the difference between a tautology and a self-contradiction.
2. Is every statement either a tautology or a self-contradiction? Why or why not?
3. Describe how to find the converse, inverse, and contrapositive of a conditional statement.
4. How can you decide if two statements are logically equivalent?
5. How can you decide if one statement is the negation of another?
6. Is a statement always logically equivalent to its converse? Explain.

Computational Exercises

For Exercises 7–16, determine which statements are tautologies, self-contradictions, or neither.

7. $(p \vee q) \vee (\sim p \wedge \sim q)$
8. $(p \to q) \wedge (p \vee q)$
9. $(p \wedge q) \wedge (\sim p \vee \sim q)$
10. $\sim p \vee (p \to q)$
11. $(p \leftrightarrow q) \vee \sim(q \leftrightarrow p)$
12. $(p \wedge q) \leftrightarrow (p \to \sim q)$
13. $(p \vee q) \wedge (\sim p \vee \sim q)$
14. $(p \wedge q) \vee (p \vee q)$
15. $(p \leftrightarrow q) \wedge (\sim p \leftrightarrow \sim q)$
16. $(p \to q) \wedge (\sim p \vee q)$

For Exercises 17–26, determine if the two statements are logically equivalent statements, negations, or neither.

17. $\sim q \to p$; $\sim p \to q$
18. $p \wedge q$; $\sim q \vee \sim p$
19. $\sim(p \vee q)$; $p \to \sim q$
20. $\sim(p \to q)$; $\sim p \wedge q$
21. $q \to p$; $\sim(p \to q)$
22. $p \vee (\sim q \wedge r)$; $(p \wedge \sim q) \vee (p \wedge r)$
23. $\sim(p \vee q)$; $\sim(\sim p \wedge \sim q)$
24. $(p \vee q) \to r$; $\sim r \to \sim(p \vee q)$
25. $(p \wedge q) \vee r$; $p \wedge (q \vee r)$
26. $p \leftrightarrow \sim q$; $(p \wedge \sim q) \vee (\sim p \wedge q)$

For Exercises 27–32, write the converse, inverse, and contrapositive of each.

27. $p \to q$
28. $\sim p \to \sim q$
29. $\sim q \to p$
30. $\sim p \to q$
31. $p \to \sim q$
32. $q \to p$

Real-World Applications

In Exercises 33–42, use De Morgan's laws to write the negation of the statement.

33. The concert is long or it is fun.
34. The soda is sweet or it is not carbonated.
35. It is not cold and I am soaked.
36. I will walk in the Race for the Cure walkathon and I will be tired.
37. I will go to the beach and I will not get sunburned.
38. The coffee is a latte or an espresso.
39. The student is a girl or the professor is not a man.
40. I will go to college and I will get a degree.
41. It is right or it is wrong.
42. Our school colors are not blue or they are not green.

For Exercises 43–49, let p = "I need to talk to my friend" and q = "I will send her a text message." Write each of the following in symbols (see Example 6).

43. If I need to talk to my friend, I will send her a text message.
44. If I will not send her a text message, I do not need to talk to my friend.
45. Sending a text message is necessary for needing to talk to my friend.
46. I will send her a text message if I need to talk to my friend.
47. Needing to talk to my friend is sufficient for sending her a text message.

48. I need to talk to my friend only if I will send her a text message.
49. I do not need to talk to my friend only if I will not send her a text message.
50. Are any of the statements in Exercises 43–49 logically equivalent?

For Exercises 51–56, write the converse, inverse, and contrapositive of the conditional statement.

51. If he graduated with a bachelor's degree in management information systems, then he will get a good job.
52. If she does not earn $5,000 this summer as a barista at the coffeehouse, then she cannot buy the green Ford Focus.
53. If the *American Idol* finale is today, then I will host a party in my dorm room.
54. If my cell phone will not charge, then I will replace the battery.
55. I will go to Nassau for spring break if I lose 10 pounds by March 1.
56. The politician will go to jail if he gets caught taking kickbacks.

Critical Thinking

57. In this section, we wrote the negation of $p \rightarrow q$ by using a disjunction. See if you can write the negation of $p \rightarrow q$ by using a conjunction.
58. Try to write the negation of the biconditional $p \leftrightarrow q$ by using only conjunctions, disjunctions, and negations.
59. Can you think of a true conditional statement about someone you know so that the converse is true as well? How about so that the converse is false?
60. Can you think of a true conditional statement about someone you know so that the inverse is true as well? How about so that the inverse is false?
61. Prove the first of De Morgan's laws by using truth tables:

$$\sim(p \vee q) \equiv \sim p \wedge \sim q$$

62. Prove the second of De Morgan's laws by using truth tables:

$$\sim(p \wedge q) \equiv \sim p \vee \sim q$$

Section 3-4 Logical Arguments

LEARNING OBJECTIVES

- ☐ 1. Define *valid argument* and *fallacy.*
- ☐ 2. Use truth tables to determine the validity of an argument.
- ☐ 3. Identify common argument forms.
- ☐ 4. Determine the validity of arguments by using common argument forms.

Common sense is a funny thing in our society: we all think we have it, and we also think that most other people don't. This thing that we call common sense is really the ability to think logically, to evaluate an argument or situation and decide what is and is not reasonable. It doesn't take a lot of imagination to picture how valuable it is to be able to think logically. We're pretty well protected by parents for our first few years of life, but after that the main tool we have to guide us through the perils of life is our brain. The more effectively that brain can analyze and evaluate information, the more successful we're likely to be. The work we've done in building the basics of symbolic logic in the first three sections of this chapter has prepared us for the real point: analyzing logical arguments objectively. That's the topic of this important section.

Valid Arguments and Fallacies

A logical argument is made up of two parts: a premise or premises and a conclusion based on those premises. We will call an argument **valid** if assuming the premises are true guarantees that the conclusion is true as well. An argument that is not valid is called **invalid** or a **fallacy.**

Let's look at an example.

Premise 1:	All students in this class will pass.
Premise 2:	Rachel is a student in this class.
Conclusion:	Rachel will pass this class.

We can easily tell that if the two premises are true, then the conclusion is true, so this is an example of a valid argument.

Sidelight LOGIC AND THE ART OF FORENSICS

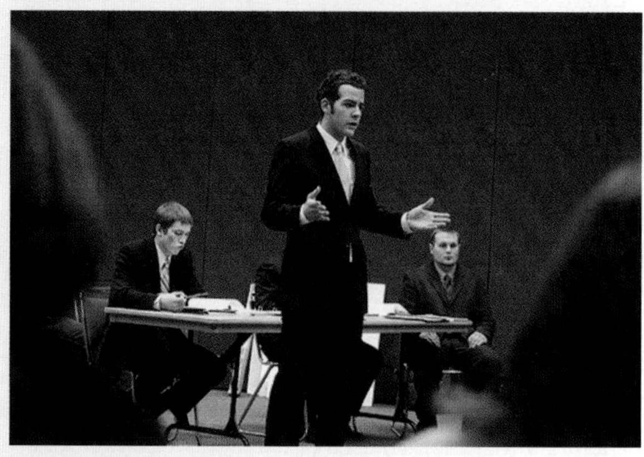

Many students find it troubling that an argument can be considered valid even if the conclusion is clearly false. But arguing in favor of something that you don't necessarily believe to be true isn't a new idea by any means—lawyers do it all the time, and it's commonly practiced in the area of formal debate, a style of intellectual competition that has its roots in ancient times.

In formal debate (also known as forensics), speakers are given a topic and asked to argue one side of a related issue. Judges determine which speakers make the most effective arguments and declare the winners accordingly. One of the most interesting aspects is that in many cases, the contestants don't know which side of the issue they will be arguing until right before the competition begins. While that aspect is intended to test the debater's flexibility and preparation, a major consequence is that opinion, and sometimes truth, is taken out of the mix, and contestants and judges must focus on the validity of arguments.

A variety of organizations sponsor national competitions in formal debate for colleges. The largest is an annual championship organized by the National Forensics Association. Students from well over 100 schools participate in a wide variety of categories. The 2008 team champions were Tennessee State University, Kansas State University, California State Long Beach, and Western Kentucky University.

It's very important at this point to understand the difference between a true statement and a conclusion to a valid argument. A statement that is known to be false can still be a valid conclusion if it follows logically from the given premises. For example, consider this argument:

Los Angeles is in California or Mexico.
Los Angeles is not in California.
Therefore, Los Angeles is in Mexico.

☑ 1. Define *valid argument* and *fallacy*.

This is a valid argument: if the two premises are true, then the conclusion, "Los Angeles is in Mexico," must be true as well. We know, however, that Los Angeles isn't actually in Mexico. That's the tricky part. In determining whether an argument is valid, *we will always assume that the premises are true*. In this case, we're assuming that the premise "Los Angeles is not in California" is true, even though in fact it is not. We will discuss this aspect of logical arguments in greater depth later in this section.

Truth Table Method

One method for determining the validity of an argument is by using truth tables. We will use the following procedure.

Procedure for Determining the Validity of Arguments

Step 1 Write the argument in symbols.

Step 2 Write the argument as a conditional statement; use a conjunction between the premises and the implication ($\Rightarrow$) for the conclusion. (*Note:* The $\Rightarrow$ is the same as $\rightarrow$ but will be used to designate an argument.)

Step 3 Set up and construct a truth table as follows:

Symbols	Premise $\wedge$ Premise $\Rightarrow$ Conclusion

Step 4 If all truth values under $\Rightarrow$ are Ts (i.e., a tautology), then the argument is valid; otherwise, it is invalid.

EXAMPLE 1 Determining the Validity of an Argument

Determine if the following argument is valid or invalid.

If a figure has three sides, then it is a triangle.
This figure is not a triangle.
____
Therefore, this figure does not have three sides.

SOLUTION

Step 1 *Write the argument in symbols.* Let p = "The figure has three sides," and let q = "The figure is a triangle."

Translated into symbols:

$p \rightarrow q$ (Premise)
$\sim q$ (Premise)
____
$\therefore \sim p$ (Conclusion)

A line is used to separate the premises from the conclusion and the three triangular dots $\therefore$ mean "therefore."

Step 2 *Write the argument as an implication* by connecting the premises with a conjunction and implying the conclusion as shown.

Premise 1		**Premise 2**		**Conclusion**
$(p \rightarrow q)$	$\wedge$	$\sim q$	$\Rightarrow$	$\sim p$

Step 3 *Construct a truth table* as shown.

p	q	$\sim p$	$\sim q$	$p \rightarrow q$	$(p \rightarrow q) \wedge \sim q$	$[(p \rightarrow q) \wedge \sim q] \Rightarrow \sim p$
T	T	F	F	T	F	T
T	F	F	T	F	F	T
F	T	T	F	T	F	T
F	F	T	T	T	T	T

Step 4 *Determine the validity of the argument.* Since all the values under the $\Rightarrow$ are true, the argument is valid.

▼ Try This One 1

Determine if the argument is valid or invalid.

I will run for student government or I will join the athletic boosters.
I did not join the athletic boosters.
____
Therefore, I will run for student government.

EXAMPLE 2 Determining the Validity of an Argument

Determine the validity of the following argument. "If a professor is rich, then he will buy an expensive automobile. The professor bought an expensive automobile. Therefore, the professor is rich."

SOLUTION

Step 1 *Write the argument in symbols.* Let p = "The professor is rich," and let q = "The professor buys an expensive automobile."

$$\begin{array}{l} p \rightarrow q \\ \underline{q \qquad} \\ \therefore p \end{array}$$

Step 2 *Write the argument as an implication.*

$$(p \rightarrow q) \wedge q \Rightarrow p$$

Step 3 *Construct a truth table* for the argument.

p	q	$p \rightarrow q$	$(p \rightarrow q) \wedge q$	$[(p \rightarrow q) \wedge q] \Rightarrow p$
T	T	T	T	T
T	F	F	F	T
F	T	T	T	F
F	F	T	F	T

Step 4 *Determine the validity of the argument.* This argument is invalid since it is not a tautology. (Remember, when the values are not all Ts, the argument is invalid.) In this case, it cannot be concluded that the professor is rich.

▼ Try This One 2

Determine the validity of the following argument. "If John blows off work to go to the playoff game, he will lose his job. John lost his job. Therefore, John blew off work and went to the playoff game."

CAUTION

Remember that in symbolic logic, whether or not the conclusion is true is not important. The main concern is whether the conclusion follows from the premises.

Consider the following two arguments.

1. Either $2 + 2 \neq 4$ or $2 + 2 = 5$

   $\underline{\quad 2 + 2 = 4 \quad}$

   $\therefore 2 + 2 = 5$

2. If $2 + 2 \neq 5$, then I passed the math quiz.

   I did not pass the quiz.

   $\therefore 2 + 2 \neq 5$

The truth tables on the next page show that the first argument is valid even though the conclusion is false, and the second argument is invalid even though the conclusion is true!

Let p be the statement "2 + 2 = 4" and q be the statement "2 + 2 = 5." Then the first argument is written as

$$\begin{array}{l} (\sim p \vee q) \\ \underline{p \qquad\quad} \\ \therefore q \end{array}$$

Truth table for argument 1

p	q	$\sim p$	$\sim p \vee q$	$(\sim p \vee q) \wedge p$	$[(\sim p \vee q) \wedge p] \Rightarrow q$
T	T	F	T	T	T
T	F	F	F	F	T
F	T	T	T	F	T
F	F	T	T	F	T

Let p be the statement "2 + 2 ≠ 5" and q be the statement "I passed the math quiz." The second argument is written as

$$\begin{array}{l} p \to q \\ \underline{\sim q \qquad} \\ \therefore p \end{array}$$

Truth table for argument 2

p	q	$\sim q$	$p \to q$	$(p \to q) \wedge \sim q$	$[(p \to q) \wedge \sim q] \Rightarrow p$
T	T	F	T	F	T
T	F	T	F	F	T
F	T	F	T	F	T
F	F	T	T	T	F

The validity of arguments that have three variables can also be determined by truth tables, as shown in Example 3.

EXAMPLE 3 Determining the Validity of an Argument

Determine the validity of the following argument.

$$\begin{array}{l} p \to r \\ p \wedge r \\ \underline{p \qquad\qquad} \\ \therefore \sim q \to p \end{array}$$

SOLUTION

Step 1 *Write the argument in symbols.* This has been done already.

Step 2 *Write the argument as an implication.* Make a conjunction of all three premises and imply the conclusion:

$$(p \to r) \wedge (q \wedge r) \wedge p \Rightarrow (\sim q \to p)$$

Step 3 *Construct a truth table.* When there are three premises, we will begin by finding the truth values for each premise and then work the conjunction from left to right as shown.

p	q	r	$\sim q$	$p \to r$	$q \wedge r$	$\sim q \to p$	$(p \to r) \wedge (q \wedge r) \wedge p$	$[(p \to r) \wedge (q \wedge r) \wedge p] \Rightarrow (\sim q \to p)$
T	T	T	F	T	T	T	T	T
T	T	F	F	F	F	T	F	T
T	F	T	T	T	F	T	F	T
T	F	F	T	F	F	T	F	T
F	T	T	F	T	T	T	F	T
F	T	F	F	T	F	T	F	T
F	F	T	T	T	F	F	F	T
F	F	F	T	T	F	F	F	T

Since the truth value for ⇒ is all Ts, the argument is valid.

2. Use truth tables to determine the validity of an argument.

▼ Try This One 3

Determine whether the following argument is valid or invalid.

$p \vee q$
$p \vee \sim r$
$\therefore q$

Common Valid Argument Forms

We have seen that truth tables can be used to test an argument for validity. But some argument forms are common enough that they are recognized by special names. When an argument fits one of these forms, we can decide if it is valid or not just by knowing the general form, rather than constructing a truth table.

We'll start with a description of some commonly used valid arguments.

1. *Law of detachment* (also known by Latin name *modus ponens*):

   $p \rightarrow q$
   p
   $\therefore q$

   Example:

   If our team wins Saturday, then they go to a bowl game.
   Our team won Saturday.
   Therefore, our team goes to a bowl game.

2. *Law of contraposition* (Latin name *modus tollens*):

   $p \rightarrow q$
   $\sim q$
   $\therefore \sim p$

   Example:

   If I try hard, I'll get an A.
   I didn't get an A.
   Therefore, I didn't try hard.

3. *Law of syllogism,* also known as *law of transitivity:*

   $p \rightarrow q$
   $q \rightarrow r$
   $\therefore p \rightarrow r$

   Example:

   If I make an illegal U-turn, I'll get a ticket.
   If I get a ticket, I'll get points on my driving record.
   Therefore, if I make an illegal U-turn, I'll get points on my driving record.

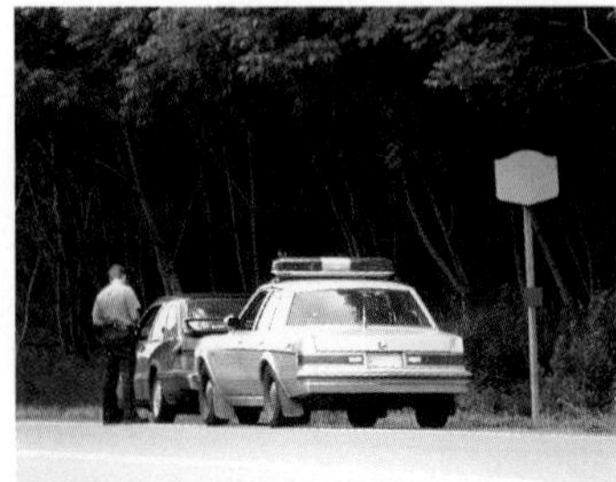

4. *Law of disjunctive syllogism:*

   $p \vee q$
   $\sim p$
   $\therefore q$

Example:

You're either brilliant or insane.
You're not brilliant.
———
Therefore, you're insane.

Common Fallacies

Next, we will list some commonly used arguments that are invalid.

Math Note

It's more common to mistakenly think that an invalid argument is valid rather than the other way around, so you should pay special attention to the common fallacies listed.

1. *Fallacy of the converse:*

   $p \rightarrow q$
   $\underline{q \quad\quad}$
   $\therefore p$

   Example:

   If it's Friday, then I will go to happy hour.
   I am at happy hour.
   ———
   Therefore, it is Friday.

   This is not valid! You can go to happy hour other days, too.

2. *Fallacy of the inverse:*

   $p \rightarrow q$
   $\underline{\sim p \quad\quad}$
   $\therefore \sim q$

   Example:

   If I exercise every day, then I will lose weight.
   I don't exercise every day.
   ———
   Therefore, I won't lose weight.

   This is also not valid. You could still lose weight without exercising *every* day.

3. *Fallacy of the inclusive or:*

   $p \vee q$
   $\underline{p \quad\quad}$
   $\therefore \sim q$

   Example:

   I'm going to take chemistry or physics.
   I'm taking chemistry.
   ———
   Therefore, I'm not taking physics.

   Remember, we've agreed that by *or* we mean *one or the other, or both*. So you could be taking both classes.

You will be asked to prove that some of these are invalid by using truth tables in the exercises.

Sidelight CIRCULAR REASONING

Circular reasoning (sometimes called "begging the question") is a sneaky type of fallacy in which the premises of an argument contain a claim that the conclusion is true, so naturally if the premises are true, so is the conclusion. But this doesn't constitute evidence that a conclusion is true. Consider the following example: A suspect in a criminal investigation tells the police detective that his statements can be trusted because his friend Sue can vouch for him. The detective asks the suspect how he knows that Sue can be trusted, and he says, "I can assure you of her honesty." Ultimately, the suspect becomes even more suspect because of circular reasoning: his argument boils down to "I am honest because I am honest."

While this example might seem blatantly silly, you'd be surprised how often people try to get away with this fallacy. A Google search for the string *circular reasoning* brings up hundreds of arguments that are thought to be circular.

EXAMPLE 4 Recognizing Common Argument Forms

Determine whether the following arguments are valid, using the given forms of valid arguments and fallacies.

(a) $p \rightarrow q$
$\underline{p}$
$\therefore q$

(b) $\sim p \rightarrow q$
$\underline{\sim q}$
$\therefore p$

(c) $\sim p \rightarrow \sim q$
$\underline{\sim q}$
$\therefore \sim p$

(d) $\sim r \rightarrow s$
$\underline{s \rightarrow t}$
$\therefore \sim r \rightarrow t$

SOLUTION

(a) This is the law of detachment, therefore a *valid* argument.
(b) This fits the law of contraposition with the statement $\sim p$ substituted in place of p, so it is valid.
(c) This fits the fallacy of the converse, using statement $\sim p$ and $\sim q$ rather than p and q, so it is an invalid argument.
(d) This is the law of syllogism, with statements $\sim r$, s, and t, so the argument is valid.

▼ Try This One 4

☑ 3. Identify common argument forms.

Determine whether the arguments are valid, using the commonly used valid arguments and fallacies.

(a) $\sim p \vee q$
$\underline{p}$
$\therefore q$

(b) $r \vee s$
$\underline{s}$
$\therefore \sim r$

(c) $\sim p \rightarrow q$
$\underline{q}$
$\therefore \sim p$

(d) $(p \wedge q) \rightarrow \sim r$
$\underline{r}$
$\therefore \sim(p \wedge q)$

EXAMPLE 5 Determining the Validity of an Argument by Using Common Argument Forms

Determine whether the following arguments are valid, using the given forms of valid arguments and fallacies.

(a) If you like dogs, you will live to be 120.
<u>You like dogs.</u>
Therefore, you will live to be 120.

(b) If the modem is connected, then you can access the Web.
<u>The modem is not connected.</u>
Therefore, you cannot access the Web.

(c) If you watch *Big Brother,* you watch reality shows.
If you watch reality shows, you have time to kill.
___
Therefore, if you have time to kill, you watch *Big Brother*.

(d) The movie *Scream* is a thriller or a comedy.
The movie *Scream* is a thriller.
___
Therefore, the movie *Scream* is not a comedy.

(e) My iPod is in my backpack or it is at my friend's house.
My iPod is not in my backpack.
___
Therefore, my iPod is at my friend's house.

SOLUTION

(a) In symbolic form this argument is $(p \rightarrow q) \wedge p \Rightarrow q$. We can see that this is the law of detachment, so the argument is *valid.*
(b) In symbolic form this argument is $(p \rightarrow q) \wedge \sim p \Rightarrow \sim q$. This is the fallacy of the inverse, so the argument is *invalid.*
(c) In symbolic form this argument is $(p \rightarrow q) \wedge (q \rightarrow r) \Rightarrow (r \rightarrow p)$. We know by the law of transitivity that if $(p \rightarrow q) \wedge (q \rightarrow r)$, then $p \rightarrow r$. The given conclusion, $r \rightarrow p$, is the converse of $p \rightarrow r$, so is not equivalent to $p \rightarrow r$ (the valid conclusion). So the argument is *invalid.*
(d) In symbolic form the argument is $(p \vee q) \wedge p \Rightarrow \sim q$. This is the fallacy of the inclusive or, so the argument is *invalid.*
(e) In symbolic form the argument is $(p \vee q) \wedge \sim p \Rightarrow q$. This is the law of disjunctive syllogism, so the argument is *valid.*

▼ Try This One 5

☑ 4. Determine the validity of arguments by using common argument forms.

Determine whether the following arguments are valid using the given forms of valid arguments and fallacies.

(a) If Elliot is a freshman, he takes English.
Elliot is a freshman.
___
Therefore, Elliot takes English.

(b) If you work hard, you will be a success.
You are not a success.
___
Therefore, you do not work hard.

(c) Jon is cheap or financially broke.
He is not financially broke.
___
Therefore, he is cheap.

(d) If Jose asks me out, I will not study Friday.
I didn't study Friday.
___
Therefore, Jose asked me out.

There are two big lessons to be learned in this section. First, sometimes an argument can appear to be legitimate superficially, but if you study it carefully, you may find out that it's not. Second, the validity of an argument is not about whether the conclusion is true or false—it's about whether the conclusion follows logically from the premises.

Answers to Try This One

1 Valid

2 Invalid

3 Invalid

4 (a) Valid (b) Invalid (c) Invalid (d) Valid

5 (a) Valid (b) Valid (c) Valid (d) Invalid

EXERCISE SET 3-4

Writing Exercises

1. Describe the structure of an argument.
2. Is it possible for an argument to be valid, yet have a false conclusion? Explain your answer.
3. Is it possible for an argument to be invalid, yet have a true conclusion? Explain your answer.
4. When you are setting up a truth table to determine the validity of an argument, what connective is used between the premises of an argument? What connective is used between the premises and the conclusion?
5. Describe what the law of syllogism says, in your own words.
6. Describe why the fallacy of the inclusive or is a fallacy.

Computational Exercises

For Exercises 7–16, using truth tables, determine whether each argument is valid.

7. $p \to q$
   $\underline{p \wedge q}$
   $\therefore p$
8. $p \vee q$
   $\underline{\sim q}$
   $\therefore p$
9. $\sim p \vee q$
   $\underline{p}$
   $\therefore p \wedge \sim q$
10. $p \leftrightarrow \sim q$
    $\underline{p \wedge \sim q}$
    $\therefore p \vee q$
11. $\sim q \vee p$
    $\underline{q}$
    $\therefore \sim p$
12. $p \vee q$
    $\underline{\sim p \wedge \sim q}$
    $\therefore p$
13. $p \leftrightarrow q$
    $\underline{\sim q}$
    $\therefore \sim p$
14. $p \vee \sim q$
    $\underline{\sim q \to p}$
    $\therefore p$
15. $p \wedge \sim q$
    $\underline{\sim r \to q}$
    $\therefore q$
16. $p \leftrightarrow q$
    $\underline{q \leftrightarrow r}$
    $\therefore p \wedge q$
17. Write the law of detachment in symbols; then prove that it is a valid argument by using a truth table.
18. Write the law of contraposition in symbols; then prove that it is a valid argument by using a truth table.
19. Write the law of syllogism in symbols; then prove that it is a valid argument by using a truth table.
20. Write the law of disjunctive syllogism in symbols; then prove that it is a valid argument by using a truth table.
21. Write the fallacy of the converse in symbols; then prove that it is not a valid argument by using a truth table.
22. Write the fallacy of the inclusive or in symbols; then prove that it is not a valid argument by using a truth table.

For Exercises 23–32, determine whether the following arguments are valid, using the given forms of valid arguments and fallacies.

23. $p \to q$
    $\underline{\sim q}$
    $\therefore \sim p$
24. $p \vee q$
    $\underline{q}$
    $\therefore \sim p$
25. $\sim p \to q$
    $\underline{\sim q}$
    $\therefore \sim p$
26. $p \vee \sim q$
    $\underline{q}$
    $\therefore p$
27. $p \to q$
    $\underline{r \to \sim q}$
    $\therefore p \to \sim r$
28. $p \to \sim q$
    $\underline{\sim q}$
    $\therefore p$
29. $\sim p \vee q$
    $\underline{\sim q}$
    $\therefore \sim p$
30. $\sim p \to q$
    $\underline{\sim q}$
    $\therefore p$
31. $p \vee \sim q$
    $\underline{q}$
    $\therefore \sim p$
32. $p \to \sim q$
    $\underline{\sim r \to q}$
    $\therefore p \to r$

Real-World Applications

For Exercises 33–42, identify p, q, and r if necessary. Then translate each argument to symbols and use a truth table to determine whether the argument is valid or invalid.

33. If it rains, then I will watch the Real World marathon.
    It did not rain.
    ∴ I did not watch the Real World marathon.
34. Ted will get a Big Mac or a Whopper with cheese.
    Ted did not get a Whopper with cheese.
    ∴ Ted got a Big Mac.
35. If Julia uses monster.com to send out her resume, she will get an interview.
    Julia got an interview.
    ∴ Julia used monster.com to send out her resume.
36. If it snows, I can go snowboarding.
    It did not snow.
    ∴ I cannot go snowboarding.
37. I will go to the party if and only if my ex-boyfriend is not going.
    My ex-boyfriend is not going to the party.
    ∴ I will go to the party.
38. If you are superstitious, then do not walk under a ladder.
    If you do not walk under a ladder, then you are superstitious.
    ∴ You are superstitious and you do not walk under a ladder.
39. Either I did not study or I passed the exam.
    I did not study.
    ∴ I failed the exam.
40. I will run the marathon if and only if I can run 30 miles by Christmas.
    I can run 30 miles by Christmas or I will not run the marathon.
    ∴ If I ran the marathon, then I was able to run 30 miles by Christmas.
41. If you eat seven pieces of pizza, then you will get sick.
    If you get sick, then you will not be able to go to the game.
    ∴ If you eat seven pieces of pizza, then you will not be able to go to the game.
42. If you back up your hard drive, then you are protected.
    Either you are protected or you are daring.
    ∴ If you are daring, you won't back up your hard drive.

For Exercises 43–50, write the argument in symbols; then decide whether the argument is valid by using the common forms of valid arguments and fallacies.

43. If the cheese doesn't melt, the nachos are ruined.
    The nachos are ruined.
    ∴ The cheese didn't melt.
44. I studied or I failed the class.
    I did not fail the class.
    ∴ I studied.
45. If I go to the student symposium on environmental issues, I will fall asleep.
    If the speaker is interesting, I will not fall asleep.
    ∴ If I go to the student symposium on environmental issues, the speaker will not be interesting.
46. If it is sunny, I will wear SPF 50 sun block.
    It is not sunny.
    ∴ I will not wear SPF 50 sun block.
47. If I get an A in this class, I will do a dance.
    I did a dance.
    ∴ I got an A in this class.
48. I will not wear a Speedo at the beach or I will be embarrassed.
    I am not embarrassed.
    ∴ I did not wear a Speedo at the beach.
49. If I see the movie at an IMAX theater, I will get dizzy.
    I see the movie at an IMAX theater.
    ∴ I get dizzy.
50. I will backpack through Europe if I get at least a 3.5 grade point average.
    I do not get at least a 3.5 grade point average.
    ∴ I will not backpack through Europe.

Critical Thinking

51. Oscar Wilde once said, "Few parents nowadays pay any regard to what their children say to them. The old-fashioned respect for the young is fast dying out." This statement can be translated to an argument, as shown next.

    If parents respected their children, then parents would listen to them.
    Parents do not listen to their children.
    ∴ Parents do not respect their children.

Using a truth table, determine whether the argument is valid or invalid.

52. Winston Churchill once said, "If you have an important point to make, don't try to be subtle or clever. Use a pile driver. Hit the point once. Then come back and hit it again. Then a third time—a tremendous wack!" This statement can be translated to an argument as shown.

    If you have an important point to make, then you should not be subtle or clever.

    You are not being subtle or clever.
    ____
    ∴ You will make your point.

    Using a truth table, determine whether the argument is valid or invalid.

53. Write an argument matching the law of syllogism that involves something about your school. Then explain why the conclusion of your argument is valid.
54. Write an example of the fallacy of the inverse that involves something about your school. Then explain why the conclusion of your argument is invalid.
55. Look up the literal translation of the Latin term *modus ponens* on the Internet, and explain how that applies to the law of detachment.
56. Look up the literal translation of the Latin term *modus tollens* on the Internet, and explain how that applies to the law of contraposition.

Section 3-5 Euler Circles

LEARNING OBJECTIVES

- ☐ 1. Define *syllogism.*
- ☐ 2. Use Euler circles to determine the validity of an argument.

Abraham Lincoln once said, "You can fool some of the people all of the time, and all of the people some of the time, but you cannot fool all of the people all of the time." Lincoln was a really smart guy—he understood the power of logical arguments and the fact that cleverly crafted phrases could be an effective tool in the art of persuasion. What's interesting about this quote from our perspective is the liberal use of the quantifiers *some* and *all.* In this section, we will study a particular type of argument that uses these quantifiers, along with *no* or *none.* A technique developed by Leonhard Euler way back in the 1700s is a useful method for analyzing these arguments and testing their validity.

Euler circles are diagrams similar to Venn diagrams. We will use them to study arguments using four types of statements. The statement types are listed in Table 3-4, and the Euler circle that illustrates each is shown in Figure 3-1 on the next page.

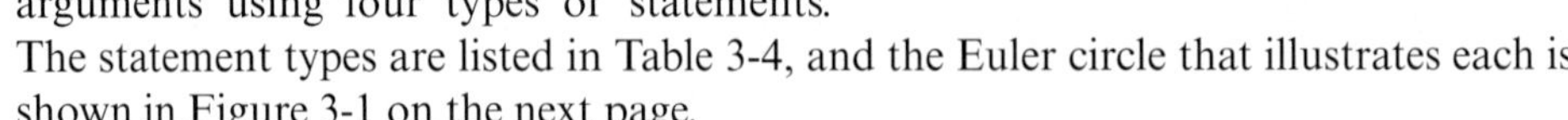

Each statement can be represented by a specific diagram. The universal affirmative "All A is B" means that every member of set A is also a member of set B. For example,

TABLE 3-4 Types of Statements Illustrated by Euler Circles

Type	General Form	Example
Universal affirmative	All A is B	All chickens have wings.
Universal negative	No A is B	No horses have wings.
Particular affirmative	Some A is B	Some horses are black.
Particular negative	Some A is not B	Some horses are not black.

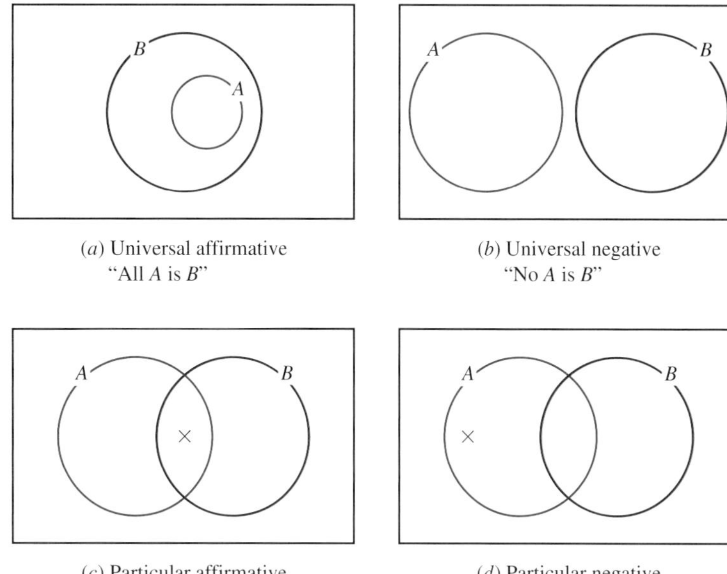

(*a*) Universal affirmative "All *A* is *B*"

(*b*) Universal negative "No *A* is *B*"

(*c*) Particular affirmative "Some *A* is *B*"

(*d*) Particular negative "Some *A* is not *B*"

Figure 3-1

Math Note

If I say that "Some horses are black," you cannot assume that "Some horses are not black." If I say, "Some horses are not black," you cannot assume that "Some horses are black."

the statement "All chickens have wings" means that the set of all chickens is a subset of the set of animals that have wings.

The universal negative "No *A* is *B*" means that no member of set *A* is a member of set *B*. In other words, set *A* and set *B* are *disjoint sets*. For example, "No horses have wings" means that the set of all horses and the set of all animals with wings are disjoint (nonintersecting).

The particular affirmative "Some *A* is *B*" means that there is at least one member of set *A* that is also a member of set *B*. For example, the statement "Some horses are black"means that there is at least one horse that is a member of the set of black animals. The × in Figure 3-1(*c*) means that there is at least one black horse.

The particular negative "Some *A* is not *B*" means that there is at least one member of set *A* that is not a member of set *B*. For example, the statement "Some horses are not black" means that there is at least one horse that does not belong to the set of black animals. The diagram for the particular negative is shown in Figure 3-1(*d*). The × is placed in circle *A* but not in circle *B*. The × in this example means that there exists at least one horse that is some color other than black.

☑ 1. Define *syllogism*.

Many of the arguments we studied in Section 3-4 consisted of two premises and a conclusion. This type of argument is called a **syllogism.** We will use Euler circles to test the validity of syllogisms involving the statement types in Table 3-4. Here's a simple example:

Premise	All cats have four legs.
Premise	Some cats are black.
Conclusion	Therefore, some four-legged animals are black.

Remember that we are not concerned with whether the conclusion is true or false, but only whether the conclusion logically follows from the premises. If yes, the argument is valid. If no, the argument is invalid.

Euler Circle Method for Testing the Validity of an Argument

To determine whether an argument is valid, diagram both premises in the same figure. If the conclusion is shown in the figure, the argument is valid.

Many times the premises can be diagrammed in several ways. If there is even one way in which the diagram contradicts the conclusion, the argument is *invalid* since the conclusion does not necessarily follow from the premises.

Examples 1, 2, and 3 show how to determine the validity of an argument by using Euler circles.

EXAMPLE 1 Using Euler Circles to Determine the Validity of an Argument

Use Euler circles to determine whether the argument is valid.

All cats have four legs.
Some cats are black.
___
Therefore, some four-legged animals are black.

SOLUTION

The first premise, "All cats have four legs," is the universal affirmative; the set of cats diagrammed as a subset of four-legged animals is shown.

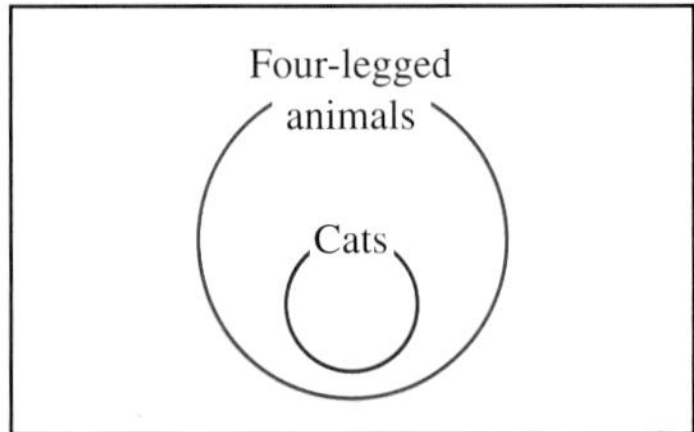

The second premise, "Some cats are black," is the particular affirmative and is shown by placing an × in the intersection of the cats' circle and the black animals' circle. The diagram for this premise is drawn on the diagram of the first premise and can be done in two ways, as shown.

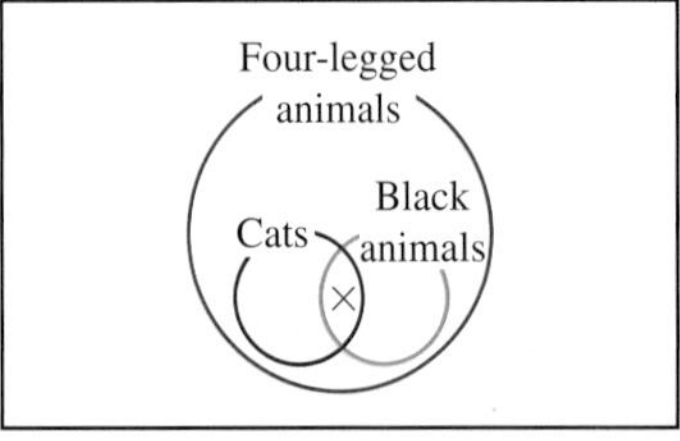

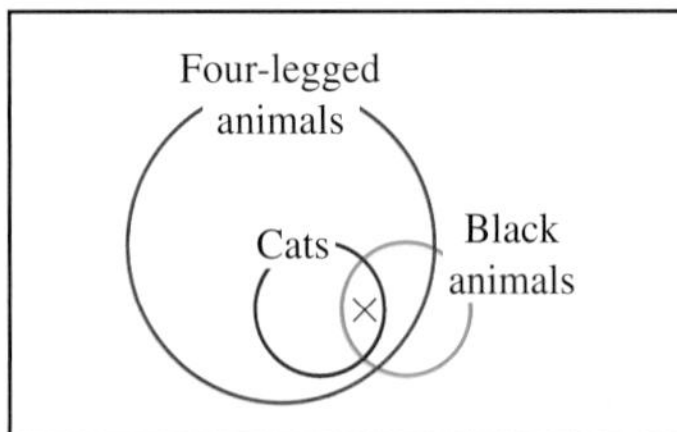

The conclusion is that some four-legged animals are black, so the diagram for the conclusion must have an × in the four-legged animals' circle and in the black animals' circle. Notice that both of the diagrams corresponding to the premises do have an × in both circles, so the conclusion matches the premises and the argument is valid. Since there is no other way to diagram the premises, the conclusion is shown to be true without a doubt.

▼ Try This One 1

Use Euler circles to determine whether the argument is valid.

All college students buy textbooks.
Some book dealers buy textbooks.
___
Therefore, some college students are book dealers.

It isn't necessary to use actual subjects such as cats, four-legged animals, etc. in syllogisms. Arguments can use letters to represent the various sets, as shown in Example 2.

EXAMPLE 2 Using Euler Circles to Determine the Validity of an Argument

Use Euler circles to determine whether the argument is valid or invalid.

Some *A* is not *B*.
All *C* is *B*.
____
∴ Some *A* is *C*.

SOLUTION

The first premise, "Some *A* is not *B*," is diagrammed as shown.

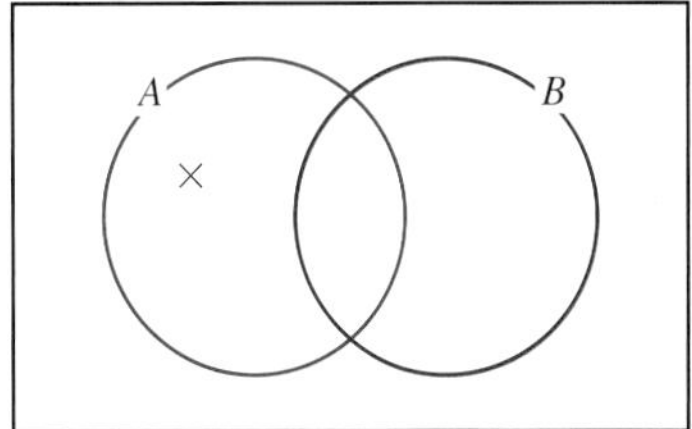

The second premise, "All *C* is *B*," is diagrammed by placing circle *C* inside circle *B*. This can be done in several ways, as shown.

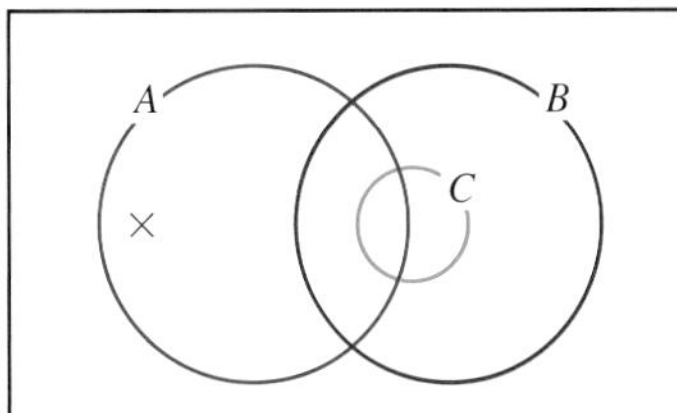

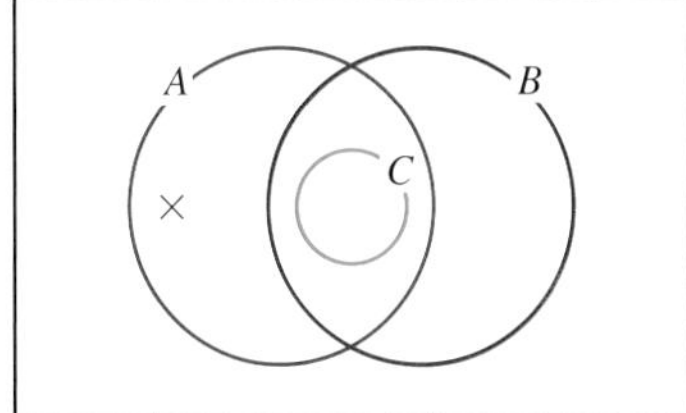

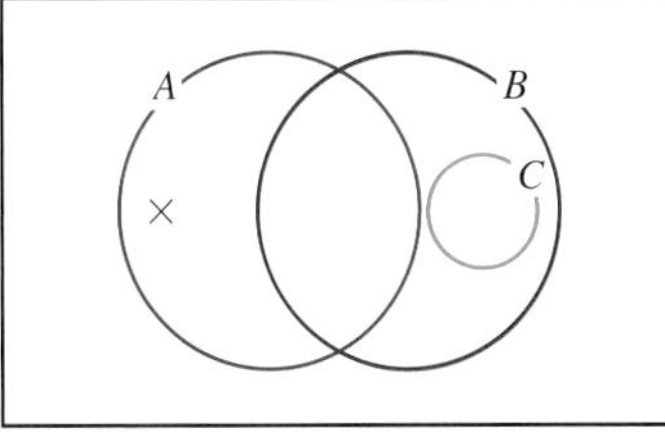

The third diagram shows that the argument is invalid. It matches both premises, but there are no members of *A* that are also in *C*, so it contradicts the conclusion "Some *A* is *C*."

▼ Try This One 2

Use Euler circles to determine whether the argument is valid.

Some *A* is *B*.
Some *A* is not *C*.
____
∴ Some *B* is not *C*.

Let's try one more specific example.

EXAMPLE 3 Using Euler Circles to Determine the Validity of an Argument

Use Euler circles to determine whether the argument is valid.

No criminal is admirable.
Some athletes are not criminals.

∴ Some admirable people are athletes.

SOLUTION

Diagram the first premise, "No criminal is admirable."

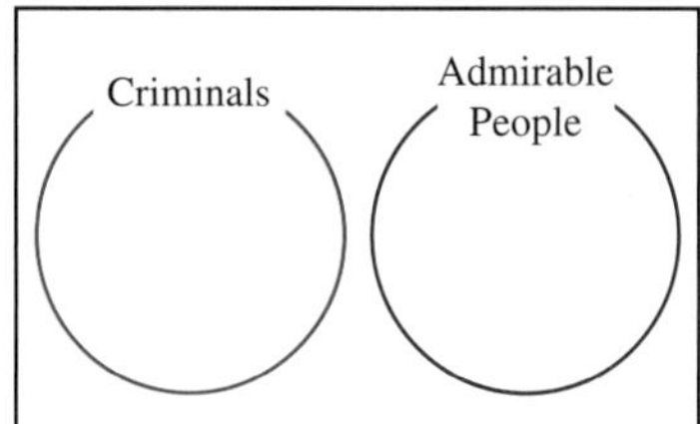

We can add the second premise, "Some athletes are not criminals," in at least two different ways:

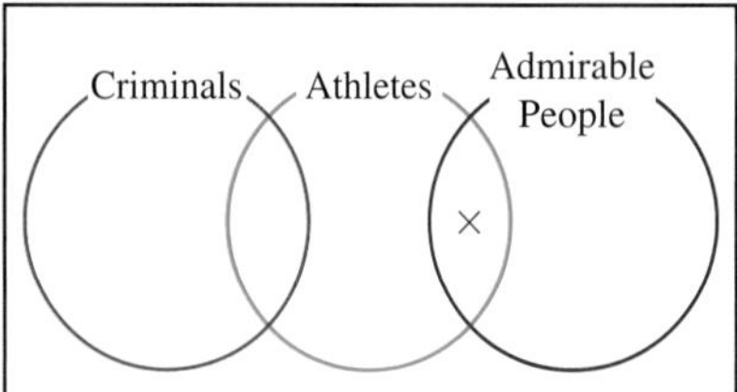

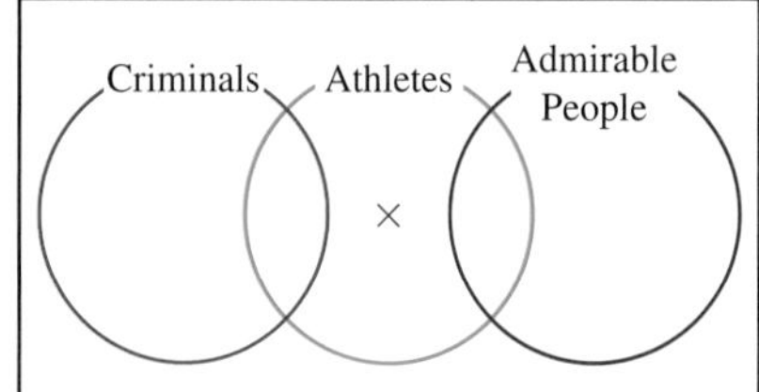

In the first diagram, the conclusion appears to be valid: some athletes are admirable. But the second diagram doesn't support that conclusion, so the argument is invalid.

2. Use Euler circles to determine the validity of an argument.

▼ Try This One 3

Use Euler circles to determine whether the argument is valid.

All dogs bark.
No animals that bark are cats.

∴ No dogs are cats.

We have now seen that for syllogisms that involve the quantifiers *all, some,* or *none,* Euler circles are an efficient way to determine the validity of the argument. We diagram both premises on the same figure, and if all possible diagrams display the conclusion, then the conclusion must be valid.

Answers to Try This One

1 Invalid

2 Invalid

3 Valid

EXERCISE SET 3-5

Writing Exercises

1. Name and give an example of each of the four types of statements that can be diagrammed with Euler circles.
2. Explain how to decide whether an argument is valid or invalid after drawing Euler circles.
3. What is a syllogism?
4. How do Euler circles differ from Venn diagrams?

Computational Exercises

For Exercises 5–14, draw an Euler circle diagram for each statement.

5. All computers are calculators.
6. No unicorns are real.
7. Some people do not go to college.
8. Some CD burners are DVD burners.
9. No math courses are easy.
10. Some fad diets do not result in weight loss.
11. Some laws in the United States are laws in Mexico.
12. All members of Mensa are smart.
13. No cheeseburgers are low in fat.
14. Some politicians are crooks.

For Exercises 15–24, determine whether each argument is valid or invalid.

15. All X is Y.
    Some Y is Z.
    ∴ Some X is Z.
16. Some A is not B.
    No B is C.
    ∴ Some A is not C.
17. Some P is Q.
    No Q is R.
    ∴ Some P is not R.
18. All S is T.
    No S is R.
    ∴ Some T is R.
19. No M is N.
    No N is O.
    ∴ Some M is not O.
20. Some U is V.
    Some U is not W.
    ∴ No W is U.
21. Some A is not B.
    No A is C.
    ∴ Some A is not C.
22. All P is Q.
    All Q is R.
    ∴ All P is R.
23. No S is T.
    No T is R.
    ∴ No S is R.
24. Some M is N.
    Some N is O.
    ∴ Some M is O.

Real-World Applications

For Exercises 25–38, use Euler circles to determine if the argument is valid.

25. All phones are communication devices.
    Some communication devices are inexpensive.
    ∴ Some phones are inexpensive.
26. Some students are overachievers.
    No overachiever is lazy.
    ∴ Some students are not lazy.
27. Some animated movies are violent.
    No kids' movies are violent.
    ∴ No kids' movies are animated.
28. Some protesters are angry.
    Some protesters are not civil.
    ∴ Some civil people are not angry.
29. Some math tutors are patient.
    No patient people are demeaning.
    ∴ Some math tutors are not demeaning.
30. Some students are hard-working.
    Some hard-working people are not successful.
    ∴ Some students are not successful.
31. Some movie stars are fake.
    No movie star is talented.
    ∴ No fake people are talented.

32. Some CEOs are women.
    Some women are tech-savvy.
    ∴ Some CEOs are not tech-savvy.

33. Some juices have antioxidants.
    Some fruits have antioxidants.
    ∴ No juices are fruits.

34. Some funny people are sad.
    No serious people are sad.
    ∴ No funny people are serious.

35. Some women have highlighted hair.
    All women watch soap operas.
    All people who watch soap operas are emotional.
    ∴ Some emotional people watch soap operas.

36. All students text message during class.
    Some students in class take notes.
    All students who take notes in class pass the test.
    ∴ Some students pass the test.

37. Some birds can talk.
    Some animals that can talk can also moo.
    All cows can moo.
    ∴ Some cows can talk.

38. All cars use gasoline.
    All things that use gasoline emit carbon dioxide.
    Some cars have four doors.
    ∴ Some things with four doors emit carbon dioxide.

Critical Thinking

For Exercises 39–42, write a conclusion so that the argument is valid. Use Euler circles.

39. All A is B.
    All B is C.
    ∴

40. No M is P.
    All S is M.
    ∴

41. All calculators can add.
    No adding machines can make breakfast.
    ∴

42. Some people are prejudiced.
    All people have brains.
    ∴

CHAPTER 3 Summary

Section	Important Terms	Important Ideas
3-1	Statement Simple statement Compound statement Connective Conjunction Disjunction Conditional Biconditional Negation	**Formal** symbolic logic uses statements. A statement is a sentence that can be determined to be true or false but not both. A simple statement contains only one idea. A compound statement is formed by joining two or more simple statements with connectives. The four basic connectives are the conjunction (which uses the word *and* and the symbol $\wedge$), the disjunction (which uses the word *or* and the symbol $\vee$), the conditional (which uses the words *if... then* and the symbol $\rightarrow$), and the biconditional (which uses the words *if and only if* and the symbol $\leftrightarrow$). The symbol for negation is $\sim$. Statements are usually written using logical symbols and letters of the alphabet to represent simple statements.
3-2	Truth table	**A truth** table can be used to determine when a compound statement is true or false. A truth table can be constructed for any logical statement.
3-3	Tautology Self-contradiction Logically equivalent statements Converse Inverse Contrapositive De Morgan's laws	**A statement** that is always true is called a tautology. A statement that is always false is called a self-contradiction. Two statements that have the same truth values are said to be logically equivalent. De Morgan's laws are used to find the negation of a conjunction or disjunction. From the conditional statement, three other statements can be made: the converse, the inverse, and the contrapositive.
3-4	Argument Premise Conclusion	**Truth** tables can be used to determine the validity of an argument. An argument consists of two or more statements called premises and a statement called the conclusion. An argument is valid if when the premises are true, the conclusion is true. Otherwise, the argument is invalid.
3-5	Syllogism Euler circles Universal affirmative Universal negative Particular affirmative Particular negative	**A mathematician** named Leonhard Euler developed a method using circles to determine the validity of an argument that is particularly effective for syllogisms involving quantifiers. This method uses four types of statements: (1) the universal affirmative, (2) the universal negative, (3) the particular affirmative, and (4) the particular negative. Euler circles are similar to Venn diagrams.

MATH IN The Art of Persuasion REVISITED

All the arguments on the list are logically invalid except the last one. Even though the claim made by the last argument is false—it did not snow in Hawaii in 2008—this is so because the first premise is false. Remember, the validity of an argument is based on whether it follows from the premises, not on its actual truth.

The first is invalid because there are sources of smoke other than fire. The second is invalid because it's possible to be happy for some reason other than having lots of money. The third is invalid because the initial statement says nothing about teams not in the SEC. The fourth is invalid because it is circular reasoning (see Sidelight in Section 3-4). The fifth is invalid because the first statement doesn't say that weapons of mass destruction are the *only* reason to go to war.

Review Exercises

Section 3-1

For Exercises 1–5, decide whether the sentence is a statement.

1. Let's go with the flow.
2. My duvet cover is indigo.
3. The girls at this school are smart.
4. Ignorance is always a choice.
5. Are we there yet?

For Exercises 6–10, decide whether each statement is simple or compound.

6. Andre is interesting and caring.
7. The monitor is blinking.
8. If it is not raining, I will go kiteboarding.
9. The book is stimulating or informative.
10. There is a silver lining behind every cloud.

For Exercises 11–20, write the negation of the statement.

11. It is scary.
12. The cell phone is out of juice.
13. The Popsicle is green.
14. No people who live in glass houses throw stones.
15. Some failing students can learn new study methods.
16. Everyone will pass the test on logic.
17. There is a printer that has no ink.
18. None of these links are broken.
19. At least one of the contestants will be voted off the island.
20. All SUVs are gas guzzlers.

For Exercises 21–25, classify the statement as a conjunction, disjunction, conditional, or biconditional.

21. In the department store the air is scented and stuffy.
22. If you dream it, you can achieve it.
23. I will go to the 7-Eleven or to the IHOP.
24. It is dangerous if and only if there is ice on the sidewalk.
25. I will slip if I walk on the icy sidewalk.

For Exercises 26–35, let p = "It is ambitious" and let q = "It is worthwhile." Write each statement in symbols.

26. It is ambitious and worthwhile.
27. If it is worthwhile, then it is ambitious.
28. It is worthwhile if and only if it is ambitious.
29. It is worthwhile and not ambitious.
30. If it is not ambitious, then it is not worthwhile.
31. It is not true that it is worthwhile and ambitious.
32. It is not true that if it is ambitious, then it is worthwhile.
33. It is not worthwhile if and only if it is not ambitious.
34. It is not true that it is not worthwhile.
35. It is neither ambitious nor worthwhile.

For Exercises 36–40, let p = "It is cool." Let q = "It is cloudy." Write each statement in words.

36. $p \vee \sim q$
37. $q \rightarrow p$
38. $p \leftrightarrow q$
39. $(p \vee q) \rightarrow p$
40. $\sim(\sim p \vee q)$

Section 3-2

For Exercises 41–48, construct a truth table for each statement.

41. $p \leftrightarrow \sim q$
42. $\sim p \rightarrow (\sim q \vee p)$
43. $(p \rightarrow q) \wedge \sim q$
44. $\sim p \vee (\sim q \rightarrow p)$
45. $\sim q \leftrightarrow (p \rightarrow q)$
46. $(p \rightarrow \sim q) \vee r$
47. $(p \vee \sim q) \wedge r$
48. $r \rightarrow (\sim p \vee q)$

For Exercises 49–52, use the truth value of each simple statement to determine the truth value of the compound statement.

49. p: January is the first month of the year.
    q: It snows in January in every state.
    Statement: $p \rightarrow q$

50. p: Gas prices reached record highs in 2008.
    q: Automobile makers started making more fuel-efficient cars.
    Statement: $p \wedge q$

51. p: Barack Obama was a Presidential candidate in 2008.
    q: Lindsay Lohan was a Presidential candidate in 2008.
    r: Obama won the Democratic nomination in 2008.
    Statement: $(p \vee q) \rightarrow r$

52. p: Attending college costs thousands of dollars.
    q: Lack of education does not lead to lower salaries.
    r: The average college graduate will make back more than they paid for school.
    Statement: $(p \wedge \sim q) \leftrightarrow r$

Section 3-3

For Exercises 53–57, determine if the statement is a tautology, self-contradiction, or neither.

53. $p \rightarrow (p \vee q)$
54. $(p \rightarrow q) \rightarrow (p \vee q)$
55. $(p \wedge \sim q) \leftrightarrow (q \wedge \sim p)$
56. $q \rightarrow (p \vee \sim p)$
57. $(\sim q \vee p) \wedge q$

For Exercises 58–60, determine whether the two statements are logically equivalent.

58. $\sim(p \rightarrow q)$; $\sim p \wedge \sim q$
59. $\sim p \vee \sim q$; $\sim(p \leftrightarrow q)$
60. $(\sim p \wedge q) \vee r$; $(\sim p \vee r) \wedge (q \vee r)$

For Exercises 61–64, use De Morgan's laws to write the negation of each statement.

61. The Internet connection is either dial-up or DSL.
62. We will increase sales or our profit margin will go down.
63. The signature is not authentic and the check is not valid.
64. It is not strenuous and I am tired.

For Exercises 65 and 66, assign a letter to each simple statement and write the compound statement in symbols.

65. I will be happy only if I get rich.
66. Having a good career is sufficient for a fulfilling life.

For Exercises 67–69, write the converse, inverse, and contrapositive of the statement.

67. If gas prices go any higher, I will start riding my bike to work.
68. If I don't pass this class, my parents will kill me.
69. The festival will move inside the student center only if it rains.

Section 3-4

For Exercises 70–73, use truth tables to determine whether each argument is valid or invalid.

70. $p \rightarrow \sim q$
$\sim q \leftrightarrow \sim p$
$\therefore p$

71. $\sim q \vee p$
$p \wedge q$
$\therefore \sim q \leftrightarrow p$

72. $\sim p \vee q$
$q \vee \sim r$
$\therefore q \rightarrow (\sim p \wedge \sim r)$

73. $\sim r \rightarrow \sim p$
$\sim q \vee \sim r$
$\therefore p \leftrightarrow q$

For Exercises 74–77, write the argument in symbols; then use a truth table to determine if the argument is valid.

74. If I go to Barnes & Noble, then I will buy the latest Stephen King novel.
I did not buy the latest Stephen King novel.
∴ I did not go to Barnes and Noble.

75. I'm going to Wal-Mart and McDonald's.
If I go to McDonald's, I will get the sweet tea.
∴ I did not get sweet tea and go to Wal-Mart.

76. If we don't hire two more workers, the union will strike.
If the union strikes, our profit will not increase.
We hired two more workers.
∴ Our profit will not increase.

77. If I don't study, I won't make honor roll.
I made the honor roll.
∴ I didn't study or I cheated.

For Exercises 78–81, use the commonly used forms of arguments from Section 3-4 to determine if the argument is valid.

78. I will drink a mineral water or a Gatorade.
I drank a Gatorade.
∴ I did not drink mineral water.

79. If it is early, I will get tickets to the comedy club.
It is not early.
∴ I will not get tickets to the comedy club.

80. If I am cold, I will wear a sweater.
If I wear a sweater, I am warm.
∴ If I am cold, I am warm.

81. If pigs fly, then I'm a monkey's uncle.
Pigs fly or birds don't sing.
Birds do sing.
∴ I'm a monkey's uncle.

Section 3-5

For Exercises 82–86, use Euler circles to determine whether the argument is valid or invalid.

82. No *A* is *B*.
Some *B* is *C*.
∴ No *A* is *C*.

83. Some *A* is not *C*.
Some *B* is not *C*.
∴ Some *A* is not *B*.

84. All money is green.
All grass is green.
∴ Grass is money.

85. No humans have three eyes.
Some Martians have three eyes.
∴ No Martians are humans.

86. Some movies are rated "R."
All R-rated movies are inappropriate for children.
∴ Some movies are inappropriate for children.

Chapter Test

For Exercises 1–4, decide whether the sentence is a statement.

1. I'm going to the karaoke club.
2. 4 + 7 = 10
3. That woman is really smart.
4. Don't let it get to you.

For Exercises 5–8, write the negation of the statement.

5. The image is uploading to my online bio.
6. All men have goatees.
7. Some students ride a bike to school.
8. No short people can dunk a basketball.

For Exercises 9–14, let p = "It is warm." Let q = "It is sunny." Write each statement in symbols.

9. It is warm and sunny.
10. If it is sunny, then it is warm.
11. It is warm if and only if it is sunny.
12. It is warm or sunny.
13. It is false that it is not warm and sunny.
14. It is not sunny, and it is not warm.

For Exercises 15–19, let p = "It is sunny." Let q = "It is warm." Write each in words.

15. $p \vee \sim q$
16. $q \rightarrow p$
17. $p \leftrightarrow q$
18. $(p \vee q) \rightarrow p$
19. $\sim(\sim p \vee q)$

For Exercises 20–24, construct a truth table for each statement.

20. $p \rightarrow \sim q$
21. $(p \rightarrow \sim q) \wedge r$
22. $(p \wedge \sim q) \vee \sim r$
23. $(\sim q \vee p) \wedge p$
24. $p \rightarrow (\sim q \vee r)$

For Exercises 25–29, determine whether each statement is a tautology, self-contradiction, or neither.

25. $(p \wedge q) \wedge \sim p$
26. $(p \vee q) \rightarrow (p \rightarrow q)$
27. $(p \vee \sim q) \leftrightarrow (p \rightarrow \sim q)$
28. $q \wedge (p \vee \sim p)$
29. $\sim(p \wedge q) \vee p$

For Exercises 30–31, determine if the two statements are logically equivalent.

30. p; $\sim(\sim p)$
31. $(p \vee q) \wedge r$; $(p \wedge r) \vee (q \wedge r)$
32. Write the converse, inverse, and contrapositive for the statement "If I exercise regularly, then I will be healthy."

For Exercises 33–34 use De Morgan's laws to write the negation of the compound statement.

33. It is not cold and it is snowing.
34. I am hungry or thirsty.

For Exercises 35–38, use truth tables to determine the validity of each argument.

35. $p \leftrightarrow \sim q$
$\underline{\sim q \rightarrow \sim p}$
$\therefore p$

36. $p \rightarrow q$
$\underline{\sim q \vee \sim r}$
$\therefore q \leftrightarrow (\sim p \wedge \sim r)$

37. $\sim q \vee p$
$\underline{p \vee q}$
$\therefore \sim q \rightarrow p$

38. $\sim p \rightarrow \sim r$
$\underline{\sim r \vee \sim q}$
$\therefore q \leftrightarrow p$

For Exercises 39–41, determine if the argument is valid or invalid by using the given forms of valid arguments and fallacies.

39. If I finish my paper early, I will have my professor proofread it.
I have my professor proofread my paper.
∴ I finish my paper early.

40. The scratch-off ticket is a winner or a loser.
The scratch-off ticket is a winner.
∴ The scratch-off ticket is not a loser.

41. If Starbuck's is not too busy, I will study there.
If Starbuck's is too busy, I will study at the library.
∴ If I do not study at Starbuck's, I will study at the library.

For Exercises 42–45, use Euler circles to determine whether the argument is valid or invalid.

42. No *B* is *A*.
Some *A* is *C*.
∴ No *B* is *C*.

43. Some *C* is not *A*.
Some *B* is not *A*.
∴ Some *C* is not *B*.

44. No good relationship has bad moments.
Fred and Suzie have a good relationship.
∴ Fred and Suzie never have bad moments.

45. Some computer mice are wireless.
All computer devices that are wireless are expensive.
∴ Some computer mice are expensive.

Projects

1. Truth tables are related to Euler circles. Arguments in the form of Euler circles can be translated into statements by using the basic connectives and the negation as follows:

   Let p be "The object belongs to set A." Let q be "The object belongs to set B."

   All A is B is equivalent to $p \rightarrow q$.
   No A is B is equivalent to $p \rightarrow \sim q$.
   Some A is B is equivalent to $p \wedge q$.
   Some A is not B is equivalent to $p \wedge \sim q$.

   Determine the validity of the next arguments by using Euler circles; translate the statements into logical statements using the basic connectives; and using truth tables, determine the validity of the arguments. Compare your answers.

   (a) No *A* is *B*.
   Some *C* is *A*.
   ∴ Some *C* is not *B*.

   (b) All *B* is *A*.
   All *C* is *A*.
   ∴ All *C* is *B*.

2. Politicians argue in favor of positions all the time. An informed voter doesn't vote for a candidate because of the candidate's party, gender, race, or how good they look on TV—an informed voter listens to the candidates' positions and evaluates them.

   Do a Google search for the text of a speech by each of the main candidates in the 2008 Presidential election. Then find at least three logical arguments within the text, write the arguments in symbols, and use truth tables or commonly used argument forms to analyze the arguments, and see if they are valid.
3. Electric circuits are designed using truth tables. A circuit consists of switches. Two switches wired in *series* can be represented as $p \wedge q$. Two switches wired in *parallel* can be represented as $p \vee q$.

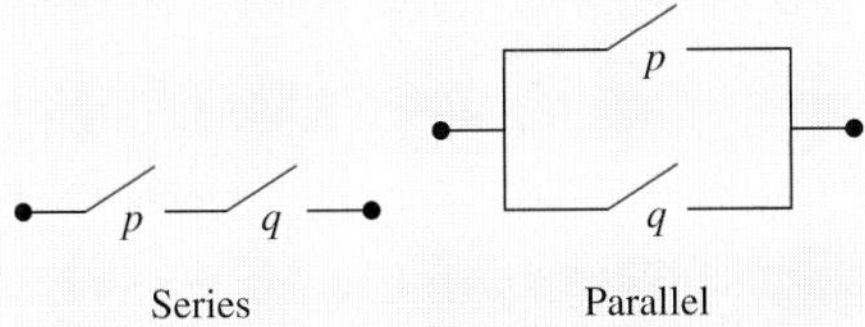

Series Parallel

In a series, circuit electricity will flow only when both switches p and q are closed. In a parallel circuit, electricity will flow when one or the other or both switches are closed. In a truth table, T represents a closed switch and F represents an open switch. So the truth table for $p \wedge q$ shows electricity flowing only when both switches are closed.

Truth table

p	q	$p \wedge q$
T	T	T
T	F	F
F	T	F
F	F	F

Circuit

p	q	$p \wedge q$
closed	closed	current
closed	open	no current
open	closed	no current
open	open	no current

Also, when switch p is closed, switch $\sim p$ will be open and vice versa, and p and $\sim p$ are different switches. Using this knowledge, design a circuit for a hall light that has switches at both ends of the hall so that the light can be turned on or off from either switch.

CHAPTER 4

Numeration Systems

Outline

4-1 Early and Modern Numeration Systems

4-2 Tools and Algorithms in Arithmetic

4-3 Base Number Systems

4-4 Operations in Base Number Systems

Summary

MATH IN Retail Sales

If you're like most college students, having endured 10 to 12 years of math class as you grew up, you probably feel like you know an awful lot about numbers. That may well be true if you're talking about the base 10 number system that you started learning even before you went to school. But you may be surprised to learn that there are many other number systems that can be used. Some of them are historical artifacts that give us a glimpse into the developmental stages of mankind's study of mathematics. But others are used commonly today. In fact, you're using one every single time you buy an item from a store, even if you don't realize it.

At first, it seems odd to many students to study different types of numbers, especially ancient number systems that are no longer in common use. But just as studying other languages gives you a deeper appreciation for the nuances of language itself, the study of other number systems provides deeper insight into the one we commonly use. The history of human thought is intimately tied to the development of language and mathematics, so studying early number systems gives us an interesting look at our species' intellectual past.

The first systems we will study use symbols to represent numbers that are completely different from the ones you're accustomed to. But we'll also study different systems that use familiar numerals like 0 and 1. One system is of particular interest in the modern world: the binary system, which uses only 0 and 1 as its digits.

The binary system is the mathematical driving force behind most of the world's computers, and is the system we had in mind when we referred to buying items. Unless you just got here from Mars, you're familiar with the UPC bar codes that tell the register at a store how much an item costs. What is the bar code really telling the register? A binary number that identifies every product for sale. The register is then programmed with the appropriate price for that item.

In Section 4-3, when we learn about the binary system, a Sidelight will help you to decode the familiar UPC codes. If all goes well, and you have a keen eye, you can use what you learn to find what products the bar codes below come from. When you know the 10-digit number that each represents, you can look that number up on the Internet to find that each code is from a product that most college students are very familiar with.

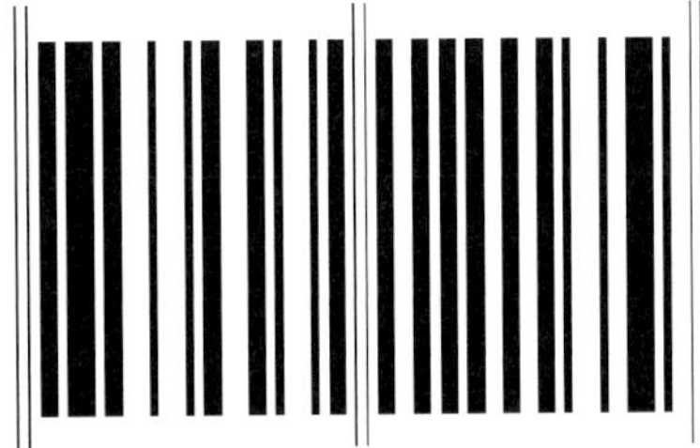

For answers, see Math in Retail Sales Revisited on page 193

Section 4-1 Early and Modern Numeration Systems

LEARNING OBJECTIVES

- ☐ 1. Define a numeration system.
- ☐ 2. Work with numbers in the Egyptian system.
- ☐ 3. Work with numbers in the Chinese system.
- ☐ 4. Identify place values in the Hindu-Arabic system.
- ☐ 5. Write Hindu-Arabic numbers in expanded notation.
- ☐ 6. Work with numbers in the Babylonian system.
- ☐ 7. Work with Roman numerals.

Are the words "number" and "numeral" synonymous? Most people answer yes to that question, but the answer is actually no. A **number** is a *concept,* or an idea, used to represent some quantity. A **numeral,** on the other hand, is a *symbol* used to represent a number. For example, there's only one concept of the number "five," but there are many different numerals (symbols) that can be used to represent the number five; 5, V, cinco, and 𝍸 are just a few.

Just as there are a wide variety of languages that developed in various parts of the world, there are a wide variety of systems for representing numbers that developed at different times and in different places. In this section we will study a few of them.

> A **numeration system** consists of a set of symbols (numerals) to represent numbers, and a set of rules for combining those symbols.

We will examine four basic types of numeration systems: tally, simple grouping, multiplicative grouping, and positional.

Tally Systems

☑ 1. Define a numeration system.

A tally system is the simplest kind of numeration system, and almost certainly the oldest. In a tally system there is only one symbol needed and a number is represented by repeating that symbol. For example, an ancient cave dweller might have drawn three stick-figure children on the wall of his cave to indicate that he had three children living there.

In modern times, tally systems are still used as a crude method of counting items. Most often, they are used to keep track of the number of occurrences of some event. For example, umpires in baseball often keep track of the runs scored by each team by making a mark for each run as it scores.

The most common symbol used in tally systems is |, which we call a stroke. Tallies are usually grouped by fives, with the fifth stroke crossing the first four, as in 𝍸.

EXAMPLE 1 Using a Tally System

An amateur golfer gets the opportunity to play with Tiger Woods, and, starstruck, his game completely falls apart. On the very first hole, it takes him six shots to reach the green, then three more to hole out. Use a tally system to represent his total number of shots on that hole.

SOLUTION

The total number of shots is nine, which we tally up as 𝍸 ||||.

▼ Try This One 1

During the unfortunate experience in Example 1, 10 people watched from the tee and 8 more watched from around the green. Use a tally system to represent the total number of people who watched.

The most obvious disadvantage to tally systems is that they make it tremendously cumbersome to represent large numbers. That's where grouping systems have an advantage.

Simple Grouping Systems

In a simple grouping system there are symbols that represent select numbers. Often, these numbers are powers of 10. To write a number in a simple grouping system, repeat the symbol representing the appropriate value(s) until the desired quantity is reached. For example, suppose in a simple grouping system the symbol Δ represents the quantity "ten" and the symbol Γ represents the quantity "one." Then to write the numeral representing the quantity "fifty-three" in this system we would use five Δ's and three Γ's as follows: ΔΔΔΔΔΓΓΓ.

The Egyptian Numeration System

One of the earliest formal numeration systems was developed by the Egyptians sometime prior to 3000 BCE. It used a system of hieroglyphics using pictures to represent numbers. These symbols are shown in Figure 4-1.

Symbol	Number	Description	Symbol	Number	Description
𓏺	1	Vertical staff	𓂭	10,000	Pointing finger
𓎆	10	Heel bone	𓆐	100,000	Burbot fish (or tadpole)
𓍢	100	Scroll	𓁨	1,000,000	Astonished person
𓆼	1,000	Lotus flower			

Figure 4-1

Sidelight MAYAN MATHEMATICS

The Mayans lived in Southeastern Mexico and the Central American countries of Guatemala, Honduras, and El Salvador. Their civilization lasted from 2000 BCE until 1700 CE, although some archaeological finds date it to earlier times. They built many temples, most on the Yucatan peninsula, similar to the pyramids of Egypt. They had a highly developed civilization that contained the elements of religion, trade, government, mathematics, and astronomy. They developed an advanced form of writing consisting of symbols similar to Egyptian hieroglyphics. They made paper from fig tree bark and wrote books that contained astronomical tables and religious ceremonies. They had two calendars. One consisted of 260 days and was used for religious purposes. The other consisted of 365 days and was based on the orbit of the earth about the sun. This calendar divided the year into 18 months with 20 days in each month and 5 days were added at the end of the year.

Their mathematical symbols consisted of three symbols. A dot represented a one. A horizontal line represented a five, and the symbol 𝋠 was used for a zero. (This was one of the first cultures to have a specific symbol for zero.)

It is interesting to examine how these symbols were used to create numbers. For example, the number 3 was written as • • •. The number 12 was written 𝋬. The numbers were written vertically, except for the ones. Using the two symbols • and —, the Mayans could write all the numbers up through 19. However, once the number 20 was reached, a space would be needed between the digits. For example, the number

𝋧

𝋮

represents 14 ones plus 7 twenties or $14 \times 1 + 7 \times 20 = 154$.

The next place value after 20 is 360, which is 18×20. This value was used because the calendar had $360 + 5$ days. The numeral

𝋭 Three hundred sixties

𝋪 Twenties

𝋩 Ones

represents $9 \times 1 + 10 \times 20 + 13 \times 360 = 4{,}889$.

The next higher place value is 18×20^2 or 7,200. The symbol for zero was used to indicate that there were no digits in a place value position of a number. For example, the number

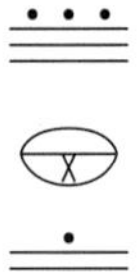

means $11 \times 1 + 0 \times 20 + 17 \times 360 = 6{,}131$. The Mayan system combines elements of two types studied in this section: grouping and positional systems.

The Egyptian system is a simple grouping system: the value of any numeral is determined by counting up the number of each symbol and multiplying the number of occurrences by the corresponding value from Figure 4-1. Then the numbers for each symbol are added, as we see in Example 2.

EXAMPLE 2 Using the Egyptian Numeration System

Find the numerical value of each Egyptian numeral.

(a)

(b)

(c)

SOLUTION

(a) The number has 4 heel bones (or 4 tens) and 3 vertical staffs (or 3 ones); this equals $4 \times 10 + 3 \times 1$ or 43.

(b) The number consists of 3 one hundred thousands, 3 ten thousands, 2 hundreds, 3 tens, and 6 ones; it equals $3 \times 100{,}000 + 3 \times 10{,}000 + 2 \times 100 + 3 \times 10 + 6 \times 1 = 300{,}000 + 30{,}000 + 200 + 30 + 6 = 330{,}236$.

(c) The number consists of 1 million, 2 ten thousands, 2 thousands, 1 hundred, 1 ten, and 3 ones; the number is $1{,}000{,}000 + 20{,}000 + 2{,}000 + 100 + 10 + 3 = 1{,}022{,}113$.

▼ Try This One 2

Find the numerical value of each Egyptian numeral.

(a)

(b)

(c)

In order to write numbers using hieroglyphic symbols, simple groupings of ones, tens, hundreds, etc. are used. For example, 28 is equal to $10 + 10 + 8$ and is equal to

EXAMPLE 3 Writing Numbers in Egyptian Notation

Write each number as an Egyptian numeral.

(a) 42 (b) 137 (c) 5,283 (d) 3,200,419

SOLUTION

(a) Forty-two can be written as $4 \times 10 + 2 \times 1$, so it consists of four tens and two ones. We would write it using four of the tens symbol (the heel bone) and two of the ones symbol (the vertical staff).

(b) Since 137 consists of 1 hundred, 3 tens, and 7 ones, it is written as

𓍢∩∩∩|||||||

(c) Since 5,283 consists of 5 thousands, 2 hundreds, 8 tens, and 3 ones, it is written as

𓆼𓆼𓆼𓆼𓆼𓍢𓍢∩∩∩∩∩∩∩∩|||

(d) Since 3,200,419 consists of 3 millions, 2 hundred thousands, 4 one hundreds, 1 ten, and 9 ones, it is written as

𓁨𓁨𓁨𓆐𓆐𓍢𓍢𓍢𓍢∩|||||||||

▼ Try This One 3

Write each number as an Egyptian numeral.

(a) 43 (b) 627 (c) 3,286

Addition and subtraction can be performed by grouping symbols, as shown in Examples 4 and 5.

EXAMPLE 4 Adding in the Egyptian System

Find the sum of 𓍢𓍢∩∩∩∩∩|||||| + 𓍢∩∩∩∩∩∩∩|||||.

SOLUTION

The answer is found by taking the total number of each symbol and converting the appropriate symbols. The total number of symbols is

𓍢𓍢𓍢 ∩∩∩∩∩∩∩∩∩∩ ∩∩ |||||||||| |

(the ten ∩ are grouped into 𓍢; the ten | are grouped into ∩)

For each 10 heel bones, replace them with a scroll and for each 10 vertical staffs, replace them with a heel bone. The final answer is

𓍢𓍢𓍢𓍢∩∩∩|

Math Note

One of the objectives of this section is to expand your numerical ability by performing calculations using numerical systems other than our own. So in problems like Examples 4 and 5, you should be using these systems in your solution. But don't forget that you can check your answer by converting the original problem into our system, performing the calculation, and seeing if it matches your result.

▼ Try This One 4

Perform the addition in Egyptian notation.

𓍢𓍢∩∩∩∩∩∩|||||||| + 𓍢𓍢𓍢𓍢∩∩∩||||

EXAMPLE 5 Subtracting in the Egyptian System

Subtract

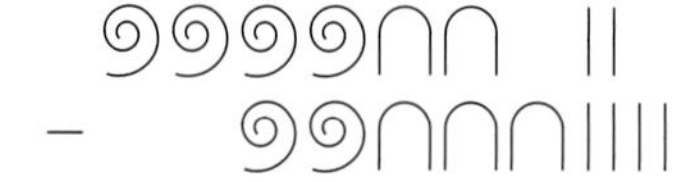

SOLUTION

In this case, we're going to have to do some rewriting before we subtract since there are more heel bones and vertical staffs in the number being subtracted. In the top number, we can convert one heel bone (10) into 10 vertical staffs, and one scroll (100) into 10 heel bones. Once this is done, the number of symbols can be subtracted as shown below, with the answer on the bottom line. You might find it helpful to cross out matching symbols in both lines.

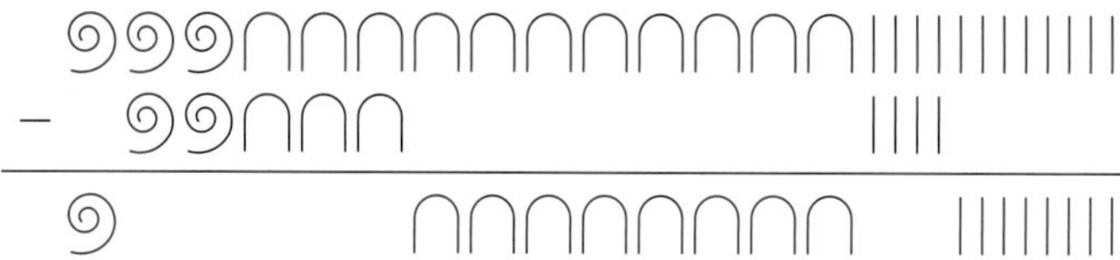

▼ Try This One 5

☑ 2. Work with numbers in the Egyptian system.

Perform the subtraction in Egyptian notation.

Multiplicative Grouping Systems

In a multiplicative grouping system, there is a symbol for each value 1 through 9 (the multipliers), and also for select other numbers (usually powers of 10 or some other common base). To write a number in a multiplicative grouping system, a multiplier is followed by the symbol representing the value of the appropriate power of 10.

For example, to write the number 53 in a multiplicative grouping system there would be two "groups" of two. The first group would consist of the multiplier representing the number five, followed by the symbol for ten ($5 \times 10 = 50$); the second group would be the multiplier representing the number three, followed by the symbol for one ($3 \times 1 = 3$). (*Note:* the symbol for one is sometimes omitted in a multiplicative system.)

EXAMPLE 6 Using a Multiplicative Grouping System

Suppose the symbols used in a multiplicative grouping system are as follows:

one	α	six	θ
two	β	seven	γ
three	χ	eight	η
four	δ	nine	ι
five	ε	ten	φ

Write the symbols that would be used to represent the number 45.

Symbol	Value
零 or 〇	0
一	1
二	2
三	3
四	4
五	5
六	6
七	7
八	8
九	9
十	10
百	100
千	1,000

Figure 4-2

SOLUTION

Forty-five consists of four 10s and five 1s. To represent four 10s, we write δφ (the multiplier 4 times the base value 10). To represent five 1s, we write εα (the multiplier 5 times the base value 1), or we could simply write ε. So the number 45 is written δφ εα, or δφ ε.

▼ Try This One 6

Using the symbols in Example 6, write the symbols that would be used to represent the number 96.

The traditional Chinese numeration system is an example of a multiplicative grouping system.

The Chinese Numeration System

The symbols used for the Chinese numeration system are shown in Figure 4-2. Because Chinese is written vertically rather than horizontally, their numbers are also represented vertically. Fifty-three would be written:

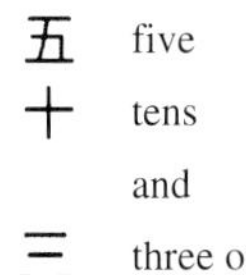

EXAMPLE 7 Using the Chinese Numeration System

Find the value of each Chinese numeral.

(a)
六
百
五
十
四

(b)
三
千
七
百
二
十
六

(c)
五
千
六
十
五

SOLUTION

Reading from the top down, we can calculate each value as below. Remember that in each group of symbols, the multiplier comes first, followed by the power of 10.

(a)
六百 } $6 \times 100 = 600$
+
五十 } $5 \times 10 = 50$
+
四 } $4 = 4$
654

(b)
三千 } $3 \times 1{,}000 = 3{,}000$
+
七百 } $7 \times 100 = 700$
+
二十 } $2 \times 10 = 20$
+
六 } $6 = 6$
3,726

(c)
五千 } $5 \times 1{,}000 = 5{,}000$
+
六十 } $6 \times 10 = 60$
+
五 } $5 = 5$
5,065

▼ Try This One 7

Find the value of each Chinese numeral.

(a) 四
百
二
十
七

(b) 六
千
七
十
五

(c) 二
十
六

EXAMPLE 8 Writing Numbers in the Chinese Numeration System

Write each number as a Chinese numeral.

(a) 65 (b) 183 (c) 8,749

SOLUTION

(a)
六 } 6 × 10
十
五 } 5

(b)
一 } 1 × 100
百
八 } 8 × 10
十
三 } 3

(c)
八 } 8 × 1,000
千
七 } 7 × 100
百
四 } 4 × 10
十
九 } 9

▼ Try This One 8

☑ 3. Work with numbers in the Chinese system.

Write each number as a Chinese numeral.

(a) 45 (b) 256 (c) 6,321

Positional Systems

In a **positional system** no multiplier is needed. The value of the symbol is understood by its position in the number. To represent a number in a positional system you simply put the numeral in an appropriate place in the number, and its value is determined by its location.

The Hindu-Arabic Numeration System

Many of the numeration systems we study in this chapter rely heavily on exponents, so a clear understanding of exponents is important.

For any number b and natural number n, we define the **exponential expression** b^n as

$$b^n = b \cdot b \cdot b \cdots b$$

where b appears as a factor n times. The number b is called the **base,** and n is called the **exponent.** We also define $b^1 = b$ for any base b, and $b^0 = 1$ for any nonzero base b.

The numeration system we use today is called the Hindu-Arabic system. (See the sidelight on page 159 for some perspective on this name.) It uses 10 symbols called **digits:** 0, 1, 2, 3, 4, 5, 6, 7, 8, and 9.

The Hindu-Arabic system is a positional system since the position of each digit indicates a specific value. The place value of each number is given as

billion	hundred million	ten million	million	hundred thousand	ten thousand	thousand	hundred	ten	one
10^9	10^8	10^7	10^6	10^5	10^4	10^3	10^2	10^1	1

The number 82,653 means there are 8 ten thousands, 2 thousands, 6 hundreds, 5 tens, and 3 ones. We say that the place value of the 6 in this numeral is hundreds.

EXAMPLE 9 Finding Place Values

In the number 153,946, what is the place value of each digit?

(a) 9 (b) 3 (c) 5 (d) 1 (e) 6

SOLUTION

(a) hundreds
(b) thousands
(c) ten thousands
(d) hundred thousands
(e) ones

▼ Try This One 9

4. Identify place values in the Hindu-Arabic system.

According to the U.S. Census Bureau, there were 8,724,560 people in New Jersey in 2006. What are the place values of the digits 5, 6, 7, and 8?

To clarify the place values, Hindu-Arabic numbers are sometimes written in **expanded notation.** An example, using the numeral 32,569, is shown below.

$$\begin{aligned} 32{,}569 &= 30{,}000 + 2{,}000 + 500 + 60 + 9 \\ &= 3 \times 10{,}000 + 2 \times 1{,}000 + 5 \times 100 + 6 \times 10 + 9 \\ &= 3 \times 10^4 + 2 \times 10^3 + 5 \times 10^2 + 6 \times 10^1 + 9 \end{aligned}$$

Since all of the place values in the Hindu-Arabic system correspond to powers of 10, the system is known as a **base 10 system.** We will study base number systems in depth in Section 4-3.

EXAMPLE 10 Writing a Base 10 Number in Expanded Notation

Write 9,034,761 in expanded notation.

SOLUTION

9,034,761 can be written as

$$\begin{aligned} & 9{,}000{,}000 + 30{,}000 + 4{,}000 + 700 + 60 + 1 \\ &= 9 \times 1{,}000{,}000 + 3 \times 10{,}000 + 4 \times 1{,}000 + 7 \times 100 + 6 \times 10 + 1 \\ &= 9 \times 10^6 + 3 \times 10^4 + 4 \times 10^3 + 7 \times 10^2 + 6 \times 10^1 + 1. \end{aligned}$$

▼ Try This One 10

☑ 5. Write Hindu-Arabic numbers in expanded notation.

Write each number in expanded notation.

(a) 573 (b) 86,471 (c) 2,201,567

So far, we have studied four types of numeration systems, but they are not the only ones that exist. The ancient Babylonian system is sort of a cross between a multiplier system and a positional system. The Roman numeration system, which is still in use today, is basically a grouping system with a twist—the use of subtraction. We'll conclude this section by studying these two systems. But remember that there are many other numeration systems that we haven't discussed.

The Babylonian Numeration System

The Babylonians had a numerical system consisting of two symbols. They are < and ▼. (These wedge-shaped symbols are known as "cuneiform.") The < represents the number of 10s, and ▼ represents the number of 1s.

EXAMPLE 11 Using the Babylonian Numeration System

What number does <<< ▼▼▼▼▼▼ represent?

SOLUTION

Since there are 3 tens and 6 ones, the number represents 36.

▼ Try This One 11

What number does <<<<< ▼▼ represent?

You might think it would be cumbersome to write large numbers in this system; however, the Babylonian system was also positional. Numbers from 1 to 59 were written using the two symbols shown in Example 11, but after the number 60, a space was left between the groups of numbers. For example, the number 2,538 was written as

and means that there are 42 sixties and 18 ones. The space separates the 60s from the ones. The value is found as follows:

$$\begin{array}{r} 42 \times 60 = 2{,}520 \\ +\ 18 \times \ \ 1 = \quad 18 \\ \hline 2{,}538 \end{array}$$

When there are three groupings of numbers, the symbols to the left of the first space represent the number of 3,600s (note that $3{,}600 = 60 \times 60$). The next group of symbols represents the number of 60s, and the final group represents the number of 1s.

A Babylonian clay tablet with numeric symbols.

EXAMPLE 12 Using the Babylonian Numeration System

Write the numbers represented.

(a) 𒌋𒌋𒌋𒌋𒌋𒁹𒁹 𒌋𒌋𒌋𒁹𒁹𒁹𒁹

(b) 𒌋𒁹𒁹 𒌋𒌋𒌋𒌋𒌋𒁹 𒌋𒌋𒁹𒁹𒁹

SOLUTION

(a) There are 52 sixties and 34 ones; so the number represents

$$\begin{array}{r} 52 \times 60 = 3{,}120 \\ +\ 34 \times \ \ 1 = \quad 34 \\ \hline 3{,}154 \end{array}$$

(b) There are twelve 3,600s, fifty-one 60s, and twenty-three 1s. The numeral represents

$$\begin{array}{r} 12 \times 3{,}600 = 43{,}200 \\ 51 \times \quad 60 = \ \ 3{,}060 \\ 23 \times \quad\ \ 1 = \quad\ \ 23 \\ \hline 46{,}283 \end{array}$$

▼ Try This One 12

Write the number represented.

(a) <<▼▼ <<<<<▼▼▼▼▼ (b) <▼ <<▼ <▼▼▼

EXAMPLE 13 Writing a Number in the Babylonian System

Write 5,217 using the Babylonian numeration system.

SOLUTION

Since the number is greater than 3,600, it must be divided by 3,600 to see how many 3,600s are contained in the number.

$$5{,}217 \div 3{,}600 = 1 \text{ remainder } 1{,}617$$

The remainder, 1,617, is then divided by 60 to see how many 60s are in 1,617.

$$1{,}617 \div 60 = 26 \text{ remainder } 57$$

So, the number 5,217 consists of

$1 \times 3{,}600 = 3{,}600$	▼
$26 \times 60 = 1{,}560$	<<▼▼▼▼▼▼
$57 \times 1 = 57$	<<<<<▼▼▼▼▼▼▼
Total = 5,217	

It can be written as

▼ <<▼▼▼▼▼▼ <<<<<▼▼▼▼▼▼▼

Math Note

The Babylonians didn't have a symbol for zero. This complicated their writings. For example, how is the number 7,200 distinguished from the number 72?

▼ Try This One 13

☑ 6. Work with numbers in the Babylonian system.

Write each number using the Babylonian numeration system.

(a) 42 (b) 384 (c) 4,278

The Roman Numeration System

The Romans used letters to represent their numbers. They are

Symbol	Number
I	1
V	5
X	10
L	50
C	100
D	500
M	1,000

The Roman system is similar to a simple grouping system, but to save space, the Romans also used the concept of subtraction. For example, 8 is written as VIII, but 9 is written as IX, meaning that 1 is subtracted from 10 to get 9. There are three rules for writing numbers in Roman numerals:

1. When a letter is repeated in sequence, its numerical value is added. For example, XXX represents 10 + 10 + 10, or 30.
2. When smaller-value letters follow larger-value letters, the numerical values of each are added. For example, LXVI represents 50 + 10 + 5 + 1, or 66.
3. When a smaller-value letter precedes a larger-value letter, the smaller value is subtracted from the larger value. For example, IV represents 5 − 1, or 4, and XC represents 100 − 10, or 90.

In addition, I can only precede V or X, X can only precede L or C, and C can only precede D or M. Then 4 is written as IV, 9 is written as IX, 40 is written as XL, 90 is written XC, 400 is written as CD, and 900 is written as CM.

Example 14 shows how to convert Roman numerals to Hindu-Arabic numerals.

EXAMPLE 14 Using Roman Numerals

Roman numerals are still in use today. For example, many clocks and watches contain Roman numerals. Can you think of other places Roman numerals are still used?

Find the value of each Roman numeral.

(a) LXVIII (b) XCIV (c) MCML (d) CCCXLVI (e) DCCCLV

SOLUTION

(a) L = 50, X = 10, V = 5, and III = 3; so LXVIII = 68.
(b) XC = 90 and IV = 4; so XCIV = 94.
(c) M = 1,000, CM = 900, L = 50; so MCML = 1,950.
(d) CCC = 300, XL = 40, V = 5, and I = 1; so CCCXLVI = 346.
(e) D = 500, CCC = 300, L = 50, V = 5; so DCCCLV = 855.

▼ Try This One 14

Convert each Roman numeral to a Hindu-Arabic numeral.

(a) XXXIX (b) MCLXIV (c) CCCXXXIII

Sidelight ROMAN AND HINDU-ARABIC NUMERALS

The Romans spread their system of numerals throughout the world as they conquered their enemies. The system was well entrenched in Europe until the 1500s, when our present system, called the Hindu-Arabic system, became widely accepted.

The present system is thought to have been invented by the Hindus before 200 BCE. It was spread throughout Europe by the Arabs, who traded with the Europeans and traveled throughout the Mediterranean region. It is interesting to note that for about 400 years, the mathematicians of early Europe were divided into two groups—those favoring the use of the Roman system and those favoring the use of the Hindu-Arabic system. The Hindu-Arabic system eventually won out, although Roman numerals are still widely used today.

Numbers can be written using Roman numerals as shown in Example 15.

EXAMPLE 15 Writing Numbers Using Roman Numerals

Math Note

For larger numbers, the Romans placed a bar over their symbols. The bar means to multiply the numerical value of the number under the bar by 1,000. For example, $\overline{\text{VII}}$ means 7 × 1,000 or 7,000, and $\overline{\text{XL}}$ means 40,000.

Write each number using Roman numerals.

(a) 19 (b) 238 (c) 1,999 (d) 840 (e) 72

SOLUTION

(a) 19 is written as 10 + 9 or XIX.
(b) 238 is written as 200 + 30 + 8 or CCXXXVIII.
(c) 1,999 is written as 1,000 + 900 + 90 + 9 or MCMXCIX.
(d) 840 is written as 500 + 300 + 40 or DCCCXL.
(e) 72 is written as 50 + 20 + 2 or LXXII.

☑ 7. Work with Roman numerals.

▼ Try This One 15

Write each number using Roman numerals.

(a) 67 (b) 192 (c) 202 (d) 960

Remember that the value of studying other number systems is that it allows us to develop a deeper understanding of the symbols we use to represent numbers, even if we always use our own system to represent numbers and perform calculations.

Answers to Try This One

1 卌 卌 卌 |||

2 (a) 456;
(b) 1,211;
(c) 1,102,041

3 (a) ∩∩∩∩|||
(b) 𓍢𓍢𓍢𓍢𓍢𓍢∩∩|||||||
(c) 𓆼𓆼𓆼𓍢𓍢∩∩∩∩∩∩∩∩||||||

4 𓍢𓍢𓍢𓍢𓍢𓍢𓍢∩||

5 𓍢𓍢𓍢𓍢𓍢𓍢𓍢∩∩∩∩∩∩|||||||||

6 ιφ θ

7 (a) 427
(b) 6,075
(c) 26

8 (a) 四十五 (b) 二百五十六 (c) 六千三百二十一

9 5: hundreds; 6: tens; 7: hundred thousands; 8: millions

10 (a) $5 \times 10^2 + 7 \times 10^1 + 3$
(b) $8 \times 10^4 + 6 \times 10^3 + 4 \times 10^2 + 7 \times 10^1 + 1$
(c) $2 \times 10^6 + 2 \times 10^5 + 1 \times 10^3 + 5 \times 10^2 + 6 \times 10^1 + 7$

11 52

12 (a) 1,375; (b) 40,873

13 (a) 𒌋𒌋𒌋𒌋𒁹𒁹
(b) 𒁹𒁹𒁹𒁹𒁹𒁹   𒌋𒌋𒁹𒁹𒁹𒁹
(c) 𒁹   𒌋𒁹   𒌋𒁹𒁹𒁹𒁹𒁹𒁹𒁹𒁹

14 (a) 39; (b) 1,164; (c) 333

15 (a) LXVII; (b) CXCII; (c) CCII; (d) CMLX

EXERCISE SET 4-1

Writing Exercises

1. Describe the difference between a number and a numeral.
2. Briefly describe how a grouping system works.
3. Briefly describe how a multiplicative grouping system works.
4. Describe what place values are and what they represent in the Hindu-Arabic numeration system.
5. Both the Roman and Egyptian systems use symbols to represent certain numbers. Explain how they differ (aside from the fact that they use different symbols).
6. Describe how to write a number in the Hindu-Arabic system in expanded notation.

Computational Exercises

For Exercises 7–16, write each number using Hindu-Arabic numerals.

7. ∩∩∩|||||
8. 𓍢𓍢𓍢𓍢∩∩∩||
9. 𓆼𓆼𓍢𓍢∩∩|||||
10. 𓆐𓆐𓍢∩∩∩||
11. 𓆼𓆼𓆼𓍢∩∩∩∩∩∩∩|||
12. 𓁨𓍢𓍢∩
13. 𓆼𓆼𓍢𓍢𓍢∩||||
14. 𓁨𓁨𓆐𓆐
15. 𓁨𓆐𓆼𓂭𓂭∩
16. 𓂭𓂭∩∩∩∩||||||||

For Exercises 17–26, write each number using Egyptian numerals.

17. 7
18. 18
19. 37
20. 52
21. 168
22. 365
23. 801
24. 955
25. 1,256
26. 8,261

For Exercises 27–32, perform the indicated operations. Write your answers as Egyptian numerals.

27. ∩∩∩∩∩||| + ∩∩∩∩∩∩∩|||
28. 𓍢𓍢∩∩|| + 𓍢∩∩∩||
29. 𓆐𓆐𓆼𓆼|| + 𓆐𓆼𓆼𓆼∩∩∩|||
30. ∩∩∩|| − ∩||||
31. 𓆼𓆼𓍢𓍢||| − 𓆼𓍢𓍢𓍢||||
32. 𓁨𓆐𓂭∩∩||| − 𓆐𓂭𓂭∩∩∩|||||

For Exercises 33–38, write each number using Hindu-Arabic numerals.

33. 一百八十九
34. 三千四百七
35. 五十二
36. 九千八百三十四
37. 七百一十三
38. 八十九

For Exercises 39–44, write each number using Chinese numerals.

39. 89
40. 567
41. 284
42. 9,857
43. 2,356
44. 21

For Exercises 45–50, use the number 3,421,578 and find the place value of the given digit.

45. 5
46. 1
47. 2
48. 3
49. 8
50. 4

For Exercises 51–60, write each number in expanded notation.

51. 86
52. 325
53. 1,812
54. 32,714
55. 6,002
56. 29,300
57. 162,873
58. 200,321,416
59. 17,531,801
60. 1,326,419

For Exercises 61–70, write each number using Hindu-Arabic numerals.

61. <▼▼
62. <<<▼▼▼▼
63. <<<<<▼
64. <<<▼▼▼▼▼
65. <▼ <<▼ <▼
66. <<▼▼ <<<< <▼▼▼
67. <<< <<▼ <<▼▼▼▼
68. <<<▼▼ <▼ <▼
69. <▼▼▼ <▼▼
70. <<<▼ <▼ ▼▼

For Exercises 71–80, write each number using Babylonian numerals.

71. 32
72. 23
73. 78
74. 156
75. 292
76. 514
77. 1,023
78. 1,776
79. 5,216
80. 8,200

For Exercises 81–90, write each number using Hindu-Arabic numerals.

81. XVII
82. XCIX
83. XLIII
84. CCXXI
85. LXXXVI
86. CCXXXIII
87. CDXVIII
88. MMCMXVII
89. CDXC
90. CMVI

For Exercises 91–100, write each number using Roman numerals.

91. 39
92. 142
93. 567
94. 893
95. 1,258
96. 3,720
97. 1,462
98. 2,170
99. 3,000
100. 2,222

Real-World Applications

Most movies use Roman numerals in the credits to indicate the date the film was made. Shown are some movies and their dates. Find the year the movie was made.

Movie	Year
101. Gone With the Wind	MCMXXXIX
102. Casablanca	MCMXLII
103. Animal House	MCMLXXVIII
104. Raiders of the Lost Ark	MCMLXXXI
105. Batman Begins	MMV
106. Shrek II	MMIV

Critical Thinking

107. Make up your own numeration system using your own symbols. Indicate whether it is a simple grouping, multiplicative grouping, or a positional system. Explain how to add and subtract in your numeration system.
108. Most clocks that use Roman numerals have four written incorrectly, as IIII rather than IV. Think of as many potential reasons for this as you can, then do an internet search to see if you can find the reason.
109. A colleague of mine once gave a quiz with the question "Why is it useful to learn about the Babylonian numeration system?," and one student answered "If you have any Babylonian friends, you could communicate with them about numbers." Explain why we laughed hysterically at that answer.
110. Which of the ancient numeration systems we studied (Egyptian, Babylonian, Roman) do you think is the most efficient? Why?

Section 4-2 Tools and Algorithms in Arithmetic

LEARNING OBJECTIVES

- ☐ 1. Multiply using the Egyptian algorithm.
- ☐ 2. Multiply using the Russian peasant method.
- ☐ 3. Multiply using the lattice method.
- ☐ 4. Multiply using Napier's bones.

Handheld calculators have been widely available since the early 1970s, which means that for most college students, it seems like they've always existed. Compared to your lifespan, 40 years seems like a long time, but compared to the history of mathematics, it's practically the blink of an eye. In order to have a greater appreciation of the modern tools we have, it's useful to look back at what people had to do before calculators and computers were invented. In this section, we'll look at a handful of extremely clever methods that were developed to perform multiplications. While you'll probably never use them in real life, studying them can give you insight into the type of innovations that helped advance human thought from the simple to the abstract, paving the way for all of the modern advances that we too often take for granted.

The Egyptian Algorithm

The Egyptian algorithm is an ancient method of multiplication that can be done by hand because it requires only doubling numbers and addition. We'll illustrate it with an example, then summarize.

EXAMPLE 1 Using the Egyptian Algorithm

Use the Egyptian algorithm to multiply 13×24.

SOLUTION

Step 1 Form two columns with 1 at the top of the first column and 24 at the top of the second column:

1	24

Step 2 Double the numbers in each column, and continue to do so until the first column contains numbers that can be added to get the other number in the product, 13:

1	24
2	48
4	96
8	192

We stop here because we can get 13 from adding 1, 4, and 8.

Step 3 Add the numbers in the second column that are next to 1, 4, and 8: $24 + 96 + 192 = 312$. This is the product of 13 and 24.

▼ Try This One 1

Use the Egyptian algorithm to multiply 22×15.

We could have put either original number at the top of the second column, but it's usually a little quicker if we put the larger number there, as we did in Example 1.

Sidelight DIVIDING WITH THE EGYPTIAN ALGORITHM

The Egyptian algorithm can be used for dividing as well. Suppose, for example, we want to divide 1,584 by 24. Make two columns headed with 1 and 24 and double as before:

1	24
2	48
4	96
8	192
16	384
32	768
64	1,536

We can stop here since the next entry would be larger than 1,584. Now find the numbers in the right column that add up to 1,584 (in this case 48 and 1,536), then add the corresponding numbers in the left column. The sum $2 + 64 = 66$ is the quotient of 1,584 and 24.

This method will not work unless the numbers divide evenly, but a modified Egyptian method will work in such cases.

☑ 1. Multiply using the Egyptian algorithm.

The Egyptian Algorithm

To multiply two numbers A and B:

1. Form two columns with the numeral one at the top of the first, and one of the numbers (we'll say B) to be multiplied at the top of the second.
2. Double the numbers in each column repeatedly until the first column contains numbers that can be added to A.
3. Add the numbers in the second column that are next to the numbers in the first column that add to A. This sum is the product of A and B.

The Russian Peasant Method

Another method for multiplying by hand is known as the Russian peasant method. As you will see from Example 2, it's similar to the Egyptian algorithm, but maybe a bit simpler in that you don't have to keep searching for numbers that add to one of the factors.

EXAMPLE 2 Using the Russian Peasant Method

Use the Russian peasant method to multiply 24×15.

SOLUTION

Step 1 Form two columns with 24 and 15 at the top.

24 15

Step 2 Divide the numbers in the first column by two (ignoring remainders), and double the numbers in the second column, until you reach one in the first column.

24	15
12	30
6	60
3	120
1	240

Step 3 Add the numbers in the second column that are next to odd numbers in the first column: $120 + 240 = 360$. This is the product of 24 and 15.

▼ Try This One 2

Use the Russian peasant method to multiply 18 × 12.

The Russian Peasant Method

To multiply two numbers A and B:

1. Form two columns with A at the top of one column and B at the top of the other.
2. Divide the numbers in the first column by two repeatedly, ignoring remainders, until you reach one. Double the numbers in the second column, with the last result next to the one in the first column.
3. Add the numbers in the second column that are next to odd numbers. The result is the product of A and B.

The Russian peasant method can be used to multiply numbers with more digits, as in Example 3.

EXAMPLE 3 Using the Russian Peasant Method

Use the Russian peasant method to multiply 103 × 19.

SOLUTION

Form the columns as described in the colored box above:

103	19
51	38
25	76
12	152
6	304
3	608
1	1,216

Now add the numbers in the second column that are next to odd numbers in the first:

$$19 + 38 + 76 + 608 + 1{,}216 = 1{,}957.$$

So 103 × 19 = 1,957.

▼ Try This One 3

☑ 2. Multiply using the Russian peasant method.

Use the Russian peasant method to multiply 210 × 21.

The Lattice Method

The lattice method for multiplication was used in both India and Persia as early as the year 1010. It was later introduced in Europe in 1202 by Leonardo of Pisa (more commonly known as Fibonacci) in his work entitled *Liber Abacii* (Book of the Abacus). The lattice method reduces multiplying large numbers into multiplying single digit numbers, as illustrated in the next two examples.

EXAMPLE 4 Using the Lattice Method

Find the product 36×568 using the lattice method.

SOLUTION

Step 1 Form a lattice as illustrated with one of the numbers to be multiplied across the top, and the other written vertically along the right side.

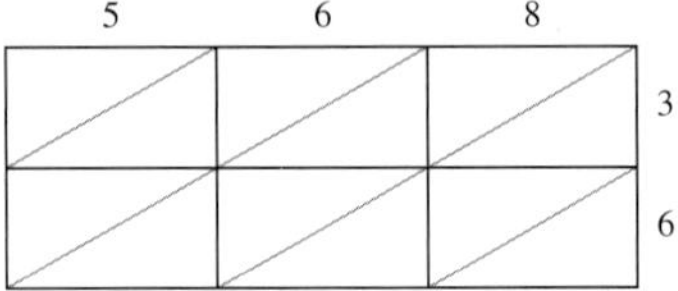

Step 2 Within each box, write the product of the numbers from the top and side that are above and next to that box. Write the first digit above the diagonal and the second below it, using zero as first digit if necessary.

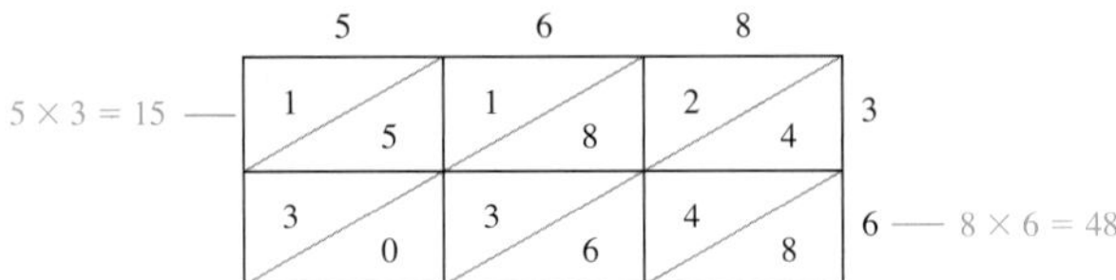

Step 3 Starting at the bottom right of the lattice, add the numbers along successive diagonals, working toward the left. If the sum along a diagonal is more than 9, write the last digit of the sum and carry the first digit to the addition along the next diagonal.

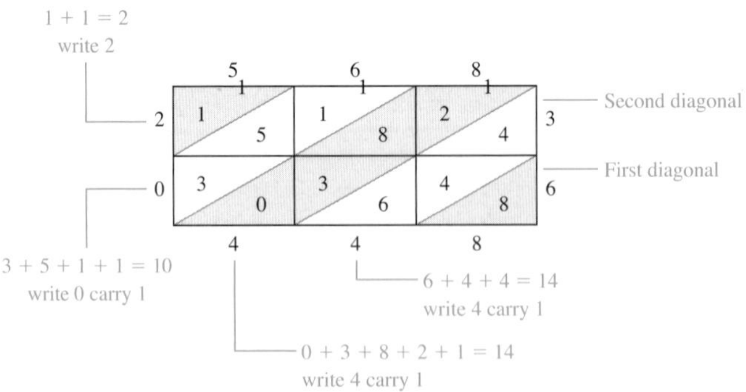

Step 4 Read the answer, starting down the left side then across the bottom:

$$36 \times 568 = 20{,}448$$

▼ Try This One 4

Find the product 53×844 using the lattice method.

EXAMPLE 5 Using the Lattice Method

Find the product 2,356 × 547 using the lattice method.

SOLUTION

Step 1 Form a lattice with one of the numbers to be multiplied across the top and the other one down the right side.

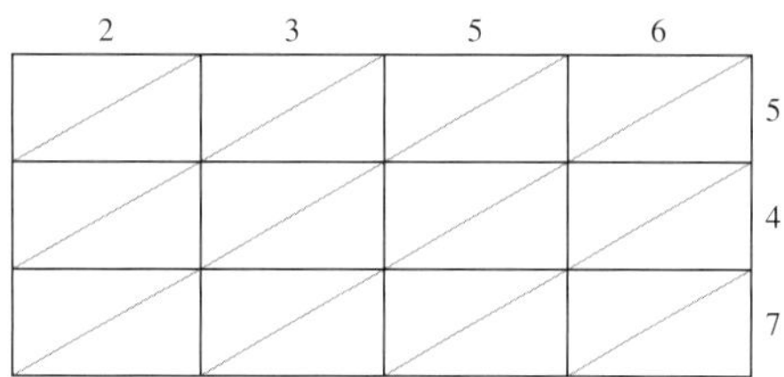

Step 2 Form the individual products in each box.

2	3	5	6	
1 / 0	1 / 5	2 / 5	3 / 0	5
0 / 8	1 / 2	2 / 0	2 / 4	4
1 / 4	2 / 1	3 / 5	4 / 2	7

Step 3 Add along the diagonals.

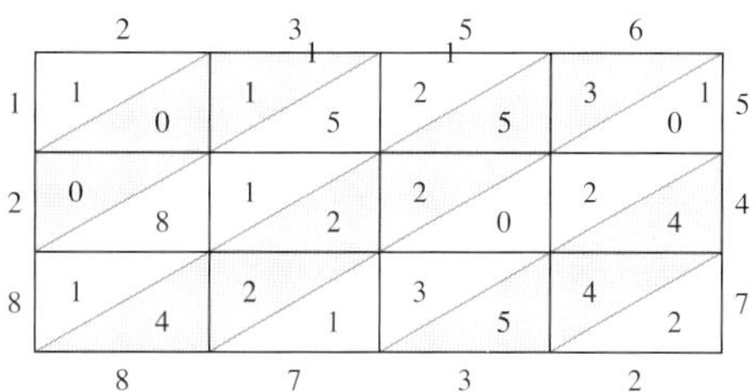

Step 4 Read the answer down the left and across the bottom.

$$2,356 \times 547 = 1,288,732$$

▼ Try This One 5

☑ 3. Multiply using the lattice method.

Use the lattice method to find the product 568 × 478.

Napier's Bones

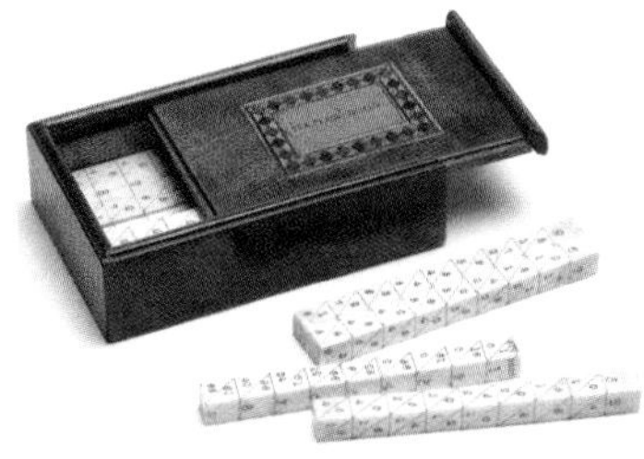

John Napier (1550–1617), a Scottish mathematician, introduced Napier's bones as a calculating tool based on the lattice method of multiplication. Napier's bones consist of a set of 11 rods: the first rod called the index and 1 rod for each digit 0–9, with multiples of each digit written on the rod in a lattice column as illustrated in Figure 4-3 on the next page.

The next example illustrates how Napier's bones are used to multiply by a single-digit number.

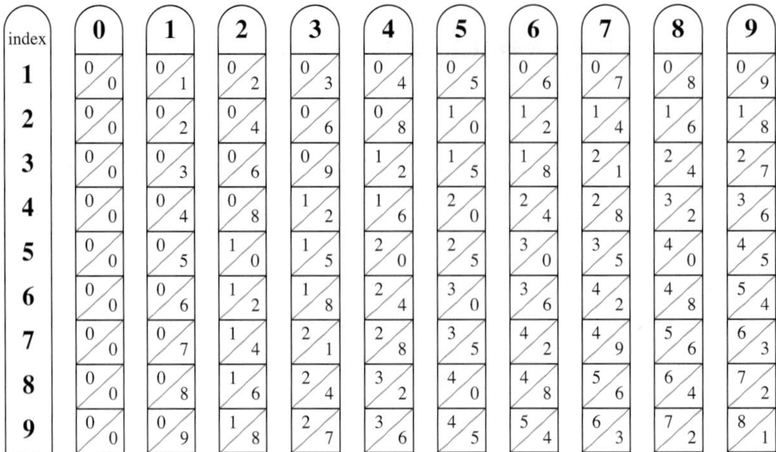

Figure 4-3

EXAMPLE 6 Using Napier's Bones

Use Napier's bones to find the product 2,745 × 8.

SOLUTION

Choose the rods labeled 2, 7, 4, and 5 and place them side by side; also, place the index to the left. Then locate the level for the multiplier 8, as shown in Figure 4-4.

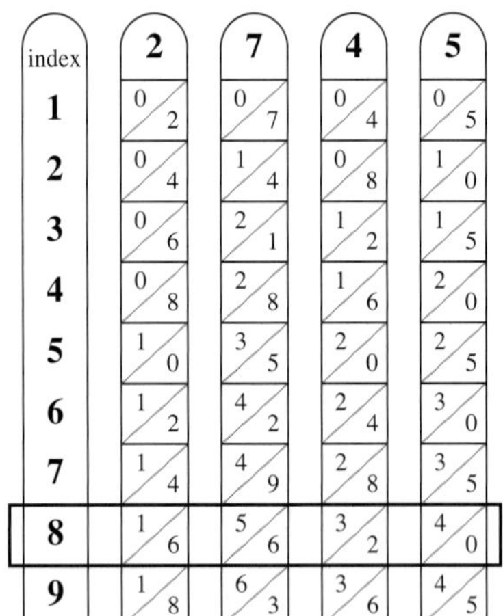

Figure 4-4

Add the numbers diagonally as in the lattice method (Figure 4-5).

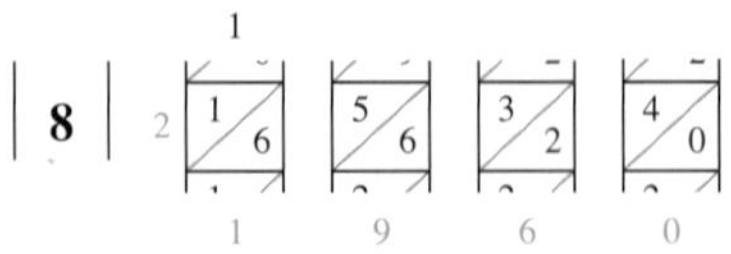

Figure 4-5

The product is 21,960.

▼ Try This One 6

Use Napier's bones to multiply 6 × 8,973.

Sidelight JOHN NAPIER (1550–1617)

John Napier was born to a wealthy family in Merchiston, Scotland, in 1550. He lived in the time of Copernicus and Kepler, an age of great discoveries (and great fame) for those involved in astronomy. Napier was a young genius, entering St. Salvator's College in St. Andrews at the age of 13, but he left after two years, spending time traveling throughout Europe and studying a variety of subjects in different countries. A man of varied talents and interests, Napier found himself especially interested in simplifying the calculations with large numbers that were so crucial to the astronomical observations of the day. He is widely credited with inventing logarithms, a computational tool still in widespread use today. Napier worked with the Englishman Henry Briggs on his logarithms, and Briggs published the first logarithm table after Napier's death in 1617. Logarithm tables maintained their status as vitally important to calculations in a variety of sciences well into the 20th century, when the advent of computers reduced them to an interesting historical artifact.

EXAMPLE 7 Using Napier's Bones

Use Napier's bones to multiply 234 × 36.

SOLUTION

Choose the 2, 3, and 4 rods and place them side by side, with the index to the left. Locate the multipliers 3 and 6.

index	2	3	4
1	0/2	0/3	0/4
2	0/4	0/6	0/8
3	0/6	0/9	1/2
4	0/8	1/2	1/6
5	1/0	1/5	2/0
6	1/2	1/8	2/4
7	1/4	2/1	2/8
8	1/6	2/4	3/2
9	1/8	2/7	3/6

Add the diagonals as in the lattice method.

3	0	0/6	0/9	1/2
6	8	1/2	1/8	2/4
		4	2	4

The product is 8,424.

▼ Try This One 7

☑ 4. Multiply using Napier's bones.

Use Napier's bones to multiply 126 × 73.

Answers to Try This One

1	330	**3**	4,410	**5**	271,504	**7**	9,198
2	216	**4**	44,732	**6**	53,838		

EXERCISE SET 4-2

Writing Exercises

1. Describe the connection between the Egyptian algorithm and the Russian peasant method. How are they different?

2. Describe the connection between the lattice method and Napier's bones. How are they different?

Computational Exercises

For Exercises 3–12, use the Egyptian algorithm to multiply.

3. 12 × 21
4. 15 × 30
5. 23 × 17
6. 29 × 15
7. 34 × 110
8. 17 × 45
9. 56 × 8
10. 11 × 13
11. 18 × 12
12. 7 × 35

For Exercises 13–22, use the Russian peasant method to multiply.

13. 11 × 23
14. 14 × 32
15. 23 × 16
16. 29 × 17
17. 34 × 11
18. 17 × 42
19. 56 × 7
20. 11 × 17
21. 18 × 15
22. 7 × 45

For Exercises 23–32, use the lattice method to multiply.

23. 23 × 456
24. 453 × 938
25. 89 × 1,874
26. 287 × 7,643
27. 876 × 903
28. 45 × 583
29. 67 × 875
30. 359 × 83
31. 568 × 359
32. 2,348 × 83,145

Construct a set of Napier's bones out of poster board or construction paper and use them to do the multiplications in Exercises 33–40.

33. 9 × 523
34. 8 × 731
35. 23 × 45
36. 71 × 52
37. 47 × 123
38. 69 × 328
39. 154 × 236
40. 211 × 416

Critical Thinking

41. Perform the multiplication 38 × 147 three different ways: using the traditional method taught in grade school, the Egyptian algorithm, and the Russian peasant method. Did you get the same answer all three times? Which method was the fastest? If you got more than one answer, use a calculator to check. Which do you think you're most likely to get the correct answer with most often?

42. Repeat Exercise 41, this time using the multiplication 140 × 276, and substituting the lattice method for the Russian peasant method.

43. If you had to perform a multiplication of two 3-digit numbers without a calculator for a million dollars on a game show, and you could only do it once, what method would you use? Why?

Section 4-3 Base Number Systems

LEARNING OBJECTIVES

- ❑ 1. Convert between base 10 and other bases.
- ❑ 2. Convert between binary, octal, and hexadecimal.

In Section 4-1, we studied a variety of numeration systems other than our own. The thing that they all have in common is that they use different numerals than the ones we're all so familiar with in the Hindu-Arabic system. In this section, you'll find out that there are numeration systems that use the numerals you're familiar with, but are still different than the Hindu-Arabic system. The key difference is that all of the digits in our system are based on powers of 10. You can also define systems based on powers of other numbers. If a system uses some of our "regular" numerals, but is based on powers other than 10, we will call it a **base number system.**

The best way to get some perspective on base number systems is to review the base 10 positional system that we use, and to be completely clear on what the significance of every digit is. A number like 453 can be expanded out as

$$\begin{aligned} 453 &= 4 \times 100 + 5 \times 10 + 3 \times 1 \\ &= 4 \times 10^2 + 5 \times 10^1 + 3 \times 10^0 \end{aligned}$$

> **Math Note**
>
> Recall that we have defined 10^0 to be 1, $10^1 = 10$, $10^2 = 10 \cdot 10$, $10^3 = 10 \cdot 10 \cdot 10$, etc. We will study exponents in depth in Section 5-6.

and we understand from experience that a 5 in the second digit from the right means five 10s. We can expand numbers in positional systems with bases other than 10 in the same way. The only difference is that the digits represent powers of some number other than 10.

Base Five System

In a base five system it is not necessary to have 10 numerals as in the Hindu-Arabic system; only five numerals (symbols) are needed. A base five number system can be formed using only the numerals 0, 1, 2, 3, and 4. Just as each digit in the Hindu-Arabic system represents a power of 10, each digit in a base five system represents a power of five. The place values for the digits in base five are:

etc.	six hundred twenty-five (5^4)	one hundred twenty-five (5^3)	twenty-five (5^2)	five (5^1)	one (5^0)

When writing numbers in base five, we use the subscript "five" to distinguish them from base 10 numbers, because a numeral like 453 in base 5 corresponds to a different number than the numeral 453 in base 10. (This is where understanding the difference between a number and a numeral is crucial!) Table 4-1 shows some base 10 numbers also written in base five.

This might look confusing, but it can be clarified using the reasoning in the comments next to the table. The numbers 1 through 4 are written the same in both systems. The number 5 can't be written in base five using the numeral 5 because the base five system only uses the digits 0, 1, 2, 3, and 4. So we have to write it as 10_{five}, meaning $1 \times 5 + 0 \times 1$, or 1 five and no ones. In the same way, the number 8 would be written in base five as 13_{five}, meaning 1 five and 3 ones.

TABLE 4-1 Base Five Numbers

Base 10 number	Corresponding base five number	
1	1_{five}	
2	2_{five}	
3	3_{five}	
4	4_{five}	
5	10_{five}	1 five and no ones
6	11_{five}	
7	12_{five}	
8	13_{five}	1 five and 3 ones
9	14_{five}	
10	20_{five}	2 fives and 0 ones
11	21_{five}	
25	100_{five}	1 twenty-five, no fives or ones
30	110_{five}	1 twenty-five, 1 five, no ones
50	200_{five}	
125	1000_{five}	
625	10000_{five}	

Sidelight MATH AND THE SEARCH FOR LIFE OUT THERE

One of mankind's greatest questions is "Are we alone in the universe?" The search for life beyond our solar system is going on every hour of every day, with giant antennas listening for radio signals from outer space. In the hit movie "Contact," a scientist recognizes that a signal she's receiving must be from an intelligent source because she recognizes that it's based on mathematics (prime numbers). In fact, most scholars feel like our best chance of communicating with an alien civilization is through the universal nature of mathematics! While almost every aspect of languages is dependent on the nature and experiences of the speaker or listener, numbers may be the one true universal constant.

If you think of it, this is intimately tied to the difference between numbers and numerals: while there may be many different ways to represent numbers, the concept of a number is always the same in any language, culture, time, or place. You may then wonder how we would decipher messages from another culture based on mathematics. A good guess is in the same way historians have been able to learn about the numeration systems of ancient cultures. The concrete nature of numbers gives us a big advantage in decoding messages, providing our best chance at finding life out there.

Converting Base Five Numbers to Base 10 Numbers

Base five numbers can be converted to base 10 numbers using the place values of the base five numbers and expanded notation. For example, the number 242_{five} can be expanded as

$$\begin{aligned} 242_{\text{five}} &= 2 \times 5^2 + 4 \times 5^1 + 2 \times 5^0 \\ &= 2 \times 25 + 4 \times 5 + 2 \times 1 \\ &= 50 \quad + 20 \quad + 2 \\ &= 72 \end{aligned}$$

EXAMPLE 1 Converting Numbers from Base Five to Base 10

Write each number in base 10.

(a) 42_{five}
(b) 134_{five}
(c) 4213_{five}

SOLUTION

The place value chart for base five is used in each case.

(a) $42_{\text{five}} = 4 \times 5^1 + 2 \times 1 = 20 + 2 = 22$

(b) $$\begin{aligned} 134_{\text{five}} &= 1 \times 5^2 + 3 \times 5 + 4 \times 1 \\ &= 1 \times 25 + 3 \times 5 + 4 \times 1 \\ &= 25 + 15 + 4 = 44 \end{aligned}$$

(c) $$\begin{aligned} 4213_{\text{five}} &= 4 \times 5^3 + 2 \times 5^2 + 1 \times 5 + 3 \times 1 \\ &= 4 \times 125 + 2 \times 25 + 1 \times 5 + 3 \times 1 \\ &= 500 + 50 + 5 + 3 = 558 \end{aligned}$$

▼ Try This One 1

Write each number in the base 10 system.

(a) 302_{five} (b) 1324_{five} (c) 40000_{five}

Converting Base 10 Numbers to Base Five Numbers

Base 10 numbers can be written in the base five system using the place values of the base five system and successive division. This method is illustrated in Examples 2 and 3.

EXAMPLE 2 Converting Numbers from Base 10 to Base Five

Write 84 in the base five system.

SOLUTION

Step 1 Identify the largest place value number (1, 5, 25, 125, etc.) that will divide into the base 10 number. In this case, it is 25.

Step 2 Divide 25 into 84, as shown.

$$\begin{array}{r} 3 \\ 25\overline{)84} \\ \underline{75} \\ 9 \end{array}$$

This tells us that there are three 25s in 84.

Step 3 Divide the remainder by the next lower place value. In this case, it is 5.

$$\begin{array}{r} 1 \\ 5\overline{)9} \\ \underline{5} \\ 4 \end{array}$$

Step 4 Continue dividing until the remainder is less than 5. In this case, it is 4, so the division process is stopped. In other words, four 1s are left. The answer, then, is 314_{five}. In 84, there are three 25s, one 5, and four 1s.

Math Note

The answer can be checked using multiplication and addition:
$3 \times 25 + 1 \times 5 + 4 \times 1$
$= 75 + 5 + 4 = 84.$

▼ Try This One 2

Write 73 in the base five system.

EXAMPLE 3 Converting Numbers from Base 10 to Base Five

Write 653 in the base five system.

SOLUTION

Step 1 Since 625 is the largest place value that will divide into 653, it is used first.

$$\begin{array}{r} 1 \\ 625\overline{)653} \\ \underline{625} \\ 28 \end{array}$$

There is one 625 in 653.

Step 2 Divide by 125.

$$125\overline{)28}\quad \begin{array}{r} 0 \\ 0 \\ \hline 28 \end{array}$$

There are no 125s in 28.

Even though 125 does not divide into the 28, the zero must be written to hold its place value in the base five number system.

Step 3 Divide by 25.

$$25\overline{)28}\quad \begin{array}{r} 1 \\ 25 \\ \hline 3 \end{array}$$

There is one 25 in 28.

Step 4 Divide by 5.

$$5\overline{)3}\quad \begin{array}{r} 0 \\ 0 \\ \hline 3 \end{array}$$

There are no 5s in 3.

The solution is 10103_{five}.

Check: $1 \times 625 + 0 \times 125 + 1 \times 25 + 0 \times 5 + 3 \times 1 = 653$.

▼ Try This One 3

Write each number in the base five system.

(a) 52 (b) 486 (c) 1,000

Other Number Bases

Once we understand the idea of alternative bases, we can define new number systems with as few as two symbols, or digits. (Remember, we only needed digits zero through four for base five numbers.) For example, a base two, or **binary system** (used extensively in computer programming) uses only two digits, 0 and 1. The place values of the digits in the base two numeration system are powers of two:

| etc.
| sixteen (2^4)
| eight (2^3)
| four (2^2)
| two (2^1)
| one (2^0)

The base eight or **octal system** consists of eight digits, 0, 1, 2, 3, 4, 5, 6, and 7. The place values of the digits in the base eight system are powers of eight:

| etc.
| four thousand ninety-six (8^4)
| five hundred twelve (8^3)
| sixty-four (8^2)
| eight (8^1)
| one (8^0)

When the base number is greater than 10, new digits must be created to make the numbers. For example, base 16 (called the **hexadecimal system**) is used in computer technology. We need 16 digits for this system; the digits in base 16 are 0, 1, 2, 3, 4, 5, 6, 7, 8, 9, A, B, C, D, E, and F, where A represents 10, B represents 11, C represents 12, etc. (We can't use 10 through 15 because they have 2 digits!) The place values of the digits in base 16 are powers of 16:

| etc.
| four thousand ninety-six (16^3)
| two hundred fifty-six (16^2)
| sixteen (16^1)
| one (16^0)

Table 4-2 shows the digits for some of the base number systems and the place values of the digits in the system. It should be pointed out that place values go on indefinitely for any base number system.

TABLE 4-2 Base Number Systems

Base two (binary system)
Digits used: 0, 1
Place values: 2^6 2^5 2^4 2^3 2^2 2^1 2^0
Numbers: 0, 1, 10 , 11, 100, 101, 110, 111, 1000, 1001, 1010, etc.

Base three
Digits used: 0, 1, 2
Place values: 3^6 3^5 3^4 3^3 3^2 3^1 3^0
Numbers: 0, 1, 2, 10, 11, 12, 20, 21, 22, 100, 101, 102, 110, etc.

Base five
Digits used: 0, 1, 2, 3, 4
Place values: 5^6 5^5 5^4 5^3 5^2 5^1 5^0
Numbers: 0, 1, 2, 3, 4, 10, 11, 12, 13, 14, 20, 21, 22, etc.

Base eight (octal system)
Digits used: 0, 1, 2, 3, 4, 5, 6, 7
Place values: 8^6 8^5 8^4 8^3 8^2 8^1 8^0
Numbers: 0, 1, 2, 3, 4, 5, 6, 7, 10, 11, 12, 13, 14, 15, 16, 17, 20, etc.

Base 10
Digits used: 0, 1, 2, 3, 4, 5, 6, 7, 8, 9
Place values: 10^6 10^5 10^4 10^3 10^2 10^1 10^0
Numbers: 0, 1, 2, 3, 4, 5, 6, 7, 8, 9, 10, 11, 12, 13, 14, 15, etc.

Base 16 (hexadecimal system)
Digits used: 0, 1, 2, 3, 4, 5, 6, 7, 8, 9, A, B, C, D, E, F
Place values: 16^6 16^5 16^4 16^3 16^2 16^1 16^0
Numbers: 0, 1, 2, 3, 4, 5, 6, 7, 8, 9, A, B, C, D, E, F, 10, 11, etc.

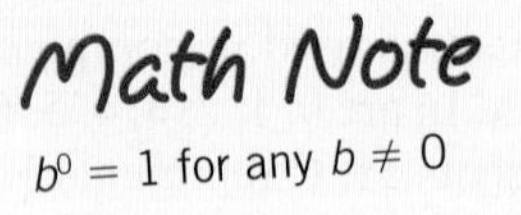

Looking at the table, several things become apparent. First, the number of symbols is equal to the base. Second, the place values of any base are

$$\cdots \quad \underline{b^6} \; \underline{b^5} \; \underline{b^4} \; \underline{b^3} \; \underline{b^2} \; \underline{b^1} \; \underline{b^0}$$

where b is the base. For example, the place values for base six are

$$\cdots \quad \underline{46{,}656} \quad \underline{7{,}776} \quad \underline{1{,}296} \quad \underline{216} \quad \underline{36} \quad \underline{6} \quad \underline{1}$$

In order to convert from numbers written in bases other than 10 to base 10 numbers, expanded notation is used. This is the same procedure used in Example 1. Example 4 shows the procedure.

EXAMPLE 4 Converting Numbers to Base 10

A microphone converts sound to a voltage signal, which in turn is converted into a binary number. Each measurement is recorded as a 16-bit number and then interpreted by an amplifier.

Write each number in base 10.

(a) 132_{six} (b) 10110_{two} (c) 1532_{eight} (d) 2102_{three} (e) $5BD8_{\text{sixteen}}$

SOLUTION

(a) The place values of the digits in base six are powers of 6:

$$\begin{aligned} 132_{\text{six}} &= 1 \times 6^2 + 3 \times 6^1 + 2 \times 1 \\ &= 1 \times 36 + 3 \times 6 + 2 \times 1 \\ &= 36 + 18 + 2 = 56 \end{aligned}$$

(b) The place values of the digits in base two are powers of 2:

$$\begin{aligned} 10110_{\text{two}} &= 1 \times 2^4 + 0 \times 2^3 + 1 \times 2^2 + 1 \times 2^1 + 0 \times 1 \\ &= 1 \times 16 + 0 \times 8 + 1 \times 4 + 1 \times 2 + 0 \times 1 \\ &= 16 + 0 + 4 + 2 + 0 = 22 \end{aligned}$$

(c) The place values of the digits in base eight are powers of 8:

$$\begin{aligned} 1532_{\text{eight}} &= 1 \times 8^3 + 5 \times 8^2 + 3 \times 8^1 + 2 \times 1 \\ &= 1 \times 512 + 5 \times 64 + 3 \times 8 + 2 \times 1 \\ &= 512 + 320 + 24 + 2 = 858 \end{aligned}$$

(d) The place values of the digits in base three are powers of 3:

$$\begin{aligned} 2102_{\text{three}} &= 2 \times 3^3 + 1 \times 3^2 + 0 \times 3^1 + 2 \times 1 \\ &= 2 \times 27 + 1 \times 9 + 0 \times 3 + 2 \times 1 \\ &= 54 + 9 + 0 + 2 = 65 \end{aligned}$$

(e) The place values of the digits in base 16 are powers of 16:

$$\begin{aligned} 5BD8_{\text{sixteen}} &= 5 \times 16^3 + 11 \times 16^2 + 13 \times 16 + 8 \times 1 \\ &= 5 \times 4{,}096 + 11 \times 256 + 13 \times 16 + 8 \times 1 \\ &= 20{,}480 + 2{,}816 + 208 + 8 = 23{,}512 \end{aligned}$$

▼ Try This One 4

Write each number in base 10.

(a) 5320_{seven} (b) 110110_{two} (c) 32021_{four} (d) $42AE_{\text{sixteen}}$

Converting Base 10 Numbers to Other Base Numbers

In Examples 2 and 3, we used division to convert base 10 numbers to base five. The same procedure can be used to convert to other base number systems as well.

EXAMPLE 5 Converting Numbers to Bases Other Than 10

(a) Write 48 in base three.
(b) Write 51 in base two.
(c) Write 19,443 in base 16.

SOLUTION

(a) **Step 1** The place values for base three are powers of three. The largest power of three less than 48 is 3^3, or 27, so we divide 48 by 27.

$$\begin{array}{r} 1 \\ 27\overline{)48} \\ \underline{27} \\ 21 \end{array}$$

Step 2 Divide the remainder by 3^2 or 9.

$$\begin{array}{r} 2 \\ 9\overline{)21} \\ \underline{18} \\ 3 \end{array}$$

Step 3 Divide the remainder by 3^1 or 3.

$$\begin{array}{r} 1 \\ 3\overline{)3} \\ \underline{3} \\ 0 \end{array}$$

So, 48 is $1 \times 3^3 + 2 \times 3^2 + 1 \times 3^1 + 0 \times 3^0$, which makes it 1210_{three}.

(b) The place values for base two are 1, 2, 4, 8, 16, 32, etc. Use successive division, as shown.

$$\begin{array}{rrrrr} 1 & 1 & 0 & 0 & 1 \\ 32\overline{)51} & 16\overline{)19} & 8\overline{)3} & 4\overline{)3} & 2\overline{)3} \\ \underline{32} & \underline{16} & \underline{0} & \underline{0} & \underline{2} \\ 19 & 3 & 3 & 3 & 1 \end{array}$$

So, $51 = 110011_{\text{two}}$.

(c) The place values in base 16 are 1, 16, 256 (16^2), 4096 (16^3), etc. Use successive division as shown. (Remember, in base 16, B plays the role of 11 and F plays the role of 15.)

$$\begin{array}{rrr} 4 & \text{B} & \text{F} \\ 4096\overline{)19443} & 256\overline{)3059} & 16\overline{)243} \\ \underline{16384} & \underline{2816} & \underline{240} \\ 3059 & 243 & 3 \end{array}$$

So $19{,}443 = 4\text{BF}3_{\text{sixteen}}$.

▼ Try This One 5

☑ 1. Convert between base 10 and other bases.

(a) Write 84 in base two.
(b) Write 258 in base six.
(c) Write 122 in base three.
(d) Write 874 in base 16.

Sidelight Bar Codes

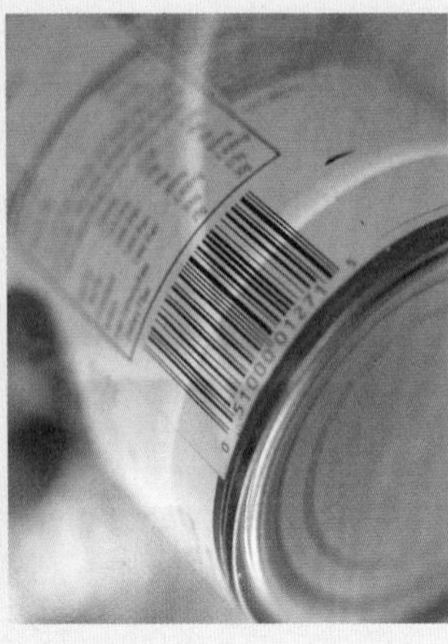

Bar codes are a series of black and white stripes that vary in width. The width of each stripe determines a binary digit that the scanner then decodes. The most familiar bar codes are the UPC codes that appear on almost every product you buy. Most UPCs have a left and right margin that tells the reader where to begin and end, a five-digit manufacturer's code, a check digit in the middle, then a five-digit product code. Once a register reads the code and knows what the product is, it computes the price that was programmed in for that product.

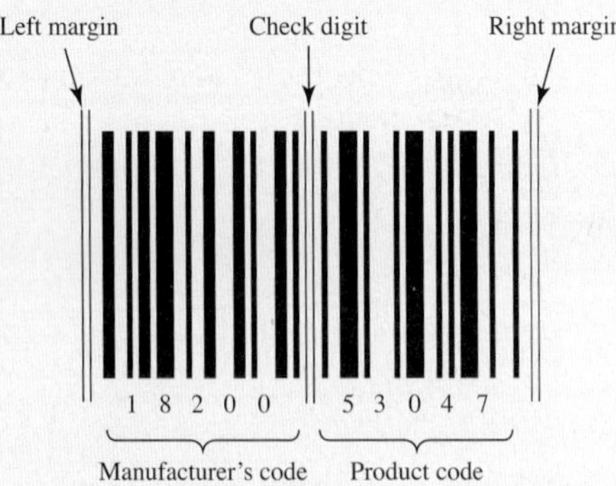

To read a UPC, you need the information in the table below. Every digit in the base 10 system is represented by a seven-digit binary number.

Digit	Manufacturer's Number	Product Number
0	0001101	1110010
1	0011001	1100110
2	0010011	1101100
3	0111101	1000010
4	0100011	1011100
5	0110001	1001110
6	0101111	1010000
7	0111011	1000100
8	0110111	1001000
9	0001011	1110100

Notice that all of the binary numbers in the manufacturer's code begin with zero, while those in the product code begin with one. This is done so that the scan can be done from left to right, or right to left. The computer can tell the correct direction by recognizing the difference in the digits. As an example of a digit, the digit five is represented by 1001110 in the product code. This number is represented with black and white bars, with a single-width white bar representing a zero and a single-width black bar representing a one.

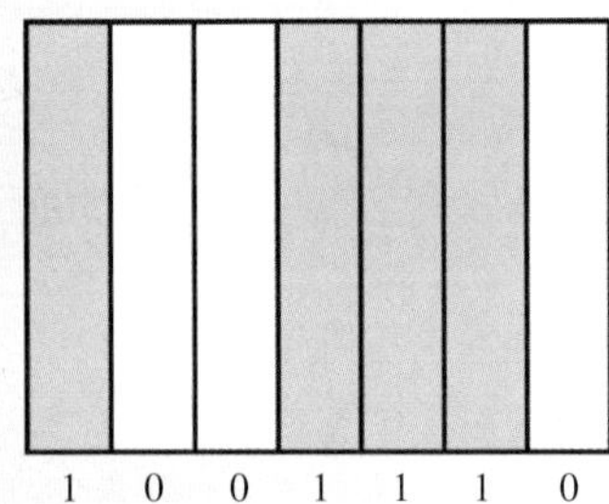

Notice how this matches the first four stripes to the right of the check digit in the UPC above (without the borders around the individual bars). Every digit in the left half of a UPC is represented by four stripes: white, then black, then white, then black. On the right half, it's black, then white, then black, then white. The key is to recognize the width of each stripe to decide how many bars it represents.

It takes a little bit of practice to distinguish bars that represent one, two, three, and four consecutive digits, but once you can do that, you can decode UPCs—just not as fast as a computer can.

Base Numbers and Computers

Computers use three bases to perform operations. They are base two, the binary system; base eight, the octal system; and base 16, the hexadecimal system.

Base two is used since it contains only two characters, 0 and 1. Electric circuits can differentiate two types of pulses, on and off. Early computers used one on-off vacuum tube to store one binary character. Since it is cheaper and faster to use the binary system, modern computers still use this system, and assembly language programmers must become proficient in the binary system.

The base eight system is also used by computer programmers. In computer language, 1 bit is used to represent one character, and 8 bits constitute a byte. It then becomes convenient to write numbers using a series of bytes (eight characters).

TABLE 4-3 Binary Equivalents for Octal and Hexadecimal Digits

Octal Digit	Binary Equivalent
0	000
1	001
2	010
3	011
4	100
5	101
6	110
7	111

Hex Digit	Binary Equivalent
0	0000
1	0001
2	0010
3	0011
4	0100
5	0101
6	0110
7	0111

Hex Digit	Binary Equivalent
8	1000
9	1001
A	1010
B	1011
C	1100
D	1101
E	1110
F	1111

The base 16 is used for several reasons. First of all, 16 characters consist of 2 bytes. Also, 16 is 2^4, which means one hexadecimal character can replace four binary characters. This increases the speed at which the computer is able to perform numerical applications since there are fewer characters for the computer to read and fewer operations to perform. Finally, large numbers can be written in base 16 with fewer characters than in base two or base 10, saving much needed space in the computer's memory.

There is a short cut for converting between binary and octal or binary and hexadecimal. To use this short cut, first notice that every octal digit can be written as a three-digit binary number, and every hexadecimal digit can be written as a four-digit binary number, as illustrated in Table 4-3.

The following examples illustrate the shortcut for converting between these bases.

EXAMPLE 6 Converting between Binary and Octal

(a) Convert the binary number 1001110110_{two} to octal.
(b) Convert the octal number 7643_{eight} to binary.

SOLUTION

(a) Starting at the rightmost digit, group the digits of the binary number into groups of three (if there are not three digits that remain at the left of the number, fill them in with zeros), then use Table 4-3 to change each group to an octal digit as follows.

001	001	110	110
1	1	6	6

So, $1001110110_{\text{two}} = 1166_{\text{eight}}$.

(b) First, convert each octal digit into a three-digit binary digit using Table 4-3, and then string them together to form a binary number.

7	6	4	3
111	110	100	011

So, $7643_{\text{eight}} = 111110100011_{\text{two}}$.

▼ Try This One 6

(a) Convert 1100101_{two} to octal.
(b) Convert 6147_{eight} to binary.

Sidelight THE BINARY SYSTEM AND COMPUTER STORAGE

Most computer users are familiar with the term megabytes, or MB—it's one of the standard measurements of file size. Our study of octal numbers can give us some insight into just how much information goes into a typical computer file. One character is called a bit (short for "binary digit"), and 8 bits make up 1 byte of storage. A kilobyte (kB) is 1,000 bytes, and a megabyte (MB) is 1,000 kilobytes, or 1 million bytes. (There is a slight difference between processor memory, where 1 MB is 1,024 kB, and disk storage, where 1 MB is 1,000 kB.)

This raises the question: exactly how many characters are in certain computer files? A typical MP3 file is in the neighborhood of 4 MB. Doing some simple multiplying, this means there are 4 million bytes, each of which contains 8 characters, so there are 32 million characters in a typical MP3. To get some perspective on how much information that is, think of it this way: if you were to try to write out those characters by hand, writing at a reasonable pace of one per second, you better pack a lunch. In fact, you better pack a little heavier than that—it would take a little bit over 370 *days* to write that many characters, and that's writing constantly with no sleep breaks! An iPod with a full 80-gigabyte hard drive (a gigabyte is 1,000 megabytes) contains 6.4×10^{11}, or 640 *trillion* characters. It's pretty hard to get any perspective on how big a number that is, but this amounts to about 910 characters for every human being on the face of the earth, and about 6 characters for every human being who has *ever* lived. All on one tiny music player.

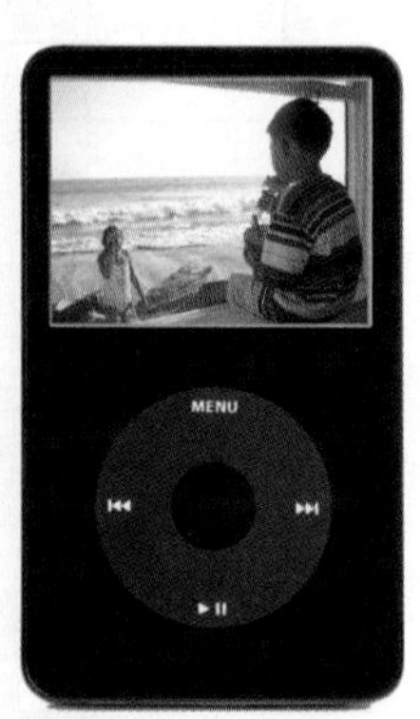

EXAMPLE 7 Converting between Binary and Hexadecimal

(a) Convert the binary number $1110011001111_{\text{two}}$ to hexadecimal.
(b) Convert the hexadecimal number $9D7A3_{\text{sixteen}}$ to binary.

SOLUTION

(a) Starting at the rightmost digit, group the binary number into groups of four (adding zeros in front as needed) and then convert each group of four to a hexadecimal digit using Table 4-3.

0001	1100	1100	1111
1	C	C	F

So $1110011001111_{\text{two}} = 1CCF_{\text{sixteen}}$.

(b) First, convert each hexadecimal number to a four-digit binary number, and then string them together to form a binary number.

9	D	7	A	3
1001	1101	0111	1010	0011

So $9D7A3_{\text{sixteen}} = 10011101011110100011_{\text{two}}$.

▼ Try This One 7

(a) Convert 100011001_{two} to hexadecimal. (b) Convert $ABCD_{\text{sixteen}}$ to binary.

☑ 2. Convert between binary, octal, and hexadecimal.

Answers to Try This One

1 (a) 77 (b) 214 (c) 2,500

2 243_{five}

3 (a) 202_{five} (b) 3421_{five} (c) 13000_{five}

4 (a) 1,876 (b) 54 (c) 905 (d) 17,070

5 (a) 1010100_{two} (b) 1110_{six} (c) 11112_{three} (d) $36A_{sixteen}$

6 (a) 145_{eight} (b) 110001100111_{two}

7 (a) $119_{sixteen}$ (b) 1010101111001101_{two}

EXERCISE SET 4-3

Writing Exercises

1. Explain why the base two system requires only two digits.
2. Explain the significance of place values in base number systems. What does each place value represent relative to the base number?

Computational Exercises

For Exercises 3–22, convert each number to base 10.

3. 1011_{two}
4. 372_{twelve}
5. 53_{six}
6. $8A21_{twelve}$
7. 99_{eleven}
8. 3451_{seven}
9. 10221_{three}
10. 110011_{two}
11. 2221_{five}
12. 5320_{eight}
13. 153_{six}
14. 11001_{two}
15. 438_{nine}
16. 2561_{eight}
17. 352_{seven}
18. 11112_{five}
19. $921E_{sixteen}$
20. 2101_{three}
21. 812_{twelve}
22. 10000001_{two}

For Exercises 23–42, write each base 10 number in the given base.

23. 31 in base two
24. 186 in base eight
25. 345 in base six
26. 266 in base three
27. 16 in base seven
28. 3,050 in base 12
29. 745 in base nine
30. 3,217 in base four
31. 22 in base two
32. 5,621 in base 11
33. 18 in base five
34. 97 in base four
35. 2,361 in base 16
36. 96 in base two
37. 18,432 in base five
38. 25,000 in base 16
39. 32 in base seven
40. 88 in base three
41. 256 in base two
42. 497 in base four

For Exercises 43–52, convert each number to base 10, then change to the specified base.

43. 134_{six} to base two
44. 1011010_{two} to base eight
45. 342_{five} to base 12
46. 4711_{nine} to base three
47. 1221_{three} to base four
48. 1521_{seven} to base three
49. 432_{eight} to base 12
50. $3AB_{twelve}$ to base six
51. 1782_{nine} to base seven
52. 3000_{four} to base 11

For Exercises 53–62, convert each binary number (a) to octal then (b) to hexadecimal. (The subscript two is omitted since it's understood that all numbers are binary).

53. 11010111010
54. 1011101
55. 110011111
56. 101100001
57. 10111011001
58. 11110110111
59. 11111111101
60. 100011000111
61. 110101101011
62. 11110100110111

For Exercises 63–72, convert each octal number to binary. (The subscript eight is omitted since it's understood that all numbers are octal).

63. 1570
64. 354
65. 6504
66. 76241
67. 620
68. 2742
69. 61
70. 432
71. 72165
72. 4731

For Exercises 73–82, convert each hexadecimal number to binary. (The subscript 16 is omitted since it's understood that all numbers are hexadecimal.)

73. 6A2B
74. 657F1
75. AB
76. 3A4
77. 362
78. 9E73
79. A5D
80. 95C1
81. 654B
82. AF3

Real-World Applications

Problems similar to base number problems occur often in everyday life. Using a procedure like the one used in Examples 2, 3, and 5, solve each.

83. Change 87 ounces to pounds and ounces (1 lb = 16 oz).
84. Change 237 ounces to quarts, pints, and ounces (1 qt = 2 pt = 32 oz).
85. Change 1,256 inches to yards, feet, and inches (1 yd = 3 ft = 36 in.).
86. Change $5.88 to quarters, dimes, nickels, and pennies using the smallest number of coins.

A process of encoding data using two symbols is called binary coding. One of the earliest codes was the Morse code. *Words were encoded using dots and dashes and transmitted by sound. If we consider a dot as a zero and a dash as a one, the various letters of the alphabet can be transformed into a binary code, as follows.*

A 01	K 101	U 001
B 1000	L 0100	V 0001
C 1010	M 11	W 011
D 100	N 10	X 1001
E 0	O 111	Y 1011
F 0010	P 0110	Z 1100
G 110	Q 1101	
H 0000	R 010	
I 00	S 000	
J 0111	T 1	

For Exercises 87–92, decode the following messages.

87. 000 1 111 0110
88. 10 111 0000 111 11 0 011 111 010 101
89. 1010 0100 01 000 000 00 000 111 0001 0 010
90. 010 0 000 1 0010 111 010 0100 001 10 1010 0000
91. 00 1 00 000 010 01 00 10 00 10 110
92. 1 001 010 10 01 010 111 001 10 100

The U.S. Postal Service uses a Postnet code on business reply forms. This code consists of a five-digit Zip code plus a four-digit extension and a check digit. This code consists of long and short vertical bars.

Digit	Bar code	Digit	Bar code
1		6	
2		7	
3		8	
4		9	
5		0	

For example, the Zip code 15131 would be encoded as follows:

1 5 1 3 1

Using the Postnet code, find the following five-digit Zip codes for Exercises 93–98.

93.
94.
95.
96.
97.
98.

For Exercises 99–104, write each Zip code using the Postnet code.

99. 26135
100. 14157
101. 18423
102. 30214
103. 11672
104. 54901

Critical Thinking

105. Think about how you would write numbers in a base one system. The result is actually a numeration system we discussed in Section 4-1. Which system? Explain.

For Exercises 106–110, use the information found in the Sidelight found on page 178. Identify the digit shown by the bar code. State whether the digit belongs to the manufacturer's number or the product number.

106. 0111101
107. 0101111
108. 1100110
109. 1110010
110. 0111011

Section 4-4 Operations in Base Number Systems

LEARNING OBJECTIVES

- ☐ 1. Add in bases other than 10.
- ☐ 2. Subtract in bases other than 10.
- ☐ 3. Multiply in bases other than 10.
- ☐ 4. Divide in bases other than 10.

Now that we are able to write numbers in other base systems, it seems reasonable to take a look at working with numbers in those systems using the basic operations of addition, subtraction, multiplication, and division. While it's definitely true that you could just convert the numbers to base 10, perform the operations the way you always have, then convert the answer back, we're going to advise you against doing that. The real value of this section is the mental exercise of adapting arithmetic skills you learned a long time ago to a new situation. This mental flexibility has a lot of benefits: it helps give a deeper understanding of the familiar operations in arithmetic and it's also good practice for problem solving, which almost always involves using some basic skills you've acquired and adapting them to a certain situation.

For the most part, arithmetic operations can be performed in other base number systems just as they are performed in base 10. For example, $4_{five} + 3_{five} = 12_{five}$. Why? For the same reason that $9 + 3 = 12$ in base 10: when adding 3 to 9, we get past the highest digit in base 10 (which is nine), so we "wrap around" and get one 10, with two 1s left over. In $4_{five} + 3_{five} = 12_{five}$, adding three gets us past the highest digit in base five (which is four), and we get one 5, with two 1s left over.

You're probably able to add very well in base 10 because you learned the sums of all of the digits back when you first started school. When working in a different base, you can build an addition table that displays the sums of all the digits in that system to stand in for those years of experience. A table for base five is shown below. (The subscript five has been omitted.) All of the values were obtained using the same reasoning we used to add 4 and 3 in base five.

+	0	1	2	3	4
0	0	1	2	3	4
1	1	2	3	4	10
2	2	3	4	10	11
3	3	4	10	11	12
4	4	10	11	12	13

Math Note

Computations like $4_{five} + 3_{five} = 12_{five}$ probably look a little bizarre to you, since your past experience tells you that 4 plus 3 is not 12. It might help to think of 12_{five} not as "twelve base five," but rather "one two base five."

In order to add using the table, find one number in the left column and the other number in the top row. Then draw a horizontal and vertical line. The intersection of the lines is the sum. For example, $2_{five} + 4_{five} = 11_{five}$, as shown.

+	0	1	2	3	(4)
0	0	1	2	3	4
1	1	2	3	4	10
(2)	2	3	4	10 →	(11)
3	3	4	10	11	12
4	4	10	11	12	13

We perform addition of numbers with more than one digit in base five the same way we do in base 10: by lining up the numbers vertically and adding one digit at a time. When a result is two digits, we will apply the first digit to the next column on the left. (You probably called this "carrying" a digit in elementary school.)

EXAMPLE 1 Adding in Base Five

Add in base five: $324_{five} + 24_{five}$.

SOLUTION

Step 1 Add the digits in the ones column, 4 and 4. $4_{five} + 4_{five} = 13_{five}$ (see the table). Write the 3 in the ones place; the digit 1 represents 1 five, so we apply it to the fives column.

$$\begin{array}{r} {}^{1} \\ 324_{five} \\ +\ 24_{five} \\ \hline 3_{five} \end{array}$$

Step 2 Add $1_{five} + 2_{five} + 2_{five} = 10_{five}$ (see the table). Write the 0 and apply the 1 to the 25s column as shown.

$$\begin{array}{r} {}^{1} \\ 324_{five} \\ +\ 24_{five} \\ \hline 03_{five} \end{array}$$

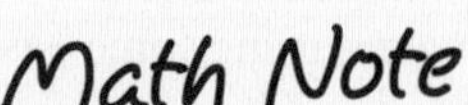

The answer can be checked by converting the numbers to base ten and seeing if the answers are equal:

$$\begin{array}{rcr} 324_{five} & = & 89_{ten} \\ +24_{five} & = & 14_{ten} \\ \hline 403_{five} & = & 103_{ten} \end{array}$$

Step 3 Add $1_{five} + 3_{five} = 4_{five}$ (see the table). Write the 4 as shown.

$$\begin{array}{r} 324_{five} \\ +\ 24_{five} \\ \hline 403_{five} \end{array}$$

The sum in base five is $324_{five} + 24_{five} = 403_{five}$.

▼ Try This One 1

Add in base five: $324_{five} + 203_{five}$.

In the remainder of the section, we will use the term "carrying" to refer to applying a second digit in a sum to the next place column to the left. But you should definitely take a moment to think about why we do this process, rather than memorize it as a "trick."

EXAMPLE 2 Adding in Base Five

Add in base five: $1244_{five} + 333_{five}$.

SOLUTION

$$\begin{array}{r} {}^{111} \\ 1244_{five} \\ +\ 333_{five} \\ \hline 2132_{five} \end{array}$$

$4_{five} + 3_{five} = 12_{five}$ Write the 2 and carry the 1.

$1_{five} + 4_{five} + 3_{five} = 13_{five}$ Write the 3 and carry the 1.

$1_{five} + 2_{five} + 3_{five} = 11_{five}$ Write the 1 and carry the 1.

$1_{five} + 1_{five} = 2_{five}$

The sum is 2132_{five}.

▼ Try This One 2

Add in base five: $4301_{five} + 2024_{five}$.

Addition can be performed in other bases as well as in base five. It might help to construct an addition table for the given base.

EXAMPLE 3 Adding in Base Two

Add in base two: $10111_{\text{two}} + 110_{\text{two}}$.

SOLUTION

The addition table for base two is

+	0	1
0	0	1
1	1	10

Then

$$\begin{array}{r} {}^{//} \\ 10111_{\text{two}} \\ +\ 110_{\text{two}} \\ \hline 11101_{\text{two}} \end{array}$$

- $1_{\text{two}} + 0_{\text{two}} = 1_{\text{two}}$
- $1_{\text{two}} + 1_{\text{two}} = 10_{\text{two}}$ Write the 0; carry 1.
- $1_{\text{two}} + 1_{\text{two}} + 1_{\text{two}} = 11_{\text{two}}$ Write the 1 and carry 1.
- $1_{\text{two}} + 0_{\text{two}} = 1_{\text{two}}$
- bring down the 1

The sum is 11101_{two}.

▼ Try This One 3

Add in base two: $11001_{\text{two}} + 1001_{\text{two}}$.

The addition table for the hexadecimal system is shown next. Remember that in base 16, the digit A corresponds to 10, B to 11, C to 12, D to 13, E to 14, and F to 15.

+	0	1	2	3	4	5	6	7	8	9	A	B	C	D	E	F
0	0	1	2	3	4	5	6	7	8	9	A	B	C	D	E	F
1	1	2	3	4	5	6	7	8	9	A	B	C	D	E	F	10
2	2	3	4	5	6	7	8	9	A	B	C	D	E	F	10	11
3	3	4	5	6	7	8	9	A	B	C	D	E	F	10	11	12
4	4	5	6	7	8	9	A	B	C	D	E	F	10	11	12	13
5	5	6	7	8	9	A	B	C	D	E	F	10	11	12	13	14
6	6	7	8	9	A	B	C	D	E	F	10	11	12	13	14	15
7	7	8	9	A	B	C	D	E	F	10	11	12	13	14	15	16
8	8	9	A	B	C	D	E	F	10	11	12	13	14	15	16	17
9	9	A	B	C	D	E	F	10	11	12	13	14	15	16	17	18
A	A	B	C	D	E	F	10	11	12	13	14	15	16	17	18	19
B	B	C	D	E	F	10	11	12	13	14	15	16	17	18	19	1A
C	C	D	E	F	10	11	12	13	14	15	16	17	18	19	1A	1B
D	D	E	F	10	11	12	13	14	15	16	17	18	19	1A	1B	1C
E	E	F	10	11	12	13	14	15	16	17	18	19	1A	1B	1C	1D
F	F	10	11	12	13	14	15	16	17	18	19	1A	1B	1C	1D	1E

EXAMPLE 4 Adding in Base 16

Add in base 16: $135E_{sixteen} + 21C_{sixteen}$.

SOLUTION

$$\begin{array}{r} \overset{1}{} \\ 135E_{sixteen} \\ +\ 21C_{sixteen} \\ \hline 157A_{sixteen} \end{array}$$

- $E_{sixteen} + C_{sixteen} = 1A_{sixteen}$ Write the A and carry the 1.
- $1_{sixteen} + 5_{sixteen} + 1_{sixteen} = 7_{sixteen}$
- $3_{sixteen} + 2_{sixteen} = 5_{sixteen}$
- Bring down the 1

The sum is $157A_{sixteen}$.

▼ Try This One 4

☑ 1. Add in bases other than 10.

Add in base 16: $8D51_{sixteen} + 947A_{sixteen}$

Now that we know how to perform addition in other bases, we should be able to subtract as well. Addition tables can help with subtraction; the addition table for base five is below. To perform a subtraction like $12_{five} - 4_{five}$, we find 4 in the first column of the table, then move across that row until we find 12. The number at the top of the column is the difference: $12_{five} - 4_{five} = 3_{five}$.

+	0	1	2	(3)	4
0	0	1	2	3	4
1	1	2	3	4	10
2	2	3	4	10	11
3	3	4	10	11	12
(4)	4	10	11	(12)	13

We will subtract numbers with more than one digit using a method similar to addition. When the digit to be subtracted is larger, as in the first step of Example 5, we'll need to "borrow" a one from the next larger place value.

EXAMPLE 5 Subtracting in Base Five

Subtract in base five:

$$\begin{array}{r} 321_{five} \\ -\ 123_{five} \\ \hline \end{array}$$

SOLUTION

Step 1 Since 3 is larger than 1, it is necessary to borrow a one from the next column; this makes the subtraction $11_{five} - 3_{five} = 3_{five}$. Change 2 in the fives column to a 1.

$$\begin{array}{rrr} & \overset{1}{} & \overset{11}{} \\ 3 & \not{2} & 1_{five} \\ -1 & 2 & 3_{five} \\ \hline & & 3_{five} \end{array}$$

Step 2 In the second column, $1_{five} - 2_{five}$ requires borrowing; change 3 in the third column to 2 and take $11_{five} - 2_{five}$ to get 4_{five}.

$$\begin{array}{r} {\scriptstyle 2\ 11} \\ \not{3}\ \not{2}\ 1_{five} \\ -1\ 2\ 3_{five} \\ \hline 4\ 3_{five} \end{array}$$

Step 3 Subtract $2_{five} - 1_{five}$ to get 1_{five}.

$$\begin{array}{r} {\scriptstyle 2} \\ 3\ 2\ 1_{five} \\ -1\ 2\ 3_{five} \\ \hline 1\ 4\ 3_{five} \end{array}$$

The difference is $321_{five} - 123_{five} = 143_{five}$.

Math Note

The answer can be checked by adding $143_{five} + 123_{five}$ and seeing if the answer is 321_{five} (which it is).

▼ Try This One 5

☑ 2. Subtract in bases other than 10.

Perform the indicated operation: $7316_{eight} - 1257_{eight}$.

The multiplication table for base five is shown next.

×	0	1	2	3	4
0	0	0	0	0	0
1	0	1	2	3	4
2	0	2	4	11	13
3	0	3	11	14	22
4	0	4	13	22	31

For example, $3_{five} \times 4_{five} = 22_{five}$ ($3 \times 4 = 12$ in base ten, which is 22_{five}).

Multiplication is done in base five using the same basic procedure you learned in base 10.

EXAMPLE 6 Multiplying in Bases Five and Two

Math Note

As always, you can check your answer by converting to base 10.

(a) Multiply in base five:

$$\begin{array}{r} 314_{five} \\ \times\ 23_{five} \\ \hline \end{array}$$

(b) Multiply in base two:

$$\begin{array}{r} 1011_{two} \\ \times\ 11_{two} \\ \hline \end{array}$$

SOLUTION

(a) First, we multiply each digit in 314_{five} by the last digit in 23_{five}.

Step 1 Multiply $4_{five} \times 3_{five} = 22_{five}$. Write the second digit and carry the first to the fives column.

$$\begin{array}{r} {\scriptstyle 2\ \ } \\ 314_{five} \\ \times\ 23_{five} \\ \hline 2_{five} \end{array}$$

Step 2 Multiply $1_{five} \times 3_{five}$ to get 3_{five} and then add the carried 2_{five} to get 10_{five}. Write the 0 and carry the 1 to the 25s column.

$$\begin{array}{r} \overset{/}{3}14_{five} \\ \times\ 23_{five} \\ \hline 02_{five} \end{array}$$

Step 3 Multiply $3_{five} \times 3_{five} = 14_{five}$ and add 1_{five} to get 20_{five}.

$$\begin{array}{r} \overset{/}{3}14_{five} \\ \times\ 23_{five} \\ \hline 2002_{five} \end{array}$$

Now, repeat, multiplying each digit in 314_{five} by the first digit in 23_{five}.

Step 4 Multiply $4_{five} \times 2_{five} = 13_{five}$. Write the 3 in the fives column and carry the 1.

$$\begin{array}{r} \overset{/}{3}14_{five} \\ \times\ 23_{five} \\ \hline 2002_{five} \\ 30\ \ \end{array}$$

Put zero in the ones column.

Step 5 Multiply $1_{five} \times 2_{five} = 2_{five}$ and add the carried 1 to get 3_{five}.

$$\begin{array}{r} 314_{five} \\ \times\ 23_{five} \\ \hline 2002_{five} \\ 330\ \ \end{array}$$

Step 6 Multiply $3_{five} \times 2_{five}$ to get 11_{five}.

$$\begin{array}{r} 314_{five} \\ \times\ 23_{five} \\ \hline 2002_{five} \\ 11330_{five} \end{array}$$

Step 7 Add the partial products in base five.

$$\begin{array}{r} 314_{five} \\ \times\ 23_{five} \\ \hline 2002_{five} \\ 11330_{five} \\ \hline 13332_{five} \end{array}$$

The product is $314_{five} \times 23_{five} = 13332_{five}$.

(b) The multiplication table for base two is

×	0	1
0	0	0
1	0	1

Using the procedure from part (a),

$$\begin{array}{r} 1011_{two} \\ \times\ \ 11_{two} \\ \hline 1011\ \ \\ 10110\ \ \\ \hline 100001_{two} \end{array}$$

(Remember to add the partial products in base two!) The product is 100001_{two}.

☑ 3. Multiply in bases other than 10.

▼ Try This One 6

(a) $\begin{array}{r} 321_{\text{four}} \\ \times\ 12_{\text{four}} \\ \hline \end{array}$

(b) $\begin{array}{r} 621_{\text{eight}} \\ \times\ 45_{\text{eight}} \\ \hline \end{array}$

Division is performed in the same way that long division is performed in base 10. The basic procedure for long division is illustrated in Example 7.

It will be helpful to look at the multiplication table for the given base in order to find the quotients.

EXAMPLE 7 Dividing in Base Five

Divide in base five:

$$3_{\text{five}}\overline{)2032_{\text{five}}}$$

SOLUTION

Step 1 Using the multiplication table for base five, we need to find a product less than or equal to 20_{five} that is divisible by 3_{five}. This is done as follows.

×	0	1	2	3	4
0	0	0	0	0	0
1	0	1	2	3	4
2	0	2	4	11	13
3	0	3	11	14	22
4	0	4	13	22	31

The number we need is 14_{five} and $3_{\text{five}} \times 3_{\text{five}} = 14_{\text{five}}$. The first digit in the quotient is 3_{five}.

$$\begin{array}{r} 3\phantom{32_{\text{five}}} \\ 3_{\text{five}}\overline{)2032_{\text{five}}} \end{array}$$

Step 2 Then multiply $3_{\text{five}} \times 3_{\text{five}} = 14_{\text{five}}$ and write the quotient under 20. Subtract $20_{\text{five}} - 14_{\text{five}}$ to get 1_{five}, then bring down the next digit.

$$\begin{array}{r} 3\phantom{32_{\text{five}}} \\ 3_{\text{five}}\overline{)2032_{\text{five}}} \\ \underline{14}\downarrow\phantom{2_{\text{five}}} \\ 13\phantom{2_{\text{five}}} \end{array}$$

Step 3 Next find a product smaller than or equal to 13_{five} in the table. It is 11_{five}. Since $3_{\text{five}} \times 2_{\text{five}} = 11_{\text{five}}$, write the 2 in the quotient. Then multiply $2_{\text{five}} \times 3_{\text{five}} = 11_{\text{five}}$. Write the 11 below the 13. Subtract $13_{\text{five}} - 11_{\text{five}}$, which is 2_{five}, and then bring down the 2.

$$\begin{array}{r} 32\phantom{2_{\text{five}}} \\ 3_{\text{five}}\overline{)2032_{\text{five}}} \\ \underline{14}\phantom{32_{\text{five}}} \\ 13\phantom{2_{\text{five}}} \\ \underline{11}\downarrow\phantom{_{\text{five}}} \\ 22\phantom{_{\text{five}}} \end{array}$$

Step 4 Find a product in the multiplication table divisible by 3_{five} that is less than or equal to 22_{five}. It is 22_{five} since $3_{five} \times 4_{five} = 22_{five}$. Write the 4 in the quotient and the 22 below the 22 in the problem. Subtract.

$$\begin{array}{r} 324 \\ 3_{five}\overline{)2032_{five}} \\ \underline{14} \\ 13 \\ \underline{11} \\ 22 \\ \underline{22} \\ 0 \end{array}$$

The remainder is 0. So $2032_{five} \div 3_{five} = 324_{five}$.

This answer can be checked by multiplication, as in Example 6.

$$\begin{array}{r} 324_{five} \\ \times \quad 3_{five} \\ \hline 2032_{five} \end{array}$$

▼ Try This One 7

☑ 4. Divide in bases other than 10.

Divide in base five:

$$4_{five}\overline{)112_{five}}$$

Hopefully, studying a variety of numerical systems in this chapter has given you some perspective on the awesome scope of the study of math by human beings. Sometimes, you need to take a step back to appreciate something that you take for granted. In this case, maybe you'll feel a little bit better about the system of numbers and operations you learned as a child by comparing it to other systems that are unfamiliar to you.

Sidelight GRACE MURRAY HOPPER (1906–1992)

Computer Genius

Grace Murray Hopper received a doctorate in mathematics from Yale and taught at Vassar College before joining the U.S. Navy in 1943. She was commissioned a lieutenant, junior grade, and was assigned to the Harvard Computer Laboratory in 1944. She worked with Dr. Howard Aiken on the first modern computer called the MARK I.

During her time at Harvard, she developed a computer known as "FLOWMATIC,"which enabled programmers to write programs in simple languages such as COBOL and have them translated into the complicated language that the computer uses.

During her lifetime, she received honorary degrees from more than 40 colleges and universities in the United States. She retired from the U.S. Navy in 1969 at the age of 60 only to return to active duty a year later. She stayed on until the mid-1980s, when she retired for the last time as a Rear Admiral. Her retirement ceremony was held aboard the *USS Constitution*, and she was recognized as "the greatest living female authority in the computer field"at that time.

Answers to Try This One

1 1032_{five}

2 11330_{five}

3 100010_{two}

4 $121CB_{\text{sixteen}}$

5 6037_{eight}

6 (a) 11112_{four} (b) 34765_{eight}

7 13_{five}

EXERCISE SET 4-4

Writing Exercises

1. Write how you would explain to a classmate that $5_{\text{six}} + 4_{\text{six}} = 13_{\text{six}}$.
2. When adding in a base number system, whenever the result of an addition in a column has two digits, we "carry" the first digit to the next column to the left. Explain what we're really doing when we do this, and why it makes sense.

Computational Exercises

3. Make an addition table for base 3.
4. Make an addition table for base 4.
5. Make a multiplication table for base 4.
6. Make a multiplication table for base 3.

For Exercises 7–32, perform the indicated operations.

7. $11_{\text{five}} + 21_{\text{five}}$
8. $44_{\text{five}} + 33_{\text{five}}$
9. $3230_{\text{four}} + 1322_{\text{four}}$
10. $8A2B_{\text{twelve}} + 191A_{\text{twelve}}$
11. $321_{\text{six}} + 1255_{\text{six}}$
12. $143_{\text{five}} + 432_{\text{five}}$
13. $4344_{\text{nine}} + 2313_{\text{nine}}$
14. $3145_{\text{seven}} + 216_{\text{seven}}$
15. $43_{\text{five}} - 12_{\text{five}}$
16. $143_{\text{five}} - 34_{\text{five}}$
17. $262_{\text{seven}} - 161_{\text{seven}}$
18. $9327_{\text{eleven}} - 7318_{\text{eleven}}$
19. $42831_{\text{nine}} - 2781_{\text{nine}}$
20. $6323_{\text{seven}} - 415_{\text{seven}}$
21. $12AB_{\text{twelve}} - 93A_{\text{twelve}}$
22. $2121_{\text{three}} - 222_{\text{three}}$
23. $52_{\text{six}} \times 4_{\text{six}}$
24. $241_{\text{seven}} \times 6_{\text{seven}}$
25. $818_{\text{nine}} \times 62_{\text{nine}}$
26. $423_{\text{five}} \times 332_{\text{five}}$
27. $AB5_{\text{twelve}} \times 42_{\text{twelve}}$
28. $5186_{\text{nine}} \times 23_{\text{nine}}$
29. $3_{\text{nine}} \overline{)1568_{\text{nine}}}$
30. $2_{\text{three}} \overline{)1202_{\text{three}}}$
31. $4_{\text{five}} \overline{)2023_{\text{five}}}$
32. $6_{\text{seven}} \overline{)1425_{\text{seven}}}$

Real-World Applications

The binary, octal, and hexidecimal systems are used extensively in computer programming; arithmetic in these systems has very real applications. For Exercises 33–48, perform the indicated operations.

33. $1001_{\text{two}} + 111_{\text{two}}$
34. $62_{\text{eight}} + 145_{\text{eight}}$
35. $3BA_{\text{sixteen}} + 49_{\text{sixteen}}$
36. $10111_{\text{two}} + 1101_{\text{two}}$
37. $1100_{\text{two}} - 11_{\text{two}}$
38. $732_{\text{eight}} - 45_{\text{eight}}$
39. $526B_{\text{sixteen}} - 4A1_{\text{sixteen}}$
40. $1000_{\text{two}} - 101_{\text{two}}$

41. $\begin{array}{r} 1010_{\text{two}} \\ \times\ 101_{\text{two}} \\ \hline \end{array}$

42. $\begin{array}{r} 54_{\text{eight}} \\ \times\ 2_{\text{eight}} \\ \hline \end{array}$

43. $\begin{array}{r} A25_{\text{sixteen}} \\ \times\ 4_{\text{sixteen}} \\ \hline \end{array}$

44. $\begin{array}{r} 326_{\text{eight}} \\ \times\ 21_{\text{eight}} \\ \hline \end{array}$

45. $11_{\text{two}}\overline{)1011_{\text{two}}}$

46. $6_{\text{eight}}\overline{)437_{\text{eight}}}$

47. $5_{\text{sixteen}}\overline{)37B1_{\text{sixteen}}}$

48. $10_{\text{two}}\overline{)11111_{\text{two}}}$

Critical Thinking

49. In a certain base number system, $5 + 6 = 13$. What is the base?
50. In a certain base number system, $15 - 6 = 6$. What is the base?

The American Standard Code for Information (ASCII) is used to encode characters of the alphabet as binary numbers. Each character is assigned an eight-digit binary number written in two groups of four digits as follows:

A–O are prefixed by 0100, and the second grouping starts with A = 0001, B = 0010, C = 0011, etc.

P–Z are prefixed by 0101, and the second grouping starts with P = 0000, Q = 0001, R = 0010, etc. For example, C = 0100 0011 and Q = 0101 0001.

For Exercises 51–54, find the letter of the alphabet corresponding to the binary code.

51. 0100 1100
52. 0101 0101
53. 0100 0111
54. 0101 1010

For Exercises 55–58, write each word in ASCII code.

55. DORM
56. PARTY
57. UNION
58. QUAD

CHAPTER 4 Summary

Section	Important Terms	Important Ideas
4-1	Number Numeral Numeration system Tally system Simple grouping system Multiplicative grouping system Digit Place value Positional system Expanded notation	**Throughout** history, people have used different numeration systems. These systems include the Egyptian, the Chinese, the Babylonian, and the Roman numeration systems. The system that is used in our world today is called the Hindu-Arabic numeration system. It uses the base 10 and 10 symbols called digits to represent numbers.
4-2	Algorithm	**There** were many methods developed for performing calculations before the development of electronic calculators. The Egyptian algorithm, the Russian peasant method, the lattice method, and Napier's bones are all procedures that can be used to multiply by hand.
4-3	Base Binary system Octal system Hexadecimal system	**Numbers** can be written using different bases. For example, the base five system has only five digits. They are 0, 1, 2, 3, and 4. The place values of the numbers written in the base five system are $5^0 = 1$, $5^1 = 5$, $5^2 = 25$, etc.
4-4		**Operations** such as addition, subtraction, multiplication, and division can be performed in other number bases the same way they are performed in base 10.

MATH IN Retail Sales REVISITED

The first UPC shown decodes as 21000 77436: this is from a staple of many college students' diet, a box of Kraft original macaroni and cheese.

The second decodes as 85909 12179: this corresponds to a 16-GB iPod Touch.

Review Exercises

Section 4-1

For Exercises 1–5, write each number using Hindu-Arabic numerals.

1. 𓁨𓍢𓍢𓎆𓎆𓏺
2. 𓂭𓆼𓆼𓎆𓎆𓏺𓏺𓏺𓏺𓏺𓏺
3. 𒌋𒁹 𒌋𒌋𒁹
4. MCXLVII
5. CDXIX
6. 𒌋𒌋𒌋𒁹 𒌋𒌋𒁹𒁹
7. 二千六百四
8. 九百五十七

For Exercises 9–16, write each number in the system given.

9. 49 in the Egyptian system
10. 896 in the Roman system
11. 88 in the Babylonian system
12. 125 in the Egyptian system
13. 503 in the Roman system
14. 8,325 in the Chinese system
15. 165 in the Babylonian system
16. 74 in the Chinese system

For Exercises 17–18, perform the indicated operation. Leave answers in the Egyptian system.

17. 𓍢𓍢𓍢𓎆𓎆𓎆𓎆𓎆||||| + 𓆼𓍢𓍢𓎆𓎆||||||
18. 𓆼𓆼𓍢𓍢𓍢𓎆𓎆||| + 𓂭𓆼𓎆𓎆|||||

Section 4-2

For Exercises 19–22, use the Egyptian algorithm to multiply.

19. 23×12
20. 13×8
21. 7×21
22. 15×16

For Exercises 23–26, multiply using the Russian peasant method.

23. 15×22
24. 12×17
25. 13×12
26. 22×45

For Exercises 27–30, use the lattice method of multiplication to find each product.

27. 23×85
28. 45×398
29. 439×833
30. 548×505

For Exercises 31–34, use Napier's bones (constructed for the exercises in Section 4-2) to find each product.

31. 31×82
32. 74×53
33. 147×95
34. 88×796

Section 4-3

For Exercises 35–44, write each number in base 10.

35. 1110111_{two}
36. 672_{eight}
37. $A03B_{twelve}$
38. 231_{four}
39. 14441_{five}
40. 2012_{three}
41. 6000_{seven}
42. 28645_{nine}
43. 555_{six}
44. $1A214_{eleven}$

For Exercises 45–54, write each number in the specified base.

45. 32 in base six
46. 105 in base 12
47. 2,001 in base nine
48. 81 in base three
49. 43 in base two
50. 213 in base eight
51. 19 in base four
52. 51 in base two
53. 343 in base seven
54. 899 in base 12

In Exercises 55–62, subscripts are omitted since the base is provided in the instructions.

For Exercises 55–58, convert each binary number to (a) octal and (b) hexadecimal.

55. 111011011
56. 10001110111
57. 1101100111
58. 111000111101

For Exercises 59 and 60, convert each octal number to binary.

59. 7324
60. 643

For Exercises 61 and 62, convert each hexadecimal number to binary.

61. A5B3
62. 9F87

Section 4-4

For Exercises 63–80, perform the indicated operation.

63. $156_{nine} + 84_{nine}$
64. $434_{five} + 341_{five}$
65. $101110_{two} + 1101_{two}$
66. $5342_{six} + 1305_{six}$
67. $6A20_{twelve} + B096_{twelve}$
68. $7267_{nine} - 354_{nine}$
69. $1010011_{two} - 100111_{two}$
70. $2120_{three} - 1212_{three}$
71. $3312_{four} - 2321_{four}$
72. $65602_{seven} - 46031_{seven}$
73. $371_{nine} \times 51_{nine}$
74. $242_{five} \times 3_{five}$
75. $1101_{two} \times 111_{two}$
76. $6A5_{sixteen} \times 8_{sixteen}$
77. $3_{five}\overline{)1242_{five}}$
78. $7_{eight}\overline{)3426_{eight}}$
79. $10_{eight}\overline{)3426_{eight}}$
80. $5_{sixteen}\overline{)324_{sixteen}}$

Chapter Test

For Exercises 1–5, write each number using Hindu-Arabic numerals.

1. 𓁨𓁨𓍢𓍢𓍢𓎆||
2. 𓂭𓂭𓆼𓆼𓆼𓆼𓎆𓎆𓎆𓎆𓎆|||||
3. 𒌋𒌋𒁹 𒌋𒁹
4. MCMLXVI
5. CDXXVI
6. 三百六十八

For Exercises 7–12, write each number in the system given.

7. 93 in the Egyptian system
8. 567 in the Roman system
9. 55 in the Babylonian system
10. 521 in the Egyptian system
11. 605 in the Roman system
12. 873 in the Chinese system

For Exercises 13–16, multiply using the given method.

13. Multiply 17×13 by the Egyptian algorithm.
14. Multiply 23×15 by the Russian peasant method.
15. Use the lattice method to multiply 364×736.
16. Use Napier's bones (constructed for the exercises in Section 4-2) to multiply 112×237.

For Exercises 17–26, write each number in base 10.

17. 341_{five}
18. 573_{eight}
19. $\text{A07B}_{\text{twelve}}$
20. 312_{four}
21. 14411_{five}
22. 21101_{three}
23. 4000_{five}
24. 1100111_{two}
25. 463_{seven}
26. $\text{1A436}_{\text{eleven}}$

For Exercises 27–36, write each number in the specified base.

27. 43 in base five
28. 183 in base 12
29. 4673 in base nine
30. 65 in base three
31. 17 in base two
32. 316 in base eight
33. 91 in base four
34. 48 in base two
35. 434 in base seven
36. 889 in base 12

For Exercises 37 and 38 convert the binary number to (a) octal and (b) hexadecimal.

37. 111011001
38. 11000111011
39. Convert 7324_{eight} to binary.
40. Convert $\text{A6D92}_{\text{sixteen}}$ to binary.

For Exercises 41–54, perform the indicated operation.

41. $\begin{array}{r} 263_{\text{nine}} \\ +\ 18_{\text{nine}} \\ \hline \end{array}$

42. $\begin{array}{r} 341_{\text{five}} \\ +213_{\text{five}} \\ \hline \end{array}$

43. $\begin{array}{r} 111010_{\text{two}} \\ +\ 1101_{\text{two}} \\ \hline \end{array}$

44. $\begin{array}{r} 2435_{\text{six}} \\ +5013_{\text{six}} \\ \hline \end{array}$

45. $\begin{array}{r} \text{5A79}_{\text{twelve}} \\ +\text{B068}_{\text{twelve}} \\ \hline \end{array}$

46. $\begin{array}{r} 6772_{\text{eight}} \\ -\ 735_{\text{eight}} \\ \hline \end{array}$

47. $\begin{array}{r} 11001010_{\text{two}} \\ -\ 110011_{\text{two}} \\ \hline \end{array}$

48. $\begin{array}{r} 2212_{\text{three}} \\ -1202_{\text{three}} \\ \hline \end{array}$

49. $\begin{array}{r} 3213_{\text{four}} \\ -2123_{\text{four}} \\ \hline \end{array}$

50. $\begin{array}{r} 20665_{\text{seven}} \\ -10364_{\text{seven}} \\ \hline \end{array}$

51. $\begin{array}{r} 254_{\text{six}} \\ \times\ 3_{\text{six}} \\ \hline \end{array}$

52. $\begin{array}{r} 413_{\text{five}} \\ \times\ 21_{\text{five}} \\ \hline \end{array}$

53. $7_{\text{eight}}\overline{)1342_{\text{eight}}}$

54. $2_{\text{three}}\overline{)1012_{\text{three}}}$

Project

In this project, we'll test various methods of multiplication, with a goal of making an individual decision on which method is most effective. We'll consider both ease of use, amount of time needed, and likelihood of getting the correct product. You will need a calculator to check the products you compute, as well as a watch to time each calculation.

(a) **Two-digit multiplication** Multiply the first two digits of your social security number and the last two digits using (i) the Egyptian algorithm, (ii) the Russian peasant method, (iii) the lattice method, (iv) Napier's bones, and (v) the method you learned in grade school. In each case, time how long it takes the method from start to finish, and make note of whether you got the right answer for each.

If you are working in a group, each group member should repeat the above steps. If you are working on your own, repeat the process, multiplying the first two digits of your phone number by the last two, then again multiplying the first two digits of your zip code by the last two.

(b) **Three-digit multiplication** Repeat question 1, but this time choose numbers this way: turn to a random page past this one in the book and note the page number, then to another page at least 100 pages further on and note that page number. Use each method of multiplication again. If you are working in a group, each group member should choose different page numbers. If you are working on your own, choose three sets of random pages.

(c) Calculate the average amount of time it took to perform all calculations with each of the five methods, and rank them in order from fastest to slowest.

(d) Calculate the percentage of problems in which the correct answer was obtained for each method, and rank them from highest to lowest.

(e) Based on your experience in using the methods, and your rankings in questions 3 and 4, write a short essay identifying the method that you think is most effective, and justify that choice.

CHAPTER 5

The Real Number System

Outline

5-1 The Natural Numbers
5-2 The Integers
5-3 The Rational Numbers
5-4 The Irrational Numbers
5-5 The Real Numbers
5-6 Exponents and Scientific Notation
5-7 Arithmetic and Geometric Sequences
Summary

MATH IN Government Spending

Most Americans have some vague idea that the government spends way more than it has, and that our nation is in debt. But very few people have any idea just how much money we're talking about. Maybe it's because ignorance is bliss, or maybe the numbers are just so staggering that not just anyone can appreciate their size.

One of the really interesting things about being a math professor is that when you tell people what you do, they almost always respond with something like "Wow, I'm not very good at math." Why are people so anxious to brag about not being good at math? Nobody has ever met an English professor and said "Wow, I stink at English . . . I'm practically illiterate!". Of course, literacy is important for success in life, but so is *numeracy*—the ability to understand and work with numbers.

This chapter is all about studying different categories of numbers, and the way we combine them using operations. It's unlikely that you'll encounter operations that you have never seen before in this chapter. Instead, we will focus on a deeper understanding of numbers and operations, much the same way you focus on *understanding* sentences rather than just reading individual words in English class.

That brings us back to our good friends in Washington. Here are some frightening numbers about the financial state of our government, followed by some questions. By the time you finish this chapter, you should be able to answer all of the questions. But more importantly, you should have a real understanding of the significance of these numbers. All of the dollar amounts are written in scientific notation, which we will study in Section 5-6. Write your answers in decimal notation, and then write how those numbers would be read aloud.

Fact: The government budget deficit for 2008 was projected to be $\$4.1 \times 10^{11}$. (This means that the government spent 410 billion dollars more than it took in for 2008.)

Fact: In 2008, the government spent $\$4.12 \times 10^{11}$ just paying *interest* on the national debt. By contrast, total federal spending for 2008 was $\$2.931 \times 10^{12}$.

Fact: The national debt, which is the amount of money that is owed by the government to various creditors, was about $\$1.06 \times 10^{13}$ at the end of 2008.

Question 1: The population of the U.S. in 2008 was about 3.06×10^{8}. How many dollars per person did the government spend total in 2008, and how many dollars per person did it spend above what it took in?

Question 2: What percentage of the budget for 2008 went to simply pay interest on the national debt?

Question 3: If the national debt were divided evenly among all citizens, how much would each person owe?

Question 4: Most congressional representatives are pretty well off. There are 435 members of the House of Representatives and 100 senators. If they got together to pay off the national debt, how much would each owe?

For answers, see Math in Government Spending Revisited on page 272

Section 5-1 The Natural Numbers

LEARNING OBJECTIVES

- ☐ 1. Find the factors of a natural number.
- ☐ 2. Identify prime and composite numbers.
- ☐ 3. Find the prime factorization of a number.
- ☐ 4. Find the greatest common factor of two or more numbers.
- ☐ 5. Find the least common multiple of two or more numbers.

Your first exposure to numbers had nothing to do with symbols written on a page—you learned about the concept of numbers by counting the things around you. Maybe you had one dog, two parents, three sisters, four stuffed animals, and five pairs of those little footy pajamas. At some point you could count to 10 (using your fingers), and maybe even 20 if you were barefoot. Eventually, you headed off to school and learned about other types of numbers—negatives, fractions, decimals—but you can still get pretty far in the material world with the counting numbers you started with.

In this section, we'll study the counting numbers in a bit more depth than you did when you used your fingers and toes. This will be a good start on our study of the real number system, because in a very real sense, the counting numbers form the building blocks of the real numbers.

Prime and Composite Numbers

The counting numbers we talked about above are often called the *natural numbers* because we use them to enumerate objects in the natural world.

> The set of **natural numbers,** also known as **counting numbers,** consists of the numbers 1, 2, 3, 4, ...

Multiplication is used extensively in studying the natural numbers, largely because of the following fact: every natural number can be written as the product of two or more natural numbers. For example,

$$12 = 3 \times 4 \quad 16 = 4 \times 4 \quad 19 = 1 \times 19 \quad 30 = 2 \times 3 \times 5$$

The natural numbers that are multiplied to get a product are called the *factors* of that product. So the statement $12 = 3 \times 4$ tells us that 3 and 4 are factors of 12. The example $16 = 4 \times 4$ shows that the factors of a number don't have to be distinct, and the example $19 = 1 \times 19$ shows that 1 is a perfectly acceptable factor.

We know that 3 and 4 are factors of 12, but they're not the only factors. We can also write 12 as 1×12 or 2×6. Now we have a list of *all* factors of 12: 1, 2, 3, 4, 6, and 12.

EXAMPLE 1 Finding Factors

Find all factors of 24.

SOLUTION

Think of all the ways you can write 24 as a product of two numbers, starting with 1 as one of the factors, and working upward.

$$1 \times 24 \qquad 2 \times 12 \qquad 3 \times 8 \qquad 4 \times 6$$

These are all of the possibilities, so the factors of 24 are 1, 2, 3, 4, 6, 8, 12, and 24.

☑ 1. Find the factors of a natural number.

Calculator Guide

We can use division on a calculator to quickly see if one number is a factor of another. For example, to decide if 8 or 12 are factors of 156:

Standard Scientific Calculator:

156 [÷] 8 [=] Result: 19.5

156 [÷] 12 [=] Result: 13

Standard Graphing Calculator:

156 [÷] 8 [Enter] Result: 19.5

156 [÷] 12 [Enter] Result: 13

Since 19.5 is not a natural number, 8 is not a factor of 156. But 12 is, because the result of the division is 13, a natural number.

▼ Try This One 1

Find all factors of 50.

Notice that if you divide 24 by any of its factors, the remainder is zero. In fact, this is a good way to test to see if one number is a factor of another. Six is a factor of 24 because $24 \div 6 = 4$ with no remainder, but 5 is not a factor of 24 because $24 \div 5$ has remainder 4. For this reason, we will sometimes call the factors of a number its **divisors.**

A natural number a is **divisible** by another natural number b if dividing a by b results in a remainder of zero. In this case, we will write $b|a$. This is also read as "b divides a."

We know from above that 6 is a factor of 24, so this means that 24 is divisible by 6, and we would write $6|24$. (This should not be confused with 6/24, which is the number 6 divided by 24.)

Most numbers have more than two factors, but not all: for example, 7 can only be written as 1×7. We give these numbers a special name.

A natural number is called **prime** if it has exactly two factors, 1 and itself.

Sidelight TESTING DIVISIBILITY

It can be very difficult to decide if a large number is divisible by some other number without a calculator. The table below lists some clever tests for divisibility by certain numbers.

Divisibility Tests

Divisible By	Test	Example
2	The last digit is even (that is 0, 2, 4, 6, or 8).	3,456,786 is divisible by 2 since the last digit, 6, is even
3	The sum of the digits is divisible by three.	387 is divisible by 3 since $3 + 8 + 7 = 18$ is divisible by 3.
4	Last two digits are divisible by 4.	234,564 is divisible by 4 since 64 is divisible by 4
5	Ends in a 0 or 5.	56,325 is divisible by 5 since it ends in 5.
6	Divisible by both 2 and 3.	534 is divisible by 6 since it's even and the sum of the digits is 12 (meaning it is divisible by both 2 and 3).
7	Double the last digit and subtract from the number formed by deleting the last digit. Repeat this until it's easy to determine if the number is divisible by 7. If so, then the original number is divisible by 7.	2,247 is divisible by 7 since $224 - 14 = 210$ is divisible by 7.
8	The last three digits are divisible by 8.	9,655,064 is divisible by 8 since 064 is divisible by 8.
9	The sum of the digits is divisible by 9.	3,258 is divisible by 9 since $3 + 2 + 5 + 8 = 18$ is divisible by 9.
10	Last digit is 0.	12,254,260 is divisible by 10 since its last digit is 0.
11	Alternate the signs of the digits and add; if this sum is divisible by 11, so is the original number.	352,715 is divisible by 11 since $3 + (-5) + 2 + (-7) + 1 + (-5) = -11$ is divisible by 11.

Not surprisingly, we also have a name for numbers with more than two factors.

A natural number is called **composite** if it has three or more factors.

EXAMPLE 2 Deciding if a Number Is Prime

Math Note

Is the number 1 prime? See Exercise 106 at the end of the section for a discussion. (Today, 1 is not considered to be prime.)

Decide whether each number is prime or composite.

(a) 25 (b) 17 (c) 12 (d) 31

SOLUTION

If we can find even one factor other than 1 and the number itself, then it is composite.

(a) Five is a factor of 25, since $5 \times 5 = 25$, so 25 is composite.
(b) The only factors of 17 are 1 and 17, so it is prime.
(c) Two is a factor of 12 (as are several other numbers), so 12 is composite.
(d) There are no factors of 31 other than 1 and 31, so it is prime.

▼ Try This One 2

☑ 2. Identify prime and composite numbers.

Decide whether each number is prime or composite.

(a) 34 (b) 29 (c) 27 (d) 10

You might wonder if there's a formula for finding all of the prime numbers. If you find one, we can guarantee you'll be famous, because people have been looking for one for well over 2,000 years. But here's a clever way to generate a list of the prime numbers up to 50. (See Figure 5-1). We write a list of all natural numbers up to 50 and cross out the ones that can't be prime. Begin by crossing out 1, since it's not considered to be prime. Any number that has 2 as a factor can't be prime (other than 2 itself), so we cross out 4, 6, 8, 10, and all the even numbers. Any number other than 3 that is divisible by 3 also can't be prime, so we cross out 3, 6, 9, 12, ... (The ones that 2 also divides are already crossed out.) We continue this process for numbers divisible by 5 and 7. At this point, all remaining numbers are prime. (In Exercise 109, we'll look at why stopping at 7 is sufficient.)

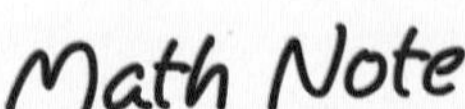

The process in Figure 5-1 is known as the Sieve of Eratosthenes, named after the Greek scholar from the third century BCE who invented it. He was also the first human to calculate the circumference of the earth to a reasonably accurate value.

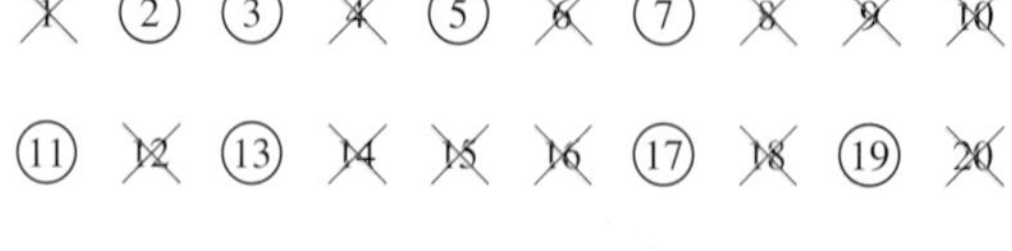

Figure 5-1

All of the prime numbers less than 50 are circled in Figure 5-1.

Sidelight TWIN PRIMES

There are four pairs of prime numbers that differ by 2 in Figure 5-1: 3 and 5, 17 and 19, 29 and 31, and 41 and 43. Pairs of prime numbers that differ by 2 are called twin primes, and mathematicians have been trying for a very long time to answer the question "Are there infinitely many pairs of twin primes?" It was proved over 2,000 years ago by Euclid that there are infinitely many prime numbers, but nobody has been able to prove that there are an infinite number of twin primes. Computer evidence makes it likely that there are, but this little conjecture has escaped proof since before the time of the Roman empire.

Prime Factorization

We know that there's more than one way to write 12 as a product: for example, we could write it as $12 = 2 \times 6$, or $12 = 3 \times 4$. But notice that some of those factors can also be written as products: $6 = 2 \times 3$ and $4 = 2 \times 2$. Now let's rewrite each multiplication statement:

$$12 = 2 \times 6 = 2 \times 2 \times 3$$
$$12 = 3 \times 4 = 3 \times 2 \times 2$$

We ended up with the same result (aside from the order, which doesn't matter), and all of the factors are prime numbers. This result indicates a very important property of natural numbers known as the fundamental theorem of arithmetic. (A theorem is a statement that has been proved to be true.)

The Fundamental Theorem of Arithmetic

Every composite number can be written as the product of prime numbers, and there's only one way to do so. (The order of the factors is always unimportant.)

The next two examples illustrate two methods for finding the prime factorization of a number. In the *tree method*, a diagram is built by finding factorizations successively.

EXAMPLE 3 Finding Prime Factorization Using the Tree Method

Find the prime factorization of 100 using the tree method.

Math Note

When using the tree method, it does not matter how you start. For example, 100 could be factored as 10×10 or 4×25. The fundamental theorem of arithmetic states that you will always end with the same factorization, $2 \times 2 \times 5 \times 5$.

SOLUTION

Start with any factorization of 100, say 2×50, then factor 50 as 5×10. Finally factor 10 as 2×5. This is shown using a tree.

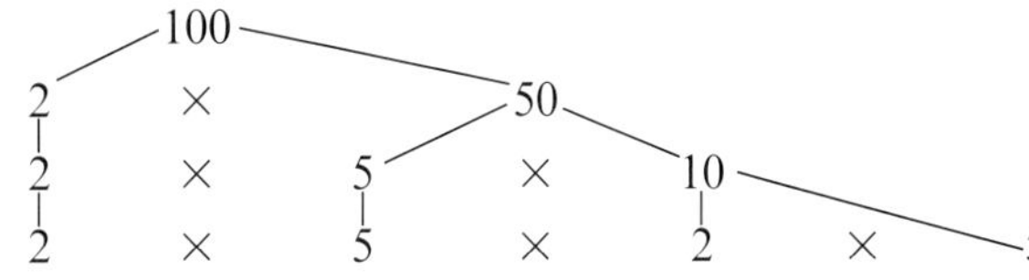

Rearrange the factors in order: $2 \times 2 \times 5 \times 5$ or $2^2 \times 5^2$.

▼ Try This One 3

Use the tree method to find the prime factorization of 360.

In the *division method*, we start with the original number and keep dividing by prime numbers until the result of the division is prime.

EXAMPLE 4 Finding Prime Factorization Using the Division Method

Find the prime factorization of 100 using the division method.

Math Note

To avoid mistakes, always start with the smallest prime factor and keep using it until you can't divide any further, and then move to the next prime number. Do not skip around.

SOLUTION

First, divide 100 by 2 and then divide the answer by 2. Continue dividing the answer until you cannot find an answer that is divisible by 2, then try to divide by 3, then 5, etc., as shown.

$$\begin{array}{r|l} & 5 \\ 5 & 25 \\ 2 & 50 \\ 2 & 100 \end{array}$$

This shows that $100 = 2 \times 2 \times 5 \times 5 = 2^2 \times 5^2$.

▼ Try This One 4

Use the division method to find the prime factorization of 360.

☑ 3. Find the prime factorization of a number.

Sidelight PERFECT NUMBERS

A *perfect number* is a number such that the sum of its proper factors is equal to the number. The proper factors of a number are all the factors of the number except the number itself. The first perfect number is 6. The factors of 6 are 1, 2, 3, and 6. The sum of the factors of 6 (excluding 6) is equal to 6: $1 + 2 + 3 = 6$. The factors of 28 are 1, 2, 4, 7, 14, and 28, and the sum of the proper factors is $1 + 2 + 4 + 7 + 14 = 28$.

Can you guess what the next perfect number is? It is 496. The sum of its proper factors is 496; $1 + 2 + 4 + 8 + 16 + 31 + 62 + 124 + 248 = 496$.

The first four perfect numbers were thought to be known by the Pythagoreans. (See the Sidelight on page 236.) The fifth perfect number (33,550,336) was found in 1461. Around 1600, Pietro Cataldi discovered the sixth and seventh perfect numbers. Around 1732, Leonhard Euler is said to have discovered the eighth perfect number. Since these perfect numbers are really large, it's difficult to verify that they are indeed perfect numbers without the aid of a computer. The eighth perfect number has 19 digits!

The 9th, 10th, 11th, and 12th perfect numbers were discovered in the 1800s. The 12th perfect number contains 77 digits and is the last perfect number to be discovered without the aid of a computer. Up to 1950, only 12 perfect numbers were known.

With the advent of modern computers and as the capacity of computers has increased, larger and larger perfect numbers have been discovered.

Today over 35 perfect numbers are known, and the search continues. All perfect numbers (except 28) that have been found to date end in 6. It has not been proved if other perfect numbers that do not end in 6 exist. It is not known if there are a finite number or an infinite number of perfect numbers and whether there are any odd perfect numbers.

Greatest Common Factors

For a variety of reasons that will become apparent later, finding factors that are common to two or more numbers is useful. Let's look at the factors of 18 and 24:

18: 1, 2, 3, 6, 9, 18
24: 1, 2, 3, 4, 6, 8, 12, 24

These two numbers have four factors in common: 1, 2, 3, and 6. Obviously, the largest of these is 6, so we will call it the *greatest common factor* of 18 and 24.

Math Note

The GCF is also sometimes called the greatest common divisor.

The **greatest common factor** of two or more numbers is the largest number that is a factor of all of the original numbers. We will use the abbreviation GCF to represent the greatest common factor.

For relatively small numbers like 18 and 24, we can list all of the factors and select the largest. But this is really inconvenient for large numbers, so we'll use a different strategy, based on prime factorization.

Procedure for Finding the Greatest Common Factor of Two or More Numbers

Step 1 Write the prime factorization of each number.

Step 2 Make a list of each prime factor that appears in all prime factorizations. For prime factors with exponents, choose the smallest power that appears in each.

Step 3 The GCF is the product of the numbers you listed in Step 2.

Example 5 illustrates the procedure.

EXAMPLE 5 Finding the Greatest Common Factor of Two Numbers

Find the GCF of 72 and 180.

Math Note

When the GCF of two numbers is 1, we call the numbers **relatively prime.** For example, 15 and 17 have no factors in common other than 1, so they are relatively prime.

SOLUTION

Step 1 Write the prime factorizations of 72 and 180:

$$72 = 2 \times 2 \times 2 \times 3 \times 3 = 2^3 \times 3^2$$
$$180 = 2 \times 2 \times 3 \times 3 \times 5 = 2^2 \times 3^2 \times 5$$

Step 2 List the common factors: 2 (with exponent 2) and 3 (with exponent 2).

Step 3 The GCF is $2^2 \times 3^2 = 36$.

▼ Try This One 5

Find the GCF of 54 and 144.

Now let's try finding the GCF of three numbers.

EXAMPLE 6 Finding the Greatest Common Factor of Three Numbers

Find the GCF of 40, 60, and 100.

SOLUTION

Step 1 Write the prime factorization of each number:

$$40 = 2 \times 2 \times 2 \times 5 = 2^3 \times 5$$
$$60 = 2 \times 2 \times 3 \times 5 = 2^2 \times 3 \times 5$$
$$100 = 2 \times 2 \times 5 \times 5 = 2^2 \times 5^2$$

Step 2 List the common factors: 2 (with exponent 2), and 5 (with exponent 1).

Step 3 The GCF is $2^2 \times 5 = 20$.

▼ Try This One 6

☑ 4. Find the greatest common factor of two or more numbers.

Find the GCF of 45, 75, and 150.

Finding the greatest common factor will come in handy when reducing fractions in Section 5-3, but it also has real-world applications.

EXAMPLE 7 Applying the GCF to Packaging Goods for Sale

An enterprising college student gets a great deal on slightly past-their-prime packets of instant coffee from a chain of coffee stores. He acquires 200 packets of decaf and 280 packets of regular coffee. The plan is to package them for resale at a college fair so that there's only one type of coffee in each box, and every box has the same number of packets. How can he do this so that each box contains the largest number of packets possible? (Of course, he wants to package all of the coffee he has.)

SOLUTION

The number of packets per box has to be a factor of both 200 and 280 so that there won't be any left over. If the boxes are to contain the largest number of packets possible, our entrepeneur will need to find the greatest common factor.

$$200 = 2 \times 2 \times 2 \times 5 \times 5 = 2^3 \times 5^2$$
$$280 = 2 \times 2 \times 2 \times 5 \times 7 = 2^3 \times 5 \times 7$$

The factors common to both numbers are 2 (with exponent 3) and 5 (with exponent 1). So the GCF is $2^3 \times 5 = 40$. That tells us that he should put 40 packets in each box. He'll have 5 boxes of decaf and 7 boxes of regular to sell.

▼ Try This One 7

The enterprising student from Example 7 later buys unsold Halloween candy for resale. If he has 360 "fun size" bags of peanut M&M's and 420 bags of plain M&M's, how should he package them to follow the same guidelines as in Example 7?

Least Common Multiples

If you multiply any natural number by 1, then 2, then 3, and so on, you will generate a list of numbers. We call this list the **multiples** of a number. For example, starting with 6, we would get $6 \times 1 = 6$, $6 \times 2 = 12$, $6 \times 3 = 18$, etc. So the multiples of 6 are 6, 12, 18, 24, 30,. . . .

Take a look at the multiples of 4 and 6:

Multiples of 4: 4, 8, 12, 16, 20, 24,. . .
Multiples of 6: 6, 12, 18, 24, 30, 36,. . .

The smallest number on both lists is 12, so we will call it the *least common multiple* of 4 and 6.

The **least common multiple (LCM)** of two or more numbers is the smallest number that is a multiple of each. If you like, you can think of it as the smallest number that is divisible by all of the numbers.

You can sometimes find the least common multiple by listing out the multiples, but for large numbers this can be very difficult. So a procedure similar to the one we used for finding the GCF can be developed for finding the LCM; it also uses prime factorization.

Procedure for Finding the Least Common Multiple of Two or More Numbers

Step 1 Write the prime factorization of each number.

Step 2 Make a list of every prime factor that appears in *any* of the prime factorizations. For prime factors with exponents, choose the largest that appears in any factorization.

Step 3 The LCM is the product of the numbers you listed in Step 2.

We'll illustrate this procedure with an example.

EXAMPLE 8 Finding the Least Common Multiple of Three Numbers

Find the LCM of 24, 30, and 42.

SOLUTION

Step 1 Write the prime factorization of each number:

$$24 = 2 \times 2 \times 2 \times 3 = 2^3 \times 3$$
$$30 = 2 \times 3 \times 5$$
$$42 = 2 \times 3 \times 7$$

Step 2 List all the factors that appear: 2 (with largest exponent 3), 3, 5, and 7.

Step 3 The LCM is $2^3 \times 3 \times 5 \times 7 = 840$.

☑ 5. Find the least common multiple of two or more numbers.

▼ Try This One 8

Find the LCM of each.

(a) 40, 50 (b) 28, 35, 49 (c) 16, 24, 32

The least common multiple is very useful in adding and subtracting fractions, but it has real-world applications as well.

EXAMPLE 9 Applying the LCM to Grocery Shopping

Have you ever noticed that many hot dogs come in packages of 10, but most hot dog buns come in packages of eight? (Who thought *that* was a good idea?) What's the smallest number of packages you can buy of each so that you end up with the same number of hot dogs and buns?

SOLUTION

The total number of hot dogs we buy will be a multiple of 10, while the number of buns will be a multiple of 8. What we need to find is the least common multiple of 8 and 10. So we write the prime factorization of each:

$$8 = 2 \times 2 \times 2 = 2^3$$
$$10 = 2 \times 5$$

The factors that appear in either list are 2 (with largest exponent 3) and 5, so the LCM is $2^3 \times 5 = 40$. That means we would need four packages of hot dogs and five packages of buns. Hope you're hungry!

▼ Try This One 9

After getting her first job out of college, Colleen vows to buy herself a new item of clothing every 15 days, and a new pair of shoes every 18 days just because she can. How long will it be until she buys both items on the same day?

Sidelight FERMAT NUMBERS

In 1650, a French lawyer who studied mathematics in his spare time made a strange observation: the first four numbers of the form $2^{2^n} + 1$ are prime. For $n = 1, 2, 3$, and 4, the numbers are $2^2 + 1 = 5$, $2^4 + 1 = 17$, $2^8 + 1 = 257$, and $2^{16} + 1 = 65{,}537$. At that time, the lawyer, named Pierre de Fermat, speculated that all such numbers were prime. Needless to say, it was not easy to decide that a number like 65,537 was prime in 1650; the next such number, for $n = 5$, is $2^{32} + 1$, which has 10 digits. It turns out that Fermat was incorrect, but it took until the advent of computers to attack the problem in real depth.

The numbers of the form $2^{2^n} + 1$ are called Fermat numbers in his honor, and mathematicians have now been able to show that the Fermat numbers for values of n from 5 to 32 are composite. Beyond that, it gets a little spotty. But we can forgive mathematicians for this limitation: the Fermat number for $n = 33$ has almost 26 billion digits!

An Alternative Method for Finding the GCF

Many students find this alternative method to be easier to execute than the one we learned earlier. You should consider trying both and seeing which you understand better.

The Divide by Primes Method for Finding the GCF of Two or More Numbers

Step 1 List the numbers horizontally.

Step 2 Find a prime number that divides all of the numbers, then divide each by that number.

Step 3 Find a prime number that divides each quotient, then divide each quotient by that prime. Continue this process until no prime divides all remaining quotients.

Step 4 The product of all the prime numbers you divide by is the GCF.

EXAMPLE 10 Using the Divide by Primes Method

Use the divide by primes method to find the GCF of 40, 60, and 100.

SOLUTION

40	60	100	Prime divisor: 2
20	30	50	Prime divisor: 2
10	15	25	Prime divisor: 5
2	3	5	

There are no further prime divisors, so the GCF is $2 \times 2 \times 5 = 20$. This matches our answer from Example 6.

▼ Try This One 10

Use the divide by primes method to find the GCF of 45, 75, and 150.

Answers to Try This One

1 1, 2, 5, 10, 25, 50.

2 (a) Composite (b) Prime (c) Composite (d) Composite

3 $360 = 2^3 \times 3^2 \times 5^2$

4 $360 = 2^3 \times 3^2 \times 5^2$

5 18

6 15

7 60 bags per box, for 6 boxes of peanut and 7 boxes of plain.

8 (a) 200 (b) 980 (c) 96

9 90 days

10 15

EXERCISE SET 5-1

Writing Exercises

1. Describe the set of natural numbers in your own words.
2. What's the difference between a prime number and a composite number?
3. Explain why every natural number other than one has at least two factors.
4. What is the prime factorization of a number? Can you find the prime factorization for every natural number?
5. Explain what the greatest common factor of a list of numbers is.
6. Explain what the last common multiple of a list of numbers is.

Computational Exercises

For Exercises 7–26, find all factors of each number.

7. 16
8. 225
9. 126
10. 54
11. 32
12. 48
13. 9
14. 10
15. 96
16. 100
17. 17
18. 19
19. 64
20. 120
21. 105
22. 365
23. 98
24. 36
25. 71
26. 47

For Exercises 27–36, find the first five multiples of each.

27. 3
28. 7
29. 10
30. 12
31. 15
32. 20
33. 17
34. 19
35. 1
36. 25

For Exercises 37–56, find the prime factorization of each.

37. 16
38. 18
39. 1,296
40. 1,960
41. 17
42. 19
43. 50
44. 64
45. 128
46. 169
47. 300
48. 500
49. 475
50. 625
51. 448
52. 77
53. 247
54. 56
55. 750
56. 825

For Exercises 57–74, find the greatest common factor of the given numbers.

57. 3, 9
58. 10, 35
59. 7, 10
60. 6, 11
61. 30, 36
62. 75, 105
63. 105, 126
64. 210, 140
65. 75, 100
66. 125, 175
67. 12, 24, 48
68. 5, 15, 25
69. 12, 18, 30
70. 42, 60, 18
71. 36, 60, 108
72. 60, 90, 84
73. 100, 225, 350
74. 42, 56, 63

For Exercises 75–92, find the least common multiple of the given numbers.

75. 5, 10
76. 12, 24
77. 7, 5
78. 6, 10
79. 18, 21
80. 25, 35
81. 50, 75
82. 60, 90
83. 70, 90
84. 195, 390
85. 4, 7, 11
86. 5, 6, 13
87. 6, 5, 4
88. 30, 18, 42
89. 12, 18, 36
90. 42, 48, 56
91. 35, 21, 40
92. 22, 33, 44

Real-World Applications

93. In one college class, there are 28 students taking the class for a grade and 20 taking it pass-fail. The instructor wants to assign groups for a project, and has two requirements: every group should contain only students with the same grading option, and all the groups should have the same number of students. What's the greatest number of students that he can put in each group?
94. The manager of a cable company that offers both digital TV and high-speed Internet is trying to arrange his service calls efficiently. He wants to divide the people awaiting service into groups based on whether they need TV or Internet service so that his techs can concentrate on one type of service each day. If there are 40 people waiting for TV service and 32 waiting for Internet service, and he wants to schedule the same number of visits each day, what's the greatest number of visits he can schedule each day?
95. Two people bike on a circular trail. One person can ride around the trail in 24 minutes, and another

person can ride around the trail in 36 minutes. If both start at the same place at the same time, when will they be at the starting place at the same time again?

96. Two clubs offer a "free day" (no admission charge) to recruit new members. The health club has a free day every 45 days. A nearby swimming club has a free day every 30 days. If today is a free day for both clubs, how long will it be until a person can again use both clubs for free on the same day?
97. There are 30 women and 36 men in a bowling league. The president wants to divide the members into all-male and all-female teams, each of the same size. Find the number of members and the number of teams for each gender.
98. A teacher has 24 colored pencils and 18 pictures to be placed in groups with the same number of pencils in each group and the same number of pictures in each group. If each group must have the same type of items, how many groups of pencils and how many groups of pictures can be made? How many of each item will be in a group?
99. An amusement park has two shuttle buses. Shuttle bus A makes six stops and shuttle bus B makes eight stops. The buses take 5 minutes to go from one stop to the next. Each bus takes a different route. If they start at 10:00 A.M. from station one, at what time will they both return to station one? Station one is at the beginning and the end of the loop and is not counted twice.
100. Four lighthouses can be seen from a boat offshore. One light blinks every 10 seconds. The second light blinks every 15 seconds. The third light blinks every 20 seconds, and the last light blinks every 30 seconds. If all four were started at the same time, how often will all the lights be on at the same time?
101. In a political science class, the teacher wants to assign students with the same political affiliation to groups in order to prepare for a debate. She finds that there are 12 Republicans, 15 Democrats, and 21 Independents. If she wants the same number of people in each group, how many groups will there be for each affiliation?
102. A bakery makes three types of muffins to be distributed to local grocery stores: chocolate chip, blueberry, and banana nut. Each morning, the bakers produce 120 chocolate chip muffins, 150 blueberry, and 90 banana nut. They want to package the muffins efficiently for shipping. The plan is to pack them in boxes so that every box has just one kind of muffin, and every box has the same number of muffins. What's the smallest number of boxes they can ship in total?
103. There are two classes of periodic cicadas that are native to the Midwest: one group emerges every 13 years and eats every tree they can find, while the other emerges every 17 years. If they both emerge in a given year, how long will it be until they both emerge together again?
104. Suppose that in one city, the baseball team has a winning season every 5 years, while the football team has a winning season every 3 years. If they both had winning records this year, how long will it be until that happens again?

Critical Thinking

105. How many prime numbers are even? Explain how you decided.
106. Until the 19th century, most mathematicians considered 1 to be a prime number. First, make a case for why it would be reasonable to consider 1 a prime number. Then review the definition of prime number on page 199 and make a case for why 1 should not be considered a prime number.
107. A German mathematician Christian Goldbach (1690–1764) made the following conjecture: Every even number greater than 2 can be expressed as the sum of two prime numbers. For example, $6 = 3 + 3$, $8 = 5 + 3$, etc. Express every even number from 4 through 20 as a sum of two prime numbers.
108. Another conjecture attributed to Goldbach is that any odd number greater than 7 can be expressed as the sum of three odd prime numbers. For example, $11 = 3 + 3 + 5$. Express every odd number from 9 through 25 as the sum of three odd prime numbers.
109. In using the sieve of Eratosthenes to generate a list of prime numbers less than 50, we were able to stop after crossing off the multiples of the prime number 7. Why was this sufficient? (*Hint:* Think about the numbers less than 50 that are multiples of 11—why are they already crossed out?). If we wanted to generate a list of all prime numbers less than 200, what's the largest prime we'd need to cross out the multiples of?
110. Fill in each blank with "less than or equal to" or "greater than or equal to," then explain your responses.

    (a) The GCF of a list of numbers is always __________ all of the numbers.

    (b) The LCM of a list of numbers is always __________ all of the numbers.

Section 5-2 The Integers

LEARNING OBJECTIVES

- ❑ 1. Define whole numbers and integers.
- ❑ 2. Find the opposite and absolute value of a number.
- ❑ 3. Compare numbers using >, <, and =.
- ❑ 4. Add and subtract integers.
- ❑ 5. Multiply and divide integers.
- ❑ 6. Perform calculations using the order of operations.

In Section 5-1, we saw that the natural numbers are practically as old as humanity itself, since they are the natural way that we count things, like family members, possessions, days, and so forth. You might wonder why we didn't include zero in our definition of natural numbers, since it's possible to have zero siblings, zero pairs of torn jeans, or zero days left until the end of the semester. Surprisingly, the number zero wasn't used until around 200 CE, which is thousands of years after natural numbers came into use.

The story for negative numbers starts even later—they entered the picture at least a hundred years later. It's interesting that numbers that we take for granted were in some sense unknown to some of the greatest mathematical minds in human history. But that's the nature of progress.

In this section, we'll extend our study of numbers to include zero and the negatives of natural numbers, which will allow us to study the basic operations of arithmetic in depth. It will also allow us to apply numbers to more real-world situations.

Definition of Integers

When we add the number zero to the natural numbers, we get a new set called the *whole numbers*.

> The set of **whole numbers** is defined as {0, 1, 2, 3, 4,...}.

☑ 1. Define whole numbers and integers.

Every natural number has an **opposite,** or additive inverse. The opposite of 2 is −2, for example. When we expand the set of whole numbers by including the negatives of the natural numbers (−1, −2, −3,...) the resulting set is called the *integers*.

> The set of **integers** is defined as {..., −3, −2, −1, 0, 1, 2, 3,...}.

Math Note

Zero is neither positive nor negative, and its opposite is itself.

One good way to study the integers is to picture them on a number line, as in Figure 5-2.

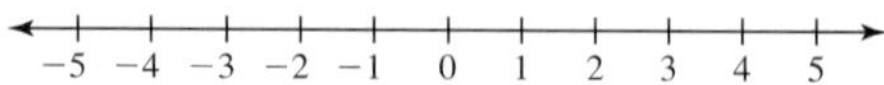

Figure 5-2

Calculator Guide

You cannot use the subtraction sign to enter negative numbers into most calculators. To enter the negative integer −20:

Standard Scientific Calculator
20 [±]

Standard Graphing Calculator
[(−)] 20

The arrows on each end indicate that the line extends infinitely far in each direction. Zero is called the *origin*. The numbers to the right of zero are the *positive integers*, and those to the left are the *negative integers*. Notice that when looking at a number line, two numbers are opposites when they are the same distance away from zero, but in opposite directions, as seen in Figure 5-3.

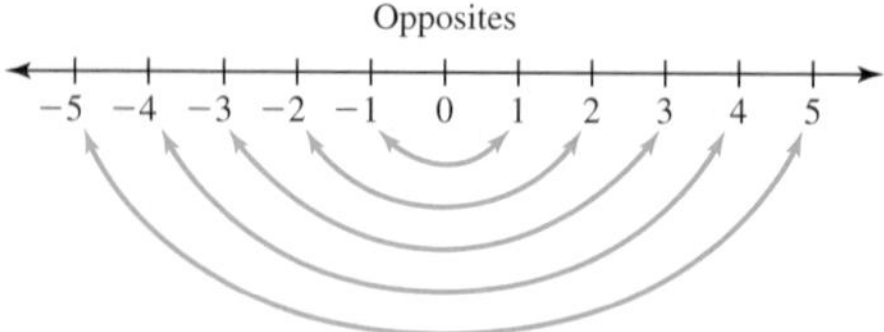

Figure 5-3

Math Note

You can write a positive sign (+) in front of a positive number if you like, but we will assume that any number with no sign in front of it is positive.

If we use the symbol x to represent any arbitrary number, its opposite would look like $-x$, or if you like, $-(x)$. For example, the opposite of 8 is $-(8)$. But this should be true for negative numbers also: the opposite of -8 (which we know is 8) should be written as $-(-8)$. This tells us that $-(-8) = 8$! (Two wrongs don't make a right, but two negatives do make a positive, at least in this case.)

Earlier, we pointed out that a number and its opposite are always the same distance away from zero. We will use the term **absolute value** to describe a number's distance from zero on a number line. Since distance is a physical quantity, the absolute value of every number except zero is positive. So the absolute value of 8 is 8, and the absolute value of -8 is also 8. The following rules describe absolute value:

Rules for Absolute Value

The absolute value of any positive number is the number itself.
The absolute value of any negative number is the opposite of that number.
The absolute value of zero is zero.

The symbol for absolute value is $|\ \ |$. For example,

$$|11| = 11 \qquad |-6| = 6$$

EXAMPLE 1 Finding Opposites and Absolute Values

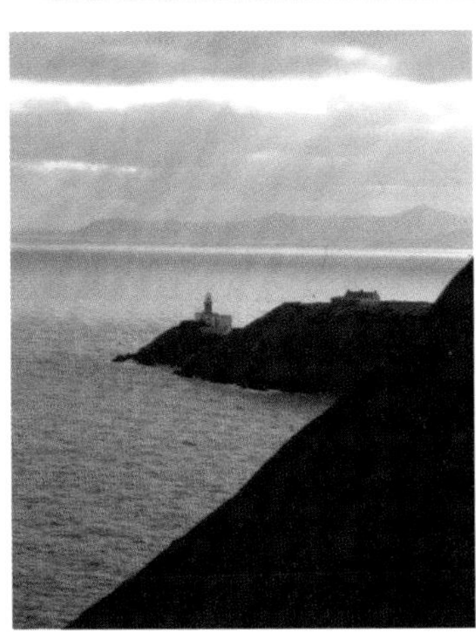

Integers are often used for applications where it makes sense to set a zero and then talk about distances to either side. For example, sea level corresponds to 0; the distance above sea level corresponds to the positive integers; and depth below sea level corresponds to negative integers.

Find each number.

(a) The opposite of 100.
(b) The opposite of -50.
(c) The opposite of 0.
(d) $|26|$
(e) $|-99|$

SOLUTION

(a) The opposite of 100 is -100.
(b) The opposite of -50 is 50 (or $-(-50)$).
(c) Zero is the only number that is its own opposite.
(d) The absolute value of any positive number is the number itself, so $|26| = 26$.
(e) The absolute value of a negative number is its opposite, so $|-99| = 99$.

▼ Try This One 1

Find each number.

(a) The opposite of -9.
(b) The opposite of 24.
(c) $|-15|$
(d) $|18|$
(e) $|0|$

☑ 2. Find the opposite and absolute value of a number.

One application of the number line is to help us decide which of two integers is larger. In comparing two integers, the one that is farther to the right is the larger number. The number 10, for example, is to the right of the number 6 on a number line, so we say that 10 is *greater than* 6, and write $10 > 6$. The number -1 is to the left of 3 on a number line, so we say that -1 is *less than* 3, and write $-1 < 3$.

EXAMPLE 2 Using the Symbols >, <, and =

Fill in the space between the two numbers with >, <, or =.

(a) 4 $\quad$ -2 (b) -20 $\quad$ 0 (c) -10 $\quad$ -7 (d) $|-4|$ $\quad$ 4

Math Note

In mathematics, three other inequality signs are also used. They are:

$\geq$ means "is greater than or equal to."

$\leq$ means "is less than or equal to."

$\neq$ means "is not equal to."

SOLUTION

(a) Since 4 is to the right of -2 on a number line, $4 > -2$.
(b) Since -20 is to the left of 0 on a number line, $-20 < 0$.
(c) Since -10 is to the left of -7 on a number line, $-10 < -7$.
(d) Since $|-4|$ is 4, the two numbers are the same; $|-4| = 4$.

▼ Try This One 2

Fill in the space between the two numbers with $>$, $<$, or $=$.

(a) -5 $\quad$ -15 (b) -3 $\quad$ 2 (c) $|6|$ $\quad$ $|-6|$ (d) 4 $\quad$ -2

☑ 3. Compare numbers using $>$, $<$, and $=$.

Addition and Subtraction of Integers

You learned how to add and subtract integers pretty early in life, but there's a chance you memorized a lot of the results. In studying these operations as an adult, we will focus on really understanding them, and we'll use a number line as a tool. To add 4 and 1, we'll start at 4, then move one unit to the right; this shows that $4 + 1 = 5$, as in Figure 5-4 (a).

A little more thought is required when negative numbers are involved. To add $1 + (-4)$, we start at 1, then move 4 units to the left (since we're adding *negative* 4). We end up at -3, so $1 + (-4) = -3$. See Figure 5-4 (b).

Math Note

The result of adding two numbers is called their *sum*.

To add $-3 + (-4)$, start at -3 on the number line, then count 4 units to the left to get -7. This shows that $-3 + (-4) = -7$. See Figure 5-4 (c).

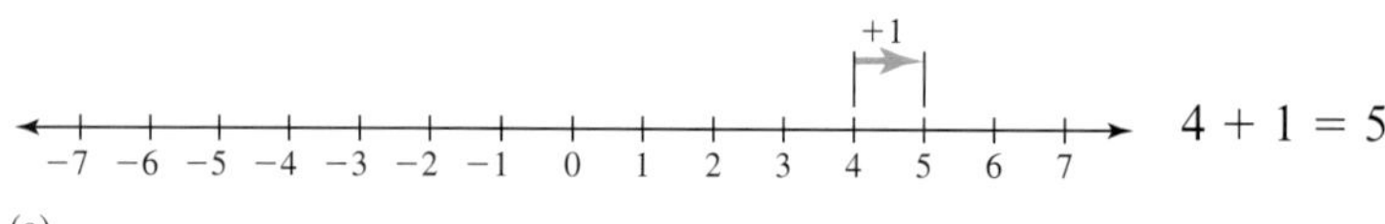

(a)

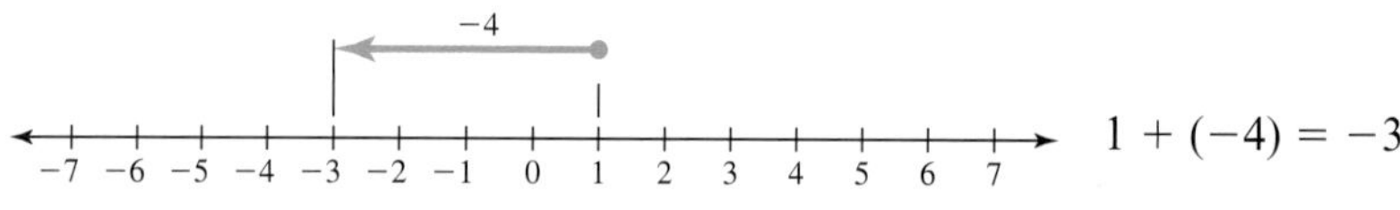

(b)

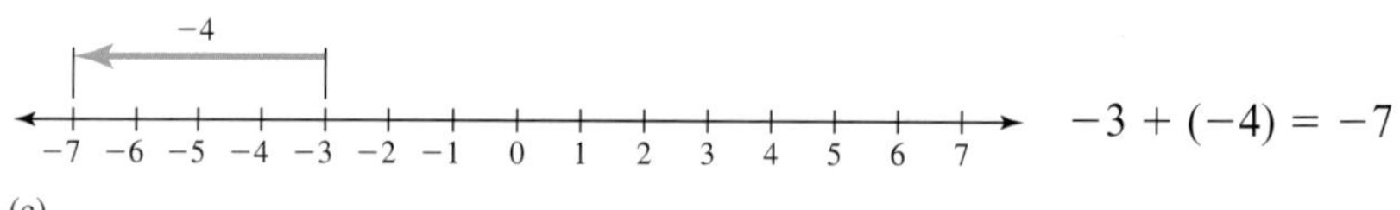

(c)

Figure 5-4

We can use these examples to summarize rules for adding integers.

Rules for Addition of Integers

Rule 1 To add two integers with the same sign, add the absolute values of the numbers and give the answer the common sign.

Rule 2 To add two integers with different signs, find the difference between the two absolute values and give the answer the sign of the number with the larger absolute value.

EXAMPLE 3 Adding Integers

Find each sum:

(a) $8 + 6$ (b) $-2 + (-8)$ (c) $-6 + 8$ (d) $6 + (-12)$

SOLUTION

(a) The two numbers have the same sign, so use Rule 1: each number is its own absolute value, and if we start at 8 and move 6 units right, we get 14. So $8 + 6 = 14$.
(b) Again, the numbers have the same sign, so use Rule 1. The sum of the absolute values is $2 + 8 = 10$, so $-2 + (-8) = -10$.
(c) This time, the numbers have opposite signs, so we use Rule 2. The difference between the absolute values is 2, and the number with the larger absolute value is positive, so $-6 + 8 = 2$.
(d) Again, the numbers have opposite signs. The difference between the absolute values is 6, and the negative number has the larger absolute value, so $6 + (-12) = -6$.

▼ Try This One 3

Find each sum.

(a) $9 + 14$ (b) $(-3) + (-8)$ (c) $(-14) + 22$ (d) $19 + (-32)$

Math Note

The rule for subtracting integers illustrates a good method of problem solving: we turned a new problem (subtraction) into one we already know how to solve (addition).

When you first learned how to subtract, it was probably taught to you in terms of taking away objects. If you had 8 cookies and that little monster that sat in front of you took 3, you had $8 - 3$ or 5 left. This works fine for natural numbers, but loses its effectiveness when negative numbers are involved. (I challenge you to take away negative 3 cookies.) So a better definition of subtraction uses what we already know about addition.

Rule for Subtracting Integers

To subtract two integers $a - b$, add the opposite of b to a. That is, $a - b = a + (-b)$.

EXAMPLE 4 Subtracting Integers

Find each difference:

(a) $9 - 6$ (b) $-2 - (-8)$ (c) $3 - (-7)$ (d) $-2 - 6$

SOLUTION

(a) Rewrite $9 - 6$ as $9 + (-6)$ since -6 is the opposite of 6. Now use Rule 2 for addition: the difference between the absolute values is 3, and the positive number has the larger absolute value, so $9 - 6 = 3$.
(b) Rewrite as $-2 + (8)$ since 8 is the opposite of -8. Now use Rule 2 for addition: the difference between absolute values is 6, and the positive number has the larger absolute value, so $-2 - (-8) = 6$.
(c) Rewrite as $3 + (7)$ since 7 is the opposite of -7. Now add to get $3 - (-7) = 10$.
(d) Rewrite as $-2 + (-6)$ since -6 is the opposite of 6. Now use Rule 1 for addition. The sum of the absolute values is 8, and we give the result a negative sign, so $-2 - 6 = -8$.

☑ 4. Add and subtract integers.

▼ Try This One 4

Find each difference.

(a) $20 - 17$
(b) $5 - 12$
(c) $-3 - 6$
(d) $-8 - (-2)$
(e) $25 - (-6)$
(f) $-14 - 8$

EXAMPLE 5 Applying Addition and Subtraction to Account Balances

Calculator Guide

To be completely honest, we don't recommend using a calculator for addition and subtraction of integers—if you don't practice working with negatives, you'll never get good at it. But in case you're interested in checking answers, like Example 4 (b):

Standard Scientific Calculator

2 [±] [−] 8 [±] [=]

Standard Graphing Calculator

[(−)] 2 [−] [(−)] 8 [Enter]

The same basic keystrokes will work for addition, multiplication, and division, with [+], [×], or [÷] in place of [−].

If you have a Paypal account, you can add money to your account, giving you a positive balance, or you can spend more than you currently have in your account, giving you a negative balance that must be paid off. A 5-day history for an active Internet shopper is given below:

Monday: Balance of +\$12
Tuesday: Balance of −\$5
Wednesday: Balance of −\$16
Thursday: Balance of −\$25
Friday: Balance of \$20

(a) Find the sum of the balances for Monday and Tuesday.
(b) Find the difference of the balances for Thursday and Friday. What does this tell you about the account activity?

SOLUTION

(a) The sum of the balances is $\$12 + (-\$5) = \$7$.
(b) The difference is $\$20 - (-\$25) = \$20 + \$25 = \$45$. This tells us that the shopper deposited \$45 into her Paypal account on Friday.

▼ Try This One 5

Find the sum of the balances for Wednesday and Thursday, and the difference of the balances from the beginning to the end of the week.

Sidelight AMICABLE NUMBERS

The pair of numbers 220 and 284 are called "amicable" or "friendly" numbers since the proper factors or divisors of 220 add up to 284 and the proper factors or divisors of 284 add up to 220.

$1 + 2 + 4 + 5 + 10 + 11 + 20 + 22 + 44 + 55 + 110 = 284$
$1 + 2 + 4 + 71 + 142 = 220$

This pair of numbers was the only pair of amicable numbers known to the Pythagoreans over 2,000 years ago. The next pair, 17,296 and 18,416, was not discovered until 1636 by Pierre de Fermat. In 1638, Descartes discovered another pair. In 1747, Euler discovered 62 pairs of amicable numbers. Today over 600 pairs of amicable numbers are known.

Another interesting note is that in 1866, a 16-year-old Italian named Nicolo Paganini found the pair of amicable numbers 1,184 and 1,210 that had been overlooked by the greatest mathematicians of the time.

Multiplication and Division of Integers

Many college students are creatures of habit. Let's suppose that every Friday afternoon, you stop by the ATM in the union to withdraw \$40 for the weekend. Six weeks into the semester, you would have withdrawn \$40 six times. You could find the total amount withdrawn in dollars using addition:

$$\underbrace{40 + 40 + 40 + 40 + 40 + 40}_{\text{40 added 6 times}} = 240$$

Maybe you looked at the calculation above and thought "Wouldn't it make more sense to just multiply 40 by 6?" If so, that's great! It means you completely understand the fact that multiplication is nothing other than a shortcut for repeated addition. (That's why we use the word "times" when describing multiplication—the phrase "six times forty" is an indication of the fact that the multiplication means to add 40 six *times*.) So we could represent the repeated addition as $6 \times 40 = 240$. We would read the left side as "six times forty" or "the **product** of six and forty."

If we multiply a positive number times a negative, the same idea works just as well. In fact, you can think of your weekly withdrawals as a negative balance on your account, in which case the overall effect on your account would be

$$-40 - 40 - 40 - 40 - 40 - 40 = -240 \text{ dollars}$$

This could also be written as $6 \times (-40) = -240$. This result suggests that when we multiply a positive number and a negative number, the result is negative. But what if we switch the order? We can interpret a multiplication like -40×6 as *subtracting* six 40 times, which would again result in -240. Again, the product of two integers with the opposite sign is negative.

What if we multiply two negatives? Think about $-3 \times (-2)$. This would mean subtracting negative two from zero three times; the result is $0 - (-2) - (-2) - (-2) = 6$. In this case, multiplying two negative numbers results in a positive number. We can summarize these rules for multiplication as follows:

Rules for Multiplying Integers

Rule 1 The product of two numbers with the same sign is positive.

Rule 2 The product of two numbers with opposite signs is negative.

EXAMPLE 6 Multiplying Integers

Find each product:

(a) $(-6) \times (-4)$
(b) $3 \times (-9)$
(c) -5×16
(d) $(-6) \times (-3) \times 8 \times (-5)$

SOLUTION

(a) The signs are the same, so the product is positive: $(-6) \times (-4) = 24$.
(b) The signs are opposite, so the product is negative: $3 \times (-9) = -27$.
(c) The signs are again opposite, so the product is negative: $-5 \times 16 = -80$.
(d) Whenever we repeat the same operation, we work from left to right: first, $(-6) \times (-3) = 18$. Next, $18 \times 8 = 144$. Finally, $144 \times (-5) = -720$.

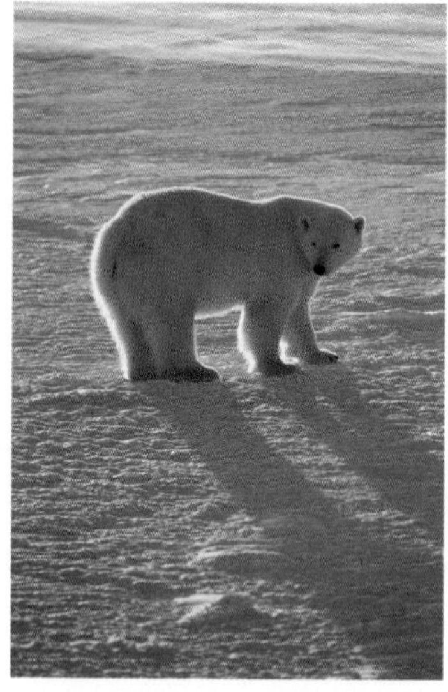

One application of negative integers is temperature. For example, if over 5 days the low temperatures in degrees Fahrenheit were $-25°$, $-21°$, $-15°$, $-23°$, $-21°$, find the average temperature by adding the integers and dividing the result by 5: $-105°/5 = -21°$.

▼ Try This One 6

Find each product:

(a) $(-16) \times 3$
(b) $(-10) \times (-8)$
(c) $12 \times (-8)$
(d) $(-2) \times (-16) \times 7 \times (-3)$

The final operation we will cover in this section is division. Division is the opposite of multiplication in the same way that subtraction is the opposite of addition. For example, $6 - 4 = 2$ because $2 + 4 = 6$. In the same way, $12 \div 2 = 6$ because $6 \times 2 = 12$. We read the operation $12 \div 2$ as "twelve divided by two" or "the **quotient** of twelve and two." Since division and multiplication are so closely related, it comes as no surprise that the rules for dividing signed numbers are the same as the rules for multiplying:

Rules for Dividing Integers

Rule 1 The quotient of two numbers with the same sign is positive.

Rule 2 The quotient of two numbers with opposite signs is negative.

EXAMPLE 7 Dividing Integers

Find each quotient:

(a) $32 \div 8$ (b) $-27 \div (-3)$ (c) $-42 \div 7$ (d) $20 \div (-5)$

SOLUTION

(a) The signs are the same, so the quotient is positive: $32 \div 8 = 4$.
(b) Again, the signs are the same, so the quotient is positive: $-27 \div (-3) = 9$.
(c) The signs are opposite, so the quotient is negative: $-42 \div 7 = -6$.
(d) Again, the signs are the opposite, so the quotient is negative: $20 \div (-5) = -4$.

☑ 5. Multiply and divide integers.

▼ Try This One 7

Find each quotient.

(a) $96 \div 6$ (b) $-48 \div (-3)$ (c) $-84 \div (-2)$ (d) $100 \div (-25)$

Order of Operations

How would you interpret the following sentence?

My professor said Josh is not doing very well.

Exactly who isn't doing well—Josh, or the professor? It depends on how the sentence is punctuated. Here are two different interpretations:

"My professor," said Josh, "is not doing very well."
My professor said, "Josh is not doing very well."

The point is that the exact same string of words in the same order can have two completely different meanings, depending on how the words are grouped. The same is true for mathematical operations. The expression $3 + 5 \times 7$ could be interpreted as:

$$(3 + 5) \times 7 \quad \text{or} \quad 3 + (5 \times 7)$$

In the first case, the result is 8×7, or 56. In the second case, the result is $3 + 35$, or 38.

Whenever different operations are combined in a calculation, there's always the potential for the result to be ambiguous. One way to avoid this is to use parentheses, as we did in the example above. But this requires using parentheses in practically *every* calculation with multiple operations. To avoid this, people long ago agreed on a set of rules to clarify the meaning of such expressions. These rules are known as the **order of operations.**

Math Note

The absolute value bars | | act like grouping symbols when there are multiple operations in a calculation. So in $3 + 2|4 - 6|$, the absolute value would be computed first.

Steps for the Order of Operations

Step 1 Perform all calculations inside grouping symbols first. Grouping symbols commonly used are parentheses ( ), brackets [ ], and braces { }.

Step 2 Evaluate all exponents.

Step 3 Perform all multiplication and division in order from left to right.

Step 4 Perform all addition and subtraction in order from left to right.

In the next four examples, we'll illustrate how to use the order of operations.

EXAMPLE 8 Using the Order of Operations

Math Note

A single dot is often used in place of a multiplication sign. In Example 8, $9 \cdot 3$ means the same thing as 9×3.

Perform the calculation: $9 \cdot 3 - (15 \div 5)$

SOLUTION

$9 \cdot 3 - (15 \div 5)$ *Divide inside parentheses*

$= 9 \cdot 3 - 3$ *Multiply before subtracting*

$= 27 - 3$ *Subtract*

$= 24$

▼ Try This One 8

Perform the calculation: $(2 + 12) - 15 \div 3$.

EXAMPLE 9 Using the Order of Operations

Math Note

Remember that exponents represent repeated multiplication: 2^3 means $2 \times 2 \times 2$.

Perform the calculation: $5 \cdot (8 - 10) + 2^3 \div 4$.

SOLUTION

$5 \cdot (8 - 10) + 2^3 \div 4$ *Subtract inside parentheses*

$= 5 \cdot (-2) + 2^3 \div 4$ *Apply the exponent*

$= 5 \cdot (-2) + 8 \div 4$ *Multiply and divide*

$= -10 + 2$ *Add*

$= -8$

▼ Try This One 9

Perform the calculation: $216 + (4 \times 5)^2 - 13 \cdot 2$

When an expression contains more than one set of grouping symbols, start by performing the operations in the innermost set and work your way outward.

EXAMPLE 10 Using the Order of Operations

Perform the calculation: $84 \div 4 - \{3 \times [10 + (15 - 2)]\}$.

SOLUTION

$84 \div 4 - \{3 \times [10 + (15 - 2)]\}$	*Subtract inside parentheses*
$= 84 \div 4 - \{3 \times [10 + 13]\}$	*Add inside brackets*
$= 84 \div 4 - \{3 \times 23\}$	*Multiply inside braces*
$= 84 \div 4 - 69$	*Divide before subtracting*
$= 21 - 69$	*Subtract*
$= -48$	

▼ Try This One 10

☑ 6. Perform calculations using the order of operations.

Perform the calculation: $9 \times 5 + \{[32 - (6 \times 4)] - 5\}$.

The real world is a pretty complicated place, so it's not a big surprise that calculations in the real world often involve more than one operation. This is why it's important to remember the order of operations.

EXAMPLE 11 Applying Order of Operations to a Checking Account

Suppose you have a checking account set up just to pay your share of the rent each month. At the start of the school year, there's \$1,100 in the account. You write a check for \$240 on the first of every month from August to May, and you deposit a student loan check for \$600 at the beginning of each semester (fall and spring). Find the ending balance in your account.

SOLUTION

This problem involves repeated additions and subtractions, so multiplication will come in handy. There are 10 months from August to May, so 10 payments of \$240. There are also two deposits of \$600. So the amount in dollars is given by

$$\begin{aligned} &1{,}100 - 10 \times 240 + 2 \times 600 \\ &= 1{,}100 - 2{,}400 + 1{,}200 \\ &= -100 \end{aligned}$$

Uh oh—you're overdrawn! Time to get a summer job. By the way, if we were to simply perform that calculation from left to right without regard for the order of operations, it would show an ending balance of \$156,961,200. Nice, but unrealistic.

▼ Try This One 11

A new company makes a profit of \$22,340 in the first year. The owner wants to keep \$10,000 of that for herself, then divide the remainder among three salespeople and two associates. How much will each salesperson and associate get?

Answers to Try This One

1 (a) 9 (b) −24 (c) 15 (d) 18 (e) 0

2 (a) > (b) < (c) = (d) >

3 (a) 23 (b) −11 (c) 8 (d) −13

4 (a) 3 (b) −7 (c) −9 (d) −6 (e) 31 (f) −22

5 Sum: −\$41; difference: \$8

6 (a) −48 (b) 80 (c) −96 (d) −672

7 (a) 16 (b) 16 (c) 42 (d) −4

8 9

9 590

10 48

11 \$2,468

EXERCISE SET 5-2

Writing Exercises

1. What is the difference between the natural numbers and the whole numbers?
2. What is the difference between the whole numbers and the integers?
3. How can we use rules for addition to help us subtract?
4. What is the connection between multiplication and addition?
5. How do you find the absolute value of a number?
6. Describe the order of operations.

Computational Exercises

For Exercises 7–16, find each.

7. |−8|
8. |−12|
9. |+10|
10. |+14|
11. The opposite of −8
12. The opposite of +27
13. The opposite of +10
14. The opposite of −16
15. The opposite of 0
16. The opposite of −9

For Exercises 17–26, insert >, <, or =.

17. 16 22
18. 8 14
19. −5 −10
20. −6 −22
21. 0 −3
22. −5 0
23. −9 +8
24. 16 −32
25. −10 −7
26. −14 +3

For Exercises 27–86, perform the indicated operation(s).

27. −6 + 5
28. −8 +4
29. 16 + (−7)
30. (−5) + (−7)
31. (−8) + (−3)
32. (−4) + 9
33. −3 + (−9)
34. −2 + 4
35. −3 + (−4) + (−6)
36. −5 + (−6) + (−8)
37. 8 −(−6)
38. 9 − 2
39. 6 − 11
40. 14 − 20
41. −3 − (−4)
42. −8 − (−10)
43. −12 − (−7)
44. −15 −9
45. −20 − 50
46. −14 − 29
47. (5)(9)
48. (6)(7)
49. (−3)(8)
50. (−12)(6)
51. 4(−9)
52. 6(−14)
53. (−3)(−14)
54. (−7)(−14)
55. (−9)(0)
56. 0(6)
57. 64 ÷ 8
58. 72 ÷ 9
59. −25 ÷ 5
60. −42 ÷ 7
61. 32 ÷ (−8)
62. 49 ÷ (−7)
63. −14 ÷ (−2)
64. −15 ÷ (−3)

65. $-90 \div (-90)$
66. $-56 \div 4$
67. $0 \div 16$
68. $0 \div (-10)$
69. $-42 \div 6 + 7$
70. $32 \div (8 \times 2)$
71. $5^3 - 2 \cdot 7$
72. $4 \cdot 3^2 - 2 \cdot 4$
73. $9 \cdot 9 - 5 \cdot 6$
74. $32 - (-6)(4)$
75. $3^3 + 5^2 - 2^4$
76. $14^2 - 5^3 + 8^2$
77. $-3[6 + (-10) - (-2)]$
78. $-5 \cdot 4 - [-3 + 8 - (-5)]$
79. $376 - 14 \cdot 3^4$
80. $82 - 9 \cdot 6 - (-2)^2$
81. $256 - 4^3 \cdot 5 + (8 \cdot 4 - 6 \cdot 4)$
82. $6^2 + 5 \cdot 9 - (-27 + 3 \cdot 2)$
83. $-56 \div 8 - \{3 \times [-10 - (4 \times 3)]\}$
84. $(96 - 70) + [(-4 \times 9) - 32 \div 8]$
85. $32 - \{-16 + 5[25 + 9^2 + (8 - 6)]\}$
86. $2\{-5 - 6[3^2 - 7 \cdot (4 + 1)]\}$

Real-World Applications

87. A student's bank balance at the beginning of the month was \$867. During the month, the student made deposits of \$83, \$562, \$37, and \$43. Also, the student made withdrawals of \$74, \$86, and \$252. What was the student's bank balance at the end of the month?
88. Pike's Peak in Colorado is 14,110 feet high, while Death Valley is 282 feet below sea level. Find the vertical distance from the top of Pike's Peak to the bottom of Death Valley.
89. The manager of a biological supply company runs a breeding facility for baby rats. At the beginning of a week, there were 1,286 baby rats. The table shows the number of new rats born each day of the week and the number that were sold. How many rats were left at the end of the week?

	Mon.	Tue.	Wed.	Thur.	Fri.	Sat.
Born	382	494	327	778	256	641
Sold	105	850	416	237	192	965

90. A large grocery store has 354 cases of canned vegetables in the storeroom. During the past month, the store removed 87 cases, 53 cases, 42 cases, and 67 cases to put on the shelves. Also, the store received two lots of 80 cases each. How many cases are in the storeroom now?
91. The table shows the fastest-growing counties in the United States from 2000 to 2007.

County	Housing Units 2000	Housing Units 2007
1. Flagler (FL)	24,452	48,454
2. Sumter (FL)	25,195	43,992
3. Paulding (GA)	29,246	50,328
4. Kendall (IL)	19,527	33,404
5. Pinal (AZ)	81,146	137,410

Find the increase in the number of housing units for each county.

92. The coldest month on average in Barrow, Alaska is February, with an average high temperature of −11.8 degrees Fahrenheit (°F). The warmest month is July, with an average high of 45.2°. Find the difference between the highest and lowest averages.
93. The 30-year average snowfall in Bismark, North Dakota, is approximately 18 inches per year. As of January 10, Bismark received approximately 10 inches. How much more snow will Bismark receive this year if this is an average year?
94. The average annual snowfall in Salt Lake City is about 59 inches. As of January 10, the recorded snowfall was 34 inches. How much more snow will Salt Lake City receive this year if this is an average year?
95. A food bank sends a truck with 400 pounds of food to each of five local food pantries every week, and receives government surplus shipments of 3,000 pounds every other week. If they have 10,000 pounds of food on hand at the beginning of July, how much will they have 6 weeks later?
96. A small college admits a freshman class of 330 students at the beginning of each year, and graduates 145 students twice a year. What will be the net change in student population over a 5-year span?

Critical Thinking

97. If x represents an arbitrary integer, can you make any general statement about the sign of $-x$? What about $-(-x)$?
98. Write a sentence in English that can have two different meanings depending on the punctuation, then write a mathematical expression whose value can be

different depending on the order in which the operations are performed.

99. When you first learned to subtract in grade school, you were probably taught to think of it as "taking away," as in "Suzie has four cookies, and Johnny takes away two of them." Explain why this isn't a reliable way to think of subtraction when working with integers.

100. Using the idea of multiplication as repeated additions or subtractions, justify the rules for multiplication that say that two numbers with the same sign always have a positive product.

Section 5-3 The Rational Numbers

LEARNING OBJECTIVES

- ❑ 1. Define rational numbers.
- ❑ 2. Convert between improper fractions and mixed numbers.
- ❑ 3. Reduce fractions to lowest terms.
- ❑ 4. Multiply and divide fractions.
- ❑ 5. Add and subtract fractions.
- ❑ 6. Write fractions in decimal form.
- ❑ 7. Write terminating and repeating decimals in fraction form.

We've seen a lot of realistic situations in this chapter that involve integers, but everything in the world doesn't come in whole pieces. As just one of many examples, Big Brown won the 2008 Kentucky Derby by $4\frac{3}{4}$ lengths, covering the $1\frac{1}{4}$-mile distance in $2{:}01\frac{4}{5}$ minutes. So many things come in fractional parts that an inability to understand and work with fractions and decimals won't necessarily leave you helpless, but you sure will miss out on a lot. This section is all about studying things that come in fractional parts by studying the numbers that can be used to represent them.

☑ 1. Define rational numbers.

Definition of Rational Numbers

The word "ratio" in math refers to a comparison of the sizes of two different quantities. For example, if there is one man for every three women in a class, we would say that the ratio of men to women is 1 to 3. Ratios are often written as fractions: in this case $\frac{1}{3}$. For that reason, numbers that can be written as fractions are called *rational numbers*.

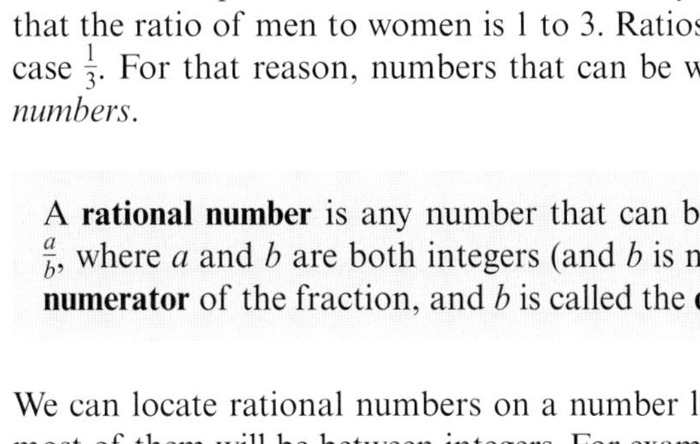

A **rational number** is any number that can be written as a fraction in the form $\frac{a}{b}$, where a and b are both integers (and b is not zero). The integer a is called the **numerator** of the fraction, and b is called the **denominator.**

We can locate rational numbers on a number line; most of them will be between integers. For example, $\frac{2}{3}$ is between 0 and 1, and $-\frac{11}{2}$ is between -5 and -6. See Figure 5-5.

All of the integers we've studied are also rational numbers because they can be written as fractions with denominator 1. For example, $3 = \frac{3}{1}$, which fits the definition of rational number. This means that a number like 3 is a natural number, a whole number, an integer, and a rational number. Of course, there are many rational numbers that are not integers, like $\frac{2}{3}$. Every rational number can also be written in decimal form. We'll study decimals later in this section.

Math Note

Rational numbers are the result of division: if you divide 11 by 2, the result is not an integer, but instead the rational number $\frac{11}{2}$.

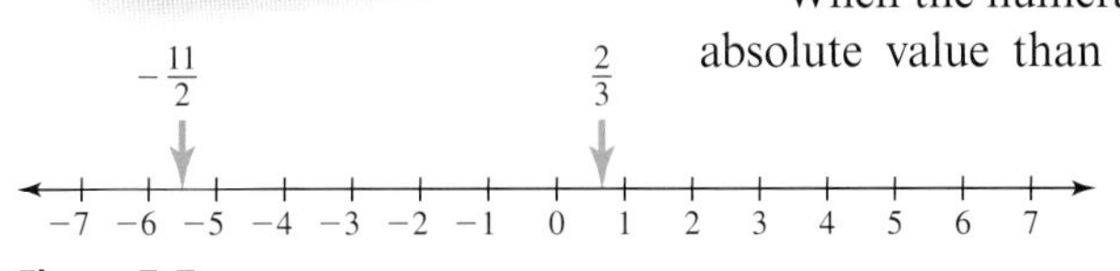

Figure 5-5

When the numerator of a fraction has a smaller absolute value than the denominator, we call the fraction a **proper fraction.** For example, $\frac{2}{3}$, $\frac{1}{2}$, and $-\frac{7}{8}$ are all proper fractions. When the absolute value of

Fractions are very commonly used in recipes.

Math Note

Notice that every proper fraction has absolute value less than one, and every improper fraction has absolute value greater than or equal to one.

the numerator is larger than or equal to that of the denominator, we call the fraction an **improper fraction.** For example, $\frac{4}{3}$, $\frac{10}{4}$, and $-\frac{11}{2}$ are all improper fractions.

A **mixed number** consists of a whole number and a fraction. For example, $3\frac{1}{2}$ and $-10\frac{3}{5}$ are mixed numbers. It's important to understand that a mixed number is really an addition without the + sign: $3\frac{1}{2}$ means the same thing as $3 + \frac{1}{2}$. (The mixed number is read as "three and one half.") We can use division to write improper fractions as integers or mixed numbers. This can be helpful in interpreting their size.

EXAMPLE 1 Writing an Improper Fraction as a Mixed Number

Write $\frac{17}{6}$ as a mixed number.

SOLUTION

First, divide 17 by 6:

$$\begin{array}{r} 2 \\ 6\overline{)17} \\ \underline{-12} \\ 5 \end{array}$$

The quotient is 2, with remainder 5. That tells us that $17 = 2 + \frac{5}{6}$. In mixed number form, we write $\frac{17}{6} = 2\frac{5}{6}$.

▼ Try This One 1

Write $\frac{23}{7}$ as a mixed number.

Since we use division to rewrite improper fractions as mixed numbers, it shouldn't come as a surprise that we can use multiplication to do the opposite—rewrite mixed numbers as improper fractions. This will be useful in performing operations on mixed numbers.

EXAMPLE 2 Writing a Mixed Number as an Improper Fraction

Write $7\frac{2}{5}$ as an improper fraction.

SOLUTION

Step 1 Multiply the whole number part (7 in this case) by the denominator (5), then add the numerator (2):

$$7 \cdot 5 + 2 = 35 + 2 = 37$$

Step 2 Write as a fraction with the result of step 1 as numerator, and the original denominator of the fractional part.

$$7\frac{2}{5} = \frac{37}{5}$$

▼ Try This One 2

Write $10\frac{1}{4}$ as an improper fraction.

☑ 2. Convert between improper fractions and mixed numbers.

Since fractions can be thought of as the division of two integers, we can take advantage of rules for division of signed numbers to address the sign of a fraction. If either the numerator or denominator is negative while the other is positive, it's like dividing two numbers with opposite signs. In this case, the result is negative. So rather than write fractions with a negative in the numerator or denominator, we will write a negative in front of the fraction to indicate that it represents a negative number. For example, all of $\frac{-3}{5}$, $\frac{3}{-5}$, and $-\frac{3}{5}$ represent the same number.

On the other hand, if both the numerator and denominator are negative, the value of the fraction is positive, so we wouldn't bother to write the negatives at all. For example, $\frac{-3}{-5} = \frac{3}{5}$.

Reducing Fractions

Suppose that you and your roommate order a pizza that you want to split evenly. When it gets delivered, you find that the pizza place didn't cut it. If you cut it into two equal pieces, each of you is getting $\frac{1}{2}$ of the pizza. But if the pizza place had cut it into 8 equal slices, you would each take 4 slices, and get $\frac{4}{8}$ of the pizza. In either case, you got half of the pizza, so it must be true that $\frac{4}{8} = \frac{1}{2}$. We could recognize this mathematically by dividing the numerator and denominator of $\frac{4}{8}$ by 4, resulting in $\frac{1}{2}$. This is known as **reducing a fraction to lowest terms.** In this case, we say that $\frac{4}{8}$ and $\frac{1}{2}$ are **equivalent fractions,** which means they represent the same number.

EXAMPLE 3 Reducing a Fraction to Lowest Terms

Reduce $\frac{18}{24}$ to lowest terms.

SOLUTION

Both the numerator and denominator can be divided by 2 with no remainder:

$$\frac{18 \div 2}{24 \div 2} = \frac{9}{12}$$

Before we congratulate ourselves, notice that there's still more that can be done: each of 9 and 12 is divisible by 3!

$$\frac{9 \div 3}{12 \div 3} = \frac{3}{4}$$

Now the fraction is in lowest terms because 3 and 4 have no common divisors. We could have accomplished this in one step by finding the greatest common factor (GCF) of 18 and 24, which is 6, then dividing the numerator and denominator by it:

$$\frac{18 \div 6}{24 \div 6} = \frac{3}{4}$$

Either method is acceptable.

▼ Try This One 3

☑ 3. Reduce fractions to lowest terms.

Reduce $\frac{56}{64}$ to lowest terms.

If we can divide both sides of a fraction by the same nonzero number, it seems reasonable that we can also multiply both sides by the same nonzero number. This is basically the opposite of reducing fractions, and it will come in handy later in the section.

Sidelight A WORD TO CANCEL FROM YOUR MATH VOCABULARY

When reducing fractions, it's common to not write out the division as we did in Example 3. Often, we represent dividing the numerator and denominator of a fraction by some number (in this case 6) like this:

$$\frac{\cancel{18}^{3}}{\cancel{24}_{4}}$$

There is nothing wrong with this approach, unless you forget that what you did was to divide numerator and denominator by the same number. Here's another way to look at it:

$$\frac{18}{24} = \frac{\cancel{6}\cdot 3}{\cancel{6}\cdot 4}$$

It's common to use the word "canceling" to refer to this process. The problem with that term is that it makes it really easy to forget the math of what you've done, and just think of it as "crossing out" the same symbol in the numerator and denominator. And here's why that's dangerous:

$$\frac{\cancel{3}+1}{\cancel{3}+3} = \frac{1}{3}$$

The original fraction, before "canceling," is $\frac{4}{6}$, or $\frac{2}{3}$. Hopefully, it's clear that the statement $\frac{2}{3} = \frac{1}{3}$ is nonsense. What went wrong here is that we didn't *divide* the numerator and denominator by 3—we *subtracted* 3 from the numerator and denominator. This is a useful example because it shows that adding or subtracting something to the numerator and denominator in this manner completely changes the value of the fraction.

We can guarantee from past experience that if you get in the habit of making this mistake, it will be *extremely* difficult to break that habit. One good way to avoid it is to resist the temptation to use the word "canceling." If you call the process of reducing fractions what it is—dividing the numerator and denominator by the same number—you'll be far less likely to fall into this common trap.

EXAMPLE 4 Rewriting a Fraction with a Larger Denominator

Change each fraction to an equivalent fraction with the indicated denominator.

(a) $\frac{3}{8} = \frac{?}{32}$ (b) $\frac{5}{4} = \frac{?}{56}$

SOLUTION

(a) To change the denominator from 8 to 32, we have to multiply by 4. So to make the fractions equivalent, we also need to multiply the numerator by 4. The equivalent fraction is

$$\frac{3\cdot 4}{8\cdot 4} = \frac{12}{32}.$$

(b) It may be a bit harder to notice what number we need to multiply 4 by to get 56, so we divide $56 \div 4 = 14$. The number we need to multiply the numerator and denominator by is 14.

$$\frac{5\cdot 14}{4\cdot 14} = \frac{70}{56}$$

▼ Try This One 4

Change each fraction to an equivalent fraction with the indicated denominator.

(a) $\frac{3}{8} = \frac{?}{24}$ (b) $\frac{7}{13} = \frac{?}{65}$ (c) $\frac{5}{9} = \frac{?}{99}$ (d) $\frac{4}{5} = \frac{?}{50}$

Multiplying and Dividing Fractions

Multiplication is the simplest operation when working with fractions.

Multiplying Fractions

To multiply two fractions, multiply the numerators and the denominators separately. That is,

$$\frac{a}{b} \cdot \frac{c}{d} = \frac{a \cdot c}{b \cdot d}.$$

EXAMPLE 5 Multiplying Fractions

Find each product, and write the answer in lowest terms.

(a) $\frac{5}{8} \cdot \frac{3}{5}$ (b) $-\frac{5}{9} \cdot \frac{3}{11}$ (c) $1\frac{3}{4} \times 2\frac{2}{5}$

SOLUTION

(a) Rather than actually multiply out the numerator and denominator, we'll write as $5 \cdot 3$ and $8 \cdot 5$, which allows us to reduce easily.

$$\frac{5}{8} \cdot \frac{3}{5} = \frac{\overset{1}{\cancel{5}} \cdot 3}{8 \cdot \underset{1}{\cancel{5}}} = \frac{3}{8}$$

(b) The product will be negative since the two fractions have opposite signs.

$$-\frac{5}{9} \cdot \frac{3}{11} = -\frac{5 \cdot \overset{1}{\cancel{3}}}{\underset{3}{\cancel{9}} \cdot 11} = -\frac{5}{33}$$

(c) Our multiplication rule doesn't apply to mixed numbers, so we should first rewrite each as an improper fraction, then multiply.

$$1\frac{3}{4} \times 2\frac{2}{5} = \frac{7}{4} \times \frac{12}{5} = \frac{7 \cdot \overset{3}{\cancel{12}}}{\underset{1}{\cancel{4}} \cdot 5} = \frac{21}{5} \quad \text{or} \quad 4\frac{1}{5}$$

▼ Try This One 5

Find each product, and write the answer in lowest terms.

(a) $\frac{2}{7} \cdot \left(-\frac{21}{8}\right)$ (b) $\frac{11}{9} \times \frac{12}{55}$ (c) $3\frac{1}{2} \cdot 2\frac{2}{3}$

If we switch the numerator and denominator of a fraction, then multiply by the original fraction, something interesting happens:

$$\frac{3}{5} \cdot \frac{5}{3} = \frac{15}{15} = 1$$

For any nonzero fraction $\frac{a}{b}$, the fraction $\frac{b}{a}$ is called the **multiplicative inverse,** or **reciprocal,** of $\frac{a}{b}$. This term is useful in dividing fractions.

Dividing Fractions

To divide two fractions, multiply the first by the reciprocal of the second. That is, $\frac{a}{b} \div \frac{c}{d} = \frac{a}{b} \cdot \frac{d}{c}$.

EXAMPLE 6 Dividing Fractions

Find each quotient, and write the answer in lowest terms.

(a) $\frac{3}{4} \div \left(-\frac{5}{8}\right)$ (b) $\frac{11}{2} \div \frac{4}{9}$

SOLUTION

(a) Multiply $\frac{3}{4}$ by the reciprocal of $-\frac{5}{8}$. The two fractions have opposite signs, so the quotient will be negative.

$$\frac{3}{4} \div \left(-\frac{5}{8}\right) = \frac{3}{4} \times \left(-\frac{8}{5}\right) = -\frac{3 \cdot \cancel{8}^{2}}{{}_{1}\cancel{4} \cdot 5} = -\frac{6}{5}$$

(b) Again, multiply the first fraction by the reciprocal of the second.

$$\frac{11}{2} \div \frac{4}{9} = \frac{11}{2} \times \frac{9}{4} = \frac{99}{8}$$

This time, the quotient cannot be reduced.

▼ Try This One 6

4. Multiply and divide fractions.

Find each quotient, and write the answer in lowest terms.

(a) $-\frac{5}{3} \div \left(-\frac{10}{9}\right)$ (b) $\frac{14}{5} \div \frac{3}{8}$

Adding and Subtracting Fractions

Based on the method we use for multiplying fractions, it's tempting to assume we do the same for addition: add the numerators and denominators separately. But there's a small problem with this approach—it doesn't work!

Think back to our earlier example of a pizza cut into eight equal pieces. Suppose you had split the pizza three ways, and you got three slices. Then you'd have $\frac{3}{8}$ of the pizza. If one of your friends decided she wasn't that hungry and gave you one more piece, you'd now have $\frac{3}{8} + \frac{1}{8}$ of the pizza. But if we add numerators and denominators separately, we get

$$\frac{3}{8} + \frac{1}{8} \stackrel{?}{=} \frac{3+1}{8+8} = \frac{4}{16}$$

This is $\frac{1}{4}$ of the pizza, which is a silly answer; it's less than the $\frac{3}{8}$ you started with!

This example shows that addition of fractions is different from multiplication. If two fractions happen to have the same denominator, there's a simple rule for adding (or subtracting).

Adding and Subtracting Fractions with a Common Denominator

To add or subtract two fractions with the same denominator, add or subtract the numerators, and keep the common denominator the same in your answer.

EXAMPLE 7 Adding and Subtracting Fractions with a Common Denominator

Find each sum or difference.

(a) $\frac{5}{12} + \frac{11}{12}$ (b) $\frac{7}{3} - \frac{4}{3}$

Math Note

From this point on, we will assume that any answer that's a fraction should be reduced to lowest terms.

SOLUTION

(a) $\frac{5}{12} + \frac{11}{12}$ *Add numerators, keep denominator 12*

$= \frac{16}{12}$ *Reduce*

$= \frac{4}{3}$

(b) $\frac{7}{3} - \frac{4}{3}$ *Subtract numerators, keep denominator 3*

$= \frac{3}{3}$ *Reduce*

$= 1$

▼ Try This One 7

Find each sum or difference:

(a) $\frac{17}{9} + \frac{4}{9}$ (b) $\frac{2}{15} - \frac{11}{15}$

To add or subtract fractions with different denominators, we'll need to rewrite the fractions so that they have the same denominator.

Steps for Adding or Subtracting Fractions with Different Denominators

Step 1 Find the least common multiple of the denominators. (This is usually called the **least common denominator,** or **LCD.**)

Step 2 Rewrite each fraction as an equivalent fraction with denominator equal to the LCD.

Step 3 Add or subtract as in Example 7.

EXAMPLE 8 Adding and Subtracting Fractions

Find each sum or difference.

(a) $\frac{1}{4} + \frac{5}{6}$ (b) $\frac{4}{9} - \frac{2}{5}$ (c) $2\frac{1}{2} + 3\frac{1}{4}$

SOLUTION

(a) The LCD of 4 and 6 is 12, so we rewrite each fraction with denominator 12, then add.

$$\frac{1}{4} + \frac{5}{6} = \frac{1 \cdot 3}{4 \cdot 3} + \frac{5 \cdot 2}{6 \cdot 2} = \frac{3}{12} + \frac{10}{12} = \frac{13}{12} \quad \text{or} \quad 1\frac{1}{12}$$

(b) The LCD of 9 and 5 is 45. Rewrite each fraction with denominator 45 and subtract:

$$\frac{4}{9} - \frac{2}{5} = \frac{4 \cdot 5}{9 \cdot 5} - \frac{2 \cdot 9}{5 \cdot 9} = \frac{20}{45} - \frac{18}{45} = \frac{2}{45}$$

(c) First, we need to rewrite the mixed numbers as improper fractions

$$2\frac{1}{2} = \frac{5}{2} \quad \text{and} \quad 3\frac{1}{4} = \frac{13}{4}$$

The LCD of 2 and 4 is 4, so we rewrite $\frac{5}{2}$ as $\frac{10}{4}$, then add.

$$\frac{10}{4} + \frac{13}{4} = \frac{23}{4} \quad \text{or} \quad 5\frac{3}{4}$$

▼ Try This One 8

☑ 5. Add and subtract fractions.

Find each sum or difference.

(a) $-\frac{3}{8} + \frac{5}{6}$

(b) $\frac{9}{10} - \frac{1}{2}$

(c) $5\frac{3}{4} - 4\frac{1}{3}$

Gas prices are often given in thousandths of a dollar.

Fractions and Decimals

You might be familiar with the decimal form of fractions, especially when using a calculator. Any fraction can be written in decimal form. (The opposite is not true—many decimals can be written as fractions, but not all). In order to work with numbers in decimal form, you need to be familiar with **place value,** which is described by Table 5-1.

TABLE 5-1 Place Value Chart for Decimals

millions	hundred thousands	ten thousands	thousands	hundreds	tens	ones		tenths	hundredths	thousandths	ten thousandths	hundred thousandths	millionths
_	_	_	_	_	_	_	.	_	_	_	_	_	_

Sidelight MUSIC AND FRACTIONS

Music is based on math, and fractions play a big role. The timing of music is described by fractions: there are half notes, quarter notes, eighth notes, and so forth. The tones of the notes themselves illustrate fractions as well. When the string of a guitar is plucked, a note is sounded, and if that string is shortened to make it $\frac{1}{2}$ as long, the same note will be heard, but a full octave higher. By increasing the length of the string, other notes can be sounded. For example, if a string $\frac{16}{15}$ as long as the string that plays note C is used, it will play the next lower note, B. If it is stretched to $\frac{6}{5}$ as long, it will sound an A note. This might sound like a pretty modern idea, but the famous Greek mathematician Pythagoras discovered the mathematical relationship between the notes C, F, and G over 2,500 years ago.

To change a fraction to decimal form, we use division.

EXAMPLE 9 Writing a Fraction in Decimal Form

Write $\frac{5}{8}$ as a decimal.

SOLUTION

Divide 5 by 8, as shown.

$$\begin{array}{r} 0.625 \\ 8\overline{)5.000} \\ \underline{-48} \\ 20 \\ \underline{-16} \\ 40 \\ \underline{-40} \\ 0 \end{array}$$

Since we got a remainder of zero, we're done: $\frac{5}{8} = 0.625$.

▼ Try This One 9

Write each fraction as a decimal.

(a) $\frac{3}{8}$ (b) $\frac{7}{20}$

Because we eventually got a remainder of zero, the decimal equivalent for $\frac{5}{8}$ ends after three digits. We call this a **terminating decimal.** Not all fractions can be converted to terminating decimals, as we'll see in Example 10.

EXAMPLE 10 Writing a Fraction in Decimal Form

Write $\frac{5}{6}$ as a decimal.

SOLUTION

$$\begin{array}{r} 0.8333\ldots \\ 6\overline{)5.0000} \\ \underline{-48} \\ 20 \\ \underline{-18} \\ 20 \\ \underline{-18} \\ 20 \\ \underline{-18} \\ 2 \end{array}$$

Notice that the pattern will keep repeating, so $\frac{5}{6} = 0.8333\ldots$.

▼ Try This One 10

☑ 6. Write fractions in decimal form.

Write each fraction as a decimal.

(a) $\frac{5}{12}$ (b) $\frac{19}{33}$

Math Note

The decimal $0.5\overline{6}$ is not the same as $0.\overline{56}$. In the first case, only the 6 repeats (0.5666...). In the second case, the 56 repeats (0.565656...).

The decimal for $\frac{5}{6}$, 0.8333..., is called a **repeating decimal.** Repeating decimals can be written by placing a line over the digits that repeat.

0.8333... is written as $0.8\overline{3}$

0.626262... is written as $0.\overline{62}$

Any terminating decimal can be written in fraction form, using the following procedure.

Procedure for Writing a Terminating Decimal as a Fraction

Step 1 Drop the decimal point and place the resulting number in the numerator of a fraction.

Step 2 Use a denominator of 10 if there was one digit to the right of the decimal point, a denominator of 100 if there were two digits to the right of the decimal point, a denominator of 1,000 if there were three digits to the right of the decimal point, and so on.

Step 3 Reduce the fraction if possible.

EXAMPLE 11 Writing a Terminating Decimal as a Fraction

Write each decimal as a fraction.

(a) 0.8 (b) 0.65 (c) 0.024

SOLUTION

(a) $0.8 = \frac{8}{10} = \frac{4}{5}$ *One digit to the right of the decimal point*

(b) $0.65 = \frac{65}{100} = \frac{13}{20}$ *Two digits to the right of the decimal point*

(c) $0.024 = \frac{24}{1000} = \frac{3}{125}$ *Three digits to the right of the decimal point*

▼ Try This One 11

Write each decimal as a fraction.

(a) 0.4 (b) 0.48 (c) 0.325

A different procedure is used to find the fractional equivalent of a repeating decimal.

Procedure for Writing a Repeating Decimal as a Fraction

Step 1 Write $n =$ the repeating decimal, and multiply both sides of that equation by 10 if one digit repeats, 100 if two digits repeat, etc.

Step 2 Now you will have two equations: subtract the first equation from the second. The repeating part of the decimal will subtract away.

Step 3 Divide both sides of the resulting equation by the number in front of n. This will be the fractional equivalent of the repeating decimal. (Reduce if necessary.)

EXAMPLE 12 Writing a Repeating Decimal as a Fraction

Change $0.\overline{8}$ to a fraction.

Math Note

The result can be checked by changing $\frac{8}{9}$ to a decimal.

SOLUTION

Step 1 Write $n = 0.\overline{8}$ and multiply both sides of the equation by 10 to get $10n = 8.\overline{8}$.

Step 2 Subtract the first equation from the second one as shown.

$$\begin{array}{r} 10n = 8.\overline{8} \\ -n = 0.\overline{8} \\ \hline 9n = 8 \end{array}$$

Step 3 Divide both sides by 9.

$$\frac{9n}{9} = \frac{8}{9}$$

$$n = \frac{8}{9}$$

▼ Try This One 12

Change $0.\overline{4}$ to a fraction.

EXAMPLE 13 Writing a Repeating Decimal as a Fraction

Change $0.\overline{63}$ to a fraction.

SOLUTION

Step 1 Write $n = 0.\overline{63}$ *Multiply both sides by 100*

$$100n = 63.\overline{63}$$

Step 2

$$\begin{array}{r} 100n = 63.\overline{63} \\ -n = 0.\overline{63} \\ \hline 99n = 63 \end{array}$$

Subtract first equation from second

Divide both sides by 99

Step 3 $\frac{99n}{99} = \frac{63}{99}$ *Reduce*

$$n = \frac{63}{99} = \frac{21}{33} = \frac{7}{11}, \quad \text{so} \quad 0.\overline{63} = \frac{7}{11}.$$

▼ Try This One 13

Change $0.\overline{56}$ to a fraction.

☑ 7. Write terminating and repeating decimals in fraction form.

We will close the section with an application of operations involving rational numbers.

EXAMPLE 14 Applying Rational Numbers to Fitness Training

In her final 2 days of preparation for a triathlon, Cat hopes to swim, run, and bike a total of 50 miles. She works out at a nearby state park with a swimming quarry and running/biking trail. One lap in the quarry is $\frac{1}{4}$ mile and the trail is $3\frac{1}{4}$ miles. The first day, Cat swims six laps, runs the trail twice, and bikes it five times. How many more miles does she need to cover the second day?

SOLUTION

The total distance she covers on the first day is

$$6 \cdot \frac{1}{4} + 2\left(3\frac{1}{4}\right) + 5\left(3\frac{1}{4}\right)$$

The mixed number $3\frac{1}{4}$ is $\frac{13}{4}$ as an improper fraction. Using order of operations, we perform the multiplications first, then add.

$$\frac{6}{1} \cdot \frac{1}{4} + \frac{2}{1} \cdot \frac{13}{4} + \frac{5}{1} \cdot \frac{13}{4} = \frac{6}{4} + \frac{26}{4} + \frac{65}{4} = \frac{97}{4} \text{ miles}$$

Now we subtract from 50 to get the distance Cat needs to cover the second day.

$$50 - \frac{97}{4} = \frac{200}{4} - \frac{97}{4} = \frac{103}{4} \text{ or } 25\frac{3}{4} \text{ miles}$$

▼ Try This One 14

If Cat swims three laps and runs and bikes four laps each on the second day, will she reach her goal? If she is over or under her goal, by how much?

Answers to Try This One

1 $3\frac{2}{7}$

2 $\frac{41}{4}$

3 $\frac{7}{8}$

4 (a) $\frac{9}{24}$ (b) $\frac{35}{65}$ (c) $\frac{55}{99}$ (d) $\frac{40}{50}$

5 (a) $-\frac{3}{4}$ (b) $\frac{4}{15}$ (c) $\frac{28}{3}$ or $9\frac{1}{3}$

6 (a) $\frac{3}{2}$ (b) $\frac{112}{15}$

7 (a) $\frac{7}{3}$ (b) $-\frac{3}{5}$

8 (a) $\frac{11}{24}$ (b) $\frac{2}{5}$ (c) $\frac{17}{12}$ or $1\frac{5}{12}$

9 (a) 0.375 (b) 0.35

10 (a) $0.41\overline{6}$ (b) $0.\overline{57}$

11 (a) $\frac{2}{5}$ (b) $\frac{12}{25}$ (c) $\frac{13}{40}$

12 $\frac{4}{9}$

13 $\frac{56}{99}$

14 Yes, with 1 mile to spare.

EXERCISE SET 5-3

Writing Exercises

1. Define *rational number* in your own words.
2. Explain why every integer is also a rational number, but the reverse is not true.
3. How can you tell if a fraction is in lowest terms?
4. Explain how to reduce a fraction to lowest terms.
5. Describe how to multiply two fractions.
6. Describe how to divide two fractions.
7. Describe how to add or subtract two fractions.
8. What is a repeating decimal? What about a terminating decimal?

Computational Exercises

For Exercises 9–18, reduce each fraction to lowest terms.

9. $\frac{7}{42}$
10. $\frac{8}{24}$
11. $\frac{42}{60}$
12. $\frac{16}{20}$
13. $\frac{30}{36}$
14. $\frac{25}{75}$
15. $\frac{91}{104}$
16. $\frac{68}{119}$
17. $\frac{420}{756}$
18. $\frac{950}{2{,}400}$

For Exercises 19–28, change each fraction to an equivalent fraction with the given denominator.

19. $\frac{5}{16} = \frac{?}{48}$
20. $\frac{15}{32} = \frac{?}{96}$
21. $\frac{19}{24} = \frac{?}{48}$
22. $\frac{5}{8} = \frac{?}{40}$
23. $\frac{7}{9} = \frac{?}{45}$
24. $\frac{3}{10} = \frac{?}{30}$
25. $\frac{11}{16} = \frac{?}{80}$
26. $\frac{3}{7} = \frac{?}{28}$
27. $\frac{1}{5} = \frac{?}{30}$
28. $\frac{5}{16} = \frac{?}{64}$

For Exercises 29–48, perform the indicated operations and reduce the answer to lowest terms.

29. $-\frac{5}{6} + \frac{2}{3}$
30. $\frac{3}{4} + \frac{7}{10}$
31. $-\frac{11}{12} - \frac{5}{8}$
32. $\frac{19}{24} - \frac{7}{18}$
33. $-\frac{5}{12} \times -\frac{7}{10}$
34. $\frac{5}{18} \times \frac{9}{25}$
35. $\frac{7}{9} \div \frac{2}{3}$
36. $-\frac{7}{24} \div \frac{23}{30}$
37. $\left(\frac{7}{16} \div \frac{3}{8}\right) \times \frac{3}{5}$
38. $-\frac{7}{8} \div \left(\frac{2}{3} \div \frac{15}{16}\right)$
39. $-\frac{11}{22} \times \left(\frac{1}{6} \times \frac{3}{4}\right)$
40. $\left(\frac{9}{10} - \frac{2}{3}\right) \times \frac{1}{2}$
41. $\left(\frac{5}{8} + \frac{3}{4}\right) \times \frac{2}{3}$
42. $-\frac{5}{6} \times \frac{7}{8}$
43. $\left(-\frac{3}{4}\right) \div \left(-\frac{5}{8}\right)$
44. $\left(-\frac{7}{8}\right) \div \left(-\frac{3}{4}\right)$
45. $\left(\frac{9}{14} \div \frac{3}{7}\right) \times \frac{1}{2}$
46. $\left(\frac{4}{5} + \frac{7}{8}\right) \div \frac{1}{9}$
47. $\left(\frac{9}{10} - \frac{2}{3}\right) \times \frac{5}{6}$
48. $\frac{3}{4} \div \left(\frac{5}{8} + \frac{1}{2}\right)$

For Exercises 49–60, change each fraction to a decimal.

49. $\frac{1}{5}$
50. $\frac{3}{10}$
51. $\frac{2}{3}$
52. $\frac{7}{5}$
53. $\frac{9}{4}$
54. $\frac{11}{9}$
55. $\frac{11}{36}$
56. $\frac{12}{7}$
57. $\frac{3}{4}$
58. $\frac{15}{8}$
59. $\frac{48}{51}$
60. $\frac{17}{24}$

For Exercises 61–70, change each decimal to a reduced fraction.

61. 0.12
62. 0.36
63. 0.375
64. 0.925
65. $0.\overline{7}$
66. $0.\overline{2}$
67. $0.\overline{54}$
68. $0.\overline{62}$
69. $0.45\overline{3}$ (Hint: If $x = 0.45\overline{3}$, what are $1{,}000x$ and $100x$?)
70. $0.27\overline{4}$ (See hint for exercise 69.)

Real-World Applications

71. Lamont and Marge are driving to a business conference. They stop for gas, and Lamont estimates that they've driven $\frac{3}{5}$ of the distance. If the total trip is 285 miles, how many miles have they driven?
72. An automobile race track in city A is $\frac{3}{8}$ of a mile long and an automobile race track in city B is $\frac{1}{6}$ of a mile long. If a driver drives 24 laps on each track, how many miles will the automobile travel?
73. A company uses $\frac{2}{7}$ of its budget for advertising. Of that, $\frac{1}{2}$ is spent on television advertisement. What part of its budget is spent on television advertisement?
74. According to the Census Bureau, 4 out of 25 men do not have health insurance. In a group of 250 men, about how many men would not have health insurance?
75. According to the U.S. Census Bureau, $\frac{1}{8}$ of the population of the United States lives in poverty. Out of

500 randomly selected people, how many would be likely to live in poverty?

76. According to the U.S. Census bureau, in 2006, $\frac{46}{125}$ citizens of the United States live in one of the five most populated states (California, Texas, New York, Florida, and Illinois). If the total population was about 300 million in 2006, how many people lived in one of those five states?
77. According to the Bureau of Labor Statistics, $\frac{13}{200}$ adults in the U.S. were unemployed in November 2008. How many adults in a representative group of 1,000 were employed at that time?
78. On a map, 1 inch represents 80 miles. If two cities are $2\frac{3}{8}$ inches apart, how far apart in miles are they?
79. An architect's rendering of a house plan shows that $\frac{1}{4}$ inch represents 1 foot. If the family room plan is 3 inches long, how long will the actual family room be?
80. A piece of wire is $\frac{3}{4}$ meter long. If it is cut into 10 pieces of equal length, how long will each piece be?
81. An estate was divided among five people. The first person received $\frac{1}{8}$ of the estate. The next two people each received $\frac{1}{5}$ of the estate. The fourth person received $\frac{1}{10}$ of the estate. What fractional part of the estate did the last person receive?
82. For a certain municipality, $\frac{2}{3}$ of the waste generated consisted of paper products, $\frac{1}{10}$ consisted of glass products, and $\frac{1}{5}$ consisted of plastic products. Eight thousand tons of waste were hauled. How much of it consisted of paper, glass, and plastic products?
83. A recipe calls for $2\frac{1}{2}$ cups of flour and $\frac{2}{3}$ cup of sugar. If a person wanted to cut the recipe in half, how many cups of flour and sugar are needed?
84. The recipe for original Toll House chocolate chip cookies calls for, among other things, $2\frac{1}{4}$ cups of flour, $\frac{3}{4}$ cup of granulated sugar, and $\frac{3}{4}$ cup of brown sugar. It makes 60 cookies. A student group wants to bake 300 cookies for a fundraiser. How much flour, sugar, and brown sugar will they need?

Exercises 85–90 refer to the 2008 presidential election.

85. Of all votes cast, $\frac{66}{125}$ were cast for Barack Obama and $\frac{229}{500}$ were cast for John McCain. What fraction of ballots were cast for someone other than McCain or Obama?
86. Among voters, $\frac{39}{100}$ identified themselves as Democrats and $\frac{8}{25}$ as Republicans. What fraction identified themselves as neither Democrats nor Republicans?
87. Among voters in the election, $\frac{37}{50}$ listed white as their race and $\frac{13}{100}$ listed black. What fraction listed some other race?
88. Of voters with a family income of over \$200,000 annually, $\frac{13}{25}$ voted for Obama and $\frac{1}{50}$ voted for someone other than the two major-party candidates. What fraction voted for McCain?
89. The table below lists the fraction of the total electorate made up by different groups based on education level.

No high school diploma	$\frac{1}{25}$
High school graduate, no college	$\frac{1}{5}$
Some college	$\frac{31}{100}$

What fraction of voters had a college degree?

90. Refer to the table in Exercise 99. In addition, $\frac{7}{25}$ of the voters finished their schooling with an undergraduate degree. What fraction went on to graduate study?

Critical Thinking

91. When adding fractions, we usually find the least common denominator and rewrite both fractions with that denominator. But do you really need the *least* common denominator? In the sum $\frac{3}{8} + \frac{5}{12}$, first add by using the least common denominator. Then add by using a common denominator that is the product of the two original denominators. Do you get the same answer? Try again for the sum $\frac{5}{6} + \frac{5}{9}$. What can you conclude? What is the advantage of finding the least common denominator?
92. We added mixed numbers by first rewriting them as improper fractions, then adding as usual. For the sum in Example 8, part *c*, instead of rewriting as improper fractions, add the whole number parts, then the fractional parts. What would you do next to find the overall answer? Did you get the same answer as in the solution to the example problem? Which method do you prefer?
93. We have seen how to find the fractional equivalent for repeating decimals where one, two, or three digits repeat. It seems reasonable, then, to conclude that we can find a fractional equivalent for any repeating decimal. Is there a decimal that has the largest possible number of repeating digits? Why or why not?
94. We know that we can find a fractional equivalent for any decimal that is repeating or terminating. Does that mean that all decimals are rational numbers? Why or why not?
95. One property of the rational numbers is that they are **dense.** This means that between any two rational numbers, you can always find another rational number. Given two arbitrary rational numbers, how would you find another rational number in between them?
96. Review the procedure for converting a mixed number to an improper fraction illustrated in Example 2. Explain why that procedure works using addition of fractions.

Section 5-4 The Irrational Numbers

LEARNING OBJECTIVES

- ❑ 1. Define irrational numbers.
- ❑ 2. Simplify radicals.
- ❑ 3. Multiply and divide square roots.
- ❑ 4. Add and subtract square roots.
- ❑ 5. Rationalize denominators.

If you're paying close attention, you may have noticed that in each of the last two sections, we've built upon the set of numbers defined in the previous section. We started with natural numbers, then included zero and the negatives of whole numbers to get the integers. Then we expanded that further, defining the rational numbers, a set that contains all of the natural numbers, whole numbers, and integers. In this section, we break that pattern, defining a new set of numbers distinct from all of those we've studied so far.

To accomplish that, we will think about numbers that are not rational—that is, cannot be written as fractions with integers in the numerator and denominator. The problem is that it's not so easy to find such numbers when using that approach. Instead, we'll focus on a key fact about rational numbers: any decimal that is either terminating or repeating can be written as a fraction. But there are decimals that neither terminate nor repeat. Here's an example: 0.0103050709011013.... We can see the pattern—a zero followed by an odd number, with the odds increasing each time. But this pattern continues forever, so the decimal is not terminating, and there's no definite string that repeats, so it's not repeating either. Numbers of this form are called *irrational numbers* because they are not rational.

A number is **irrational** if it can be written as a decimal that neither terminates nor repeats.

Even though this definition may be new to you, there's one big class of irrational numbers that you probably have some experience working with: the square roots. For example, the number $\sqrt{2} \approx 1.41421356237\ldots$ is irrational.

The symbol $\sqrt{2}$ is read as "the square root of two," and the symbol around 2 is called a *radical* sign. Here's a definition of square roots:

The **square root** of a number a, symbolized $\sqrt{a}$, is the nonnegative number you have to multiply by itself (or square) to get a.

Math Note

Notice that we used the "approximately equal to" sign ($\approx$) rather than the "equal to" sign ($=$). This is because we can't write the exact value of an irrational number in decimal form, because there are infinitely many digits.

It's not particularly easy to prove conclusively that $\sqrt{2}$ is an irrational number, but here's a way of understanding why that's likely to be the case.

You can approximate $\sqrt{2}$ by finding a number that when squared gives an answer close to 2. For example, $(1.4)^2 = 1.96$ and $(1.5)^2 = 2.25$. So the square root of 2 is a number between 1.4 and 1.5. You could guess 1.41, but $(1.41)^2 = 1.9881$. Although $(1.41)^2$ is closer to 2 than $(1.4)^2$, it's still too small. Now try 1.42; $(1.42)^2 = 2.0164$. This number is too large. So $\sqrt{2}$ is between 1.41 and 1.42.

You can continue the process as shown.

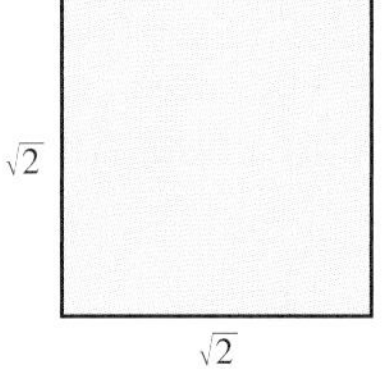

A square with an area of 2 square units would have sides $\sqrt{2}$ units long.

$$1 < \sqrt{2} < 2$$
$$1.4 < \sqrt{2} < 1.5$$
$$1.41 < \sqrt{2} < 1.42$$
$$1.414 < \sqrt{2} < 1.415$$
$$1.4142 < \sqrt{2} < 1.4143$$

etc.

Notice that squaring the number on the left side of each inequality gives a number slightly smaller than 2 and squaring the number on the right side of each inequality gives a number slightly larger than 2. You could continue this process forever, but you would never get a number that when squared would give an answer that is exactly equal to 2. (This fact was proved by Euclid over 3,000 years ago.)

Sidelight PYTHAGORAS

Almost everybody remembers the Pythagorean theorem—maybe not what it actually says, but that they studied it at one time or another in elementary school, middle school, or high school.

Pythagoras was born on the island of Samos about 570 B.C.E. He studied under another famous mathematician, Thales. Pythagoras' early life is only conjecture. He founded a society in 509 B.C.E. that was devoted to the study of mathematics. The society taught, "Numbers ruled the universe."

In order to be accepted into the society, the candidates had to undergo several degrees of initiation. For the first 5 years, the initiates to the society were not allowed to speak, wear wool, eat meat or beans, or touch a white rooster. After an initiate completed the initiation, he or she was taught the truth about mathematics.

The Pythagoreans discovered the mathematical relationship between the length of a string of a musical instrument and its pitch. They also studied prime numbers, perfect numbers, and amicable numbers.

In addition, the Pythagoreans studied geometry. They were interested only in the study of geometric forms such as triangles, squares, and circles. They were not concerned with the measurement of lines and the numerical nature of geometry.

They did not recognize zero or the negative numbers since they could not find any physical meaning for them.

Finally, they attributed mystical powers to numbers. For example, they considered odd numbers to be masculine and even numbers to be feminine. They also believed that the future could be predicted by numbers and a person's character could be revealed by numbers.

The members of the society were sworn to secrecy; one member, Hippasus, spoke of the discovery of irrational numbers, which the Pythagoreans refused to accept. Later when he perished in a shipwreck, the Pythagoreans believed that this was a sign that the discovery was "alagon" (unutterable)!

At the time the citizens of the community where the Pythagoreans lived became suspicious of the secret society, and when Pythagoras started to speak in public, he drew large crowds. The distrust of some then became a riot; the citizens sacked and burned the society's buildings.

Pythagoras and what was left of his followers fled to Tarentum to start a new school. His reputation, however, preceded him, and he was forced to flee again to Metapontium. At Metapontium, another riot broke out, and this time, Pythagoras was killed.

Calculator Guide

You can find decimal approximations of irrational numbers using a calculator. To approximate $\sqrt{2}$:

Standard Scientific Calculator

[2] [√] or [2] [2nd] [x²]

Standard Graphing Calculator

[2nd] [x²] [2]

It turns out that $\sqrt{2}$ is not alone when it comes to square roots that are irrational numbers. One thing that is easy to determine is that not *every* square root is irrational. Since $2^2 = 4$, $\sqrt{4} = 2$, so $\sqrt{4}$ is an integer, and consequently a rational number. The same is true for $\sqrt{9}$, $\sqrt{16}$, $\sqrt{25}$, $\sqrt{36}$, and so on. In each case, the number under the radical is the square of an integer. We will call such a number a **perfect square,** and observe that the square root of a perfect square is always an integer. But for any other number, the square root happens to be irrational.

In case you're wondering, there are plenty of irrational numbers that are not square roots. The most famous of them is probably π (pi), which is defined to be the distance around a circle (the circumference) divided by the diameter.

☑ 1. Define irrational numbers.

Simplifying Radicals

There is a simple property of square roots, commonly known as the product rule, that will help us in working with irrational numbers that are defined by square roots.

The Product Rule For Square Roots

For any two positive numbers a and b, $\sqrt{ab} = \sqrt{a} \cdot \sqrt{b}$.

Sidelight A SLICE OF PI

You would more than likely be amazed, if not appalled, at the tremendous amount of effort that has gone into the study of the number pi. The earliest mentions of pi come from about 1900 BCE, when it was approximated as 25/8 in Babylon and 256/81 in Egypt. As hard as it may be to believe, almost 4,000 years later mathematicians are still working on better and better approximations! The value of pi is about 3.14, but that's a tremendously crude approximation by modern standards. In fact, the number of digits computed as of the end of 2008 was 1.24 *trillion*. It's almost hard to imagine how many digits that is. If you recited two digits each second, it would take you almost 20 *years* with no breaks to read that many digits.

One of the more interesting aspects of pi is that the digits appear to be totally random. Remember, it's been calculated out to over a trillion digits, and nobody has found any pattern whatsoever. In fact, it's been speculated that the digits are completely randomly distributed, which would mean that *every possible string of numbers* can be found somewhere in the decimal expansion of pi! Your birthdate? It's in there. Your cell phone number? It's in there. Social security number? In there. All assuming, of course, that the digits really are randomly distributed. There is a Web page at http://www.angio.net/pi/piquery.html where you can search for any string of numbers in the first 200 million digits of pi.

Notes:

- The product rule is also true if a and/or b are equal to zero, but it's not particularly useful in that case.
- This is called the *product* rule for a good reason—a similar formula is not true for *all* operations. Most notably, $\sqrt{a+b}$ is *not equal to* $\sqrt{a} + \sqrt{b}$! See Exercises 65–68.

One way to use the product rule is to simplify square roots. We will call a square root *simplified* if the number inside the radical has no factors that are perfect squares. The procedure for simplifying square roots is illustrated in Example 1.

EXAMPLE 1 Simplifying Radicals

Simplify each radical.

(a) $\sqrt{40}$
(b) $\sqrt{200}$
(c) $\sqrt{26}$

SOLUTION

(a) Notice that 40 can be written as $4 \cdot 10$, and that 4 is a perfect square. Also, 4 is the largest factor of 40 that is a perfect square. Using the product rule, we can write

$$\begin{aligned} \sqrt{40} &= \sqrt{4 \cdot 10} && \text{Use the product rule} \\ &= \sqrt{4} \cdot \sqrt{10} && \sqrt{4} = 2 \\ &= 2\sqrt{10} \end{aligned}$$

(b) The largest perfect square factor of 200 is 100.

$$\begin{aligned} \sqrt{200} &= \sqrt{100 \cdot 2} && \text{Use the product rule} \\ &= \sqrt{100} \cdot \sqrt{2} && \sqrt{100} = 10 \\ &= 10\sqrt{2} \end{aligned}$$

(c) The prime factorization of 26 is $2 \cdot 13$. It has no perfect square factors, so $\sqrt{26}$ is already simplified.

2. Simplify radicals.

▼ Try This One 1

Simplify each radical.

(a) $\sqrt{75}$ (b) $\sqrt{56}$ (c) $\sqrt{74}$

The product rule can also be used to multiply two square roots.

EXAMPLE 2 Multiplying Square Roots

Find each product.

(a) $\sqrt{6} \cdot \sqrt{2}$ (b) $\sqrt{5} \cdot \sqrt{20}$ (c) $\sqrt{2} \cdot \sqrt{3} \cdot \sqrt{3}$

Math Note

In the same way that we assume any fractional answer should be reduced, from now on we'll assume that any square root answer should be simplified.

SOLUTION

(a) Using the product rule (reading it from right to left), we can write the product as a single square root by multiplying the numbers under the radical.

$$\sqrt{6} \cdot \sqrt{2} = \sqrt{6 \cdot 2} = \sqrt{12}$$

Now we can simplify like we did in Example 1.

$$\sqrt{12} = \sqrt{4 \cdot 3} = \sqrt{4} \cdot \sqrt{3} = 2\sqrt{3}$$

(b) Again, we can multiply the two numbers under the radical.

$$\sqrt{5} \cdot \sqrt{20} = \sqrt{5 \cdot 20} = \sqrt{100}$$

This time, it's easier to simplify: $\sqrt{100}$ is 10 because $10^2 = 100$. This shows that the product of two irrational numbers is not always irrational.

(c) The product rule doesn't specifically tell us how to multiply three radicals, but we can use it in stages (multiply the first two, then multiply the result by the third) to show that we can multiply the three numbers inside the radical.

$$\sqrt{2} \cdot \sqrt{3} \cdot \sqrt{3} = \sqrt{2 \cdot 3 \cdot 3} = \sqrt{18}$$

Now we can simplify.

$$\sqrt{18} = \sqrt{9 \cdot 2} = \sqrt{9} \cdot \sqrt{2} = 3\sqrt{2}$$

In this case, it would have been simpler to notice that the product of the second and third factors, $\sqrt{3} \cdot \sqrt{3}$, is 3, in which case we would have obtained the same answer, but much more quickly.

▼ Try This One 2

Find each product.

(a) $\sqrt{35} \cdot \sqrt{5}$ (b) $\sqrt{18} \cdot \sqrt{6}$ (c) $\sqrt{42} \cdot \sqrt{15} \cdot \sqrt{15}$

Since multiplication and division are strongly related, it's not a big surprise that there is a quotient rule for square roots, similar to the product rule. It is used to divide two square roots, and to simplify square roots of fractions.

The Quotient Rule for Square Roots

For any two positive numbers a and b, $\sqrt{\frac{a}{b}} = \frac{\sqrt{a}}{\sqrt{b}}$.

Note: The quotient rule is also true if $a = 0$, but not if $b = 0$, as this would make the denominator zero.

EXAMPLE 3 Using the Quotient Rule

Find each quotient.

(a) $\frac{\sqrt{27}}{\sqrt{3}}$ (b) $\frac{\sqrt{60}}{\sqrt{5}}$

SOLUTION

In each case, we'll use the quotient rule to write the quotient as a single square root, dividing the numbers underneath the radical, then simplify.

(a) $\frac{\sqrt{27}}{\sqrt{3}} = \sqrt{\frac{27}{3}} = \sqrt{9} = 3$

(b) $\frac{\sqrt{60}}{\sqrt{5}} = \sqrt{\frac{60}{5}} = \sqrt{12} = \sqrt{4 \cdot 3} = \sqrt{4} \cdot \sqrt{3} = 2\sqrt{3}$

▼ Try This One 3

☑ 3. Multiply and divide square roots.

Find each quotient.

(a) $\frac{\sqrt{80}}{\sqrt{10}}$ (b) $\frac{\sqrt{48}}{\sqrt{3}}$

Adding and subtracting square roots is very different than multiplying and dividing them. This is because there is no sum or difference rule for square roots. Instead, we will need to define a few new terms. In an expression like $3\sqrt{5}$, the number under the radical is known as the **radicand,** and the number in front of the radical is called the **coefficient.** When the radicands of two or more different square roots are the same, we call them **like radicals.** For example, $4\sqrt{3}$, $2\sqrt{3}$, and $-\sqrt{3}$ are like radicals because all have the same radicand, 3. On the other hand, $5\sqrt{5}$ and $5\sqrt{3}$ are not like radicals because they have different radicands. Only like radicals can be added or subtracted.

Addition and Subtraction of Like Radicals

To add or subtract like radicals, add or subtract their coefficients and keep the radical the same. In symbols,

$$a\sqrt{c} + b\sqrt{c} = (a + b)\sqrt{c} \text{ and}$$

$$a\sqrt{c} - b\sqrt{c} = (a - b)\sqrt{c}.$$

EXAMPLE 4 Adding Square Roots

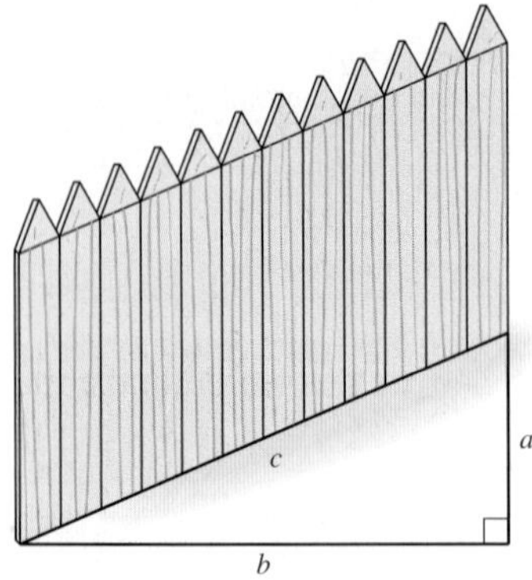

A familiar use of radicals is determining sides of right triangles with the Pythagorean theorem, $a^2 + b^2 + c^2$. If the sides of the garden a and b are 10 ft and 24 ft, respectively, then $10^2 + 24^2 = c^2$, and $c = \sqrt{10^2 + 24^2}$, or 26 ft.

Find the sum: $6\sqrt{3} + 8\sqrt{3} + 5\sqrt{3}$.

SOLUTION

Since all the radicals are like radicals, the sum can be found by adding the coefficients of the radicals.

$$6\sqrt{3} + 8\sqrt{3} + 5\sqrt{3} = (6 + 8 + 5)\sqrt{3} = 19\sqrt{3}$$

▼ Try This One 4

Find the sum: $7\sqrt{5} + 8\sqrt{5} + 9\sqrt{5}$.

If you have a hard time visualizing the addition of coefficients in Example 4, it might help to mentally replace the number $\sqrt{3}$ with some objects, like MP3s. In that case, we'd have 6 MP3s + 8 MP3s + 5 MP3s, which would be 19 MP3s. This will work for subtraction as well.

EXAMPLE 5 Subtracting Square Roots

Find the difference: $3\sqrt{10} - 7\sqrt{10}$.

SOLUTION

$$3\sqrt{10} - 7\sqrt{10} = (3 - 7)\sqrt{10} = -4\sqrt{10}$$

▼ Try This One 5

Find the difference: $12\sqrt{11} - 20\sqrt{11}$.

In life, things are not always as they appear. Based on our definition of like radicals, it seems reasonable to conclude that $\sqrt{2}$ and $\sqrt{8}$ are not like, since their radicands are different. But as we'll see in Example 6, simplifying can expose like radicals in disguise.

EXAMPLE 6 Adding and Subtracting Square Roots

Perform the indicated operations.

$$4\sqrt{2} - 3\sqrt{8} + 5\sqrt{32}$$

SOLUTION

Simplify each radical first, and then add or subtract like radicals as shown.

$$-3\sqrt{8} = -3\sqrt{4 \cdot 2} = -3\sqrt{4} \cdot \sqrt{2} = -3 \cdot 2\sqrt{2} = -6\sqrt{2}$$

$$5\sqrt{32} = 5\sqrt{16 \cdot 2} = 5\sqrt{16} \cdot \sqrt{2} = 5 \cdot 4\sqrt{2} = 20\sqrt{2}$$

Now we see that we have like radicals.

$$4\sqrt{2} - 6\sqrt{2} + 20\sqrt{2} = (4 - 6 + 20)\sqrt{2} = 18\sqrt{2}$$

Math Note

The expressions $5\sqrt{2} + 6\sqrt{3}$ cannot be added since they are not like radicals.

▼ Try This One 6

4. Add and subtract square roots.

Perform the indicated operations.

(a) $2\sqrt{8} + 5\sqrt{50}$ (b) $6\sqrt{28} + 4\sqrt{112} - 2\sqrt{12}$

Another method used to simplify radical expressions is called **rationalizing the denominator.**

When a radical expression contains a square root sign in the denominator of a fraction, it can be simplified by multiplying the numerator and denominator by a radical expression that will make the radicand in the denominator a perfect square. This is called rationalizing the denominator. Examples 7 and 8 illustrate the process.

EXAMPLE 7 Rationalizing Denominators

Simplify each radical expression.

(a) $\frac{18}{\sqrt{3}}$ (b) $\frac{6}{\sqrt{18}}$

SOLUTION

(a) If we multiply the numerator and denominator by $\sqrt{3}$, the denominator will become $\sqrt{9}$, which is of course 3.

$$\frac{18}{\sqrt{3}} = \frac{18}{\sqrt{3}} \cdot \frac{\sqrt{3}}{\sqrt{3}} = \frac{18\sqrt{3}}{\sqrt{9}} = \frac{18\sqrt{3}}{3} = 6\sqrt{3}$$

(b) We could mimic what happened in part (a) and multiply the numerator and denominator by $\sqrt{18}$, but it will be easier to simplify if we multiply instead by $\sqrt{2}$. This will make the denominator $\sqrt{36}$, which is 6.

$$\frac{6}{\sqrt{18}} = \frac{6}{\sqrt{18}} \cdot \frac{\sqrt{2}}{\sqrt{2}} = \frac{6\sqrt{2}}{\sqrt{36}} = \frac{6\sqrt{2}}{6} = \sqrt{2}$$

▼ Try This One 7

Simplify each radical expression.

(a) $\frac{20}{\sqrt{5}}$ (b) $\frac{3}{\sqrt{12}}$

When the radicand of a square root is a fraction, we can use the quotient rule to apply the root to the numerator and denominator separately, then simplify by rationalizing the denominator.

EXAMPLE 8 Simplifying the Square Root of a Fraction

Simplify $\sqrt{\frac{5}{6}}$.

SOLUTION

Apply the quotient rule to split into two separate roots:

$$\frac{\sqrt{5}}{\sqrt{6}}$$

Math Note

Radical expressions such as $\frac{\sqrt{30}}{6}$ can also be written as $\left(\frac{1}{6}\right)\sqrt{30}$. A radical expression like $\frac{2\sqrt{3}}{5}$ can also be written as $\frac{2}{5}\sqrt{3}$.

Now multiply the numerator and denominator by $\sqrt{6}$ to rationalize the denominator:

$$\frac{\sqrt{5}}{\sqrt{6}} \cdot \frac{\sqrt{6}}{\sqrt{6}} = \frac{\sqrt{30}}{\sqrt{36}} = \frac{\sqrt{30}}{6}$$

▼ Try This One 8

Simplify each radical expression.

(a) $\sqrt{\frac{5}{18}}$ (b) $\sqrt{\frac{75}{8}}$

☑ 5. Rationalize denominators.

In this section, we focused on operations on certain types of irrational numbers—square roots. But don't forget that there are many other irrational numbers, among them the multiples of π. When working with irrational numbers, one of the key features you should always keep in mind is that decimal forms are always approximations because by definition, the irrational numbers have decimal forms with infinitely many digits.

Answers to Try This One

1 (a) $5\sqrt{3}$ (b) $2\sqrt{14}$ (c) $\sqrt{74}$

2 (a) $5\sqrt{7}$ (b) $6\sqrt{3}$ (c) $15\sqrt{42}$

3 (a) $2\sqrt{2}$ (b) 4

4 $24\sqrt{5}$

5 $-8\sqrt{11}$

6 (a) $29\sqrt{2}$ (b) $28\sqrt{7} - 4\sqrt{3}$

7 (a) $4\sqrt{5}$ (b) $\frac{\sqrt{3}}{2}$

8 (a) $\frac{\sqrt{10}}{6}$ (b) $\frac{5\sqrt{6}}{4}$

EXERCISE SET 5-4

Writing Exercises

1. How can you tell the difference between an irrational number and a rational number?
2. Is every square root an irrational number? Explain.
3. Are there irrational numbers other than square roots? Explain.
4. Describe two ways that we use the product rule for square roots.
5. How can you tell if a square root is simplified or not?
6. Explain how to rationalize the denominator of a square root.

Computational Exercises

For Exercises 7–12, state whether each number is rational or irrational.

7. $\sqrt{49}$
8. $\sqrt{37}$
9. 0.232332333…
10. $\frac{5}{6}$
11. π
12. 0

For Exercises 13–18, without using a calculator, name two integers that the given square root is between on a number line.

13. $\sqrt{11}$
14. $\sqrt{28}$
15. $\sqrt{100}$
16. $\sqrt{75}$
17. $\sqrt{200}$
18. $\sqrt{160}$

For Exercises 19–28, simplify the radical.

19. $\sqrt{24}$
20. $\sqrt{27}$
21. $\sqrt{80}$
22. $\sqrt{175}$
23. $\sqrt{30}$
24. $\sqrt{42}$
25. $10\sqrt{20}$
26. $4\sqrt{8}$
27. $3\sqrt{700}$
28. $2\sqrt{162}$

For Exercises 29–56, perform the operations and simplify the answer.

29. $\sqrt{2} \cdot \sqrt{10}$
30. $\sqrt{15} \cdot \sqrt{6}$
31. $\sqrt{18} \cdot \sqrt{15}$
32. $\sqrt{5} \cdot \sqrt{25}$
33. $2\sqrt{6} \cdot 3\sqrt{8}$
34. $6\sqrt{15} \cdot 2\sqrt{5}$
35. $\dfrac{\sqrt{60}}{\sqrt{2}}$
36. $\dfrac{\sqrt{42}}{\sqrt{6}}$
37. $\dfrac{\sqrt{64}}{\sqrt{8}}$
38. $\dfrac{\sqrt{15}}{\sqrt{3}}$
39. $2\sqrt{7} + 10\sqrt{7}$
40. $50\sqrt{11} + 11\sqrt{11}$
41. $8\sqrt{3} - 15\sqrt{3}$
42. $5\sqrt{7} - 11\sqrt{7}$
43. $2\sqrt{3} + 5\sqrt{3} - 9\sqrt{3}$
44. $8\sqrt{5} - 6\sqrt{5} - 7\sqrt{5}$
45. $\sqrt{320} - \sqrt{80}$
46. $\sqrt{125} + \sqrt{20}$
47. $6\sqrt{5} - 3\sqrt{80}$
48. $13\sqrt{90} + 5\sqrt{40}$
49. $6\sqrt{72} - 9\sqrt{8}$
50. $5\sqrt{10} + 2\sqrt{40}$
51. $3\sqrt{2} - \sqrt{8} + 4\sqrt{12}$
52. $10\sqrt{20} - 20\sqrt{10} + 6\sqrt{5}$
53. $5\sqrt{40} + 4\sqrt{50} - 6\sqrt{32}$
54. $8\sqrt{12} - 9\sqrt{20} + \sqrt{75}$
55. $\sqrt{5}(\sqrt{75} + \sqrt{12})$
56. $\sqrt{3}(\sqrt{48} - \sqrt{27})$

For Exercises 57–64, rationalize the denominator and simplify.

57. $\dfrac{1}{\sqrt{5}}$
58. $\dfrac{3}{\sqrt{8}}$
59. $\dfrac{3}{\sqrt{6}}$
60. $\dfrac{10}{\sqrt{20}}$
61. $\sqrt{\dfrac{3}{28}}$
62. $\sqrt{\dfrac{1}{9}}$
63. $\sqrt{\dfrac{2}{3}}$
64. $\sqrt{\dfrac{7}{8}}$

Exercises 65–68 will help you to understand why $\sqrt{a+b}$ is not equal to $\sqrt{a} + \sqrt{b}$.

65. Find $\sqrt{9}$, $\sqrt{16}$, and $\sqrt{25}$. Is $\sqrt{9} + \sqrt{16} = \sqrt{25}$?
66. Find $\sqrt{25}$, $\sqrt{144}$, and $\sqrt{169}$. Is $\sqrt{25} + \sqrt{144} = \sqrt{169}$?
67. Use a calculator to approximate $\sqrt{40}$ and $\sqrt{80}$ to the nearest hundredth. Is $\sqrt{40} + \sqrt{40} = \sqrt{80}$?
68. Use a calculator to approximate $\sqrt{60}$ and $\sqrt{120}$ to the nearest hundredth. Is $\sqrt{60} + \sqrt{60} = \sqrt{120}$?

Real-World Applications

If an object is dropped from a given height h, the time in seconds it takes to reach the ground is given by the formula

$$t = \sqrt{\frac{2h}{32}}$$

(This formula ignores air resistance.) Use this formula for Exercises 69–74.

69. How long will it take an object to reach the ground when dropped from a 144-foot building?
70. How long will it take an object to reach the ground when dropped from a 256-foot bridge?
71. The tallest observation deck in the United States is the Sears Tower in Chicago, at a height of 1,353 feet. Use a calculator to find how long it will take an object to reach the ground when dropped from that height.
72. The tallest observation deck in the world is the Shanghai World Financial Center in China, at a height of 1,555 feet. Use a calculator to find how long it will take an object to reach the ground when dropped from that height. Round to the nearest tenth of a second.
73. How much longer will it take an object dropped from 200 feet to reach the ground than one dropped from 100 feet? Is the answer surprising? Why?
74. How much longer will it take an object dropped from 400 feet to reach the ground than one dropped from 100 feet? Is the answer surprising? Why?

Use this information for Exercises 75–78: the voltage of an electric circuit can be found by the formula

$$V = \sqrt{P \cdot r}$$

where V = volts, P = power in watts, and r = resistance in ohms.

75. Find the voltage when $P = 80$ watts and $r = 5$ ohms.
76. Find the voltage when $P = 360$ watts and $r = 10$ ohms.
77. Use a calculator to approximate to one decimal place the voltage in a circuit with 1,000 watts of power and 20 ohms of resistance.
78. Use a calculator to approximate to one decimal place the voltage in circuit with 1,200 watts of power and 25 ohms of resistance.

From the time of their invention in 1656 until the 1930s, pendulum clocks were the most accurate in the world, and they are still in use today. The time it takes for a pendulum to complete one swing and return to its starting point is called its period. The period of a pendulum is important in using one as a timekeeping device, and it can be calculated using the formula

$$t = 2\pi\sqrt{\frac{l}{32}}$$

where t is the period in seconds and l is the length in feet of the pendulum arm.

79. Find the period of a pendulum whose arm length is 128 feet.
80. Find the period of a pendulum whose arm length is 64 feet.
81. Using a calculator and trial-and-error, find the approximate arm length of a pendulum with a period of one second.
82. Using a calculator and trial-and-error, find the approximate length of a pendulum with a period of 5 seconds.

Critical Thinking

83. When two integers are multiplied, the result is always another integer. We say that the integers are **closed under multiplication.** Are the irrational numbers closed under multiplication? What about the rational numbers? Explain.
84. Use several examples to decide whether $\sqrt{a-b} = \sqrt{a} - \sqrt{b}$ for positive numbers a and b.
85. Based on the way we defined square roots in this section, why can't you compute the square root of a negative number?
86. In the same way that we define square roots in terms of squares, we can define **cube roots** in terms of cubes. So the cube root of a number a, denoted $\sqrt[3]{a}$, is the number whose cube, or third power, is a. Do you think that cube roots of integers are irrational numbers? Always? Sometimes? Never? Discuss your answer.
87. Can you find a pair of numbers a and b for which $\sqrt{a+b} = \sqrt{a} + \sqrt{b}$? Make a conjecture as to what condition must be satisfied in order for $\sqrt{a} + \sqrt{b}$ to equal $\sqrt{a+b}$.

Section 5-5 The Real Numbers

LEARNING OBJECTIVES

- ☐ 1. Define the real numbers.
- ☐ 2. Identify properties of the real numbers.

The Russian art of nesting dolls, or matryoshka, dates back to the 1890s. A series of hollow carved dolls with similar proportions fits one inside the other, with all of them fitting inside the largest. This is a pretty good (but not perfect) model for the sets of numbers that we've built while working through this chapter. With the exception of the irrational numbers, we have a group of sets that live inside one another: natural numbers, whole numbers, integers, rational numbers.

In this section, we'll study a larger set of numbers that contains all of the other sets we've studied so far, including the irrationals. This set is called the *real numbers,* and they make up the set of numbers that are used throughout the remainder of the book.

☑ 1. Define the real numbers.

The set of **real numbers** consists of the union of the set of rational numbers and the set of irrational numbers.

Using set notation, {real numbers} = {rational numbers} ∪ {irrational numbers}.

Figures 5-6 and 5-7 on the next page are two diagrams that demonstrate the structure of the real numbers. Figure 5-6 traces the way we built the real numbers step-by-step through the first four sections of this chapter, always including a new group of numbers to build a larger set.

Another way to illustrate the structure of the real number system is shown in Figure 5-7. The set of natural numbers is a subset of the set of whole numbers. The set of whole numbers is a subset of the set of integers. The set of integers is a subset of the

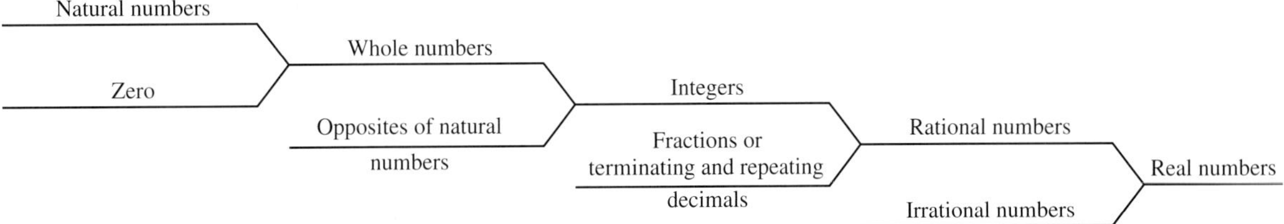

Figure 5-6

set of rational numbers. The set of rational numbers and the set of irrational numbers are subsets of the set of real numbers. But the set of irrational numbers is completely outside the set of rational numbers. One good way to make sure you understand the definitions of the various sets of numbers is to classify numbers according to the sets that they are in. For example, the number 2 is a natural number, a whole number, an integer, a rational number, and a real number. The number -5 is an integer, a rational number, and a real number. The number $\sqrt{3}$ is an irrational number and a real number.

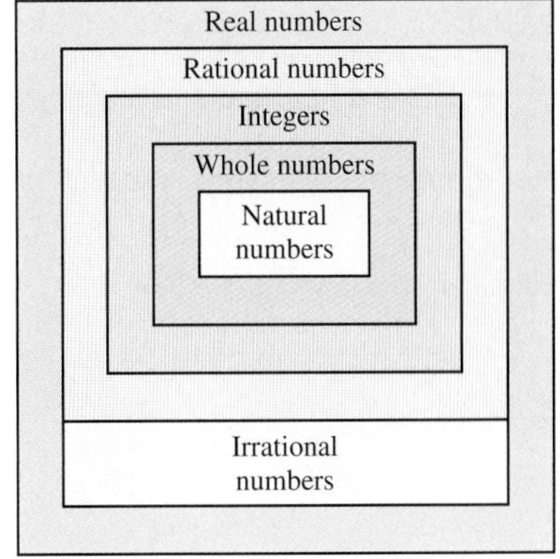

Figure 5-7

EXAMPLE 1 Classifying Numbers

Classify each number according to type.

(a) 0
(b) $\sqrt{15}$
(c) $-\dfrac{3}{4}$
(d) $0.8\overline{6}$
(e) $\sqrt{25}$

SOLUTION

(a) Zero is a whole number, and all whole numbers are integers, rational numbers, and real numbers.

(b) Since 15 is not a perfect square, $\sqrt{15}$ is an irrational number. Since all irrational numbers are real numbers, $\sqrt{15}$ is also a real number.

(c) The number $-\frac{3}{4}$ is a rational number. Since all rational numbers are real numbers, it is also a real number.

(d) The number $0.8\overline{6}$ is a repeating decimal; so it is a rational number and a real number.

(e) At first glance, you might think $\sqrt{25}$ is an irrational number because of the radical sign, but $\sqrt{25} = 5$, and 5 is a natural number, a whole number, an integer, a rational number, and a real number.

▼ Try This One 1

Classify each number according to type.

(a) π
(b) $-9.\overline{2}$
(c) $\sqrt{6}$
(d) -10
(e) $\sqrt{36}$

Sidelight KEEPING IT REAL

Why are the real numbers called the real numbers? Probably because they can all represent the size or measure of some real object. Integers represent tons of real quantities—anything that can be counted, sizes in whole-number units like 10 feet, and things like temperatures, elevations, and debts, which can be negative. Rational numbers very commonly describe measurements and sizes: a standard piece of paper is $8\frac{1}{2} \times 11$ inches, for example. We can even think of physical quantities that are represented by irrational numbers. The floor of a room that measures 10 feet by 10 feet is $\sqrt{200}$ feet from one corner to the opposite corner, and a soccer ball, which has diameter 10 inches, has a volume of $\frac{500}{3}\pi$ cubic inches.

But what about numbers that are not real? Think about a number like $\sqrt{-4}$. Based on our definition of square root, we would need to find a number whose square is -4. But whenever you multiply a number by itself, the result is positive because (obviously) in this case the two factors have the same sign. So there is no real number whose square is -4, and we define $\sqrt{-4}$ to be an imaginary number. Surprisingly, imaginary numbers actually do have applications in the real world! They play a role in electrical engineering and quantum mechanics, among other areas.

Properties of Real Numbers

You might recall that when you multiply two numbers, the order in which they are multiplied doesn't matter. For example, 3×2 and 2×3 are both equal to 6. In other words, $3 \times 2 = 2 \times 3$. But when you subtract two numbers, order is very important. For example $5 - 3 = 2$, but $3 - 5 = -2$. This tells us that $5 - 3 \neq 3 - 5$.

The fact that two numbers can be multiplied in any order is called the *commutative property of multiplication.* Subtraction is *not* commutative. This is just one example of a property of real numbers. By **property,** we mean a fact that is always true, regardless of the specific numbers chosen. We will study 11 basic properties of the real numbers.

For any set of numbers, the *closure property* for a given operation says that if two numbers from the set are combined using that operation, the result is another number in that set. For example, the set of natural numbers is closed under addition because the sum of two natural numbers is always a natural number. But the natural numbers are not closed under subtraction. Both 5 and 10 are natural numbers, but 5 – 10 is not, since it's negative.

The set of real numbers is closed under addition and multiplication since the sum or product of any two real numbers will always be a real number. This provides our first two properties of real numbers.

The closure property of addition: For any two real numbers a and b, the sum $a + b$ is also a real number.

The closure property of multiplication: For any two real numbers a and b, the product $a \cdot b$ is also a real number.

The *commutative property* for an operation states that for any two numbers, the order in which the operation is performed on the numbers doesn't matter. Addition of real numbers is commutative since $a + b = b + a$. For example, $6 + 7 = 7 + 6$, and $5 + 10 = 10 + 5$.

Multiplication of real numbers is also commutative. For example, $8 \times 6 = 6 \times 8$, and $15 \times 3 = 3 \times 15$.

The commutative property of addition: For any real numbers a and b, $a + b = b + a$.

The commutative property of multiplication: For any real numbers a and b, $a \times b = b \times a$.

Sidelight MATH AND BODY COMPOSITION

As obesity becomes more and more of a health issue in the United States, an index that compares height and weight has become widely used to evaluate body composition. The body mass index, or BMI, was actually developed in the mid-1800s by a Belgian scholar who dabbled in astronomy, mathematics, statistics, and sociology, but it's probably used today more than ever.

To determine your BMI, (1) square your height in inches, (2) divide your weight in pounds by that square, then (3) multiply the result by 703. The following guidelines have been created by the World Health Organization:

Under 18.5	Underweight
18.5 – 24.9	Normal
25 – 29.9	Overweight
30 – 39.9	Obese
40 and over	Severely obese

Here's a sample calculation for a person who is 5 foot 7 inches tall and weighs 135 pounds.

$$5'7'' = 67''$$
$$67^2 = 4{,}489$$
$$135 \div 4{,}489 \approx 0.0300$$
$$0.0300 \times 703 \approx 21.1$$

You will need this formula to determine the BMIs for some well-known athletes in exercises 45–52.

Remember, to show that a conjecture is not true, all that's needed is one counterexample. So we can easily see that subtraction and division are not commutative—choosing any two nonzero numbers will work. For example, $6 - 3$ and $3 - 6$ are not equal; $6 \div 3$ and $3 \div 6$ are also not equal.

Our next property involves operations with three numbers. Think about the sum $4 + 5 + 8$. If we use parentheses to group the 4 and the 5, we get $(4 + 5) + 8 = 9 + 8 = 17$. If instead we group the 5 and the 8, we get $4 + (5 + 8) = 4 + 13 = 17$. This is different than the commutative property—we didn't change the order, but rather the grouping. When grouping numbers differently in this manner doesn't affect the result, we say that an operation is *associative*. Both addition and multiplication are associative.

The associative property of addition: For any real numbers a, b, and c, $(a + b) + c = a + (b + c)$.

The associative property of multiplication: For any real numbers a, b, and c, $(a \times b) \times c = a \times (b \times c)$.

Subtraction and division are not associative, as we will see in the exercises.

You surely know that any number multiplied by one is the original number. We could say that 1 identifies any number when multiplying; in math terms, we call 1 the *identity* for multiplication. The identity for addition is zero because any number added to zero is the original number.

Math Note

Recall that the expression $-a$ doesn't mean that the expression is negative; it depends on the value of a. If a is positive, then $-a$ is negative. But if a is negative, then the expression $-a$ is positive.

The identity property of addition: The sum of any real number a and zero is the original number a. The number zero is called the identity for addition.

The identity property of multiplication. The product of any real number a and one is the original number a. One is called the identity for multiplication.

Numbers in a set can have *inverses* for a specific operation. If an operation is performed on a number and its inverse, the answer will be the identity for that operation.

For addition, 2 and -2 are *additive inverses* since $2 + (-2) = 0$. (Note that 0 is the identity for addition.) In algebra, the inverses for addition are called *opposites*. Every real number has an additive inverse (or opposite). The additive inverse of 0 is 0. The additive inverse for a number a is designated by $-a$.

Inverse property of addition: For any real number a, there exists a real number $-a$ such that $a + (-a) = 0$ and $-a + a = 0$. The number $-a$ is called the additive inverse or opposite of a.

For multiplication, 6 and $\frac{1}{6}$ are *multiplicative inverses* since $6 \times \frac{1}{6} = 1$. (Recall that 1 is the identity for multiplication.) The multiplicative inverse of a number is also called its *reciprocal*. Every real number a except zero has a multiplicative inverse, denoted $\frac{1}{a}$.

Inverse property of multiplication. For any real number a except zero, there exists a real number $\frac{1}{a}$ such that $a \times \frac{1}{a} = 1$ and $\frac{1}{a} \times a = 1$. The number $\frac{1}{a}$ is called the multiplicative inverse, or reciprocal, of a.

Math Note

Remember that the reciprocal of a fraction is obtained by "flipping the fraction upside down." For example, the reciprocal of $\frac{2}{3}$ is $\frac{3}{2}$.

Make sure you don't confuse inverses for addition with inverses for multiplication. To find an additive inverse for a number, change its sign. To find the multiplicative inverse for a number, use its reciprocal. The additive inverse of $-\frac{2}{3}$ is $+\frac{2}{3}$. The multiplicative inverse of $-\frac{2}{3}$ is $-\frac{3}{2}$. The sign does not change.

The *distributive property of multiplication over addition* states that when a number is multiplied by a sum, the number can first be multiplied by each number in the sum, and then the addition can be performed. For example,

$$\begin{aligned} 5(6+3) &= 5 \cdot 6 + 5 \cdot 3 \\ 5(9) &= 30 + 15 \\ 45 &= 45 \end{aligned}$$

Stated formally,

The distributive property of multiplication over addition: For any real numbers a, b, and c, $a \cdot (b + c) = a \cdot b + a \cdot c$.

This property will be referred to simply as the distributive property since addition cannot be distributed over multiplication, as we will see in the exercises.

EXAMPLE 2 Identifying Properties of Real Numbers

Identify the property illustrated by each calculation.

(a) $6 + 5 = 5 + 6$
(b) $0 + 3 = 3$
(c) $4(3 + 9) = 12 + 36$
(d) $\frac{2}{5} \times \frac{5}{2} = 1$
(e) $(6 \cdot 7) \cdot 3 = 6 \cdot (7 \cdot 3)$

SOLUTION

(a) Commutative property of addition
(b) Identity property of addition
(c) Distributive property
(d) Inverse property of multiplication
(e) Associative property of multiplication

▼ Try This One 2

☑ 2. Identify properties of the real numbers.

Identify the property illustrated by each calculation.

(a) $-2(3 + 5) = -6 - 10$
(b) $127 \times 1 = 127$
(c) $18 + (-18) = 0$
(d) $(-5 \cdot 1 + 3 \cdot 2) + 4 \cdot 7 = -5 \cdot 1 + (3 \cdot 2 + 4 \cdot 7)$
(e) $\frac{2}{3} \cdot \frac{4}{5} = \frac{4}{5} \cdot \frac{2}{3}$

The properties illustrated in Example 2 are straightforward. Identifying the properties in Example 3 is a little more challenging.

EXAMPLE 3 Identifying Properties of Real Numbers

Identify the property illustrated by each calculation.

(a) $(6 + 5) + 3 = (5 + 6) + 3$ (b) $1 \times 7 = 7 \times 1$ (c) $5 \cdot (6 + 2) = (6 + 2) \cdot 5$

SOLUTION

(a) This is not the associative property of addition since different numbers would have to be in parentheses. Notice that only the order of the 6 and 5 was changed. This is an example of the commutative property of addition, $6 + 5 = 5 + 6$.
(b) This is not an example of the identity property of multiplication. That property states that $1 \times 7 = 7$. This is an example of the commutative property of multiplication since $a \cdot b = b \cdot a$.
(c) This is not an example of the distributive property. The distributive property states that $5 \cdot (6 + 2) = 5 \cdot 6 + 5 \cdot 2$. Here the order of the multiplication was changed. So this is an example of the commutative property of multiplication.

▼ Try This One 3

Identify the property illustrated by each calculation.

(a) $4 + (2 + 3) = 4 + (3 + 2)$
(b) $5 \cdot (6 + 1) = 5 \cdot (1 + 6)$
(c) $3 \cdot \frac{1}{3} = \frac{1}{3} \cdot 3$

The 11 properties of real numbers that we studied in this section are summarized in Table 5-2.

TABLE 5-2 Properties of the Real Number System

Name	Property	Example
(For any real numbers a, b, and c)		
Closure property of addition	$a + b$ is a real number	$8 + (-3)$ is a real number
Closure property of multiplication	$a \times b$ is a real number	-5×8 is a real number
Commutative property of addition	$a + b = b + a$	$9 + 8 = 8 + 9$
Commutative property of multiplication	$a \cdot b = b \cdot a$	$6 \cdot 8 = 8 \cdot 6$
Associative property of addition	$(a + b) + c = a + (b + c)$	$(12 + 7) + 3 = 12 + (7 + 3)$
Associative property of multiplication	$(a \cdot b) \cdot c = a \cdot (b \cdot c)$	$(5 \cdot 3) \cdot 2 = 5 \cdot (3 \cdot 2)$
Identity property of addition	$0 + a = a$	$0 + 14 = 14$
Identity property of multiplication	$1 \times a = a$	$1 \times (-3) = -3$
Inverse property of addition	$a + (-a) = 0$	$6 + (-6) = 0$
Inverse property of multiplication	$a \cdot \frac{1}{a} = 1, a \neq 0$	$4 \cdot \frac{1}{4} = 1$
Distributive property	$a(b + c) = a \cdot b + a \cdot c$	$6(2 + 5) = 6 \cdot 2 + 6 \cdot 5$

Answers to Try This One

1 (a) Irrational and real
(b) Rational and real
(c) Irrational and real
(d) Integer, rational, and real
(e) Natural number, whole number, integer, rational, and real

2 (a) Distributive property
(b) Identity property of multiplication
(c) Inverse property of addition
(d) Associative property of addition
(e) Commutative property of multiplication

3 (a) Commutative property of addition
(b) Commutative property of addition
(c) Commutative property of multiplication

EXERCISE SET 5-5

Writing Exercises

1. Describe the set of real numbers in your own words.
2. Explain what is meant by the term "property of real numbers."
3. Explain what commutative properties refer to. Which operations are commutative?
4. Explain what associative properties refer to. Which operations are associative?
5. What does it mean for a set to be closed under a certain operation?
6. Explain why zero is the identity element for addition, while 1 is the identity element for multiplication.

Computational Exercises

For Exercises 7–22, classify each number by using one or more of the categories—natural, whole, integer, rational, irrational, real.

7. -5
8. 18
9. $\frac{3}{4}$
10. $-\frac{2}{3}$
11. 6.25
12. -18.376
13. $-\sqrt{6}$
14. $-\sqrt{18}$
15. 0.03030030003...
16. $-\pi$
17. 2.8
18. 13.6
19. 33
20. -17
21. $\sqrt{9}$
22. $\sqrt{100}$

For Exercises 23–38, name the property illustrated.

23. $4 + 8$ is a real number
24. $6 \cdot 1 = 6$
25. $17 + 6 = 6 + 17$
26. $-5 \times (3 + 4) = -5(3) + (-5)(4)$
27. $4 \times 8 = 8 \times 4$
28. $6 + (-6) = 0$
29. $4 \cdot (\sqrt{5} + \sqrt{11}) = 4\sqrt{5} + 4\sqrt{11}$
30. $\sqrt{3} + \sqrt{4}$ is a real number
31. $\frac{5}{8} \cdot \frac{8}{5} = 1$
32. $-5 + (+5) = 0$
33. $-6 \cdot (2 + 3) = -6 \cdot (3 + 2)$
34. $(16 + 3) + 5 = 16 + (3 + 5)$
35. $5 + 0 = 5$
36. $-6 + (+6) = (+6) + (-6)$
37. $\frac{3}{4} \times \frac{4}{3} = \frac{4}{3} \times \frac{3}{4}$
38. $(8 \times 4) \times 2 = (4 \times 8) \times 2$

For Exercises 39–44, determine under which operations (addition, subtraction, multiplication, division) the system is closed.

39. Natural numbers
40. Whole numbers
41. Integers
42. Rational numbers
43. Irrational numbers
44. Real numbers

Real-World Applications

For Exercises 45–52, use the body mass index formula found in the sidelight on page 247 to find the BMI for the given athlete, then use the table in the sidelight to classify him or her according to World Health Organization standards.

45. Lebron James, basketball player: 6′8″, 250 lb
46. Yao Ming, basketball player: 7′6″, 310 lb
47. Michael Phelps, Olympic swimmer: 6′4″, 195 lb
48. Lance Armstrong, bicycle racer: 5′10″, 165 lb
49. Nastia Liukin, Olympic gymnast: 5′3″, 99 lb
50. Julie Krone, professional jockey: 4′10$\frac{1}{2}$″, 105 lb
51. Warren Sapp, football player: 6′2″, 300 lb
52. Jonathan Ogden, football player: 6′9″, 345 lb

For Exercises 53–56, decide whether the two procedures described are commutative or not, and explain your answer.

53. Putting on your shoes and socks.
54. Going to the grocery store and the laundromat.
55. Doing your homework and working out.
56. Washing and drying your dirty clothes.

For Exercises 57–60, decide if the three procedures described are associative or not, and explain your answer.

57. Driving to the bank and McDonalds, and walking your dog.
58. Riding an exercise bike, talking on the phone, and watching TV.
59. Graduating from college, buying a house, and getting married.
60. Getting on a plane, flying to Mexico, and renting a car.

Critical Thinking

For Exercises 61–64, complete the two operations and compare the results. Then describe the result in terms of a property from this section.

61. $(12 - 8) - 15$ $12 - (8 - 15)$
62. $100 \div (25 \div 5)$ $(100 \div 25) \div 5$
63. $10 + (3 \cdot 5)$ $(10 + 3) \cdot (10 + 5)$
64. $8(20 - 4)$ $8 \cdot 20 - 8 \cdot 4$

Section 5-6 Exponents and Scientific Notation

LEARNING OBJECTIVES

- ❑ 1. Define integer exponents.
- ❑ 2. Use rules for exponents.
- ❑ 3. Convert between scientific and decimal notation.
- ❑ 4. Perform operations with numbers in scientific notation.
- ❑ 5. Use scientific notation in applied problems.

It wasn't that long ago that a million dollars was an almost unimaginable amount of money. The first million dollar lottery was held in New York in 1970, and at that time, winning a million dollars instantly made someone ultra-rich.

But times have changed. In the 21st century, a lot of lottery players don't even bother to buy a ticket when the jackpot is "only" a million dollars! The amount of money made by CEOs, entertainers, and athletes continues to grow at an amazing clip, and even the average Joe can make a salary that would have been considered a king's ransom 50 years ago. As the numbers that describe our monetary system get larger and larger, the ability to work with and interpret really large numbers will become more and more important. In this section, we'll study a way to work with really large (and really small) numbers. It starts with an understanding of exponents.

In December 2008, pitcher C. C. Sabathia signed a free-agent contract worth $161 million.

Exponents are a concise method of representing repeated multiplications. For example, $5 \cdot 5 \cdot 5 \cdot 5$ can be written as 5^4, the exponent 4 describing how many times to multiply 5 by itself. In general:

For any positive integer n,

$$a^n = \underbrace{a \cdot a \cdot a \cdot a \cdot \cdots \cdot a}_{n \text{ factors}}$$

where a is called the **base** and n is called the **exponent.**

The expression a^n is read as "a to the nth power." When the exponent is 2, such as 5^2, it can be read as "five squared" or "five to the second power." When the exponent is one, it is usually not written; that is, $a^1 = a$.

When an exponent is negative, it is defined as follows: For any positive integer n,

$$a^{-n} = \frac{1}{a^n}$$

For example, $5^{-3} = \frac{1}{5^3}$ or $\frac{1}{125}$. Finally, when the exponent is zero, it is defined as follows.

For any nonzero number a, $a^0 = 1$.

That is, any nonzero number to the 0 power is equal to one. For example, $6^0 = 1$.

EXAMPLE 1 Evaluating Expressions with Exponents

Evaluate each expression.

(a) 6^3 (b) 3^{-4} (c) 9^0

SOLUTION

(a) $6^3 = 6 \cdot 6 \cdot 6 = 216$

(b) $3^{-4} = \frac{1}{3^4} = \frac{1}{3 \cdot 3 \cdot 3 \cdot 3} = \frac{1}{81}$

(c) $9^0 = 1$

▼ Try This One 1

☑ 1. Define integer exponents.

Evaluate each expression.

(a) 3^6 (b) 2^{-6} (c) 7^0

There are three rules for working with exponents that help us greatly in performing calculations. Think about the expression $4^2 \cdot 4^3$. Using the definition of exponents, we can write this as $(4 \cdot 4) \cdot (4 \cdot 4 \cdot 4)$, which is the same as 4^5. The net result is that we added the exponents: there were two factors from 4^2 and three factors from 4^3, for a total of five factors. This is an example of an exponent rule.

Rules for Exponents

For any nonzero real number a and any integers m and n:

1. (The product rule) $a^m \cdot a^n = a^{m+n}$

2. (The quotient rule) $\frac{a^m}{a^n} = a^{m-n}$
3. (The power rule) $(a^m)^n = a^{m \cdot n}$

Most students find it easier to understand and use these rules when they're described in words rather than symbols.

Math Note

Neither the product rule nor the quotient rule says *anything* about multiplying or dividing exponential expressions with *different bases!*

1. **Product rule.** When two expressions with the same base are multiplied, you keep the base unchanged and add the exponents.
2. **Quotient rule.** When two expressions with the same base are divided, you keep the base unchanged and subtract the exponents.
3. **Power rule.** When an expression with an exponent is raised to a power, you keep the base unchanged and multiply the two exponents.

Example 2 illustrates the use of these rules in simplifying and performing calculations.

EXAMPLE 2 Using Rules For Exponents

Simplify the expression using rules for exponents, and then evaluate the resulting expression.

(a) $3^3 \cdot 3^5$ (b) $\frac{5^4}{5^6}$ (c) $(2^3)^4$

Calculator Guide

To find 3^8 for Example 2 using a calculator:

Standard Scientific Calculator

3 [y^x] or [x^y] 8 [=]

Standard Graphing Calculator

3 [^] 8 [Enter]

SOLUTION

(a) The bases are the same and we're multiplying, so we add exponents (product rule).

$$3^3 \cdot 3^5 = 3^{3+5} = 3^8 \text{ which is } 6{,}561$$

(b) The bases are the same and we're dividing, so we subtract the exponents (quotient rule)

$$\frac{5^4}{5^6} = 5^{4-6} = 5^{-2} \quad \text{which is} \quad \frac{1}{5^2} \quad \text{or} \quad \frac{1}{25}$$

(c) When raising an exponential expression to a power, multiply the exponents (power rule).

$$(2^3)^4 = 2^{3 \cdot 4} = 2^{12} \quad \text{which is} \quad 4{,}096$$

▼ Try This One 2

☑ 2. Use rules for exponents.

Simplify the expression using rules for exponents, and then evaluate the resulting expression.

(a) $6^2 \cdot 6^4$ (b) $\frac{10^8}{10^{11}}$ (c) $(4^3)^2$

Scientific Notation

Now we turn our attention to working with really large and really small numbers. We will express numbers in terms of powers of 10, because this allows us to write a lot of digits with just a few symbols:

Powers of 10

	$10^0 = 1$	$10^{-1} = 0.1$	
	$10^1 = 10$	$10^{-2} = 0.01$	
	$10^2 = 100$	$10^{-3} = 0.001$	*Thousandth*
Thousand	$10^3 = 1{,}000$	$10^{-4} = 0.0001$	
	$10^4 = 10{,}000$	$10^{-5} = 0.00001$	
	$10^5 = 100{,}000$	$10^{-6} = 0.000001$	*Millionth*
Million	$10^6 = 1{,}000{,}000$	$10^{-7} = 0.0000001$	
	$10^7 = 10{,}000{,}000$	$10^{-8} = 0.00000001$	
	$10^8 = 100{,}000{,}000$		
Billion	$10^9 = 1{,}000{,}000{,}000$		

Math Note

Because it's very common to work with really large or really small numbers in science, the system for writing them concisely is called scientific notation.

For example, the earth is about 93,000,000 (93 million) miles from the sun. Another way to write this number is 9.3 times 10 million, or 9.3×10^7. A number is in **scientific notation** when it's written as a decimal with one digit to the left of the decimal point times some power of 10. The rules for writing a number in scientific notation are below.

Procedure for Writing a Number in Scientific Notation

Step 1 Move the decimal point either right or left so that there is exactly one digit to the left of the decimal point.

Step 2 Write the resulting number times 10 to some power:

(a) If the decimal point was moved to the left, the exponent of 10 is the number of places the decimal point was moved.

(b) If the decimal point was moved right, the exponent of 10 is the negative of the number of places the decimal point was moved.

EXAMPLE 3 Writing Numbers in Scientific Notation

Write each number in scientific notation.

(a) 3,572,000,000 (b) 0.000087

Math Note

If you have trouble remembering whether to use a positive or negative exponent, think of it this way: a power of 10 with a positive exponent, like 10^8, is a really large number, so large numbers get positive exponents in scientific notation, and small numbers get negative exponents.

SOLUTION

(a) Move the decimal point 9 places to the left so that it falls between the 3 and 5.

3572000000
987654321

In scientific notation, $3{,}572{,}000{,}000 = 3.572 \times 10^9$.

(b) Move the point to the right 5 places so that it will fall between the 8 and 7.

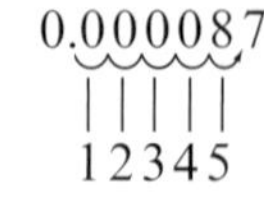

In scientific notation, $0.000087 = 8.7 \times 10^{-5}$.

▼ Try This One 3

Write each number in scientific notation.

(a) 516,000,000 (b) 0.000162

As we'll see shortly, writing numbers in scientific notation is very useful in performing calculations, but you should also be able to convert from scientific to decimal notation. This sometimes makes it easier to get perspective on the size of the numbers involved. To do so, we just reverse the procedure above.

Rules for Converting a Number in Scientific Notation to Decimal Notation

(a) If the exponent of the power of 10 is positive, move the decimal point to the right the same number of places as the exponent. You may need to put in zeros as placeholders when you run out of digits.

(b) If the exponent is negative, move the decimal point to the left the same number of places as the exponent, again putting in zeros as needed.

EXAMPLE 4 Converting from Scientific to Decimal Notation

Write each number in decimal notation.

(a) 4.192×10^8 (b) 6.37×10^{-8}

SOLUTION

Calculator Guide

Here's how to enter 4.192×10^8 into a calculator:

Standard Scientific Calculator

4.192 [EE] 8 [=]

Standard Graphing Calculator

4.192 [EE] or [2nd] [,] [Enter]

(a) Since the exponent is positive, move the decimal point 8 places to the right.

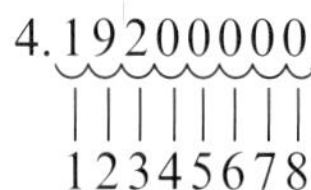

In decimal notation, $4.192 \times 10^8 = 419{,}200{,}000$.

(b) Since the exponent is negative, move the decimal point 8 places to the left.

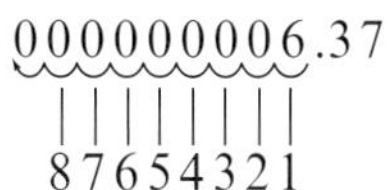

In decimal notation, $6.37 \times 10^{-8} = 0.0000000637$.

▼ Try This One 4

Write each number in decimal notation.

(a) 9.61×10^{10} (b) 2.77×10^{-6}

☑ 3. Convert between scientific and decimal notation.

Operations with Numbers in Scientific Notation

One of the biggest reasons that scientific notation is useful is that it makes multiplication and division of very large and very small numbers very easy. Suppose you are asked to multiply 2×10^7 and 4.2×10^{10}. In decimal form, this would be challenging, to say the least. But let's write it out and see what happens:

$$(2 \times 10^7) \times (4.2 \times 10^{10}) = (2 \times 4.2) \times (10^7 \times 10^{10}) = 8.4 \times 10^{17}$$

Commutative Property of Multiplication

Multiplying Numbers in Scientific Notation

To multiply two numbers in scientific notation, multiply the numbers to the left of the powers of 10 and add the exponents of the 10s. (You may have to rewrite to put the result in proper scientific notation.)

EXAMPLE 5 Multiplying Numbers in Scientific Notation

Calculator Guide

Many calculators will display very small or large answers in scientific notation, even if the original numbers were entered in decimal notation. For example, when multiplying 3 million and 5 million entered in decimal form, the calculator displays:

Standard Scientific Calculator

1.5e+13

Standard Graphing Calculator

1.5E13

This represents 1.5×10^{13}.

Find each product. Write your answer in scientific notation.

(a) $(3 \times 10^5)(2.2 \times 10^3)$ (b) $(5 \times 10^2)(7 \times 10^3)$

SOLUTION

(a) Multiply 3×2.2 to get 6.6, and add exponents $5 + 3$ to get 8.

$$(3 \times 10^5)(2.2 \times 10^3) = 6.6 \times 10^8$$

(b) This time, when we multiply 5×7, we get a two-digit number, so our answer won't be in proper scientific notation:

$$(5 \times 10^2)(7 \times 10^3) = 35 \times 10^5$$

We can remedy this by moving the decimal point one place left, which raises the exponent by 1:

$$35 \times 10^5 = 3.5 \times 10^6$$

To make sure we chose the right exponent, notice that in changing 35 to 3.5, we made it smaller, so we needed to make the exponent larger.

▼ Try This One 5

Find each product. Write your answer in scientific notation.

(a) $(6 \times 10^5)(1 \times 10^4)$ (b) $(5 \times 10^3)(3.1 \times 10^6)$

The same rule for multiplication works just fine when the exponents are negative.

EXAMPLE 6 Multiplying Numbers in Scientific Notation

Find the product. Round the decimal part of your answer to two decimal places.

(a) $(3.25 \times 10^{-4})(5.1 \times 10^{-3})$ (b) $(8.6 \times 10^3)(9.7 \times 10^{-6})$

Math Note

The part of a number in scientific notation that comes before the power of 10 is often called the *decimal part* of the number, as we did in Example 6.

SOLUTION

(a) $(3.25 \times 10^{-4})(5.1 \times 10^{-3}) = 16.575 \times 10^{-7} \approx 1.66 \times 10^{-6}$

Be careful when moving the decimal point! Adding 1 to -7 makes it -6.

(b) $(8.6 \times 10^{3})(9.7 \times 10^{-6}) = 83.42 \times 10^{-3} \approx 8.34 \times 10^{-2}$

▼ Try This One 6

Find the product. Write each answer in scientific notation, rounding the decimal part of the answer to two decimal places.

(a) $(4.2 \times 10^{-4})(2 \times 10^{-2})$
(b) $(7.3 \times 10^{-9})(5.1 \times 10^{6})$
(c) $(8.32 \times 10^{-3})(6.48 \times 10^{5})$

The rule for dividing numbers in scientific notation is similar to the one for multiplying. In Exercise 102, we'll see why it makes perfect sense.

Dividing Numbers in Scientific Notation

To divide two numbers in scientific notation, divide the numbers to the left of the power of 10 and subtract the exponents of the 10s. (You may have to rewrite to put the result in proper scientific notation.)

EXAMPLE 7 Dividing Numbers in Scientific Notation

Find each quotient. Write your answer in scientific notation.

(a) $\dfrac{9 \times 10^{6}}{3 \times 10^{4}}$ (b) $\dfrac{9 \times 10^{-3}}{2 \times 10^{5}}$ (c) $\dfrac{1.2 \times 10^{-5}}{4.8 \times 10^{-4}}$

SOLUTION

In each case, we'll divide the decimal parts, then subtract the exponents to find the appropriate power of 10.

(a) $\dfrac{9 \times 10^{6}}{3 \times 10^{4}} = \dfrac{9}{3} \times 10^{6-4} = 3 \times 10^{2}$

(b) $\dfrac{9 \times 10^{-3}}{2 \times 10^{5}} = \dfrac{9}{2} \times 10^{-3-5} = 4.5 \times 10^{-8}$

(c) $\dfrac{1.2 \times 10^{-5}}{4.8 \times 10^{-4}} = \dfrac{1.2}{4.8} \times 10^{-5-(-4)} = 0.25 \times 10^{-5+4} = 0.25 \times 10^{-1}$

This time, the answer needs to be rewritten to put it in proper scientific notation. The decimal point will be moved one place to the right, so we need to subtract one from the exponent.

$$0.25 \times 10^{-1} = 2.5 \times 10^{-2}$$

▼ Try This One 7

☑ 4. Perform operations with numbers in scientific notation.

Find each quotient. Round the decimal part of the answer to two decimal places.

(a) $\dfrac{3.2 \times 10^{7}}{8 \times 10^{4}}$ (b) $\dfrac{9.6 \times 10^{-5}}{1.3 \times 10^{-8}}$ (c) $\dfrac{2.2 \times 10^{-6}}{4.8 \times 10^{5}}$

Applications of Scientific Notation

5. Use scientific notation in applied problems.

As we pointed out earlier, many areas in science and economics require the use of very large or very small numbers. This is when scientific notation really comes in handy. Many calculators display only 8 to 12 digits, so numbers with more digits can only be entered in scientific notation.

EXAMPLE 8 Applying Scientific Notation to Astronomy

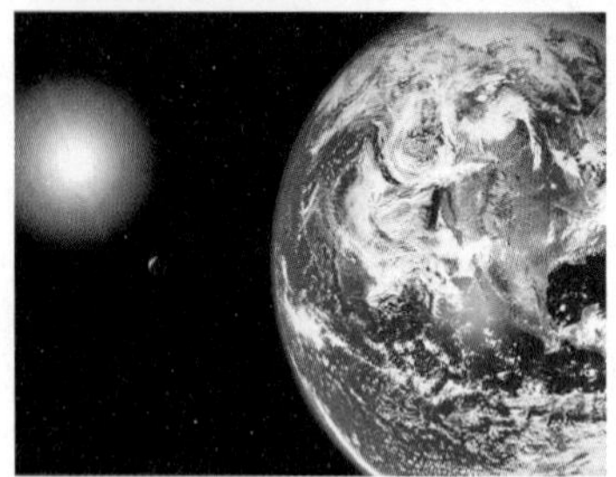

The earth's orbit around the sun is not circular, but it's reasonably close to a circle with radius 9.3×10^7 miles. How far does the earth travel in one year? (The formula for circumference of a circle is $C = 2\pi r$.)

SOLUTION

The distance the earth travels in one year is one full turn around a circle with radius 9.3×10^7 miles. Using the formula for circumference, and approximating π with 3.14,

$$\begin{aligned} C &\approx 2(3.14)(9.3 \times 10^7) \\ &\approx 58.4 \times 10^7 \\ &= 5.84 \times 10^8 \text{ miles} \end{aligned}$$

In decimal form, this is 584,000,000, or 584 million miles!

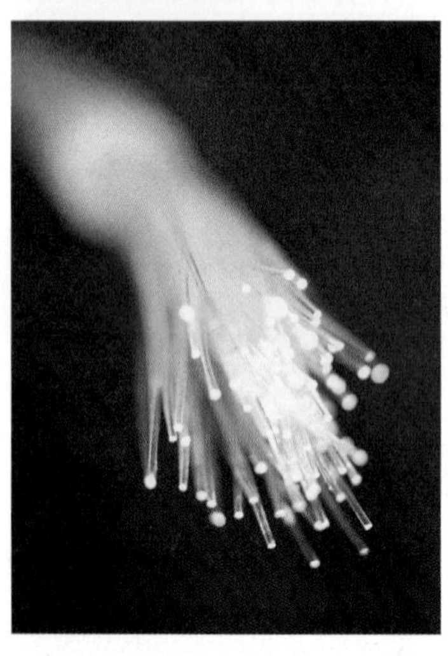

▼ Try This One 8

The speed of light in glass is about 6.56×10^8 feet per second, and the circumference of the earth is about 1.31×10^8 feet. Using the formula

$$\text{Time} = \frac{\text{distance}}{\text{speed}}$$

find how long it takes light to circle the globe in a glass fiber-optic cable.

Answers to Try This One

1 (a) 729
(b) $\frac{1}{64}$
(c) 1

2 (a) $6^6 = 46{,}656$
(b) $10^{-3} = \frac{1}{1{,}000}$
(c) $4^6 = 4{,}096$

3 (a) 5.16×10^8
(b) 1.62×10^{-4}

4 (a) 96,100,000,000
(b) 0.00000277

5 (a) 6×10^9
(b) 1.55×10^{10}

6 (a) 8.4×10^{-6}
(b) 3.72×10^{-2}
(c) 5.39×10^3

7 (a) 4×10^2
(b) 7.38×10^3
(c) 4.58×10^{-12}

8 About 0.2 seconds

EXERCISE SET 5-6

Writing Exercises

1. Describe how exponents are defined using multiplication.
2. What is scientific notation? How do you convert a number from decimal to scientific notation?
3. Explain how to convert a number from scientific to decimal notation.
4. What are some of the advantages of using scientific notation?
5. Describe in your own words how to multiply two numbers in scientific notation.
6. Describe in your own words how to divide two numbers in scientific notation.

Computational Exercises

For Exercises 7–16, evaluate each.

7. 3^5
8. 6^4
9. 8^0
10. 9^0
11. $(-5)^0$
12. $(-4)^0$
13. 3^{-5}
14. 6^{-4}
15. 2^{-6}
16. 7^{-2}

For Exercises 17–36, simplify the expression using rules for exponents, then evaluate the resulting expression.

17. $3^4 \cdot 3^2$
18. $5^3 \cdot 5^3$
19. $4^4 \cdot 4^3$
20. $2^6 \cdot 2^4$
21. $\frac{3^4}{3^2}$
22. $\frac{6^5}{6^3}$
23. $\frac{2^5}{2^4}$
24. $\frac{8^3}{8}$
25. $(5^2)^3$
26. $(4^4)^2$
27. $3^2 \cdot 3^{-4}$
28. $4^{-3} \cdot 4^5$
29. $5^{-3} \cdot 5^{-2}$
30. $6^3 \cdot 6^{-3}$
31. $\frac{2^5}{2^7}$
32. $\frac{3^4}{3^6}$
33. $\frac{4^4}{4^7}$
34. $\frac{5^2}{5^5}$
35. $\frac{7^2}{7^3}$
36. $\frac{8^2}{8^4}$

For Exercises 37–50, write each number in scientific notation.

37. 625,000,000
38. 9,910,000
39. 0.0073
40. 0.261
41. 528,000,000,000
42. 2,220,000
43. 0.00000618
44. 0.0000000077
45. 43,200
46. 56,000
47. 0.0814
48. 0.0011
49. 32,000,000,000,000
50. 43,500,000

For Exercises 51–64, write each number in decimal notation.

51. 5.9×10^4
52. 6.28×10^6
53. 3.75×10^{-5}
54. 9×10^{-10}
55. 2.4×10^3
56. 7.72×10^5
57. 3×10^{-6}
58. 4×10^{-9}
59. 1×10^3
60. 2.26×10^4
61. 8.02×10^9
62. 1×10^{-4}
63. 7×10^{12}
64. 1.33×10^2

For Exercises 65–78, perform the indicated operations. Write the answers in scientific notation, rounding to two decimal places. Then write the rounded answer in decimal notation as well.

65. $(3 \times 10^4)(2 \times 10^6)$
66. $(5 \times 10^3)(8 \times 10^5)$
67. $(6.2 \times 10^{-2})(4.3 \times 10^{-6})$
68. $(1.7 \times 10^{-5})(3.8 \times 10^{-6})$
69. $(4 \times 10^4)(2.2 \times 10^{-7})$
70. $(2.2 \times 10^5)(3.6 \times 10^{-4})$
71. $(5 \times 10^{-2})(3 \times 10^{-8})$
72. $(4.3 \times 10^5)(2.2 \times 10^{-6})$
73. $\frac{5 \times 10^4}{2.5 \times 10^2}$
74. $\frac{9 \times 10^6}{3 \times 10^2}$
75. $\frac{4.2 \times 10^{-2}}{7 \times 10^{-3}}$
76. $\frac{6.4 \times 10^8}{8 \times 10^{-2}}$
77. $\frac{6.6 \times 10^3}{1.1 \times 10^5}$
78. $\frac{3 \times 10^7}{1.5 \times 10^{-5}}$

For Exercises 79–84, write each number in scientific notation and perform the indicated operations. Leave answers in scientific notation. Round to two decimal places.

79. (63,000,000)(41,000,000)
80. (52,000)(3,000,000)
81. $\frac{600{,}000{,}000}{25{,}000{,}000}$
82. $\frac{32{,}000{,}000}{64{,}000{,}000}$
83. (0.00000025)(0.000004)
84. $\frac{0.0000036}{0.0009}$

Real-World Applications

85. Light travels through air at 1.86×10^5 miles per second. How many miles does light travel in 10 minutes? Write your answer in decimal notation.
86. There are about 1×10^{14} cells in the human body. About how many body cells are there in the United States? (Use 300 million as the population of the United States.)
87. The mass of a proton is about 1.67×10^{-24} grams. There are 1.27×10^{26} molecules in a gallon of water, and 10 protons in one molecule of water. Also, 1 gallon of water weighs 3,778 grams. Based on this information, what percentage of the weight of water comes from protons?

88. It has been estimated that there are 1×10^{20} grains of sand on the beach at Coney Island, New York. By one estimate, the beach has an area of 3.3 million square feet. How many grains of sand are there on average per square foot of beach?

For Exercises 89–92, use the following information. The number of miles that light travels in one year is called a light-year. One light-year is equal to 5.88×10^{12} miles.

89. The nearest star (other than the sun) to the earth is Proxima Centauri, at 4.2 light-years away. How far is this in miles?
90. The star Gruis is 280 light-years from the earth. How far is this in miles?
91. The planet Pluto (if it's actually a planet, which is being debated) is 4,681 million miles from the earth. Write this number in scientific notation. How far is that in light-years?
92. The average distance from the earth to the sun is 93 million miles. Write this number in scientific notation. Use the number of miles in a light-year to find how many minutes it takes light to reach the earth from the sun.
93. One atom is 1×10^{-8} centimeters in length. How many atoms could be laid end-to-end on a meter stick (which is 100 cm in length)?
94. One grain of spruce pollen weighs about 7×10^{-5} grams. How many grains are there in a pound of spruce pollen? (One pound is about 454 grams.)
95. The planet Venus is about 67 million miles from the sun. Assuming that its orbit is approximately circular, how far does it travel in one year? (One Venus year, not one earth year.)
96. Each red blood cell contains 250 million molecules of hemoglobin. There are about 25 trillion red blood cells in the average human. How many molecules of hemoglobin are there in an average human?
97. The largest oil company in the United States, Exxon Mobil, announced a profit of 4.06×10^{10} dollars in 2007. How much on average did the company earn every day in that year?
98. Refer to Exercise 97. If the CEO of Exxon Mobil had gone stark raving mad and decided to equally distribute the company's 2007 profit to the 300 million people in the United States, how much would each person have received?
99. In December of 2008, the New York Yankees signed three free-agent players, committing a total of 4.235×10^{8} dollars to these three contracts. The team was expected to have an average ticket price of \$74 in the 2009 season. How many tickets would the team have to sell to cover the cost of those three contracts?
100. At the end of the college football season, there are five bowl games that are part of the Bowl Championship Series. ESPN signed a contract to televise these five games from 2010 through 2014 at a cost of 1.25×10^{8} dollars per year. In 2008, an estimated 62 million viewers watched the games. If viewership continues at that rate, how much is ESPN spending per viewer?

Critical Thinking

101. Read the second column of the table below from top to bottom. Using inductive reasoning, complete the missing entries in the table. Then explain how that justifies our definitions of zero and negative exponents.

Exponent (n)	2^n
5	32
4	16
3	8
2	4
1	2
0	?
−1	?
−2	?

102. The quotient rule for exponents is $\frac{a^m}{a^n} = a^{m-n}$, where a is any nonzero number, and m and n are integers. Use the definition of exponents from the beginning of the section and reducing fractions to justify this rule. Does your justification require any restrictions on m or n? Explain.
103. The power rule for exponents is $(a^m)^n = a^{m \cdot n}$, where a is any nonzero number, and m and n are integers. Use the definition of exponents from the beginning of the section and multiplication to justify this rule. Does your justification require any restrictions on m or n? Explain.
104. We have already developed rules for multiplying and dividing numbers in scientific notation. In this exercise, we'll develop a power rule for scientific notation.

    (a) Using the definition of exponent from the beginning of this section and our rule for multiplying numbers in scientific notation, compute $(2 \times 10^4)^3$.
    (b) Repeat for $(a \times 10^n)^3$.
    (c) Use the results of (a) and (b) to verbally describe a rule for raising a number in scientific notation to a power. Does your rule work for zero and negative powers?

Section 5-7 Arithmetic and Geometric Sequences

LEARNING OBJECTIVES

- ☐ 1. Define arithmetic sequence.
- ☐ 2. Find a particular term of an arithmetic sequence.
- ☐ 3. Add terms of an arithmetic sequence.
- ☐ 4. Define geometric sequence.
- ☐ 5. Find a particular term of a geometric sequence.
- ☐ 6. Add terms of a geometric sequence.

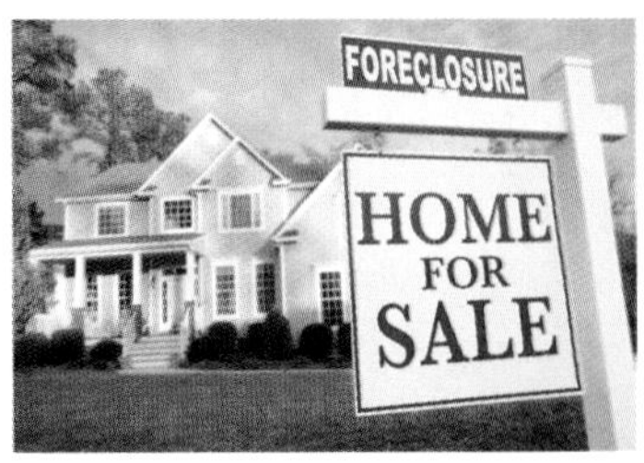

For most of the first decade of the 21st century, if you had a job and a pulse, you could get a home loan. Then the banks started finding out that giving loans to people that couldn't afford them wasn't such a smart idea. The housing market collapsed, millions of people lost their homes, and major financial institutions went belly-up. By the beginning of 2009, it was very difficult to get a home loan unless you had a significant down payment. This forced a lot of people to rent housing and try to save money for a home.

Suppose that this is the predicament a young couple find themselves in. They determine that they can afford to sock away \$400 each month. They don't trust banks, so they plan to put \$400 in cash in their cookie jar each month, which, by the way, is not the brightest idea we've ever heard. The amount they save would grow like this as the months pass: \$400, \$800, \$1,200, \$1,600, \$2,000.... This is a special type of *sequence* that we will study in this section.

A **sequence** is a list of numbers in a definite order. Each number in the sequence is called a **term** of the sequence.

Note: In many cases, there is a rule, or formula, that determines the terms of a sequence. But not always.

Arithmetic Sequences

In the sequence defined by saving money for a down payment above, every term after the first can be obtained by adding 400 to the previous term. This is an example of an **arithmetic sequence.** A sequence is arithmetic when every term after the first is obtained by adding the same fixed number to the previous term. This amount is called the **common difference**.

Examples of Arithmetic Sequences	Common Difference
1, 3, 5, 7, 9, 11,...	2
2, 7, 12, 17, 22, 27,...	5
100, 97, 94, 91, 88, 85,...	−3
$\frac{1}{2}, \frac{2}{2}, \frac{3}{2}, \frac{4}{2}, \frac{5}{2}, \frac{6}{2}, \ldots$	$\frac{1}{2}$

EXAMPLE 1 Finding the Terms of an Arithmetic Sequence

Write the first five terms of an arithmetic sequence with

(a) First term 9 and common difference 7.
(b) First term $\frac{1}{16}$ and common difference $-\frac{1}{8}$.

SOLUTION

(a) We begin at 9, and add 7 to obtain each successive term:
9, 16, 23, 30, 37.

(b) This time we will start at $\frac{1}{16}$ and *subtract* $\frac{1}{8}$ to obtain each successive term. It will be helpful to write $\frac{1}{8}$ as $\frac{2}{16}$ so we can perform the subtractions easily.

$$\frac{1}{16}, -\frac{1}{16}, -\frac{3}{16}, -\frac{5}{16}, -\frac{7}{16}$$

▼ Try This One 1

1. Define arithmetic sequence.

Write the first five terms of an arithmetic sequence with

(a) First term 6 and common difference 10.
(b) First term $\frac{3}{8}$ and common difference $-\frac{1}{2}$.

We often use the letter a with a subscript to represent the terms of a generic sequence. The symbol a_1 represents the first term, a_2 the second term, a_3 the third, and so on. So an arbitrary sequence looks like $a_1, a_2, a_3, a_4, a_5, \ldots$. We call a_n the nth term of the sequence.

When a sequence is arithmetic with first term a_1 and common difference d, we can find a formula to quickly calculate any term of the sequence without having to find all of the terms that come before it. In this case, the sequence looks like

$$a_1, \quad a_1 + d, \quad a_1 + d + d, \quad a_1 + d + d + d, \ldots$$

Doing some simplifying, we find that

$$a_1 = a_1 \qquad a_2 = a_1 + d \qquad a_3 = a_1 + 2d \qquad a_4 = a_1 + 3d$$

Every term a_n looks like a_1 plus some coefficient times d, and the coefficient is one less than the subscript of that term in the sequence. This gives us a formula:

Finding the nth Term of an Arithmetic Sequence

The nth term of an arithmetic sequence can be found using the formula

$$a_n = a_1 + (n - 1)d$$

where n is the number of the term you want to find, a_1 is the first term, and d is the common difference.

EXAMPLE 2 Finding a Particular Term of an Arithmetic Sequence

Find the eighth term of an arithmetic sequence with first term 5 and common difference 9.

SOLUTION

Substitute in the formula using $a_1 = 5$, $n = 8$, and $d = 9$.

$$\begin{aligned} a_n &= a_1 + (n - 1)d \\ a_8 &= 5 + (8 - 1)(9) \\ &= 5 + (7)(9) \\ &= 5 + 63 \\ &= 68 \end{aligned}$$

The eighth term is 68.

Math Note

The answer can be checked by writing eight terms of the sequence as shown:

5, 14, 23, 32, 41, 50, 59, 68
↑

▼ Try This One 2

Find the 10th term of an arithmetic sequence with first term 2 and common difference −5.

EXAMPLE 3 Finding a Particular Term of an Arithmetic Sequence

The trombone plays a different note when the musician changes the length of the instrument while setting up a vibrating column of air through the mouthpiece. For a given length of trombone L, it will resonate for notes with wavelengths

$$\frac{2L}{1}, \frac{2L}{2}, \frac{2L}{3}, \frac{2L}{4}, \frac{2L}{5}, \ldots$$

(This is true of other brass and woodwind instruments as well.) What is the nth term of this sequence?

☑ 2. Find a particular term of an arithmetic sequence.

Find the 10th term of the arithmetic sequence with first term $\frac{1}{7}$ and common difference $\frac{2}{5}$.

SOLUTION

Substitute in the formula using $a_1 = \frac{1}{7}$, $n = 10$, and $d = \frac{2}{5}$.

$$\begin{aligned} a_n &= a_1 + (n-1)d \\ a_{10} &= \frac{1}{7} + (10-1)\left(\frac{2}{5}\right) \\ &= \frac{1}{7} + (9)\left(\frac{2}{5}\right) \\ &= \frac{5}{35} + \frac{126}{35} \\ &= \frac{131}{35} \end{aligned}$$

▼ Try This One 3

Find the requested term of the arithmetic sequence with the given first term and common difference.

(a) First term $\frac{1}{9}$, common difference $\frac{1}{3}$: find the seventh term.

(b) First term $-\frac{5}{6}$, common difference $-\frac{1}{10}$: find the sixth term.

Sometimes it's helpful to find the sum of the terms of a sequence. When the sequence is arithmetic, we can use the formula below.

Finding the Sum of an Arithmetic Sequence

The sum of the first n terms of an arithmetic sequence is given by the formula

$$S_n = \frac{n(a_1 + a_n)}{2}$$

where a_1 is the first term of the sequence and a_n is the nth term.

EXAMPLE 4 Finding the Sum of an Arithmetic Sequence

Find the sum of the first 20 terms of the sequence.

$$2, 4, 6, 8, 10, 12, \ldots$$

SOLUTION

First, we need to find the 20th term. Using the formula shown previously,

$$a_{20} = a_1 + (n-1)d$$

where $a_1 = 2$, $n = 20$, and $d = 2$.

$$\begin{aligned} a_{20} &= 2 + (20-1)(2) \\ &= 2 + 19(2) \\ &= 40 \end{aligned}$$

Next, substitute into the formula for finding the sum.

$$S_n = \frac{n(a_1 + a_n)}{2}$$
$$= \frac{20(2 + 40)}{2}$$
$$= 420$$

The sum of the first 20 even numbers is 420.

▼ Try This One 4

☑ 3. Add terms of an arithmetic sequence.

Find the sum of the first 12 terms of each sequence.

(a) 5, 12, 19, 26, 33,...
(b) −1, −3, −5, −7, −9,...
(c) $\frac{1}{5}, \frac{2}{5}, \frac{3}{5}, \frac{4}{5}, \ldots$

Geometric Sequences

Suppose that 1 month after starting to save $400 each month toward a house, the wife in our young couple wins American Idol, and the money starts rolling in. Instead of adding $400 to their savings every month, now they're able to double the total each month. The amount would now look like $400, $800, $1,600, $3,200, $6,400,.... This is an example of a **geometric sequence.** A sequence is geometric when every term after the first is obtained by multiplying the previous term by the same fixed nonzero number. This multiplier is called the **common ratio.**

Examples of Geometric Sequences	Common Ratio
1, 3, 9, 27, 81, 243,...	$r = 3$
2, 10, 50, 250, 1,250,...	$r = 5$
5, −10, 20, −40, 80,...	$r = -2$
$1, \frac{1}{4}, \frac{1}{16}, \frac{1}{64}, \frac{1}{256}, \ldots$	$r = \frac{1}{4}$

Sidelight THE FIBONACCI SEQUENCE

Sometime around the year 1200, an Italian mathematician named Leonardo Pisano, who went by the name Fibonacci, discovered the sequence 1, 1, 2, 3, 5, 8, 13, 21, 34,.... Do you see the pattern? After the first two terms, each successive term comes from adding the two previous terms. Not exactly earth-shattering, right? The truly amazing thing about this innocent little list of numbers is that numbers from this sequence are found all over the place in nature. The number of petals that many flowers have correspond to Fibonacci numbers. Lilies and irises have 3 petals, buttercups have 5, delphiniums have 8, and marigolds have 13. Different varieties of daisies have 21, 34, 55, or 89 petals. The seeds on the head of a sunflower spiral out from the center in clockwise and counterclockwise directions: some sunflowers have 21 spirals in one direction and 34 in the other. A giant sunflower has 89 in one direction and 144 in the other, and these are Fibonacci numbers as well.

Speaking of spirals, the beautiful shell of the chambered nautilus, pictured here, is a smooth spiral that can be built using a series of squares with areas matching—you guessed it—the Fibonacci sequence.

Even today, nearly a thousand years later, mathematicians still study the Fibonacci sequence. In fact, there is a society devoted to the study of the mathematics of Fibonacci that publishes a newsletter four times a year.

EXAMPLE 5 Finding the Terms of a Geometric Sequence

Write the first five terms of a geometric sequence with first term 4 and common ratio -3.

SOLUTION

Start with first term 4, then multiply by -3 to obtain each successive term:

$$4, -12, 36, -108, 324$$

▼ Try This One 5

☑ 4. Define geometric sequence.

Write the first five terms of a geometric sequence with first term $\frac{1}{2}$ and common ratio 4.

We can also find a formula to quickly calculate any term of a geometric sequence without having to find all of the terms that come before it. A geometric sequence with first term a_1 and common ratio r looks like

$$a_1, \quad a_1 \cdot r, \quad a_1 \cdot r \cdot r, \quad a_1 \cdot r \cdot r \cdot r, \ldots$$

Doing some simplifying, we find that

$$a_1 = a_1 \qquad a_2 = a_1 \cdot r \qquad a_3 = a_1 \cdot r^2 \qquad a_4 = a_1 \cdot r^3$$

Every term a_n looks like a_1 times some power of r, and the power is one less than the subscript of that term in the sequence. This gives us a formula:

Finding the nth Term of a Geometric Sequence

The nth term of a geometric sequence can be found using the formula

$$a_n = a_1 \cdot r^{n-1}$$

where n is the number of the term you want to find, a_1 is the first term, and r is the common ratio.

EXAMPLE 6 Finding a Particular Term of a Geometric Sequence

Find the fifth term of a geometric sequence with first term $\frac{1}{2}$ and common ratio 6.

Math Note

This can be checked by writing the first five terms of the sequence. They are $\frac{1}{2}$, 3, 18, 108, 648.

SOLUTION

Substitute in the formula using $a_1 = \frac{1}{2}$, $r = 6$, and $n = 5$.

$$a_n = a_1 r^{n-1}$$

$$a_5 = \tfrac{1}{2}(6)^{5-1}$$

$$= \tfrac{1}{2}(6^4) \quad \text{Don't forget order of operations!}$$

$$= 648$$

The fifth term is 648.

Sidelight SEQUENCES AND THE PLANETS

In the late 1700s, the world of scientists became very excited when two German astronomers discovered a mathematical sequence that actually predicted the average distance the then-known planets were from the sun. This distance is measured in what are called astronomical units. One astronomical unit (AU) is equal to the average distance the Earth is located from the sun (about 93 million miles).

The sequence, called the Titius-Bode law (named for its discoverers in 1777), is 0, 3, 6, 12, 24, 48, 96, 192, ... When 4 is added to each number and the sum is divided by 10, the result gives the approximate distance in AUs each planet is from the sun as shown.

Planet	Sequence	AU
Mercury	$(0 + 4) \div 10$	0.4
Venus	$(3 + 4) \div 10$	0.7
Earth	$(6 + 4) \div 10$	1.0
Mars	$(12 + 4) \div 10$	1.6
___	$(24 + 4) \div 10$	2.8
Jupiter	$(48 + 4) \div 10$	5.2
Saturn	$(96 + 4) \div 10$	10.0
___	$(192 + 4) \div 10$	19.6

Between Mars and Jupiter, no planet exists, and it looks like the sequence breaks down. However, an asteroid belt is located between Mars and Jupiter, and some astronomers thought that this was once a planet.

More amazing was the fact that in 1781 William Herchel discovered the planet Uranus, which is located at 19.2 AU from the sun!

Unfortunately, the next two planets that were discovered did not fit the sequence's pattern. Neptune's location is 30.1 AU, and Pluto's location is 39.5 AU.

▼ Try This One 6

☑ 5. Find a particular term of a geometric sequence.

Find the ninth term of a geometric sequence with first term 3 and common ratio (-2).

There is also a formula for adding terms of a geometric sequence.

Finding the Sum of a Geometric Sequence

The sum of the first n terms of a geometric sequence is $S_n = \dfrac{a_1(1 - r^n)}{1 - r}$ where a_1 is the first term and r is the common ratio.

EXAMPLE 7 Finding the Sum of a Geometric Sequence

Math Note

The solution can be checked by finding the sum of the first six terms directly:

$8 + (-4) + 2 + (-1) + \frac{1}{2} + \left(-\frac{1}{4}\right) = 5\frac{1}{4}$

Find the sum of the first six terms of a geometric sequence with first term 8 and common ratio $-\frac{1}{2}$.

SOLUTION

Using $a_1 = 8$, $r = -\frac{1}{2}$, and $n = 6$, substitute into the formula

$$S_n = \frac{a_1(1 - r^n)}{1 - r}$$

$$S_6 = \frac{8\left[1 - \left(-\frac{1}{2}\right)^6\right]}{\left[1 - \left(-\frac{1}{2}\right)\right]}$$ *A calculator is a big help here.*

$$= 5.25 \text{ or } 5\tfrac{1}{4}$$

The sum is 5.25.

▼ Try This One 7

☑ 6. Add terms of a geometric sequence.

Find a decimal approximation for the sum of the first seven terms of a geometric sequence with first term 6 and common ratio $\frac{1}{4}$.

Applications of Sequences

EXAMPLE 8 Applying Arithmetic Sequences to Saving for a Home

We began the section with an example in which a couple are saving money for a down payment on a house. Suppose that your plan is similar, but you decide to put away \$150 the first month, \$175 the second, \$200 the third, and so on. Would you save more or less in the first 2 years than if you put away \$400 each month?

SOLUTION

If we list the amount saved each month, we get an arithmetic sequence with first term 150 and common difference 25. To find the amount after 2 years, we need to find the sum of the first 24 terms, or S_{24}, for this sequence. But first we need to know a_{24} since it's part of the formula for S_{24}.

$$a_n = a_1 + (n - 1)d$$

$$a_{24} = 150 + (24 - 1)25 = 150 + 23 \cdot 25 = 725$$

Now we use the formula for the sum, S_n.

$$S_n = \frac{n(a_1 + a_n)}{2}$$

$$S_{24} = \frac{24(150 + 725)}{2} = 12(875) = \$10{,}500$$

If you had put away \$400 each month, the total after 2 years would simply be $24 \times 400 = \$9{,}600$. The first plan saves more money in 2 years.

▼ Try This One 8

If you are offered a 1-year job where you get paid just \$20 the first week, but \$20 is added to your pay each week, how much money would you earn in 1 year?

EXAMPLE 9 Applying Geometric Sequences to Salary

When a worker earns a 4% raise each year, to find his or her salary in any given year, you multiply the salary from the year before by 1.04. If you take a job with a starting annual salary of \$40,000 and earn a 4% raise each year, how much money would you earn in your first 10 years?

SOLUTION

Since the salary gets multiplied by 1.04 each year, the annual salaries form a geometric sequence with first term 40,000 and common ratio 1.04. We can use the formula for the sum of a geometric sequence with $n = 10$:

$$S_n = \frac{a_1(1 - r^n)}{1 - r}$$

$$S_{10} = \frac{40{,}000(1 - 1.04^{10})}{1 - 1.04}$$

$$= \$480{,}244.28$$

▼ Try This One 9

For the job in Example 9, if you're offered the option of starting at $30,000 and taking an 8% raise each year, would that earn you more or less in the first 10 years? By how much?

Answers to Try This One

1 (a) 6, 16, 26, 36, 46
(b) $\frac{3}{8}, -\frac{1}{8}, -\frac{5}{8}, -\frac{9}{8}, -\frac{13}{8}$

2 −43

3 (a) $\frac{19}{9}$ (b) $-\frac{4}{3}$

4 (a) 522 (b) −144 (c) $\frac{78}{5}$

5 $\frac{1}{2}$, 2, 8, 32, 128

6 768

7 About 7.9995

8 $27,560

9 $45,647.41 less

EXERCISE SET 5-7

Writing Exercises

1. If you're given a list of numbers, explain how to tell if the list is an arithmetic sequence.
2. If you're given a list of numbers, explain how to tell if the list is a geometric sequence.
3. For an arbitrary sequence, explain what the symbol a_n represents.
4. For an arbitrary sequence, explain what the symbol S_n represents.
5. Give an example of a real-world quantity that might be described by an arithmetic sequence, and justify your answer.
6. Give an example of a real-world quantity that might be described by a geometric sequence, and justify your answer.

Computational Exercises

For the arithmetic sequence in Exercises 7–22, find each:

(a) the first five terms
(b) the common difference
(c) the 12th term
(d) the sum of the first 12 terms

7. $a_1 = 1, d = 6$
8. $a_1 = 10, d = 5$
9. $a_1 = -9, d = -3$
10. $a_1 = -15, d = -2$
11. $a_1 = \frac{1}{4}, d = \frac{3}{8}$
12. $a_1 = \frac{3}{7}, d = \frac{1}{7}$
13. $a_1 = 4, d = -\frac{1}{3}$
14. $a_1 = \frac{5}{2}, d = -\frac{1}{4}$
15. $5, 13, 21, 29, 37, \ldots$
16. $2, 12, 22, 32, 42, \ldots$
17. $50, 48, 46, 44, 42, \ldots$
18. $12, 7, 2, -3, -8, \ldots$
19. $\frac{1}{8}, \frac{19}{24}, \frac{35}{24}, \frac{17}{8}, \frac{67}{24}, \ldots$
20. $\frac{1}{2}, \frac{9}{10}, \frac{13}{10}, \frac{17}{10}, \frac{21}{10}, \ldots$
21. $0.6, 1.6, 2.6, 3.6, 4.6, \ldots$
22. $0.3, 0.7, 1.1, 1.5, 1.9, \ldots$

For the geometric sequence in Exercises 23–38, find each:

(a) the first five terms
(b) the common ratio
(c) the 12th term
(d) the sum of the first 12 terms

23. $a_1 = 12, r = 2$
24. $a_1 = 8, r = 3$
25. $a_1 = -5, r = \frac{1}{4}$
26. $a_1 = -9, r = \frac{2}{3}$
27. $a_1 = \frac{1}{6}, r = -6$
28. $a_1 = \frac{3}{7}, r = -3$
29. $a_1 = 100, r = -\frac{1}{4}$
30. $a_1 = 10, r = -\frac{1}{10}$
31. $4, 12, 36, 108, 324, \ldots$
32. $6, 12, 24, 48, 96, \ldots$
33. $\frac{1}{2}, \frac{1}{4}, \frac{1}{8}, \frac{1}{16}, \frac{1}{32}, \ldots$
34. $\frac{2}{3}, \frac{2}{9}, \frac{2}{27}, \frac{2}{81}, \frac{2}{243}, \ldots$
35. $-3, 15, -75, 375, -1875, \ldots$
36. $-3, 12, -48, 192, -768, \ldots$
37. $1, 3, 9, 27, 81, \ldots$
38. $8, 2, \frac{1}{2}, \frac{1}{8}, \frac{1}{32}, \ldots$

For Exercises 39–46, determine whether each sequence is an arithmetic sequence, a geometric sequence, or neither.

39. $5, -15, 45, -135, 405, \ldots$
40. $42, 35, 28, 21, 14, \ldots$
41. $2, 4, 8, 14, 22, \ldots$
42. $2, 4, 12, 48, 240, \ldots$
43. $6, 2, -2, -6, -10, \ldots$
44. $\frac{1}{10}, \frac{3}{40}, \frac{9}{160}, \frac{27}{640}, \frac{81}{2560}, \ldots$
45. $\frac{5}{8}, \frac{1}{8}, -\frac{3}{8}, -\frac{7}{8}, -\frac{11}{8}, \ldots$
46. $4, -3, \frac{9}{16}, -\frac{27}{64}, \frac{81}{256}, \ldots$

Real-World Applications

47. A new car that cost $27,000 originally depreciates in value by $3,500 in the first year, $3,000 in the second year, $2,500 in the third year, and so on.
    (a) How much value does it lose in the seventh year?
    (b) How much is the car worth at the end of the seventh year?
48. A large piece of machinery at a factory originally cost $50,000. It depreciates by $1,800 the first year, $1,750 the second year, $1,700 the third year, and so on.
    (a) What is the amount of depreciation in the fifth year?
    (b) What is the value of the machinery after the fifth year?
49. At one university, the first parking violation in the union lot results in a $5 fine. The second violation is an $8 fine, the third $11, and so on. One particular student has the fines sent home to her parents, who inform her that if she racks up over $100 in fines, they're selling her car on eBay. If she has been given eight tickets, will she be able to keep her car?
50. A company decided to fine its workers for parking violations on its property. The first offense carries a fine of $25, the second offense is $30, the third offense is $35, and so on. What is the fine for the eighth offense?
51. A contractor is hired to build a cell phone tower, and its price depends on the height of the tower. The contractor charges $1,000 for the first 10 feet, and after that the price for each successive 10-foot section is $250 more than the previous section. What would be the cost of a 90-foot tower?
52. At one plant that produces auto parts, management had to lay off 25 workers in the first quarter of 2009, 30 workers in the second quarter, 35 in the third, and continued this pattern through 2010. If the plant originally had 700 workers, how many remained at the end of 2010?
53. A bungee jumper reaches the bottom of her jump and rebounds 80 feet upward on the first bounce. Each successive bounce is $\frac{1}{2}$ as much as the previous. If she bounces a total of 10 times before coming to a stop, what is the total height of all bounces?
54. A ball that rebounds $\frac{7}{8}$ as high as it bounced on the previous bounce is dropped from a height of 8 feet. How high does it bounce on the fourth bounce and how far has it traveled after the fourth bounce?
55. A desperate gambler named Phil borrows $1,200 from Vito the loan shark. Vito informs him that he will be charging 10% interest each month, and Phil must pay off his debt in 1 year or his thumbs will be broken. How much cash does Phil have to cough up at the end of 1 year to save his thumbs?
56. A student deposited $500 in a savings account that pays 5% annual interest. At the end of 10 years, how much money will be in the savings account?
57. A contestant on a quiz show gets $1,000 for answering the first question correctly, $2,000 more for answering the second question correctly, $4,000 more for the third question, $8,000 for the fourth, and so on. The contestants can stop at any time, but if they get a question wrong, they lose it all. Monica's goal is to make enough money to pay cash for a $225,000 house. How many questions does she have to answer correctly?

58. An eccentric business owner gives a new employee a choice: he can work for \$4,000 per month, or he can get 1 cent the first day, 2 cents the second, 4 cents the third, 8 cents the fourth, and so on. The employee, completely offended at the thought of working all day for a measly penny, chooses the \$4,000 without a second thought. If there are 22 workdays in an average month, how much money per month did his haste cost him?
59. Suppose that you're offered two jobs: both have a starting salary of \$30,000 per year; the first includes a raise of \$2,000 each year, and the second a raise of 4% each year. Which job will pay more in the 10th year, and by how much?
60. Your financial advisor calls, offering you two can't-miss propositions if you can scrape together an initial investment of \$5,000. The first investment will increase in value by \$1,000 per year, every year. The second will increase at the rate of 10% every year. Which is the better investment if you plan to keep the money invested for 12 years?

Critical Thinking

When a geometric sequence has infinitely many terms, if the common ratio is between −1 and 1, you can still find the sum of all the terms. The formula for the sum in this case is $S = \frac{a_1}{1-r}$. *Use this formula for Exercises 61 and 62.*

61. A repeating decimal between −1 and 1 can be written as the sum of an infinite geometric sequence as shown below.

$$0.333\ldots = \frac{3}{10} + \frac{3}{100} + \frac{3}{1{,}000} + \cdots$$

Find a_1 and r, then find the sum of all of the terms using the sum formula above.

62. Refer to Exercise 61. Write the repeating decimal 0.151515... as the sum of an infinite geometric sequence and find the sum using the given sum formula.
63. Consider the following situation: in a mythical post-apocalyptic world, there are just two people left to repopulate. Fortunately, one is a man and one is a woman, and they're both of reproductive age. Do you think that a sequence describing the population is more likely to be arithmetic or geometric? Explain your answer.

CHAPTER 5 Summary

Section	Important Terms	Important Ideas
5-1	Natural number Factor Divisor Divisible Multiples Prime number Prime factorization Composite number Fundamental theorem of arithmetic Greatest common factor Relatively prime Least common multiple	**The set** of natural numbers is {1, 2, 3, 4, 5, . . . }. It consists of the set of prime numbers, composite numbers, and the number 1, which is neither a prime number nor a composite number. Every composite number can be factored as a product of prime numbers (called a prime factorization) in only one way. For two or more numbers, one can find the greatest common factor (GCF) and the least common multiple (LCM).
5-2	Whole number Integer Opposite Absolute value Order of operations	**The set** of whole numbers is {0, 1, 2, 3, . . . }. The set of integers is { . . . −3, −2, −1, 0, 1, 2, 3, . . . }. Each integer has an opposite. The absolute value of any integer except 0 is positive. The absolute value of 0 is 0. When performing a calculation with multiple operations, the usual order of operations has to be observed.
5-3	Rational number Proper fraction Improper fraction Mixed number Lowest terms Reciprocal Least common denominator Place value Terminating decimal Repeating decimal	**Rational** numbers can be written either as fractions or decimals. The decimals are either terminating or repeating. A fraction is in lowest terms when the numerator and denominator have no common factors. The rules for performing operations on rational numbers are given in this section.
5-4	Irrational number Radical Perfect square Like radicals Rationalizing the denominator	**Irrational** numbers are nonterminating nonrepeating decimals. Square roots of numbers that are not perfect squares are irrational numbers. Rules for performing operations on square roots are given in this section.
5-5	Real number Property Identity	**A real number** is either rational or irrational. We studied 11 properties of the real numbers. They are the closure properties for addition and multiplication, the commutative properties for addition and multiplication, the associative properties for addition and multiplication, the identity properties for addition and multiplication, the inverse properties for addition and multiplication, and the distributive property.

5-6	Exponential notation Base Exponent Scientific notation	**In order** to write very large or very small real numbers without a string of zeros, mathematicians and scientists use scientific notation. Using powers of 10, scientific notation simplifies operations such as multiplication and division of large or small numbers.
5-7	Sequence Arithmetic sequence Common difference Geometric sequence Common ratio	**A sequence** of numbers is a list of numbers in a definite order. We studied two basic types of sequences. They are arithmetic sequences and geometric sequences. Many real-world problems in math can be solved using sequences.

MATH IN Government Spending REVISITED

Question 1: Divide the total spending, 2.931×10^{12}, by the population, 3.06×10^{8} to get 9.58×10^{3}. This means the government spent $9,580 per person. Next divide the deficit, 4.1×10^{11} by the population to get 1.34×10^{3}. The government spent $1,340 per person more than it took in.

Question 2: Divide the amount spent on interest, 4.12×10^{11}, by the total spending, 2.931×10^{12}, to get 0.141. This means that 14.1% of every dollar spent went just to pay interest on the national debt.

Question 3: Divide the national debt, 1.06×10^{13}, by the population to get 3.46×10^{4}. This tells us that it would take $34,600 dollars from every man, woman, and child in the United States to pay off the national debt.

Question 4: Divide the national debt by 535 to get 1.98×10^{10}. This means that every representative in congress would have to contribute about $19.8 billion to pay off the national debt. Nobody is *that* well off!

Review Exercises

Section 5-1

For Exercises 1–6, find all factors of each.

1. 78
2. 81
3. 45
4. 38
5. 140
6. 324

For Exercises 7–10, find the first five multiples of each.

7. 4
8. 32
9. 9
10. 60

For Exercises 11–16, find the prime factorization for each.

11. 96
12. 44
13. 250
14. 720
15. 600
16. 75

For Exercises 17–22, find the GCF and LCM.

17. 6, 10
18. 18, 20
19. 35, 40
20. 50, 75
21. 60, 80, 100
22. 27, 54, 72
23. An investor takes out two certificates of deposit. One matures every 18 months, the other every 22 months. He decides to continue rolling them over until they both mature at the same time. How long will it be until that happens?

Section 5-2

For Exercises 24–33, perform the indicated operations.

24. $-6 + 24$
25. $18 - 32$
26. $5(-9)$
27. $32 \div (-8)$
28. $6 + (-2) - (-3)$
29. $6 \cdot 8 - (-2)^2$
30. $4 \cdot 3 \div (-3) + (-2)$
31. $100 - \{[6 + (2 \cdot 3) - 5] + 4\}$
32. $\{8 \cdot 7^3 - 55[(3 + 4) - 6]\} + 20$
33. $(-5)^3 + (-7)^2 - 3^4$

34. Ty has a separate checking account set aside for housing, utilities, and car payment. For fall semester, the account starts with \$2,400. During the 5-month semester, Ty writes checks each month for rent (\$340), utilities (\$45), and car payment (\$170). His parents make three deposits of \$350 during the semester. How much money is left at the end of the semester?

Section 5-3

For Exercises 35–38, reduce each fraction to lowest terms.

35. $\frac{75}{95}$ 36. $\frac{56}{64}$ 37. $\frac{48}{60}$ 38. $\frac{24}{30}$

For Exercises 39–50, perform the indicated operations. Write your answer in lowest terms.

39. $\frac{1}{8} + \frac{5}{6}$
40. $\frac{3}{10} - \frac{2}{5} + \frac{1}{4}$
41. $\frac{5}{9} \times \frac{3}{7}$
42. $\frac{15}{16} \div \left(-\frac{21}{40}\right)$
43. $\frac{1}{2} \div \left(\frac{2}{3} + \frac{3}{4}\right)$
44. $\frac{9}{10} \times \left(\frac{5}{6} - \frac{1}{8}\right)$
45. $\frac{2}{3}\left(\frac{3}{4} + \frac{1}{2} - \frac{1}{6}\right)$
46. $1\frac{7}{8} - \left(\frac{3}{4}\right)^2$
47. $-\frac{6}{7}\left(\frac{1}{2} + 2\frac{1}{3}\right)$
48. $\frac{9}{10} + \left(-\frac{2}{5}\right)\left(-\frac{1}{4}\right)$
49. $\frac{5}{8} - \frac{2}{3}\left(-1 + \frac{2}{5}\right)$
50. $\frac{1}{2} - \frac{3}{4} - \frac{7}{8} \cdot \frac{1}{6}$

For Exercises 51–54, change each fraction to a decimal.

51. $\frac{9}{10}$ 52. $\frac{5}{16}$ 53. $\frac{6}{7}$ 54. $\frac{1}{9}$

For Exercises 55–58, change each decimal to a reduced fraction.

55. 0.6875
56. 0.22
57. $0.2\overline{5}$
58. $0.\overline{45}$

59. In the National Football League, $\frac{3}{8}$ of the teams make the playoffs; in Major League Baseball, $\frac{4}{15}$ make the playoffs, and in the National Basketball Association and the National Hockey League, $\frac{8}{15}$ do. There are 32 teams in the NFL, and 30 teams in MLB, NBA, and NHL. How many teams make the playoffs in each sport?

Section 5-4

For Exercises 60–65, simplify each.

60. $\sqrt{48}$
61. $\sqrt{112}$
62. $\frac{7}{\sqrt{5}}$
63. $\frac{5}{\sqrt{20}}$
64. $\sqrt{\frac{3}{8}}$
65. $\sqrt{\frac{5}{12}}$

For Exercises 66–73, perform the indicated operations.

66. $\sqrt{20} + 2\sqrt{75} - 3\sqrt{5}$
67. $\sqrt{18} - 5\sqrt{2} + 4\sqrt{72}$
68. $\sqrt{27} \cdot \sqrt{63}$
69. $\sqrt{40} \cdot \sqrt{30}$
70. $\frac{\sqrt{20}}{\sqrt{5}}$
71. $\frac{\sqrt{96}}{\sqrt{16}}$
72. $\sqrt{6}(\sqrt{2} + \sqrt{5})$
73. $\sqrt{42}(\sqrt{14} - \sqrt{6})$

Section 5-5

For Exercises 74–79, classify each number as natural, whole, integer, rational, irrational, and/or real.

74. $-\frac{5}{16}$
75. 0.86
76. $0.3\overline{7}$
77. $\sqrt{15}$
78. 0
79. 16

For Exercises 80–83, state which property of the real numbers is being illustrated.

80. $8 \cdot \frac{1}{8} = 1$
81. $3 + 5 = 5 + 3$
82. $6 + 5$ is a real number
83. $2(3 + 8) = 2 \cdot 3 + 2 \cdot 8$

Section 5-6

For Exercises 84–93, evaluate each.

84. 4^5
85. 2^0
86. $(-3)^0$
87. 3^{-4}
88. 6^{-5}
89. $7^2 \cdot 7^4$
90. $\frac{5^6}{5^2}$
91. $(3^4)^2$
92. $2^3 \cdot 2^{-5}$
93. $6^{-2} \cdot 6^{-3}$

For Exercises 94–97, write each number in scientific notation. Round to two decimal places.

94. 3826
95. 25,946,000,000
96. 0.00000327
97. 0.00048

For Exercises 98–101, write each number in decimal notation.

98. 5.8×10^{11}
99. 2.33×10^9
100. 6.27×10^{-4}
101. 8.8×10^{-6}

For Exercises 102–105, perform the indicated operations and write the answers in scientific notation.

102. $(2 \times 10^4)(4.6 \times 10^{-6})$
103. $(3.2 \times 10^{-5})(8.9 \times 10^{-7})$
104. $\frac{4.8 \times 10^4}{2.4 \times 10^{-6}}$
105. $\frac{1.8 \times 10^{-5}}{3 \times 10^2}$
106. According to numerous websites, the average American drinks the equivalent of 5.97×10^2 cans of soda per year. According to the U.S. Census Bureau, the population of the United States at the end of 2008 was about 3.06×10^8. How many cans of soda were consumed in the United States in 2008? Write your answer in both scientific and decimal notation, then write how the decimal answer would be read aloud.
107. The speed of sound in air is about 1.126×10^3 feet per second. When the volcano Krakatoa erupted on August 26, 1883, the blast could be heard 3,000 miles away. How long would it have taken for the sound to reach that far?

Section 5-7

For Exercises 108–111, write the first six terms of the arithmetic sequence. Find the ninth term and the sum of the first nine terms.

108. $a_1 = 8, d = 10$
109. $a_1 = 4, d = -3$
110. $a_1 = -13, d = -5$
111. $a_1 = -\frac{1}{5}, d = \frac{1}{2}$

For Exercises 112–115, write the first six terms of the geometric sequence. Find the ninth term and the sum of the first nine terms.

112. $a_1 = 7.5, r = 2$
113. $a_1 = -3, r = 3$
114. $a_1 = \frac{1}{9}, r = \frac{1}{4}$
115. $a_1 = -\frac{2}{5}, r = -\frac{1}{2}$

116. The number of people without health insurance in the United States is increasing by 1 million people per year. If there were about 46 million U.S. residents without health insurance in 2007, find the approximate number of people without health insurance in 2010.
117. The net profit of a small company is increasing by 5% each year. If the net profit for this year is $20,000, find the projected profit for the sixth year of operation and the total amount of money the company can be expected to make in the next 6 years.

Chapter Test

For Exercises 1–10, classify each number as natural, whole, integer, rational, irrational, and/or real.

1. -27
2. 8.6
3. $\frac{5}{9}$
4. $0.6\overline{2}$
5. $\sqrt{50}$
6. π
7. 0
8. $-\frac{13}{20}$
9. $-\sqrt{25}$
10. $\sqrt{\frac{49}{25}}$

For Exercises 11–14, find the GCF and LCM of each group of numbers.

11. 42, 56
12. 36, 45
13. 150, 175, 200
14. 80, 110, 120

For Exercises 15–20, reduce each fraction to lowest terms.

15. $\frac{15}{35}$
16. $\frac{81}{108}$
17. $\frac{112}{175}$
18. $\frac{64}{128}$
19. $\frac{49}{70}$
20. $\frac{98}{128}$

For Exercises 21–22, simplify the radical.

21. $\sqrt{48}$
22. $\sqrt{243}$

For Exercises 23–32, perform the indicated operations.

23. $-5 \cdot (-6) + 3 \cdot 2$
24. $18 - 3^2 - 4^2 + 6 \div 3$
25. $\left(\frac{5}{6} \cdot \frac{3}{4}\right) \div \frac{2}{3}$
26. $\left(\frac{1}{7} + \frac{1}{9}\right) - \frac{2}{3} \cdot \frac{3}{4}$
27. $-6 + \frac{1}{4} \div \frac{2}{3} + \sqrt{81}$
28. $[4 + (2 \times 3) - 6^2] + 18$
29. $\sqrt{27} + \sqrt{3}(2\sqrt{2} - 1)$
30. $\frac{16}{\sqrt{32}}$
31. $\frac{\sqrt{45}}{\sqrt{5}}$
32. $2\sqrt{50} - 3\sqrt{32}$

For Exercises 33–36, change each decimal into a reduced fraction.

33. 0.875
34. 0.64
35. $0.\overline{2}$
36. $0.\overline{35}$

For Exercises 37–42, state the property illustrated.

37. $0 + 15 = 15 + 0$
38. 6×7 is a real number
39. $0 + (-2) = -2$
40. $\frac{1}{5} \cdot 5 = 1$
41. $(4 \times 6) \times 10 = 4 \times (6 \times 10)$
42. $6(5 + 7) = 6 \cdot 5 + 6 \cdot 7$

For Exercises 43–47, evaluate each.

43. 8^4
44. 7^{-3}
45. 6^0
46. $4^3 \cdot 4^5$
47. $5^{-3} \cdot 5^{-2}$
48. Write 52,000,000 in scientific notation.
49. Write 0.00236 in scientific notation.
50. Write 9.77×10^3 in decimal notation.
51. Write -6×10^{-5} in decimal notation.
52. $(5.2 \times 10^8)(3 \times 10^{-5}) =$ ______ in scientific notation.
53. Divide $\frac{2.1 \times 10^9}{7 \times 10^5}$.
54. Write the first seven terms, the 20th term, and

the sum of the first 20 terms for the arithmetic sequence where $a_1 = 1$ and $d = 2.5$.

55. Write the first seven terms, the 15th term, and the sum of the first 15 terms for the geometric sequence where $a_1 = \frac{3}{4}$ and $r = -\frac{1}{6}$.

56. A runner decides to train for a marathon by increasing the distance she runs by $\frac{1}{2}$ mile each week. If she can run 15 miles now, how long will it take her to run 26 miles?

57. A gambler decides to double his bet each time he wins. If his first bet is $20 and he wins five times in a row, how much did he bet on the fifth game? Find the total amount he bet.

Projects

1. In Section 5-4, we pointed out that $\sqrt{2}$ is an irrational number, but that fact is certainly not obvious. Search the internet for a proof of this fact, then explain the proof step-by-step.

2. To add together the first hundred natural numbers, you can think of them as an arithmetic sequence with first term and common difference both equal to one. In this project, we'll develop a different method.

   (a) Find the sum using the formula for the sum of an arithmetic sequence.
   (b) Write out the sum of the first 100 natural numbers. You don't have to write out all 100 numbers, but write at least the first five and the last five with an ellipsis (. . .) between.
   (c) Write the same sum underneath the first one, but in the opposite order. What property of real numbers can you use to conclude that the two sums are equal?
   (d) Add the two sums together one term at a time. You should now have a sum that's easy to compute using the fact that multiplication is repeated addition by the same number.
   (e) The sum of those two lists is twice the number we're looking for, so divide by two. Did you get the same answer as in part (a)?
   (f) Repeat steps (b) through (e), but this time find the sum of the first n natural numbers, $1 + 2 + 3 + \cdots + n$. The result should be a formula with variable n.
   (g) Use your answer from part (f) with $n = 100$. Do you again get the same answer as in part (a)?
   (h) Now compare your answer from part (f) to the formula for the sum of an arithmetic sequence with first term and common difference both equal to one. Do you get the same formula?

CHAPTER 6

Topics in Algebra

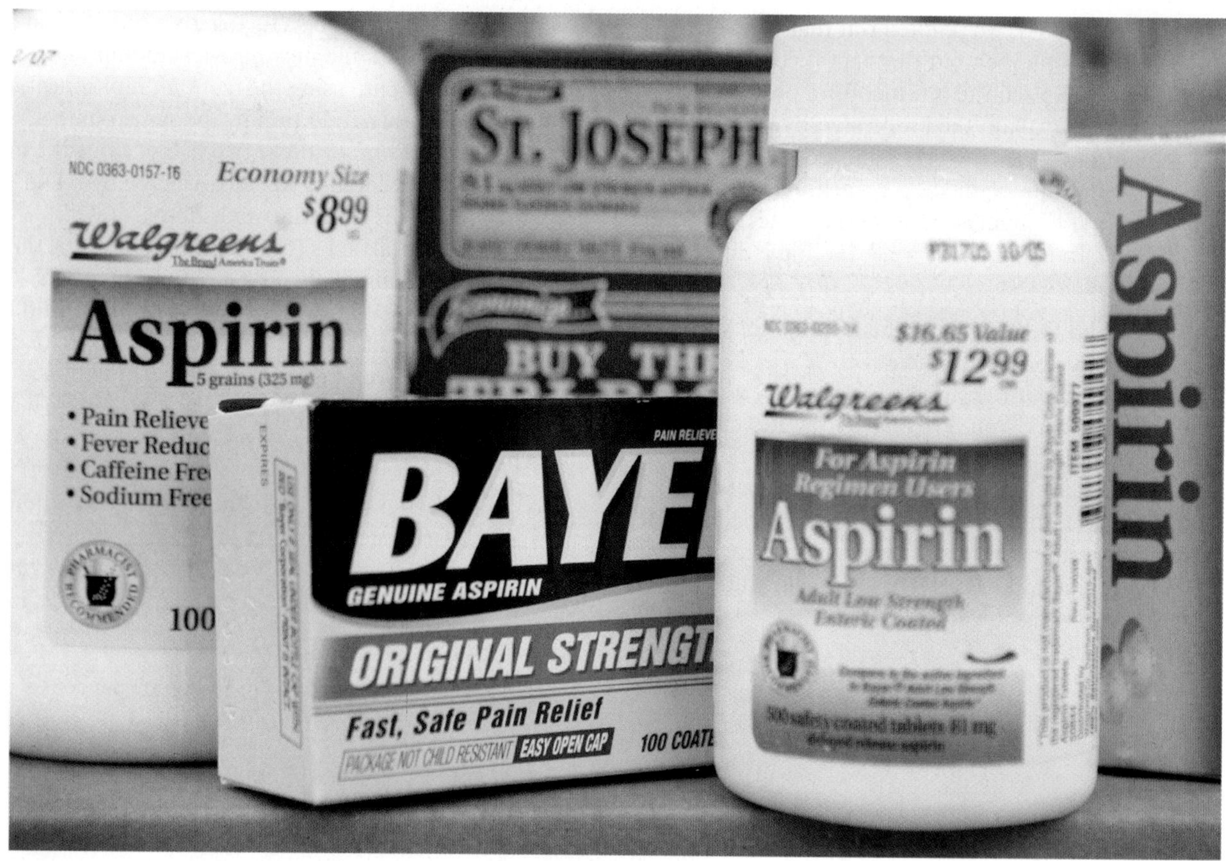

Outline

6-1 The Fundamentals of Algebra
6-2 Solving Linear Equations
6-3 Applications of Linear Equations
6-4 Ratio, Proportion, and Variation
6-5 Solving Linear Inequalities
6-6 Solving Quadratic Equations
Summary

MATH IN Drug Administration

Have you ever looked at the dosage information on a bottle of aspirin and thought "It just doesn't seem reasonable to recommend the same dosage for all adults"? People come in all shapes and sizes, and the effect of a certain dosage is in large part dependent on the size of the individual. If a 105-pound woman and her 230-pound husband both take two aspirin the morning after their wedding reception, she is in effect getting more than twice as much medicine as he is.

Because there's a lot of variation in the world, the study of algebra was created, based on a brilliantly simple idea: using a symbol, rather than a number, to represent a quantity that can change. Since the word "vary" is a synonym for change, we call such a symbol a *variable*. The use of variables is the thing that distinguishes algebra from arithmetic and makes it extremely useful to describe phenomena in a world in which very few things stay the same for very long. An understanding of the basics of algebra is, in a very real sense, the gateway to higher mathematics and its applications. Simply put, you can only go so far with just arithmetic. To model real situations, expressions involving variables are almost always required.

In this chapter, you will be introduced to the fundamental ideas of algebra: variables, expressions, equations, and inequalities. We're setting the groundwork for a lot of the math that will follow in the book, but we will always keep an eye on how the fundamentals of algebra can be applied to real-world situations, like dosage calculations. The wrong dose of aspirin might upset your stomach, but more serious drugs carry with them more serious consequences. In many cases, an incorrect dosage could lead to death. Using the skills you learn in the chapter, you should be able to answer the questions below.

Suppose that a new pain-killing drug is being tested for safety and effectiveness in a variety of people. The recommended dosage for an average 170-pound man is 400 mg. The manufacturer claims that the minimum effective dose for a person of that size is 250 mg, and that anything over 1,500 mg could be lethal. Two of the patients in the test are a 275-pound college football player recovering from a major injury, and a 65-pound girl being treated for sickle-cell disease. According to the manufacturer, dosages for this drug are proportional to body weight.

- What would the recommended dosage be for each patient?
- If the medications are mixed up and the recommended dosage for the football player is given to the child, is she in danger of dying?
- Would the child's recommended dosage be at all effective for the football player?

For answers, see Math in Drug Administration Revisited on page 341

Section 6-1 The Fundamentals of Algebra

LEARNING OBJECTIVES

- ☐ 1. Identify terms and coefficients.
- ☐ 2. Simplify algebraic expressions.
- ☐ 3. Evaluate algebraic expressions.
- ☐ 4. Apply evaluating expressions to real-world situations.

In the movie "The Shawshank Redemption," there's a character who is released from prison after being locked away for a very long time. The thing that jumped out at him after going back out into the world was how quickly everything and everyone seemed to move. In a letter back to his friends in prison, he wrote "The world went and got itself in a big damn hurry."

Things do change very quickly in our modern world, and those who can't adapt to change get left behind. We have seen that arithmetic is a very valuable tool in the real world, but working only with numbers has one main limitation: a number is what it is, and it can't change. That might be the world's shortest explanation of the value of algebra—in the study of algebra we use *variables* to represent quantities that change (or *vary*). That gives us the flexibility to model far more things in our changing world than when we strictly use fixed numbers. So it's only reasonable that our coverage of the fundamentals of algebra begins with a look at variables.

Basic Definitions

Algebra is a branch of mathematics that generalizes the concepts of arithmetic by applying them to quantities that are allowed to vary.

A **variable,** usually represented by a letter, is a quantity that can change. It represents unknown values in a situation.

For example, we can write the formula distance = rate × time as $d = rt$. Then the letters d, r, and t are all variables. This allows us to describe a relationship between these quantities not just for a specific distance, rate, and time, but a wide variety of them. The right side of the equation $d = rt$ is an example of an algebraic expression.

An **algebraic expression** is a combination of variables, numbers, operation symbols, and grouping symbols.

Math Note

An algebraic expression does *not* have an equal sign in it; $d = rt$ is an *equation,* not an expression.

Some examples of algebraic expressions are

$$3x + 2 \quad 8x^2 \quad 111 \quad \frac{9}{5}C + 32 \quad 7(2y^2 - 5)$$

Algebraic expressions are made up of one or more *terms.* Terms are the pieces in an expression that are separated by addition or subtraction signs. In the expression $8x^2 + 6x - 3$, each of $8x^2$, $6x$, and -3 is a term. The expression 111 has just one term, namely 111.

Every term has a **numerical coefficient,** or just coefficient. This is the number part of a term, like the 8 in $8x^2$. Terms also may or may not have variables in them; the term -3 in the preceding paragraph doesn't have a variable. For the term $-17x^2y$, -17 is the coefficient, and x and y are the variables.

EXAMPLE 1 Identifying Terms and Coefficients

Math Note

In the expression in Example 1, the negative sign in front of 3 is considered to be part of the term. We could write the expression as $-8y + \frac{5}{2}x^3 + xy + (-3)$.

Identify the terms of the algebraic expression, and the coefficient for each term.

$$-8y + \frac{5}{2}x^3 + xy - 3$$

SOLUTION

The expression has four terms: $-8y$, which has coefficient -8; $\frac{5}{2}x^3$, which has coefficient $\frac{5}{2}$; xy, which has coefficient 1; and -3, which has coefficient -3. (When no number appears in a term, as in xy, the coefficient is 1; we just don't need to write it.)

☑ 1. Identify terms and coefficients.

▼ Try This One 1

Identify the terms of the algebraic expression, and the coefficient for each term.

$$3x^2y - \frac{1}{2}y^3 + \sqrt{3}x$$

The Distributive Property

The distributive property, which we studied in Chapter 5, is used very often in working with algebraic expressions. Using new terminology from this section, the distributive property tells us that when an expression with more than one term in parentheses is multiplied by a term outside the parentheses, the term outside can be distributed to each term inside. This is illustrated in Example 2.

EXAMPLE 2 Using the Distributive Property

Math Note

When the term being distributed is negative, notice that the sign of *every* original term inside the parentheses changes. Not doing so is one of the most commonly made mistakes in all of math, so be especially careful not to make it!

Use the distributive property to multiply out the parentheses.

(a) $5(3x + 7)$ (b) $-3(6A - 7B + 10)$

SOLUTION

(a) Distributing the 5 to each term inside the parentheses, we get

$$5(3x + 7) = 5 \cdot 3x + 5 \cdot 7 = (5 \cdot 3)x + 35 = 15x + 35$$

Note the use of the associative property in deciding that $5 \cdot 3x$ is $15x$.

(b) This time we have to be careful because the term we're distributing is negative.

$$-3(6A - 7B + 10) = -3 \cdot 6A - (-3) \cdot 7B + (-3) \cdot 10 = -18A + 21B - 30$$

▼ Try This One 2

Use the distributive property to multiply out the parentheses.

(a) $7(4x - 20)$ (b) $-5(3x - 7y + 18)$

Simplifying Algebraic Expressions

When two terms have the same variables with the same exponents, we will call them **like terms.** For example, $3x$ and $-5x$ are like terms, but $3x$ and $-5x^2$ are not. Table 6-1 shows some examples of like and unlike terms.

TABLE 6-1 Like Terms and Unlike Terms

Like Terms		Unlike Terms	
$6x$	$-10x$	$6x$	$-10x^2$
$8x^3$	$6x^3$	$8x^3$	$6y^3$
$2x^2y$	$-5x^2y$	$2x^2y$	$-5xy^2$
5	12	x	5

Like terms can be added or subtracted by using the reverse of the distributive property. For example,

$$3x + 5x = (3 + 5)x = 8x$$
$$-2x^2y + 3x^2y + 9x^2y = (-2 + 3 + 9)x^2y = 10x^2y$$

Unlike terms cannot be added or subtracted since the distributive property does not apply.

In other words, to add or subtract like terms (i.e., combine like terms), add or subtract the numerical coefficients of the like terms. Unlike terms cannot be combined by addition or subtraction.

EXAMPLE 3 Combining Like Terms

Combine like terms for each, if possible.

(a) $9x - 20x$ (b) $3x^2 + 8x^2 - 2x^2$ (c) $6x + 8x^2$

SOLUTION

(a) $9x - 20x = (9 - 20)x = -11x$ (Found by subtracting $9 - 20$)
(b) $3x^2 + 8x^2 - 2x^2 = (3 + 8 - 2)x^2 = 9x^2$ (Found by adding and subtracting $3 + 8 - 2$)
(c) $6x + 8x^2$ (These terms cannot be combined since they are not like terms.)

▼ Try This One 3

Combine like terms for each, if possible.

(a) $12y^2 - 18y^2$ (b) $7xy + 9xy - 11xy$ (c) $10z^2 + 10$

Algebraic expressions can have any number of terms. The phrase "to simplify an expression" means to find any terms that are like and combine them as in Example 2. For example, in the expression $-6x + 3y - 12 + 8y - 2x + 10$, $-6x$ and $-2x$ are like terms that combine to give $-8x$; $3y$ and $8y$ are like terms that combine to give $11y$; and -12 and 10 are like terms that combine to give -2. So the simplified version of the expression is $-8x + 11y - 2$. (The order of terms is not important.)

EXAMPLE 4 Combining Like Terms

Simplify each expression.

(a) $9x - 7y + 18 - 27 + 6y - 10x$
(b) $3x^3 + 4x^2 - 6x + 10 - 7x^2 + 4x^3 + 2x - 6$

SOLUTION

(a) Combine like terms:

$$9x - 10x = -x$$
$$-7y + 6y = -y$$
$$18 - 27 = -9$$

The answer is $-x - y - 9$.

(b) Combine like terms:

$$3x^3 + 4x^3 = 7x^3$$
$$4x^2 - 7x^2 = -3x^2$$
$$-6x + 2x = -4x$$
$$10 - 6 = 4$$

The answer is $7x^3 - 3x^2 - 4x + 4$.

▼ Try This One 4

Simplify each expression.

(a) $2x - 6 + 3y - 7x + 8y - 12$
(b) $9y^3 + 7y - 2y^2 + 6 - 8 + 8y^3 - 7y + 12y^2$
(c) $3a - 2b + 4c - 7a + 3b + 2c$

When an algebraic expression has parentheses, it can be simplified by combining the two skills we've already practiced: multiplying out the parentheses using the distributive property, and combining like terms.

EXAMPLE 5 Simplifying an Algebraic Expression

Simplify the expression $8(3x^2 + 5) + 3(2 - x) - (5x^2 + x)$.

Math Note

A negative sign outside parentheses changes the sign of every term inside. For example, $-(3x - 2) = -3x + 2$.

SOLUTION

First, we multiply out the parentheses. We can think of the last set as $-1(5x^2 + x)$.

$$8(3x^2 + 5) = 24x^2 + 40$$
$$3(2 - x) = 6 - 3x$$
$$-1(5x^2 + x) = -5x^2 - x$$

Now we combine like terms:

$$24x^2 - 5x^2 - 3x - x + 40 + 6 = 19x^2 - 4x + 46$$

▼ Try This One 5

☑ 2. Simplify algebraic expressions.

Simplify each expression.

(a) $5(7y + 10) - (4y + 8)$ (b) $2z + 5 - 3(4 - 2z^2) + 10(z^2 + z)$

Evaluating Algebraic Expressions

Algebraic expressions almost always contain variables, which can be any number. But when we substitute numbers in for the variables, the result is an arithmetic problem. Finding the value of this problem is called *evaluating* the expression. This is illustrated in Example 6.

Sidelight A BRIEF HISTORY OF ALGEBRA

Algebra had its beginnings when people attempted to solve mathematical riddles. One of the earliest books which contained algebraic problems (i.e., riddles) was a collection of 85 problems copied by an Egyptian priest, Ahmes, around 1650 BCE. This manuscript later became known as the *Rhind Papyrus.*

The next important development in algebraic thinking came around 250 CE, when a famous mathematician named Diophantus wrote a book called *Arithmetica,* which contained about 130 algebraic problems and a number of algebraic principles called theorems. *Arithmetica* contained problems that were solved using first-degree and second-degree equations in one unknown. Diophantus studied mathematics in Alexandria in northern Egypt where Euclid and Hypatia had also lived.

Diophantus is known as the "Father of Algebra" because he was the first mathematician to use symbols to represent mathematical concepts. Prior to Diophantus, all mathematical concepts were written out in words.

A Persian mathematician, Al-khwārizmī, wrote a book on algebra, and when it was translated into Latin, the word "al-jabr,"which later became "algebra,"was used in the title of the translation. The word means the science of reduction and cancellation.

One of the first mathematicians to realize the existence of negative numbers was Leonardo de Pisa, called Fibonacci (1170–1250 CE), who used them to solve financial problems. The use of the plus sign (+) and minus sign (−) first appeared in print in an arithmetic textbook written by Johann Widman in 1489. Prior to this, plus and minus signs were used by merchants to mark bales or barrels that were weighed at warehouses and compared to a standard weight. If the barrels were heavier than the standard weight, they were marked with a plus sign. If they weighed less than the standard weight, they were marked with a minus sign.

In 1494, a Franciscan friar, Luca Pacioli, published a book called *Summa de Arithmetica,* which contained all the known algebra to date. A French mathematician, François Vieta (1540–1603), was the first person to use letters to represent quantities. He used vowels for variables or unknowns and consonants for constants or knowns.

René Descartes in 1637 used a variation of Vieta's symbolization. He used the first letters of the alphabet to represent constants or knowns and the letters at the end of the alphabet to represent variables or unknowns. When an equation had one unknown, Descartes used the letter "x" to represent it. Descartes also used the letters "x" and "y" to represent the coordinates of a point.

You probably had no idea that the algebra you studied in high school was the product of 3,500 years of study!

EXAMPLE 6 Evaluating an Expression

Evaluate $9x - 3$ when $x = 5$.

SOLUTION

Substitute 5 for the variable x, then perform the calculation:

$$\begin{aligned} 9x - 3 &= 9(5) - 3 && \text{Multiply first.} \\ &= 45 - 3 && \text{Subtract.} \\ &= 42 \end{aligned}$$

The value of $9x - 3$ when $x = 5$ is 42.

▼ Try This One 6

Evaluate each expression.

(a) $9x - 17$ when $x = 3$. (b) $2x^2 - 3x + 5$ when $x = -10$.

When an expression contains more than one variable, it's evaluated in the same way.

EXAMPLE 7 Evaluating an Expression with Two Variables

Evaluate $5x^2 - 7y + 2$ when $x = -3$ and $y = 6$.

SOLUTION

Substitute -3 for x and 6 for y in the expression, and then simplify.

$$\begin{aligned} 5x^2 - 7y + 2 &= 5(-3)^2 - 7(6) + 2 && \textit{Exponent first.} \\ &= 5(9) - 7(6) + 2 && \textit{Multiply twice.} \\ &= 45 - 42 + 2 && \textit{Add and Subtract.} \\ &= 5 \end{aligned}$$

The value of $5x^2 - 7y + 2$ when $x = -3$ and $y = 6$ is 5.

▼ Try This One 7

☑ 3. Evaluate algebraic expressions.

Evaluate $6x + 8y - 15$ when $x = -5$ and $y = 7$

Algebraic expressions have a seemingly endless supply of applications in the real world. In the remainder of the section, we'll look at just a few examples.

EXAMPLE 8 Finding Commission on Sales

A salesperson at a popular clothing store gets a \$600 monthly salary and a 10% commission on everything she sells. The expression $0.10x + 600$ describes the amount of money she earns each month, where x represents the dollar amount of sales. If she had net sales of \$13,240 in July, how much did she earn?

SOLUTION

Since we were told that x represents monthly sales, and that her sales for July were \$13,240, we substitute 13,240 in for x, then simplify.

$$\begin{aligned} 0.10(13{,}240) + 600 &= 1{,}324 + 600 \\ &= 1{,}924 \end{aligned}$$

The salesperson earned \$1,924 in July.

Calculator Guide

Graphing calculators can be used to evaluate expressions. To perform the evaluation in Example 8, use the following keystrokes:

[Y=] 0.10 [X, T, θ, n] [+] 600

[2nd] [Window], then use the arrows and [ENTER] to select "Ask" next to "Indpnt." Then do [2nd] [Graph] 13,240 [ENTER].

▼ Try This One 8

The salesperson in Example 8 is considering a job at a different store that pays a \$400 salary per month, but a 15% commission, so that the amount she would earn is given by $0.15x + 400$, where x is the dollar amount of sales. Would she earn more or less on her July sales? By how much?

EXAMPLE 9 Computing Total Cost Including Tax

The state of Florida has a sales tax of 6%.

(a) Write an expression for the total cost of an item purchased in Florida, including sales tax. The variable should represent the cost of the item before tax.
(b) John bought an iPhone at a store in Hollywood, Florida. The price was \$349. What was the total cost John paid, including tax?

SOLUTION

(a) We'll use the variable x to represent the cost of the item before tax. The sales tax will be 6% of x, or $0.06x$. That makes the total cost $x + 0.06x$, which simplifies to $1.06x$.
(b) Substitute 349 in for x: $1.06(349) = \$369.94$.

▼ Try This One 9

The sales tax in Georgia is only 4%. How much money would John have saved if he bought the iPhone at the same chain store in Georgia instead of Florida?

EXAMPLE 10 Computing a Discount Price

While shopping at her favorite department store, Carmen found a dress she's been hoping to buy on the 40%-off clearance rack.

(a) Write an algebraic expression representing the new price of the dress before tax, then one for the total price including tax. Use 6% as the sales tax rate.
(b) If the original price of the dress was \$59, find the discounted price, and the total amount that Sally would pay including tax.

SOLUTION

(a) Let x represent the original price of the dress. Then the amount of the discount is 40% of x, or $0.4x$. So the sale price is $x - 0.4x$, which simplifies to $0.6x$. With a sales tax rate of 6%, the total cost will be

$$\underbrace{0.6x}_{\text{Discount price}} + \underbrace{0.06(0.6x)}_{\text{6\% Sales tax}} = 0.6x + 0.036x = 0.636x$$

(b) In our expressions in part (a), the variable x represented the original price of the dress, so we substitute 59 in for x. The discount price is $0.6(59) = \$35.40$, and the total cost including tax is $0.636(59) = \$37.52$.

▼ Try This One 10

Write an expression for the discounted price of items in a store that is having a going-out-of-business sale, with all items 70% off. Then use your expression to find the discount price for a golf club that was originally \$180 and a dozen golf balls that were originally \$24.99.

Formulas are used in almost any area where numbers can describe some quantity. A formula is an equation, usually with more than one variable, that allows you to calculate some quantity of interest. For example, you probably recognize the equation $A = lw$ as the formula for calculating the area of a rectangle. You can use the same procedure used to evaluate expressions to get information from a formula, as in Example 11.

EXAMPLE 11 Using the Formula for Distance

The distance in miles an automobile travels is given by the formula $d = rt$, where r is the rate (or speed) in miles per hour and t is the time in hours. How far will an automobile travel in 6 hours at a rate of 55 miles per hour?

SOLUTION

In the formula $d = rt$, substitute 55 for r and 6 for t and evaluate.

$$d = rt$$
$$d = 55(6)$$
$$d = 330$$

The automobile will travel 330 miles in 6 hours.

Math Note

Formulas can be evaluated using units in addition to numbers. For example, in the previous problem, if the units were included, the formula would look like this:

$$d = \frac{55\text{ miles}}{\text{hour}} \cdot \frac{6\text{ hours}}{1}$$
$$= 330\text{ miles}$$

The hours divide out, leaving behind miles.

☑ 4. Apply evaluating expressions to real-world situations.

▼ Try This One 11

In Canada and in other countries where the metric system is used, the temperature is given in Celsius instead of Fahrenheit. If the temperature in Montreal is 36°C, find the corresponding Fahrenheit temperature. The formula is $F = \frac{9}{5}C + 32$.

Answers to Try This One

1

Term	Coefficient
$3x^2y$	3
$-\frac{1}{2}y^3$	$-\frac{1}{2}$
$\sqrt{3}x$	$\sqrt{3}$

2 (a) $28x - 140$
(b) $-15x + 35y - 90$

3 (a) $-6y^2$
(b) $5xy$
(c) Cannot be combined

4 (a) $-5x + 11y - 18$
(b) $17y^3 + 10y^2 - 2$
(c) $-4a + b + 6c$

5 (a) $31y + 42$
(b) $16z^2 + 12z - 7$

6 (a) 10
(b) 235

7 11

8 \$462 more

9 \$6.98

10 $0.3x$; \$61.50

11 96.8°F

EXERCISE SET 6-1

Writing Exercises

1. Explain in your own words what a variable is.
2. Describe the difference between an algebraic expression and an equation.
3. Explain the connection between terms, coefficients, and algebraic expressions.
4. What does it mean to evaluate an algebraic expression? What do you need to be given in order to do so?
5. What does it mean to simplify an expression?
6. How does the distributive property come into play when simplifying expressions?

Computational Exercises

For Exercises 7–40, simplify the expression.

7. $5x + 12x - 6x$
8. $3x^2 + 8x^2 - 15x^2$
9. $4y - 10y - 12y$
10. $8A - 15A + 2A$
11. $3p + 2q - 7 + 6p - 3q - 10$
12. $5x - 8y + 9 + 4y - 27 + 2x$
13. $8x^2 + 6x - 10 + 15 - 7x + 3x^2$
14. $-9x^2 - 2x - 7 + 3 - 5x + 21x$
15. $5(6x - 7)$
16. $9(3x + 8)$
17. $-4(12x - 10)$
18. $-8(4m + 7)$
19. $3(2x + 6) - 5x + 9$
20. $4(6x - 3) - 7 - 10x$
21. $-7(3x + 8) - 5x + 6$
22. $-10(4x + 11) - 15x + 19$
23. $3x + 7 + 5(x - 6)$
24. $9b + 12 + 8(2b + 3)$
25. $4x - 17 - 5(x - 6)$
26. $14x + 9 - 6(3x - 2)$
27. $4x - 2 - (x + 3)$
28. $3(x - 2) - (x + 4)$
29. $3(4 - x^2) - 2(5x + 7) + x^2$
30. $-2(3y^2 + y) + 5y^2 - 7(y + 2)$
31. $7x^2y + 8xy^2 - 9 + 10xy^2 - 11x^2y$
32. $-2ab + 5a^2b + 8ab - 11b^2 + 14a^2b$
33. $\frac{5}{2}(2m - 7) + 8m + \frac{17}{2}$
34. $4k - \frac{2}{3} + \frac{5}{3}(3k + 4)$
35. $\frac{15}{4}yz + \frac{1}{4}(3 - yz) - \frac{3}{2}$
36. $\frac{5}{8}ab + \frac{11}{8}(2 - ab) + \frac{9}{4}$
37. $1.4x + 7.1 - 5.3x + 2.9$
38. $3.2x - 11.9 + 0.8x + 6$
39. $0.5(3 + tz) - 0.25(tz + 6)$
40. $-0.4(rs - 5) + 0.3(2 + tz)$

For Exercises 41–64, evaluate the expression for the given value or values of the variable.

41. $5x - 7$ when $x = 18$
42. $-3x + 8$ when $x = 5$
43. $-2c + 10$ when $c = -3$
44. $4w + 9$ when $w = -12$
45. $3x^2 + 2x - 6$ when $x = 5$
46. $8x^2 - 7x + 4$ when $x = 16$
47. $9r^2 - 5r - 10$ when $r = -7$
48. $14x^2 - 6x + 30$ when $x = -7$
49. $5x + 18y + 10$ when $x = 8$ and $y = 3$
50. $9a + 18b - 5$ when $a = 7$ and $b = 2$
51. $3x^2 - 2y^2 + 6x$ when $x = -8$ and $y = 2$
52. $5x^2 - 7x + 2y^2$ when $x = -1$ and $y = 5$
53. $13y^2 - 6x^2 + 7y - 6x + 1$ when $x = -5$ and $y = 9$
54. $5x^2 - 4x + 3 - 2y$ when $x = 7$ and $y = -3$
55. $9x^2 + 7y^2 + 6x + 2y + 5$ when $x = 1$ and $y = 5$
56. $10y^3 + 10y^2 + 7x - 6$ when $x = -3$ and $y = 10$
57. $x + \frac{3y}{2}$ when $x = 8$ and $y = 6$
58. $3x - \frac{7y}{6}$ when $x = 7$ and $y = 3$
59. $8x^2 - \frac{5}{2y}$ when $x = 4$ and $y = 6$
60. $6x^2 - \frac{10}{3y}$ when $x = -5$ and $y = 15$
61. $5.6y + 11x$ when $x = -2$ and $y = 3$
62. $3 + 8.1t - 7z$ when $t = 2$ and $z = -5$
63. $8st + 5s^2 - 3t^2$ when $t = -2.5$ and $s = 0.5$
64. $x^2 + y^2 - 2xy$ when $x = 1.5$ and $y = -0.5$

For Exercises 65–74, evaluate each formula.

65. $A = 2\pi rh$ when $\pi \approx 3.14$, $r = 6$ in., and $h = 10$ in.
66. $P = 2l + 2w$ when $l = 10$ feet and $w = 5$ feet
67. $A = P(1 + rt)$ when $P = \$5{,}000$, $r = 0.07$ per year, and $t = 3$ years
68. $S = \frac{1}{2}gt^2$ when $g = 32$ ft/sec^2 and $t = 20$ seconds
69. $V = \frac{4}{3}\pi r^3$ when $r = 4$ mm and $\pi \approx 3.14$
70. $S = 4\pi r^2$ when $r = 7$ and $\pi \approx 3.14$
71. $FV = P(1 + r)^n$ when $P = \$20{,}000$, $r = 0.06$, and $n = 8$
72. $v = V + gt$ when $V = 50$, $g = 32$, and $t = 8$
73. $SA = 2\pi rh + 2\pi r^2$ when $\pi \approx 3.14$, $r = 6$, and $h = 12$
74. $T = 2\pi\sqrt{\frac{L}{g}}$ when $\pi \approx 3.14$, $L = 10$, and $g = 32$

Real-World Applications

75. The simple interest (I) on a certain amount of money (p) that is invested at a specified interest rate (r) for a specific period of time (t) can be found by the formula $I = prt$. Find the interest on a principal of \$500 invested at 5% yearly for 4 years.
76. For a particular occupation, a person's hourly income can be estimated by using this expression: $11.2 + 1.88x + 0.547y$, where x is the number of years of experience on the job and y is the number of years of higher education completed. Find the income of a person who has completed 4 years of college and has worked for the company for 5 years.
77. A manufacturer found that the number of defective items can be estimated by the expression $2.2x - 1.08y + 9.6$, where x is the number of hours an employee worked on a shift and y is the total number of items produced. Find the number of defective items produced for a person who has worked 9 hours and produced 24 items.
78. A real estate agent found that the value of a seaside mansion in thousands of dollars can be estimated by $7.56x - 0.266y + 44.9$, where x is the number of acres the mansion sits on and y is the number of rooms in the mansion. Predict the value of a mansion that has 371 acres with 14 rooms.
79. Find the Celsius temperature that corresponds to a Fahrenheit temperature of 50° using the formula $C = \frac{5}{9}(F - 32)$.
80. Find the electric current, I, delivered by battery cells connected in a series given by the formula $I = \frac{nE}{R} + nr$ when $n = 4$, $E = 2$ volts, $R = 12$ ohms, and $r = 0.2$ ohms.
81. The kinetic energy (KE) in ergs of an object is given by the formula $\text{KE} = \frac{mv^2}{2}$ where m is the mass of the object and v is the velocity. Find the kinetic energy when $m = 30$ g and $v = 200$ cm/s.
82. The heat energy from electricity is given by the formula $E = 0.238I^2\,Rt$. Find E when $I = 25$ amps, $R = 12$ ohms, and $t = 175$ s.
83. The resistance of an electrical conductor is given by the formula $R = \frac{kl}{d^2}$. Find R when $k = 8$, $l = 100$ feet, and $d = 30$ mil.
84. The volume of a sphere is given by the formula $V = \frac{4}{3}\pi r^3$. Find V when $\pi \approx 3.14$ and $r = 4$ inches.
85. The future value (FV) of a compound interest investment (P) at a specific interest rate (r) for a specific number of periods, n, is found by the formula $FV = P(1 + r)^n$. Find the future value of \$9,000 invested at 8% compounded annually for 6 years.
86. Use Einstein's mass-energy equivalence formula $E = mc^2$ to find how many joules of energy E are equivalent to an object with a mass $m = 2$ kg that is moving at the speed of light $c = 300{,}000$ km/s.
87. Finding the distance between two points in a coordinate system has a wide variety of applications in navigation and many other fields. Find the distance from one point to another, where the distance formula is given by $d = \sqrt{(x_2 - x_1)^2 + (y_2 - y_1)^2}$, and $x_1 = 5$, $x_2 = 7$, $y_1 = 3$, and $y_2 = 6$.
88. The effective interest rate, which is the actual interest rate earned after interest is compounded N times a year, is given by the formula $r_{\text{eff}} = \left(1 + \frac{r}{N}\right)^N - 1$, and then that decimal is converted to a percentage. Find the effective interest rate for an account where the given interest rate is $r = 0.0625$ and the interest is compounded monthly ($N = 12$).
89. The height of an object that is thrown upward with an initial velocity of 48 ft/s from a height of 1,500 ft is given by the equation $h = -16t^2 + 48t + 1{,}500$, where h is the height in feet after t seconds. What is the height in feet of the object after 5 seconds?
90. The profit equation for a company can be found by taking its revenue (money coming in) and subtracting its costs (money going out). The revenue equation for a certain company is $R = 3.25x$ and the cost equation is $C = 1.15x + 500$, where x is the number of units sold. Find a formula for the profit equation and then find the profit made for selling 1,000 units.

Critical Thinking

91. The amount of medication a person receives is sometimes based on what is called *body surface area* (BSA) in square meters (m^2). The formula for BSA is BSA = (weight in kg) × (height in cm/3,600). If an order of medication is 50 mg/m^2, how much medication should be given to a person who weighs 88 kg and has a height of 150 cm?
92. A track is 110 yards long and 50 yards wide and has the shape in the diagram below. If the perimeter of a half circle is $P = \pi r$, where $r = 25$ yards (half the distance across the circle), find the distance around the track.

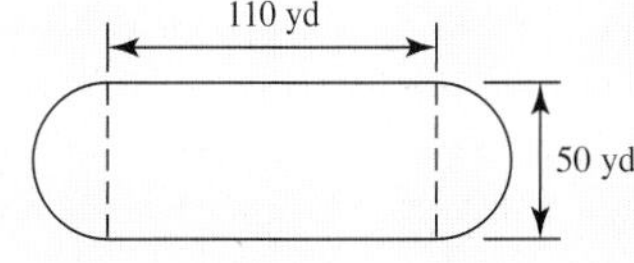

Section 6-2 Solving Linear Equations

LEARNING OBJECTIVES

- ❑ 1. Decide if a number is a solution of an equation.
- ❑ 2. Identify linear equations.
- ❑ 3. Solve general linear equations.
- ❑ 4. Solve linear equations containing fractions.
- ❑ 5. Solve formulas for one specific variable.
- ❑ 6. Determine if an equation is an identity or a contradiction.

In Section 6-1, we saw many situations in which algebraic expressions can help us find information in the real world. But sometimes there's a little more work involved. In Example 9, we found the total cost of an iPhone including tax. But what if we knew the total cost, and wanted to find the price before tax? In that case, the formula we set up would look like $1.06x = 369.94$, and we would be interested in finding the value of x that makes the equation true.

The truth is that very often when you are using a formula to find some quantity of interest, the quantity you want to find is going to be mixed into the equation, like the original price in the example above. This requires the ability to solve equations, which is probably the most useful skill in algebra.

Basic Definitions

An **equation** is a statement that two algebraic expressions are equal.

It's absolutely crucial that you understand the difference between an equation and an expression, because there are some things that can be done to equations, but not expressions. Table 6-2 reviews the distinction between equations and expressions.

For most equations, if you substitute in some values of the variable the equation will be a true statement, and for other values it will be false.

A **solution** of an equation is a value of the variable that makes the equation a true statement when substituted into the equation. **Solving** an equation means finding *every* solution of the equation. We call the set of all solutions the **solution set,** or simply the **solution** of an equation.

Math Note

Equations can have any number of solutions. Some have one, some have two, some have many, and some have none at all.

For example, $x = 2$ is one solution of the equation $x^2 - 4 = 0$, because $(2)^2 - 4 = 0$ is a true statement. But $x = 2$ is not **the** solution, because $x = -2$ is a solution as well. The solution set is actually $\{-2, 2\}$.

TABLE 6-2 Expressions and Equations

Expressions	Equations
$5x + 7$	$5x + 7 = 15$
$3x + 4y + 7$	$3x + 4y + 7 = 10$
$8 + 5$	$8 + 5 = 13x$
$4(x + 6)$	$4(x + 6) = 20x - 7$

EXAMPLE 1 Identifying Solutions of an Equation

Determine if the given value is a solution of the equation.

(a) $4(x - 1) = 8; x = 2$
(b) $x + 7 = 2x - 1; x = 8$
(c) $2y^2 = 200; y = -10$

SOLUTION

(a) Substituting 2 in for x we get:
$$4(2 - 1) = 8$$
$$4(1) = 8$$
$$4 = 8$$
This is not true, so $x = 2$ is not a solution of the equation.

(b) Substituting 8 in for x we get:
$$8 + 7 = 2(8) - 1$$
$$15 = 16 - 1$$
$$15 = 15$$
This is true, so $x = 8$ is a solution of the equation.

(c) Substituting in -10 for y, we get:
$$2(-10)^2 = 200$$
$$2 \cdot 100 = 200$$
$$200 = 200$$
This is true, so $y = -10$ is a solution of the equation.

▼ Try This One 1

☑ 1. Decide if a number is a solution of an equation.

Determine if the given value is a solution of the equation.

(a) $3(t - 4) = 5t;\ t = -6$
(b) $x^2 + x = 12;\ x = 3$
(c) $7y + 3 = 10y - 4;\ y = 2$

We will begin our study of solving equations by working with a special type. *Linear equations* are equations in which the variable appears only to the first power. Here's a formal definition:

A **linear equation in one variable** is an equation that can be written in the form $Ax + B = 0$, where A and B are real numbers, and A is not zero.

For example, $2x + 3 = 7$ is a linear equation. Certainly, it fits our informal definition because the only variable that appears has exponent 1. It also fits the formal definition because it can be rewritten as $2x - 4 = 0$ (we will see shortly how to do so). In that form, it fits the formal definition with $A = 2$ and $B = -4$.

EXAMPLE 2 Identifying Linear Equations

Determine which of the equations below are linear equations.

(a) $2(x - 3) = 3x + 2$
(b) $y^2 - 4 = 0$
(c) $\frac{1}{2}t - 3^2 = 10$
(d) $x(x + 3) = 4$

SOLUTIONS

(a) This is a linear equation: the variable appears only to the first power.
(b) This is not a linear equation: the variable has exponent 2.
(c) This is a linear equation. Don't let the fraction, or the exponent on 3 fool you! The variable has exponent 1.
(d) This is not a linear equation. It looks like it at first, because both xs have exponent 1. But the left side can be simplified to $x^2 + 3x$, so we see that x has exponent two.

2. Identify linear equations.

▼ Try This One 2

Determine which of the equations are linear equations.

(a) $3z + 7 = z^3$
(b) $7x - 4 = 11$
(c) $19(2y - 7) = 5(3y + 1)$
(d) $2x = x(4 - x)$

Solving Basic Linear Equations

When two equations have the same solution set, we call them **equivalent equations.** The equations $x - 3 = 0$ and $x = 3$ are equivalent because in each case, the only number that makes them true is $x = 3$. Our strategy for solving linear equations will be to perform one or more steps that transforms the original equation into an equivalent one, until we get to the point where the solution set is obvious, like the equation $x = 3$.

We'll use two main tools to accomplish this. The first involves addition and subtraction.

The Addition and Subtraction Properties of Equality

You can add or subtract the same real number or algebraic expression to both sides of an equation without changing the solution set. In symbols, if $a = b$, then $a + c = b + c$ and $a - c = b - c$.

In Example 3, we'll see how to use these properties to solve simple linear equations.

EXAMPLE 3 Solving Linear Equations Using the Addition and Subtraction Properties

Math Note

When solving equations, make sure you add or subtract on *both sides* of the equation! Adding or subtracting on one side only will almost always change the solutions.

Solve each equation, and check your answer.

(a) $x - 5 = 9$
(b) $y + 30 = 110$

SOLUTION

(a) The goal is to transform the equation into one whose solution is obvious. We can do this by isolating the variable, x, on the left side. To do so, we will add 5 to both sides.

$$x - 5 = 9$$
$$x - 5 + 5 = 9 + 5$$
$$x = 14$$

The solution set is now obvious: it's $\{14\}$. One of the great things about solving equations is that you can easily check your answer by substituting back into the original equation:

Check: $14 - 5 = 9$
$9 = 9$ ✓ *The equation is true, so 14 is a solution.*

(b) This time, to isolate the variable, we subtract 30 from both sides.

$$y + 30 = 110$$
$$y + 30 - 30 = 110 - 30$$
$$y = 80$$

The solution set is $\{80\}$.

Check: $80 + 30 = 110$
$110 = 110$ ✔

▼ Try This One 3

Solve each equation, and check your answer.

(a) $y + 25 = 70$ (b) $x - 13 = 20$

When solving an equation where the coefficient of the variable is not 1, we'll need one of the next two properties.

The Multiplication and Division Properties of Equality

You can multiply or divide both sides of an equation by the same *nonzero* real number without changing the solution set. In symbols, if $a = b$, then $a \cdot c = b \cdot c$ and $\frac{a}{c} = \frac{b}{c}$ as long as $c \neq 0$.

EXAMPLE 4 Solving Linear Equations Using the Multiplication and Division Properties

Solve each equation, and check your answer.

(a) $\frac{t}{6} = 3$

(b) $5x = 30$

Math Note

The properties we're using to solve equations don't say you can do *anything you want* to both sides of an equation. They say you can add, subtract, multiply, or divide by the same number on both sides (except zero when multiplying or dividing).

SOLUTION

(a) This time, to isolate the variable, we need to multiply both sides of the equation by 6.

$$\frac{t}{6} = 3 \quad \text{Multiply both sides by 6.}$$
$$\frac{t}{6} \cdot 6 = 3 \cdot 6 \quad \text{Simplify.}$$
$$t = 18$$

The solution set is $\{18\}$.

Check: $\frac{18}{6} = 3$
$3 = 3$ ✔

(b) The variable is multiplied by 5, so we can isolate x by dividing both sides by 5 (or multiplying both sides by $\frac{1}{5}$).

$$5x = 30 \quad \textit{Divide both sides by 5.}$$
$$\frac{5x}{5} = \frac{30}{5} \quad \textit{Simplify.}$$
$$x = 6$$

Check: $5 \cdot 6 = 30$
$30 = 30$ ✔

▼ Try This One 4

Solve each equation, and check your answer.

(a) $11z = 121$ (b) $\frac{x}{7} = -4$

Solving General Linear Equations

So far, we've seen how to solve simple linear equations. Any linear equation can be solved using the procedure listed below.

Procedure for Solving Linear Equations

Step 1 Simplify the expressions on both sides of the equation by distributing and combining like terms.

Step 2 Use the addition and/or subtraction property of equality to move all the variable terms to one side of the equation and all the constant terms to the opposite side of the equation.

Step 3 Combine like terms.

Step 4 Use the multiplication or division property of equality to eliminate the numerical coefficient and solve for the variable.

The next three examples illustrate our general procedure.

EXAMPLE 5 Solving a General Linear Equation

Solve the equation $5x + 9 = 29$, and check your answer.

SOLUTION

$$5x + 9 = 29 \quad \textit{Subtract 9 from both sides of the equation.}$$
$$5x + 9 - 9 = 29 - 9 \quad \textit{Combine like terms.}$$
$$5x = 20 \quad \textit{Divide both sides by 5.}$$
$$\frac{5x}{5} = \frac{20}{5}$$
$x = 4$, or the solution set is $\{4\}$

Check:

$$5x + 9 = 29$$
$$5(4) + 9 \stackrel{?}{=} 29$$
$$29 = 29 \checkmark$$

▼ Try This One 5

Solve the equation $3x - 25 = 8$, and check your answer.

EXAMPLE 6 Solving a General Linear Equation

Solve $6x - 10 = 4x + 8$, and check your answer.

SOLUTION

$6x - 10 = 4x + 8$	*Subtract 4x from both sides of the equation.*
$6x - 4x - 10 = 4x - 4x + 8$	*Combine like terms.*
$2x - 10 = 8$	*Add 10 to both sides of the equation.*
$2x - 10 + 10 = 8 + 10$	*Combine like terms.*
$2x = 18$	*Divide both sides by 2.*
$\frac{2x}{2} = \frac{18}{2}$	
$x = 9$, or the solution set is $\{9\}$	

Check:

$$6x - 10 = 4x + 8$$
$$6(9) - 10 \stackrel{?}{=} 4(9) + 8$$
$$54 - 10 \stackrel{?}{=} 36 + 8$$
$$44 = 44 \checkmark$$

▼ Try This One 6

Solve the equation $8x - 27 = 3x + 33$, and check your answer.

EXAMPLE 7 Solving a General Linear Equation

Solve $3(2x + 5) - 10 = 3x - 10$, and check your answer.

SOLUTION

First, we simplify the left side, then proceed as usual.

$3(2x + 5) - 10 = 3x - 10$	*Multiply out parentheses.*
$6x + 15 - 10 = 3x - 10$	*Combine like terms.*

$6x + 5 = 3x - 10$	*Subtract 3x from both sides of the equation.*
$6x - 3x + 5 = 3x - 3x - 10$	*Combine like terms.*
$3x + 5 = -10$	*Subtract 5 from both sides of the equation.*
$3x + 5 - 5 = -10 - 5$	*Combine like terms.*
$3x = -15$	*Divide both sides by 3.*
$\frac{3x}{3} = \frac{-15}{3}$	
$x = -5$, or the solution set is $\{-5\}$	

Math Note

It's very important that, when checking your solution, you substitute back into the *original equation*. If you use one of the later equations you wrote and made a mistake in the very first step, a wrong answer might appear to be a correct answer.

Check:

$$3(2x + 5) - 10 = 3x - 10$$
$$3(2 \cdot (-5) + 5) - 10 \stackrel{?}{=} 3(-5) - 10$$
$$3(-10 + 5) - 10 \stackrel{?}{=} -15 - 10$$
$$3(-5) - 10 \stackrel{?}{=} -25$$
$$-15 - 10 = -25$$
$$-25 = -25 \checkmark$$

▼ Try This One 7

3. Solve general linear equations.

Solve each equation, and check your answer.

(a) $2(x - 7) + 5 = 3x - 10$ (b) $-5(2x - 8) + 6 = 4x - 32$

Solving Equations Containing Fractions

Suppose you were given a choice on a test between solving these two equations:

$$6x + 4 = x - 7 \qquad \frac{3x}{2} + 1 = \frac{x}{4} - \frac{7}{4}$$

If you're like pretty much everyone on the planet, you'd choose the first. It just looks easier because there are no fractions. Actually, these are the same equation in different forms. The good news is that there's a simple procedure that will turn *any* equation with fractions into one with no fractions at all. You just need to find the least common denominator of all fractions that appear in the equation, and multiply every single term on each side of the equation by the LCD. If there are any fractions left after doing so, you made a mistake! In the remaining examples in this section, we'll leave the checking to you.

EXAMPLE 8 Solving a Linear Equation Containing Fractions

Solve the equation $\frac{2x}{3} + \frac{x}{5} = \frac{26}{3}$.

SOLUTION

The least common denominator of fractions with denominators 3 and 5 is 15. So the first step is to multiply every term in the equation by 15.

$$\frac{2x}{3} + \frac{x}{5} = \frac{26}{3}$$ *Multiply every term on each side by 15.*

$$15 \cdot \frac{2x}{3} + 15 \cdot \frac{x}{5} = 15 \cdot \frac{26}{3}$$ *Simplify fractions.*

$$5 \cdot 2x + 3 \cdot x = 5 \cdot 26$$ *Multiply.*

$$10x + 3x = 130$$ *Combine like terms.*

$$13x = 130$$ *Divide both sides by 13.*

$$\frac{13x}{13} = \frac{130}{13}$$ *Simplify.*

$x = 10$, or the solution set is $\{10\}$

▼ Try This One 8

Solve the equation $\frac{3x}{2} + 1 = \frac{x}{4} - \frac{7}{4}$

EXAMPLE 9 Solving a Linear Equation Containing Fractions

Solve the equation $\frac{2x - 3}{3} + \frac{5x}{2} = \frac{x}{2} - 4$.

SOLUTION

The least common denominator is 6, so we begin by multiplying every term on each side by 6.

Math Note

Make sure you multiply *every* term on each side by the LCD, not just the ones that are fractions.

$$6 \cdot \frac{2x - 3}{3} + 6 \cdot \frac{5x}{2} = 6 \cdot \frac{x}{2} - 6 \cdot 4$$ *Simplify fractions.*

$$2(2x - 3) + 3 \cdot 5x = 3x - 24$$ *Multiply.*

$$4x - 6 + 15x = 3x - 24$$ *Combine like terms.*

$$19x - 6 = 3x - 24$$ *Subtract 3x from both sides.*

$$19x - 6 - 3x = 3x - 24 - 3x$$ *Combine like terms.*

$$16x - 6 = -24$$ *Add 6 to both sides.*

$$16x - 6 + 6 = -24 + 6$$ *Simplify.*

$$16x = -18$$ *Divide both sides by 16.*

$$\frac{16x}{16} = \frac{-18}{16}$$ *Simplify.*

$x = -\frac{18}{16} = -\frac{9}{8}$, or the solution set is $\left\{-\frac{9}{8}\right\}$

▼ Try This One 9

☑ 4. Solve linear equations containing fractions.

Solve the equation $\frac{2(x - 3)}{5} - \frac{x}{3} = \frac{3x - 1}{5} - 2$.

Solving a Formula for a Specific Variable

Most formulas contain more than one variable, and it's often helpful to rearrange a formula so that the quantity you want to find is isolated on one side. For example, the formula $I = \frac{V}{R}$ is used in electronics to find the current I in a circuit when you know the voltage (V) and the resistance (R). But what if you know the current, and need to find the resistance? You can solve the equation for variable R, treating the other variables as numbers. The result is a formula that tells you the resistance.

EXAMPLE 10 Solving a Formula in Electronics for One Variable

Solve the formula $I = \dfrac{V}{R}$ for R.

SOLUTION

$$I = \frac{V}{R} \qquad \textit{Multiply both sides by } R.$$

$$I \cdot R = \frac{V}{\cancel{R}} \cdot \cancel{R}$$

$$IR = V \qquad \textit{Divide both sides by } I.$$

$$\frac{\cancel{I}R}{\cancel{I}} = \frac{V}{I}$$

$$R = \frac{V}{I}$$

▼ Try This One 10

The formula for the area of a trapezoid is $A = \frac{1}{2} b(h_1 + h_2)$. Solve this formula for b.

EXAMPLE 11 Finding a Formula for Temperature in Celsius

The formula $F = \frac{9}{5}C + 32$ gives the Fahrenheit equivalent for a temperature in Celsius. Transform this into a formula for calculating the Celsius temperature C.

SOLUTION

$$F = \frac{9}{5}C + 32 \qquad \textit{Multiply both sides by 5.}$$

$$5F = 5 \cdot \frac{9}{5}C + 5 \cdot 32 \qquad \textit{Simplify.}$$

$$5F = 9C + 160 \qquad \textit{Subtract 160 from both sides.}$$

$$5F - 160 = 9C + 160 - 160 \qquad \textit{Combine like terms.}$$

$$5F - 160 = 9C \qquad \textit{Divide both sides by 9.}$$

$$\frac{5}{9}F - \frac{160}{9} = \frac{9C}{9} \qquad \textit{Simplify.}$$

$$\frac{5}{9}F - \frac{160}{9} = C$$

The formula for Celsius temperature is $C = \frac{5}{9}F - \frac{160}{9}$

▼ Try This One 11

5. Solve formulas for one specific variable.

The formula $A = P(1 + rt)$ can be used to calculate the amount in an interest-bearing account. Transform this into a formula for calculating the interest rate r.

Contradictions and Identities

Some equations have no solution. For example, it's easy to see that the equation $x = x + 1$ has no solution since there is no real number such that adding one results in the original number. Such an equation is called a *contradiction*. If the normal procedures for solving a linear equation were applied, we would get

$$x = x + 1 \quad \text{Subtract } x \text{ from both sides}$$
$$x - x = x - x + 1 \quad \text{Simplify.}$$
$$0 = 1$$

Since the final equation is always a false statement, the same is true for original equation. Formally,

A **contradiction** is an equation with no solution.

Informally, a contradiction is an equation that when solved results in a false statement. The solution set of a contradiction is the empty set, { } or $\varnothing$.

On the other hand, there are equations in one variable with infinitely many solutions. Such equations are called *identities*.

An **identity** is an equation that is true for any value of the variable for which both sides are defined.

When you solve an equation that is an identity, the final equation will be a statement that is always true as in the following example:

$$2(5x + 8) - 10 = 10x + 6 \quad \text{Distribute.}$$
$$10x + 16 - 10 = 10x + 6 \quad \text{Simplify.}$$
$$10x + 6 = 10x + 6 \quad \text{Subtract } 10x \text{ from both sides.}$$
$$10x - 10x + 6 = 10x - 10x + 6 \quad \text{Simplify.}$$
$$6 = 6$$

In this case every real number is a solution to the equation so the solution set is $\{x \mid x \text{ is a real number}\}$.

In summary, if the variable is eliminated when solving an equation and the resulting equation is false then the equation is a contradiction and the solution set is the empty set. If the variable is eliminated and the resulting equation is true, then the equation is an identity and the solution set is the set of all real numbers.

EXAMPLE 12 Recognizing Identities and Contradictions

Indicate whether the equation is an identity or a contradiction, and give the solution set.

(a) $3(x - 6) + 2x = 5x - 18$ (b) $6x - 4 + 2x = 8x - 10$

SOLUTION

(a) $3(x - 6) + 2x = 5x - 18$

$3x - 18 + 2x = 5x - 18$

$5x - 18 = 5x - 18$

$-18 = -18$

Since the resulting equation is true, it is an identity and the solution set is $\{x \mid x \text{ is a real number}\}$.

(b) $6x - 4 + 2x = 8x - 10$

$8x - 4 = 8x - 10$

$8x - 8x - 4 = 8x - 8x - 10$

$-4 = -10$

Since the resulting equation is false, it's a contradiction, and the solution set is ∅.

▼ Try This One 12

6. Determine if an equation is an identity or a contradiction.

Indicate whether the equation is an identity or a contradiction, and give the solution set.

(a) $13x - 6 = 2(5x + 4) + 3x$ (b) $5(x + 6) - 5x = 30$

Answers to Try This One

1 (a) Yes (b) Yes (c) No

2 (b) and (c) are linear

3 (a) {45} (b) {33}

4 (a) {11} (b) {−28}

5 {11}

6 {12}

7 (a) {1} (b) $\left\{\frac{39}{7}\right\}$

8 $\left\{-\frac{11}{5}\right\}$

9 $\left\{\frac{15}{8}\right\}$

10 $b = \frac{2A}{h_1 + h_2}$

11 $r = \frac{A}{Pt} - \frac{1}{t}$ or $r = \frac{1}{t}\left(\frac{A}{P} - 1\right)$

12 (a) contradiction; ∅
(b) identity; $\{x \mid x \text{ is a real number}\}$

EXERCISE SET 6-2

Writing Exercises

1. How can you tell when an equation is a linear equation?
2. Explain what it means to solve an equation. Why is that not the same thing as "find a solution"?
3. List the steps for solving a general linear equation in your own words.
4. What's the best way to begin solving an equation that contains fractions?
5. When does an equation have an empty solution set?
6. Why is it sometimes useful to solve a formula for one of the variables?

Computational Exercises

In Exercises 7–14, state whether the equation is a linear equation in one variable or not.

7. $3x(2 - x) = x - 3$
8. $4x^2 = 5(x - 2)$
9. $7x = 3(x + 3)$
10. $9x - 7 + 2x = 3$
11. $\frac{5}{x} + x = 3x$
12. $\frac{2}{5}x - \frac{3}{x} = 4$
13. $7(x - 3) = 2x$
14. $\frac{1}{3}x = 7$

In Exercises 15–20, determine whether the given value is a solution to the equation.

15. $x^2 = 1; x = -1$
16. $3x - 2 = 7; x = 3$
17. $4(x - 1) = 3x; x = -3$
18. $\frac{2}{3}x = 3(x - 2); x = 3$
19. $3(x + 2)^2 = 12; x = 0$
20. $2(x + 3) - 4 = 3x + 3; x = -1$

For Exercises 21–74, solve each equation.

21. $x + 6 = 32$
22. $7 + x = 43$
23. $x - 5 = 54$
24. $36 = x - 9$
25. $-5 = y - 2$
26. $12 = y + 18$
27. $9x = 27$
28. $6x = 42$
29. $-3z = 36$
30. $-42 = -7z$
31. $6x + 12 = 48$
32. $10x - 30 = -5$
33. $-5x + 25 = -55$
34. $-3x + 18 = 42$
35. $2t + 10 = 4t - 30$
36. $5t - 6 = 2t - 24$
37. $x = 6x - 55$
38. $9x - 18 = 7x + 4$
39. $-2x = 15 - 5x$
40. $5x + 8 = 10x - 32$
41. $-6x + 15 = 4x - 25$
42. $9 - 2x = 7 - x$
43. $2(y + 8) = 40$
44. $2(y - 6) = 2$
45. $3(x + 2) = 26$
46. $7(x - 3) = 42$
47. $6 + 3(x - 5) = 2(x - 3)$
48. $-2(4x - 7) = 3x - 8$
49. $6 + 7(m - 3) = 2m + 10$
50. $5(9 - n) = 4(n + 6)$
51. $12(x - 2) - 10(x + 7) = 14$
52. $-2x + 3 + 4(x - 6) = 18$
53. $6(-x + 5) = 2(x + 8) - 10$
54. $-\frac{3}{7}x = 21$
55. $\frac{5}{6}x = 30$
56. $-\frac{1}{4}x = 2$
57. $\frac{5}{8}y = 40$
58. $-\frac{1}{2}y = 25$
59. $\frac{3}{4}x + 2 = 21$
60. $\frac{1}{8}x - 10 = -16$
61. $\frac{5z}{6} + \frac{z}{3} = 30$
62. $\frac{3t}{4} + \frac{7t}{2} = 18$
63. $\frac{7x}{3} + \frac{4x}{2} = 28$
64. $\frac{4x}{6} + \frac{x}{5} = \frac{2}{3} - \frac{x}{5}$
65. $\frac{7x}{3} + 5 = \frac{4x}{8} + 10$
66. $\frac{3x}{2} + \frac{1}{2} = \frac{4x}{5} + \frac{3}{5}$
67. $\frac{x}{6} + \frac{3x}{2} = \frac{4}{5} - \frac{2x}{15}$
68. $\frac{5}{12} - \frac{1}{3}x = \frac{2}{3}x - \frac{7}{6}$
69. $\frac{9x}{5} - \frac{2x}{7} = \frac{3}{5} - \frac{6}{7}$
70. $\frac{6x}{8} + \frac{4x}{2} - 5 = \frac{3x}{4}$
71. $\frac{3x - 2}{3} + 4 = \frac{x}{4}$
72. $\frac{x + 3}{2} + \frac{3}{4} = \frac{3x}{2} - 1$
73. $\frac{2}{5} = \frac{2(x - 1)}{3} + \frac{x}{5}$
74. $\frac{x}{3} + \frac{5(x - 1)}{4} = \frac{3x}{4} - 1$

For Exercises 75–80, solve each equation for the specified variable.

75. $3x + 8 = 2y + 4$ for y
76. $5y = 3x + 2$ for x
77. $2 + 5x - 7y = 18$ for x
78. $5y - 3x + 2 = 10$ for y
79. $7x + 2y = 9$ for y
80. $3y + 6 = 2x + 8$ for x

For Exercises 81–88, indicate whether the equation is an identity or a contradiction and give the solution set.

81. $8x - 5 + 2x = 10x - 10 + 5$
82. $3x + 7 - x = 2x + 21$
83. $5(x - 3) + 2 = 5x - 8$
84. $4(x + 2) + 6 = 2x + 2x + 14$
85. $3x - 2 = 5(x - 3) - 2x$
86. $4x + 3(x - 2) = 7(x - 2) + 8$
87. $x + 6 = 3(x + 2) - 2x$
88. $\frac{2}{3}x = \frac{4x - 2}{6}$

Real-World Applications

89. The electrical resistance for a conductor can be found by the formula $R = \frac{KL}{d^2}$. Solve the formula for L.
90. The illumination of a light can be found by the formula $I = \frac{C}{D^2}$. Solve the formula for C.
91. The volume of a cylinder can be found by the formula $V = \pi r^2 h$. Solve the formula for h.
92. The formula for the perimeter of a rectangle is $P = 2l + 2w$. Solve the formula for w.
93. The formula for the volume of a rectangular solid is $V = lwh$. Solve the formula for h.
94. The formula for converting mass to energy is $E = mc^2$. Solve the formula for m.
95. The formula for the distance traveled during an acceleration period is $d = \frac{1}{2}at^2$. Solve the formula for a.
96. The formula for earned run average in baseball is $E = \frac{R}{I} \times 9$. Solve the formula for I.
97. The formula for slugging percentage in baseball is $P = \frac{S + 2D + 3T + 4H}{A}$. Solve the formula for H.
98. The formula for the area of a triangle is $A = \frac{1}{2}bh$. Solve the formula for h.
99. The formula for the average a of two numbers b and c is $a = \frac{b + c}{2}$. Solve for b.
100. The centripetal force of an object can be found by using the formula $F = \frac{mv^2}{r}$. Solve the formula for r.

Exercises 101–104 use the following formula: the force in newtons exerted on an object is given by the formula $F = ma$, where m is the mass in kg and a is acceleration in m/s^2.

101. Find the acceleration for a force $F = 100$ newtons and a mass of 20 kg.
102. Find the acceleration for a force $F = 350$ newtons and a mass of 50 kg.
103. Find the mass for a force of $F = 120$ newtons and an acceleration of 30 m/s^2.
104. Find the mass for a force of $F = 450$ newtons and an acceleration of 75 m/s^2.

Exercises 105–108 use the following formula: the cost of a cell phone plan is given by the formula $C = 0.035m + 45$, where C is the cost per month, m is the number of minutes used per month, and \$45 is the basic monthly charge.

105. What is the cost if you use 300 minutes per month?
106. What is the cost if you use 1,000 minutes per month?
107. How many minutes would you get to talk per month if you want the total bill to be \$100?
108. How many minutes would you get to talk per month if you want the total bill to be \$150?

Exercises 109–112 use the following formula: the momentum of a person who weighs 68 kg and is riding a roller coaster is given by the formula $p = 68v$, where p is the momentum in kg m/s and v is the velocity in m/s.

109. What is the momentum of the person when the velocity of the roller coaster is 40 m/s?
110. What is the momentum of the person when the velocity of the roller coaster is 78 m/s?
111. What is the velocity if the momentum is 100 kg-m/s?
112. What is the velocity if the momentum is 350 kg-m/s?

Exercises 113–116 use the following formula: the grade for Steve's English class can be calculated if he takes the sum of his three essay scores and divides that total by 3, as in the formula $G = \frac{E1 + E2 + E3}{3}$. Say he got an 80 on the first essay, E1, and he got a 95 on the second essay, E2.

113. What is his grade G if he gets a 70 on his third essay?
114. What is his grade G if he gets an 87 on his third essay?
115. What does he need to score on his third essay to get an overall grade of 80?
116. What does he need to score on his third essay to get an overall grade of 90?

Exercises 117–120 use the following formula: the power of a circuit can be computed by the formula $P = \frac{V^2}{R}$, where P is the power in watts, V is the voltage, and R is the resistance in ohms.

117. For a 25-volt circuit, find the power for a resistance of 10 ohms.
118. For a 50-volt circuit, find the power for a resistance of 25 ohms.
119. For a 30-volt circuit, find the resistance in ohms for 100 watts of power.
120. For a 20-volt circuit, find the resistance in ohms for 250 watts of power.

Critical Thinking

121. Fact: You can solve linear equations using only the addition and multiplication properties of equality. Explain why.
122. Explain why it's incorrect to multiply both sides of an equation by zero when you are solving.
123. Review your answer to Exercise 122, then explain why you can't multiply or divide both sides of an equation by an expression containing the variable.
124. Describe a method for easily writing an equation that is an identity and an equation that is a contradiction.
125. For the equation in Example 8, try to solve without multiplying both sides by the LCD. Instead, get a common denominator and perform the addition on the left side, then solve using the usual procedure. Did you get the same answer? Explain why multiplying both sides by the LCD is a much more efficient way to solve equations with fractions.

Section 6-3 Applications of Linear Equations

LEARNING OBJECTIVES

- ❑ 1. Translate verbal expressions into mathematical symbols.
- ❑ 2. Solve real-world problems using linear equations.

Two of the most common questions that students ask when learning about solving equations are "When would I ever *use* this?" and "Who cares what x is?" (If you want to drive your professor up the wall, ask these questions at least once a day.) Nobody will claim that you will solve multiple equations every day once you leave the hallowed halls of college. But the fact is that solving equations is a topic that is widely applied to almost every area of study. And even if it weren't, the problem-solving skills that you learn and hone while solving applied problems are among the best "brain exercise" you can get. And what could possibly be more useful to your education than training your brain to work better?

So in this section, we will solve problems that relate to real-world issues. In most cases, the plan is to write an equation that describes a situation, then solve that equation to find some quantity of interest. And even if you don't find the situations applicable to your life, keep the bigger picture in mind: you'll be practicing useful problem-solving skills with every question.

In the box below, we've outlined a general strategy for attacking word problems using algebra. If you look close enough, you'll notice that our strategy is based on Polya's problem-solving strategy that we studied in Chapter 1, with a few elaborations matching this specific type of problem.

A General Procedure for Solving Word Problems Using Equations

Step 1 Read the problem carefully, but *don't read it all at once without doing anything!* As you're reading, write down any information provided by the problem that seems relevant. This will at least get you started. Make sure you carefully note what it is you're being asked to find. Draw a diagram if the situation calls for one.

Step 2 Assign a variable to an unknown quantity in the problem. Most of the time, the variable should represent the quantity you're being asked to find.

Step 3 Write an equation based on the information given in the problem. Remember, an equation is a statement that two quantities are equal, so keep an eye out for statements in the problem indicating two different ways to express the same quantity.

Step 4 Solve the equation.

Step 5 Make sure that you answer the question! The best approach is to write your answer in sentence form.

Step 6 Check to see if your solution makes sense based on the original wording of the problem.

The step that almost everyone finds most challenging is step 3. In order to write an equation that describes a situation, you have to translate verbal statements into mathematical symbols. For example, the verbal statement "six more than three times some number" can be written in symbols as "$6 + 3x$" or "$3x + 6$." A careful read of Table 6-3 will help get you started on these types of translations. The table is also very useful to refer back to as you work on problems.

TABLE 6-3 Common Phrases That Represent Operations

Phrase	Symbols
Phrases that represent addition	
6 more than a number	$6 + x$
A number increased by 8	$x + 8$
5 added to a number	$5 + x$
The sum of a number and 17	$x + 17$
Phrases that represent subtraction	
18 decreased by a number	$18 - x$
6.5 less than a number	$x - 6.5$
3 subtracted from a number	$x - 3$
The difference between a number and 5	$x - 5$
Phrases that represent multiplication	
8 times a number	$8x$
Twice a number	$2x$
A number multiplied by 4	$4x$
The product of a number and 19	$19x$
$\frac{2}{3}$ of a number	$\frac{2}{3}x$
Phrases that represent division	
A number divided by 5	$x \div 5$
35 divided by a number	$35 \div x$
The quotient of a number and 6	$x \div 6$

EXAMPLE 1 Translating Verbal Statements into Symbols

Translate each verbal statement into symbols.

(a) 14 times a number
(b) A number divided by 7
(c) 10 more than the product of 8 and a number
(d) 3 less than 4 times a number
(e) 6 times the sum of a number and 18

SOLUTION

(a) Using variable x to represent the unspecified number, we can write this as $14x$.
(b) $\frac{x}{7}$
(c) $10 + 8x$
(d) It might help to reword this as 3 subtracted from 4 times a number: $4x - 3$
(e) Parentheses are required here because the multiplication is 6 times the sum: $6(x + 18)$

Math Note

The actual letter you choose for a variable is unimportant. You can choose any letter you like.

▼ Try This One 1

Translate each verbal statement into symbols.

(a) 100 divided by a number
(b) 5 more than the product of a number and 7
(c) The difference between 25 and a number
(d) The product of 8 and the difference of a number and 4

☑ 1. Translate verbal expressions into mathematical symbols.

Solving Word Problems

We will begin our study of solving word problems with a basic translation problem. It's not terribly realistic, but gives you a start on the basic steps for solving.

EXAMPLE 2 Solving a Basic Translation Problem

If 8 times a number plus 3 is 27, find the number.

> **Math Note**
>
> When translating a statement into an equation, the word "is" usually indicates where the equal sign should go.

SOLUTION

Step 1 Write the relevant information:

8 times a number plus 3 is 27

Identify what we're asked to find: that unknown number.

Step 2 Use variable x to represent the unknown number.

Step 3 Translate the relevant information into an equation:

8 times a number plus 3 is 27

$$8x \qquad + \ 3 = 27$$

Step 4 Solve the equation:

$$8x + 3 = 27 \qquad \textit{Subtract 3 from both sides.}$$
$$8x + 3 - 3 = 27 - 3 \qquad \textit{Simplify.}$$
$$8x = 24 \qquad \textit{Divide both sides by 8.}$$
$$\frac{8x}{8} = \frac{24}{8} \qquad \textit{Simplify.}$$
$$x = 3$$

Step 5 Answer the question: the requested number is 3.

Step 6 Check: 8 times 3 is 24, and when you add 3, you get 27. This matches the description of the problem.

▼ Try This One 2

Ten less than twice a number is 42. Find the number.

In the next example, see if you can recognize the similarity to the abstract problem in Example 2.

EXAMPLE 3 A Problem Involving Contract Negotiations

Two basketball teams are interested in signing a free-agent player. An inside source informs the general manager of one team that the other has made an offer, and the player's agent said "Double that and add an extra million per year, and you're in our league." According to a published report, the player is seeking a contract of $18 million per year. What was the rival team's offer?

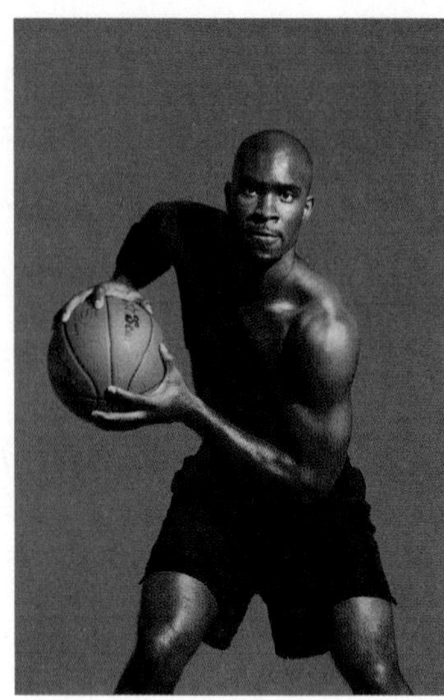

SOLUTION

Step 1 Relevant information: Twice the offer plus 1 million is 18 million. We're asked to find the offer.

Step 2 Use the variable x to represent the offer. Since the numbers are in millions, we'll let x stand for the offer in million dollar units—that will keep the arithmetic simpler.

Step 3 Translate the relevant information into an equation:

Twice the offer plus one million is 18 million

$$2x \quad + \quad 1 \quad = \quad 18$$

Step 4 Solve the equation:

$$2x + 1 = 18 \qquad \text{Subtract 1 from both sides.}$$
$$2x + 1 - 1 = 18 - 1 \qquad \text{Simplify.}$$
$$2x = 17 \qquad \text{Divide both sides by 2.}$$
$$x = \frac{17}{2} \text{ or } 8.5$$

Step 5 Answer the question: The team's offer was \$8.5 million. Only \$8.5 million? That's an insult!

Step 6 Check: Doubling \$8.5 million gives \$17 million, and adding 1 million more makes it \$18 million as required.

Math Note

Remember, we're using million dollar units, so 1 million and 18 million are represented by 1 and 18.

▼ Try This One 3

A teacher with a fondness for joking around with students tells one that if he doubled his score on the last test and subtracted 12%, he would have just barely gotten an A. The syllabus says that the minimum cutoff for A is 92%. What was the student's score?

Sometimes when there are two unknowns in a word problem, one unknown can be represented in terms of the other. For example, if I know that one number is 5 more than another number, then the first number can be represented by x and the second number can be represented by $x + 5$. Example 4 uses this idea.

EXAMPLE 4 An Application to Home Improvement

Pat and Ron are planning to build a deck off the back of their house, and they buy some plans from the Internet. The plans can be customized to the required deck height, which in this case will be 92 inches. They call for support posts of two different heights. The taller ones are 8 inches longer than the shorter ones, and the plans say that the sum of the lengths should be the height of the deck. How long should the support posts be cut?

Math Note

We could also call the length of the longer posts x; in that case, the shorter ones would have length $x - 8$.

SOLUTION

Step 1 Relevant information: The posts are 8 inches different in length, and the lengths should add to 92 inches.

Step 2 We'll call the length of the shorter posts x. The other posts are 8 inches longer, so they must be $x + 8$.

Step 3 Translate the relevant information into an equation:

Length of shorter post + length of longer post is 92

$$x \quad + \quad x + 8 \quad = 92$$

Step 4 Solve the equation:

$$\begin{aligned} x + x + 8 &= 92 \\ 2x + 8 &= 92 \\ 2x &= 84 \\ x &= 42 \end{aligned}$$

Step 5 Answer the question—this is where it becomes really important to consider the original question. We were asked to find two lengths, so $x = 42$ doesn't answer the question. We found that the length of the shorter post is 42 inches, and the longer is 8 inches longer, so the answer is 50 inches.

Step 6 Check: The two lengths are definitely separated by 8 inches and 42 inches + 50 inches = 92 inches, as required.

▼ Try This One 4

The railings for the deck in Example 4 have three different lengths of board. The shortest is 10 inches less than the next shortest, which is 14 inches shorter than the longest. Their combined length is supposed to match the overall length of the deck, which in this case is 20 feet. How long should each piece be?

EXAMPLE 5 An Application Involving Money

After a busy homecoming weekend, the tip jar at an off-campus bar is stuffed full of quarters and dollar bills. It turns out that there are 3 times as many quarters as dollar bills, and the jar contains \$140. How many quarters and how many dollar bills are there?

SOLUTION

Step 1 Relevant information: Three times as many quarters as dollar bills, and the total value is \$140.

Step 2 Use variable d to represent the number of dollar bills. Then $3d$ is the number of quarters (because there are three times as many).

Step 3 Translate the relevant information into an equation: the value in dollars of the quarters is the number of quarters ($3d$) times \$0.25. The value of the dollar bills is the number of them (d).

Total value is \$140

$$0.25(3d) + d = 140$$

Step 4 Solve the equation:

$$\begin{aligned} 0.25(3d) + d &= 140 && \textit{Multiply.} \\ 0.75d + d &= 140 && \textit{Combine like terms.} \\ 1.75d &= 140 && \textit{Divide both sides by 1.75.} \\ d &= \frac{140}{1.75} = 80 \end{aligned}$$

Math Note

In this case, we could have use a variable like q to represent the number of quarters, but then the number of dollar bills would be $\frac{q}{3}$, and we'd be introducing fractions. So it's simpler to use d = the number of dollar bills.

Step 5 Answer the question: There are 80 dollar bills in the jar, and 240 quarters (because there are 3 times as many).

Step 6 Check: 240 quarters is \$60; plus \$80 in dollar bills gives us \$140 as required.

▼ Try This One 5

☑ 2. Solve real-world problems using linear equations.

Corrine collected \$137 in tips during a Friday evening shift waiting tables, split among one- and five-dollar bills. She got 53 more singles than fives. How many of each did she get?

As you work the exercises, keep in mind that one of the main goals is to practice organized thinking and problem-solving skills. You shouldn't worry too much about how realistic or interesting you think the situations are—focus on the process of setting up and solving the equations.

Answers to Try This One

1 (a) $\frac{100}{x}$ (b) $5 + 7x$ (c) $25 - x$ (d) $8(x - 4)$

2 26

3 52% (ouch!)

4 $68\frac{2}{3}$ in., $78\frac{2}{3}$ in., and $92\frac{2}{3}$ in.

5 67 singles and 14 fives

EXERCISE SET 6-3

Writing Exercises

1. Write some reasons why it's a bad idea to read an entire word problem without writing down any information.
2. Write some reasons why it's a great idea to write your answer to a word problem in the form of a sentence.
3. What's the difference between checking your answer when solving an equation, and checking your answer when solving a word problem?
4. How do you choose what the variable in a word problem should represent?

Computational Exercises

For Exercises 5–24, write each phrase in symbols.

5. 3 less than a number
6. A number decreased by 17
7. A number increased by 9
8. 6 increased by a number
9. 11 decreased by a number
10. 8 more than a number
11. 9 less than a number
12. 6 subtracted from a number
13. 7 minus a number
14. 7 times a number
15. A number multiplied by 8
16. One-half a number
17. 5 more than 3 times a number
18. The quotient of 3 times a number and 6
19. 3 more than 5 times a number
20. 4 times a number
21. Double a number
22. 4 less than 6 times a number
23. The quotient of a number and 14
24. A number divided by 8

For Exercises 25–34, solve each.

25. Six times a certain number plus the number is equal to 56. Find the number.
26. The sum of a number and the number plus 2 is equal to 20. Find the number.
27. Twice a number is 32 less than 4 times the number. Find the number.
28. The larger of two numbers is 10 more than the smaller number. The sum of the number is 42. Find the numbers.
29. The difference of two numbers is 6. The sum of the numbers is 28. Find the numbers.
30. Five times a number is equal to the number increased by 12. Find the number.
31. Twice a number is 24 less than 4 times the number. Find the number.
32. The difference between one-half a number and the number is 8. Find the number.
33. Twelve more than a number is divided by 2. The result is 20. Find the number.
34. Eighteen less than a number is tripled, and the result is 10 more than the numbers. Find the numbers.

Real-World Applications

35. A math class containing 57 students was divided into two sections. One section has three more students than the other. How many students were in each section?
36. For a certain year, the combined revenues for PepsiCo and Coca-Cola were $47 billion. If the revenue for PepsiCo was $11 billion more than Coca-Cola, how much was the revenue of each company?
37. The cost, including sales tax, of a Ford Focus SE is $16,655.78. If the sales tax is 6%, find the cost of the car before the tax was added.
38. Pete is 3 times as old as Bill. The sum of their ages is 48. How old is each?
39. An electric bill for September is $2.32 less than the electric bill for October. If the total bill for the 2 months was $119.56, find the bill for each month.
40. If a person invested half of her money at 8% and half at 6% and received $210 interest, find the total amount of money invested.
41. A basketball team played 32 games and won 4 more games than it lost. Find the number of games the team won.
42. The enrollment of students in evening classes at a local university decreased by 6% between the years of 2007 and 2008. If the total number of students attending evening classes in both years was 16,983, find how many students enrolled in evening classes in each of those years.
43. If a television set is marked $\frac{1}{3}$ off and sells for $180, what was the original price?
44. A carpenter wanted to cut a 6-foot board into three pieces so that each piece is 6 inches longer than the preceding one. Find the length of each piece.
45. The Halloween Association reported that last year, Americans spent $0.68 billion more on candy than costumes for Halloween. If the total spent by Americans for both items was $3.18 billion, how much did Americans spend on each item?
46. In a charity triathlon, Mark ran half the distance and swam a quarter of the distance. When he took a quick break to get a drink of Gatorade, he was just starting to bike the remaining 15 miles. What was the total distance of the race?
47. A ball bounces half as far as it did on the preceding bounce and traveled a total distance of 12 feet on two bounces. How far did it go on the first bounce?
48. A person sold her house for $82,000 and made a profit of 20%. How much did she pay for her home?
49. A father left $\frac{1}{2}$ of his estate to his son, $\frac{1}{3}$ of his estate to his granddaughter, and the remaining $6,000 to charity. What was his total estate?
50. In 2009, there were 95 female officials in Congress, and there were 61 more female members of the House of Representatives than female senators. Find the number of females in each house of Congress.
51. While shopping on BlueFly.com, Juanita notices a special where if she buys two items, the third will be half off. She buys one item, then another item that is half of that amount, and then a third item that is a quarter of the original item amount. The discount she is given is half off the cheapest item. She ends up spending $65 on the order (neglect taxes). What is the price of the first item she bought?
52. There were five winning lottery tickets for a total jackpot of $24 million. Three of the winners won twice as much as the other two. How much did each of the two that won the least get?
53. In Mary's purse, there are $3.15 worth of nickels and dimes. There are 5 times as many nickels as dimes. The vending machine is only taking dimes, and Mary needs 10 dimes for her purchase. Does Mary have enough dimes?
54. If the perimeter of a triangular flower bed is 15 feet with two sides the same length and the third side 3 feet longer, what are the measures of the three sides of the flower bed?
55. Last semester, Marcus's tuition bill was 12% cheaper than this semester's tuition bill of $640. How much did Marcus pay for tuition last semester?
56. Jane and her two friends will rent an apartment for $875 a month, but Jane will pay double what each

friend does because she will have her own bedroom. How much will Jane pay a month?

57. In the recent election for Student Congress President, there were a total of 487 votes cast for the two candidates. The winner received 75 more votes than the loser did. How many votes did each receive?

58. Three sisters inherited \$100,000 from a rich uncle. The uncle's favorite niece got twice as much as his second favorite niece and the second favorite niece got twice as much as the least favorite niece. How much did the favorite niece get?

Critical Thinking

59. The temperature and the wind combine to cause body surfaces to lose heat. Meteorologists call this effect "the wind chill factor." For example, if the actual temperature outside is 10°F and the wind speed is 20 miles per hour, it will feel like it is −20°F outside, so −20°F is called the wind chill. When it is 25°F outside and the wind speed is 40 miles per hour, it will feel like it is −35°F outside. From the information given, write a linear equation for determining the wind chill using the actual temperature and the wind speed.

60. Suppose your roommate brags that he made \$250 in singles and fives one night waiting tables, and that he collected four times as many one-dollar bills as five-dollar bills. How can you tell that he's not telling the exact truth?

Section 6-4 Ratio, Proportion, and Variation

LEARNING OBJECTIVES

- ☐ 1. Write ratios in fraction form.
- ☐ 2. Solve proportions.
- ☐ 3. Solve real-world problems using proportions.
- ☐ 4. Solve real-world problems using direct variation.
- ☐ 5. Solve real-world problems using inverse variation.

The most obvious way to compare the sizes of two numbers is to subtract them. But is that the *best* way? Suppose you're comparing the cost of an item at two different stores, and you find that the item is a dollar more at Target than at Wal-Mart. If that item is a bottle of Coke, and the prices are \$1 and \$2, that dollar difference is significant. But if the item is a 50-inch flat-screen TV and the prices are \$1,201 and \$1,200, it's essentially the same price. If you *divide* the prices rather than subtract them, however, something interesting happens:

$$\text{Coke: } \frac{\$2}{\$1} = 2$$

$$\text{TV: } \frac{\$1{,}201}{\$1{,}200} = 1.0008$$

The point is that the most meaningful way to compare the sizes of two numbers is to divide them, forming what we called a *ratio* in Section 5-3.

Ratios

A **ratio** is a comparison of two quantities using division.

For example, in the 2008 presidential election, 52% of Independents voted for Obama and 44% voted for McCain, so we would say that the ratio of Independent Obama voters to Independent McCain voters was 52 to 44.

For two nonzero numbers, a and b, the **ratio of a to b** is written as $a{:}b$ (read a to b) or $\frac{a}{b}$.

Ratios can be written using a colon or a fraction as shown in the definition, but in mathematics, the fraction is used more often.

EXAMPLE 1 Writing Ratios

Math Note

To set up a correct ratio, whatever number comes first in the ratio statement should be placed in the numerator of the fraction and whatever number comes second in the ratio statement should be placed in the denominator of the fraction.

According to the Sporting Goods Manufacturers' Association, 95.1 million Americans participate in recreational swimming, 56.2 million Americans participate in recreational biking, 52.6 million Americans participate in bowling, and 44.5 million Americans participate in freshwater fishing. Find each:

(a) The ratio of recreational swimmers to recreational bikers
(b) The ratio of people who fish to people who bowl

SOLUTION

(a) $\dfrac{\text{Number of swimmers}}{\text{Number of bikers}} = \dfrac{95.1}{56.2}$

(b) $\dfrac{\text{Number of people who fish}}{\text{Number of people who bowl}} = \dfrac{44.5}{52.6}$

▼ Try This One 1

From 1969 through 1977, there were 24 teams in Major League Baseball, and only 4 made the playoffs each year. Now, there are 30 teams, and 8 make the playoffs. Find the ratio of teams making the playoffs to those not making the playoffs in 1969 and today.

Since ratios can be expressed as fractions, they can be simplified by reducing the fraction. For example, the ratio of 10 to 15 is written 10:15 or $\frac{10}{15}$ and the fraction $\frac{10}{15}$ can be reduced to $\frac{2}{3}$. So the ratio 10:15 is the same as 2:3.

Sometimes ratios can be written as fractions where the numerator and the denominator have the same units of measure. In such cases, it is necessary to make the units match as shown in Example 2.

EXAMPLE 2 Writing a Ratio Involving Units

Find the ratio of 18 inches to 2 feet.

SOLUTION

It's tempting to simply write $\frac{18}{2}$, but this is deceiving—it makes it seem like 18 inches is 9 times as much as 2 feet, which is of course silly. Instead, to make the ratio meaningful, we want the units to be the same. Since 1 foot is 12 inches, 2 feet is 24 inches. So the ratio is

$$\frac{18 \text{ inches}}{24 \text{ inches}}$$

Now we can reduce. The unit inches divides out, and we're left with

$$\frac{18}{24} = \frac{3 \cdot 6}{4 \cdot 6} = \frac{3}{4}$$

The ratio of 18 inches to 2 feet is $\frac{3}{4}$.

▼ Try This One 2

Find the ratio of 40 ounces to 2 pounds. (There are 16 ounces in 1 pound.)

☑ 1. Write ratios in fraction form.

Proportions

When two ratios are equal, they can be written as a *proportion*.

A **proportion** is a statement of equality of two ratios.

For example, the ratio of 4:7 and 8:14 can be written as a proportion as shown.

$$\frac{4}{7} = \frac{8}{14}$$

Two fractions, $\frac{a}{b}$ and $\frac{c}{d}$, are equal if $ad = bc$. (This will be shown in Exercise 53.) The product of the numerator of one fraction and the denominator of the other fraction is called a cross product. For example, $\frac{3}{4} = \frac{6}{8}$ since $3 \cdot 8 = 4 \cdot 6$, or $24 = 24$.

This tells us that two ratios form a proportion if the cross products of their numerators and denominators are equal. For example, the two ratios $\frac{5}{6}$ and $\frac{15}{18}$ can be written as a proportion since

$$\frac{5}{6} \times \frac{15}{18}$$ *This is called cross multiplying.*

$$5 \cdot 18 = 6 \cdot 15$$

$$90 = 90$$

So we can write 5:6 = 15:18, or $\frac{5}{6} = \frac{15}{18}$.

EXAMPLE 3 Deciding if a Proportion Is True

Decide if each proportion is true or false.

(a) $\frac{3}{5} = \frac{9}{15}$ (b) $\frac{5}{3} = \frac{7}{2}$ (c) $\frac{14}{16} = \frac{7}{8}$

SOLUTION

In each case, we will cross multiply and see if the two products are equal.

(a) $3 \cdot 15 = 45$; $5 \cdot 9 = 45$ The proportion is true.
(b) $5 \cdot 2 = 10$; $3 \cdot 7 = 21$ The proportion is false.
(c) $14 \cdot 8 = 112$; $16 \cdot 7 = 112$ The proportion is true.

▼ Try This One 3

Decide if each proportion is true or false.

(a) $\frac{2}{9} = \frac{6}{25}$ (b) $\frac{5}{2} = \frac{25}{4}$ (c) $\frac{11}{2} = \frac{55}{10}$

If there is an unknown value in a proportion, we can solve for the unknown value by cross multiplying as shown in Examples 4 and 5.

EXAMPLE 4 Solving a Proportion

Solve the proportion for x.

$$\frac{12}{48} = \frac{3}{x}$$

The height of the people and the statue are in proportion. If we know the ratio of the heights and the height of the people, we can find the height of the statue.

SOLUTION

We begin by cross multiplying.

$$\frac{12}{48} \times \frac{3}{x} \quad \textit{Cross multiply.}$$

$$12x = 3 \cdot 48$$

$$12x = 144 \quad \textit{Divide both sides by 12.}$$

$$\frac{12x}{12} = \frac{144}{12} \quad \textit{Simplify.}$$

$$x = 12$$

Check:

$$\frac{12}{48} \stackrel{?}{=} \frac{3}{12}$$

$$\frac{1}{4} \stackrel{?}{=} \frac{1}{4} \quad ✓$$

▼ Try This One 4

Solve the proportion: $\frac{x}{7} = \frac{22}{25}$

EXAMPLE 5 Solving a Proportion

Solve the proportion. $\frac{x-5}{10} = \frac{x+2}{20}$

SOLUTION

$$\frac{x-5}{10} \times \frac{x+2}{20} \quad \textit{Cross multiply.}$$

$$20(x-5) = 10(x+2) \quad \textit{Multiply out parentheses.}$$

$$20x - 100 = 10x + 20 \quad \textit{Subtract 10x from both sides.}$$

$$10x - 100 = 20 \quad \textit{Add 100 to both sides.}$$

$$10x = 120 \quad \textit{Divide both sides by 10.}$$

$$x = 12$$

Check:

$$\frac{x-5}{10} = \frac{x+2}{20}$$

$$\frac{12-5}{10} \stackrel{?}{=} \frac{12+2}{20}$$

$$\frac{7}{10} \stackrel{?}{=} \frac{14}{20}$$

$$\frac{7}{10} = \frac{7}{10} \quad ✓$$

▼ Try This One 5

Solve the proportion: $\frac{x+6}{15} = \frac{x-2}{5}$

☑ 2. Solve proportions.

Applications of Proportions

Proportions have been around for a really long time in one form or another (see Sidelight on page 313) because they're very useful in solving real-world problems. In the next example, we will illustrate a problem-solving procedure that works well for problems where ratios are provided.

EXAMPLE 6 Applying Proportions to Fuel Consumption

While on a spring break trip, a group of friends burns 12 gallons of gas in the first 228 miles, then stops to refuel. If they have 380 miles yet to drive, and the SUV has a 21-gallon tank, can they make it without refueling again?

SOLUTION

Step 1 *Identify the ratio statement.* The ratio the problem gives us is 12 gallons of gas to drive 228 miles.

Step 2 *Write the ratio as a fraction.* The ratio is $\frac{12 \text{ gallons}}{228 \text{ miles}}$.

Step 3 *Set up the proportion.* We need to find the number of gallons of gas needed to drive 380 miles, so we'll call that x. The ratio we already have is gallons compared to miles, so the second ratio in our proportion should be as well. We have x gallons, and 380 miles, so the proportion is

$$\frac{12 \text{ gallons}}{228 \text{ miles}} = \frac{x \text{ gallons}}{380 \text{ miles}}$$

When setting up a proportion, be sure to put like quantities in the numerators and like quantities in the denominators. In Example 6, gallons were placed in the numerators and miles in the denominators.

Step 4 Solve the proportion.

$$\frac{12}{228} = \frac{x}{380}$$

$$\frac{12}{228} \times \frac{x}{380}$$

$$228x = 12 \cdot 380$$

$$228x = 4{,}560$$

$$\frac{228x}{228} = \frac{4{,}560}{228}$$

$$x = 20$$

Step 5 *Answer the question.* The SUV will burn 20 gallons of gas to make the last 380 miles, so they can make it without stopping.

▼ Try This One 6

In 2009, roughly 13 of every 100 people in the United States were African-American. A marketing company wants to select a group of 250 people that accurately reflects the racial makeup of the country. How many African-Americans should be included?

In order to decide that a certain species is endangered, biologists have to know how many individuals are in a population. But how do they do that? It's actually an interesting application of proportions illustrated in Example 7.

EXAMPLE 7 Applying Proportions to Wildlife Population

As part of a research project, a biology class plans to estimate the number of fish living in a lake thought to be polluted. They catch a sample of 35 fish, tag them, and release them back into the lake. A week later, they catch 80 fish and find that 5 of them are tagged. About how many fish live in the lake?

SOLUTION

Step 1 *Identify the ratio statement.* Five of 80 fish caught were tagged.

Step 2 *Write the ratio as a fraction.* $\frac{5 \text{ tagged}}{80 \text{ total}}$

Step 3 *Set up the proportion.* We want to know the number of fish in the lake, so call that x. The comparison in the lake overall is $\frac{35 \text{ tagged}}{x \text{ total}}$, so the proportion is

$$\frac{5 \text{ tagged}}{80 \text{ total}} = \frac{35 \text{ tagged}}{x \text{ total}}$$

Step 4 *Solve the proportion.*

$$\frac{5}{80} = \frac{35}{x}$$

$$5x = 35 \cdot 80$$

$$5x = 2{,}800$$

$$x = \frac{2{,}800}{5} = 560$$

Step 5 *Answer the question.* There are approximately 560 fish in the lake.

▼ Try This One 7

☑ 3. Solve real-world problems using proportions.

A student can type 18 pages in 10 minutes. At the same rate, how long will it take the student to type 72 pages?

Sidelight RATIOS, PROPORTIONS, AND ARCHITECTURE

Architects often use the term "proportion" as a way to describe comparisons between the sizes of architectural features, like the length compared to the width of a window or a doorway. Some proportions are commonly found in nature, and others aren't. Some proportions just tend to be pleasing to most people, and some don't. This idea goes *at least* back to the ancient Egyptians, but it was taken to the extreme by the Greeks, who defined a particular ratio, now known as the Golden Ratio, which is approximately 1:1.618. They felt that this particular ratio was found often in nature, and they considered it to be the most pleasing to the eye. As a result, the Greeks used it extensively in architecture. When examining ancient Greek structures, archaeologists have found that almost every part of their buildings used the Golden Ratio in some way.

It turns out that the Golden Ratio is related to the Fibonacci sequence we looked at in a sidelight in Section 5-7. The ratio of any term in the Fibonacci sequence to the previous term approaches 1.618 as you go further and further out in the sequence.

The Golden Ratio is still widely used in modern architecture. If you've ever noticed that rectangular features of buildings often seem to be the same shape, it's because they are—a rectangle with its sides in the ratio 1:1.618.

Variation

Two quantities are often related in such a way that if one goes up, the other does too, and if one goes down, the other goes down as well. For example, if you have a job that pays \$95 a day, the amount you make goes up or down depending on how many days you work. This is an example of what is called **direct variation.** In this case, we can write a ratio statement based on the pay: $\frac{\$95}{1 \text{ day}}$. We could then use this to write an equation that describes your total pay depending on how many days you work: $y = \frac{\$95}{1 \text{ day}} \cdot x$ days, or just $y = 95x$.

A quantity y is said to **vary directly** with x if there is some nonzero constant k so that $y = kx$. The constant k is called the **constant of proportionality.**

EXAMPLE 8 Using Direct Variation to Find Wages

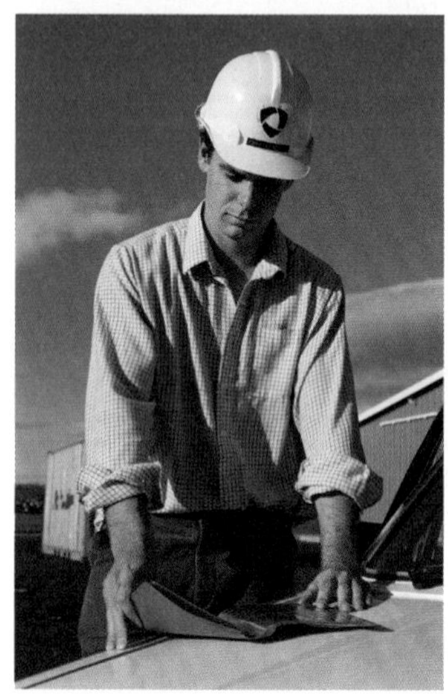

Suppose you earn \$95 per day. Write a variation equation that describes total pay in terms of days worked, and use it to find your total pay if you work 6 days and if you work 15 days.

SOLUTION

Let y = the total amount earned
x = the number of days you work
k = \$95 per day (as we saw above)

Then $y = 95x$ is the variation equation.
For $x = 6$ days: $y = 95 \cdot 6 = \$570$.
For $x = 15$ days: $y = 95 \cdot 15 = \$1{,}425$.

▼ Try This One 8

A 6-month-old Labrador puppy gets $4\frac{1}{2}$ cups of food per day. Write a variation equation that describes how many cups of food she eats in terms of days, then use it to find how much she eats in 6 days and in 2 weeks.

Sometimes it is possible to find the value of k based on certain information and then solve the problem. The next example shows this.

EXAMPLE 9 Using Direct Variation to Find a Weight

The weight of a certain type of cable varies directly with its length. If 20 feet of cable weighs 4 pounds, find k and determine the weight of 75 feet of cable.

SOLUTION

Step 1 Write the equation of variation.

$y = kx$ where y = the weight
x = length of cable in feet
k = the constant

$$4 \text{ lb} = k \cdot 20 \text{ ft}$$

Step 2 Solve for k.

$$4 \text{ lb} = k \cdot 20 \text{ ft}$$
$$\frac{4 \text{ lb}}{20 \text{ ft}} = \frac{k \cdot \cancel{20 \text{ ft}}}{\cancel{20 \text{ ft}}}$$
$$k = 0.2 \text{ lb/ft}$$

Now we know that the equation of variation can be written as $y = 0.2x$.

Step 3 Solve the problem for the new values of x and y using $k = 0.2$.

$$y = 0.2x \qquad \text{Substitute } x = 75$$
$$y = 0.2 \cdot 75$$
$$y = 15 \text{ pounds}$$

So 75 feet of cable will weigh 15 pounds.

▼ Try This One 9

☑ 4. Solve real-world problems using direct variation.

The weight (in pounds) of a hollow statue varies directly with the square of its height (in feet); i.e., $y = kx^2$, where y = the weight and x = the height. If a statue that is 4 feet tall weighs 2 pounds, find the weight of a statue that is 6 feet tall.

A quantity can also vary *inversely* with another quantity. For example, the time it takes to drive a certain distance to a vacation home varies inversely with the rate of speed of the automobile. That is, if a person drives on average 55 miles per hour as opposed to 40 miles per hour, the person will get there in less time. The higher the speed, the lower the time. This is called **inverse variation.**

A quantity y is said to **vary inversely** with x if there is some nonzero constant k such that $y = \frac{k}{x}$.

EXAMPLE 10 Using Inverse Variation to Find Driving Time

The time it takes to drive a certain distance varies inversely with the speed, and the constant of proportionality is the distance. A family has a vacation cabin that is 378 miles from their residence. Write a variation equation describing driving time in terms of speed. Then use it to find the time it takes to drive that distance if they take the freeway and average 60 miles per hour, and if they take the scenic route and average 35 miles per hour.

SOLUTION

Let y = the time it takes to drive the distance
x = the average speed
k = 378 miles (the distance)

Then the variation equation is $y = \frac{k}{x}$ or $y = \frac{378}{x}$.

If they average 60 miles per hour:

$$y = \frac{378}{60} = 6.3 \text{ hours}$$

If they average 35 miles per hour:

$$y = \frac{378}{35} = 10.8 \text{ hours}$$

I vote for the freeway.

▼ Try This One 10

A student lives 120 miles from the college she attends. Write a variation equation describing the time it takes to drive home for a weekend in terms of speed. Then use it to find driving time if they average 72 miles per hour and manage to not get pulled over by the state patrol.

EXAMPLE 11 Applying Inverse Variation to Construction

In construction, the strength of a support beam varies inversely with the cube of its length. If a 12-foot beam can support 1,800 pounds, how many pounds can a 15-foot beam support?

SOLUTION

Step 1 *Write the variation equation.* Let

y = strength of the beam in pounds it can support
x = length of the beam
k = the constant of proportionality

The variation equation is $y = \frac{k}{x^3}$, since y varies inversely with the cube of x.

Step 2 Find k.

$$y = \frac{k}{x^3} \quad \text{Substitute } y = 1{,}800 \text{ and } x = 12$$

$$1{,}800 = \frac{k}{12^3}$$

$$1{,}800 = \frac{k}{1{,}728}$$

$$k = 1{,}800 \cdot 1{,}728 = 3{,}110{,}400$$

Step 3 *Substitute in the given value for x.* In this case, it is 15.

$$y = \frac{3{,}110{,}400}{x^3}$$

$$y = \frac{3{,}110{,}400}{15^3} = 921.6$$

A 15-foot beam can support 921.6 pounds.

☑ 5. Solve real-world problems using inverse variation.

▼ Try This One 11

If the temperature of a gas is held constant, the pressure the gas exerts on a container varies inversely with its volume. If a gas has a volume of 38 cubic inches and exerts a pressure of 8 pounds per square inch, find the volume when the pressure is 64 pounds per square inch.

Answers to Try This One

1 In 1969: $\frac{4}{20}$. Today: $\frac{8}{22}$.

2 $\frac{5}{4}$

3 (a) False (b) False (c) True

4 $x = \frac{154}{25}$

5 $x = 6$

6 Either 32 or 33

7 40 minutes

8 $y = \frac{9}{2}x$. In 6 days: 27 cups. In 2 weeks: 63 cups.

9 4.5 lb

10 $y = \frac{120}{x}$; about 1.7 hours

11 4.75 cubic inches

EXERCISE SET 6-4

Writing Exercises

1. Write a real-world example of a ratio.
2. Write a real-world example of two quantities that vary directly.
3. Write a real-world example of two quantities that vary inversely.
4. Describe the procedure for solving a proportion.

Computational Exercises

For Exercises 5–14, write each ratio statement as a fraction and reduce to lowest terms if possible.

5. 18 to 28
6. 5 to 12
7. 14:32
8. 40:75
9. 12 cents to 15 cents
10. 18 inches to 42 inches
11. 3 weeks to 8 weeks
12. 2 pounds to 12 ounces
13. 5 feet to 30 inches
14. 12 years to 2 decades

For Exercises 15–24, solve each proportion.

15. $\frac{3}{x} = \frac{14}{45}$
16. $\frac{x}{2} = \frac{18}{6}$
17. $\frac{5}{6} = \frac{x}{42}$
18. $\frac{9}{8} = \frac{45}{x}$
19. $\frac{x-6}{12} = \frac{1}{3}$
20. $\frac{x+3}{5} = \frac{35}{25}$
21. $\frac{2}{x-3} = \frac{5}{x+8}$
22. $\frac{4}{x-3} = \frac{16}{x-2}$
23. $\frac{x-3}{4} = \frac{x+6}{20}$
24. $\frac{x}{10} = \frac{x-2}{20}$

Real-World Applications

25. The Information Resources Institute reports that one out of every five people who buy ice cream buys vanilla ice cream. If a store sells 75 ice cream cones in one day, about how many will be vanilla?
26. The U.S. Department of Agriculture reported that 57 out of every 100 milk drinkers drink skim milk. If a storeowner orders 25 gallons of milk, how many should be skim?
27. Under normal conditions, 1.5 feet of snow will melt into 2 inches of water. After a recent snowstorm, there were 3.5 feet of snow. How many inches of water will there be when the snow melts?
28. The Travel Industry Association of America reports that 4 out of every 35 people who travel do so by air. If there are 180 students who are traveling for spring break, how many of them will fly?
29. A gallon of paint will cover 640 square feet of wall space. If a person has to paint a room whose walls measure 2,560 square feet, how many gallons of paint will the person need?

30. The American Dietetic Association reported that 31 out of every 100 people want to lose weight. If 384 students were surveyed at random in the student union, how many would want to lose weight?
31. The U.S. Census Bureau reported that 9 out of every 20 joggers are female. On a trail, there were 220 joggers on July 4. Approximately how many were female?
32. If a person drives 4,000 miles in 8 months, how many miles will that person drive every 2 years?
33. Out of every 50 iPods sold by a discount website, 2 were returned as defective. If the website sold 1,000 iPods this holiday season, how many should they expect will be returned as defective?
34. The American Dietetic Association states that 11 out of every 25 people do not eat breakfast. If there are 175 students in a large lecture hall, about how many of them did not eat breakfast?
35. At South Campus, the student-faculty ratio is about 16 to 1. If 128 students enroll in Statistics 101, how many sections should be offered?
36. If a 10-foot pole casts a shadow of 4 feet, how tall is a tree whose shadow is 7 feet?
37. A small college has 1,200 students and 80 professors. The college is planning to increase enrollment to 1,500 students next year. How many new professors should be hired, assuming they want to maintain the same ratio?
38. The taxes on a house assessed at $64,000 are $1,600 a year. If the assessment is raised to $80,000 and the tax rate did not change, how much would the taxes be now?
39. If five small cans of paint can cover 20 square feet of wood siding, how many cans of paint should be purchased to cover $13\frac{1}{3}$ square feet of siding?
40. If you need a minimum of 27 correct out of 30 to get an A on a test, how many correct answers would you need to get an A on an 80-point test assuming that the same ratio is used for scoring purposes?
41. The amount of simple interest on a specific amount of money varies directly with the time the money is kept in a savings account when the interest rate is constant. Find the amount of interest on a $5,000 savings account, if the interest rate is 6%, and the money has been invested for 4 years.
42. The number of tickets purchased for a prize varies directly with the amount of the prize. For a prize of $1,000, 250 tickets are purchased. Find the approximate number of tickets that will be purchased on a prize worth $5,000.

Use the following information for Exercises 43–46. If everyone had the same body proportions, your weight in pounds would vary directly with the cube of your height in feet. According to Wikipedia, the most recent statistics available in 2009 indicated that the average height and weight for an adult male in the United States is 5 feet 9.4 inches and 191 lb. Use this information to write a variation equation, then use it to find the weight that each of the following famous athletes would be if they had the same body type as the average male.

43. Basketball player Lebron James: 6′8″ (Actual weight is 260 lb.)
44. Basketball player Yao Ming: 7′6″ (Actual weight is 310 lb.)
45. Jockey Pat Day: 4′11″ (Actual weight is 105 lb.)
46. Football player Tom Brady: 6′4″ (Actual weight is 225 lb.)
47. Under certain conditions, the pressure of a gas varies inversely with its volume. If a gas with a volume of 20 cubic inches has 36 pounds of pressure, find the amount of pressure 30 cubic inches of gas is under.
48. The strength of a beam varies inversely with the square of its length. If a 10-foot beam can support 500 pounds, how many pounds can a 12-foot beam support?
49. In karate, the force needed to break a board varies inversely with the length of the board. If it takes 5 lb of force to break a board that is 3 feet long, how many pounds of force will it take to break a board that is 5 feet long?
50. The weight of a body varies inversely with the square of the distance from the center of the earth. If the radius of the earth is 4,000 miles, how much would a 150-lb woman weigh 500 miles above the surface of the earth?
51. The time to complete a project is inversely proportional to the number of people who are working on the project. A class project can be completed by 3 students in 20 days. In order to finish the project in 5 days, how many more students should the group add?
52. The intensity of sound varies inversely as the square of the distance from the source. A sound with an intensity of 300 decibels is heard from 2 feet away from a speaker. What is the intensity of the sound 5 feet away from the same speaker?

Critical Thinking

53. Starting with the equation $\frac{a}{b} = \frac{c}{d}$, find the LCD and multiply both sides by it. Make sure you simplify fractions. This proves something that's very important in working with proportions. What is it?
54. Write the generic variation equations for direct and inverse proportionality. Now fill in the blanks in the next two sentences, and explain how you can deduce your answer from the equations.

When a quantity goes up in proportion to another going up, this is ________________ variation. When a quantity goes down in proportion to another going up, this is ___________________ variation.

Stores are required by law to display unit prices. A unit price is the ratio of the total price to the number of units. For example, if a 40-pound bag of sand costs \$5.00, then the unit price per pound would be

$$\begin{aligned} \textit{Unit price} &= \frac{\textit{Price}}{\textit{Number of pounds}} \\ &= \frac{\$5.00}{\textit{40 pounds}} \\ &= \$0.125 \end{aligned}$$

or \$0.12$\frac{1}{2}$ per pound.

For Exercises 55–59, find the unit price and then decide which is a better buy. These prices were obtained from actual foods.

55. Flour: 10 pounds for \$3.39 or 25 pounds for \$7.49
56. Candy: 20 ounces for \$1.50 or 24 ounces for \$1.75
57. Potato sticks: 7 ounces for \$1.99 or 1.5 ounces for \$0.50
58. Cookies: 7 ounces for \$0.99 or 14 ounces for \$1.50
59. Coffee: 11.5 ounces for \$2.75 or 34.5 ounces for \$7.49

Section 6-5 Solving Linear Inequalities

LEARNING OBJECTIVES

- ❑ 1. Graph solution sets for simple inequalities.
- ❑ 2. Solve linear inequalities in one variable.
- ❑ 3. Solve three-part linear inequalities.
- ❑ 4. Solve real-world problems using inequalities.

We have seen that solving linear equations is a very useful tool in solving problems from a wide variety of areas. But think about the following situation: you've finished school and are looking for a real job. As you nervously wait for your big interview, you decide not to be too demanding, but that the minimum compensation you're willing to accept in salary and benefits is \$40,000 per year. Nobody has ever gone into an interview and said "I want 40k and I won't accept a penny more!" Of course, you would be perfectly happy with any amount over \$40,000 as well.

This is where inequalities can be more useful than equations. Rather than describing your salary requirement as $S = 40{,}000$, it would be more sensible to describe it as $S \geq 40{,}000$. Inequalities are custom-built to describe situations where a range of possible outcomes is acceptable. The main goal of this section is to learn how to solve linear inequalities. But first, we should get comfortable with the terminology and the notation that we'll be using.

Recall that a linear equation in one variable was defined to be an equation of the form $Ax + B = 0$. The statement $3x + 4 < 0$ is an example of a linear inequality in one variable.

A **linear inequality in one variable** is a statement that can be written in any of the following four forms: $Ax + B < 0$, $Ax + B > 0$, $Ax + B \leq 0$, or $Ax + B \geq 0$. In each case, B can be any real number, and A can be any real number except zero.

Just like equations, the simple way to recognize a linear inequality is that the variable appears only to the first power. To **solve** a linear inequality means to find the set of all numbers that make the inequality a true statement when substituted in for the variable. That set is called the **solution set** for the inequality. Finding the graphs of solution sets to very basic inequalities will help us become familiar with the concept of inequalities.

Graphing Inequalities on a Number Line

For an inequality of the form $x \geq 5$, the solution set is obvious: every real number that's 5 or larger. We can represent this solution set on a number line like this:

To show that the number 5 is included in the solution set, a *solid* or *closed circle* (•) is used. On the other hand, the solution set for the inequality $x > 5$ includes all real numbers greater than 5; however, the actual number 5 is not in the solution set. The graph for the solution set of $x > 5$ looks like this:

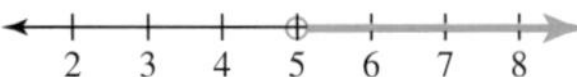

To show that the number 5 is not included in the solution set, an *open circle* (○) is used.

If the inequality is of the form $x \leq 5$, the solution set is every real number that's 5 or smaller, represented like this:

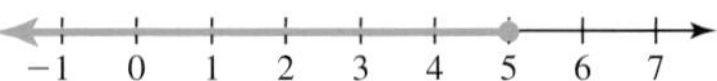

Again, the closed circle indicates that 5 is included. For $x < 5$, an open circle indicates that only the numbers to the left of 5 on the number line are in the solution set:

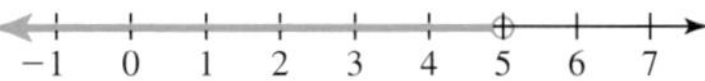

If the set you want to describe has only one boundary, like the set of salaries that are $40,000 or greater, inequalities like these are fine. But what if the range of numbers has a boundary on each end, like "I plan to work between two and four evenings a week to earn some spending money"? We could write this as two separate inequalities, $x \geq 2$ *and* $x \leq 4$, but it's more efficient to write it this way: $2 \leq x \leq 4$. This is the way we'll represent an interval between two numbers. The graph for this solution set looks like:

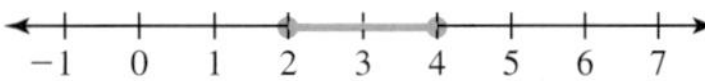

A summary of simple inequalities and the graphs of their solution sets is shown in Table 6-4.

TABLE 6-4 **Graphs of Solution Sets for Linear Inequalities**

For any real numbers *a* and *b*

Inequality	Graph	Solution set
$x \geq a$		$\{x \mid x \geq a\}$
$x > a$		$\{x \mid x > a\}$
$x \leq a$		$\{x \mid x \leq a\}$
$x < a$		$\{x \mid x < a\}$
$a \leq x \leq b$		$\{x \mid a \leq x \leq b\}$
$a < x < b$		$\{x \mid a < x < b\}$
$a < x \leq b$		$\{x \mid a < x \leq b\}$
$a \leq x < b$		$\{x \mid a \leq a < b\}$

EXAMPLE 1 Graphing Solution Sets for Simple Inequalities

Graph the solution set for each inequality.

(a) $x \leq 10$ (b) $y > -4$ (c) $-30 < x \leq 50$

SOLUTIONS

(a) −2 0 2 4 6 8 10 12 14

(b) −6 −5 −4 −3 −2 −1 0 1 2

(c) −40 −30 −20 −10 0 10 20 30 40 50 60

▼ Try This One 1

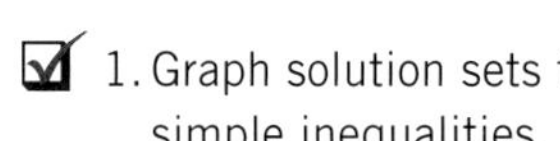

1. Graph solution sets for simple inequalities.

Graph the solution set for each inequality.

(a) $y < -2$ (b) $x \geq 20$ (c) $-9 \leq t < -4$

Solving Linear Inequalities in One Variable

Since we're now really good at solving linear equations in one variable, the obvious question is "Can we solve inequalities the same way?" The answer, sadly, is no. But all is not lost—a better answer might be "almost." We will be able to use the technique we used for solving linear equations, with one important difference.

Our procedure for solving equations was based on the addition, subtraction, multiplication, and division properties of equality. So we begin with similar properties for inequalities.

The Addition and Subtraction Properties for Inequalities

You can add or subtract the same real number or algebraic expression to both sides of an inequality without changing the solution set. In symbols, if $a < b$, then $a + c < b + c$, $a - c < b - c$, and the same is true for the inequality symbols $\leq$, $>$, and $\geq$.

This is the analog of half of what we needed to solve linear equations. The other half was the multiplication and division properties of equality, and that's where things get interesting.

Consider the inequality $3 < 5$, which is most definitely a true statement. If we multiply both sides by some positive number, like 4, we get $4 \cdot 3 < 5 \cdot 3$, or $12 < 15$. This is still true—so far, so good. But if we multiply by *negative* 4, we get $-4 \cdot 3 < -4 \cdot 5$, or $-12 < -20$. This is completely false. But if we *reverse* the inequality symbol, we get $-12 > -20$, which *is* true. This hints at the multiplication and division properties we need.

The Multiplication and Division Properties for Inequalities

- If you multiply or divide both sides of an inequality by the same *positive* real number, it does not change the solution set. In symbols, if $a < b$ and $c > 0$, then $a \cdot c < b \cdot c$, $\frac{a}{c} < \frac{b}{c}$, and the same is true for the inequality symbols $\leq$, $>$, and $\geq$.
- If you multiply or divide both sides of an inequality by the same *negative* real number, the direction of the inequality symbol is reversed. In symbols, if $a < b$ and $c < 0$, then $a \cdot c > b \cdot c$, $\frac{a}{c} > \frac{b}{c}$, and the analogous result is true for the inequality symbols $\leq$, $>$, and $\geq$.

The bottom line is that this gives us a simple procedure for solving linear inequalities:

Math Note

Reversing an inequality means that $\geq$ becomes $\leq$, $>$ becomes $<$, $\leq$ becomes $\geq$, and $<$ becomes $>$.

Procedure for Solving Linear Inequalities

To solve a linear inequality, proceed in the same way you solve a linear equation except that when multiplying or dividing by a *negative* number, you have to reverse the inequality symbol.

EXAMPLE 2 Solving a Linear Inequality

Solve and graph the solution set for $5x - 9 \geq 21$.

SOLUTION

$$5x - 9 \geq 21 \quad \text{Add 9 to both sides.}$$
$$5x - 9 + 9 \geq 21 + 9$$
$$5x \geq 30 \quad \text{Divide both sides by 5.}$$
$$\frac{5x}{5} \geq \frac{30}{5}$$
$$x \geq 6$$

The solution set is $\{x \mid x \geq 6\}$. The graph of the solution set is

4 5 6 7

Math Note

An inequality can't be checked for the exact answer like an equation can; however, an approximate check can be made by selecting some number in the solution set for x, substituting it into the inequality and seeing if the inequality is true. In Example 2, try $x = 8$. (You can choose any value for x as long as it is 6 or larger.)

$$5x - 9 \geq 21$$
$$5(8) - 9 \overset{?}{\geq} 21$$
$$40 - 9 \overset{?}{\geq} 21$$
$$31 \geq 21 \quad \text{True}$$

▼ Try This One 2

Solve and graph the solution set for $7x + 29 \geq 1$.

EXAMPLE 3 Solving a Linear Inequality

Solve and graph the solution set for $16 - 3x > 40$.

SOLUTION

$$16 - 3x > 40 \quad \text{Subtract 16 from both sides.}$$
$$16 - 16 - 3x > 40 - 16$$
$$-3x > 24 \quad \text{Divide both sides by -3 and reverse the inequality sign.}$$
$$\frac{-3x}{-3} < \frac{24}{-3}$$
$$x < -8$$

The solution set is $\{x \mid x < -8\}$ and the graph of the solution set is

−9 −8 −7 −6

▼ Try This One 3

Solve and graph the solution set for $14 - 7x < 56$.

Checking answers is always a good idea, but it's an especially good idea to check your answer when you reverse the inequality sign. Also, be careful about the number you choose to test. For example, if your answer is $x \le 8$, testing $x = 8$ would be a bad choice because it will make the inequality true even if $x \ge 8$ is the right solution set! Checking with $x = 6$ would be a much better choice.

EXAMPLE 4 Solving a Linear Inequality

Solve and graph the solution set for $4(x + 3) < 2x - 26$.

SOLUTION

$$4(x + 3) < 2x - 26 \quad \textit{Multiply out parentheses.}$$
$$4x + 12 < 2x - 26 \quad \textit{Subtract 2x from both sides.}$$
$$4x - 2x + 12 < 2x - 2x - 26$$
$$2x + 12 < -26 \quad \textit{Subtract 12 from both sides.}$$
$$2x + 12 - 12 < -26 - 12$$
$$2x < -38 \quad \textit{Divide both sides by 2.}$$
$$\frac{2x}{2} < \frac{-38}{2}$$
$$x < -19$$

The solution set is $\{x \mid x < -19\}$ and the graph is

−20 −19 −18

▼ Try This One 4

☑ 2. Solve linear inequalities in one variable.

Solve and graph the solution set for $6(2x - 7) \le 3x + 8$.

When solving a three-part inequality, the goal is to isolate the variable in the middle. Make sure that whatever you do to one part of the inequality, you do to all three.

EXAMPLE 5 Solving a Three-Part Linear Inequality

Solve and graph the solution set for $-4 < 3 - 2y \le 9$.

SOLUTION

$$-4 < 3 - 2y \le 9 \quad \textit{Subtract 3 from all three parts.}$$
$$-4 - 3 < 3 - 2y - 3 \le 9 - 3$$
$$-7 < -2y \le 6 \quad \textit{Divide all three parts by −2; reverse both inequality symbols.}$$
$$\frac{-7}{-2} > y \ge \frac{6}{-2}$$
$$\frac{7}{2} > y \ge -3 \quad \text{or} \quad -3 \le y < \frac{7}{2}$$

The solution set is $\left\{y \mid -3 \le y < \frac{7}{2}\right\}$ and the graph is

−4 −3 −2 −1 0 1 2 3 $\frac{7}{2}$ 4

▼ Try This One 5

✓ 3. Solve three-part linear inequalities.

Solve and graph the solution set for $-10 \le 8 - x < 4$.

Applications of Inequalities

As we indicated at the beginning of the chapter, there are many situations where we are interested in some quantity being at least a certain amount, or at most a certain amount. In such situations, a couple of which are illustrated in Examples 6 and 7, inequalities are a natural fit. We'll proceed using the same problem-solving steps we used in Section 6-3. Table 6-5 is a handy reference for translating common phrases into inequality form.

TABLE 6-5 Common Phrases Used in Inequality Word Problems

>	<
Greater than	Less than
Above	Below
Higher than	Lower than
Longer than	Shorter than
Larger than	Smaller than
Increased	Decreased
≥	**≤**
Greater than or equal to	Less than or equal to
At least	Is at most
Not less than	Not more than

EXAMPLE 6 Applying Inequalities to Vacation Planning

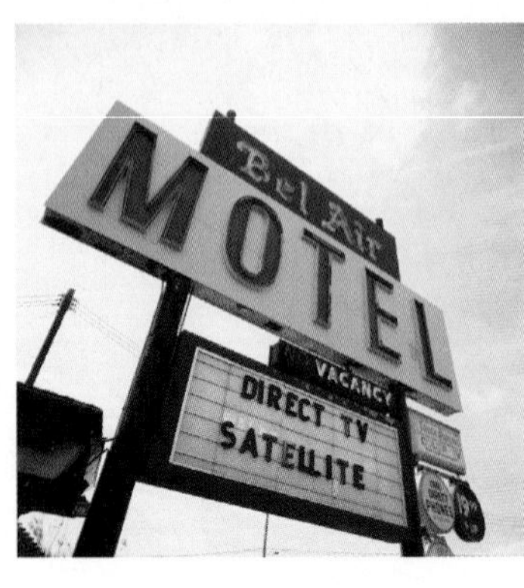

With the stress of finals behind you, you decide to plan a vacation to relax a little bit. After poking around on the Internet, you find a room in the area you want to visit for \$65 per night. Some quick estimating leads you to conclude that you'll need at least \$250 for gas, food, beverages, and entertainment expenses. Upon checking your bank balance, you decide that you can afford to spend at most \$600 on the trip. How many nights can you stay?

SOLUTION

Step 1 Relevant information: Lodging is \$65 per night, other expenses are \$250, maximum you can spend is \$600. We're asked to find the number of nights.

Step 2 Use variable n to represent the number of nights.

Step 3 Translate the relevant information into an inequality.

\$65 × number of nights + other expenses is no more than \$600

$$65 \times \quad n \quad + \quad 250 \quad \le \quad 600$$

Step 4 Solve the inequality.

$$65n + 250 \le 600$$
$$65n \le 350$$
$$n \le \frac{350}{65} \approx 5.4$$

Step 5 Answer the question. Staying 5.4 nights doesn't make sense, so you could stay at most 5 nights.

Step 6 Check: Will 5 nights work?

$$65 \cdot 5 + 250 \stackrel{?}{\le} 600$$
$$325 + 250 \stackrel{?}{\le} 600$$
$$575 \le 600 \quad \textit{True}$$

▼ Try This One 6

Philip doesn't want to look like a cheapskate, so he plans to spend at least \$200 on his girlfriend for Christmas. He's already bought her a necklace for \$135, and plans to buy some \$15 iTunes gift cards as well. What's the smallest number he can buy?

EXAMPLE 7 Applying Inequalities to the Cost of Buying Food

Mike is planning to buy lunch for himself and some coworkers. He decides to buy cheeseburgers and fries from the value menu—the burgers are \$1 each, and the fries cost \$0.80. He also needs to pay 5% of the total in sales tax. What is the largest number of items he can buy if he wants to buy the same number of burgers as fries, and he only has \$10 to spend?

SOLUTION

Step 1 Relevant information: Burgers are \$1 each, fries are \$0.80 each; 5% sales tax; same number of burgers and fries; maximum cost \$10.

Step 2 We can use variable x to represent both the number of burgers and the number of fries because those numbers are equal.

Step 3 $\$1 \cdot x$ is the total cost of burgers, and $\$0.80x$ is the total cost of fries. The tax is 5% of their sum, which is $0.05(x + 0.80x)$ dollars. That makes the total cost $x + 0.80x + 0.05(x + 0.80x)$, which must be less than or equal to 10. The inequality is

$$\underset{\text{Cost of burgers}}{x} + \underset{\text{Cost of fries}}{0.80x} + \underset{\text{Tax}}{0.05(x + 0.80x)} \le 10$$

Step 4

$$x + 0.80x + 0.05(x + 0.80x) \le 10$$
$$x + 0.80x + 0.05x + 0.04x \le 10$$
$$1.89x \le 10$$
$$x \le \frac{10}{1.89} \approx 5.29$$

Step 5 Mike can't buy 5.29 burgers, so we round down to 5. He can buy at most 5 burgers and 5 fries.

4. Solve real-world problems using inequalities.

Step 6 Use $x = 5$ to check:

5 hamburgers cost $5 \times \$1.00 =$	\$5.00
5 fries cost $5 \times \$0.80 =$	+ 4.00
Total cost of food	9.00
Tax: $0.05 \times 9.00 = 0.45$	+ 0.45
Total bill	9.45

▼ Try This One 7

Maureen plans to buy a computer and monitor. She wants to spend at most \$1,000. She finds that there is a 10% rebate on all computers purchased at a certain store. If the monitor costs \$100 (no rebate), what is the most she can pay for the computer? (Ignore the sales tax.)

Answers to Try This One

1 (a) −7 −6 −5 −4 −3 −2 −1 0 1

(b) −10 0 10 20 30 40 50 60 70

(c) −10 −9 −8 −7 −6 −5 −4 −3 −2

2 $\{x \mid x \geq -4\}$; −6 −5 −4 −3 −2 −1 0 1 2

3 $\{x \mid x > -6\}$ −6 0

4 $\left\{x \mid x \leq 5\frac{5}{9}\right\}$ 4 5 $5\frac{5}{9}$ 6

5 $\{x \mid 4 < x \leq 18\}$; 4 6 8 10 12 14 16 18 20

6 4

7 \$1,000

EXERCISE SET 6-5

Writing Exercises

1. What is the difference between the solution set of an inequality and the solution set of an equation?
2. What's the biggest difference between solving linear inequalities and solving linear equations?
3. Write a real-world situation where an inequality would be useful.
4. What does it mean to graph the solution set of an inequality?

Computational Exercises

For Exercises 5–14, show the solutions using a graph.

5. $x \geq 3$
6. $x < -2$
7. $y < -4$
8. $y \geq 0$
9. $x \leq -9$
10. $x \leq 1$
11. $-3 < t < 7$
12. $4 \leq t \leq 10$
13. $2 < x \leq 5$
14. $-3 \leq x < 0$

For Exercises 15–56, solve each inequality and graph the solution set on a number line.

15. $x + 6 < 11$
16. $x - 2 \leq 15$
17. $x - 7 \leq 23$
18. $x + 9 > 20$
19. $3y \geq 18$
20. $5y < 30$
21. $7 - x > 42$
22. $9 - x \leq 20$

23. $\frac{2}{3}x < 18$
24. $\frac{3}{4}x \geq 36$
25. $-10t < 30$
26. $-25t \geq 100$
27. $2z + 8 \geq 32$
28. $5z - 6 < 39$
29. $-3x + 12 \leq 36$
30. $5 - 2x > 25$
31. $6(x - 12) \leq 54$
32. $-3(2x + 7) < -16$
33. $-5(3 - y) \leq 27$
34. $9(4y - 1) > 71$
35. $16 - 5x \geq 22$
36. $5 - 3x < 25$
37. $3(x + 1) - 10 < 2x + 7$
38. $4(n - 8) - 2n < -22$
39. $9 - 5(n + 6) \geq 32$
40. $18 - 6(x + 2) < 41$
41. $6(2x + 3) \geq 5(2x - 15)$
42. $-x \geq -15$
43. $6x - 7 \geq 5(x - 2) - 17$
44. $3x + 6 < -8x + 7$
45. $5.4 - 3.1x \geq 8 + 5.9x$
46. $2z - 11.9 < 6.3z - 4.2$
47. $-3(1.5t + 4.1) + 3 \leq 2t - 4.7$
48. $9.8 - 7(6.5x + 4) > 3.2 - x$
49. $3 \leq x + 10 < 18$
50. $-5 < 11 + y \leq -1$
51. $-8 < 3y + 1 < 10$
52. $3 \leq 2x - 7 \leq 11$
53. $-5 < -4z \leq 20$
54. $3 \leq -2x < 4$
55. $0 < 5(3 - 2x) < 25$
56. $-30 \leq -2(2t + 4) \leq 44$

Real-World Applications

The list shown here represents the average number of tornados per year for the given states. Let x represent the average number of tornados per year. For Exercises 57–62, write the name or names of the states that are described by the solution to the inequality.

State	Average number of tornados per year
Texas	168
Florida	79
Kansas	75
Colorado	58
West Virginia	30
Missouri	26
Pennsylvania	22
California	14
Maryland	11
Arizona	5
Delaware	2
Maine	1

Source: *USA TODAY*

57. $x \geq 58$
58. $x < 22$
59. $26 \leq x < 58$
60. $2 < x \leq 22$
61. $11 \leq x \leq 30$
62. $30 < x < 168$
63. Mary plans to purchase a used car. She wants to spend at most \$8,000. The sales tax rate in her state is 7%. Title and license plate fee is \$120. What is the maximum amount she can spend for a car?
64. Bill has three test grades of 95, 84, and 85 so far. If the final examination, still to come, counts for two test scores, what is the lowest he can score on the final exam and still get an A for the course? He needs at least 450 points for an A.
65. In order to get a C for her sociology course, Betsy needs at least a 70% average. On exam 1 she scored 78% and on exam 2 she scored 68%. What is the lowest score she can get on the last exam?
66. A husband and wife want to sell their house and make at least a 10% profit. The real estate agent's commission is 7% and closing costs are \$1,000. If they paid \$150,000 for their home, what is the minimum price they should ask for their house?
67. Kiki has up to \$100 to purchase five prizes for the raffle for the Spring Fling Dance. The sales tax is 6% and each prize should cost the same amount. What is the most she can spend on each prize?
68. In Beth's Greek mythology course, quizzes are worth 25% of the grade, tests are worth 60% of the grade, and the final exam is worth 15% of the grade. If Beth has an 82% quiz average and a 76% test average, what grade will she need on the final exam to get a B in the course with at least an 80% overall average?
69. When renting a car for her trip, Janice has the option of paying \$0.32 a mile plus putting \$25 down (option A) or she can pay \$0.40 a mile (option B). What is the least amount of miles Janice can drive to make option A the cheaper option?
70. Farrah has a \$20 bill and needs to buy three birthday cards for her friends. With the leftover money, she'd like to be able to buy as many holiday cards as possible. If each birthday card costs \$1.99 and each holiday card costs \$0.59, how many holiday cards can she purchase? Neglect sales tax.
71. Alana's new cell phone plan has a monthly access fee of \$18.83 with a flat rate of \$0.033 per minute. If she wants to keep her monthly bill under \$50, what is the maximum number of minutes she can talk on the phone each month?
72. Dave works at the mall part-time for \$9 per hour and notices that 20% of his check is deducted for taxes. If he wants to take home at least \$200 per week, how many hours does he have to work?
73. Shelly has saved \$10,000 to put down on her first house. She wants to make a down payment of 8% of the purchase price of the house and she was told it would cost 5% of the purchase price for closing costs and legal fees. What is the maximum price of house she can afford?
74. Amy spends \$300 a month on bills, \$600 a month on food and gas, \$350 for her car, and \$125 for her car insurance. If she makes \$2,250 a month but wants to put

10% of the remaining money into her savings account, what is the most she can spend on her monthly rent?

75. Dawn and Monica want to build a set of shelves for their dorm room. The width of the shelves needs to be 4 times the height of the shelves, and the set of shelves must have four horizontal pieces and two vertical pieces. They can get up to 72 feet of wood to make the entire shelving unit. What is the tallest height they can make the shelving unit?
76. Hideo can select from two salary plans. Plan A will pay him \$20 per hour plus a \$100 weekly bonus and Plan B will pay him \$25 per hour. How many hours should Hideo work so that he earns the most from Plan B?
77. On his first four tests, Alon got 65%, 72%, 85%, and 90%. He wants a B in the course, which would mean his overall average would have to be between 80% and 89%. What range of scores can Alon get on his last test so he can get a B in the course?
78. On Molly's new diet, she is supposed to eat a lunch that is between 300 and 500 calories. She selects a yogurt for 100 calories and a half a bagel for 135 calories. How many calories could a third item contain so she remains on her diet?
79. Adrian spends a third of his week sleeping, 10 hours a week in class, 25 hours a week studying, 5 hours a week traveling, and 20 hours a week socializing. He also works part time at the campus café. Altogether, Adrian spends between 136 and 146 hours a week for all of those activities. How many hours a week does Adrian work?
80. Hannah wants to invest a certain amount of money into various accounts. She wants to invest 20% into Fund A and 15% into Fund B and then take half of the remainder and put it into Fund C. The amount left is between \$1,000 and \$1,500. How much money did Hannah start with?

Critical Thinking

81. It turns out that it's possible to solve linear inequalities without ever having to worry about changing the direction of an inequality symbol. Why? How can you arrange the procedure for solving linear inequalities to make sure of it?

Section 6-6 Solving Quadratic Equations

LEARNING OBJECTIVES

- ❑ 1. Identify the standard form of a quadratic equation.
- ❑ 2. Multiply binomials using FOIL.
- ❑ 3. Factor trinomials.
- ❑ 4. Solve quadratic equations using factoring.
- ❑ 5. Solve quadratic equations using the quadratic formula.
- ❑ 6. Solve real-world problems using quadratic equations.

The main theme of our study of linear equations and inequalities is that they can help us to solve a wide variety of real-world problems. But there are many other types of equations, and not every problem that applies to solving equations can be solved with linear equations. For example, the motion of objects (including bungee jumpers) through the air can be described by equations where the highest power of the variable is 2. We call such equations *quadratic,* and studying them in this section will greatly expand the types of problems we can solve using equations.

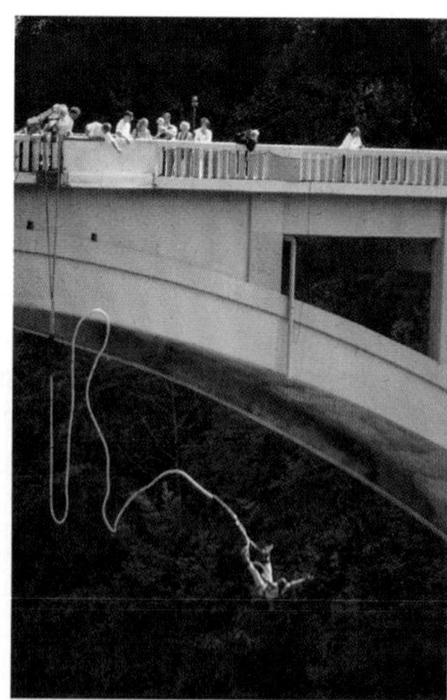

Quadratic Equations

A **quadratic equation** is any equation that can be written in the form $ax^2 + bx + c = 0$, where a, b, and c are real numbers, and a is not zero. When written this way, we say that a quadratic equation is in **standard form.**

To write a quadratic equation in standard form, we arrange it so that zero is on the right side of the equation, the term with exponent 2 comes first on the left side, followed by the term with exponent 1 (if there is one) and then the constant (numeric) term. This is illustrated in Example 1.

EXAMPLE 1 Writing a Quadratic Equation in Standard Form

Write each equation in standard form and identify a, b, and c.

(a) $7 + 9x^2 = 3x$
(b) $4x - 15 = 3x^2$
(c) $5x^2 = 25$

SOLUTION

(a) $7 + 9x^2 = 3x$ *Subtract 3x from both sides.*
$9x^2 - 3x + 7 = 0$

Now we use the definition of standard form: a is the coefficient of x^2, b is the coefficient of x, and c is the constant term, so we get $a = 9$, $b = -3$, and $c = 7$.

(b) $4x - 15 = 3x^2$ *Subtract $3x^2$ from both sides.*
$-3x^2 + 4x - 15 = 0$
$a = -3, b = 4, c = -15$

Notice that if we wanted to, we could multiply both sides of the equation in standard form by -1, giving us the equivalent equation $3x^2 - 4x + 15 = 0$. In this case, $a = 3$, $b = -4$, and $c = 15$.

(c) $5x^2 = 25$ *Subtract 25 from both sides.*
$5x^2 - 25 = 0$
$a = 5, b = 0, c = -25$

Math Note

All quadratic equations with variable x must have an x^2 term; however, they do not have to have an x term or a constant term. For the equation $x^2 - 6 = 0$, $a = 1$, $b = 0$, and $c = -6$. For the equation $2x^2 - 10x = 0$, $a = 2$, $b = -10$, and $c = 0$.

▼ Try This One 1

☑ 1. Identify the standard form of a quadratic equation.

Write each quadratic equation in standard form and identify a, b, and c.

(a) $6 + 8x - x^2 = 0$
(b) $-6x^2 = 5$
(c) $4x + 5x^2 = 0$

Multiplying Binomials

Before we tackle solving quadratic equations, we need some background information. A **binomial** is an algebraic expression with two terms in which any variable has a whole number exponent. Some examples of binomials are

$$x - 5 \qquad 2x + 3 \qquad -6x^2 + 4$$

There is a clever method for multiplying two binomials known as the FOIL method. **F** represents the product of the *first* terms of the binomial. **O** represents the product of the *outer* terms. **I** represents the product of the *inner* terms. **L** represents the product of the *last* terms of the binomials.

The product of two binomials using the FOIL method is

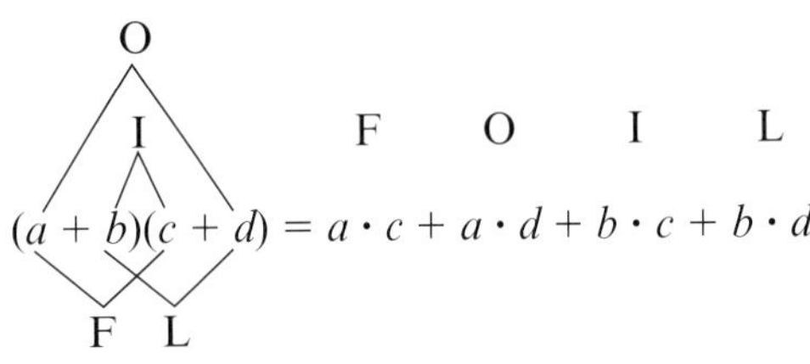

Examples 2 and 3 show how to multiply binomials using the FOIL method.

EXAMPLE 2 Multiplying Binomials Using FOIL

Multiply $(x - 8)(x + 3)$.

Math Note

Always look for like terms that can be combined after finding the four products in the FOIL method.

SOLUTION

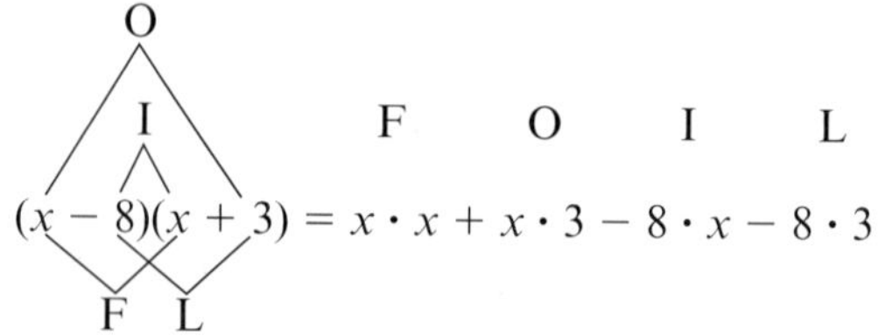

Multiply the *first* terms: $x \cdot x = x^2$
Multiply the *outer* terms: $x \cdot 3 = 3x$
Multiply the *inner* terms: $-8 \cdot x = -8x$
Multiply the *last* terms: $(-8) \cdot (+3) = -24$

These two are like terms.

Notice that $3x$ and $-8x$ are like terms and when combined equal $-5x$. So the product is $x^2 - 5x - 24$.

▼ Try This One 2

Multiply $(x + 7)(x + 9)$

EXAMPLE 3 Multiplying Binomials Using FOIL

Multiply $(2x - 5)(3x - 8)$.

SOLUTION

O, I, F O I L, F L

$$(2x - 5)(3x - 8) = 2x \cdot 3x - 2x \cdot 8 - 5 \cdot 3x - 5(-8)$$

$$= 6x^2 - 16x - 15x + 40 \quad \textit{Combine like terms.}$$

$$= 6x^2 - 31x + 40$$

The product is $6x^2 - 31x + 40$.

▼ Try This One 3

☑ 2. Multiply binomials using FOIL.

Find the product of each.

(a) $(4x - 9)(2x + 5)$
(b) $(3x - 8)(5x - 2)$

Factoring Trinomials

When a quadratic equation is written in standard form, there are three terms on the left side if none of the coefficients are zero. Three term expressions like this are called **trinomials,** and you may have noticed that the result of each example of using FOIL in Examples 2 and 3 fit that form. It's important that you're really good at using FOIL, because one of our methods for solving quadratic equations will involve performing FOIL in reverse, a process we call *factoring*.

In Chapter 5, we called the process of writing a number as a product **factoring.** The same can often be done to algebraic expressions. Take another look at the result of Example 2:

$$(x - 8)(x + 3) = x^2 - 5x - 24$$

If we read this equation from right to left, it's an example of factoring, because we start with the trinomial $x^2 - 5x - 24$ and write it as a product of two factors: $x - 8$ and $x + 3$.

The key to being able to factor trinomials is recognizing where each term in the trinomial in the equation above comes from. The x^2 comes from multiplying the first terms of the two factors, x and x. The constant term, -24, comes from multiplying the last two numbers of the two factors, -8 and 3. And the x term comes from combining the products of the outsides and insides. In essence, we will reverse the process of FOIL to factor. It's easiest to describe the process using some examples.

EXAMPLE 4 Factoring a Trinomial with $a = 1$

Factor $x^2 + 10x + 16$

SOLUTION

Step 1 The factored form will be the product of two binomials; the first term of each is x (since the product of the firsts must be x^2).

$$x^2 + 10x + 16 = (x \quad)(x \quad)$$

Step 2 Write all pairs of factors of the constant term. One of these pairs must be the product of the last terms inside the parentheses. In this case, they are $1 \cdot 16$, $2 \cdot 8$, and $4 \cdot 4$.

Step 3 Set up the factorization with each possible pair of last terms, found in Step 2. Then find the product of outside and inside terms for each possibility.

$1 \cdot 16$: $(x \quad 1)(x \quad 16)$	Outsides: $16x$	Insides: x
$2 \cdot 8$: $(x \quad 2)(x \quad 8)$	Outsides: $8x$	Insides: $2x$
$4 \cdot 4$: $(x \quad 4)(x \quad 4)$	Outsides: $4x$	Insides: $4x$

Step 4 Find a sum or difference of the outsides and insides that yields the x term in the trinomial. In this case, we need $10x$, and $8x + 2x = 10x$. That tells us that the two last terms we need are $+8$ and $+2$, and the trinomial factors as

$$x^2 + 10x + 16 = (x + 2)(x + 8)$$

Check, using FOIL:

$$(x + 2)(x + 8) = x^2 + 8x + 2x + 16 = x^2 + 10x + 16$$

▼ Try This One 4

Factor $x^2 + 13x + 36$.

EXAMPLE 5 Factoring a Trinomial with $a = 1$

Factor $x^2 - 7x + 12$.

SOLUTION

Step 1 Write as the product of two binomials with first term x.

$$x^2 - 7x + 12 = (x \quad)(x \quad)$$

Step 2 The pairs of factors of the constant term are $1 \cdot 12$, $2 \cdot 6$, and $3 \cdot 4$.

Step 3 Possible factorizations:

$(x \quad 1)(x \quad 12)$	Outsides: $12x$	Insides: x
$(x \quad 2)(x \quad 6)$	Outsides: $6x$	Insides: $2x$
$(x \quad 3)(x \quad 4)$	Outsides: $4x$	Insides: $3x$

Step 4 The pair of outsides and insides that yields middle term $-7x$ is $4x$ and $3x$, if both are negative: $-4x - 3x = -7x$. Also, $(-4)(-3) = 12$, which is the correct constant term. So the factors are

$$x^2 - 7x + 12 = (x - 3)(x - 4)$$

Check:

$$(x - 3)(x - 4) = x^2 - 4x - 3x + 12 = x^2 - 7x + 12$$

Math Note

Notice that if the constant term is positive and the x term is negative, both factors of the constant term have to be negative.

▼ Try This One 5

Factor $x^2 - 12x + 20$

EXAMPLE 6 Factoring a Trinomial with $a = 1$

Factor $x^2 - 2x - 8$.

SOLUTION

Step 1 Write as the product of two binomials with first term x.

$$x^2 - 2x - 8 = (x \quad)(x \quad)$$

Step 2 The pairs of factors of the constant term are $1 \cdot 8$ and $2 \cdot 4$.

Step 3 Possible factorizations:

$(x \quad 1)(x \quad 8)$	Outsides: $8x$	Insides: x
$(x \quad 2)(x \quad 4)$	Outsides: $4x$	Insides: $2x$

Math Note

Notice that if the constant term is negative, one factor of the constant is negative, and the other is positive.

Step 4 The pair of outsides and insides that yields middle term $-2x$ is $-4x$ and $2x$: $-4x + 2x = -2x$. Also, $(-4)(2) = -8$, which is the correct constant term. So the factors are

$$x^2 - 2x - 8 = (x + 2)(x - 4)$$

Check: $(x + 2)(x - 4) = x^2 - 4x + 2x - 8 = x^2 - 2x - 8$

▼ Try This One 6

Factor each trinomial:

(a) $x^2 - 3x - 40$ (b) $x^2 + x - 30$

Factoring Trinomials with $a \neq 1$

Step one was the same in each factoring example up to this point because the product of the first terms was always just $x \cdot x$. But that will not be the case if the coefficient of x^2 isn't 1, as in the next two examples. The good news, though, is that the same procedure still works.

EXAMPLE 7 Factoring a Trinomial with $a \neq 1$

Factor $3x^2 - 4x - 15$.

SOLUTION

Step 1 Write as the product of two binomials; the product of the first terms has to be $3x^2$, so they are $3x$ and x.

$$3x^2 - 4x - 15 = (3x \quad)(x \quad)$$

Step 2 The pairs of factors of the constant term are $1 \cdot 15$ and $3 \cdot 5$.

Step 3 Possible factorizations (important note: this time the order matters, so we have to write the possible combinations in both orders):

$(3x \quad 1)(x \quad 15)$	Outsides: $45x$	Insides: x
$(3x \quad 15)(x \quad 1)$	Outsides: $3x$	Insides: $15x$
$(3x \quad 3)(x \quad 5)$	Outsides: $15x$	Insides: $3x$
$(3x \quad 5)(x \quad 3)$	Outsides: $9x$	Insides: $5x$

Step 4 The pair of outsides and insides that yields middle term $-4x$ is $-9x$ and $5x$: $-9x + 5x = -4x$. This comes from factors 5 and -3, and $(5)(-3) = -15$, which is the correct constant term. So the factors are

$$3x^2 - 4x - 15 = (3x + 5)(x - 3)$$

Check: $(3x + 5)(x - 3) = 3x^2 - 9x + 5x - 15 = 3x^2 - 4x - 15$

Math Note

With enough practice, you'll eventually be able to try the possibilities mentally, rather than on paper. But we suggest you write them out until you're really comfortable with the process.

▼ Try This One 7

Factor $3x^2 - 11x + 6$.

The most challenging problems are ones where the coefficient of x^2 has more than one pair of factors, as in Example 8.

EXAMPLE 8 Factoring a Trinomial with $a \neq 1$

Factor $14x^2 - 33x + 10$.

SOLUTION

Step 1 Write as the product of two binomials; the product of the first terms has to be $14x^2$, so they can be $14x$ and x, or $7x$ and $2x$.

$$14x^2 - 33x + 10 = (14x \quad)(x \quad) \quad \text{or} \quad (7x \quad)(2x \quad)$$

Step 2 The pairs of factors of the constant term are $1 \cdot 10$ and $2 \cdot 5$.

Step 3 We'll start trying the combinations, starting with $14x$ and x as first terms:

$(14x \quad 10)(x \quad 1)$	Outsides: $14x$	Insides: $10x$
$(14x \quad 1)(x \quad 10)$	Outsides: $140x$	Insides: x
$(14x \quad 5)(x \quad 2)$	Outsides: $28x$	Insides: $5x$

Step 4 We haven't written all the combinations, but at this point we can stop. With factors $(14x - 5)$ and $(x - 2)$, we'll get outsides $-28x$ and insides $-5x$, which gives us the $-33x$ we need.

$$14x^2 - 33x + 10 = (14x - 5)(x - 2)$$

Check:

$$(14x - 5)(x - 2) = 14x^2 - 28x - 5x + 10 = 14x^2 - 33x + 10$$

Math Note

If you exhaust all of the possible combinations and don't find one that works, then the trinomial cannot be factored using integers, and we call it **prime**.

▼ Try This One 8

☑ 3. Factor trinomials.

Factor $6x^2 + 25x + 21$.

Solving Quadratic Equations Using Factoring

We come now to the main point of factoring: it's easy to find the solutions to a quadratic equation if the left side is factored. Here's why: consider the equation $(x - 3)(x + 2) = 0$. This is a statement that the product of two numbers, one named $x - 3$, and the other $x + 2$, is zero. But the only way a product can be zero is if one of the two factors is zero. (If you don't believe it, try to find two nonzero numbers with product zero!)

So from $(x - 3)(x + 2) = 0$, we conclude that either $x - 3 = 0$ or $x + 2 = 0$. This leaves us with two easy linear equations to solve, giving us $x = 3$ or $x = -2$. Now we have a general procedure, which will be demonstrated in Examples 9 and 10:

Math Note

The fact that the product of two numbers can only be zero if at least one of the two is zero is called the **zero product property.**

Procedure for Solving Quadratic Equations by Factoring

Step 1 Write the quadratic equation in standard form.

Step 2 Factor the left side.

Step 3 Set both factors equal to zero.

Step 4 Solve each equation for x.

EXAMPLE 9 Solving a Quadratic Equation Using Factoring

Solve $x^2 - 13x = -36$.

SOLUTION

Step 1 $x^2 - 13x + 36 = 0$ *Write in standard form.*

Step 2 $(x - 9)(x - 4) = 0$ *Factor the left side.*

Step 3 $x - 9 = 0$ or $x - 4 = 0$ *Set both factors equal to zero.*

Step 4 *Solve each equation for x.*

$x - 9 = 0$	$x - 4 = 0$
$x - 9 + 9 = 0 + 9$	$x - 4 + 4 = 0 + 4$
$x = 9$	$x = 4$

The solution set is {9, 4}.

The solutions can be checked by substituting the values into the original equation, as shown.

For $x = 9$,	For $x = 4$,
$x^2 - 13x = -36$	$x^2 - 13x = -36$
$9^2 - 13(9) \stackrel{?}{=} -36$	$4^2 - 13(4) \stackrel{?}{=} -36$
$81 - 117 \stackrel{?}{=} -36$	$16 - 52 \stackrel{?}{=} -36$
$-36 = -36$	$-36 = -36$

▼ Try This One 9

Solve $x^2 + 6 = -7x$

EXAMPLE 10 Solving a Quadratic Equation Using Factoring

Solve $6x^2 - 6 = -5x$.

SOLUTION

Step 1 $6x^2 + 5x - 6 = 0$ *Write in standard form.*

Step 2 $(3x - 2)(2x + 3) = 0$ *Factor.*

Step 3 $3x - 2 = 0 \quad 2x + 3 = 0$ *Set each factor equal to 0.*

Step 4 *Solve each equation.*

$3x - 2 = 0$	$2x + 3 = 0$
$3x - 2 + 2 = 0 + 2$	$2x + 3 - 3 = 0 - 3$
$3x = 2$	$2x = -3$
$\frac{3x}{3} = \frac{2}{3}$	$\frac{2x}{2} = \frac{-3}{2}$
$x = \frac{2}{3}$	$x = -\frac{3}{2}$

The solution set is $\left\{\frac{2}{3}, -\frac{3}{2}\right\}$. We'll leave the check to you.

▼ Try This One 10

☑ 4. Solve quadratic equations using factoring.

Solve $6x^2 - 11x = 2$ using factoring.

Solving Quadratic Equations Using the Quadratic Formula

Factoring is a very useful tool for solving quadratic equations, but it's not the only one we need. There are plenty of trinomials that can't be factored using integers, and plenty of others that can be factored, but the size of the coefficients makes it impractical. Just when things look bad, along comes a formula to save the day, providing solutions to quadratic equations where factoring doesn't help.

The formula

$$x = \frac{-b \pm \sqrt{b^2 - 4ac}}{2a}$$

is called the **quadratic formula** and can be used to solve any quadratic equation written in standard form, $ax^2 + bx + c = 0$, $a \neq 0$.

The good thing about this formula is that there's no algebraic manipulation involved: you simply substitute in the appropriate values for a, b, and c, then use order of operations to simplify the solutions.

EXAMPLE 11 Using the Quadratic Formula

Solve $2x^2 - x - 8 = 0$ using the quadratic formula.

SOLUTION

Identify a, b, and c. (The equation is in standard form.)

$$a = 2, b = -1, \text{ and } c = -8$$

Substitute into the quadratic formula.

$$x = \frac{-b \pm \sqrt{b^2 - 4ac}}{2a}$$

$$= \frac{-(-1) \pm \sqrt{(-1)^2 - 4(2)(-8)}}{2(2)}$$

$$= \frac{1 \pm \sqrt{1 + 64}}{4}$$

$$x = \frac{1 + \sqrt{65}}{4} \quad \text{or} \quad x = \frac{1 - \sqrt{65}}{4} \quad \textit{Exact solutions.}$$

$$\approx 2.265 \qquad \approx -1.766 \quad \textit{Decimal approximations.}$$

$$\{2.265, -1.766\}$$

Math Note

Don't forget to include negative signs if appropriate when identifying a, b, and c in the quadratic formula.

▼ Try This One 11

Solve $3x^2 - 5x + 1 = 0$ using the quadratic formula.

When solutions found using the quadratic formula are left in radical form, they are exact; decimal equivalents are usually approximate values. The exact solutions can sometimes be simplified; for example,

$$\frac{9 \pm \sqrt{27}}{6} = \frac{9 \pm \sqrt{9 \cdot 3}}{6} = \frac{9 \pm 3\sqrt{3}}{6} = \frac{\overset{1}{\cancel{3}}(3 \pm \sqrt{3})}{\underset{2}{\cancel{6}}} = \frac{3 \pm \sqrt{3}}{2}$$

First $\sqrt{27}$ was simplified to $3\sqrt{3}$ and a 3 was factored out of the numerator. Finally, a factor of 3 was divided out of the numerator and denominator.

EXAMPLE 12 Using the Quadratic Formula

Solve $5 = 5y^2 + 8y$ using the quadratic formula.

SOLUTION

Before identifying a, b, and c, we need to rewrite the equation in standard form by subtracting $5y^2$ and $8y$ from both sides.

$$-5y^2 - 8y + 5 = 0 \qquad a = -5,\ b = -8,\ c = 5.$$

Make sure to use variable y in the solution!

$$y = \frac{-b \pm \sqrt{b^2 - 4ac}}{2a}$$

$$= \frac{-(-8) \pm \sqrt{(-8)^2 - 4(-5)(5)}}{2(-5)}$$

$$= \frac{8 \pm \sqrt{64 + 100}}{-10}$$

$$= \frac{8 \pm \sqrt{164}}{-10} \qquad \sqrt{164} = \sqrt{4} \cdot \sqrt{41} = 2\sqrt{41}$$

$$= \frac{8 \pm 2\sqrt{41}}{-10} \qquad \text{Factor 2 out of numerator.}$$

$$= \frac{2(4 \pm \sqrt{41})}{-10} \qquad \text{Divide numerator and denominator by 2.}$$

$$= \frac{4 + \sqrt{41}}{-5}, \frac{4 - \sqrt{41}}{-5}$$

The solution set is $\left\{\frac{4 + \sqrt{41}}{-5}, \frac{4 - \sqrt{41}}{-5}\right\}$. To two decimal places, the decimal approximations are $\{-2.08, 0.48\}$.

Calculator Guide

To find decimal approximations for $\frac{4 + \sqrt{41}}{-5}$:

Standard Scientific Calculator

(4 + 41 √) ÷ 5 ± =

Standard Graphing Calculator

(4 + √ 41)) ÷ (−) 5 ENTER

The second close parenthesis after 41 is needed if your calculator automatically puts in an open parenthesis when you push the √ button.

Try This One 12

5. Solve quadratic equations using the quadratic formula.

Solve each quadratic equation using the quadratic formula.

(a) $3x^2 - 3x = 1$ (b) $x = x^2 - 13$

When you apply the quadratic formula and the number under the root ends up being negative, the equation has no real solutions. We won't study such equations in this book, but they do have applications. Which brings us to our next topic.

Applications of Quadratic Equations

Quadratic equations can be used to solve many real-world problems using the problem-solving procedure we developed in Section 6-3.

EXAMPLE 13 Applying Quadratic Equations to Woodworking

The plans for a make-it-yourself picnic table call for the length to be 2 feet more than the width. If you want a table with 15 square feet of area, what dimensions should you choose?

SOLUTION

Let x = the width of the table. The length is 2 feet more, so $x + 2$ is the length. The area of a rectangle is length times width, so we can set up an equation:

Length	times	Width	is	Area
$(x + 2)$	$\times$	x	$=$	15

Now we solve:

$(x + 2)x = 15$ *Distribute.*

$x^2 + 2x = 15$ *Subtract 15 from both sides.*

$x^2 + 2x - 15 = 0$ *Factor.*

$(x + 5)(x - 3) = 0$ *Set each factor equal to zero.*

$x + 5 = 0 \quad | \quad x - 3 = 0$ *Solve each equation.*

$x = -5 \quad | \quad x = 3$

Even though we got two solutions, only one makes sense in the problem – you can't build a table that is −5 feet wide. So the width is 3 feet and the length is 5 feet.

Check:

$$5 \text{ ft} \times 3 \text{ ft} = 15 \text{ sq ft} \checkmark$$

▼ Try This One 13

☑ 6. Solve real-world problems using quadratic equations.

The formula for the distance that an object falls freely to the ground is $d = rt + 16t^2$, where:

d is the distance it falls (in feet);

r is the rate at which the object starts to fall; and

t is the number of seconds the object falls.

How long will it take an object that is dropped from the top of the Sony Building to hit the ground? The Sony Building, 1 Madison Square Plaza, New York City, is 576 feet tall. (*Hint:* Since the object is dropped, $r = 0$.)

Answers to Try This One

1 (a) $-x^2 + 8x + 6 = 0$; $a = -1$; $b = 8$; $c = 6$ or $x^2 - 8x - 6 = 0$; $a = 1$; $b = -8$; $c = -6$

(b) $-6x^2 - 5 = 0$; $a = -6$; $b = 0$; $c = -5$ or $6x^2 + 5 = 0$; $a = 6$; $b = 0$; $c = 5$

(c) $5x^2 + 4x = 0$; $a = 5$; $b = 4$; $c = 0$

2 $x^2 + 16x + 63$

3 (a) $8x^2 + 2x - 45$ (b) $15x^2 - 46x + 16$

4 $(x + 9)(x + 4)$

5 $(x - 10)(x - 2)$

6 (a) $(x - 8)(x + 5)$ (b) $(x + 6)(x - 5)$

7 $(3x - 2)(x - 3)$

8 $(6x + 7)(x + 3)$

9 $\{-1, -6\}$

10 $\left\{2, -\frac{1}{6}\right\}$

11 $\frac{5 \pm \sqrt{13}}{6} \approx \{1.43, 0.23\}$

12 (a) $\frac{3 \pm \sqrt{21}}{6} \approx \{1.26, -0.26\}$

(b) $\frac{1 \pm \sqrt{53}}{2} \approx \{4.14, -3.14\}$

13 6 seconds

EXERCISE SET 6-6

Writing Exercises

1. Describe the process we use to multiply two binomial expressions.
2. What does it mean to factor an expression?
3. Describe the steps we use to factor a trinomial.
4. Why is factoring useful in solving quadratic equations?
5. Since we first learned to solve quadratic equations using factoring, why do we need the quadratic formula as well?
6. If you had to choose only one of solving by factoring and solving using the quadratic formula, which would you choose? Why?

Computational Exercises

For Exercises 7–12, write the equation in standard form and identify a, b, and c.

7. $2x + 3x^2 = 5$
8. $4 - 2x^2 = 7x$
9. $10 = 30x^2$
10. $-8 = -4x^2$
11. $-5x = 2x^2$
12. $100x = -50x^2$

For Exercises 13–22, use the FOIL method to multiply the two binomials.

13. $(x + 7)(x + 9)$
14. $(x - 8)(x - 12)$
15. $(y - 7)(y - 10)$
16. $(y + 4)(y + 2)$
17. $(x - 15)(x + 8)$
18. $(x + 10)(x - 3)$
19. $(2x - 7)(7x - 9)$
20. $(4x - 1)(4x - 1)$
21. $(5z + 7)(3z - 8)$
22. $(2t - 5)(3t + 8)$

For Exercises 23–46, solve each quadratic equation by factoring.

23. $x^2 + 5x + 6 = 0$
24. $x^2 + 9x + 20 = 0$
25. $x^2 + x - 12 = 0$
26. $x^2 - 3x - 10 = 0$
27. $t^2 - 14t = 51$
28. $y^2 - y = 20$
29. $x^2 + 24x = 81$
30. $x^2 - 12x = 64$
31. $y^2 + 15 = 8y$
32. $z^2 + 20 - 12z = 0$
33. $2x^2 - x - 21 = 0$
34. $5x^2 + 27x - 18 = 0$
35. $6x^2 - x - 12 = 0$
36. $4x^2 + 13x - 12 = 0$
37. $6m^2 - 12 = m$
38. $10k^2 + 21k = 10$
39. $7x - 6 = -5x^2$
40. $5x^2 - 18 = 27x$
41. $12 - x = 6x^2$
42. $6x^2 + 6 = 13x$
43. $\frac{1}{2}x^2 + \frac{1}{2}x - 3 = 0$
44. $\frac{1}{3}x^2 + 2x + \frac{5}{3} = 0$
45. $\frac{9}{4}y = -1 - \frac{1}{2}y^2$
46. $1 = \frac{1}{2}t^2 - \frac{7}{6}t$

For Exercises 47–64, solve each quadratic equation by using the quadratic formula.

47. $3x^2 + x - 1 = 0$
48. $4x^2 - 7x = 2$
49. $2x^2 - 5x = 12$
50. $x^2 + 5x - 12 = 0$
51. $3y^2 + 5y + 1 = 0$
52. $5y^2 + 2y = 3$
53. $x^2 - 8x - 9 = 0$
54. $6x^2 + x = 35$
55. $x^2 + 5x = 3$
56. $6x - 1 = 4x^2$
57. $\frac{1}{3}y^2 = 7y - 10$
58. $\frac{5}{2} + y = 4y^2$
59. $\frac{7}{5}x - \frac{1}{2}x^2 = -3$
60. $-1 + \frac{11}{3}x^2 = \frac{1}{5}x$
61. $0.2x^2 + 1.2 = -1.3x$
62. $4.5x = x^2 + 1.1$
63. $2.3 - 1.6z = z^2$
64. $2.3t^2 - 4.5 = 6.7t$
65. The product of two consecutive even integers is 288. Find the numbers. (*Hint:* Consecutive even integers can be written as x and $x + 2$.)
66. The product of two consecutive integers is 156. Find the numbers. (*Hint:* Consecutive integers can be written as x and $x + 1$.)

Real-World Applications

67. How long will it take an object to hit the ground if it is dropped from a height of 1,296 feet? (Use $d = rt + 16t^2$.) Assume the object initially is held still.
68. If the height of a triangular clothing tag is 6 inches longer than its base and the area of the tag is 8 square inches, find the lengths of the base and height. Use the formula $A = \frac{1}{2}bh$.
69. Gary made two square pizza boxes, one that has a side 6 inches longer than the other. The sum of the areas of the bases of both boxes is 116 inches. What are the lengths of the sides of each pizza box? (Use the formula $A = s^2$.)
70. In one city, a new bus route and a new subway stop get added each month. There are five more subway stops than bus routes, and the product of the current number of each is 3 times what it was 5 months ago. How many of each are there?
71. To form part of a garden sculpture, an artist cuts a 20-foot wire into two pieces, and then bends each piece to form a square. The product of the lengths of one side of each square is 6 feet. What is the area of the larger square that is formed? (Use the formula $A = s^2$.)
72. A rectangular banner advertising a homecoming dance is 8 feet long and 6 feet wide. The committee wants to increase visibility by making the banner larger. If each dimension is increased by the same number of feet, the area of the new rectangle formed is 32 square feet more than the area of the original rectangle. By how many feet was each dimension increased?

73. The profit for a company that produces custom gift baskets can be modeled by the equation $P = -2x^2 + 20x + 2{,}400$, where x is the number of gift baskets produced per month. How many gift baskets does the company have to sell to break even (no profit but no loss)?
74. A rectangular piece of cardboard has a length that is quadruple its width. When 2-inch squares are cut out of the four corners of the cardboard and the flaps are folded to form a box without a top, the box has a volume of 32 cubic inches. What is the length and width of the piece of cardboard? (The volume of a rectangular box is the product of length, width, and height.)
75. Lucy puts a rug that measures 105 square feet in her family room so that there is a 1½-foot border of wood flooring around the room. The length of the room is triple the width of the room. What are the dimensions of Lucy's family room?
76. A hamburger wrapper is thrown from a sixth floor dorm window, and its motion can be modeled by the equation $h = -3t^2 + 3t + 60$, where h is the height in feet and t is the time in seconds. How long will it take for the wrapper to hit the ground?
77. A rectangular piece of land that is 30 feet by 20 feet is being sectioned off in the middle of campus to create a September 11 memorial monument. The developers would like to create a monument in the center that is 144 square feet, with a walkway border that is a uniform width. How wide should the walkway border be?
78. Dom and Sally conduct an experiment for extra credit in their math class. Standing on a 160-foot tall building, Sally drops her pencil over the edge at the same time Dom throws his pencil straight down with an initial velocity of 48 m/s. The motion of Sally's pencil can be modeled by the equation $h = -16t^2 + 160$ and the motion of Dom's pencil can be modeled by the equation $h = -16t^2 - 48t + 160$, where h is the height of the pencil and t is the time in seconds. By how many seconds does Dom's pencil beat Sally's to the ground below?
79. The distance it takes a particular type of car to stop is a function of the speed it is traveling (in miles per hour) when the brakes are applied and can be modeled by the equation $y = 0.07x^2 - 0.514x + 23$. When the distance this car takes to stop is 40 feet, find the speed the car is traveling when the brakes are applied.
80. The surface area A of a cylinder is given by the formula $A = 2\pi rh + 2\pi r^2$. Fizzy-Up, the manufacturer of a leading energy drink, uses 60π square inches of aluminum sheets for each can. If the height of each can is 7 inches, what is the radius of the can?

The Pythagorean theorem is a formula describing the relationship among the side lengths of a right triangle (a triangle with a 90 degree angle). If a and b represent the two shorter sides, and c is the longest side, then $a^2 + b^2 = c^2$. Use this formula for Exercises 81–84.

81. A bus leaves from campus headed east at 35 miles per hour and another bus leaves from campus headed south at 25 miles per hour. How long will it take for the buses to be 215 miles apart?
82. When a 20-foot ladder is leaned against a perpendicular wall, the distance from the base of the wall to the bottom of the ladder is 7 feet less than the height the ladder reaches up the wall. How far away from the wall is the bottom?
83. A kite is flying on 584 feet of string. Its vertical distance from the ground is 12 feet more than its horizontal distance from the person flying the kite. Assuming the string is being held at ground level, find the horizontal distance from the person and the vertical distance from the ground.
84. The roads Akani has to drive from home to school, school to work, and then work to home form a right triangle. She drives 7 more miles from school to work than from home to school. The distance from work to home is 13 miles. What is the round-trip distance Akani drives from home to school to work to home?

Critical Thinking

85. It can be shown using algebra that $2 = 1$. Incorrect algebra, that is. Look at the "proof" and find the error.
Let $x = 1$, then multiply both sides by x:
$x^2 = x$, then subtract one from each side:
$x^2 - 1 = x - 1$, then factor the left side:
$(x + 1)(x - 1) = x - 1$; divide both sides by $x - 1$:
$$(x+1)\frac{(x-1)}{(x-1)} = \frac{(x-1)}{(x-1)}$$
$$x + 1 = 1;$$
substitute 1 for x:
$$1 + 1 = 1$$
$$2 = 1$$
86. Early in Section 6-6, we saw that the generic formula for multiplying two binomials using FOIL is
$$(a + b)(c + d) = a \cdot c + a \cdot d + b \cdot c + b \cdot d$$
Show that this is really just two consecutive applications of the distributive property. (*Hint:* Begin by treating the expression $(a + b)$ as if it were a single number.
87. Pick out three quadratic equations that were solved in this section using factoring, then rework them using the quadratic formula. What can you say about the expression under the radical sign in the quadratic formula, $b^2 - 4ac$, whenever the equation can be solved using factoring?

CHAPTER 6 Summary

Section	Important Terms	Important Ideas
6-1	Variable Algebraic expression Distributive property Like terms Evaluate Formula	**Algebra** involves the use of expressions and equations. Expressions can be simplified by using the distributive property and combining like terms. Expressions can be evaluated by substituting the values for the variables and using the order of operations to simplify the expression. Many real-world problems can be solved by using specific formulas that apply to a given situation.
6-2	Equation Solution Linear equation Solution set Equivalent equations Contradiction Identity	**An equation** is a statement that two algebraic expressions are equal. To solve an equation means to find all values of the variable that make the equation a true statement. We solve linear equations using the addition, subtraction, multiplication, and division properties of equality. The solutions to an equation can be checked by substituting them back in for the variable.
6-3		**Many** real-world problems can be solved by writing an appropriate equation that describes a problem and then solving the equation.
6-4	Ratio Proportion Cross multiply Direct variation Inverse variation	**Two** quantities can be compared using a ratio. Two equal ratios constitute a proportion. Proportions can be used to solve many real-world problems.
6-5	Linear inequality Three-part inequality	**Inequalities** are similar to equations but use an inequality sign instead of an equal sign. Linear inequalities can be solved by using the same principles as equations with one exception: If you multiply or divide both sides of the inequality by a negative number, the inequality sign must be reversed.
6-6	Quadratic equation Standard form Binomial FOIL method Trinomial Factoring Quadratic formula	**An equation** is called a quadratic equation when the largest exponent of the variable is 2. Some quadratic equations can be solved by factoring. When the trinomial cannot be factored, the quadratic formula can be used. Quadratic equations can also be used to solve many real-world problems.

MATH IN Drug Administration REVISITED

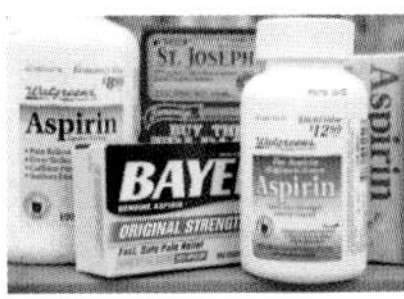

Using proportions, comparing the amount of drug in milligrams to the weight of the patient in pounds, you can find that the recommended dosage for the football player is 647 mg, and for the child is 153 mg. The potentially lethal dose for the child is 573 mg, so the player's dose could kill her. The minimum dose for the football player is 404 mg, so the child's dose would do him absolutely no good.

Review Exercises

Section 6-1

For Exercises 1–8, simplify each algebraic expression.

1. $6x + 3y - 10 + 2y - 8x + 3$
2. $4x - 9 - 2x + 7y - 3y + 16$
3. $5(x - 6) + 2(x - 3)$
4. $-9(2x + 4) - 3(x - 2)$
5. $2x + 7(x - 3) + 4x$
6. $6x + 7 - 3(2x - 8) + 3x$
7. $3 - (x + 2) + 2x$
8. $3(2x - 1) - (2 - x)$

For Exercises 9–14, evaluate each algebraic expression or formula.

9. $2x^2 + 5x - 3$ when $x = 6$
10. $3x - 5 + x^2$ when $x = -5$
11. $6(x - 8) - 10$ when $x = -2$
12. $2(x - 4) + 3x$ when $x = 9$
13. $d = rt$ when $r = 8$ and $t = 15$
14. $A = P(1 + rt)$ when $P = 3{,}000$, $r = 0.08$, and $t = 5$

Section 6-2

For Exercises 15–27, solve each equation.

15. $4x + 8 = -32$
16. $5x + 6 = 36$
17. $8y - 3 = 6y + 37$
18. $2x - 10 = 7x + 55$
19. $5(x + 9) = -20$
20. $3(z - 6) = 33$
21. $6(t + 8) - 4t = 3t - 19$
22. $9(x - 7) - 5x = 2x + 10$
23. $5x + 3 = 16x + 47$
24. $9(2x - 4) = 15x - 27$
25. $\frac{3x}{2} + 2 = \frac{5x}{3}$
26. $\frac{y - 2}{3} + 2y = \frac{5}{3}$
27. $\frac{2x + 1}{4} + \frac{x}{8} = \frac{5}{4}$

For Exercises 28-35, solve for the specified variable.

28. $3x + 3y = 12$; solve for y
29. $y = 3x - 2$; solve for x
30. $P = a + b + c$; solve for c
31. $I = prt$; solve for r
32. $\frac{1}{x} = \frac{1}{y} - \frac{1}{z}$; solve for y
33. $C = 2\pi r$; solve for r
34. $P = 2L + 2W$; solve for W
35. $A = \frac{1}{2}bh$; solve for h

Section 6-3

For Exercises 36–39, write each statement in symbols.

36. 8 times a number decreased by 4
37. the product of 6 and two times a number
38. 3 added to four times a number
39. a number increased by 5
40. Cindy is making the drive home for Thanksgiving, which takes 8 hours round trip. With the holiday traffic, she averages 40 miles per hour on the way home and 50 miles per hour back to school. Find the time it took her to get home and the time it took her to get back to school. (Use the formula distance = rate × time).
41. The health food store sells mixed soy nuts for \$1.20 a pound and Asian trail mix for \$1.80 a pound. If the store sold 2 more pounds of Asian trail mix than soy nuts and the total sales were \$21.60 for these items for the day, how many pounds of each did the store sell?
42. Tickets for a school play sold for \$8.00, \$10.00, and \$12.00. Twice as many \$8.00 tickets were sold as \$10.00 tickets, and 10 more of the \$12.00 tickets than the \$10.00 tickets were sold. If the total revenue was \$3,122, how many of each denomination were sold?
43. On a train excursion trip, the adult fare was \$12.00 and the child's fare was \$6.00. The number of passengers was 400, and the total revenue for the trip was \$4,020. How many adults were there on the train?
44. In a department store Juaquin notices a jacket that he wanted to buy on the "60% off" rack. If the original price of the jacket was \$75 before sales tax, what is the sale price of the jacket before sales tax?
45. Latasha has 4 times as many dimes as she does nickels and twice as many quarters as nickels. She has 35 coins in all and a total of \$4.75. How many coins of each type does she have?
46. At a New Year's Eve party in a certain restaurant the cost was \$60 per couple. If more than 50 couples attend, the restaurant agreed to drop the price by 50 cents a couple for each couple in excess of 50. If 76 couples attended the party, what was the cost per couple?

Section 6-4

For Exercises 47–50, write each ratio as a fraction.

47. 82 miles to 15 gallons of gasoline
48. 16 ounces cost \$2.37
49. 4 months to 2 years
50. 18 minutes to 2 hours

For Exercises 51–54, solve each proportion for x.

51. $\frac{2}{x} = \frac{14}{63}$
52. $\frac{16}{5} = \frac{x}{2.5}$
53. $\frac{8}{24} = \frac{24}{x}$
54. $\frac{5}{11} = \frac{6.2}{x}$
55. If you burn 300 calories when exercising for 12 minutes, how many calories will you burn when exercising for 30 minutes?
56. The U.S. Center for Disease Control reported that 4 out of 10 people with incomes between \$15,000 and \$24,999 exercise regularly. About how many people exercise regularly in a group of 85 people who are in that income bracket?
57. In his will, a man's estate was divided according to a ratio of three parts for his wife and two parts for his son. If his estate amounted to \$18,000, how much did each receive?
58. A professor states that if a student misses a unit test (worth 30 points), he will use the score on the final exam (100 points), proportionally reduced, for the score on the unit test. If a student scored 85 on the final exam, what would be the student's score on the unit test?
59. The cost of building a deck varies directly with the area of the deck. If a 6-foot by 9-foot deck costs \$2,160, find the cost of building a 9-foot by 12-foot deck.
60. The amount of paint needed to paint a spherical object varies directly with the square of the diameter. If 3 pints of paint are needed to paint a spherical object with a diameter of 36 inches, how much paint must be purchased to paint a spherical object with a diameter of 60 inches?
61. The amount of amperage in amps of electricity passing through a wire varies inversely with the resistance in ohms of the wire when the potential remains the same. If the resistance is 20 ohms when the amperage is 10 amps, find the amperage when the resistance is 45 ohms.
62. The cost of producing an item varies inversely with the square root of the number of items produced (i.e., $y = \frac{k}{\sqrt{x}}$, where y = the cost of the item and x = the number of items produced). Find the cost of producing 1,600 items if the cost of producing 900 items is \$600.

Section 6-5

For Exercises 63–68, solve each inequality.

63. $7x + 10 > 80$
64. $3x + 6 \le 2x - 14$
65. $4 - 5y < -31$
66. $4(x - 6) > 3(x - 15)$
67. $6t - 3 \ge 5(2t + 18)$
68. $2x + 7 \le 6(2x + 9) - 20$
69. Membership in a health club is \$75 to join and \$32.50 per month. How many months can a person sign up for if the person wishes to spend at most \$1,000?
70. The graduation committee has at most \$200 to spend on decorations for the stage at the commencement ceremony. How many potted plants can be purchased if each plant costs \$12.95 plus 6% sales tax?

Section 6-6

For Exercises 71–76, solve each equation by factoring.

71. $x^2 - 6 = x$
72. $x^2 + 11x - 26 = 0$
73. $z^2 - 4z - 21 = 0$
74. $2x^2 + 5x = 3$
75. $2 = y + 3y^2$
76. $2x^2 + 9 = 9x$

For Exercises 77–82, solve each equation using the quadratic formula.

77. $x^2 - 5x = 7$
78. $5t^2 - 7t - 4 = 0$
79. $8x^2 + 14x + 4 = 0$
80. $9x^2 - 12x = 7$
81. $4k^2 - 5 = -14k$
82. $5x^2 + 5 = 12x$
83. If the product of two consecutive numbers is 132, find the numbers.
84. How long will it take a dropped object to fall a distance of 1,024 feet? Use $d = rt + 16t^2$.

Chapter Test

For Exercises 1 and 2, simplify.

1. $3x - 7y + 2x - 3y + 5$
2. $5(x - 6) + 2x - 10$

For Exercises 3 and 4, evaluate.

3. $3x^2 - 2x + 6$ when $x = -5$
4. $E = 0.2381I^2Rt$ when $I = 30$, $R = 5$, and $t = 80$

For Exercises 5 and 6, solve the equation.

5. $3x - 5(2x + 10) = -59$
6. $2(y - 6) = 7 + 4(y + 16)$
7. Solve: $F = \frac{mv^2}{r}$ for r
8. Solve for y: $3x + 2y = 10$

For Exercises 9 and 10, solve the equation.

9. $4 - 3x \geq x + 10$
10. $2(y - 3) < 5y + 12$

For Exercises 11 and 12, solve the proportion.

11. $\frac{x}{9} = \frac{16}{36}$
12. $\frac{3}{7} = \frac{15}{x}$

For Exercises 13 and 14, find the product.

13. $(x - 8)(2x + 3)$
14. $(3x - 5)(4x - 7)$

For Exercises 15–17, solve by factoring.

15. $x^2 - 14x - 51 = 0$
16. $t^2 + 12t - 40 = -27$
17. $6x^2 + x = 12$

For Exercises 18 and 19, solve using the quadratic formula.

18. $3x^2 - x - 1 = 0$
19. $5y^2 + 2y = 3$
20. A person has invested part of $5,000 in stocks paying a 4% dividend and the rest in stocks paying a 6% dividend. If the total of the dividends was $270, how much did the person invest in each stock?
21. A person invested part of $3,000 at 6% interest and the rest at 8% interest. If the interest was $190, how much money was invested at each rate?
22. Ava purchased two jump drives for $57. How many jump drives can be purchased for $228?
23. If you can bike 2 miles in $12\frac{1}{2}$ minutes, how many hours will it take to bike 210 miles without counting rest stops?
24. The number of vibrations per second of a metal string varies directly with the square root of the tension when all other factors remain unchanged. Find the number of vibrations per second of a string under a tension of 64 pounds when a string under a tension of 25 pounds makes 125 vibrations per second.
25. The number of hours it takes to do a certain job varies inversely with the number of people working on the job. If it takes 12 people 8 hours to build a front porch on a Habitat for Humanity house, how many hours will it take 8 people to build a porch?
26. The product of two consecutive even numbers is 624. Find the numbers. (*Hint:* Consecutive even numbers can be represented by x and $x + 2$.)
27. A board game is played on a rectangular board whose area is 96 square inches. If the length is 4 inches longer than the width, find its dimensions.

Projects

1. We studied two methods for solving quadratic equations in Chapter 6: factoring and the quadratic formula. The biggest advantage of the quadratic formula is that it works on *every* quadratic equation, not just the ones that factor. But is it always the best choice?

   (a) Solve the equation $x^2 + x - 6 = 0$ using factoring. Use a stopwatch or timer to time how long it takes, then check your solution.

   (b) Solve the same equation, this time using the quadratic formula. Again, time yourself, and note whether you got the correct solution.

   (c) Repeat parts *a* and *b* for each of the equations below.

   $x^2 - 3x - 4 = 0$
   $2x^2 - x - 15 = 0$
   $3x^2 - 20x - 7 = 0$

   (d) Compute your average time for each method on all four equations, then calculate the percentage of problems that you got right using each method. Which method overall would you say was more efficient, and why?

   (e) Repeat parts (a) through (d) for the four equations below.

   $x^2 - 2x - 80 = 0$
   $4x^2 + 26x - 48 = 0$
   $6x^2 - 55x - 50 = 0$
   $2x^2 + 12x - 144 = 0$

   (f) Was your decision on which method is more efficient different for the second group of equations? Write a general conclusion about the value of each method based on all of your results.

2. An interesting method for solving quadratic equations came from India. The steps are

   1. Move the constant term to the right side of the equation.
   2. Multiply each term in the equation by four times the coefficient of the x^2 term.
   3. Square the coefficient of the original x term and add it to both sides of the equation.
   4. Take the square root of both sides.
   5. Set the left side of the equation equal to the positive square root of the number on the right side and solve for x.

6. Set the left side of the equation equal to the negative square root of the number on the right side of the equation and solve for x.

Example: Solve $x^2 + 3x - 10 = 0$.

$$x^2 + 3x = 10$$
$$4x^2 + 12x = 40$$
$$4x^2 + 12x + 9 = 40 + 9$$
$$4x^2 + 12x + 9 = 49$$
$$2x + 3 = \pm 7$$

$2x + 3 = 7$	$2x + 3 = -7$
$2x = 4$	$2x = -10$
$x = 2$	$x = -5$

Try these.

(a) $x^2 - 2x - 13 = 0$
(b) $4x^2 - 4x + 3 = 0$
(c) $x^2 + 12x - 64 = 0$
(d) $2x^2 - 3x - 5 = 0$
(e) Do you find this method to be more or less effective than the ones we studied in Chapter 6?
(f) Do you think this method will work on any quadratic equation? Why or why not?

CHAPTER 7

Additional Topics in Algebra

Outline

7-1 The Rectangular Coordinate System and Linear Equations in Two Variables
7-2 Systems of Linear Equations
7-3 Solving Systems of Linear Equations Using Matrices
7-4 Linear Inequalities
7-5 Linear Programming
7-6 Functions
7-7 Linear, Quadratic, and Exponential Functions
Summary

MATH IN The Stock Market

There was a time when the fluctuations of the stock market were exclusively of concern to the rich. Most people weren't directly invested in the market, and felt like it didn't affect their lives very much. But those days are long gone. Today, most people's retirement savings, at the very least, are invested in stocks and mutual funds. This means that value changes in the market are of more interest to more Americans than ever before, and the ability to monitor those changes is certainly a relevant skill.

Now that we're well acquainted with using variables to represent quantities that can change, we will start to think about relationships between two changing quantities. These types of relationships are all around you if you look for them. The total weight of the clothing you put on in the morning depends on the temperature. The number of minutes you spend getting ready depends on what time you woke up. How much money you spend on gas is affected by the mileage your car gets and how far away you live from work or school. There are three examples before your day really even gets started! And, of course, the value of the stock market changes as the days pass, so it is related to time.

These relationships can often be described by equations with two variables, or with functions, which are the main topics of this chapter. Equations and functions are used to model anything that can be described by numbers, so being able to work with them allows us to study our world in depth. In particular, an understanding of graphs is a tremendously useful real-world skill. The graph on the facing page shows the fluctuations of the Dow Jones Industrial Average for a 22-month period from 2007 to 2009. Looking at a formula, or the raw numbers, would be useful in obtaining specific information, like the Dow Jones average on a given date. But for interpreting overall trends and making general predictions, nothing beats a graph.

The equation $y = -23.088x^2 + 276.21x + 12{,}575$ can be used to model the monthly Dow Jones Industrial Average for the period from March 2007 to January 2009, where the variable x represents months after March 2007. In this chapter, you'll learn how to work with equations of this form, and also get some insight into how they are obtained. By the end of the chapter, you should be able to answer the following questions.

1. According to the model, what was the Dow Jones Industrial Average for March 2007?
2. When did the Dow Jones Industrial Average reach its highest value in that period, and what was the average for that month?
3. What does this model predict the Dow Jones Industrial Average would be for February 2009? Use the Internet to find the actual average for February 2009 and see how accurate the prediction was.

For answers, see Math in the Stock Market Revisited on page 414

Section 7-1 The Rectangular Coordinate System and Linear Equations in Two Variables

LEARNING OBJECTIVES

- ☐ 1. Plot points in a rectangular coordinate system.
- ☐ 2. Graph linear equations.
- ☐ 3. Find the intercepts of a linear equation.
- ☐ 4. Find the slope of a line.
- ☐ 5. Graph linear equations in slope-intercept form.
- ☐ 6. Graph horizontal and vertical lines.
- ☐ 7. Find linear equations that describe real-world situations.

The table below lists the U.S. federal budget surplus (positive number) or deficit (negative number) for each year from 2000 to 2008, according to the Congressional Budget Office. The amounts are in billions of dollars.

Year	'00	'01	'02	'03	'04	'05	'06	'07	'08
Surplus/Deficit	236	128	−158	−378	−413	−318	−248	−161	−455

If you study the table carefully, you can see that the budget went from a surplus to a deficit, which kept getting larger until 2004. Then that trend reversed, with the deficit shrinking, until a drastic change in the other direction occurred in 2008.

Now let's look at the same information in graphic form:

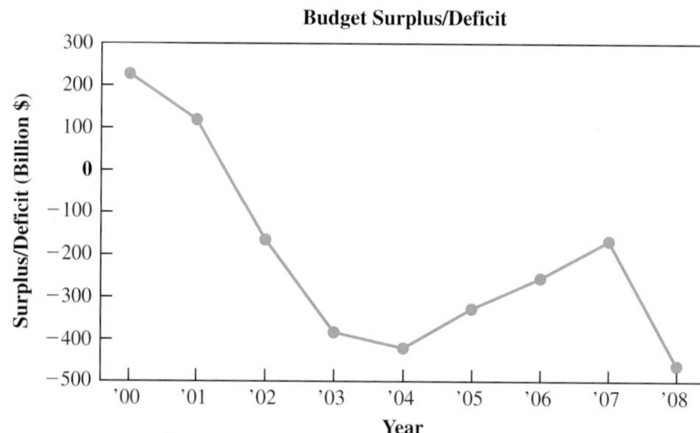

Can you see how much easier it is to see the trends? This is the big advantage of being able to draw graphs in mathematics. Graphing data almost always makes it easier to understand its significance than simply looking at raw numbers.

In this section, we will discuss the basics of graphing and study a particular kind of graph that we can use to explore real-world situations.

The Rectangular Coordinate System

The foundation of graphing in mathematics is a system for locating data points using a pair of perpendicular number lines. We call each one an **axis.** The horizontal line is called the ***x* axis,** and the vertical line is called the ***y* axis.** The point where the two intersect is called the **origin.** Collectively, they form what is known as a **rectangular coordinate system,** sometimes called the **Cartesian plane.** The two axes divide the plane into four regions called **quadrants.** They are numbered using Roman numerals I, II, III, and IV as shown in Figure 7-1.

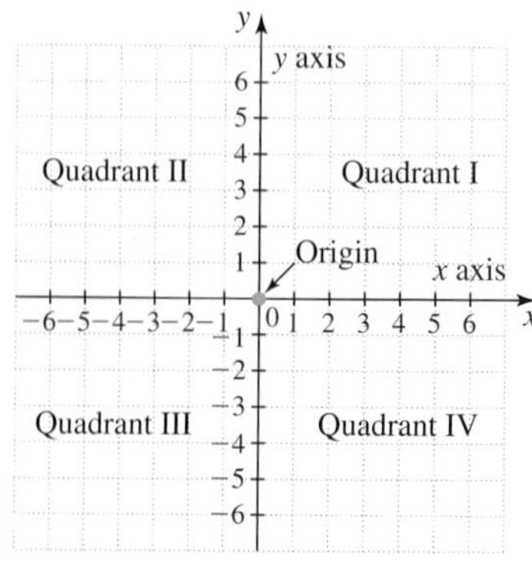

Figure 7-1

When visualizing data, it's often natural to have two quantities paired together, like the year 2000 and the number 236 in the table above. We can associate pairs of numbers in a rectangular coordinate system with points by locating each number on one of the two number lines that make up the coordinate

Sidelight DESCARTES AND ANALYTIC GEOMETRY

The rectangular coordinate system is often called the Cartesian plane in honor of the 17th-century philosopher René Descartes. Descartes devised a method for combining arithmetic, algebra, and geometry into a single subject, now called "analytic geometry." In 1637, Descartes published a book entitled *Discourse on the Method of Rightly Conducting the Reason and Seeking for Truth in the Sciences,* in which he advocated the position that all knowledge should be devised using mathematical reasoning. Today the book is considered an important work of philosophy. Interestingly, the part that became the study of analytic geometry, still taught to every high school student almost 400 years later, was an appendix in his book. To be fair, though, it was a 106-page appendix!

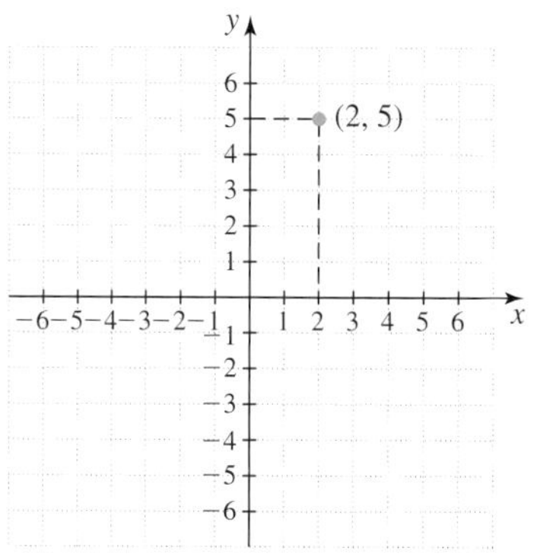

Figure 7-2

system. We call the two numbers **coordinates** of a point, and write them as (x, y), where the first number describes a number on the x axis and the second describes a number on the y axis. The coordinates of the origin are (0, 0). A point P whose x coordinate is 2 and whose y coordinate is 5 is written as $P = (2, 5)$. It is plotted by starting at the origin and moving two units right and five units up, as shown in Figure 7-2.

Example 1 illustrates the process of plotting points.

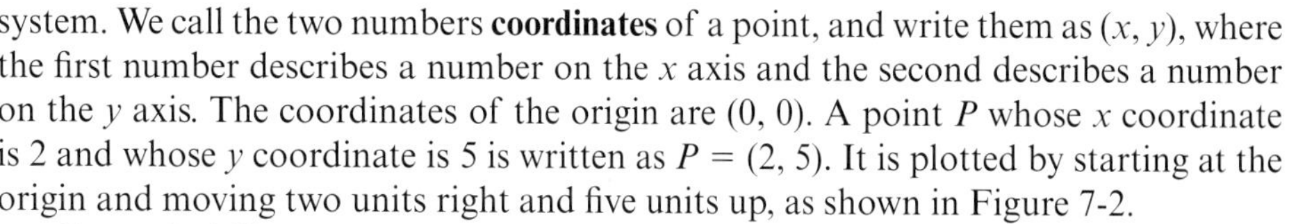

EXAMPLE 1 Plotting Points

Plot the points (5, −3), (0, 4), (−3, −2), (−2, 0), and (2, 6).

SOLUTION

To plot each point, start at the origin and move left or right according to the x value, and then up or down according to the y value.

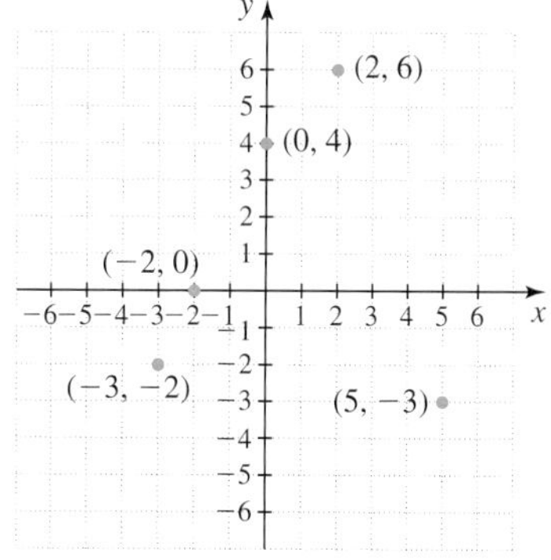

Math Note

To plot a point with a negative x coordinate, start at the origin and move left. For a negative y coordinate, move down.

▼ Try This One 1

Plot the points whose coordinates are (5, 3), (−1, 5), (0, −5), (−3, 0), and (−4, −2).

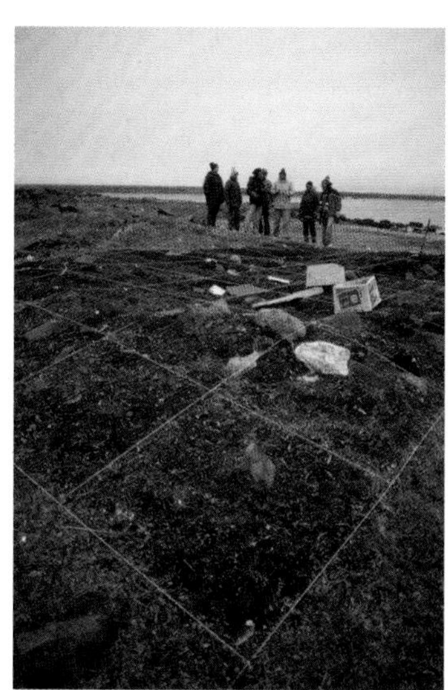

Archaeological digs use a rectangular coordinate system to track where objects are found.

Given a point on the plane, its coordinates can be found by drawing a vertical line back to the x axis and a horizontal line back to the y axis. For example, the coordinates of point C shown in Figure 7-3 are (−3, 4).

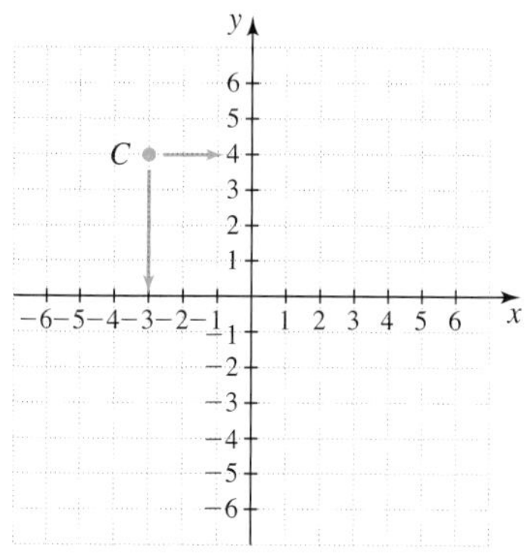

Figure 7-3

EXAMPLE 2 Finding the Coordinates of Points

Math Note

If both coordinates of a point are positive, the point is located in quadrant I. If x is negative and y is positive, the point is located in quadrant II. If x and y are both negative, the point is located in quadrant III. If x is positive and y is negative, the point is in quadrant IV. If $x = 0$, the point is on the y axis, and if $y = 0$, the point is located on the x axis.

Find the coordinates of each point shown on the plane below.

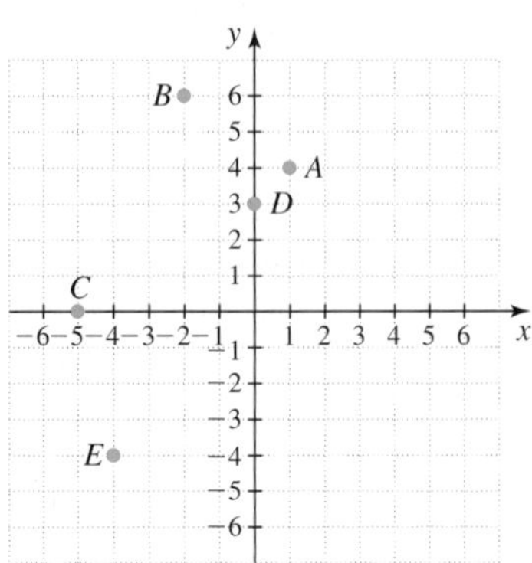

SOLUTION

$A = (1, 4)$ $\quad$ $B = (-2, 6)$ $\quad$ $C = (-5, 0)$ $\quad$ $D = (0, 3)$ $\quad$ $E = (-4, -4)$

▼ Try This One 2

1. Plot points in a rectangular coordinate system.

Find the coordinates of the points shown.

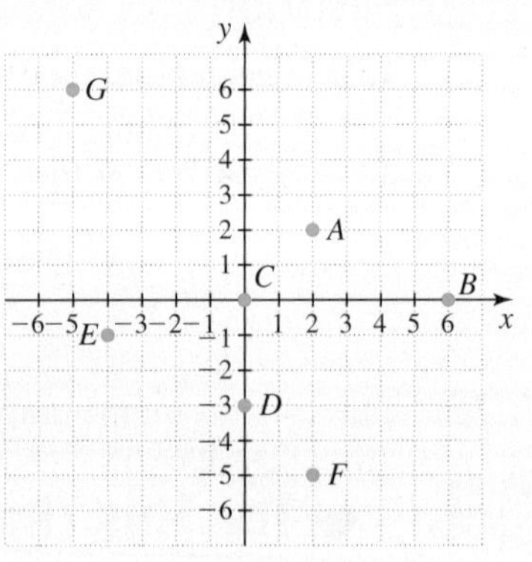

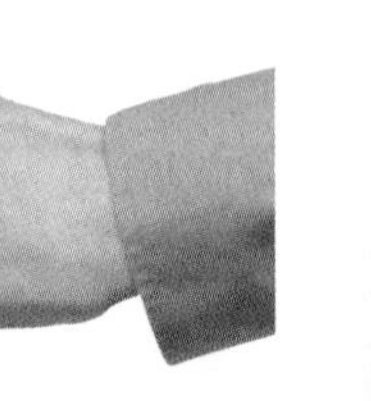

The interface between a mouse and a computer uses a coordinate system to match the cursor's motion to the mouse's motion.

Linear Equations in Two Variables

An equation of the form $ax + by = c$, where a, b, and c are real numbers, is called a **linear equation in two variables.** Let's look at the example $3x + y = 6$. If we choose a pair of numbers to substitute in for x and y, the resulting equation will be either true or false. For example, for $x = 1$ and $y = 3$, the equation becomes $3(1) + 3 = 6$, which is a true statement. We call the pair (1, 3) a **solution** to the equation, and say the pair **satisfies** the equation. But for $x = 3$ and $y = 1$, the equation is $3(3) + 1 = 6$, which is false.

Listed below are a handful of pairs that satisfy the equation $3x + y = 6$.

$$(0, 6) \qquad (1, 3) \qquad (2, 0) \qquad (3, -3) \qquad (4, -6)$$

If we plot the points corresponding to these pairs, something interesting happens (see Figure 7-4).

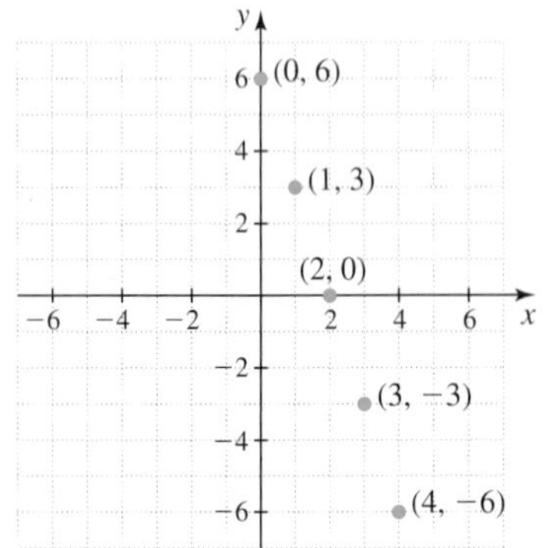

Figure 7-4

Notice that all of the points appear to line up in a straight line pattern. This is not a coincidence, and in fact is why the equation is called *linear*. If we connect these points with a line (Figure 7-5), the result is called the **graph of the equation.** The graph is a geometric representation of *every* pair of numbers that is a solution to the equation.

In Example 3, we illustrate the process of drawing the graph of a linear equation in two variables.

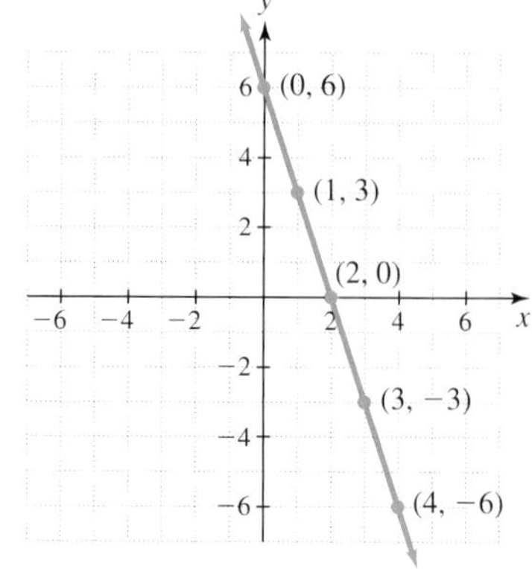

Figure 7-5

EXAMPLE 3 Graphing a Linear Equation in Two Variables

Graph $x + 2y = 5$.

SOLUTION

Only two points are *necessary* to find the graph of a line, but it's a good idea to find three to make sure that you haven't made a mistake. To find pairs of numbers that make the equation true, we will choose some numbers to substitute in for x, then solve the resulting equation to find the associated y. In this case, we chose $x = -1$, $x = 1$, and $x = 5$, but any three will do.

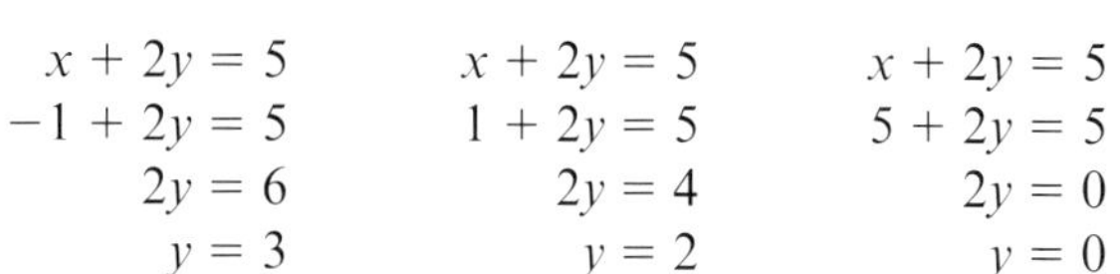

$x + 2y = 5$	$x + 2y = 5$	$x + 2y = 5$
$-1 + 2y = 5$	$1 + 2y = 5$	$5 + 2y = 5$
$2y = 6$	$2y = 4$	$2y = 0$
$y = 3$	$y = 2$	$y = 0$

Three points on the graph are $(-1, 3)$, $(1, 2)$, and $(5, 0)$. We plot those three points and draw a straight line through them.

Calculator Guide

The table feature on a graphing calculator can be used to quickly find points on a graph. First, the equation must be solved for y. In Example 3, $x + 2y = 5$ becomes $y = \frac{(5 - x)}{2}$. The keystrokes:

[Y=] [(] 5 [−] [X, T, θ, n] [)] [÷] 2 [2nd] [Window], then use the arrows and [Enter] to select "Ask" next to "Indpnt." Then do [2nd] [Graph] [(−)] 1 [Enter], 1 [Enter], and 5 [Enter]. A typical result is shown below.

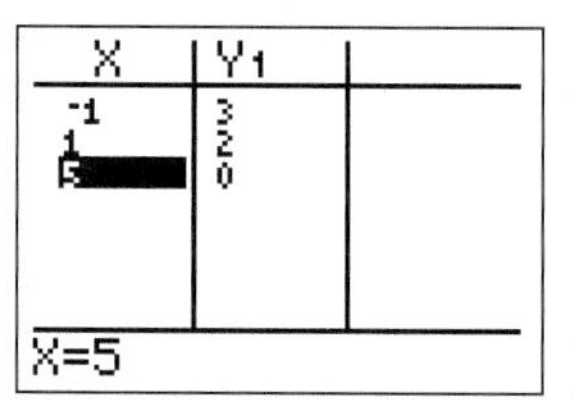

▼ Try This One 3

Graph $2x - y = 10$.

☑ 2. Graph linear equations.

Intercepts

It's often convenient to find two special points on a graph. The point where a graph crosses the x axis is called the **x intercept.** The point where a graph crosses the y axis is called the **y intercept.** Every point on the x axis has y coordinate zero, and every point on the y axis has x coordinate zero, so we get the following rules.

Finding Intercepts

To find the x intercept, substitute 0 for y and solve the equation for x.
To find the y intercept, substitute 0 for x and solve the equation for y.

EXAMPLE 4 Finding Intercepts

Find the intercepts of $2x - 3y = 6$, and use them to draw the graph.

Math Note

Intercepts are *points on the graph*, not numbers. Make sure you give both coordinates, not just the result of solving the equation in the procedure.

SOLUTION

To find the x intercept, let $y = 0$ and solve for x.

$$\begin{aligned} 2x - 3y &= 6 \\ 2x - 3(0) &= 6 \\ 2x &= 6 \\ x &= 3 \end{aligned}$$

The x intercept has the coordinates (3, 0).

To find the y intercept, let $x = 0$ and solve for y.

$$\begin{aligned} 2x - 3y &= 6 \\ 2(0) - 3y &= 6 \\ -3y &= 6 \\ y &= -2 \end{aligned}$$

The y intercept has the coordinates $(0, -2)$.

Now we plot the points (3, 0) and $(0, -2)$, and draw a straight line through them. (It would still be a good idea to find one additional point to check your work. If the three points don't line up, there must be a mistake.)

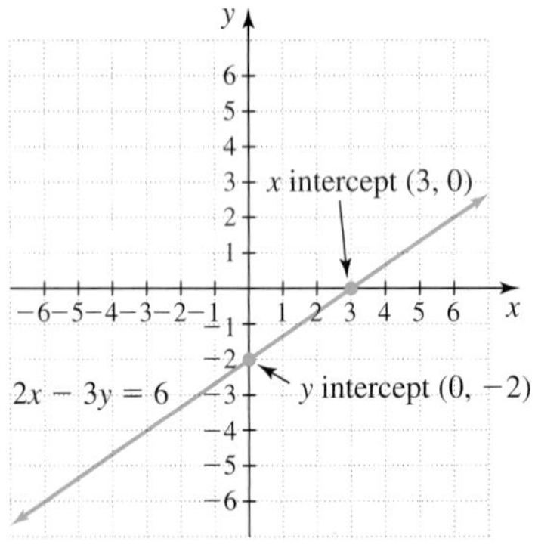

▼ Try This One 4

For each equation, find the intercepts and use them to draw the graph.

(a) $x + 5y = 10$ (b) $4x - 3y = 12$

☑ 3. Find the intercepts of a linear equation.

Slope

The slope of a line on a plane is analogous to the slope of a road. It is the "steepness" of the road or line. Consider the two roads shown in Figure 7-6.

The "slope" of a road can be defined as the "rise" (vertical height) divided by the "run" (horizontal distance) or as the change in y with respect to the change in x. In road A, we have

$$\frac{30\text{ ft}}{50\text{ ft}} = 0.6$$

That is, for every 50 feet horizontally the road rises a height of 30 feet. Road B has a slope of

$$\frac{10\text{ ft}}{50\text{ ft}} = 0.2$$

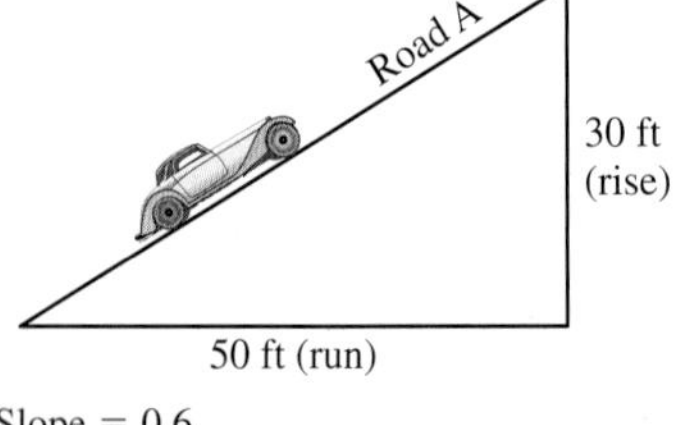

Slope = 0.6

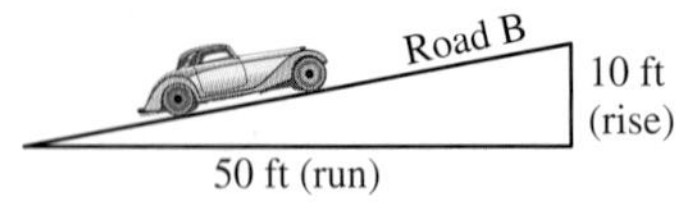

Slope = 0.2

Figure 7-6

Since the slope of road A is larger than the slope of road B, we say that road A is steeper than road B.

On the Cartesian plane, slope is defined as follows:

The **slope** of a line (designated by m) is

$$m = \frac{y_2 - y_1}{x_2 - x_1}$$

where (x_1, y_1) and (x_2, y_2) are two points on the line.

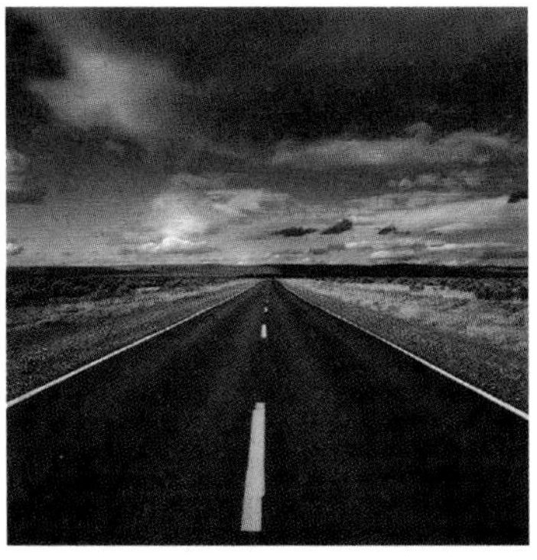
The slope of this road is 0, because the rise is 0.

In words, the slope of a line can be determined by subtracting the y coordinates (the vertical height) of two points and dividing that difference by the difference obtained from subtracting the x coordinates (the horizontal distance) of the same two points. See Figure 7-7.

Example 5 shows the procedure for finding the slope of a line given two points.

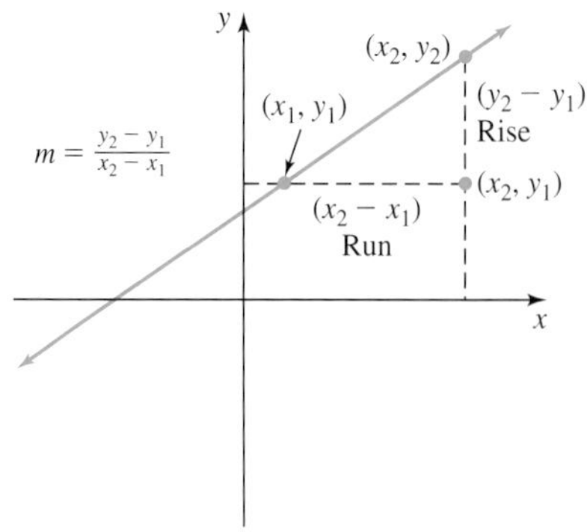

Figure 7-7

EXAMPLE 5 Finding the Slope of a Line

Find the slope of a line passing through the points (2, 3) and (5, 8).

SOLUTION

Designate the two points as follows

$$\begin{array}{ccc} (2,\ 3) & \text{and} & (5,\ 8) \\ \downarrow\ \downarrow & & \downarrow\ \downarrow \\ (x_1, y_1) & & (x_2, y_2) \end{array}$$

Substitute into the formula

$$m = \frac{y_2 - y_1}{x_2 - x_1} = \frac{8 - 3}{5 - 2} = \frac{5}{3}$$

The slope of the line is $\frac{5}{3}$. That means the line is rising 5 feet vertically for every 3 feet horizontally.

Math Note

If the line goes "uphill" from left to right, the slope will be positive. If a line goes "downhill" from left to right, the slope will be negative. The slope of a vertical line is *undefined*. The slope of a horizontal line is 0 (see sample graphs at the bottom of the page).

▼ Try This One 5

Find the slope of a line passing through the points (−1, 4) and (2, −8).

When finding slope, it doesn't matter which of the two points you choose to call (x_1, y_1) and which you call (x_2, y_2). But the order of the subtraction in the numerator and denominator has to be consistent—if you subtract $y_2 - y_1$ in the numerator, you have to subtract $x_2 - x_1$ in the denominator.

If you know the equation of a line, you can find the slope by first finding two points like we did in graphing, then using the slope formula.

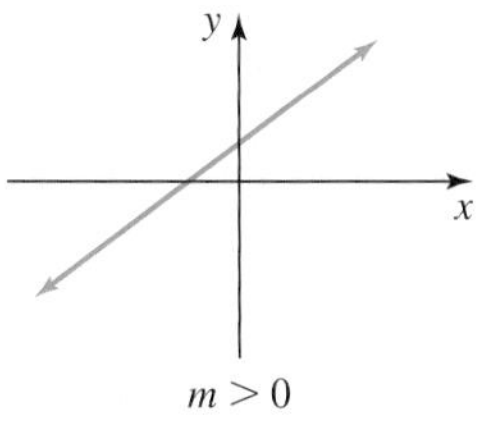

$m > 0$

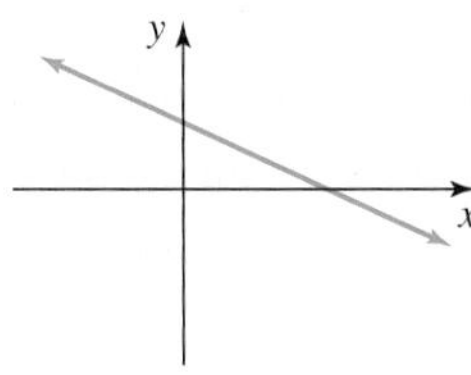

$m < 0$

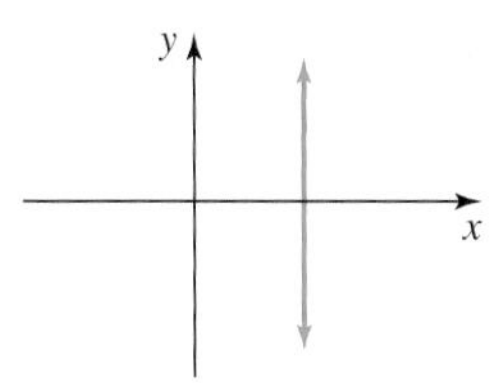

m is undefined

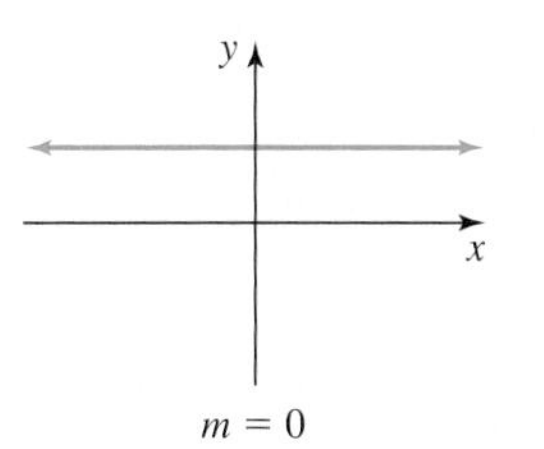

$m = 0$

EXAMPLE 6 Finding Slope Given the Equation of a Line

Math Note

It doesn't matter which two different points are used to find the slope of a line.

Find the slope of the line $5x - 3y = 15$.

SOLUTION

Find the coordinates of any two points on the line. In this case, we choose the intercepts, which are (3, 0) and (0, −5). Then substitute into the slope formula.

$$m = \frac{y_2 - y_1}{x_2 - x_1} = \frac{-5 - 0}{0 - 3} = \frac{-5}{-3} = \frac{5}{3}$$

The slope of the line $5x - 3y = 15$ is $\frac{5}{3}$.

▼ Try This One 6

☑ 4. Find the slope of a line.

Find the slope of the line $-2x + 4y = 8$.

If we start with the equation $5x - 3y = 15$ from Example 6 and solve the equation for y, the result suggests a useful fact.

$$\begin{aligned} 5x - 3y &= 15 \\ -3y &= -5x + 15 \\ y &= \frac{5}{3}x - 5 \end{aligned}$$

Notice that the coefficient of x is 5/3, which is the same as the slope of the line, as found in Example 6. This is not a coincidence. Also, notice that in the final equation, the y intercept is easy to find: substituting in $x = 0$ gives us the equation $y = -5$. So solving the equation for y was useful because we can get important information about the line very easily when the equation is in that form.

The **slope-intercept form** for an equation in two variables is $y = mx + b$, where m is the slope and $(0, b)$ is the y intercept.

The graph of a line in slope-intercept form can be drawn by using the y intercept as a point, then plotting the "rise," which is the numerator of the slope in fraction form, and then the run, which is the denominator, as shown in Example 7.

EXAMPLE 7 Using Slope-Intercept Form to Draw a Graph

Math Note

When the slope of a line is negative, like $-\frac{2}{3}$, start at the y intercept and move 2 units *down* and 3 units to the right to get the second point.

Graph the line $y = \frac{5}{3}x - 6$.

SOLUTION

The slope is 5/3 and the y intercept is (0, −6). Starting at the point (0, −6), we move vertically upward 5 units for the rise, and move horizontally 3 units right for the run. That gives us second point (3, −1). Then draw a line through these points. To check, notice that (3, −1) satisfies the equation.

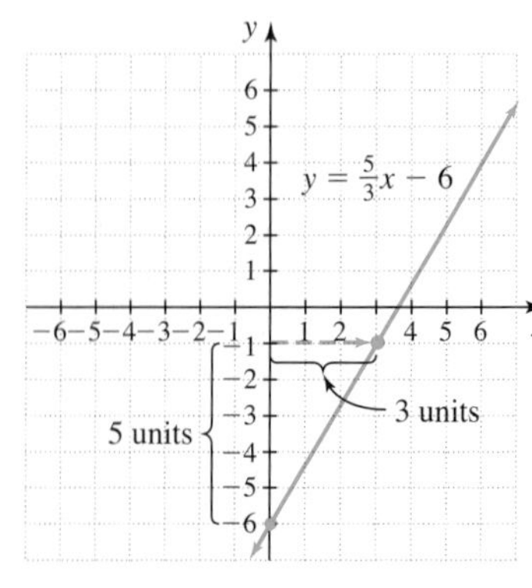

▼ Try This One 7

☑ 5. Graph linear equations in slope-intercept form.

Graph the line $y = \frac{2}{5}x - 2$.

Math Note

The graph of $x = 0$ is the y axis, and the graph of $y = 0$ is the x axis.

Horizontal and Vertical Lines

Think about what the equation $y = 3$ says in words: that the y coordinate is always 3. This is a line whose height is always 3, which is a horizontal line. Similarly, an equation like $x = -6$ is a vertical line with every point having x coordinate -6.

EXAMPLE 8 Graphing Vertical and Horizontal Lines

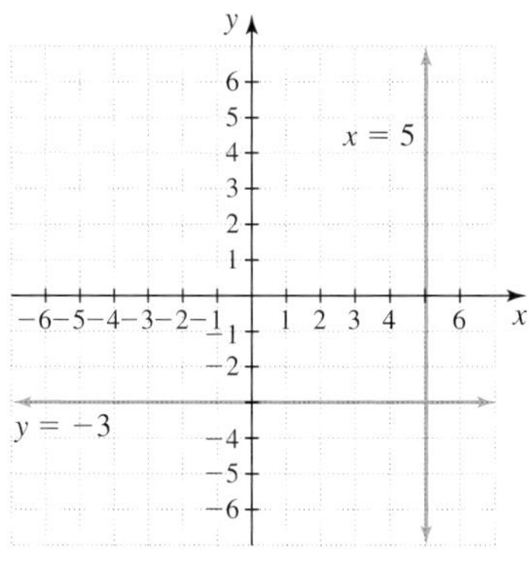

Graph each line.

(a) $x = 5$ (b) $y = -3$

SOLUTION

(a) The graph of $x = 5$ is a vertical line with every point having x coordinate 5. We draw it so that it passes through 5 on the x axis.
(b) The graph of $y = -3$ is a horizontal line with every point having y coordinate -3. We draw it so that it passes through -3 on the y axis.

▼ Try This One 8

☑ 6. Graph horizontal and vertical lines.

Graph each line.

(a) $x = -7$ (b) $y = 1$

Applications of Linear Equations in Two Variables

A wide variety of real-world situations can be modeled using linear equations in two variables. These equations are usually written in slope-intercept form, $y = mx + b$. We will close the section with two examples.

EXAMPLE 9 Finding a Linear Equation Describing Cab Fare

The standard fare for a taxi in one city is \$5.50, plus \$0.30 per mile. Write a linear equation that describes the cost of a cab ride in terms of the length of the ride in miles. Then use your equation to find the cost of a 6-mile ride, an 8.5-mile ride, and a 12-mile ride.

SOLUTION

The first quantity that varies in this situation is the length of the trip, so we will assign variable x to number of miles. The corresponding quantity that changes is the cost, so we will let y = the cost of the ride. Since each mile costs \$0.30, the total mileage cost is $0.30x$. Adding the upfront cost of \$5.50, the total cost is given by $y = 0.30x + 5.50$.

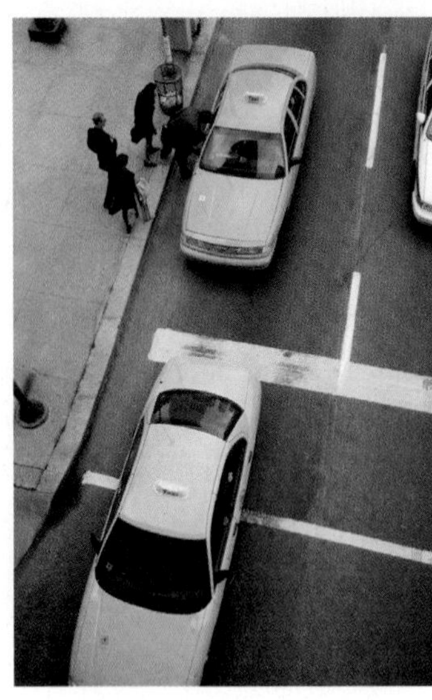

For $x = 6$ miles,

$$y = 0.30(6) + 5.50$$
$$y = \$7.30$$

For $x = 8.5$ miles,

$$y = 0.30(8.5) + 5.50$$
$$y = \$8.05$$

For $x = 12$ miles,

$$y = 0.30(12) + 5.50$$
$$y = \$9.10$$

▼ Try This One 9

The cost of a medium cheese pizza at Mario's Campus Pizzeria is \$6.75, and each additional topping costs \$0.35. Write a linear equation that describes the cost of a pizza in terms of the number of toppings. Then use your equation to find the cost of a pizza with three toppings and one with five toppings.

In Example 9, the slope of the equation describing the cab ride was 0.30. This also happens to be the rate at which the cost changes as the mileage increases. This is a very useful observation in applying linear equations to real-world situations.

Slope and Rate of Change

The slope of any line tells us the rate at which *y* changes with respect to *x*.

EXAMPLE 10 Finding a Linear Equation Describing Distance

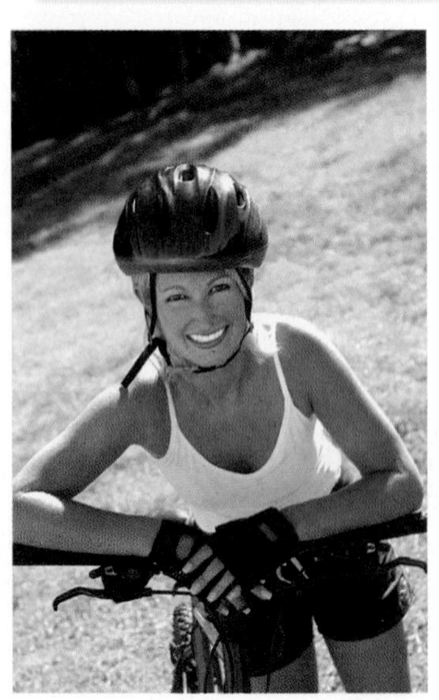

After a brisk bike ride, you take a break and set out for home. Let's say you start out 15 miles from home and decide to relax on the way home and ride at 9 miles per hour. Write a linear equation that describes your distance from home in terms of hours, and use it to find how long it will take you to reach home.

SOLUTION

In this case, we know two key pieces of information: at time zero (when you start out for home) the distance is 15, and the rate at which that distance is changing is -9 miles per hour (negative because the distance is decreasing). The rate is the slope of a line describing distance, and the distance when time is zero is the y intercept.

Let y = distance and x = hours after starting for home. Then the equation is

$$y = -9x + 15.$$

You reach home when the distance (y) is zero, so substitute in $y = 0$ and solve for x:

$$0 = -9x + 15$$
$$9x = 15$$
$$x = \frac{15}{9} = \frac{5}{3}$$

It will take $1\frac{2}{3}$ hours, or 1 hour and 40 minutes to get home.

7. Find linear equations that describe real-world situations.

Try This One 10

A contestant on The Biggest Loser starts out weighing 345 pounds and loses weight steadily at the rate of 7 pounds per week. Write a linear equation that describes his weight in terms of weeks, and use it to find how long it will take him to reach his goal of 175 pounds.

Answers to Try This One

1

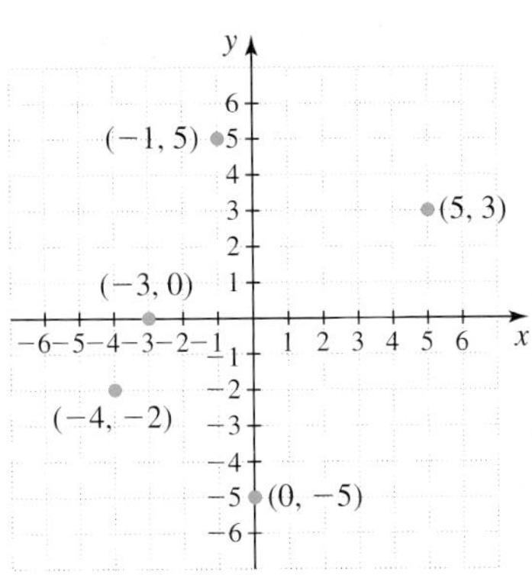

2 $A = (2, 2)$; $B = (6, 0)$; $C = (0, 0)$; $D = (0, -3)$; $E = (-4, -1)$; $F = (2, -5)$; $G = (-5, 6)$

3

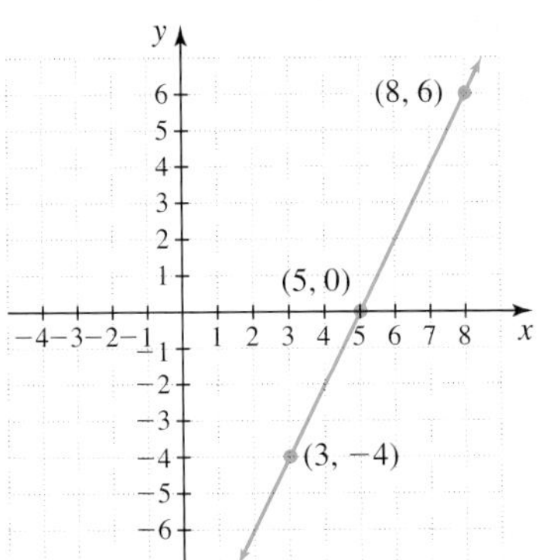

4 (a) and (b)

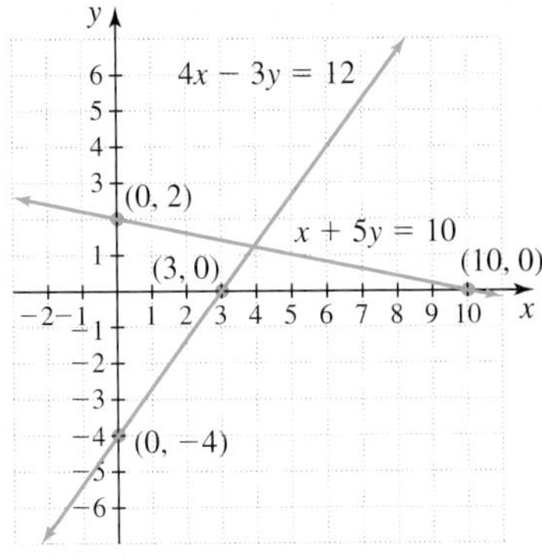

5 $m = -4$

6 $m = \frac{1}{2}$

7

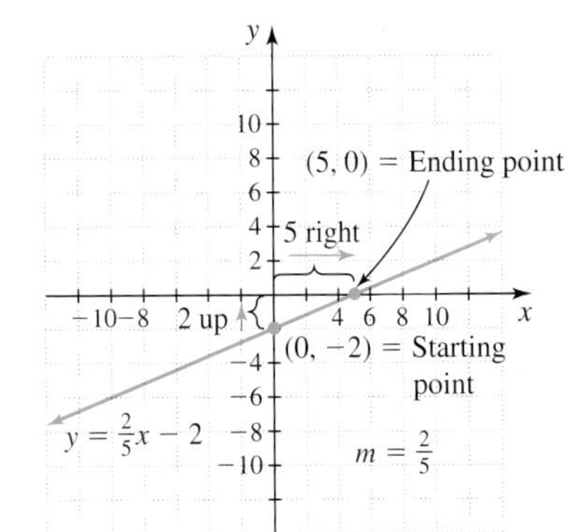

8

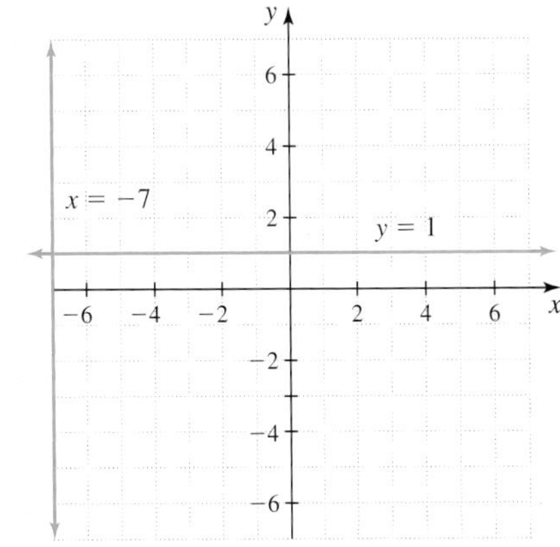

9 $y = 6.75x + 0.35x$; 3 toppings: \$7.80; 5 toppings: \$8.50

10 $y = -7x + 345$; about 24.3 weeks

EXERCISE SET 7-1

Writing Exercises

1. Explain how the quadrant where a point is located can be determined by looking at the signs of the coordinates.
2. Explain what the slope of a line means.
3. Explain how to find the intercepts of a line.
4. Explain how to determine if a line is vertical by looking at its equation.
5. Explain how to determine if a line is horizontal by looking at its equation.
6. Explain how to find the slope of a line without finding two points on the line. Assume the line is neither vertical nor horizontal.

Computational Exercises

For Exercises 7–20, plot each point on the Cartesian plane.

7. $(-2, -5)$
8. $(-3, -8)$
9. $(-6, 4)$
10. $(3, -7)$
11. $(6, 0)$
12. $(-4, 0)$
13. $(0, 3)$
14. $(0, -2)$
15. $(5.6, -3.2)$
16. $(-4.8, 7.3)$
17. $(-6, -10)$
18. $(-4, -9)$
19. $(0, 0)$
20. $(5, 5)$

For Exercises 21–30, draw the graph for each equation by first finding at least two points on the line.

21. $5x + y = 20$
22. $x + 4y = 24$
23. $3x - y = 15$
24. $2x - y = 10$
25. $4x + 7y = 28$
26. $3x - 8y = 24$
27. $2x + 7y = -12$
28. $5x - 3y = -18$
29. $6x - 4y = 28$
30. $2x - 7y = 28$

For Exercises 31–38, find the slope of the line passing through the two points.

31. $(-3, -2), (6, 7)$
32. $(4, 0), (3, -5)$
33. $(2, 10), (4, 9)$
34. $(6, 3.5), (4.2, 6)$
35. $(-4, -5), (-9, -2)$
36. $(3.8, -1.2), (2.2, 3.1)$
37. $(2, -6.1), (3.4, -2.8)$
38. $(4, 0), (0, 7)$

For Exercises 39–46, find the coordinates of the x intercept and the y intercept for each line.

39. $3x + 4y = 24$
40. $-2x + 7y = -28$
41. $-5x - 6y = 30$
42. $x + 6y = 10$
43. $2x - y = 18$
44. $9x + 4y = -36$
45. $5x - 2y = 15$
46. $9x - 7y = 18$

For Exercises 47–54, write the equation in slope-intercept form, then find the slope and the y intercept. Finally, draw the graph of the line.

47. $7x + 5y = 35$
48. $-2x + 7y = 14$
49. $x - 4y = 16$
50. $4x - 8y = 15$
51. $8x - 3y = 24$
52. $3x - 7y = 14$
53. $2x - y = 19$
54. $3x - 9y = 20$

For Exercises 55–58, draw the graph for each equation.

55. $x = -3$
56. $y = 2$
57. $y = 6$
58. $x = 7$

Real-World Applications

59. A newspaper advertisement costs $6.50 per week to run the ad plus a setup charge of $50.00. Find the cost of running the ad for
(a) 3 weeks. (b) 5 weeks. (c) 10 weeks.
60. A painter's labor charges are $50 plus $40 per room to paint the interior of a house. Find the cost of painting
(a) a five-room house. (c) a nine-room house.
(b) a seven-room house.
61. The cost of renting an automobile is $40 per day plus $1.10 per mile. Find the cost of renting the automobile for 1 day if it is driven
(a) 63 miles. (b) 42 miles. (c) 127 miles.
62. The number of e-mails, in billions, sent per day in the United States is approximated by the equation $y = 10x + 190$, where x is the number of years from now. Find the number of emails that will be sent 5 years from now, assuming this trend continues.
63. The number in thousands of civilian staff in the military can be approximated by the equation $y = 10.5x$, where x is the number of years from now. Find the number of civilians in the military 3 years from now.
64. The percentage of alcohol-related traffic deaths can be approximated by the equation $y = 1.2x$, where x is the number of years from now. What will be the percentage of alcohol-related traffic deaths 2 years from now?
65. The number in millions of Americans over age 65 can be approximated by the equation $y = 0.5x + 35.3$, where x is the number of years from now. Find the number of people over 65 living 5 years from now.

66. The percentage of the population of the United States with less than 12 years of school can be approximated by the equation $y = 18 - 1.1x$, where x represents the number of years from now. Find the percentage of the population with less than 12 years of school 4 years from now.
67. The amount of \$10 in 2000 has the same buying power as \$12.32 in 2008. If something cost \$10 in 2000, its cost x years from 2000 can be approximated by the equation $y = 10 + 0.29x$. Find the cost of an item in 2016 if it cost \$10 in 2000.
68. The number of women who use a computer at work in the United States can be modeled by the equation $y = 13{,}219x + 24{,}813$, where x is the number of years from now. How many women will be using a computer at work 10 years from now if this trend continues?
69. The cost of reserving a particular domain name is \$25 up front plus \$4.99 per month. Find the cost of reserving that domain name for
(a) 3 months. (b) 6 months. (c) 2 years.
70. The number of full-time students enrolled in college can be modeled by the equation $y = 42{,}000x + 1{,}558{,}000$, where x is the number of years from now. How many students will be enrolled full-time in college 9 years from now if this trend continues?
71. The homecoming committee is blowing up balloons for the big dance on Saturday. Forty balloons have already been inflated, and now the next shift comes in and is able to blow up 20 balloons per hour. Assuming they continue at this rate, write a linear equation that describes the total number of balloons blown up in terms of hours and use the equation to find out how many hours it will take them to reach their goal of having 500 balloons.
72. It costs Damage Inc. \$1.73 for each Super Slingshot it produces, and the initial cost to the company for production, marketing, and paying workers is \$12,000. Write a linear equation that describes the total cost of producing x slingshots, and use your equation to find out how much it will cost to produce 1,000 slingshots.
73. Sandi weighs 160 lb, but she has started a new diet plan that promises she will lose 3 lb per month every month until she reaches her goal weight. Write a linear equation that describes how much Sandi weighs each month, and then use your equation to find out how many months it will take her to get to 130 lb.
74. The state lottery starts at \$2 million, and then for each \$1 ticket purchased, it adds \$0.50 to the pot. Write a linear equation that describes how much money is in the pot in terms of number of tickets sold and then use your equation to find out how much will be in the pot if 250,000 tickets are sold.
75. There are 350 known diseases within a community and doctors can eliminate 5 of them per year. Assuming no new diseases are introduced to the community, write a linear equation that represents the number of diseases in terms of years from now. Use your equation to predict how many years it will take for the community to be free of all diseases.
76. There are currently 1,250 known computer viruses floating around on the Internet and 12 new viruses are introduced every day. Assuming the number of viruses keeps growing at this rate, write a linear equation that represents how many viruses there are in terms of days from now, and use your equation to find out how many days it will take for there to be 10,000 viruses.

Critical Thinking

77. Show why the slope of a vertical line is said to be undefined. (*Hint:* Select two points on a vertical line and calculate the slope.)
78. Why is the slope of a horizontal line zero?
79. Find the formula for the distance between two points on a line. (*Hint:* Use the Pythagorean theorem. See Exercises 81–84 in Section 6-6.)
80. Find the formulas for the coordinates of a midpoint when given two points on a line. (A midpoint is halfway between the two given points.)

Section 7-2 Systems of Linear Equations

One of the most important ideas in Chapter 6 is that linear equations can be used to solve real-world problems. In that chapter, we looked at equations that have one variable. But there are a lot of problems where more than one quantity might change. For example, suppose you're helping to organize a tailgate event for a student group, and your job is to buy cups and plates, staying within a certain budget. In this situation, there are two quantities that can vary—the number of cups and the number of plates.

LEARNING OBJECTIVES

- ☐ 1. Solve systems of linear equations graphically.
- ☐ 2. Identify inconsistent and dependent systems.
- ☐ 3. Solve systems of linear equations by substitution.
- ☐ 4. Solve systems of linear equations by the addition/subtraction method.
- ☐ 5. Solve real-world problems using systems of linear equations.

So in this section we'll look at equations with more than one variable. It turns out that in order to find a numeric answer when solving equations with two variables, you need two equations also. (With only one, you can solve for one variable, but your solution will always have the other variable in it.) A pair of linear equations with the same two variables is called a **system of two linear equations.**

A **system of two linear equations in two variables** x and y is a pair of equations that can be written in the form

$$a_1x + b_1y = c_1$$
$$a_2x + b_2y = c_2$$

An example of a linear system is

$$2x + y = 7$$
$$x - 4y = -10$$

Solving a system of two linear equations means finding all pairs of numbers that satisfy both equations in the system. For the system above, the pair (2, 3) is a solution because substituting in $x = 2$ and $y = 3$ makes both equations true. We will study three methods for finding solutions to systems of linear equations.

Solving a Linear System Graphically

We know that the graph of any linear equation with two variables is a straight line. We also know that the coordinates of every point on the line satisfy the equation. So a pair of numbers that satisfies *both* equations in a system must correspond to a point that is on *both* lines.

Solving a System of Equations Graphically

Step 1 Draw the graphs of both equations on the same Cartesian plane.

Step 2 Find the point or points of intersection, if there are any.

EXAMPLE 1 Solving a System Graphically

Math Note

The solution can be checked by substituting the solution into both equations and seeing if both equations are true:

$x + 3y = 8$	$2x - y = 9$
$5 + 3(1) \stackrel{?}{=} 8$	$2(5) - 1 \stackrel{?}{=} 9$
$5 + 3 \stackrel{?}{=} 8$	$10 - 1 \stackrel{?}{=} 9$
$8 = 8$	$9 = 9$

Solve the system graphically:

$$x + 3y = 8$$
$$2x - y = 9$$

SOLUTION

Step 1 Draw the graphs for both equations on the Cartesian plane.

$x + 3y = 8$

x	y
8	0
2	2
−1	3

$2x - y = 9$

x	y
1	−7
4	−1
5	1

Find three points as we did in Section 7-1.

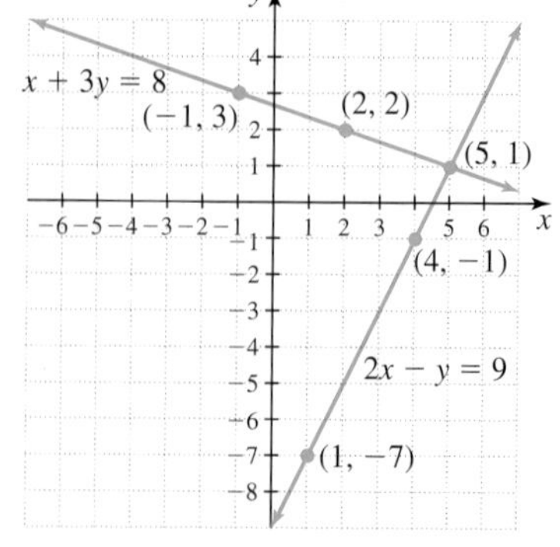

Step 2 Find the point of intersection of the two lines. In this case, it is (5, 1), so the solution set is {(5, 1)}.

▼ Try This One 1

☑ 1. Solve systems of linear equations graphically.

Solve the system graphically:

$$\begin{aligned} x - 4y &= 8 \\ 2x - y &= -5 \end{aligned}$$

There are three possibilities to consider when you are finding the solution to a system of two linear equations in two variables.

1. *The lines intersect at a single point.* In this case, there is only one solution, and that is the point of intersection of the lines. See Figure 7-8(a). In this case, the system is said to be **consistent** and **independent.**
2. *The lines are parallel.* In this case, there would be no solution, or the solution would be the empty set, ∅, since parallel lines never intersect. See Figure 7-8(b). In this case, the system is said to be **inconsistent** and independent.
3. *The lines coincide.* In this case, the graph of both equations is the same line; so any point on the line will satisfy both equations. There are infinitely many solutions. In this case, the system is said to be consistent and **dependent.** The solution set is written as $\{(x, y) \mid ax + by = c\}$, where $ax + by = c$ is either one of the two equations. See Figure 7-8(c).

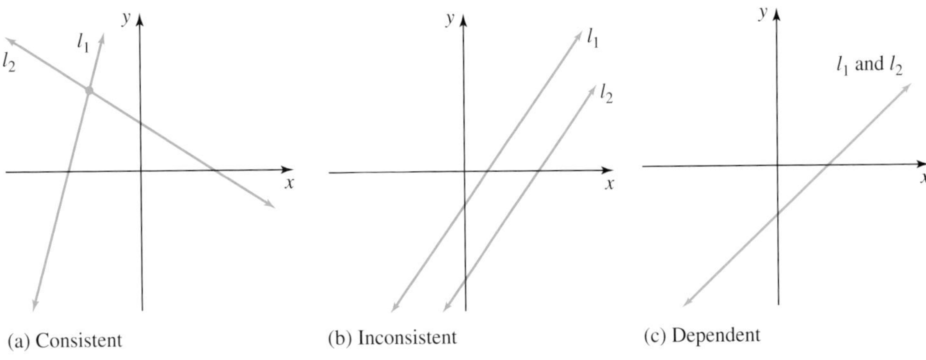

Figure 7-8

EXAMPLE 2 Identifying an Inconsistent System Graphically

Solve the system graphically:

$$\begin{aligned} 3x + y &= 12 \\ 6x + 2y &= 18 \end{aligned}$$

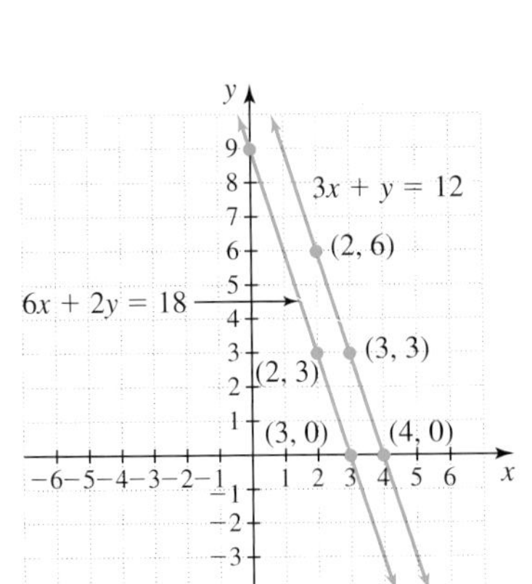

SOLUTION

Step 1 Graph both equations.

$3x + y = 12$

x	y
4	0
3	3
2	6

$6x + 2y = 18$

x	y
3	0
0	9
2	3

Find three points on each line.

Step 2 Find the point or points of intersection. In this case, the lines are parallel, and never intersect. The solution set is ∅, and the system is inconsistent.

▼ Try This One 2

Solve the system graphically:

$$3x + 4y = 12$$
$$6x + 8y = 18$$

EXAMPLE 3 Identifying a Dependent System Graphically

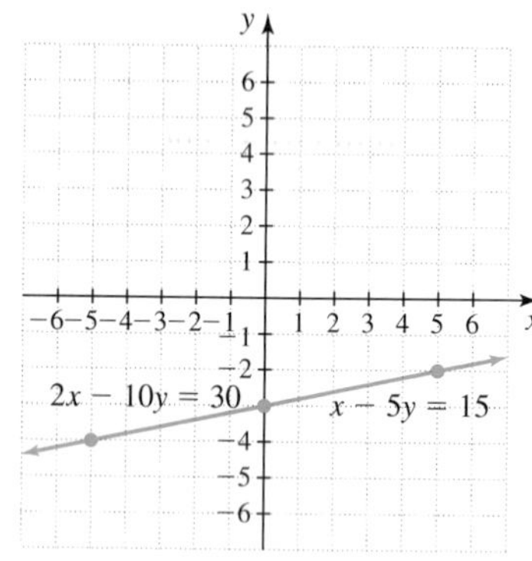

Solve the system graphically:

$$x - 5y = 15$$
$$2x - 10y = 30$$

SOLUTION

Step 1 Graph both equations.

$x - 5y = 15$

x	y
−5	−4
5	−2
0	−3

$2x - 10y = 30$

x	y
−5	−4
0	−3
5	−2

Find three points on each line.

Step 2 Find the point or points of intersection. In this case, the lines coincide; so the solution set is any point on either line. We can write this as $\{(x, y) \mid x - 5y = 15\}$. The system is dependent.

☑ 2. Identify inconsistent and dependent systems.

▼ Try This One 3

Solve the system graphically:

$$2x - y = 9$$
$$6x - 3y = 27$$

EXAMPLE 4 An Application of the Graphical Method

Mary leaves home at 9 A.M. and starts hiking around a lake at 3 miles per hour. At 10 A.M., her boyfriend Dave follows her, jogging at 5 miles per hour. When and how far from home do they meet?

SOLUTION

There are two variables for each person: the amount of time passed, and how far they've gone. We will choose 10 A.M. as our base time, since they're both moving after that point. Let x = the time in hours after 10 A.M., and let y = the distance traveled.

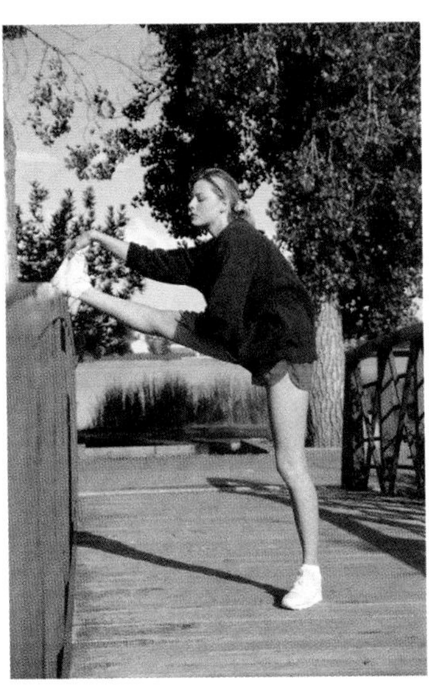

Mary: At 10 A.M., she's been hiking for an hour, and has gone 3 miles (her speed is 3 miles/hour). Her rate of change (slope) is 3, so Mary's distance is $y = 3x + 3$.

Dave: At 10 A.M., his distance is zero, and his rate of change is 5, so Dave's distance is $y = 5x$.

Now we have the system

$$y = 3x + 3$$
$$y = 5x$$

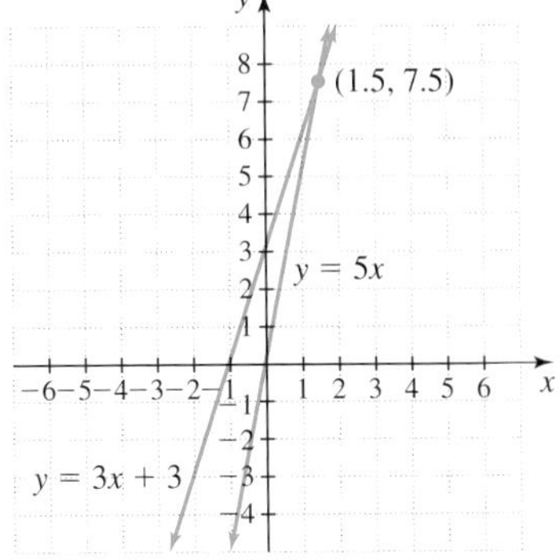

The graphs of the two equations shown to the right intersect at (1.5, 7.5), so Dave and Mary are at the same place at the same time 1.5 hours after 10 A.M., which is 11:30. And at that time, they are 7.5 miles from home.

▼ Try This One 4

At the beginning of 2010, a company had plants in two different cities. The plant in Sheboygan started with 60 employees, and added new employees at the rate of 6 per month. The plant in Kalamazoo started with 120 employees and added 3 new workers per month. How long will it take for the two plants to have the same number of employees, and how many employees will they have?

Solving a Linear System by Substitution

There is one big limitation to solving systems graphically: it can be very imprecise. Even in the best of circumstances, with very accurate graphs, it can be hard to tell exactly what the solution is. Fortunately, there are algebraic methods that will provide exact solutions. The first one we will study is known as **substitution.**

Math Note

You can use either equation and either variable, so look at the equations and decide which equation to use, and which variable is easier to solve for. To check your answer, substitute the values for x and y into the original equations and see if they are true.

Procedure for Solving a System of Equations by Substitution

Step 1 Pick one equation and solve it for one variable (either x or y) in terms of the other variable (see Math Note).

Step 2 Substitute the expression that you found in step 1 into the other equation for the variable you solved for.

Step 3 Solve the resulting equation for the unknown (it now has only one variable).

Step 4 Pick one of the original equations, substitute the value found in step 3 for the variable, and solve to find the value of the other variable.

Example 5 shows this procedure.

EXAMPLE 5 Solving a System by Substitution

Solve the following system by substitution.

$$x + 3y = 8$$
$$2x - y = 9$$

SOLUTION

Step 1 It will be easy to solve the first equation for x, so we start there.

$$x + 3y = 8 \qquad \text{Subtract } 3y \text{ from both sides.}$$
$$x = 8 - 3y$$

Step 2 Substitute the expression for x (i.e., $8 - 3y$) into the second equation.

$$2x - y = 9$$
$$2(8 - 3y) - y = 9$$

Step 3 Solve the equation for y.

$$2(8 - 3y) - y = 9 \qquad \text{Distribute.}$$
$$16 - 6y - y = 9 \qquad \text{Subtract 16 from both sides, and combine like terms.}$$
$$-7y = 9 - 16 \qquad \text{Simplify.}$$
$$-7y = -7 \qquad \text{Divide both sides by } -7.$$
$$\frac{-7y}{-7} = \frac{-7}{-7}$$
$$y = 1$$

Step 4 Substitute $y = 1$ into either equation and solve for x.

$$x + 3y = 8 \qquad \text{Substitute } y = 1.$$
$$x + 3(1) = 8 \qquad \text{Multiply.}$$
$$x + 3 = 8 \qquad \text{Subtract 3 from both sides.}$$
$$x = 5$$

The solution set is $\{(5, 1)\}$. Note that this is the same solution obtained by graphing, as shown in Example 1.

Math Note

When the solution set to a system of equations is a single point, like $\{(5, 1)\}$ in Example 5, it's common to write the solution in the form (5, 1) without the set notation brackets. This is the way we'll write such solutions in the remainder of Chapter 7.

▼ Try This One 5

Solve the system by substitution.

$$5x + y = 8$$
$$x + 2y = 7$$

EXAMPLE 6 Solving a System by Substitution

Solve the following system by substitution.

$$3x - 4y = 10$$
$$2x + 3y = 1$$

SOLUTION

Step 1 Neither of the variables in either equation is easier to solve for than the other, so we randomly choose to solve the first equation for x.

$$3x - 4y = 10 \qquad \text{Add } 4y \text{ to both sides.}$$
$$3x = 10 + 4y \qquad \text{Divide both sides by 3.}$$
$$x = \frac{10}{3} + \frac{4y}{3}$$

Step 2 Substitute the expression for x into the other equation.

$$2x + 3y = 1 \qquad \text{Substitute } x = \frac{10}{3} + \frac{4y}{3}.$$

$$2\left(\frac{10}{3} + \frac{4y}{3}\right) + 3y = 1 \qquad \text{Distribute.}$$

Step 3 Solve for y.

$$\frac{20}{3} + \frac{8y}{3} + 3y = 1 \qquad \text{Multiply every term by 3 to clear fractions.}$$

$$\frac{3}{1} \cdot \frac{20}{3} + \frac{3}{1} \cdot \frac{8y}{3} + 3 \cdot 3y = 3 \cdot 1 \qquad \text{Simplify.}$$

$$20 + 8y + 9y = 3 \qquad \text{Subtract 20 from both sides, and combine like terms.}$$

$$17y = -17 \qquad \text{Divide both sides by 17.}$$

$$\frac{17y}{17} = -\frac{17}{17} \qquad \text{Simplify.}$$

$$y = -1$$

Step 4 Substitute $y = -1$ into one equation and solve for x.

$$3x - 4y = 10 \qquad \text{Substitute } y = -1.$$

$$3x - 4(-1) = 10 \qquad \text{Multiply.}$$

$$3x + 4 = 10 \qquad \text{Subtract 4 from both sides.}$$

$$3x = 6 \qquad \text{Divide both sides by 3.}$$

$$\frac{3x}{3} = \frac{6}{3} \qquad \text{Simplify.}$$

$$x = 2$$

The solution is $(2, -1)$.

Check:

$$3x - 4y = 10 \qquad 2x + 3y = 1$$

$$3(2) - 4(-1) \stackrel{?}{=} 10 \qquad 2(2) + 3(-1) \stackrel{?}{=} 1$$

$$6 + 4 \stackrel{?}{=} 10 \qquad 4 - 3 \stackrel{?}{=} 1$$

$$10 = 10 \qquad 1 = 1$$

▼ Try This One 6

☑ 3. Solve systems of linear equations by substitution.

Solve the system by substitution.

$$3x - 5y = 29$$
$$2x + 4y = -10$$

Solving a System by Addition/Subtraction (Elimination)

Another algebraic method that is used to solve a system of linear equations in two variables is called the **addition/subtraction method** or the *elimination* method. The steps are shown next.

Procedure for Solving a System of Linear Equations Using the Addition/Subtraction Method

Step 1 If necessary, rewrite the equations so they are in the form $ax + by = c$.

Step 2 Multiply one or both of the equations by a number so that the two coefficients of either x or y have the same absolute value.

Step 3 Either add or subtract the two equations so that one of the variables is eliminated. Add if the matching coefficients from Step 2 are negatives of each other, and subtract if they are equal.

Step 4 Solve the resulting single-variable equation for the variable.

Step 5 Substitute the value of the variable from Step 4 into either of the original equations, and solve for the other variable.

The process will be a lot clearer when illustrated with some examples.

EXAMPLE 7 Solving a System by Addition/Subtraction

Solve the following system by the addition/subtraction method:

$$2x - 3y = -6$$
$$x = 7 - y$$

SOLUTION

Step 1 Write both equations in the form $ax + by = c$. (Add y to both sides of the second equation.)

$$2x - 3y = -6$$
$$x + y = 7$$

Step 2 Multiply the second equation by 2 in order to make the coefficients of the x terms equal.

$$2x - 3y = -6$$
$$2(x + y) = 2 \cdot 7 \qquad \textit{Distribute.}$$

which gives us

$$2x - 3y = -6$$
$$2x + 2y = 14$$

Step 3 Subtract the second equation from the first equation to eliminate the x variable.

$$\begin{array}{r} 2x - 3y = -6 \\ -2x - 2y = -14 \\ \hline -5y = -20 \end{array} \qquad \textit{Subtract every term!}$$

Step 4 Solve the equation for y.

$$-5y = -20 \qquad \textit{Divide both sides by } -5.$$
$$\frac{-5y}{-5} = \frac{-20}{-5} \qquad \textit{Simplify.}$$
$$y = 4$$

Step 5 Pick one equation and substitute 4 for y, then solve for x.

$$2x - 3y = -6 \quad \text{Substitute } y = 4.$$
$$2x - 3(4) = -6 \quad \text{Multiply.}$$
$$2x - 12 = -6 \quad \text{Add 12 to both sides.}$$
$$2x = 6 \quad \text{Divide both sides by 2.}$$
$$\frac{2x}{2} = \frac{6}{2} \quad \text{Simplify.}$$
$$x = 3$$

The solution is (3, 4).

Check:

$$2x - 3y = -6 \qquad x = 7 - y$$
$$2(3) - 3(4) \stackrel{?}{=} -6 \qquad 3 \stackrel{?}{=} 7 - 4$$
$$6 - 12 \stackrel{?}{=} -6 \qquad 3 = 3$$
$$-6 = -6$$

▼ Try This One 7

Solve the system by the addition/subtraction method:

$$5x = -23 - y$$
$$2x + 3y = -4$$

Sometimes it's necessary to multiply each equation by a different number in order to eliminate one of the variables.

EXAMPLE 8 Solving a System by Addition/Subtraction

Solve the system by the addition/subtraction method:

$$5x - 2y = 10$$
$$3x + 5y + 56 = 0$$

SOLUTION

Step 1 Write the equations in the form $ax + by = c$. (Subtract 56 from both sides of the second equation.)

$$5x - 2y = 10$$
$$3x + 5y = -56$$

Step 2 Multiply the first equation by 5 and the second equation by 2 to make the coefficients of the y terms equal in absolute value.

$$5 \cdot (5x - 2y) = 5 \cdot 10$$
$$2 \cdot (3x + 5y) = 2 \cdot (-56)$$

which gives us

$$25x - 10y = 50$$
$$6x + 10y = -112$$

Step 3 Add the equations.

$$\begin{aligned} 25x - 10y &= 50 \\ 6x + 10y &= -112 \\ \hline 31x \qquad &= -62 \end{aligned}$$

Step 4 Solve the resulting equation for x.

$$\frac{31x}{31} = \frac{-62}{31} \quad \text{Divide both sides by 31.}$$

$$x = -2$$

Step 5 Substitute $x = -2$ into one of the equations and solve for y.

$$\begin{aligned} 5x - 2y &= 10 && \text{Substitute } x = -2. \\ 5(-2) - 2y &= 10 && \text{Multiply.} \\ -10 - 2y &= 10 && \text{Add 10 to both sides.} \\ -2y &= 20 && \text{Divide both sides by } -2. \\ \frac{-2y}{-2} &= \frac{20}{-2} && \text{Simplify.} \\ y &= -10 \end{aligned}$$

The solution is $(-2, -10)$.

Check:

$$\begin{aligned} 5x - 2y &= 10 \\ 5(-2) - 2(-10) &\stackrel{?}{=} 10 \\ -10 + 20 &\stackrel{?}{=} 10 \\ 10 &= 10 \end{aligned} \qquad \begin{aligned} 3x + 5y + 56 &= 0 \\ 3(-2) + 5(-10) + 56 &\stackrel{?}{=} 0 \\ -6 - 50 + 56 &\stackrel{?}{=} 0 \\ 0 &= 0 \end{aligned}$$

☑ 4. Solve systems of linear equations by the addition/subtraction method.

▼ Try This One 8

Solve the system by the addition/subtraction method:

$$3x - 4y - 4 = 0$$
$$4x = 3y + 3$$

Recall that a system of linear equations can also be inconsistent (meaning the lines are parallel) or dependent (meaning the lines coincide). When you try to solve an inconsistent or dependent system by substitution or addition/subtraction, both variables will be eliminated. *If the resulting equation is false, the system is inconsistent, and if the resulting equation is true, the system is dependent.*

EXAMPLE 9 Identifying an Inconsistent System

Solve the system:

$$3x + y = 12$$
$$6x + 2y = 15$$

SOLUTION

Solving by substitution, we get

$$\begin{aligned} 3x + y &= 12 && \text{Subtract } 3x \text{ from both sides.} \\ y &= 12 - 3x && \text{Substitute into second equation.} \\ 6x + 2(12 - 3x) &= 15 && \text{Distribute.} \\ 6x + 24 - 6x &= 15 && \text{Combine like terms.} \\ 24 &= 15 \end{aligned}$$

Since the resulting equation, $24 = 15$, is false, the system is inconsistent. The lines are parallel, and the solution set is $\varnothing$.

▼ Try This One 9

Solve the system:

$$\begin{aligned} 6x + 3y &= 4 \\ -2x - y &= 1 \end{aligned}$$

EXAMPLE 10 Identifying a Dependent System

Solve the system:

$$\begin{aligned} 5x + y &= 9 \\ -20x - 4y &= -36 \end{aligned}$$

SOLUTION

This time, for variety, we'll use the addition/subtraction method.

First, we multiply the first equation by 4.

$$\begin{aligned} 4(5x + y) &= 4 \cdot 9 && \text{Distribute.} \\ -20x - 4y &= -36 \end{aligned}$$

This gives us

$$\begin{aligned} 20x + 4y &= 36 \\ -20x - 4y &= -36 \end{aligned}$$

Now we add the equations:

$$\begin{aligned} 20x + 4y &= 36 \\ \underline{-20x - 4y} &\underline{= -36} \\ 0 &= 0 \end{aligned}$$

Since the variables are all eliminated and the resulting equation is true, the system is dependent. The solution set is $\{(x, y) \mid 5x + y = 9\}$.

▼ Try This One 10

Solve the system:

$$\begin{aligned} 4x &= 2y + 9 \\ 6y &= 12x - 27 \end{aligned}$$

Applications of Linear Systems

As we pointed out at the beginning of the section, there are a lot of real-world problems with more than one variable, so systems of equations can be very useful in that regard. In this section, though, we're limiting the discussion to two equations with two variables.

EXAMPLE 11 Applying Systems of Equations to Finance

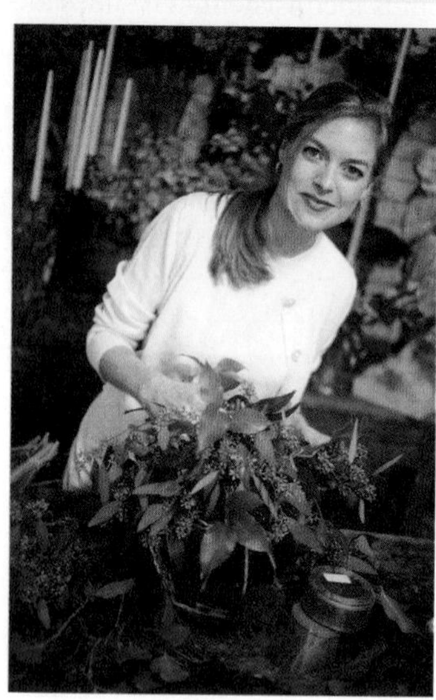

A small business owner plans to set up an investment plan that will generate enough income to pay the property taxes on her shop, a total of \$2,200 per year. She has \$24,000 to invest, and divides that amount between two accounts. One pays 8% annually. The other is a bit riskier, but it pays 10% annually unless it collapses. How much should be invested in each account to raise the required \$2,200 in annual income?

SOLUTION

Let x = the amount of money invested at 8%. Let y = the amount of money invested at 10%. The first equation is $x + y = \$24{,}000$ since this is the total amount of money invested. The second equation is $0.08x + 0.10y = \$2{,}200$, since this is the amount of interest earned from 8% of the investment amount x and 10% of the investment amount y.

The system is

$$\begin{aligned} x + y &= 24{,}000 \\ 0.08x + 0.10y &= 2{,}200 \end{aligned}$$

Solving the system by substitution:

$$\begin{aligned}
x + y &= 24{,}000 && \textit{Subtract } x \textit{ from both sides.} \\
y &= 24{,}000 - x && \textit{Substitute into the second equation.} \\
0.08 + 0.10y &= 2{,}200 && \\
0.08x + 0.10(24{,}000 - x) &= 2{,}200 && \textit{Distribute.} \\
0.08x + 2{,}400 - 0.1x &= 2{,}200 && \textit{Combine like terms.} \\
-0.02x + 2{,}400 &= 2{,}200 && \textit{Subtract 2,400 from both sides.} \\
-0.02x + 2{,}400 - 2{,}400 &= 2{,}200 - 2{,}400 && \textit{Simplify.} \\
-0.02x &= -200 && \textit{Divide both sides by } -0.02. \\
\frac{-0.02x}{-0.02} &= \frac{-200}{-0.02} && \textit{Simplify.} \\
x &= \$10{,}000 &&
\end{aligned}$$

\$10,000 should be invested at 8%. This leaves \$14,000 invested at 10%.

▼ Try This One 11

A new campus is being formed in a university system, and it was decided that the first class admitted will consist of 400 students. The school anticipates a 20% annual increase in male students, and a 30% annual increase in female students. It also anticipates being able to accommodate 495 students next year. How many men and women should they admit now?

EXAMPLE 12 Applying Systems of Equations to Unit Cost

A campus bookstore received two shipments from Apple Computer over the last month. The first contained six iPods and eight MacBooks, and the cost to the bookstore was \$6,840. The second shipment was three iPods and five MacBooks, at a wholesale cost of \$4,170. The bookstore manager is unable to find the itemized invoice, and accounting needs to know how much each individual item cost. What were the individual costs?

SOLUTION

We are asked to find two quantities: the cost of one iPod and the cost of one MacBook. Let x = the iPod cost and y = the MacBook cost. Then $6x + 8y$ is the cost of the first shipment, and $3x + 5y$ is the cost of the second. This gives us a system:

$$6x + 8y = 6{,}840$$
$$3x + 5y = 4{,}170$$

We will solve using addition/subtraction. Multiply the second equation by 2:

$$2(3x + 5y) = 2 \cdot 4{,}170$$
$$6x + 10y = 8{,}340$$

Now subtract the second equation from the first, then solve for y:

$$\begin{aligned} 6x + 8y &= 6{,}840 \\ -6x - 10y &= -8{,}340 \\ \hline -2y &= -1{,}500 \quad \text{Divide both sides by } -2. \\ y &= 750 \end{aligned}$$

Substitute $y = 750$ into one of the original equations and solve for x:

$$\begin{aligned} 3x + 5(750) &= 4{,}170 \quad \text{Multiply.} \\ 3x + 3{,}750 &= 4{,}170 \quad \text{Subtract 3,750 from both sides.} \\ 3x &= 420 \quad \text{Divide both sides by 3.} \\ x &= 140 \end{aligned}$$

The bookstore paid \$140 for each iPod and \$750 for each MacBook.

▼ Try This One 12

☑ 5. Solve real-world problems using systems of linear equations.

Pablo bought three bags of chips and two bottles of soda and paid \$9.15. His friend Michael bought two bags of chips and four bottles of soda and paid \$8.74. What is the price of a bottle of soda? What is the price of a bag of chips?

Answers to Try This One

1 $\{(-4, -3)\}$

2 No solution; the lines are parallel.

3 $\{(x, y) \mid 2x - y = 9\}$

4 In 20 months, each will have 180 employees.

5 $\{(1, 3)\}$

6 $(3, -4)$

7 $(-5, 2)$

8 $(0, -1)$

9 No solution

10 $\{(x, y) \mid 4x - 2y = 9\}$

11 250 men and 150 women

12 A bag of chips is \$2.39 and a bottle of soda is \$0.99.

EXERCISE SET 7-2

Writing Exercises

1. What does it mean to be a solution to a system of two equations?
2. Describe how to solve a system of linear equations graphically.
3. Explain why a point common to both lines is a solution to a system of linear equations.
4. Explain why algebraic methods for solving systems of equations are important.
5. What types of real-world problems can be solved using systems of equations?
6. Based on the graphical method, how many solutions are possible for any system of two linear equations?

Computational Exercises

For Exercises 7–12, decide whether or not the given points are solutions to the system of equations.

7. $3x - y = 5$
   $x + 2y = 11$ (3, 4), (5, 10)
8. $2x + 4y = 4$
   $x - 3y = -8$ (4, −1), (−2, 2)
9. $6x + 7y = 24$
   $-2x + y = 2$ (11, −6), (1/2, 3)
10. $7x - 3y = 12$
    $-4x - 6y = -12$ (2, 2/3), (3, 3)
11. $4x = 6y - 190$
    $5x + 5y = 75$ (−10, 25), (5, 35)
12. $-10x + 15y = 230$
    $10y = -20x - 60$ (4, 18), (−8, 10)

For Exercises 13–20, solve each system by graphing.

13. $2x - y = 14$
    $x = y + 7$
14. $x - 3y = 6$
    $4x + 3y = 9$
15. $2x - y = -4$
    $x + y = -2$
16. $x + y = 2$
    $3x - 2y = -9$
17. $4x + 3y = 2$
    $3x + 5y = -4$
18. $x + 2y = 0$
    $2x - y = 0$
19. $x - 2y = 10$
    $x + y = 4$
20. $3x - 5y = -2$
    $10y - 6x = 4$

For Exercises 21–28, solve each system by substitution.

21. $x + y = 0$
    $x - 2y + 9 = 0$
22. $x = 2 - y$
    $3x - 2y = 1$
23. $x - 3y = 7$
    $4x + 3y = 13$
24. $x + 2y = 11$
    $2x - y = 7$
25. $4x + 3y = 24$
    $3x + 5y = 22$
26. $3x = 5y + 16$
    $5y - 3x = -16$
27. $2x - 3y = 1$
    $x = y + 2$
28. $2x + 3y = -1$
    $x = y - 13$

For Exercises 29–36, solve each system by addition/subtraction.

29. $3x = 2y + 10$
    $9y = 3x - 7$
30. $5x - y = 10$
    $x - 2y = 18$
31. $3x + 4y = 2$
    $y = x - 3$
32. $x = 3y - 6$
    $x - 3y = 12$
33. $5x - 2y = 11$
    $15x - 6y = 33$
34. $-x = -2y + 11$
    $x + 3y = 14$
35. $x = 3y - 2$
    $5x - 4y = 12$
36. $2x - 3y = -2$
    $5x = 2y - 9$

For Exercises 37–44, solve each system by any method; state whether the system is consistent, inconsistent, or dependent; and give the solution.

37. $4x + y = 2$
    $7x + 3y = 1$
38. $8x - 2y = -2$
    $3x - 5y = 4$
39. $x - 6y = 19$
    $2x + 7y = 0$
40. $4x - y = 11$
    $7x + 3y = 14$
41. $3x - 5y = 5$
    $5x - 7y = 1$
42. $x = 3y + 2$
    $x + 3y = 14$
43. $4x - y = 11$
    $7x + 3y = 8$
44. $4x - y = 3$
    $8x - 2y = 6$

Real-World Applications

45. Lane has two jobs, Job A and Job B, and two savings accounts, S1 and S2. Every week, he puts 10% of his check from Job A into S1 and 20% of his check from Job A into S2. He then puts 30% of his check from Job B into S1 and 40% of his check from Job B into S2. Every week, the deposit into S1 is \$127 and the deposit into S2 is \$184. The rest of his salaries from both jobs go into his checking account. How much does Lane get paid at each job per week and how much gets deposited into his checking account?
46. A grocer mixes coffee that sells for \$6.40 a pound with coffee that sells for \$10.80 a pound. If she wants to have 30 pounds of coffee to sell at \$9.04 a pound, how many pounds of each would she have to use?
47. Adult tickets for a tourist train ride cost \$12.00 and child tickets cost \$8.00. If there were 40 passengers

on the train and the net revenue was \$420.00, how many adults and how many children rode the train?

48. Two groups of students bought airline tickets to go to a conference. One group bought three tickets from Airline 1 and two tickets from Airline 2, and their total was \$800 before taxes and fees. The other group bought one ticket from Airline 1 and four tickets from Airline 2, and their total was \$850 before taxes and fees. How much does a ticket cost for each airline?
49. In preparation for the week's reality chef competition, the producer went to one market and got boneless chicken for \$4.53 per pound and filet mignon for \$8.28 per pound. She then went to another market since the first one was now out of the meat she needed, and she got the same amount of chicken and filet mignon, but at this market, the chicken was \$5.72 per pound and the filet mignon was \$7.54 per pound. At the first market, she spent \$144.66 and at the second market she spent \$147.68. How many pounds of chicken and filet mignon did she buy at both places?
50. A candy store owner mixes spearmint candy selling for \$0.99 a pound with cinnamon candy selling for \$0.89 a pound. If there are 10 pounds of the mixture selling for \$0.94 per pound, how many pounds of each type were mixed together?
51. At a fast food restaurant, three chicken sandwiches and two large orders of French fries cost \$8.87, and five chicken sandwiches and four large orders of French fries cost \$15.55. How much does each item cost?
52. A person wants to invest \$2,400.00, part at 9% and part at 6%. If the total interest desired at the end of the year is \$189.00, how much should be invested at each rate?
53. At a school concert, student tickets cost \$5.00 and general admission tickets for nonstudents were \$8.00. If the total revenue was \$3,034 and 500 tickets were sold, how many students attended the concert? How many general admission tickets were sold?
54. At a flea market, Joe bought some VCR tapes and CDs. The CDs sold for \$5.00 each, and the VCR tapes sold for \$3.00 each. If the total cost of the items was \$78.00 and the total number of products sold was 18, find how many of each item Joe bought.
55. At a thrift store, all tops are the same price and all pairs of pants are the same price. Stacy buys three tops and five pairs of pants for \$42 and her friend Lacy buys one top and two pairs of pants for \$16. How much does the thrift store charge for a top and for a pair of pants?
56. On a game show, two numbers are selected at random from a box. Twice the smaller number less the larger number yields −7. Four times the smaller number and twice the larger number yields 54. What are the two numbers pulled from the box?
57. For an upcoming concert, one lawn seat and one general admission seat costs \$38. A group of five bought the only tickets left, and they got three general admission tickets and two lawn seats for \$98. How much does a general admission ticket cost?
58. On one website, each music video is a certain price and each movie is a certain price. Stan bought 10 music videos and 3 movies and spent \$105. Steve went to the same website and bought 12 music videos and 2 movies and spent \$94. How much does each item cost?
59. The Student Government Association sold carnations for \$5 each for Valentine's Day and they sold green bagels for \$3 each for St. Paddy's Day. They sold 800 items for \$2,800. How many carnations were sold and how many bagels were sold?
60. A cabinetmaker needs 200 feet of two kinds of wood. Poplar sells for \$10 per 8-foot board and mahogany sells for \$12 per 8-foot board. Before taxes, the total amount he spent to get the proper type and amount of wood is \$288.75. How many feet does the cabinet maker need of each type of wood?
61. At a bookstore, Janie selected three paperbacks that were all the same price and two hardcover books that were each the same price, and she thought she'd be spending \$55.96 for the entire purchase before taxes. When she got to the register, she found out that the paperbacks were all 10% off and the hardcover books were all 5% off, so she ended up spending \$51.96 before taxes. What is the sale price of the paperbacks and what is the sale price of the hardcover books?
62. On a 7-hour trip home for winter break, Marta was forced to drive on a stretch of highway that had a lot of construction. At first, she had to go 65 miles per hour for a certain amount of time, but then she was able to go 75 miles per hour for the rest of the trip, and in total she drove 485 miles. Assuming she made no stops, how long did she drive 65 miles per hour and how long did she drive 75 miles per hour?

Critical Thinking

63. Write a system of equations that has (3, −5) as a solution. (There are many possible answers.)
64. A system of three equations with three unknowns can be solved by selecting a pair of equations and eliminating one variable and then selecting another pair and eliminating the same variable. The resulting two equations will have two variables. This system can be solved by methods shown in this chapter. Solve this system:

$$\begin{aligned} 3x + 2y - 2z &= 1 \\ x + 3y + z &= 10 \\ 2x - 4y + z &= -3 \end{aligned}$$

65. Which method is more likely to result in having to work with fractional expressions, substitution or addition/subtraction? Explain.

66. (a) Solve the system below graphically.

$$\begin{aligned} 6x - 4y &= 8 \\ -12x + 12y &= -19 \end{aligned}$$

(b) Now solve the system using the addition/subtraction method. Did you get the same solution? What does this tell you about the limitations of the graphical method?

Section 7-3 Solving Systems of Linear Equations Using Matrices

LEARNING OBJECTIVES

- ☐ 1. Identify matrices in row echelon form.
- ☐ 2. Solve systems of two linear equations using matrices.
- ☐ 3. Solve systems of three linear equations using matrices.

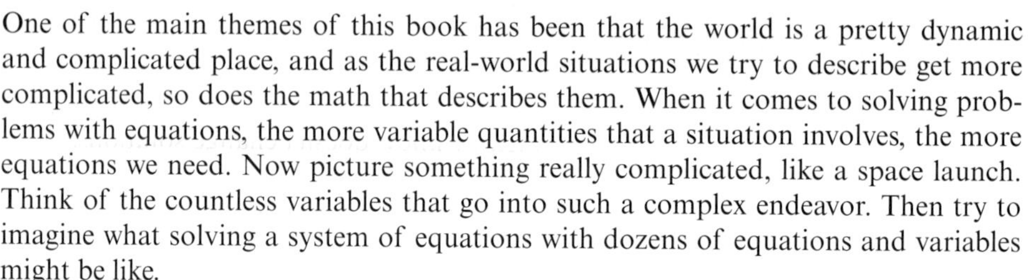

One of the main themes of this book has been that the world is a pretty dynamic and complicated place, and as the real-world situations we try to describe get more complicated, so does the math that describes them. When it comes to solving problems with equations, the more variable quantities that a situation involves, the more equations we need. Now picture something really complicated, like a space launch. Think of the countless variables that go into such a complex endeavor. Then try to imagine what solving a system of equations with dozens of equations and variables might be like.

In Section 7-2, we developed two algebraic procedures for solving systems of linear equations, and used those procedures to solve systems of two equations. As the number of equations and variables gets larger, not surprisingly, the algebra gets more complex. It turns out that the addition/subtraction method for solving systems is mostly based on what the coefficients and constants are—the variables sort of "come along for the ride."

In this section, we will study an alternative approach to solving systems that temporarily leaves out the variables and focuses on the numbers. This makes larger systems more manageable, and also paves the way for computers to get involved. We begin by defining the tool we will use to solve systems in this section.

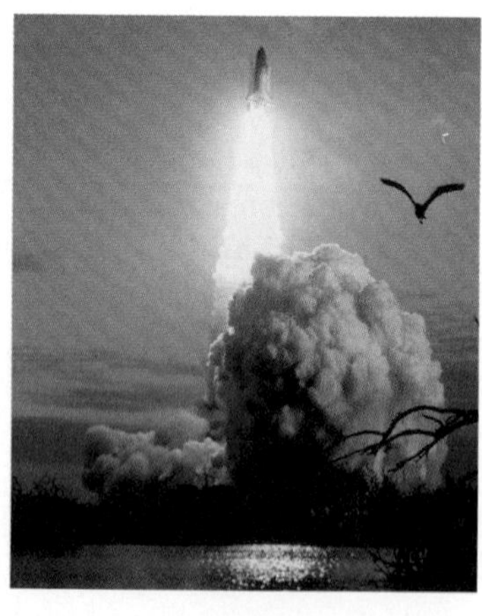

A **matrix** (plural **matrices**) is a rectangular array of numbers enclosed in brackets, like the following:

$$\begin{bmatrix} 2 & -13 & 4 & 0 \\ -5 & 31 & 27 & 3 \\ 12 & 4 & -3 & 0 \end{bmatrix}$$

The individual numbers in a matrix are called **entries.**

Since the above matrix has 3 rows (horizontal) and 4 columns (vertical), we say that it is a 3×4 matrix. In general, a matrix with m rows and n columns is called an **$m \times n$ matrix.**

Gaussian Elimination

The Gaussian elimination method, published by Carl Friedrich Gauss in 1809, is an algorithm for obtaining solutions to systems of equations. It is based on the addition/subtraction method we studied in the last section, but uses matrices rather than equations.

First, we will need to define the augmented matrix for a system of equations. An **augmented matrix** for a system of equations is the matrix of coefficients and constants of the system of equations. For example, given the system of equations below, its augmented matrix is shown to the right.

Math Note

A helpful hint to remember which are the rows and which are the columns: think of the columns that hold up a building. They are vertical, so columns are vertical in a matrix.

System of linear equations

$$\begin{aligned} 3x + 2y - z &= -6 \\ 2x - 3y + 4z &= 10 \\ x - 3y + z &= 2 \end{aligned}$$

Augmented matrix

$$\left[\begin{array}{ccc|c} 3 & 2 & -1 & -6 \\ 2 & -3 & 4 & 10 \\ 1 & -3 & 1 & 2 \end{array}\right]$$

Notice that a vertical bar is used to separate the coefficients from the constants.

Since the rows in the matrix represent the equations in the system, we can perform certain row operations that correspond to operations performed on the equations. When row operations are performed, the resulting matrix represents an equivalent system of equations.

The row operations we will need are:

1. Exchanging any two rows in a matrix. This simply corresponds to rearranging the order in which equations in a system are listed, so doesn't change the solutions.
2. Multiplying or dividing any row in a matrix by a nonzero constant. This corresponds to multiplying or dividing both sides of an equation by a nonzero number, which we know doesn't change solutions.
3. Adding a constant multiple of any row to another row. This is a combination of multiplying both sides of an equation by a constant, and adding equations together as in the addition/subtraction method.

In order to solve a system of equations using Gaussian elimination, we will need to use the above operations to transform a matrix into *row echelon form.*

A matrix is in **row echelon form** if the following properties hold.

1. Any rows of all zeros are below all rows with at least one nonzero entry.
2. The first nonzero number in any row, called the **leading coefficient,** is always to the right of the leading coefficient in the row above it.
3. All entries below a leading coefficient, if there are any, are zero.

EXAMPLE 1 Identifying Row Echelon Form

Decide whether each matrix is in row echelon form.

(a) $\begin{bmatrix} 1 & -3 & 2 & 5 \\ 0 & 2 & -3 & 12 \\ 0 & 0 & 3 & 3 \end{bmatrix}$ (b) $\begin{bmatrix} 1 & -4 & 3 & 7 \\ 0 & 0 & 0 & 0 \\ 0 & 0 & 2 & 6 \end{bmatrix}$

(c) $\begin{bmatrix} 4 & -5 & -2 & 9 \\ 0 & 2 & 3 & 8 \\ 2 & 0 & 3 & 3 \end{bmatrix}$ (d) $\begin{bmatrix} 2 & -4 & 10 & 15 \\ 0 & 3 & 14 & 5 \\ 0 & 0 & 0 & 0 \end{bmatrix}$

SOLUTION

(a) This matrix is in row echelon form.

(b) This one is not. It violates rule 1, because the second row has all zeros, but the third does not.

(c) This is not in row echelon form. It violates both rules 2 and 3: the leading coefficient in row 3 is to the left of the one in row 2, and the first entry in row 3 is below a leading coefficient and is not zero.

(d) This matrix is in row echelon form.

▼ Try This One 1

1. Identify matrices in row echelon form.

Decide whether each matrix is in row echelon form.

(a) $\left[\begin{array}{ccc|c} 1 & 7 & -3 & 13 \\ 0 & 0 & 3 & 8 \\ 0 & 2 & 0 & 3 \end{array}\right]$ (b) $\left[\begin{array}{ccc|c} 1 & 2 & 3 & 4 \\ 0 & 5 & 6 & 7 \\ 0 & 0 & 8 & 9 \end{array}\right]$

(c) $\left[\begin{array}{ccc|c} 2 & 0 & -3 & 10 \\ 0 & 0 & -11 & 100 \\ 0 & 0 & 0 & 0 \end{array}\right]$ (d) $\left[\begin{array}{ccc|c} 6 & 8 & -10 & 12 \\ 1 & 2 & -6 & -8 \\ 0 & 5 & 2 & 4 \end{array}\right]$

We will use the following steps to solve a system of equations using Gaussian elimination.

Procedure for Solving A System of Linear Equations Using Gaussian Elimination

1. Write the augmented matrix for the system.
2. Using the row operations listed on page 375, rewrite the matrix so that it is in row echelon form.
3. Convert back to equation form, then solve for each of the variables.

The procedure is illustrated in Examples 2 through 4.

EXAMPLE 2 Solving a System of Two Equations Using Gaussian Elimination

Solve the following system using Gaussian elimination:

$$\begin{aligned} x - 2y &= 8 \\ 4x + 3y &= -1 \end{aligned}$$

SOLUTION

First write the augmented matrix for the system

$$\left[\begin{array}{cc|c} 1 & -2 & 8 \\ 4 & 3 & -1 \end{array}\right]$$

To put the matrix into row echelon form, we need to get a zero in the second row, first column. We can accomplish this with row operation 3, adding -4 times row 1 to row 2, which we will abbreviate $-4R_1 + R_2 \rightarrow R_2$.

$$\left[\begin{array}{cc|c} 1 & -2 & 8 \\ 4 & 3 & -1 \end{array}\right] -4R_1 + R_2 \rightarrow R_2 \left[\begin{array}{cc|c} 1 & -2 & 8 \\ -4 \cdot 1 + 4 & -4(-2) + 3 & -4 \cdot 8 + -1 \end{array}\right]$$

$$= \left[\begin{array}{cc|c} 1 & -2 & 8 \\ 0 & 11 & -33 \end{array}\right]$$

Now the matrix is in row echelon form. The next step is to restore the variables, changing the new augmented matrix back into equation form.

$$\left[\begin{array}{cc|c} 1 & -2 & 8 \\ 0 & 11 & -33 \end{array}\right] \Rightarrow \begin{aligned} x - 2y &= 8 \\ 11y &= -33 \end{aligned}$$

Math Note

Remember, the vertical line in an augmented matrix reminds you where the equal signs go when converting back to equation form. The entries in the first column in Example 2 correspond to coefficients of *x*, and the entries in the second column correspond to coefficients of *y*.

Notice that the second equation is now easy to solve for y. We divide both sides by 11 and find that $y = -3$. We can substitute this value back into the first equation to find x.

$x - 2y = 8$	*Substitute $y = -3$.*
$x - 2(-3) = 8$	*Multiply.*
$x + 6 = 8$	*Subtract 6 from both sides.*
$x = 2$	

The solution to the system is $(2, -3)$.

☑ 2. Solve systems of two linear equations using matrices.

▼ Try This One 2

Solve the system using Gaussian elimination:

$$\begin{aligned} 3x - 5y &= 22 \\ 9x + 4y &= 28 \end{aligned}$$

EXAMPLE 3 Solving a System of Three Equations Using Gaussian Elimination

Solve the system using Gaussian elimination:

$$\begin{aligned} 2x + y - 3z &= 12 \\ x - 2y + 4z &= -15 \\ 3x - y + 2z &= -5 \end{aligned}$$

SOLUTION

First write the augmented matrix for the system:

$$\left[\begin{array}{rrr|r} 2 & 1 & -3 & 12 \\ 1 & -2 & 4 & -15 \\ 3 & -1 & 2 & -5 \end{array}\right]$$

In Example 2, it wasn't difficult to get zero below the first leading coefficient because that coefficient was 1. We can accomplish that here by switching rows 1 and 2 (row operation 1).

$$\left[\begin{array}{rrr|r} 2 & 1 & -3 & 12 \\ 1 & -2 & 4 & -15 \\ 3 & -1 & 2 & -5 \end{array}\right] R_1 \leftrightarrow R_2 \left[\begin{array}{rrr|r} 1 & -2 & 4 & -15 \\ 2 & 1 & -3 & 12 \\ 3 & -1 & 2 & -5 \end{array}\right]$$

Now it's easier to get zeros in the first column below the one, using the row operations shown.

$$\left[\begin{array}{rrr|r} 1 & -2 & 4 & -15 \\ 2 & 1 & -3 & 12 \\ 3 & -1 & 2 & -5 \end{array}\right] \begin{array}{l} \\ -2R_1 + R_2 \rightarrow R_2 \\ -3R_1 + R_3 \rightarrow R_3 \end{array} \left[\begin{array}{rrr|r} 1 & -2 & 4 & -15 \\ 0 & 5 & -11 & 42 \\ 0 & 5 & -10 & 40 \end{array}\right]$$

Next, get a zero below the 5 in the second row.

$$\left[\begin{array}{rrr|r} 1 & -2 & 4 & -15 \\ 0 & 5 & -11 & 42 \\ 0 & 5 & -10 & 40 \end{array}\right] \begin{array}{l} \\ \\ -1R_2 + R_3 \rightarrow R_3 \end{array} \left[\begin{array}{rrr|r} 1 & -2 & 4 & -15 \\ 0 & 5 & -11 & 42 \\ 0 & 0 & 1 & -2 \end{array}\right]$$

Now we restore the variables:

$$\left[\begin{array}{ccc|c} 1 & -2 & 4 & -15 \\ 0 & 5 & -11 & 42 \\ 0 & 0 & 1 & -2 \end{array}\right] \quad \Rightarrow \quad \begin{aligned} x - 2y + 4z &= -15 \\ 5y - 11z &= 42 \\ z &= -2 \end{aligned}$$

Substitute $z = -2$ into the second equation and solve for y:

$$\begin{aligned} 5y - 11z &= 42 && \text{Substitute } z = -2. \\ 5y - 11(-2) &= 42 && \text{Multiply.} \\ 5y + 22 &= 42 && \text{Subtract 22 from both sides.} \\ 5y &= 20 && \text{Divide both sides by 5.} \\ y &= 4 \end{aligned}$$

Finally, substitute $z = -2$ and $y = 4$ into the first equation and solve for x:

$$\begin{aligned} x - 2y + 4z &= -15 && \text{Substitute } z = -2 \text{ and } y = 4. \\ x - 2(4) + 4(-2) &= -15 && \text{Multiply.} \\ x - 8 - 8 &= -15 && \text{Add 16 to both sides.} \\ x &= 1 \end{aligned}$$

We write the solution as $(1, 4, -2)$.

▼ Try This One 3

Solve the system using Gaussian elimination. (*Hint:* In the third equation, the coefficient of y is zero.)

$$\begin{aligned} x - 2y + 4z &= -25 \\ -2x + y - 2z &= 11 \\ 3x \qquad + z &= -2 \end{aligned}$$

In some cases, a fourth row operation can help us to avoid fractions.

Row Operation 4: Adding a multiple of one row to a multiple of another row.

This operation will be used in Example 4.

Sidelight JUST HOW OLD *IS* THIS METHOD?

If you're really observant, you may have noticed that we said that Carl Friedrich Gauss *published* this method for solving systems—we didn't say he *discovered* it. Gauss has been dead for over 150 years, but he's current news compared to the method that bears his name. In fact, the method can be traced back to a Chinese text called *Jiuzhang Suanshu* or *The Nine Chapters on the Mathematical Art.* The first reference to the book by this name is dated to the year 179, but parts of it were written as early as 150 BCE!

But let's not accuse our friend Gauss of academic dishonesty: the Chinese texts were unknown in Europe until far later than Gauss' time. But still, crediting Gauss for developing Gaussian elimination is sort of like crediting Columbus for discovering America. There were people who knew of its existence a long time earlier—they just didn't happen to be European.

EXAMPLE 4 Solving a System of Three Equations Using Gaussian Elimination

Solve the system using Gaussian elimination:

$$\begin{aligned} 2x - 3y + 7z &= 11 \\ 3x + 5y - 2z &= -2 \\ 4x - y - 3z &= -19 \end{aligned}$$

SOLUTION

First write the augmented matrix for the system:

$$\left[\begin{array}{ccc|c} 2 & -3 & 7 & 11 \\ 3 & 5 & -2 & -2 \\ 4 & -1 & -3 & -19 \end{array}\right]$$

Next reduce the matrix to row echelon form. Start by getting zeros in the first column below the 2.

$$\left[\begin{array}{ccc|c} 2 & -3 & 7 & 11 \\ 3 & 5 & -2 & -2 \\ 4 & -1 & -3 & -19 \end{array}\right] \begin{array}{l} \\ -3R_1 + 2R_2 \to R_2 \\ -2R_1 + R_3 \to R_3 \end{array} \left[\begin{array}{ccc|c} 2 & -3 & 7 & 11 \\ 0 & 19 & -25 & -37 \\ 0 & 5 & -17 & -41 \end{array}\right]$$

Math Note

A calculator is definitely helpful when the numbers start to get large, like the ones in Example 4.

Next, we need a zero in row 3 below the 19.

$$\left[\begin{array}{ccc|c} 2 & -3 & 7 & 11 \\ 0 & 19 & -25 & -37 \\ 0 & 5 & -17 & -41 \end{array}\right] \begin{array}{l} \\ \\ -5R_2 + 19R_3 \to R_3 \end{array} \left[\begin{array}{ccc|c} 2 & -3 & 7 & 11 \\ 0 & 19 & -25 & -37 \\ 0 & 0 & -198 & -594 \end{array}\right]$$

The last row gives us $-198z = -594$, or $z = 3$. Back substituting to find y and z (details omitted), we get $y = 2$ and $x = -2$. The solution is $(-2, 2, 3)$.

▼ Try This One 4

☑ 3. Solve systems of three linear equations using matrices.

Solve the system using Gaussian elimination:

$$\begin{aligned} x - 7y + 3z &= -13 \\ 2x - 3y - z &= 3 \\ 4x + 2y + z &= 19 \end{aligned}$$

In a dependent system the row echelon form of the matrix will have at least one row of all zeros. In an inconsistent system there will be a row where the coefficients are zero but the constant is not.

The system represented by the augmented matrix below (which is in row echelon form) is a dependent system.

$$\left[\begin{array}{ccc|c} 2 & -3 & 10 & 15 \\ 0 & 3 & 14 & 5 \\ 0 & 0 & 0 & 0 \end{array}\right]$$

There would be infinitely many solutions to the system. Since the bottom row is all zeros, there is no specific solution for z; z can be any number. We would start by letting z equal some constant t. So, if $z = t$, then we can back substitute and find that $y = \frac{5 - 14t}{3}$ or $\frac{5}{3} - \frac{14}{3}t$, and that $x = 10 - 12t$. So the solutions to the system are ordered triples of the form $(10 - 12t, \frac{5 - 14t}{3}, t)$, where t is any real number.

The system whose reduced augmented matrix is given below represents an inconsistent system.

$$\left[\begin{array}{ccc|c} 4 & -5 & -2 & 9 \\ 0 & 2 & 3 & 8 \\ 0 & 0 & 0 & 3 \end{array}\right]$$

This row represents $0x + 0y + 0z = 3$, which is never true.

The system, therefore, has no solution.

Sidelight COMPUTERS AND GAUSSIAN ELIMINATION

One of the big advantages of solving systems of equations with matrices is that computers can be programmed to find the row echelon form of a matrix. And, within reason, computers don't care how large the numbers are, or how many rows there are in the matrix. That makes them an ideal tool for solving systems of equations that have a very large number of variables. How many? Currently, supercomputers are able to solve systems of equations with over 60,000 variables and equations! It would take an awfully complicated situation to have more variables than that.

Answers to Try This One

1 (a) No (b) Yes (c) Yes (d) No

2 (4, −2)

3 (1, 3, −5)

4 (4, 2, −1)

EXERCISE SET 7-3

Writing Exercises

1. Explain what a matrix is in your own words.
2. How do we use a matrix to represent a system of equations?
3. In the procedure for solving a system using Gaussian elimination, we sometimes use the term "back substitute." What does that mean?
4. Describe how to tell if a matrix is in row echelon form.

Computational Exercises

For Exercises 5–12, decide if the matrix is in row echelon form. If it is not, explain why.

5. $\left[\begin{array}{cc|c} 1 & 3 & 5 \\ 0 & 0 & 2 \end{array}\right]$

6. $\left[\begin{array}{cc|c} -5 & 0 & 7 \\ 0 & 0 & 0 \end{array}\right]$

7. $\left[\begin{array}{ccc|c} 2 & 4 & 6 & 8 \\ 0 & 0 & 3 & 5 \\ 0 & 1 & -7 & -10 \end{array}\right]$

8. $\left[\begin{array}{ccc|c} -3 & 2 & -4 & -5 \\ 0 & 0 & 0 & 0 \\ 0 & 0 & 1 & 8 \end{array}\right]$

9. $\left[\begin{array}{ccc|c} 1 & 0 & 0 & -15 \\ 0 & 1 & 0 & 3 \\ 0 & 0 & 0 & 4 \end{array}\right]$

10. $\left[\begin{array}{ccc|c} \frac{3}{2} & -1 & 9 & 6 \\ 1 & 1 & \frac{5}{2} & 2 \\ 0 & 2 & 4 & -\frac{4}{3} \end{array}\right]$

11. $\left[\begin{array}{ccc|c} 0 & 1 & -5 & \frac{11}{2} \\ 0 & 2 & \frac{7}{3} & -6 \\ 0 & 0 & 0 & 0 \end{array}\right]$

12. $\left[\begin{array}{ccc|c} 1 & 0 & 0 & -50 \\ 0 & 1 & 0 & 12 \\ 0 & 0 & 1 & 18 \end{array}\right]$

For problems 13–24, use Gaussian elimination to solve the system of equations.

13. $x - 3y = -1$
    $2x + 5y = 9$
14. $2x + 5y = 0$
    $4x - 3y = -26$
15. $7x - 2y = 7$
    $5x + y = 22$
16. $6x - 10y = -16$
    $3x + 2y = -1$
17. $3x - y + 2z = 9$
    $x - 2y + 3z = 7$
    $2x + y + 7z = 10$
18. $2x - 3y + z = 11$
    $4x - 2y + 5z = 24$
    $6x + 2y - 3z = 10$

19. $7x + 2y - 3z = -6$
$2x - 3y + 5z = -2$
$x + 4y - 6z = 1$

20. $3x + 5y - z = -1$
$6x + y + 2z = 17$
$3x - 4y + 3z = 18$

21. $3x - 2y + z = -3$
$2x + 3y - 3z = 5$
$5x + y - 2z = 2$

22. $4x - 3y + 2z = -1$
$3x + 2y - z = 7$
$2x + y + z = 7$

23. $3x - 10y + 4z = -11$
$5x - 2y + 3z = 0$
$4x + 3y - 7z = 14$

24. $5x + y - 3z = 6$
$3x + 2y - 3z = 5$
$4x - 2y + 4z = 5$

Real-World Applications

Use matrices to solve the system of equations for each problem.

25. A movie theater sells adult tickets for \$7, child tickets for \$4, and senior citizen tickets for \$6. For one movie the theater sold 39 tickets for a total of \$200. The number of child tickets was 1 more than 3 times the number of senior citizen tickets. How many tickets of each kind did they sell?
26. Sue has a sock drawer where she collects nickels, dimes, and quarters. The total number of coins in her drawer is 37 with a total value of \$4.00. The number of quarters plus the number of dimes is 7 more than the number of nickels. How many coins of each type does she have?
27. For the upcoming Coldplay concert, one group bought three lawn tickets, five general admission tickets, and two ground floor tickets and spent \$820. Another group bought two lawn tickets, four general admission tickets, and one ground floor ticket and spent \$550. A third group bought three ground floor tickets and spent \$450. How much is each type of ticket to the concert?
28. At a local Target store, one mother bought three binders, two graphing calculators, and ten backpacks for her kids for school and spent \$290 before taxes. Another mother bought one binder and three graphing calculators and spent \$260 before taxes. A third mother bought one graphing calculator and three binders and spent \$101 before taxes. Assuming each mother bought the same three items at the same prices, how much does each item cost?
29. Frank wants to invest \$2,000 into three funds. The amount he puts into Fund B and Fund C is triple the amount he puts into Fund A. Twice what he puts into Fund B and a third of what he puts into Fund C is 5 times what he puts into Fund A. How much does he invest into each fund?
30. During inventory at the student bookstore, a worker notes there are only college algebra, liberal arts math, and intermediate algebra books left on the shelf, and he writes down that there are 48 books altogether. A college algebra book is \$105, a liberal arts math book is \$85, and an intermediate algebra book is \$120, and there is \$4,590 worth of books on the shelves. Twice the number of college algebra books is 6 more than the number of intermediate algebra books. How many of each type of book are left?
31. A football team played 16 games last season, and of course they either won, lost, or tied each game. If the team won just three more games, it would be the same as the number of games they tied and twice the number they lost. The number of games they lost or tied is 2 less than the number they won. How many games did they win, lose and tie?
32. When Celia went to a Las Vegas Night fundraiser, she brought \$40 with her and played blackjack, Texas hold ‘em poker, and roulette. Twice the amount she spent on blackjack was 60% of her money, and half the amount she spent playing roulette was 10% of her money. How much did she spend on each game?
33. During the last Olympics, there were 650 people in the arena watching the woman’s gymnastics finals. Since no one there could watch all the events, some watched the beam competition, others watched the floor exercise, and others watched the uneven bars. The vault event was to be held the next night, so no one watched that event. The number who watched the floor exercise and twice the number who watched the uneven bars totaled 8 times the number who watched the beam competition. Twice the number who watched the beam competition and the number who watched the floor exercise together was the same as twice those who watched the uneven bars. How many watched each of the three events?
34. Mary wanted to buy a mixture of nuts to put on the tables at her daughter’s wedding, so she went to the bulk food store, where she bought 15 lb of three types of nuts. She bought cashews for \$1.19 per pound, almonds for \$2.20 per pound, and pecans for \$1.59 per pound, and her total before taxes was \$26.12. The number of pounds of almonds and pecans was the same as twice the number of cashews. How much of each type of nut did Mary buy?

Critical Thinking

35. Re-solve the system in Example 2, this time using the addition/subtraction method from Section 7-2. Then compare the steps in each method. How do they compare? Which method do you think is easier?

36. The first three row operations introduced in this section can be used to put most matrices into a form called *reduced row echelon* form. (This is the form that most computer programs use to solve systems of equations.) The reduced row echelon form for the matrix in Try This One 4 is

$$\left[\begin{array}{ccc|c} 1 & 0 & 0 & 4 \\ 0 & 1 & 0 & 2 \\ 0 & 0 & 1 & -1 \end{array}\right]$$

(a) Explain why the reduced row echelon form is helpful in solving systems.
(b) Use the first three row operations from this section to transform the original augmented matrix for Try This One 4 into the matrix to the left. (*Hint:* First, get the first column to look like the one in the matrix to the left. Then get a 1 in the second row second column and proceed from there.)

Section 7-4 Linear Inequalities

LEARNING OBJECTIVES

- ☐ 1. Graph linear inequalities in two variables.
- ☐ 2. Graph a system of linear inequalities.
- ☐ 3. Model a real-world situation with a system of linear inequalities.

When we introduced inequalities in Chapter 6, we pointed out that many real-world problems don't have just a single solution, but rather a range of reasonable outcomes. When more complicated situations require two variables, the same is true. For example, in Section 7-2, we referred to buying cups and plates for a tailgate outing, where the number of each is variable. No reasonable person would say "Go buy cups and plates with this \$30, and you better spend every penny of it!" Of course, any amount \$30 or less would be fine.

So in this section, we will study systems of linear inequalities. The graphical method we used to solve systems of linear equations will come in very handy, so we begin by examining what the graph of a linear inequality in two variables should look like.

We know that an equation of the form $ax + by = c$ is called a linear equation in two variables, and that its graph is a straight line. When the equal sign is replaced by $>$, $<$, $\geq$, or $\leq$, the equation becomes a **linear inequality in two variables.** Some examples of linear inequalities are shown below.

$$2x + y > 6 \qquad 3x - 8y \leq 20 \qquad x - y \geq 3$$

We can use our ability to graph linear equations to develop a procedure for graphing linear inequalities.

Math Note

The origin is usually a good choice of test point because the coordinates, (0, 0), are easy to substitute in. But if the line looks like it might pass through the origin, you should pick another point that's definitely not on the line.

Procedure for Graphing a Linear Inequality in Two Variables

Step 1 Replace the inequality symbol with an equal sign and graph the line.

(a) Use a dashed line if the inequality is either $<$ or $>$ (a strict inequality) to indicate that the points on the line are not included in the solution.

(b) Use a solid line if the inequality is either $\leq$ or $\geq$ to indicate that the points on the line are included in the solution.

Step 2 Pick any point not on the line and substitute its coordinates into the inequality to see if the resulting statement is true or false.

Step 3 The line divides the plane into two half planes.

(a) If the test point makes the inequality true, shade the half plane containing that point.

(b) If the test point makes the inequality false, shade the half plane not containing that point.

The shaded region is the solution set of the inequality.

Examples 1 and 2 demonstrate the procedure.

EXAMPLE 1 Graphing a Linear Inequality in Two Variables

Graph $x - y \geq 6$.

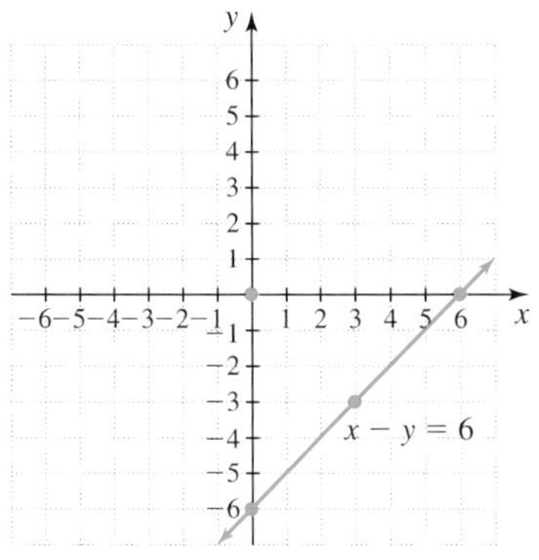

SOLUTION

Step 1 Graph the line $x - y = 6$. It will be easy to find the intercepts, and we'll choose one other point to be safe.

x	y
0	−6
6	0
3	−3

Since the sign is $\geq$, equality is included. That means the points on the line satisfy the inequality, so we draw a solid line through the three points.

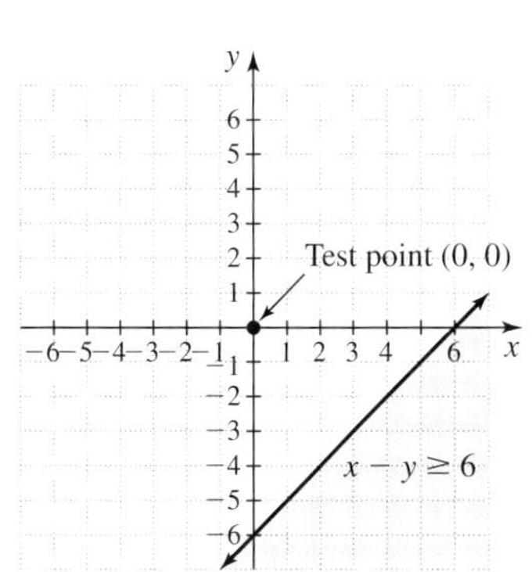

Step 2 Pick a test point not on the line, substitute into the equation, and see if a true or false statement results.

For (0, 0):

$$x - y \geq 6$$
$$0 - 0 \overset{?}{\geq} 6$$
$$0 \geq 6 \quad \textit{False.}$$

Step 3 The point (0, 0) is above the line, and is not in the solution set for the inequality, so we shade the half-plane below the line.

▼ Try This One 1

Graph $3x + 2y \leq 12$.

EXAMPLE 2 Graphing a Linear Inequality in Two Variables

Graph $2x - 4y < 12$.

SOLUTION

Step 1 Graph the line $2x + 4y = 12$. It is again easy to find the intercepts, and we'll choose one other point to be safe.

x	y
0	3
6	0
−6	6

Since the sign is <, equality is not included. That means the points on the line do not satisfy the inequality, and we draw a dashed line through the three points.

Step 2 The origin is definitely not on the line, so we use it as our test point:

$$2x + 4y < 12$$
$$2(0) + 4(0) \overset{?}{<} 12$$
$$0 < 12 \quad \textit{True.}$$

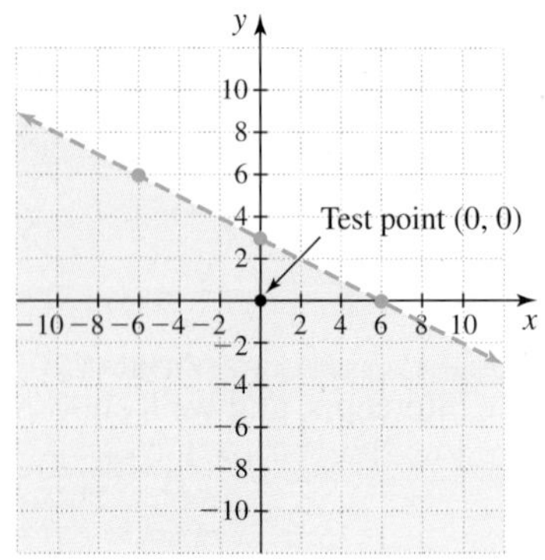

Step 3 The point (0, 0) is below the line, and is in the solution set for the inequality, so we shade the half-plane below the line.

▼ Try This One 2

Graph $2x - y < 8$.

Systems of Linear Inequalities

☑ 1. Graph linear inequalities in two variables.

Systems of linear equations were solved graphically by graphing each line and finding the point where the lines intersect. **Systems of linear inequalities** can be solved graphically by graphing each inequality and finding the intersection of the shaded regions.

EXAMPLE 3 Solving a System of Linear Inequalities

Solve the system graphically.

$$x + 3y \geq 6$$
$$2x - y < 10$$

SOLUTION

Begin by graphing the first inequality, $x + 3y \geq 6$, as in Example 1. The intercepts are (6, 0) and (0, 2), and we draw a solid line through those points since equality is included. Pick (0, 0) as a test point: $0 + 3(0) \geq 6$ is false, so we shade the half-plane above the line, which does not contain (0, 0).

Next we add the second inequality to the graph, starting with the points (5, 0) and (4, −2). The line is dashed because equality is not included in $2x - y < 10$. We can again use (0, 0) as a test point: $2(0) - 0 < 10$ is a true statement, so we shade the region to the left of the dashed line, which contains (0, 0).

The solution to the system is the intersection of the two shaded regions, shown below right. The points on the solid line are included, but those on the dashed line are not. (Note the open circle where the lines intersect.)

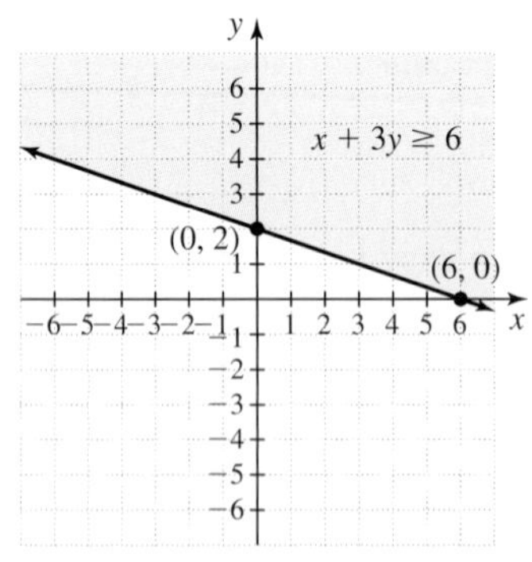

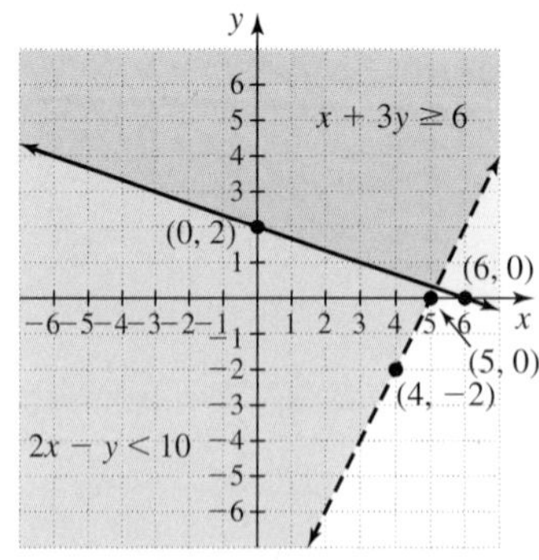

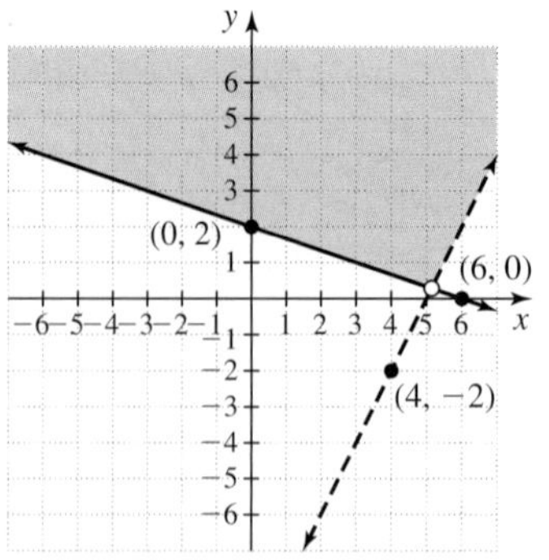

▼ Try This One 3

Solve the system graphically.

$$4x - 3y \geq 12$$
$$5x - 2y < 10$$

EXAMPLE 4 Solving a System of Linear Inequalities

Solve the system graphically.

$$x > -3$$
$$y \leq 10$$

Math Note

The solution to a system can be roughly checked by selecting a point on the graph contained in the intersection of the shaded regions, then substituting the coordinates of the point into the two inequalities. If both inequalities are true, then the area shaded is possibly correct.

SOLUTION

With a little bit of thought, the individual inequalities can be graphed without plotting points.

The inequality $x > -3$ represents the set of all points whose x coordinates are larger than -3. This is all points to the right of the dashed vertical line $x = -3$, shaded in blue.

The inequality $y \leq 2$ is the set of all points whose height (y coordinate) is less than or equal to 2. This is all points on or under the solid horizontal line $y = 2$, shaded in pink.

The solution set to the inequality is the intersection of those regions, shaded in purple.

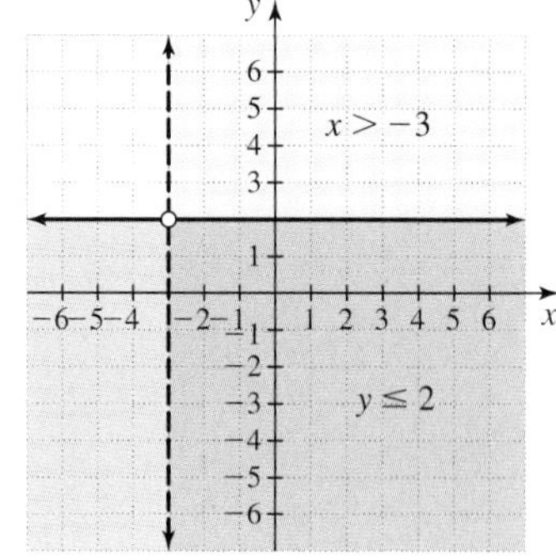

▼ Try This One 4

☑ 2. Graph a system of linear inequalities.

Solve the system graphically.

$$x < 4$$
$$y \geq 1$$

Multivariable problems with a range of reasonable solutions are a natural fit for systems of inequalities, as we will see in Example 5.

EXAMPLE 5 Applying Systems of Inequalities to Purchasing

You are dispatched to buy cups and plates for a homecoming tailgate event. The plates you choose are \$0.40 each, and the cups are \$0.25 each. You were told that any more than \$30 is coming out of your own pocket (unacceptable, obviously), and that you have to get at least 40 plates. Draw a graph that represents acceptable numbers of plates and cups.

SOLUTION

The two quantities that can vary are the number of plates purchased, and the number of cups. We will let

x = the number of plates
y = the number of cups

The cost of plates will be $0.40x$ (40 cents times number of plates). The cost of cups will be $0.25y$ (25 cents times the number of cups). The total cost is the sum, and it has to be \$30 or less.

$$0.40x + 0.25y \leq 30$$

We need at least 40 plates, so $x \geq 40$. Finally, the number of cups certainly can't be negative, so $y \geq 0$. Now we have a system:

$$\begin{aligned} 0.40x + 0.25y &\leq 30 \\ x &\geq 40 \\ y &\geq 0 \end{aligned}$$

To graph the first inequality, find the intercepts and choose (0, 0) as test point:

x	y
0	120
75	0

Test point (0, 0): $0.40(0) + 0.25(0) \le 30$. True.

The solid line is on the first graph below, with the region containing (0, 0) shaded.

The inequality $x \ge 40$ represents all points on or to the right of the vertical line $x = 40$ (second graph below). The inequality $y \ge 0$ is all points on or above the horizontal line $y = 0$, which is the x axis (third graph).

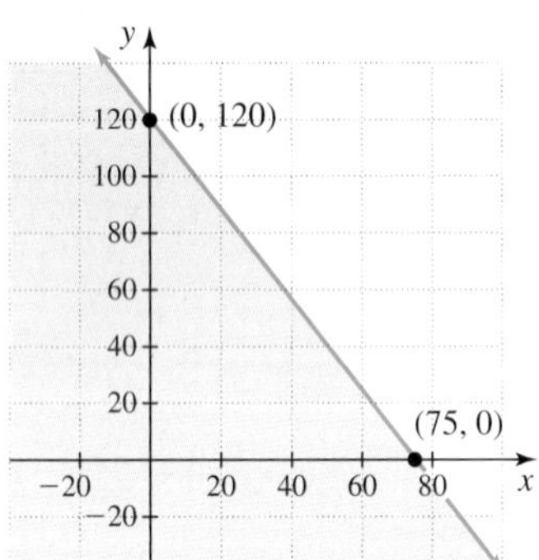

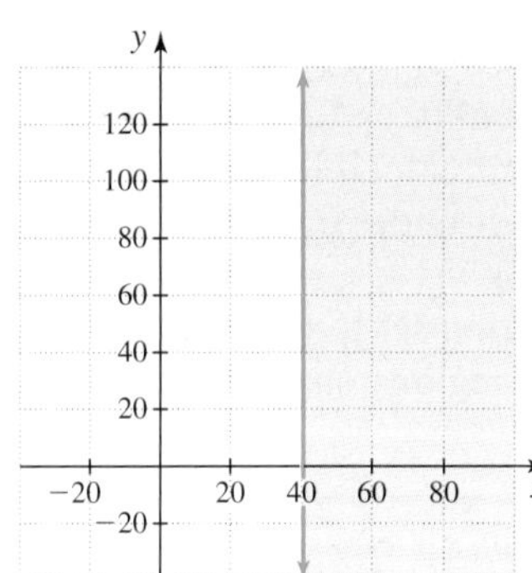

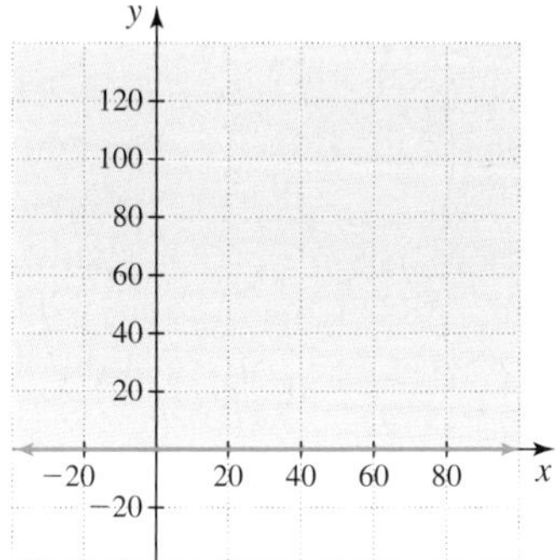

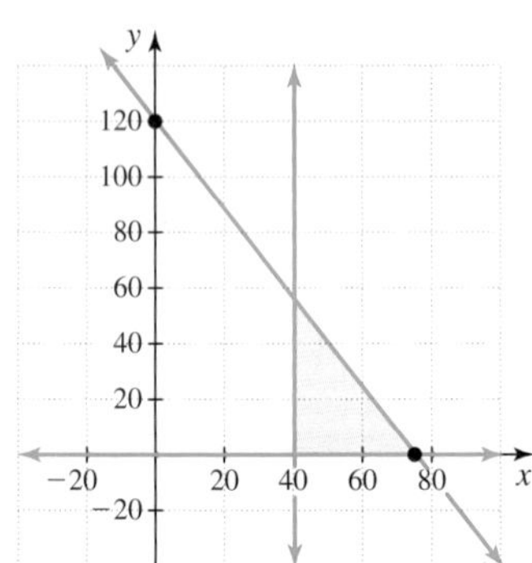

The intersection of the shaded regions is the triangular region to the left. Any combination of plates (x coordinate) and cups (y coordinate) from points in that region is an acceptable solution. For example, the point (40, 40) is in the region, so 40 cups and 40 plates will do.

3. Model a real-world situation with a system of linear inequalities.

▼ Try This One 5

To quickly build a home, a construction company plans to hire a large crew of carpenters at \$240 per day and bricklayers at \$320 per day. The total daily budget for these workers is \$5,000, and the foreman insists on having at least four of each. Draw a graph that represents acceptable numbers of each type of worker.

Answers to Try This One

1

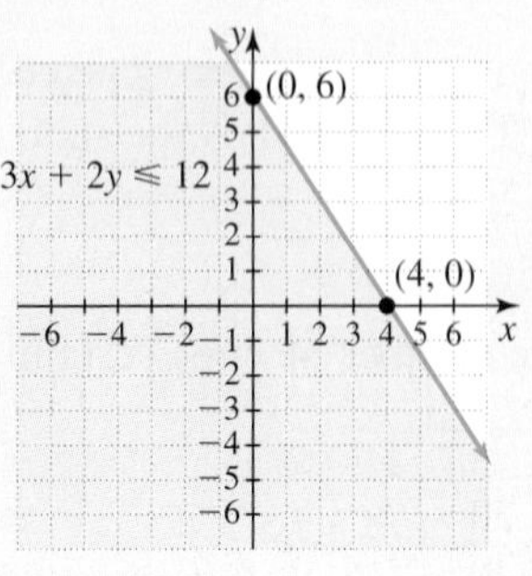

2

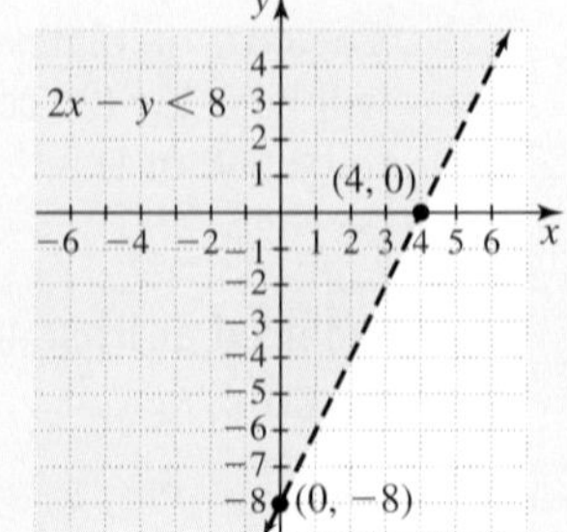

3

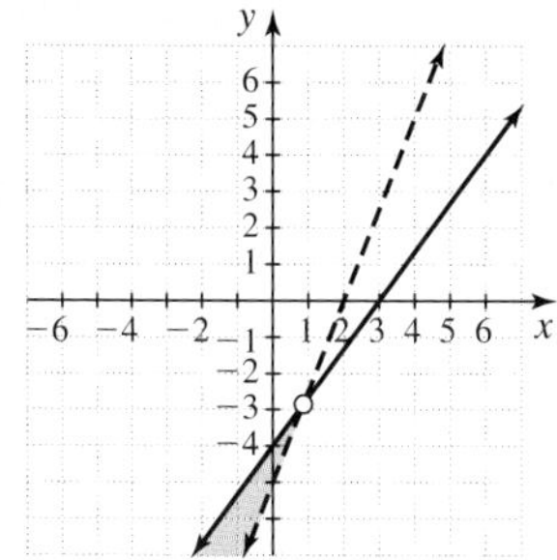

4

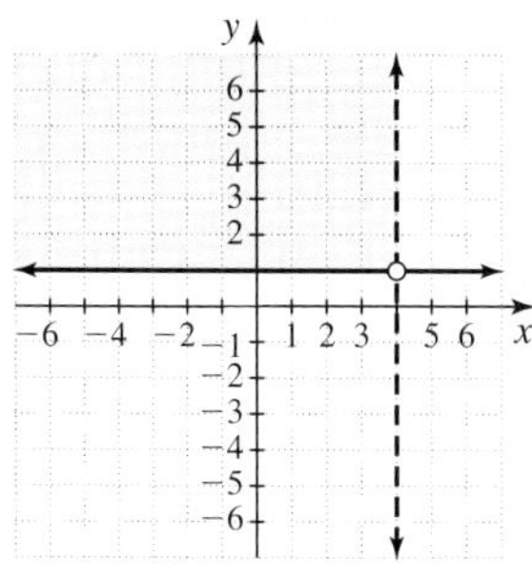

5 The shaded region is all acceptable values; x is the number of carpenters and y is the number of bricklayers.

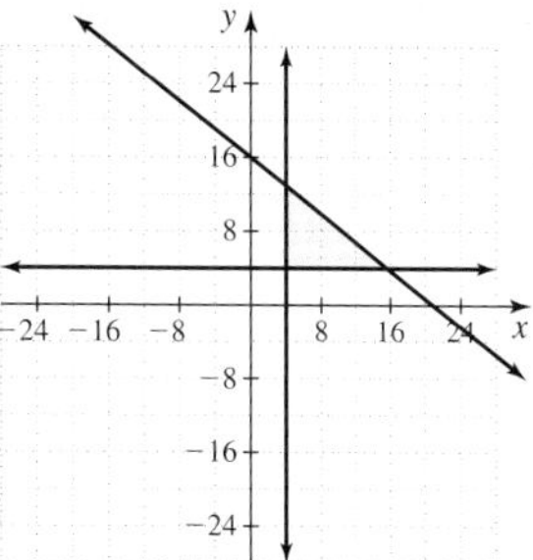

EXERCISE SET 7-4

Writing Exercises

1. Explain the difference between using a solid line and a dashed line in graphing a linear inequality in two variables. Focus on why, not when.
2. Describe how we use a test point to decide which part of the plane to shade when graphing an inequality.
3. Explain how to quickly graph the inequality $x \geq -3$.
4. Explain how to quickly graph the inequality $y < 5$.

Computational Exercises

For Exercises 5–14, graph each linear inequality.

5. $2x - y \geq 5$
6. $x + 4y < -3$
7. $3x - 4y \leq 12$
8. $12x - 3y > 10$
9. $x \geq 5$
10. $y < -6$
11. $y \leq 0$
12. $x > -3$
13. $3x + y < -6$
14. $-2x + y \geq 0$

For Exercises 15–24, find the solution set for the system of linear inequalities.

15. $2x - y \leq 6$
 $x + y > 3$
16. $x - 3y < 10$
 $3x - 2y \geq 5$
17. $x \geq -2$
 $y < -3$
18. $x < -3$
 $y \geq 0$
19. $5x - 3y \geq 15$
 $3x - 8y \geq 12$
20. $x - y < 6$
 $2x - y \geq 3$
21. $x + y \geq 7$
 $x - y < -5$
22. $3x - y \leq -6$
 $2x + y > 6$
23. $4x - 2y < 6$
 $8x - 4y > -3$
24. $x + 7y < -5$
 $7x - y \leq 10$

Real-World Applications

25. Eileen is taking her liberal arts math test, and she has up to 60 minutes to complete up to 12 problems. She can choose from computation problems and word problems. She can do computation problems in 4 minutes and word problems in 5 minutes. The directions say she must do at least 5 word problems.

Draw a graph that represents the number of computation and word problems she can do to meet the requirements of the test.

26. Stephanie is trying a new diet that requires she consume at least 35 grams of protein and no more than 24 grams of fat each day. On her way to the gym, she stops to get a grilled chicken wrap and baked pita chips. The wrap has 5 grams of protein per ounce and 6 grams of fat per ounce. The pita chips have 7 grams of protein per ounce and 3 grams of fat per ounce. Draw a graph that represents the number of ounces of each food she can eat during lunch and still stay on her diet.
27. Sammy starts an eBay store where she will only sell two items. To make a profit, she determined she must sell at least 15 tubes of body lotion each week, but she can only get up to 60 tubes from her supplier per week. She must also sell at least 20 loofah sponges each week, but she can only get up to 100 sponges from her supplier per week. Draw a graph that represents the total number of tubes of lotion and loofah sponges she can sell each week to be profitable.
28. Lucette does not have any storage in her dorm room and does not have anywhere to put away her clothes and personal items. She has up to \$70 to spend on modular storage units, and there is only 36 square feet available to put them. She goes to Wal-Mart to buy the storage units, and she sees two types of modular units that would be perfect. Model A costs \$7 per unit, and it takes up 6 square feet of space. Model B costs \$10 per unit and takes 4 square feet of space. Assuming she buys at least one of each, draw a graph to represent how many of Model A and Model B storage units Lucette can purchase.
29. At a local university, there is a unique degree program that requires students to take both graduate and undergraduate courses at the same time. To complete the degree, a student must take between 4 and 8 undergraduate courses and between 6 and 11 graduate courses. Draw a graph that represents the number of graduate and undergraduate courses a student can take to fulfill the requirements of the degree.
30. A building company is creating a new subdivision out of two types of modular homes. They have 100 units of wood and 180 units of brick available for constructing these homes. The Ski Lodge model requires 10 units of wood and 5 units of brick. The Golf View model requires 25 units of wood and 9 units of brick. Draw a graph that represents the number of each model the builders can construct with the given materials.
31. A textile company must produce at least 36 pairs of jeans and 64 T-shirts a day to meet consumer demand. Two factories produce the jeans, but they have different equipment, so their production levels vary. A worker at one factory can produce 3 pairs of jeans and 8 T-shirts a day and a worker at the other factory can produce 6 pairs of jeans and 4 T-shirts a day. Draw a graph that represents how many workers they need at each factory to make enough jeans and T-shirts a day to meet the demand.
32. A cabinetmaking company produces two types of kitchen cabinets. Each cabinet must go through a cabinet shop to be built and a finishing shop for staining before it can be sent to a customer. The company must produce at least 6 cabinets a day to fulfill customer orders. The cabinet shop has enough employees available for up to 12 hours a day and the finishing shop has enough workers for up to 15 hours a day. The cherry-finish cabinets must go through the cabinet shop for 1 hour and the finishing shop for 3 hours before they are ready for shipping. The walnut-finish cabinets must go through the cabinet shop for 2 hours and the finishing shop for 5 hours before they are ready for shipping. Draw a graph that represents the number of each type of cabinet that can be produced to meet both the company's and customers' requirements.
33. A law firm specializes in civil and malpractice suits. A civil suit requires 3 months of prep time and 4 expert witnesses. A malpractice suit requires 4 months of prep time and 6 expert witnesses. In a 12-month stretch of time, with only 24 expert witnesses available, the firm decides they will do at least as many civil suits as malpractice suits. Draw a graph that represents the possible number of suits of each type the firm will pursue.
34. An airline is trying to decide if flying from Miami to Tampa is profitable enough to continue this route. For each daily flight, they must have at least 6 first-class passengers and 15 coach passengers to make a profit. They cannot have more than 47 passengers on board due to capacity regulations. Draw a graph representing the number of first-class and coach passengers the airline can have to still be profitable.

Critical Thinking

35. When is the point of intersection of the lines included in a solution of a system of linear inequalities?
36. Can the solution set for a system of linear inequalities be the empty set? Explain your answer.
37. Solve this system:

$$x + y > 6$$
$$x + y < -3$$

Section 7-5 Linear Programming

LEARNING OBJECTIVE

☐ 1. Use linear programming to solve real-world problems.

When applying systems of linear inequalities to situations in Section 7-4, we found an entire set of possible solutions. In this section, we'll take that a little further, finding the best possible solution to certain problems. To do so, we will learn a method known as **linear programming.** Linear programming was developed to handle military logistical problems such as efficiently transporting equipment and personnel during wartime. Today the techniques of linear programming are used by business and industry to make decisions and find cost-effective solutions to many problems.

The idea is to write a formula describing some quantity of interest, like profit or cost. We call this formula the **objective function.** (Functions are the main topic of the next section. For now, it's okay to think of a function as a formula that describes some quantity of interest.) The objective function must be linear, meaning that all variables appear only to the first power. The goal is to find values of the variables that either **maximize** (make as large as possible) or **minimize** (make as small as possible) the objective function.

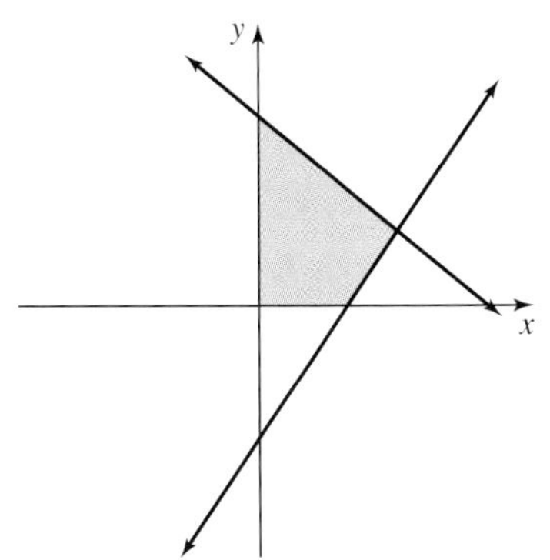

A typical polygonal region in a linear programming problem. The vertices are where the lines intersect.

For example, suppose the objective function $P = 3x + 8y$ represents a company's profit from manufacturing x footballs and y basketballs. It would be reasonable to want to maximize the profit. In most real-world problems, there are limitations on values of the variable, which we call **constraints.** A constraint in this case might be that the company can make at most 200 footballs in an 8-hour day; we could represent this with the inequality $x \le 200$.

The constraints of the problem will form a system of linear inequalities, and in most cases the graph of that system will be a **polygonal region.** (A polygon is a closed figure with at least three straight sides.) Each corner point of the polygonal region is called a **vertex** (plural "vertices"). Every vertex has a pair of coordinates that represent a potential solution to the problem at hand. Advanced mathematical techniques have shown that the maximum or minimum value of an objective function subject to the constraints always occurs at one of the vertices. So our job will be to find the vertices of the region and evaluate the objective function for each, picking out the largest or smallest value.

In Example 1, we demonstrate a step-by-step procedure for solving a linear programming problem.

EXAMPLE 1 Solving a Linear Programming Problem

The owner of a small business manufactures playground sets and playhouses. The process involves two steps. First, the lumber must be cut and drilled, and second, the product must be assembled. For a playground set, it takes one worker 2 hours to cut and drill the lumber and another worker 1 hour to assemble the set. For the playhouse, it takes one worker 1 hour to cut and drill the lumber and another worker 1.5 hours to assemble the playhouse. Workers work at most 8 hours per day. The owner makes \$100 profit on every playground set that is sold and \$75 profit on every playhouse that is sold. Using linear programming techniques, how many of each product should be manufactured in 1 day in order to maximize profit?

SOLUTION

Step 1 Write the objective function.

Since the owner makes a profit of \$100 on each playground set sold and a profit of \$75 on each playhouse sold, the objective function can be written as

$$P = 100x + 75y$$

where x = the number of playground sets sold, y = the number of playhouses sold, and P = the profit made.

Step 2 Write the constraints.

Listing the information in a table makes it somewhat easier to find the constraints.

Item	Cut (hours)	Assemble (hours)	Profit
Playground set	2	1	\$100
Playhouse	1	1.5	\$ 75
Time limit	8	8	

Since it takes 2 hours to cut lumber for one playground set, $2 \cdot x$ is the number of hours needed to cut lumber for x playground sets. It only takes 1 hour to cut lumber for a playhouse, so $1 \cdot y$ is the number of hours needed to cut lumber for y playhouses. The cutter works at most 8 hours, so we have our first constraint:

$$2x + y \le 8$$

Math Note

The shaded region in a linear programming problem is often called the **feasible region** because it represents all values of the variables that are possible solutions based on the constraints.

Since it takes 1 hour to assemble a playground set and 1.5 hours to assemble a playhouse, the second constraint is

$$x + 1.5y \le 8 \quad \text{Maximum 8 hours on assembly.}$$

We can't make a negative number of playground sets or playhouses, so, there are two additional constraints:

$$x \ge 0$$
$$y \ge 0$$

Step 3 Graph the linear system made up of the constraints.

$$2x + y \le 8 \qquad x \ge 0$$
$$x + 1.5y \le 8 \qquad y \ge 0$$

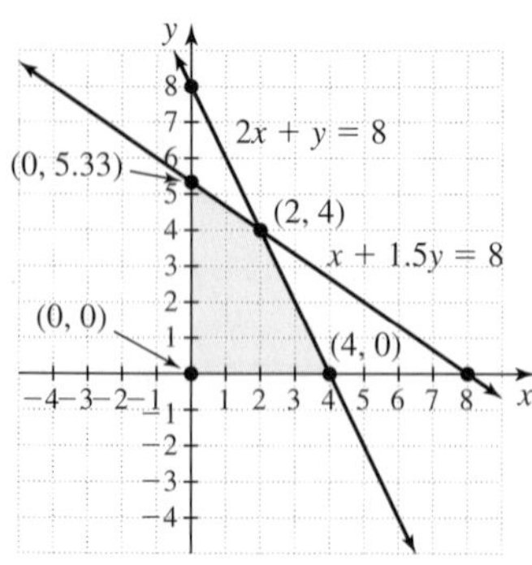

First, we find the intercepts for the lines in the first two constraints.

$2x + y = 8$

x	y
0	8
4	0

$x + 1.5y = 8$

x	y
0	5.33
8	0

The graph is to the left. Note that the two lines meet at the point (2, 4).

Step 4 Find the vertices of the polygonal region. The vertices of the polygon are (0, 0), (0, 5.33), (2, 4), (4, 0).

The vertex (0, 0) is the origin. The vertex (2, 4) was found by determining the coordinates of the intersection of the two lines $2x + y = 8$ and $x + 1.5y = 8$. The vertex (0, 5.33) was found by substituting 0 for x and solving for y in the equation $x + 1.5y = 8$. Finally, the vertex (4, 0) was found by substituting 0 for y and solving for x in the equation $2x + y = 8$.

Step 5 Substitute the vertices into the objective function and find the maximum value. (Recall that the maximum or minimum values of the objective function will occur at a vertex of the polygonal region.)

	$P = 100x + 75y$
(0, 0)	$P = 100(0) + 75(0) = 0 + 0 = 0$
(0, 5.33)	$P = 100(0) + 75(5.33) = 0 + 399.75 = 399.75$
(2, 4)	$P = 100(2) + 75(4) = 200 + 300 = 500$
(4, 0)	$P = 100(4) + 75(0) = 400 + 0 = 400$

The vertex (2, 4) produces the maximum value 500. The solution is to make two playground sets and four playhouses for a maximum profit of \$500 per day.

▼ Try This One 1

A craftsperson makes leather purses and cloth purses. She can make at most 20 purses per week. She plans to make at least 5 leather purses and no more than 15 cloth purses per week. Her profit on a leather purse is \$5.00, and her profit on a cloth purse is \$10.00. How many of each type of purse should she make to maximize her profit?

The procedure for solving problems using linear programming is summarized next.

Procedure for Using Linear Programming

Step 1 Write the objective function.

Step 2 Write the constraints.

Step 3 Graph the constraints.

Step 4 Find the vertices of the polygonal region.

Step 5 Substitute the coordinates of the vertices into the objective function and find the maximum or minimum value.

(*Note*: The solutions will not always be integers.)

EXAMPLE 2 Solving a Linear Programming Problem

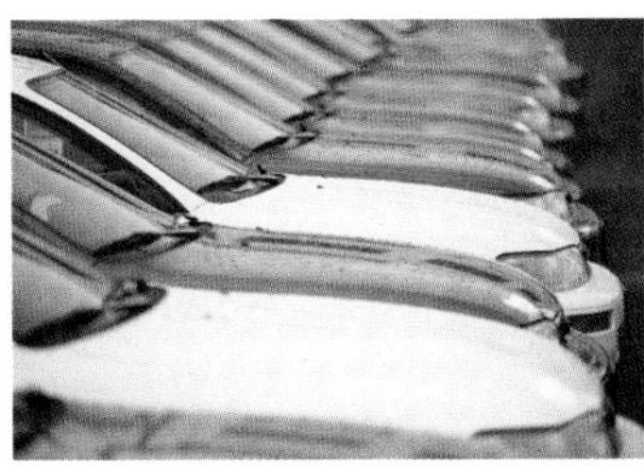

An automobile dealer has room for no more than 100 cars on his lot. The dealer sells two models, convertibles and sedans, and he sells at least 3 times as many sedans as convertibles. If he makes a profit of \$1,000 on a convertible and a profit of \$1,500 on a sedan, how many of each automobile should he have on his lot in order to maximize his profit?

SOLUTION

Step 1 Write the objective function.

Since the dealer makes a profit of \$1,000 on each convertible sold and \$1,500 on each sedan sold, the objective function is

$$P = 1{,}000x + 1{,}500y$$

where x = the number of convertibles sold, y = the number of sedans sold, and P = profit made.

Math Note

If the intersection point isn't clear from the graph, you could find it by solving the system of equations $x + y = 100$ and $y = 3x$ algebraically.

Step 2 Write the constraints.

$$x + y \leq 100 \quad \text{At most 100 cars.}$$
$$3x \leq y \quad \text{At least 3 times as many sedans.}$$

In addition, x and y can't be negative:

$$x \geq 0$$
$$y \geq 0$$

Step 3 Graph the system.

$$x + y \leq 100 \qquad x \geq 0$$
$$3x \leq y \qquad y \geq 0$$

Find points for the lines $x + y = 100$ and $y = 3x$:

$x + y = 100$

x	y
0	100
100	0

$y = 3x$

x	y
0	0
25	75

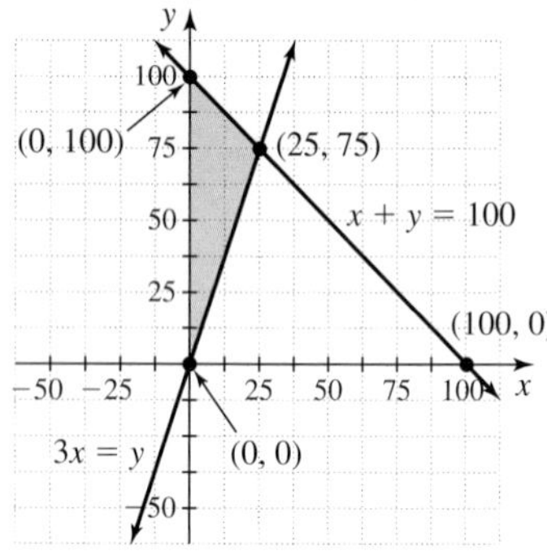

The two lines meet at (25, 75). The region is shown to the left.

Step 4 Find the vertices of the polygonal region. They are (0, 0), (0, 100), and (25, 75). The vertex (0, 0) is the origin. The vertex (25, 75) was found by finding the coordinates of the point of the intersection of the two lines $x + y = 100$ and $3x = y$. The vertex (0, 100) was found by substituting $x = 0$ and solving for y in the equation $x + y = 100$.

Step 5 Substitute into the objective function and find the maximum value.

$$P = 1{,}000x + 1{,}500y$$
$$(0, 0) \quad P = 1{,}000(0) + 1{,}500(0) = 0 + 0 = 0$$
$$(0, 100) \quad P = 1{,}000(0) + 1{,}500(100) = 0 + 150{,}000 = 150{,}000$$
$$(25, 75) \quad P = 1{,}000(25) + 1{,}500(75) = 25{,}000 + 112{,}500 = 137{,}500$$

The maximum profit occurs at the point (0, 100), so the dealer should stock no convertibles and 100 sedans to maximize his profit.

☑ 1. Use linear programming to solve real-world problems.

▼ Try This One 2

A professor offers students a choice of how they want to design the grading structure for a sociology course. Students can earn 20 points for a community service project and 15 points for a research paper. Everyone in the class is required to write at least two papers, and students can get credit for at most five service projects. If one student plans to complete no more than eight assignments total, how many of each assignment should she do to maximize her points?

Answers to Try This One

1 Five leather and fifteen cloth

2 Five community service projects and three research papers

EXERCISE SET 7-5

Writing Exercises

1. What is the goal of a linear programming problem?
2. What is an objective function?
3. What are constraints and how do we represent them mathematically?
4. Why are the vertices of a polygonal region important in linear programming?

Computational Exercises

For Exercises 5–14, write each constraint or objective function.

5. A company that makes calculators can manufacture at most 50 calculators in one day. Write the constraint.
6. The profit on a scientific calculator is \$10.00, and the profit on a graphing calculator is \$12.50. Write the objective function for the profit as a function of the number of scientific calculators sold and the number of graphing calculators sold.
7. A company plans to manufacture twice as many pens as pencils. Write the constraint.
8. The manager of a store plans to purchase at least 20 more desktop computers than laptop computers. Write the constraint.
9. The profit on a desktop computer is \$85.00, and the profit on a laptop computer is \$130.00. Write the profit function as a function of the number of desktop computers and the number of laptop computers sold.
10. A food store owner buys turkeys and hams. The owner plans to purchase at most a total of 50 products. Write a constraint.
11. The profit on each turkey sold is \$4.00, and the profit on each ham sold is \$3.00. Write the profit function as a function of the number of turkeys sold and the number of hams sold.
12. The food store owner plans to purchase at most 10 more turkeys than hams. Write the constraint.
13. It takes a person 2 hours to assemble a VCR and 3 hours to assemble a CD player. Each employee can work at most 6 hours a day. Write a constraint for the problem.
14. The profit on a CD player is \$32.00, and the profit on a VCR is \$45.00. Write a profit function as a function of the number of CD players sold and the number of VCRs sold.

For Exercises 15–20, evaluate the profit function for each vertex and determine which vertex gives the maximum profit.

15. $P = \$30x + \$40y$; (0, 0), (5, 6), (10, 0), (0, 8)
16. $P = \$85x + \$132y$; (0, 0), (9, 0), (0, 3), (7, 2)
17. $P = \$15x + \$5y$; (0, 0), (4, 8), (0, 6), (7, 0)
18. $P = \$5x + \$20y$; (0, 0), (2, 7), (0, 10), (9, 0)
19. $P = \$120x + \$340y$; (0, 0), (20, 50), (0, 62), (43, 0)
20. $P = \$92x + \$45y$; (0, 0), (20, 13), (0, 18), (25, 0)

Real-World Applications

21. A store owner sells televisions and VCRs, but he can stock at most 30 of such appliances in any combination. The profit on the sale of a television set is \$40, and the profit on a VCR is \$60. The owner plans to carry at least twice as many television sets as VCRs. How many of each should the owner stock to maximize the profit?
22. A small bakery makes cakes and pies every business day. It can make a total of 30 items of which at least 5 must be cakes and at least 10 must be pies for its restaurant customers. The profit on each cake is \$1.50, and the profit on each pie is \$2.00. How many of each should be made in order to maximize the profit?
23. A sorority is knitting hats and scarves as a fundraiser for a local charity. The members can produce as many as 50 hats and scarves in the amount of time they have, and yarn costs \$4 for a hat and \$3 for a scarf. Their budget is \$360 for the project. When they sell the hats and scarves, they raise \$5 on each hat and \$4 on each scarf. Assuming they can sell as many of each as they want, how many hats and scarves should be made to maximize the amount they raise?
24. A toy maker makes wooden toy train sets and wooden toy monster trucks. It takes 5 hours to cut and assemble a train set and 2 hours to paint it by hand. It takes 3 hours to cut and assemble a truck and 1 hour to paint it by hand. When he sells a train set, he makes a profit of \$26, and when he sells a truck, he makes a profit of \$15. How many of each should he make within a 40-hour week, if he does not want any unfinished toys at the end of the week? (Assume he wants the maximum profit.)
25. A neighborhood has come together to make a community garden out of a 30-square-foot plot of land between two houses and in it, they grow organic tomatoes and squash. It costs them \$10 to grow a square foot of the tomatoes and \$15 to grow a square foot of the squash each month. They are able to raise \$360 each month to cover the costs of growing the vegetables. The profit for each square foot of tomatoes is \$15 and the profit for each square foot of squash is \$20.

How many square feet of each vegetable should be grown in order to maximize the community's profit?

26. A school theater has 120 seats. The theater manager plans to reserve some seats for students and some seats for nonstudents (i.e., the general public). Students pay \$4 per ticket, and nonstudents pay \$8 per ticket. The school has a rule that at least twice as many seats for students have to be reserved as seats for nonstudents. In order to maximize the profit, how many student seats should be reserved?
27. A local animal shelter receives at least four cats and seven dogs a day. The shelter has the capacity to receive up to 30 cats and dogs a day. If the shelter can raise donations for the animals at a rate of \$3 per cat and \$5 per dog per day, how many cats and dogs should the shelter accept to maximize the amount of donations per day?
28. A boutique franchise buys its hair products from either Biolage or Rusk. Biolage offers a package of four bottles of glaze and three bottles of mousse and Rusk offers a package of two bottles of glaze and six bottles of mousse. Biolage charges \$5 per product if the boutique buys it as part of the package, and Rusk similarly charges \$6 per product. The boutique wants to buy at least 2,000 bottles of glaze and at least 2,400 bottles of mousse to stock all of its locations. Since it regularly buys from Biolage, it struck a deal that at least 25% of the total number of products it buys from both vendors will be from Biolage. How many of products does it buy from each vendor to minimize its cost?
29. A company sells two types of computers, one a laptop and the other a desktop model. It takes workers 4 hours to build a laptop and 2 hours to build a desktop computer. In a given month, the engineers can work up to 800 hours according to their union contract, and they must produce at least 200 computers, which is the company's minimum production level. Due to consumer demand, they must build between 100 and 150 laptops a month. The company brings in a profit of \$300 for each laptop and \$200 for each desktop computer. How many of each type of computer should the company build each month to maximize its profits?
30. An acting troupe wants to rent out the campus amphitheater for its summer Shakespeare production. They can sell no more than 1,000 general admission tickets and up to 500 student tickets according to school policy. They must sell at least 100 student tickets according to student government rules. To make enough money, they must sell at least 800 tickets. The school charges \$2 per student ticket and \$3 per general admission ticket as its fee for renting the amphitheater. How many of each ticket must be sold to minimize the cost of renting the theater and still make enough money?

Critical Thinking

31. Consider a scenario where you would have to use linear programming to solve a problem in your field of study. Write the constraints and a profit function. Then solve the problem as shown in this section.
32. In most linear programming problems, we get constraints from the fact that the variables have to be nonnegative. Think of a real-world situation in which the variables have constraints, but can be negative.

Section 7-6 Functions

LEARNING OBJECTIVES

- ☐ 1. Identify functions.
- ☐ 2. Write functions in function notation.
- ☐ 3. Evaluate functions.
- ☐ 4. Find the domain and range of functions.
- ☐ 5. Determine if a graph represents a function.

Each of the following statements came up in a Google search for the string "is a function of":

> "Intelligence is a function of experience."
> "Health is a function of proper nutrition."
> "Freedom is a function of economics."
> "What we wear each day is a function of atmospheric conditions."

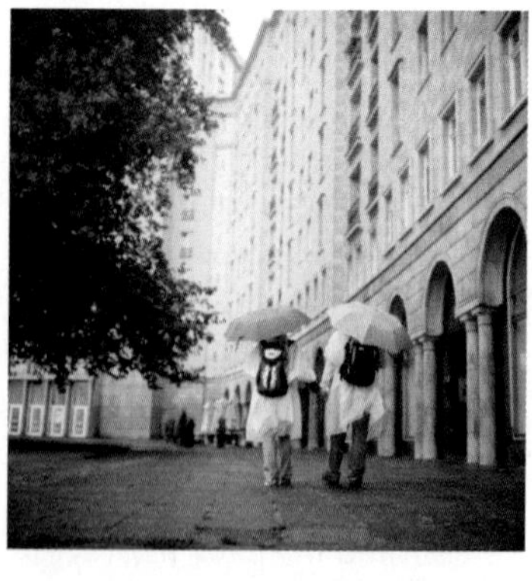

These statements illustrate the simple idea behind one of the most important concepts in all of math, the function. In every case, the statement is telling us that one thing depends on another, and this is the essence of functions.

But first, let's take a look at a related idea, the relation.

A **relation** is a rule matching up two sets of objects. Relations are often represented by sets of ordered pairs.

The following are examples of relations:

$$A = \{(2, 3), (-2, 1), (12, -3), (2, -1)\}$$
$$B = \{(9, 0), (-3, 1), (3, 9), (-1, 5), (2, 3)\}$$
$$C = \{(x, y) \mid 3x + 5y = 7\}$$

Often, equations are used to represent sets of ordered pairs, as in relation C above. Usually we will simplify the notation by simply writing the equation; it is understood that the equation represents a relation between two sets, with x representing elements from one set, and y representing elements from another.

A function is a special type of relation.

A **function** is a relation in which each x coordinate gets paired with exactly one y coordinate. In other words, the first coordinate is never repeated with a different second coordinate.

Relation A above is not a function because the points (2, 3) and (2, −1) have the same first coordinate. Relation B is a function because no first coordinate is repeated. Relation C is also a function (although it's not as clear as the other two) because for each value of x that you substitute into the equation, there is only one possible value of y that corresponds to it.

EXAMPLE 1 Identifying Functions Defined by Equations

Which of the following equations represent functions? (Assume that x represents the first coordinate.)

(a) $y = x^2$ (b) $3x^2 + y - 2x = 5$ (c) $x = |y|$

SOLUTION

(a) This is a function. Every number has only one square, so every value of x has only one associated y.
(b) This is also a function. Every x will again have only one associated y.
(c) This is not a function. Positive values of x will correspond to two possible values of y. For example, if $x = 2$, y can be 2 or −2.

▼ Try This One 1

☑ 1. Identify functions.

Which of the following equations represent functions? (Assume that x represents the first coordinate.)

(a) $x = y^2$ (b) $y = |x|$ (c) $x^2 + y^2 = 4$

Math Note

The function notation $f(x)$ is *not* a product, and should *never* be read as "f times x." It designates that f is a function with variable x.

Function Notation

The equation $y = x^2$ represents a function that relates variables x and y. We call x the **independent variable** and y the **dependent variable** because its value depends on the choice of x. Another way to write the same function is $f(x) = x^2$. This is known as **function notation,** and is read aloud as "f of x equals x squared." This is the notation most commonly used to describe functions.

Functions can also be called by names other than f. Letters like $f, g, h,$ and k are commonly used to represent functions, but a letter that is more representative can be used. For example, the circumference of a circle is a function of the radius (meaning that it depends on the value of the radius), so we could use the letter C to represent this function:

$$C(r) = 2\pi r.$$

To write an equation in function notation, solve for y in terms of x, then change the letter y to the symbol $f(x)$.

EXAMPLE 2 Writing a Function in Function Notation

Write $3x - 2y = 6$ in function notation.

SOLUTION

We need to solve the equation for y, then replace y with $f(x)$.

$$3x - 2y = 6 \quad \text{Subtract } 3x \text{ from both sides.}$$
$$-2y = -3x + 6 \quad \text{Divide both sides by } -2.$$
$$y = \frac{3}{2}x - 3 \quad \text{Replace } y \text{ with } f(x).$$
$$f(x) = \frac{3}{2}x - 3$$

▼ Try This One 2

☑ 2. Write functions in function notation.

Write $10y + 30 = 5x$ in function notation.

When a function is written as $f(x)$, $f(2)$ means to find the value of the function when $x = 2$. This is known as **evaluating** a function. We call $x = 2$ the **input,** and the resulting value of the function the **output.**

EXAMPLE 3 Evaluating a Function

Math Note

When finding $f(-2)$, make sure you replace *all* occurrences of the variable x with (-2). The parentheses are almost always a good idea.

Let $f(x) = x^2 + 3x - 5$. Find $f(3)$, $f(-2)$, and $f(0)$.

SOLUTION

$$f(3) = (3)^2 + 3(3) - 5 = 9 + 9 - 5 = 13$$
$$f(-2) = (-2)^2 + 3(-2) - 5 = 4 - 6 - 5 = -7$$
$$f(0) = (0)^2 + 3(0) - 5 = 0 + 0 - 5 = -5$$

▼ Try This One 3

Let $f(x) = 3x^2 - 2x + 5$. Find $f(-1)$, $f(2)$, and $f(3)$.

☑ 3. Evaluate functions.

The **domain** of a function is the set of all values of the independent variable x that result in real number values for y. The **range** of a function is the set of all possible y values.

EXAMPLE 4 Finding the Domain and Range of a Function

Find the domain and range of each function:

(a) $f(x) = x^2$ (b) $f(x) = \sqrt{x}$ (c) $f(x) = \dfrac{3x - 2}{x + 1}$

SOLUTION

(a) There are no restrictions on what values x can be; therefore the domain is all real numbers. Since x^2 is never negative, the range is $\{y \mid y \geq 0\}$.

(b) Since the square root of a negative number is undefined, x cannot be negative. Therefore, the domain is $\{x \mid x > 0\}$. Since $\sqrt{x}$ is never negative the range is $\{y \mid y \geq 0\}$.

(c) Since the denominator of a fraction cannot be zero, we must exclude $x = -1$. Every other x value will result in a real number output, so the domain is all real numbers except -1, which we write as $\{x \mid x \neq -1\}$. The range is not obvious, but notice that an output of 3 would make the equation

$$3 = \frac{3x - 2}{x + 1}$$

Multiplying both sides by $x + 1$, we get the contradiction $3x + 3 = 3x - 2$, so the range is $\{y \mid y \neq 3\}$.

▼ Try This One 4

☑ 4. Find the domain and range of functions.

Find the domain and range of the following functions:

(a) $f(x) = |x|$ (b) $f(x) = \sqrt{x - 2}$ (c) $f(x) = \dfrac{2x - 1}{x - 3}$

The Vertical Line Test

As we have seen, an equation in two variables sometimes represents a function, and sometimes does not. If we know the graph of an equation, there's a simple way to determine whether or not the equation represents a function. We know that a relation is not a function if any x value corresponds to more than one output. Consider the graph of an equation shown in Figure 7-9(a). The two points labeled have the same x coordinate, so the equation is not a function. The vertical line through those two points indicates that if any vertical line crosses a graph more than once, the graph does not represent a function. In Figure 7-9(b), there is no vertical line that crosses the graph more than once, so it is the graph of a function.

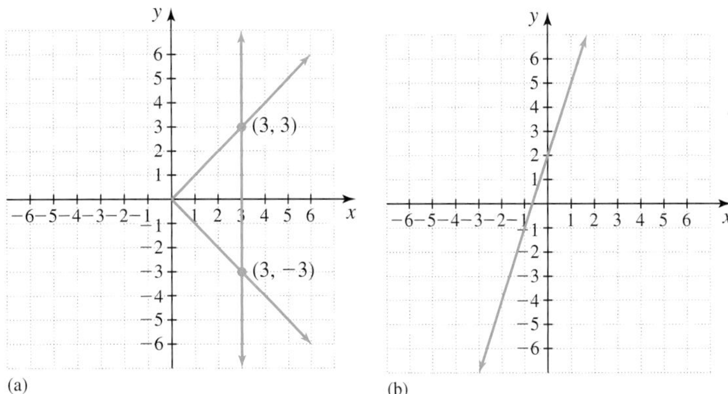

Figure 7-9

The Vertical Line Test for Functions

If no vertical line can intersect the graph of a relation at more than one point, then the relation is a function.

EXAMPLE 5 Using the Vertical Line Test

Use the vertical line test to determine whether each relation graphed represents a function.

(a)

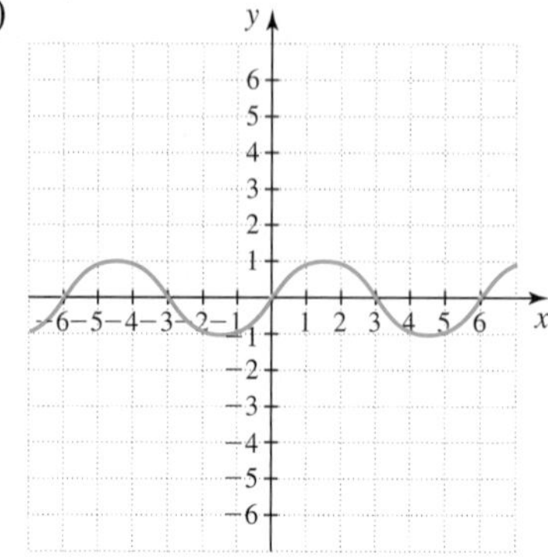

(b)

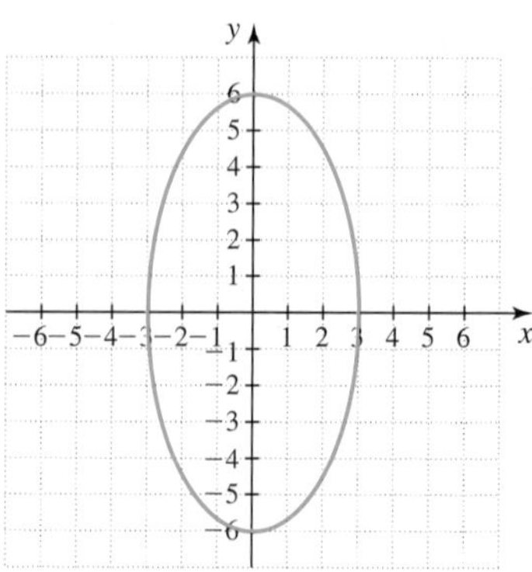

SOLUTION

(a) This is a function since any vertical line will cross the graph only once.

(b) This is not a function since there is a vertical line that crosses the graph in more than one place, as seen to the left.

▼ Try This One 5

☑ 5. Determine if a graph represents a function.

Use the vertical line test to determine whether each relation graphed represents a function.

(a)

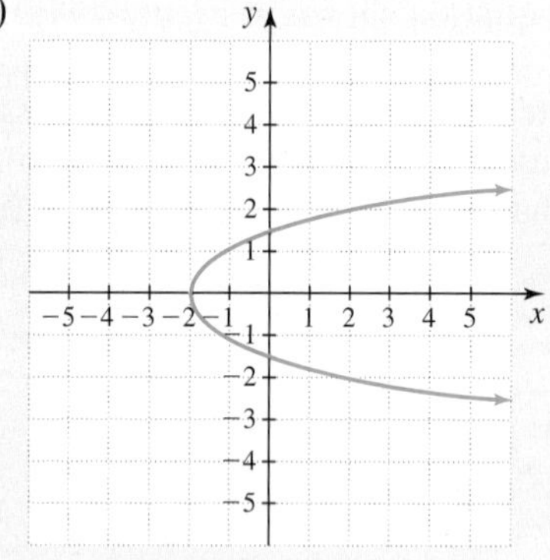

(b)

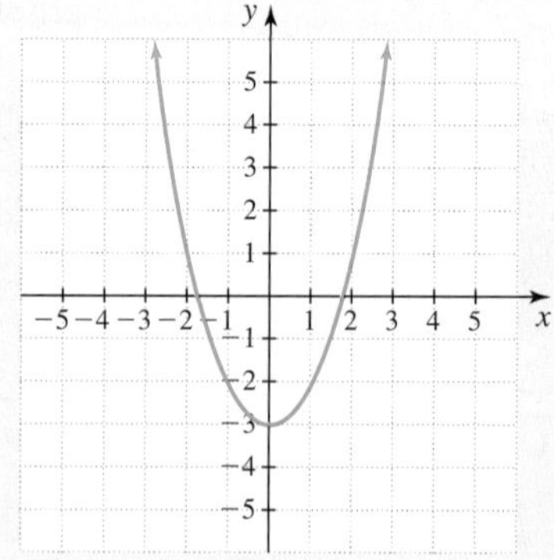

Answers to Try This One

1 (b) is a function.

2 $f(x) = \frac{1}{2}x - 3$

3 10; 13; 26

4 (a) Domain: All real numbers. Range: $\{y \mid y \geq 0\}$.

(b) Domain: $\{x \mid x \geq 2\}$. Range: $\{y \mid y \geq 0\}$.

(c) Domain: $\{x \mid x \neq 3\}$. Range: $\{y \mid y \neq 2\}$.

5 (b) represents a function.

EXERCISE SET 7-6

Writing Exercises

1. What is the difference between a function and a relation?
2. What is the domain of a function?
3. What is the range of a function?
4. What does it mean to evaluate a function?
5. Describe the vertical line test and why it works.
6. Write how the expression $f(x) = 3x - 7$ would be read aloud.

Computational Exercises

For Exercises 7–16, state whether or not the relation is a function. Assume that x represents the first coordinate.

7. $\{(5, 8), (6, 9), (7, 10), (8, 11)\}$
8. $\{(0, 1), (0, 2), (0, 3), (0, 4), (0, 5)\}$
9. $\{(6, 11), (7, 11), (8, 11), (9, 11)\}$
10. $\{(-2, 4), (-1, 5), (0, 3), (1, 8), (-2, 2)\}$
11. $3x + 2y = 6$
12. $-6x = y + 5$
13. $y = 5x$
14. $x = 2y^2$
15. $|x - 3| = 2y$
16. $3x = |2y|$

For Exercises 17–24, evaluate the function for the indicated values.

17. $f(x) = 3x + 8; f(3), f(-2)$
18. $f(x) = -2x + 5; f(-5), f(2)$
19. $f(x) = 4x - 8; f(10), f(-10)$
20. $f(x) = 6x^2 - 2x + 5; f(2), f(-4)$
21. $f(x) = 8x^2 + 3x; f(0), f(6)$
22. $f(x) = -3x^2 + 5; f(1.5), f(-2.1)$
23. $f(x) = x^2 + 4x + 7; f(-3.6), f(4.5)$
24. $f(x) = -x^2 + 6x - 3; f(0), f(-20)$

For Exercises 25–34, find the domain and range of each function.

25. $f(x) = (x - 1)^2$
26. $f(x) = \sqrt{3 - x}$
27. $f(x) = \frac{x}{x - 2}$
28. $y = \frac{2}{3}x + 3$
29. $y = 3 + \sqrt{x}$
30. $f(x) = -\sqrt{x}$
31. $2x - y = 7$
32. $y = \frac{2x + 1}{x}$
33. $f(x) = -\sqrt{x + 1} - 1$
34. $f(x) = 2(x + 3)^2$

For Exercises 35–38, use the vertical line test to determine if the graph represents a function.

35.

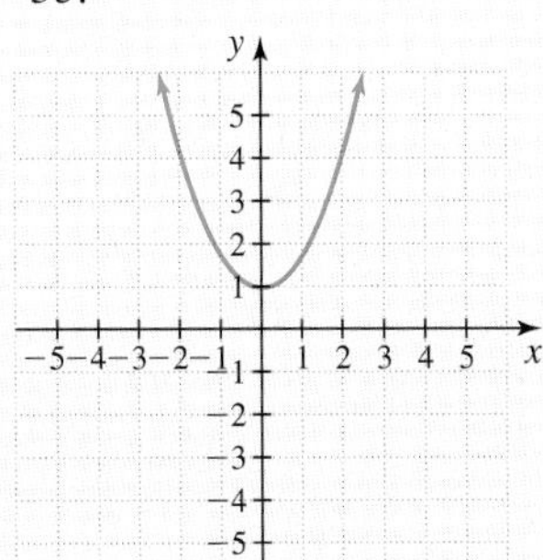

37.

36.

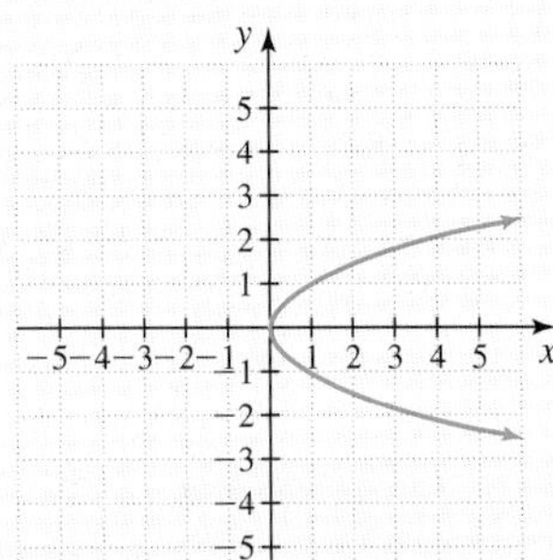

38.

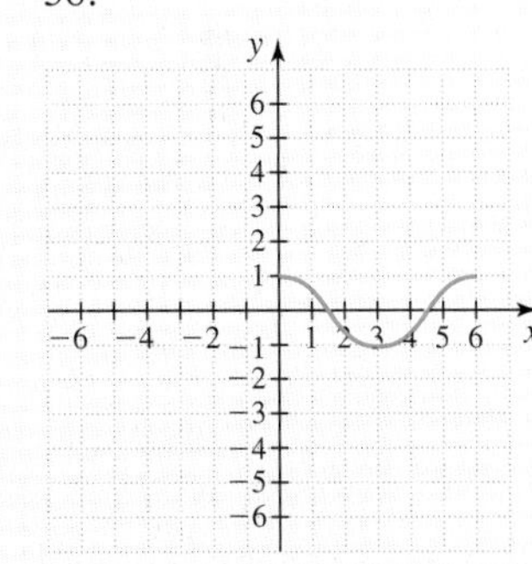

Real-World Applications

39. A waitress made \$2.52 per hour plus \$78 in tips on a Friday evening shift. The amount of money she makes for the night is given by the function $P(t) = 2.52t + 78$, where t is the number of hours she worked. How much would a waitress make for a 4-hour shift if the pay rate is the same? An 8-hour shift?
40. A realtor makes a commission of 6% of the purchase price of a house plus a third of the \$3,000 closing costs. The amount he makes on a house is given by the function $A(s) = 1{,}000 + 0.06s$, where s is the selling price of the house. How much more would he make on a \$300,000 dollar house than a \$120,000 house?
41. A company started out spending \$578 to get up and running, and then it planned on charging \$25 for its product. The total amount of profit (which is money brought in minus cost) is given by the function $p(x) = 25x - 578$. If the company sells 18 units, will it make enough to cover its start-up costs? How much will it make if it sells 120 units?
42. On a history test, Miley got a quarter of the true/false questions wrong, two multiple choice questions wrong, and every other question right. The function $w(q) = \frac{1}{4}q + 2$ describes the number of questions she got wrong in terms of the number of true/false questions on the test. If there were 20 true/false

questions on the test, how many total questions did Miley get wrong?

43. A particular car rental agency charges \$20 a day to rent a car. Write the cost of renting a car as a function of the number of days rented. How much does it cost to rent the car for 5 days?
44. A salesperson earns a base income of \$250 a week plus 10% commission on the dollar value of each sale. Write a function to represent the amount of his weekly income based on dollar value of sales. If his sales for the week totaled \$5,000, what is his income for the week?
45. At the big sale after Thanksgiving, Zoe spent triple what Molly spent less \$10. Write what Zoe spent as a function of what Molly spent. If Molly spent \$45, how much did Zoe spend?
46. Macy spent \$234.56 at the bookstore this semester on her textbooks. The average price per book is found by taking the total amount and dividing by the number of books purchased. Write the average price per book as a function of the total number of books purchased. If Macy bought six books, what was her average price per book?
47. On a road trip, Frank and Mandy were already 80 miles into their trip before they realized the rest of the distance could be covered in the time they had if they drove 65 mph the rest of the way. Write the total distance for their trip as a function of time in hours. If their total trip is 275 miles, how many more hours will they drive?
48. A company has endorsed a team of runners in the citywide marathon and says it will pay the runners \$0.40 per mile they run plus donate a fixed amount of \$500 to the charity the runners are sponsoring. Write the total amount the company will pay as a function of the total number of miles the runners travel during the marathon. If the company pays \$600 and there were 10 runners who all ran the same distance, how far did each runner travel?
49. While Jenny is studying for class, she realizes that reading her math book and taking notes as her professor suggested takes her triple the time it used to take her to just read without taking notes. However, she also notices her grades rise dramatically, so she decides this is a good idea. Write the time it takes Jenny to read and take notes as a function of the time it takes her to only read her book. If it took Jenny 45 minutes to read a section of her book without taking notes, how long would it take her to read the same section while taking notes?
50. When Marcy is walking, she notes that she is only burning 80% of the calories she would be if she were jogging. Write the amount of calories Marcy burns walking as a function of the amount of calories Marcy would burn by jogging. If she burns 120 calories jogging, how much would she burn by walking?
51. The Brighton soccer team won twice as many as the Westview soccer team plus four more games. Write the number of games the Brighton team won as a function of the games Westview won. If the Brighton team won 12 games, how many did the Westview team win?
52. Andrea got triple the number of problems right on the test that Mark did less five problems. Write the amount Andrea got right as a function of the problems Mark got right. If Mark got six problems right, how many did Andrea get right?
53. A shoe store was having a sale. Customers could get 20% off their entire purchase if they bought at least two pairs of shoes. Write the sale price that a customer who bought at least two pairs of shoes would pay as a function of the price before the discount. If Cindy bought three pairs of shoes totaling \$119.78, what would the sale price be?
54. A company has determined its price per unit is \$15.40 divided by the product of 0.07 and the number of units it sells. Write the price per unit as a function of the number of units it sells. How much will the price per unit be if it sells 2,000 units?
55. Before class, Mike ate a third of the cookies in his bag, and then he ate half the cookies remaining, and then later on the bus, he ate four more cookies. Write the amount of cookies Mike ate as a function of the number of cookies in the bag. If there were 30 cookies in the bag, how many did Mike eat?
56. At the grocery store, the cashier rang up a total of \$140.28 for groceries before taxes. The customer told her child to get the sodas she forgot, and they cost \$1.59 each. Write the total amount of the grocery bill as a function of the number of sodas the customer buys. If the customer buys four sodas, how much will her new total be after a 6% sales tax has been applied?
57. At a poker tournament, the dealer makes 75% of the amount the players must contribute to play plus \$48 in tips for the night. Write the amount the dealer makes if there are 10 players at the table as a function of the amount the players must contribute to play. If the dealer made \$198 for the night, how much did each player have to pay to play the game?
58. At a dieter's meeting, Sandy weighed 90% of her original weight less the 5 pounds she'd lost that month. Write Sandy's current weight as a function of her original weight. If Sandy started at 180 lb, how much does she weigh now?

Critical Thinking

59. For the function $f(x) = x^2 + 3x + 1$, find $f(3)$, $f(-3)$, $f(x + 3)$, and $f(x - 3)$.
60. Think of four real-life situations you've encountered today that could be modeled by a function.

Section 7-7 Linear, Quadratic, and Exponential Functions

LEARNING OBJECTIVES

- ❑ 1. Graph linear functions.
- ❑ 2. Graph quadratic functions.
- ❑ 3. Apply quadratic functions to real-world problems.
- ❑ 4. Graph exponential functions.
- ❑ 5. Apply exponential functions to real-world problems.

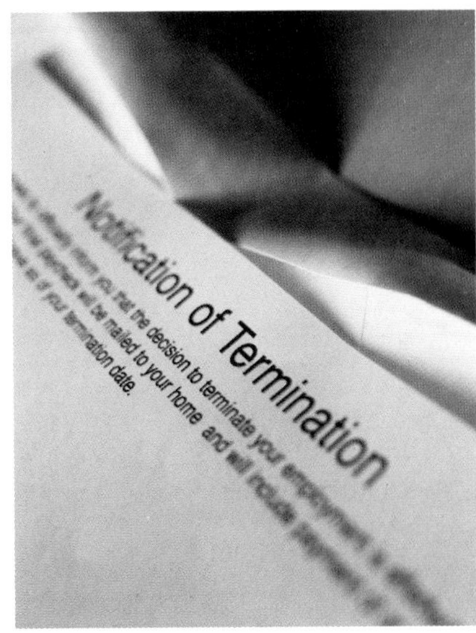

It's no secret that 2008 was a very tough year for the economy, particularly in the area of employment. Almost 2.6 million Americans lost their jobs in 2008, the highest total in over 60 years. The function $f(x) = 0.217x + 4.40$ can be used to model the unemployment rate for 2008, with f representing the percentage of workers who are unemployed and x representing the month of the year. For example, $f(10) = 6.57$; this tells us that in October (the 10th month), the unemployment rate was 6.57%.

This example illustrates the main reason why functions are so useful: they can be used to model a practically limitless variety of real-world phenomena. Anything that can be quantified can be modeled with a function, from the number of wins by your favorite football team to important economic data. In this section we will study three types of functions that are often used to model real-world situations.

Linear Functions

A **linear function** is a function of the form $f(x) = ax + b$, where a and b are real numbers.

We studied linear equations extensively in Chapter 6, and much of what we learned carries over to the study of linear functions. The graph of every linear function is a straight line. In the function $f(x) = ax + b$, a is the slope of the graph, and $(0, b)$ is the y intercept. In Example 1, we'll look at two different quick methods for graphing a linear function. You should decide which method you understand best and use that one.

EXAMPLE 1 Graphing Linear Functions

Math Note

Remember that if the slope is negative, move downward from left to right.

Graph each linear function.

(a) $f(x) = 3x - 2$

(b) $f(x) = \frac{2}{3}x + 3$

SOLUTION

(a) In this method, we'll take advantage of knowing the slope, which in this case is 3, and the y intercept, which is $(0, -2)$. Plot the point $(0, -2)$, and then use a rise of 3 and a run of 1 to find a second point. Then we draw the line connecting those points.

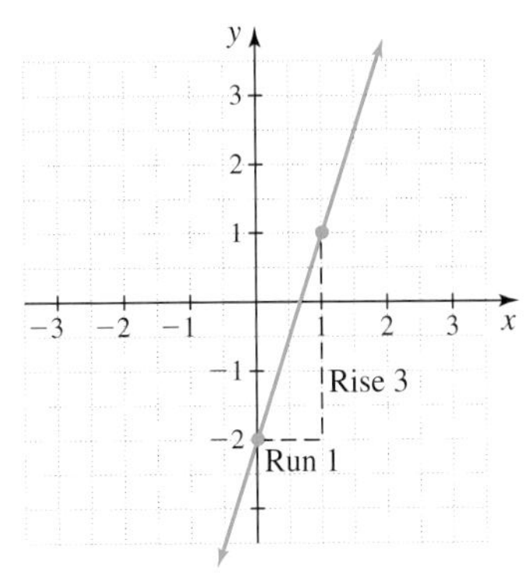

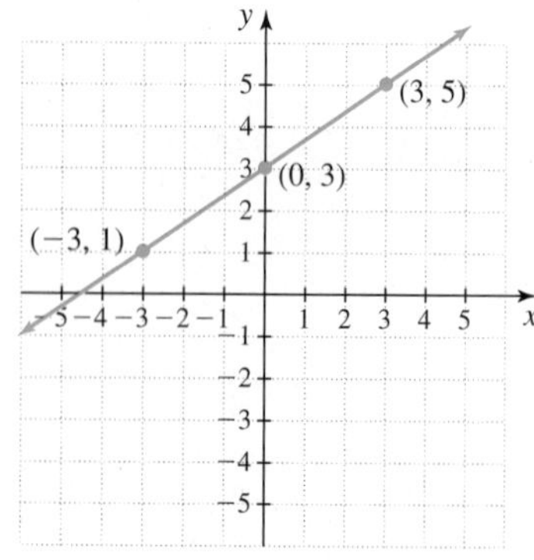

(b) In the second method, we'll evaluate the function for three x values then plot the associated points, and again draw a line connecting the points.

$$f(-3) = \frac{2}{3}(-3) + 3 = -2 + 3 = 1$$
$$f(0) = 3$$
$$f(3) = \frac{2}{3}(3) + 3 = 2 + 3 = 5$$

▼ Try This One 1

Graph each linear function.

(a) $f(x) = 2x - 4$ (b) $f(x) = -\frac{3}{4}x + 1$

Math Note

Notice that in Example 1(b) we chose x values that made it easy to evaluate the function by eliminating the fraction.

☑ 1. Graph linear functions.

Quadratic Functions

Linear functions are naturally limited as models in that the graph can never change direction. If a quantity first gets larger, then starts to get smaller, we would need a function with a graph that changes direction to model it. The quadratic functions, which are related to the quadratic equations we studied in Chapter 6, fit the bill nicely. Their graphs are shaped like the time-lapse photo of a ball in flight in the photo to the left.

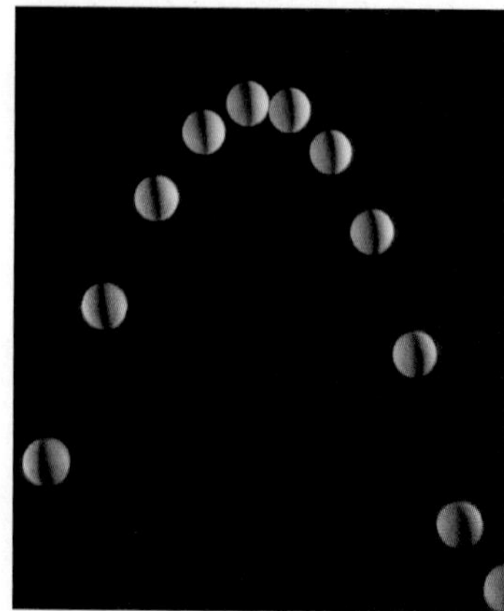

Parabolas often occur in problems about the motion of a falling object.

A function of the form $f(x) = ax^2 + bx + c$, where a, b, and c are real numbers and $a \neq 0$, is called a **quadratic function.** The graph of a quadratic function is called a **parabola.**

When a parabola decreases in height, reaches a low point, and then increases, we say that it *opens upward.* This occurs when a, the coefficient of x^2, is positive. When a parabola increases in height, reaches a high point, then decreases, we say that it *opens downward.* This occurs when a is negative.

The point where a parabola changes direction is called the **vertex.** For parabolas that open upward, the vertex is the lowest point. For parabolas that open downward, the vertex is the highest point. Every parabola has two halves that are mirror images of each other, with the divider being a vertical line through the vertex. We call this line the **axis of symmetry.** It's not part of the parabola, but a guide to help us draw one. We say that a parabola is *symmetric* about its axis of symmetry. Two representative parabolas are shown in Figure 7-10 below.

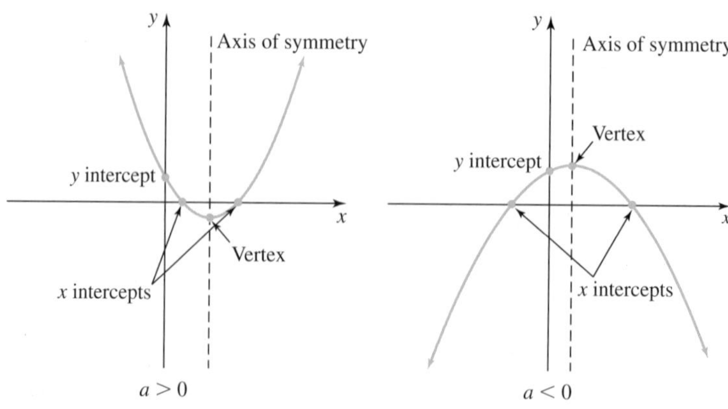

Figure 7-10

Math Note

You can check to make sure your parabola opens in the correct direction by noting the sign of a. If $a > 0$, it should open upward, and if $a < 0$, it should open downward.

Procedure For Graphing a Parabola

Step 1 Identify a, b, and c; then use the formula $x = \frac{-b}{2a}$ to find the x coordinate of the vertex. The y coordinate can then be found by substituting the x value into the function.

Step 2 Find the y intercept by evaluating the function for $x = 0$.

Step 3 Find the x intercepts, if any, by substituting 0 for $f(x)$ and solving for x, using either factoring or the quadratic formula.

Step 4 If there are no x intercepts, find at least one other point on each side of the vertex to help determine the shape.

Step 5 Plot all the points you found, then connect them with a smooth curve.

EXAMPLE 2 Graphing a Quadratic Function

Graph the function

$$f(x) = x^2 - 6x + 5$$

SOLUTION

Step 1 In this case, $a = 1$, $b = -6$, and $c = 5$. We can begin by finding the x coordinate of the vertex:

$$x = \frac{-b}{2a} = \frac{-(-6)}{2(1)} = \frac{6}{2} = 3$$

Evaluate $f(3)$ to find the y coordinate:

$$f(3) = 3^2 - 6(3) + 5 = 9 - 18 + 5 = -4$$

The vertex is $(3, -4)$.

Math Note

Since the axis of symmetry is a vertical line and passes through the vertex, its equation is $x = 3$.

Step 2 Find the y intercept by evaluating $f(0)$.

$$f(0) = 0^2 - 6(0) + 5 = 5$$

The y intercept is $(0, 5)$.

Step 3 Find the x intercepts by substituting 0 for $f(x)$ and solving.

$$0 = x^2 - 6x + 5 \quad \textit{Factor.}$$
$$0 = (x - 5)(x - 1) \quad \textit{Set each factor} = 0.$$

$0 = x - 5$	$0 = x - 1$
$5 = x$ or	$1 = x$ or
$x = 5$	$x = 1$

The x intercepts are $(5, 0)$ and $(1, 0)$.

Math Note

If the equation cannot be solved by factoring, use the quadratic formula. If $b^2 - 4ac$ is negative, the parabola has no x intercepts since the square root of a negative number is not a real number.

Step 4 We already have at least one point on each side of the vertex, so this should be enough to draw the graph.

Step 5 Plot all of the points we found: the vertex $(3, -4)$, and intercepts $(0, 5)$, $(5, 0)$, and $(1, 0)$. Then connect them with a smooth curve, making sure to change direction at the vertex.

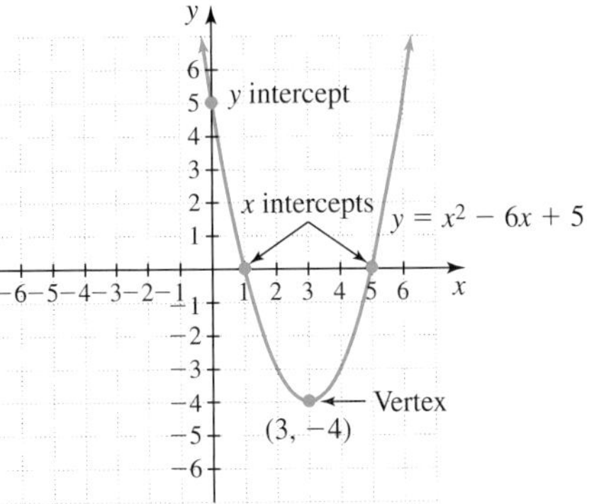

▼ Try This One 2

2. Graph quadratic functions.

Graph $f(x) = x^2 - 3x - 10$.

Applications of Quadratic Functions

The next two examples illustrate some of the ways that quadratic functions can be used to solve real-world problems.

EXAMPLE 3 Modeling Demographic Data with a Quadratic Function

The percentage of the population in the United States that was foreign-born can be modeled by the function $P(x) = 0.0033x^2 - 0.420x + 19.5$, where x is the number of years after 1900.

(a) When did the percentage reach its low point?
(b) What was the percentage in 2008?
(c) If the model accurately predicts future trends, what will the percentage be in 2020?

SOLUTION

(a) The graph of this function is a parabola that opens upward ($a = 0.0033$ is positive), so the lowest point occurs at the vertex.

$$x = \frac{-b}{2a} = \frac{-(-0.420)}{2(0.0033)} = \frac{0.420}{0.0066} \approx 64$$

The vertex occurs at about $x = 64$, or 64 years after 1900. The low point for the percentage was in 1964.

(b) The year 2008 is 108 years after 1900, and corresponds to $x = 108$.

$$P(108) = 0.0033(108)^2 - 0.420(108) + 19.5 \approx 12.6$$

In 2008, about 12.6% of the population was foreign-born.

(c) The year 2020 corresponds to $x = 120$.

$$P(120) = 0.0033(120)^2 - 0.420(120) + 19.5 \approx 16.6$$

The model predicts that 16.6% of the population will be foreign-born in 2020.

▼ Try This One 3

The percentage of cars sold in the United States that are imported can be modeled by the formula $P(x) = 0.159x^2 - 2.62x + 29.0$, where x is years after 1988. Find the year when the percentage was lowest, and use the model to estimate the percentage in 2008.

In some situations, a quadratic function can be written based on a description.

EXAMPLE 4 Fencing a Yard Efficiently

A family buys a new puppy, and plans to fence in a rectangular portion of their backyard for an exercise area. They buy 60 feet of fencing, and plan to put the fence against the house so that fencing is needed on only three sides. Find a quadratic function that describes the area enclosed, and use it to find the dimensions that will enclose the largest area.

SOLUTION

In this situation, a diagram will be very helpful. If we let x = the length of the sides touching the house, then the side parallel to the house will have length $60 - 2x$ (60 feet total minus two sides of length x).

House

x x

$60 - 2x$

The area of a rectangle is length times width, so in this case, we get

$$A(x) = x(60 - 2x)$$

To find a, b, and c, we should multiply out the parentheses:

$$A(x) = 60x - 2x^2 \text{ or } A(x) = -2x^2 + 60x \quad a = -2,\ b = 60,\ c = 0.$$

The maximum area will occur at the vertex of the parabola:

$$x = \frac{-b}{2a} = \frac{-60}{2(-2)} = \frac{-60}{-4} = 15$$

The sides touching the house should be 15 feet long. The side parallel to the house should be $60 - 2(15) = 30$ feet long.

▼ Try This One 4

 3. Apply quadratic functions to real-world problems.

Suppose the family in Example 4 decides to move the dog pen away from the house, so that fence is needed on all four sides. What dimensions will provide the largest area?

Exponential Functions

Some of the functions we've worked with so far have exponents, but in every case the exponent has been a number, while the base has been a variable. The next class of functions we will study reverses that—the variable appears in an exponent. These functions are very useful in modeling real-world situations like investments and loans, depreciation, growth and decay, and population changes.

An **exponential function** has the form $f(x) = a^x$, where a is *a* positive real number, but not 1.

Examples of exponential functions are

$$f(x) = 2^x \qquad f(x) = 10^x$$
$$f(x) = \left(\frac{1}{3}\right)^x \qquad f(x) = (0.5)^x$$

The graph of an exponential function has two forms.

1. When $a > 1$, the function increases as x increases, as in Figure 7-11a.
2. When $0 < a < 1$, the function decreases as x increases, as in Figure 7-11b.

For any acceptable value of a, the expression $a^0 = 1$, so every exponential function of the form $f(x) = a^x$ has y intercept (0, 1). Also, the graph approaches the x axis in one direction but never touches it. When this happens, we say that the x axis is a **horizontal asymptote** of the graph.

Since all of the exponential functions have the same basic shape, we can draw the graph by plotting a handful of points and drawing a curve similar to the ones in Figure 7-11.

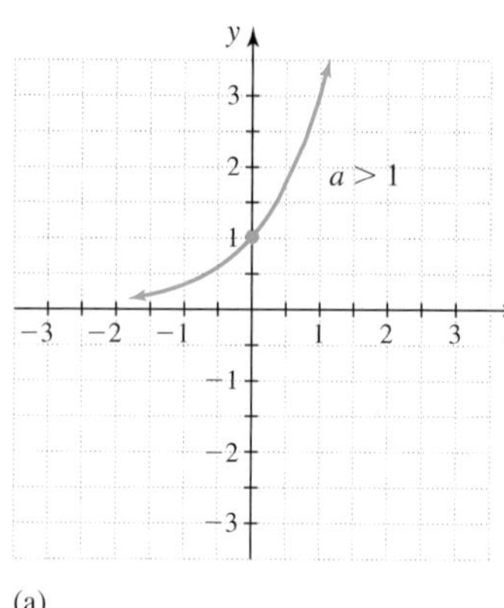

(a)

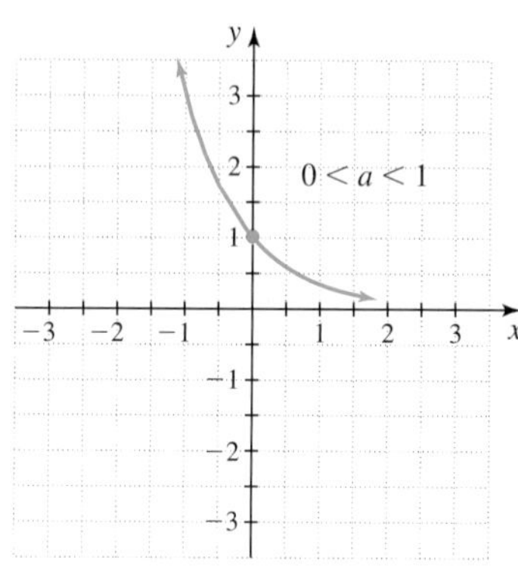

(b)

Figure 7-11

EXAMPLE 5 Graphing Exponential Functions

Draw the graph of each function.

(a) $f(x) = 2^x$ (b) $g(x) = 3^{x+1}$

SOLUTION

(a) Pick several numbers for x and find $f(x)$:

$$x = -2: f(-2) = 2^{-2} = \frac{1}{2^2} = \frac{1}{4}$$
$$x = -1: f(-1) = 2^{-1} = \frac{1}{2}$$
$$x = 0: f(0) = 2^0 = 1$$
$$x = 1: f(1) = 2^1 = 2$$
$$x = 2: f(2) = 2^2 = 4$$

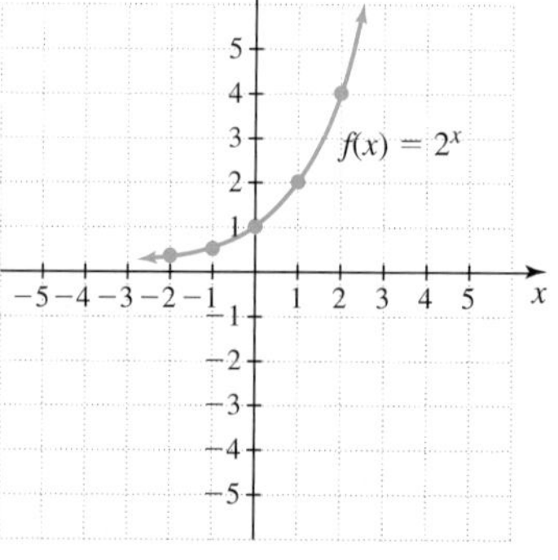

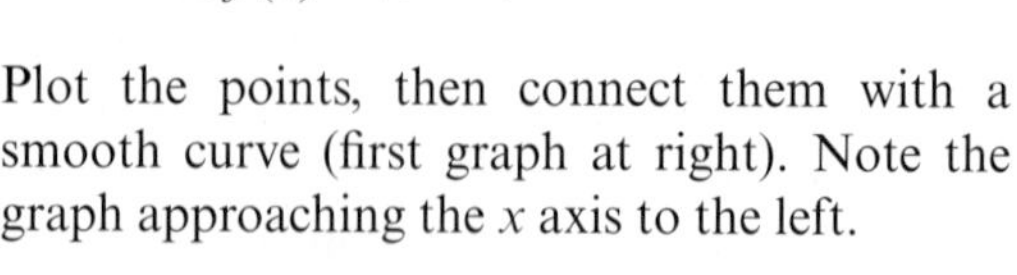
Plot the points, then connect them with a smooth curve (first graph at right). Note the graph approaching the x axis to the left.

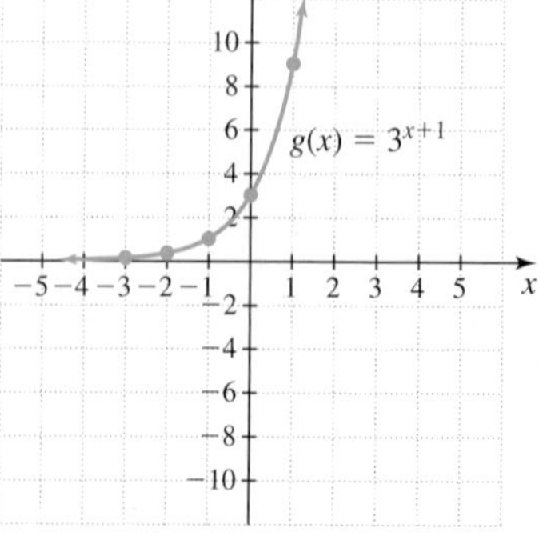

(b) Pick several numbers for x and find $g(x)$:

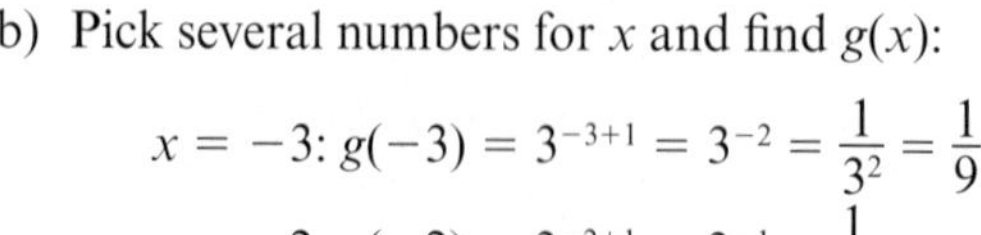

$$x = -3: g(-3) = 3^{-3+1} = 3^{-2} = \frac{1}{3^2} = \frac{1}{9}$$
$$x = -2: g(-2) = 3^{-2+1} = 3^{-1} = \frac{1}{3}$$
$$x = -1: g(-1) = 3^{-1+1} = 3^0 = 1$$
$$x = 0: g(0) = 3^{0+1} = 3^1 = 3$$
$$x = 1: g(1) = 3^{1+1} = 3^2 = 9$$

Plot the points, then connect them with a smooth curve (second graph above).

▼ Try This One 5

Draw the graph of $f(x) = 2^{x-2}$.

EXAMPLE 6 Graphing an Exponential Function

Draw the graph for $f(x) = \left(\frac{1}{2}\right)^x$.

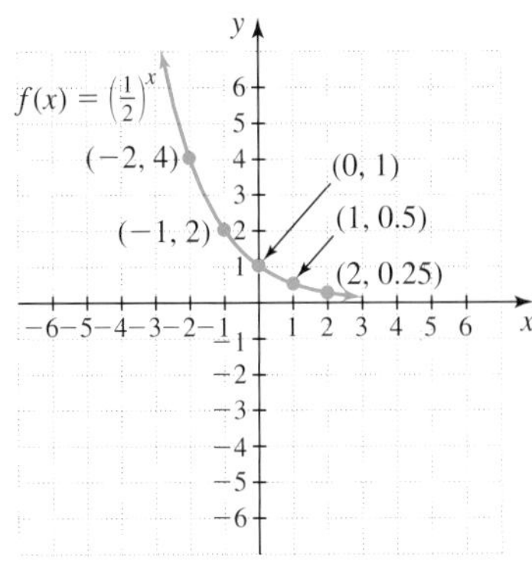

SOLUTION

Pick several numbers for x and find $f(x)$.

For $x = -2, f(-2) = \left(\frac{1}{2}\right)^{-2} = 4$.

For $x = -1, f(-1) = \left(\frac{1}{2}\right)^{-1} = 2$.

For $x = 0, f(0) = \left(\frac{1}{2}\right)^{0} = 1$.

For $x = 1, f(1) = \left(\frac{1}{2}\right)^{1} = 0.5$.

For $x = 2, f(2) = \left(\frac{1}{2}\right)^{2} = 0.25$.

Plot the points and draw the graph.

▼ Try This One 6

☑ 4. Graph exponential functions.

Draw the graph for the function $f(x) = \left(\frac{1}{3}\right)^x$.

We mentioned earlier that exponential functions are useful for modeling a variety of situations. The last three examples in the section illustrate some of these applications.

Sidelight WHERE DO MODELING FUNCTIONS COME FROM?

In some of the examples in this section, a function that models some real-world quantity was provided. This leads to an obvious question: where do these functions come from? How do you find a formula that models some actual data? This was a much harder question 20 or so years ago, before the rise of graphing calculators.

The process of finding a certain type of function to model a data set is known as regression. It originated with a method to find a straight line that best fits certain data, which was discovered in the early 1800s. The advent of computers changed everything, though, as other types of models can be found using techniques that would be at best extremely tedious by hand. Today's graphing calculators come preprogrammed with a variety of options that can be used to model a wide variety of data, not just data that happen to resemble a straight line when you plot the data points. The screen shot to the right, from a TI-84, shows some of the options available. Of particular interest to us are LinReg, which finds a linear function; QuadReg, which finds a quadratic function; and ExpReg, which finds an exponential function. Each of those choices was used to find at least one function mentioned in this section.

```
EDIT CALC TESTS
4:LinReg(ax+b)
5:QuadReg
6:CubicReg
7:QuartReg
8:LinReg(a+bx)
9:LnReg
0↓ExpReg
```

There are a number of excellent resources available on the Internet to help you learn to use this extremely valuable feature of a graphing calculator.

EXAMPLE 7 Exponential Population Growth

Riverside County, California, is one of the five fastest-growing counties in the United States. One estimate uses the function $f(t) = A_0(1.04)^t$ to model the population, where A_0 is the population at time $t = 0$ and t is the time in years. The population in 2007 reached 2,000,000 for the first time. Use the model to predict the population in 2027.

SOLUTION

Let $A_0 = 2{,}000{,}000$ and $t = 20$ years (since 2027 is 20 years after our base year, 2007). Then

$$f(20) = 2{,}000{,}000(1.04)^{20} = 4{,}382{,}246$$

The population is predicted to be 4,382,246 in 2027.

▼ Try This One 7

Another estimate was released in early 2009 with a less optimistic projection for growth due to economic factors. This predicts a model of $f(x) = A_0(1.02)^t$, where A_0 is the population at time $t = 0$, and t is time in years. The county had grown to 2,163,000 by 2009. What does this model predict the population will be in 2027?

EXAMPLE 8 Computing the Value of an Investment

It is often said that people would save more if they understood how compound interest worked. Use the formula in Example 8 to solve the following: At age 25, you invest your savings of \$10,000 at 8% interest compounded quarterly. How much do you have when you are 65?

Interest on a savings account can be compounded annually, semiannually, quarterly, or daily. The formula for compound interest is given by

$$A = P\left(1 + \frac{r}{n}\right)^{nt}$$

where A is the amount of money, which includes the principal plus the earned interest, P is the principal (amount initially invested), r is the yearly interest rate in decimal form, n is the number of times the interest is compounded a year, and t is the time in years that the principal has been invested. If \$10,000 is invested at 8% per year compounded quarterly, find the value of the investment (amount) after 40 years.

SOLUTION

$P = \$10{,}000$
$r = 8\% = 0.08$
$t = 40$ years
$n = 4$, since the interest is compounded quarterly, or four times a year.

$$\begin{aligned} A &= P\left(1 + \frac{r}{n}\right)^{nt} \\ &= \$10{,}000\left(1 + \frac{0.08}{4}\right)^{4\cdot 40} \\ &= \$237{,}699.07 \end{aligned}$$

At the end of 40 years, the \$10,000 will grow to \$237,699.07!

▼ Try This One 8

Find the value of a \$5,000 investment compounded semiannually (i.e., twice a year) for 2 years at 6% interest.

EXAMPLE 9 Carbon Dating

Carbon-14 is a radioactive isotope found in all living things. It begins to decay when an organism dies, and scientists can use the proportion remaining to estimate the age of objects derived from living matter, like bones, wooden tools, or textiles. The function $f(x) = A_0 2^{-0.0175x}$ describes the amount of carbon-14 in a sample, where A_0 is the original amount and x is the number of centuries (100 years) since the carbon-14 began decaying. An archaeologist claims that a recently unearthed wooden bowl is 4,000 years old. What percentage of the original carbon-14 must remain for him to make that claim?

SOLUTION

We need to evaluate the given function for $x = 40$, since 4,000 years is 40 centuries.

$$f(x) = A_0 2^{-0.0175(40)} \approx A_0(0.616)$$

The amount is 0.616 times what it was when it started decaying, so there is about 61.6% of the carbon-14 left.

▼ Try This One 9

☑ 5. Apply exponential functions to real-world problems.

If another sample from the dig site has 72% of the carbon-14 remaining, can it be over 3,000 years old? Why or why not?

Answers to Try This One

1 (a)

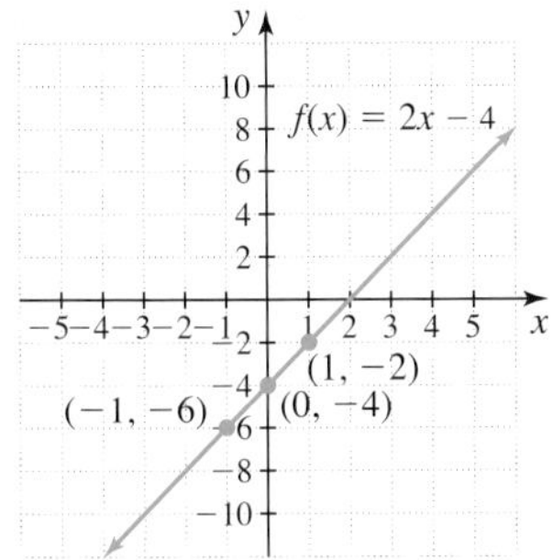

(b)

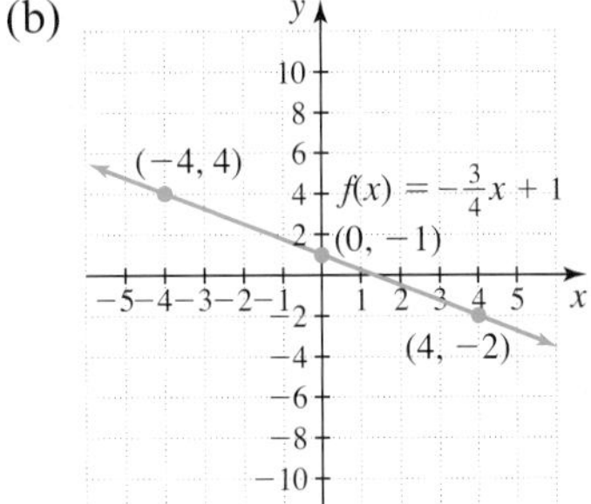

2

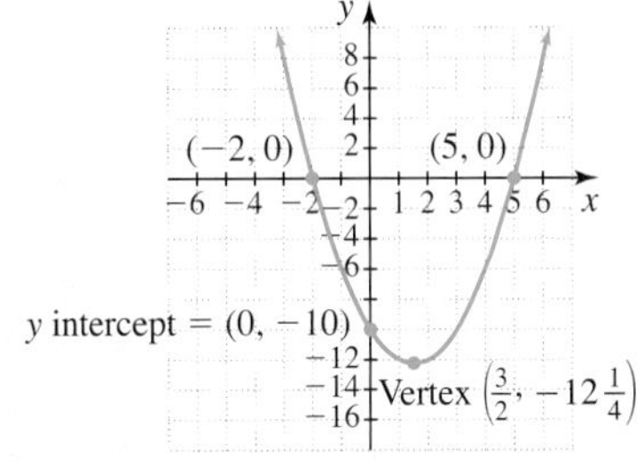

3 1996; 40.2% in 2008

4 15 feet by 15 feet

5

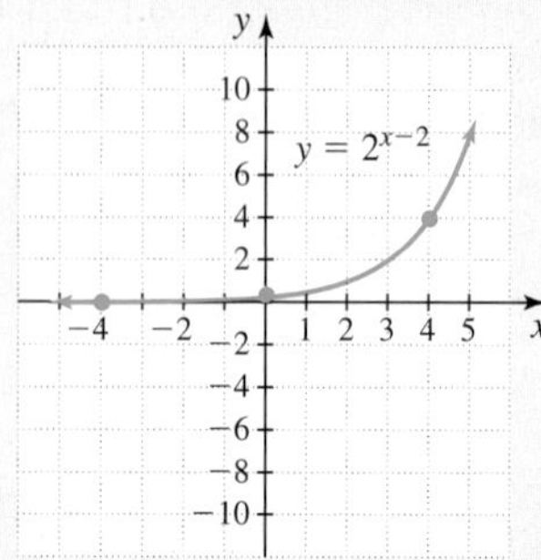

6

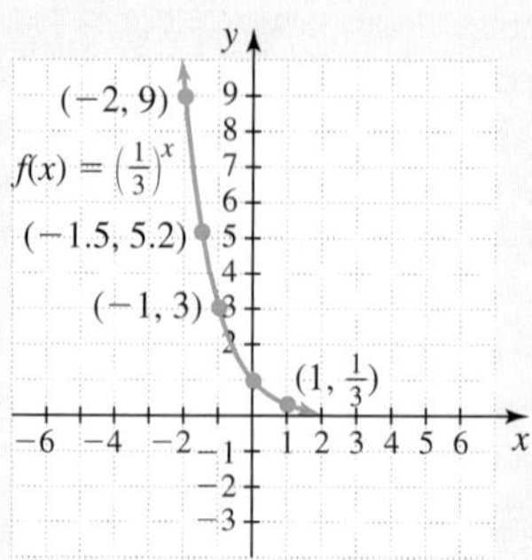

7 3,089,297

8 $5,627.54

9 No. At 3,000 years, less than 70% would remain.

EXERCISE SET 7-7

Writing Exercises

1. In the standard linear function $f(x) = ax + b$, what is the significance of the constants a and b?
2. Describe in your own words what a parabola is.
3. How can you tell whether a parabola opens up or down?
4. Explain how to find the maximum or minimum of a quadratic function.
5. What is the major difference between exponential functions and quadratic functions?
6. What does it mean to say that a function models some quantity?
7. What type of quantity is a quadratic function likely to be a reasonable model for?
8. Name some quantities that exponential functions are used to model.

Computational Exercises

For Exercises 9–16, graph each linear function.

9. $f(x) = 7x - 8$
10. $f(x) = 6x$
11. $f(x) = -3x + 2$
12. $f(x) = -5x + 1$
13. $f(x) = \frac{2}{3}x - 4$
14. $f(x) = -\frac{1}{4}x + 2$
15. $f(x) = -\frac{3}{5}x + 1$
16. $f(x) = \frac{3}{4}x + 2$

For Exercises 17–24, graph each parabola.

17. $f(x) = x^2$
18. $f(x) = x^2 - 6x$
19. $f(x) = x^2 + 6x + 9$
20. $f(x) = 4x^2 - 4x + 1$
21. $f(x) = -x^2 + 12x - 36$
22. $f(x) = -2x^2 + 3x + 4$
23. $f(x) = -10x^2 + 20x$
24. $f(x) = -3x^2 + 5x + 2$

For Exercises 25–30, graph each exponential function.

25. $f(x) = 5^x$
26. $f(x) = 3^{x+1}$
27. $f(x) = \left(\frac{1}{2}\right)^{x-2}$
28. $f(x) = \left(\frac{1}{4}\right)^{x+1}$
29. $f(x) = 2^{-x}$
30. $f(x) = 3^{-x}$

Real-World Applications

31. A tour boat operator found that when the price charged for a scenic boat tour was $25, five hundred customers were willing to pay that price for the tour. When the price was changed to $20, the number of customers willing to pay that price increased to 600.
   (a) Write the number of customers as a linear function of the price. Use independent variable p and dependent variable C.
   (b) Graph your linear function, labeling the axes with correct variables. Does it make sense to

extend the graph into the second and fourth quadrants? Why or why not?

(c) If the price was \$45, how many customers would pay for the tour?

(d) According to your model, if 300 customers paid for the tour, what price would they pay?

(e) What is the p-intercept? What does it mean?

(f) What is the C-intercept? What does it mean?

32. In 2000, a house was appraised at a value of \$256,000. In 2008, with the declining real estate market, the same house was valued at \$200,000. Let $t = 0$ correspond to the year 2000.

(a) Write the value of the house as a linear function of the year.

(b) Graph your function, labeling axes with correct variables. Start your graph with the year 2000.

(c) According to this trend, in what year would the house have a value of \$116,000?

(d) According to this trend, what will the house's value be in the year 2028? (Recall that $t = 0$ is 2000).

33. On a road trip, Shelly and her friends drove 140 miles in 2 hours. They drove at the same speed the entire way and did not take a break, so in 5 hours, they had gone 350 miles.

(a) Write the distance they traveled as a linear function of the time in hours.

(b) Graph your function, labeling the axes with correct variables. Start your graph at $t = 0$, which is when they started their trip.

(c) If they drove for 3 more hours at the same speed without stopping, how far would they have traveled?

(d) How long were they driving when they were 210 miles into their trip?

34. In 2005, sales for Stacy's Super Slingshots were \$32,000. In 2008, the novelty of the slingshots wore off, and sales had dwindled to \$20,000. Let $t = 0$ represent the year 2000.

(a) Write the sales (s) as a linear function of time (t) in years.

(b) Graph your function, labeling the axes with correct variables. Start your graph at $t = 0$.

(c) What is the s-intercept? What does this intercept mean?

(d) What is the t-intercept? What does this intercept mean?

(e) According to this trend, how much did Stacy's Super Slingshots have in sales in 1997?

(f) How much will sales be in 2011?

35. If a ball is thrown vertically upward from a height of 6 feet with an initial velocity of 60 feet per second, then the height of the ball t seconds after it is thrown is given by $f(t) = -16t^2 + 60t + 6$. Find the maximum height that the ball will attain and find the number of seconds that will elapse from the moment the ball is thrown to the moment it hits the ground.

36. A vendor sells boxes of computer paper. The amount of revenue made per week is given by the function $f(x) = 12x - 0.1x^2$, where x is the number of boxes sold. How many boxes should be sold if the vendor is to make the maximum possible profit?

37. A stone is dropped from a bridge, and 3 seconds later a splash is heard. How high is the bridge? Use $f(x) = 16x^2$, where $f(x)$ is the distance the stone falls in x seconds.

38. The profit for a company that produces discount sneakers can be modeled by the quadratic function $P(x) = 4x^2 - 32x + 210$, where x is the number of pairs of sneakers sold.

(a) For how many pairs of sneakers did the profit reach its lowest point?

(b) What was the lowest profit made by the company?

(c) What is the profit for selling 18 pairs of sneakers?

(d) How many pairs of sneakers must be sold to have a profit of \$162?

39. The cost (in hundreds of dollars) for Jake's Snowplowing business can be modeled by the quadratic function $C(t) = -t^2 + 12t$, where t is the time in months. Let $t = 0$ correspond to the start of a new year.

(a) In what month does the snowplowing business have the most cost? Why do you think that is?

(b) What is the most cost the company will have to spend to stay in business?

(c) In March, what will the costs be to run the business?

(d) In which month(s) will the cost be \$3,200? (*Hint:* Let $C = 32$.)

40. The revenue for jet ski rentals can be modeled by $R(x) = -x^2 - 6x + 432$, where x represents the number of \$0.50 price hikes.

(a) What is the most revenue and how many price hikes should occur to obtain the most revenue? What is the meaning of your answer?

(b) If you charge \$2.00 more, how much revenue will you make?

(c) If you charge \$5.00 less, how much revenue will you make?

41. The length of a picture is $2\frac{1}{2}$ times its width.

(a) Find an equation for the area as a function of the width.

(b) If the width is 10 inches, what is the length and what is the area?

(c) If the area is 40 square inches, what is the length and width of the picture?

(d) If the width was doubled, what would be the new area function?

42. Find the length of the sides of a gutter consisting of three sides that can be made from a piece of aluminum that is 16 inches wide in order for it to carry the maximum capacity of water. (The shape of the gutter is ⊔.)

43. The population growth of a certain geographic region is defined by the function $f(t) = A_0(1.4)^t$, where A_0 is the present population and t is the time in decades. If the present population is 4,000,000, find the population in 5 years.

44. In a biology experiment, there were 300,000 cells present initially and then the number started decreasing every second. The number of cells present can be modeled by the function $f(t) = 300{,}000(5)^{-0.4307t}$, where t is the time in seconds. Find how many cells remain after 10 seconds.

45. The monthly mortgage payment on a house that was financed for \$100,000 at a 7.5% interest rate can be calculated using the function

$$P(t) = \frac{(0.075)(100{,}000)}{1 - \left(1 + \frac{0.075}{12}\right)^{-12t}} \div 12$$

where t is the time the mortgage was financed in years. What is the monthly payment if the house was financed for 30 years?

46. You invest \$1 at 100% interest ($r = 1.00$) into an account, and the interest is compounded n times per year for one year. Use the formula

$$A = P\left(1 + \frac{r}{n}\right)^{nt}$$

to make a chart of the amount you would make in one year for compounding the interest once a year, twice a year, four times a year, every month, and every day.

47. A baby that weighs 7 lb at birth may increase his weight by 11% per month. Use the function $f(t) = K(1 + r)^t$, where K is the initial value, r is the decimal value of the percentage of increase ($r = 0.11$) and t is the time in months. How much would the baby weigh 6 months from birth?

Critical Thinking

48. What is the difference between the function $f(x) = 2^{-x}$ and the function $f(x) = \left(\frac{1}{2}\right)^x$?

49. Look at the linear model for the unemployment rate at the very beginning of this section. Explain the significance of the constants 0.217 and 4.40, not in mathematical terms, but in terms of unemployment.

50. Suppose that you know the following about the graph of some quadratic function $f(x)$: the vertex is $(4, -10)$ and one of the x intercepts is $(-1, 0)$. What is the other one?

For Exercises 51 and 52, the equation represents a relation that is not a function. Graph each relation by plotting points and describe the nature of the graph.

51. $x = y^2 - 6y - 7$

52. $x = -y^2 + 2y + 4$

CHAPTER 7 Summary

Section	Important Terms	Important Ideas
7-1	Rectangular coordinate system Cartesian plane x axis y axis Origin Quadrants Coordinates x intercept y intercept Slope Slope-intercept form	**The rectangular** coordinate system (also called the Cartesian plane) consists of two number lines: one vertical axis called the y axis and one horizontal axis called the x axis. The axes divide the plane into four quadrants. A point is located on the plane by its coordinates, which consist of an ordered pair of numbers (x, y). The graph of an equation is the set of all points in the plane whose coordinates make the equation true when substituted in for the variable. A line on the plane can be represented by an equation of the form $ax + by = c$. The slope of a line is defined as the rise divided by the run. The slope describes the rate at which the y coordinate is changing. The slope of a horizontal line is zero. The slope of a vertical line is undefined. The point where a line crosses the x axis is called the x intercept. The point where a line crosses the y axis is called the y intercept.
7-2	System of linear equations Consistent system Independent system Inconsistent system Dependent system Substitution method Addition/subtraction method	**A system** of linear equations consists of two or more linear equations in two variables. The solution to a system of linear equations is a value of x and a value of y that, when substituted into both equations, makes them true equations. Systems can be solved graphically by finding the point of intersection of the lines, or algebraically using the substitution method or the addition/subtraction method. If a system has only one solution, it is said to be consistent, and the solution is the coordinates of the point of intersection of the lines. If the lines are parallel, the system is said to be inconsistent, and the solution is the empty set. If the lines coincide, the system is said to be dependent and every point on the line is a solution.
7-3	Matrix Augmented matrix Row operations Row echelon form Gaussian elimination	**Matrices** provide an alternative method for solving systems of linear equations. An augmented matrix is written, which represents the coefficients of the system. Then row operations are performed until a matrix is obtained that represents an equivalent system of equations, but one that is much easier to solve.
7-4	Linear inequality Half plane Test point System of linear inequalities	**A system** of linear inequalities consists of two or more linear inequalities in two variables. The solution to a system of linear inequalities is found by graphing the linear inequalities and finding the intersection of the half planes.
7-5	Linear programming Constraint Polygonal region Objective function Maximize Minimize Vertex	**Linear** programming is one real application that uses systems of linear inequalities. Solving a problem using linear programming consists of representing the constraints as linear inequalities and the relationship of the variables using an objective function. When the inequalities are graphed, a polygonal region is formed and the coordinates of the vertices are substituted into the objective function. A decision is then made on the maximum or minimum values of the objective function.

7-6	Relation Function Function notation Evaluate Domain Range Vertical line test	**A relation** is a rule matching up two sets of objects. Relations are often represented by sets of ordered pairs. A relation is called a function if every element in the domain (first coordinates) corresponds to exactly one element in the range (second coordinates). Many functions are represented by equations relating two variables. The vertical line test can be used to determine if a graph is the graph of a function.
7-7	Linear function Quadratic function Parabola Vertex Axis of symmetry Exponential function Asymptote	**A linear** function has the form $f(x) = ax + b$. The graph is a straight line with slope a and y intercept $(0, b)$. A quadratic function has the form $f(x) = ax^2 + bx + c$. The graph is a parabola that opens up if $a > 0$ and down if $a < 0$. Exponential functions have the variable in an exponent. Exponential functions have many applications to finance, population studies, archaeology, and other areas.

MATH IN The Stock Market REVISITED

1. Evaluating for $x = 0$, we get 12,575 as the Dow Jones Industrial Average for March 2007.
2. The vertex occurs very close to $x = 6$. This represents 6 months after March 2007, which is September 2007. The corresponding value from the equation for $x = 6$ is $y = 13{,}401$. Notice that this is lower than shown on the graph, which peaks above 14,000. This happens often in approximating real-world data with a function. Those spikes above 14,000 were aberrational and don't fit the general pattern of the data.
3. For $x = 23$, which is February 2009, the model predicts an average of 6,714. The actual average for February was around 7,100, so as bad as things were, the function thought it would be even worse!

Review Exercises

Section 7-1

For Exercises 1–6, draw the graph for each line.

1. $4x - y = 8$
2. $y = 2x + 6$
3. $x = -5$
4. $y = 8$
5. $y = -3x + 12$
6. $x - 6y = 11$

For Exercises 7–12, find the slope of the line containing the two given points.

7. $(-8, 6), (4, 3)$
8. $(3, 8), (-2, 6)$
9. $(-5, -6), (-3, 8)$
10. $(2, 6), (2, 10)$
11. $(5, 9), (-3, 9)$
12. $(-3, -8), (6, -2)$

For Exercises 13–16, write the equation in slope-intercept form and find the slope, x intercept, and y intercept, and graph the line.

13. $3x + y = 12$
14. $-2x + 8y = 15$
15. $4x - 7y = 28$
16. $x - 3y = 9$

17. After collecting money for a local charity from their respective classes, Mary had \$16.72 more than Betty. Together, they collected \$32.00. How much did each collect?
18. Marcus invested a sum of money at 5% annual interest and twice as much at 8%. At the end of the year, his total interest earned was \$89. How much did he have in each investment?

Section 7-2

For Exercises 19–22, solve each system graphically.

19. $x - 2y = 6$
$2x - y = 18$
20. $x + y = 9$
$x - y = 1$
21. $2x + y = 12$
$5x - 2y = 21$
22. $4x - y = 10$
$8x - 2y = 20$

For Exercises 23–26, solve each system by substitution.

23. $x + 6y = 4$
$x - y = 11$
24. $3x - y = 10$
$x - 2y = 5$
25. $x - y = 3$
$x - 3y = -3$
26. $3x + 2y = 10$
$2x - 3y = 11$

For Exercises 27–30, solve each system by addition/subtraction.

27. $3x - y = 12$
$x + 2y = 4$

28. $5x - 6y = 15$
$10x - 12y = 21$

29. $4x - 2y = 11$
$5x - y = 10$

30. $6x - y = 15$
$2x + 5y = 21$

31. Three pounds of coffee and 4 pounds of tea cost \$14.50. Five pounds of coffee and 2 pounds of tea cost \$16.00. What is the cost per pound of the coffee and the tea?

32. A merchant has invested money in the partial ownership of two stores. Last year, store 1 earned 10% of the money invested in it while store 2 lost 8% of the amount invested in it. The net total gain was \$1,300.00. This year, the respective quantities were +6%, +5%, and \$2,250.00. Find the merchant's investment for each store.

Section 7-3

For Exercises 33–34, write the augmented matrix corresponding to the system of equations. Do not solve the system.

33. $2x - y + 4z = 10$
$-3x - 4y - 5z = 8$
$x + y + z = -8$

34. $x - y + z = -4$
$2y + z = 7$
$3x + y = 11$

For Exercises 35–38, solve the system using Gaussian elimination.

35. $x + 5y = 7$
$-4x - 2y = 26$

36. $3x - 4y = -16$
$2x + 6y = 11$

37. $x - 2y - z = 6$
$2x - 3y + 2z = 33$
$-2x + y - 8z = -65$

38. $-5x + y + z = 0$
$2x + 2y + 2z = -12$
$x - 2y + 3z = -6$

Section 7-4

For Exercises 39–42, graph each inequality.

39. $x - 19y \geq 5$
40. $2x + 7y < 14$
41. $-3x + 10y \leq -15$
42. $x > -5$

For Exercises 43–46, find the solution set to each system by graphing.

43. $5x - y > 10$
$2x + 3y \leq 12$

44. $6x + 2y \leq 15$
$3x - y > 6$

45. $12x + 3y \geq 24$
$-x + y \leq 5$

46. $3x + y \geq -4$
$x + y < 0$

Section 7-5

47. A manufacturer produces children's scooters in two models. Model A takes 6 hours to manufacture and 2 hours to paint. Model B takes 5 hours to manufacture and 1 hour to paint. The assembly plant works 160 hours and the painting department works 100 hours. If the profit for model A is \$20.00 and for model B is \$10.00, how many models of each should be produced to maximize the profit?

48. An appliance store owner stocks two types of microwaves and has warehouse space for at most 36 units. The owner keeps at least twice as many of model I as model II. If the profit on model I is \$24.00 and the profit on model II is \$36.00, how many of each model should the owner carry to maximize profit?

Section 7-6

For Exercises 49–50, state whether or not the relation is a function.

49. $\{(2, 5), (5, -7), (6, -10)\}$
50. $\{(-1, 5), (2, 6), (-1, 3)\}$

For Exercises 51–54, find the domain and range for each relation and state whether or not the relation is a function.

51. $y = \sqrt{3 - x}$
52. $3x + 2y = 6$
53. $y = \dfrac{2}{x - 3}$
54. $y^2 = x$

For Exercises 55–58, evaluate each function for the specific value.

55. $f(x) = 5x - 12 \quad f(5)$
56. $f(x) = -2x + 10 \quad f(-3)$
57. $f(x) = x^2 + 7x + 10 \quad f(-10)$
58. $f(x) = 2x^2 - 3x \quad f(8)$

Section 7-7

For Exercises 59–62, graph each linear function.

59. $f(x) = -2x + 5$
60. $f(x) = 3x - 3$
61. $f(x) = \frac{1}{3}x - 2$
62. $f(x) = -\frac{3}{4}x + 4$

For Exercises 63–66, graph each parabola.

63. $f(x) = x^2 + 10x + 25$
64. $f(x) = 3x^2 - 4x - 4$
65. $f(x) = -x^2 + 25$
66. $f(x) = -6x^2 + 12x$

For Exercises 67–70, graph each exponential function.

67. $f(x) = -3^x$
68. $f(x) = 6^x$
69. $f(x) = \left(\frac{1}{3}\right)^x$

70. $f(x) = -\left(\frac{1}{2}\right)^x$
71. A ball is thrown upward from a height of 4 feet with a velocity of 80 feet per second. The function $f(x) = -16x^2 + v_0x + h_0$ describes the height after x seconds, where v_0 is the initial velocity and h_0 is the initial height. Find the maximum height reached by the ball.
72. A radioactive isotope decays according to the function $F(x) = A_0 2^{-0.5x}$ where A_0 is the initial amount and x is the number of days since it started decaying. If there were 200 pounds initially, how much remains after 10 days?

Chapter Test

For Exercises 1–6, draw the graph for each.

1. $3x - 5y = -15$
2. $y = 8x - 24$
3. $x = 0$
4. $y = 0$
5. $2x + 3y = -8$
6. $-4x - 5y = 20$

For Exercises 7 and 8, find the slope of the line containing the two points.

7. $(-5, -10), (2, 7)$
8. $(3, -6), (4, 1)$

For Exercises 9 and 10, write each equation in slope-intercept form and find the slope, x intercept, and y intercept and graph the line.

9. $x + 5y = 20$
10. $2x - 11y = 22$

For Exercises 11–16, solve by any method you choose.

11. $x - 7y = 10$
 $2x + y = 5$
12. $4x + y = 15$
 $x - y = 8$
13. $3x + 4y = 19$
 $9x + 12y = 57$
14. $5x - y = 18$
 $2y = 10x + 36$
15. $x + 6y \geq 10$
 $6x - y \leq 23$
16. $x + y > 10$
 $x - y < 0$

For Exercises 17 and 18, solve the system using Gaussian elimination.

17. $x + y + z = 4$
 $3x - y + 2z = 19$
 $-x + y - z = -10$
18. $3x - y - z = 9$
 $x + 4y + z = 13$
 $5x - 2y - 2z = 14$

For Exercises 19 and 20, state whether or not each relation is a function.

19. $\{(-4, 6), (-10, 18), (12, 5)\}$
20. $\{(3, 7), (4, 8), (5, 9), (3, 6)\}$

For Exercises 21 and 22, evaluate each function for the specific value.

21. $f(x) = -3x + 10$; find $f(-15)$
22. $f(x) = 2x^2 + 6x - 5$; find $f(3)$

For Exercises 23–26, graph each.

23. $f(x) = x^2 - 9x + 14$
24. $f(x) = 2x^2 + x - 3$
25. $f(x) = 2^{1.5x}$
26. $f(x) = -3^{-0.8x}$
27. A college algebra class of 57 students was divided into two sections. There were 5 more students in one section than in the other section. How many students were in each section?
28. A newsstand sells a certain number of *Daily News* newspapers for $0.35 and *Tribune* newspapers for $0.50. The gross amount made on both papers was $38.70. If the stand would sell twice as many *Tribunes* and half as many *Daily News* papers, the gross amount made would be $44.85. Find the number of each paper sold.
29. The wages per hour for three cooks and eight food servers is $66.50. The cooks receive $2.00 an hour more than the food servers. How much does each earn per hour?
30. To make some extra money for the holidays, a college student makes Christmas wreaths and decorated snowmen to sell at a local craft shop. It takes her 30 minutes to assemble a wreath and 20 minutes to decorate it. It takes her 40 minutes to assemble a snowman and 20 minutes to decorate it. She makes a profit of $8 on a wreath and $10 on a snowman. If she can work at most 24 hours on these crafts, how many of each should she make to maximize her profit?
31. A contractor needs to make a rectangular drain consisting of three sides from a piece of aluminum that is 24 inches wide. Find the length of each side if it needs to carry the maximum capacity of water. The drain is ⊔ shaped.
32. Find the compound interest on a $4,000 investment held for 5 years, if the rate is 8.5% compounded semiannually. Use $A = P(1 + \frac{r}{n})^{nt}$.

Projects

1. Using the Internet, find population statistics for your state for the beginning of every decade from 1950 to 2000. Then either use the regression feature on your calculator or do an Internet search for a website that computes regression equations. Find a linear regression model for the population, a quadratic regression model, and an exponential regression model. Then use each of those models to calculate the current population of your state, and find the most recent estimate you can. Which of the models gave the best estimate? Based on the graphs we studied in this chapter, explain why the best model did so based on what the graph of that type of function tends to look like.
2. A function is linear when it increases or decreases at a steady rate. Make a list of at least five real-world quantities that you suspect might be modeled well by a linear function. Then do some research, finding values for the quantities on your list and plotting them on a graph. Rank your list from best to worst in terms of how linear the graphs appear to be.

CHAPTER 8

Consumer Mathematics

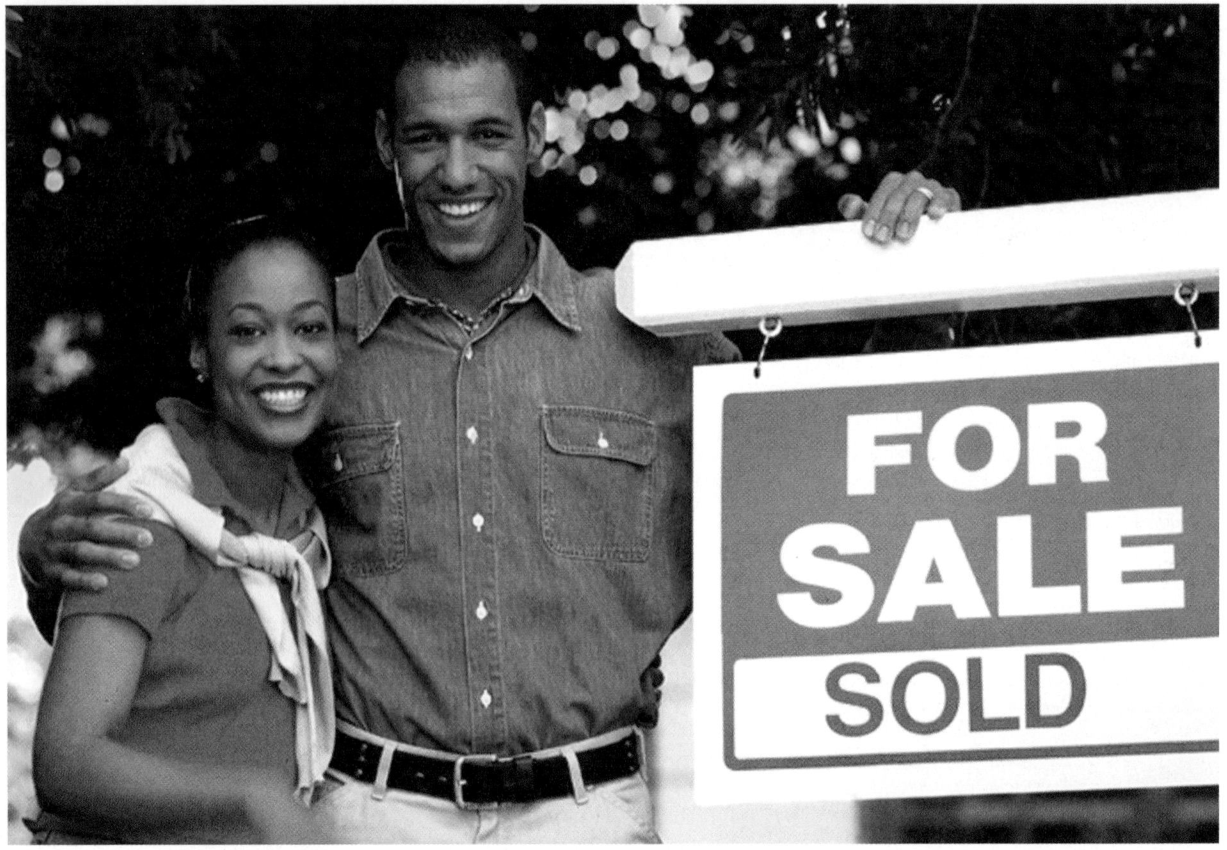

Outline

8-1 Percents
8-2 Simple Interest
8-3 Compound Interest
8-4 Installment Buying
8-5 Home Ownership
8-6 Stocks and Bonds
Summary

MATH IN Home Buying

Home buying shows have become very popular in the last couple of years. It's interesting to watch people looking for just the right home, and maybe more interesting to find out what homes are selling for in other parts of the country. The downside as a viewer is that it can be pretty intimidating to find out just how much houses cost. Almost everyone wonders at some point "Will I ever be able to afford a home of my own?"

This might lead you to come to a larger realization: long ago, survival depended on overcoming physical challenges that modern humans can only imagine, but today, survival is more about navigating the waters of the modern financial system. There are over 300 million people in the United States, and at any given time, probably half of them are trying to figure out a way to separate you from your hard-earned money. That's why this chapter is about some aspects of our financial system that are particularly important to the average adult. Success in this chapter will help you to be a well-informed consumer, which makes it less likely that those 150 million people will succeed in getting your cash.

We begin with a thorough look at percents, which play a big role in almost all areas of consumer mathematics. Then we'll study loans and investments, two topics of particular interest in the current financial climate. The goal is to help you take the guesswork out of financial planning, so that rather than hoping you're doing the right things for your future, you can be sure that you are.

Now to return to the original question: how much can you afford to pay for a home? Many financial experts suggest that typical home buyers can afford to pay 28% of their gross monthly income on housing. Obviously, this determines how much house you can afford. Below is a description of a hypothetical young couple. Using the math you learn in this chapter, you should be able to help them determine the price range they can afford.

Troy and Lisa have a combined annual income of $72,000, and while renting for the first 4 years of their marriage, have managed to save $15,000 for a down payment. How much can they afford to pay for a house if they qualify for a 30-year mortgage with an interest rate of 6.0%? Assume that 5% of income will go to property tax and insurance on the home.

For answers, see Math in Home Buying Revisited on page 475

Section 8-1 Percents

LEARNING OBJECTIVES

- ☐ 1. Convert between percent, decimal, and fraction form.
- ☐ 2. Perform calculations involving percents.
- ☐ 3. Solve real-world problems involving percents.
- ☐ 4. Find percent increase or decrease.
- ☐ 5. Evaluate the validity of claims based on percents.

Have you ever been shopping, come across a clearance rack that said something like "40% off lowest ticketed price," and had to ask someone what the discounted price of a certain item would be? If so, you're certainly not alone. Any math teacher will tell you that people ask them questions like that all of the time.

Percents are a math topic with a huge number of applications to everyday life, so they should be high on our priority list if we're trying to learn about math in our world. In this section, we'll learn about percents: what they really mean, how to use them in calculations, and even how they can be deceptively misused.

The word "percent" can be translated literally as "per hundred."

Percent means hundredths, or per hundred. That is, $1\% = \frac{1}{100}$.

For example, the International Mass Retail Association reported that 20% of adults plan to buy Valentine's Day cards next year. Of those who plan to buy them, 36% of women and 26% of men plan to buy romantic cards. This tells us that 20 out of every 100 adults plan to buy a card. Also, 26 out of 100 men who buy a card will choose a romantic one, and 36 out of 100 women will do so.

Percent Conversions

To work with percents in calculations, we will need to convert them to either decimal or fractional form. But to interpret the answers to calculations, we will want to convert them back into percents.

Converting Percents to Decimals

In order to change a percent to a decimal, drop the % sign and move the decimal point two places to the left.

EXAMPLE 1 Changing Percents to Decimals

Change each percent to a decimal.

(a) 84% (c) 37.5%
(b) 5% (d) 172%

SOLUTION

(Drop the % sign and move the decimal point two places to the left.)

(a) 84% = 84.% = 0.84% = 0.84 *Put decimal point in if necessary.*

(b) 5% = 5.% = 0.05% = 0.05

(c) 37.5% = 0.375% = 0.375

(d) 172% = 172.% = 1.72% = 1.72

Math Note

Moving the decimal point two places to the left is the same as dividing by 100.

▼ Try This One 1

Change each percent to a decimal.

(a) 62.5% (b) 3% (c) 250%

Converting Percents to Fractions

A percent can be converted to a fraction by dropping the percent sign and using the percent number as the numerator of a fraction whose denominator is 100.

Be sure to reduce fractions to lowest terms when possible.

EXAMPLE 2 Changing Percents to Fractions

Change each percent to a fraction.

(a) 42% (c) $37\frac{1}{2}\%$
(b) 6% (d) 15.8%

SOLUTION

(a) $42\% = \frac{42}{100} = \frac{21 \cdot \cancel{2}}{50 \cdot \cancel{2}} = \frac{21}{50}$

(b) $6\% = \frac{6}{100} = \frac{3 \cdot \cancel{2}}{50 \cdot \cancel{2}} = \frac{3}{50}$

(c) When converting fractional or decimal percents, it might be helpful to multiply by some number over itself to clear fractions or decimals. In this case, multiplying by $\frac{2}{2}$ is helpful:

$$\frac{37\frac{1}{2}}{100} \times \frac{2}{2} = \frac{75}{200} = \frac{\cancel{25} \cdot 3}{\cancel{25} \cdot 8} = \frac{3}{8}$$

(d) $15.8\% = \frac{15.8}{100} = \frac{15.8}{100} \cdot \frac{10}{10} = \frac{158}{1{,}000} = \frac{79 \cdot \cancel{2}}{500 \cdot \cancel{2}} = \frac{79}{500}$

▼ Try This One 2

Change each percent to a fraction.

(a) 90% (b) 16.5% (c) 130%

Converting a Decimal to a Percent

To change a decimal to a percent, move the decimal point two places to the right and add a percent sign.

EXAMPLE 3 Changing Decimals to Percents

Change each decimal to a percent.

(a) 0.74 (b) 0.05 (c) 0.327 (d) 5.463

Math Note

Moving the decimal point two places to the right is the same as multiplying by 100.

SOLUTION

(a) $0.74 = 074.\% = 74\%$ (c) $1.327 = 132.7\% = 132.7\%$

(b) $0.05 = 005.\% = 5\%$ (d) $5.463 = 546.3\% = 546.3\%$

▼ Try This One 3

Change each decimal to a percent.

(a) 0.974 (b) 0.04 (c) 3.75

Changing a Fraction to a Percent

To change a fraction to a percent, first change the fraction to a decimal, and then change the decimal to a percent.

EXAMPLE 4 Changing Fractions to Percents

Convert each fraction to a percent.

(a) $\frac{7}{8}$ (b) $\frac{3}{4}$ (c) $\frac{5}{6}$ (d) $1\frac{1}{2}$

Math Note

Recall that to change a fraction to a decimal, we divide the numerator by the denominator.

SOLUTION

(a) $\frac{7}{8} = 7 \div 8 = 0.875 = 87.5\%$

(b) $\frac{3}{4} = 3 \div 4 = 0.75 = 75\%$

(c) $\frac{5}{6} = 0.83\overline{3} = 83.\overline{3}\%$

(d) $1\frac{1}{2} = 1.5 = 150\%$

Calculator Guide

Calculators come in very handy when converting fractions to percents. For $\frac{5}{6}$:

Standard Scientific Calculator

5 [÷] 6 [=]

Standard Graphing Calculator

5 [÷] 6 [Enter]

In each case, the display will show something like 0.83333333333. This needs to be interpreted as the repeating decimal $0.8\overline{3}$.

▼ Try This One 4

Change each fraction to a percent.

(a) $\frac{5}{16}$ (b) $\frac{7}{9}$ (c) $1\frac{3}{4}$

Problems Involving Percents

The most common calculations involving percents involve finding a percentage of some quantity. To understand how to do so, consider the following example. You probably know that 50% of 10 is 5. Let's rewrite that statement, then turn it into a calculation:

50% of 10 is 5

$0.5 \times 10 = 5$

When writing a percentage statement in symbols, the word "of" becomes multiplication, and the word "is" becomes an equal sign. Also, we must change the percent into decimal or fractional form. The next three examples show how to use this procedure to set up calculations.

☑ 1. Convert between percent, decimal, and fraction form.

EXAMPLE 5 Finding a Certain Percentage of a Whole

In a class of 66 students, 32% got a B on the first exam. How many students got a B?

SOLUTION

First write 32% in decimal form, as 0.32. The question is "what is 32% of 66?" which we translate into symbols:

32% of 66 is _____

$0.32 \times 66 = 21.12$

We can't have 0.12 students, so we interpret the answer as 21 students got a B.

▼ Try This One 5

In another section of the same course, 19% of the 54 students got a B. How many students got a B?

CAUTION

In calculations with percents, it's very common for the result to contain digits after a decimal point, like 21.12 in Example 5. You should always think about what quantity your answer represents to decide if it's appropriate to round to the nearest whole number.

EXAMPLE 6 Finding a Percentage from a Portion

Of these 5 kittens, 2 are mostly gray. What percent are gray?

Of 60 runners who started a 5k race, 45 finished the race in under 40 minutes. What percent is that?

SOLUTION

Write the statement in the form we discussed above:

45 is what percent of 60?

$45 = \quad x \quad \times 60$ *x is the percent in decimal form.*

This is the equation $60x = 45$, which we solve for x.

$$60x = 45 \qquad \text{Divide both sides by 60.}$$

$$\frac{60x}{60} = \frac{45}{60}$$

$$x = \frac{45}{60} = 0.75$$

The decimal 0.75 corresponds to 75%, so 75% finished in under 40 minutes.

▼ Try This One 6

In 2008, 21 out of 50 states had a population of 5 million or higher. What percent is that?

Sidelight MONEY, BANKS, AND CREDIT CARDS

Long before money existed, people bartered for goods and services. For example, if you needed the roof of your hut fixed, you might pay the repair person two chickens. The first known coins were made over 2,500 years ago in Western Turkey. They consisted of a mixture of gold and silver and were stamped to guarantee uniformity. These coins were first accepted by the merchants of the area. Also around that time, coins were made in India and China.

Paper money was first made in China about 1,400 years ago; however, when Marco Polo brought the idea to Europe, it was rejected by the people. It wasn't until the 1600s that banks in Europe began to issue paper money to their depositors and borrowers.

In the United States, early settlers used tobacco, beaver skins, and foreign coins as currency. A popular coin was the Spanish dollar, called "pieces of eight." For purchases of less than one dollar, the coin was cut into eight pieces. Each piece was called a "bit" and was worth $12\frac{1}{2}$ cents. Hence 25 cents became known as "two bits," 50 cents as "four bits," etc.

During the Revolutionary War, the Continental Congress authorized the printing of paper money to pay war debts. The government printed more money than it could back up with gold and silver, and the dollar became virtually worthless. The phrase "not worth a Continental dollar" is still used today.

The U.S. Mint opened on April 2, 1792, in Washington, D.C., to mint coins. Gold was used for the $10.00, $5.00, and $2.50 coins. Silver was used for the $1.00, $0.50, $0.25, $0.10, and the $0.05 coins, and copper was used for the 1 cent and $\frac{1}{2}$ cent.

Most paper money was issued by state banks until 1863 when Congress established national banks to issue currency notes. In 1913, the Federal Reserve System was established to issue notes, which became our standard currency.

Banking began in ancient Babylon about 2000 BCE when people kept their money in temples. They thought that if the temples were robbed, the gods would punish the robbers. In medieval times, money was kept in vaults in castles and protected by the armies of the nobles. The first bank was established in 1148 in Genoa, Italy. It was called the Bank of San Giorgio. The first bank in the United States was established in 1781 in Philadelphia and was called the Bank of North America.

Credit cards were first issued by large hotels in the early 1900s. These cards were considered to be prestigious and were issued only to customers who spent a lot of money at the hotel. Department stores and gasoline companies began to issue credit cards around 1915. During World War II, the United States forbade the use of credit cards. Banks began to issue credit cards in the 1950s. Finally, in the late 1960s, banks agreed to sponsor credit cards such as Master Card, Visa, etc.

EXAMPLE 7 Finding a Whole Amount Based on a Percentage

A medium-sized company reported that it had to cut its work force back to 70% of what it was last year. If it has 63 workers now, how many did it have a year ago?

SOLUTION

Convert 70% to a decimal: 70% = 0.70. Now write as a question and translate to symbols:

$$\begin{array}{cccc} 70\% \text{ of} & \text{what number} & \text{is} & 63? \\ 0.70 \quad \times & x & = & 63 \end{array}$$

This gives us the equation $0.70x = 63$, which we solve for x.

$$0.70x = 63 \qquad \textit{Divide both sides by 0.70.}$$

$$\frac{0.70x}{0.70} = \frac{63}{0.70}$$

$$x = 90$$

The company had 90 workers a year ago.

▼ Try This One 7

2. Perform calculations involving percents.

After a really rotten year in 2008, a baseball team won 120% as many games in 2009, which was 84 games. How many games did they win in 2008?

Applications of Percents

Many aspects of consumer mathematics deal with finding parts of a whole. For example, you may want to leave a 15% tip, you may want to figure out a 33% markup, or you may want to calculate an 8% commission on sales. Some applications are shown in Examples 8 through 10.

EXAMPLE 8 Calculating Sales Tax

Math Note

In Example 8, you can find the total by multiplying the cost by 1.07 (i.e., 107%). 1.07($89.95) = $96.25 (rounded).

The sales tax in Allegheny County, Pennsylvania, is 7%. What is the tax on a calculator that costs $89.95? What is the total amount paid?

SOLUTION

Find 7% of $89.95: Write 7% as 0.07 in decimal form, then multiply.

$$0.07 \times \$89.95 = \$6.30 \text{ (rounded)}$$

The sales tax is $6.30. The total amount paid is

$$\$89.95 + \$6.30 = \$96.25$$

▼ Try This One 8

The sales tax in Atlanta, Georgia, is 5%. Find the amount of tax and the total cost of a portable DVD player on sale for $149.

EXAMPLE 9 Calculating Cost of Sale from Commission

A real estate agent receives a 7% commission on all home sales. How expensive was the home if she received a commission of $5,775.00?

SOLUTION

In this case, the problem can be written as $5,775.00 is 7% of what number?

$$\$5{,}775 \text{ is } 7\% \text{ of}$$

$$5{,}775 = 0.07 \times x \qquad \textit{Divide both sides by 0.07.}$$

$$\frac{5{,}775}{0.07} = \frac{0.07 \times x}{0.07}$$

$$82{,}500 = x$$

The home was purchased for $82,500.00.

▼ Try This One 9

3. Solve real-world problems involving percents.

A sales clerk receives a 9% commission on all sales. Find the total sales the clerk made if his commission was $486.00.

Sometimes it is useful to find the percent increase or the percent decrease in a specific situation. In this case, we can use the following method.

Procedure for Finding Percent Increase or Decrease

Step 1 Find the amount of the increase or the decrease.

Step 2 Make a fraction as shown:

$$\frac{\text{Amount of increase}}{\text{Original amount}} \quad \text{or} \quad \frac{\text{Amount of decrease}}{\text{Original amount}}$$

Step 3 Change the fraction to a percent.

EXAMPLE 10 Finding a Percent Change

A large latte at the Caffeine Connection sells at a regular price of \$3.50. Today it is on sale for \$3.00. Find the percent decrease in the price.

SOLUTION

The original price is \$3.50.

Step 1 Find the amount of decrease. \$3.50 − \$3.00 = \$0.50.

Step 2 Make a fraction as shown

$$\frac{\text{Amount of decrease}}{\text{Original price}} = \frac{\$0.50}{\$3.50}$$

Step 3 Change the fraction to a percent.

$$\frac{0.50}{3.50} \approx 0.1428 = 14.3\% \text{ (rounded)}$$

The decrease in price is 14.3%.

▼ Try This One 10

In 2008 the population of the town of Oak Creek was 23,258. In 2009, the population increased to 23,632. Find the percent increase in the population.

☑ 4. Find percent increase or decrease.

Percent increase or decrease is often misused, sometimes intentionally, sometimes not. In Example 11, we'll look at a common deceptive use of percents in advertising.

EXAMPLE 11 Recognizing Misuse of Percents in Advertising

A department store advertised that certain merchandise was reduced 25%. Also, an additional 10% discount card would be given to the first 200 people who entered the store on a specific day. The advertisement then stated that this amounted to a 35% reduction in the price of an item. Is the advertiser being honest?

SOLUTION

Let's say that an item was originally priced at \$50.00. (Note: any price can be used.) First find the discount amount.

$$\begin{aligned}\text{Discount} &= \text{rate} \times \text{selling price}\\ &= 25\% \times \$50.00\\ &= 0.25 \times \$50.00\\ &= \$12.50\end{aligned}$$

Then find the reduced price.

$$\begin{aligned}\text{Reduced price} &= \text{original price} - \text{discount}\\ &= \$50.00 - \$12.50\\ &= \$37.50\end{aligned}$$

Next find 10% of the reduced price.

$$\begin{aligned}\text{Discount} &= \text{rate} \times \text{reduced price}\\ &= 10\% \times \$37.50\\ &= \$3.75\end{aligned}$$

Find the second reduced price.

$$\begin{aligned}\text{Reduced price} &= \$37.50 - \$3.75\\ &= \$33.75\end{aligned}$$

Now find the percent of the total reduction.

$$\frac{\text{Reduction}}{\text{Original price}} = \frac{\$12.50 + \$3.75}{\$50.00} = \frac{16.25}{50.00} = 0.0325 = 32.5\%$$

The total percent of the reduction was 32.5%, and not 35% as advertised.

▼ Try This One 11

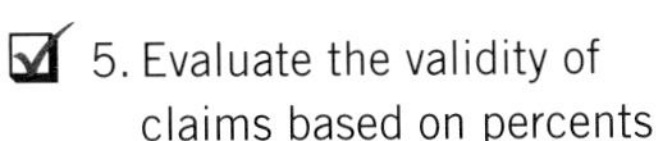

5. Evaluate the validity of claims based on percents.

A department store offered a 20% discount on all television sets. They also stated that the fist 50 customers would receive an additional 5% discount. Find the total percent discount. You can use any selling price for the televisions.

Answers to Try This One

1 (a) 0.625 (b) 0.03 (c) 2.50

2 (a) $\frac{9}{10}$ (b) $\frac{33}{200}$ (c) $\frac{13}{10}$

3 (a) 97.4% (b) 4% (c) 375%

4 (a) 31.25% (b) $77.\overline{7}\%$ (c) 175%

5 10

6 42%

7 70

8 Tax: $7.45; total cost: $156.45

9 $5,400

10 1.6%

11 24%

EXERCISE SET 8-1

Writing Exercises

1. What exactly does the word "percent" mean?
2. Explain how to change percents into decimal and fraction form.
3. Explain how to change decimals and fractions into percent form.
4. Explain how the word "of" plays an important role in calculations involving percents.
5. How do you find the percent increase or decrease of a quantity?
6. Is it possible to have more than 100% of a quantity? Explain.

Computational Exercises

For Exercises 7–18, express each as a percent.

7. 0.63
8. 0.87
9. 0.025
10. 0.0872
11. 1.56
12. 3.875
13. $\frac{1}{5}$
14. $\frac{5}{8}$
15. $\frac{2}{3}$
16. $\frac{1}{6}$
17. $1\frac{1}{4}$
18. $2\frac{3}{8}$

For Exercises 19–26, express each as a decimal.

19. 18%
20. 23%
21. 6%
22. 2%
23. 62.5%
24. 75.6%
25. 320%
26. 275%

For Exercises 27–36, express each as a fraction or mixed number.

27. 24%
28. 36%
29. 9%
30. 4%
31. 236%
32. 520%
33. $\frac{1}{2}$%
34. $12\frac{1}{2}$%
35. $16\frac{2}{3}$%
36. $4\frac{1}{6}$%

Real-World Applications

37. Find the sales tax and total cost of a laser printer that costs $299.99. The tax rate is 5%.
38. Find the sales tax and total cost of an espresso machine that costs $59.95. The tax rate is 7%.
39. Find the sales tax and total cost of a Sony Playstation that costs $149.99. The tax rate is 6%.
40. Find the sales tax and total cost of a wireless mouse that costs $19.99. The tax rate is 4.5%.
41. A diamond ring was reduced from $999.99 to $399.99. Find the percent of the reduction in the price.
42. An MP3 player was reduced from $109.99 to $99.99. Find the percent reduction in price.
43. A 20-inch flat panel computer monitor is on sale for $249.99. It was reduced $80.00 from the original price. Find the percent reduction in price.
44. The sale price of a spring break vacation package was $179.99, and the travel agent said by booking early, you saved $20. Find the percent reduction in price.
45. A luggage set was selling for $159.99, and the ad states that it has now been reduced 40%. Find the sale price.
46. If a sales clerk receives a 7% commission on all sales, find the commission the clerk receives on the sale of a computer system costing $1,799.99.
47. If the cost of a gas grill is $199.99 and it is on sale for 25% off, find the sale price.
48. If the commission for selling a 52-inch high-definition television set is 12%, find the commission on a television set that costs $2,499.99.
49. Milo receives a commission of 6% on all sales. If his commission on a sale was $75.36, find the cost of the item he sold.
50. The sales tax in Pennsylvania is 6%. If the tax on an item is $96, find the cost of the item.
51. For a certain year, 19% of all books sold were self-help books. If a bookstore sold 12,872 books, about how many were self-help books?
52. You saved $200 on your new laptop because you bought it online. If this was a 25% savings from the original price, find the original cost of the laptop.
53. The average teachers' and superintendents' salaries in a school district in western Pennsylvania was $50,480. Five years later, the new average was $54,747. Find the percent increase.
54. In 2000 there were 97 million cell phone subscribers. Four years later, there were 169.5 million cell phone subscribers. Find the percent increase.
55. In the 1992–1993 school year in Pennsylvania, there were 20 teachers' strikes. In the 2006–2007 school year, there were 8 teacher's strikes. Find the percent decrease.
56. In 1998, the winning competitor in the Nathan's hot dog eating contest ate 19 hot dogs. In 2008, the winner ate 59 hot dogs. Find the percent increase.
57. The website forsalebyowner.com reported that total real estate commissions in 2007 were $55 billion, an increase of $19 million over the year 2000. What was the percent increase?
58. In the 2004–2005 school year, the average cost of tuition, fees, room, and board for a public 4-year university was $12,127 and in the 2007–2008 school year, the average cost was $13,589. Find the percent increase.

Critical Thinking

59. A store has a sale with 30% off every item. When you enter the store, you receive a coupon that states that you receive an additional 20% off. Is this equal to a 50% discount? Explain your answer.
60. You purchase a stock at $100 per share. It drops 30% the next day; however, a week later, it increases in value by 30%. If you sell it, will you break even? Explain your answer.
61. Suppose a friend planning a shopping spree on the day after Thanksgiving tells you he plans to buy a 65-inch plasma TV, and you say "There's no way you can afford that!" He then tells you that the store is offering 50% off any one item, and he has an Internet coupon good for 50% off any price, even a discounted one. So that's 100% off, and he'll get it for free! Explain why your friend will come home very disappointed.
62. A store that used to sell a grill for $90 now offers it at $60, and advertises "33% off our best-selling grill!" An amusement park used to have 60 rides, and now boasts 90 rides, claiming "50% more rides this year!" Which one of them is lying?

Section 8-2 Simple Interest

LEARNING OBJECTIVES

- ❑ 1. Compute simple interest and future value.
- ❑ 2. Compute principal, rate, or time.
- ❑ 3. Compute interest using the Banker's rule.
- ❑ 4. Compute the true rate for a discounted loan.

The topic of the next two sections is of interest to anyone who plans to buy a house or a car, have a credit card, invest money, have a savings account—in short, pretty much everyone. This interesting topic is interest—a description of how fees are calculated when money is borrowed, and how your money grows when you save. Unless you don't mind being separated from your hard-earned money, this is a topic you should be eager to understand well.

Interest is a fee paid for the use of money. For example, if you borrow money from a bank to buy a car, you must not only pay back the amount of money that you borrowed, but also an additional amount, called the interest, for the use of the bank's money. On the other hand, if you deposit money in a savings account, the bank will pay you interest for saving money since it will be using your money to provide loans, mortgages, etc. to people who are borrowing money. The stated rate of interest is generally given as a yearly percentage of the amount borrowed or deposited.

There are two kinds of interest. *Simple interest* is a one-time percent of an amount of money. *Compound interest* is a percentage of an original amount, as well as a percentage of the new amount including previously calculated interest. We will study simple interest in this section, and compound interest in the next.

Simple Interest

In order to compute simple interest, we will need three pieces of information: the *principal,* the *rate,* and the *time.*

Math Note

Remember: P (principal) is the beginning amount borrowed or invested, and A (future value) is the final amount repaid or accumulated.

Interest (I) is the fee charged for the use of money.
Principal (P) is the amount of money borrowed or placed into a savings account.
Rate (r) is the percent of the principal paid for having money loaned, or earned for investing money. Unless indicated otherwise, rates are given as a percent for a term of 1 year.
Time (t) or **term** is the length of time that the money is being borrowed or invested. When the rate is given as a percent per year, time has to be written in years.
Future value (A) is the amount of the loan or investment plus the interest paid or earned.

The basic formulas for computing simple interest use principal, rate, and time as follows:

Math Note

The formula $A = P(1 + rt)$ can be used to find future value without explicitly computing the interest first.

Formulas for Computing Simple Interest and Future Value

1. Interest = principal × rate × time:

$$I = Prt$$

2. Future value = principal + interest:

$$A = P + I \quad \text{or} \quad A = P(1 + rt)$$

EXAMPLE 1 Computing Simple Interest

Find the simple interest on a loan of \$3,600.00 for 3 years at a rate of 8% per year.

SOLUTION

Change the rate to a decimal and substitute into the formula $I = Prt$:

$$8\% = 0.08$$
$$\begin{aligned} I &= Prt \\ &= (\$3{,}600.00)(0.08)(3) \\ &= \$864.00 \end{aligned}$$

The interest on the loan is \$864.00.

▼ Try This One 1

Find the simple interest on a \$12,000 loan for 5 years at 7%.

EXAMPLE 2 Finding Future Value

Find the future value for the loan in Example 1.

SOLUTION

Substitute into the formula $A = P + I$

$$\begin{aligned} A &= P + I \\ &= \$3{,}600.00 + \$864.00 \\ &= \$4{,}464.00 \end{aligned}$$

The total amount of money to be paid back is \$4,464.00.

ALTERNATE SOLUTION

Substitute into the formula $A = P(1 + rt)$

$$\begin{aligned} A &= P(1 + rt) \\ &= \$3{,}600(1 + 0.08 \cdot 3) \\ &= \$4{,}464.00 \end{aligned}$$

▼ Try This One 2

Find the future value of the loan in Try This One 1.

Since rates are typically given in terms of percent per year, when the time of a loan or investment is given in months, we need to divide it by 12 to convert to years.

EXAMPLE 3 Computing Simple Interest for a Term in Months

To meet payroll during a down period, United Ceramics Inc. needed to borrow \$2,000.00 at 4% simple interest for 3 months. Find the interest.

SOLUTION

Change 3 months to years by dividing by 12, and change the rate to a decimal. Substitute in the formula $I = Prt$.

$$I = (\$2{,}000.00)(0.04)\left(\frac{3}{12}\right) \quad 4\% = 0.04$$

$$= \$20.00$$

The interest is \$20.00.

▼ Try This One 3

Marta needs some quick cash for books at the beginning of spring semester, so she borrows \$600 at 11% simple interest for 2 months. How much interest will she pay?

☑ 1. Compute simple interest and future value.

Often, a simple interest loan is paid off in monthly installments. To find the monthly payment, divide the future value of the loan by the number of months in the term of the loan.

EXAMPLE 4 Computing Monthly Payments

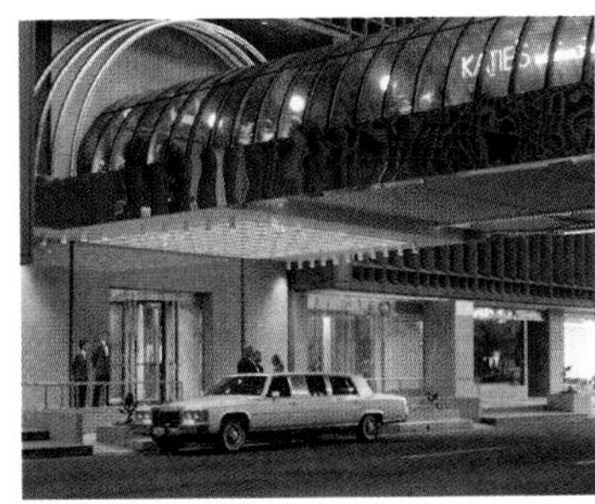

Admiral Chauffeur Services borrowed \$600.00 at 9% simple interest for $1\frac{1}{2}$ years to repair a limousine. Find the interest, future value, and the monthly payment.

SOLUTION

Step 1 Find the interest.

$$I = Prt$$

$$= (\$600.00)(0.09)\left(1\frac{1}{2}\right) \quad 9\% = 0.09$$

$$= \$81$$

The interest is \$81.00.

Step 2 Find the future value of the loan.

$$A = P + I$$

$$= \$600.00 + \$81.00$$

$$= \$681.00$$

Step 3 Divide the future value of the loan by the number of months. Since $1\frac{1}{2}$ years = 18 months, divide \$681.00 by 18 to get \$37.83. The monthly payment is \$37.83.

▼ Try This One 4

The Lookout Restaurant took out a loan for \$5,000.00. The simple interest rate was 6.5%, and the term of the loan was 3 years. Find the interest, future value, and monthly payment.

Finding the Principal, Rate, and Time

In addition to finding the interest and future value for a loan or investment, we can find the principal, the rate, and the time period by substituting into the formula $I = Prt$ and solving for the unknown.

Examples 5–7 show how to find the principal, rate, and time.

EXAMPLE 5 Computing Principal

Calculator Guide

In the calculation for Example 5, order of operations is very important.

Standard Scientific Calculator

93.5 [÷] [(] .055 [×] 2 [)] [=]

Standard Graphing Calculator

93.5 [÷] [(] .055 [×] 2 [)] [Enter]

Phillips Health and Beauty Spa is replacing one of its workstations. The interest on a loan secured by the spa was \$93.50. The money was borrowed at 5.5% simple interest for 2 years. Find the principal.

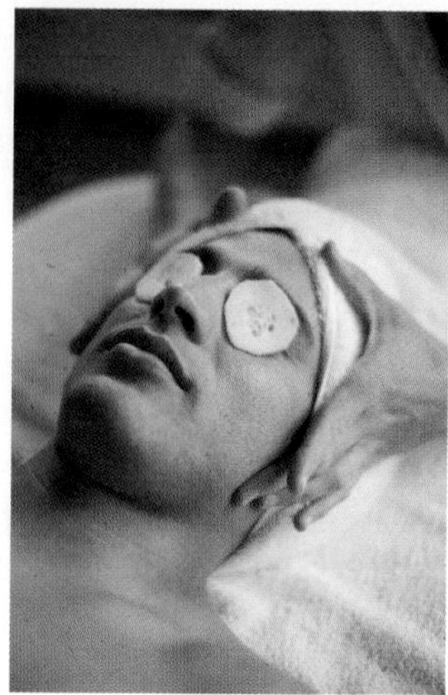

SOLUTION

$$I = \$93.50,\ r = 5.5\% = 0.055, \text{ and } t = 2$$

$$I = Prt$$

$$\$93.50 = P(0.055)(2) \quad \text{Divide both sides by } (0.055)(2).$$

$$\frac{\$93.50}{(0.055)(2)} = \frac{P(0.055)(2)}{(0.055)(2)}$$

$$P = \$850$$

The amount of the loan was \$850.00.

▼ Try This One 5

Find the principal on a savings account that paid \$76.50 in simple interest at 6% over 3 years.

The same formulas can be used for investments as well. Example 6 shows this.

EXAMPLE 6 Computing Interest Rate

R & S Furnace Company invested \$15,250.00 for 10 years and received \$9,150.00 in simple interest. What was the rate that the investment paid?

SOLUTION

$$P = \$15{,}250,\ t = 10, \text{ and } I = \$9{,}150$$

$$I = Prt$$

$$\$9{,}150 = (\$15{,}250)(r)(10) \quad \text{Divide both sides by } (\$15{,}250)(10).$$

$$\frac{\$9{,}150}{(\$15{,}250)(10)} = \frac{(\$15{,}250)(r)(10)}{(\$15{,}250)(10)}$$

$$0.06 = r$$

$$r = 0.06 \text{ or } 6\%$$

The interest paid on the investment was 6%.

▼ Try This One 6

If you invest \$8,000 for 30 months and receive \$1,000 in simple interest, what was the rate?

CAUTION

Be sure to change the decimal to a percent since rates are given in percents.

EXAMPLE 7 Computing the Term of a Loan

Judi and Laura borrowed \$4,500.00 at $8\frac{3}{4}\%$ to put in a hot tub. They had to pay \$2,756.25 interest. Find the term of the loan.

SOLUTION

$$P = \$4{,}500,\ r = 8\tfrac{3}{4}\% = 0.0875, \text{ and } I = \$2{,}756.25$$

$$I = Prt$$

$$\$2{,}756.25 = (\$4{,}500)(0.0875)t \qquad \text{Divide both sides by } (\$4{,}500)(0.0875).$$

$$\frac{\$2{,}756.25}{(\$4{,}500)(0.0875)} = \frac{(\$4{,}500)(0.0875)t}{(\$4{,}500)(0.0875)}$$

$$7 = t$$

The term of the loan was 7 years.

▼ Try This One 7

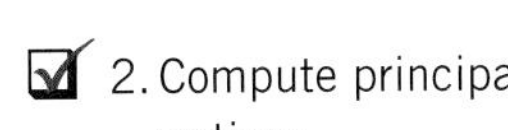

2. Compute principal, rate, or time.

A pawn shop offers to finance a guitar costing \$750 at 4% simple interest. The total interest charged will be \$150. What is the term of the loan?

The Banker's Rule

Simple interest for short term loans is sometimes computed in days. For example, the term of a loan may be 90 days. In this case, the time would be $\frac{90 \text{ days}}{365 \text{ days}} = \frac{90}{365}$ since there are 365 days in a year. However, many lending institutions use what is called the *Banker's rule*. The Banker's rule treats every month like it has 30 days, so it uses 360 days in a year. They claim that the computations are easier to do. When a lending institution uses 360 days instead of 365, how does that affect the amount of interest? For example, on a \$5,000 loan at 8% for 90 days, the interest would be

$$I = Prt$$

$$= (\$5{,}000)(0.08)\left(\frac{90}{365}\right) \qquad 8\% = 0.08;\ 90 \text{ days is } \frac{90}{365} \text{ year.}$$

$$= \$98.63$$

Using the Banker's rule, the interest is

$$I = Prt$$

$$= (\$5{,}000)(0.08)\left(\frac{90}{360}\right) \qquad \text{We used a 360 day year here.}$$

$$= \$100.00$$

We can see why this is called the Banker's rule and not the customer's rule!

Sidelight AN OUTDATED RULE

The Banker's rule is very old, and there was a time when it made sense. Originally, interest on savings and loans had to be calculated by hand, and standardizing every month to 30 days did in fact make the calculations a lot simpler. But those types of calculations haven't been done by hand for over 50 years, and in the age of computers, it's just plain silly to worry about how difficult it is to compute the interest. That's exactly why computers are called computers—they're really good at computing things! So why do some lenders still use the Banker's rule? Because they can, probably, and it makes them more money.

EXAMPLE 8 Using the Banker's Rule

Find the simple interest on a $1,800 loan at 6% for 120 days. Use the Banker's rule.

SOLUTION

$$P = \$1{,}800,\ r = 6\% = 0.06,\ t = \frac{120}{360}$$

$$I = Prt$$

$$= (\$1{,}800)(0.06)\left(\frac{120}{360}\right) = \$36$$

The interest using the Banker's rule is $36.

▼ Try This One 8

3. Compute interest using the Banker's rule.

Find the simple interest on a $2,200 loan at 7% interest for 100 days. Use the Banker's rule.

Discounted Loans

Sometimes the interest on a loan is paid upfront by deducting the amount of the interest from the amount the bank gives you. This type of loan is called a **discounted loan.** The interest that is deducted from the amount you receive is called the **discount.** Example 9 illustrates how it works.

EXAMPLE 9 Finding the True Rate of a Discounted Loan

A student obtained a 2-year $4,000 loan for college tuition. The rate was 9% simple interest and the loan was a discounted loan.

(a) Find the discount.
(b) Find the amount of money the student received.
(c) Find the true interest rate.

SOLUTION

(a) The discount is the total interest for the loan.

$$P = \$4{,}000,\ r = 9\%,\ t = 2 \text{ years}$$

$$I = Prt$$

$$= (\$4{,}000)(0.09)(2) \quad 9\% = 0.09$$

$$= \$720$$

The discount is $720.

(b) The student received $4,000 − $720 = $3,280.
(c) The true interest rate is calculated by finding the rate on a $3,280 loan with $720 interest.

$$I = Prt$$

$$\$720 = (\$3{,}280)r(2) \quad \textit{Multiply.}$$

$$\$720 = \$6{,}560r \quad \textit{Divide both sides by \$6,560.}$$

$$r = \frac{\$720}{\$6{,}560} = 0.1098 \text{ (rounded)}$$

The true interest rate is approximately 10.98%.

Sidelight WHO GETS THE DISCOUNT?

Everyone likes getting discounts, so "discounted loan" sounds great, right? This is just another way that lenders can take advantage of customers. Since you're paying the interest up front, out of the amount you're borrowing, the effect is that you're borrowing less money and paying the same amount of interest. As Example 9 shows, the actual interest rate you end up paying is quite a bit higher than the rate you're quoted. The only thing they're really "discounting" is your ability to make smart financial decisions.

4. Compute the true rate for a discounted loan.

▼ Try This One 9

Mary Dixon obtained a $5,000 discounted loan for 3 years at 6% simple interest.

(a) Find the discount.
(b) Find the amount of money Mary received.
(c) Find the true interest rate.

Answers to Try This One

1 $4,200

2 $16,200

3 $11

4 $975, $5,975, $165.97

5 $425

6 5%

7 5 years

8 $42.78

9 (a) $900
(b) $4,100
(c) 7.32%

EXERCISE SET 8-2

Writing Exercises

1. Describe what interest is, and how simple interest is calculated.
2. What do the terms "principal" and "future value" refer to?
3. What is meant by the term of a loan?
4. How is the rate of a loan or a savings account typically described?
5. What is the Banker's rule? Does it help borrowers or lenders?
6. What does it mean for a loan to be discounted?

Computational Exercises

For Exercises 7–26, find the missing value.

	Principal	Rate	Time	Simple Interest
7.	$12,000	6%	2 years	______
8.	$25,000	8.5%	6 months	______
9.	$1,800	10%	______	$360
10.	$600	4%	______	$72
11.	$4,300	______	6 years	$1,290
12.	$200	______	3 years	$45
13.	______	9%	4 years	$354.60

	Principal	Rate	Time	Simple Interest
14.	______	15%	7 years	$65,625
15.	$500	______	2.5 years	$40
16.	$1,250	5%	______	$375
17.	$900	$9\frac{1}{2}\%$	18 months	______
18.	$420	______	30 months	$31
19.	$660	12%	______	$514.80
20.	$1,975	7.2%	$3\frac{1}{2}$ years	______
21.	$14,285	______	6 years	$8,571
22.	$650	15%	______	$877.50
23.	$325	______	8 years	$156
24.	$15,000	11%	______	$742.50
25.	$700	$6\frac{3}{4}\%$	______	$141.75
26.	$135	7%	6.5 years	______

For Exercises 27–32, find the future value of the loan

27. $P = \$800$, $r = 4\%$, $t = 5$ years
28. $P = \$15{,}000$, $r = 9\%$, $t = 7$ years
29. $P = \$700$, $r = 10\%$, $t = 6$ years
30. $P = \$1{,}250$, $r = 8\frac{1}{2}\%$, $t = 6$ months
31. $P = \$475$, $r = 13\%$, $t = 3$ months
32. $P = \$3{,}360$, $r = 5.5\%$, $t = 2$ years

For Exercises 33–38, find the interest on each loan using the Banker's rule.

33. $P = \$220$, $r = 3\%$, $t = 90$ days
34. $P = \$3{,}800$, $r = 9\%$, $t = 60$ days
35. $P = \$500$, $r = 14\%$, $t = 180$ days
36. $P = \$760$, $r = 6.2\%$, $t = 270$ days
37. $P = \$1{,}100$, $r = 8\frac{1}{4}\%$, $t = 105$ days
38. $P = \$420$, $r = 12\%$, $t = 150$ days

For Exercises 39–44, the loans are discounted. For each exercise, find (a) the discount, (b) the amount of money received, and (c) the true interest rate.

39. $P = \$3{,}000$, $r = 8\%$, $t = 3$ years
40. $P = \$1{,}750$, $r = 4\frac{1}{2}\%$, $t = 6$ years
41. $P = \$22{,}000$, $r = 5\%$, $t = 4$ years
42. $P = \$600$, $r = 2\%$, $t = 10$ years
43. $P = \$780$, $r = 6\%$, $t = 9$ years
44. $P = \$1{,}850$, $r = 3\%$, $t = 5$ years

Real-World Applications

For Exercises 45–64, assume that all interest is simple interest.

45. In addition to working and her family's contribution, Jane had to borrow $8,000 over the course of 6 years to complete her education. The interest is $4,046.40. Find the rate.
46. Fred started a new e-Commerce business and borrowed $15,000 for 12 years to get the business up and running. The interest is $18,000. Find the rate.
47. To take advantage of a going-out-of-business sale, the College Corner Furniture Store had to borrow some money. It paid back a total of $150,000 on a 6-month loan at 12%. Find the principal.
48. To purchase two new copiers, the campus bookstore paid $1,350 interest on a 9% loan for 3 years. Find the principal.
49. To train employees to use new equipment, Williams Muffler Repair had to borrow $4,500.00 at $9\frac{1}{2}\%$. The company paid $1,282.50 in interest. Find the term of the loan.
50. Berger Car Rental borrowed $8,650.00 at 6.8% interest to cover the increasing cost of auto insurance. Find the term of the loan if the interest is $441.15.
51. To pay for new supplies, Jiffy Photo Company borrowed $9,325.00 at 8% and paid $3,170.50 in interest. Find the term of the loan.
52. Mary Beck earned $216 interest on a savings account at 8% over 2 years. Find the principal.
53. John White has savings of $4,300, which earned $9\frac{3}{4}\%$ interest for 5 years. Find the interest.
54. Ed Bland had savings of $816 invested at $4\frac{1}{2}\%$ for 3 years. Find the interest.
55. Matt's Appliance Store borrowed $6,200 for 3 years for repairs. The rate was 6%. Find the future value of the loan.
56. Adrienne's Pub borrowed $12,000 for 6 years at 9%. Find the future value of the loan.
57. Sally borrowed $600 for 90 days at 4%. Find the interest using the Banker's rule.
58. Harry borrowed $950 for 120 days at $6\frac{3}{4}\%$. Find the interest using the Banker's rule.
59. The West Penn Finance group secured an $18,000 discounted loan to remodel its offices. The rate was 5% and the term was 6 years. Find
    (a) The discount.
    (b) The amount of money the group received.
    (c) The true interest rate.
60. The University Center obtained a $20,000 discounted loan for 3 years to remodel the student game room. The rate was 9%. Find
    (a) The discount.
    (b) The amount of money the center received.
    (c) The true interest rate.
61. Susan would like to buy a new car. Which loan would have the higher interest amount: a personal loan of $10,000 at 9% for 6 years or an auto loan of $10,000 at 8% for 60 months? Why?
62. Sea Drift Motel is converting its rooms into privately owned condominiums. The interest on a $1,000,000, 20-year construction loan is $98,000. What is the rate of interest? Does the rate seem unreasonable?

63. The Laurel Township Fire Department must decide whether to purchase a new tanker truck or repair the one they now use. For a new truck loan, the interest rate on \$25,000 is 18% for a 10-year period; to repair the existing truck, the department must borrow \$18,000 at $12\frac{1}{2}\%$ for 8 years. Which loan is less expensive?

64. A local miniature golf course owner must recarpet his fairways and replace the golf clubs his customers use. A bank will lend the owner the necessary \$7,800 at 9.5% interest over 48 months. A savings and loan company will lend the owner \$7,800 at 8.5% interest for 54 months. Which loan will be less expensive for the golf course owner to assume?

Critical Thinking

65. Suppose that you have a choice of two loans: one at 5% simple interest for 6 years, and one at 6% simple interest for five years. Which will result in the smaller future value? Does it depend on the principal?

66. When a loan is discounted, is it better or worse for the borrower if the term is longer? Try some specific examples and make a conjecture.

Section 8-3 Compound Interest

LEARNING OBJECTIVES

- ☐ 1. Compute compound interest.
- ☐ 2. Compute the effective interest rate of an investment.
- ☐ 3. Compare the effective rate of two investments.
- ☐ 4. Find the future value of an annuity.
- ☐ 5. Compute the periodic payment needed to meet an investment goal.

If you think about simple interest over a long period of time, it doesn't sound like a great deal for the investor. Suppose you put \$1,000 into an account that pays 5% simple interest, and you keep it untouched for 30 years. Each year, you're getting 5% of \$1,000 in interest. But for all that time, the bank could have been increasing your money through loans and investments, and they're still paying interest only on the original amount.

This is where compound interest comes into play. It seems more fair for the bank to pay interest on the actual value of the account each year, not just the original amount. When interest is computed on the principal *and* any previously earned interest, it is called **compound interest.** Let's look at an example that compares simple and compound interest.

EXAMPLE 1 Comparing Simple and Compound Interest

Suppose that \$5,000 is invested for 3 years at 8%.

(a) Find the amount of simple interest.
(b) Find the compound interest if interest is calculated once per year.

SOLUTION

(a) Using the formula $I = Prt$ with $P = \$5{,}000$, $r = 0.08$, and $t = 3$, we get

$$I = \$5{,}000 \times 0.08 \times 3 = \$1{,}200$$ *Simple interest over 3 years.*

(b) **First year** For the first year, we have $P = \$5{,}000$, $r = 0.08$ and $t = 1$:

$$I = Prt = \$5{,}000 \times 0.08 \times 1 = \$400$$

The interest for the first year is \$400.
Second year At the beginning of the second year, the account now contains \$5,400, so we use this as principal for the second year. The rate and time remain the same.

$$I = Prt = \$5{,}400 \times 0.08 \times 1 = \$432$$

The interest for the second year is \$432.
Third year The principal is now \$5,400 + \$432 = \$5,832.

$$I = Prt = \$5{,}832 \times 0.08 \times 1 = \$466.56$$

The interest for the third year is \$466.56, and the total interest for three years is \$400 + \$432 + \$466.56 = \$1,298.56. This is almost a hundred dollars more than with simple interest.

▼ Try This One 1

For an investment of \$100,000 at 6% interest for 4 years, find (a) the simple interest, and (b) the compound interest if interest is calculated once per year.

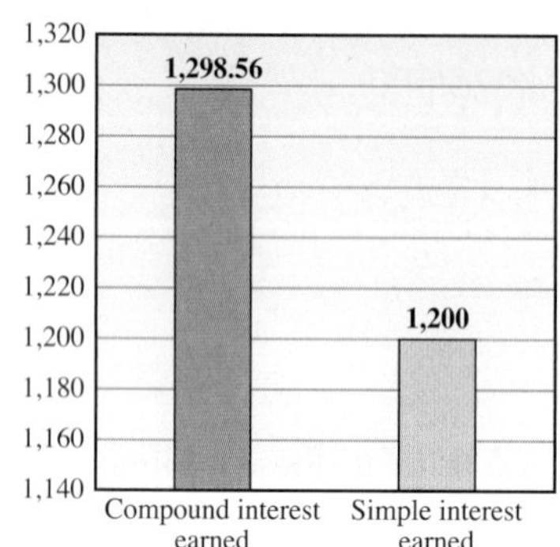

For the investment in Example 1, a simple-interest savings account would earn \$1,200.00 interest while a compound-interest savings account would earn \$1,298.56, almost \$100.00 (\$98.56) more over a 3-year period.

When interest is calculated once each year, we say that it is **compounded yearly.** In many cases, interest is computed at more frequent intervals than that. It can be compounded **semiannually** (twice a year), **quarterly** (4 times a year), **monthly** (12 times a year), or even **daily** (every day). In Example 1, we had to do three separate calculations to find the interest after 3 years. If the interest were instead compounded monthly, we would have needed 36 calculations! Needless to say, such calculations can get pretty cumbersome pretty quickly. Instead, using an approach similar to the steps in Example 1, we can develop a formula to perform the calculation in one step.

Math Note

When interest is compounded yearly, $n = 1$; semiannually, $n = 2$; quarterly, $n = 4$; and daily, $n = 365$.

Formula for Computing Compound Interest

$$A = P\left(1 + \frac{r}{n}\right)^{nt}$$

where A is the future value (principal + interest)
r is the yearly interest rate in decimal form
n is the number of times per year the interest is compounded
t is term of the investment in years

EXAMPLE 2 Computing Compound Interest

Find the interest on \$7,000.00 compounded quarterly at 3% for 5 years.

SOLUTION

Quarterly means 4 times a year, so $n = 4$.

$$P = \$7{,}000.00,\ r = 3\% = 0.03,\ t = 5$$

$$A = P\left(1 + \frac{r}{n}\right)^{nt}$$

$$= \$7{,}000.00\left(1 + \frac{0.03}{4}\right)^{4\cdot 5}$$

$$= \$7{,}000(1.0075)^{20} \quad \textit{See Calculator Guide.}$$

$$= \$8{,}128.29$$

To find the interest, subtract the principal from the future value.

$$I = \$8{,}128.29 - \$7{,}000.00$$

$$= \$1{,}128.29$$

The interest is \$1,128.29.

Calculator Guide

To compute the future value in Example 2:

Standard Scientific Calculator

7000 [×] [(] 1.0075 [)] [y^x] 20 [=]

Standard Graphing Calculator

7000 [(] 1.0075 [)] [^] 20 [Enter]

▼ Try This One 2

Find the interest on \$600.00 compounded semiannually at 4.5% for 6 years.

EXAMPLE 3 Computing Compound Interest

Calculator Guide

The future value in Example 3 can be computed in one step using a graphing calculator, but should be done in stages with a scientific calculator:

Standard Scientific Calculator

1 [+] .05 [÷] 365 [=] [y^x]
2190 [=] [×] 11000

(Note that 365 × 6 must be calculated first.)

Standard Graphing Calculator

11000 [(] 1 [+] .05 [÷]
365 [)] [^] [(] 365 [×]
6 [)] [Enter]

Find the interest on \$11,000 compounded daily at 5% for 6 years. Assume a 365-day year.

SOLUTION

$$P = \$11{,}000,\ r = 5\% = 0.05,\ n = 365,\ t = 6$$

$$A = P\left(1 + \frac{r}{n}\right)^{nt}$$

$$= \$11{,}000\left(1 + \frac{0.05}{365}\right)^{365\cdot 6} \qquad \textit{See Calculator Guide.}$$

$$= \$14{,}848.14$$

To find the interest, subtract the principal from the future value.

$$I = \$14{,}848.14 - \$11{,}000$$

$$= \$3{,}848.14$$

The interest is \$3,848.14.

☑ 1. Compute compound interest.

▼ Try This One 3

Find the interest on \$50,000 compounded weekly at 7% for 20 years. Assume a 52-week year.

In Example 2, we simplified $1 + \frac{0.03}{4}$ to 1.0075 before calculating, but didn't do the same for $1 + \frac{0.05}{365}$ in Example 3. This is because the second number has a much longer decimal expansion, and would likely have required rounding. This affects the accuracy considerably. If you rework the calculation in Example 3, but round $1 + \frac{0.05}{365}$ to four decimal places, the result is \$2,692.99, which is off by over a thousand dollars!

Effective Rate

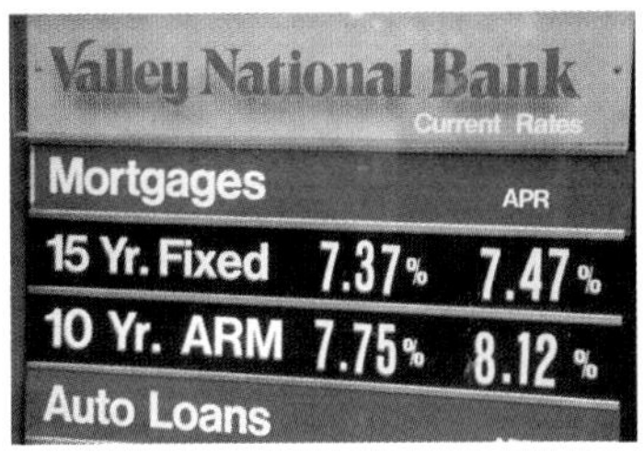

APR on a loan is similar to the effective rate on an investment, but may also factor in upfront fees.

As the number of times per year that interest is compounded goes up, the amount of interest does as well, but not by as much as you might think. For a \$1,000 investment at 4% interest for 10 years, the difference between compounding yearly and compounding daily is only about \$11. (See Sidelight on page 443). Still, because of this relatively small difference, when interest is compounded more than once per year, the interest earned on a savings account is actually a bit higher than the stated rate.

For example, consider a savings account with a principal of \$5,000.00 and an interest rate of 4% compounded semiannually. The interest is compounded twice a year at 2%. Using the formula shown in Example 2, the actual interest for one year is \$202.00. The stated rate is 4%, but the actual rate can be found by dividing \$202.00 by \$5,000.00, and it is 4.04%. This rate is called the *effective rate* or *annual yield.*

The **effective rate** (also known as the **annual yield**) is the simple interest rate which would yield the same future value over 1 year as the compound interest rate.

The next formula can be used to calculate the effective interest rate.

Formula for Effective Interest Rate

$$E = \left(1 + \frac{r}{n}\right)^n - 1$$

where

E = effective rate
n = number of periods per year the interest is calculated
r = interest rate per year (i.e., stated rate)

The stated rate is also called the **nominal rate.**

Example 4 illustrates the use of this formula.

EXAMPLE 4 Finding Effective Interest Rate

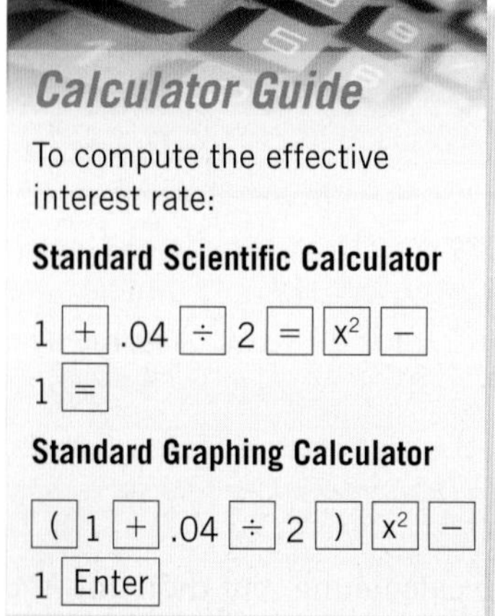
Calculator Guide

To compute the effective interest rate:

Standard Scientific Calculator

1 [+] .04 [÷] 2 [=] [x²] [−] 1 [=]

Standard Graphing Calculator

[(] 1 [+] .04 [÷] 2 [)] [x²] [−] 1 [Enter]

Find the effective interest rate when the stated rate is 4% and the interest is compounded semiannually.

SOLUTION

Let $r = 0.04$ (rate is 4%) and $n = 2$ (compounded semiannually) and then substitute into the formula.

$$E = \left(1 + \frac{r}{n}\right)^n - 1$$

$$= \left(1 + \frac{0.04}{2}\right)^2 - 1 \quad \textit{See Calculator Guide.}$$

$$= 0.0404 = 4.04\%$$

The effective rate is 4.04%. (This is the same rate that we calculated previously.)

▼ Try This One 4

Find the effective rate for a stated interest rate of 8% compounded quarterly.

☑ 2. Compute the effective interest rate of an investment.

The effective rates of two savings accounts can be used to determine which account would be a better investment. The next example shows how to do this.

EXAMPLE 5 Comparing the Effective Rate of Two Investments

Which savings account is a better investment: 6% compounded quarterly or 6.2% compounded semiannually?

SOLUTION

Find the effective rates of both accounts and compare them.

6% quarterly	**6.2% semiannually**
$r = 0.06, n = 4$	$r = 0.062, n = 2$
$E = \left(1 + \frac{r}{n}\right)^n - 1$	$E = \left(1 + \frac{r}{n}\right)^n - 1$
$= \left(1 + \frac{0.06}{4}\right)^4 - 1$	$= \left(1 + \frac{0.062}{2}\right)^2 - 1$
$\approx 0.0614 = 6.14\%$	$\approx 0.063 = 6.3\%$

The 6.2% semiannual investment gives an effective rate of 6.3%, while 6% quarterly gives an effective rate of 6.14%, so 6.2% semiannually is a better investment.

▼ Try this One 5

☑ 3. Compare the effective rate of two investments.

Which is a better investment, 3% compounded monthly or 3.25% compounded semiannually?

Annuities

An **annuity** is a savings investment for which an individual or business makes the same payment each period (i.e., annually, semiannually, or quarterly) into a compound-interest account where the interest does not change during the term of the investment. For example, an individual may pay $500 annually for 3 years into an account that yields 6% interest compounded annually. The total amount accumulated (payments plus interest) is called the **future value** of the annuity. Annuities are set up by individuals to pay for college expenses, vacations, or retirement. Annuities are set up by businesses to pay future expenses such as purchasing new equipment, expanding their business, etc. The payments are made at the end of each period.

Example 6 shows how an annuity works.

EXAMPLE 6 Finding Future Value of an Annuity

Find the future value of an annuity where a $500 payment is made annually for 3 years at 6%.

SOLUTION

The interest rate is 6% and the payment is $500 each year for 3 years.

I. End of the first year $500 (payment)
II. End of the second year

The $500 collected 6% interest and a $500 payment is made; the value of the annuity at the end of the second year is

$500(0.06) = $ 30	Interest.
$500	Principal paid at end of first year.
+$500	Payment at the end of the second year.
$1,030	

III. End of the third year

During the third year, the \$1,030 earns 6% interest and a payment of \$500 is made at the end of the third year. The annuity is worth

\$1,030(0.06) =	\$ 61.80	Interest.
	\$1,030.00	Principal at end of second year.
	+\$500.00	Payment at end of third year.
	\$1,591.80	

The future value of the annuity at the end of the three years is \$1,591.80.

▼ Try This One 6

Find the future value of an annuity where a \$300 payment is made annually for 4 years at 9%.

Finding the future value of an annuity by the method shown in Example 6 could become quite lengthy if the annuity is taken out for a long term. For example, if payments were made quarterly for 10 years on an annuity, 40 calculations (4 × 10) would be required to compute the future value. In order to avoid this situation, the next formula can be used for calculating the future value of an annuity.

Formula for Finding the Future Value of an Annuity

$$A = \frac{R\left[\left(1 + \frac{r}{n}\right)^{nt} - 1\right]}{\frac{r}{n}}$$

where A is the future value of the annuity
R is the regular periodic payment
r is the annual interest rate
n is the number of payments made per year
t is the term of the annuity in years

CAUTION

You should be especially careful in using the annuity formula because it uses both capital R and lowercase r. The lowercase represents what it always does—annual interest rate. Think of the capital R as standing for "Regular" or "Recurring" payment.

EXAMPLE 7 Finding the Future Value of an Annuity

Find the future value of an annuity when the payment is \$800 semiannually, the interest rate is 5% compounded semiannually, and the term is 4 years.

SOLUTION

$R = \$800$

$r = 5\% = 0.05$

$n = 2$ (semiannual)

$t = 4$

The calculations in the annuity formula are a little complicated, so we'll be especially careful to work in stages.

$$A = \frac{R\left[\left(1 + \frac{r}{n}\right)^{nt} - 1\right]}{\frac{r}{n}}$$

$$A = \frac{\$800\left[\left(1 + \frac{0.05}{2}\right)^{2 \cdot 4} - 1\right]}{\frac{0.05}{2}} = \frac{\$800\left[(1.025)^{8} - 1\right]}{0.025} = \$6{,}988.89$$

The future value of the annuity at the end of 4 years is $6,988.89.

▼ Try This One 7

4. Find the future value of an annuity.

Find the future value of an annuity when the payment is $275 quarterly, the interest rate is 6.5% compounded quarterly, and the term is 4 years.

In many cases, an investor may have a specific goal in mind for future value, and would want to find the regular payment that would be necessary to reach that goal. In this case, we can solve the annuity formula for R to calculate the payment.

Formula for Finding Regular Annuity Payments Needed to Reach a Goal

$$R = \frac{A\left(\frac{r}{n}\right)}{\left[\left(1 + \frac{r}{n}\right)^{nt} - 1\right]}$$

All variables represent the same values as the previous annuity formula.

Sidelight DOUBLING YOUR MONEY

People will often ask, "If I put my money in a savings account, how long will it take me to double it?" The answer, of course, depends on the interest rate. Bankers use a simple rule to find the answer in what is called the *rule of 72.*

All that is necessary to find the answer is to divide 72 by the current interest rate. For example, if the interest rate is 3%, then it will take you 72 ÷ 3 or 24 years. If the interest rate is 10%, it will take you 72 ÷ 10 or approximately 7.2 years.

There are several things to consider, though. First, the answer is only an approximation. Second, it is assumed that the interest is compounded yearly. When interest is compounded semiannually, quarterly, or daily, the answer found by the rule is less precise. Other simple rules can be used to find how to triple your money, quadruple your money, etc.

As you know, the higher the interest rate, the faster your money will grow. The growth rate of your money also depends on how often the interest rate is compounded. Usually, interest is compounded yearly (once a year), semiannually (two times a year), or quarterly (four times a year). There are two other ways to compound interest. They are daily (365 times a year) and continuously (all the time). You might think that interest compounded continuously will make you a lot of money, but in reality, continuous interest pays only slightly more money than daily interest. The comparison is shown in the chart.

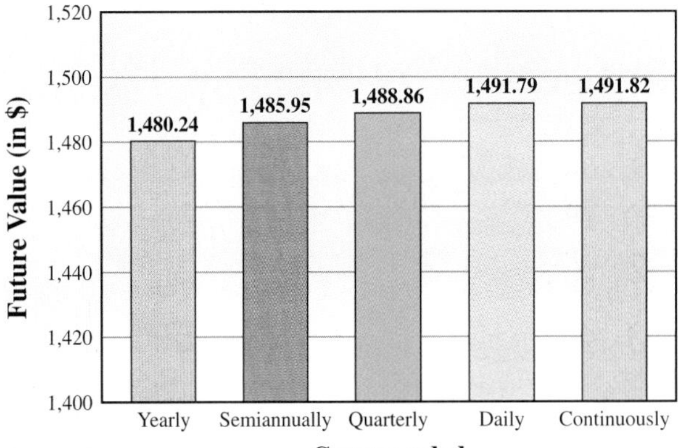

Compounded	Future Value
Yearly	$1,480.24
Semiannually	$1,485.95
Quarterly	$1,488.86
Daily	$1,491.79
Continuously	$1,491.82

As you can see, you will earn only about 3 cents more from continuous interest than you would earn with daily compounding.

EXAMPLE 8 Finding the Monthly Payment for an Annuity

Suppose you've always dreamed of opening your own tattoo parlor, and decide it's time to do something about it. A financial planner estimates that you would need a \$35,000 initial investment to start the business, and you plan to save that amount over the course of 5 years by investing in an annuity that pays 7.5% compounded weekly. How much would you need to invest each week?

SOLUTION

We know the following values: $A = 35{,}000$, $r = 7.5\% = 0.075$, $t = 5$, and $n = 52$.

$$R = \frac{A\left(\frac{r}{n}\right)}{\left[\left(1 + \frac{r}{n}\right)^{nt} - 1\right]}$$

$$R = \frac{35{,}000\left(\frac{0.075}{52}\right)}{\left(1 + \frac{0.075}{52}\right)^{52 \cdot 5} - 1}$$

Reduce error by not rounding $\frac{0.075}{52}$.

$$= \frac{\frac{2{,}625}{52}}{\left(1 + \frac{0.075}{52}\right)^{260} - 1} = 111.04$$

A payment of \$111.04 per week would be necessary to save the required \$35,000.

▼ Try This One 8

5. Compute the periodic payment needed to meet an investment goal.

Find the monthly payment needed to save a \$6,000 down payment for a car in an annuity that pays 5.5% interest compounded monthly over 2 years.

Answers to Try This One

1 (a) \$24,000
(b) \$26,247.70

2 \$183.63

3 \$152,569.20

4 8.24%

5 3.25% semiannually

6 \$1,371.94

7 \$4,979.15

8 \$237.07

EXERCISE SET 8-3

Writing Exercises

1. Describe the difference between simple interest and compound interest.
2. What does it mean to say that interest is compounded quarterly? What about compounded monthly?
3. What is the effective rate of an investment?
4. Describe how an annuity works.

Computational Exercises

For Exercises 5–14, find the compound interest and future value for each.

	Principal	Rate	Compounded	Time
5.	\$825	4%	Annually	10 years
6.	\$3,250	2%	Annually	5 years
7.	\$75	3%	Semiannually	6 years
8.	\$1,550	5%	Semiannually	7 years
9.	\$625	8%	Quarterly	12 years
10.	\$2,575	4%	Quarterly	2 years
11.	\$1,995	5%	Semiannually	6 years
12.	\$460	6%	Quarterly	7 years
13.	\$750	9%	Daily	1 year
14.	\$3,500	11%	Daily	$4\frac{1}{2}$ years

For Exercises 15–24, find the future value of each annuity.

	Payment	Rate	Compounded	Time
15.	\$200	9%	Annually	9 years
16.	\$750	3%	Annually	12 years
17.	\$1,250	6.8%	Semiannually	3 years
18.	\$375	5%	Quarterly	4 years
19.	\$1,530	4.5%	Quarterly	8 years

	Principal	Rate	Compounded	Time
20.	\$1,750	10%	Annually	12 years
21.	\$1,425	2%	Semiannually	6 years
22.	\$3,500	6%	Semiannually	2 years
23.	\$240	4%	Quarterly	7 years
24.	\$175	7%	Annually	5 years

For Exercises 25–28, find the effective interest rate.

25. Rate: 6% Compounded: Quarterly
26. Rate: 10% Compounded: Semiannually
27. Rate: 6.5% Compounded: Quarterly
28. Rate: 9.55% Compounded: Semiannually

For Exercises 29–32, determine which is the better investment.

29. 4.5% compounded semiannually or 4.25% compounded quarterly.
30. 7% compounded monthly or 7.2% compounded semiannually.
31. 3% compounded daily or 3.1% compounded quarterly.
32. 5.74% compounded semiannually or 5.6% compounded daily.

Real-World Applications

33. A couple decides to set aside \$5,000 in a savings account for a second honeymoon trip. It is compounded quarterly for 10 years at 9%. Find the amount of money they will have in 10 years.
34. In order to pay for college, the parents of a child invest \$20,000 in a bond that pays 8% interest compounded semiannually. How much money will there be in 18 years?
35. A 25-year-old plans to retire at age 50. She decided to invest an inheritance of \$60,000 at 7% interest compounded semiannually. How much will she have at age 50?
36. To pay for new machinery in 5 years, a company owner invests \$10,000 at $7\frac{1}{2}$% compounded quarterly. How much money will be available in 5 years?
37. A husband and wife plan to save money for their daughter's college education in 4 years. They decide to purchase an annuity with a semiannual payment earning 7.5% compounded semiannually. Find the future value of the annuity in 4 years if the semiannual payment is \$2,250.
38. A business owner decided to purchase an annuity to pay for new copy machines in 3 years. The payment is \$600 quarterly at 8% interest compounded quarterly. Find the future value of the annuity in 3 years.
39. Find the future value of an annuity if you invest \$200 quarterly for 20 years at 5% interest compounded quarterly.
40. The Washingtons decide to save money for a vacation in 2 years. They purchase an annuity with semiannual payments of \$200 at 9% interest compounded semiannually. Find the amount of money they can spend for the vacation.
41. The owner of the Campus Café plans to open a second location on the satellite campus in 5 years. She purchases an annuity that pays 10.5% interest compounded annually. If the payment is \$4,000 a year, find the future value of the annuity in 5 years.
42. Titan Thigh, the owner of Work-it-out Fitness, wants to buy an annuity that pays 4% interest compounded quarterly for 4 years. If the quarterly payment is \$160, find the future value of the annuity.
43. In order to plan for their retirement, a married couple decides to purchase an annuity that pays 8% interest compounded semiannually. If the semiannual payment is \$2,000, how much will they have saved in 10 years?
44. The Wash-n-Surf, an Internet café and laundromat, plans to replace two wash and surf hubs in 3 years. If the owner purchases an annuity at 6% interest compounded annually with an annual payment of \$800, find the value of the annuity in 3 years.
45. Petty Marine Co. has a long-term plan to expand to a second location that's actually near some water, so they want to start a monthly annuity to save \$150,000 in capital over 5 years. The best rate they can find is 7%. Find the monthly payment.

46. Suppose you plan to work right after you graduate, but still save money for grad school. You decide to save \$10,000 before starting, and find a weekly annuity that pays 6.5% interest for 4 years. How much will you need to pay each week?
47. The Massive Chemical Corporation starts an annuity to pay for the huge government penalties they expect in 10 years when a pending case finally gets litigated. Their lead attorney informs them that they can expect a \$4,000,000 fine. An investment house offers 11% interest on annuities of that size, compounded semiannually. What will the semiannual payment be on this annuity?
48. A 25-year-old decides that her goal is to retire at age 50 with at least \$2,000,000 in savings. The company investment annuity offers 7.4% annual returns, compounded monthly. What amount will she need to invest each month?

Critical Thinking

49. A person deposits \$5,000 into an account paying 4% compounded semiannually. Two years later the person deposited \$2,000 into the same account. How much money was there at the end of 5 years?
50. Sam deposited \$3,500 into a savings account paying 3% compounded quarterly. Two years later he withdrew \$800. How much money was in the account at the end of 6 years?
51. Akish deposited \$900 into a savings account paying 2% compounded quarterly. Two years later, she deposited \$400 into the same account. One year after that, she withdrew \$200. How much money was in the account at the end of 6 years?
52. Find the future value of a \$15,000 investment at 6% compounded monthly for 5 years. Then calculate the monthly payment that would be needed to accumulate that amount in a 6% monthly annuity. Finally, calculate the total amount that would be paid into the annuity. What can you conclude about the value of a lump-sum investment versus an annuity?

Section 8-4 Installment Buying

LEARNING OBJECTIVES

- ☐ 1. Find amount financed, total installment price, and finance charge for a fixed installment loan.
- ☐ 2. Use a table to find APR for a loan.
- ☐ 3. Compute unearned interest and payoff amount for a loan paid off early.
- ☐ 4. Compute credit card finance charges using the unpaid balance method.
- ☐ 5. Compute credit card finance charges using the average daily balance method.

A lot of people celebrate their college graduation by getting a new car. Unless you're very wealthy, you won't be plunking down a stack of crisp hundred dollar bills to make that happen—you'll need to get a loan. In this case, you'll be doing what is called **installment buying.** This is when an item is purchased and the buyer pays for it by making periodic partial payments, or installments.

There are natural advantages and disadvantages to installment buying. The most obvious advantage is that it allows you to buy an item that you don't have enough money to pay for, and use it while you're raising that money. The most obvious disadvantage is that you pay interest on the amount borrowed, so you end up paying more for the item—in some cases a *lot* more.

Fixed Installment Loans

A **fixed installment loan** is a loan that is repaid in equal payments. Sometimes the buyer will pay part of the cost at the time of purchase. This is known as a **down payment.** The other terms used to describe installment loans are defined next.

The **amount financed** is the amount a borrower will pay interest on.

Amount financed = Price of item − Down payment

The **total installment price** is the total amount of money the buyer will ultimately pay.

Total installment price = Sum of all payments + Down payment

The **finance charge** is the interest charged for borrowing the amount financed.

Finance charge = Total installment price − Price of item

EXAMPLE 1 Calculating Information About a Car Loan

Cat bought a 2-year old Santa Fe for \$12,260. Her down payment was \$3,000, and she will have to pay \$231.50 for 48 months. Find the amount financed, the total installment price, and the finance charge.

SOLUTION

Using the formulas in the box on page 446:

$$\begin{aligned}\text{Amount financed} &= \text{Cash price} - \text{Down payment}\\ &= \$12{,}260 - \$3{,}000\\ &= \$9{,}260\end{aligned}$$

Since she paid \$231.50 for 48 months and her down payment was \$3,000,

$$\begin{aligned}\text{Total installment price} &= \text{Total of monthly payments} + \text{Down payments}\\ &= 48 \times \$231.50 + \$3{,}000 \quad \text{48 payments at \$231.50.}\\ &= \$14{,}112.00\end{aligned}$$

Now we can find the finance charge:

$$\begin{aligned}\text{Finance charge} &= \text{Total installment price} - \text{Cash price}\\ &= \$14{,}112.00 - \$12{,}260.00\\ &= \$1{,}852.00\end{aligned}$$

The amount financed was \$9,260.00; the total installment price was \$14,112.00, and the finance charge was \$1,852.00.

▼ Try This One 1

If you buy a used car for \$8,200 with a down payment of \$1,000 and 36 monthly payments of \$270, find the amount financed, the total installment price, and the finance charge.

EXAMPLE 2 Computing a Monthly Payment

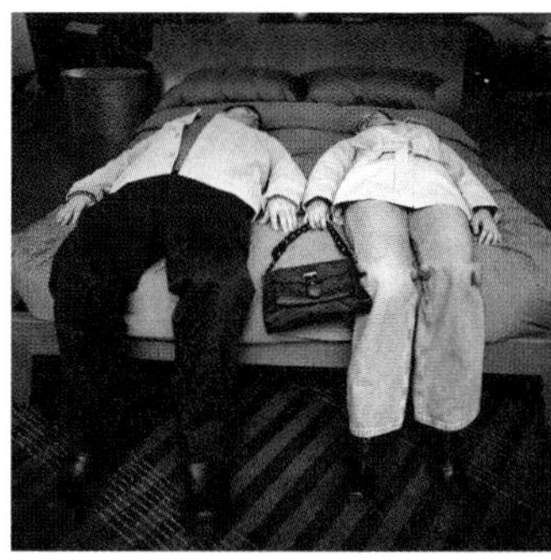

After a big promotion, a young couple bought \$9,000 worth of furniture. The down payment was \$1,000. The balance was financed for 3 years at 8% simple interest per year.

(a) Find the amount financed.
(b) Find the finance charge (interest).
(c) Find the total installment price.
(d) Find the monthly payment.

SOLUTION

(a) Amount financed = Price of item – Down payment

$$= \$9{,}000 - \$1{,}000 = \$8{,}000$$

(b) To find the finance charge, we use the simple interest formula:

$$\begin{aligned}I &= Prt\\ &= \$8{,}000 \times 0.08 \times 3 \quad 8\% = 0.08\\ &= \$1{,}920\end{aligned}$$

Math Note

The monthly payment is the total installment price minus the down payment, divided by the number of payments.

(c) In this case, the total installment price is simply the cost of the furniture plus the finance charge:

$$\text{Total installment price} = \$9{,}000 + \$1{,}920 = \$10{,}920$$

(d) To calculate the monthly payment, divide the amount financed plus the finance charge (\$8,000 + \$1,920) by the number of payments:

$$\text{Monthly payment} = \$9{,}920 \div 36 = \$275.56$$

In summary, the amount financed is \$8,000, the finance charge is \$1,920, the total installment price is \$10,920, and the monthly payment is \$275.56.

☑ 1. Find amount financed, total installment price, and finance charge for a fixed installment loan.

▼ Try This One 2

A graphic design pro buys a new iMac for \$1,499 with a \$200 down payment, and gets manufacturer financing for 5 years at 12% simple interest. Find (a) the amount financed, (b) the finance charge, (c) the total installment price, and (d) the monthly payment.

Annual Percentage Rate

Many lenders add upfront fees to a loan and then spread them over the life of the loan. This has the effect of making the actual interest rate that a borrower pays higher than the quoted rate. Because this can get confusing, lenders are required by law to disclose an **annual percentage rate,** or **APR,** that reflects the true interest charged. This allows consumers to compare loans with different terms. The mathematical procedures for computing APR are extremely complicated, so tables have been compiled that help you to estimate APR for a loan. A partial APR table is shown in Table 8-1. There are three steps required to find an APR from the table; they are listed at the top of the next page.

TABLE 8-1 APR Table

Number of Payments	Annual Percentage Rate												
	6.0%	6.5%	7.0%	7.5%	8.0%	8.5%	9.0%	9.5%	10.0%	10.5%	11.0%	11.5%	12.0%
	(Finance charge per \$100 of amount financed)												
6	\$1.76	\$1.90	\$2.05	\$2.20	\$2.35	\$2.49	\$2.64	\$2.79	\$2.94	\$3.08	\$3.23	\$3.38	\$3.53
12	3.28	3.56	3.83	4.11	4.39	4.66	4.94	5.22	5.50	5.78	6.06	6.34	6.62
18	4.82	5.22	5.63	6.04	6.45	6.86	7.28	7.69	8.10	8.52	8.93	9.35	9.77
24	6.37	6.91	7.45	8.00	8.54	9.09	9.64	10.19	10.75	11.30	11.86	12.42	12.98
30	7.94	8.61	9.30	9.98	10.66	11.35	12.04	12.74	13.43	14.13	14.83	15.54	16.24
36	9.52	10.34	11.16	11.98	12.81	13.64	14.48	15.32	16.16	17.01	17.86	18.71	19.57
48	12.73	13.83	14.94	16.06	17.18	18.31	19.45	20.59	21.74	22.90	24.06	25.23	26.40
60	16.00	17.40	18.81	20.23	21.66	23.10	24.55	26.01	27.48	28.96	30.45	31.96	33.47

Using the APR Table

Step 1 Find the finance charge per $100 borrowed using the formula

$$\frac{\text{Finance charge}}{\text{Amount financed}} \times \$100$$

Step 2 Find the row in the table marked with the number of payments and move to the right until you find the amount closest to the number from Step 1.

Step 3 The APR (to the nearest half percent) is at the top of the corresponding column.

EXAMPLE 3 Finding APR

Burk Carter purchased a color laser printer for $600.00. He made a down payment of $50.00 and financed the rest for 2 years with a monthly payment of $24.75. Find the APR.

Math Note

There are many websites with calculators for finding APR without having to use a table. A Google search will find dozens of them, in fact.

SOLUTION

Step 1 Find the finance charge per $100.00. The total amount he will pay is $24.75 per month × 24 payments, or $594.00. Since he financed $550.00, the finance charge is $594.00 − $550.00 = $44.00.

$$\begin{aligned}\text{Finance charge per }\$100.00 &= \frac{\text{Finance charge}}{\text{Amount financed}} \times \$100\\ &= \frac{\$44}{\$550} \times \$100\\ &= \$8.00\end{aligned}$$

Step 2 Find the row for 24 payments and move across the row until you find the number closest to $8.00. In this case, it is exactly $8.00.

Step 3 Move to the top of the column to get the APR. It is 7.5%.

▼ Try This One 3

☑ 2. Use a table to find APR for a loan.

Darla Connor purchased a Mercury Mariner for $19,900. Her down payment was $3,000. She financed the balance at $623 per month for 30 months. Find the APR.

One way to save money on a fixed installment loan is to pay it off early. This will allow a buyer to avoid paying the entire finance charge. The amount of the finance charge that is saved when a loan is paid off early is called **unearned interest.** There are two methods for calculating unearned interest, the **actuarial method** and the **rule of 78.** The actuarial method uses the APR table, and the following formula:

Math Note

The value h is found using Table 8-1: it is the entry in the row with the number of remaining payments and the column matching the loan's APR.

The Actuarial Method

$$u = \frac{kRh}{100 + h}$$

where u = unearned interest
k = number of payments remaining, excluding the current one
R = monthly payment
h = finance charge per \$100 for a loan with the same APR and k monthly payments

EXAMPLE 4 Using the Actuarial Method

Our friend Burk from Example 3 decides to use part of his tax refund to pay off the full amount of his laser printer with his 12th payment. Find the unearned interest and the payoff amount.

SOLUTION

To use the formula for the actuarial method, we'll need values for k, R, and h. Half of the original 24 payments will remain, so $k = 12$. From Example 3, the monthly payment is \$24.75 and the APR is 7.5%. Using Table 8-1, we find the row for 12 payments and the column for 7.5%; the intersection shows \$4.11, so $h = \$4.11$. Now we substitute those values into the formula:

$$u = \frac{kRh}{100 + h} \qquad k = 12,\ R = 24.75,\ h = 4.11$$

$$= \frac{(12)(24.75)(4.11)}{100 + 4.11} \qquad \text{Multiply numerator and denominator separately.}$$

$$= \frac{1{,}220.67}{104.11} \approx 11.72$$

The unearned interest is \$11.72.

The payoff amount is the amount remaining on the loan minus unearned interest. At this point, Burk has made 11 payments, so there would be 13 remaining if he were not paying the loan off early.

$$\text{Payoff amount} = 13 \times \$24.75 - \$11.72$$
$$= \$310.03$$

With a payment of \$310.03, Burk is the proud owner of a laser printer.

▼ Try This One 4

The buyer in Try This One 3 decided to pay off her car loan in 24 months instead of 30. Use the actuarial method to find the unearned interest and the payoff amount.

The next example will show how to find the unearned interest and payoff amount of a fixed installment loan using the rule of 78.

Sidelight MONEY FOR NOTHING?

Since most credit card companies either use the unpaid balance method, or don't charge interest on new purchases, if you pay off the full balance each month, you never pay any interest. In essence, you're getting to use the bank's money for nothing! But don't shed any tears for the poor banks. They charge the merchant a fee for each transaction. Also, the freedom that comes from buying stuff and walking out the door without handing over any money makes it extremely difficult for people to spend only what they can afford, especially young people using credit cards for the first time. Consider the following statistics:

- 76% of college undergrads have at least one credit card.
- 55% of credit card users keep a balance on their card.
- Total credit card debt in the United States in 2008 reached $951.7 *billion* dollars.
- The average credit card debt per borrower was $5,710.

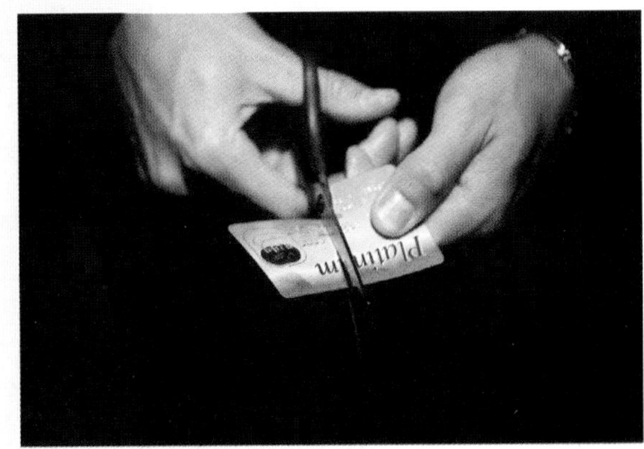

If you find yourself spending more than you can pay off, and having to keep a balance on your card, the best approach is probably the one pictured above.

EXAMPLE 7 Computing a Credit Card Finance Charge

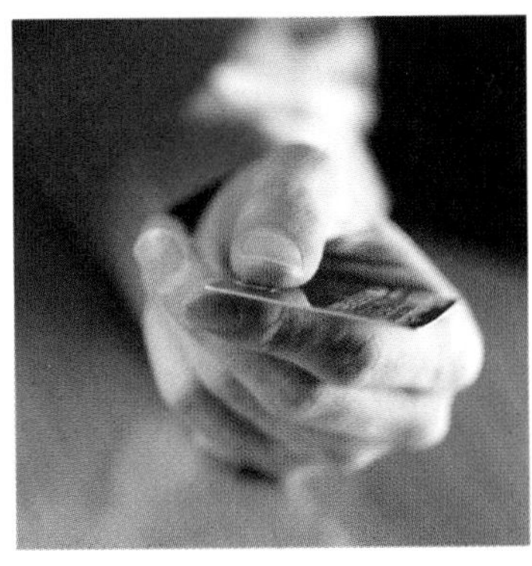

Betty's credit card statement showed the following transactions during the month of August.

August 1	Previous balance	$165.50
August 7	Purchases	59.95
August 12	Purchases	23.75
August 18	Payment	75.00
August 24	Purchases	107.43

(a) Find the average daily balance, the finance charge for the month, and the new balance on September 1. The interest rate is 1.5% per month on the average daily balance.

SOLUTION

Step 1 Find the balance as of each transaction.

August 1	$165.50
August 7	$165.50 + $59.95 = $225.45
August 12	$225.45 + $23.75 = $249.20
August 18	$249.20 − $75.00 = $174.20
August 24	$174.20 + $107.43 = $281.63

Step 2 Find the number of days for each balance.

Date	Balance	Days	Calculations
August 1	$165.50	6	(7 − 1 = 6)
August 7	$225.45	5	(12 − 7 = 5)
August 12	$249.20	6	(18 − 12 = 6)
August 18	$174.20	6	(24 − 18 = 6)
August 24	$281.63	8	(31 − 24 + 1 = 8)

Math Note

Since the transaction period starts on August 24 and ends on the last day of the month, which must be included, we had to add 1 to the last period of days in step 2. The total number of days has to equal the number of days in the given month.

Step 3 Multiply each balance by the number of days, and add these products.

Date	Balance	Days	Calculations
August 1	\$165.50	6	\$165.50(6) = \$993.00
August 7	\$225.45	5	\$225.45(5) = \$1,127.25
August 12	\$249.20	6	\$249.20(6) = \$1,495.20
August 18	\$174.20	6	\$174.20(6) = \$1,045.20
August 24	\$281.63	8	\$281.63(8) = \$2,253.04
		31	\$6,913.69

Step 4 Divide the total by the number of days in the month to get the average daily balance.

$$\text{Average daily balance} = \frac{\$6{,}913.69}{31} \approx \$223.02$$

The average daily balance is \$223.02.

Step 5 Find the finance charge. Multiply the average daily balance by the rate, which is 1.5%, or 0.015.

$$\text{Finance charge} = \$223.02 \times 0.015 \approx \$3.35.$$

Step 6 Find the new balance. Add the finance charge to the balance as of the last transaction.

$$\text{New balance:} \quad \$281.63 + \$3.35 = \$284.98$$

The average daily balance is \$223.02. The finance charge is \$3.35, and the new balance is \$284.98.

The procedure for finding the average daily balance is summarized next.

Procedure for the Average Daily Balance Method

Step 1 Find the balance as of each transaction.

Step 2 Find the number of days for each balance.

Step 3 Multiply the balances by the number of days and find the sum.

Step 4 Divide the sum by the number of days in the month.

Step 5 Find the finance charge (multiply the average daily balance by the monthly rate).

Step 6 Find the new balance (add the finance charge to the balance as of the last transaction).

5. Compute credit card finance charges using the average daily balance method.

▼ Try This One 7

A credit card statement for the month of November showed the following transactions.

November 1	Previous balance	\$937.25
November 4	Purchases	\$531.62
November 13	Payment	\$400.00
November 20	Purchases	\$89.95
November 28	Payment	\$100.00

(a) Find the average daily balance.
(b) Find the finance charge. The interest rate is 1.9% per month on the average daily balance.
(c) Find the new balance on December 1.

It's definitely a good idea to check with your credit card company to see what method it uses for computing finance charges. If it uses average daily balance, you will end up paying interest even if you pay off the full amount each month. There's a lot of competition out there, so if that's the case, you can probably find another card with a better deal.

Answers to Try This One

1 Amount financed, $7,200; total installment price, $10,720; finance charge, $2,520

2 (a) $1,299 (b) $779.40 (c) $2,278.40 (d) $34.64

3 8.0%

4 Unearned interest, $85.83; Payoff amount, $4,275.17

5 $88

6 Finance charge = $16.93; balance on Feb. 1 = $1,046.29

7 (a) $1,198.69 (b) $22.78 (c) $1,081.60

EXERCISE SET 8-4

Writing Exercises

1. Explain what is meant by the term "installment loan."
2. What is a finance charge?
3. What is the difference between the purchase price of an item and the amount financed?
4. What is the difference between closed-ended credit and open-ended credit?
5. What is the annual percentage rate (APR) for a loan? Why is it typically different than the stated interest rate?
6. What is the difference between the unpaid balance method and the average daily balance method for computing interest? Which is better for the consumer?

Real-World Applications

7. Mary Lee purchased a stove for $460. She made a down payment of $60 and paid $42 a month for 10 months. Find the total installment price of the stove.
8. Martin Dennis purchased a wide screen television for $1,720. He made a down payment of 15% and paid the balance over 18 months. The finance charge was 4% of the amount financed. Find the down payment and the installment price of the television and the monthly payment.
9. Joy Lansung purchased a microwave oven for $375.00. She made a down payment of 15% and financed the rest for 12 months with payments of $27.25. Find the down payment and the total installment price of the oven.
10. Mary Scherer purchased a wristwatch for $845.00. She made a down payment of $95.00 and financed the rest with six monthly payments of $128.75. Find the total installment price.
11. Hang Yo bought a treadmill for $925. He made a 20% down payment and financed the rest over 18 months. Find the monthly payment if the finance charge was 5% of the amount financed.
12. Stacie Howard purchased a water softener for $550. She made a down payment of $75 and paid the balance off in 12 monthly payments. Find the monthly payment if the interest rate was 11%.
13. Richard Johnston bought a 2010 Hummer for $39,905. He made a down payment of $15,000 and paid $614 monthly for 4 years. Find the APR.
14. Jennifer Siegel bought a new Ford Focus for $14,400. She made a down payment of $4,000 and made monthly payments of $319 for 3 years. Find the APR.
15. Erin LaRochelle bought four comforter sets for $900.00. She made a down payment of $100.00 and paid off the balance in 12 monthly payments of $71.05. Find the APR.

16. Ben King bought his wife a turquoise bracelet, earrings, and pendant for her birthday. He paid $1,125 for the set and had a down payment of $175. He paid the balance with 12 monthly payments of $84. Find the APR.
17. Matt Brawn bought a diamond engagement ring for $2,560. His down payment was $600, and he made 24 monthly payments of $90. Find the APR.
18. Mary Sinclair purchased a preowned Corvette for $27,900.00. She made a down payment of $8,000.00 and financed the rest over 5 years with monthly payments of $389.50. Find the APR.
19. In Exercise 13, Richard was able to pay off his loan at the end of 30 months. Using the actuarial method, find the unearned interest and payoff amount.
20. In Exercise 14, Jennifer decided to pay off her loan at the end of 2 years. Using the actuarial method, find the unearned interest and payoff amount.
21. In Exercise 15, Erin decided to pay off her loan at the end of 6 months. Using the actuarial method, find the unearned interest and the payoff amount.
22. In Exercise 16, Ben was able to pay off his loan at the end of 6 months. Using the actuarial method, find the unearned interest and the payoff amount.
23. In Exercise 17, Matt was able to pay off his loan at the end of 18 months. Using the actuarial method, find the unearned interest and the payoff amount.
24. In Exercise 18, Mary decided to pay off her loan at the end of 3 years. Using the actuarial method, find the unearned interest and the payoff amount.

For Exercises 25–30, use the rule of 78.

25. A $4,200.00 loan is to be paid off in 36 monthly payments of $141.17. The borrower decides to pay off the loan after 20 payments have been made. Find the amount of interest saved.
26. Fred borrowed $150.00 for 1 year. His payments are $13.75 per month. If he decides to pay off the loan after 6 months, find the amount of interest that he will save.
27. Greentree Limousine Service borrowed $200.00 to repair a limousine. The loan was to be paid off in 18 monthly installments of $13.28. After a good season, they decide to pay off the loan early. If they pay off the loan after 10 payments, how much interest do they save?
28. Household Lighting Company borrowed $600.00 to purchase items from another store that was going out of business. The loan required 24 monthly payments of $29.50. After 18 payments were made, the company decided to pay off the loan. How much interest was saved?
29. Lydia needed to have her roof repaired. She borrowed $950.00 for 10 months. The monthly payments were $99.75 each. After seven payments, she decided to pay off the balance of the loan. How much interest did she save?
30. The owners of Scottdale Village Inn decided to remodel the dining room at a cost of $3,250.00. They borrowed the money for 1 year and repaid it in monthly payments of $292.50. After eight payments were made, the owners decided to pay off the loan. Find the interest saved.
31. For the month of January, Juan had an unpaid balance on a credit card statement of $832.50 at the beginning of the month and made purchases of $675.00. A payment of $400.00 was made during the month. If the interest rate was 2% per month on the unpaid balance, find the finance charge and the new balance on February 1.
32. For the month of July, the unpaid balance on Sue's credit card statement was $1,131.63 at the beginning of the billing cycle. She made purchases of $512.58. She also made a payment of $750.00 during the month. If the interest rate was 1.75% per month on the unpaid balance, determine the finance charge and the new balance on the first day of the August billing cycle.
33. Sam is redecorating his apartment. On the first day of his credit card billing cycle, his balance was $2,364.79. He has recently made purchases totaling $1,964.32. He was able to make a payment of $1,000.00 during this billing cycle. If his interest rate is 1.67% per month on the unpaid balance, what is the finance charge and what will Sam's new balance be on the first day of the next billing cycle?
34. Janine has recently accepted a position with an upscale clothing store. On the first day of her March credit card billing cycle, her unpaid balance was $678.34. She has made clothing purchases totaling $3,479.03. She was able to make one payment of $525.00 during the billing cycle. If the interest rate is 2.25% per month on the unpaid balance, find the finance charge and the new balance on the first day of the April billing cycle.
35. Joe's credit card statement on the first day of the May billing cycle shows a balance of $986.53. During this billing cycle, he charged $186.50 to his account and made a payment of $775.00. At 1.35% interest per month on the unpaid balance, what is the finance charge? Also, find the balance on the first day of the next billing cycle.
36. Frank's credit card statement shows a balance of $638.19 on the first day of the billing cycle. If he makes a payment of $475.00 and charges $317.98 during this billing period, what will his finance charge be (the interest rate is 1.50% of the unpaid balance per month)? What will his beginning balance be at the start of the next billing cycle?
37. Mary's credit card statement showed these transactions during September:

September 1	Previous balance	$627.75
September 10	Purchase	$87.95
September 15	Payment	$200.00
September 27	Purchases	$146.22

(a) Find the average daily balance.
(b) Find the finance charge for the month. The interest rate is 1.2% per month on the average daily balance.
(c) Find the new balance on October 1.

38. Pablo's credit card statement showed these transactions during March:

March 1	Previous balance	\$2,162.56
March 3	Payment	\$800.00
March 10	Purchases	\$329.27
March 21	Payment	\$500.00
March 29	Purchases	\$197.26

(a) Find the average daily balance.
(b) Find the finance charge for the month. The interest rate is 2% per month on the average daily balance.
(c) Find the new balance on April 1.

39. Mike's credit card statement showed these transactions during the month of June:

June 1	Previous balance	\$157.95
June 5	Purchases	\$287.62
June 20	Payment	\$100.00

(a) Find the average daily balance.
(b) Find the finance charge for the month. The interest rate is 1.4% per month on the average daily balance.
(c) Find the new balance on July 1.

40. Charmaine's credit card statement showed these transactions during the month of December:

December 1	Previous balance	\$1,325.65
December 15	Purchases	\$287.62
December 16	Purchases	\$439.16
December 22	Payment	\$700.00

(a) Find the average daily balance.
(b) Find the finance charge for the month. The interest rate is 2% per month on the average daily balance.
(c) Find the new balance on January 1.

41. Ruth's credit card statement showed these transactions for the month of July:

July 1	Previous balance	\$65.00
July 2	Purchases	\$720.25
July 8	Payment	\$500.00
July 17	Payment	\$100.00
July 28	Purchases	\$343.97

(a) Find the average daily balance.
(b) Find the finance charge for the month. The interest rate is 1.1% per month.
(c) Find the new balance on August 1.

42. Tamera's credit card statement showed these transactions for the month of September:

September 1	Previous balance	\$50.00
September 13	Purchases	\$260.88
September 17	Payment	\$100.00
September 19	Purchases	\$324.15

(a) Find the average daily balance.
(b) Find the finance charge for the month. The interest rate is 1.9% per month on the average daily balance.
(c) Find the new balance on October 1.

Critical Thinking

43. Find the principal on a loan at 8% for 4 years when the monthly payments are \$100 per month.
44. A couple borrowed \$800 for 1 year at 12% interest. Payments were made monthly. After eight payments were made, they decided to pay it off. Find the interest that was saved if it was computed equally over 12 months. Then find the interest saved using the rule of 78. Explain which is a better deal for the borrower.

Section 8-5 Home Ownership

For many people, the day they buy their first home is one of the proudest days of their life—and one of the scariest. There is nothing that compares to the feeling of looking at a house and knowing that it's all yours. But the buying process is tremendously intimidating. There are dozens of documents to sign, and the sheer numbers involved are enough to make almost everyone wonder if they're making a colossal mistake.

For the very few among us who pay cash for a house, it's not as big an issue. But almost everyone will secure a large loan to buy a home, and the most common home loans are paid over a 30-year span. That's a major commitment, and one that nobody

LEARNING OBJECTIVES

- ❑ 1. Find a monthly mortgage payment using a payment table.
- ❑ 2. Find the total interest on a home loan.
- ❑ 3. Compare two mortgages with different lengths.
- ❑ 4. Find a monthly mortgage payment using a formula.
- ❑ 5. Make an amortization schedule for a home loan.

should enter into without an understanding of the mathematics that go into the process. In this section, we will study that math, hopefully helping you to become a well-informed home buyer.

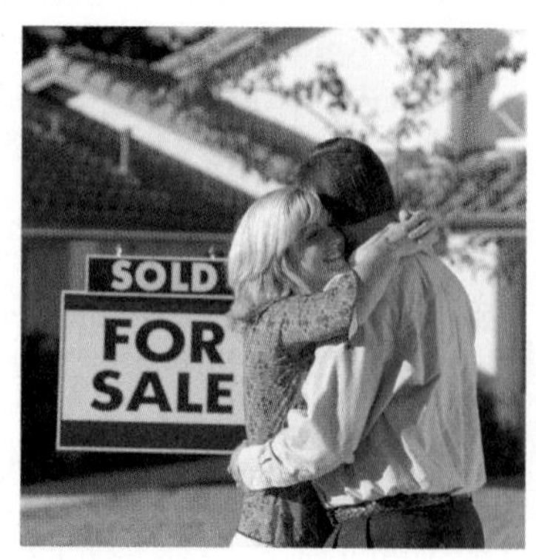

Mortgages

A **mortgage** is a long-term loan where the lender has the right to seize the property purchased if the payments are not made. Homes are the most common items bought using mortgages. The most common mortgage term is 30 years, but they are widely available in terms from 15 to as many as 50 years.

There are several types of mortgages. **A fixed-rate mortgage** means that the rate of interest remains the same for the entire term of the loan. The payments (usually monthly) stay the same. An **adjustable-rate mortgage** means that the rate of interest may fluctuate (i.e., increase and decrease) during the period of the loan. Some lending institutions will allow you to make **graduated payments.** This means that even though the interest does not change for the period of the loan, you can make smaller payments in the first few years and larger payments at the end of the loan period.

Finding Monthly Payments and Total Interest

One way to find the monthly payments for a fixed-rate mortgage is to use a table like Table 8-2 below. The table displays the monthly payment required for each $1,000 of a mortgage.

TABLE 8-2 Monthly Payment per $1,000 of Mortgage (Includes Principal and Interest)

	Number of Years					
Rate (%)	15	20	25	30	35	40
3.5	$7.15	$5.80	$5.01	$4.49	$4.13	$3.87
4	7.40	6.06	5.28	4.77	4.43	4.18
4.5	7.65	6.33	5.56	5.07	4.73	4.50
5	7.91	6.60	5.85	5.37	5.05	4.82
5.5	8.17	6.88	6.14	5.68	5.37	5.16
6	8.44	7.16	6.44	6.00	5.70	5.50
6.5	8.71	7.46	6.75	6.32	6.04	5.85
7	8.99	7.75	7.07	6.65	6.39	6.21
7.5	9.27	8.06	7.39	6.99	6.74	6.58
8	9.56	8.36	7.72	7.34	7.10	6.95
8.5	9.85	8.68	8.05	7.69	7.47	7.33
9	10.14	9.00	8.39	8.05	7.84	7.71
9.5	10.44	9.32	8.74	8.41	8.22	8.10
10	10.75	9.65	9.09	8.78	8.60	8.49
10.5	11.05	9.98	9.44	9.15	8.98	8.89
11	11.37	10.32	9.80	9.52	9.37	9.28

Math Note

The word "mortgage" comes from a combination of Old French words "mort" (dead) and "gage" (pledge). It is believed the intent was that the debtor pledged the property to secure the loan, and if he or she failed to pay, the property was taken, and was therefore "dead" to the debtor.

In Example 1, we'll find the monthly payment on a mortgage using the table and the procedure that follows:

Math Note

A Google search for the phrase "mortgage calculator" results in over 20 million hits! There are thousands of pages available that can be used to quickly calculate monthly payments on a mortgage. But doing the calculations here will help you to become very familiar with the terms involved in mortgages.

Procedure for Finding the Monthly Payment for a Fixed-Rate Mortgage

Step 1 Find the down payment.

Step 2 Subtract the down payment from the cost of the home to find the principal of the mortgage.

Step 3 Divide the principal by 1,000.

Step 4 Find the number in the table that corresponds to the interest rate and the term of the mortgage.

Step 5 Multiply that number by the number obtained in step 3 to get the monthly payment.

EXAMPLE 1 Finding Monthly Mortgage Payments

Math Note

You can reduce steps 1 and 2 to a single calculation. In Example 1, a 20% down payment leaves 80% to be financed: 80% of $174,900 = $139,920.

The Petteys family plans to buy a home for $174,900, and have been offered a 30-year mortgage with a rate of 5.5% if they make a 20% down payment. What will their monthly payment be with this loan?

SOLUTION

Step 1 Find the down payment.

$$20\% \text{ of } \$174,900 = 0.20 \times \$174,900 = \$34,980$$

Step 2 Subtract the down payment from the cost of the home to get the principal.

$$\$174,900 - \$34,980 = \$139,920$$

Step 3 Divide by 1,000.

$$\frac{\$139,920}{1,000} = 139.92$$

Step 4 Find the value in Table 8-2 for a 30-year mortgage at 5.5%. It is $5.68.

Step 5 Multiply the value from step 3, 139.92, by $5.68.

$$139.92 \times \$5.68 \approx \$794.75$$

The monthly payment is $794.75.

▼ Try This One 1

☑ 1. Find a monthly mortgage payment using a payment table.

The Trissel family agreed on a price of $229,500 for a home. Their company credit union offers a 5.0% 20-year loan with 15% down. Calculate the monthly payment.

It's an eye-opening experience to calculate the total interest on a mortgage. To do so, multiply the monthly payments by the total number of payments and then subtract the principal.

EXAMPLE 2 Finding Total Interest on a Mortgage

Find the total amount of interest the Petteys family would pay if they take the loan in Example 1.

SOLUTION

On a 30-year mortgage, there are $30 \times 12 = 360$ payments. We found that the monthly payment would be \$794.75.

$$\$794.75 \times 360 = \$286{,}110$$

This is the total of payments. We subtract the amount financed from Example 1:

$$\$286{,}110 - \$139{,}920 = \$146{,}190 \quad \textit{Interest on the loan.}$$

The interest paid exceeds the principal of the loan by over \$6,000!

▼ Try This One 2

☑ 2. Find the total interest on a home loan.

Find the total interest paid on the loan in Try This One 1.

Not surprisingly, the length of a loan has a profound effect on how much interest is paid. In the next example, we'll weigh the amount of extra monthly payment required versus the amount of interest saved.

EXAMPLE 3 Comparing Mortgages with Different Terms

Suppose that the Petteys family from Examples 1 and 2 is also offered a 15-year mortgage with the same rate and down payment. Find the difference in monthly payment and interest paid between the 15- and 30-year mortgages.

SOLUTION

We essentially need to rework Examples 1 and 2 with a 15-year mortgage, then compare the results. Fortunately, some of the work we did carries over. We know that the principal is \$139,920, and the principal divided by 1,000 is 139.92. This time we use the 15-year column and 5.5% row in Table 8-2 to get \$8.17. Now we multiply that by 139.92:

$$139.92 \times \$8.17 \approx \$1{,}143.15 \quad \textit{Monthly payment with 15-year term.}$$

The difference in monthly payments is

$$\$1{,}143.15 - \$794.75 = \$348.40 \quad \textit{\$794.75 was payment for 30 years.}$$

With a monthly payment of \$1,143.14 for 15 years (which is 180 months) the total payments are

$$\$1{,}143.15 \times 180 = \$\ 205{,}767.00$$

and the interest paid is

$$\$205{,}767 - \$139{,}920 = \$65{,}847$$

The interest paid on the 30-year mortgage was \$146,190:

$$\$146{,}190 - \$65{,}847 = \$80{,}343$$

If the Petteys family can manage an extra \$348.40 a month, they will save over \$80,000 in interest!

▼ Try This One 3

☑ 3. Compare two mortgages with different lengths.

If the Trissel family from Try This One 1 chooses a 15-year mortgage instead of 20, find the increase in monthly payment and total interest saved.

There is a formula for computing monthly payments on a mortgage that can be used in place of Table 8-2. It can also be used for interest rates or terms not included in the table.

Formula for Computing Monthly Payments on a Mortgage

$$R = \frac{P\left(\frac{r}{n}\right)}{1 - \left(1 + \frac{r}{n}\right)^{-nt}}$$

R = regular monthly payment

P = amount financed, or principal

r = rate written as a decimal

n = number of payments per year

t = number of years

CAUTION

Obviously, the payment formula is pretty complicated. We will have to be extra careful in doing the calculations and take it in stages.

Sidelight THE GREAT MORTGAGE CRISIS

Throughout much of the first decade of the 21st century, homeowners were very happy people. Housing prices were rising at an almost unprecedented rate, and people watched the value of their investment soar. New homes were being built everywhere you looked, and lending institutions were practically climbing over each other to hand out home loans.

Then a funny thing happened—the housing market got oversaturated and prices started to fall. At the same time, people who took adjustable-rate mortgages to buy larger houses had their rates go up and couldn't make their payments anymore. Foreclosures (when the lending institution takes back a home) started to rise, causing even more houses to go on the market. Soon, the whole house of cards came crashing down, taking the U.S. economy with it.

There is plenty of blame to go around, but put in its simplest terms, the blame is to be shared equally between home buyers and lenders. Millions of people bought homes they couldn't afford with adjustable-rate mortgages, and the lenders gave out loans to millions of people who couldn't afford them. The result was an economic bust that had cost over 7 million jobs by mid-2009. One lesson has been learned from this mess—lenders will be far less likely to put people in homes that they can't afford from now on.

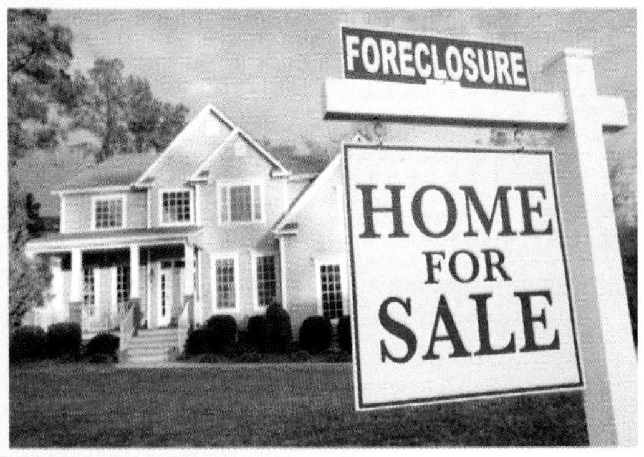

EXAMPLE 4 Finding a Monthly Payment Using the Formula

Calculator Guide

The keystrokes for the final calculation in Example 4 are:

Standard Scientific Calculator

8500 [÷] [(] 1 [−] 1.005 [y^x] 360 [±] [)] [=]

Standard Graphing Calculator

8500 [÷] [(] 1 [−] 1.005 [^] [(−)] 360 [)] [Enter]

After one hit single, a young singer unwisely decides that she needs a $2.2 million dollar mansion. With some of the proceeds from her CD, she puts down $500,000, leaving $1,700,000 to finance at 6% for 30 years. Find her monthly payment.

SOLUTION

In the formula above, use $P = 1{,}700{,}000$, $r = 0.06$, $n = 12$, and $t = 30$.

$$R = \frac{P\left(\frac{r}{n}\right)}{1 - \left(1 + \frac{r}{n}\right)^{-nt}}$$

$$R = \frac{1{,}700{,}000\left(\frac{0.06}{12}\right)}{1 - \left(1 + \frac{0.06}{12}\right)^{-12 \cdot 30}}$$

Multiply in numerator. $1 + \frac{0.06}{12} = 1.005$ (in some cases, rounding might by necessary here); $-12 \cdot 30 = -360$.

$$= \frac{8{,}500}{1 - (1.005)^{-360}}$$

See Calculator Guide.

$$\approx \$10{,}192.36$$

The monthly payment is $10,192.36, and the singer better hope her next CD does well, too.

▼ Try This One 4

Use the payment formula to find the monthly payment on the one-hit-wonder pop star's second house, a $120,000 mortgage at 5.2% for 15 years.

☑ 4. Find a monthly mortgage payment using a formula.

Computing an Amortization Schedule

After securing a mortgage, the lending institution will prepare an **amortization schedule.** This schedule shows what part of the monthly payment is paid on the principal and what part of the monthly payment is paid in interest.

In order to prepare an amortization schedule, the next procedure can be used.

Math Note

Be sure to subtract any down payment from the cost of the home before beginning the amortization table.

Procedure for Computing an Amortization Schedule

Step 1 Find the interest for the first month. Use $I = Prt$, where $t = \frac{1}{12}$. Enter this value in a column labeled Interest.

Step 2 Subtract the interest from the monthly payment to get the amount paid on the principal. Enter this amount in a column labeled Payment on Principal.

Step 3 Subtract the amount of the payment on principal found in step 2 from the principal to get the balance of the loan. Enter this in a column labeled Balance of Loan.

Step 4 Repeat the steps using the amount of the balance found in step 3 for the new principal.

EXAMPLE 5 Preparing an Amortization Schedule

Compute the first two months of an amortization schedule for the loan in Example 1.

Math Note

Preparing an amortization schedule makes it easy to see why more interest is paid earlier in a fixed installment loan, and more principal is paid later.

SOLUTION

The value of the mortage is \$139,920, the interest rate is 5.5%, and the monthly payment is \$794.75.

Step 1 Find the interest for month 1.

$$\begin{aligned} I &= Prt \\ &= \$139{,}920 \times 0.055 \times \frac{1}{12} \\ &= \$641.30 \end{aligned}$$

Enter this in a column labeled Interest.

Step 2 Subtract the interest from the monthly payment.

$$\$794.75 - \$641.30 = \$153.45$$

This goes into the Payment on Principal column.

Step 3 Subtract principal payment from principal.

$$\$139{,}920 - \$153.45 = \$139{,}766.55$$

This goes into the Balance of Loan column. Now we repeat steps 1–3 using the balance of \$139,766.55.

Step 4 $I = Prt$

$$\begin{aligned} &= \$139{,}766.55 \times 0.055 \times \frac{1}{12} \\ &= \$640.60 \end{aligned}$$

Step 5 \$794.75 − \$640.60 = \$154.15

Step 6 \$139,766.55 − \$154.15 = \$139,612.40

The first 2 months of the amortization schedule are:

Payment Number	Interest	Payment on Principal	Balance of Loan
1	\$641.30	\$153.45	\$139,766.55
2	\$640.60	\$154.15	\$139,612.40

▼ Try This One 5

Make an amortization schedule for the first 3 months for the loan in Try This One 1.

☑ 5. Make an amortization schedule for a home loan.

Answers to Try This One

1 $1,287.50

2 $113,925

3 Increase in payment = $255.54; interest saved = $31,252.80

4 $961.50

5

Payment Number	Interest	Payment on Principal	Balance of Loan
1	$812.81	$474.69	$194,600.31
2	$810.83	$476.67	$194,123.64
3	$808.85	$478.65	$193,644.99

EXERCISE SET 8-5

Writing Exercises

1. What specifically makes a loan a mortgage?
2. Explain how to find the total interest paid on a mortgage.
3. What are the advantages and disadvantages of getting a mortgage with a shorter term?
4. What is an amortization schedule?

Real-World Applications

Use Table 8-2 for Exercises 5–12.

5. A house sells for $145,000 and a 15% down payment is made. A mortgage was secured at 7% for 25 years.
   (a) Find the down payment.
   (b) Find the amount of the mortgage.
   (c) Find the monthly payment.
   (d) Find the total interest paid.
6. A house sells for $182,500 and a 5% down payment is made. A mortgage is secured at 7% for 15 years.
   (a) Find the down payment.
   (b) Find the amount of the mortgage.
   (c) Find the monthly payment.
   (d) Find the total interest paid.
7. A building sells for $200,000 and a 40% down payment is made. A 30-year mortgage at 6% is obtained.
   (a) Find the down payment.
   (b) Find the amount of the mortgage.
   (c) Find the monthly payment.
   (d) Find the total interest paid.
8. An ice cream store sells for $125,000 and a 12% down payment is made. A 25-year mortgage at 7.5% is obtained.
   (a) Find the down payment.
   (b) Find the amount of the mortgage.
   (c) Find the monthly payment.
   (d) Find the total interest paid.
9. An auto parts store sells for $325,000 and a 10% down payment is made. A 40-year mortgage at 7.5% is obtained.
   (a) Find the down payment.
   (b) Find the amount of the mortgage.
   (c) Find the monthly payment.
   (d) Find the total interest paid.
10. A beauty shop sells for $175,000 and a 22% down payment is made. A 20-year mortgage at 6.5% is obtained.
   (a) Find the down payment.
   (b) Find the amount of the mortgage.
   (c) Find the monthly payment.
   (d) Find the total interest paid.

11. A computer store sells for $1,200,000. The buyer made a 30% down payment and secured a 20-year mortgage on the balance at 5.5%.
    (a) Find the down payment.
    (b) Find the amount of the mortgage.
    (c) Find the monthly payment.
    (d) Find the total interest paid.
12. A grocery store sells for $550,000 and a 25% down payment is made. A 40-year mortgage at 6% is obtained.
    (a) Find the down payment.
    (b) Find the amount of the mortgage.
    (c) Find the monthly payment.
    (d) Find the total interest paid.

For Exercises 13–20, use the monthly payment formula on page 461.

13. A house is purchased for $232,000 with a 15% down payment. A mortgage is secured at 6.75% for 20 years. Find the monthly payment.
14. A building sells for $330,000 with a down payment of 25%. Find the monthly payment on a 25-year mortgage at 5.3%.
15. A home is purchased for $163,000 with a 12% down payment. Find the monthly payment if the mortgage is 6.75% for 18 years.
16. A house sells for $289,000. The buyer made a 20% down payment and financed the balance with a 5.75% loan for 15 years. Find the monthly payment.
17. A store was purchased for $725,000 and the buyer made a 10% down payment. The balance was financed with a 6.35% loan for 27 years. Find the monthly payment.
18. A house was purchased for $162,000.00 with a 5% down payment. The balance was financed at 5.6% for 18 years. Find the monthly payment.
19. A pizza parlor was purchased for $327,000 with no down payment and a 6.7% loan for 10 years. Find the monthly payment.
20. A supermarket building was purchased for $375,000. The down payment was 10%. The balance was financed at 7.2% for 15 years. Find the monthly payment.
21. Compute an amortization schedule for the first 3 months for the house purchased in Exercise 5 of this section.
22. Compute an amortization schedule for the first 3 months for the house purchased in Exercise 6 of this section.
23. Compute an amortization schedule for the first 3 months for the computer store purchased in Exercise 11 of this section.
24. Compute an amortization schedule for the first 3 months for the grocery store purchased in Exercise 12 of this section.

Critical Thinking

25. You decide to buy a $180,000 home. If you make a 25% down payment, you can get a 20-year mortgage at 9%, but if you can make a 10% down payment, you can get a 25-year mortgage at 7%. Which is the better option for you?
26. Which mortgage would cost you less, a 30-year mortgage at 6.5% or a 15-year mortgage at 10%?

Section 8-6 Stocks and Bonds

We have seen that the magic of compound interest allows your money to grow considerably over long periods of time. However, interest rates on basic savings accounts are usually quite low, so if you rely on savings alone to build a nest egg, you better hope that you live a very, very long life indeed. Most successful investors grow their money much more quickly using the stock market. In this section, we'll learn about the basics of stocks and bonds, and how to get information about the performance of stocks that you might be interested in.

When a company files legal papers to become a corporation, it is able to issue **stock.** If an investor purchases shares of stock, he or she becomes a part owner of the company; for example, if a company issues 1,000 shares of stock and an investor

LEARNING OBJECTIVES

- ❑ 1. Find information from a stock listing.
- ❑ 2. Compute the P/E ratio for a stock.
- ❑ 3. Compute the total cost of a stock purchase.
- ❑ 4. Compute the profit or loss from a stock sale.

purchases 250 shares, then the investor owns one-quarter of the company. The investor is called a **shareholder.**

When a company makes money, it distributes part of the profit to its shareholders. This money is called a **dividend.** The stockholder receives a sum of money based on the number of shares of the stock that he or she owns. Sometimes if a company does not make a profit or its owners or managers decide to reinvest the money into the company, no dividends are paid.

Besides issuing stock, a company can also issue **bonds.** Usually bonds are issued to raise money for the company for start-up costs or special projects. A person who purchases a bond is really lending money to the company. The company, in turn, repays the owner of the bond its **face value** plus interest. As a general rule, bonds are a safer investment than stocks, but stocks have greater growth potential.

Stocks can be bought and sold on a **stock exchange.** The price of a stock varies from day to day (even from minute to minute) depending on the amount that investors are willing to pay for it. This can be affected by the profitability of the company, the economy, scandals, even global political concerns. Investors buy or sell stock through a **stockbroker.** Traditionally, this was an individual working for a brokerage firm, but it has become common for people to use online brokers, in which the investor initiates the buying and selling of stocks. In either case, the brokerage charges a fee, called a **commission,** for the service of having their representatives buy or sell the stock at an exchange. Bonds can also be bought and sold like stock.

Investors often own a combination of stocks and bonds. The set of all stocks and bonds owned is called an investor's **portfolio.** Sometimes a group of investors hire a manager to handle their investments. The manager invests in stocks and bonds, follows the activities of companies, and buys and sells in an attempt to achieve maximum profit for the group. This type of investment is called a **mutual fund.**

Stocks

In order to get information about a certain stock, you can refer to a stock table. These tables can be found in newspapers and online. The listings vary somewhat depending on the source. In this case, a stock listing for a company called Computer Programming and Systems, Inc. will be used as an example.

52 weeks								
HI	LO	STOCK	DIV	YLD%	P/E	VOL (1,000s)	CLOSE	NET CHG
31.00	17.07	CPSI	1.44	6.5%	16.8	1,244	22.25	+0.40

The first two columns give the highest and lowest selling prices for one share of stock in this company during the past 52 weeks. In this case, they are \$31.00 and \$17.07 respectively. The column labeled STOCK contains the letters CPSI. This is the symbol the company uses for trading. The column labeled DIV is the dividend per share that was paid to shareholders last year. In this case, it was \$1.44 per share. The column labeled YLD% is the annual percentage yield: It is the dividend per share divided by the current price. In this case, it is 6.5%. This percent can be compared to other stocks as a measure of performance. The P/E column is the price-to-earnings ratio. It is the ratio of yesterday's closing price of the stock (found in the CLOSE column) to its annual earnings per share. In this case, the closing price of the stock, \$22.25, is a bit less than 17 times the annual earnings per share. This concept will be explained in more detail after the first example.

The column labeled VOL (1,000s) means the number of shares in thousands that were traded yesterday. In this case, $1,244 \times 1,000 = 1,244,000$ shares were traded as of closing time. The column labeled NET CHG is the change in the price of the stock between the day before yesterday and yesterday at closing time. In this case, the value of the stock increased \$0.40. This tells us that the value of the stock the day before yesterday was \$22.25 − \$0.40 = \$21.85. Since the net change was positive, a + appears in the column. When . . . appears in this column, it means that there is no change.

EXAMPLE 1 Reading a Stock Listing

The following is a stock listing for the Terex Corporation. Use the listing to answer the questions.

52 weeks								
HI	LO	STOCK	DIV	YLD%	P/E	VOL (1,000s)	CLOSE	NET CHG
35	20.97	TEX	0.24	1.0	25	7,143	24.51	−0.06

(a) What was the highest price that the stock sold for during the past 52 weeks?
(b) What was the lowest price that the stock sold for during the past 52 weeks?
(c) What was the amount of the dividend per share that TEX paid last year?
(d) If you owned 250 shares of stock, how much did you make in dividends last year?
(e) How many shares were traded yesterday?
(f) What was the closing price per share the day before yesterday?

SOLUTION

(a) \$35.00 *Found in the "HI" column.*

(b) \$20.97 *Found in the "LO" column.*

(c) \$0.24 *Found in the DIV column.*

(d) $250 \times \$0.24 = \$60$ *\$0.24 dividend per share x 250 shares.*

(e) $7{,}143 \times 1{,}000 = 7{,}143{,}000$ *Number in VOL column x 1,000.*

(f) $\$24.51 + 0.06 = \$24.57$ *Closing price of \$24.51 is 0.06 below previous day.*

▼ Try This One 1

☑ 1. Find information from a stock listing.

The following is a stock listing for Wabtec Corporation. Using the listing to answer the questions.

52 weeks								
HI	LO	STOCK	DIV	YLD%	P/E	VOL (1,000s)	CLOSE	NET CHG
22.70	16.53	WAB	0.04	0.2%	27	4,870	21.97	+0.16

(a) What was the highest price that the stock sold for during the past 52 weeks?
(b) What was the lowest price that the stock sold for during the past 52 weeks?
(c) What was the amount of the dividend per share that Wabtec paid last year?
(d) If you owned 432 shares of stock, how much did you make in dividends last year?
(e) How many shares were traded yesterday?
(f) What was the closing price per share the day before yesterday?

CAUTION

When finding the number of shares traded, don't forget to multiply by the units given for VOL in the table. In Example 1, it is thousands.

P/E Ratio

The P/E ratio of a stock is a comparison of the current selling price to the company's earnings per share.

Math Note

The abbreviation P/E is used to remind you that this is a ratio of Price to Earnings.

Formula for the P/E ratio

$$\text{P/E ratio} = \frac{\text{Yesterday's closing price}}{\text{Annual earnings per share}}$$

The annual earnings per share is found by dividing a company's total earnings by the number of shares that are owned by the stockholders for the last year. The annual earnings per share for a stock is found by subtracting expenses, taxes, losses, etc. from the gross revenues. These figures can be found in a company's annual reports.

EXAMPLE 2 Computing a P/E Ratio

If the annual earnings per share for Terex is \$0.98, find the P/E ratio.

SOLUTION

$$\text{P/E ratio} = \frac{\text{Yesterday's closing price}}{\text{Annual earnings per share}}$$ *Closing price from Example 1.*

$$= \frac{\$24.51}{\$0.98} = 25 \text{ (rounded)}$$

▼ Try This One 2

☑ 2. Compute the P/E ratio for a stock.

The most recent annual earnings per share for Wabtec, the company in Try This One 1, was \$2.62. Find the updated P/E ratio for Wabtec.

Our answer to Example 2 means that the price of a share of stock is 25 times the company's annual earnings per share. If you divide \$1.00 by the P/E ratio 25, you get 0.04, which means that for every dollar you invest in the company by purchasing its stock, the company makes 4¢. This however, does not mean that the company pays a dividend of 4¢. The dividends paid are determined by the board of directors of the company, and they may want to use some of the profits for other purposes, such as expansion.

Another way of looking at the P/E ratio is that you are paying the company \$1.00 so it can make 4¢. Now if the P/E ratio for another company's stock is 20, then \$1.00 ÷ 20 = 5¢. This means that you are paying the company \$1.00 so that it can earn 5¢. Which is better? Obviously the investment in the second company is better. So in general, the lower the P/E ratio is, the better the investment, but there are many other factors to consider. Also remember that since the price of a company's stock is constantly changing, the P/E ratio also changes.

Knowing the price per share of stock and the P/E ratio, you can find the annual earnings per share for the last 12 months by using the following formula:

Formula for Annual Earnings per Share

$$\text{Annual earnings per share} = \frac{\text{Yesterday's closing price}}{\text{P/E ratio}}$$

Math Note

This formula is obtained from solving the P/E ratio formula for earnings.

EXAMPLE 3 Computing Annual Earnings per Share

If the closing price for Kellogg's stock was \$44.23 and the P/E ratio is 15, find the annual earnings per share for last year.

SOLUTION

$$\text{Annual earnings per share} = \frac{\text{Yesterday's closing price}}{\text{P/E ratio}}$$

$$= \frac{\$44.23}{15} \approx \$2.95$$

The annual earnings per share for Kellogg's is \$2.95.

▼ Try This One 3

Find the annual earnings per share for a stock if yesterday's closing price was \$62.43 and the P/E ratio is 8.6.

The current yield for a stock can be calculated by using the following formula:

Formula for Current Yield for a Stock

$$\text{Current stock yield} = \frac{\text{Annual dividend per share}}{\text{Closing price of stock}}$$

EXAMPLE 4 Computing Yield for a Stock

For the CPSI stock from page 466, the annual percent yield is 6.5%. Verify the current yield by using the preceding formula.

SOLUTION

The dividend per share is \$1.44 and the closing price is \$22.25:

$$\text{Current yield} = \frac{\text{Annual dividend per share}}{\text{Closing price of stock}}$$

$$= \frac{\$1.44}{\$22.25} = 0.065 \text{ (rounded)} = 6.5\%$$

After rounding, the figure agrees with the 6.5% shown in the listing.

▼ Try This One 4

Find the current yield for a stock if the annual dividend per share is \$1.51 and the closing price of the stock is \$30.15.

There are two ways to make money from stocks: buy shares of a stock that pays dividends, or buy stock at a low price and sell it at a higher price. But of course, you can't just go buy stock at the corner store—you need to use a brokerage firm, placing an order which is then carried out by representatives at the stock exchange. In exchange for that service, the broker charges a commission, which varies among brokers. Brokers can also make recommendations concerning what stocks to buy and sell, which further justifies their commissions.

The amount that an investor receives from the sale of a stock is called the **proceeds.** The proceeds are equal to the amount of the sale minus the broker's commission. The next two examples illustrate the buying and selling of stocks.

EXAMPLE 5 Finding the Total Cost of Buying Stock

Shares of Apple Computer (AAPL) closed at \$12.89 on April 1, 2004. Suppose that an investor bought 600 shares at that price using a broker that charged a 2% commission. Find the amount of commission and the total cost to the investor.

SOLUTION

Step 1 Find the purchase price.

$$600 \text{ shares} \times \$12.89 = \$7{,}734.00$$

Step 2 Find the broker's commission.

$$\begin{aligned} 2\% \text{ of purchase price} &= 0.02 \times \$7{,}734.00 \\ &= \$154.68 \end{aligned}$$

Step 3 Add the commission to the purchase price.

$$\$7{,}734.00 + \$154.68 = \$7{,}888.68$$

The investor paid a total of \$7,888.68 for the transaction.

▼ Try This One 5

☑ 3. Compute the total cost of a stock purchase.

Apple closed at \$38.45 on January 3, 2005. If 250 shares were bought at that price through a broker with a 1.6% commission, find the commission and total cost to the investor.

EXAMPLE 6 Finding the Amount Made from Selling Stock

On May 1, 2008, shares of Apple stock reached \$192.24. If the investor in Example 5 sold all of his Apple stock at that point, and the broker also charges a 2% commission on sales, find the commission, proceeds, and the amount of profit made by the investor.

SOLUTION

Step 1 Find the total amount of the sale.

$$600 \text{ shares} \times \$192.24 = \$115{,}344.00$$

Step 2 Find the commission.

$$2\% \text{ of } \$115{,}344.00 = 0.02 \times \$115{,}344.00 = \$2{,}306.88$$

Step 3 Subtract the commission amount from the total amount of the sale to get the proceeds.

$$\$115{,}344.00 - \$2{,}306.88 = \$113{,}037.12$$

Step 4 The profit is the proceeds minus the total cost from Example 5.

$$\$113{,}037.12 - \$7{,}888.68 = \$105{,}148.44$$

Math Note

Stockbrokers charge commission for both buying and selling your stock, so you end up paying on both ends when investing in stock.

▼ Try This One 6

When the market tanked at the end of 2008, Apple stock reached a low of \$78.20 on January 20, 2009. If the investor in Try This One 5 sold all of his Apple stock that day, and the broker charges 1.6% for sales, find the commission, proceeds, and profit made.

☑ 4. Compute the profit or loss from a stock sale.

Bonds

When an investor purchases bonds, the investor is lending money to the company that issues the bonds. Bonds can also be issued by local, state, and federal governments. Bonds have a **face value,** which is usually \$1,000, and a fixed interest rate. Bonds can be sold just like stocks and the prices vary with the market conditions. Bonds also have a **maturity date,** which is the date that they come due. Bonds are listed in tables that are similar to stock tables. Brokers also earn commissions for buying and selling bonds. Investment procedures for bonds are similar to those for stocks.

Mutual Funds

Many times investors purchase a group of stocks and bonds called a **mutual fund.** Mutual funds are managed by professional managers and include money from other investors. The manager follows the markets and makes the decisions of when to buy or sell the stocks and bonds. Mutual funds usually consist of a large number of small investments in companies. This way, if a single stock does not perform well, only a small amount of money is lost. Sometimes mutual funds can be high return but also high risk.

Ratings for mutual funds can be found on most financial websites, and in business publications like the *Wall Street Journal*. They are rated either from A to F, or from 5 to 1, with 5 being the best. Often, two separate ratings are given. An overall rating compares the fund to all other stock funds. A category rating compares a fund to other funds that have similar holdings. For example, there are funds that invest strictly in smaller businesses, and it makes sense to compare those funds to others like them, as well as to the market as a whole.

Answers to Try This One

1 (a) \$22.70
(b) \$16.53
(c) \$0.04
(d) \$17.28
(e) 4,870,000
(f) \$21.81

2 8.4 (rounded)

3 \$7.26

4 5.0%

5 Commission = \$153.80, total cost = \$9,766.30

6 Commission = \$312.80, proceeds = \$19,237.20, profit = \$9,470.90

EXERCISE SET 8-6

Writing Exercises

1. Explain in your own words what stock is.
2. What's the difference between stocks and bonds?
3. What is a mutual fund?
4. What is meant by the term P/E ratio?
5. What does a stockbroker do?
6. When selling stock, what's the difference between sale price, proceeds, and profit or loss?

Real-World Applications

Use the following information about Sunoco stock for Exercises 7–16.

52 weeks								
HI	LO	STOCK	DIV	YLD%	P/E	VOL (1,000s)	CLOSE	NET CHG
97.25	57.50	SUN	1.23	1.6	7	4,626	62.06	+0.77

7. What is the highest price that the stock sold for during the last 52 weeks?
8. What was the lowest price that the stock sold for during the last 52 weeks?
9. What was the amount of the dividend per share that the company paid last year?
10. If you own 175 shares, how much in dividends did you make last year?
11. How many shares were traded yesterday?
12. What was the closing price of the stock yesterday?
13. Find the annual earnings per share.
14. If you purchase 480 shares of Sunoco stock at \$62.06 per share and the broker's commission is 1.5%, find the total cost of the purchase.
15. If an investor had 623 shares of Sunoco stock last year and the dividend per share was \$1.23 last year, how much did the investor receive?
16. What was the closing price per share of stock the day before yesterday?

Use the following information about Wabtec stock for Exercises 17–26.

52 weeks								
HI	LO	STOCK	DIV	YLD%	P/E	VOL (1,000s)	CLOSE	NET CHG
40.08	24.75	WAB	0.04	0.1	20	345	29.79	+0.39

17. What was the highest price that the stock sold for during the last 52 weeks?
18. What was the lowest price that the stock sold for during the last 52 weeks?

19. What was the amount of the dividend per share that the company paid last year?
20. If you own 357 shares, how much in dividends did you make last year?
21. How many shares were traded yesterday?
22. What was the closing price of the stock yesterday?
23. Find the annual earnings per share.
24. If you purchase 1,247 shares of stock at the closing price and the broker's commission is 2.6%, find the total cost of the purchase.
25. If an investor owned 1,562 shares of Wabtec and the dividend per share was $0.10, how much income did the investor receive?
26. What was the closing price of the stock the day before yesterday?

Use the following information about Wal-Mart stock for Exercises 27–36.

52 weeks								
HI	LO	STOCK	DIV	YLD%	P/E	VOL (1,000s)	CLOSE	NET CHG
50.87	42.31	WMT	0.67	1.4	19	9,662	48.12	−0.10

27. What was the highest price that the stock sold for during the last 52 weeks?
28. What was the lowest price that the stock sold for during the last 52 weeks?
29. What was the amount of the dividend per share that the company paid last year?
30. If you own 682 shares, how much in dividends did you make last year?
31. How many shares were traded yesterday?
32. What was the closing price of the stock yesterday?
33. Find the annual earnings per share.
34. If you purchase 842 shares of Wal-Mart stock at $52.67 per share and the broker's commission is 2%, find the total cost of the purchase.
35. If an investor had 1,225 shares of Wal-Mart stock and the dividend per share was $ 0.67 last year, how much did the investor make?
36. What was the closing price of the stock the day before yesterday?
37. If the closing price of a stock is $21.92 and the annual earnings per share is $0.88, find the P/E ratio.
38. If the closing price of a stock is $6.65 and the annual earnings per share is $0.35, find the P/E ratio.
39. If the closing price of a stock is $24.19 and the annual earnings per share is $1.61, find the P/E ratio.
40. If the closing price of annual DirecTV stock is $20.18 and the annual earnings per share is $1.06, find the P/E ratio.
41. If the closing price of Gaither stock is $18.53 and the P/E ratio is 55, find the annual earnings per share.
42. If the closing price of Jacob Energy is $75.66 and the P/E ratio is 25, find the annual earnings per share.
43. If the closing price of Marine Max is $25.76 and the P/E ratio is 13, find the annual earnings per share.
44. If the closing price of Omnicare is $43.73 and the P/E ratio is 30, find the annual earnings per share.
45. An investor purchased 800 shares of stock for $63.25 per share and sold them later for $65.28 per share. The broker's commission was 2% of the purchase price and 2% of the selling price. Find the amount the investor made on the stock.
46. An investor purchased 200 shares of a stock at $93.75 per share and sold it later at $89.50 per share. The broker's commission on the purchase and sale of the stock is 2.5%. Find the amount of money the investor lost on the sale.
47. An investor purchased 550 shares of stock at $51.60 per share. She later sold it at $49.70. The broker's commission on the purchase was 2% and 1.5% on the sale. Find the amount of money the investor lost on the stock.
48. An investor purchased 670 shares of a stock at $73.20 per share. Then he sold the stock at $82.35. If the broker's commission was 2.5% on the purchase and sale of the stocks, how much money did the investor make on the transaction?

Critical Thinking

49. Compare the two investments below and decide which would have been the better choice.

    Investment 1: $10,000 was invested in a 24-month CD that earned 5.1% annual interest compounded daily.

    Investment 2: 1,400 shares of stock in the Lybarger Aviation Company were bought at $7.11 per share using a brokerage with a 0.75% commision rate on both buying and selling stock. Over the 2 years the stock was held, it paid a dividend of $0.48 per share in the first year and $0.36 per share in the second year. The stock was sold through the same brokerage for $7.95 per share.

CHAPTER 8 Summary

Section	Important Terms	Important Ideas
8-1	Percent Percent increase Percent decrease	**Percent** means "per hundred," or "hundredths." So 45% means 45 per hundred. In order to do calculations with percents, they must be changed to fractions or decimals. The word "of" is important in calculations with percents: the phrase "40% of 80 is 32" translates to the equation $0.40 \times 80 = 32$. This allows us to set up many percent calculations.
8-2	Interest Simple interest Principal Rate Term Future value Banker's rule Discounted loan	**When you** borrow money, you pay a fee for its use. This fee is called interest. Likewise, when you put money into a savings account, the bank pays interest for the use of your money. Simple interest is interest computed only as a percentage of the principal. The formula $I = Prt$ is used to compute simple interest. Future value is the sum of the principal and any interest earned.
8-3	Compound interest Effective rate Annual yield Annuity	**Compound interest** is interest calculated on both the principal and any interest previously earned. Compound interest investments earn more interest than simple interest investments at the same rate. Since the actual rate is higher when interest is compounded more than once per year, the true rate is called the effective rate or annual yield. An annuity is a savings plan where an individual or business makes the same payment each period into a compound interest account where the rate remains the same for the term of the annuity.
8-4	Fixed installment loan Finance charge Down payment Total installment price Annual percentage rate (APR) Payoff amount Actuarial method Rule of 78 Closed-ended credit Open-ended credit Unpaid balance method Average daily balance method	**A fixed** installment loan is a loan that is repaid in equal (usually monthly) payments. A down payment is a cash payment made on the purchase. Many times a finance charge is added to the amount financed. The total installment price is found by summing the monthly payments and adding the down payment. Because you pay back some of the principal each month, you do not have the full use of the money for the term of the loan. This means that the actual interest is higher than the stated interest rate. This actual interest rate is called the annual percentage rate and can be computed approximately by the constant ratio formula. When an installment loan is paid off early, the amount of interest saved can be determined by the rule of 78, or the actuarial method. Credit card companies also charge interest. There are two ways the companies compute interest. One method is computing interest on the unpaid balance. In this case you are charged interest only on last month's balance. The other method is called the average daily balance. Here the interest is computed on the average balance on all of the days of the month. This includes any purchases and payments made during the month.

8-5	Mortgage Fixed-rate mortgage Adjustable-rate mortgage Graduated payments Amortization schedule	**Because homes** cost so much, most people need a loan to buy one. A loan for which the property being bought is used as security against not making payments is called a mortgage. A table listing the amount of each payment going to pay interest, the amount toward principal, and the balance of the loan is called an amortization schedule.
8-6	Stock Shareholder Dividend Bond Face value Stock exchange Stockbroker Commission Portfolio Yield P/E ratio Proceeds Mutual fund	**Investors** can purchase stocks and bonds. A stock is a share of ownership in a company. A bond is actually a loan to a company. A mutual fund is a combination of stocks and bonds that is managed by a professional investor. Newspapers and websites show information about stocks and bonds by using tables. The tables show the 52-week high price and low price of the stocks and bonds. The table also shows the yield, the P/E ratio, the dividend, the volume of sales, the closing price, and the net change of a stock. P/E ratio is a comparison between the share price of a stock and the company's earnings per share.

MATH IN Home Buying REVISITED

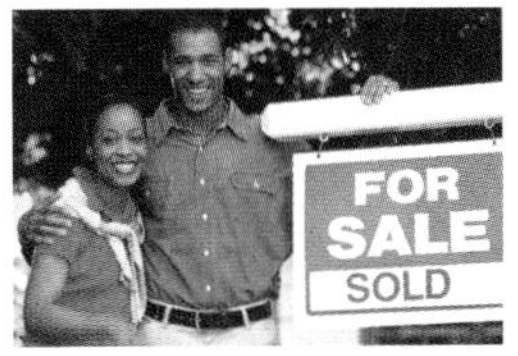

Since 5% of income goes to taxes and insurance, that leaves 23% for the monthly payment. The income is $72,000 per year, which is 72,000/12, or $6,000 per month.

$$23\% \text{ of } \$6{,}000 = 0.23 \times \$6{,}000 = \$1{,}380$$

This is the monthly payment our friends can afford. Now we look in the payment chart (Table 8-2) in Section 8-4, and find the entry corresponding to 6.0% and 30 years: it is $6.00. So the payment should equal 6.0 times the mortgage amount in thousands:

$$6.0x = 1{,}380$$
$$x = \frac{1{,}380}{6.0} = 230$$

Now we multiply by 1,000 because this is the mortgage amount in thousands:

$$230 \times 1{,}000 = 230{,}000$$

This is the mortgage amount, and we add the $15,000 down payment to find that our couple can buy a $245,000 house. (We essentially did Example 1 in Section 8-4 backward.)

As a side note, 20 years ago, a 30-year mortgage could have had a rate of 9% or more. At 9% rather than 6%, the amount of home this couple could afford goes down to $186,429!

Review Exercises

Section 8-1

For Exercises 1–10, find the missing value.

	Fraction	Decimal	Percent
1.	$\frac{7}{8}$	_____	_____
2.	_____	0.54	_____
3.	_____	_____	80%
4.	$\frac{5}{12}$	_____	_____
5.	_____	_____	185%
6.	_____	0.06	_____
7.	$5\frac{3}{4}$	_____	_____

	Fraction	Decimal	Percent
8.	______	1.55	______
9.	______	______	45.5%
10.	$\frac{3}{8}$	______	______

11. Find 72% of 96.
12. 18 is what percent of 60?
13. 25% of what number is 275?
14. If the sales tax is 5% on a calculator, find the tax and the total cost if the calculator is $19.95.
15. If the sales tax on a coffee table is $3.60, find the cost of the table if the tax rate is 6%.
16. Marcia received a commission of $2,275 for selling a small home. If she receives a 7% commission, find the price of the home.
17. In 2000, households received 3.4 credit card offers per month on average. In 2005 the average was 5.9. Find the percent increase.
18. In 2001, there were 3,147 adolescents under 18 being held in state prisons. In 2005, the number was 2,226. Find the percent decrease.

Section 8-2

For Exercises 19–26, find the missing value.

	Principal	Rate	Time	Simple Interest
19.	$4,300	9%	6 years	______
20.	$16,000	______	3 years	$1,920
21.	$875	12%	______	$262.50
22.	$50	6%	18 months	______
23.	$230	______	6.5 years	$104.65
24.	______	3%	5 years	$63.75
25.	______	14%	2 years	$385
26.	$785.00	______	12 years	$1,130.40

27. Ace Auto Parts borrowed $6,000 at 6% for 5 years to enlarge its display area. Find the simple interest and future value of the loan.
28. Sam's Sound Shack borrowed $13,450 at 8% for 15 years to remodel its existing store. Find the simple interest and future value of the loan.
29. Julie earned $60.48 in simple interest on a savings account balance of $4,320.00 over a 12-month period. Find the rate of interest.
30. John has an opportunity to buy a new boat. He has to borrow $5,300 at 11% simple interest for 36 months. Find the monthly payment.
31. Find the simple interest on a $2,300 loan at 5% for 80 days. Use the Banker's rule.
32. Find the simple interest on a $8,750 loan at 8.5% for 100 days. Use the Banker's rule.
33. David obtained a 3-year, $6,000 discounted loan at 6%. Find the discount and the amount of money David received.
34. Marla obtained a 4-year $9,250 discounted loan at 12%. Find the discount and the amount of money Marla received.

Section 8-3

For Exercises 35–38, find the compound interest and future value.

	Principal	Rate	Compounded	Time
35.	$1,775	5%	annually	6 years
36.	$200	4%	semiannually	10 years
37.	$45	8%	quarterly	3 years
38.	$21,000	6%	quarterly	7 years

39. Find the effective rate when the stated rate is 12% and the interest is computed quarterly.
40. The Evergreen Landscaping Company will need to purchase a new backhoe in 7 years. The owner purchases an annuity that pays 8.3% interest compounded semiannually. If the semiannual payment is $4,000, find the future value of the annuity in 7 years.
41. Mike and Marie plan to take an African vacation in 3 years. In order to save money for the trip, they purchase an annuity that pays 3% interest compounded quarterly. Find the future value of the annuity in 3 years if their quarterly payment is $650.

Section 8-4

42. Brad Johnson purchased a washer and dryer for a total of $854. He made a 25% down payment and financed the rest with 12 monthly payments of $54.30. Find the total installment price and the finance charge.
43. Mary Cartworth purchased a four-piece luggage set for $750. She made a down payment of 15% and was charged 6% interest. Find the total installment price and the monthly payment if she paid it off in 8 months.
44. Judy Harper purchased a Cobalt for $10,900. Her down payment was $1,000. She paid the balance with monthly payments of $310 for 3 years. Find the APR.
45. Max Dunbar bought a used BMW for $20,500 on www.Autotrader.com. His down payment was $6,000. He paid off the balance with monthly payments of $311 for 5 years. Find the APR.

46. Mike purchased a home for $149,500. He made a down payment of $8,000 and financed the remainder with an 8.5% loan for 25 years. Find his monthly payment.
47. In Exercise 44, Judy decided to pay off her loan at the end of 24 months. Use the actuarial method and find the unearned interest and the payoff amount.
48. In Exercise 45, Max was able to pay off his loan at the end of 3 years. Use the actuarial method and find the unearned interest and the payoff amount.
49. A loan for $1,500.00 is to be paid back in 30 monthly installments of $61.25. The borrower decides to pay off the balance after 24 payments have been made. Find the amount of interest saved. Use the rule of 78.
50. For the month of February, Pete had an unpaid balance on his credit card of $563.25 at the beginning of the month. He had purchases of $563.25 and made a payment of $350.00 during the month. Find the finance charge if the interest rate is 1.75% per month on the unpaid balance and find the new balance on March 1.
51. Sid's Used Cars had these transactions on its credit card statement:

April 1	Unpaid balance	$5,628.00
April 10	Purchases	$2,134.60
April 22	Payment	$ 900.00
April 28	Purchases	$ 437.80

Find the finance charge if the interest rate is 1.8% on the average daily balance and find the new balance for May 1.

Section 8-5

52. A home was purchased for $145,000 with a 20% down payment. The mortgage rate was 8.5% and the term of the mortgage was 25 years.
    (a) Find the amount of the down payment.
    (b) Find the amount of the mortgage.
    (c) Find the monthly payment.
    (d) Compute an amortization schedule for the first 2 months.
53. A business sold for $252,000. The down payment was 8%. The buyer financed the balance at 8.25% for 25 years. Find the monthly payment on the mortgage.

Section 8-6

Use the table shown for Exercises 54–59.

52 weeks								
HI	LO	STOCK	DIV	YLD%	P/E	VOL (1,000s)	CLOSE	NET CHG
34.28	27.09	TRBCQ	0.72	2.2	30	5,528	32.79	−0.25

54. What was the high price and low price of the stock for the last 52 weeks?
55. If you own 475 shares of this stock, how much was the dividend you received?
56. How many shares of the stock were sold yesterday?
57. What was the closing price of the stock the day before yesterday?
58. Find the annual earnings per share of the stock.
59. An investor purchased 90 shares of stock for $86.43 per share and later sold it for $92.27 per share. How much did she make on the stock if the broker's fee was 2% on the purchase and the sale of the stock? Ignore the dividends.

Chapter Test

1. Change $\frac{5}{16}$ to a percent.
2. Write 0.63 as a percent.
3. Write 28% as a fraction in lowest terms.
4. Change 16.7% to a decimal.
5. 32 is what percent of 40?
6. Find 87.5% of 48.
7. 45% of what number is 135?
8. Find the sales tax on a toaster oven that sells for $29.95. The tax rate is 8%.
9. If a salesperson receives a 15% commission on all merchandise sold, find the amount sold if his commission is $385.20.
10. On the first day of math class, 28 students were present. The next day, 7 more students enrolled in the class because the other section was canceled. Find the percent increase in enrollment.
11. Find the simple interest on $1,350 at 12% for 3 years.

12. Find the rate for a principal of \$200 invested for 15 years if the simple interest earned is \$150.
13. Ron's Detailing Service borrowed \$435 at 3.75% for 6 months to purchase new equipment. Find the simple interest and future value of the loan, and the monthly payment.
14. The Express Delivery borrowed \$1,535 at 4.5% for 3 months to purchase safety equipment for its employees. Find the simple interest and future value of the loan, and the monthly payment.
15. Benson Electric borrowed \$1,800 at 12% for 1 year from a local bank. Find the simple interest and future value of the loan, and the monthly payment.
16. Find the simple interest on a \$5,000 loan at 4% for 60 days. Use the Banker's rule.
17. Latoya obtained a 6-year \$12,650 discounted loan at 7.5%. Find the discount and the amount of money Latoya received.
18. Find the interest and future value for a principal of \$500 invested at 6.5% compounded semiannually for 4 years.
19. Find the interest and future value on a principal of \$9,750 invested at 10% compounded quarterly for 6 years.
20. In order to purchase a motorcycle, Jayden borrowed \$12,000 at 9.5% for 4 years. Find his monthly payment.
21. Find the effective rate when the stated interest rate is 8% and the interest is compounded semiannually.
22. In order to open a new branch of her business in 3 years, the owner of Quick Fit Fitness Center purchases an annuity that pays 4.5% interest compounded semiannually. If her semiannual payment is \$3,000, find the future value of the annuity in 3 years.
23. Sara bought furniture for her first apartment at a cost of \$935. She made a down payment of 30% and financed the rest for 6 months at 10% interest. Find the total installment charge and the monthly payment.
24. Bart Johnston purchased a Mazda for \$15,000 and had a down payment of \$2,000. He financed the balance at \$305 per month for 48 months. Find the APR.
25. In Exercise 24, Bart was able to pay off his loan at the end of 24 months. Using the actuarial method, find the unearned interest and the payoff amount.
26. A loan for \$2,200 is to be paid off in 24 monthly installments of \$111.85. The borrower decides to pay off the loan after 20 payments have been made. Find the amount of interest saved, using the rule of 78.
27. For the month of November, Harry had an unpaid balance of \$1,250 on his credit card. During the month, he made purchases of \$560 and a payment of \$800. Find the finance charge if the interest rate is 1.6% per month on the unpaid balance and find the new balance on December 1.
28. Rhonda's credit card statement for the month of May shows these transactions.

May 1	Unpaid balance	\$474.00
May 11	Payment	\$300.00
May 20	Purchases	\$ 86.50
May 25	Purchases	\$120.00

Find the finance charge if the interest rate is 2% on the average daily balance and find the new balance on June 1.

29. Tamara borrowed \$800.00 for tuition. She is to pay it back in 12 monthly installments of \$70.70. Find the annual percentage rate.
30. A home is purchased for \$180,000 with a 5% down payment. The mortgage rate is 6% and the term is 30 years.
    (a) Find the amount of the down payment.
    (b) Find the amount of the mortgage.
    (c) Find the monthly payment.
    (d) Compute an amortization schedule for the first 2 months.
31. A group of business people purchased a sporting goods store for \$475,000. They made a 15% down payment and obtained a mortgage at 5.75% for 20 years. Find the monthly payment.

Use the following table for Exercises 32–36.

52 weeks								
HI	LO	STOCK	DIV	YLD%	P/E	VOL (1,000s)	CLOSE	NET CHG
36.98	23.17	CAT	0.20	2.7	12	1,501	27.45	+0.80

32. What were the 52-week high and low prices of the stock?
33. If you own 300 shares of stock, how much money in dividends did you receive?

34. How many shares of the stock were sold yesterday?
35. What was the closing price of the stock the day before yesterday?
36. Find the annual earnings per share of the stock.

Projects

1. Compare the investments below to decide which you think is the best. Consider such things as total profit, length of time, and amount of money needed up front.
   (a) $20,000 placed into a savings account at 3.8% compounded monthly for 10 years.
   (b) A 10-year annuity that pays 4.5% interest with monthly payments of $170.
   (c) Buying 700 shares of stock at $14.30 per share; selling 200 shares 4 years later at $25.10, and the rest 3 years after that at $28.05. The brokerage charges 1% commission on both buying and selling.
   (d) Buying a $150,000 house with $20,000 down and financing the rest with a 15-year mortgage at 5%. Then selling the house and paying off the balance of the loan in 8 years at a selling price, after commission, of $192,000. (*Hint:* You will need to calculate the monthly payment, then compute unearned interest and payoff amount.)

2. You have $1,000 to invest. Investigate the advantages and disadvantages of each type of investment.

   (a) Checking account
   (b) Money market account
   (c) Passbook savings account
   (d) Certificate of deposit

   Write a short paper indicating which type of account you have chosen and why you chose that account.

3. There are many fees involved in buying or selling a home. Some of these include an appraisal fee, survey fee, etc. Consult a real estate agency to see what is necessary to purchase a home in your area and write a short paper on the necessary closing costs.

CHAPTER 9

Measurement

Outline

9-1 Measures of Length: Converting Units and the Metric System

9-2 Measures of Area, Volume, and Capacity

9-3 Measures of Weight and Temperature

Summary

MATH IN Travel

To measure something means to assign a number that represents its size. In fact, measurement might be the most common use of numbers in everyday life. Numbers are used to measure heights, weights, distances, grades, weather, sizes of homes, capacities of bottles and cans, and much more. Even in our monetary system, we are using numbers to measure sizes: little ones, like the cost of a candy bar, and big ones, like an annual salary.

The one thing that every measurement has in common is that it's good for absolutely nothing unless there are units attached to the number. If a friend asks you how far you live from campus and you answer "three," at best they'll look at you funny and cautiously back away. An answer like that is meaningless because without units, there's no context to describe what the number represents. And that's entirely because there are many different units that can be used to describe similar measurements. That number 3 could mean 3 miles, kilometers, blocks, houses, even parsecs if you come from a galaxy far, far away.

In this chapter, we will study the different ways that things are measured in our world, with a special focus on the different units that are used. You probably have at least a passing familiarity with the metric system, since measurements in units like liters and kilometers are becoming more common. When you gather information of interest to you, you don't always get to choose the units that the information is provided in, so the ability to convert measurements from one unit to another is a particularly useful skill in the information age.

One obvious application of this skill comes into play when traveling outside the United States. Most countries use the metric system almost exclusively, meaning that to understand various bits of information, you need to be able to interpret measurements in an unfamiliar system. Suppose that you and your sweetie are planning a dream week in Aruba: the following questions might be of interest. By the time you complete this chapter, you'll be able to answer the questions, making you a more well-informed traveler.

1. A check of a tourist web site informs you that the temperature is expected to range from 24° to 32° Celsius. How should you pack?
2. The airline that will fly you to Aruba allows a maximum of 30 kilograms for each piece of luggage. If your empty suitcase weighs 4 pounds, how many pounds can you pack into it?
3. The resort you will be staying at is 4.2 kilometers from the airport and 0.6 kilometers from downtown. Can you walk downtown? And how long will it take to drive your rental car from the airport at 35 miles per hour?
4. At the gas station closest to the airport, gas costs \$0.43 per liter. If you use ¾ of the car's 16-gallon tank, how much will it cost to refill?

For answers, see Math in Travel Revisited on page 506

Section 9-1 Measures of Length: Converting Units and the Metric System

LEARNING OBJECTIVES

- ☐ 1. Convert measurements of length in the English system.
- ☐ 2. Convert measurements in the metric system.
- ☐ 3. Convert between English and metric units of length.

Units of length are among the oldest and most common units used to measure objects. They are used to measure everything from your waistline to the distance between the earth and far away planets. In this section, we will study two systems for measuring length, and become expert at converting from one unit to another. This skill will serve you well in any topic where different units of measure are used.

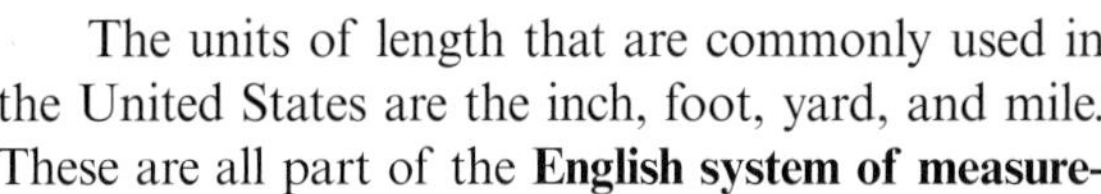

The units of length that are commonly used in the United States are the inch, foot, yard, and mile. These are all part of the **English system of measurement,** which is very old indeed. (See Sidelight on page 483 for some historical perspective.) The basic conversion factors between units in the English system are shown in Table 9-1.

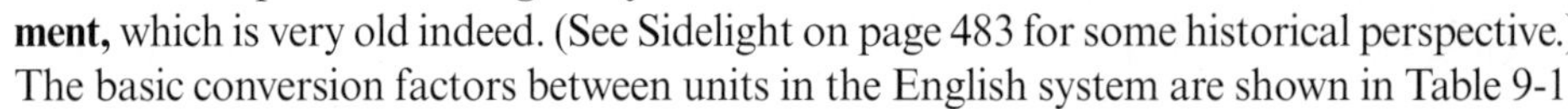

There are several ways to convert from one unit of measure to another. Many people use the "common sense" method. For example, to convert 6 yards to feet, multiply 6×3 feet $= 18$ feet, since there are 3 feet in 1 yard. To convert 564 inches to feet, divide 564 inches by $12 = 47$ feet, since there are 12 inches in 1 foot.

A more systematic, and reliable, method for converting units is known as **dimensional analysis.** It is based on one simple idea: if measurements in two different units represent the same length, like 1 yard and 3 feet, then a fraction formed with the two measurements is equal to 1. For example,

$$\frac{1 \text{ yd}}{3 \text{ ft}} = 1 \quad \text{and} \quad \frac{3 \text{ ft}}{1 \text{ yd}} = 1$$

We call a fraction of this type a **unit fraction.** Our procedure for converting units will be to multiply by one or more unit fractions in such a way that the units we don't want can be divided out, leaving behind the units that we do want. Since all we're really doing is multiplying by 1, we won't be changing the actual value of the measurement—just the units it's measured in. The procedure is illustrated in Examples 1 to 4 below.

TABLE 9-1 Units of Length in the English System

12 inches (in.) = 1 foot (ft)
3 feet = 1 yard (yd)
36 inches = 1 yard
5,280 feet = 1 mile (mi)
1,760 yards = 1 mile

EXAMPLE 1 Converting Yards to Feet

Convert 6 yards to feet.

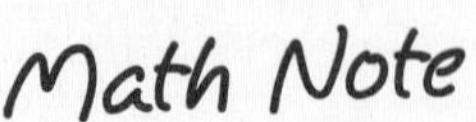

Math Note

You might find it helpful to write the original measurement as a fraction with denominator 1 as in Example 1, but it isn't necessary.

SOLUTION

The goal is to eliminate the units of yards, and leave behind feet. So we would like to use a unit fraction that has yards in the denominator (to divide out yards) and feet in the numerator. That would be the second one shown above $\left(\frac{3 \text{ ft}}{1 \text{ yd}}\right)$:

$$\frac{6 \cancel{\text{yd}}}{1} \cdot \frac{3 \text{ ft}}{1 \cancel{\text{yd}}} = 18 \text{ ft}$$

The yards divided out, leaving behind feet in the numerator.

▼ Try This One 1

Convert 81 feet to yards.

EXAMPLE 2 Converting Feet to Miles

The height of Mount Everest is 29,035 feet. How many miles is that?

SOLUTION

This time, we want a unit fraction with feet in the denominator and miles in the numerator: that is, $\frac{1\text{ mi}}{5{,}280\text{ ft}}$.

$$\frac{29{,}035\,\cancel{\text{ft}}}{1}\cdot\frac{1\text{ mi}}{5{,}280\,\cancel{\text{ft}}}=\frac{29{,}035\text{ mi}}{5{,}280}\approx 5.5\text{ mi}$$

▼ Try This One 2

The maximum cruising height of a Boeing 737 is about 7.77 miles. How many feet is that?

Sometimes we will need to multiply by more than one unit fraction to make a conversion.

Sidelight A BRIEF HISTORY OF MEASURING LENGTHS

Nobody knows how and when human beings started measuring physical objects, but historians seem to be in general agreement that the earliest recorded measurements were lengths. In particular, they were based on parts of the human body. If you search the Internet for early units of measurement, the most common one you will find is the **cubit,** first used by the Egyptians around 5,000 years ago. The cubit was equal to the distance from a person's elbow to the outstretched middle finger. (If you try to measure something that way, make sure to stretch out your other fingers as well.) Other early units were the **palm** (the distance across the base of a person's four fingers) and the **digit** (the thickness of a person's middle finger).

This leads us to believe that one of two things was true: either everyone was exactly the same size in ancient Egypt, or their units of measurement were pretty imprecise. Things got a little more specific sometime around the year 1100 CE, when King Henry I of England declared that the distance from the tip of his nose to the end of his thumb would be known as a **yard.** (At least this distance was based on one specific person's physical size.) The oldest known yardstick still in existence, believed to have been made in 1445, is accurate to a modern yard within about $\frac{1}{300}$th of an inch.

A couple hundred years later, King Edward I declared that one third of a yard should be called a **foot.** In 1595, Queen Elizabeth changed the length of a mile from the Roman tradition of 5,000 feet to the current 5,280 feet so that it was exactly 8 furlongs. (Perhaps she was a big fan of horse racing, which still uses furlongs.) Eventually, the English system of measurement became standard throughout the world. Of course, these units are still used in the United States today. The real irony? England has been officially using the metric system since 1965. In fact, only three countries in the world have not officially adopted the metric system, which we'll study later in this section: the United States, Liberia, and Myanmar.

EXAMPLE 3 Converting Measurements Using Two Unit Fractions

The distance by air from JFK airport to LaGuardia airport in New York is about 11 miles. How many inches is that?

SOLUTION

Looking at Table 9-1, there is no conversion factor for miles to inches. But we can go from miles to feet $\left(\text{using } \frac{5{,}280 \text{ ft}}{1 \text{ mi}}\right)$ then from feet to inches $\left(\text{using } \frac{12 \text{ in.}}{1 \text{ ft}}\right)$:

$$\frac{11 \cancel{\text{mi}}}{1} \cdot \frac{5{,}280 \cancel{\text{ft}}}{\cancel{\text{mi}}} \cdot \frac{12 \text{ in.}}{1 \cancel{\text{ft}}} = 11 \times 5{,}280 \times 12 \text{ in.} = 696{,}960 \text{ in.}$$

Math Note

Notice that we put the units we wanted to eliminate in the denominator, and the units we wanted to keep in the numerator.

▼ Try This One 3

A football field is 100 yards long. How many inches is that?

Dimensional analysis can also be used to convert more complicated units, like those for speed, as shown in Example 4.

EXAMPLE 4 Converting Units for Speed

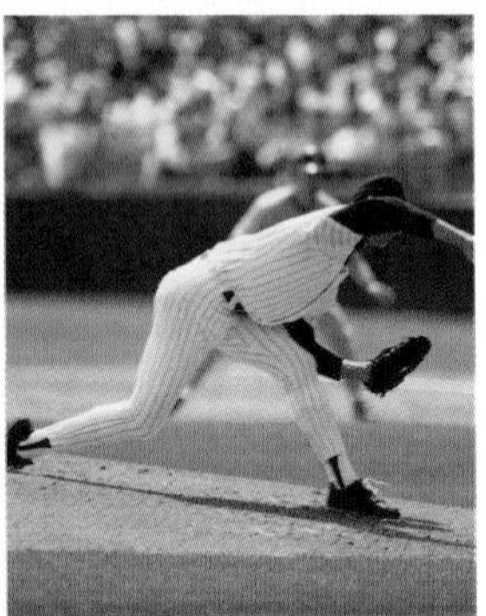

A decent major league pitcher can throw a baseball 90 miles per hour. How fast is that in feet per second?

SOLUTION

Notice that we can write 90 miles per hour as a fraction: $\frac{90 \text{ mi}}{1 \text{ hr}}$. Now we can multiply by unit fractions that convert miles to feet $\left(\frac{5{,}280 \text{ ft}}{1 \text{ mi}}\right)$, hours to minutes $\left(\frac{1 \text{ h}}{60 \text{ min}}\right)$, and minutes to seconds $\left(\frac{1 \text{ min}}{60 \text{ s}}\right)$.

$$\frac{90 \cancel{\text{mi}}}{1 \cancel{\text{h}}} \cdot \frac{5{,}280 \text{ ft}}{1 \cancel{\text{mi}}} \cdot \frac{1 \cancel{\text{h}}}{60 \cancel{\text{min}}} \cdot \frac{1 \cancel{\text{min}}}{60 \text{ s}} = \frac{90 \times 5{,}280 \text{ ft}}{60 \times 60 \text{ s}} = 132 \frac{\text{ft}}{\text{s}}$$

CAUTION

In this case, we wanted seconds in the denominator, so we had to be extra careful to put the units we wanted to eliminate in the numerator, and the ones we wanted to keep in the denominator.

▼ Try This One 4

The speed of sound at sea level and room temperature is about 1,126 feet per second. How fast is that in miles per hour?

 1. Convert measurements of length in the English system.

The Metric System

If you grew up in the United States, you're probably perfectly happy with the English system of measurement. You've been using it your whole life, after all. But the truth

Sidelight THE LUNAR WEIGHT-LOSS PLAN

Since weight is a measure of gravitational pull, your weight is less in locations where gravity is not as strong. The effect is negligible pretty much anywhere on earth, although you would weigh about 0.3% less at the top of Mt. Everest than you would at sea level. But once you leave earth, weight can vary widely. The force of gravity on the moon, for example, is about one-sixth of what it is on earth. So by heading to the moon, you would instantly shed five-sixths of your weight! Of course, you would still be the same size (i.e., your mass won't change), but the scale would seem like a much friendlier place to most of us.

Math Note

Technically speaking, the gram is a unit of mass, not weight, and the two are slightly different. The weight of an object is the measure of the gravitational pull on it, so your weight can vary depending on where you are. (See Sidelight above.) Mass is a constant measure of the physical size of an object regardless of location. Since we all live on the earth, we'll consider mass and weight to be synonymous.

is that it's not that great of a system because there's no rhyme or reason to it. Why is 12 inches a foot? Why is 5,280 feet a mile? They just are. Somebody decided, and that was that. The metric system, on the other hand, makes a bit more sense because the different units of measure are all related by powers of 10. We'll see exactly what that means after introducing the base units. The metric system uses three basic units of measure: the *meter,* the *liter,* and the *gram.*

Length in the metric system is measured using **meters.** One meter is a bit more than a yard, about 39.4 inches. The symbol for meter is "m."

The basic unit for capacity in the metric system is the **liter,** which is a little bit more than a quart. (You're probably familiar with 2-liter bottles of soda—that's close to a half gallon.) The symbol for liter is "L."

The basic unit for measuring weight in the metric system is the **gram.** This is a very small unit—a nickel weighs about 5 grams. The symbol for gram is "g."

Now back to other units in the metric system. All units other than the three basic units are based on multiples of the basic units, with those multiples all being powers of 10. For example, a standard unit of weight for people in the metric system is the kilogram: this is equal to 1,000 grams. A commonly used unit of length for smaller measurements is the centimeter, which is $\frac{1}{100}$ meter. It's the prefix in front of the basic unit that determines the size. The most common metric prefixes are listed in Table 9-2.

TABLE 9-2 Metric Prefixes

Prefix	Symbol	Meaning
kilo	k	1,000 units
hecto	h	100 units
deka	da	10 units
	m, L, g	1 unit
deci	d	$\frac{1}{10}$ of a unit
centi	c	$\frac{1}{100}$ of a unit
milli	m	$\frac{1}{1,000}$ of a unit

The relationship between units in the metric system is a lot like the units used in our monetary system. Table 9-3 demonstrates this comparison for the base unit of meter.

TABLE 9-3 Relationship between Metric Units and U.S. Currency

Metric Unit	Meaning	Monetary Comparison
kilometer	1,000 meters	$1,000
hectometer	100 meters	100
dekameter	10 meters	10
meter (base unit)	1 meter	dollar (base unit)
decimeter	0.1 meter	0.10 (10 cents)
centimeter	0.01 meter	0.01 (1 cent)
millimeter	0.001 meter	0.001 (1/10 cent)

There are several ways to do conversions in the metric system. The easiest way is to multiply or divide by powers of 10. The basic rules are given next.

Conversion When You Use the Metric System

To change a larger unit to a smaller unit in the metric system, multiply by 10^n, where n is the number of steps that you move down in Table 9-4.

To change a smaller unit to a larger unit in the metric system, divide by 10^n, where n is the number of steps that you move up in Table 9-4.

TABLE 9-4 Metric Unit Conversions

divide	kilo hecto deka base unit deci centi milli	multiply	*Units get smaller as you read down the chart.*

EXAMPLE 5 Converting Metric Units

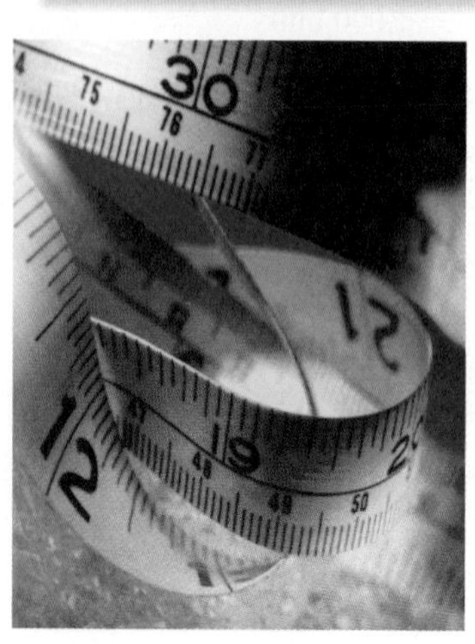

Convert 42.5 kilometers to centimeters.

SOLUTION

The prefix "kilo" is higher in Table 9-4 than "centi," so we're converting to smaller units, which means we need to multiply. There are 5 steps from "kilo" to "centi" in the table, so the power of 10 we multiply by is 10^5. This can be quickly accomplished by moving the decimal point 5 places to the right:

$$42.5 \text{ km} \times 10^5 = 42.50000 = 4{,}250{,}000 \text{ cm}$$

Many rulers and tape measures are marked with both inches and centimeters.

▼ Try This One 5

Convert 170 dekagrams to milligrams.

EXAMPLE 6 Converting Metric Units

Convert 1,253.7 milligrams to grams.

SOLUTION

The gram is the base unit of weight, and the prefix "milli" is below the base unit in the table, so we are converting to a larger unit and must divide. There are 3 steps from "milli" to the base unit in the table, so we divide by 10^3. This can be quickly done by moving the decimal point 3 places to the left:

$$1{,}253.7 \text{ mg} \div 10^3 = 1\underbrace{253}.7 = 1.2537 \text{ grams}$$

▼ Try This One 6

Convert 173.6 deciliters to dekaliters.

Conversion in the metric system can also be done using dimensional analysis. We will use Table 9-2 to convert first to the base unit, then the unit we want.

EXAMPLE 7 Converting in the Metric System with Dimensional Analysis

Math Note

You will be able to do metric conversion a lot more efficiently if you take a few minutes to learn which prefixes indicate units larger than the base unit, and which indicate units smaller than the base unit.

Convert 42.5 kilometers to centimeters using dimensional analysis.

SOLUTION

According to Table 9-2, 1 kilometer is 1,000 meters, and 1 meter is 100 centimeters.

$$\frac{42.5 \cancel{\text{km}}}{1} \cdot \frac{1{,}000 \cancel{\text{m}}}{1 \cancel{\text{km}}} \cdot \frac{100 \text{ cm}}{1 \cancel{\text{m}}} = 42.5 \times 100{,}000 = 4{,}250{,}000 \text{ cm}$$

This agrees with our answer to Example 5.

▼ Try This One 7

Convert 170 dekagrams to milligrams using dimensional analysis.

☑ 2. Convert measurements in the metric system.

Because most of the world uses the metric system while the United States uses the English system, it is often necessary to convert between the two systems. We will use dimensional analysis, along with the table below, which provides conversion factors for units of length.

English and Metric Equivalents for Length

1 in. = 2.54 cm	1 mm ≈ 0.03937 in.
1 ft = 30.48 cm = 0.3048 m	1 cm ≈ 0.3937 in.
1 yd = 91.44 cm = 0.9144 m	1 m ≈ 1.0936 yd = 3.2808 ft
1 mi ≈ 1.61 km	1 km ≈ 0.6214 mi

Many road signs now list mileages in both miles and kilometers.

EXAMPLE 8 Converting Between English and Metric Lengths

Convert each.

(a) 135 feet to centimeters (b) 87 centimeters to inches
(c) 213.36 millimeters to feet

SOLUTION

(a) To convert feet to centimeters use the unit fraction $\frac{30.48 \text{ cm}}{1 \text{ ft}}$.

$$135 \text{ ft} \cdot \frac{30.48 \text{ cm}}{1 \text{ ft}} = 135 \cdot 30.48 \text{ cm} = 4{,}114.8 \text{ cm}$$

(b) To convert centimeters to inches use the unit fraction $\frac{1 \text{ in.}}{2.54 \text{ cm}}$.

$$87 \text{ cm} \cdot \frac{1 \text{ in.}}{2.54 \text{ cm}} = \frac{87 \text{ in.}}{2.54} = 34.25 \text{ in. (rounded)}$$

(c) To convert from millimeters to feet, we first convert millimeters to centimeters using the unit fraction $\frac{1 \text{ cm}}{10 \text{ mm}}$, then convert centimeters to feet using the unit fraction $\frac{1 \text{ ft}}{30.48 \text{ cm}}$.

$$213.36 \text{ mm} \cdot \frac{1 \text{ cm}}{10 \text{ mm}} \cdot \frac{1 \text{ ft}}{30.48 \text{ cm}} = \frac{213.36 \text{ ft}}{(10 \cdot 30.48)} = 0.7 \text{ ft}$$

Alternatively, we could convert millimeters to inches, then inches to feet:

$$213.36 \text{ mm} \cdot \frac{0.03937 \text{ in.}}{1 \text{ mm}} \cdot \frac{1 \text{ ft}}{12 \text{ in.}} \approx \frac{213.36 \cdot 0.0397}{12} \text{ ft} \approx 0.7 \text{ ft}$$

The results are close, but not exactly the same. The first worked out to be exactly 0.7, while the second came out to 0.6999986.

Math Note

Because some of the English-metric conversions are approximations due to rounding, the converted measurements are often approximations as well. This accounts for the fact that the two methods shown for part (c) lead to slightly different results.

▼ Try This One 8

☑ 3. Convert between English and metric units of length.

Convert each of the following. Round to two decimal places.

(a) 237 feet to meters (b) 128 cm to inches (c) 23,342 mm to feet

Answers to Try This One

1 27 yd
2 41,025.6 ft
3 3,600 in.
4 About 767.7 mi/h
5 1,700,000 mg
6 1.736 daL
7 1,700,000 mg
8 (a) 72.24 m (b) 50.39 in. (c) 76.58 ft

EXERCISE SET 9-1

Writing Exercises

1. What are the basic units of length in the English measurement system, and how are they related?
2. What are the basic units of length, capacity, and weight in the metric system?
3. What importance do prefixes play in different units of measure in the metric system?
4. Describe how to use dimensional analysis to convert units of measure.

Computational Exercises

For Exercises 5–16, convert each using dimensional analysis. round to 2 decimal places if necessary.

5. 12 feet to yards
6. 36 inches to feet
7. 17 feet to yards
8. 10,345 feet to miles
9. 6 yards to feet
10. 5 miles to yards
11. 21 inches to feet
12. 10 yards to feet

13. 18 inches to yards
14. 5,237,834 inches to miles
15. 876 inches to yards
16. 182 feet to yards

For Exercises 17–56, convert each measurement to the specified unit.

17. 8 m = ______ cm
18. 0.25 m = ______ dm
19. 12 dam = ______ m
20. 24 hm = ______ dam
21. 0.6 km = ______ hm
22. 30 cm = ______ dm
23. 90 m = ______ dm
24. 18,426 mm = ______ m
25. 375.6 cm = ______ m
26. 63 m = ______ km
27. 405.3 m = ______ km
28. 0.6 dam = ______ hm
29. 12 km = ______ cm
30. 50,000 cm = ______ km
31. 1.85 km = ______ mm
32. 650,000 mm = ______ km
33. 12.62 km = ______ dm
34. 39 m = ______ cm
35. 8 hm = ______ km
36. 540 dm = ______ cm
37. 5 meters = ______ inches
38. 14 yards = ______ meters
39. 16 inches = ______ millimeters
40. 50 meters = ______ yards
41. 235 feet = ______ dekameter
42. 563 decimeters = ______ inches
43. 1,350 meters = ______ feet
44. 4,375 dekameters = ______ feet
45. 0.6 inch = ______ millimeters
46. 256 kilometers = ______ miles
47. 0.06 hectometers = ______ feet
48. 54 inches = ______ centimeters
49. 1,345 feet = ______ decimeters
50. 44,000 millimeters = ______ yards
51. 2.35 kilometers = ______ miles
52. 837 miles = ______ kilometers
53. 42 decimeters = ______ yards
54. 75 centimeters = ______ inches
55. 333 inches = ______ meters
56. 1,256 kilometers = ______ inches

Real-World Applications

57. Steve sprinted the last 5 yards of a race. How many feet did he sprint?
58. A professional basketball court is 94 feet long. How many inches is that?
59. On a baseball diamond, the bases are 90 feet apart. How many yards does a batter run in reaching second base?
60. A football player completed a 45-yard pass. How many inches did he throw the ball?
61. A traffic reporter announced a backup on I-95 of 1,500 yards. How far is that in miles?
62. Sarah wanted to buy a new queen-sized mattress. She only has an area that measures 5 ft × 6 ft to put the mattress. The mattress size is 60 in. × 80 in. Will the mattress fit? Convert the measurements to feet to determine this.
63. Celia used a tape measure to section off areas for her vegetable garden. She measured the section for tomatoes to be 126 inches long. How many yards long is the section for Celia's tomatoes?
64. What are the dimensions of a 6 dm by 5 dm piece of poster board in centimeters? In millimeters?
65. To make a batch of an energy drink, Gina used 2.3 grams of citric acid. How many milligrams of citric acid did she use?
66. If a millipede has 1,000 legs, how many legs would a decipede have?
67. Sanford's gym coach made him run 20 laps on a 700-m track. How many kilometers did he run?
68. How many miles will a runner cover in a 10-km race?
69. Frank drives 65 miles per hour on a freeway in Detroit. When he crosses into Windsor, Ontario, he resumes the same speed. How fast does Frank drive in kilometers per hour?
70. Mary brings a 20-inch by 24-inch picture to the frame shop, but the frame she wants only lists measurements in terms of centimeters. What are the measurements of her picture in centimeters?
71. A bowling lane from the foul line to the pins is 62 feet $10\frac{3}{16}$ inches long. How many meters is a bowling lane?
72. A 5-carat round-cut diamond measures half an inch in diameter. How many millimeters is the diameter of the diamond?
73. The new Mars Pathfinder postage stamp measures 1.5 × 3 inches. What are the stamp's measurements in millimeters?
74. Light travels at 186,283 miles every second. How many feet per hour does light travel?
75. The earth is on average 92,900,000 miles from the sun. In July, the distance is 94,400,000 miles, the farthest of any month. What is the distance in feet of the difference of the distance in July from the average?
76. To build a loft in his dorm room, Jay got 8-foot planks from the lumber yard. To get the right size, he had to cut off a quarter of each plank. How many inches did he cut off each?
77. A college is building a new basketball court that is 94 feet long and 50 feet wide, but the wood for the

court comes only in 10-meter planks that are 8 centimeters wide. How many planks are needed to surface the court? (*Hint:* Round each measurement up to estimate the number of planks needed.)

78. If three servings of ice cream have 20 grams of fat, how many milligrams of fat does one serving have?
79. A popular pain-relieving drug has 200 mg of its active ingredient in one capsule. If a bottle has 100 capsules, how many grams of the active ingredient are in the bottle?
80. The campus fitness center is installing an Olympic-sized swimming pool that measures 164 ft × 82 ft. The builders want to put 6-inch tiles around the border. How many tiles will they need?
81. If one U.S. dollar was worth 1.22 Canadian dollars when Sal went to a debate tournament in Toronto, and he had $400 in U.S. currency as spending money, how much was his money worth in Canadian dollars?
82. While in Chile, where $1 in U.S. currency was worth 610.045 in Chilean pesos, you won 10,000,000 Chilean pesos. Did you plan to eat a nice dinner out, buy a new car, or buy a new house? Explain your answer.
83. On her summer abroad in France, Jane bought a pair of shoes for 54.23 euros. The store owner only had francs to give her as change. She gave him 60 euros. How much did he give her back in francs? At the time, 1 euro was worth 6.55957 francs.
84. In August of 2008, one dollar could buy 9.86 Mexican pesos. In early February of 2009, that same dollar could buy 14.49 Mexican pesos. If you paid 750 pesos per night at a hotel in Cancun in August 2008, and your friend paid the same rate the following February, how much less per night in dollars did he spend?

Critical Thinking

85. Suppose that four world-class sprinters compare times after running different distances. The first ran the 100-yard dash in 9.12 seconds; the second ran the 100-meter dash in 9.72 seconds; the third ran the 220-yard dash in 19.75 seconds; the fourth ran the 200-meter dash in 19.71 seconds. Which is the most impressive feat? Explain your reasoning.
86. What units would 7,000 cm have to be converted to so that the result has no zeros?
87. What units would 0.006 m have to be converted to so that the result has no zeros?

Section 9-2 Measures of Area, Volume, and Capacity

LEARNING OBJECTIVES

- ☐ 1. Convert units of area.
- ☐ 2. Convert units of volume.

The units of length we studied in Section 9-1 are useful for measuring many things, but they have a major limitation: length is a one-dimensional measurement, and we live in a three-dimensional world. It's equally important to be able to measure in two dimensions, like the size of a lot for a new home, and in three dimensions, like the capacity of a swimming pool. So in this section, we will study area (two dimensions) and volume and capacity (both three dimensions).

Conversions of Area

Units of length are also called **linear units,** because they are used to measure along a line. The units used to measure the size of an object in two dimensions are called **square units.** They are based on a very simple idea: one square unit is defined to be the area of a square that is one unit long on each side. For example, a square inch is the area of a square that is 1 inch on each side, as shown in Figure 9-1.

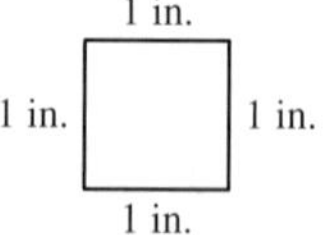

Figure 9-1 One square inch

The area of other figures is then defined to be the number of square units that will fit inside the figure. For example, the rectangle in Figure 9-2 is 3 inches by 5 inches, and we can see that 15 one-inch by one-inch squares will fit inside. So its area is 15 square inches, which we abbreviate as 15 in.^{2}.

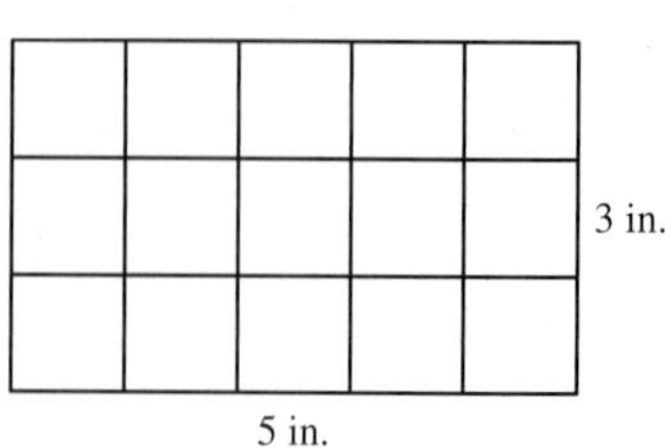

Figure 9-2 15 in.^{2}

Finding the area of various geometric shapes will be covered in Chapter 10. In this section we will focus on conversions between different units of area.

Because the units used in measuring area are square units, the conversions will be the squares of the conversions in linear measure. For example, we know that 1 ft = 12 in., so $(1 \text{ ft})^2 = (12 \text{ in.})^2$, or $1 \text{ ft}^2 = 144 \text{ in.}^2$. The conversions for area are summarized in Table 9-5.

TABLE 9-5 English and Metric Equivalents for Area

English Conversions for Area	English/Metric Conversions for Area
$1 \text{ ft}^2 = 144 \text{ in.}^2$	$1 \text{ in.}^2 \approx 6.45 \text{ cm}^2$
$1 \text{ yd}^2 = 9 \text{ ft}^2 = 1{,}296 \text{ in.}^2$	$1 \text{ ft}^2 \approx 0.093 \text{ m}^2$
$1 \text{ mi}^2 = 27{,}878{,}400 \text{ ft}^2 = 3{,}097{,}600 \text{ yd}^2$	$1 \text{ yd}^2 \approx 0.836 \text{ m}^2$
$1 \text{ acre} = 43{,}560 \text{ ft}^2$	$1 \text{ mi}^2 \approx 2.59 \text{ km}^2$
$1 \text{ mi}^2 = 640 \text{ acres}$	$1 \text{ acre} \approx 4{,}047 \text{ m}^2$

Math Note

The acre is a very old measure of area, traditionally thought to be the area that could be plowed by one man behind one ox in one day.

The next examples will show how to use dimensional analysis to convert measures of area.

EXAMPLE 1 Converting Square Inches to Square Feet

A 52-in. wide-screen TV has a screen area of 1,064 in.². How many square feet is that?

SOLUTION

We want to replace in.² with ft², so we multiply by the unit fraction $\frac{1 \text{ ft}^2}{144 \text{ in.}^2}$.

$$\frac{1{,}064 \text{ in.}^2}{1} \cdot \frac{1 \text{ ft}^2}{144 \text{ in.}^2} \approx 7.39 \text{ ft}^2$$

▼ Try This One 1

A 32-in. wide-screen TV has screen area 3.05 ft². How many square inches is that?

EXAMPLE 2 Converting Acres to Square Miles

Yellowstone National Park has an area of 2,219,789 acres. How large is it in square miles?

SOLUTION

From Table 9-5, the unit fraction we need is $\frac{1 \text{ mi}^2}{640 \text{ acres}}$.

$$\frac{2{,}219{,}789 \text{ acres}}{1} \cdot \frac{1 \text{ mi}^2}{640 \text{ acres}} \approx 3{,}468.42 \text{ mi}^2$$

▼ Try This One 2

The state of Delaware has an area of 2,490 square miles. How many acres is that?

EXAMPLE 3 Converting Square Yards to Acres

An average of 2,191,781 square yards of carpet is discarded into landfills every day in the United States. How many acres would that carpet cover?

SOLUTION

There is no direct conversion between square yards and acres in Table 9-5, so we will convert to square feet (using $\frac{9 \text{ ft}^2}{1 \text{ yd}^2}$) and then to acres (using $\frac{1 \text{ acre}}{43{,}560 \text{ ft}^2}$).

$$\frac{2{,}191{,}781 \cancel{\text{yd}^2}}{1} \cdot \frac{9 \cancel{\text{ft}^2}}{1 \cancel{\text{yd}^2}} \cdot \frac{1 \text{ acre}}{43{,}560 \cancel{\text{ft}^2}} \approx 452.85 \text{ acres}$$

Math Note

We could also have used Table 9-5 to convert square yards to square miles, then square miles to acres.

▼ Try This One 3

An American football field is about 1.32 acres, counting the end zones. Promoters are planning to unfurl a giant flag covering the entire field at a playoff game. How many square yards of fabric will be needed?

Because both the metric and English systems are widely used today, it's useful to be able to convert between English and metric units of area.

EXAMPLE 4 Converting Area from English to Metric Units

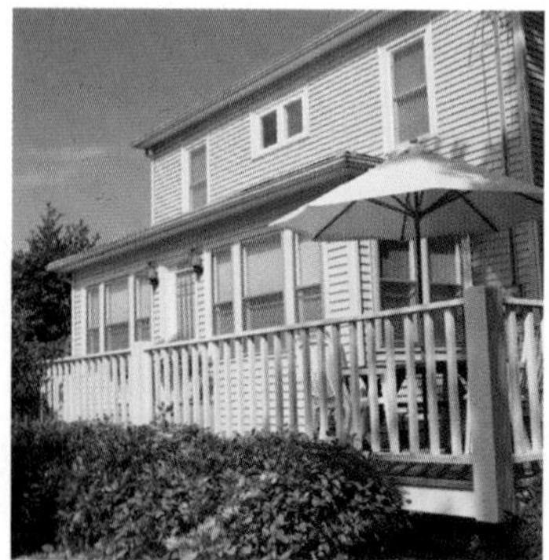

The plans for building a deck call for 28 square meters of deck boards. John has already bought 300 square feet. Does he have enough?

SOLUTION

John needs to convert 28 square meters to square feet to see how much wood is required. This is done using the unit fraction $\frac{1 \text{ ft}^2}{0.093 \text{ m}^2}$.

$$28 \cancel{\text{m}^2} \cdot \frac{1 \text{ ft}^2}{0.093 \cancel{\text{m}^2}} \approx 301.1 \text{ ft}^2$$

The 300 ft^2 that John bought is not quite enough.

▼ Try This One 4

Fussy University prohibits any displays on bulletin boards that have an area greater than 300 square inches. Does a poster with an area of 1,700 square centimeters meet this restriction?

EXAMPLE 5 Converting Area from Metric to English Units

Prince Edward County in Ontario, Canada, has an area of 1,050.1 square kilometers. Is it larger or smaller than Butler County, Ohio, which is 470 square miles?

SOLUTION

To find how many square miles in 1,050.1 square kilometers, multiply by the unit fraction $\frac{1\text{ mi}^2}{2.59\text{ km}^2}$.

$$1{,}050.1\ \cancel{\text{km}^2}\cdot\frac{1\text{ mi}^2}{2.59\ \cancel{\text{km}^2}}\approx 405.4\text{ mi}^2$$

Butler County is about 64.6 square miles larger.

▼ Try This One 5

The campus of the University of Guelph in Ontario has an area of 4.1 km^2. How many square miles is that?

The next example shows a very common and very useful application of area conversions for homeowners.

EXAMPLE 6 Finding the Cost of Carpeting a Room

Jane plans to have new carpet installed in her living room. She measures and finds that the room is 300 ft^2. The carpet she wants sells for $14 per square yard installed. How much will it cost to have the room carpeted?

SOLUTION

We first convert square feet to square yards, then use the unit fraction $\frac{\$14}{1\text{ yd}^2}$ to find the cost.

$$300\ \cancel{\text{ft}^2}\cdot\frac{1\ \cancel{\text{yd}^2}}{9\ \cancel{\text{ft}^2}}\cdot\frac{\$14}{1\ \cancel{\text{yd}^2}}=\$466.67$$

▼ Try This One 6

 1. Convert units of area.

Find the cost of carpeting Jane's basement, which is 520 square feet, with cheaper carpet costing $9.50 per square yard.

Conversions of Volume and Capacity

Just as the measurements of area are based on the two-dimensional square, measurements of volume are based on the three-dimensional analog, the cube. So volume is measured in cubic units.

In the English system, this is typically cubic inches (in.^{3}), cubic feet (ft^3), and cubic yards (yd^3). For example, 1 cubic foot consists of a cube whose measure is 12 inches on each side. Recall that to find the volume of a cube, we multiply the length times the width times the height. Since all these measures are the same for a cube, 1 cubic foot = 12 inches × 12 inches × 12 inches or 1,728 cubic inches. Also, 1 cubic yard = 3 feet × 3 feet × 3 feet or 27 cubic feet. See Figure 9-3.

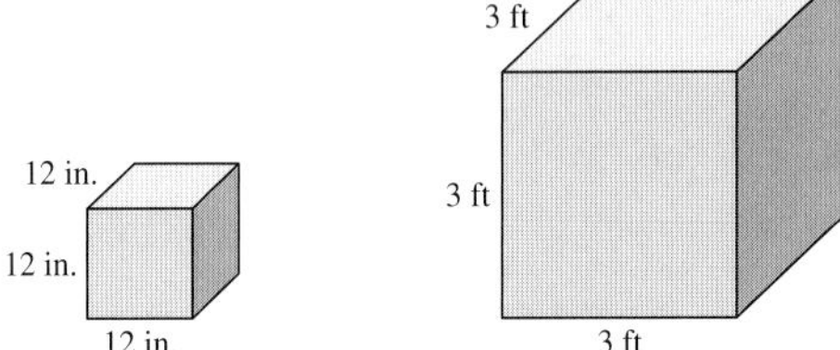

Figure 9-3

Measures of volume also include measures of capacity. The capacity of a container is equal to the amount of fluid the container can hold. In the English system, capacity is measured in fluid ounces, pints, quarts, and gallons. The conversion factors for capacity are shown in Table 9-6.

TABLE 9-6 Units of Capacity in the English System

1 pint (pt) = 16 fluid ounces (oz)
1 quart (qt) = 2 pints (pt)
1 gallon (gal) = 4 quarts (qt)

Since volume and capacity are both measures of size in three dimensions, we can compare the units.

A cubic foot (ft^3) of water is about 7.48 gallons, and a cubic yard of water is about 202 gallons. If a gallon of water is poured into a container, it would take up about 231 cubic inches. Finally, a cubic foot of freshwater weighs about 62.5 pounds and a cubic foot of seawater weighs about 64 pounds. These measures are summarized in Table 9-7.

TABLE 9-7 Conversion Factors for Capacity in the English System

1 cubic foot $\approx$ 7.48 gal
1 cubic yard $\approx$ 202 gal
1 gal. $\approx$ 231 $in.^3$
1 ft^3 freshwater $\approx$ 62.5 lb
1 ft^3 seawater $\approx$ 64 lb

EXAMPLE 7 Converting Cubic Feet to Gallons

If a water tank at an aquarium has a volume of 3,000 cubic feet, how many gallons of water will the tank hold?

SOLUTION

From Table 9-6, 1 cubic foot = 7.48 gallons.
Use dimensional analysis as shown.

$$3{,}000\ \cancel{ft^3} \cdot \frac{7.48\text{ gal}}{1\ \cancel{ft^3}} = 22{,}440 \text{ gallons}$$

▼ Try This One 7

How many cubic feet of water are in a 100-gallon freshwater fish tank?

EXAMPLE 8 Finding Weight of a Liquid

If the tank in the previous example contains freshwater, find its weight.

SOLUTION

1 cubic foot of freshwater weighs 62.5 pounds, so $\frac{62.5\text{ lb}}{1\text{ ft}^3}$ is a unit fraction.

$$3{,}000\text{ cubic feet} = \frac{3{,}000\ \cancel{\text{ft}^3}}{1} \cdot \frac{62.5\text{ pounds}}{1\ \cancel{\text{ft}^3}} = 187{,}500\text{ pounds}$$

▼ Try This One 8

What is the weight of the water in the freshwater fish tank in Try This One 7?

EXAMPLE 9 Computing the Weight of Water in Bottles

The team managers for a college football team carry quart bottles of water for the players to drink during breaks at practice. The bottles are stored in carrying racks that weigh 6 pounds and hold 16 bottles. How heavy (in pounds) is a rack when full? (Ignore the weight of the empty bottles).

SOLUTION

First we will need to know how much a quart of freshwater weighs.

$$1\ \cancel{\text{qt}} \cdot \frac{1\ \cancel{\text{gal}}}{4\ \cancel{\text{qt}}} \cdot \frac{1\ \cancel{\text{ft}^3}}{7.48\ \cancel{\text{gal}}} \cdot \frac{62.5\text{ lb}}{1\ \cancel{\text{ft}^3}} = 2.1\text{ lb}$$

So 16 bottles weigh $16 \times 2.1 = 33.6$ lb; adding the 6 pounds for the rack, a full load weighs 39.6 pounds.

▼ Try This One 9

A pallet of 1-pint water bottles delivered to a grocery store contains 288 bottles. How much does that much water weigh in pounds?

In the metric system, a cubic centimeter consists of a cube whose measure is 1 centimeter on each side. Since there are 100 centimeters in 1 meter, 1 cubic meter is equal to 100 cm × 100 cm × 100 cm or 1,000,000 cubic centimeters. See Figure 9-4.

The base unit for volume in the metric system is the liter. One liter is defined to be 1,000 cubic centimeters (see Figure 9-5). Cubic centimeters are abbreviated cm^3 or cc.

1 cm

1 cm 1 cm

1 cubic centimeter

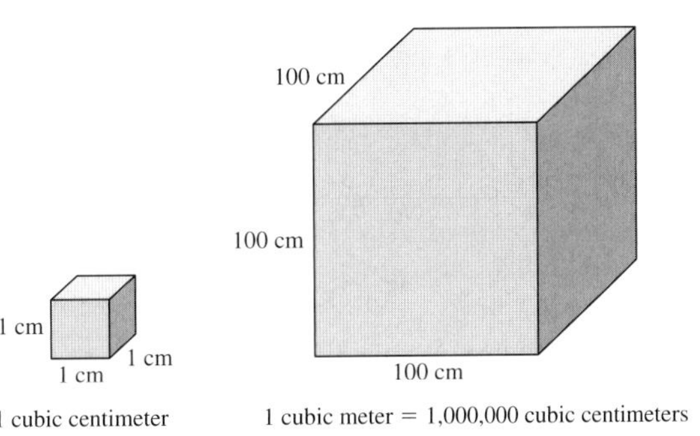

1 cubic meter = 1,000,000 cubic centimeters

Figure 9-4

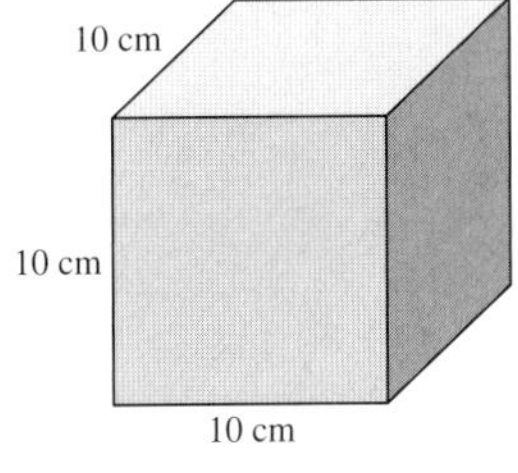

1 liter = 1,000 cubic centimeters

Figure 9-5

This provides a simple and interesting connection between two of the base units in the metric system, meters and liters: 1 cubic centimeter is the same as 1 milliliter. Comparisons of other volume units are shown in Table 9-8.

TABLE 9-8 Conversion Factors for Capacity in the Metric System

$1 \text{ cm}^3 = 1 \text{ mL}$
$1 \text{ dm}^3 = 1{,}000 \text{ cm}^3 = 1 \text{ L}$
$1 \text{ m}^3 = 1{,}000 \text{ dm}^3 = 1 \text{ kL}$
$1 \text{ L} \approx 1.06 \text{ quarts}$

EXAMPLE 10 Converting Metric Units of Volume

A typical backyard in-ground pool has a volume of about 82.5 cubic meters. How many milliliters of water are needed to fill such a pool?

SOLUTION

There's a conversion in Table 9-8 from mL to cm^3, so we first convert to cm^3, then to mL. Since 100 cm = 1 m, $(100 \text{ cm})^3 = (1 \text{ m})^3$; that is, $1 \text{ m}^3 = 1{,}000{,}000 \text{ cm}^3$.

$$82.5 \cancel{\text{m}^3} \cdot \frac{1{,}000{,}000 \cancel{\text{cm}^3}}{1 \cancel{\text{m}^3}} \cdot \frac{1 \text{ mL}}{1 \cancel{\text{cm}^3}} = 82{,}500{,}000 \text{ mL}$$

▼ Try This One 10

On average, 2,476,000 cubic centimeters of Coke are consumed worldwide every second. How many kiloliters is that?

The last equation in Table 9-8 provides a conversion between English and metric units of volume.

EXAMPLE 11 Converting Liters to Gallons

How many gallons of Mountain Dew are in one 2-liter bottle?

SOLUTION

We know that 1 liter is about 1.06 quarts, and that 1 gallon is 4 quarts:

$$2 \cancel{\text{L}} \cdot \frac{1.06 \cancel{\text{qt}}}{1 \cancel{\text{L}}} \cdot \frac{1 \text{ gal}}{4 \cancel{\text{qt}}} \approx 0.53 \text{ gal}$$

▼ Try This One 11

☑ 2. Convert units of volume.

How many liters of gas does an SUV with a 21-gallon tank hold?

Answers to Try This One

1 439.2 in.^{2}

2 1,593,600 acres

3 6,388.8 yd^2

4 Yes, it is about 263.6 in.^{2}.

5 1.58 mi^2

6 \$548.89

7 13.37 ft^3

8 835.625 lb

9 300.8 lb

10 2.476 kL

11 79.25 L

EXERCISE SET 9-2

Writing Exercises

1. Explain what "square units" mean. Use examples like "square inches" and "square miles."
2. If 1 yd = 3 ft, how come 1 yd$^2 \neq$ 3 ft^2?
3. What is the connection between the base units of length and capacity in the metric system?
4. Describe three things in your house, dorm, or apartment that could be measured using linear units, and explain why those units would be appropriate.
5. Repeat Question 4 for square units.
6. Repeat Question 4 for cubic units.

Computational Exercises

For Exercises 7–76, convert each measurement to the specified equivalent unit.

7. 3,456 in.^{2} = ________ ft^2
8. 245,678 ft^2 = ________ acres
9. 4,653 acres = ________ mi^2
10. 45,678,231 yd^2 = ________ mi^2
11. 12 yd^2 = ________ ft^2
12. 7 mi^2 = ________ acres
13. 345 yd^2 = ________ acres
14. 77,234,587 in.^{2} = ________ acres
15. 654,543 in.^{2} = ________ yd^2
16. 10 yd^2 = ________ in.^{2}
17. 18 in.^{2} = ________ cm^2
18. 52 ft^2 = ________ m^2
19. 40 m^2 = ________ yd^2
20. 72 mi^2 = ________ km^2
21. 32 acres = ________ km^2
22. 43 in.^{2} = ________ cm^2
23. 18 ft^2 = ________ dm^2
24. 93 cm^2 = ________ in.^{2}
25. 3 yd^2 = ________ dm^2
26. 15.6 m^2 = ________ ft^2
27. 103 km^2 = ________ acres
28. 80,000 cm^2 = ________ in.^{2}
29. 42 dm^2 = ________ in.^{2}
30. 19.4 cm^2 = ________ ft^2
31. 1,875 in.^{2} = ________ dm^2
32. 35 yd^2 = ________ m^2
33. 5,326 mm^2 = ________ in.^{2}
34. 152 acres = ________ km^2
35. 777 dm^2 = ________ ft^2
36. 1,000,000 m^2 = ________ yd^2
37. 3 cubic feet = ________ fluid ounces
38. 6 cubic yards = ________ gallons
39. 400 gallons = ________ cubic feet
40. 3,724 gallons = ________ cubic yards
41. 12,561 fluid ounces = ________ gallons
42. 12,000 cubic inches = ________ cubic feet
43. 22,000 cubic feet = ________ gallons
44. 32 pints = ________ cubic inches
45. 8 cubic yards = ________ gallons
46. 4,532 cubic inches = ________ gallons
47. 4.5 liters = ________ cubic centimeters
48. 97 milliliters = ________ cubic meters
49. 28.5 cubic centimeters = ________ milliliters
50. 140 cubic decimeters = ________ deciliters
51. 433 milliliters = ________ cubic centimeters
52. 87,250 cubic centimeters = ________ liters
53. 32 liters = ________ cubic centimeters
54. 437 centiliters = ________ cubic meters
55. 32 centiliters = ________ cubic meters
56. 1.6 liters = ________ cubic centimeters
57. 500 mL = ________ L
58. 80 hl = ________ L
59. 92 L = ________ dL
60. 48 daL = ________ hL
61. 8 dL = ________ cL
62. 42 L = ________ mL

63. 6.7 hL = _______ daL
64. 92 L = _______ hL
65. 12 gallons = _______ L
66. 1.9 L = _____ gallons
67. 19 L = _____ gallons
68. 155 gallons = _____ L
69. 19 quarts = _______ mL
70. 1,500 mL = _______ quarts
71. 21,050 mL = _______ quarts
72. 27 quarts = _______ mL
73. 175 kL = _______ gallons
74. 940 pints = _______ kL
75. 11,460 ounces = _______ kL
76. 84.1 kL = _______ gallons
77. Find the weight of 500 cubic feet of freshwater.
78. Find the weight of 23 cubic feet of saltwater.
79. Find the volume in cubic feet of 200 pounds of freshwater.
80. Find the volume in cubic feet of 755 pounds of saltwater.

Real-World Applications

81. Becky wants to paint the walls in her bedroom. A gallon of paint will cover 400 ft^2. She measures the surface to be painted and it is 69,120 in.^{2}. How many gallons of paint should she buy?
82. Jose has a 15-m^2 garden and plans to buy some mulch to cover his garden. On the bag of mulch it says that one bag covers 20 ft^2. If each bag costs $2, how much will it cost to mulch his garden?
83. Dan wants to pour a concrete patio in his backyard. He measures the length, width, and thickness of the patio and finds that the amount of concrete he needs is 200 ft^3. When he called the concrete company they asked how many cubic yards of concrete he wanted. Help Dan answer the question.
84. Chloe drank two 20-oz bottles of Diet Pepsi while she was studying. She poured the soda into the only clean glass she could find, which was a 1-pint beer glass. How many glasses did she fill?
85. A new campus was built on a 1,000-acre plot of land. How many square miles does it cover?
86. A new extension center is being built on 5 acres of campus north of the engineering building. How many square yards is the extension center being built on?
87. A pump added 5 gallons of water to the pool to level it after a diver displaced too much water during a cannonball jump. How many cubic feet of water were added?
88. A bottle containing a quarter of a cubic meter of a potent drug is half empty. Nurses give doses of 100 cc at a time. How many doses are left?
89. Sally wants to tile the backsplash in her kitchen. She measures and finds that the area she plans to cover is 32 square feet. She orders custom tiles that measure 8 square inches. How many tiles should she order?
90. A store has twenty 1-quart containers of milk, but the manager wants to pour the milk into gallon containers to sell at a higher price. How many gallon containers can the manager fill?
91. Five cubic feet of seawater have flooded into a boat through a leak in the side. If the boat itself weighs 200 lb, how much does it weigh with the seawater?
92. Gianna determines she needs 10,000 square inches of material to make a slipcover for her sofa. At the fabric store, the material is sold by the square yard. How much of the fabric she chose should she tell the clerk to cut?
93. Shelly buys eight 20-oz containers of Gatorade for a race, but the bikers' water bottles are quart-sized. If she fills each biker's bottle completely, how many bikers will have Gatorade for the race?
94. A dump truck with 1,500 gallons of soil arrives on campus to fill in the new planters on the quad. Each planter needs 2 cubic yards of soil. How many planters can be filled?
95. A search party covered 6 square miles of territory looking for a prized statue that fell from a cargo plane. How many square kilometers did they search?
96. A Jacuzzi was filled with 200 gallons of water. The Jacuzzi itself weighed 50 lb. If a person who weighs 120 lb and a person who weighs 180 lb get into the Jacuzzi, how much is the total weight including the two people, the tub, and the water?

Critical Thinking

97. The capacity of a dump truck is listed by the manufacturer as 10 yd^3. A construction site requires the removal of 200 m^3 of dirt, and the contractor has three identical trucks. How many trips will each need to make?
98. After the excavation at the construction site in Exercise 97, a period of bad weather delays the construction. When the weather clears up, the crew chief finds that the excavated dirt has been replaced with water that now needs to be pumped out of the hole dug for the foundation. The contractor has two industrial pumps that can each pump water out at the rate of 750 gallons per minute. How long will the pumps need to empty the site?
99. Traditionally, carpet and tile have been priced by the square yard, but in recent years, more and more stores are listing prices by the square foot. Why do you suppose they made this change?

Section 9-3 Measures of Weight and Temperature

LEARNING OBJECTIVES

- ❑ 1. Converting weights in the English system.
- ❑ 2. Converting weights in the metric system.
- ❑ 3. Converting weights between English and metric units.
- ❑ 4. Converting temperatures between the Fahrenheit and Celsius scales.

According to the Nintendo Corporation, the weight limit for the fitness board that can be used with its Wii game system is 150 kg. If you're like the average American, you have no idea exactly how heavy you can be and still safely use the board. This, of course, is because you are so accustomed to weights being measured in pounds. But like any other form of measurement, there are a variety of units that can be used for weight. We will study them in this section, then conclude with a quick look at measuring temperatures.

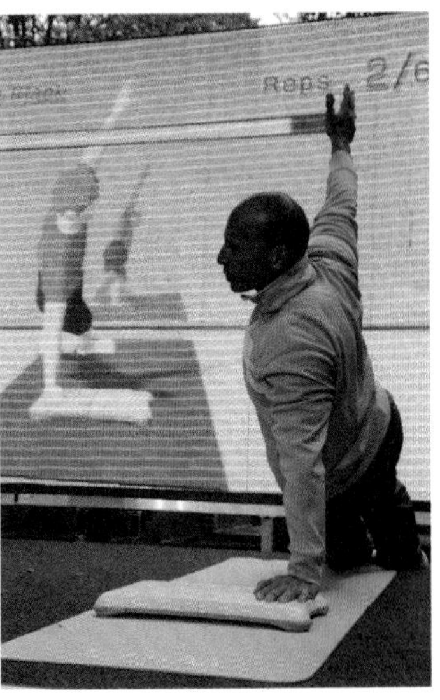

Conversions for Weight

In Section 9-1, we learned that weight and mass are not quite the same thing. But in this section, we'll assume that all the objects we're weighing are earthbound, so that mass and weight are equivalent.

In the English system, weight is most often measured in ounces, pounds, and tons. In the metric system, the base unit for weight is the gram, but this is a very small unit—a dollar bill weighs about a gram, for example. So a more common unit is the kilogram, which is 1,000 grams. There is also a metric equivalent of the ton: a metric ton, sometimes written "tonne," is 1,000 kg. The basic conversions for weight, along with the symbols used to represent units, are shown in Table 9-9.

TABLE 9-9 Units of Weight

English Units	Metric Units	English/Metric Conversions
16 ounces (oz) = 1 pound (lb)	1 kilogram (kg) = 1,000 grams (g)	1 kg ≈ 2.2 lb
2,000 lb = 1 ton (T)	1,000 kg = 1 metric ton (t)	1 oz ≈ 28 g

Conversions within the metric system follow the same procedures we learned for converting lengths in Section 9-1. In the English system, our old friend dimensional analysis will do the work for us.

EXAMPLE 1 Converting Weights in the English System

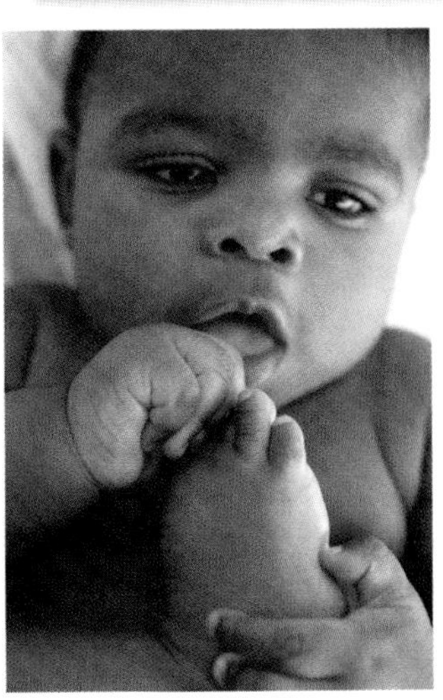

According to the Guinness Book of World Records, the largest healthy baby ever delivered weighed 360 ounces. How much did he weigh in pounds?

SOLUTION

The unit fraction we need is $\frac{1 \text{ lb}}{16 \text{ oz}}$:

$$360 \cancel{\text{oz}} \cdot \frac{1 \text{ lb}}{16 \cancel{\text{oz}}} = 22.5 \text{ lb}$$

This is almost exactly 3 times as large as an average newborn boy!

▼ Try This One 1

How many ounces is a 12-pound bowling ball?

EXAMPLE 2 Converting Weights in the English System

An empty Boeing 747 weighs about 175 tons. How much does it weigh in ounces?

SOLUTION

We will need to first convert tons to pounds, then pounds to ounces.

$$175\,\cancel{T} \cdot \frac{2{,}000\,\cancel{lb}}{1\,\cancel{T}} \cdot \frac{16\text{ oz}}{1\,\cancel{lb}} = 5{,}600{,}000\text{ oz}$$

▼ Try This One 2

☑ 1. Converting weights in the English system.

An average male African elephant weighs about 176,000 ounces. How many tons does he weigh?

EXAMPLE 3 Converting Metric Units of Weight

Convert each of the following:

(a) 150 grams of protein powder to milligrams
(b) 23 grams of salt to kilograms
(c) 3 kilograms of chicken to milligrams

SOLUTION

For reference, here is the order of the metric prefixes from Table 9-4 from largest to smallest: kilo, hecto, deka, base unit, deci, centi, milli.

Math Note

Recall that when converting metric units, we multiply or divide by powers of 10. When moving three steps in Table 9-4, we multiply or divide by 10^3, or 1,000. When moving six steps, we multiply or divide by 10^6, or 1,000,000.

(a) The prefix milli is three steps below the base unit, so multiply by 1,000 (or move the decimal three places right).

$$150\text{ g} = 150.000\text{ g} = 150{,}000\text{ mg}$$

(b) The prefix kilo is three units above the base unit, so divide by 1,000 (or move the decimal three places left).

$$23\text{ g} = 0023\text{ g} = 0.023\text{ kg}$$

(c) The prefix milli is six steps below the prefix kilo, so multiply by 1,000,000 (which is easy in this case, without moving the decimal).

$$3\text{ kg} \times 1{,}000{,}000 = 3{,}000{,}000\text{ mg}$$

▼ Try This One 3

☑ 2. Converting weights in the metric system.

Convert each of the following:

(a) A 10-kilogram dumbbell to grams
(b) A 250-milligram aspirin to grams
(c) 400 milligrams of a steroid to kilograms

One of the most common uses of weight conversion is converting between English and metric units, as in Example 4.

EXAMPLE 4 Converting between English and Metric Units of Weight

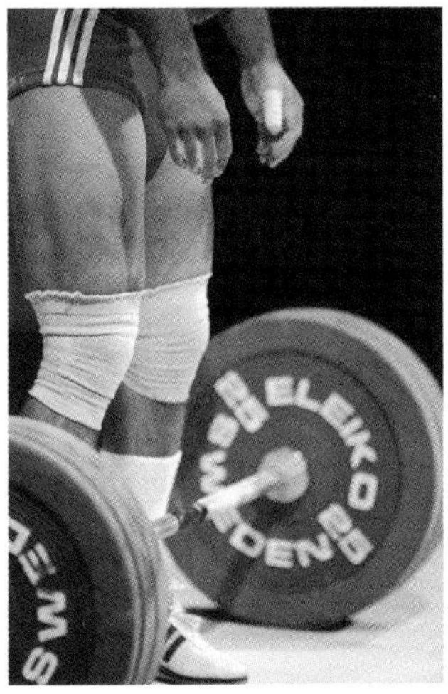

(a) The heaviest amount ever successfully lifted in Olympic competition was in the clean and jerk competition in Athens in 2004: 263.5 kg. How many pounds is that?

(b) An average adult hummingbird weighs one-eighth of an ounce. How many grams does it weigh?

SOLUTION

(a) We can use the unit fraction $\frac{2.2\text{ lb}}{1\text{ kg}}$:

$$263.5\ \cancel{\text{kg}} \cdot \frac{2.2\text{ lb}}{1\ \cancel{\text{kg}}} \approx 579.7\text{ lb}$$

(b) This time, the unit fraction $\frac{28\text{ g}}{1\text{ oz}}$ is helpful:

$$\frac{1}{8}\ \cancel{\text{oz}} \cdot \frac{28\text{ g}}{1\ \cancel{\text{oz}}} \approx = \frac{28}{8}\text{ g} \approx 3.5\text{ g}$$

Math Note

Since 1 kg ≈ 2.2 lb, for a very quick (and very imprecise) idea of the equivalent in pounds for a weight in kilograms, double the weight and add a bit more.

▼ Try This One 4

(a) One of the weight classes in Olympic boxing is 81 kg. How heavy is that in pounds?

(b) An average golden retriever puppy weighs $12\frac{1}{2}$ ounces at birth. How many grams does it weigh?

EXAMPLE 5 Assessing Weight Loss

Megan decides that she absolutely will not go to Florida for spring break unless she loses 15 pounds in the next 3 months. She finds an advertisement online for a British weight loss system that guarantees loss of 3 kg per month. If that claim is true, will she reach her goal?

SOLUTION

We can do this in one calculation by multiplying by the unit fraction $\frac{2.2\text{ lb}}{1\text{ kg}}$ to eliminate kg, then multiplying by 3 months:

$$\frac{3\ \cancel{\text{kg}}}{1\ \cancel{\text{month}}} \cdot \frac{2.2\text{ lb}}{1\ \cancel{\text{kg}}} \cdot 3\ \cancel{\text{months}} \approx 19.8\text{ lb}$$

If the claim is accurate (a *very* big if), Megan should start saving money for her trip.

▼ Try This One 5

☑ 3. Converting weights between English and metric units.

Bart's football coach tells him that he'll start next year if he gains 10 pounds in the 14 weeks between spring practice and summer drills. He hires a personal trainer who promises him a gain of 1 kg every 3 weeks. Will he make his goal?

Sidelight IT'S ABSOLUTELY FREEZING!

You're probably perfectly satisfied with the Fahrenheit scale, and when you mix in Celsius, that seems more than sufficient for measuring temperatures. Physicists and chemists would tend to disagree with you, however. In their view, temperature is a measure of how much heat is contained in some medium, so negative temperatures just don't seem right. In 1948, a third temperature scale was developed to address this issue. It is known as the **Kelvin** scale, in honor of the British physicist who introduced it. Zero degrees Celsius might be a bit warm for our friend the polar bear, and zero degrees Fahrenheit is more to his liking. But zero Kelvins (the terminology used, rather than "zero degrees Kelvin") would be far too cold even for him.

On the Kelvin scale, one degree corresponds to the same amount of temperature difference as the Celsius scale. The difference is the temperature that corresponds to zero degrees. On the Kelvin scale, this temperature is known as **absolute zero,** which is the coldest possible temperature. Heat comes from an interaction of molecules, and at absolute zero, there is no heat because molecules stop moving. This temperature corresponds to –459.67°F and –237.15°C, and because one Kelvin is the same amount of heat as one degree Celsius, you can convert temperatures between those two scales by simply adding or subtracting 237.15°.

It is theoretically impossible to cool anything to absolute zero, but in 2000, Finnish scientists reported reaching a low of 1 ten-billionth of a Kelvin in the lab.

Conversions for Temperature

Temperature is measured in the English system using degrees **Fahrenheit.** On this scale, water freezes at 32° and boils at 212°. The average temperature of the human body is 98.6°. In the metric system, temperatures are measured in degrees **Celsius** (also known as **centigrade**). Like all metric measurements, the Celsius scale is based on powers of 10 in some form: water freezes at 0° and boils at 100°. The average temperature of the human body is 37°.

One degree Celsius corresponds to $\frac{9}{5}$ degrees Fahrenheit, and there are relatively simple formulas for converting between the two.

Fahrenheit–Celsius Conversions

To convert Celsius to Fahrenheit:

$$F = \frac{9}{5}C + 32$$

To convert Fahrenheit to Celsius:

$$C = \frac{5}{9}(F - 32)$$

EXAMPLE 6 Converting Celsius to Fahrenheit

In preparing for vacation, Randy and Catalina check the Internet and find that the average temperature over the next week at their destination is predicted to be 28° Celsius. Should they pack coats or bathing suits?

SOLUTION

Substitute 28 for C in the first formula above:

$$\begin{aligned} F &= \frac{9}{5}C + 32 \\ &= \frac{9}{5} \cdot 28 + 32 = 82.4° \end{aligned}$$

Swimsuits and sunblock it is!

▼ Try This One 6

The average high temperature for March in Moose Jaw, Saskatchewan, is –1° Celsius. Find the temperature on the Fahrenheit scale.

EXAMPLE 7 Converting Fahrenheit to Celsius

A frozen pizza needs a 450° Fahrenheit oven to cook. Suppose you bought a cheap oven on eBay, and found that it had been scavenged from Romania, so the temperatures on the dial are Celsius. At what temperature should you set the oven?

SOLUTION

Substitute $F = 450$ into the second conversion formula:

$$C = \frac{5}{9}(F - 32)$$

$$C = \frac{5}{9}(450 - 32) \approx 232.2°$$

You should set it close to 230°.

▼ Try This One 7

☑ 4. Converting temperatures between the Fahrenheit and Celsius scales.

The pizza box recommends 225° Fahrenheit for rewarming. Where should the dial be set for rewarming?

Answers to Try This One

1 192 oz

2 $5\frac{1}{2}$ T

3 (a) 10,000 g (b) 0.250 g (c) 0.000400 kg

4 (a) ≈178.2 lb (b) ≈350 g

5 Yes, by about 0.3 lb

6 30.2°F

7 About 110°C (225°F ≈ 107.2°C)

EXERCISE SET 9-3

Writing Exercises

1. Even though mass and weight are technically different, we usually treat them as equivalent. Why is that reasonable?
2. Metric measurements are based on powers of 10. How does the Celsius scale fit that pattern?

Computational Exercises

For Exercises 3–12, convert to the indicated unit.

3. 33 oz to pounds
4. 48 oz to pounds
5. 2.3 lb to ounces
6. 3.1 lb to ounces
7. 4.2 T to pounds
8. 3.8 T to pounds
9. 3,500 lb to tons
10. 2,350 lb to tons
11. 34,567 oz to tons
12. 2.1 T to ounces

For Exercises 13–32, convert each metric weight to the indicated unit.

13. 9 dg = ______ cg
14. 6 mg = ______ cg
15. 44 kg = ______ hg
16. 16.34 hg = ______ g
17. 18 dg = ______ cg
18. 27 cg = ______ g
19. 71 cg = ______ dg
20. 215 g = ______ dag
21. 5 g = ______ dg
22. 32 kg = ______ hg
23. 0.325 g = ______ mg
24. 3,217 cg = ______ g
25. 4,325 kg = ______ g
26. 5 dag = ______ kg
27. 86 mg = ______ g
28. 24 hg = ______ g
29. 400 g = ______ kg
30. 6.6 kg = ______ g
31. 5,632 g = ______ hg
32. 150 mg = ______ cg

For Exercises 35–52, convert each unit to the specified equivalent unit.

33. 120 grams = ______ ounces
34. $15\frac{3}{4}$ ounces = ______ grams
35. 4,823 centigrams = ______ pounds
36. 27 metric tons = ______ pounds
37. 3 tons = ______ hectograms
38. 14 decigrams = ______ ounces
39. 357,201 pounds = ______ metric tons
40. 13 pounds = ______ decigrams
41. 5.75 tons = ______ metric tons
42. 16.3 metric tons = ______ tons
43. 213 ounces = ______ decigrams
44. 64 pounds = ______ dekagrams
45. 815 dekagrams = ______ ounces
46. 37 tons = ______ metric tons
47. 183 ounces = ______ decigrams
48. 550 hectograms = ______ ounces
49. 27 pounds = ______ decigrams
50. 14,625 milligrams = ______ pounds
51. 41 pounds = ______ grams
52. 42 grams = ______ ounces

For Exercises 53–62, convert each Celsius temperature to an equivalent Fahrenheit temperature.

53. 14°C
54. 27°C
55. 55°C
56. 100°C
57. 150°C
58. –5°C
59. –18°C
60. –20°C
61. –33°C
62. –50°C

For Exercises 63–72, convert each Fahrenheit temperature to an equivalent Celsius temperature.

63. 5°F
64. 27°F
65. 32°F
66. 158°F
67. 100°F
68. –3°F
69. –10°F
70. –22°F
71. –14°F
72. 212°F

Real-World Applications

73. Cindy and Juan are the proud parents of a 100-oz baby. Juan's parents think this sounds like a rather large baby! Convert the baby's weight to pounds and decide if Juan's parents are correct.
74. Billy's pickup truck is capable of carrying $\frac{3}{4}$ ton. Billy is picking up a load of rock that weighs 1,700 pounds. Will his pickup truck be able to handle the load?
75. The maximum weight a certain elevator can withstand is 4,400 kilograms. Ten large men, weighing about 210 pounds each, step into the elevator. Will the elevator hold them?
76. On the scale at the doctor's office Susan weighed 68 kilograms. The healthy weight for a person her height is 130 lb. Does Susan need to go on a diet?
77. Olivia Sanchez is going to visit Europe on vacation; she checked the local weather of the country she's visiting and found that the average temperature this time of year is about 35°C. This temperature sounds pretty cold to the American-born Olivia, but she decides before packing she should convert this to °F. After converting, will Olivia pack a coat?
78. For a chicken Marsala recipe, Janine needed 24 oz of boneless, skinless chicken breasts. At the supermarket, the chicken was measured in pounds. How many pounds did she need to buy?
79. Five cars weighing 2,540 lb each are put onto a ferry. The ferry has a maximum weight capacity of 6 tons. Will the ferry be able to hold all the cars without sinking into the bay?
80. A chemistry teacher has 750 grams of a substance and he wants to separate the substance into 3-oz jars. How many 3-oz jars can he fill?
81. A thermometer in the window of your car reads 20 degrees Celsius, but the other side where the Fahrenheit scale is located has faded in the sun. What is the Fahrenheit temperature?
82. On a summer trip to France, you catch a virus and feel like you have a fever. The only thermometer they sell at the store shows you have a 38.8 degrees Celsius temperature. What is your Fahrenheit temperature?
83. The border patrol seizes 5 lb of an illegal substance at a check point. How many grams of the substance did they seize?
84. While building a new playground at the city recreation center, a truck pours 3,000 kg of sand into the foundation. The specs require 2.5 metric tons of sand. Did the truck pour the right amount into the foundation?
85. Shirl lost 3.6 lb of fat on her diet. How many ounces of fat did she lose?
86. During a dinner at Outback Steakhouse, a family of six eats four 12-oz filet mignons and two 20-oz porterhouse steaks. How many pounds of steak did the family consume during the dinner?

87. An elevator says the maximum capacity is 1,000 kg. Twelve people are already on the elevator, their weight totaling 2,000 lb. Another man who weighs 260 lb wants to get on. Should they let him onto the elevator?
88. To mold silver into a ring, Tawny has to heat it up to its boiling point, which is 3,924 degrees Fahrenheit. Her heating element can only heat to 2,000 degrees Celsius before it will cool down for safety reasons. Can she make the ring?
89. The cement Yuan bought at the hardware store weighs 52 lb per cubic foot. To pour his driveway, the specifications require at least a weight of 25 kg per cubic foot. Did Yuan buy the right type of cement?
90. A 1.7-ton truck crosses the border into Canada. The scale measures in kilograms. How much does the scale say the car weighs?
91. The sugar fructose melts at between 103 and 105 degrees Celsius. Julia heats up the fructose on a burner to 222.8 degrees Fahrenheit. Is the sugar melted yet?
92. A lab tech had a fifth of a kilogram left of a substance and wanted to divide it into 50 bags for the next day's experiment. Each bag needed 4,500 mg of the substance. Would the lab tech have enough?
93. On a see-saw, three Americans weighing a total of 450 lb are sitting on one side. Two Europeans weighing a total of 137 kg sit on the other side. What would a person have to weigh in kilograms to balance out the see-saw?

CHAPTER 9 Summary

Section	Important Terms	Important Ideas
9-1	English system of measurement Dimensional analysis Unit fractions The metric system Base unit	**Unit** fractions such as $\frac{1 \text{ ft}}{12 \text{ in.}}$ are used to convert between different units of measure. Conversions within the metric system can be done by moving the decimal the appropriate number of places, based on the prefix of the units. Conversions between metric and English units of measure can be done with dimensional analysis.
9-2	Linear units Square units Cubic units	**Area** is measured in square units. Volume is measured in cubic units. Squaring/cubing the conversions for linear measure will give the conversions necessary for measure of area/volume.
9-3	Pounds Ounces Tons Grams Fahrenheit Celsius	**Pounds,** ounces, and tons are units commonly used to measure weight in the English system. The base unit for weight in the metric system is the gram. Dimensional analysis is used to convert between units of weight. Degrees Fahrenheit are the units for temperature in the English system, and degrees Celsius are the units in the metric system. There are simple formulas to convert from one to the other.

MATH IN Travel REVISITED

1. The Fahrenheit equivalents are 75.2° and 89.6°, so beach wear is the way to go.
2. Thirty kilograms is about 66 pounds, so you can pack 62 pounds of clothes and personal items.
3. Walking to downtown shouldn't be a problem—0.6 kilometers is about 0.37 miles. The distance from the airport is about 2.6 miles, which will take just about 4½ minutes at 35 miles per hour.
4. You used 12 gallons, which is about 45.3 liters. At 43 cents per liter, the refill costs $19.48.

Review Exercises

Section 9-1

For Exercises 1–16, convert each of the following using dimensional analysis.

1. 4 yards to feet
2. 23,564 feet to miles
3. 21 inches to feet
4. 356 inches to yards
5. 8,245,264 inches to miles
6. 10 m to cm
7. 0.36 m to dm
8. 13.54 km to dm
9. 34,500 mm to m
10. 0.56 dam to hm
11. 7 m to inches
12. 35 m to yards
13. 85 cm to inches
14. 23 yd to dm
15. 345 feet to dm
16. 1,235,543 in. to km
17. How many meters are run in a 100-yard dash?
18. Traffic is backed up for 10 km on a major highway. How many miles is the backup?
19. The 2009 Chevrolet Corvette has a top speed of 205 miles per hour. How fast is that in kilometers per hour? In feet per second?
20. It is about 3,600 meters from the Lincoln Memorial to the United States Capitol. How long would it take to walk that distance at a speed of 3 miles per hour?
21. The Eiffel Tower stands 1,986 feet tall. The Chrysler Building in New York City is 319 meters. Which is taller?

Section 9-2

For Exercises 22–41, convert each of the following to the indicated units.

22. 5,698 in.^{2} to square feet
23. 543 yd^2 to acres
24. 2 mi^2 to square feet
25. 456,321 in.^{2} to square yards
26. 5,643 acres to square miles
27. 23 in.^{2} to square centimeters
28. 23 ft^2 to square decimeters
29. 123 acres to square kilometers
30. 32.5 cm^2 to square feet
31. 324 km^2 to square feet
32. 7 ft^3 to fluid ounces
33. 42,000 ft^3 to cubic inches
34. 6 yd^3 to gallons
35. 3.5 L to cubic centimeters
36. 45 pints to cubic inches
37. 300 L to mL
38. 673 hL to liters
39. 54,457 mL to kL
40. 231 cL to L
41. 45,672 dL to kL
42. A nurse is instructed to give a patient 500 cc's (cubic centimeters) of a drug. The only thing she has to measure with is marked off in milliliters, so she measures out 500,000 mL and gives it to the patient (her reasoning, be it right or not, is that milli is 10 times centi). Is the patient in danger?
43. The instructions on a dechlorinator for a fish tank says to put in five drops per gallon. Geoff has a fish tank that holds 2.27 cubic feet of water. How many drops should he use?
44. The football field and surrounding sideline area at Enormous State U. covers an area of 79,000 square feet. The athletic department decides to replace the natural grass with Field Turf, a popular synthetic alternative. The cost of installing field turf is $119 per square yard. How much will the project cost?
45. Juan is considering two properties to buy for investment purposes. One is 2 acres and will cost $41,000. The other is 9,500 square meters and will cost $46,000. Which costs less per square foot?
46. Ron's Lawnscaping Artists charge $45 for one fertilizer treatment on a 5,000-square-foot lawn. How much should they charge for a lawn that is 5,000 square meters?

Section 9-3

For Exercises 47–62, convert each of the following to the indicated units.

47. 65 oz to pounds
48. 2.4 T to pounds
49. 34,500 lb to tons
50. 1.3 T to ounces
51. 567,376 oz to tons
52. 7 dg to cg
53. 0.457 g to mg
54. 5.6 kg to g
55. 4 dag to kg
56. 34,345 g to hg
57. 130 g to ounces
58. 4,536 cg to pounds
59. 45.5 oz to grams
60. 156 oz to decigrams
61. 23,456 mg to pounds
62. 456 hg to ounces

For Exercises 63–66, convert each Celsius temperature to Fahrenheit.

63. 13°C
64. 23°C
65. –5°C
66. –12°C

For Exercises 67–70, convert each Fahrenheit temperature to Celsius.

67. 67°F
68. 92°F
69. –21°F
70. –9°F
71. How many pounds is a baby who weighs 3,200 g?
72. Heinrich's temperature is 38.5°C. A person is running a fever if his or her temperature is above 100°F. Does Heinrich have a fever?
73. If the high temperature for a day was 8° Celsius higher than the low temperature for that day, and the low temperature was 52° F, what was the high temperature in degrees Fahrenheit?
74. A contestant on the TV Show "The Biggest Loser" started out weighing 346 pounds and lost 58 kg on the show. What was the contestant's ending weight?
75. On the American side of Niagara Falls, a home improvement center sells landscaping rock for $65 per ton. On the Canadian side, a different business sells a similar stone for $58 per metric ton. Which is the better buy?

Chapter Test

Convert each of the following:

1. 132 in. to yards
2. 8,665 ft to miles
3. 0.123 m to centimeters
4. 2 hm to decimeters
5. 34 cm to inches
6. 12 ft to meters
7. 3,475 in.^{2} to square feet
8. 3,235 acres to square miles
9. 10 cm^2 to square inches
10. 12 ft^2 to square meters
11. 100 cm^3 to cubic inches
12. 15,154 in.^{3} to cubic meters
13. 3 yd^3 to gallons
14. 34 pints to cubic inches
15. 365 oz to pounds
16. 34,675 oz to tons
17. 245,231 oz to tons
18. 5 dg to cg
19. 250 g to ounces
20. 2,854 cg to pounds
21. 13°C to °F
22. 85°F to °C
23. A marathon is 26 miles. How many kilometers is a marathon?
24. How many pounds does a 88-kg man weigh?
25. Tyron has a choice between two carpets that he likes equally well. One costs $12.50 per square yard; the other costs $1.70 per square foot. Which is the better buy?
26. How many feet is a 5-km race?
27. It is about 353 km from Toronto to Ottawa. How long would it take to make that drive averaging 65 miles per hour?
28. The Ironman Triathlon is a race in which participants swim 2.4 miles, bike 112 miles, and run 26.2 miles. Find the total distance covered in kilometers.
29. A carpet installer brags that he can install Berber and pad at the rate of 50 square feet per hour. Convert that rate to square yards per minute.
30. The famous rotunda of the Capitol building has a floor area of about 7,238 square feet. A marble restoration company charges $40 per square yard for its services. How much would it cost to restore the entire floor in the rotunda?
31. A young family is looking to buy a house and decides they need at least a ¾-acre lot so there's room for the kids and their dog Moose to play. Their agent plans to show them a house with a 30,000-sq-ft lot. Will she be wasting her time?
32. The ill-fated ship *Titanic* weighed 46,000 tons. How many kilograms did it weigh?
33. The warmest month in Aruba is September, with an average high of 32.8° Celsius. The coolest month is January, with an average low of 24.4°C. What is the difference between these temperatures in degrees Fahrenheit?

Project

While the metric system has never been officially adopted by the United States government, at various times the government and other entities have made attempts to make use of the metric system more common. Using a library or the Internet as a resource:

a. Make a list of the top 20 countries in the world in terms of population, and find when each of those countries officially adopted the metric system.
b. Research the metric system in Great Britain. (While it has been officially adopted there, many people still cling to the English system. See if you can find any information on governmental attempts to encourage more people to abandon the English system.)
c. Write about any initiatives you can find in the United States to make the metric system either official, or at least more commonly used, both today and historically.
d. Make a list of difficulties that you think our society would encounter if the government decided to switch to the metric system for all measurements overnight.

CHAPTER 10

Geometry

Outline

10-1 Points, Lines, Planes, and Angles
10-2 Triangles
10-3 Polygons and Perimeter
10-4 Areas of Polygons and Circles
10-5 Volume and Surface Area
10-6 Right Triangle Trigonometry
10-7 A Brief Survey of Non-Euclidean and Transformational Geometries
Summary

MATH IN Home Improvement

It probably won't surprise you to learn that one of the most common questions that math students ask is "How can I actually use this stuff?" Of course, how math is used in our world is the main theme of this book, and this chapter fits that framework especially well. The ideas presented in this chapter are commonly used in everyday things like working around the home, so we'll present some actual projects that geometry was used for in the home of one of the authors.

Geometry is one of those topics that most people recognize when they see it, but might have a hard time writing an actual definition for. If you look up a mathematical definition on the Web, it will refer to specific figures, like points, lines, triangles, and so forth. A more casual definition might say that geometry is the study of physical shapes and objects. Since we live in a physical world, in a very real sense, geometry is the study of the world around us.

If you really look for simple geometric figures, you can see them almost everywhere. The roads or sidewalks you took to get where you are right now are kind of like lines. The desk or table you're working at is probably rectangular, and the pen or pencil you're writing with is basically a cylinder. One way to think of the value of geometry in studying our world is that most of the things we encounter are in some way made up of basic geometric figures. This makes geometry one of the easiest areas of math to apply to the real world.

In each of the projects described below, one or more of the techniques you will learn in this chapter was used to solve a problem. By the time you finish the chapter, you should be able to solve the problems yourself.

1. The diagram shows the measurements of all countertops in our kitchen. The two styles we liked cost \$42 per square foot and \$60 per square foot. How much did we save by going with the less expensive style?

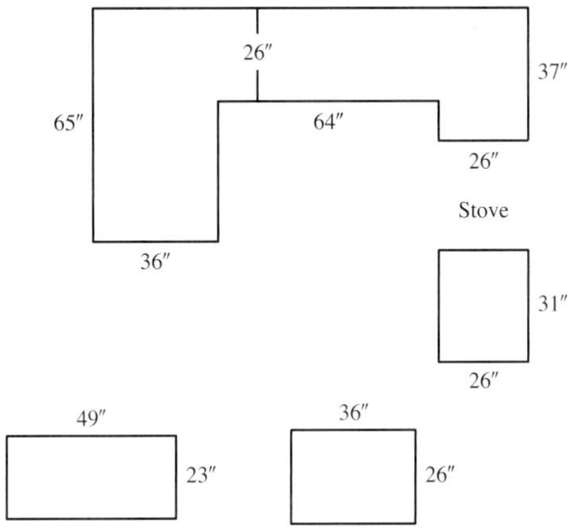

2. When finishing the basement, we needed trim along the new stairwell. From the landing at the top of the stairs in the top right photo on the facing page, the stairs extend 57 inches horizontally, and drop 42 inches. I wanted to cut the trim running along the stairs so that the bottom was parallel to the floor. At what angle did it need to be cut?
3. We built a bar out of 28 glass blocks arranged in four rows of seven each. Each block is $7\frac{3}{4}$ inches on a side, with $\frac{1}{4}$-inch mortar joints in between. We wanted to put a 3-inch trim around the outside of the blocks. How many square feet of wood did we need for the trim?
4. The boards for a cornbag toss game are 4 feet long. The bottom edge has a height of 4 inches, and the top edge has a height of 12 inches. The legs are attached so that they are perpendicular to the top. What angle did the bottom of the legs need to be cut at so that they sit flat on the ground?

For answers, see Math in Home Improvement Revisited on page 566

Section 10-1 Points, Lines, Planes, and Angles

LEARNING OBJECTIVES

- ☐ 1. Write names for angles.
- ☐ 2. Use complementary and supplementary angles to find angle measure.
- ☐ 3. Use vertical angles to find angle measure.
- ☐ 4. Find measures of angles formed by a transversal.

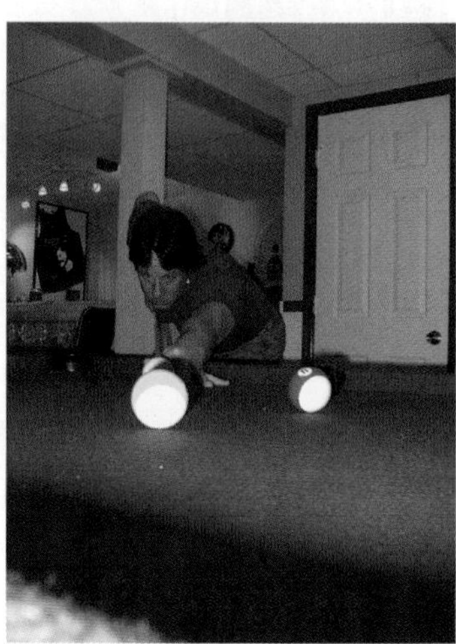

If you've ever watched really good billiards players, you probably noticed that they make bank shots look very easy. But if you try them yourself, you find they're not so easy at all. The secret to bank shots, and really to being a good pool player at all, is angles. Understanding the relationship between the angle at which two balls hit and the angles at which they move after impact is the most important part of the game.

There are many, many applications in which understanding angles is key, so our main focus in this section will be an understanding of angles. To begin, we will need to familiarize ourselves with some background information that will form the fundamentals of our entire study of geometry.

Points, Lines, and Planes

The most basic geometric figures we will study are points, lines, and planes. A **point** is easiest thought of as a location, like a particular spot on this page. We represent points with dots, but in actuality a point has no length, width, or thickness. (We call it *dimensionless.*) A **line** is a set of connected points that has an infinite length, but no width. We draw representations of lines, but again, in actuality a line cannot be seen because it has no thickness. We will assume that lines are straight, meaning that they follow the shortest path between any two points on the line. This means that only two points are needed to describe an entire line. A **plane** is a two-dimensional flat surface that is infinite in length and width, but has no thickness. You might find it helpful to think of a plane as an infinitely thin piece of paper that extends infinitely far in each direction. Examples of the way we represent these basic figures are shown in Figure 10-1.

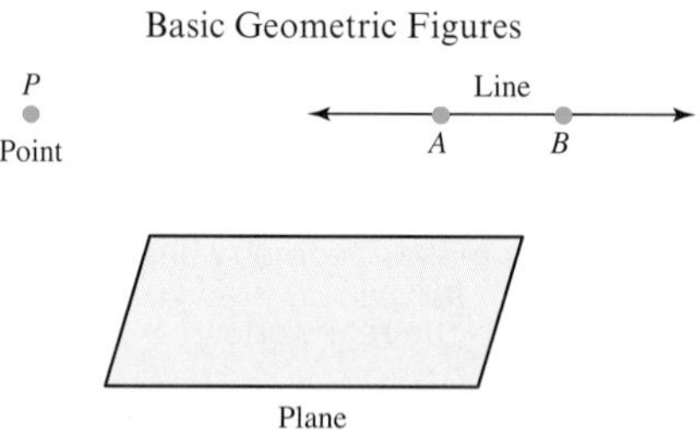

Figure 10-1

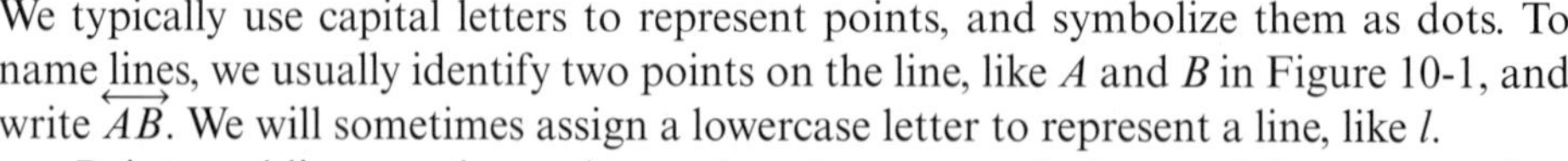

We typically use capital letters to represent points, and symbolize them as dots. To name lines, we usually identify two points on the line, like A and B in Figure 10-1, and write $\overleftrightarrow{AB}$. We will sometimes assign a lowercase letter to represent a line, like l.

Points and lines can be used to make other geometric figures. A **line segment** is a finite portion of a line consisting of two distinct points, called **endpoints,** and all of the points on a line between them. The line segment connecting two points A and B is written as $\overline{AB}$.

Any point on a line separates the line into two halves, which we call **half lines.** A half-line beginning at point A and continuing through point B is written as $\overset{\circ\rightarrow}{AB}$. The open circle over A indicates that the point A is not included. When the endpoint of a half line is included, we call the resulting figure a **ray.** A ray with endpoint A that continues through point B is written as $\overrightarrow{AB}$. The figures described in the last few paragraphs are summarized in Table 10-1.

Rays can be represented by rays of light starting at a source and continuing outward.

TABLE 10-1 Lines and Portions of Lines

Name	Figure	Symbol
Line	A B	$\overleftrightarrow{AB}$
Line segment	A B	$\overline{AB}$
Half line	A B	$\overset{\circ\!\rightarrow}{AB}$
Ray	A B	$\overrightarrow{AB}$

Rays are used to define angles, which are the most important figures in this section.

An **angle** is a figure formed by two rays with a common endpoint. The rays are called the **sides** of the angle, and the endpoint is called the **vertex.**

Math Note

In using three letters to name an angle, the vertex is always the letter in the middle.

Some angles are represented in Figure 10-2. The symbol for angle is $\measuredangle$, and there are a number of ways to name angles. The angle in Figure 10-2a could be called $\measuredangle ABC$, $\measuredangle CBA$, $\measuredangle B$, or $\measuredangle 1$.

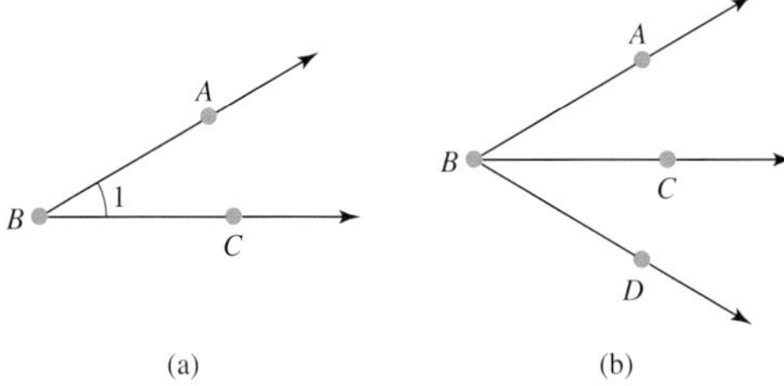

Figure 10-2

CAUTION

You should only use a single letter to denote an angle if there's no question as to the angle represented. In Figure 10-2b, $\measuredangle B$ is ambiguous, because there are three different angles with vertex at point B.

EXAMPLE 1 Naming Angles

Name the angle shown here in four different ways.

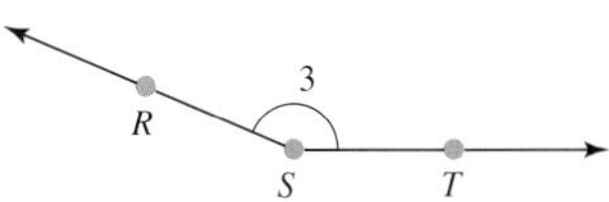

SOLUTION

$\measuredangle RST$, $\measuredangle TSR$, $\measuredangle S$, and $\measuredangle 3$.

▼ Try This One 1

☑ 1. Write names for angles.

Write three different ways to represent the bottom angle in the diagram.

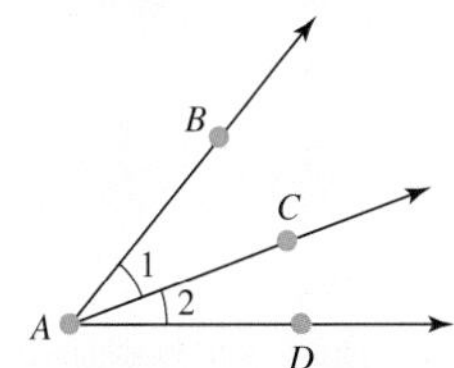

One way to measure an angle is in **degrees,** symbolized by °. One degree is defined to be $\frac{1}{360}$ of a complete rotation.

The instrument that is used to measure an angle is called a **protractor.** Figure 10-3 shows how to measure an angle using a protractor. The center of the base of the protractor is placed at the vertex of the angle, and the bottom of the protractor is placed on one side of the angle. The angle measure is marked where the other side falls on the scale. The angle shown in Figure 10-3 has a measure of 40°. The symbol for the measure of an angle is $m\measuredangle$; so we would write $m\measuredangle ABC = 40°$.

Angles can be classified by their measures.

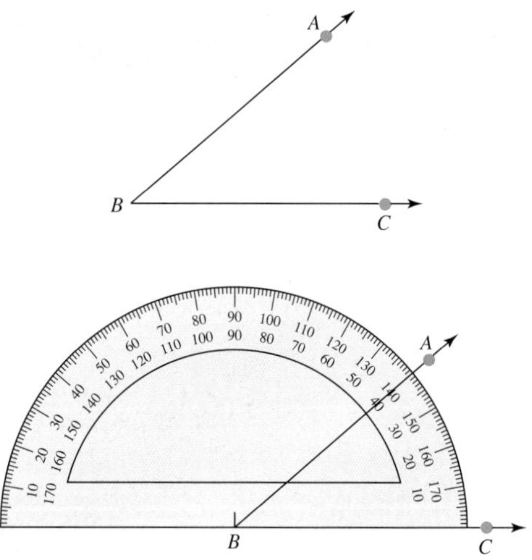

Figure 10-3

An **acute angle** has a measure between 0° and 90°.
A **right angle** has a measure of 90°.
An **obtuse angle** has a measure between 90° and 180°.
A **straight angle** has a measure of 180°.
See Figure 10-4.

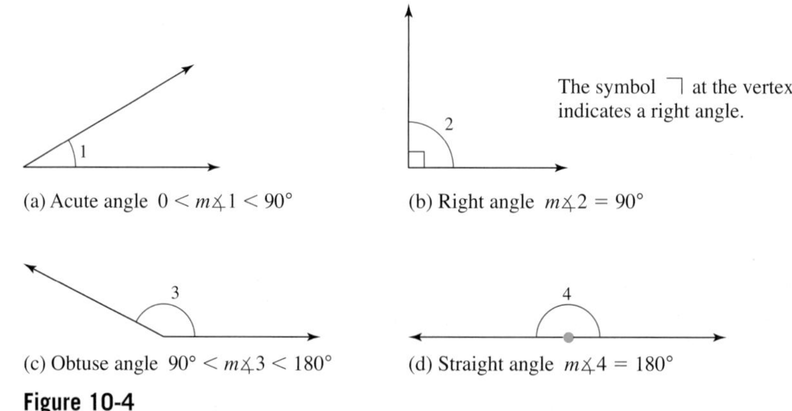

(a) Acute angle $0 < m\measuredangle 1 < 90°$

(b) Right angle $m\measuredangle 2 = 90°$

(c) Obtuse angle $90° < m\measuredangle 3 < 180°$

(d) Straight angle $m\measuredangle 4 = 180°$

Figure 10-4

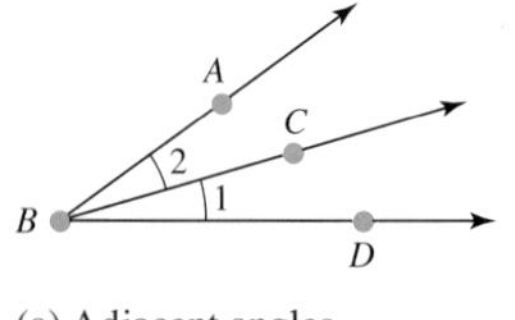
(a) Adjacent angles

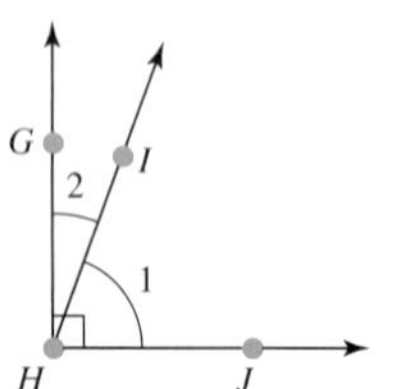
(b) Complementary angles $m\measuredangle 1 + m\measuredangle 2 = 90°$

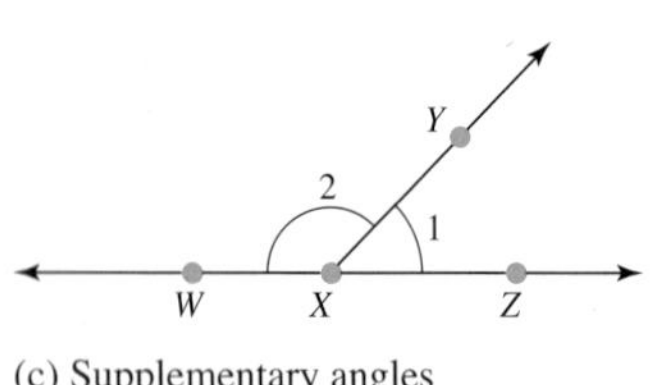
(c) Supplementary angles $m\measuredangle 1 + m\measuredangle 2 = 180°$

Figure 10-5

Math Note

Two angles are complementary when together they form a right angle. Two angles are supplementary when together they form a straight angle.

Pairs of Angles

Pairs of angles have various names depending on how they are related.

Two angles are called *adjacent angles* if they have a common vertex and a common side. Figure 10-5a shows a pair of adjacent angles, $\measuredangle ABC$ and $\measuredangle CBD$. The common vertex is B and the common side is $\overrightarrow{BC}$.

Two angles are said to be **complementary** if the sum of their measures is 90°. Figure 10-5b shows two complementary angles. The sum of the measures of $\measuredangle GHI$ and $\measuredangle IHJ$ is 90°.

Two angles are said to be **supplementary** if the sum of their measures is equal to 180°. Figure 10-5c shows two supplementary angles. The sum of the measures of $\measuredangle WXY$ and $\measuredangle YXZ$ is 180°.

EXAMPLE 2 Using Complementary Angles

If $\measuredangle FEG$ and $\measuredangle GED$ are complementary and $m\measuredangle FEG$ is 28°, find $m\measuredangle GED$.

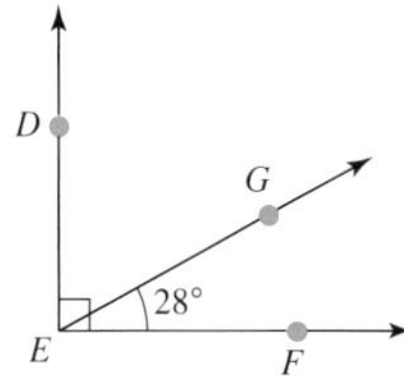

SOLUTION

Since the two angles are complementary, the sum of their measures is 90°, so $m\measuredangle GED + m\measuredangle FEG = 90°$, and solving we get:

$$\begin{aligned} m\measuredangle GED &= 90° - m\measuredangle FEG \\ &= 90° - 28° \\ &= 62° \end{aligned}$$

We say that the **complement** of an angle with measure 28° has measure 62°.

Math Note

It's not always *necessary* to draw a diagram for a geometry problem, but it is always *helpful.*

▼ Try This One 2

Find the measure of the complement of an angle with measure 41°.

EXAMPLE 3 Using Supplementary Angles

If $\measuredangle RQS$ and $\measuredangle SQP$ are supplementary and $m\measuredangle RQS = 135°$, find the measure of $\measuredangle SQP$.

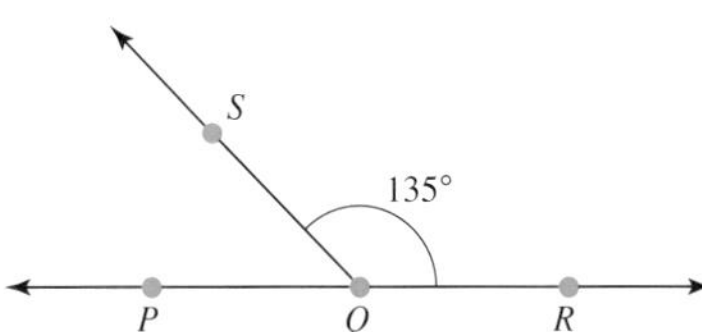

SOLUTION

Since $\measuredangle RQS$ and $\measuredangle SQP$ are supplementary, the sum of their measures is 180°, so $m\measuredangle SQP + m\measuredangle RQS = 180$, and solving we get

$$\begin{aligned} m\measuredangle SQP &= 180° - m\measuredangle RQS \\ &= 180° - 135° \\ &= 45° \end{aligned}$$

We say that the **supplement** of an angle with measure 135° has measure 45°.

▼ Try This One 3

Find the measure of the supplement of an angle with measure 74°.

EXAMPLE 4 Using Supplementary Angles

If two adjacent angles are supplementary and one angle is 3 times as large as the other, find the measure of each.

SOLUTION

Let $x =$ the measure of the smaller angle. The larger is three times as big, so $3x =$ the measure of the larger. The angles are supplementary, so their measures add to 180°. This gives us an equation:

$$x + 3x = 180° \quad \text{Smaller angle plus larger angle is } 180°.$$
$$4x = 180°$$
$$\frac{4x}{4} = \frac{180°}{4}$$
$$x = 45°$$

The smaller angle has measure 45°, and the larger has measure $3 \times 45°$, or 135°.

2. Use complementary and supplementary angles to find angle measure.

▼ Try This One 4

Two angles are complementary, and the smaller has measure 17° less than the larger. Find the measure of each.

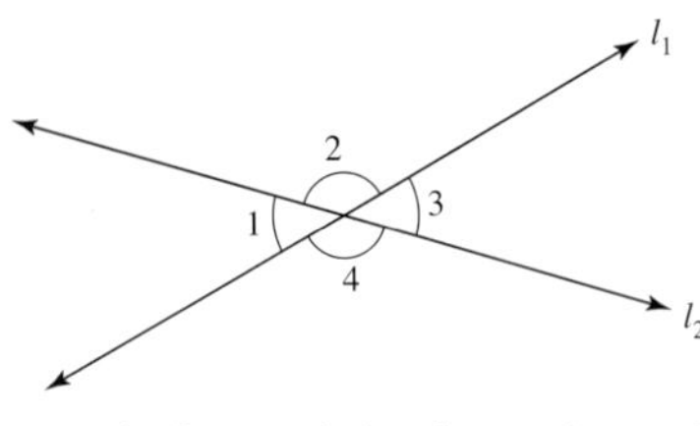

∡1 and ∡3 are vertical angles; $m∡1 = m∡3$
∡2 and ∡4 are vertical angles; $m∡2 = m∡4$

Figure 10-6

Just as two intersecting streets have four corners at which you can cross, when two lines intersect, four angles are formed, as we see in Figure 10-6. Angles 1 and 3 are called **vertical angles,** as are angles 2 and 4. It's easy to believe from the diagram that *two vertical angles have the same measure.*

If the measure of one of the four angles is known, then the measure of the other three angles can be found. This is shown in Example 5.

Sidelight THE FATHER OF GEOMETRY

The geometry that we are studying in most of this chapter is known as Euclidean geometry, in honor of Euclid, who was a professor of mathematics at the University of Alexandria sometime around 300 BCE. That particular university was formed to serve as the center of all of the knowledge that had been accumulated to that point. By all indications, it wasn't terribly different from a modern university—it had lecture rooms, laboratories, gardens, museums, and a library with over 600,000 papyrus scrolls. There is no record of off-campus bars, but it probably had those, too.

The bible is the only book ever written that has been more widely translated, edited, and read than Euclid's famous *Elements*, which consists of 13 books on geometry, number theory, and algebra. The book remained for the most part unchanged for over 2,000 years, but in the 1800s some logical flaws were corrected by mathematicians of the time. All of the geometry we study in the first six sections of this chapter has its roots in the works of Euclid. In Section 10-7, however, we will take just a quick look at some other geometries that don't adhere to Euclid's framework.

EXAMPLE 5 Using Vertical Angles

Find $m\measuredangle 2$, $m\measuredangle 3$, and $m\measuredangle 4$ when $m\measuredangle 1 = 40°$.

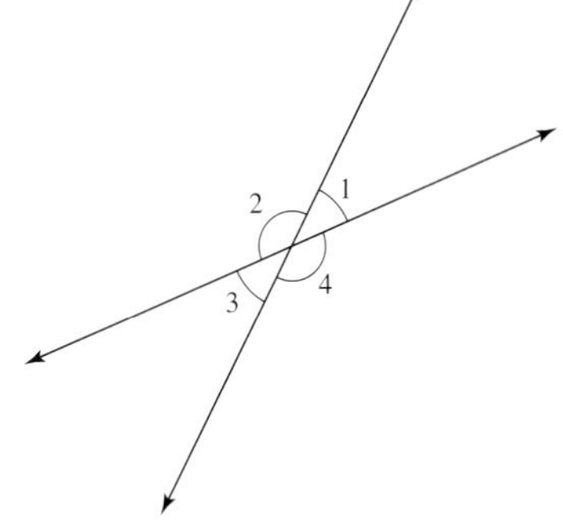

SOLUTION

Since $\measuredangle 1$ and $\measuredangle 3$ are vertical angles and $m\measuredangle 1 = 40°$, $m\measuredangle 3 = 40°$. Since $\measuredangle 1$ and $\measuredangle 2$ form a straight angle (180°), $m\measuredangle 1 + m\measuredangle 2 = 180°$, and

$$\begin{aligned} m\measuredangle 2 &= 180° - m\measuredangle 1 \\ &= 180° - 40° \\ &= 140° \end{aligned}$$

Finally, since $\measuredangle 2$ and $\measuredangle 4$ are vertical angles, $m\measuredangle 4 = 140°$.

Math Note

When two angles are supplementary, you can always find the measure of one by subtracting the measure of the other from 180°.

▼ Try This One 5

☑ 3. Use vertical angles to find angle measure.

Find $m\measuredangle 2$, $m\measuredangle 3$, and $m\measuredangle 4$ when $m\measuredangle 1 = 75°$.

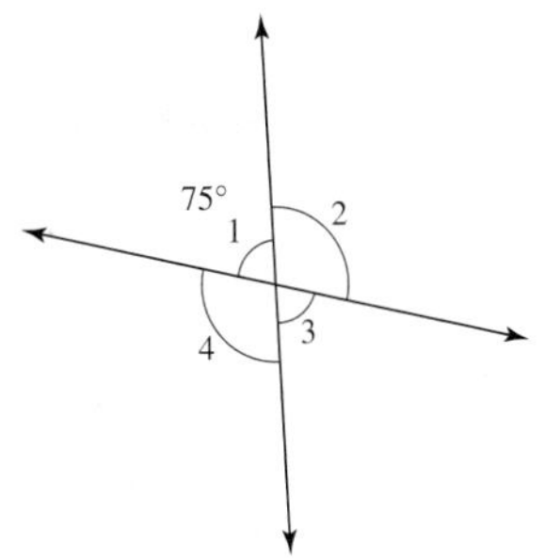

Two lines in the same plane are called **parallel** if they never intersect. You might find it helpful to think of them as lines that go in the exact same direction. If two lines l_1 and l_2 are parallel, we write $l_1 \parallel l_2$. When two parallel lines are intersected by a third line, we call the third line a **transversal.** As you can see in Figure 10-7 on the next page, eight angles are formed. The angles between the parallel lines (angles 3–6) are called **interior angles,** and the ones outside the parallel lines (angles 1, 2, 7, and 8) are called **exterior angles.** The box below Figure 10-7 defines special relationships among these angles that allow us to find the measure of all of the angles if we know just one.

You can find parallel roads and transversals on almost any city map.

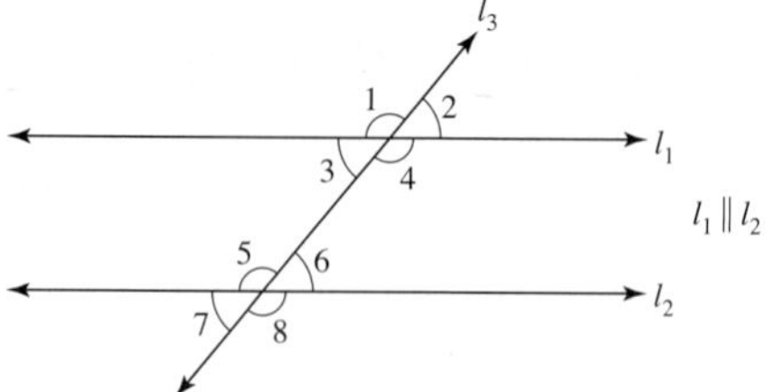

Pairs of alternate interior angles
$\measuredangle 3$ and $\measuredangle 6$ $\quad m\measuredangle 3 = m\measuredangle 6$
$\measuredangle 4$ and $\measuredangle 5$ $\quad m\measuredangle 4 = m\measuredangle 5$

Pairs of alternate exterior angles
$\measuredangle 1$ and $\measuredangle 8$ $\quad m\measuredangle 1 = m\measuredangle 8$
$\measuredangle 2$ and $\measuredangle 7$ $\quad m\measuredangle 2 = m\measuredangle 7$

Pairs of corresponding angles
$\measuredangle 1$ and $\measuredangle 5$ $\quad m\measuredangle 1 = m\measuredangle 5$
$\measuredangle 2$ and $\measuredangle 6$ $\quad m\measuredangle 2 = m\measuredangle 6$
$\measuredangle 3$ and $\measuredangle 7$ $\quad m\measuredangle 3 = m\measuredangle 7$
$\measuredangle 4$ and $\measuredangle 8$ $\quad m\measuredangle 4 = m\measuredangle 8$

Figure 10-7

Alternate interior angles are the angles formed between two parallel lines on the opposite sides of the transversal that intersects the two lines. Alternate interior angles have equal measures.

Alternate exterior angles are the opposite exterior angles formed by the transversal that intersects two parallel lines. Alternate exterior angles have equal measures.

Corresponding angles consist of one exterior and one interior angle with no common vertex on the same side of the transversal that intersects two parallel lines. Corresponding angles have equal measures.

EXAMPLE 6 Finding Angles Formed by a Transversal

Find the measures of all the angles shown when the measure of $\measuredangle 2$ is 50°.

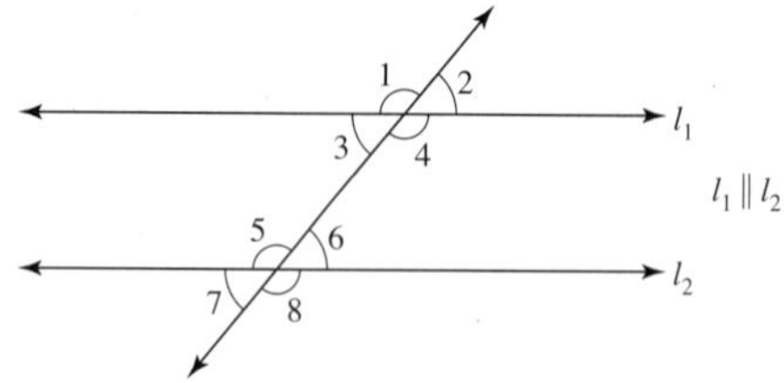

SOLUTION

First, let's identify the angles that have the same measure as $\measuredangle 2$. They are $\measuredangle 3$ (vertical angles), $\measuredangle 6$ (corresponding angles), and $\measuredangle 7$ (alternate exterior angles). Mark all of these as 50° on the diagram (Figure 10-8).

Since $\measuredangle 1$ and $\measuredangle 2$ are supplementary, $m\measuredangle 1 = 180° - 50° = 130°$. This allows us to find the remaining angles: $\measuredangle 4$ is a vertical angle with $\measuredangle 1$, so it has measure 130° as well. Now $\measuredangle 8$ is a corresponding angle with $\measuredangle 4$, and $\measuredangle 5$ is an alternate interior angle with $\measuredangle 4$, which means they both have measure 130° as well.

Figure 10-8

☑ 4. Find measures of angles formed by a transversal.

▼ Try This One 6

Find the measures of all the angles shown when the measure of ∡1 is 165°.

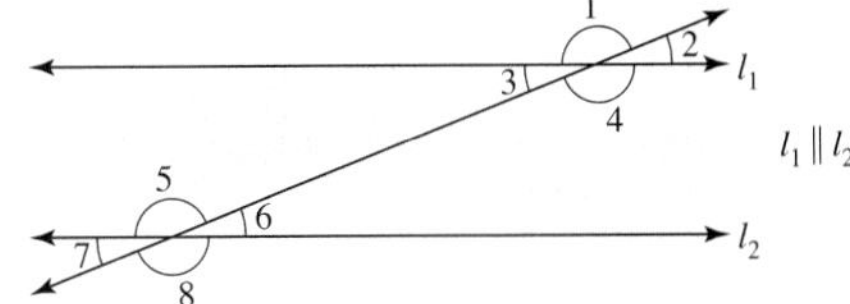

Answers to Try This One

1 ∡2, ∡*CAD*, ∡*DAC*

2 49°

3 106°

4 $36\frac{1}{2}°$ and $53\frac{1}{2}°$

5 $m∡2 = 105°$, $m∡3 = 75°$, $m∡4 = 105°$

6 $m∡2$, ∡3, ∡6, $∡7 = 15°$; $m∡4$, ∡5, $∡8 = 165°$

EXERCISE SET 10-1

Writing Exercises

1. Explain why you can't actually draw a point or a line, just figures that represent them.
2. What is the difference between a half line and a ray?
3. Describe the four different ways that we name angles.
4. Explain why the two lines below are not parallel even though they don't meet in the diagram.
5. Describe how to find the complement and supplement of an angle.
6. When parallel lines are intersected by a transversal, there are four types of pairs of angles with equal measures formed. Describe each, using a diagram to illustrate.

Computational Exercises

For Exercises 7–12, identify and name each figure.

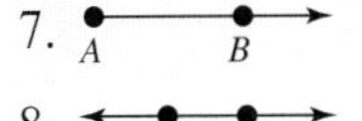

10. • P

11. T U

12. E F

For Exercises 13–14, name each angle in four different ways.

13.

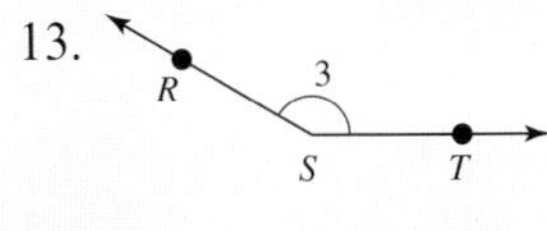

14.

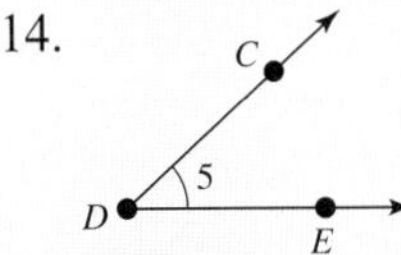

For Exercises 15–18, classify each angle as acute, right, obtuse, or straight.

15.

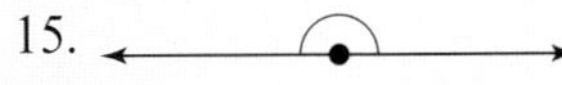

16.

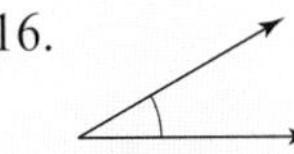

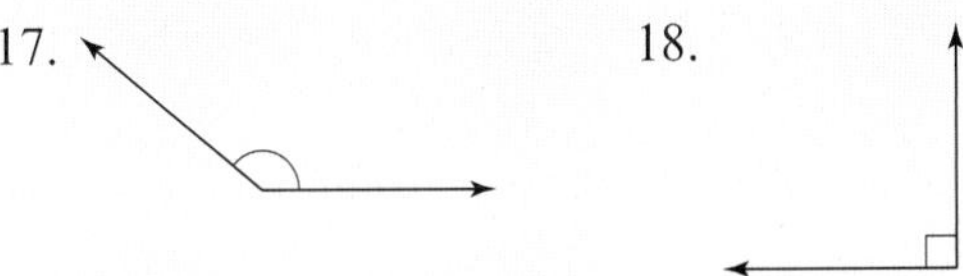

For Exercises 19–26, identify each pair of angles as alternate interior, alternate exterior, corresponding, or vertical.

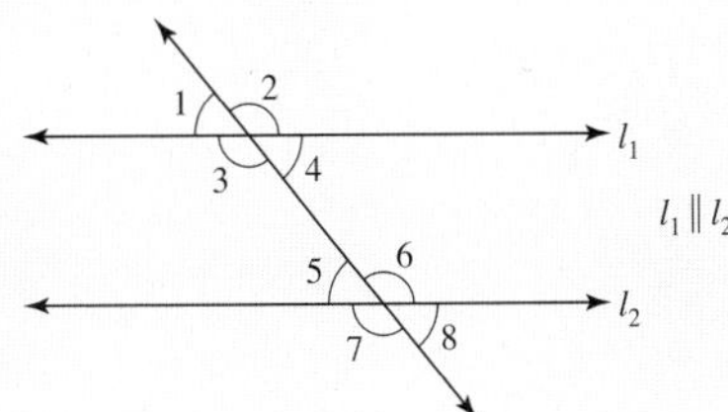

19. ∡1 and ∡4
20. ∡3 and ∡6
21. ∡2 and ∡6
22. ∡5 and ∡8
23. ∡1 and ∡5
24. ∡2 and ∡7
25. ∡1 and ∡8
26. ∡4 and ∡8

For Exercises 27–32, find the measure of the complement of each angle.

27. 8°
28. 24°
29. 32°
30. 56°
31. 78°
32. 84°

For Exercises 33–38, find the measure of the supplement for each angle.

33. 156°
34. 90°
35. 62°
36. 143°
37. 120°
38. 38°

For Exercises 39–42, find the measures of ∡1, ∡2, and ∡3.

39.

40.

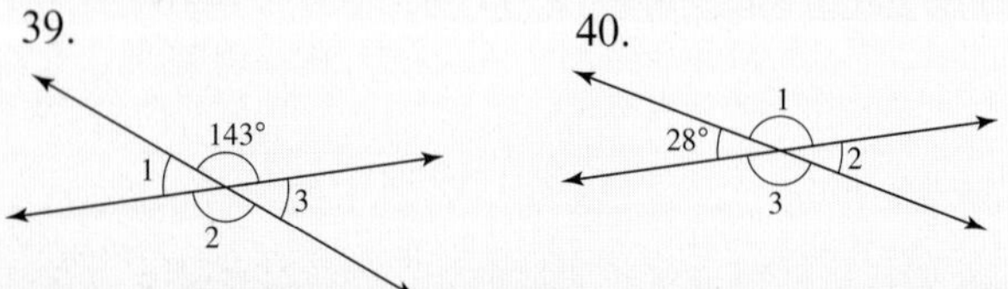

41.

42.

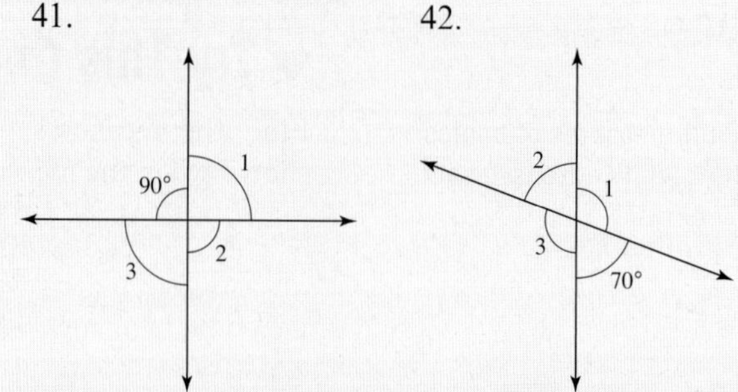

For Exercises 43 and 44, find the measure of ∡1 through ∡7.

43.

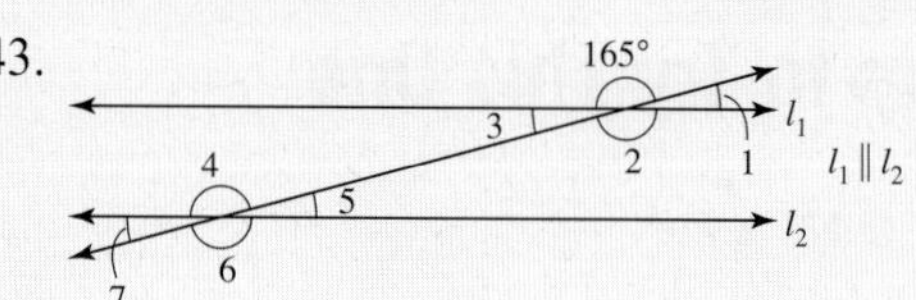

44.

1 2 3 120° 4 5 6 7 l_1 l_2 $l_1 \parallel l_2$

Real-World Applications

For Exercises 45–48, identify the measure of the angle made by the hands of a clock at these times:

45. 3 o'clock
46. 6 o'clock
47. 2 o'clock
48. 4 o'clock

Exercises 49–56 use the following description: Euclid Avenue and Prospect Avenue are parallel, both running west to east, with Euclid north of Prospect. East Ninth Street is a transversal running northwest to southeast. The angle made with Euclid on the northwest corner is 76°. Find what angle a motorist turns at when making the following turns.

49. Driving west on Euclid and turning northwest on E. Ninth
50. Driving east on Prospect and turning northwest on E. Ninth
51. Driving southeast on E. Ninth and turning east on Euclid
52. Driving northwest on E. Ninth and turning east on Euclid
53. Driving east on Euclid and turning southeast on E. Ninth
54. Driving west on Prospect and turning northwest on E. Ninth
55. Driving northwest on E. Ninth and turning west on Prospect
56. Driving southeast on E. Ninth and turning west on Prospect

Critical Thinking

57. Find the measure of ∡3 and ∡4 if $m\measuredangle 1 = m\measuredangle 2$ and line l_1 is parallel to line l_2.

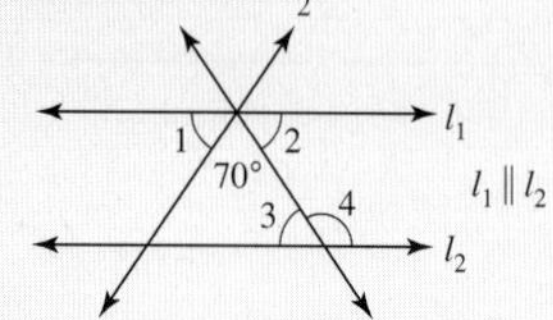

58. Find $m\measuredangle 1$ if line l_1 is parallel to line l_2.

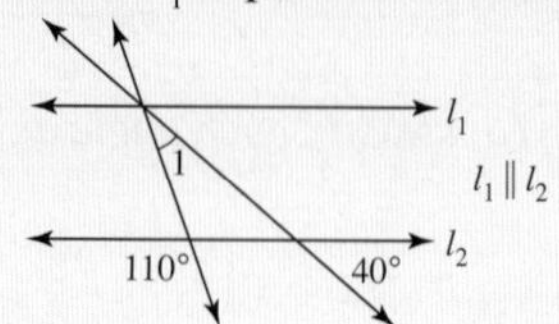

Section 10-2 Triangles

LEARNING OBJECTIVES

- ❑ 1. Identify types of triangles.
- ❑ 2. Find one missing angle in a triangle.
- ❑ 3. Use the Pythagorean theorem to find side lengths.
- ❑ 4. Use similar triangles to find side lengths.

If you look for them, you can find familiar geometric shapes in a surprising number of places. Architects in particular are fascinated by creating complex designs out of basic shapes. The pyramid structure at the famous Louvre in Paris is one of the most well-known examples, blending an assortment of simple shapes into an architectural masterpiece.

As we continue our study of geometry in our world, we use points, lines, rays, and angles to form more complex figures, beginning with the triangle. Although they are very simple figures, triangles have a surprising number of applications to real-world situations.

A geometric figure is said to be **closed** when you can start at one point, trace the entire figure, and finish at the point you started without lifting your pen or pencil off the paper. One such figure is the triangle:

A **triangle** is a closed geometric figure that has three sides and three angles. The three sides of the triangle are line segments, and the points where the sides intersect are called the **vertices** (plural of "vertex").

One way we can name a triangle is according to its vertices, using the symbol $\triangle$. For example, we could call the triangle in Figure 10-9 $\triangle ABC$. (In this case, the order of the vertices doesn't matter.)

A

Vertices

Sides

B

C

Figure 10-9

Types of Triangles

Special names are given to certain types of triangles based on either the lengths of their sides or the measures of their angles. For sides:

An **isosceles triangle** has two sides with the same length.

An **equilateral triangle** has three sides with the same length.

A **scalene triangle** has sides that are three different lengths.

Examples are shown in Figure 10-10a.

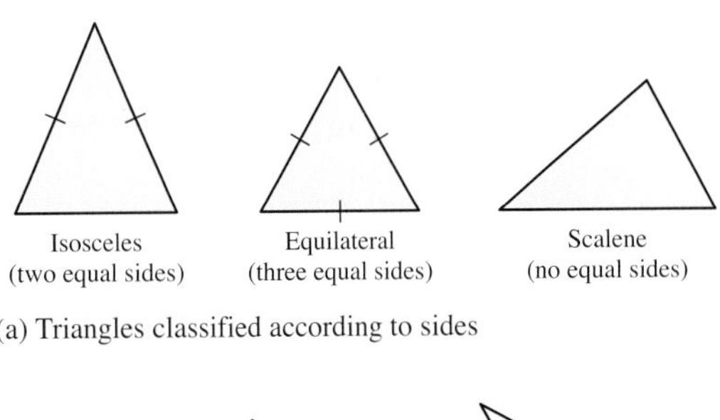

(a) Triangles classified according to sides

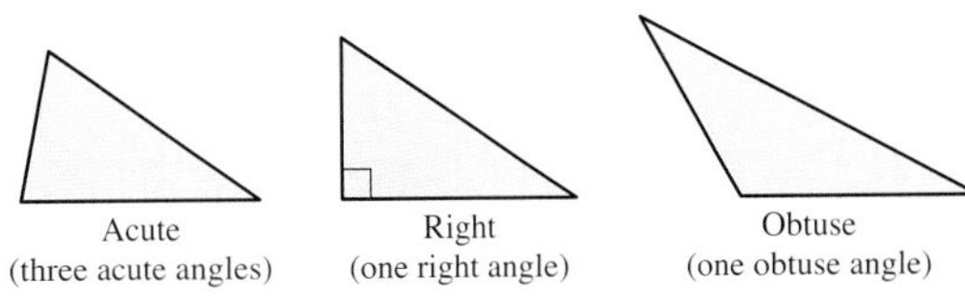

(b) Triangles classified according to angles

Figure 10-10

Math Note

Because an isosceles triangle has two equal sides, it also has two equal angles. The equilateral triangle has three equal angles. The dash marks on the sides of the triangles indicate the sides have the same length.

For angles:

An **acute triangle** has three acute angles (less than 90°).

A **right triangle** has one right angle (90°).

An **obtuse triangle** has one obtuse angle (greater than 90° and less than 180°).

Examples are shown in Figure 10-10b.

EXAMPLE 1 Identifying Types of Triangles

Identify the type of triangle.

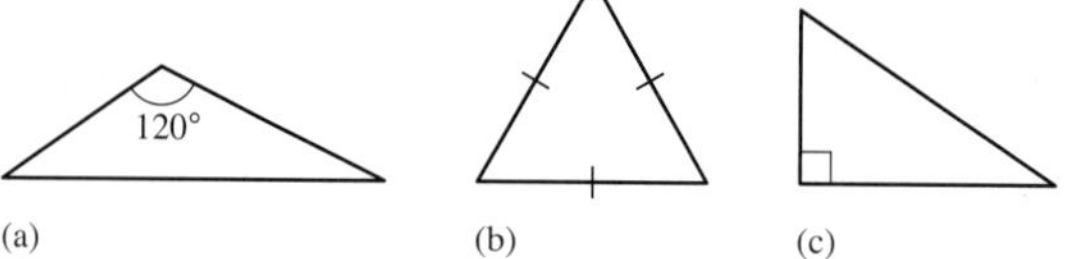

SOLUTION

(a) Obtuse triangle since one angle is greater than 90°.
(b) Equilateral triangle since all sides are equal, or acute triangle since all angles are acute.
(c) Right triangle since one angle is 90°.

▼ Try This One 1

☑ 1. Identify types of triangles.

Identify the type of triangle.

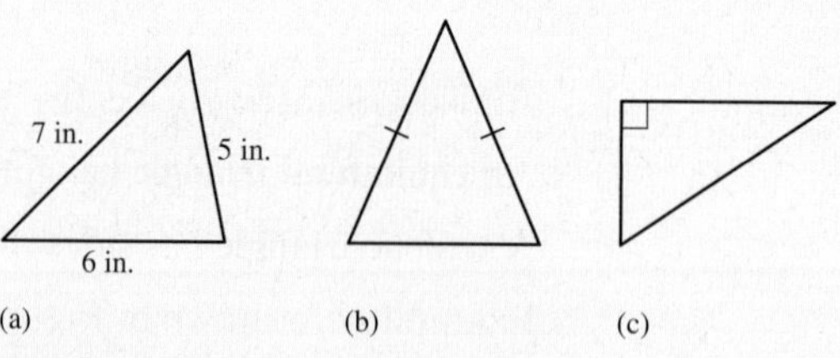

It is very useful to know the following fact:

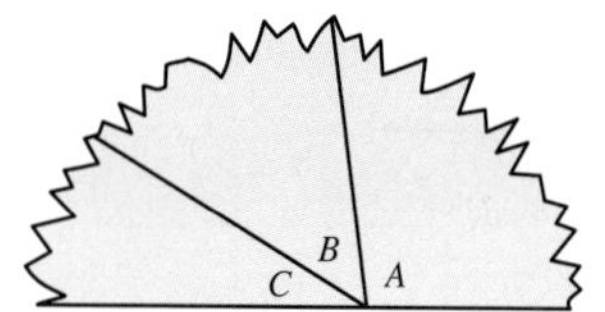

Figure 10-11

The Sum of Angle Measures in a Triangle

In any triangle, the measures of the three angles add to 180°.

We will outline a proof of this fact in Exercises 47–51, but you can see that it's true with a simple demonstration. Draw any triangle on a piece of paper, then tear it into three pieces, one containing each angle. Now arrange the three angles next to each other, as shown in Figure 10-11: you will find that when the vertices of the three angles are placed together, they always form a straight line (or a 180° angle).

EXAMPLE 2 Finding an Angle in a Triangle

Find the measure of angle C in the triangle.

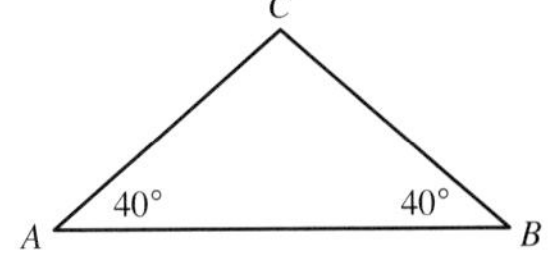

SOLUTION

Since the sum of the measures of the angles of a triangle is 180°,

$$
\begin{aligned}
m\measuredangle A + m\measuredangle B + m\measuredangle C &= 180^\circ && m\measuredangle A \text{ and } m\measuredangle B = 40^\circ. \\
40^\circ + 40^\circ + m\measuredangle C &= 180^\circ && \text{Add on left side.} \\
80^\circ + m\measuredangle C &= 180^\circ && \text{Subtract } 80^\circ \text{ from both sides.} \\
m\measuredangle C &= 180^\circ - 80^\circ \\
&= 100^\circ
\end{aligned}
$$

The measure of angle C is 100°.

▼ Try This One 2

☑ 2. Find one missing angle in a triangle.

Find the measure of angle B.

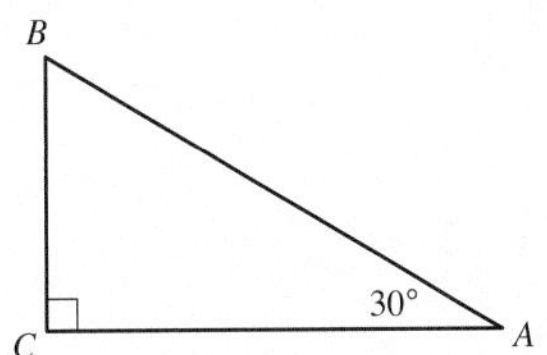

The Pythagorean Theorem

Right triangles are very special because there is a simple relationship among their sides. This is the famous Pythagorean theorem, which is attributed to the Greek mathematician Pythagoras around 500 BCE, although it was actually known much earlier in some form (see Sidelight on the next page). In any case, a **theorem** is a fact that has been proven true using deductive reasoning. This particular theorem allows us to find the third side of a right triangle if we know two sides. The side across from the right angle is called the **hypotenuse,** and the other two sides are called the **legs.**

The Pythagorean Theorem

The sum of the squares of the lengths of the two legs in a right triangle always equals the square of the length of the hypotenuse. If we use a and b to represent the lengths of the legs and c to represent the length of the hypotenuse, as in the figure, then

$$a^2 + b^2 = c^2$$

In addition, if the sides of any triangle satisfy this equation, the triangle is a right triangle.

CAUTION Remember, the Pythagorean theorem only holds true if the triangle is a right triangle!

Sidelight PYTHAGORAS AND THE PYRAMIDS

The great pyramids in Egypt were completed about 2,000 years before Pythagoras was born, but the theorem that now bears his name played an important role in their construction. Three whole numbers that satisfy the equation in the Pythagorean theorem are called a **Pythagorean triple.** For example, $3^2 + 4^2 = 5^2$, so 3, 4, and 5 form a Pythagorean triple. So if a triangle has legs of length 3 units and 4 units, and a hypotenuse of length 5 units, it must be a right triangle.

This brings us back to the pyramid builders. Right angles are tremendously important in building structures, especially massive structures like the Pyramids. If all of the angles are off by just a little bit, the result is a crooked pyramid. To ensure that angles were right angles, the Egyptians used a rope divided into even sections of 3, 4, and 5 units long with knots. If the rope wouldn't stretch into a taut triangle with sides of length 3 and 4 against an edge that was supposed to be a right angle, then they would know that the angle was off.

EXAMPLE 3 Using the Pythagorean Theorem

Calculator Guide

To find the hypotenuse in Example 3:

Standard Scientific Calculator:

10 [x²] [+] 24 [x²] [=] [√]

Standard Graphing Calculator:

[2nd] [x²] 10 [x²] + 24 [x²] [)]

[Enter]

Note: [2nd] [x²] is the square root operator: if your calculator does not automatically put in a (, you will need to do so manually.

For the right triangle shown, find the length of the hypotenuse (i.e., side c).

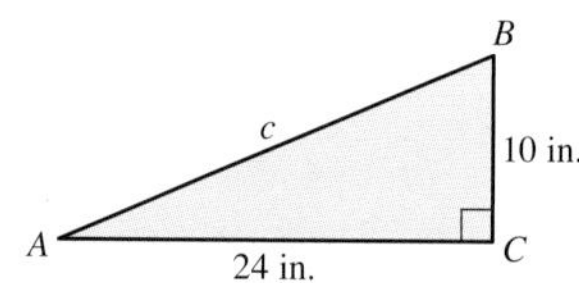

SOLUTION

$$c^2 = a^2 + b^2 \quad \text{The two legs are 10 in. and 24 in.}$$
$$= 10^2 + 24^2$$
$$= 100 + 576$$
$$= 676 \quad \text{Apply square root to both sides.}$$
$$c = \sqrt{676} \quad c \text{ is a length, so ignore the negative solution.}$$
$$= 26$$

The hypotenuse is 26 inches long.

▼ Try This One 3

If the hypotenuse of a right triangle is 13 feet long and one leg is 5 feet long, find the length of the other leg.

EXAMPLE 4 Applying the Pythagorean Theorem

To build the frame for the roof of a shed, the carpenter must cut a 2 × 4 to fit on the diagonal. If the length of the horizontal beam is 12 feet and the height is 3 feet, find the length of the diagonal beam.

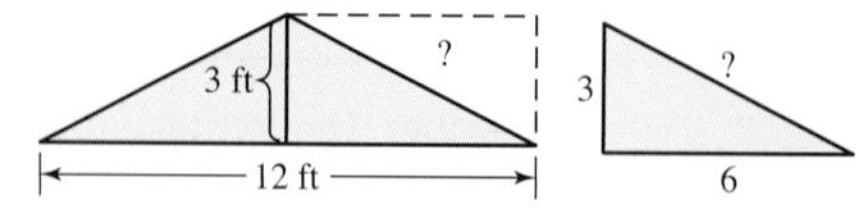

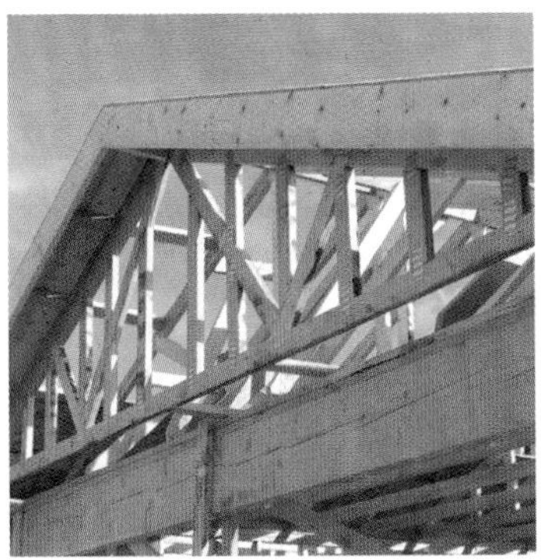

Triangles play a very important role in roof construction.

SOLUTION

The frame forms two right triangles with legs of 3 feet and 6 feet. The length of the diagonal beam can be found using the Pythagorean theorem.

$$\begin{aligned} c^2 &= a^2 + b^2 && \text{Use } a = 3,\ b = 6. \\ &= 3^2 + 6^2 && \text{Square each.} \\ &= 9 + 36 && \text{Add.} \\ &= 45 && \text{Apply square root to both sides.} \\ c &= \sqrt{45} \\ &\approx 6.7 \text{ feet} \end{aligned}$$

The length of the beam should be about 6.7 feet.

▼Try This One 4

☑ 3. Use the Pythagorean theorem to find side lengths.

The rectangular frame for a large sign is 10 feet long and 8 feet high. Find the length of a diagonal beam that will be used for bracing. (*Hint:* Draw a diagram.)

Similar Triangles

When two triangles have the same shape but not necessarily the same size, they are called **similar triangles.** Consider the triangles in Figure 10-12. Since the measure of angle A and the measure of angle A' are equal, they are called *corresponding angles.* Likewise, angle B and angle B' are corresponding angles since they have the same measure. Finally, angle C and angle C' are corresponding angles. When two triangles are similar, then their corresponding angles will have the same measure. Also, if all of the corresponding angles have equal measure, we know that two triangles are similar.

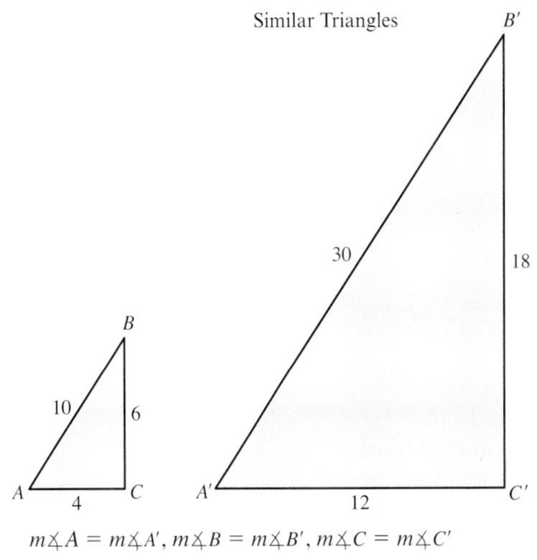

$$\frac{\text{Length of side } AB}{\text{Length of side } A'B'} = \frac{\text{Length of side } BC}{\text{Length of side } B'C'} = \frac{\text{Length of side } AC}{\text{Length of side } A'C'}$$

Figure 10-12

Math Note

Actually, since the sum of all three angle measures in a triangle is 180°, if two triangles have two corresponding angles equal, the third angle must be equal as well, and they are similar triangles.

The sides that are opposite the corresponding angles are called *corresponding sides.* When two triangles are similar, the ratios of the corresponding sides are equal. For the two triangles shown in Figure 10-12,

$$\frac{\text{Length of side } AB}{\text{Length of side } A'B'} = \frac{10}{30} = \frac{1}{3}$$

$$\frac{\text{Length of side } AC}{\text{Length of side } A'C'} = \frac{4}{12} = \frac{1}{3}$$

$$\frac{\text{Length of side } BC}{\text{Length of side } B'C'} = \frac{6}{18} = \frac{1}{3}$$

Since these three ratios are all equal, we can describe the situation by saying that the lengths of the corresponding sides are in proportion. In summary:

Similar Triangle Relationships

If triangle ABC is similar to triangle $A'B'C'$ then

$$\frac{\text{Length of side } AB}{\text{Length of side } A'B'} = \frac{\text{Length of side } AC}{\text{Length of side } A'C'} = \frac{\text{Length of side } BC}{\text{Length of side } B'C'}$$

EXAMPLE 5 Using Similar Triangles

If the two triangles below are similar, find the length of side $B'C'$.

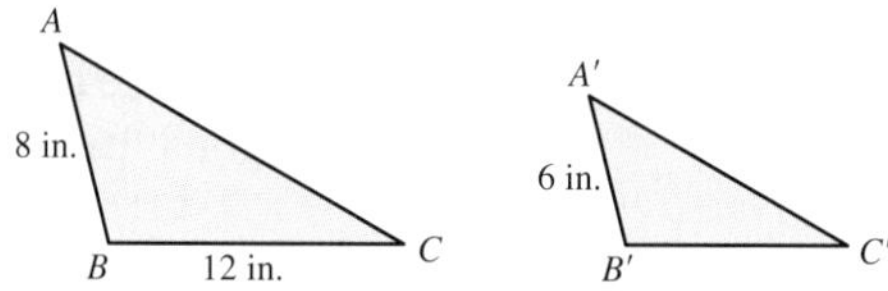

SOLUTION

To begin, we will use the variable x to represent the length we're asked to find. Now we want to write ratios of corresponding sides. Sides AB and $A'B'$ are corresponding, and sides BC and $B'C'$ are as well. So the ratios formed from these corresponding sides will be equal:

$$\frac{\text{Length of side } AB}{\text{Length of side } A'B'} = \frac{\text{Length of side } BC}{\text{Length of side } B'C'}$$

Now we substitute in the lengths and solve for x.

$$\frac{8}{6} = \frac{12}{x} \quad \textit{Cross multiply.}$$

$$8x = 72 \quad \textit{Divide both sides by 8.}$$

$$x = 9$$

The length of side $B'C'$ is 9 inches.

Math Note

In the proportion $\frac{8}{6} = \frac{12}{x}$, both numerators are sides from one triangle, and both denominators are sides from the other. Both lengths on the left side are left sides of the triangles, and both lengths on the right side are bottom sides. This can help assure you've set up the proportion correctly.

▼ Try This One 5

The two triangles shown are similar. Find the length of side AC.

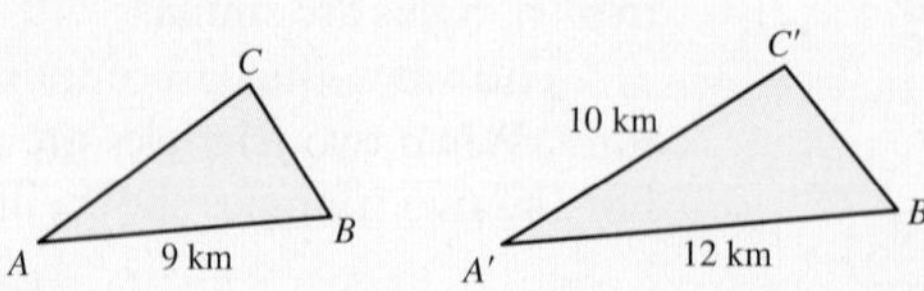

One clever use of similar triangles is to measure the height of objects that are difficult to measure directly. Example 6 shows how to use this form of indirect measurement.

EXAMPLE 6 Using Similar Triangles in Measurement

If a tree casts a shadow 12 feet long and at the same time a person who is 5 feet 10 inches tall casts a shadow of 5 feet, find the height of the tree.

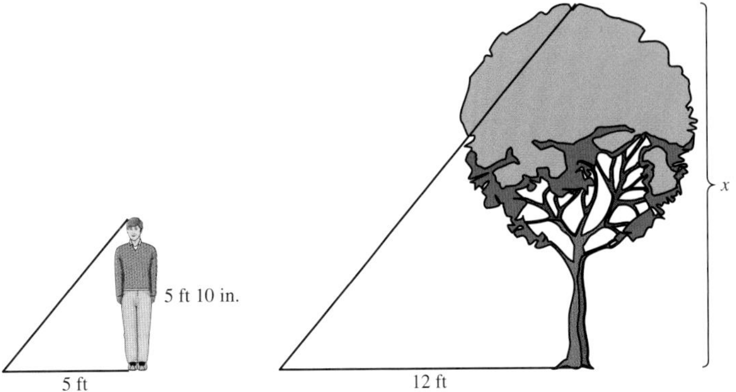

SOLUTION

Both triangles in the diagram are right triangles, and since the sun makes the same angle with the ground in both diagrams, the triangles have two corresponding angles equal, and must be similar. That means we can set up and solve a proportion:

$$\frac{\text{Height of person}}{\text{Length of person's shadow}} = \frac{\text{Height of tree}}{\text{Length of tree's shadow}}$$

$$\frac{5\frac{10}{12}}{5} = \frac{x}{12} \quad \text{Cross multiply.}$$

$$5x = 70 \quad \text{Divide both sides by 5.}$$

$$x = 14 \text{ feet}$$

The height of the tree is 14 feet.

▼ Try This One 6

☑ 4. Use similar triangles to find side lengths.

Find the length of a pole if it casts a 20-foot shadow at the same time that a man 6 feet tall casts a 15-foot shadow.

Answers to Try This One

1 (a) Scalene and acute
(b) Isosceles and acute
(c) Right

2 60°

3 12 feet

4 12.8 feet

5 7.5 km

6 8 feet

EXERCISE SET 10-2

Writing Exercises

1. Describe the three ways that triangles can be classified based on lengths of sides.
2. Describe the three ways that triangles can be classified based on measures of angles.
3. Explain why the following statement is incorrect: the sum of the squares of the lengths of two sides in a triangle is equal to the square of the length of the third side.
4. What does it mean for two triangles to be similar?
5. Explain how to find the measure of the third angle in a triangle if you know the measures of the other two.
6. Explain how to find the length of the third side of a right triangle when you know the lengths of the other two sides.
7. What does it mean for a geometric figure to be closed? Is a plane a closed figure?
8. Explain why two triangles have to be similar if they have two pairs of corresponding angles that are equal.

Computational Exercises

For Exercises 9-14, classify each triangle.

9. 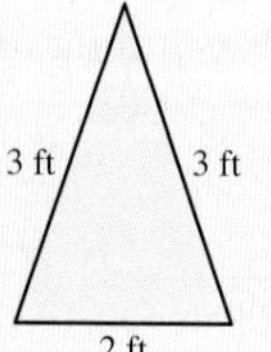

10. 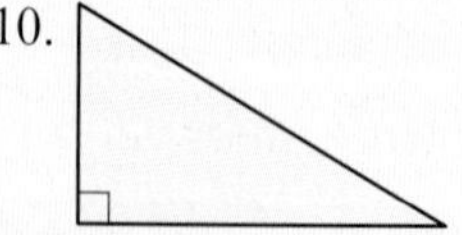

11. 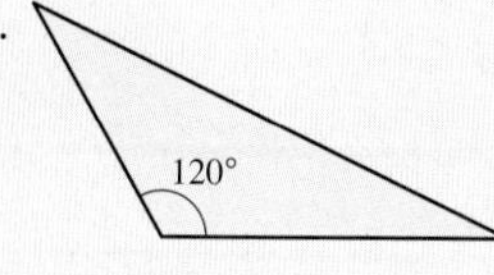

12. 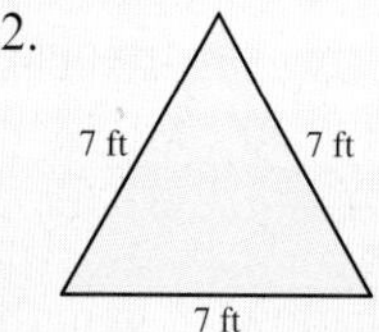

13. 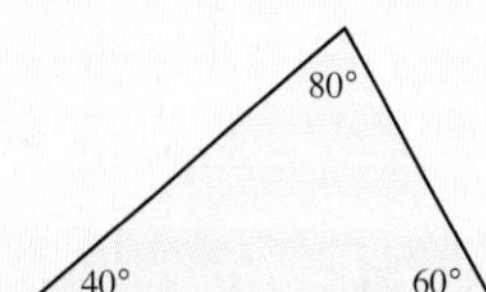

14. 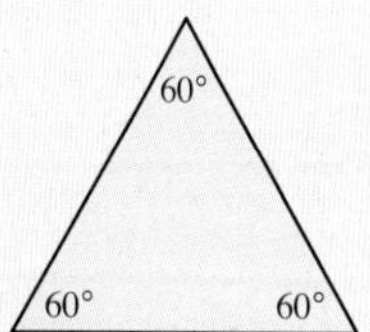

For Exercises 15–20, find the measure of angle C.

15. 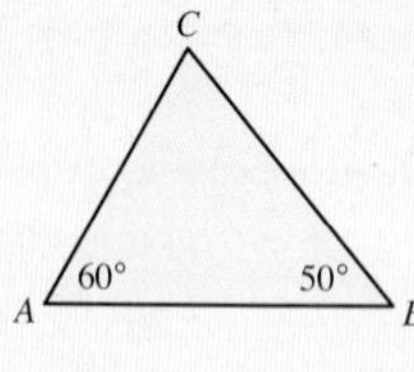

16. 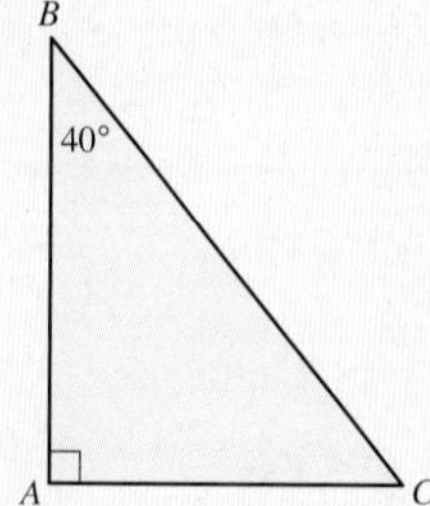

17. 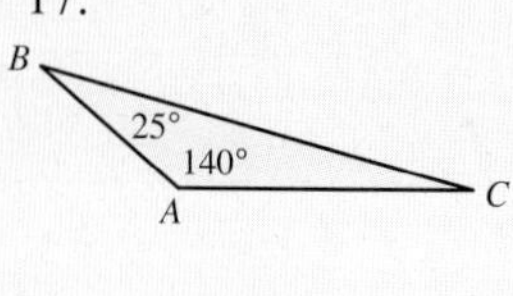

18. 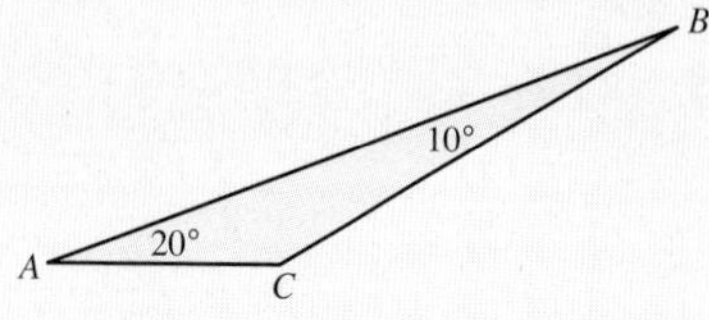

19. 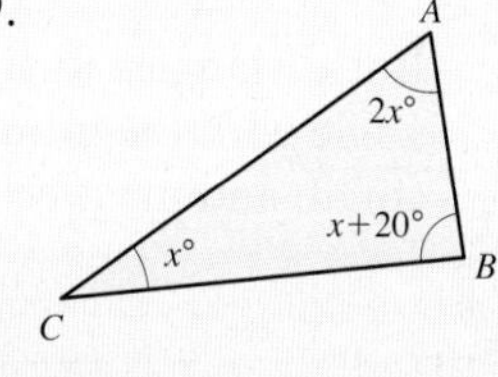

20. 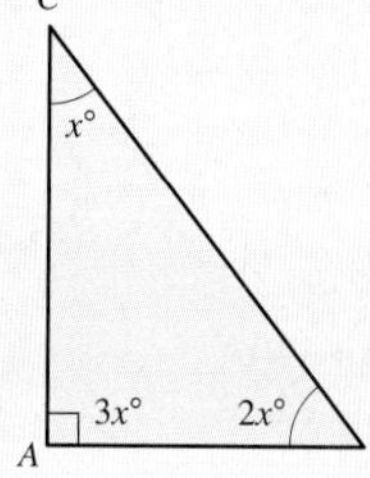

For Exercises 21–26, use the Pythagorean theorem to find the measure of side x.

21. 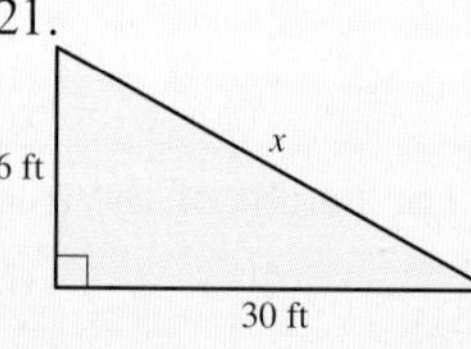

22. 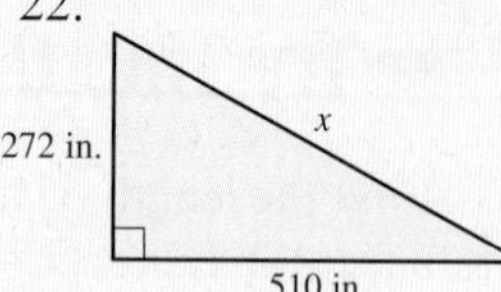

23. 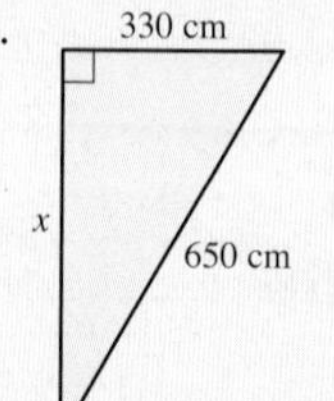

24. 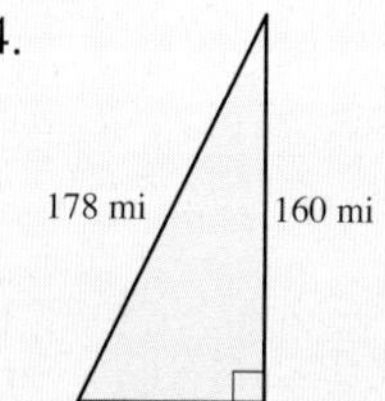

25. 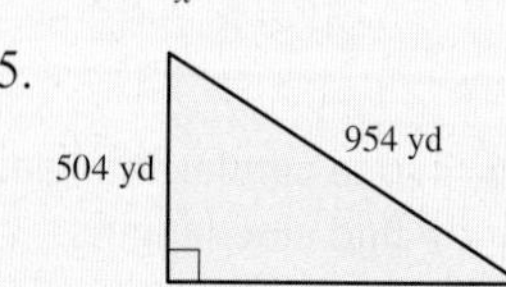

26. 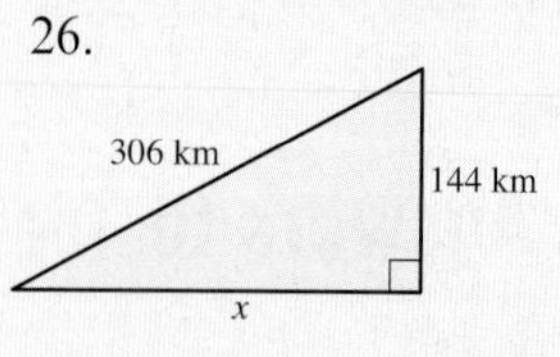

For Exercises 27–30, the two triangles drawn are similar. Find the length of side x.

27.  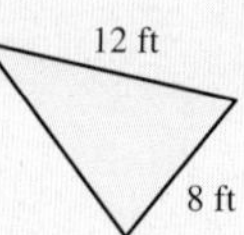

28.

29.

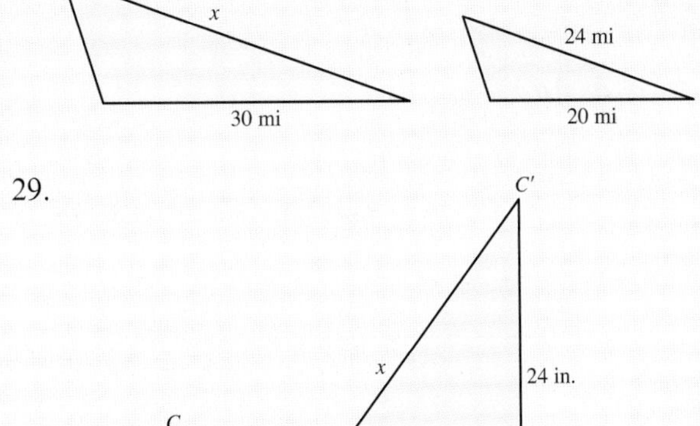

30.

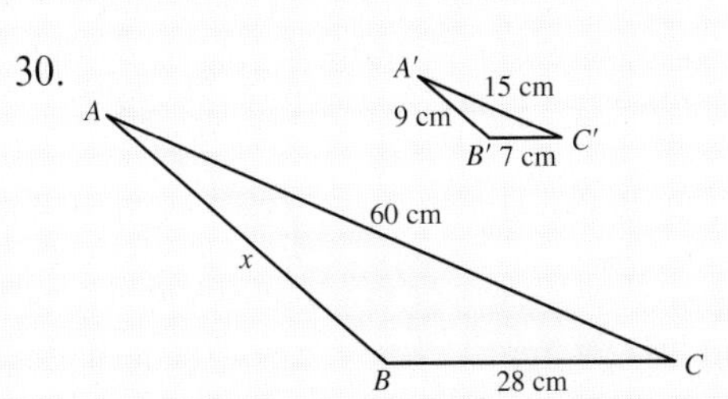

For Exercises 31–36, explain why the two triangles are similar, then find the length x.

31. $m\measuredangle A = 35°$; $m\measuredangle B = 35°$

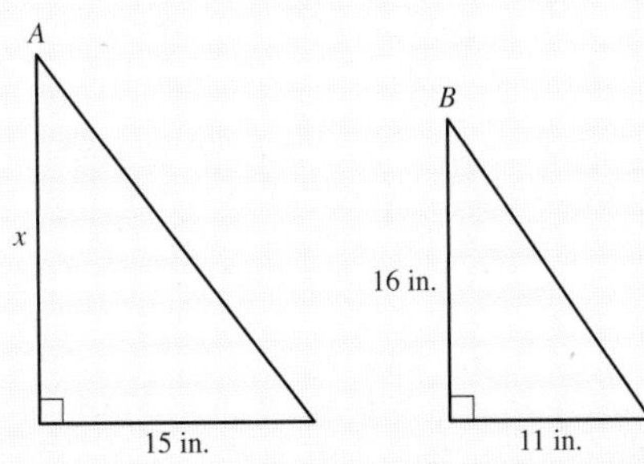

32. $m\measuredangle C = 49°$; $m\measuredangle D = 41°$

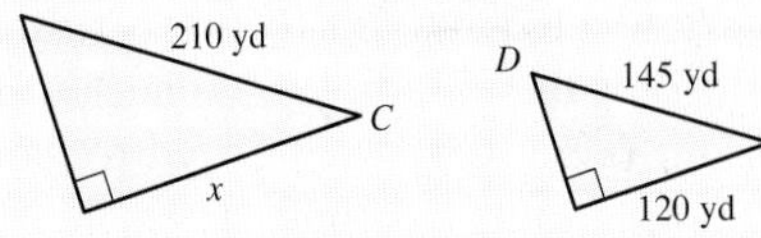

33.

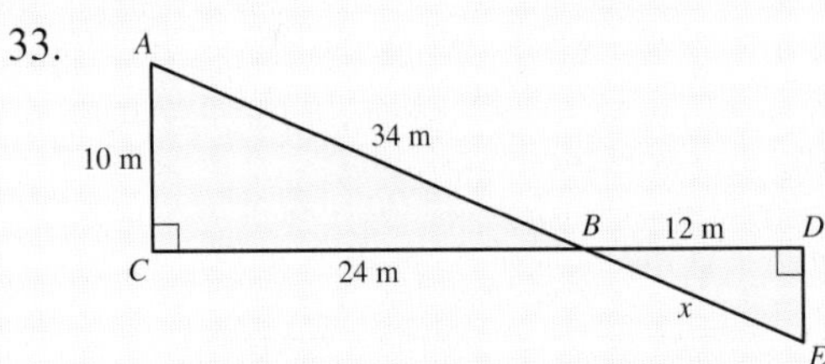

34.

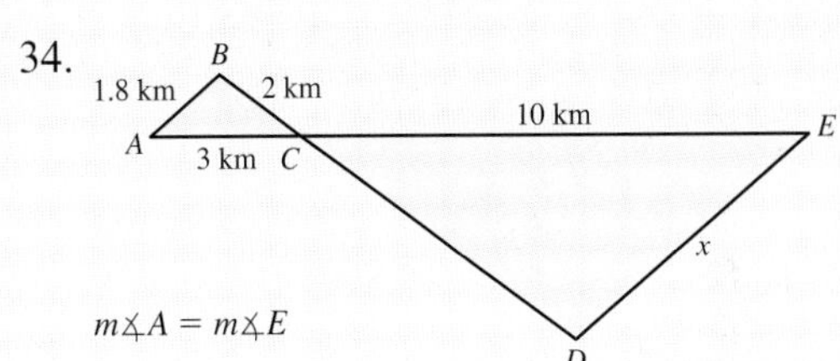

35.

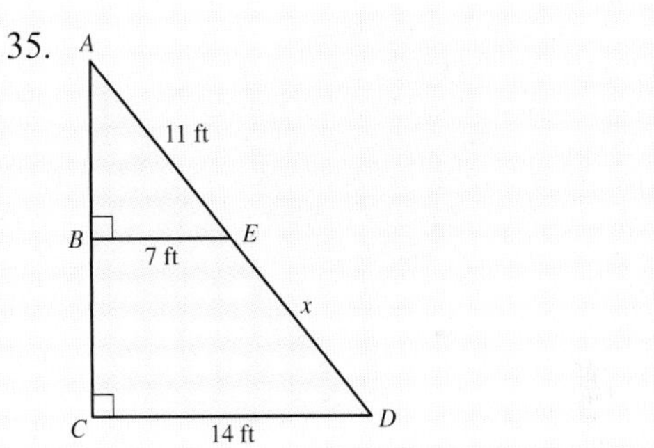

36.

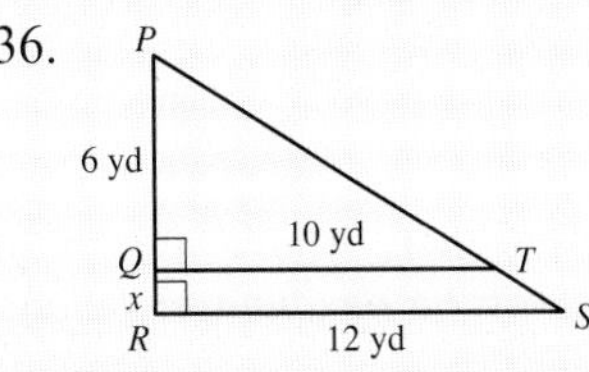

Real-World Applications

37. A baseball diamond is really a square with the bases at the corners. While playing intramural softball, Jane ran 60 feet from home plate to first base. The next batter hit the ball from home plate directly to second base. What distance did he hit the ball?

38. Television screens are sized according to the length of a diagonal across the screen, from one corner to another. The screen of a 52 in. widescreen TV is 24.5 in. high. How wide is it?

39. For the Campus Triathlon, the athletes start at point *A*, swim directly across Siegel Lake to point *B*, then cycle to point *C*, a distance of 7 miles. They then discard their bikes and run 2 miles to point *D*, then turn to run 4 miles to point *E*, then go under the tunnel another 2 miles back to point *C*. They then finish the race back on their bicycles for 7 miles to the finish line at point *A*. What is the distance they swam across Siegel Lake if the measure of angle *A* is the same as the measure of angle *E*?

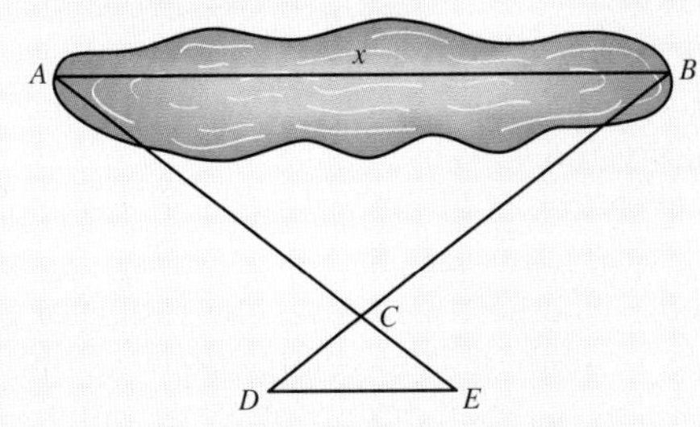

40. At the local mall, Suzette and her friends ate lunch at the food court, located at point A. Afterward, they walked 60 yd to Best Buy at point C to check out the latest CDs. From there, they went 25 yd to Macy's at point B to sample some perfume. After, they continued walking 150 yd to point D, where they saw the latest Batman movie. After the show, they found Suzette's car at point E. How far away from the movie theater was Suzette's car?

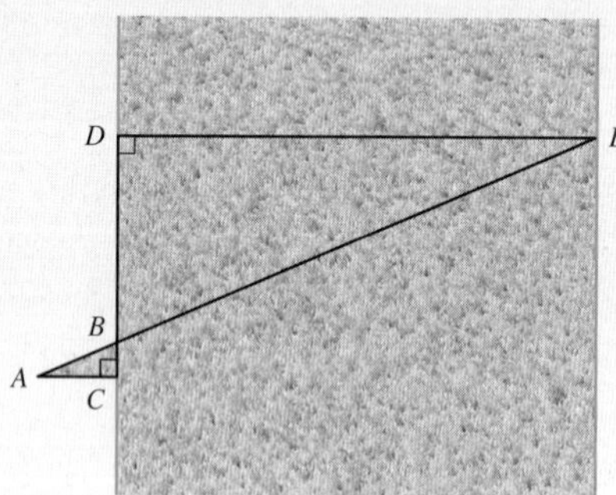

41. Find the height of the tree.

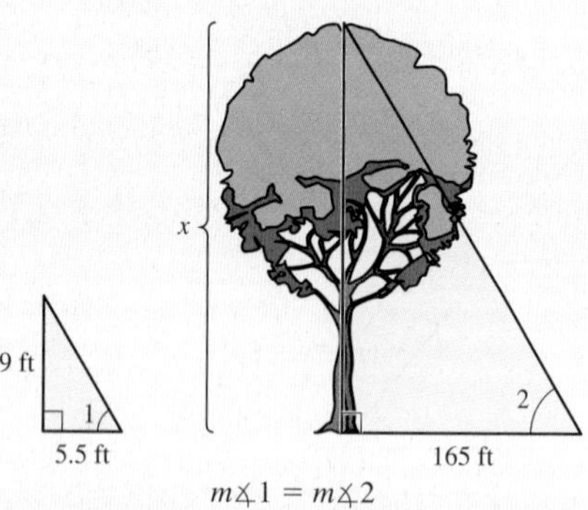

$m\measuredangle 1 = m\measuredangle 2$

42. Find the height of the tower.

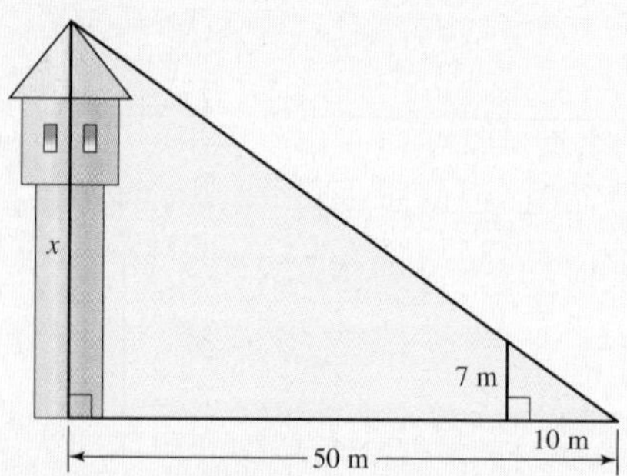

43. To build a loft in their dorm room, Kevin and Neil had to figure out how long to cut a beam so that the beam would run diagonally from a point on the floor 6 feet away from the wall to the top of the 8-foot-high wall. How long should they cut the beam?
44. How high up on a wall is the top of a 20-foot ladder if its bottom is 6 feet from the base of the wall?
45. A plane flies 175 miles north then 120 miles due east. How far diagonally is the plane from its starting point?
46. Two cell phone towers are 20 feet tall and 38 feet tall and they sit 30 feet apart. How far is it from the top of one tower to the other?

Critical Thinking

In Exercises 47–51, we will prove that the measures of the angles in a triangle sum to 180°, using the diagram below.

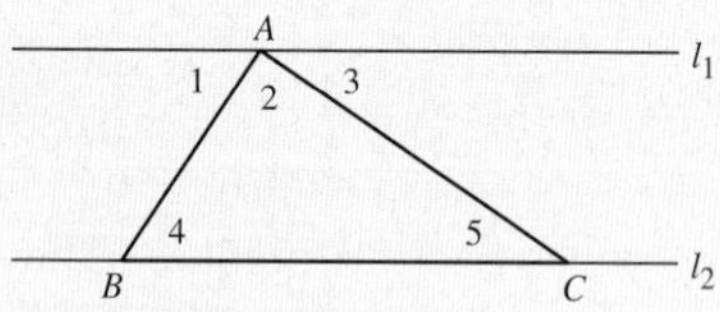

Lines l_1 and l_2 are parallel.

47. What is $m\measuredangle 1 + m\measuredangle 2 + m\measuredangle 3$? Why?
48. What is the relationship between the measures of angles 3 and 5? Why?
49. What is the relationship between the measures of angles 1 and 4? Why?
50. Using Exercises 47–49, what can you conclude about $m\measuredangle 2 + m\measuredangle 4 + m\measuredangle 5$?
51. Discuss whether this proves the result for every triangle, or if there is anything special about this diagram that makes it apply only to certain triangles.
52. Explain how the Pythagorean theorem guarantees that the hypotenuse is the longest side in any right triangle.
53. The length of a side in a triangle depends on the measure of the angle across from it—larger angles means longer sides. This observation, combined with Exercise 52, shows that in a right triangle, the right angle is the largest angle. How else can you show this is true?
54. In the right triangle below, it can be shown that the lengths of the segments have these properties:

$$\frac{\text{Length of side } AD}{\text{Length of side } CD} = \frac{\text{Length of side } CD}{\text{Length of side } DB}$$

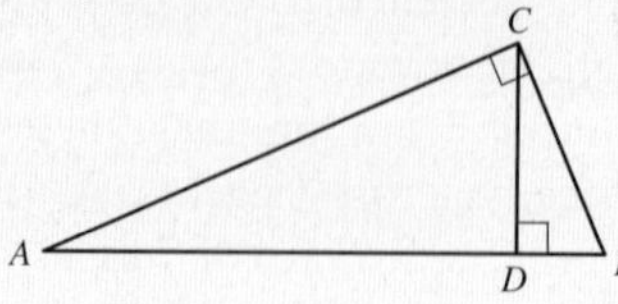

Show the proportion is true by using similar triangles.

Section 10-3 Polygons and Perimeter

LEARNING OBJECTIVES

- ☐ 1. Find the sum of angle measures of a polygon.
- ☐ 2. Find the angle measures of a regular polygon.
- ☐ 3. Find the perimeter of a polygon.

High-rise buildings capture our imagination and get most of the attention, but did you know that as recently as 2008, the largest building in the United States had only five floors? The Pentagon in suburban Washington, D.C., has a floor area of over $6\frac{1}{2}$ million square feet, and is one of the world's finest examples of the importance of geometry in architecture.

Triangles, as we've seen, are very useful in many settings. But of course they are limited to three sides. In this section, we will study closed figures with more than three sides, including the five-sided figure known as the pentagon.

Polygons

Closed geometric figures whose sides are line segments are classified according to the number of sides. These figures are called *polygons.* Table 10-2 shows the number of sides and some of the shapes of these polygons: triangle, **quadrilateral, pentagon, hexagon, heptagon, octagon, nonagon, decagon, dodecagon,** and **icosagon.**

In Section 10-2, we saw that the sum of the measures of the angles of a triangle is equal to 180°. The sum of the measures of the angles of any polygon can be found by using the next formula.

The sum of the measures of the angles of a polygon with n sides is $(n - 2)180°$.

TABLE 10-2 Basic Polygons

Math Note

The word "polygon" comes from the Greek word "polygonos," meaning "many-angled."

Name	Number of sides
Triangle	3
Quadrilateral	4
Pentagon	5
Hexagon	6
Heptagon	7
Octagon	8
Nonagon	9
Decagon	10
Dodecagon	12
Icosagon	20

Many shapes found in nature are polygons.

EXAMPLE 1 Finding the Sum of Angle Measures of a Polygon

Find the sum of the measures of the angles of a heptagon.

SOLUTION

According to Table 10-2, a heptagon has seven sides, so the sum of the measures of the angles of the heptagon is

$$\begin{aligned}(n-2)180^\circ &= (7-2)180^\circ \quad \text{Use } n = 7.\\ &= 5 \cdot 180^\circ \\ &= 900^\circ\end{aligned}$$

The sum of the measures of the angles of a heptagon is 900°.

1. Find the sum of angle measures of a polygon.

▼ Try This One 1

Find the sum of the measures of the angles of an icosagon.

Types of Quadrilaterals

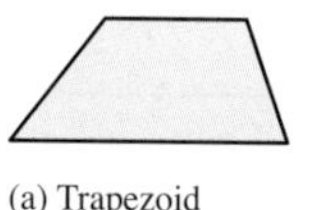

(a) Trapezoid

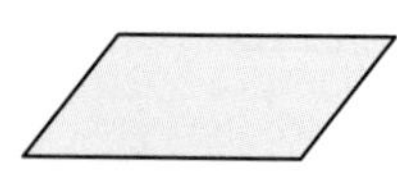

(b) Parallelogram

(c) Rectangle

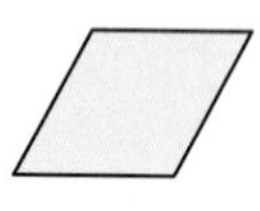

(d) Rhombus

(e) Square

Figure 10-13

Quadrilaterals

Just as there are special names for certain types of triangles, there are names for certain types of quadrilaterals as well. (Recall that a quadrilateral is a polygon with four sides.)

A **trapezoid** is a quadrilateral that has exactly two parallel sides. See Figure 10-13a.

A **parallelogram** is a quadrilateral in which opposite sides are parallel and equal in measure. See Figure 10-13b.

A **rectangle** is a parallelogram with four right angles. See Figure 10-13c.

A **rhombus** is a parallelogram in which all sides are equal in length. See Figure 10-13d.

A **square** is a rhombus with four right angles. See Figure 10-13e.

The next diagram explains the relationships of the quadrilaterals.

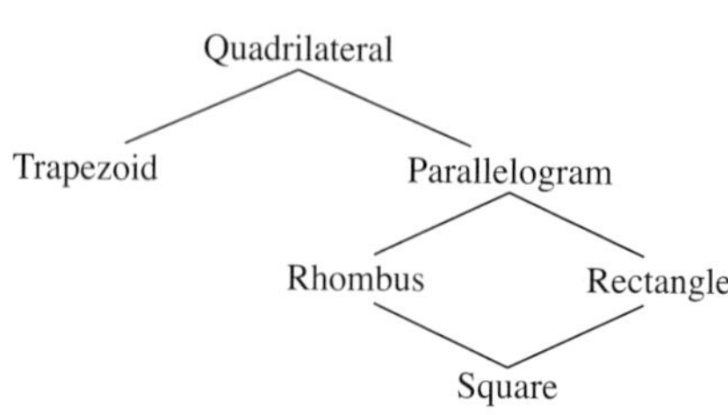

Looking at the relationships, you can see that a square is also a rectangle and a rhombus. A rhombus and a rectangle are also parallelograms.

Most of the polygons drawn in Table10-2 are actually special polygons: all of the sides have the same length, and all of the angles are equal in measure. We call such a figure a **regular polygon.** The most common examples of regular polygons are squares and equilateral triangles.

EXAMPLE 2 Finding Angle Measure for a Regular Polygon

☑ 2. Find the angle measures of a regular polygon.

Find the measure of each angle of a regular hexagon.

SOLUTION

First, find the sum of the measures of the angles for a hexagon. The formula is $(n - 2) \cdot 180°$, where n is the number of sides. Since a hexagon has six sides, the sum of the measures of the angles is $(6 - 2) \cdot 180° = 720°$. Next, divide the sum by 6 since a hexagon has six angles: $720 \div 6 = 120°$.

Each angle of a regular hexagon has a measure of 120°.

▼ Try This One 2

Find the measure of each angle of a regular pentagon.

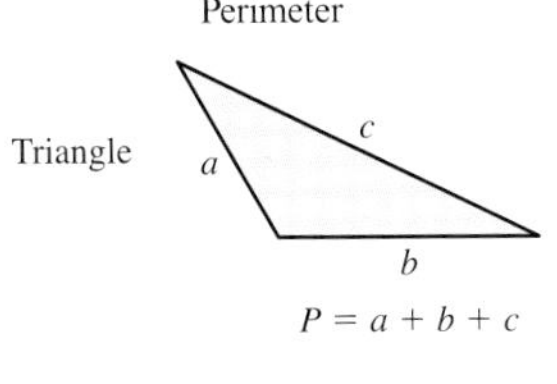
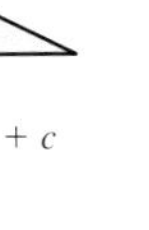

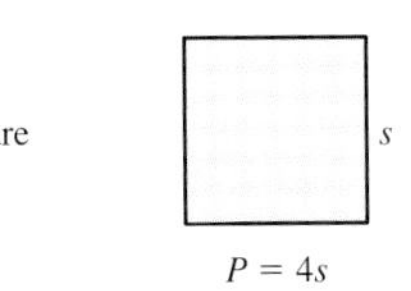

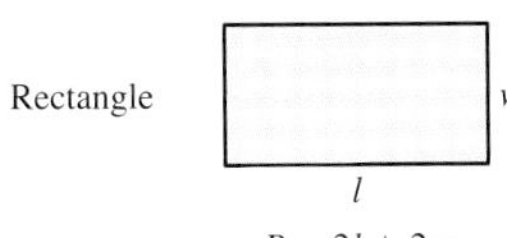

Figure 10-14

Perimeter

The **perimeter** of a polygon is the sum of the lengths of its sides. The perimeter of a triangle with sides of length a, b, and c is simply $P = a + b + c$. For a square with side length s, the perimeter is $P = s + s + s + s$, or $P = 4s$. For a rectangle with length l and width w, there are two sides that are l units long, and two that are w units long, so the perimeter is $P = l + l + w + w$, or $P = 2l + 2w$. See Figure 10-14.

EXAMPLE 3 Finding the Perimeter of a Rectangle

The Houser family finds their dream home perfect in every way except one: the backyard is not fenced in, and their dog Bunch needs room to roam. The rectangular portion they plan to enclose is 95 feet wide and 70 feet long. How much fence will they need to enclose the yard on all four sides?

SOLUTION

The amount of fence needed is the perimeter of the rectangle.

$$\begin{aligned} P &= 2l + 2w \\ &= 2(70) + 2(95) \\ &= 140 + 190 = 330 \end{aligned}$$

The Housers need 330 feet of fence.

▼ Try This One 3

The intramural field at Fiesta University is a large square measuring 600 yards on a side. Find the perimeter.

Sidelight A TRIUMPH OF GEOMETRY

Completed in 1943, the Pentagon is a marvel of architectural design, packing an incredible amount of floor space into a five-story building. Although it covers an area of just 29 acres, the total floor space is more than 152 acres. Over 25,000 people work there on an average day; this would make it the fourth-largest city in Alaska. There are 17.5 *miles* of corridors in the building; at an average walking pace, it would take almost 6 hours to walk all of them. And yet because of the geometric design, it takes at most 7 minutes to walk from any point in the building to any other.

EXAMPLE 4 Finding the Perimeter of a Polygon

The length of each outside wall of the Pentagon is 921 feet. Suppose that a sentry must walk the outside wall six times during his 4-hour shift. How many miles does he walk in one shift?

SOLUTION

A pentagon has five sides, and each has length 921 feet, so the sum of the lengths of the sides is $5 \times 921 = 4{,}605$ feet. In walking the perimeter six times, the sentry covers $6 \times 4{,}605 = 27{,}630$ feet. Now we convert to miles:

$$\frac{27{,}630 \text{ \sout{feet}}}{1} \times \frac{1 \text{ mi}}{5{,}280 \text{ \sout{feet}}} \approx 5.23 \text{ miles}$$

▼ Try This One 4

☑ 3. Find the perimeter of a polygon.

The running path at a state park is a right triangle with legs 0.7 mile and 1.3 miles. What's the total length around the path?

Answers to Try This One

1 3,240°

2 108°

3 2,400 yards

4 About 3.5 miles

EXERCISE SET 10-3

Writing Exercises

1. Is a circle a polygon? Why or why not?
2. How can you find the sum of the angle measures for a polygon?
3. What makes a polygon regular?
4. How can you find the measure of the angles for a regular polygon?

5. What is the perimeter of a polygon?

6. Is the perimeter of a pentagon five times the length of one side? Why or why not?

Computational Exercises

For Exercises 7–12, identify each polygon and find the sum of the measures of the angles.

7.

8.

9.

10.

11.

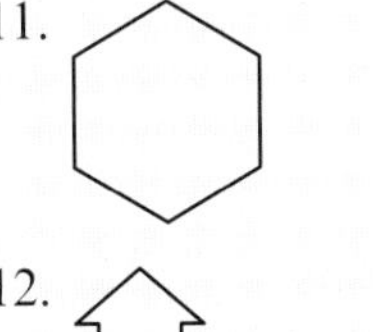

12.

For Exercises 13–16, identify each quadrilateral.

13.

14.

15.

16.

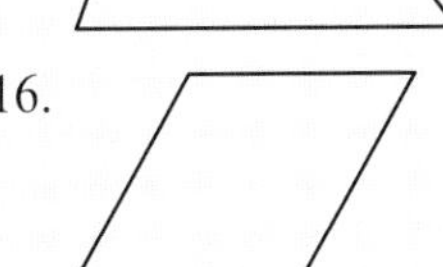

For Exercises 17–20, find the perimeter.

17.

22 yd

16 yd

18.

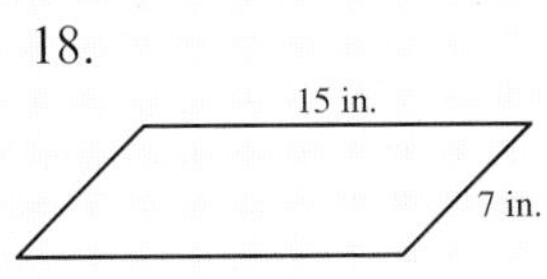

19.

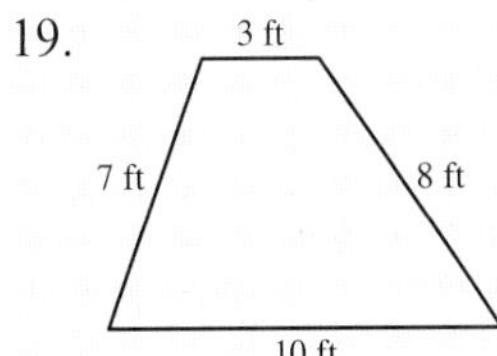

20.

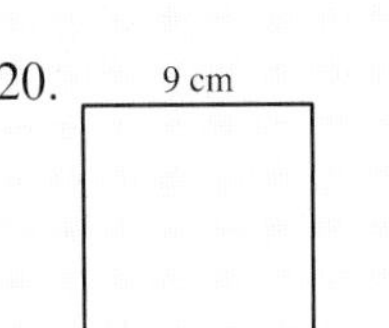

21.

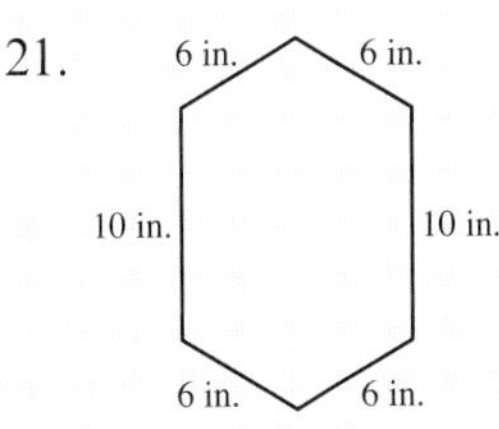

22.

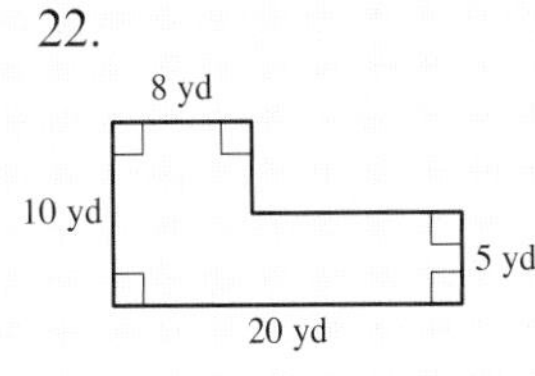

23.

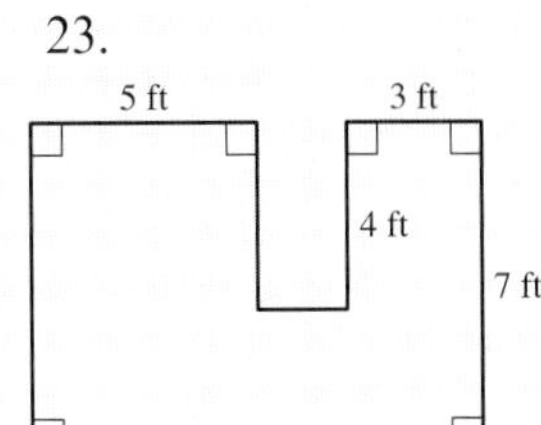

24.

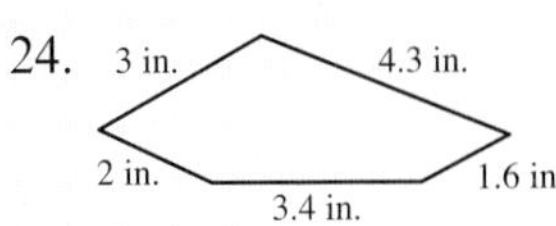

25.

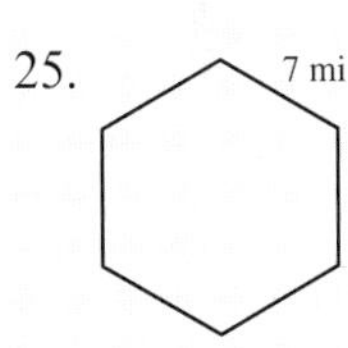

26.

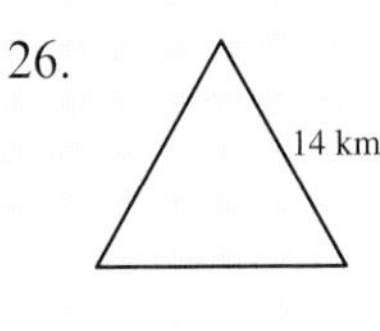

Real-World Applications

27. At least how far does a major league player run when he hits a home run? The baseball diamond is a square with sides 90 feet.
28. How many feet of fence are needed to fence in a rectangular plot 24 feet by 18 feet if an opening of 3 feet is left for a gate?
29. How many feet of hedges will be needed to enclose a triangular display at an amusement park if the sides measure 62 feet, 85 feet, and 94 feet?
30. Jim and Joe have built a rectangular stage for their band that measures 40 feet by 56 feet. They want to put fiber optic lighting around the border that comes in 8-foot sections for $24.00 each. Find the cost of the lighting for their stage.
31. How much molding in length will be needed to frame an 11×14-inch picture if there is to be a 2-inch mat around the picture?
32. A carpenter needs to put baseboard around the room shown. How many feet are needed? Each door is 30 inches wide.

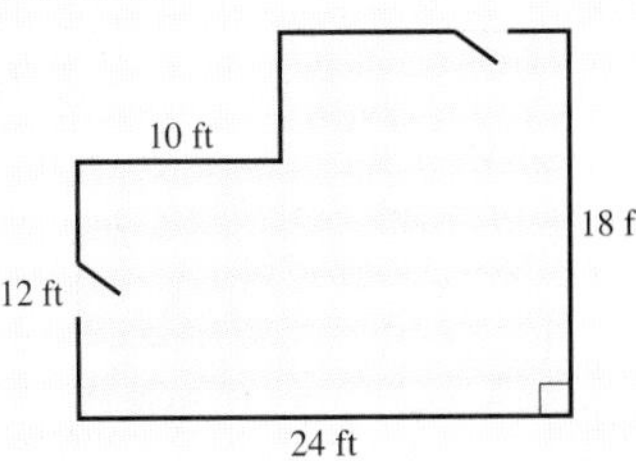

33. How many times must a person walk around a football field in order to walk a mile? The dimensions of a football field are 360 feet by 160 feet. One mile is 5,280 feet.
34. How many times must a person walk around a soccer field in order to walk a mile? The dimensions of a soccer field are 345 feet by 223 feet.

35. Ultimate Fighting matches take place in a steel cage that is a regular octagon with sides 12 feet 3 inches. Jed and Bubba decide to build an Ultimate Fighting cage in Jed's backyard, using fence they can buy at Home Depot for $4 per foot. How much will it cost to build the cage?
36. For a LiveStrong walkathon, Cat is walking her dogs Macleod and Tessa around the perimeter of a rectangular park that is 0.4 miles by 0.7 miles. Her sponsors have pledged a total of $22.50 per mile. If Cat and the girls do four laps around the park, how much money will they raise?
37. Refer to Example 4. If the sentry was told he needed to complete two circuits around the perimeter of the Pentagon in an hour, how fast would he have to go in miles per hour?
38. Refer to Exercise 36. If Cat completed her four laps in 2 hours, what was her average speed in miles per hour?

Critical Thinking

39. For the triangle shown, $\measuredangle BCD$ is called an exterior angle. If you know the measures of $\measuredangle A$ and $\measuredangle B$, explain how the measure of $\measuredangle BCD$ can be found.

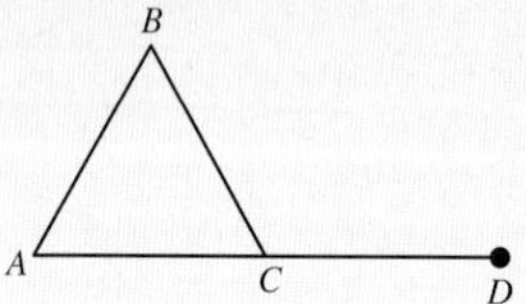

40. What is the measure of each exterior angle of a regular pentagon?
41. Make a table with the number of sides in a regular polygon in one column and the measure of each angle in the other (see Example 2). Start with 8 sides and increment by 2 until you reach 20 sides. Do you see any pattern? Where do you think that pattern is headed as the number of sides gets larger and larger?

Section 10-4 Areas of Polygons and Circles

LEARNING OBJECTIVES

- ❑ 1. Find areas of rectangles and parallelograms.
- ❑ 2. Find areas of triangles and trapezoids.
- ❑ 3. Find circumferences and areas of circles.

In the 21st century, more and more people are taking on home improvement projects that would have been done only by professionals 20 years ago. Stores like Lowe's and Home Depot have become regular stops for homeowners looking to personalize their little corner of the world.

Suppose that you plan to install ceramic tile in your kitchen to cover up that ridiculous yellow linoleum—what were the previous owners thinking? Tiles come in many different sizes, and while planning the job, you want to make sure you buy enough tile. But you also don't want to buy TOO much. A calculation of area is just the ticket to make sure the job is well planned and ultimately successful. In this section we will learn how to find the area of various geometric figures.

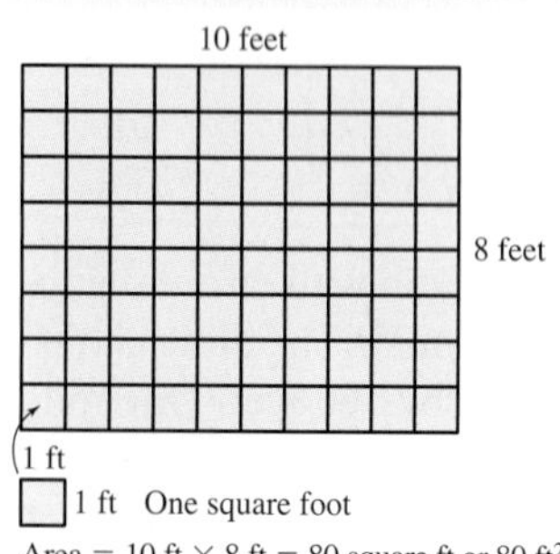

Figure 10-15

Areas of Polygons

We already know that the area of a geometric figure is a measure of the region bounded by its sides. We also know that area is measured in square units, like square feet or square meters. Let's say that your kitchen is a rectangle measuring 8 feet by 10 feet, and that the tiles you've picked out are 12 inches on a side. Then each tile covers exactly one square foot of space. Figure 10-15 is a diagram of the kitchen.

Notice that the floor is divided into a grid by marking off 1-foot units along each side. This shows that it would take 80 tiles to cover the entire floor, so the area is 80 square feet. It is no coincidence that this number is the product of the length and width—this gives us our first area formula:

Area Formulas for Rectangles and Squares

The area of a rectangle is the product of the length and width. If l is the length and w is the width, then

$$A = lw$$

In a square, the length and width are equal, so if s is the length of the sides,

$$A = s^2$$

EXAMPLE 1 Finding the Cost of Installing Carpet

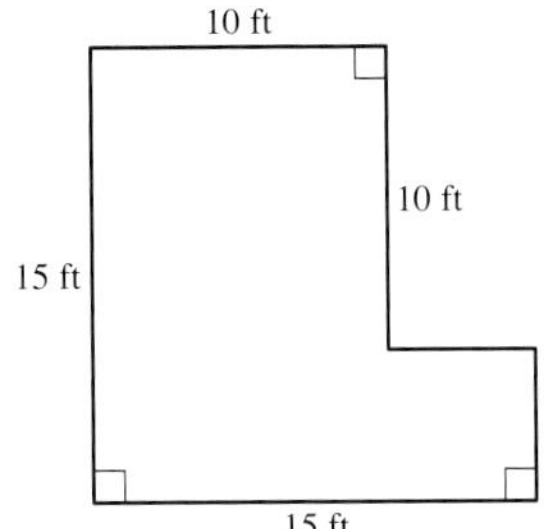

A couple plans to carpet an L-shaped living room, as shown to the left. Find the total cost of the carpet if it is priced at \$25.00 per square yard.

SOLUTION

With a little bit of ingenuity, we can divide the room into two figures we know the area of: a rectangle and a square.

$$A = lw \qquad A = s^2$$
$$= 15 \cdot 10 \qquad = 5^2$$
$$= 150 \text{ square feet} \qquad = 25 \text{ square feet}$$

The total area is 150 square feet + 25 square feet = 175 square feet.

Now we can use dimensional analysis to finish the calculation:

$$175 \text{ ft}^2 \times \frac{1 \text{ yd}^2}{9 \text{ ft}^2} \times \frac{\$25}{1 \text{ yd}^2} = \$486.11$$

It will cost \$486.11 to carpet the room.

▼ Try This One 1

A homeowner plans to install sod around his new house, as shown.

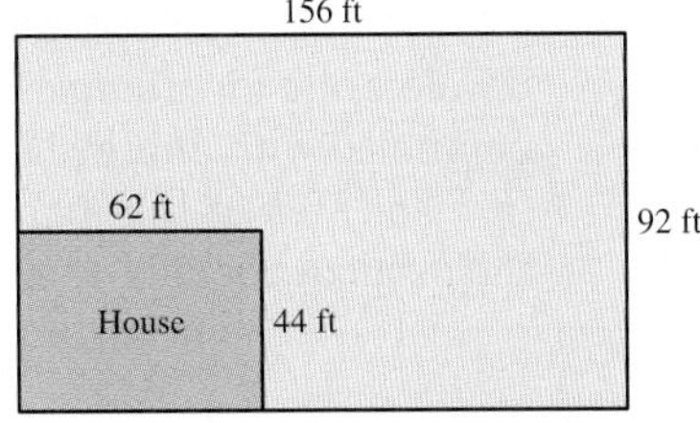

If sod costs \$3.98 per square yard, find the total cost.

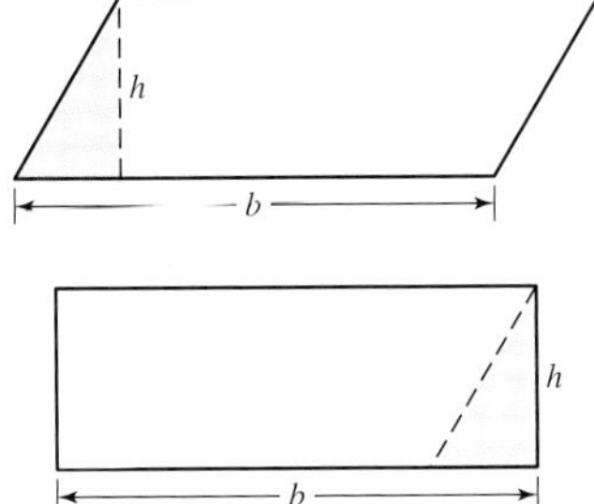

Figure 10-16

Once we know the formula for the area of a rectangle, we can use it to develop formulas for other polygons. We'll start with the parallelogram, shown in Figure 10-16.

The key dimensions here are the base (length of the bottom side) and height (vertical distance from the bottom side to the top side). The trick is to "cut" the bottom left corner piece off and attach it to the right side. This turns the parallelogram into a rectangle, which we can find the area of using length times width.

Area Formula for Parallelograms

The area of a parallelogram is the product of the base and the height. If b is the length of the base and h is the height,

$$A = bh$$

CAUTION

Be careful when working with parallelograms! The base is the length of the bottom (or top) side, but the height is *not* the length of the left or right side. It is the vertical distance from the bottom side to the top side.

EXAMPLE 2 Finding the Area of a Parallelogram

Find the area of the parallelogram:

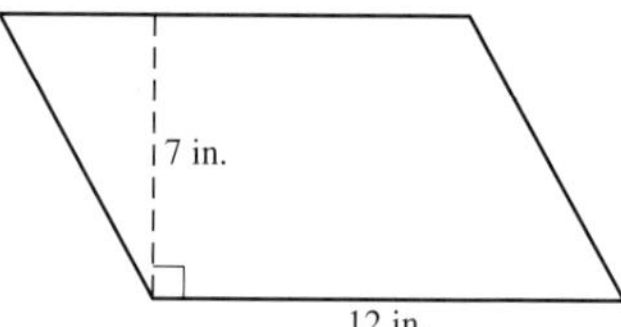

SOLUTION

The base is 12 inches and the height 7 inches, so the area is

$$\begin{aligned} A &= bh \\ &= 12 \text{ in.} \times 7 \text{ in.} = 84 \text{ in.}^2 \end{aligned}$$

▼ Try This One 2

☑ 1. Find areas of rectangles and parallelograms.

Find the area of the parallelogram:

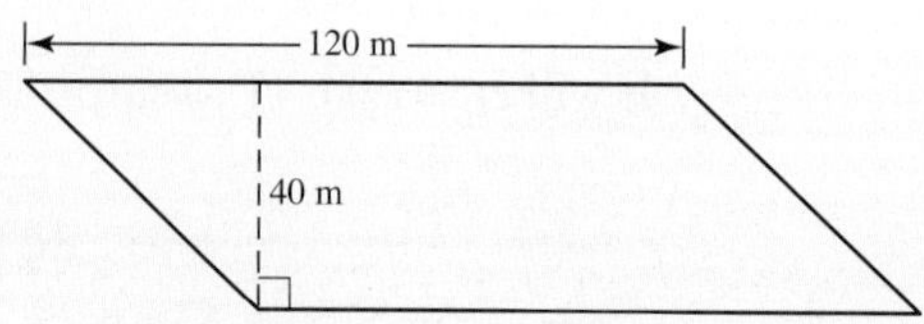

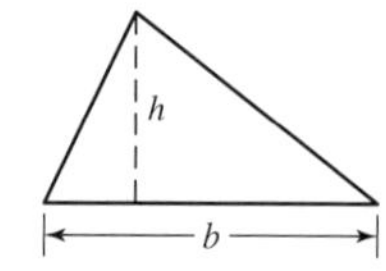

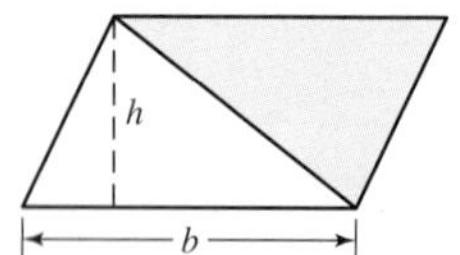

Figure 10-17

Now that we can find the area of a parallelogram, we can use that formula to develop one for triangles. In Figure 10-17, we see a triangle with base b and height h combined with another identical triangle to form a parallelogram.

The area of that parallelogram is $A = bh$, and since it is built from two copies of the original triangle, the area of the triangle is half as much.

Area Formula for Triangles

The area of a triangle is half the base times the height. If b is the length of the base and h is the height, then

$$A = \frac{1}{2}bh$$

CAUTION

Again, be careful! The height of a triangle is not usually the length of one of the sides.

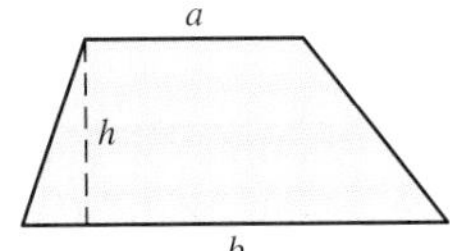

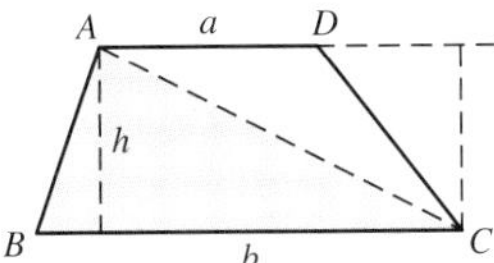

The area of $\triangle ABC = \frac{1}{2}bh$

The area of $\triangle ADC = \frac{1}{2}ah$

Figure 10-18

Finally, we can use triangles to find the area of a trapezoid. In the first drawing in Figure 10-18, we see the relevant dimensions: the height h, and the lengths of the top and bottom sides, a and b. We can cut the trapezoid diagonally to form two triangles, shown in the bottom drawing of Figure 10-18. The top triangle has area $A = \frac{1}{2}ah$, while the bottom triangle has area $A = \frac{1}{2}bh$. The area of the trapezoid is the sum of these two areas:

$$A = \frac{1}{2}ah + \frac{1}{2}bh \qquad \text{Factor out } 1/2\ h.$$
$$= \frac{1}{2}h(a + b)$$

How would you find the area of the front of this house to estimate the amount of paint needed to paint it?

Math Note

When finding the area of a trapezoid, make sure that *a* and *b* are the lengths of the parallel sides.

Area Formula for Trapezoids

The area of a trapezoid with parallel sides *a* and *b* and height *h* is

$$A = \frac{1}{2}h(a + b)$$

EXAMPLE 3 Finding the Area of a Triangle

Find the area of the triangle shown.

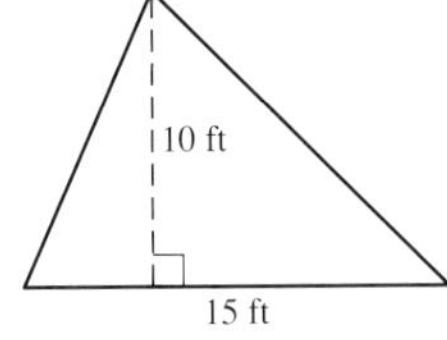

SOLUTION

The base is 15 ft and the height is 10 ft:

$$A = \frac{1}{2}bh = \frac{1}{2}(15)(10) = 75 \text{ square feet}$$

The area of the triangle is 75 square feet.

▼ Try This One 3

☑ 2. Find areas of triangles and trapezoids.

Find the area of the trapezoid shown.

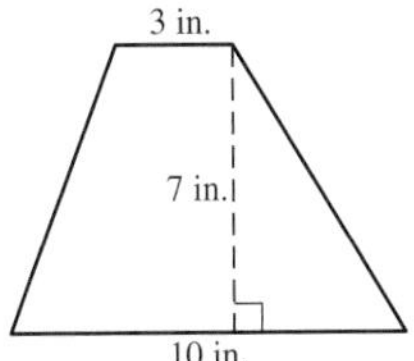

Sidelight CALCULUS AND AREA

We now have a variety of formulas for finding area, but how would you find area of a figure that's curved or oddly shaped, like the front of the building shown here? This is one of the major topics of calculus. In calculus, we learn how to approximate areas, and in many cases find exact areas of figures when we don't have basic formulas for them.

It's all based on a really simple idea: we're good at computing areas of rectangles, so to approximate areas of other shapes, we'll fit little rectangles inside the shape, and sum the areas of the rectangles. This is not terribly different than the way we started with the simple rectangle formula in this section and used it to build other area formulas.

Circles

By definition, polygons have sides that are line segments. The most common geometric figure that doesn't fit that criterion is the circle—no part of a circle is straight.

> **Math Note**
>
> The center of a circle is inside the circle, but is not actually part of the circle.

A **circle** is the set of all points in a plane that are the same distance from a fixed point, which we call the **center** of the circle.

> **Math Note**
>
> An interesting fact: manhole covers are circular because a circle is the only shape where it's completely impossible for the lid to fall in the hole.

Based on the definition of circle, a line segment from any point on a circle to the center is always the same length. We call this length the **radius** of the circle. A line segment starting at a point on a circle, going through the center, and ending at a point on the opposite side is called a **diameter** of the circle, and its length is twice the radius. If we use r to represent radius and d to represent diameter,

$$d = 2r \quad \text{and} \quad r = \frac{d}{2}$$

The distance around the outside of a circle is called the **circumference** (C) of the circle. (This is analogous to the perimeter of a polygon.) The key parts of a circle are illustrated in Figure 10-19.

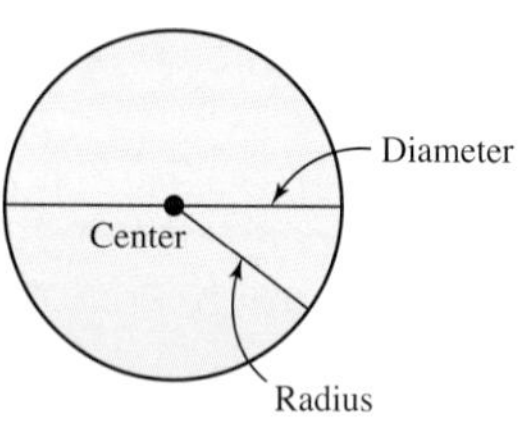

Figure 10-19

Thousands of years ago, people started to realize that if you divide the circumference of any circle by its diameter, the result is always the same number. Eventually, this number was given a special name, the Greek letter π (pronounced **pi**). We can't write an exact value of pi, because it is irrational, so its decimal expansion is infinitely long. But 3.14 is commonly used as an approximation. (See the Sidelight on page 237 for a discussion of pi.)

The fact that circumference divided by diameter always equals pi gives us a formula for the circumference of a circle.

Circumference Formula for Circles

The circumference of a circle is π times the diameter, or 2π times the radius:

$$C = \pi d \quad \text{or} \quad C = 2\pi r$$

EXAMPLE 4 Finding Distance Around a Track

A dirt track is set up for amateur auto racing. It consists of a rectangle with half-circles on the ends, as shown in Figure 10-20. The track is 300 yards wide, and 700 yards from end to end. What is the distance around the track?

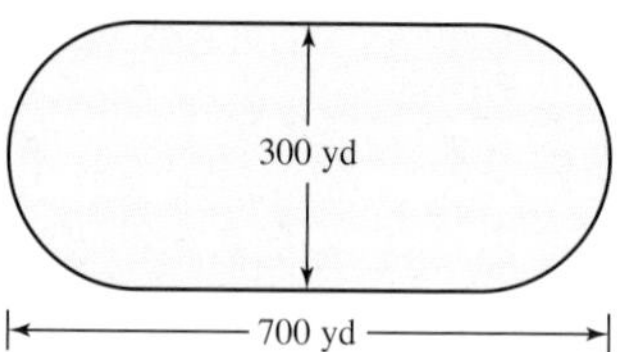

Figure 10-20

> **Calculator Guide**
>
> As a substitute for using the approximation $\pi \approx 3.14$, most calculators have π built in. To find $\pi \cdot 300$ in Example 4:
>
> **Standard Scientific Calculator**
>
> [π] [×] 300 [=]
>
> **Standard Graphing Calculator**
>
> [2nd] [^] [×] 300 [Enter]

SOLUTION

From the diagram, we can see that the diameter of the circular ends is 300 yards, so the circumference is $C = \pi d = \pi(300) \approx 942$ yards. (This is the total length of the curved portion, since the two half circles make one full circle.) The length of each straightaway is the total length of the track (700 yards) minus twice the radius of the circular ends, which is 150 yards (see Fig. 10-21). So each straightaway is $700 - 2(150) = 400$ yards. Now we can find the total length:

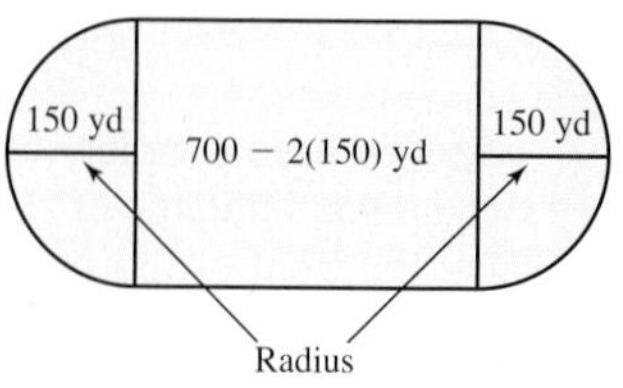

Figure 10-21

$$L = 942 + 400 + 400 = 1{,}742 \text{ yards}$$

▼ Try This One 4

The entrance to Joe's living room is an arch consisting of a rectangle that is 7 feet high and 4 feet wide with a half-circle on top. How many feet of trim would be needed to go around the outside of the arch, not including the bottom?

Through exhaustive experiments involving comparing areas of circles to other known figures, the Greeks found that the ratio of a circle's area to the square of its radius is π. This provides our next area formula:

Area Formula for Circles

The area of a circle is pi times the square of the radius: $A = \pi r^2$

EXAMPLE 5 Finding Area Enclosed by a Track

Find the area enclosed by the track in Example 4.

SOLUTION

The area is the sum of the area of a rectangle and the area of a circle (again because the two half-circles form one full circle).

Rectangle	**Circle**
$A = lw$	$A = \pi r^2$
$= 400 \cdot 300$	$= \pi \cdot 150^2$
$= 120{,}000$	$\approx 70{,}650$

The combined area is about $120{,}000 + 70{,}650 = 190{,}650$ square yards.

▼ Try This One 5

☑ 3. Find circumferences and areas of circles.

Find the area of the doorway in Try This One 4

Sidelight MATH AND THE ORBITS OF COMETS

For most of our history, humans have been fascinated and mystified by the appearance of comets. But starting in the 16th century, astronomers were able to use math to unlock the secrets of comets and were able to predict their appearances.

The most famous such comet is Halley's Comet. It has been appearing for millions of years, but its occasional appearances remained a mystery until British astronomer Sir Edmond Halley began studying it in 1704. He noticed that earlier records showed that a comet had appeared in the same region of the sky in 1456, 1531, 1607, and 1682. He was the first person to suggest that it was the same comet appearing every 75 or 76 years, and correctly predicted that it would reappear in 1758. (Sadly, he didn't live to see his prediction come true.)

When a circle is stretched in one direction, the resulting figure is called an ellipse. Scientists have determined that the paths comets take through the solar system are elliptical in shape. Armed with this knowledge and some data, they can use math to calculate the path, speed, and appearance dates of some comets.

Answers to Try This One

1 \$5,140.39

2 4,800 square meters

3 45.5 square inches

4 20.3 feet

5 34.3 square feet

EXERCISE SET 10-4

Writing Exercises

1. How can we use the formula for the area of a rectangle to find the area of a parallelogram?
2. How can we use the formula for the area of a parallelogram to find the area of a triangle?
3. Explain the difference between perimeter and circumference.
4. What is the connection between the number π and the dimensions of a circle?

Computational Exercises

For Exercises 5–16, find the area of each figure.

5. 17 in.

6.

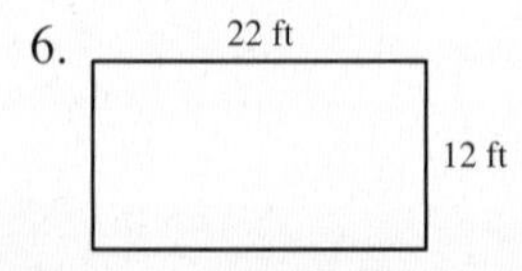

7.

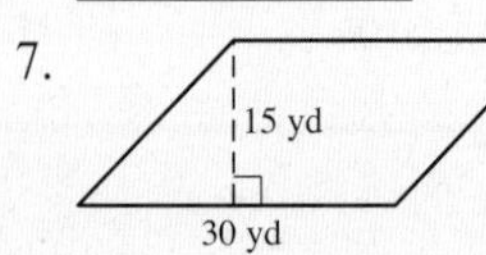

8.

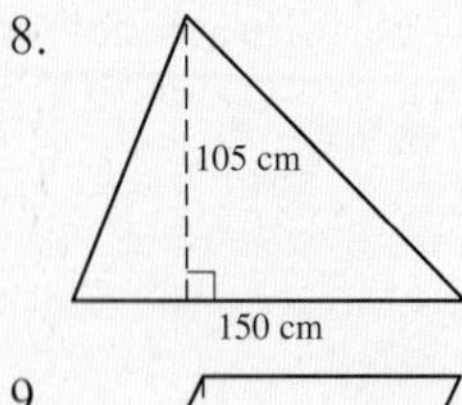

9.

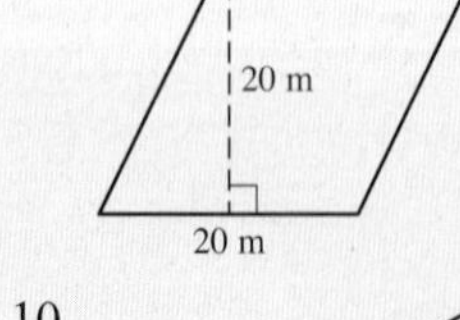

10.

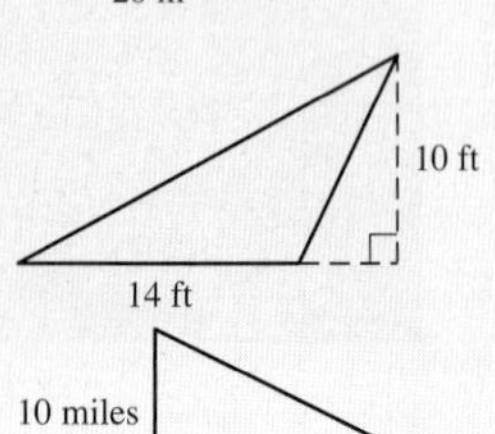

11. 10 miles; 21 miles

12. 9 in.; 9 in.

13. 20 in.; 17 in.; 35 in.

14. 13 km; 19 km; 24 km

15.

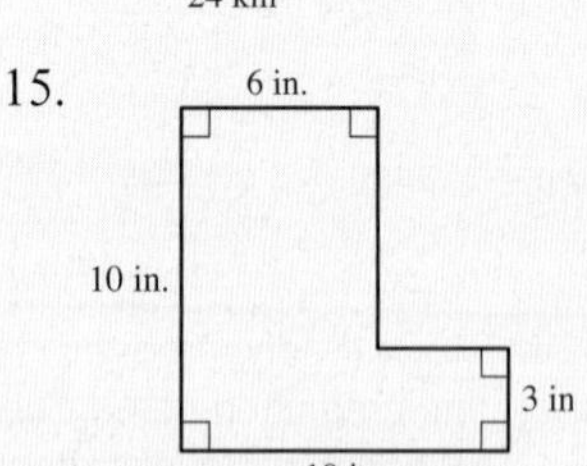

16.

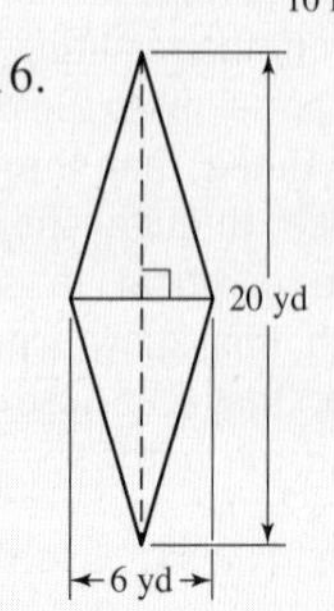

For all calculations involving π, answers may vary slightly depending on whether you use $\pi = 3.14$ or the π key on a calculator.

For Exercises 17–22, find the circumference and the area of each circle.

17.

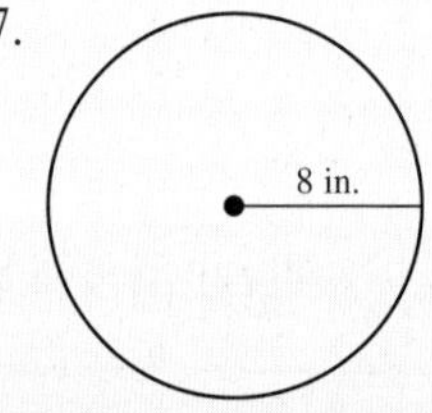

18. 10 ft

19.

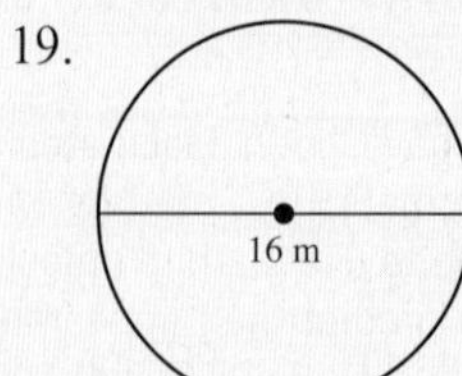

20.

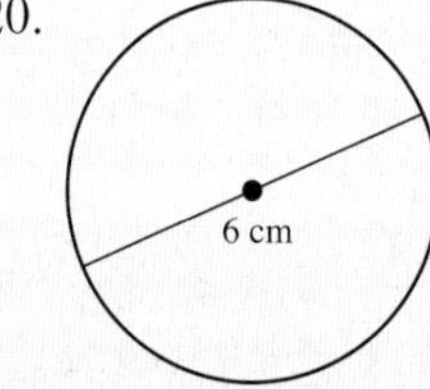

21. 21 km

22.

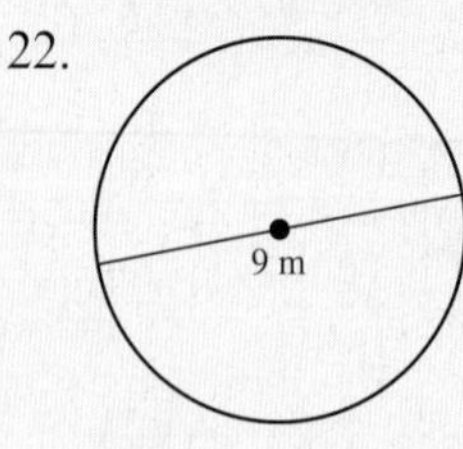

Real-World Applications

23. How many square yards of carpeting are needed to cover a square dorm room that measures 10 feet on a side?
24. Find the cost of coating a rectangular driveway that measures 11 feet by 21 feet, at \$2.50 per square foot.
25. Dawn printed off several 3-inch by 5-inch pictures from her digital camera onto a sheet of glossy photo paper measuring 24 by 25 inches. How many can she paste onto a 600-square-inch poster board if the images are aligned side by side so no white space is left?
26. Find the amount and the cost of artificial turf needed to cover a football field that measures 360 feet by 160 feet. The cost of the turf is \$20.00 per square foot.
27. A stage floor shaped like a trapezoid with bases of 60 feet and 75 feet and a height of 40 feet is to be covered with plywood for a play. What would be the cost for the plywood if it sells for \$0.60 a square foot?
28. Derek and Amir have started a lawn care service for summer money and have run into a challenge. They are to plant new sod on a lawn that is in the shape of a trapezoid, with the bases measuring 21 feet and 32 feet and a vertical height from one base to the other of 10 feet. Find the cost of the sod if each square yard sells for \$7.00.
29. A triangular-shaped shelf in a student's dormitory room that is placed in a corner has a base of 5 feet and a height of 3 feet. Find the area of the shelf.
30. How many square yards of fabric are needed to make a triangular team banner whose base is 6 feet and whose height is 6 feet for a tailgating party?
31. Janelle and Sandra are planting various colors of flowers in a circular shape on the front lawn of their university to display the school's new logo. The circle has a diameter of 12 feet and the flowers cost \$15.00 per square yard. Find the cost of the flowers.
32. How much more pizza do you get in a large pizza that has a diameter of 15 inches than you do in a small pizza that has a diameter of 10 inches?
33. A college radio station can broadcast over an area with a radius of 60 miles. How much area is covered?
34. Find the distance around the inside lane of a track that has the dimensions shown below.

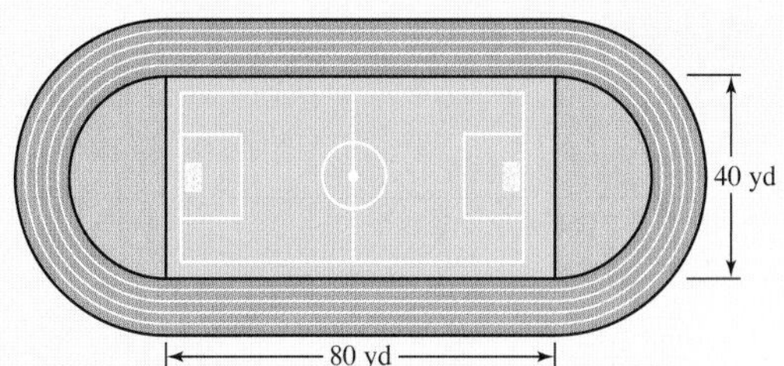

35. For a Homecoming Game Gala, the cheerleading squad wants to sell ice cream to earn money for new equipment. They plan to make a sign in the shape of an ice cream cone with the dimensions given below. Find the area of the sign so they know how much plywood to buy.

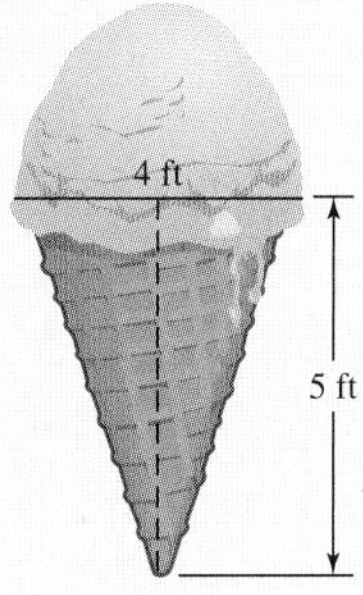

36. Find the area of the walkway (24 inches wide) that will be installed around a circular jacuzzi with a diameter of 6 feet.

Critical Thinking

37. Under what conditions will the area of a circle be the same as its circumference?
38. In the formula for the area of a triangle, $A = \frac{1}{2}bh$, b is the length of one side of the triangle, but h typically is not. Under what circumstances is h the length of one side of the triangle?
39. Review the definition of a circle on page 540. How can you use that definition to draw a perfect circle using things you can find lying around the house? (Tracing a circular object doesn't count!)
40. The area of a triangle can be found if the measures of the three sides are known. This formula was discovered about 100 BCE by a Greek mathematician known as Heron. Heron's formula is

$$A = \sqrt{s(s-a)(s-b)(s-c)}$$

where $s = \frac{1}{2}(a + b + c)$ and a, b, and c are the measures of the lengths of the sides of the triangle. Using the formula, find the area of a triangle if the sides are 5 in., 12 in., and 13 in.
41. The triangle in Problem 40 is also a right triangle. Find the area using the formula $A = \frac{1}{2}bh$ and see if you get the same answer.

Section 10-5 Volume and Surface Area

LEARNING OBJECTIVES

- ❑ 1. Find the volumes of solid figures.
- ❑ 2. Find the surface areas of solid figures.

Owning a swimming pool sounds pretty great, especially during those hot summer weekends. But a lot of work goes into maintaining a pool, and it takes a while to learn everything you need to know. A surprising amount of math goes into it, too. Once you fill up your pool with sparkling clear water, it won't stay that way very long if you don't add chemicals. But how much should you add? The amount of chemicals you need depends on the capacity of the pool, and that's where the math comes in.

In Section 9-2, we learned that capacity is a measure of volume, which is the amount of space enclosed by a three-dimensional object. We know that volume is measured in cubic units, like cubic inches or cubic meters, and we also have conversions to write volume in terms of familiar capacity units like gallons or liters.

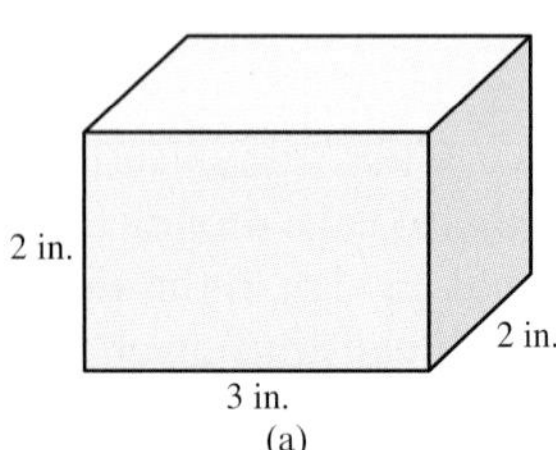

(a)

A **polyhedron** is a three-dimensional figure bounded on all sides by polygons. The simplest polyhedra are rectangular solids and cubes, which are bounded on all sides by rectangles or squares. A rectangular solid is shown in Figure 10-22a. It has three dimensions: length (3 in.), width (2 in.), and height (2 in.). In Figure 10-22b, we see that you could build the solid out of 12 cubes that are 1 inch on all sides. That tells us that the volume of the rectangular solid is 12 cubic inches.

This volume can be obtained by multiplying the length, width, and height, giving us our first volume formula.

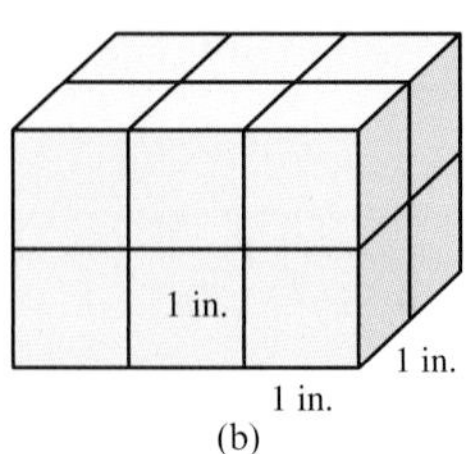

(b)

Figure 10-22

Volume Formulas for Rectangular Solids and Cubes

The volume of a rectangular solid is the product of the length, width, and height. For a cube, all three dimensions are equal, so the volume is the length of a side raised to the third power (cubed).

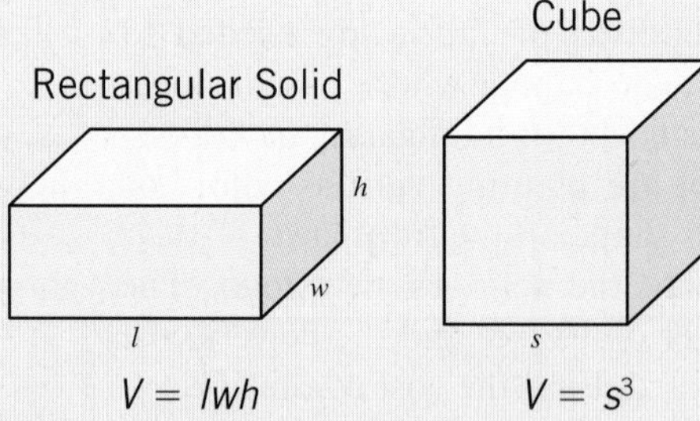

EXAMPLE 1 Finding the Volume of a Rectangular Solid

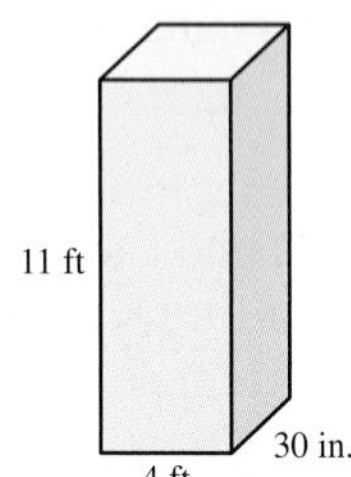

Figure 10-23

Find the volume of the rectangular solid pictured in Figure 10-23.

SOLUTION

The length is 4 feet, the width is 30 inches, and the height is 11 feet. First, we need to rewrite 30 inches in terms of feet:

$$30\,\cancel{\text{in.}} \times \frac{1\text{ ft}}{12\,\cancel{\text{in.}}} = \frac{30}{12}\text{ ft} = 2.5\text{ ft}$$

Now we use the volume formula with $l = 4$, $w = 2.5$, and $h = 11$:

$$V = lwh = 4 \times 2.5 \times 11 = 110\text{ ft}^3$$

▼ Try This One 1

Find the volume of a rectangular solid with length 10 meters, width 12 meters, and height 50 centimeters.

EXAMPLE 2 Finding the Volume of a Swimming Pool

In order to figure out how much chlorine to use, Judi needs to know the capacity of her pool. The pool is a rectangle 18 feet wide and 36 feet long, and has an average depth of 4 feet. How many gallons does it hold? (There are 7.48 gallons in 1 cubic foot.)

SOLUTION

With an average depth of 4 feet, we can think of the pool as a rectangular solid with $l = 36$ ft, $w = 18$ ft, and $h = 4$ ft.

$$V = lwh = 36 \times 18 \times 4 = 2{,}592 \text{ ft}^3$$

Now we use dimensional analysis to convert to gallons:

$$2{,}592 \cancel{\text{ft}^3} \times \frac{7.48 \text{ gal}}{1 \cancel{\text{ft}^3}} \approx 19{,}388 \text{ gal}$$

The capacity of the pool is about 19,388 gallons.

▼ Try This One 2

A rectangular dunking tank at a carnival is 4 feet long and 5 feet wide, and is filled to a depth of 4 feet. How many gallons of water were needed to fill it?

Notice that the volume of a rectangular solid can be thought of as the area of the base (length times width) multiplied by the height. The same is true for our next solid figure, the cylinder. A **right circular cylinder** is a figure with a circular top and bottom, and with sides that make a right angle with the top and bottom. (Think of a soda can.)

Volume Formula for a Right Circular Cylinder

The volume of a right circular cylinder is given by the formula

$$V = \pi r^2 h$$

where r is the radius of the circular ends and h is the height.

Cylinder

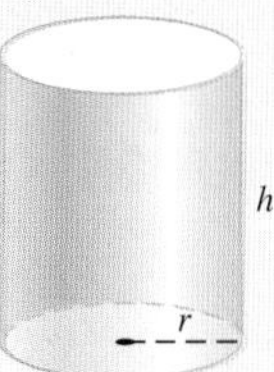

$V = \pi r^2 h$

Math Note

Since the area of a circle with radius r is πr^2, the volume of a right circular cylinder is the area of the base times the height, just like a rectangular solid.

EXAMPLE 3 Finding the Volume of a Cylinder

How many cubic inches does a soup can hold if it is a right circular cylinder with height 4 inches and radius 1.5 inches?

SOLUTION

Using $r = 1.5$ in. and $h = 4$ in., we get

$$V = \pi r^2 h = \pi \cdot 1.5^2 \cdot 4 = 9\pi \approx 28.3 \text{ in.}^3$$

▼ Try This One 3

Find the volume of an oil storage tank that is a right circular cylinder with height 40 feet and radius 15 feet.

Figure 10-24

A cylinder is not a polyhedron because the edges are not polygons, but our next figure is. A **pyramid** is a polyhedron whose base is a polygon and whose sides are triangles. An example with a square base is shown in Figure 10-24.

Volume Formula for a Pyramid

The volume of a pyramid is given by the formula

$$V = \frac{1}{3}Bh$$

where B is the area of the base and h is the height.

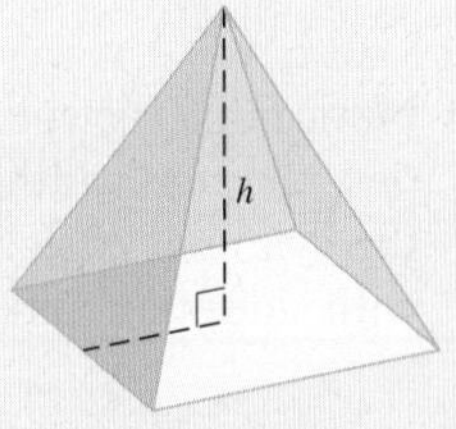

$V = \frac{1}{3}Bh$, where B is the area of the base

EXAMPLE 4 Finding the Volume of a Pyramid

The Great Pyramid at Giza was built by the Egyptians roughly 4,570 years ago. It has a square base measuring 230.6 meters on a side and is 138.8 meters high. Find its volume.

SOLUTION

Since the base is a square that is 230.6 meters on a side, its area is 230.6^2, or 53,176.36 square meters. Using the volume formula with $B = 53{,}176.36$ and $h = 138.8$, we get

$$V = \frac{1}{3} \cdot 53{,}176.36 \cdot 138.8 \approx 2{,}460{,}293 \text{ m}^2$$

▼ Try This One 4

The smallest of the three great pyramids is the Pyramid of Menkaure. Its base measures 339 feet on each side, and it is 215 feet high. Find the volume.

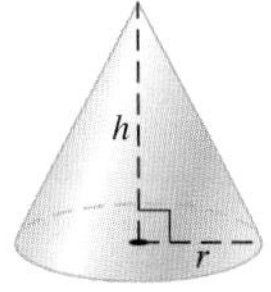

Figure 10-25

The next figure we will discuss is the **right circular cone**. This is similar to a pyramid with a circular base, as shown in Figure 10-25.

Not surprisingly, the volume formula matches the pyramid formula: one-third times the area of the base times the height. But the base is always a circle with area πr^2, so we get the following:

Volume Formula for a Right Circular Cone

The volume of a right circular cone is given by

$$V = \frac{1}{3}\pi r^2 h$$

where r is the radius of the circular bottom and h is the height.

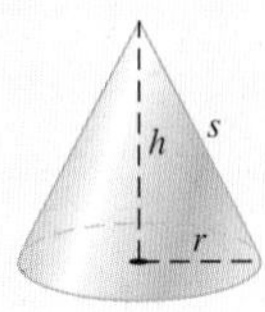

$V = \frac{1}{3}\pi r^2 h$

EXAMPLE 5 Finding the Volume of a Cone

The cups attached to a water cooler on a golf course are right circular cones with radius 1.5 inches and height 3 inches. How many ounces of water do they hold? (One ounce is about 1.8 cubic inches.)

SOLUTION

Using $r = 1.5$ and $h = 3$, we get

$$V = \frac{1}{3}\pi r^2 h = \frac{1}{3}\pi(1.5 \text{ in.})^2(3 \text{ in.}) \approx 7.1 \text{ in.}^3$$

Now we convert to ounces:

$$7.1 \cancel{\text{in.}^3} \times \frac{1 \text{ oz}}{1.8 \cancel{\text{in.}^3}} \approx 3.9 \text{ oz}$$

The cups hold about 3.9 ounces of water.

▼ Try This One 5

How many ounces of melted ice cream does an ice cream cone hold if it has a radius of 1.1 inches and a height of 5 inches?

Many of the games we play use solid objects that are circular in three dimensions—basketballs, baseballs, golf balls, pool balls, etc. The geometric name for such an object is a **sphere.** Just as a circle is the set of all points in a plane that are the same distance from a fixed point, a sphere is the set of all points in space that are the same distance away from a fixed point. And we use the same terminology: the fixed point is called the **center** of the sphere, and the distance is called the **radius.** (Twice the radius is called the **diameter.**)

Math Note

The formula for the volume of a sphere is easily developed using calculus.

Volume Formula for a Sphere

The volume of a sphere is given by the formula

$$V = \frac{4}{3}\pi r^3$$

where r is the radius of the sphere.

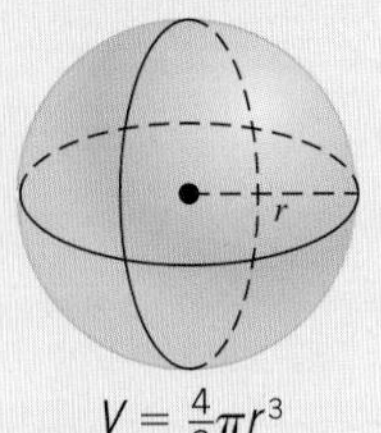

EXAMPLE 6 Finding the Volume of a Sphere

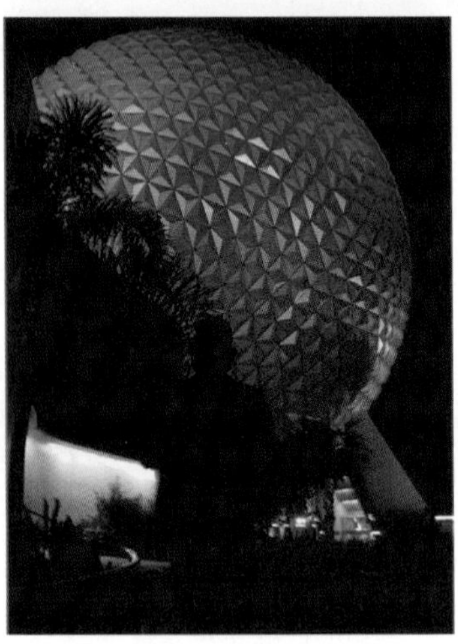

The famous ball at Epcot center in Orlando has a diameter of 164 feet. Find its volume.

SOLUTION

To find the volume, we need the radius. The diameter is twice the radius, so the radius in this case is half of 164 feet, or 82 feet.

$$V = \frac{4}{3}\pi r^3 = \frac{4}{3}\pi(82)^3 \approx 2{,}309{,}565 \text{ ft}^3$$

The volume of the ball is about 2,309,565 cubic feet.

▼ Try This One 6

An official NBA basketball is a sphere with diameter 9.4 inches. What is the volume?

☑ 1. Find the volumes of solid figures.

Math Note

In the case of the polyhedra (cube and rectangular solid) the surface area is obtained by simply adding the areas of the polygons that make up its sides.

Surface Area

The area of the outer surface of a three-dimensional figure is called the **surface area.** For example, consider a cube. A cube has six faces (think of a standard die, numbered 1 through 6). Each of them is a square with area s^2, where s is the length of a side. So the surface area is $6s^2$.

The surface areas of some of the figures we've studied are listed in Table 10-3.

TABLE 10-3 Surface Area Formulas for Solid Figures

All variables represent the same dimensions as in the volume formulas. SA represents surface area.

Cube	$\text{SA} = 6s^2$
Rectangular solid	$\text{SA} = 2lw + 2lh + 2wh$
Right circular cylinder	$\text{SA} = 2\pi r^2 + 2\pi rh$
Sphere	$\text{SA} = 4\pi r^2$

The relation between surface area and volume is important when determining how something will absorb or lose heat. Ice absorbs heat through its surface. When you break up a large piece of ice, the volume stays the same, but the surface area is greatly increased. So small ice cubes melt quickly, and large blocks of ice last much longer.

EXAMPLE 7 Finding the Surface Area of a Cylinder

How many square inches of sheet metal are needed to form the soup can in Example 3?

SOLUTION

The radius of the can is 1.5 inches, and the height is 4 inches. Using the formula from Table 10-3:

$$\begin{aligned} SA &= 2\pi r^2 + 2\pi rh \\ &= 2\pi(1.5)^2 + 2\pi(1.5)(4) \\ &= 4.5\pi + 12\pi \\ &= 16.5\pi \approx 51.8 \text{ in.}^2 \end{aligned}$$

▼ Try This One 7

☑ 2. Find the surface areas of solid figures.

How many square inches of leather are needed to cover an NBA basketball? (See Try This One 6 for dimensions, and ignore the seams on the ball.)

Answers to Try This One

1 60 cubic meters

2 598.4 gallons

3 About 28,274 cubic feet

4 8,236,005 cubic feet

5 About 3.5 ounces

6 About 3,479 cubic inches

7 About 1,110.4 square inches

EXERCISE SET 10-5

Writing Exercises

1. Explain why the volume of a rectangular solid is the product of length, width, and height.
2. Explain why the volume of a cube is the length of a side raised to the third power.

3. What is the difference between volume and surface area?
4. Describe the connection between the volume formulas $V = lwh$ and $V = \pi r^2 h$.
5. Explain why the surface area of a cube is $6s^2$, where s is the length of a side.
6. Is the height of a pyramid the length of one of the sides? Why or why not?

Computational Exercises

For Exercises 7–18, find the volume of each figure.

7.
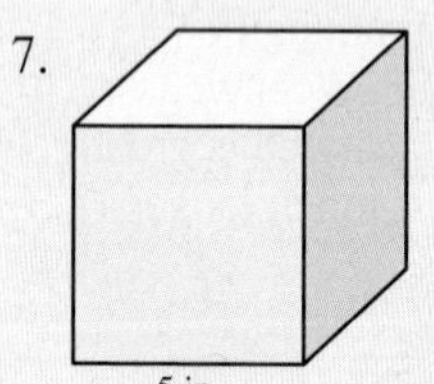

8.
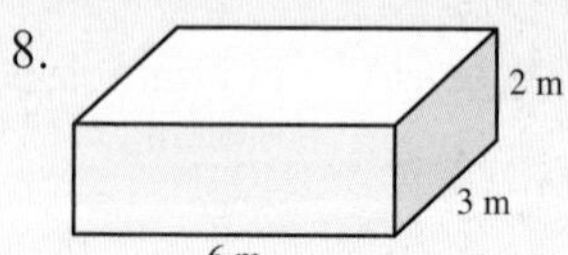

9.
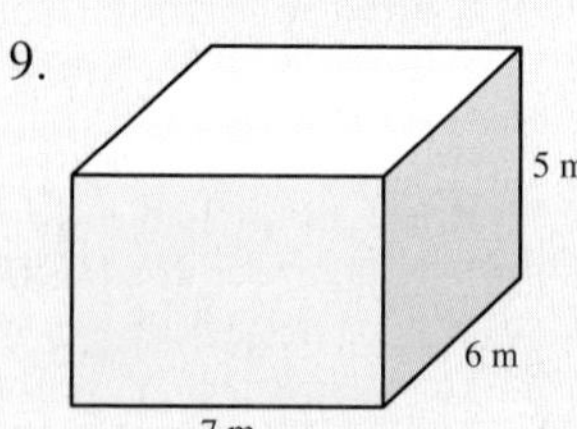

10.
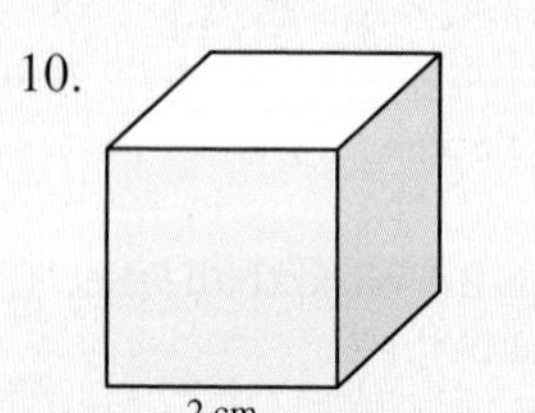

11.
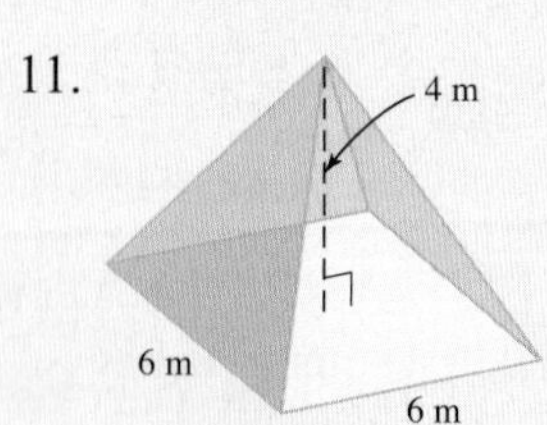

12.
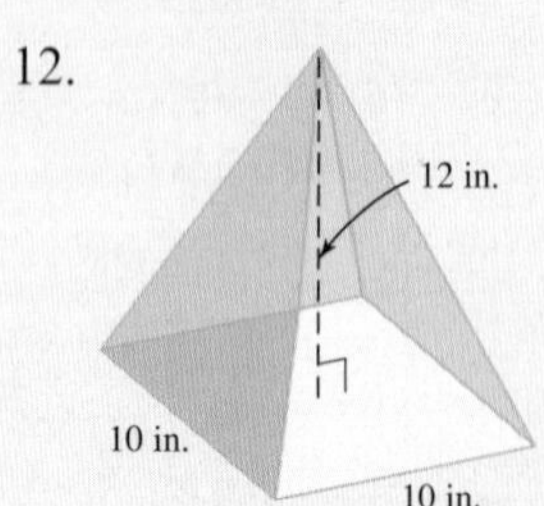

13.
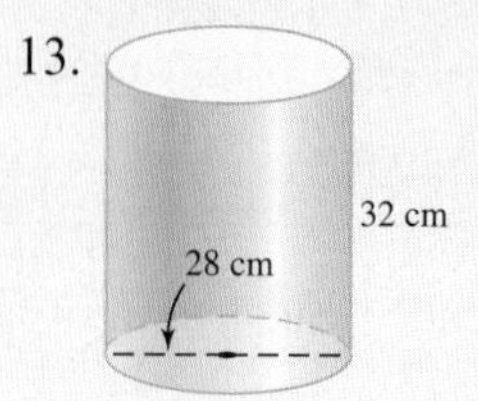

14.
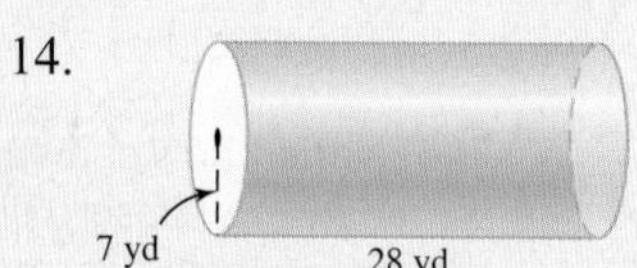

15.
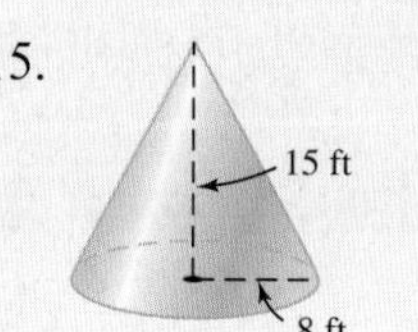

16.
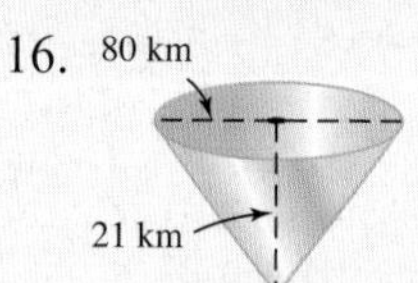

17.
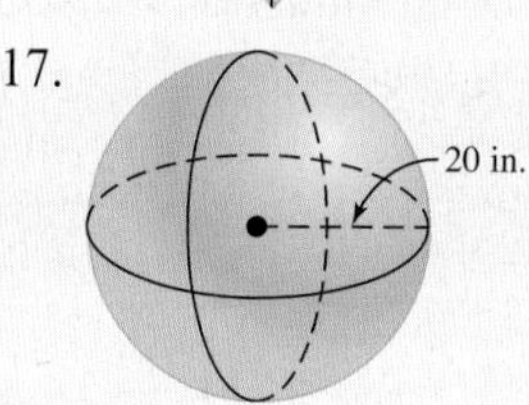

18.
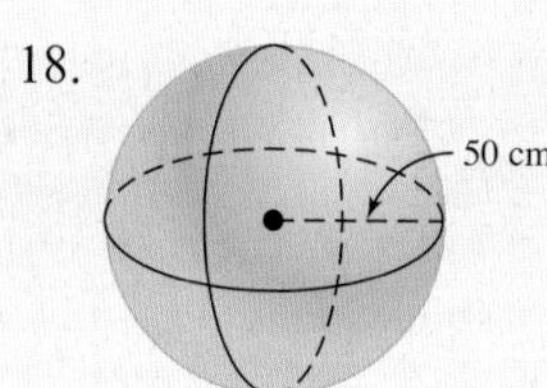

For Exercises 19 and 20, find the volume of the solid figure not including the hole cutout.

19.
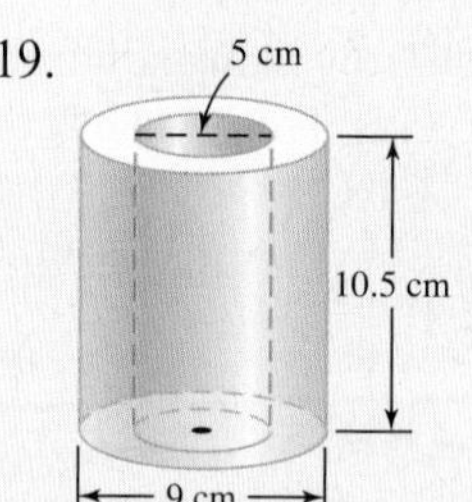

20.
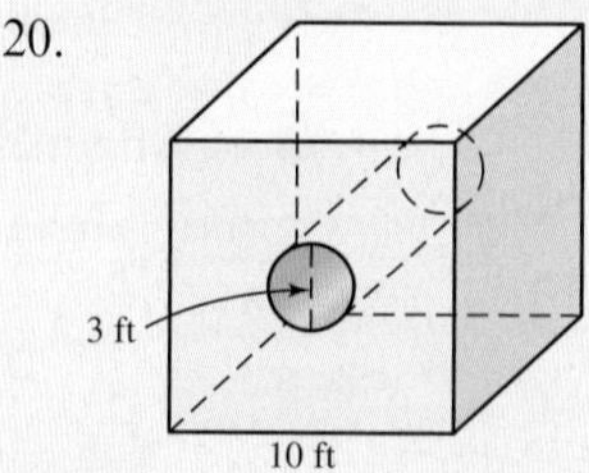

21. Find the surface area of the figure in Exercise 7.
22. Find the surface area of the figure in Exercise 8.
23. Find the surface area of the figure in Exercise 9.
24. Find the surface area of the figure in Exercise 10.
25. Find the surface area of the figure in Exercise 13.
26. Find the surface area of the figure in Exercise 14.
27. Find the surface area of the figure in Exercise 17.
28. Find the surface area of the figure in Exercise 18.

Real-World Applications

29. How many cubic feet of dirt must be removed to build an in-ground swimming pool whose dimensions are 18 feet in length, 12 feet in width, and 3 feet in depth?
30. At the College Tech Expo, Jamal wants to build a simple display table in the shape of a cube with each side measuring 6 feet. How many square feet of plywood will be needed?
31. A pyramid found in Central America has a square base measuring 932 feet on each side and a height of 657 feet. Find the volume of the pyramid.
32. For her class in entrepreneurship, Jenny designed a pop-up tent made for dogs that is in the shape of a pyramid with a square base measuring 6 feet on a side with a height of 4 feet. Find the volume of the tent.
33. A cylindrical-shaped gasoline tank has a 22-inch diameter and is 36 inches long. Find its volume.
34. Find the surface area of a can of Zoom Energy Drink with a diameter of 8 cm and a height of 8.5 cm.
35. Find the volume of a cone-shaped funnel whose base diameter is 7 inches and whose height is 11 inches.
36. Find the volume of a cone-shaped Christmas tree that is 6 feet tall and has a base diameter of 4 feet.
37. The diameter of Mars is 4,200 miles. How many square miles is the surface of Mars?
38. The diameter of the moon is 3,474 kilometers. If the moon were made of green cheese, how many cubic kilometers of cheese would there be?
39. How many cubic inches of packing peanuts can be held by a rectangular box that is 1 foot long, 8 inches wide, and 2 feet tall?
40. A storage shed maker charges $2.50 per cubic foot. Find the price of a custom shed measuring 4 yards long, 5 yards wide, and 8 feet tall.
41. What is the minimum number of square inches of cardboard required to make a rectangular box that is 1 foot long, 2 feet wide, and 7 inches high?
42. What is the minimum number of square centimeters of wrapping paper needed to wrap a rectangular box that is 40 cm wide, 60 cm long, and 85 mm high?
43. A cereal box is required to have length 9 inches and height 13 inches to fit in the space allotted by major grocery chains. To hold the weight of this particular cereal required, it needs to have a volume of 400 cubic inches. How wide should it be?
44. A motor oil manufacturer is designing a new cylindrical can for its economy size. They want it to be 23 cm tall so it fits store shelves, and they want it to hold 2 liters of oil. What should the radius be? (Recall that 1 cubic centimeter = 1 milliliter.)

For Exercises 45–48, suppose that a particular brand of paint claims that under normal conditions, each gallon will cover 350 square feet. The paint is sold in 5-gallon and 1-gallon cans, with the 5-gallon costing the same as 4 1-gallon cans. Find how many cans of each should be bought for each situation to keep the cost as low as possible.

45. To paint the walls and flat roof of a rectangular building that is 40 feet long, 30 feet wide, and 12 feet high.
46. To paint forty 55-gallon drums that are being reconditioned for sale, including the top and bottom. Each drum has a height of 35 inches and a diameter of 24 inches.
47. To paint the outside walls of a round above-ground swimming pool with a diameter of 25 feet and a height of 4 feet.
48. To paint a giant papier-mâché model of the moon, which is a sphere with a diameter of 26 feet.

Critical Thinking

49. Twelve rubber balls with 3-inch diameters are placed in a box with dimensions 12 inches × 9 inches × 3 inches. Find the volume of the space that is left over.
50. A rain gutter (cross section shown) is 24 feet long. How much water will it hold when it is full? One cubic foot of water is equal to 7.48 gallons.

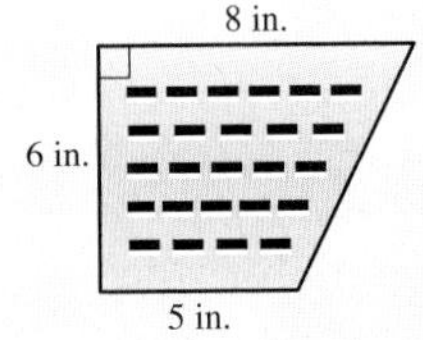

51. Find the surface area of the pyramid in Exercise 11. (*Hint:* Find the area of each side first. You will need the Pythagorean theorem.)
52. By finding the area of the top, bottom, and sides separately, prove the formula for the surface area of a cylinder provided in Table 10-3. (*Hint:* What would the sides look like if you unrolled the cylinder?)
53. Find the surface area of a silo that is a cylinder with radius 9 feet and height 22 feet, with a dome-shaped top that is half of a sphere.

Section 10-6 Right Triangle Trigonometry

LEARNING OBJECTIVES

- ☐ 1. Find basic trigonometric ratios.
- ☐ 2. Use trigonometric ratios to find sides of a right triangle.
- ☐ 3. Use trigonometric ratios to find angles of a right triangle.
- ☐ 4. Solve problems using trigonometric ratios.

We have seen that triangles can be used to solve many practical problems, but there are still many problems involving triangles that we can't solve using what we've learned so far. Here's a simple example. Like pretty much everything you buy today, ladders come with a variety of safety warnings designed to keep the manufacturer from getting sued. Suppose you buy a 12-foot ladder to do some work around the house, and the safety warning tells you it should not be placed at an angle steeper than 65° with the ground. Assuming that you don't have a protractor handy, how can you decide if it's safe to place the bottom 4 feet from the wall?

In this section we will study the basics of **trigonometry,** a very old subject whose name literally means "angle measurement." In trigonometry, we use relationships among the sides and angles of triangles to solve problems. This is in some ways like the work we did earlier with similar triangles, but is more widely applicable.

The big restriction we will work with when studying the basics of trigonometry is that all triangles have to be right triangles. In this section, we will use capital letters A, B, and C to represent the angles, and lowercase letters a, b, and c to represent the lengths of the sides. **We will always use C to represent the right angle.** In each case, the letter of an angle matches the letter of the side across from it. For example, a will represent the length of the side across from angle A. (See Figure 10-26.)

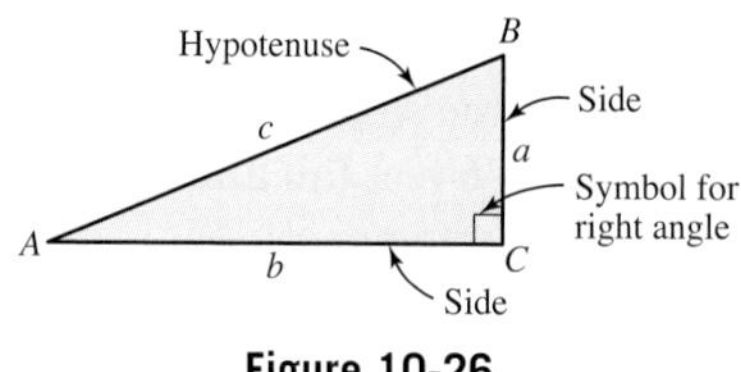

Figure 10-26

Math Note

Remember, in a right triangle, the side across from the right angle is called the hypotenuse. It's always the longest side.

The Trigonometric Ratios

There are three basic trigonometric ratios. They are called the **sine** (abbreviated sin), the **cosine** (abbreviated cos), and the **tangent** (abbreviated tan). The trigonometric ratios are defined as follows:

The Trigonometric Ratios

$$\sin A = \frac{\text{Length of side opposite angle } A}{\text{Length of hypotenuse}} = \frac{a}{c}$$

$$\cos A = \frac{\text{Length of side adjacent to angle } A}{\text{Length of hypotenuse}} = \frac{b}{c}$$

$$\tan A = \frac{\text{Length of side opposite angle } A}{\text{Length of side adjacent to angle } A} = \frac{a}{b}$$

Math Note

Generations of students have remembered the trigonometric ratios using this simple mnemonic device: SOHCAHTOA (which most students pronounce "sow-cah-tow-ah"). This stands for Sine is Opposite over Hypotenuse, Cosine is Adjacent over Hypotenuse, Tangent is Opposite over Adjacent.

EXAMPLE 1 Finding Basic Trigonometric Ratios

For the triangle shown in Figure 10-27, find sin B, cos B, and tan B.

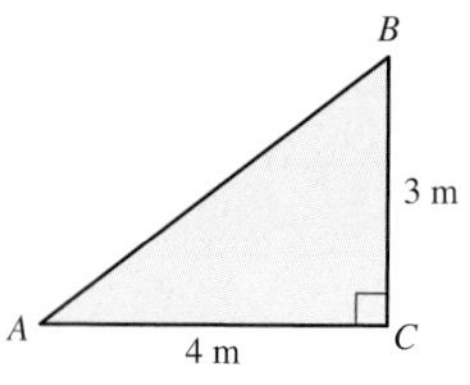

Figure 10-27

SOLUTION

Since two of the three ratios involve the length of the hypotenuse, we need to find that first, using the Pythagorean

Theorem. (Since the hypotenuse is across from angle C, we label its length c.)

$$c^2 = a^2 + b^2$$
$$= 4^2 + 3^2 = 16 + 9 = 25 \quad \textit{Apply the square root to both sides.}$$
$$c = \sqrt{25} = 5 \text{ m}$$

Now we can use the trigonometric ratios from the definitions above; the hypotenuse is 5 m, the side opposite B is 4 m, and the side adjacent to B is 3 m.

$$\sin B = \frac{\text{Opposite}}{\text{Hypotenuse}} = \frac{4 \text{ m}}{5 \text{ m}} = \frac{4}{5}$$

$$\cos B = \frac{\text{Adjacent}}{\text{Hypotenuse}} = \frac{3 \text{ m}}{5 \text{ m}} = \frac{3}{5}$$

$$\tan B = \frac{\text{Opposite}}{\text{Adjacent}} = \frac{4 \text{ m}}{3 \text{ m}} = \frac{4}{3}$$

Math Note

Notice that the units divide out when you compute the trigonometric ratios, so the units of length are unimportant in defining the ratios.

▼ Try This One 1

☑ 1. Find basic trigonometric ratios.

Find sin A, cos A, and tan A for the triangle below.

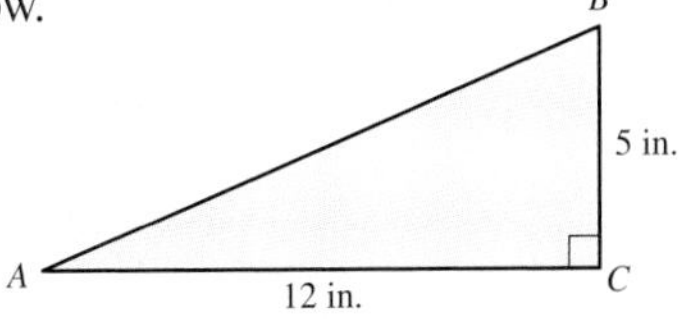

Calculator Guide

To use a calculator in solving problems involving trigonometric ratios, you will have to make sure your calculator is set to degree mode:

Standard Scientific Calculator

Press Mode, Deg, or Rad until you see "Deg" in the display window.

Standard Graphing Calculator

Press Mode, use arrow keys to select "DEGREE," then exit by pressing 2nd Mode

The trigonometric ratios would not be very useful if they gave different values for the same measure angle in different triangles—certainly we want sin 30° to be the same regardless of the size of the triangle the 30° angle came from. Fortunately, this is the case, because of what we learned about similar triangles. In Figure 10-28, we see two different right triangles with a 30° angle. Because the triangles are similar, the ratios of corresponding sides are equal, so sin 30° is $\frac{1}{2}$ in both triangles.

One situation that trigonometry is often used for is to find the lengths of sides of a triangle. If we know one side and one of the acute angles, we can find either of the remaining sides, as in Example 2.

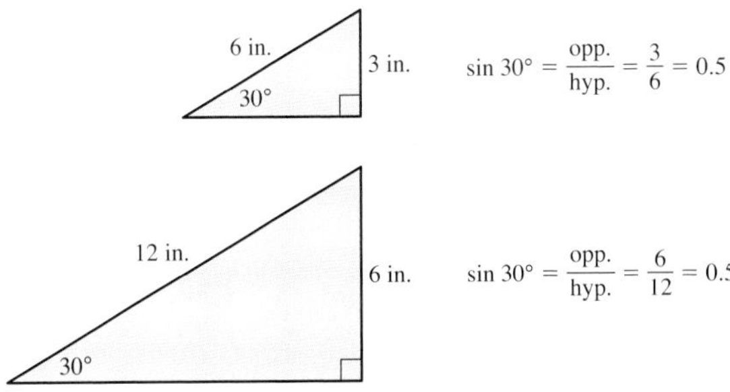

Figure 10-28

EXAMPLE 2 Finding a Side of a Triangle Using Tangent

In the right triangle ABC, find the length of side a when $m\measuredangle A = 30°$ and $b = 200$ feet.

SOLUTION

Step 1 Draw and label the figure.

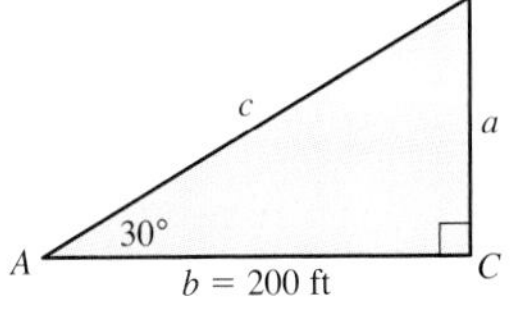

Step 2 Choose an appropriate trigonometric ratio and substitute values into it. In this case, we know the side

adjacent to angle A, and are asked to find the side opposite, so tangent is a good choice.

$$\tan A = \frac{\text{Opposite}}{\text{Adjacent}} = \frac{a}{b}$$

$$\tan 30° = \frac{a}{200}$$

Step 3 Solve the resulting equation for a.

$$\tan 30° = \frac{a}{200} \qquad \textit{Multiply both sides by 200.}$$

$$a = 200 \tan 30°$$

To evaluate this answer, we use a calculator in degree mode (see Calculator Guide on previous page) and press 200 [×] 30 [Tan] [=] (standard scientific calculator), or 200 [Tan] 30 [Enter] (standard graphing calculator). The result in either case, rounded to two decimal places, is 115.47, so the length of side a is 115.47 feet.

> ### Math Note
> There is usually more than one way to find the length of a side using the trigonometric ratios. In Example 2, we could have noticed that $m\measuredangle B = 60°$ and used the tangent of B to set up an equation.

▼ Try This One 2

In the right triangle ABC, find the length of side b if $m\measuredangle B = 58°$ and $a = 12$ inches.

EXAMPLE 3 Finding a Side of a Triangle Using Cosine

In the right triangle ABC, find the measure of side c when $m\measuredangle B = 72°$ and the length of side a is 24 feet.

SOLUTION

Step 1 Draw and label the figure.

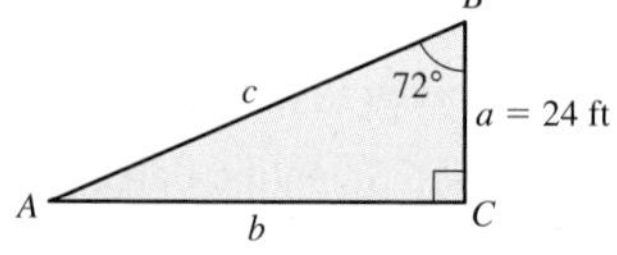

Step 2 We know the side adjacent to B and want to find the hypotenuse, so cosine is a good choice.

$$\cos B = \frac{a}{c}$$

$$\cos 72° = \frac{24}{c}$$

Step 3 Solve for c.

$$\cos 72° = \frac{24}{c} \qquad \textit{Multiply both sides by c.}$$

$$c \cdot \cos 72° = 24 \qquad \textit{Divide both sides by cos 72°.}$$

$$c = \frac{24}{\cos 72°} \approx 77.67 \text{ ft (to two decimal places)}$$

> ### Calculator Guide
> Keystrokes for finding c in Example 3:
>
> **Standard Scientific Calculator**
>
> 24 [÷] 72 [Cos] [=]
>
> **Standard Graphing Calculator**
>
> 24 [÷] [Cos] 72 [Enter]

▼ Try This One 3

In the right triangle ABC, find the measure of side b when $m\measuredangle A = 53°$ and the hypotenuse (side c) is 18 cm.

☑ 2. Use trigonometric ratios to find sides of a right triangle.

Trigonometric ratios can be used to find angles as well, provided that we know at least two sides of a right triangle. To accomplish this, we will need to use special routines programmed into a calculator known as **inverse trigonometric functions.** They are accessed using the [2nd] key on most calculators, as illustrated in Example 4.

EXAMPLE 4 Finding an Angle Using Trigonometric Ratios

In the right triangle ABC, side c (the hypotenuse) measures 25 inches and side b measures 24 inches. Find the measure of angle B.

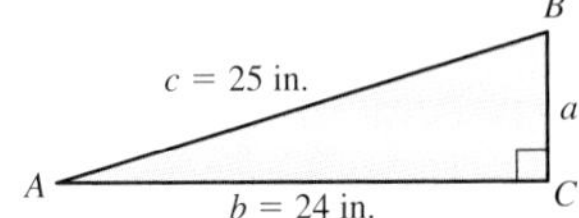

SOLUTION

Step 1 Draw and label the figure.

Step 2 We know the hypotenuse and the side opposite angle B, so sine is a good choice.

$$\sin B = \frac{\text{Opposite}}{\text{Hypotenuse}} = \frac{24}{25} = 0.96$$

Step 3 Solve for B. To accomplish this, we will need to access the inverse sine feature on a calculator: .96 [2nd] [Sin] (standard scientific calculator), or [2nd] [Sin] .96 [Enter] (standard graphing calculator). The result, rounded to two decimal places, is 73.74, so $m\measuredangle B \approx 73.74°$.

Math Note

The symbol $\sin^{-1}$ is used to represent the inverse of sine. Likewise, $\cos^{-1}$ and $\tan^{-1}$ represent the inverses of cosine and tangent, respectively.

▼ Try This One 4

Find the measure of angle A for the right triangle ABC if the measure of side a is 42 inches and the measure of side b is 18 inches.

☑ 3. Use trigonometric ratios to find angles of a right triangle.

In the remainder of the section, we will demonstrate just a few of the many, many real-world problems that can be solved using trigonometry. To begin, we return to the ladder question that began the section.

EXAMPLE 5 An Application of Trigonometry to Home Improvement

The safety label on a 12-foot ladder says that the ladder should not be placed at an angle steeper than 65° with the ground. What is the closest safe distance between the base of the ladder and the wall?

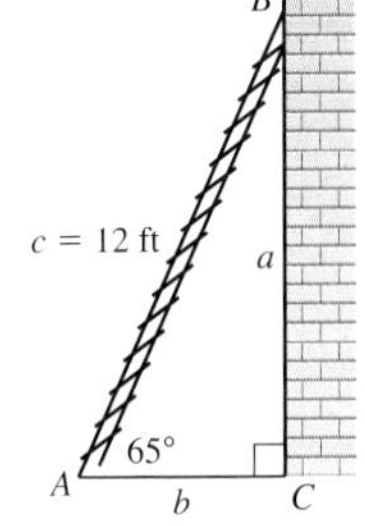

SOLUTION

Step 1 Draw and label the figure.

Step 2 Choose the appropriate formula and substitute the values for the variables. Since we need to find the length of side b and we are given the measures of angle A and side c, cosine is a good choice.

$$\cos A = \frac{\text{Adjacent}}{\text{Hypotenuse}} = \frac{b}{c}$$

$$\cos 65° = \frac{b}{12}$$

Step 3 Solve for b.

$$\cos 65° = \frac{b}{12}$$

$$b = 12 \cos 65° \approx 5.07 \text{ feet}$$

The bottom of the ladder can't be any closer than 5.07 feet from the wall.

Calculator Guide

Keystrokes for finding b:

Standard Scientific Calculator

12 [×] 65 [Cos] [=]

Standard Graphing Calculator

12 [×] [Cos] 65 [Enter]

▼ Try This One 5

Being afraid of heights, a homeowner determines that he's unwilling to place the ladder in Example 5 any steeper than a 50° angle with the ground. How far up the wall will the ladder reach at that angle?

Trigonometry is used extensively in surveying.

Angle of Elevation and Angle of Depression

Many applications of right angle trigonometry use what is called the *angle of elevation* or the *angle of depression*.

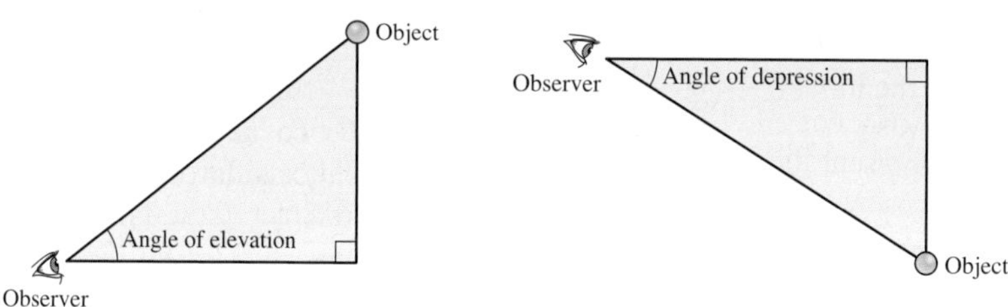

Figure 10-29

The **angle of elevation** of an object is the measure of the angle from a horizontal line at the point of an observer upward to the line of sight to the object. The angle of **depression** is the measure of an angle from a horizontal line at the point of an observer downward to the line of sight to the object. See Figure 10-29.

EXAMPLE 6 Finding the Height of a Tall Object Using Trigonometry

In order to find the height of a building, an observer who is 6 feet tall measures the angle of elevation from his head to the top of the building to be 32° when he is standing at a point 200 feet from the building. How tall is the building?

SOLUTION

Step 1 Draw and label the figure.

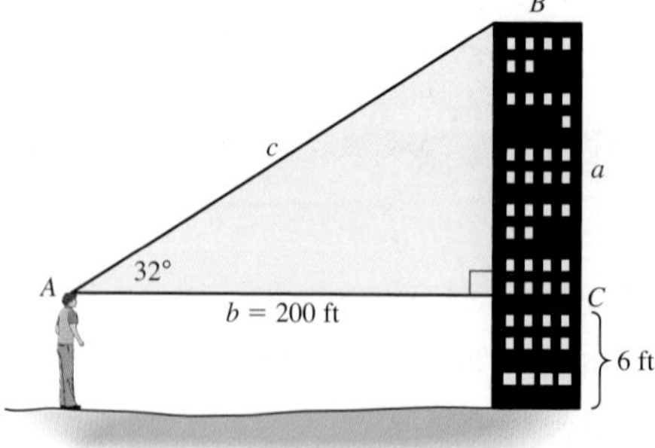

Step 2 Choose the appropriate formula and substitute the values for the variables. Since we are given the measure of angle A and the measure of side b, we can use tangent.

$$\tan A = \frac{\text{Opposite}}{\text{Adjacent}} = \frac{a}{b}$$

$$\tan 32° = \frac{a}{200}$$

Step 3 Solve for a.

$$\tan 32° = \frac{a}{200} \quad \textit{Multiply both sides by 200.}$$

$$a = 200 \tan 32° \approx 125 \text{ feet (to the nearest foot)}$$

Looking back at the diagram, we see that the height of the building is side a (125 feet) plus 6 feet, so the building is 131 feet tall.

Calculator Guide

Keystrokes for finding a:

Standard Scientific Calculator

200 [×] 32 [Tan] [=]

Standard Graphing Calculator

200 [×] [Tan] 32 [Enter]

▼ Try This One 6

A hiker standing on top of a 150-foot cliff sights a boat at an angle of depression of 24°. How far is the boat from the base of the cliff? (The hiker's eye level is 5.6 feet above the top of the cliff.)

EXAMPLE 7 Finding an Angle of Depression

A local photographer gets a hot tip that a famous starlet is sunbathing on a boat cruising down a river. He sets up his camera on a bridge that is 22 feet above the water and 60 feet from a bend in the river, hoping to get a shot as the boat comes into view. At what angle of depression should he aim the camera?

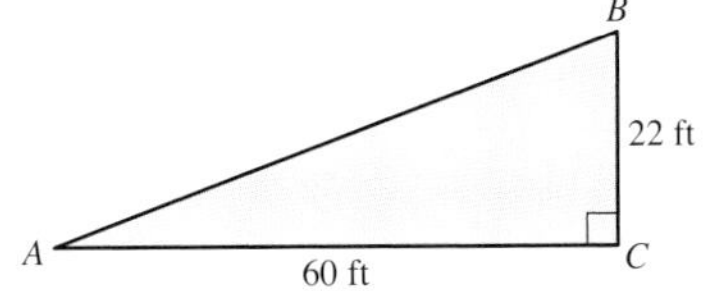

SOLUTION

Step 1 Draw and label the figure (in the margin).

Step 2 We are looking for angle B, and have the opposite and adjacent sides, so we choose tangent.

$$\tan B = \frac{\text{Opposite}}{\text{Adjacent}} = \frac{60}{22}$$

Step 3 Solve for B using the inverse tangent feature on a calculator (see Calculator Guide).

$$B \approx 69.86°$$

The angle of depression should be a bit less than 70°.

> **Calculator Guide**
>
> Keystrokes for finding angle B:
>
> **Standard Scientific Calculator**
>
> 60 [÷] 22 [=] [2nd] [Tan]
>
> **Standard Graphing Calculator**
>
> [2nd] [Tan] 60 [÷] 22 [Enter]

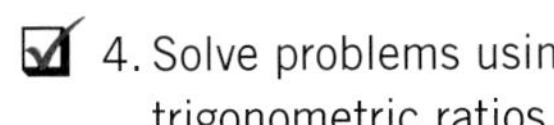
4. Solve problems using trigonometric ratios.

▼ Try This One 7

Marie hopes to photograph the rising sun as it just peeks over the top of a 20-foot-tall house across the street, 120 feet from where she is sitting on the ground. At what angle of elevation should she aim the camera?

CAUTION

When using trig ratios to solve applied problems, never forget that there must be a right triangle in the diagram, or the trig ratios do not apply.

Answers to Try This One

1 $\sin A = \frac{5}{13}$; $\cos A = \frac{12}{13}$; $\tan A = \frac{5}{12}$

2 19.20 inches

3 10.83 cm

4 66.80°

5 9.19 feet

6 349.48 feet

7 9.46°

EXERCISE SET 10-6

Writing Exercises

1. Describe each of the trigonometric ratios in terms of sides of a right triangle.
2. Given a right triangle ABC with right angle C, explain how we would label the lengths of the sides using the letters we used in this section.
3. Why can we be sure that sin 30° always has the same value regardless of the size of the right triangle containing the 30° angle?
4. Describe what is meant by an angle of elevation and an angle of depression.

Computational Exercises

For Exercises 5–10, use the given right triangle to find cos A, sin A, and tan A.

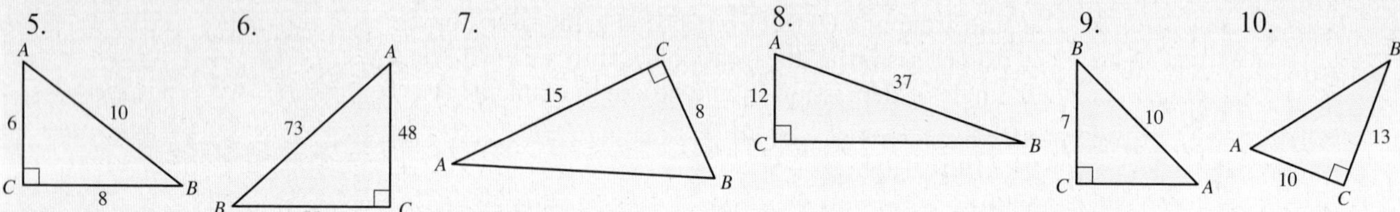

Exercises 11–30 refer to right triangles labeled like the one below. Use trigonometric ratios to find each measure.

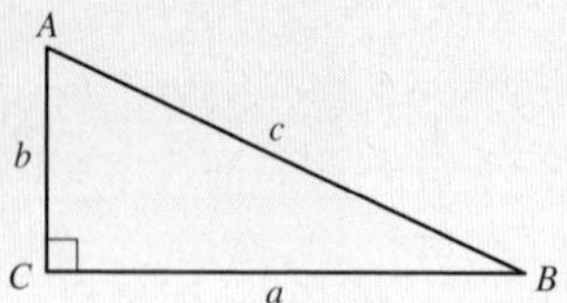

11. Side a if $m\angle B = 72^\circ$ and side $c = 300$ cm
12. Side b if $m\angle A = 41^\circ$ and side $c = 200$ yd
13. Side c if $m\angle A = 76^\circ$ and side $b = 18.6$ in.
14. Side a if $m\angle A = 30^\circ$ and side $c = 40$ km
15. Side b if $m\angle B = 8^\circ$ and side $c = 10$ mm
16. Side c if $m\angle B = 62^\circ$ and side $a = 313$ mi
17. Side a if $m\angle A = 60^\circ$ and side $b = 71$ in.
18. Side b if $m\angle B = 85^\circ$ and side $a = 110$ yd
19. Side c if $m\angle A = 28^\circ$ and side $a = 872$ ft
20. Side a if $m\angle A = 22^\circ$ and side $b = 27$ ft
21. Side b if $m\angle B = 53^\circ$ and side $c = 97$ mi
22. Side c if $m\angle A = 15^\circ$ and side $b = 1{,}250$ ft
23. $m\angle A$ if side $b = 183$ ft and side $c = 275$ ft
24. $m\angle B$ if side $b = 104$ yd and side $c = 132$ yd
25. $m\angle A$ if side $b = 18$ mi and side $c = 36$ mi
26. $m\angle B$ if side $a = 529$ mi and side $c = 1{,}000$ mi
27. $m\angle A$ if side $a = 306$ in. and side $b = 560$ in.
28. $m\angle B$ if side $a = 1{,}428$ ft and side $c = 1{,}800$ ft
29. $m\angle B$ if side $a = 413$ ft and side $b = 410$ ft
30. $m\angle A$ if side $a = 532$ yd and side $c = 780$ yd

Real-World Applications

31. An airplane is flying at an altitude of 6,780 feet and sights the angle of depression to a control tower at the next airport to be 16°. Find the horizontal distance the plane is from the control tower. (Disregard the height of the tower.)
32. If the angle of elevation from ground level to the top of a building is 38° and the observer is 500 feet from the building, find the height of the building.
33. How high is a satellite radio tower if the angle of elevation from the ground is 82 degrees and the observation point is 700 meters from the base of the tower?
34. The front of a 6-foot-long treadmill is 4 inches higher than the back. At what angle is it inclined?
35. One common use of trigonometry is to measure inaccessible objects. Based on the measurements in the diagram, how wide is the lake?

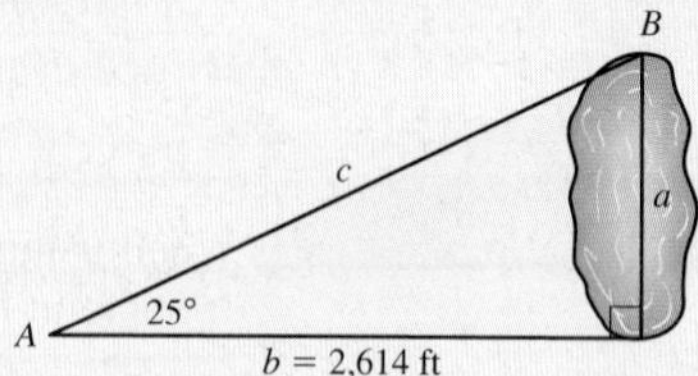

36. During a recent intramural game of Capture the Flag, Pete noticed the flag was hidden atop a light pole. Pete was standing 105 feet from the pole and he estimated that the angle of elevation from his feet to the top of the pole was 40 degrees. How far must he climb up the pole to capture the flag?

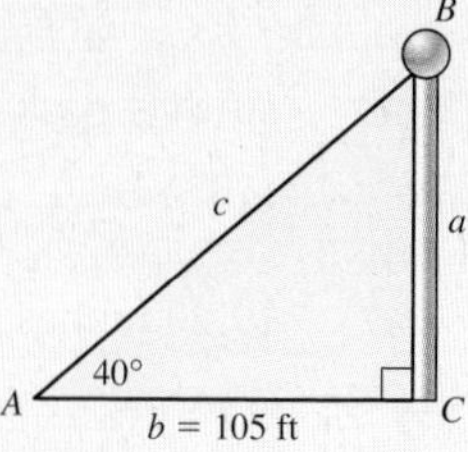

37. If a two-story building that is 20 feet tall casts a shadow 8 feet long, what is the angle of elevation of the sun at that time of day?
38. Trey locked his keys in his house again and desperately needs to get his term paper from his desk on the second floor of his house. He grabs a 25-foot ladder and leans it 63 degrees against the house to the bottom of the second floor window. How high is the window from the ground?
39. In a moment of frustration, Kelly tosses her math book out the window of her dorm room. (Legal disclaimer: don't try this at home.) The window is 138 feet from the ground, and the book was thrown at a 21 degree angle of depression. Find the horizontal distance from where the book landed to Kelly's dorm. (Assume that the book traveled a straight-line path.)
40. Archaeologists find a lost city with a map room leading them to the treasure of a long-dead civilization. When the instructions carved into a stone tablet are translated, they find that that location of the treasure room is found by following the shadow of a 10-foot staff to a location on the map at 2 P.M. on the day of the winter solstice. They check with meteorologists at a nearby university and find that the angle of elevation of the sun at that location on that day and time is 44°. How far away from the staff is the location of the treasure on the map?
41. Patty sits on a bench with her laptop 2,456 feet from the base of the WiFi antenna and the angle of elevation from her laptop, which is 2.5 feet from the ground, to the top of the antenna is 58 degrees. How tall is the antenna?
42. Sara takes an elevator 349 feet up to the science wing of the lecture hall and then walks 1,000 feet to her classroom. What is the angle of elevation from where Sara got on the elevator to the classroom?
43. On a military training exercise, a group of artillerymen are instructed to train their weapon on a target that is $\frac{3}{4}$ mile away and 80 feet high. At what angle of elevation should they set the weapon?
44. While finishing his basement, Dave finds that the stairs have a horizontal span of 12 feet while going down 8.5 feet. At what angle should the hand railing be placed to match the incline of the stairs?
45. A business jet is instructed to make a straight-line approach and landing, starting at an altitude of 4,500 feet and a distance of 7 miles from the runway. At what angle of depression should the pilot fly?
46. A sniper sets up on a roof across the street from a hostage situation. The roof is 36 feet high, and the laser distance finder on his rifle indicates that his straight-line distance from the target is 90 yards. At what angle of depression should he aim his rifle to train in on the target?

Critical Thinking

47. From the top of a building 300 feet high, the angle of elevation to a plane is 33° and the angle of depression to an automobile directly below the plane is 24°. Find the height of the plane and the distance the automobile is from the base of the building.
48. The angle of elevation to the top of a tree sighted from ground level is 18°. If the observer moves 80 feet closer, the angle of elevation from ground level to the top of the tree is 38°. Find the height of the tree.
49. Suppose that there is no trigonometric ratio known as tangent. Can you solve the problem in Example 2 without it? Describe the procedure you would use.
50. How can you solve the problem in Example 3 without using the cosine ratio?

Section 10-7 A Brief Survey of Non-Euclidean and Transformational Geometries

The Euclidean geometry that we have studied so far in this chapter helps us to solve many real-world problems, and because of this, many people would say that Euclidean geometry is the geometry that describes our physical world. But does it really? Suppose you hop in your car and drive 500 miles due south. In your perception, you have driven a straight-line path. But in fact, you were driving along a curved path, matching the surface of the earth. This is why when you look out over the ocean, you can't see the other side—the water literally curves out of your sight at some point, which we call the horizon.

LEARNING OBJECTIVES

- ☐ 1. Identify the basics of elliptic geometry.
- ☐ 2. Identify the basics of hyperbolic geometry.
- ☐ 3. Create fractal figures.
- ☐ 4. Create tessellations.

After thousands of years of accepting Euclid's geometry as the only one, in the 1800s mathematicians and scientists started to question whether the principles of Euclidean geometry really were universal, or if maybe there were other systems in which Euclid's principles didn't apply. And so the area of non-Euclidean geometry was born. There are entire courses taught on non-Euclidean geometries, so in this section we will simply take a quick look at some of them.

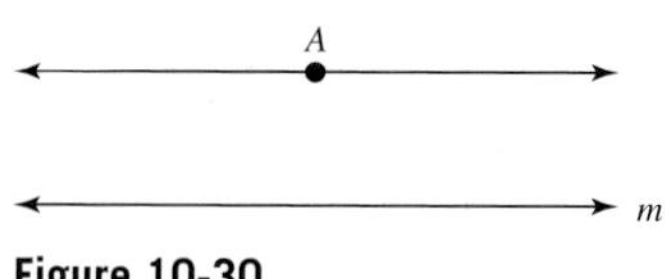

Figure 10-30

A **postulate** is a fundamental assumption underlying an area of study that cannot be proved, but rather is assumed to be true. It turns out that Euclidean geometry has at its foundation the **parallel postulate,** which says that given any line m and a point A not on that line, there is exactly one line parallel to m through A. See Figure 10-30.

Mathematicians spent about 2,000 years trying to prove the parallel postulate without success. Many of these attempts tried proof by contradiction: it was assumed that the parallel postulate didn't always hold, and an attempt was made to find something that contradicted the known theorems of geometry. Along the way, though, an interesting thing happened: new geometries were discovered that didn't follow the rules set out by Euclid. One such system is elliptic geometry.

Math Note

There are no parallel lines at all in elliptic geometry because all of the lines are great circles. To see that any two lines must meet, picture the lines of longitude on a globe. They look parallel near the equator, but meet at the poles.

Elliptic Geometry

☑ 1. Identify the basics of elliptic geometry.

Elliptic geometry was developed by the German mathematician Bernhard Riemann (1826–1866). While Euclidean geometry is based on planes, elliptic geometry is based on spheres. In elliptic geometry, the lines are finite in length, and are defined to be *great circles* of a sphere. A **great circle** is a circle on the sphere that has the same center as the sphere. See Figure 10-31.

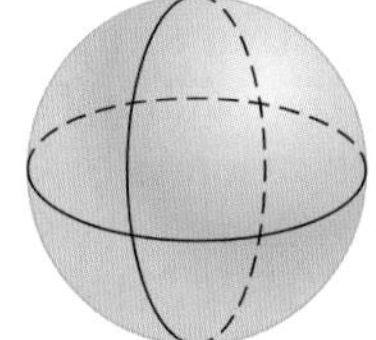

Figure 10-31 Great circles on a sphere

One interesting aspect of elliptic geometry is that the sum of the angles of a triangle is greater than 180° (see Figure 10-32). The sum of the measures of the angles varies with the area of the triangle, and approaches 180° as the area of the triangle decreases.

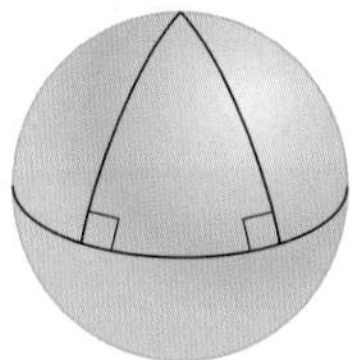

Figure 10-32 A triangle in elliptic geometry

If you find it hard to imagine an elliptic geometry, consider the fact that you live on one! That's why that 500-mile path you drove felt like a line to you even though it was curved—when you live on a sphere, lines follow the contour of the sphere.

Hyperbolic Geometry

Hyperbolic geometry was developed independently by the Russian mathematician Nikolay Lobachevsky (1792–1856) and the Hungarian mathematician János Bolyai (1802–1860).

Hyperbolic geometry is based on the assumption that, given a line m and a point not on the line, there are an infinite number of lines parallel to m. The shapes in hyperbolic geometry are drawn not on planes, but on a funnel-shaped surface known as a **pseudosphere.** A line on a pseudosphere is pictured in Figure 10-33.

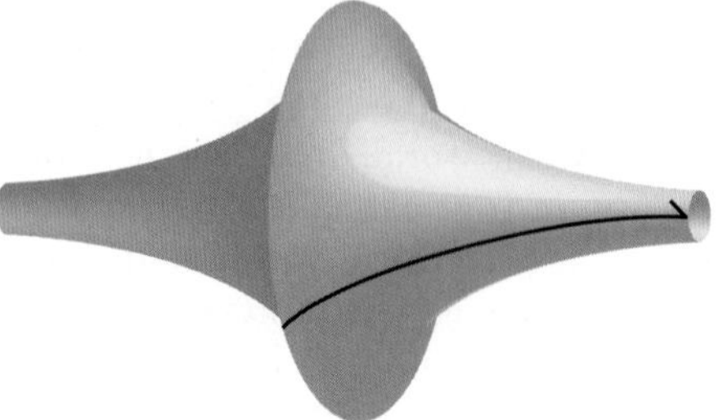

Figure 10-33 A line in hyperbolic geometry

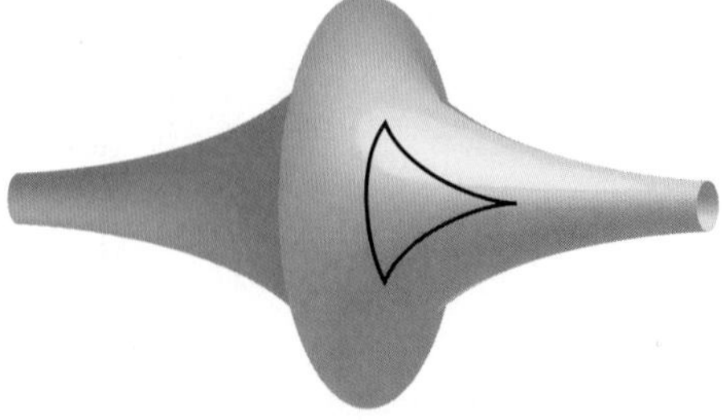

Figure 10-34 A triangle in hyperbolic geometry

☑ 2. Identify the basics of hyperbolic geometry.

In hyperbolic geometry, the shortest distance between two points is a curved line, the sides of a triangle are arcs, and the sum of the angles of a triangle is less than 180°. See Figure 10-34 on the previous page.

The remaining types of geometries we will study are called **transformational geometries** because they are based on transforming basic geometric shapes.

Fractal Geometry

☑ 3. Create fractal figures.

One type of transformational geometry is called **fractal geometry.** This geometry was developed by Benoit Mandelbrot (1924–). Fractal geometry makes geometric figures by using shapes that are repeated. See Figure 10-35 for an example. In this case, an equilateral triangle was used repeatedly. This figure is called the Sierpinski triangle.

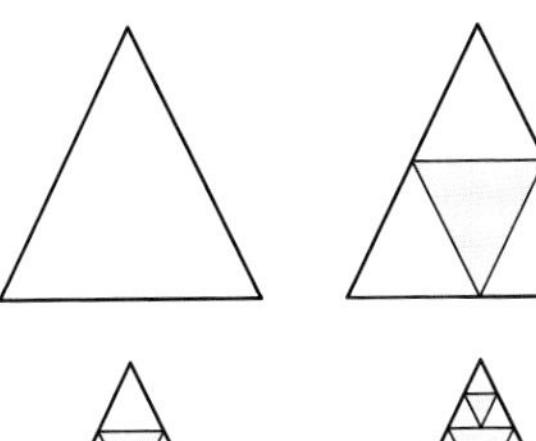

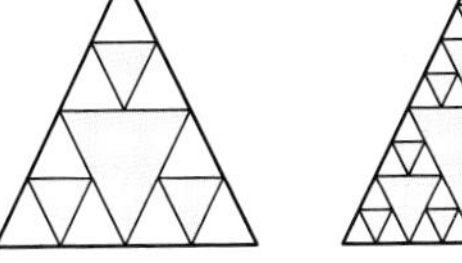

Figure 10-35

The repeating process is called **iteration.** Computers can generate fractal figures very easily and quickly. Many nature forms, such as ferns, flowers, mountains, etc. can be duplicated, or at least simulated, using fractals.

Let's look at a method for creating another fractal figure.

Step 1 Start with an equilateral triangle.

Step 2 Replace each – side with ⁄\⁄\.

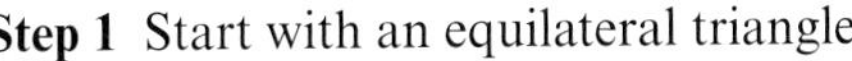

Step 3 Repeat step 2 in the new figure.

Step 4 Continue the process as long as you wish. See Figure 10-36 for several iterations of the process. The figure is called the Koch snowflake.

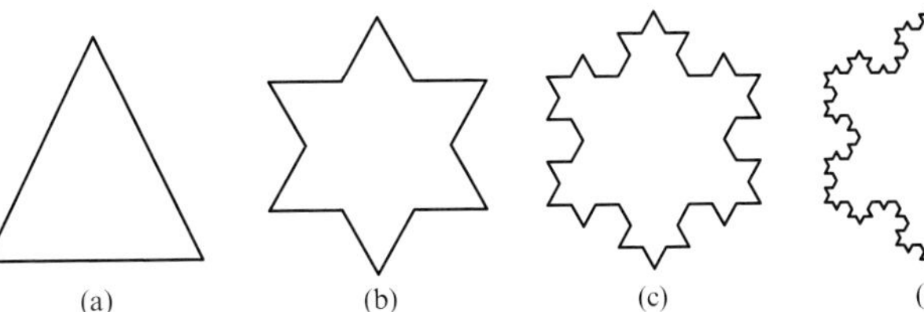

Figure 10-36

Sidelight THE MOBIUS STRIP

An unusual mathematical shape can be created from a flat strip of paper. After cutting a strip, give it a half twist and tape the ends together to make a closed ring as shown.

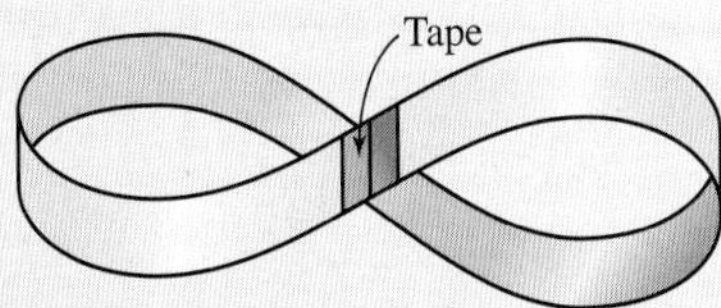

The shape you have made is called a Mobius strip, named for the German mathematician Augustus Ferdinand Mobius (1790–1868) who created it. What is unusual about the Mobius strip is that it has only one side. You can draw a line through the center of the strip, starting anywhere and ending where you started without crossing over an edge. You can color the entire strip with one color by starting anywhere and ending where you started without crossing over an edge.

The Mobius strip has more unusual properties. Take a scissors and cut through the center of the strip following the line you drew as shown.

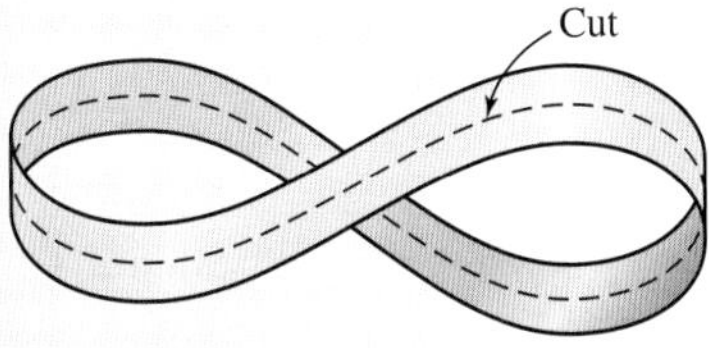

You will not get two strips as you might suspect, but you will have one long continuous two-sided strip.

Finally, make a new strip and cut it one-third of the way from the edge. You will find that you can make two complete trips around the strip using one continuous cut. The result of this cut will be two distinct strips: one will be a two-sided strip, and the other will be another one-sided Mobius strip.

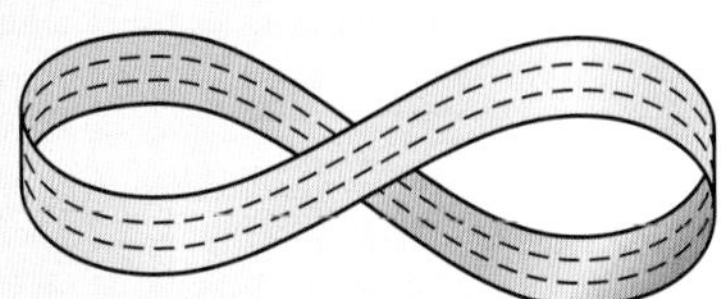

The Mobius strip is a shape that belongs to a recently developed branch of mathematics called *topology.* Topology is a kind of geometry in which solid shapes are bent or stretched into different shapes. The properties of the shape that remain unchanged in the new shape are then studied.

Fractals have been used to study natural phenomenon such as seashores, mountains, trees, galaxies, and weather patterns. In addition, a radio antenna using a fractal pattern has been found more efficient than the traditional standard telescoping metal antenna.

Tessellations

☑ 4. Create tessellations.

An interesting aspect of transformational geometry is called a *tessellation.* The formal mathematical study of tessellations dates to the 1600s, but the concept itself is almost surely one of the oldest in all of math. For thousands and thousands of years, civilizations have used tessellation in art and architecture.

A **tessellation,** also called a **tiling,** is a pattern that uses the same geometric shape to cover a plane without any gaps.

Math Note

Not every regular polygon can be used to create a tessellation. For example, it can be shown that regular pentagons alone cannot create a tessellation.

Figure 10-37 shows several tessellations. These figures are made by using regular polygons. (Recall that a regular polygon is a polygon in which all sides have the same length and all interior angles have the same measure.) The tessellation shown in Figure 10-37a is made from equilateral triangles. The one shown in Figure 10-37b is made from squares, and the one shown in Figure 10-37c is made from regular hexagons.

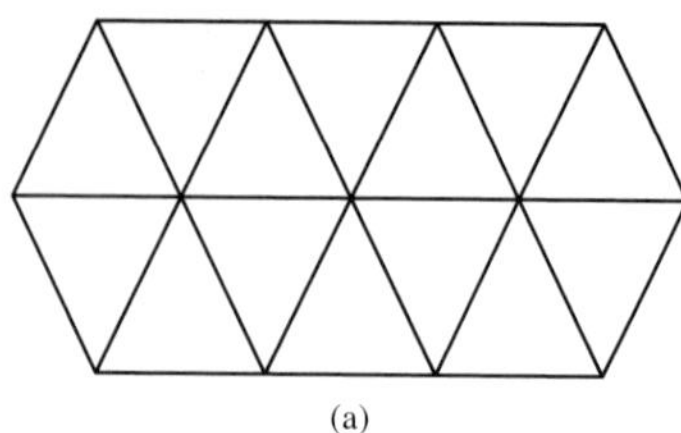
(a)

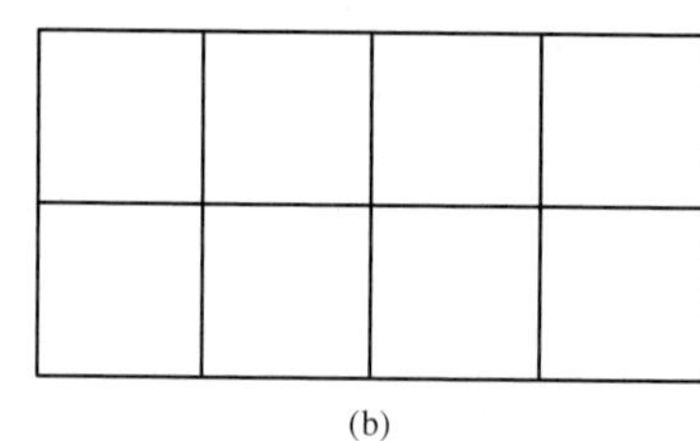
(b)

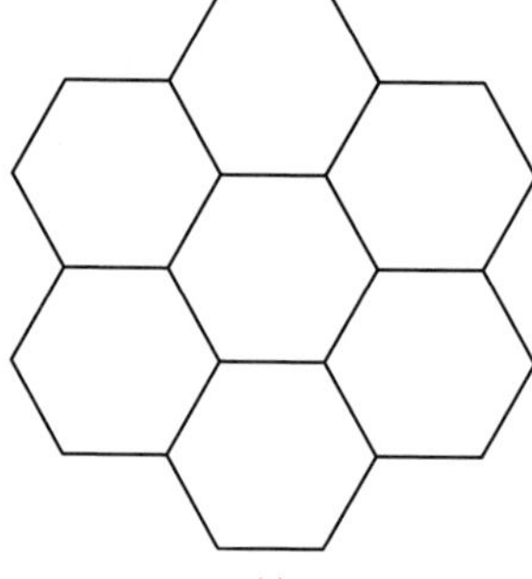
(c)

Figure 10-37

A tessellation can be made by following these steps.

Step 1 Draw a regular geometric polygon. In this case, we'll use a square.
Step 2 Cut the square from top to bottom using any design. See Figure 10-38a.

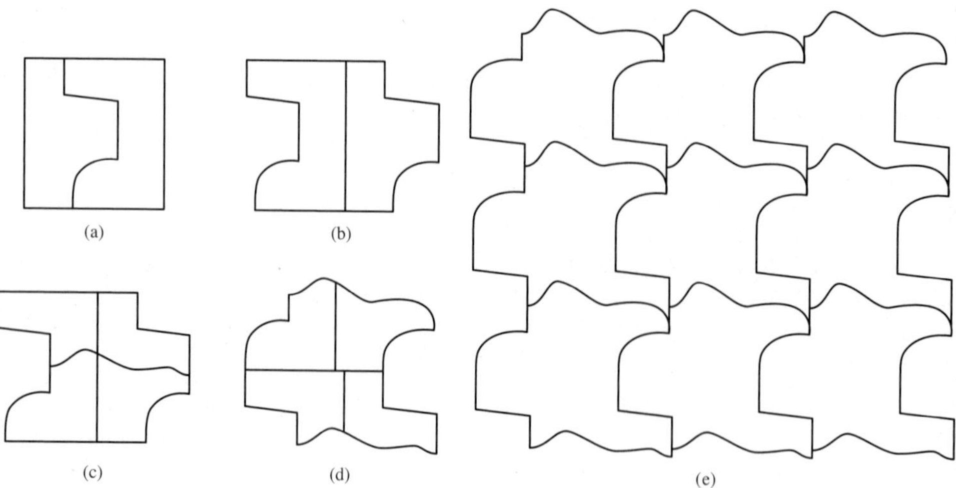

Figure 10-38

Sidelight M. C. ESCHER

A Dutch graphic artist, Maurits Cornelius Escher (1898–1972), used tessellations in his artwork and lithography. Escher had no formal training in mathematics; however, he was inspired by the beauty of Moorish artwork in the Alhambra Palace in Granada, Spain. Escher then read articles on symmetry, non-Euclidean geometry, and other areas of mathematics associated with tessellations. He is recognized for his works containing infinite repeating geometric patterns and created paintings of objects that could be drawn in two dimensions but were impossible to create in three dimensions.

Step 3 Move the left piece to the right side. Tape the pieces together. See Figure 10-38b.
Step 4 Cut the piece from left to right using any design. See Figure 10-38c.
Step 5 Move the top piece to the bottom piece and tape the pieces together. See Figure 10-38d.
Step 6 Trace the figure on another sheet of paper and then move it so that the sides line up like puzzle pieces.
Step 7 Continue the process as many times as you wish until a tessellation figure is complete. See Figure 10-38e.

Tessellations are very common in architectural design, but are most commonly used in mosaic designs of floors, walls, and windows. The most famous user of tessellations is without question the Dutch artist M. C. Escher, profiled in the sidelight.

EXERCISE SET 10-7

Writing Exercises

1. What is the surface on which shapes are drawn in Euclidean geometry?
2. What is the surface on which shapes are drawn in elliptic geometry?
3. How are lines represented in elliptic geometry?
4. Why are there no parallel lines in elliptic geometry?
5. Describe the shortest path between two points in hyperbolic geometry.
6. What is a pseudosphere?
7. Describe a fractal figure.
8. What is a tessellation? How are tesselations used?

Computational Exercises

9. Use a square and the procedure described on pages 562–563 to create a tessellation.
10. Use a regular hexagon and the procedure described on pages 562–563 to create a tessellation.
11. Using the letter H and putting an H at each corner, create a fractal-like figure.
12. Develop a fractal by replacing each _______ side in a square with __⌐¬__.

Critical Thinking

13. Try to form a tessellation using only regular pentagons, then explain why it's not possible.
14. Repeat Exercise 13 using regular octagons.
15. Try to develop a rule for determining whether or not a given geometric shape can tessellate with itself.

CHAPTER 10 Summary

Section	Important Terms	Important Ideas
10-1	Point Line Plane Line segment Half line Ray Angle Vertex Degree Acute angle Right angle Obtuse angle Straight angle Complementary angles Supplementary angles Vertical angles Parallel lines Transversal Alternate interior angles Alternate exterior angles Corresponding angles	**The principals** of geometry have long been useful in helping people understand the physical world. The basic geometric figures are the point, the line, and the plane. From these figures, other figures such as segments, rays, and half lines can be made. When two rays have a common endpoint, they form an angle. Angles can be measured in degrees using a protractor. When lines intersect, several different types of angles are formed. They are vertical, corresponding, alternate interior, and alternate exterior. Two angles are called complementary angles if the sum of their measures is equal to 90°. Two angles are called supplementary if the sum of their measures is equal to 180°. Pairs of corresponding angles have the same measure, as do pairs of alternate interior angles, and pairs of alternate exterior angles.
10-2	Triangle Sides Vertices Isosceles triangle Equilateral triangle Scalene triangle Acute triangle Obtuse triangle Right triangle Hypotenuse Legs Pythagorean theorem Similar triangles	**A closed** geometric figure with three sides is called a triangle. A triangle can be classified according to the lengths of its sides or according to the measures of its angles. For a right triangle, the Pythagorean theorem states that $c^2 = a^2 + b^2$, where c is the length of the hypotenuse and a and b are the lengths of the legs. Two triangles with the same shape are called similar triangles. Corresponding sides of similar triangles are in proportion. This allows us to solve problems involving finding the lengths of sides.
10-3	Polygon Quadrilateral Pentagon Hexagon Heptagon Octagon Nonagon Decagon Dodecagon Icosagon Trapezoid Parallelogram	**A polygon** is classified according to the number of sides. A quadrilateral has four sides; a pentagon has five sides, etc. The sum of the measures of the angles of a polygon with n sides is $(n - 2)180°$. Special quadrilaterals such as the trapezoid, parallelogram, rhombus, and square are used in this chapter. A regular polygon is a polygon in which all sides are the same length and all angles have the same measure. The distance around the outside of a polygon is called the perimeter.

	Rectangle Rhombus Square Regular polygon Perimeter	
10-4	Area Circle Center Radius Diameter Circumference π (pi)	**The measure** of the portion of the plane enclosed by a geometric figure is called the area of the geometric figure. Area formulas can be developed for many familiar geometric figures. A circle is a closed geometric figure in which all the points are the same distance from a fixed point called the center. A segment connecting the center with any point on the circle is called a radius. A segment connecting two points on the circle and passing through the center of the circle is called a diameter. The circumference of a circle is the distance around the outside of the circle.
10-5	Volume Rectangular solid Cube Right circular cylinder Pyramid Cone Sphere Surface area	**Geometry** is also the study of solid figures. Some of the familiar solid figures are the cube, the rectangular solid, the pyramid, the cone, the cylinder, and the sphere. The volume of a solid geometric figure is the amount of space that is enclosed by the surfaces of the figure. The surface area of a solid figure is the area of the faces or surfaces of the figure.
10-6	Trigonometry Sine Cosine Tangent Inverse trigonometric functions Angle of elevation Angle of depression	**Trigonometry** is the study of the relationship between the angles and the sides of a triangle. The trigonometric ratios of sine, cosine, and tangent are used with a right triangle. The concepts of right triangle trigonometry can be used to solve many problems in navigation measurement, engineering, physics, and many other areas.
10-7	Elliptic geometry Hyperbolic geometry Fractals Tessellations	**Euclidean geometry** is based on the parallel postulate, which says that given any line m and a point not on it, there is exactly one line parallel to m through that point. In elliptic geometry, there are no parallel lines, and shapes are drawn on a sphere. In hyperbolic geometry, there are infinitely many lines parallel to any given line through a point, and shapes are drawn on a pseudosphere. Fractal figures are formed by starting with a basic shape and repeating the shape indefinitely in a pattern. Tessellations are formed when the same geometric shapes are used repeatedly to cover a plane without any gaps.

MATH IN **Home Improvement REVISITED**

1. The U-shaped section can be divided into three rectangles, giving a total of six rectangles we can find the area of using length times width. The combined area is 7,835 square inches, which is about 54.4 square feet. The cost of the style at \$42 per square foot is $54.4 \times \$42 = \$2{,}284.80$. The cost of the style at \$60 is $54.4 \times \$60 = \$3{,}264$. The difference is \$979.20.

2. Draw a right triangle diagram with the given measurements:

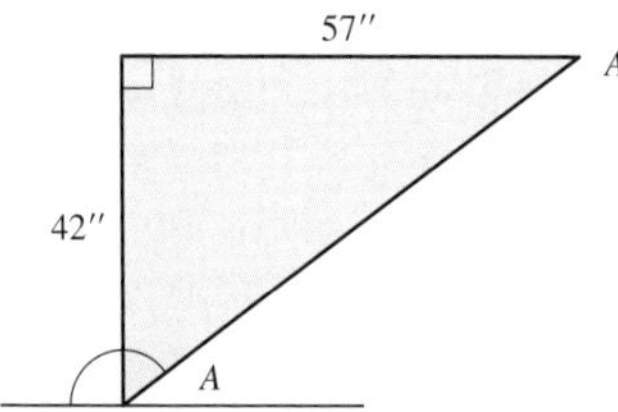

Angle A of the triangle is found using the equation $\tan A = \frac{42}{57}$. Solving for A, we get 36.4°. The angle marked A along the bottom is also 36.4° because of alternate interior angles, and the angle marked with the arc, which the trim should be cut at, is the supplement of 36.4°, or 143.6°.

3. The width of the glass block part is $55\frac{3}{4}$" (7 blocks at $7\frac{3}{4}$" each, and 6 mortar joints at $\frac{1}{4}$" each). The height is $31\frac{3}{4}$" (4 blocks and 3 mortar joints). The diagram shows the trim:

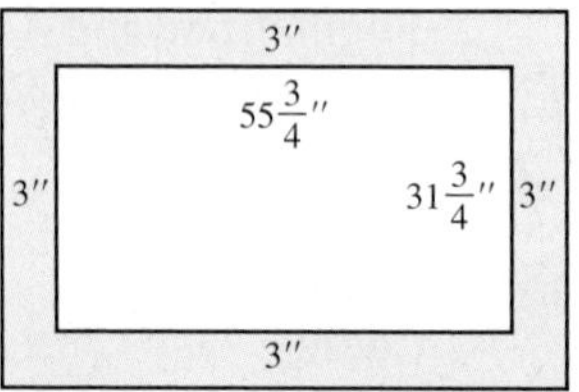

The area of the trim can be found either by subtracting the area of the small rectangle from the area of the large rectangle, or by breaking the trim into four rectangles. In either case, the result is 561 square inches of trim needed, which is about 3.9 square feet.

4. This is another triangle diagram:

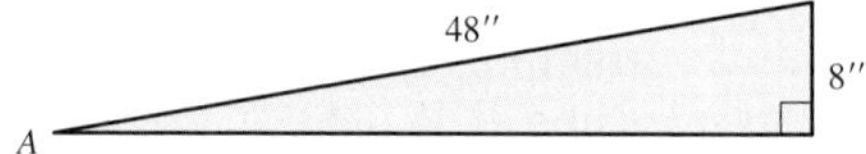

The 8″ side comes from the difference between the front edge and back edge heights. Solving the equation $\sin A = \frac{8}{48}$, we find that angle A measures 9.6°. If the legs were perpendicular to the hypotenuse and cut at a 90° angle, the bottom would make the same 9.6° angle with the ground (see below, left drawing). So to make it sit flat, it needs to be cut 9.6° short of 90°, or 80.4°.

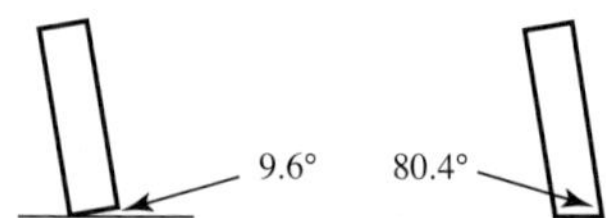

None of these were made up—they are the actual calculations done by the author while doing the projects.

Review Exercises

Mixed Review

For Exercises 1–10, identify each figure.

1. A B

2. R S

3. C D

4.

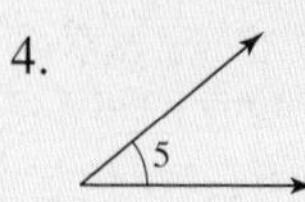

5.

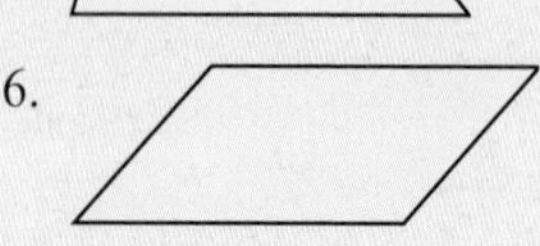

6.

7.

8.

9.

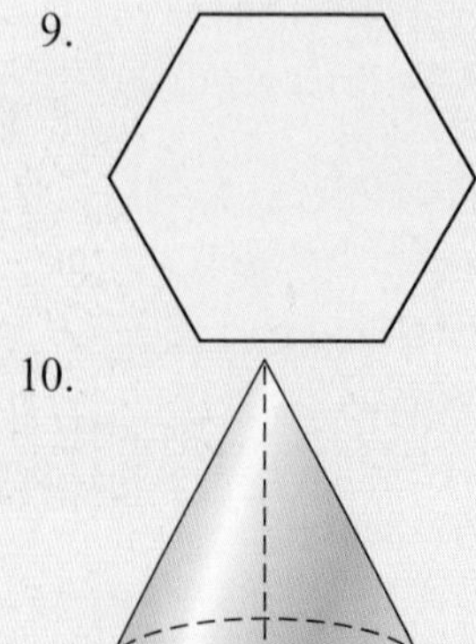

10.

Section 10-1

For Exercises 11–16, identify each type of angle or pairs of angles.

11. 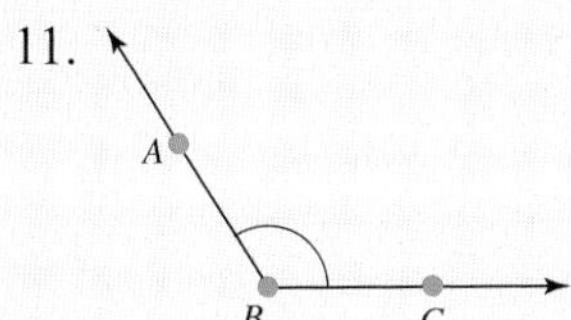

12. 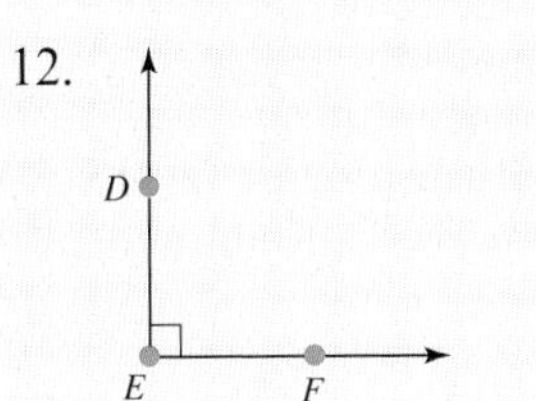

13. 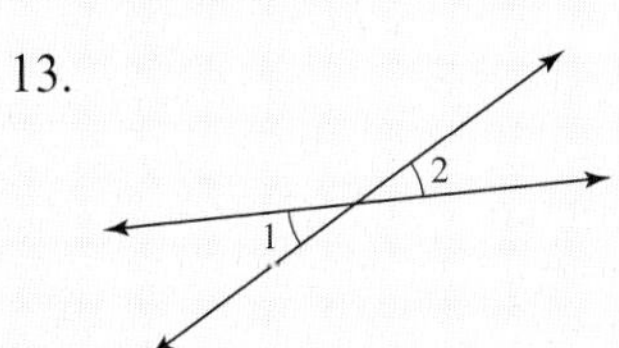

$\measuredangle 1$ and $\measuredangle 2$

14. 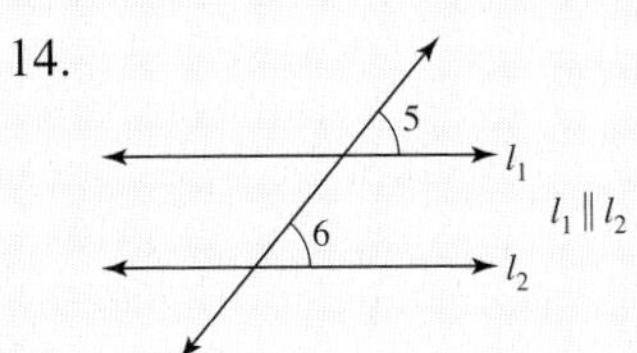

$l_1 \parallel l_2$
$\measuredangle 5$ and $\measuredangle 6$

15. 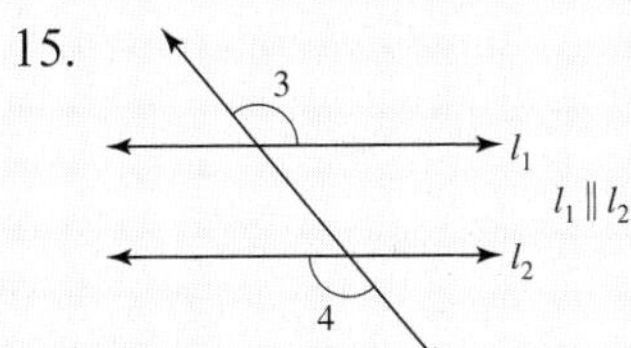

$l_1 \parallel l_2$
$\measuredangle 3$ and $\measuredangle 4$

16. 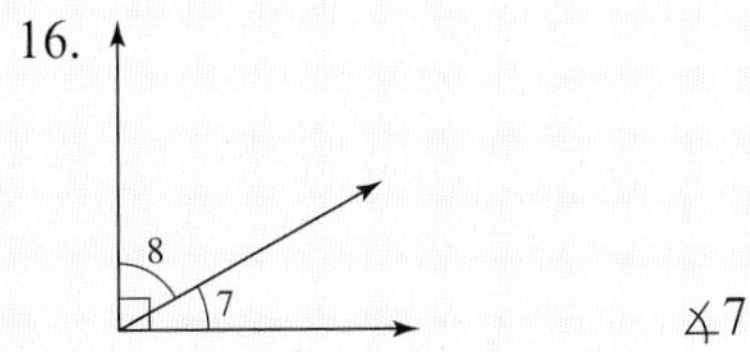

$\measuredangle 7$ and $\measuredangle 8$

17. Find the complement of each angle.
(a) 27° (b) 88°
18. Find the supplement of each angle.
(a) 172° (b) 13°
19. Find the measures of $\measuredangle 1$, $\measuredangle 2$, and $\measuredangle 3$.

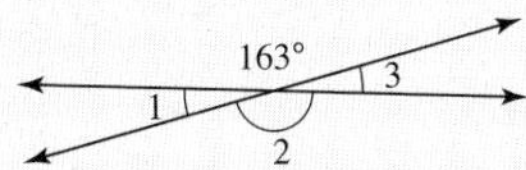

20. Find the measures of $\measuredangle 1$ through $\measuredangle 7$.

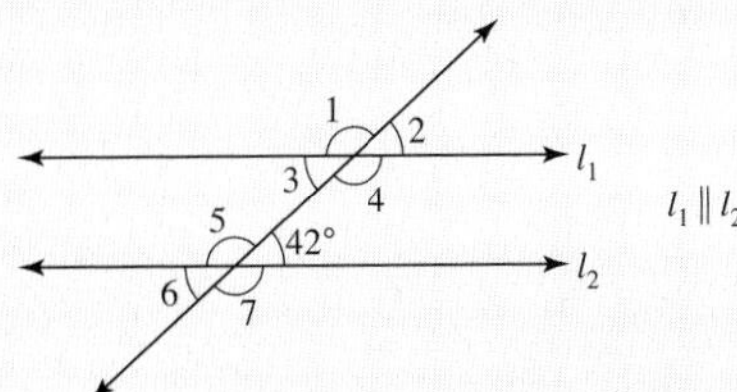

Section 10-2

21. Find the measure of the third angle of a triangle if the measures of the other two angles are 95° and 42°.
22. Find the measure of the hypotenuse of a right triangle if the lengths of the two legs are 8 inches and 15 inches.
23. Find the height of a water tower if its shadow is 24 feet when a 6-foot fence pole casts a shadow of 2 feet.

In Exercises 24–27, classify each triangle by angles (acute, obtuse, or right) and by sides (equilateral, isosceles, or scalene).

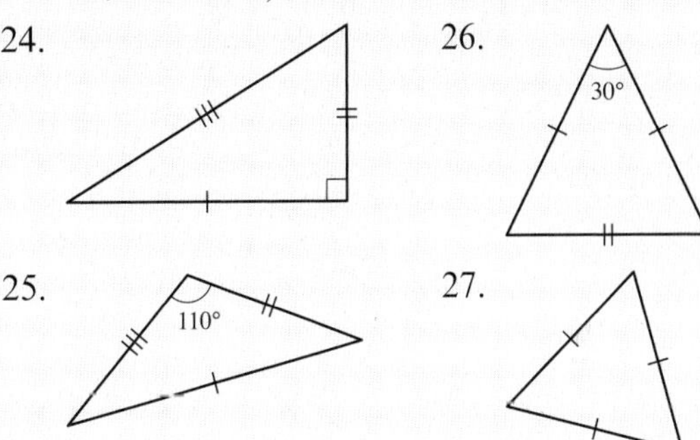

In Exercises 28–29, use the Pythagorean theorem to find the length of the missing side.

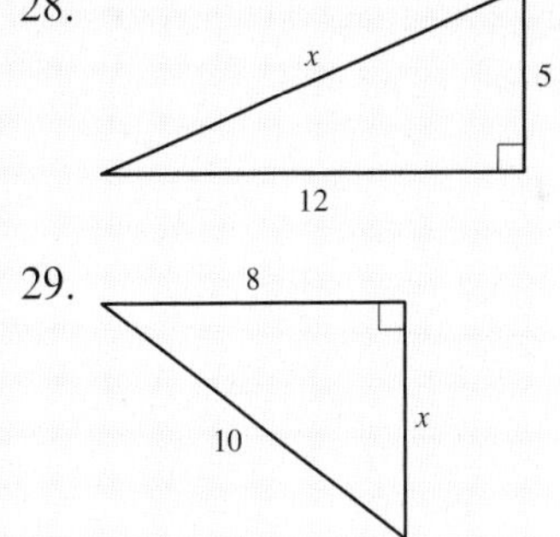

In Exercises 30–31, use the proportional property of similar triangles to find the missing sides (assume the two triangles in each exercise are similar).

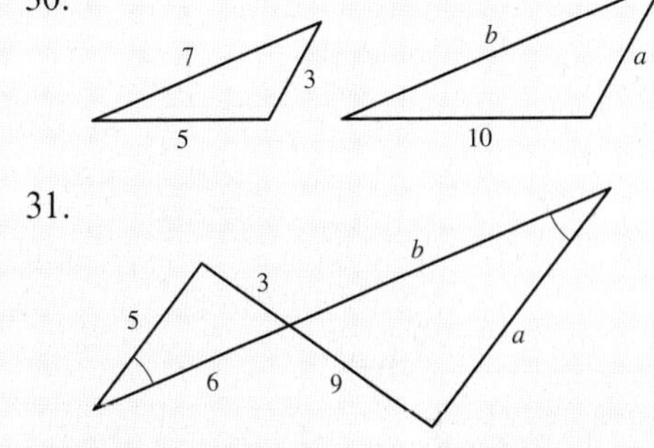

32. If a 6-foot-tall man casts a 10-foot-long shadow, find the height of a building that casts a 150-foot-long shadow at the same time.
33. If a 12-foot ladder reaches 9 feet, 6 inches up a wall, how far up would a 20-foot ladder reach when placed at the same angle?

34. If the 20-foot ladder from Question 33 is placed with its base 9 feet from the wall, how high up does it reach?
35. A series of guy wires hold a radio antenna tower in place. If the tower is 140 feet high and the wires are attached to a point on the ground 90 feet from the base of the tower, how long are the wires?
36. While doing conditioning drills for spring training, a baseball player is supposed to run the length and width of a rectangular field that is 500 yards long and 800 feet wide. If he cheats and runs straight across on a diagonal, how much less distance does he run?

Section 10-3

In Exercises 37 and 38, identify the figure, then find the sum of the measures of the angles.

37.

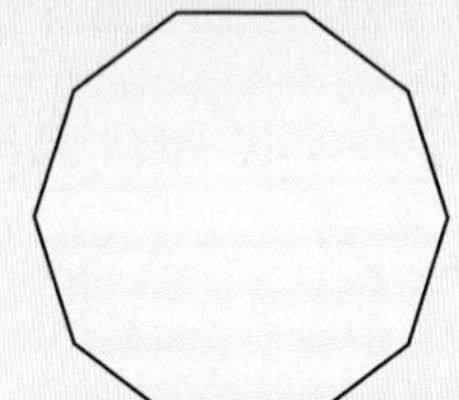

38.

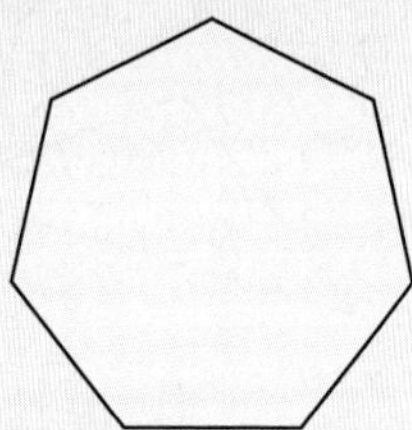

In Exercises 39–42, find the perimeter.

39. Find the perimeter of the figure shown.

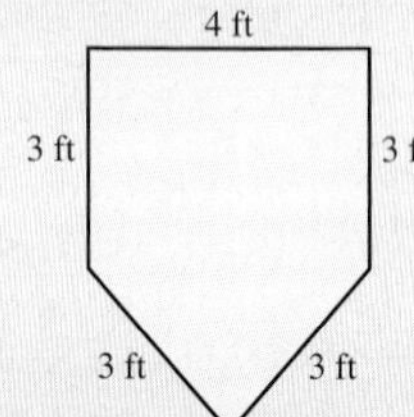

40.

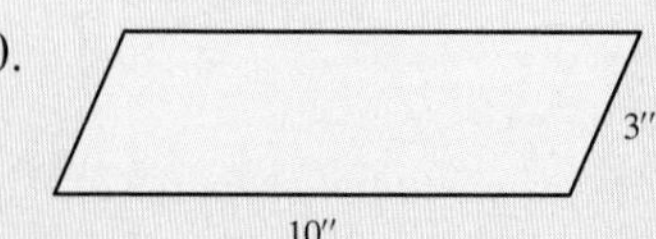

41.

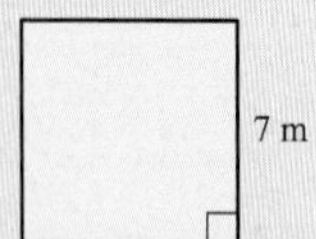

42.

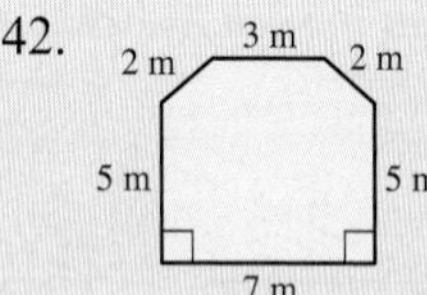

Section 10-4

For Exercises 43–46, find the area of each figure.

43.

44.

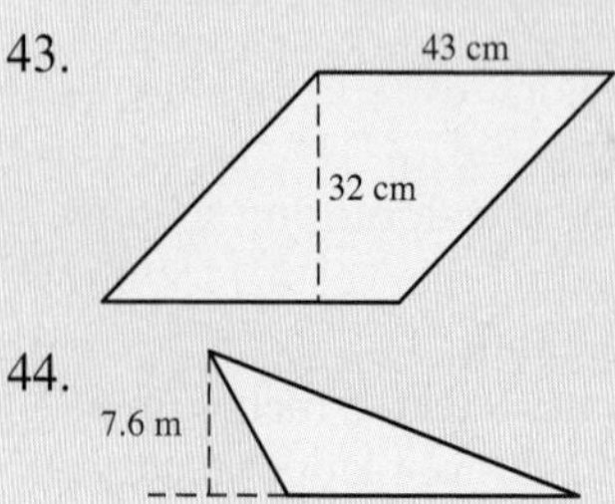

45.

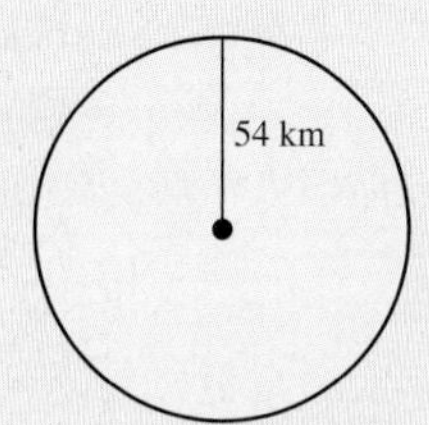

46. 12 ft; 8 ft; 17 ft

47. Find the circumference and area of a circle whose diameter is 16.5 yards.
48. The main gathering area of a newly constructed museum of modern art is shaped like a parallelogram. The north and south sides are 40 feet in length, and those two sides are 65 feet apart. Find how much it would cost to have the room tiled with marble that costs $12 per square foot installed.
49. Solar covers for swimming pools are made of a special plastic material, similar to bubble wrap, that helps to hold in heat. One supplier sells solar covers by the square foot, charging 22 cents per square foot. How much would a solar cover cost for a circular pool with a diameter of 17 feet?
50. The manufacturer of a new lawn tractor promises that in average use, it can cut 130 square yards of grass per minute. It currently takes the landscaper at East Central State an hour and 20 minutes to cut the football field, which is a rectangle with length 120 yards and width 53 yards. How much time would the new tractor save him?

Section 10-5

For Exercises 51–54, find the volume.

51.

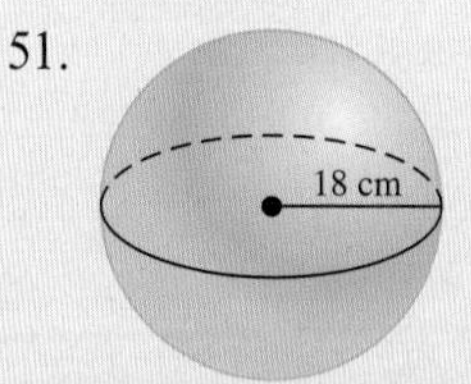

52.

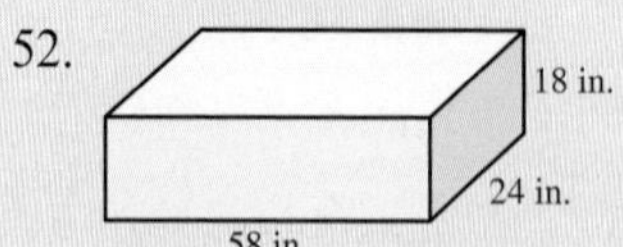

53.

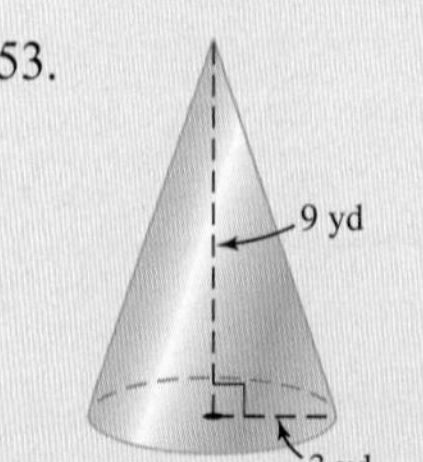

54.

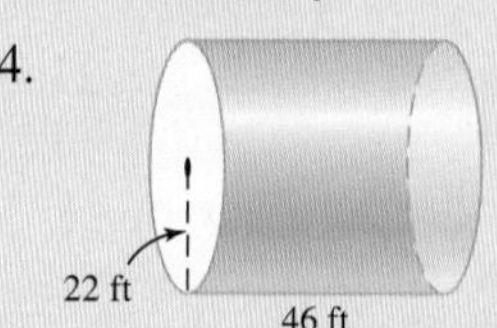

55. Find the surface area of the figure in Exercise 52.
56. Find the surface area of the figure in Exercise 54.
57. If the diameter of a bicycle wheel is 26 inches, how many revolutions will the wheel make if the rider rides 1 mile (1 mile = 5,280 feet)?
58. Find how many square inches of fabric are needed to make the kite shown.

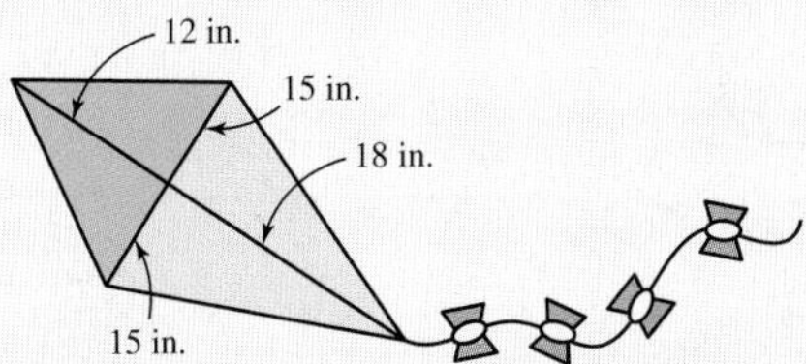

59. Find the volume of a ball if the diameter is 4.2 cm.
60. Find the surface area for the walls of the building shown. (Include openings but do not include the area of the roof.)

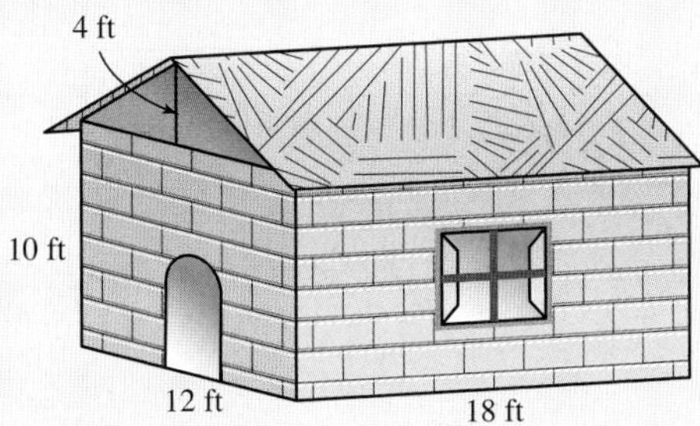

Section 10-6

61. For the triangle shown, find sin B, cos B, and tan B.

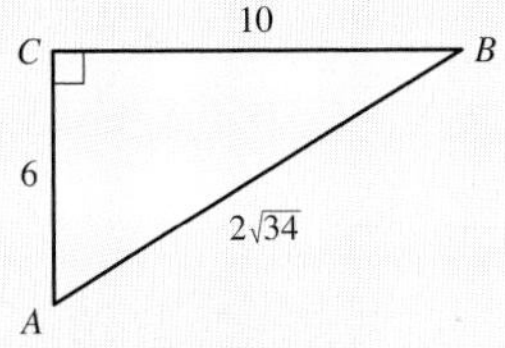

In Exercises 62–65, find the measure of the requested side or angle in triangle ABC. Note that C is the right angle.

62. Find b if $c = 27$ inches and $m\angle B = 49°$.
63. Find a if $b = 12$ yards and $m\angle A = 11°$.
64. Find $m\angle A$ if $c = 11$ meters and $b = 7.5$ meters.
65. Find $m\angle B$ if $a = 100$ miles and $b = 127$ miles.
66. If a tree 32 feet tall casts a shadow of 40 feet, find the angle of elevation of the sun.
67. A pole is leaning against a wall and is 15 feet from the base of the wall. If the angle of elevation from the ground to the base of the pole is 63°, find the length of the pole.
68. The ramp for a motorcycle jump is a straight, flat surface 130 feet long. The beginning is at ground level, and the end is 27 feet high. What angle does the ramp make with the ground?
69. A plane takes off in a straight line and flies for a half-hour at 160 miles per hour. At that point, the pilots are informed by air traffic control that they have flown 17° off course. They are instructed to change direction and fly at the same speed along a perpendicular path back to the line they should have flown on, then turn and continue to the destination. How long will it take them to get back on course?

Section 10-7

70. Starting with a square, create a tessellation using the procedure described in Section 10-7.
71. Make a fractal figure by starting with a square piece of paper 8 inches on a side, then drawing a square that is 3 inches on a side in each corner. Continue the process for at least four iterations, each time drawing squares in each corner that are $\frac{3}{8}$ the length of the square they are inside.
72. Describe lines in elliptic geometry in terms of a globe.

Chapter Test

1. Find the complement of an angle whose measure is 73°.
2. Find the supplement of an angle whose measure is 149°.
3. Find the measures of $\angle 1$, $\angle 2$, and $\angle 3$.

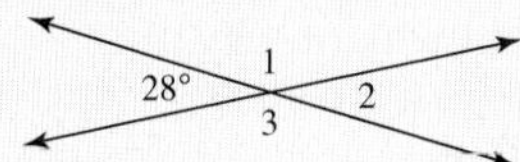

4. Find the measures of $\angle 1$ through $\angle 7$.

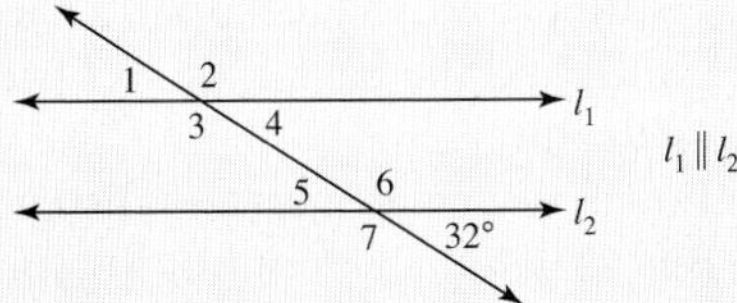

5. If two angles of a triangle have a measure of 85° and 47°, respectively, find the measure of the third angle.
6. If the hypotenuse of a right triangle is 29 inches and the length of one leg is 21 inches, find the length of the other leg.
7. Find the measure of side x. The two triangles are similar.

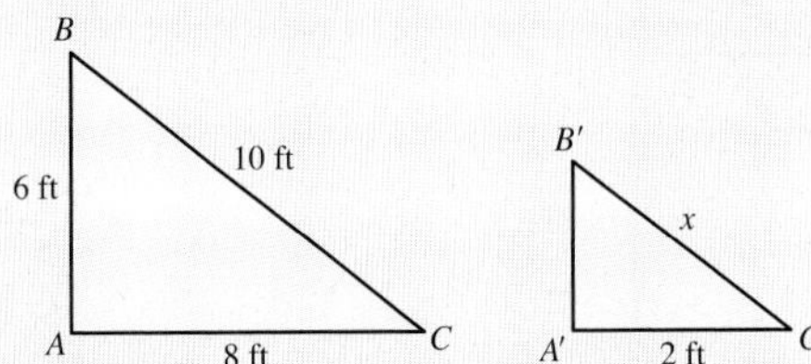

8. One way to measure the height of a tree is to cut it down and use a tape measure. But that's pretty drastic... another is to measure the length of its shadow, then measure the length of your shadow at the same time. Suppose the shadow of the tree is 13'11" long at a time when a person who is 5'10" tall casts a shadow 2'8" long. How tall is the tree?
9. Find the sum of the angles of an octagon.
10. Maureen is building a flowerbed in her backyard in the shape below. She wants to put three layers of landscape timbers along the border. If the timbers come in 8-foot lengths, how many will she need to buy?

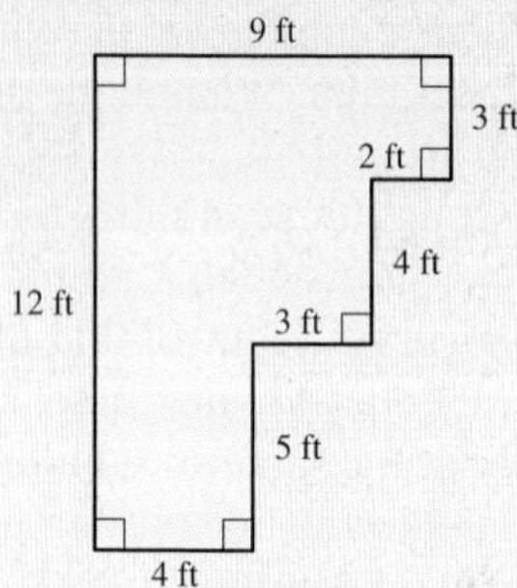

For Exercises 11-15, find the area.

11.

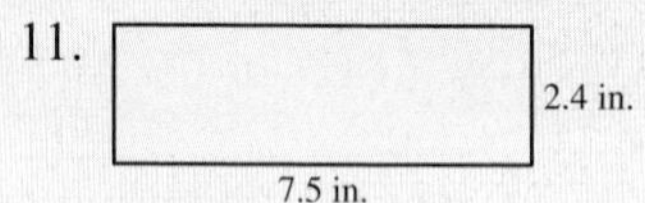

12.

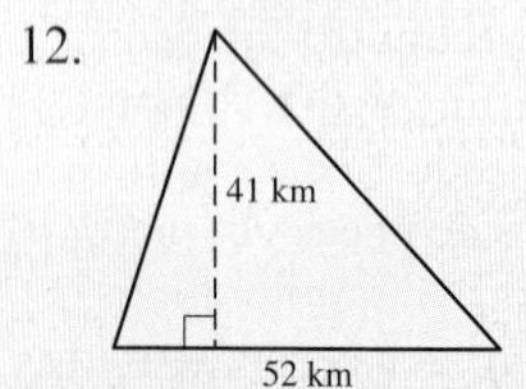

13.

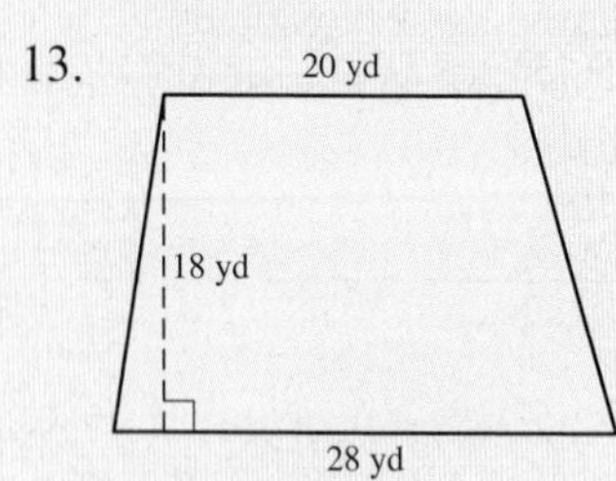

14.

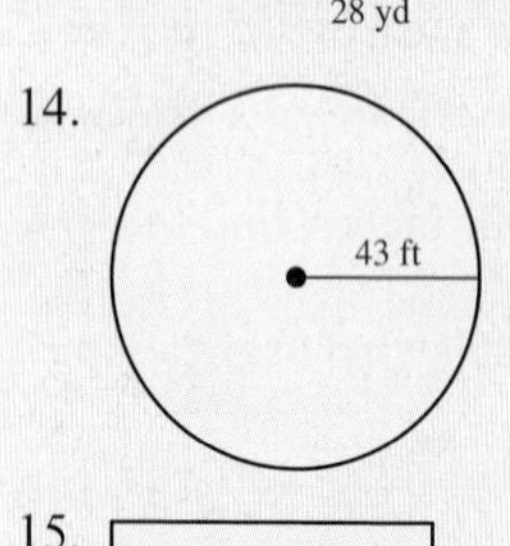

15.

16. Find the circumference of a circle whose radius is 13 cm.

For Exercises 17–22, find the volume of each.

17.

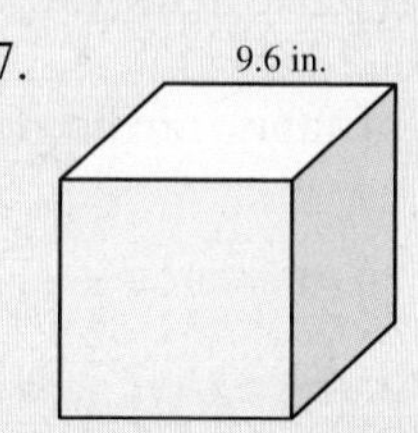

18.

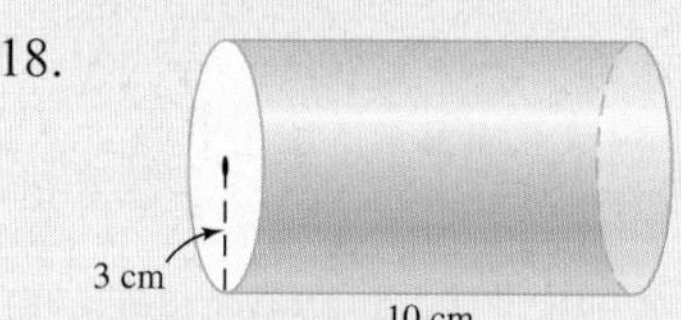

19. 21.

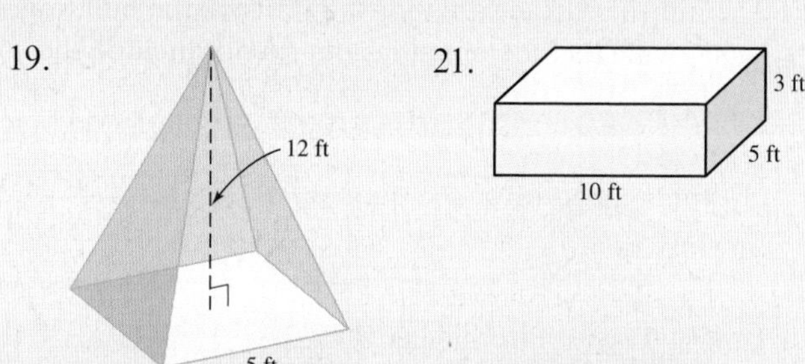

20. 22.

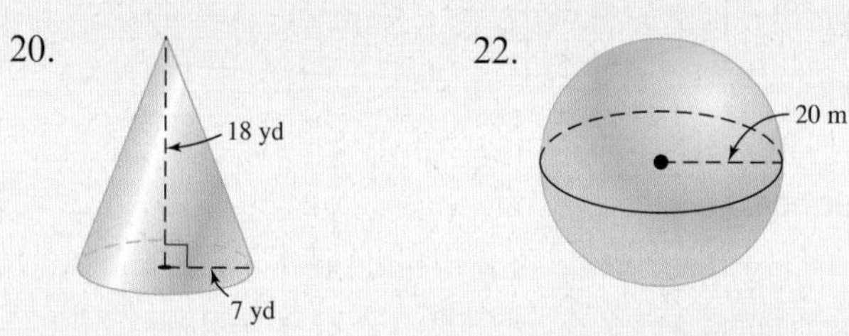

For Exercises 23–25, find the surface area of each figure.

23. The figure in Exercise 17.

24. 25.

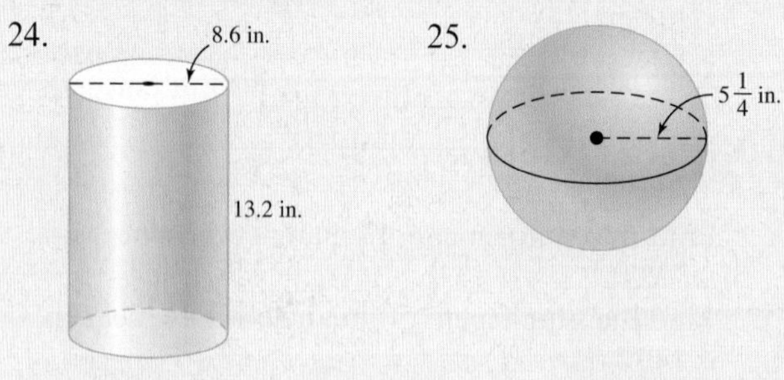

26. Are lines of latitude on a globe examples of lines in elliptic geometry? Explain why or why not.
27. Discuss whether or not a brick wall is a tessellation.
28. A city lot is shaped like a trapezoid with the front and back sides parallel and 157 feet apart. The front is 32 feet wide, and the back is 48 feet wide. Find the area of the lot.
29. Starting from home, you drive a certain distance south to Best Buy to get a new monitor for your

computer. You then drive 5 miles west to return that awful shirt your mother bought you for your birthday. You then drive 25 miles home. How far is it from your house to Best Buy?

30. You create a masterpiece in your art studio and it is selected for the student art exhibition. Your painting is 70 inches by 100 inches, and you choose framing that costs \$4.35 per yard. Find the cost of the framing.
31. During your chemistry lab, you fill a cone-shaped beaker that has a 1-inch diameter and a 2-inch height with lead, which weighs 0.41 pounds per cubic inch. Find the weight of the lead in the beaker.
32. To install a satellite television transmitter, the technician digs a cylindrical hole that is 3 feet in diameter with a depth of 5 feet. How much dirt in pounds must be removed if 1 cubic foot of dirt weighs 98 pounds?
33. A bank sets up a security camera on top of a building across the street from their entrance. The camera is controlled by a computer, and the head of security needs to input an angle of depression for proper aim. If the camera is at a height of 46 feet and the base of the building across the street is 70 feet from the bank entrance, what angle should the manager input?

Projects

1. In a Sidelight feature back in Section 5-3, we referred to a website (http://www.angio.net/pi/piquery.html) where you can search for strings of numbers within the first 200 million digits of pi. Go to the site and search for the following:

   (a) Your birthday, in the form MMDDYYY
   (b) Your phone number, without area code
   (c) Your phone number with the area code
   (d) The first 8 natural numbers consecutively
   (e) The first 9 natural numbers consecutively
   (f) The first 6 even natural numbers consecutively
   (g) The first 7 odd natural numbers consecutively

   Based on these results, what can you say about the likelihood of finding a given string based on the number of digits in that string? Try to make some general conjectures about how likely you are to find a given string based on the number of digits, and try several numbers to test out your conjectures.

2. The numbers 3, 4, and 5 are called a Pythagorean triple, because a right triangle with sides 3, 4, and 5 units satisfies the Pythagorean theorem. That is, $3^2 + 4^2 = 5^2$. The numbers 5, 12, and 13 form another Pythagorean triple.

   (a) Do 6, 8, and 10 form a Pythagorean triple?
   (b) Do 9, 12, and 15 form a Pythagorean triple?
   (c) Do 10, 24, and 26 form a Pythagorean triple?
   (d) Based on your answers for parts a through c, can you make any definite statements about how many Pythagorean triples there are?
   (e) Based on parts a through c, write a general formula or formulas for finding Pythagorean triples.
   (f) Can you find a Pythagorean triple where the numbers are not multiples of either 3, 4, 5 or 5, 12, 13?
   (g) Search the Internet for the string "Pythagorean triple" and find how many Pythagorean triples there are where all three numbers are less than 100.
   (h) Research the question "are Pythagorean triples more common, or more rare, among three-digit numbers than among two-digit numbers?" In general, are they harder or easier to find as the lengths get larger?

3. Gather a variety of objects that you could pour a liquid into: glasses, buckets, pitchers, coolers, aquariums, sinks, or anything else that you could calculate (or estimate) the volume of, using the volume formulas we learned in Section 10-5.

   (a) Calculate the volume of each in both cubic inches and cubic centimeters by taking measurements and using volume formulas. If the containers are not exactly regular, like a bucket that is wider at the top than the bottom, make your best estimate of the volume.
   (b) Convert the measurements in cubic inches to gallons using the conversions from Chapter 9. Convert the measurements in cubic centimeters to liters also.
   (c) Calculate the volume of each container in gallons and liters by pouring measured amounts of water into it using a measuring cup, pitcher, or beaker.
   (d) Compare the measurements done with water to your calculations. How well did you do?

CHAPTER 11

Probability and Counting Techniques

Outline

11-1 The Fundamental Counting Principle and Permutations
11-2 Combinations
11-3 Basic Concepts of Probability
11-4 Tree Diagrams, Tables, and Sample Spaces
11-5 Probability Using Permutations and Combinations
11-6 Odds and Expectation
11-7 The Addition Rules for Probability
11-8 The Multiplication Rules and Conditional Probability
11-9 The Binomial Distribution
Summary

MATH IN Gambling

The fact that you're reading this sentence means that you're probably taking a math class right now. But maybe not . . . you could be an instructor evaluating the book, or maybe an editor looking for mistakes (unsuccessfully, we hope). Still, I would be willing to bet that you're taking a math class. The word "probably" indicates a certain likelihood of something happening, and that basic idea is the topic of this chapter. We call the study of the likelihood of events occurring *probability*.

Probability is one of the most useful concepts in math because being able to anticipate the likelihood of events can be useful in so many different areas. Games of chance, business and investing, sports, and weather forecasting are just a few samples from a seemingly limitless list of applications. What are the chances of your team winning the championship? Should you take an umbrella to the golf course today? Will stock in a company you're keeping an eye on go up or down? Is that new job offer a good opportunity, or a disaster waiting to happen? Every day you make decisions regarding possible events that are governed at least in part by chance. The more you know about the likelihood of events, the more informed your decisions are likely to be.

In order to compute probabilities of events occurring, we're going to need to know all of the possible outcomes for an event. For example, in deciding whether or not to bring an umbrella, there are only two outcomes: it will either rain, or it won't. But in considering whether or not to take a job, there are a wide variety of possible outcomes, some good, some not so good. Since counting up the number of possible outcomes is important in probability, we will begin our study with a look at methods for counting. (Don't worry—that's not quite as elementary as it sounds!) Then we'll be ready to tackle the basic concepts of probability. Along the way, we'll learn about odds and the expected value of a probability experiment, and this is where gambling comes into play.

A handful of gambling scenarios is provided below. In each case, find the expected value (that is, the average amount a person would win or lose) if placing the bet 100 times. Then rank the scenarios from best to worst in terms of your likelihood of winning or losing money.

1. At a church fair, you bet $1 and roll two dice. If the sum is 2, 3, 11, or 12, you get back your dollar plus four more. On any other roll, you lose.
2. In a casino, you bet $1 on 33 at a roulette table. There are 38 possible numbers that can come up. If you win, you get your dollar back, plus 35 more.
3. You buy a $1 ticket to a multistate lottery. If you match all six numbers, including the Mega Ball, you win the $20 million jackpot. If not, you lose. There are 175,711,536 possible combinations, and only one of them will be a winner.
4. You bet $1 on flipping a coin with your roommate. Heads, you win, tails, your roommate wins.

For answers, see Math in Gambling Revisited on page 640

Section 11-1 The Fundamental Counting Principle and Permutations

LEARNING OBJECTIVES

- ❑ 1. Use the fundamental counting principle.
- ❑ 2. Calculate the value of factorial expressions.
- ❑ 3. Find the number of permutations of n objects.
- ❑ 4. Find the number of permutations of n objects taken r at a time.
- ❑ 5. Find the number of permutations when some objects are alike.

Many problems in probability and statistics require knowing the total number of ways a sequence of events can occur. Suppose that as part of an exciting new job, you're responsible for designing a new license plate for your state. You want it to look cool, of course. But if you want to keep your job, you better make sure that the sequence of letters and numbers you choose guarantees that there are enough different combinations so that every registered vehicle has a different plate number.

In this chapter, we'll study three basic rules for counting the number of outcomes for a sequence of events. The first is the *fundamental counting principle*.

The Fundamental Counting Principle

After getting that new job, you naturally want a new apartment, and furniture to go along with it. The hip furniture boutique around the corner has the couch you want in either leather or microsuede, and each comes in your choice of four colors. How many different couches do you have to choose from?

We'll illustrate the situation with a *tree diagram,* which displays all the possible combinations.

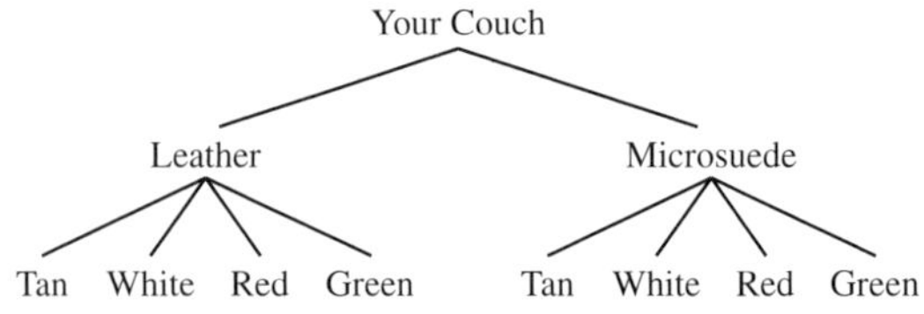

This shows us that the total number of couches possible is eight. This number can also be found by multiplying the number of materials (2) by the number of colors for each (4).

Now what if each couch can also come with or without an end recliner? Each of the eight choices in our diagram would have two more possibilities beneath it.

This would give us a total of 16 couches, which is $2 \cdot 4 \cdot 2$, or the product of the number of choices at each stage. This illustrates our first key counting principle.

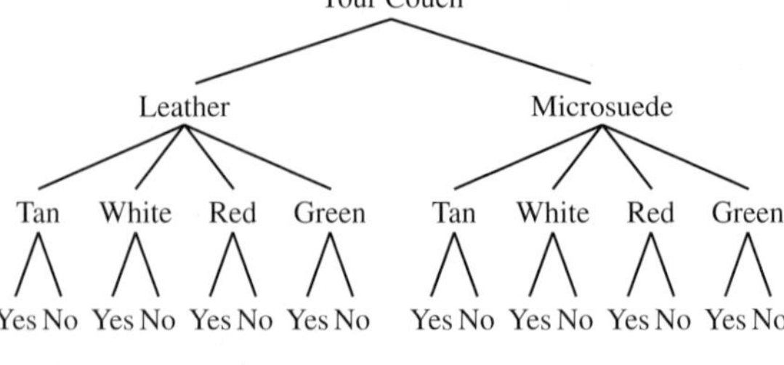

Math Note

The occurrence of the first event in no way affects the occurrence of the second event, which in turn, does not affect the occurrence of the third event, etc.

The Fundamental Counting Principle

In a sequence of n events in which the first event can occur in k_1 ways and the second event can occur in k_2 ways and the third event can occur in k_3 ways and so on, the total number of ways the sequence can occur is

$$k_1 \cdot k_2 \cdot k_3 \cdot \cdots \cdot k_n$$

EXAMPLE 1 Using the Fundamental Counting Principle

There are four blood types: A, B, AB, and O. Blood is also either Rh^+ or Rh^-. If a local blood bank labels donations according to type, Rh factor, and gender of the donor, how many different ways can a blood sample be labeled?

SOLUTION

There are four possibilities for blood type, two for Rh factor, and two for gender of the donor. Using the fundamental counting principle, there are

$$4 \cdot 2 \cdot 2 = 16$$

different ways that blood could be labeled.

▼ Try This One 1

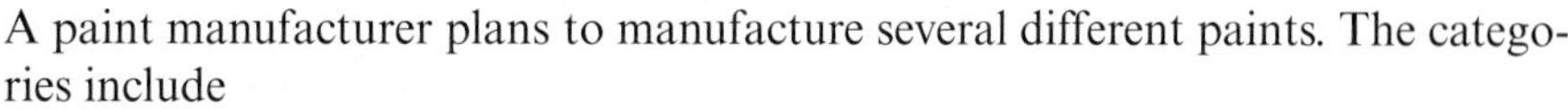

A paint manufacturer plans to manufacture several different paints. The categories include

Color	Red, blue, white, black, green, brown, yellow
Type	Latex, oil
Texture	Flat, semigloss, high gloss
Use	Outdoor, indoor

How many different kinds of paint can be made if a person can select one color, one type, one texture, and one use?

When determining the number of different ways a sequence of events can occur, we'll need to know whether or not repetitions are permitted. The next example shows the difference between the two situations.

EXAMPLE 2 Using the Fundamental Counting Principle with Repetition

How many three-digit codes are possible if repetition is not permitted?

(a) The letters A, B, C, D, and E are to be used in a four-letter ID card. How many different cards are possible if letters are allowed to be repeated?
(b) How many cards are possible if each letter can only be used once?

SOLUTION

(a) There are four spaces to fill and five choices for each. The fundamental counting principle gives us

$$5 \cdot 5 \cdot 5 \cdot 5 = 5^4 = 625$$

(b) The first letter can still be chosen in five ways. But with no repetition allowed, there are only four choices for the second letter, three for the third, and two for the last. The number of potential cards is

$$5 \cdot 4 \cdot 3 \cdot 2 = 120$$

Hopefully, this ID card is for a pretty small organization.

▼ Try This One 2

The lock on a storage facility is controlled by a keypad containing digits 1 through 5.

(a) How many three-digit codes are possible if digits can be repeated?
(b) How many three-digit codes are possible if digits cannot be repeated?

 1. Use the fundamental counting principle.

Factorial Notation

Calculator Guide

Most graphing calculators, and many scientific calculators have factorials programmed into them. To quickly compute 5!:

Standard Scientific Calculator

5 [x!]

Standard Graphing Calculator

5 [MATH]; use right arrow to choose PRB, then press [4] [ENTER].

The next couple of counting techniques we'll learn use **factorial notation.** The symbol for a factorial is the exclamation mark (!). In general, $n!$ means to multiply the whole numbers from n down to 1. For example,

$$1! = 1 = 1$$
$$2! = 2 \cdot 1 = 2$$
$$3! = 3 \cdot 2 \cdot 1 = 6$$
$$4! = 4 \cdot 3 \cdot 2 \cdot 1 = 24$$
$$5! = 5 \cdot 4 \cdot 3 \cdot 2 \cdot 1 = 120$$

The formal definition of factorial notation is given next.

For any natural number n

$$n! = n(n-1)(n-2)(n-3)\cdots 3 \cdot 2 \cdot 1$$

$n!$ is read as "***n* factorial.**"
0! is defined as 1. (This might seem strange, but will be explained later.)

Some of the formulas we'll be working with require division of factorials. This will be simple if we make two key observations:

- $\frac{n!}{n!}$ is always 1. For example, $\frac{3!}{3!} = \frac{3 \cdot 2 \cdot 1}{3 \cdot 2 \cdot 1} = \frac{6}{6} = 1$
- You can write factorials without writing all of the factors down to 1. For example, $5! = 5 \cdot 4 \cdot 3 \cdot 2 \cdot 1$, but we can also write this as $5 \cdot 4!$, or $5 \cdot 4 \cdot 3!$, etc.

In Example 3, we'll use these ideas to simplify our calculations.

EXAMPLE 3 Evaluating Factorial Expressions

Evaluate each expression:

(a) 8! (b) $\frac{12!}{10!}$

SOLUTION

(a) $8! = 8 \cdot 7 \cdot 6 \cdot 5 \cdot 4 \cdot 3 \cdot 2 \cdot 1 = 40{,}320$
(b) First, write 12! as $12 \cdot 11 \cdot 10!$, then note that $\frac{10!}{10!} = 1$.

$$\frac{12!}{10!} = \frac{12 \cdot 11 \cdot 10!}{10!} = 12 \cdot 11 = 132$$

▼ Try This One 3

2. Calculate the value of factorial expressions.

Evaluate each expression:

(a) 6! (b) $\frac{9!}{4!}$

CAUTION

You cannot divide factorial expressions by reducing fractions. In Example 3(b), $\frac{12!}{10!}$ is *not* equal to $\frac{6!}{5!}$.

Permutations

The second rule that we can use to find the total number of outcomes for a sequence of events is the *permutation rule*.

An arrangement of n distinct objects in a specific order is called a **permutation** of the objects.

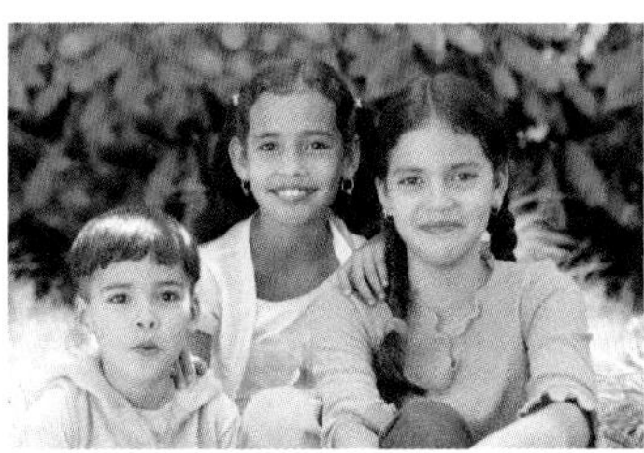

For example, suppose that a photographer wants to arrange three people, Carmen, Juan, and Christina, in a specific order for a portrait. There are six ways he could arrange them:

Carmen	Juan	Christina	Carmen	Christina	Juan
Juan	Carmen	Christina	Juan	Christina	Carmen
Christina	Juan	Carmen	Christina	Carmen	Juan

We could use the fundamental counting principle to see that there are six possible permutations as well:

3	·	2	·	1	=	3!
Choices for 1st position		Choices for 2nd position		Choices for 3rd position		

In general,

The number of permutations of n distinct objects using all of the objects is $n!$.

Example 4 illustrates this formula.

EXAMPLE 4 Calculating the Number of Permutations

In seven of the 10 years from 1998–2007, the five major league baseball teams in the American League East division finished in the exact same order: New York, Boston, Toronto, Baltimore, Tampa Bay. Just how unusual is this? Find the number of possible finishing orders for these five teams.

SOLUTION

This is a permutation problem since we're deciding on the number of ways to arrange five distinct objects. There are

$$5! = 120$$

possible finishing orders.

▼ Try This One 4

In how many different orders can the 11 basketball teams in the Big Ten conference finish? (Seriously, there are 11 teams in the Big Ten.)

☑ 3. Find the number of permutations of n objects.

So far, in calculating permutations, we've used all of the available objects. But what if only some of them are selected?

EXAMPLE 5 Solving a Permutation Problem

How many different ways can a pledge class with 20 members choose a president, vice president, and Greek Council representative? (No pledge can hold two offices.)

SOLUTION

There are 20 choices for president, 19 remaining candidates for vice president, and 18 members left to choose from for Greek Council rep. So there are $20 \cdot 19 \cdot 18 = 6{,}840$ different ways to assign these three offices.

▼ Try This One 5

How many ways can a manager and assistant manager be selected from a department consisting of 10 employees?

In Example 5, a certain number of objects (people in this case) have been chosen from a larger pool. The order of selection is important, and no repetition is allowed. (No pledge can be both president and vice president—power trip!) We call such an arrangement of objects a **permutation of *n* objects taken *r* at a time.** In the pledge problem, n is 20 and r is 3. We will use the symbol ${}_nP_r$ to represent this type of permutation.

We solved Example 5 using the fundamental counting principle, but the result suggests the formula below:

Permutation of *n* Objects Taken *r* at a Time

The arrangement of n objects in a specific order using r of those objects is called a **permutation of *n* objects taken *r* at a time.** It is written as ${}_nP_r$, and is calculated using the formula

$${}_nP_r = \frac{n!}{(n-r)!}$$

EXAMPLE 6 Solving a Permutation Problem

Calculator Guide

Most graphing calculators have the permutation formula built in. To find ${}_{10}P_5$ on a standard graphing calculator: 10 [MATH], then use right arrow to select PRB, and press [2] to select ${}_nP_r$, and finally 5 [ENTER].

How many five-digit zip codes are there with no repeated digits?

SOLUTION

This is a permutation problem because five numbers are taken from 10 possible digits, with order important and no repetition. In this case, $n = 10$ and $r = 5$.

$${}_{10}P_5 = \frac{10!}{(10-5)!} = \frac{10!}{5!} = \frac{10 \cdot 9 \cdot 8 \cdot 7 \cdot 6 \cdot 5!}{5!} = 30{,}240$$

☑ 4. Find the number of permutations of n objects taken r at a time.

▼ Try This One 6

How many six-letter passwords are there that use only lowercase letters with no letter repeated?

We now have two permutation formulas: there are $n!$ permutations of n objects using all of them, and $\frac{n!}{(n-r)!}$ using only r of the objects. We should probably make sure that these formulas are consistent with each other. To check, we'll use the second formula to calculate the number of permutations when n objects are taken n at a time. The result should be $n!$ (to match the first formula). Let's see.

If we arrange five people in order for a group picture, we can think of it as a permutation of the five people chosen five at a time; that is,

$${}_5P_5 = \frac{5!}{(5-5)!} = \frac{5!}{0!}$$

This will agree with the first formula if we agree to define 0! as 1. Now we know why $0! = 1$, and we know that our two permutation formulas are consistent. This is good news, because it means we only need to remember the second permutation rule.

Problems involving permutation without duplicate objects can be solved using the fundamental counting rule; however, not all problems that can be solved with the fundamental counting rule can be solved using permutations.

When some of the objects are the same, a different permutation rule is used. Suppose that to solve a word puzzle, you need to find the number of permutations of the letters in the word *moon*. First label the letters as M, O_1, O_2, and N. This would be 4!, or 24, permutations. But since the O's without the subscripts are the same, the permutation M, O_1, O_2, N would be the same as M, O_2, O_1, N. The duplicates are eliminated by dividing 4! by the number of ways to arrange the O's (2!) to get 12. This leads to the next rule.

Permutation Rule When Objects Are Alike

The number of permutations of n objects in which k_1 objects are alike, k_2 objects are alike, etc. is

$$\frac{n!}{k_1!k_2! \ldots k_p!}$$

where $k_1 + k_2 + \cdots + k_p = n$

EXAMPLE 7 Solving a Permutation Problem with Like Objects

How many different passwords can be made using all of the letters in the word *Mississippi*?

SOLUTION

The letters can be rearranged as M IIII SSSS PP. Then $n = 11$, $k_1 = 1$, $k_2 = 4$, $k_3 = 4$, and $k_4 = 2$.

Using our newest formula, there are

$$\frac{11!}{1!4!4!2!} = 34{,}650$$

different passwords.

▼ Try This One 7

☑ 5. Find the number of permutations when some objects are alike.

Find the number of different passwords using all of the letters in the word *Massachusetts*.

In summary, the formulas for permutations are used when we're looking for the number of ways to arrange objects when the order matters, and once an object is used, it can't be used again.

Sidelight WIN A MILLION OR BE STRUCK BY LIGHTNING?

Do you think you would be more likely to win a large lottery and become a millionaire or more likely to be struck by lightning? The answer is that you would be quite a bit more likely to be struck by lightning.

An article in the Associated Press noted that researchers have found that the chance of winning $1 million or more is about 1 in 1.9 million. The chances of winning $1 million in a recent Pennsylvania lottery were 1 in 9.6 million. The chances of winning a $10 million prize in Publisher's Clearinghouse Sweepstakes were 1 in 200 million. In contrast, the chances of being struck by lightning are about 1 in 600,000. In other words, a person is at least 3 times more likely to be struck by lightning than to win $1 million.

One way to guarantee winning a lottery is to buy all possible combinations of the winning numbers. In 1992, an Australian investment group purchased 5 million of the 7 million possible combinations of the lottery numbers in a Virginia State Lottery. Because of the time, they could not purchase the other two million tickets. However, they were able to purchase the winning number and won $27 million. Their profit was about $22 million. Not bad!

States have written laws to prevent this from happening today, and they are devising lottery games with many more possibilities so that it would be impossible to purchase all the possible tickets to win.

The consequence is that your odds of winning are now even lower! There's an old joke among statisticians: lotteries are a tax on people who are bad at math.

Answers to Try This One

1 84

2 (a) 125 (b) 60

3 (a) 720 (b) 15,120

4 39, 916, 800

5 90

6 165, 765, 600

7 64, 864, 800

EXERCISE SET 11-1

Writing Exercises

1. Explain the fundamental counting principle in your own words.
2. What do we mean by the phrase "a permutation of n distinct objects"?
3. How does a permutation of n objects differ from a permutation of n objects taken r at a time?
4. Explain what the symbols $n!$ and ${}_nP_r$ represent, and explain how to compute each for given values of n and r.

Computational Exercises

Evaluate each.

5. $10!$
6. $5!$
7. $9!$
8. $1!$
9. $0!$
10. $4!$
11. $(3!)(2!)$
12. $(7!)(2!)$
13. $\frac{8!}{5!}$
14. $\frac{9!}{7!}$
15. $\frac{7!}{2!3!}$
16. $\frac{10!}{2!5!}$
17. $\frac{150!}{148!}$
18. $\frac{200!}{197!}$
19. ${}_8P_2$
20. ${}_7P_5$
21. ${}_{12}P_{12}$
22. ${}_5P_3$
23. ${}_6P_6$
24. ${}_6P_0$
25. ${}_8P_0$
26. ${}_8P_8$
27. ${}_{11}P_3$
28. ${}_6P_2$

Real-World Applications

29. How many different four-letter permutations can be formed from the letters in the word *decagon*?
30. Out of a group of eight students serving on the Student Government Association, how many different ways can a president, a vice president, and a treasurer be selected?
31. How many different ID cards can be made if there are six digits on a card and no digit can be used more than once?
32. How many different ways can seven types of laser printer be displayed on a shelf in a computer store?
33. How many different ways can four Super Bowl raffle tickets be selected from 50 tickets if each ticket wins a different prize?
34. How many different ways can a psychology student select five subjects from a pool of 20 subjects and assign each one to a different experiment?
35. How many website graphics can be created by using at least three of five different bitmap images?
36. A chemistry lab group has seven experiments to choose from and five members in the group. How many different ways can the experiments be assigned if only one experiment is assigned to each group member?
37. A radio DJ has a choice of seven songs in the queue. He must select three different songs to play in a certain order after the commercial break. How many ways can he select the three different songs?
38. A professor has five different tasks to assign, one to each of her five teaching assistants. In how many different ways could she make the assignments?
39. A nursing student has six different patients assigned to her. How many different ways can she see each one exactly once in the same day?
40. A store owner has 50 items to advertise, and she can select one different item each week for the next 6 weeks to put on special. How many different ways can the selection be made?
41. Out of 17 contacts in her cell phone, how many different ways can Shana set the first four speed-dial contacts?
42. How many different ways can you visit four different stores in a shopping mall?
43. How many ways can a research company select three geographic areas from a list of six geographic areas to test market its product? One area will be selected and tested in September, a different area in October, and a third area will be used in November.
44. How many different code words can be made from the symbols *, *, *, @, @, $, #, #, #, # if each word has 10 symbols?
45. How many different permutations of the letters in the word *Alabama* are there?
46. A radio station must run commercial A three times, commercial B twice, and commercial C once. How many different ways can this be done?
47. How many desktop wallpaper designs can be created from two identical red squares, three identical white squares, and two identical blue squares?
48. How many different ways can two identical television sets, three identical DVD players, and one computer be arranged on a shelf?
49. How many ways can five DVD-ROMs, three zip discs, and two flash drives be arranged in a laptop case?
50. There are 14 teams in the Atlantic Ten Conference, and 12 qualify for the conference basketball tournament. Those 12 teams are seeded 1 through 12 for the conference tournament. In how many ways could this be done?
51. There are 16 teams in the Big East Conference. How many different ways can the top four seeds for the conference basketball tournament be selected?
52. A major league baseball team has 25 players on the active roster. How many choices does a manager have for batting order, listing the nine starters from 1 through 9?
53. A professional football team has 53 players on the active roster. How many choices of 22 starting players does a coach have without regard for their positions?
54. For the new fall season, a network president has 11 shows in development, and six openings in the prime time schedule. In how many ways can she arrange new shows to fit into the schedule?
55. For Ozzfest 2008, there were nine bands scheduled to perform on the main stage. In how many different orders could the bands have been scheduled to perform? How many are possible if Ozzy Osbourne is always scheduled to perform last?

Critical Thinking

56. A campus pizzeria offers regular crust, thin crust, or pan pizzas. You can get either white or red sauce. The owner is kind of eccentric, and only sells pizzas with one topping, chosen from pepperoni, sausage, ham, onions, and ground beef. First, use a tree diagram like the one on page 574 to diagram out all possible choices of pizza. Then show that using the fundamental counting principle yields the same answer.

Section 11-2 Combinations

LEARNING OBJECTIVES

- ❑ 1. Distinguish between combinations and permutations.
- ❑ 2. Find the number of combinations of n objects taken r at a time.
- ❑ 3. Use the combination rule in conjunction with the fundamental counting principle.

Suppose that after waiting in line overnight, you manage to snag the last three tickets to a big concert. Sweet! The bad news is that you can only take two of your four housemates. How many different ways can you choose the two friends that get to go?

This sounds a little bit like a permutation problem, but there's a key difference: the order in which you choose two friends doesn't make any difference. They either get to go or they don't. So choosing Ruth and Ama is exactly the same as choosing Ama and Ruth. When order matters in a selection, we call it a permutation, but when order is not important we call it a *combination*.

A selection of objects without regard to order is called a **combination.**

In Example 1, we'll examine the difference between permutations and combinations.

EXAMPLE 1 Comparing Permutations and Combinations

Given four housemates, Ruth, Elaine, Ama, and Jasmine, list the permutations and combinations when you are selecting two of them.

SOLUTION

We'll start with permutations, then eliminate those that have the same two people listed.

Permutations

Ruth Elaine	Elaine Ruth	Ruth Ama
Ama Ruth	Ruth Jasmine	Jasmine Ruth
Elaine Ama	Ama Elaine	Elaine Jasmine
Jasmine Elaine	Ama Jasmine	Jasmine Ama

Combinations

Ruth Elaine	Ruth Ama	Ruth Jasmine
Elaine Ama	Elaine Jasmine	Ama Jasmine

There are 12 permutations, but only 6 combinations.

▼ Try This One 1

If you are choosing two business classes from three choices, list all the permutations and combinations.

It will be very valuable in our study of counting and probability to be able to decide if a given selection is a permutation or a combination.

EXAMPLE 2 Identifying Permutations and Combinations

Decide if each selection is a permutation or a combination.

(a) From a class of 25 students, a group of 5 is chosen to give a presentation.
(b) A starting pitcher and catcher are picked from a 12-person intramural softball team.

SOLUTION

(a) This is a combination because there are no distinct roles for the 5 group members, so order is not important.
(b) This is a permutation because each selected person has a distinct position, so order matters.

▼ Try This One 2

☑ 1. Distinguish between combinations and permutations.

Decide if each selection is a permutation or a combination.

(a) A 5-digit passcode is chosen from the numbers 0 through 9.
(b) A gardener picks 4 vegetable plants for his garden from 10 choices.

Recall that the number of combinations in Example 1 was half as great as the number of permutations. This is because once two people are selected, there are two ways to arrange them. So the number of combinations of n objects chosen r at a time should be the number of permutations divided by the number of arrangements of r objects. But we know that is $r!$. This gives us a formula for combinations.

Math Note

Notice that the combination formula is the permutation formula from Section 11-1 with an extra factor of $r!$ in the denominator.

The Combination Rule

The number of combinations of n objects taken r at a time is denoted by $_nC_r$, and is given by the formula

$$_nC_r = \frac{n!}{(n-r)!r!}$$

EXAMPLE 3 Using the Combination Rule

How many combinations of four objects are there taken two at a time?

Calculator Guide

Most graphing calculators have the combination formula built in. To find $_4C_2$ on a standard graphing calculator:

4 [MATH], then use right arrow to select PRB, and press [3] to select $_nC_r$, and finally 2 [Enter].

SOLUTION

Since this is a combination problem, the answer is

$$_4C_2 = \frac{4!}{(4-2)!2!} = \frac{4!}{2!2!} = \frac{4 \cdot 3 \cdot \cancel{2!}}{2 \cdot 1 \cdot \cancel{2!}} = 6$$

This matches our result from Example 1.

▼ Try This One 3

How many combinations of eight objects are there taken five at a time?

EXAMPLE 4 An Application of Combinations

Math Note

Some people use the terminology "10 choose 3" to describe the combination ${}_{10}C_3$, and it is sometimes represented using the notation $\binom{10}{3}$.

While studying abroad one semester, Tran is required to visit 10 different cities. He plans to visit 3 of the 10 over a long weekend. How many different ways can he choose the 3 to visit? Assume that distance is not a factor.

SOLUTION

The problem doesn't say anything about the order in which they'll be visited, so this is a combination problem.

$$ {}_{10}C_3 = \frac{10!}{(10-3)!3!} = \frac{10!}{7!3!} = 120 $$

▼ Try This One 4

☑ 2. Find the number of combinations of n objects taken r at a time.

An instructor posts a list of eight group projects to her website. Every group is required to do four projects at some point during the semester. How many different ways can a group choose the four projects they want to do?

In some cases, the combination rule is used in conjunction with the fundamental counting principle. Examples 5 and 6 illustrate some specific situations.

EXAMPLE 5 Choosing a Committee

At one school, the student government consists of seven women and five men. How many different committees can be chosen with three women and two men?

SOLUTION

First, we will choose three women from the seven candidates. This can be done in ${}_7C_3 = 35$ ways. Then we will choose two men from the five candidates in ${}_5C_2 = 10$ ways. Using the fundamental counting principle, there are $35 \cdot 10 = 350$ possible committees.

▼ Try This One 5

On an exam, a student must select 2 essay questions from 6 essay questions and 10 multiple choice questions from 20 multiple choice questions to answer. How many different ways can the student select questions to answer?

EXAMPLE 6 Designing a Calendar

To raise money for a charity event, a sorority plans to sell a calendar featuring tasteful pictures of some of the more attractive professors on campus. They will need to choose six models from a pool of finalists that includes nine women and six men. How many possible choices are there if they want to feature at least four women?

SOLUTION

Since we need to include at least four women, there are three possible compositions: four women and two men, five women and one man, or six women and no men.

Four women and two men:

$${}_9C_4 \cdot {}_6C_2 = \frac{9!}{(9-4)!\,4!} \cdot \frac{6!}{(6-2)!\,2!} = 126 \cdot 15 = 1{,}890$$

Five women and one man:

$${}_9C_5 \cdot {}_6C_1 = \frac{9!}{(9-5)!\,5!} \cdot \frac{6!}{(6-1)!\,1!} = 126 \cdot 6 = 756$$

Six women and no men:

$${}_9C_6 \cdot {}_6C_0 = \frac{9!}{(9-6)!\,6!} \cdot \frac{6!}{(6-0)!\,0!} = 84 \cdot 1 = 84$$

The total number of possibilities is 1,890 + 756 + 84 = 2,730.

▼ Try This One 6

☑ 3. Use the combination rule in conjunction with the fundamental counting principle.

A four-person crew for the international space station is to be chosen from a candidate pool of 10 Americans and 12 Russians. How many different crews are possible if there must be at least two Russians?

Table 11-1 summarizes all of the counting rules from sections 11-1 and 11-2. It's important to know the formulas, but it's far more important to understand the situations that each formula is needed for.

TABLE 11-1 Summary of Counting Rules

Rule	Definition	Formula
Fundamental counting principle	The number of ways a sequence of n events can occur if the first event can occur in k_1 ways, the second event can occur in k_2 ways, etc. (Events are unaffected by the others.)	$k_1 \cdot k_2 \cdot k_3 \cdots k_n$
Permutation rule	The number or permutations of n objects taking r objects at a time. (Order is important.)	$\frac{n!}{(n-r)!}$
Permutation rule for duplicate objects	The number of permutations in which k_1 objects are alike, k_2 objects are alike, etc.	$\frac{n!}{k_1!\,k_2!, \cdots, k_p!}$
Combination rule	The number of combinations of r objects taken from n objects. (Order is not important.)	$\frac{n!}{(n-r)!r!}$

Answers to Try This One

1 Call the classes A, B, and C:

Permutations
A B B A A C
C A B C C B

Combinations
A B A C B C

2 (a) Permutation
(b) Combination

3 56

4 70

5 2,771,340

6 5,665

EXERCISE SET 11-2

Writing Exercises

1. What is meant by the term combination?
2. What is the difference between a permutation and a combination?
3. Describe a real-life situation in which it would be appropriate to use combinations to count possibilities.
4. Describe a situation related to the one in Exercise 3 in which it would be appropriate to use permutations to count possibilities.

Computational Exercises

For Exercises 5–14, evaluate each expression.

5. ${}_5C_2$
6. ${}_8C_3$
7. ${}_7C_4$
8. ${}_6C_2$
9. ${}_6C_4$
10. ${}_3C_0$
11. ${}_3C_3$
12. ${}_9C_7$
13. ${}_{12}C_2$
14. ${}_4C_3$

For Exercises 15–20, find both the number of combinations and the number of permutations for the given number of objects.

15. 8 objects taken 5 at a time
16. 5 objects taken 3 at a time
17. 6 objects taken 2 at a time
18. 10 objects taken 6 at a time
19. 9 objects taken 9 at a time
20. 12 objects taken 1 at a time

For Exercises 21–28, decide whether the selection described is a combination or a permutation.

21. Ten fans at a concert are chosen to go backstage after the show.
22. From a list of 20 dishes he knows how to cook, Maurice chooses different dishes for breakfast, lunch, and dinner on his girlfriend's birthday.
23. A state elects a governor and lieutenant governor from a pool of eight candidates.
24. A state elects two senators from a pool of 12 candidates.
25. Lupe chooses an eight-letter password from the letters of the alphabet.
26. Of the six optional community service projects in a service learning course, Haylee picks three of them.
27. When looking for a new car, you read about ten different models and choose four that you would like to test drive.
28. Mark looks over the novels on his bookshelf and lists his five favorites ranked 1 through 5.

Real-World Applications

29. How many different ways can five cards be selected from a standard deck of 52 cards?
30. How many ways are there to select three math help websites from a list that contains six different websites?
31. How many ways can a student select five questions from an exam containing nine questions? How many ways are there if he must answer the first question and the last question?
32. How many ways can four finalists for a job be selected from ten interviewees?
33. If a person can select three presents from 10 presents under a Christmas tree, how many different combinations are there?
34. How many different possible tests can be made from a test bank of 20 questions if the test consists of 5 questions? (Ignore the order of questions.)
35. The general manager of a fast-food restaurant chain must select 6 restaurants from 11 for a promotional program. How many different possible ways can this selection be done?
36. How many ways can 3 cars and 4 trucks be selected from 8 cars and 11 trucks to be tested for a safety inspection?
37. During the tryouts for the school jazz band, there were four trumpet players, twelve guitar players, and seven saxophonists. How many ways can the jazz band be chosen so there are two trumpet players, five guitarists, and three saxophonists?
38. There are seven men and five women in line at a Salsa dance club. The bouncer can only admit two more men and two more women. How many ways can he choose from those in line? How many ways

can he choose if instead he is told he can admit four people and at least two must be women?

39. Coca-Cola comes in two low-calorie varieties: Diet Coke and Coke Zero. If a promoter has 10 cans of each, how many ways can she select 3 cans of each for a taste test at the local mall?
40. At the movies, Shana wants to get snacks for her friends. How many ways can she select three types of candy and two types of soda from the eight types of candy and five types of soda available?
41. Steve wants to download new music into his iPod from iTunes. How many ways can Steve select two rock songs, three alternative songs, and three rap songs from a list of eight rock songs, six alternative songs, and ten rap songs?
42. How many ways can 2 men and 2 women be selected for a debate tournament if there are 10 male finalists and 12 female finalists?
43. The state narcotics bureau must form a 5-member investigative team. If it has 25 agents to choose from, how many different possible teams can be formed?
44. How many different ways can a computer programmer select 3 jobs from a possible 15?
45. The Environmental Protection Agency must investigate nine nuclear reactors for complaints of radioactive pollution. How many different ways can a representative select five of these to investigate this week?
46. How many ways can a person select 8 DVDs from 10 DVDs?
47. How many ways can 20 students be chosen for the express line at registration if there are 30 students to choose from?
48. An advertising manager decides to have an ad campaign in which eight special items will be hidden at various locations in a shopping mall. If he has 17 locations to pick from, how many different possible combinations can he choose?

Section 11-3 Basic Concepts of Probability

LEARNING OBJECTIVES

- ☐ 1. Compute classical probabilities.
- ☐ 2. Compute empirical probabilities.

Walking into a casino without knowing anything about probability is kind of like going to a stick fight without a stick—you're likely to take a beating. Casinos aren't in the business of losing money, and the games are designed so that most people lose more than they win. But an understanding of what is likely to happen in a given situation can give you an advantage over other players, giving you a better chance of walking out the door with some cash in your pockets.

The study of probability originated in an effort to understand games of chance, like those that use coins, dice, and playing cards. Generally speaking, probability is simply a number that describes how likely an event is to occur. We will use games of chance to illustrate the ideas, but will eventually see that probability has many applications beyond simple games. In this section, we'll examine the basic concepts involved in studying probability.

Sample Spaces

Processes such as flipping a coin, rolling a die, or drawing a card from a deck are called *probability experiments*.

A **probability experiment** is a process that leads to well-defined results called outcomes. An **outcome** is the result of a single trial of a probability experiment.

Some examples of a trial are flipping a coin once, rolling a single die, and drawing one card from a deck. When a coin is tossed, there are two possible outcomes: heads or tails. When rolling a single die, there are six possible outcomes: 1, 2, 3, 4, 5, or 6.

In a probability experiment, we can predict what outcomes are possible, but we cannot predict with certainty which one will occur. We say that the outcomes occur at random. In any experiment, the set of all possible outcomes is called the *sample space*.

A **sample space** is the set of all possible outcomes of a probability experiment.

Some sample spaces for various probability experiments are shown here.

Experiment	**Sample Space**
Toss one coin	{head, tail}
Roll a die	{1, 2, 3, 4, 5, 6}
Answer a true-false question	{true, false}
Toss two coins	{head/head, tail/tail, head/tail, tail/head}

Experiment: draw a card. Sample space: 52 cards. Event: drawing an ace.

It is important to realize that when two coins are tossed, there are *four* possible outcomes. Consider tossing a quarter and a dime at the same time. Both coins could fall heads up. Both coins could fall tails up. The quarter could fall heads up and the dime could fall tails up and, finally, the quarter could fall tails up and the dime could fall heads up. The situation is the same even if the coins are indistinguishable.

In finding probabilities, it is sometimes necessary to consider several outcomes of a probability experiment. For example, when a die is rolled, we may want to consider obtaining an odd number, i.e., 1, 3, or 5. Getting an odd number when rolling a die is an example of an event.

An **event** is any subset of the sample space of a probability experiment.

Classical Probability

Now we're ready to specifically define what is meant by probability. The first type we will study is called *classical probability,* because it was the first type of probability to be studied in the 17th and 18th centuries. In classical probability, we study all of the possible outcomes in a sample space and determine the probability, or likelihood, of an event occurring without actually performing experiments.

There is one key assumption we make in classical probability: that every outcome in a sample space is equally likely. For example, when a single die is rolled, we assume that each number is equally likely to come up. When a card is chosen from a deck of 52 cards, we assume that each card has the same probability of being drawn.

Math Note

Because classical probabilities are based on theory, not experiments, they are also called theoretical probabilities.

Formula for Classical Probability

Let E be an event that is a subset of the sample space S. We will write $n(E)$ to represent the number of outcomes in E, and $n(S)$ to represent the number of outcomes in S. The probability of the event E is defined to be

$$P(E) = \frac{n(E)}{n(S)} = \frac{\text{Number of outcomes in } E}{\text{Number of outcomes in } S}$$

In Example 1, we'll compute some simple probabilities using the formula above.

EXAMPLE 1 Computing Classical Probabilities

A single die is rolled. Find the probability of getting

(a) A 2.
(b) A number less than 5.
(c) An odd number.

SOLUTION

In this case, since the sample space is 1, 2, 3, 4, 5, and 6, there are six outcomes: $n(S) = 6$.

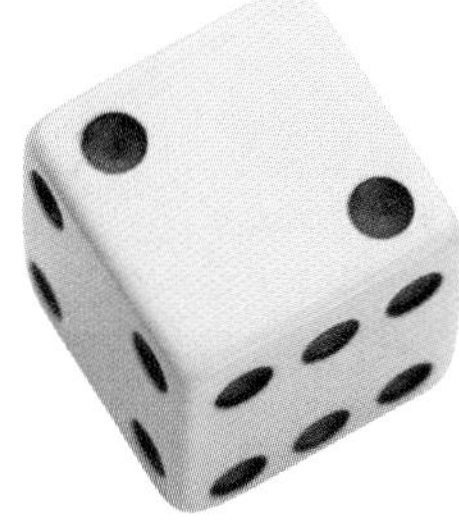

A die roll has six outcomes. If E = roll a 2, then $P(E) = \frac{1}{6}$. If E = roll an even number, then $P(E) = \frac{3}{6} = \frac{1}{2}$.

(a) There is one possible outcome that gives a 2, so $P(2) = \frac{1}{6}$.

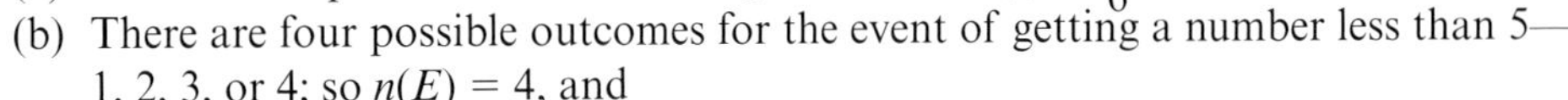

(b) There are four possible outcomes for the event of getting a number less than 5—1, 2, 3, or 4; so $n(E) = 4$, and

$$P(\text{a number less than 5}) = \frac{n(E)}{n(S)} = \frac{4}{6} = \frac{2}{3}$$

(c) There are three possible outcomes for the event of getting an odd number: 1, 3, or 5; so $n(E) = 3$, and

$$P(\text{odd number}) = \frac{n(E)}{n(S)} = \frac{3}{6} = \frac{1}{2}$$

▼ Try This One 1

Each number from one to twelve is written on a card and placed in a box. If a card is selected at random, find the probability that the number on the card is

(a) A 7.
(b) An odd number.
(c) A number less than four.
(d) A number greater than seven.

EXAMPLE 2 Computing Classical Probabilities

Two coins are tossed. Find the probability of getting

(a) Two heads.
(b) At least one head.
(c) At most one head.

Math Note

A good problem-solving strategy to use is to make a list of all possible outcomes in the sample space before computing the probabilities of events.

SOLUTION

The sample space is {HH, HT, TH, TT}; therefore, $n(S) = 4$.

(a) There is only one way to get two heads: HH. So

$$P(\text{two heads}) = \frac{n(E)}{n(S)} = \frac{1}{4}$$

(b) "At least one head" means one or more heads; i.e., one head or two heads. There are three ways to get at least one head: HT, TH, and HH. So $n(E) = 3$, and

$$P(\text{at least one head}) = \frac{n(E)}{n(S)} = \frac{3}{4}$$

(c) "At most one head" means no heads or one head: TT, TH, HT. So $n(E) = 3$, and

$$P(\text{at most one head}) = \frac{n(E)}{n(S)} = \frac{3}{4}$$

▼ Try This One 2

☑ 1. Compute classical probabilities.

Suppose that in a certain game, it is equally likely that you will win, lose, or tie. Find the probability of

(a) Losing twice in a row.
(b) Winning at least once in two tries.
(c) Having the same outcome twice in a row.

Math Note

The probability of getting tails when tossing a single coin could be given as any of ½, 0.5, or 50%.

At this point, we can make a series of simple but important observations about probabilities. Both $n(E)$ and $n(S)$ have to be zero or positive, so *probability is never negative*. Second, since S represents all possible outcomes, and E is a subset of S, $n(E)$ is always less than or equal to $n(S)$. This means that *probability is never greater than 1*. Probabilities are usually expressed as fractions or decimals between (and including) zero and one. But sometimes we will express probabilities in percent form when it seems appropriate for a given situation.

A 10% chance of rain is not the same as no chance of rain. You would expect it to rain on roughly 1 in 10 days when a 10% chance of rain was forecast.

When an event cannot possibly occur, the probability is zero, and *when an event is certain to occur, the probability is one*. For example, the probability of getting 8 when rolling a single die is zero because none of the outcomes in the sample space result in 8. The probability of getting either heads or tails when flipping a single coin is 1 because that particular event is satisfied by every outcome in the sample space.

The sum of the probabilities of all the outcomes in the sample space will always be one. For example, when a die is rolled, each of the six outcomes has a probability of $\frac{1}{6}$, and the sum of the probabilities of the six outcomes will be one.

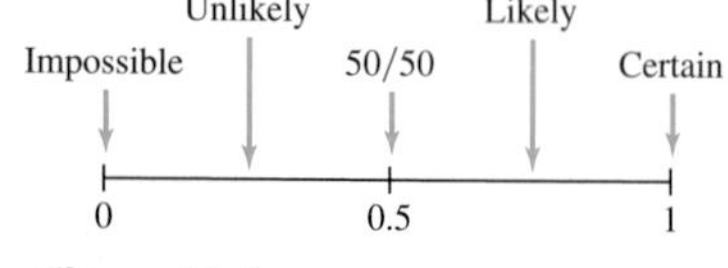

Figure 11-1

When the probability of an event is close to zero, the event is very unlikely to occur. When the probability of an event is close to $\frac{1}{2}$ or 0.5, the event has approximately a 50% chance of occurring. When the probability of an event is near one, the event is almost certain to occur. See Figure 11-1.

In addition to finding the probability that an event will occur, it is sometimes useful to find the probability that the event will *not* occur. For example, if a die is rolled, the probability that a 4 will not occur, symbolized by $P(4')$, is $\frac{5}{6}$ since there are five ways that a 4 will not occur; i.e., 1, 2, 3, 5, 6. The solution can also be found by finding the probability of getting a 4 and subtracting it from one; the sum of the probabilities of all outcomes in the sample space is one, so

$$P(4') = 1 - P(4) = 1 - \frac{1}{6} = \frac{5}{6}$$

The rule can be generalized as follows

For any event, E, $P(E') = 1 - P(E)$, where E' is the event "E does not occur."

Probability and Sets

The theory of probability is related to the theory of sets discussed in Chapter 2. For a given probability experiment, the sample space can be considered the universal set, and an event E can be considered as a subset of the universal set.

For example, when rolling a die, the sample space is $\{1, 2, 3, 4, 5, 6\}$; so the universal set is $U = \{1, 2, 3, 4, 5, 6\}$. Let the event E be getting an odd number; i.e., 1, 3, or 5. In sets, $E = \{1, 3, 5\}$. A Venn diagram can now be drawn illustrating this example. See Figure 11-2.

$E \cup E' = U$
$P(E) + P(E') = 1$

Figure 11-2

Notice that set E contains 1, 3, and 5 while the numbers 2, 4, and 6 are in the universal set but not in E. So $E' = \{2, 4, 6\}$. Now recall from set theory that $E \cup E' = U$. As stated previously, the sum of the probabilities of the outcomes in the sample space is one, so $P(U) = 1$. E' represents the elements in U but not in E, so $P(E) + P(E') = 1$. Subtracting $P(E)$ from both sides, we get $P(E') = 1 - P(E)$. Additional relationships between probability theory and set theory will be shown in other sections of this chapter.

Empirical Probability

The second type of probability we will study is computed using experimental data, rather than counting equally likely outcomes. For example, suppose 100 games into the season, your favorite baseball team has won 60 games and lost 40. You might reasonably guess that since they've won 60 of their 100 games so far, the probability of them winning any given game is about 60/100, or 0.6. This type of probability is called **empirical probability,** and is based on *observed frequencies*—that is, the number of times a particular event has occurred out of a certain number of trials. In this case, the observed frequency of wins is 60, the observed frequency of losses is 40, and the total number of trials is 60 + 40 = 100.

Sidelight YOU BET YOUR LIFE!

You probably think of gambling as betting money at a casino or on a sporting event, but people gamble all the time in many ways. In fact, people bet their lives every day by engaging in unhealthy activities like smoking, using drugs, eating a high-fat diet, and even driving too fast. Maybe people don't care about the risks involved in these activities because they don't understand the concept of probability. On the other hand, people tend to fear things that are far less likely to harm them, like flying, because the occasional negative consequence is sensationalized in the press.

In his book *Probabilities in Everyday Life* (Ivy Books, 1986), author John D. McGrevey states:

> When people have been asked to estimate the frequency of death from various causes, the most overestimated causes are those involving pregnancy, tornados, floods, fire, and homicide. The most underestimated categories include death from diseases such as diabetes, stroke, tuberculosis, asthma, and stomach cancer (although cancer in general is overestimated).

Which do you think is safer: flying across the United States on a commercial airline, or driving cross country? According to our friend McGrevey, the probability of being killed on any given airline flight is about 1/1,000,000, while the probability of being killed on a transcontinental automobile trip is just 1/8,000. That means that driving across the country is 125 times more dangerous than flying!

Math Note

The total number of trials is the sum of all observed frequencies.

Formula for Empirical Probability

$$P(E) = \frac{\text{Observed frequency of the specific event } (f)}{\text{Total number of trials } (n)} = \frac{f}{n}$$

In this coin toss, the empirical probability of heads was $\frac{6}{10}$, or $\frac{3}{5}$. With more tosses, you would expect $P(\text{heads})$ to approach $\frac{1}{2}$.

The information in the baseball problem can be written in the form of a *frequency distribution* that consists of classes and frequencies for the classes, as shown below:

Result (Class)	Observed Frequency
Win	60
Lose	40
Total	100

This technique is often helpful in working out empirical probabilities.

EXAMPLE 3 Computing an Empirical Probability

In a sample of 50 people, 21 had type O blood, 22 had type A blood, 5 had type B blood, and 2 had type AB blood. Set up a frequency distribution and find these probabilities for a person selected at random from the sample.

(a) The person has type O blood.
(b) The person has type A or type B blood.
(c) The person has neither type A nor type O blood.
(d) The person does not have type AB blood.

Source: Based on American Red Cross figures

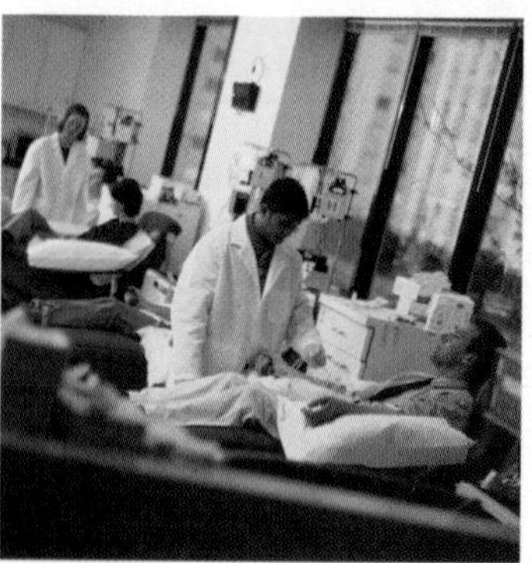

Type O negative blood is the least likely to react badly with a patient's blood. If it is not possible to test the patient's blood before a transfusion, then O negative is used.

SOLUTION

Type (class)	Observed Frequency
A	22
B	5
AB	2
O	21
Total	50

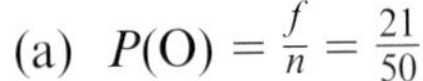

(a) $P(\text{O}) = \frac{f}{n} = \frac{21}{50}$
(b) The frequency of A or B is $22 + 5 = 27$. $P(\text{A or B}) = \frac{27}{50}$
(c) Neither A nor O means either B or AB; the frequency of either B or AB is $5 + 2 = 7$. $P(\text{neither A nor O}) = \frac{7}{50}$
(d) The probability of not AB is found by subtracting the probability of AB from 1.

$$P(\text{not AB}) = 1 - P(\text{AB}) = 1 - \frac{2}{50} = \frac{48}{50} = \frac{24}{25}$$

Math Note

The solution to part c could also be done as $P(\text{neither A nor O}) = 1 - P(\text{A or O})$:

$$1 - \left(\frac{22}{50} + \frac{21}{50}\right) = 1 - \frac{43}{50} = \frac{7}{50}.$$

☑ 2. Compute empirical probabilities.

▼ Try This One 3

A bag of Hershey's Assorted Miniatures contains 18 Hershey Milk Chocolate bars, 9 Mr. Goodbars, 9 Krackel bars, and 8 Hershey's Special Dark Chocolate bars. If a bar is selected at random from the bag, find the probability that it is

(a) A Mr. Goodbar.
(b) A Krackel or a Special Dark Chocolate bar.
(c) Not a Milk Chocolate bar.

It is important to understand the relationship between classical probability and empirical probability in certain situations. In classical probability, the probability of rolling a 3 when a die is thrown is found by looking at the sample space, and is $\frac{1}{6}$. To find the probability of getting a 3 when a die is thrown using empirical probability, you would actually toss a die a specific number of times and count the number of times a 3 was obtained; then divide that number by the number of times the die was rolled. For example, suppose a die was rolled 60 times, and a 3 occurred 12 times. Then the empirical probability of getting a 3 would be $\frac{12}{60} = \frac{1}{5}$. Most of the time, the probability obtained from empirical methods will differ from that obtained using classical probability. The question is, then, "How many times should I roll the die when using empirical probability?" There is no specific answer except to say that the more times the die is tossed, the closer the results obtained from empirical probability will be to those of classical probability.

In summary, then, classical probability uses sample spaces and assumes the outcomes are equally likely. Empirical probability uses observed frequencies and the total of the number of frequencies.

Sidelight PROBABILITY AND YOUR FEARS

All of us at one time or another have thought about dying. Some people have fears of dying in a plane crash or dying from a heart attack. In the sidelight on page 591, it was explained that it is safer to fly across the United States than to drive. Statisticians who work for insurance companies (called actuaries) also calculate probabilities for dying from other causes. For example, based on deaths in the United States, the risks of dying from various other causes are shown.

Motor vehicle accident	1 in 7,000
Shot by a gun	1 in 10,000
Accident while walking across the street	1 in 60,000
Lightning strike	1 in 3 million
Shark attack	1 in 100 million

The death risk for various diseases is much higher as shown.

Heart attack	1 in 400
Cancer	1 in 600
Stroke	1 in 2,000

As you can see, the chances of dying from diseases are much higher than dying from accidents.

Answers to Try This One

1 (a) 1/12 (b) 1/2 (c) 1/4 (d) 5/12

2 (a) 1/9 (b) 5/9 (c) 1/3

3 (a) 9/44 (b) 17/44 (c) 13/22

EXERCISE SET 11-3

Writing Exercises

1. Define in your own words what the probability of an event means.
2. What is a sample space?
3. What is the difference between an outcome and an event?
4. What is the range of numbers that can represent probabilities? Why?
5. What is the probability of an event that can't occur? Explain why.
6. What is the probability of an event that is certain to occur? Explain why.
7. Explain the difference between classical and empirical probability.
8. Describe how to find the empirical probability of an event after conducting an experiment.

Computational Exercises

For Exercises 9–16, decide whether or not the given number could represent a probability.

9. $\frac{3}{4}$
10. 0.75
11. $-\frac{1}{2}$
12. 0
13. $\frac{41}{40}$
14. $\frac{40}{41}$
15. 72%
16. 111%

For Exercises 17–20, decide whether the probability described is classical or empirical.

17. At one school, 59% of the students having lunch in the union are women, so the probability of a randomly selected student from the campus phone directory being male is 0.41.
18. A pool table has 15 balls labeled 1 through 15. The probability of a ball made on the break having a number less than 6 is $\frac{1}{3}$.
19. The probability of a randomly selected state beginning with the letter A is $\frac{2}{25}$.
20. While at a casino, Catalina won 10 of the first 15 hands of blackjack she played, so she has a $\frac{2}{3}$ chance of winning the next hand.

Real-World Applications

21. If a die is rolled one time, find the probability of
    (a) Getting a 4.
    (b) Getting an even number.
    (c) Getting a number greater than 4.
    (d) Getting a number less than 7.
    (e) Getting a number greater than 0.
    (f) Getting a number greater than 3 or an odd number.
    (g) Getting a number greater than 3 and an odd number.
22. A couple has two children. Find the probability that
    (a) Both children are girls.
    (b) At least one child is a girl.
    (c) Both children are of the same gender.
23. On the *Price Is Right* game show, a contestant spins a wheel with numbers 1 through 7, with equally sized regions for each of these numbers. If the contestant spins once, what is the probability that the number is
    (a) A 6.
    (b) An even number.
    (c) A number greater than 4.
    (d) A number less than 8.
    (e) A number greater than 7.

24. A list contains the names of five anthropology students, two sociology students, and three psychology students. If one name is selected at random to assist in the professor's new study, find the probability that the chosen student is
    (a) An anthropology student.
    (b) A psychology student.
    (c) An anthropology student or a sociology student.
    (d) Not a psychology student.
    (e) Not an anthropology student.
25. On the shelf at the gaming store, there are five Sony Playstations and four Nintendo Wii consoles left. If one gaming system is selected at random, find the probability that the system is a Wii console.
26. If there are only 50 lottery tickets for the Big Game, one of which is a winning ticket, and you buy 7 of those tickets at random, what is the probability that you will win the super jackpot?
27. In a math class of seven women and nine men, if one person is selected at random to come to the board to show the solution to a problem, what is the probability that the student is a man?
28. A recent survey reported that 67% of Americans approve of human embryonic stem cell research. If an American is selected at random, find the probability that he or she will disagree or have no opinion on the issue.
29. The Federal Bureau of Investigation reported that in 2007, there were 7,624 incidents of hate crimes that had a bias motivation based on race, religion, sexual orientation, nationality, or disability. Of these incidents, 6,965 were committed by a "known offender," where one or more attributes is known about the offender. If one incident is selected at random, find the probability that the incident was committed by a known offender.
30. Thirty-nine of 50 states are currently under court order to alleviate overcrowding and poor conditions in one or more of their prisons. If a state is selected at random, find the probability that it is currently under such a court order.

    Source: *Harper's Index*
31. In a survey, 16% of male college students said they lied to get a woman to go out on a date with them. If a male college student is chosen at random, find the probability that he does not lie to get a date with a woman.
32. In the lost and found box at the campus security office, there were nine BlackBerry SmartPhones and seven Apple iPhones. If a phone is selected at random, find the probability that it is
    (a) A BlackBerry SmartPhone.
    (b) An Apple iPhone.

Exercises 33–35 refer to a standard deck of playing cards. If you are unfamiliar with playing cards, see the description on page 600.

33. During a game of Texas Hold'em poker, each of four players is dealt two cards, then the dealer "burns" a card (puts it face down), then deals the "flop" (three cards face up). He then burns another card, then flips over the "turn" card (one card face up). One player needs a spade on the "turn" to make a flush. No one else has a spade, and he has two in his hand and there are two on the flop. If neither of the burn cards are spades, what's the probability the turn card will be a spade?
34. During a game of Gin Rummy, Sven needs the eight of diamonds to make a straight in his hand. He and the other player have been dealt 10 cards each, the other player does not have the card he wants, and all other cards are in the deck. What is the probability that the next card picked from the deck is the eight of diamonds?
35. During a game of Blackjack, three players are dealt two cards each, and the dealer has two cards. No one has a card that is worth 10 or 11, which would be a 10, a face card, or an ace. What is the probability that the next player dealt a card would get a card worth 10 or 11?
36. A survey on campus revealed that 68% of the students felt that a new attendance policy was unfair. If a student is randomly asked to give an opinion of the new attendance policy, find the probability that the student will either think it's fair or have no opinion.
37. A survey of 25 students in line during registration revealed that 3 were math majors, 10 were history majors, 2 were psychology majors, 7 were biology majors, and the rest were undecided. If the clerk calls a name from the same line of students at random, find the probability the student would be either a history or biology major.
38. On a bookshelf, there are five Tess Gerritsen novels, three Stephen King novels, and seven John Grisham novels. If Sheri selects one while turned away to talk to her friend, find the probability she chose a Tess Gerritsen novel.
39. In a class of 35 students, 22 passed the first exam. If a student is chosen at random from the class, find the probability he failed the first exam.
40. On her way out the door to class, Li dumped five energy bars and three candy bars of the same size into her bag. During class, she got hungry, and she grabbed a bar without looking. It was an energy bar. Her friend then asked for something to eat, and she grabbed another bar without looking. Find the probability that she grabbed a candy bar.
41. On a 10-question true/false test, there are seven false questions and the rest are true. If Marcus answered the first eight questions correctly, and five of them were false, find the probability that when he answers true for the next question, his answer will be correct.

42. According to www.namestatistics.com, the five most popular male names and their percentages are as follows:

Name	Percentage
James	3.318%
John	3.271%
Robert	3.143%
Michael	2.629%
William	2.451%

(a) If Mary meets a man at a party, find the probability his name is one of the most popular five male names in the country.
(b) If Jane goes to the grocery store and the clerk is a man, find the probability his name would be John or Robert.
(c) If Bob and Sue rent a new apartment and the landlord is a man, find the probability his name would be in the top three most popular male names.

43. Many people blame Wall Street greed for causing the economic crisis that gripped the nation in late 2008. In February 2009, the Harris Poll surveyed 1,010 Americans, using the statement that people on Wall Street are "as honest and moral as other people." The number giving each response is summarized in the table below:

Agree	707
Disagree	263
Not sure	40

Based on these results, if you ask a randomly selected American this question,
(a) What is the probability that a randomly selected American thinks that people on Wall Street are less honest than other people?
(b) What is the probability that he or she either agrees or disagrees?

44. In December 2008, 500 men and 500 women were surveyed by Omnitel about their opinion on whether federal government bailout money should be used to help homeowners in default. The number giving each response is shown below.

Response	Men	Women
Yes	195	235
No	290	220
Not sure	15	45

(a) If a person who participated in the survey is selected at random, what is the probability that he or she answered no?
(b) What is the probability that the person is a man who answered either yes or no?
(c) Based on the data from the survey, if you had stopped a random woman on the street in December 2008 and asked her opinion, what is the probability that she would have said that bailout money should not be used to help homeowners in default?

45. In a survey conducted by Bank of America, college graduates were asked how much money they typically donate to their alma mater each year. The responses are summarized below:

Nothing	58%
Something, but less than $500	32%
$500 or more	10%

Based on these results:
(a) What is the probability that a randomly selected college graduate gives at least something in a typical year?
(b) What is the probability that a randomly selected college graduate gives less than $500 in a typical year?

46. Jockey International surveyed men to find out how old their oldest pair of underwear is. The results are summarized below:

Less than 1 year	17%
1–4 years	59%
5–9 years	15%
10–19 years	7%
20 or more years	2%

Based on these results:
(a) What is the probability that a randomly selected man has a pair of underwear that is older then 4 years? 9 years?
(b) What is the probability that a randomly selected man has no pair of underwear more than a year old?
(c) What is the probability that a randomly selected man has no underwear more than 9 years old?

47. A student passed a fountain every day on the way to class and tossed in his good luck penny with a wish. He then retrieved it because he couldn't bear to part with the penny. Out of 100 consecutive tosses of this same penny, he noticed the penny landed on tails 73 times. Can he conclude his good luck penny is unbalanced?

48. A professor stated that with a certain method of teaching statistics, a student has about a 50% chance of passing her class. She justified this by saying, "Either a student will pass or he won't pass." Comment on this statement.

49. The students at a university are classified by a 0 for freshman, a 1 for sophomores, a 2 for juniors, a 3 for seniors, and a 4 for graduate students. There are two extra scholarships to assign, so an administrator randomly selects from a box with only the numbers 0, 1, 2, 3, and 4 to choose the class of the first recipient. She then puts the number back into the box and randomly selects a number for the class of the second recipient. Find the sample space, and then find the probability of the following events:
(a) An odd number is chosen first and an even number is chosen second. (*Note:* 0 is considered an even number.)

(b) The sum of the two numbers selected is greater than 4.
(c) For both selections, an even number is drawn.
(d) The sum of the two numbers selected is odd.
(e) The same number is drawn twice.

50. Only six students attended a school charity event, so each of their names was placed into three boxes for the three raffles for the event. What is the probability that the same student's name will be drawn from each box?

Section 11-4 Tree Diagrams, Tables, and Sample Spaces

LEARNING OBJECTIVES

- ❑ 1. Use tree diagrams to find sample spaces and compute probabilities.
- ❑ 2. Use tables to find sample spaces and compute probabilities.

For centuries, people have tried a wide variety of techniques, some of them pretty bizarre, to try and influence the gender of their children. The truth is, without the aid of cutting-edge science, you don't get to choose. But that doesn't stop many young couples from planning the type of family they hope to have. Suppose that one couple would like to have three children, but they definitely want to have at least one boy and one girl. What is the probability that they'll get their wish without having to go beyond three kids?

When working with classical probabilities, we know that we need to decide on the sample space for an event, and then find how many individual outcomes are in that event. When situations start to get complicated, it might not always be apparent how to do so. That's where tree diagrams and tables can help.

Tree Diagrams

A **tree diagram** consists of branches corresponding to the outcomes of two or more probability experiments that are done in sequence.

Math Note

Recall that we used tree diagrams to illustrate the fundamental counting principle back in Section 11-1.

When constructing a tree diagram, use branches emanating from a single point to show the outcomes for the first experiment, and then show the outcomes for the second experiment using branches emanating from each branch that was used for the first experiment, etc.

In Example 1, we'll use a tree diagram to find the sample space for our hopeful young couple.

EXAMPLE 1 Using a Tree Diagram to Find a Sample Space

Use a tree diagram to find the sample space for the genders of three children in a family.

SOLUTION

There are two possibilities for the first child, boy or girl, two for the second, boy or girl, and two for the third, boy or girl. So the tree diagram can be drawn as shown in Figure 11-3.

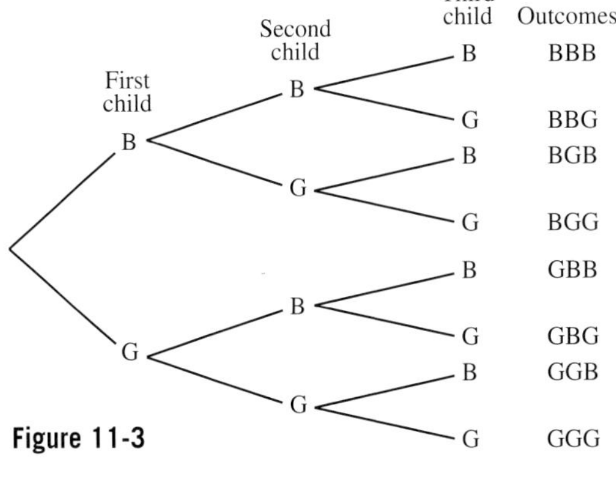

Figure 11-3

Math Note

The genders of the children are usually listed in their birth order. For example, the outcome GGB means the firstborn was a girl, the second a girl, and the third a boy.

After a tree diagram is drawn, the outcomes can be found by tracing through all of the branches. In this case, the sample space would be {BBB, BBG, BGB, BGG, GBB, GBG, GGB, GGG}.

▼ Try This One 1

A soda machine dispenses both Coke and Pepsi products, in both 12-ounce cans and 20-ounce bottles. For each brand, it has a regular cola, diet cola, and lemon-lime drink. Use a tree diagram to find the sample space for all the drinks dispensed.

Once a tree diagram is drawn and the sample space is found, you can compute the probabilities for various events.

EXAMPLE 2 Computing a Probability

If a family has three children, find the probability that all three children are the same gender; that is, all boys or all girls. (Assume that all outcomes are equally likely.)

SOLUTION

The sample space shown in Example 1 has eight outcomes, and there are two possible ways to have three children of the same gender, BBB or GGG. So, the probability of the three children being of the same gender is $\frac{2}{8}$ or $\frac{1}{4}$.

▼ Try This One 2

Suppose the soda machine from Try This One 1 goes berserk and starts dispensing drinks randomly. If you want a diet cola, what is the probability that you'll get one?

EXAMPLE 3 Using a Tree Diagram to Compute Probabilities

A coin is tossed, and then a die is rolled. Use a tree diagram to find the probability of getting heads on the coin and an even number on the die.

SOLUTION

First, we'll use a tree diagram to find the sample space. The coin will land on either heads or tails, and there are six outcomes for the die: 1, 2, 3, 4, 5, or 6. The tree diagram is shown in Figure 11-4.

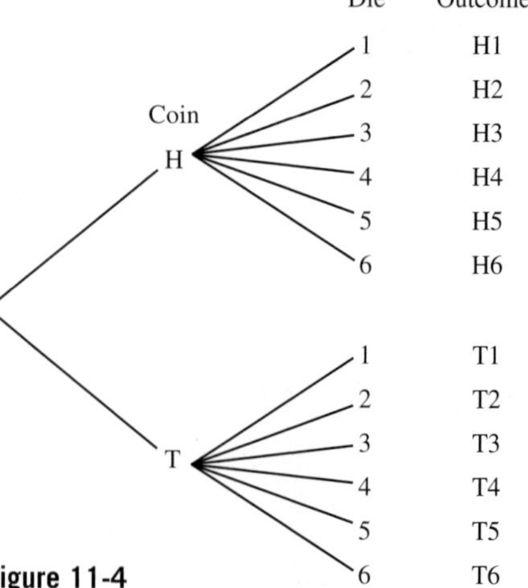

Figure 11-4

The sample space is {H1, H2, H3, H4, H5, H6, T1, T2, T3, T4, T5, T6}.

The total number of outcomes for the experiment is 12. The number of ways to get a head on the coin and an even number on the die is 3: H2, H4, or H6. So, the probability of getting a head and an even number when a coin is tossed and a die is rolled is $\frac{3}{12}$, or $\frac{1}{4}$.

▼ Try This One 3

☑ 1. Use tree diagrams to find sample spaces and compute probabilities.

In order to collect information for a student survey, a researcher classifies students according to eye color (blue, brown, green), gender (male, female), and class rank (freshman, sophomore). A folder for each classification is then made up (e.g., freshman/female/green eyes). Find the sample space for the folders using a tree diagram. If a folder is selected at random, find the probability that

(a) It includes students with blue eyes.
(b) It includes students who are female.
(c) It includes students who are male freshmen.

In constructing tree diagrams, not all branches have to be the same length. For example, suppose two players, Alice and Diego, play chess, and the first one to win two games wins the tournament. The tree diagram would be like the one shown in Figure 11-5. For any game, A means that Alice wins, and D means that Diego wins.

Notice that if Alice wins the first two games, the tournament is over. So the first branch is shorter than the second one. But if Alice wins the first game and Diego wins the second game, they need to play a third game in order to decide who wins the tournament. Similar reasoning can be applied to the rest of the branches.

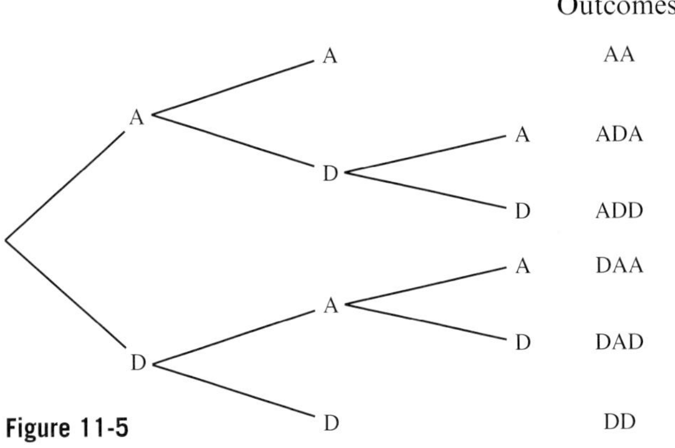

Figure 11-5

Sidelight MATH AND THE DRAFT LOTTERY

Today there are many state lotteries that offer patrons a chance to win money. At one time, there was a national lottery not to win money but to see who would be drafted for military service. During the Vietnam War, men were drafted based on their age and their birthday. The oldest men were drafted first. Men born in the last months of the year could avoid the draft since they were younger in months and days than those born earlier in the year.

In 1970, the United States decided to change the system and draft men based on a lottery. The days of the year were numbered one through 366 (leap year), placed in a barrel, and mixed up. Then numbers corresponding to birth dates were selected one at a time and men were drafted on the basis of their draft number.

It was assumed that the drawing was random. However, mathematicians found that the drawing was not random since those born later in the year had a higher chance of being drafted than those born earlier in the year. It was concluded that the capsules containing the numbers were not thoroughly mixed and that the numbers placed in the barrel last were on the top of the pile. They were selected first.

CAUTION

When the branches of a tree diagram are not equal in length, the outcomes are not equally likely even if the probability of every individual trial is $\frac{1}{2}$. In the tournament example above, if each player is equally likely to win any given game, the outcomes AA and DD both have probability $\frac{1}{4}$, while the outcomes ADA, ADD, DAA, and DAD all have probability $\frac{1}{8}$.

Tables

Another way of determining a sample space is by making a **table.** Consider the sample space of selecting a card from a standard deck of 52 cards. (The cards are assumed to be shuffled to make sure that the selection occurs at random.) There are four suits—hearts, diamonds, spades, and clubs, and 13 cards of each suit consisting of the denominations ace (A), 2, 3, 4, 5, 6, 7, 8, 9, 10, and 3 picture or face cards—jack (J), queen (Q), and king (K). The sample space is shown in Figure 11-6.

	A	2	3	4	5	6	7	8	9	10	J	Q	K
♥	A♥	2♥	3♥	4♥	5♥	6♥	7♥	8♥	9♥	10♥	J♥	Q♥	K♥
♦	A♦	2♦	3♦	4♦	5♦	6♦	7♦	8♦	9♦	10♦	J♦	Q♦	K♦
♠	A♠	2♠	3♠	4♠	5♠	6♠	7♠	8♠	9♠	10♠	J♠	Q♠	K♠
♣	A♣	2♣	3♣	4♣	5♣	6♣	7♣	8♣	9♣	10♣	J♣	Q♣	K♣

Figure 11-6

EXAMPLE 4 Using a Table to Compute Probabilities

A card is drawn from an ordinary deck. Use the sample space shown in Figure 11-6 to find the probabilities of getting

(a) A jack. (b) The 6 of clubs. (c) A 3 or a diamond.

SOLUTION

(a) There are four jacks and 52 possible outcomes, so

$$P(\text{jack}) = \frac{4}{52} = \frac{1}{13}$$

(b) Since there is only one 6 of clubs, the probability of getting a 6 of clubs is

$$P(\text{6 of clubs}) = \frac{1}{52}$$

(c) There are four 3s and 13 diamonds, but the 3 of diamonds is counted twice in this listing. So, there are 16 possibilities of drawing a 3 or a diamond, and

$$P(\text{3 or diamond}) = \frac{16}{52} = \frac{4}{13}$$

▼ Try This One 4

A single card is drawn at random from a well-shuffled deck. Using the sample space shown in Figure 11-6, find the probability that the card is

(a) An ace.
(b) A face card.
(c) A club.
(d) A 4 or a heart.
(e) A 6 and a spade.

Math Note

The sample space for rolling two dice is the same regardless of the color of the dice. Color helps distinguish between (2, 4) and (4, 2).

When two dice are rolled, how many outcomes are in the sample space? Assume for the purpose of illustration that one die is red and the other is blue. The sample space for the red die is {1, 2, 3, 4, 5, 6}, and the sample space for the blue die is also {1, 2, 3, 4, 5, 6}. Now if the outcomes for each die are combined, the sample space will consist of 36 outcomes, as shown in Figure 11-7. Note that the color indicates which die the number comes from.

Red die \ Blue die	1	2	3	4	5	6
1	(1, 1)	(1, 2)	(1, 3)	(1, 4)	(1, 5)	(1, 6)
2	(2, 1)	(2, 2)	(2, 3)	(2, 4)	(2, 5)	(2, 6)
3	(3, 1)	(3, 2)	(3, 3)	(3, 4)	(3, 5)	(3, 6)
4	(4, 1)	(4, 2)	(4, 3)	(4, 4)	(4, 5)	(4, 6)
5	(5, 1)	(5, 2)	(5, 3)	(5, 4)	(5, 5)	(5, 6)
6	(6, 1)	(6, 2)	(6, 3)	(6, 4)	(6, 5)	(6, 6)

Figure 11-7

The outcome (3, 6) means the 3 was obtained on the red die and a 6 was obtained on the blue die. The sum of the spots on the faces then would be $3 + 6 = 9$.

EXAMPLE 5 Using a Table to Compute Probabilities

When two dice are rolled, find the probability of getting

(a) A sum of 8.
(b) Doubles.
(c) A sum less than 5.

SOLUTION

Using the sample space shown in Figure 11-7, there are 36 possible outcomes.

(a) There are five ways to get a sum of 8: (2, 6), (3, 5), (4, 4), (5, 3), and (6, 2). So $n(E) = 5$, $n(S) = 36$, and

$$P(\text{sum of } 8) = \frac{n(E)}{n(S)} = \frac{5}{36}$$

(b) There are six ways to get doubles: (1, 1), (2, 2), (3, 3), (4, 4), (5, 5), and (6, 6). So $n(E) = 6$, $n(S) = 36$, and

$$P(\text{doubles}) = \frac{n(E)}{n(S)} = \frac{1}{6}$$

(c) A sum less than 5 means a sum of 4 or 3 or 2. The number of ways this can occur is 6, as shown.

Sum of 4: (1, 3), (2, 2), (3, 1)
Sum of 3: (1, 2), (2, 1)
Sum of 2: (1, 1)
$n(E) = 6$, $n(S) = 36$, and so

$$P(\text{sum less than } 5) = \frac{6}{36} = \frac{1}{6}$$

▼ Try This One 5

☑ 2. Use tables to find sample spaces and compute probabilities.

Two dice are rolled. Use the sample space shown in Figure 11-7 to find the probability of

(a) Getting a sum of 9.
(b) Getting a sum that is an even number.
(c) Getting a sum greater than 6.

Answers to Try This One

1 Sample Space: {Coke, can, cola; Coke, can, diet cola; Coke, can, lemon-lime; Coke, bottle, cola; Coke, bottle, diet cola; Coke, bottle, lemon-lime; Pepsi, can, cola; Pepsi, can, diet cola; Pepsi, can, lemon-lime; Pepsi, bottle, cola; Pepsi, bottle, diet cola; Pepsi, bottle, lemon-lime}.

2 $\frac{1}{3}$

3 Sample space: {blue, male, freshman; blue, male, sophomore; blue, female, freshman; blue, female, sophomore; brown, male, freshman; brown, male, sophomore; brown, female, freshman; brown, female, sophomore; green, male, freshman; green, male, sophomore; green, female, freshman; green, female, sophomore}.

(a) $\frac{1}{3}$ (b) $\frac{1}{2}$ (c) $\frac{1}{4}$

4 (a) $\frac{1}{13}$ (b) $\frac{3}{13}$ (c) $\frac{1}{4}$ (d) $\frac{4}{13}$ (e) $\frac{1}{52}$

5 (a) $\frac{1}{9}$ (b) $\frac{1}{2}$ (c) $\frac{7}{12}$

EXERCISE SET 11-4

Writing Exercises

1. Explain how to draw a tree diagram, and how tree diagrams help to find sample spaces.

2. Think of an example, other than the one provided in the section, where a tree diagram would have branches of different lengths.

Real-World Applications

3. There are three computers on a table that are either Dells or Gateways. Using a sample space similar to the one in Example 1, find the probability that
   (a) There will be exactly two Dells.
   (b) All three will be Gateways.
   (c) There will be at least one Dell.

4. There was a die and a quarter sitting on a table, and Klutzy Kramer knocked both off the table at the same time when he passed by. Using a sample space similar to Example 3, find the probability that as each landed on the floor,
   (a) There was a head on the coin and an odd number on the die.
   (b) There was a head on the coin and a prime number on the die.
   (c) There was a tail on the coin and a number less than 5 on the die.

5. After the incident in Exercise 4, the professor got smart and dropped the die and the quarter in a box to prevent Klutzy Kramer from knocking them onto the floor again. Using a sample space similar to Example 3, find the probability that as each landed in the box,
   (a) There was a tail on the coin and a number greater than 1 on the die.
   (b) There was a head on the coin and an even number on the die.
   (c) There was a tail on the coin and a number divisible by 3 on the die.

6. Sara came across a new website that featured a "live" fortune-teller. She typed in three questions and found (shockingly) that the fortune-teller could only answer yes or no to her questions. Draw a tree diagram for all possible answers for her three questions. Find the probability that
   (a) All answers will be yes or all answers will be no.
   (b) The answers will alternate (i.e., Yes–No–Yes or No–Yes–No).
   (c) Exactly two answers will be yes.

7. After a late night studying, Ebony decides to grab a latte before class so she can stay awake through the

lecture. She has only a one-dollar bill, a five-dollar bill, and a ten-dollar bill in her wallet. She pulls one out and looks at it, but then she puts it back. Distracted by a flyer for a new campus organization, she randomly hands a bill from her wallet to the clerk. Draw a tree diagram to determine the sample space and find the probability that

(a) Both bills have the same value.
(b) The second bill is larger than the first bill.
(c) The value of each of the two bills is even.
(d) The value of exactly one of the bills is odd.
(e) The sum of the values of both bills is less than $10.

8. When Frank threw his jeans into the washer after getting them muddy during an impromptu mud-wrestling match at the fraternity house, four coins fell out onto the floor. Find the probability that

(a) Exactly three coins land heads up.
(b) All coins land tails up.
(c) Two or more coins land heads up.
(d) No more than two coins land tails up.
(e) At least one coin lands tails up.

9. Mark and Raul play three rounds of Mortal Kombat. They are equal in ability. Draw a tree diagram to determine the sample space and find the probability that

(a) Either Mark or Raul win all three games.
(b) Either Mark or Raul win two out of three games.
(c) Mark wins only two games in a row.
(d) Raul wins the first game, loses the second game, and wins the third game.

10. Sofia is in line at the Simpsons ride at Universal Studios and notices a box containing a Bart keychain, a Homer keychain, a Marge keychain, and a Lisa keychain. She selects a keychain from the box and then puts it back. She then blindly selects another keychain from the box. Draw a tree diagram to determine the sample space and find the probability that

(a) Both key chains are the same character.
(b) The key chains are different characters.
(c) At least one is a Marge keychain.
(d) Both are Bart keychains.
(e) One keychain is Marge and the other is Homer.

11. The pool table in the student lounge is broken, and when Carrie puts her quarter into the slot, only the balls numbered 1 through 5 come down the chute. She selects one without looking and puts it on the table. She then selects another without looking. Draw a tree diagram to determine the sample space and find the probability that

(a) The sum of the numbers of the balls is odd.
(b) The number on the second ball is larger than the number on the first ball.
(c) The sum of the numbers on the balls is greater than 4.

12. Akira wants to purchase his first new car and can select one option from each category:

Model	Engine Type	Color
Ford Focus	Hybrid	Burnt copper
Honda Civic	E-85	Cobalt blue
Toyota Corolla		Metallic green

Draw a tree diagram and find the sample space for all possible choices. Find the probability that the car, if chosen at random,

(a) Has an E-85 engine.
(b) Is a burnt copper Ford Focus.
(c) Is a metallic green hybrid Honda Civic.
(d) Is cobalt blue.
(e) Is either a metallic green or cobalt blue Toyota Corolla.

13. Kimberly goes onto Dell's website to order a custom-made Tablet PC. She can select one option from each category:

Operating System	Processor	Wireless Options
Windows Vista Ultimate	Intel Core Duo	Dell Wireless Mini Card
Windows Vista Home Premium	Intel Celeron 540	Bluetooth 2.0
Windows Vista Home Basic		

Draw a tree diagram to determine the sample space and find the probability that her Tablet PC, if chosen at random, will have

(a) An Intel Celeron 540 Processor.
(b) A Windows Vista Home Premium operating system and an Intel Core 2 Duo Processor.
(c) An Intel Core 2 Duo Processor with a Bluetooth 2.0 wireless option.

14. To choose the order of bands for the finals of the campus Battle of the Bands competition, Freddy puts a penny, a nickel, a dime, a quarter, and a half-dollar into five separate envelopes and has one band choose an envelope. Another band then chooses from the envelopes remaining. Draw a tree diagram to determine the sample space and find the probability that

(a) The amount of the first coin is less than the amount of the second coin.
(b) Neither coin is a quarter.
(c) One coin is a penny and the other coin is a nickel or a dime.
(d) The sum of the amounts of both coins is even.
(e) The sum of the amounts of both coins is less than $0.40.

15. At the beginning of a magic trick, The Great Mancini shuffles an ordinary deck of 52 cards as shown in Figure 11-6, and has the nearest person in the audience draw a single card. Using the sample space for drawing a single card from this deck, find the probability that the contestant got
    (a) A 10.
    (b) A club.
    (c) The ace of hearts.
    (d) A 3 or a 5.
    (e) A 6 or a spade.
    (f) A queen or a club.
    (g) A diamond or a club.
    (h) A red king.
    (i) A black card or an 8.
    (j) A red 10.
16. After a rowdy game of 52 pickup, Lauren cleans up the mess of 52 cards from a standard deck from the floor. Use the sample space for drawing a single card from an ordinary deck as shown in Figure 11-6 to find the probability that the first card Lauren picked up was
    (a) The 6 of clubs.
    (b) A black card.
    (c) A queen.
    (d) A black 10.
    (e) A red card or a 3.
    (f) A club and a 6.
    (g) A 2 or an ace.
    (h) A club, diamond, or spade.
    (i) A diamond face card.
    (j) A red ace.
17. In between classes, Jade plays a game of online Monopoly on her laptop. Using the sample space for rolling two dice shown in Figure 11-7, find the probability that when Jade rolls the two dice, she gets a
    (a) Sum of 5.
    (b) Sum of 7 or 11.
    (c) Sum greater than 9.
    (d) Sum less than or equal to 5.
    (e) Three on one die or on both dice.
    (f) Sum that is odd.
    (g) A prime number on one or both dice.
    (h) A sum greater than 1.
18. During a charity Las Vegas Casino Night, Rosie plays craps and gets to roll the dice. Using the sample space for rolling two dice as shown in Figure 11-7, find the probability she rolled a
    (a) Sum of 8.
    (b) Sum that is prime.
    (c) Five on one or both dice.
    (d) Sum greater than or equal to 7.
    (e) Sum that is less than 3.
    (f) Sum greater than 12.
    (g) Six on one die and 3 on the other die.

Critical Thinking

19. An online math quiz made up of true-false questions starts every student out with two questions. If they get both wrong, they're done. If they get at least one right, they get a third question. Suppose that a student guesses on all questions. Draw a tree diagram and find the sample space for all possible results. Then explain why every outcome is not equally likely.
20. To get past security at a club, potential patrons must answer a series of four questions. If they get any of the first three wrong, they get no more questions and don't get in. Draw a tree diagram and find the sample space for all possible results. Then explain why all of the outcomes are not equally likely.
21. Consider the sample space when three dice are rolled. How many different outcomes would there be?
22. When three dice are rolled, how many ways can a sum of 6 be obtained?
23. When three dice are rolled, find the probability of getting a sum of 6.

Section 11-5 Probability Using Permutations and Combinations

Sometimes a friendly game of poker can become less friendly when someone seems just a bit too lucky. Suppose one player in such a game gets dealt all four aces in one hand. Would you suspect that something fishy was going on? What is the probability of that happening?

In Section 11-4, we used tree diagrams and tables to find sample spaces and the number of outcomes in certain events. This was a pretty good strategy, but when the number of possibilities gets larger, diagrams can get out of hand. Fortunately, we know about counting techniques that are tailor-made for answering questions about

probability. If our job is to find how many ways something can happen, the combination and permutation rules from Section 11-2 will be our best friends.

Our general game plan will be to use these rules to find the number of outcomes that satisfy a certain event, as well as the total number of outcomes in the sample space. Then we can divide the first number by the second to obtain the probability of the event occurring.

LEARNING OBJECTIVES

- ☐ 1. Compute probabilities using combinations.
- ☐ 2. Compute probabilities using permutations.

EXAMPLE 1 Using Combinations to Compute Probability

Stacy has the option of selecting three books to read for a humanities course. The suggested book list consists of 10 biographies and five current events books. She decides to select the three books at random. Find the probability that all three books selected will be current events books.

SOLUTION

Since there are five current events books and Stacy will need to select three of them, then there are ${}_5C_3$ ways of doing this.

$${}_5C_3 = \frac{5!}{(5-3)!3!} = \frac{5 \cdot 4 \cdot \cancel{3!}}{2 \cdot 1 \cdot \cancel{3!}} = 10$$ *10 ways to choose 3 current events books.*

The total number of outcomes in the sample space is ${}_{15}C_3$ since she has to select three books from 15 books.

$${}_{15}C_3 = \frac{15!}{(15-3)!3!} = \frac{15!}{12!3!} = \frac{15 \cdot 14 \cdot 13 \cdot \cancel{12!}}{\cancel{12!} \cdot 3 \cdot 2 \cdot 1} = 455$$ *455 total choices.*

The probability of selecting three current events books is

$$\frac{10}{455} = \frac{2}{91} \approx 0.022$$

▼ Try This One 1

There are 12 women and 8 men in a seminar course. If the professor chooses five-person groups at random, what is the probability that the first group chosen will consist of all women?

Now back to that poker game . . .

EXAMPLE 2 Using Combinations to Compute Probability

What is the probability of getting 4 aces when drawing 5 cards from a standard deck of 52 cards?

SOLUTION

First, we'll figure out how many five-card hands have four aces. Since there are only four aces in the deck, there's only one way to get all four of them in your hand. At that point there are 48 cards left for the other card in the hand. Using the fundamental counting principle, there are $1 \cdot 48$ ways to be dealt four aces.

Math Note

The probability in Example 2 means that you would expect to draw four aces on average once every 54,145 tries.

The total number of hands is the combinations of 5 cards chosen from 52, or ${}_{52}C_5$.

$${}_{52}C_5 = \frac{52!}{(52-5)!5!} = \frac{52!}{47!5!} = \frac{52 \cdot 51 \cdot 50 \cdot 49 \cdot 48 \cdot \cancel{47!}}{\cancel{47!} \cdot 5 \cdot 4 \cdot 3 \cdot 2 \cdot 1} = 2{,}598{,}960$$

The probability of getting four aces is

$$\frac{48}{2{,}598{,}960} = \frac{1}{54{,}145} \approx 0.0000185$$

We hate to accuse your friend of cheating, but to say the least, it's unlikely that she would draw a hand with 4 aces.

▼ Try This One 2

☑ 1. Compute probabilities using combinations.

Suppose the deck of cards in Example 2 has all 32 cards with numbers less than 10 removed, so that only 10s, jacks, queens, kings, and aces remain. Now what is the probability of getting 4 aces when drawing 5 cards?

Example 3 uses the permutation rule.

EXAMPLE 3 Using Permutations to Compute Probability

Permutations determine the number of combinations that can open a combination lock.

A combination lock has 40 numbers on it, from zero to 39. Find the probability that if the combination to unlock it consists of three numbers, it will contain the numbers 1, 2, and 3 in some order. Assume that numbers cannot be repeated in the combination. (It's interesting to note that a combination lock should really be called a permutation lock since the order of the numbers is important when you are unlocking the lock.)

SOLUTION

The number of combinations for the lock containing 1, 2, and 3 is ${}_3P_3$.

$${}_3P_3 = \frac{3!}{(3-3)!} = \frac{3!}{0!} = \frac{3 \cdot 2 \cdot 1}{1} = 6$$

The total number of combinations is a permutation of the 40 numbers taken 3 at a time, or ${}_{40}P_3$.

$${}_{40}P_3 = \frac{40!}{(40-3!)} = \frac{40!}{37!} = \frac{40 \cdot 39 \cdot 38 \cdot \cancel{37!}}{\cancel{37!}} = 59{,}280$$

The probability of the combination containing 1, 2, and 3 is $\frac{6}{59{,}280} \approx 0.000101$

▼ Try This One 3

☑ 2. Compute probabilities using permutations.

A different "permutation" lock has letters from A through L on it, and the combination consists of four letters with no repeats. What is the probability that the combination is I, J, K, and L in some order?

In Example 4, we will again need to use the fundamental counting principle.

EXAMPLE 4 Using Combinations to Compute Probability

A store has six different fitness magazines and three different news magazines. If a customer buys three magazines at random, find the probability that the customer will pick two fitness magazines and one news magazine.

SOLUTION

There are ${}_6C_2$ or 15 ways to select two fitness magazines from six fitness magazines, as shown:

$$ {}_6C_2 = \frac{6!}{(6-2)!2!} = \frac{6!}{4!2!} = \frac{6 \cdot 5 \cdot \cancel{4!}}{\cancel{4!} \cdot 2 \cdot 1} = 15 $$

There are ${}_3C_1$ or three ways to select one magazine from three news magazines:

$$ {}_3C_1 = \frac{3!}{(3-1)!1!} = \frac{3!}{2! \cdot 1!} = \frac{3 \cdot \cancel{2!}}{\cancel{2!} \cdot 1} = 3 $$

Using the fundamental counting principle, there are $15 \cdot 3$ or 45 ways to select two fitness magazines *and* one news magazine.

Next, there are ${}_9C_3$ or 84 ways to select three magazines from nine magazines:

$$ {}_9C_3 = \frac{9!}{(9-3)!3!} = \frac{9!}{6!3!} = \frac{9 \cdot 8 \cdot 7 \cdot \cancel{6!}}{\cancel{6!} \cdot 3 \cdot 2 \cdot 1} = 84 $$

The probability of selecting two fitness magazines and one news magazine is

$$ \frac{45}{84} \approx 0.536 $$

▼ Try This One 4

A box contains 24 imported cell phones and four of them are defective. If three phones are selected at random, find the probability that

(a) Exactly two are defective.
(b) None are defective.
(c) All three are defective.

Now we see why Chapter 11 began with a study of permutations and combinations: a large variety of probability problems can be solved using permutations and combinations in conjunction with the probability rules.

Sidelight THE CLASSICAL BIRTHDAY PROBLEM

What do you think the chances are that in a classroom with 23 students, two students have the same birthday (day and month)? Most people think that the probability would be very low since there are 365 days in a year. You may be surprised to find out that it is greater than 0.5, or 50%! Furthermore, as the number of people increases, the probability becomes even greater than 0.5 very rapidly. In a room of 30 students, there is a greater than 70% chance that two students have the same birthday. If you have 50 students in the room, the probability jumps to 97%!

The problem can be solved by using probability and permutation rules. It must be assumed that all birthdays are equally likely, but this assumption will have little effect on the answers. The way to solve the problem is to find the probability that no two people have the same birthday, and then subtract this probability from one. In other words, *P*(two students have the same birthday) = 1 − *P*(all students have different birthdays).

For example, suppose that there were only three students in a room. Then the probability that each would have a different birthday is

$$ \frac{365}{365} \cdot \frac{364}{365} \cdot \frac{363}{365} = \frac{{}_{365}P_3}{365^3} \approx 0.992 $$

So the probability that at least two of the three students have the same birthday is

$$ 1 - 0.992 = 0.008 $$

In general, in a room with *k* people, the probability that at least two people have the same birthday is

$$ 1 - \frac{{}_{365}P_k}{365^k} $$

In a room with 23 students, then, the probability that at least two students will have the same birthday is

$$ 1 - \frac{{}_{365}P_{23}}{365^{23}} \approx 0.507 \text{ or } 50.7\% $$

It is interesting to note that two presidents, James K. Polk and Warren G. Harding, were both born on November 2. Also, John Adams and Thomas Jefferson both died on July 4. The unusual thing about this is that they died on the same day of the same year, July 4, 1826.

Answers to Try This One

1 $33/646 \approx 0.051$

2 $1/969 \approx 0.00103$

3 $1/495 \approx 0.002$

4 (a) $15/253 \approx 0.059$

(b) $285/506 \approx 0.56$

(c) $1/506 \approx 0.002$

EXERCISE SET 11-5

Writing Exercises

1. Explain how combinations and permutations are useful in computing probabilities.
2. What is the biggest advantage of combinations and permutations over tree diagrams when computing probability?

Real-World Applications

3. A student-faculty government committee of 4 people is to be formed from 20 student volunteers and 5 faculty volunteers. Find the probability that the committee will consist of the following, assuming the selection is made at random:
   (a) All faculty members.
   (b) Two students and two faculty members.
   (c) All students.
   (d) One faculty member and three students.
4. In a company there are seven executives: four women and three men. Three are selected to attend a management seminar. Find these probabilities.
   (a) All three selected will be women.
   (b) All three selected will be men.
   (c) Two men and one woman will be selected.
   (d) One man and two women will be selected.
5. A city council consists of 10 members. Four are Republicans, three are Democrats, and three are Independents. If a committee of three is to be selected, find the probability of selecting
   (a) All Republicans. (b) All Democrats.
   (c) One of each party.
   (d) Two Democrats and one Independent.
   (e) One Independent and two Republicans.
6. In a class of 18 students, there are 11 men and seven women. Four students are selected to present a demonstration on the use of graphing calculators. Find the probability that the group consists of
   (a) All men.
   (b) All women.
   (c) Three men and one woman.
   (d) One man and three women.
   (e) Two men and two women.
7. Fred needs to print his term paper. He pulls down a box of twelve ink cartridges and recalls that three of them have no more ink. If he selects four cartridges from the box, find the probability that
   (a) No cartridge has ink.
   (b) One cartridge has no ink.
   (c) Three cartridges have no ink.
8. There are 50 tickets sold for a raffle for the Student Art Auction, and there are two prizes to be awarded. If Dionte buys two tickets, find the probability that he will win both prizes.
9. An engineering company has four openings and the applicant pool consists of six database administrators and eight network engineers. If the hiring is done without regard for the specific qualifications of the applicants, find the probability that the four hired will be
   (a) All network engineers.
   (b) Two database administrators and two network engineers.
   (c) All database administrators.
   (d) Three database administrators and one network engineer.
   (e) One database administrator and three network engineers.
10. At the coffee kiosk in the mall, there are eight coffee drinks, five tea drinks, and two smoothies. Nora wants to buy three drinks for herself and her friends. Find the probability that she buys
    (a) All coffee drinks.
    (b) Two smoothie drinks.
    (c) All tea drinks.
    (d) One of each type of drink.
    (e) Two coffee drinks and one tea drink.

11. Find the probability of getting any triple-digit number where all the digits are the same in a lottery game that consists of selecting a three-digit number.
12. Binh is choosing from eight YouTube videos and nine online role-playing games to link to his MySpace page. If he randomly picks seven links total, find the probability that Binh chooses three YouTube videos and four online role-playing games.
13. Drew is on a fitness kick, so he's walking around with four protein bars and eight lowfat cereal bars in his backpack. If he reaches in and randomly grabs five bars, find the probability that he gets two protein bars and three cereal bars.
14. To win a state lottery, a person must select 5 numbers from 40 numbers. Find the probability of winning if a person buys one ticket. (*Note:* The numbers can be selected in any order.)
15. A five-digit identification card is made. Find the probability that the card will contain the digits 0, 1, 2, 3, and 4 in any order.
16. The combination lock in Example 3 has 40 numbers from zero to 39, and a combination consists of 3 numbers in a specific order with no repeats. Find the probability that the combination consists only of even numbers.

Exercises 17–22 refer to poker hands consisting of 5 cards dealt at random from a standard deck of 52 cards. Find the probability of getting each hand.

17. A full house (three of one denomination and two of another)
18. A flush (five cards of the same suit)
19. Three of a kind (three of one denomination)
20. Four of a kind (four of the same denomination)
21. A royal flush (ten, jack, queen, king, and ace of the same suit)
22. A straight flush (five cards of the same suit that are consecutive in denomination)

Critical Thinking

23. At a carnival game, the player pays a dollar, then flips a quarter, rolls two dice, and draws two cards from a standard deck. If the result is tails, 12, and a pair of aces, he wins $1,000. What is the probability of this happening?

Section 11-6 Odds and Expectation

LEARNING OBJECTIVES

- ❑ 1. Compute the odds in favor of and against an outcome.
- ❑ 2. Compute odds from probability.
- ❑ 3. Compute probability from odds.
- ❑ 4. Compute expected value.

The Pittsburgh Steelers won the Super Bowl on February 1, 2009, and by the time the ink dried on newspapers reporting the victory, oddsmakers in Las Vegas had listed the odds against the Steelers winning it again in 2010 as 9 to 1. But what exactly does that mean? The term "odds" is used all the time in describing the likelihood of something happening, but a lot of people don't understand exactly what a given set of odds means.

Odds are used by casinos, racetracks, and other gambling establishments to determine the payoffs when bets are made or lottery tickets are purchased. They're also used by insurance companies in determining the amount to charge for premiums. The formulas for computing odds are similar to the formula we've been using for classical probability, and shortly we will see a strong connection between the two concepts.

If an event E has a favorable outcomes and b unfavorable outcomes, then

1. The odds in favor of event E occurring $= \frac{a}{b}$
2. The odds against event E occurring $= \frac{b}{a}$

CAUTION

Do not confuse the formula for odds in favor of an event with the formula for classical probability. The probability formula is the number of favorable outcomes over the *total* number of outcomes, while the one for odds is the number of favorable outcomes over the number of *unfavorable* outcomes.

Odds can be expressed as a fraction or a ratio. For example, the odds against Pittsburgh repeating as Super Bowl champion could be listed as $\frac{9}{1}$, 9:1, or 9 to 1. In common usage, the phrase “the odds of” really means “the odds against”; if we are told that the odds of rolling a 12 with two dice are 35 to 1, it means that the odds against rolling 12 are 35 to 1. So by setting Pittsburgh’s odds at 9:1, the oddsmakers are predicting that if the season were played 10 times, Pittsburgh would win the Super Bowl 1 time, and not win it 9 times.

EXAMPLE 1 Computing Odds

A card is drawn from a standard deck of 52 cards.

Math Note

Notice that if the odds in favor of an event occurring are $a{:}b$, the odds against it occurring are $b{:}a$.

(a) Find the odds in favor of getting an ace.
(b) Find the odds against getting an ace.

SOLUTION

(a) In a deck of cards there are 52 cards and there are 4 aces, so $a = 4$ and $b = 52 - 4 = 48$. (In other words, there are 48 cards that are not aces.)
The odds in favor of an ace $= \frac{4}{48} = \frac{1}{12}$

(b) The odds against an ace $= \frac{48}{4} = \frac{12}{1}$.
The odds in favor of an ace are 1:12 and the odds against an ace are 12:1.

▼ Try This One 1

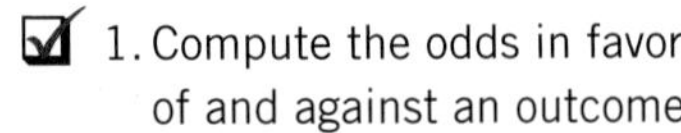
1. Compute the odds in favor of and against an outcome.

What are the odds in favor of rolling a prime number sum with a roll of two dice? What are the odds against? (Figure 11-7 on page 601 will help.)

When an event E has a favorable and b unfavorable outcomes, there are $a + b$ total outcomes, and the probability of E is

$$P(E) = \frac{a}{a+b}$$

The probability of E not occurring is

$$1 - P(E), \quad \text{or} \quad \frac{b}{a+b}$$

If we divide these two probabilities, we get

$$\frac{P(E)}{1 - P(E)} = \frac{\frac{a}{a+b}}{\frac{b}{a+b}} = \frac{a}{a+b} \cdot \frac{a+b}{b} = \frac{a}{b}$$

The result is the odds in favor of event E. This gives us a strong connection between probability and odds.

Formulas for Odds in Terms of Probability

$$\text{Odds in favor} = \frac{P(E)}{1 - P(E)}$$

$$\text{Odds against} = \frac{P(E')}{1 - P(E')}$$

where $P(E)$ is the probability that event E occurs and $P(E')$ is the probability that the event E does not occur.

Math Note

Recall that $P(E') = 1 - P(E)$.

EXAMPLE 2 Finding Odds from Probability

The probability of getting exactly one pair in a five-card poker hand is 0.423. Find the odds in favor of getting exactly one pair, and the odds against.

SOLUTION

This is a direct application of the formula relating probability and odds. The odds in favor of getting exactly one pair are

$$\frac{P(\text{getting exactly one pair})}{1 - P(\text{getting exactly one pair})} = \frac{0.423}{1 - 0.423} = \frac{0.423}{0.577}$$

We can convert this into fraction form by multiplying both the numerator and denominator by 100.

$$\frac{0.423}{0.577} \cdot \frac{100}{100} = \frac{423}{577}$$

So the odds in favor of getting exactly one pair are 423:577, and the odds against are 577:423.

▼ Try This One 2

☑ 2. Compute odds from probability.

According to the American Cancer Society, the probability of an American female developing some type of cancer at some point in her life is about $\frac{1}{3}$. Find the odds in favor of and against an American woman developing cancer.

Math Note

When the odds are 1:1, a game is said to be fair. That is, both parties have an equal chance of winning or losing.

In Example 2, the odds for an event were found when the probabilities were known. As we saw on page 610, when the odds of an event are given, the probability of an event can be found.

Formula for Probability in Terms of Odds

If the odds in favor of an event E are a:b, then the probability that the event will occur is

$$P(E) = \frac{a}{a + b}$$

EXAMPLE 3 Finding Probability from Odds

According to the National Safety Council, the odds of dying due to injury at some point in your life are about 10:237. Find the probability of dying from injury.

SOLUTION

In the formula for converting to probability, the odds in favor are a:b. In this case, those odds are 10:237, so $a = 10$ and $b = 237$.

$$P(\text{dying from injury}) = \frac{10}{10 + 237} = \frac{10}{247} \approx 0.040$$

▼ Try This One 3

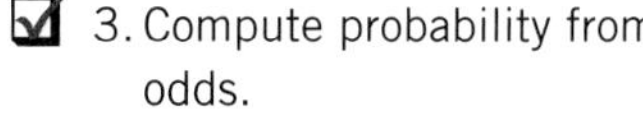

☑ 3. Compute probability from odds.

When two dice are rolled, the odds in favor of getting a sum of 9 are 1:8. Find the probability of not getting a sum of 9 when two dice are rolled.

Sidelight ODDS AND BETTING

In almost any gambling enterprise, from Vegas to a church raffle, odds are used to determine what the payouts will be. Let's use the game of roulette as an example. There are 38 partitions on a roulette wheel, with a small ball that is equally likely to land in any of them. Players make various bets on where the ball will land. If you simply bet on a particular number, the odds against you are 37:1. If you win, the casino will pay 35 times what you bet. So if you bet a dollar on each spin, on average, in 38 spins, you would lose 37 times and win once, meaning you'd be two dollars in the hole. That doesn't sound like the recipe for the casino making a lot of money, but when you consider the large number of people in a typical casino, and the fact that they're placing multiple bets at frequent intervals, it starts to make more sense.

This is where it becomes clear that state lotteries are an awful bet. In the Ohio lottery's "Classic Lotto" game, for example, the odds in favor of matching 5 of 6 numbers are 1:54,021. So to make the game completely fair, the payout should be in the neighborhood of $54,000 on a $1 ticket. The actual payout? Just $1,500! Any gambler or statistician will tell you that lotteries offer the worst odds of just about any game of chance.

Expected Value

Another concept related to odds and probability is **expectation,** or **expected value.** Expected value is used to determine the result that would be expected over the long term in some sort of gamble. It is used not only for games of chance, but in areas like insurance, management, engineering, and others. *The key element is that the events in question must have numerical outcomes.*

For example, rolling a die has a numerical outcome (1 through 6), and expected value can be used to determine what the average long-run result is likely to be. (We'll find out in Example 4.)

To find expected value, multiply the numerical result of each outcome by the corresponding probability of the outcome, then add those products.

Expected Value

The expected value for the outcomes of a probability experiment is

$$E = X_1 \cdot P(X_1) + X_2 \cdot P(X_2) + \cdots + X_n \cdot P(X_n)$$

where the X's correspond to the numerical outcomes and the $P(X)$'s are the corresponding probabilities of the outcomes.

EXAMPLE 4 Computing Expected Value

Math Note

In Example 4, a die cannot show 3.5 spots: It can only show 1, 2, 3, 4, 5, or 6 spots. However, in this case, the expected value would be the long run average—that is, if you add up the total number of spots, then divide by the number of times the die is rolled, the average would be close to 3.5.

When a single die is rolled, find the expected value of the outcome.

SOLUTION

Since each numerical outcome, 1 through 6, has a probability of $\frac{1}{6}$, the expected value is

$$E = 1 \cdot \frac{1}{6} + 2 \cdot \frac{1}{6} + 3 \cdot \frac{1}{6} + 4 \cdot \frac{1}{6} + 5 \cdot \frac{1}{6} + 6 \cdot \frac{1}{6} = \frac{21}{6} = 3.5$$

▼ Try This One 4

If seven cards are numbered with integers from −2 to 4, then placed into a box and picked out at random, find the expected value.

In gambling games, the expected value is found by multiplying the amount won, or net gain, and the amount lost by the corresponding probabilities and then finding the sum.

EXAMPLE 5 Computing Expected Value

One thousand tickets are sold at $1 each for a color television valued at $350. What is the expected value if a person purchases one ticket?

SOLUTION

We begin with two notes. First, for a win, the net gain is $349, since the person does not get the cost of the ticket ($1) back. Second, for a loss, the gain is represented by a negative number, in this case, −$1.

The problem can then be set up as follows:

	Win	Lose
Gain, *X*	$349	−$1
Probability, *P*(*X*)	$\frac{1}{1{,}000}$	$\frac{999}{1{,}000}$

The solution, then, is

$$E(X) = \$349 \cdot \frac{1}{1{,}000} + (-\$1) \cdot \frac{999}{1{,}000} = -\$0.65$$

▼ Try This One 5

With his house in foreclosure, a homeowner comes up with a plan to salvage the situation: he sells 10,000 raffle tickets at $50 each for the home, which is valued at $200,000. Find the expected value from buying one ticket.

Note that the expectation in Example 5 is −$0.65. This does not mean that a person would ever actually lose 65 cents, since the person can only win a television set valued at $350 or lose $1 on the ticket. What this expectation means is that the average of the losses is $0.65 for each of the 1,000 ticket holders. Here is another way of looking at this situation: If a person purchased one ticket each week over a long period of time, the average loss would be $0.65 per ticket, since theoretically, on average, that person would win the television set once for each 1,000 tickets purchased.

Sidelight MATH AND SLOT MACHINES

Today most slot machines are run electronically, much the same as video games are. However, early slot machines were mechanical in nature. The first slot machines were invented by the Fey Manufacturing Company of San Francisco in 1895. They consisted of three large wheels, which were spun when the handle on the side of the machine was pulled. In order to control the number of wins and the payoffs, each wheel contains 20 symbols. However, the number of the same symbols is not the same on each wheel. For example, there may be two oranges on wheel 1, six oranges on wheel 2, and no oranges on wheel 3. When a person gets two orange symbols, he may think that he almost won; i.e., 2 out of 3. However, since there is no orange symbol on the third wheel, the probability of getting three oranges is zero! The higher the probability of getting three oranges, the lower the payoff. Using probability theory, the owner of the machines can determine his long-run profit.

EXAMPLE 6 Computing Expected Value

A stock you bought two years ago with high hopes is now selling for less than you paid, and things look grim for the company. Do you sell, or hold on and hope it will come back to the original price before you sell?

A model economists use for such situations is a game no one wants to play: Suppose you have a choice: lose $100, or take a $\frac{50}{50}$ chance between losing nothing, and losing $300. Which do you choose? Find the expected value for each strategy over 10 trials. (See Exercise 24.)

4. Compute expected value.

One thousand tickets are sold at $1 each for four prizes of $100, $50, $25, and $10. What is the expected value if a person purchases two tickets?

SOLUTION

Find the expected value if the person purchases one ticket.

Gain, x	$99	$49	$24	$9	−$1
Probability, $P(x)$	$\frac{1}{1{,}000}$	$\frac{1}{1{,}000}$	$\frac{1}{1{,}000}$	$\frac{1}{1{,}000}$	$\frac{996}{1{,}000}$

$$E(x) = \$99 \cdot \frac{1}{1{,}000} + \$49 \cdot \frac{1}{1{,}000} + \$24 \cdot \frac{1}{1{,}000} + \$9 \cdot \frac{1}{1{,}000} - \$1 \cdot \frac{996}{1{,}000} = -\$0.815$$

Now multiply by 2 since two tickets were purchased.

$$-\$0.815(2) = -\$1.63$$

▼ Try This One 6

A small ski resort loses $70,000 per season when it does not snow very much and makes $250,000 profit when it does snow a lot. The probability of it snowing at least 75 inches (i.e., a good season) is 40%. Find the expectation for the profit.

In gambling games, if the expected value of the gain is 0, the game is said to be fair. If the expected value of the gain of a game is positive, then the game is in favor of the player. That is, the player has a better-than-even chance of winning. If the expected value of the gain is negative, then the game is said to be in favor of the house. That is, in the long run, the players will lose money. Can you guess what the sign of the expected value is for every game in a casino?

Math Note

In American roulette, there are 22 different types of bet you can place, but it is interesting to note that all but one of them have the exact same expected value: −$0.053 on a $1 bet. (Betting on 0, 00, 1, 2, and 3, called a five-number bet, is worse, at −$0.079.)

Answers to Try This One

1 In favor: 5:7; against: 7:5
2 In favor: 1:2; against: 2:1
3 $\frac{8}{9}$
4 1
5 −$30
6 $58,000

EXERCISE SET 11-6

Writing Exercises

1. Explain the difference between the odds in favor of an event and the odds against an event.
2. Explain the numerical relationship between the odds in favor of an event and the odds against an event.

3. Explain the meaning of odds in a gambling game.
4. Explain how to find the probability of an event occurring when given the odds in favor of an event.
5. Explain what is meant by the expected value of an event.
6. Why does every game in a casino have a negative expected value for the player?

Real-World Applications

7. In planning a gambling booth for a charity festival, Antoine needs to know the odds of various combinations in order to decide on payouts that will be high enough that people want to play, but low enough that the charity will make money. If the player rolls two dice, find the odds
   (a) In favor of getting a sum of 10.
   (b) In favor of getting a sum of 12.
   (c) Against getting a sum of 7.
   (d) Against getting a sum of 3.
   (e) In favor of getting doubles.
8. If the game in Exercise 7 has players who roll only one die, find the odds
   (a) In favor of getting a 3.
   (b) In favor of getting a 6.
   (c) Against getting an odd number.
   (d) Against getting an even number.
   (e) In favor of getting a prime number.
9. Steve shuffled a deck of 52 cards and asked Sally to draw one card to start a magic trick. Find the odds
   (a) In favor of getting a queen.
   (b) In favor of getting a face card.
   (c) Against getting a club.
   (d) In favor of getting an ace.
   (e) In favor of getting a black card.
10. While cleaning her dorm room, Monica mistakenly pushed three coins off the desk. When they land on the chair below, find the odds
    (a) In favor of getting exactly three heads.
    (b) In favor of getting exactly three tails.
    (c) Against getting exactly two heads.
    (d) Against getting exactly one tail.
    (e) In favor of getting at least one tail.
11. Your friends have taken bets on whether you will pass this class. (Sounds like maybe you need new friends.) Find the probability that you will pass given these odds:
    (a) 7:4 in favor of you passing
    (b) 2:5 against you passing
    (c) 3:1 in favor of you passing
    (d) 1:4 against you passing
12. Find the probability that you will win a Wii bowling tournament given these odds:
    (a) 3:4 in favor of you winning
    (b) 1:7 against you winning
    (c) 5:4 in favor of you winning
    (d) 6:5 in favor of you winning
13. If the odds against a horse winning a race are 9:5, find the probability that the horse will win the race.
14. A person rolls two dice and wins if he or she throws doubles. What are the odds in favor of the event? What are the odds against the event?
15. A cash prize of $5,000 is to be awarded at a fund-raiser. If 2,500 tickets are sold at $5 each, find the expected value.
16. You start your shift as a cashier with your drawer containing ten $1 bills, five $2 bills, three $5 bills, one $10 bill, and one $100 bill. Find the expectation if one bill is selected.
17. In a scratch-off game, if you scratch the two dice on the ticket and get doubles, you win $5. For the game to be fair, how much should you pay to play the game?
18. At this year's State Fair, there was a dice rolling game. If you rolled two dice and got a sum of 2 or 12, you won $20. If you rolled a 7, you won $5. Any other roll was a loss. It cost $3 to play one game with one roll of the dice. What is the expectation of the game?
19. Melinda buys one raffle ticket at the Spring Fling since there is one $1,000 prize, one $500 prize, and five $100 prizes. There were a total of 1,000 tickets sold at $3 each. What is Melinda's expectation?
20. If Melinda buys two tickets to the raffle in Exercise 19, what is her expectation?
21. For a daily lottery, a person selects any three-digit number from 000 to 999. If a person plays for $1, the person can win $500. Find the expectation. In the same daily lottery, if a person boxes a number, the person can win $80. Find the expectation if the number 123 is played for $1 and boxed. (When a number is "boxed," it can win when the digits occur in any order.)
22. If a 60-year-old buys a $1,000 life insurance policy at a cost of $60 and has a probability of 0.972 of living to age 61, find the expectation of the policy until the person reaches 61.

Critical Thinking

23. Consider the following problem: You stop on the street between errands to engage in a shell game with a street vendor. The vendor shows you a two-headed penny under one shell, a two-tailed penny under the second shell and a fair penny (one head and one tail) under the third shell. He shuffles the shells around and then

you choose a shell. He shows you that under the shell is a penny with the head side up. He is willing to bet you $5 that it is the two-headed penny. He says it cannot be the two-tailed penny because a head is showing. Therefore, he says there is a 50-50 chance of it being the two-headed coin. Should you take the bet?

24. Stuck in a bad situation, you're given a choice: lose $100, or take a 50-50 chance between losing nothing and losing $300. Choose the option that sounds better to you, then find the expected value of each option over 10 trials.
25. Since expected value only applies to numerical outcomes, it takes some ingenuity to use it for flipping a coin.
    (a) Choose any two numbers at random, assigning one to heads and another to tails. Then find the expected value of flipping the coin.
    (b) Repeat part (a) for two different numbers. What can you conclude?
26. Chevalier de Mere, a famous gambler, won money when he bet unsuspecting patrons that in four rolls of a die, he could get at least one 6, but he lost money when he bet that in 24 rolls of two dice, he could get a double 6. Using the probability rules, find the probability of each event and explain why he won the majority of the time on the first game but lost the majority of the time when playing the second game.
27. A roulette wheel has 38 numbers: 1 through 36, 0, and 00. A ball is rolled, and it falls into one of the 38 slots, giving a winning number. If a player bets $1 on a number and wins, the player gets $35 plus his $1 back. Otherwise, he loses the $1 he bet. Show that the expected value of the game matches that shown in the Math Note on page 614.

Section 11-7 The Addition Rules for Probability

LEARNING OBJECTIVES

- ☐ 1. Decide if two events are mutually exclusive.
- ☐ 2. Use the addition rule for mutually exclusive events.
- ☐ 3. Use the addition rule for events that are not mutually exclusive.

Many interesting problems in probability involve finding the probability of more than one event. Usually, some careful thought is required. For example, when the U.S. House of Representatives is in session, suppose that a political commentator feels like a certain bill is most likely to be supported by women and Democrats. She would likely be interested in the probability that a member of the House is either a woman or a Democrat. She could find the number of women representatives, and the number of Democratic representatives easily, but then what? Should she add those numbers and divide by the total number of representatives? What about those who are both women *and* Democrats? They would get counted twice.

The situation would be simpler if the commentator were interested in the probability of a representative being either a Republican or an Independent. The key difference is that any individual has to be one or the other. These two events are called *mutually exclusive,* which indicates that either one or the other must occur, but not both.

Two events are **mutually exclusive** if they cannot both occur at the same time. That is, the events have no outcomes in common.

EXAMPLE 1 Deciding if Two Events Are Mutually Exclusive

In drawing cards from a standard deck, determine whether the two events are mutually exclusive or not.

(a) Drawing a 4, drawing a 6. (b) Drawing a 4, drawing a heart.

SOLUTION

(a) Every card has just one denomination, so a card can't be both a 4 and a 6. The events are mutually exclusive.
(b) You could draw the 4 of hearts, which is one outcome satisfying both events. The events are not mutually exclusive.

▼ Try This One 1

☑ 1. Decide if two events are mutually exclusive.

If student government picks students at random to win free books for a semester, determine whether the two events are mutually exclusive or not.

(a) The winner is a sophomore or a business major.
(b) The winner is a junior or a senior.

If you select one of these T-shirts at random in the morning, the probability that it is blue or red is $\frac{1}{5} + \frac{1}{5}$.

The probability of two or more events occurring can be determined by using the **addition rules.** The first addition rule is used when the events are mutually exclusive.

Addition Rule 1

When two events A and B are mutually exclusive, the probability that A or B will occur is

$$P(A \text{ or } B) = P(A) + P(B)$$

In Exercise 2, you will examine why this formula makes sense.

EXAMPLE 2 Using Addition Rule 1

A restaurant has three pieces of apple pie, five pieces of cherry pie, and four pieces of pumpkin pie in its dessert case. If a customer selects at random one kind of pie for dessert, find the probability that it will be either cherry or pumpkin.

Math Note

Sometimes it's a bad idea to reduce fractions in the individual probabilities when using the addition rule: you'll need a common denominator to add them. Your final answer should always be reduced if possible, though.

SOLUTION

The events are mutually exclusive. Since there is a total of 12 pieces of pie, five of which are cherry and four of which are pumpkin,

$$P(\text{cherry or pumpkin}) = P(\text{cherry}) + P(\text{pumpkin})$$
$$= \frac{5}{12} + \frac{4}{12} = \frac{9}{12} = \frac{3}{4}$$

▼ Try This One 2

A liberal arts math class contains 7 freshmen, 11 sophomores, 5 juniors, and 2 seniors. If the professor randomly chooses one to present a homework problem at the board, find the probability that it's either a junior or senior.

EXAMPLE 3 Using Addition Rule 1

A card is drawn from a standard deck. Find the probability of getting an ace or a queen.

SOLUTION

The events are mutually exclusive. There are four aces and four queens; therefore,

$$P(\text{ace or queen}) = P(\text{ace}) + P(\text{queen})$$
$$= \frac{4}{52} + \frac{4}{52} = \frac{8}{52} = \frac{2}{13}$$

▼ Try This One 3

At a political rally, there are 20 Republicans, 13 Democrats, and 6 Independents. If a person is selected at random, find the probability that he or she is either a Democrat or an Independent.

The addition rule for mutually exclusive events can be extended to three or more events as shown in Example 4.

EXAMPLE 4 Using the Addition Rule with Three Events

A card is drawn from a deck. Find the probability that it is either a club, a diamond, or a heart.

SOLUTION

In the deck of 52 cards there are 13 clubs, 13 diamonds, and 13 hearts, and any card can be only one of those suits. So

$$P(\text{club, diamond, or heart}) = P(\text{club}) + P(\text{diamond}) + P(\text{heart})$$
$$= \frac{13}{52} + \frac{13}{52} + \frac{13}{52} = \frac{39}{52} = \frac{3}{4}$$

▼ Try This One 4

☑ 2. Use the addition rule for mutually exclusive events.

In rolling two dice, find the probability that the sum is 2, 3, or 4.

In Example 4, we used an extended version of addition rule 1:

$$P(A \text{ or } B \text{ or } C) = P(A) + P(B) + P(C)$$

When two events are not mutually exclusive, any outcomes that are common to two events are counted twice. We account for this by subtracting the probability of both events occurring, which results in addition rule 2.

Math Note

Addition rule 2 can actually be used when events are mutually exclusive, too—in that case, $P(A \text{ and } B)$ will always be zero, and you'll get addition rule 1 back. But it's still important to distinguish between the two situations.

Addition Rule 2

When two events A and B are not mutually exclusive, the probability that A or B will occur is

$$P(A \text{ or } B) = P(A) + P(B) - P(A \text{ and } B)$$

EXAMPLE 5 Using Addition Rule 2

A single card is drawn from a standard deck of cards. Find the probability that it is a king or a club.

SOLUTION

In this case, there are 4 kings and 13 clubs. However, the king of clubs has been counted twice since the two events are not mutually exclusive. When finding the probability of getting a king or a club, the probability of getting the king of clubs (i.e., a king and a club) must be subtracted, as shown.

$$\begin{aligned} P(\text{king or club}) &= P(\text{king}) + P(\text{club}) - P(\text{king and club}) \\ &= \frac{4}{52} + \frac{13}{52} - \frac{1}{52} = \frac{16}{52} = \frac{4}{13} \end{aligned}$$

▼ Try This One 5

A card is drawn from an ordinary deck. Find the probability that it is a heart or a face card.

EXAMPLE 6 Using Addition Rule 2

Two dice are rolled. Find the probability of getting doubles or a sum of 6.

SOLUTION

Using the sample space shown in Section 11-2, there are six ways to get doubles: (1, 1), (2, 2), (3, 3), (4, 4), (5, 5), (6, 6). So $P(\text{doubles}) = \frac{6}{36}$. There are five ways to get a sum of 6: (1, 5), (2, 4), (3, 3), (4, 2), (5, 1). So $P(\text{sum of } 6) = \frac{5}{36}$. Notice that there is one way of getting doubles and a sum of 6, so $P(\text{doubles and a sum of } 6) = \frac{1}{36}$. Finally,

$$\begin{aligned} P(\text{doubles or a sum of } 6) &= P(\text{doubles}) + P(\text{sum of } 6) - (\text{doubles and sum of } 6) \\ &= \frac{6}{36} + \frac{5}{36} - \frac{1}{36} = \frac{10}{36} = \frac{5}{18} \end{aligned}$$

▼ Try This One 6

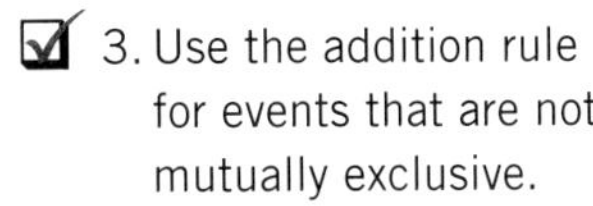
3. Use the addition rule for events that are not mutually exclusive.

When two dice are rolled, find the probability that both numbers are more than three, or that they differ by exactly two.

In many cases, the information in probability problems can be arranged in table form in order to make it easier to compute the probabilities for various events. Example 7 uses this technique.

EXAMPLE 7 Using a Table and Addition Rule 2

In a hospital there are eight nurses and five physicians. Seven nurses and three physicians are females. If a staff person is selected, find the probability that the subject is a nurse or a male.

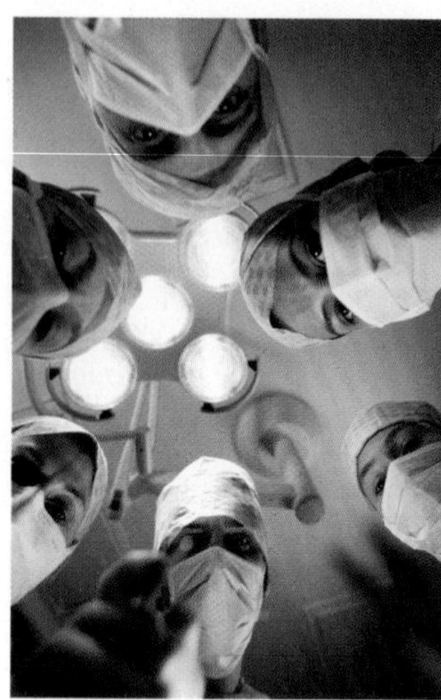

SOLUTION

The sample space can be written in table form.

Staff	Females	Males	Total
Nurses	7	1	8
Physicians	3	2	5
Total	10	3	13

Looking at the table, we can see that there are 8 nurses and 3 males, and there is one person who is both a male and a nurse. The probability is

$$P(\text{nurse or male}) = P(\text{nurse}) + P(\text{male}) - P(\text{male and a nurse})$$
$$= \frac{8}{13} + \frac{3}{13} - \frac{1}{13} = \frac{10}{13}$$

▼ Try This One 7

In a class of students, there are 15 freshmen and 10 sophomores. Six of the freshmen are males and four of the sophomores are males. If a student is selected at random, find the probability that the student is a sophomore or a male.

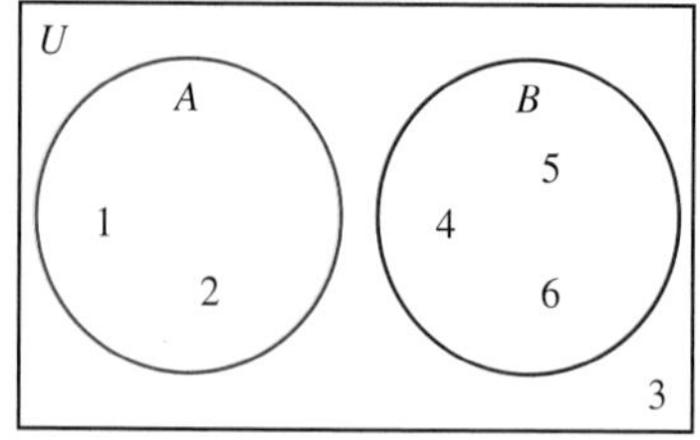

(a) Mutually exclusive events
$A \cap B = \varnothing$
$P(A \cup B) = P(A) + P(B)$

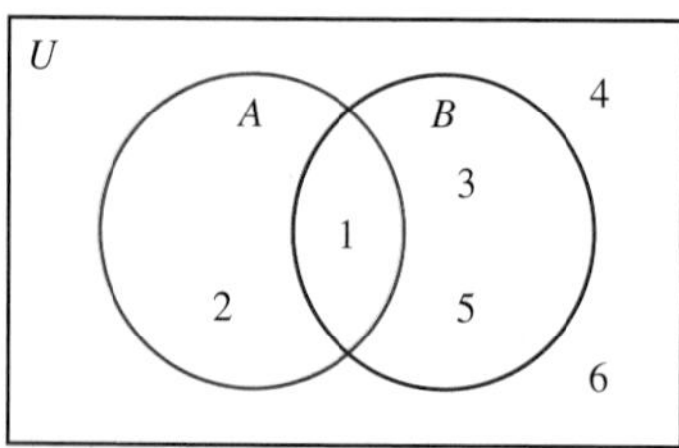

(b) Events not mutually exclusive
$A \cap B \neq \varnothing$
$P(A \cup B) = P(A) + P(B) - P(A \cap B)$

Figure 11-8

Probability and Sets

The addition rules for probability can be represented by Venn diagrams. First, consider two mutually exclusive events. For example, we will use the probability experiment of rolling a die. Let A be the event of getting a number less than 3 and B be the event of getting a number greater than 3. In set notation, $U = \{1, 2, 3, 4, 5, 6\}$, $A = \{1, 2\}$, and $B = \{4, 5, 6\}$. The Venn diagram for mutually exclusive events is shown in Figure 11-8(a). Notice that $A \cup B = \{1, 2\} \cup \{4, 5, 6\} = \{1, 2, 4, 5, 6\}$; now $P(A \cup B) = P(A) + P(B)$ or in probability language, $P(A \text{ or } B) = P(A) + P(B)$. Since $P(A) = \frac{2}{6}$, and $P(B) = \frac{3}{6}$, then $P(A \text{ or } B) = \frac{5}{6}$.

Next, consider two events that are not mutually exclusive. Let A be the event of getting a number less than 3 and B be the event of getting an odd number. The sample space for A is $\{1, 2\}$. The sample space for B is $\{1, 3, 5\}$. The Venn diagram for this example is shown in Figure 11-8(b). We can see that 4 of the 6 outcomes are in $A \cup B$, so $P(A \cup B) = \frac{4}{6}$. Also, there are 2 outcomes in A, 3 in B, and 1 in $A \cap B$, so $P(A) = \frac{2}{6}$, $P(B) = \frac{3}{6}$, and $P(A \cap B) = \frac{1}{6}$. Then

$$P(A \cup B) = P(A) + P(B) - P(A \cap B)$$
$$= \frac{2}{6} + \frac{3}{6} - \frac{1}{6} = \frac{4}{6} = \frac{2}{3}$$

In probability notation, this is equivalent to $P(A \text{ or } B) = P(A) + P(B) - P(A \text{ and } B)$.

Answers to Try This One

1. (a) Not mutually exclusive
   (b) Mutually exclusive
2. $\frac{7}{25}$
3. $\frac{19}{39}$
4. $\frac{1}{6}$
5. $\frac{11}{26}$
6. $\frac{5}{12}$
7. $\frac{16}{25}$

EXERCISE SET 11-7

Writing Exercises

1. Explain how to tell if two events are mutually exclusive or not.

For Exercises 2–8, determine whether or not the events are mutually exclusive, and explain your answer.

2. Roll a die: Get an even number, or get a number less than 3.
3. Roll a die: Get a prime number (2, 3, 5), or get an odd number.
4. Roll a die: Get a number greater than 3, or get a number less than 3.
5. Select a student in your class: the student has blond hair, or the student has blue eyes.
6. Select a student in your college: the student is a sophomore, or the student is a business major.
7. Select any course: It is a calculus course, or it is an English course.
8. Select a registered voter: The voter is a Republican, or the voter is a Democrat.

Real-World Applications

9. A young couple who plan to marry decide to select a month for their wedding. Find the probability that it will be April or May. Assume that all months have an equal opportunity of being selected.
10. At the animal shelter where Miguel volunteers on the weekends, there were two Siamese cats, four tabby cats, sixteen mixed-breed dogs, and four iguanas in their cages. If a customer picks any of these animals at random, find the probability that the animal is either a mixed-breed dog or a Siamese cat.
11. When Milo went to register for classes at the last minute, the only classes left to take were seven math courses, five computer science courses, three statistics courses, and four science courses. He shut his eyes and picked one at random. Find the probability that Milo selected either a science course or a math course. (The probability that Milo will ever graduate using this strategy is an interesting question, too.)
12. When Shana looked at the favorites list on her boyfriend's MP3 player, there were 10 songs she liked, 7 songs she'd heard but did not like, and 5 songs she hadn't heard. She let the MP3 player select a song at random. What is the probability that the selected song was one she liked or one she'd heard but did not like?
13. The bookstore has eight red school jerseys, nine blue school jerseys, and three black school jerseys to choose from. If a student selects one at random, find the probability that it is

    (a) A red or a blue jersey.
    (b) A red or a black jersey.
    (c) A blue or a black jersey.
14. In my Tae Kwon Do class, there are two black belts, three red belts, five blue belts, three green belts, and six yellow belts. If the sensei selects a student at random to lead the warm-up, find the probability that the person is

    (a) Either a green belt or a yellow belt.
    (b) Either a black belt or a blue belt.
    (c) Either a black belt, a green belt, or a yellow belt.
15. On a small college campus, there are five English professors, four mathematics professors, two science professors, three psychology professors, and three history professors. If a professor is selected at random, find the probability that the professor is

    (a) An English or psychology professor.
    (b) A mathematics or science professor.
    (c) A history, science, or mathematics professor.
    (d) An English, mathematics, or history professor.
16. A deck of cards is randomly dealt by the computer during a game of Spider Solitaire. Find the probability the first card dealt is

    (a) A 4 or a diamond.
    (b) A club or a diamond.
    (c) A jack or a black card.
17. In a statistics class there are 18 juniors and 10 seniors; 6 of the seniors are females, and 12 of the juniors are males. If a student is selected at random, find the probability of selecting

    (a) A junior or a female.
    (b) A senior or a female.
    (c) A junior or a senior.
18. A cell phone company gets a good deal on 400 new Motorola RAZR phones. Of them, 250 have digital cameras and 150 have touch screens. Of the RAZR phones that have digital cameras, 140 are lime green and the rest are metallic orange. Of the RAZR phones that have touch screens, 80 are lime green and the rest are metallic orange. If a RAZR phone is selected at random, find the probability that it will

    (a) Be lime green or have a digital camera.
    (b) Be metallic orange or have a touch screen.
19. BlueFly.com, an online designer clothing marketplace, purchases items from Kenneth Cole, Michael

Kors, and Vera Wang. The most recent purchases are shown here:

Product	Kenneth Cole	Michael Kors	Vera Wang
Dresses	24	18	12
Jeans	13	36	15

If one item is selected at random, find these probabilities:

(a) It was purchased from Kenneth Cole or is a dress.
(b) It was purchased from Michael Kors or Vera Wang.
(c) It is a pair of jeans or it was purchased from Kenneth Cole.

20. In a recent campus survey, the following data were obtained in response to the question "Do you think there should be harsher penalties for underage drinking on campus?"

	Yes	No	No opinion
Males	72	81	5
Females	103	68	7

If a person is selected at random, find these probabilities:

(a) The person has no opinion.
(b) The person is a male or is against the issue.
(c) The person is a female or favors the issue.

21. A grocery store employs cashiers, stock clerks, and deli personnel. The distribution of employees according to marital status is shown next.

Marital status	Cashiers	Stock clerks	Deli personnel
Married	8	12	3
Not married	5	15	2

If an employee is selected at random, find these probabilities:

(a) The employee is a stock clerk or married.
(b) The employee is not married.
(c) The employee is a cashier or is not married.

22. Students were surveyed on campus about their study habits. Some said they study in the morning, others study during the day between classes, and others study at night. Some students always study in a group and others always study alone. The distribution is shown below:

How students study	Morning	Between classes	Evening
Study in a group	2	3	1
Study alone	3	4	2

If a student who was surveyed is selected at random, find these probabilities:

(a) The student studies in the evening.
(b) The student studies in the morning or in a group.
(c) The student studies in the evening or studies alone.

23. Three cable channels (95, 97, and 103) air quiz shows, comedies, and dramas. The numbers of shows aired are shown here.

Type of show	Channel 95	Channel 97	Channel 103
Quiz show	5	2	1
Comedy	3	2	8
Drama	4	4	2

If a show is selected at random, find these probabilities:

(a) The show is a quiz show or it is shown on Channel 97.
(b) The show is a drama or a comedy.
(c) The show is shown on Channel 103 or it is a drama.

24. A local postal carrier distributed first-class letters, advertisements, and magazines. For a certain day, she distributed the following number of each type of item.

Delivered to	First-class letters	Ads	Magazines
Home	325	406	203
Business	732	1,021	97

If an item of mail is selected at random, find these probabilities:

(a) The item went to a home.
(b) The item was an ad or it went to a business.
(c) The item was a first-class letter or it went to a home.

25. As part of her major in microbiology, Juanita spent 3 weeks studying the spread of a disease in the jungles of South America. When she returned, she found that she had many e-mails sent to her home account and her school account, and that the e-mails were either spam, school announcements, or messages from friends as follows:

Delivered to	School announcements	Spam	Messages from friends
Home	325	406	203
School	732	1,021	97

If an e-mail is selected at random, find these probabilities:

(a) The e-mail was sent to her home.
(b) The e-mail was a school announcement or it was sent to her school account.
(c) The e-mail was spam or it was sent to her home account.

26. During a windstorm, the two large dice atop a new casino in Las Vegas crash into the pool below. Find the probabilities the dice landed with

(a) A sum of 6, 7, or 8.
(b) Doubles or a sum of 4 or 6.
(c) A sum greater than 9 or less than 4 or equal to 7.

27. Before a Walk for the Cure 10-mile walk, participants could choose a T-shirt from a box with six red shirts, two green shirts, one blue shirt, and one white shirt. When the first participant randomly selects a shirt from the box, what is the probability she will select a red shirt or a white shirt?

28. Colby tosses the contents of his jeans onto a table before putting them into the washer. Among the things thrown are three dice. Find the probability that the dice landed so that
    (a) All three have the same number showing.
    (b) The three numbers sum to 5.

Critical Thinking

29. The probability that a customer selects a pizza with mushrooms or pepperoni is 0.55, and the probability that the customer selects mushrooms only is 0.32. If the probability that he or she selects pepperoni only is 0.17, find the probability of the customer selecting both items.
30. In building new homes, a contractor finds that the probability of a buyer selecting a two-car garage is 0.70 and of selecting a one-car garage is 0.20. Find the probability that the buyer will select no garage. The builder does not build houses with garages for three or more cars.
31. In Exercise 30, find the probability that the buyer will not want a two-car garage.
32. In this exercise, we will attempt to justify addition rule 1. When two dice are rolled, we will find the probability of getting a sum of 5 or 10.
    (a) Explain why the two events are mutually exclusive.
    (b) How many total outcomes are possible when two dice are rolled?
    (c) How many outcomes result in a sum of either 5 or 10?
    (d) Find the probability of rolling a sum of 5 or 10 without the addition rule, using your answers from parts b and c.
    (e) Find the probability of rolling a sum of 5.
    (f) Find the probability of rolling a sum of 10.
    (g) Add your answers from parts e and f, then explain why the calculations from parts d and g will always be the same for any pair of mutually exclusive events.

Section 11-8 The Multiplication Rules and Conditional Probability

LEARNING OBJECTIVES

- ☐ 1. Find the probability of two or more independent events all occurring.
- ☐ 2. Find the probability of two or more dependent events all occurring.
- ☐ 3. Find conditional probabilities.

Based on data collected from 1960–2000, the probability of it raining on any given day during March in Daytona Beach, Florida, is about 1/5. So you head to Daytona for spring break, and you're more than a little disappointed when it rains the first 2 days. Trying to remain upbeat, you say "With only a $\frac{1}{5}$ chance of rain, what are the chances of it raining 3 days in a row?"

That's actually a much more interesting question than you think. One reason is that many people would say that it's less likely to rain the third day because it already rained 2 days in a row, but that's not the case. Assuming that there's not a massive storm system lasting for 3 days, what happened on Saturday has no effect on whether or not it rains on Monday, so the probability of rain on the third day is $\frac{1}{5}$ regardless of what happened the first 2 days.

Returning to the original question, the probability of it raining 3 days in a row, we see that a rule for finding the probability of consecutive events occurring would be helpful. As we will see, the probability of consecutive events depends on whether the first event affects the second.

Two events *A* and *B* are **independent** if the fact that *A* occurs has no effect on the probability of *B* occurring.

Based on our description above, rain on Saturday and rain on Monday are independent events.

Each day a professor selects a student at random to work out a problem from the homework. If the student is selected regardless of who already went, the events are independent. If the student is selected from the pool of students who haven't gone yet, the events are dependent.

Here are other examples of independent events:

Rolling a die and getting a 6, and then rolling a second die and getting a 3.

Drawing a card from a deck and getting a queen, replacing it, and drawing a second card and getting a queen.

On the other hand, when the occurrence of the first event changes the probability of the occurrence of the second event, the two events are said to be *dependent*. For example, suppose a card is drawn from a deck and *not* replaced, and then a second card is drawn. The probability for the second card is changed since the sample space contains only 51 cards when the first card is not replaced.

Two events A and B are **dependent** if the outcome of A has some effect on the probability of B occurring.

Here are some examples of dependent events:

Drawing a card from a deck, not replacing it, and then drawing a second card.

Selecting a lottery ball from a tumbler, not replacing it, and then drawing a second ball.

Parking in a no-parking zone and getting a parking ticket.

When a coin is flipped twice, the outcomes of the first and second flips are independent, and in each case, the probability of getting tails is $\frac{1}{2}$. The sample space for the two flips is {HH, TT, HT, TH}, and the probability of getting tails twice is $\frac{1}{4}$. Notice that $\frac{1}{4} = \frac{1}{2} \cdot \frac{1}{2}$. This shows that the probability of two consecutive tails is the product of the probabilities of getting tails in each individual trial. This result is our first *multiplication rule*.

Multiplication Rule 1

When two events A and B are independent, the probability of both occurring is

$$P(A \text{ and } B) = P(A) \cdot P(B)$$

EXAMPLE 1 Using Multiplication Rule 1

A coin is flipped and a die is rolled. Find the probability of getting heads on the coin and a 4 on the die.

Math Note

The problem in Example 1 could be solved using standard sample space methods, but as outcomes get more complicated, it can be extremely difficult to list sample spaces. That's why the multiplication rules are so helpful.

SOLUTION

The two events, the coin landing on heads and the die showing 4, are independent. We know from earlier examples that the probability of heads is $\frac{1}{2}$ and the probability of getting 4 with one die is $\frac{1}{6}$. Using the multiplication rule:

$$P(\text{heads and } 4) = P(\text{heads}) \cdot P(4) = \frac{1}{2} \cdot \frac{1}{6} = \frac{1}{12}$$

▼ Try This One 1

If the probability of your alarm not going off is $\frac{1}{20}$, and the probability of getting a ticket on your way to work is $\frac{1}{200}$, find the probability that both will happen.

EXAMPLE 2 Using Multiplication Rule 1

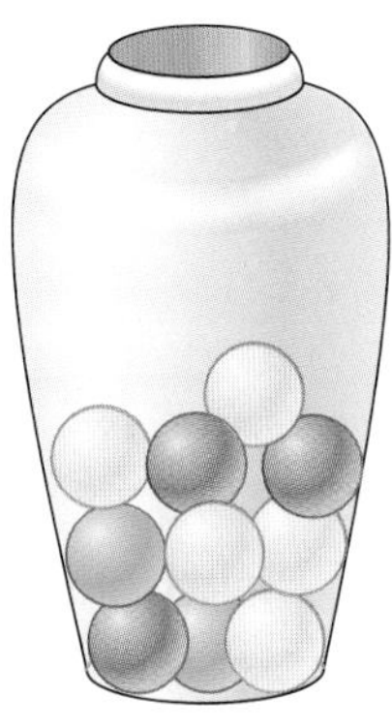

As part of a psychology experiment on perception and memory, colored balls are picked from an urn. The urn contains three red balls, two green balls, and five white balls. A ball is selected and its color is noted. Then it is replaced. A second ball is selected and its color is noted. Find the probability of each of these.

(a) Selecting two green balls.
(b) Selecting a green ball and then a white ball.
(c) Selecting a red ball and then a green ball.

SOLUTION

Remember, selection is done with replacement, which makes the events independent.

(a) The probability of selecting a green ball on each trial is $\frac{2}{10}$, so

$$P(\text{green and green}) = P(\text{green}) \cdot P(\text{green}) = \frac{2}{10} \cdot \frac{2}{10} = \frac{4}{100} = \frac{1}{25}$$

(b) The probability of selecting a green ball is $\frac{2}{10}$ and the probability of selecting a white ball is $\frac{5}{10}$, so

$$P(\text{green and white}) = P(\text{green}) \cdot P(\text{white}) = \frac{2}{10} \cdot \frac{5}{10} = \frac{10}{100} = \frac{1}{10}$$

(c) The probability of selecting a red ball is $\frac{3}{10}$ and the probability of selecting a green ball is $\frac{2}{10}$, so

$$P(\text{red and green}) = P(\text{red}) \cdot P(\text{green}) = \frac{3}{10} \cdot \frac{2}{10} = \frac{6}{100} = \frac{3}{50}$$

▼ Try This One 2

As part of a card trick, a card is drawn from a deck and replaced; then a second card is drawn. Find the probability of getting a queen and then an ace.

EXAMPLE 3 Finding Probabilities for Three Independent Events

Math Note

Multiplication rule 1 can be extended to three or more independent events using the formula

$$P(A_1 \text{ and } A_2 \text{ and } A_3 \cdots \text{ and } A_n) = P(A_1) \cdot P(A_2) \cdot P(A_3) \cdots P(A_n)$$

Three cards are drawn from a deck. After each card is drawn, its denomination and suit are noted and it is replaced before the next card is drawn. Find the probability of getting

(a) Three kings.
(b) Three clubs.

SOLUTION

(a) Since there are four kings, the probability of getting a king on each draw is $\frac{4}{52}$ or $\frac{1}{13}$. The probability of getting three kings is

$$\frac{1}{13} \cdot \frac{1}{13} \cdot \frac{1}{13} = \frac{1}{2{,}197}$$

(b) Since there are 13 clubs, the probability of getting a club is $\frac{13}{52}$ or $\frac{1}{4}$. The probability of getting three clubs in a row is

$$\frac{1}{4} \cdot \frac{1}{4} \cdot \frac{1}{4} = \frac{1}{64}$$

▼ Try This One 3

Given that the probability of rain on any given day in March in Daytona Beach is $\frac{1}{5}$, find the probability that

(a) It rains three straight days in March.
(b) It rains on March 10 and 12, but not March 11.

Assume that weather on any day is independent of the others.

When a few subjects are selected from a large number of subjects in a sample space, and the subjects are not replaced, the probability of the event occurring changes so slightly that, for the most part, it is considered to remain the same. Example 4 illustrates this concept.

EXAMPLE 4 Using Multiplication Rule 1 with Large Samples

According to a Gallup Poll in 2007, 47 percent of all parents in America with children under 18 feel frequent stress. If three parents are chosen randomly, find the probability that all three will say they feel frequent stress.

SOLUTION

Let S denote the event that the chosen parent feels stress.

$$P(S \text{ and } S \text{ and } S) = P(S) \cdot P(S) \cdot P(S) = (0.47)(0.47)(0.47) \approx 0.104$$

▼ Try This One 4

☑ 1. Find the probability of two or more independent events all occurring.

According to a report from the National Institute of Justice, 64% of males arrested for any reason in 2000 tested positive for illegal drugs. If four prisoners at random were chosen, find the probability that all four test positive.

CAUTION

Even though the three events in Example 4 are not independent, because parents were not replaced after being chosen, there are millions and millions of parents, so having one or two fewer in the sample has a negligible effect. Make sure to carefully consider the situation before using multiplication rule 1.

When we are interested in finding the probability of consecutive events that are dependent, we can still use the multiplication rule, but with a minor modification. For example, let's say we draw two cards at random from a standard deck. The probability of getting an ace on the first draw is $\frac{4}{52}$. But if the first card is not put back into the deck, that changes the probability of drawing another ace—it's now $\frac{3}{51}$. Using the multiplication rule, the probability of both events occurring is

$$\frac{4}{52} \cdot \frac{3}{51} = \frac{12}{2,652} = \frac{1}{221}.$$

We can summarize this procedure as multiplication rule 2.

Multiplication Rule 2

When two events are dependent, the probability of both occurring is

$$P(A \text{ and } B) = P(A) \cdot P(B \text{ given that } A \text{ has already occurred})$$

EXAMPLE 5 Using Multiplication Rule 2

An appliance store gets a shipment of 25 plasma TVs, and 3 of them are defective. If two of the TVs are chosen at random, find the probability that both are defective. (The first TV is not replaced after it's tested.)

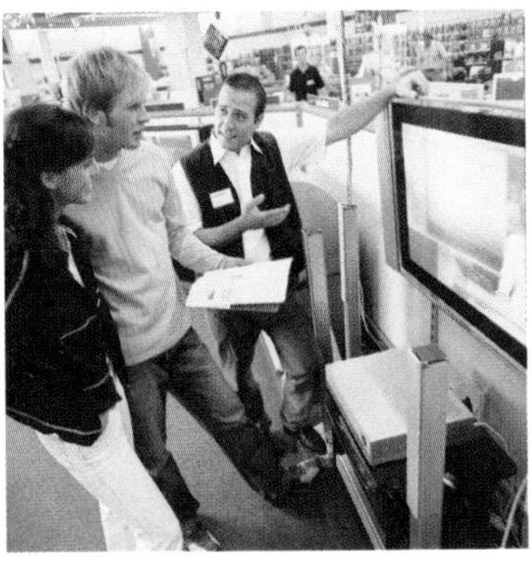

SOLUTION

Since there are 3 defective TVs out of a total of 25, the probability of the first being defective is $\frac{3}{25}$. After the first one is found to be defective and not replaced, there are 2 defective sets left out of 24, so the probability of the second being defective given that the first one is defective is $\frac{2}{24}$. Using multiplication rule 2, the probability that both are defective is

$$P(\text{1st defective and 2nd defective}) = P(\text{1st}) \cdot P(\text{2nd given 1st})$$
$$= \frac{3}{25} \cdot \frac{2}{24} = \frac{6}{600} = \frac{1}{100}$$

▼ Try This One 5

The 2009 NCAA men's basketball tournament field had (among 64 teams) 7 teams from the Big East conference, 7 from the Big Ten, 6 from the Big 12, and 3 from the Southeastern conference. If you were randomly assigned two teams in a dorm pool, find the probability that

(a) Both were from the Big East.
(b) The first was from the Southeastern conference and the second was from the Big 12.

Multiplication rule 2 can be extended to three or more events as shown in Example 6.

EXAMPLE 6 Using Multiplication Rule 2 with Three Events

Three cards are drawn from an ordinary deck and not replaced. Find the probability of

(a) Getting three jacks.
(b) Getting an ace, a king, and a queen in order.
(c) Getting a club, a spade, and a heart in order.
(d) Getting three clubs.

SOLUTION

(a) $P(\text{three jacks}) = \frac{4}{52} \cdot \frac{3}{51} \cdot \frac{2}{50} = \frac{24}{132{,}600} = \frac{1}{5{,}525}$

(b) $P(\text{ace and king and queen}) = \frac{4}{52} \cdot \frac{4}{51} \cdot \frac{4}{50} = \frac{64}{132{,}600} = \frac{8}{16{,}575}$

(c) $P(\text{club and spade and heart}) = \frac{13}{52} \cdot \frac{13}{51} \cdot \frac{13}{50} = \frac{2{,}197}{132{,}600} = \frac{169}{10{,}200}$

(d) $P(\text{three clubs}) = \frac{13}{52} \cdot \frac{12}{51} \cdot \frac{11}{50} = \frac{1{,}716}{132{,}600} = \frac{11}{850}$

☑ 2. Find the probability of two or more dependent events all occurring.

▼ Try This One 6

When drawing four cards from a deck, what is the probability that

(a) All four are aces? (b) All four are clubs?

Math Note

We use the term conditional probability because the condition that A has already occurred affects the probability of B.

Conditional Probability

We know that to find the probability of two dependent events occurring, it's important to find the probability of the second event occurring given that the first has already occurred. We call this the **conditional probability** of event B occurring given that event A has occurred, and denote it $P(B \mid A)$.

Now that we have a symbol to represent the probability of event B given that A has occurred, we can rewrite multiplication rule 2, and solve the equation for $P(B \mid A)$:

$$P(A \text{ and } B) = P(A) \cdot P(B \mid A)$$

$$P(B \mid A) = \frac{P(A \text{ and } B)}{P(A)}$$

This gives us a formula for conditional probability.

Formula for Conditional Probability

The probability that the second event B occurs given that the first event A has occurred can be found by dividing the probability that both events occurred by the probability that the first event has occurred. The formula is

$$P(B \mid A) = \frac{P(A \text{ and } B)}{P(A)}$$

Example 7 illustrates the use of this rule.

EXAMPLE 7 Finding a Conditional Probability

Military strategies (and other types of strategies) use conditional probability. If A occurs, how likely is B? How likely is B if A doesn't occur?

Suppose that your professor goes stark raving mad and chooses your final grade from A, B, C, D, F, or Incomplete totally at random. Find the probability of getting an A given that you get a letter grade higher than D.

SOLUTION

We are asked to find $P(\text{A} \mid \text{letter grade higher than D})$.

Method 1 Knowing that you got a letter grade higher than D reduces the sample space to {A, B, C}, which has three outcomes. One of them is an A, so $P(\text{A} \mid \text{letter grade higher than D}) = \frac{1}{3}$.

Method 2 With the full sample space of {A, B, C, D, F, I}, $P(\text{A}) = \frac{1}{6}$, and $P(\text{letter grade higher than D}) = \frac{3}{6}$. Using the formula for conditional probability,

$$P(\text{A} \mid \text{letter grade higher than D}) = \frac{\frac{1}{6}}{\frac{3}{6}} = \frac{1}{6} \cdot \frac{6}{3} = \frac{1}{3}$$

▼ Try This One 7

A group of buses is rented for an away game, and they are numbered from 1–8. All students are randomly assigned to a bus. What is the probability that you get an even–numbered bus given that your bus number is less than 6?

EXAMPLE 8 Finding a Conditional Probability

Hate crimes are defined to be crimes in which the victim is targeted because of one or more personal characteristics, such as race, religion, or sexual orientation. The table below lists the motivation for certain crimes as reported by the FBI for 2007.

Motivation	Crimes against persons	Crimes against property	Crimes against society
Race	3,031	1,686	7
Religion	421	1,054	2
Sexual orientation	1,039	418	3
Total	4,491	3,158	12

(a) Find the probability that a hate crime was racially motivated given that it was a crime against persons.
(b) Find the probability that a hate crime was against property given that it was motivated by the victim's sexual orientation.

SOLUTION

(a) Since we are interested only in crimes against persons, we only need to look at that column. There were 4,491 such crimes total, and 3,031 were racially motivated, so the probability is

$$\frac{3{,}031}{4{,}491} \approx 0.675$$

(b) This time we are given that the crime was motivated by sexual orientation, so we only need to look at that row. There were 1,460 such crimes total, and 418 were against property, so the probability is

$$\frac{418}{1{,}460} \approx 0.286$$

▼ Try This One 8

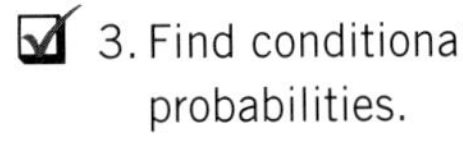
3. Find conditional probabilities.

Based on the data in the above table, find the probability that

(a) A crime was motivated by either race or religion given that it was a crime against society.
(b) A crime was against persons given that it was motivated by religion or sexual orientation.

Answers to Try This One

1 $\frac{1}{4,000}$

2 $\frac{1}{169}$

3 (a) $\frac{1}{125}$ (b) $\frac{4}{125}$

4 Approximately 0.168

5 (a) $\frac{1}{96}$ (b) $\frac{1}{224}$

6 (a) $\frac{1}{270,725}$ (b) $\frac{11}{4,165}$

7 $\frac{2}{5}$

8 (a) $\frac{3}{4}$ (b) $\frac{1,460}{2,937} \approx 0.497$

EXERCISE SET 11-8

Writing Exercises

1. What is the difference between independent and dependent events? Give an example of each.
2. What is meant by the term conditional probability?
3. Describe two methods for computing conditional probability.
4. Explain why the probability of two consecutive events occurring couldn't possibly be the sum of the two individual probabilities.

In Exercises 5–12, decide whether the events are independent or dependent, and explain your answer.

5. Tossing a coin and drawing a card from a deck.
6. Drawing a ball from an urn, not replacing it, and then drawing a second ball.
7. Getting a raise in salary and purchasing a new car.
8. Driving on ice and having an accident.
9. Having a large shoe size and having a high IQ.
10. A father being left-handed and a daughter being left-handed.
11. Smoking excessively and having lung cancer.
12. Eating an excessive amount of ice cream and smoking an excessive amount of cigarettes.

Real-World Applications

13. If 18% of students on campus said they were voting for Homer Simpson for the next President of the United States, find the probability that if three students were randomly selected on campus and asked whom they were voting for in the presidential election, all would say they were voting for Homer Simpson.
14. A national study of patients who were overweight found that 56% also have elevated blood pressure. If two overweight patients are selected, find the probability that both have elevated blood pressure.
15. According to the National Highway Traffic Safety Administration, 83% of Americans used seat belts regularly in 2008. If four people are selected at random, find the probability that all four regularly use seat belts.
16. A computer salesperson at Best Buy claims that she has a 20% chance of selling a computer when helping out a customer. If this is true and she talks to four customers before lunch, find the probability that all four will buy computers.
17. If 25% of Michael's graduating class are not U.S. citizens, find the probability that two randomly selected graduating students are not U.S. citizens.
18. If two people are selected at random, what is the probability that they were both born in December?
19. If two people are selected at random, find the probability that they were born in the same month.
20. If three people are selected, find the probability that all three were born in March.
21. If half of Americans believe that the federal government should take "primary responsibility" for eliminating poverty, find the probability that three randomly selected Americans will agree that it is the federal government's responsibility to eliminate poverty.
22. What is the probability that a husband, wife, and daughter have the same birthday?
23. A telecommunications company has six satellites, two of which are sending a weak signal. If two are selected at random without replacement find the probability that both are sending a weak signal.

24. In Exercise 23, find the probability that one satellite sends a strong signal and the other sends a weak signal.
25. The U.S. Department of Justice reported that in 2005, 50.8% of all murders are committed with a handgun. If three murder cases are selected at random, find the probability that a handgun was used in all three.
26. In a department store there are 120 customers, 90 of whom will buy at least one item. If five customers are selected at random, one by one, find the probability that all will buy at least one item.
27. During a game of online hearts, three cards are dealt, one at a time without replacement, from a shuffled, ordinary deck of cards. Find these probabilities:

    (a) All are jacks.
    (b) All are clubs.
    (c) All are red cards.
28. In a group of eight Olympic track stars, five are hurdlers. If three are selected at random without replacement, find the probability they are all hurdlers.
29. In Exercise 28, find the probability none of them are hurdlers.
30. In a class consisting of 15 men and 12 women, two different homework papers were selected at random. Find the probability that both papers belonged to women.
31. While practicing juggling with a die, a quarter, and a spoon, Ken drops the quarter first and then the die and is left holding the spoon. Find the probability of getting a 3 on the die given that the coin landed heads up.
32. Juan draws a black card from an ordinary deck as the first card for a game of Gin Rummy. What is the probability the card was a king?
33. At a carnival gambling booth, two dice are rolled, and a sum of seven wins double your money. Find the probability that the sum is 7 if it is given that one of the numbers was a 6.
34. Frank flips a coin to see who will pick up the pizza from Pizza Hut, and then Sun rolls a die to see how much the person picking up the pizza has to pay toward the meal. Find the probability that the coin landed on tails and the number on the die was odd.
35. A computer randomly deals an ordinary deck of cards for a game of FreeCell, and the first card dealt is a face card. Find the probability that the card was also a diamond.
36. During a backgammon game, Kelly rolled the two dice on the board. Find the probability that the sum obtained was greater than 8 given that the number on one die was a 6.

Use this information for Exercises 37–40.
Three red cards are numbered 1, 2, and 3. Three black cards are numbered 4, 5, and 6.
The cards are placed in a box and one card is selected at random.

37. Find the probability that a red card was selected given that the number on the card was an odd number.
38. Find the probability that a number less than 5 was selected given that the card was a black card.
39. Find the probability that a number less than 5 was selected given that the card was red.
40. Find the probability that a black card was selected given that the number on the card was an even number.

Use the following information for Exercises 41–44.
A survey of 200 college students shows the average number of minutes that people talk on their cell phones each month.

	Less than 600	600–799	800–999	1,000 or more
Men	56	18	10	16
Women	61	18	13	8

If a person is selected at random, find these probabilities:

41. The person talked less than 600 minutes if it is known that the person was a woman.
42. The person talked 1,000 or more minutes if it is known that the person was a man.
43. The person was a woman if it is known that the person talked between 600 and 799 minutes.
44. The person was a man if it is known that the person talked between 600 and 999 minutes.

Critical Thinking

45. A magician is carrying his box of tricks and out spill three dice. What is the probability that the three dice sum to 9?
46. In a rush to pay for her latte, Shelly dropped her change purse and five coins fell out. What is the probability that as the coins fell to the floor, there was at least one head?
47. During a comedy show, the comedian selected three people at random from the audience. What is the probability that the three people will have the same birthday? (Ignore the year.)
48. Suppose that I roll two dice and tell you that the result is definitely even, then offer you 3 to 1 odds that the total isn't 4. Draw a Venn diagram to illustrate the outcomes, then use the diagram to compute the conditional probability. Finally, decide if the given odds are in your favor or not.

Section 11-9 The Binomial Distribution

LEARNING OBJECTIVES

- ☐ 1. Identify binomial experiments.
- ☐ 2. Compute probabilities of outcomes in a binomial experiment.
- ☐ 3. Construct a probability distribution.

Many probability problems involve situations that have only two outcomes. When a baby is born, it will be either male or female. When a coin is tossed, it will land either heads or tails. When the New York Yankees play, they either win or they lose. That cute girl that sits behind you in class will either go out with you or she won't.

Other situations can be reduced to two outcomes. For example, medical procedures can be classified as either successful or unsuccessful. An answer to a multiple choice exam question can be classified as correct or incorrect even though there may be four answer choices.

Situations like these are called *binomial experiments.*

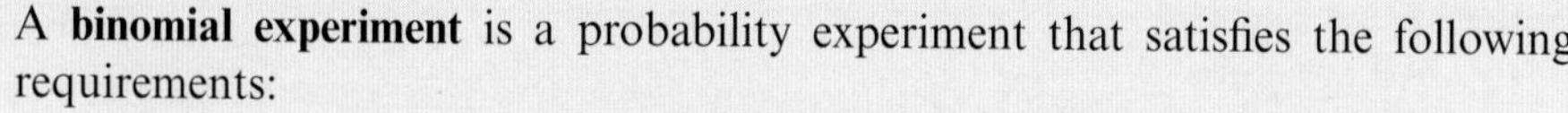

A **binomial experiment** is a probability experiment that satisfies the following requirements:

1. Each trial can have only two outcomes, or outcomes that can be reduced to two outcomes. These outcomes can be considered as either a success or a failure.
2. The outcomes must be independent of each other.
3. There must be a fixed number of trials.
4. The probability of a success must remain the same for all trials of the experiment.

EXAMPLE 1 Deciding if an Experiment is Binomial

Decide whether or not each is a binomial experiment.

(a) Drawing a card from a deck and seeing what suit it is
(b) Answering a question on a true-false test
(c) Asking 100 people whether or not they smoke
(d) Selecting cards at random from a deck without replacement and deciding if they are red or black cards

SOLUTION

(a) No, since there are four outcomes: heart, diamond, spade, or club.
(b) Yes, there are only two outcomes: correct and incorrect.
(c) Yes, there are only two outcomes: yes or no.
(d) No, since the cards are not being replaced, the probability changes on each draw.

▼ Try This One 1

☑ 1. Identify binomial experiments.

Decide whether or not each experiment is a binomial experiment.

(a) Selecting a colored ball with replacement from an urn with three balls, each of which is a different color
(b) Selecting a number from a bingo machine
(c) Selecting a card at random from a deck with replacement and noting its color, red or black
(d) Tossing a die and getting a 3

Consider the following experiment. A box contains four colored balls—one each of red, black, white, and green. A ball is randomly selected and its color is recorded. It is then replaced and a second ball is randomly selected and its color is recorded. The tree diagram for this experiment is shown in Figure 11-9.

The probability of selecting two green balls is $\frac{1}{16}$ since there is only one way to select two green balls, namely (G, G), and there are 16 total possible outcomes in the sample space. The probability of selecting exactly one green is $\frac{6}{16}$ or $\frac{3}{8}$ since there are 6 outcomes that contain one green ball: (R, G), (B, G), (W, G), (G, R), (G, B), and (G, W). The probability of selecting no green balls is $\frac{9}{16}$ since there are 9 outcomes that contain no green balls.

A table for the probabilities can be shown.

Outcome	Probability
0 green balls	$\frac{9}{16}$
1 green ball	$\frac{3}{8} = \frac{6}{16}$
2 green balls	$\frac{1}{16}$
Sum	$\frac{16}{16} = 1$

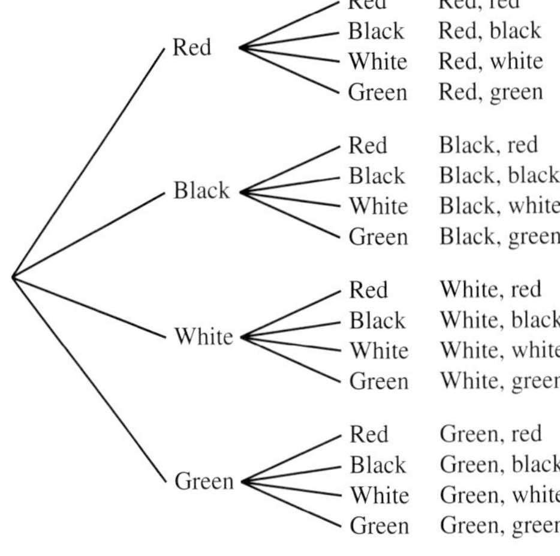

Figure 11-9

The experiment can be considered a binomial experiment since it meets the four conditions of a binomial experiment:

1. Each trial has only two outcomes. The ball chosen is either green or it is not green.
2. The outcomes are independent of each other: each ball is replaced before the second ball is selected.
3. There is a fixed number of trials. In this case, there are two draws.
4. The probability of a success remains the same in each case. In this case, since there is only one green ball and a total of four balls, $P(S) = \frac{1}{4}$ and $P(F) = 1 - \frac{1}{4} = \frac{3}{4}$. (Getting a green ball is considered a success. Getting any other colored ball is considered a failure.)

The probability of a given outcome for a binomial experiment can be found using the binomial probability formula.

The Binomial Probability Formula

The probability of exactly x successes in n trials is

$$P(x) = {}_nC_x \cdot p^x \cdot q^{n-x}$$

where p = the probability of a success

q = the probability of a failure ($q = 1 - p$)

Since p is the probability of success on any given trial, using the multiplication rule, p^x is the probability of x successes. The same reasoning shows that q^{n-x} is the probability of $n - x$ failures. We also need to multiply by the number of different ways we can get x successes in n trials, which is ${}_nC_x$: the number of combinations of n objects taken x at a time.

In our example of drawing balls from a box, let's find the probability of getting one green ball in two draws. In this case, $n = 2$, $x = 1$, and $n - x = 1$. The probability of success (p) is $\frac{1}{4}$, and the probability of failure (q) is $\frac{3}{4}$. Using the binomial probability formula,

$$\begin{aligned} P(1 \text{ green ball}) &= {}_nC_x \cdot p^x \cdot q^{n-x} \\ &= {}_2C_1 \cdot \left(\frac{1}{4}\right)^1 \cdot \left(\frac{3}{4}\right)^1 \\ &= \frac{2!}{(2-1)!1!} \cdot \frac{1}{4} \cdot \frac{3}{4} \\ &= 2 \cdot \frac{3}{16} = \frac{3}{8} \end{aligned}$$

This matches the probability found using a tree diagram.

EXAMPLE 2 Using the Binomial Probability Formula

Suppose that the morning after your birthday, you remember that you have a 20 question true or false quiz in your early class. Uh oh! Completely unprepared and a little woozy, you decide to guess on every question. What is the probability that you'll get 16 out of 20 right?

SOLUTION

Math Note

Notice that because there are 20 trials in Example 2, it would be cumbersome (to say the very least!) to use a tree diagram to find the probability.

Because the questions are true or false, and you're arbitrarily guessing, the probability of getting any given question right is $\frac{1}{2}$. The number of trials is $n = 20$, and the number of successes is $x = 16$. The probability of success is $p = \frac{1}{2}$, and the probability of failure is then $q = 1 - \frac{1}{2}$. Substituting these values into the binomial probability formula, we get

$$\begin{aligned} P(16 \text{ right}) &= {}_nC_x \cdot p^x \cdot q^{n-x} \\ &= {}_{20}C_{16} \cdot \left(\frac{1}{2}\right)^{16} \cdot \left(\frac{1}{2}\right)^{20-16} \\ &= \frac{20!}{(20-16)!16!} \cdot \left(\frac{1}{2}\right)^{16} \cdot \left(\frac{1}{2}\right)^4 \\ &= 4{,}845 \cdot \frac{1}{1{,}048{,}576} \approx 0.0046 \end{aligned}$$

In short, your chances are not good.

▼ Try This One 2

What is the probability of getting exactly 6 questions right when guessing on a 10 question true or false quiz?

EXAMPLE 3 Using the Binomial Probability Formula

A box contains three red balls and two yellow balls. Five balls are selected with each ball being replaced before the next ball is selected. Find the probability of getting

(a) Exactly two yellow balls
(b) Fewer than two yellow balls

SOLUTION

(a) Use the binomial probability formula with $n = 5$, $x = 2$, $p = \frac{2}{5}$, and $q = \frac{3}{5}$.

$$P(2 \text{ yellow}) = {}_5C_2\left(\frac{2}{5}\right)^2\left(\frac{3}{5}\right)^3$$
$$= \frac{5!}{(5-2)!2!} \cdot \frac{4}{25} \cdot \frac{27}{125}$$
$$= 0.3456$$

(b) Fewer than two yellow balls means 0 or 1 yellow ball, so we will find each probability and add the answers.

$$P(0 \text{ yellow}) = {}_5C_0\left(\frac{2}{5}\right)^0\left(\frac{3}{5}\right)^5$$
$$= \frac{5!}{(5-0)!0!}(1)\left(\frac{243}{3{,}125}\right)$$
$$= 0.07776$$

$$P(1 \text{ yellow}) = {}_5C_1\left(\frac{2}{5}\right)^1\left(\frac{3}{5}\right)^4$$
$$= \frac{5!}{(5-1)!1!}\left(\frac{2}{5}\right)\left(\frac{81}{625}\right)$$
$$= 0.2592$$

$$P(0 \text{ or } 1 \text{ yellow}) = 0.07776 + 0.2592 = 0.33696$$

▼ Try This One 3

For the experiment in Example 3, find the probability of getting

(a) Exactly one red ball

(b) More than three red balls

When several subjects are selected without replacement from a large group of subjects, theoretically the selections are not independent and the binary probability formula doesn't apply. However, since the individual probabilities of a subject selected change only slightly, binomial probabilities can be used. This is shown in the next example.

EXAMPLE 4 Using the Binary Probability Formula with a Large Sample

Forty-two percent of individuals surveyed said that a man should always pay for the first date. If 10 people are randomly selected, find the probability that 3 randomly selected people will agree that a man should pay.

SOLUTION

Since the population is large, the binomial distribution can be used even though there's no replacement. In this case, $n = 10$, $x = 3$, $n - x = 7$, $p = 0.42$ and $q = 1 - 0.42 = 0.58$.

$$P(x) = {}_nC_x \cdot p^x \cdot q^{n-x}$$
$$P(3 \text{ people agree}) = {}_{10}C_3(0.42)^3(0.58)^7$$
$$= \frac{10!}{(10-7)!3!}(0.42)^3(0.58)^7$$
$$\approx 120(0.074)(0.022)$$
$$\approx 0.196$$

2. Compute probabilities of outcomes in a binomial experiment.

▼ Try This One 4

In a large community, it was determined that 44% of the residents use the public library at least once a year. If 10 people are selected, find the probability that exactly 2 people have used the library during the last year.

To describe all of the outcomes in a probability experiment, we can build a *probability distribution.*

A **probability distribution** consists of a list of all outcomes and the corresponding probabilities for a probability experiment.

EXAMPLE 5 Constructing a Probability Distribution

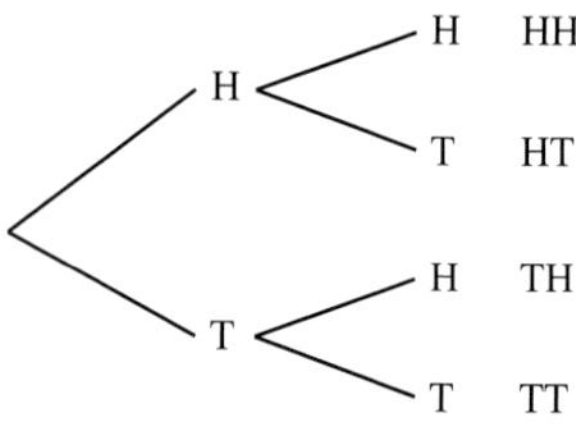

Figure 11-10

Construct a probability distribution for the possible number of tails when you flip two coins.

SOLUTION

When two coins are tossed, the outcomes can be shown using a tree diagram. See Figure 11-10.

The sample space is {HH, HT, TH, TT}. Each outcome has a probability of $\frac{1}{4}$. Notice that the outcome "No tails" is HH, and $P(\text{HH}) = \frac{1}{4}$. The outcome of one tail consists of HT and TH, and $P(\text{HT, TH}) = \frac{1}{4} + \frac{1}{4} = \frac{1}{2}$. The outcome of two tails is TT and $P(\text{TT}) = \frac{1}{4}$. Now a probability distribution can be constructed by considering the outcomes as the number of tails. The probability distribution is

Number of tails, *x*	0	1	2
Probability, *P*(*x*)	$\frac{1}{4}$	$\frac{1}{2}$	$\frac{1}{4}$

Math Note

In Example 5, the number of trials was small enough that a tree diagram was more efficient than using the binomial probability formula three times.

3. Construct a probability distribution.

▼ Try This One 5

Three cards are numbered 1, 2, 3 and placed into a bag. Another bag is set up the same way, then one card is drawn from each bag and the numbers are added. Find all possible totals, and construct a probability distribution for the experiment.

Answers to Try This One

1 (a) Yes (b) No (c) Yes (d) Yes

2 $\frac{105}{512} \approx 0.205$

3 (a) 0.0768 (b) 0.33696

4 ≈ 0.084

5

Total	2	3	4	5	6
Probability	$\frac{1}{9}$	$\frac{2}{9}$	$\frac{1}{3}$	$\frac{2}{9}$	$\frac{1}{9}$

EXERCISE SET 11-9

Writing Exercises

1. What are the four requirements for a probability experiment to be a binomial experiment?
2. Give a brief description of how an experiment with several outcomes can be reduced to one with two outcomes.
3. What is a probability distribution?
4. In a binomial experiment, given the probability of any trial being a success, explain how to find the probability of a failure.

Computational Exercises

Find the probability of each using the binary probability formula.

5. $n = 3, p = 0.40, x = 2$
6. $n = 5, p = 0.80, x = 3$
7. $n = 10, p = 0.66, x = 4$
8. $n = 5, p = 0.20, x = 8$
9. $n = 20, p = 0.93, x = 12$
10. $n = 13, p = 0.16, x = 7$
11. $n = 9, p = 0.33, x = 5$
12. $n = 6, p = 0.58, x = 2$
13. $n = 4, p = 0.72, x = 0$
14. $n = 7, p = 0.25, x = 1$

Real-World Applications

15. During a game of Yahtzee, a player needs three sixes. She rolls the same die three times. What is the probability she got three sixes?
16. A cooler held three energy drinks. One was a Monster, another was a Red Bull, and the third was Atomic X. Simon selects an energy drink at random, puts it back, then selects another, puts it back and then selects a third drink. Find the probability that
    (a) No Red Bull drinks are selected.
    (b) Exactly one Atomic X drink is selected.
    (c) Exactly three Monster drinks are selected.
17. Of all identity thefts, 27% are solved in 1 day or less. If 10 cases are selected, find the probability that exactly 5 are solved in 1 day or less.
18. Of all people who do banking, 16% prefer to use ATMs. Of 20 people who are banking customers, find the probability that exactly 4 prefer to use the ATM.
19. According to the Bureau of Labor Statistics, among married-couple families where both the husband and wife have earnings, 26% of wives earned more than their husbands. If 18 wives are selected, find the probability that exactly 6 of them earn more than their husbands.
20. It is reported that 41% of people surveyed said that they eat meals while driving. If 12 drivers are selected, find the probability that exactly 6 will say that they eat meals while driving.
21. An instructor gives a 10-question true-false exam. If a student does not study and guesses, find the probability of getting exactly five correct answers.
22. A multiple-choice quiz consists of five questions. Each question has four responses, and only one of the answers is correct. Find the probability of guessing and getting exactly three correct answers.
23. Approximately 45% of people who eat at fast food places choose McDonald's. If 10 randomly selected people are surveyed, find the probability that 5 selected McDonald's.
24. About 30% of people who listen to commercial radio change the station in 1 to 2 minutes after the commercials begin. If six people are randomly selected, find the probability that two will change the station within 1 to 2 minutes after the commercials begin.
25. A survey shows that 65% of workers ages 42 to 60 would select excellent retirement benefits over a high salary when seeking a job. If 15 people are randomly selected, find the probability that 9 would select excellent benefits over a high salary.
26. About 22% of people selected Yahoo for instant messaging services during a selected year. If six randomly selected people are asked what instant messaging service they use, find the probability that one will say that he or she uses Yahoo.
    (Source: Nielson/net Ratings.)
27. A survey done by Zoomerang found that 30% of the people who were planning to remodel were planning to remodel the family room or den. If 12 people planning to remodel are selected, find the probability that 5 are planning to remodel the family room or den.
28. Approximately 20% of all college students in the United States do not have health insurance. If 10 college students are randomly selected in the United States, find the probability that 3 will not have health insurance.

29. A survey found that 55% of people said that a diet was harder to stick to than a budget. (Source: Kelton Research for Medifast.) If 20 people are randomly selected, find the probability that 10 will say that a diet is harder to stick to than a budget.
30. A survey done by Harris found that 40% of the people surveyed said that their children will have between \$5,001 and \$20,000 in debt when they graduate from college. If 25 people are selected, find the probability that 14 will say that their children will have a debt between \$5,001 and \$20,000 when they graduate from college.
31. A survey done by Harris International QUICK-QUERY found the 10% of women had a fear of flying. If 50 women are selected, find the probability that 6 will have a fear of flying.
32. In a Harris survey, 40% of the people surveyed said that they received error messages from doing online transactions. If 18 people are randomly selected, find the probability that 5 or 6 people would have received error messages while doing online transactions.
33. A survey found that 33% of people earning between \$30,000 and \$75,000 said that they were very happy. If 6 people are selected at random, find the probability that at most 2 people who earn between \$30,000 and \$75,000 would consider themselves very happy.
34. In a recent survey, 2% of the people selected stated that they would keep their current job if they won a multi-million-dollar lottery. If 20 people are selected, find the probability that 3, 4, or 5 of them would keep their job.

Critical Thinking

35. Using coins, cards, colored balls, etc., design a binomial experiment.

CHAPTER 11 Summary

Section	Important Terms	Important Ideas
11-1	Fundamental counting principle Factorial notation Permutation Permutation rule Tree diagram	**In order** to determine the total number of outcomes for a sequence of events, the fundamental counting rule or the permutation rule can be used. When the order or arrangement of the objects in a sequence of events is important, then the result is called a permutation of the objects.
11-2	Combination Combination rule	**When the** order of the objects is not important, then the result is called a combination.
11-3	Probability experiment Outcome Sample space Event Classical probability Empirical probability Observed frequency	**Flipping coins,** drawing cards from a deck, and rolling a die are examples of probability experiments. The set of all possible outcomes of a probability experiment is called a sample space. The two types of probability are classical and empirical. Classical probability uses sample spaces. Empirical probability uses frequency distributions and is based on observation. Probability is a number that represents how likely it is that something will occur. Probability can be zero, one, or any number in between. When the probability of an event is close to zero, the event is highly unlikely to occur. When the probability of an event is near one, the event is almost certain to occur.
11-4		**When sample** spaces are difficult to identify, it is often helpful to use tree diagrams or tables to identify all possible outcomes for a probability experiment.
11-5		**Probabilities** of events can be found by using the fundamental counting principle, the permutation rule, or the combination rule, depending on the situation.
11-6	Odds Expectation (expected value)	**In order** to determine payoffs, gambling establishments give odds. There are two ways to compute odds for a game of chance. They are "odds in favor of an event" and "odds against the event." Another concept related to probability and odds is the concept of expectation or expected value. Expected value is used to determine what happens over the long run. It is found by multiplying the outcomes of a probability experiment by their corresponding probabilities and then finding the sum of the products.
11-7	Mutually exclusive events Addition rules	**Two events** are said to be mutually exclusive if they cannot occur at the same time. To find the probability of one event or another event occurring, one of two addition rules can be used, depending on whether or not the events are mutually exclusive.

11-8	Multiplication rules Independent events Dependent events Conditional probability	**Events can** be classified as independent or dependent. Events are said to be independent if the occurrence of the first event does not affect the probability of the occurrence of the next event. If the probability of the second event occurring is changed by the occurrence of the first event, then the events are dependent. To find the probability of one event and another event occurring, one of the two multiplication rules can be used, depending on whether the two events are independent or dependent. If the probability of an event *B* occurring is affected by an event *A* occurring, then we say that a condition has been imposed on the event and the probability of event *B* occurring given that *A* has occurred is called a conditional probability.
11-9	Binomial experiment Probability distribution	**Many probability** experiments have two outcomes or can be reduced to two outcomes. If the trials are independent, fixed in number, and have the same probability of a success, then the experiment can be considered a binomial experiment. These problems can be solved using the binomial probability formula. For some probability experiments, a probability distribution can be constructed, listing all outcomes with the corresponding probabilities.

MATH IN Gambling REVISITED

1. Using the table for rolling two dice on page 601, the probability of rolling 2, 3, 11, or 12 is $\frac{1}{6}$, meaning the probability of losing is $\frac{5}{6}$. So the two outcomes are $+\$4$ with probability $\frac{1}{6}$, and $-\$1$ with probability $\frac{5}{6}$. The expected value of each trial is then

$$+4 \cdot \frac{1}{6} + (-1) \cdot \frac{5}{6} = -\frac{1}{6}$$

In 100 trials, you would be expected to lose $\$100(\frac{1}{6})$, or \$16.67.

2. The probability of winning \$35 is $\frac{1}{38}$, and the probability of losing \$1 is $\frac{37}{38}$. The expected value of each spin is

$$+35 \cdot \frac{1}{38} + (-1) \cdot \frac{37}{38} = -\frac{2}{38} = -\frac{1}{19}$$

In 100 trials, you would expect to lose $\$100(\frac{1}{19})$, or \$5.26.

3. Using the same idea as scenarios 1 and 2, the expected value of each ticket is

$$+20{,}000{,}000 \cdot \frac{1}{175{,}711{,}536} + (-1) \cdot \frac{175{,}711{,}535}{175{,}711{,}536}$$
$$\approx -\$0.8862$$

Buying 100 tickets, you would be expected to lose $\$100(-0.8862)$, or \$88.62.

4. You would win \$1 with probability $\frac{1}{2}$, and lose \$1 with probability $\frac{1}{2}$, so the expected value of each flip is

$$1 \cdot \frac{1}{2} + (-1) \cdot \frac{1}{2} = 0$$

In 100 flips you would be expected to break even.

The best, by far is flipping a coin, followed by roulette, the church fair dice game, and, bringing up the rear by quite a bit, the multistate lottery. Notice that under the best of circumstances, you are likely to break even!

Review Exercises

Section 11-1

1. Compute $\frac{14!}{11!}$.
2. Compute ${}_{12}P_6$.
3. An automobile license plate consists of three letters followed by four digits. How many different plates can be made if repetitions are allowed? If repetitions are allowed in the letters but not in the digits?
4. How many different arrangements of the letters in the word *bread* are there?
5. How many different arrangements of the letters in the word *cheese* are there?
6. How many different three-digit odd numbers use only the digits 0, 1, 2, 3, 4?

Section 11-2

7. Compute ${}_9C_6$.
8. Find both the number of combinations and the number of permutations of 10 objects taken 4 at a time.
9. Describe the difference between combinations and permutations.
10. How many different three-digit combinations can be made by using the numbers 1, 3, 5, 7, and 9 without repetitions if the "right" combination can open a safe? Does a combination lock really use combinations?
11. How many two-card pairs (i.e., the same rank) are there in a standard deck?
12. How many ways can five different television programs be selected from 12 programs?
13. A quiz consists of six multiple-choice questions. Each question has three possible answer choices. How many different answer keys can be made?
14. How many different ways can a buyer select four television models from a possible choice of six models?

Section 11-3

15. Which of the following numbers could represent a probability?

    (a) $\frac{3}{2}$

    (b) $\frac{2}{3}$

    (c) 0.1

    (d) $-\frac{1}{2}$

    (e) 80%

16. When a die is rolled, find the probability of getting

    (a) A 5. (b) A 6. (c) A number less than 5.

17. When a card is drawn from a deck, find the probability of getting

    (a) A heart.
    (b) A 7 and a club.
    (c) A 7 or a club.
    (d) A jack.
    (e) A black card.

18. In a survey conducted at the food court in the local mall, 20 people preferred Panda Express, 16 preferred Sbarro Pizza, and 9 preferred Subway for lunch. If a person is selected at random, find the probability that he or she prefers Sbarro Pizza.
19. If a die is rolled one time, find these probabilities:

    (a) Getting a 7
    (b) Getting an odd number
    (c) Getting a number less than 3

20. In a recent survey in a college dorm that has 1,500 rooms, 850 have an Xbox 360. If a room in this dorm is randomly selected, find the probability that it has an Xbox 360.
21. During a Midnight Madness sale at Old Navy, 16 white cargo pants, 3 khaki cargo pants, 9 tan cargo pants, and 7 black cargo pants were sold. If a customer who made a purchase during the sale is selected at random, find the probability that he or she bought

    (a) A pair of tan cargo pants.
    (b) A pair of black cargo pants or a pair of white cargo pants.
    (c) A pair of khaki, tan, or black cargo pants.
    (d) A pair of cargo pants that was not white.

22. In a survey of college students who reported they used a search engine that day, 16 reported they used Google, 4 used MSN, 3 used Yahoo! Search, and 7 used Ask Jeeves. If a student from the survey is selected at random and the student used one search engine that day, find the probability he or she used

    (a) Yahoo! Search.
    (b) MSN or Ask Jeeves.
    (c) Google or Yahoo! Search or Ask Jeeves.
    (d) A search engine that was not MSN.

23. When two dice are rolled, find the probability of getting

    (a) A sum of 5 or 6.
    (b) A sum greater than 9.
    (c) A sum less than 4 or greater than 9.
    (d) A sum that is divisible by 4.
    (e) A sum of 14.
    (f) A sum less than 13.

24. Two dice are rolled. Find the probability of getting a sum of 8 if the number on one die is a 5.

Section 11-4

25. A person rolls an eight-sided die and then flips a coin. Draw a tree diagram and find the sample space.
26. A student can select one of three courses at 8:00 A.M.: English, mathematics, or chemistry. The student can select either psychology or sociology at 11:00 A.M. Finally, the student can select either world history or economics at 1:00 P.M. (a) Draw a tree diagram and find all the different ways the student can make a schedule. (b) Repeat part (a), but include the condition that the student will take classes only at 8 A.M. and 11 A.M. if his first class is chemistry.
27. As an experiment in probability, a two-question multiple choice quiz is given at the beginning of class, but the answers are all written in Hebrew, which none of the students can read. This forces everyone to guess. Each question has choices A, B, C, D, and E. Construct a table that displays the sample space, then use the table to find the probability that both questions in a randomly selected quiz were answered with D or E (either D-E, or E-D).

Section 11-5

28. A card is selected from a deck. Find the probability that it is a diamond given that it is a red card.
29. A person has six bond accounts, three stock accounts, and two mutual fund accounts. If three investments are selected at random, find the probability that one of each type of account is selected.
30. A newspaper advertises five different movies, three plays, and two baseball games. If a couple selects three activities at random, find the probability they will attend two plays and one movie.
31. In putting together the music lineup for an outdoor spring festival, Fast Eddie can choose from 4 student bands and 12 nonstudent bands. There are five time slots for bands; the first at 3 P.M., the others on the hour until 7 P.M. If Eddie chooses the bands randomly, find the probability that no student bands will be picked.

Section 11-6

32. Find the odds for an event E when $P(E) = \frac{1}{4}$.
33. Find the odds against an event E when $P(E) = \frac{5}{6}$.
34. Find the probability of an event when the odds for the event are 6:4.
35. The table lists five outcomes for a probability experiment with the corresponding probabilities. Find the expected value.

Outcome	5	10	15	20	25
Probability	0.5	0.2	0.1	0.1	0.1

36. Ishi has a penny, a nickel, a dime, a quarter, and a half-dollar in her wallet. One falls out onto the counter as she is paying for gas. Find the expected value of the event.
37. A person selects a card from a deck. If it is a red card, he wins \$1. If it is a black card between and including 2 and 10, he wins \$5. If it is a black face card, he wins \$10, and if it is a black ace, he wins \$100. Find the expectation of the game.

Section 11-7

In Exercises 38–40, decide if the two events are mutually exclusive.

38. You meet someone while out; she gives you her phone number or her email address.
39. You complete a course and either pass or fail.
40. You spend a weekend in Las Vegas; you either win money, lose money, or break even.
41. If one of the 50 states is selected at random to be the site of a new nuclear power plant,
    (a) Find the probability that the state either borders Canada or Mexico.
    (b) Find the probability that the state begins with either A or ends with S.
42. During Halloween, Seth collected six Snickers bars, three Twix bars, two packages of M&M's, and two boxes of Nerds while trick-or-treating in the dorm. If one candy is selected from his candy stash, find the probability of getting a package of M&M's or a Twix bar.

Section 11-8

In Exercises 43–46, decide if the two events are independent.

43. Missing 3 straight days of class and failing the next test.
44. Missing 3 straight days of class and getting overloaded with spam e-mails.
45. Drawing an ace from a standard deck, then drawing a second ace.
46. Drawing an ace from a standard deck, then replacing that card, shuffling, and drawing another ace.
47. In a family of three children, find the probability that all the children will be girls if it is known that at least one of the children is a girl.
48. A Gallup Poll found that 78% of Americans worry about the quality and healthfulness of their diet. If five people are selected at random, find the probability that all five worry about the quality and healthfulness of their diet.

Source: *The Book of Odds*, Michael D. Shook and Robert C. Shook (New York: Penguin Putnam, Inc., 1991), p. 33.

49. Twenty-five percent of the engineering graduates of a university received a starting salary of \$50,000 or more. If three of the graduates are selected at random, find the probability that all have a starting salary of \$50,000 or more.
50. Three cards are drawn from an ordinary deck *without* replacement. Find the probability of getting
    (a) All black cards.
    (b) All spades.
    (c) All queens.
51. A coin is tossed and a card is drawn from a deck. Find the probability of getting
    (a) A head and a 6.
    (b) A tail and a red card.
    (c) A head and a club.
52. The data in the table below are based on a nationwide exit poll taken on November 4, 2008. The numbers are based on an average sample of 1,000 voters. The rows represent the political affiliation of voters, and the columns represent how they voted.

	Obama	McCain	Other
Democrat	347	39	4
Republican	32	285	3
Independent	151	128	11

    (a) Find the probability that a randomly selected person voted for Obama given that he or she was not Republican.
    (b) Find the probability that a randomly selected person voted for either Obama or McCain given that he or she was Independent.
    (c) Find the probability that a randomly selected person was a Republican given that he or she voted for McCain.

Section 11-9

53. Use the binomial probability formula to find the probability of five successes in six trials when the probability of success on each trial is 1/3.
54. A survey found that 24% of families eat at home as a family five times a week. If 10 families are selected, find the probability that exactly 3 will say that they eat at home as a family five times a week.
55. According to a survey, 45% of teenagers said that they have seen passengers in an automobile encouraging the driver to speed. If 16 teens are selected, find the probability that exactly 6 will say that they have seen passengers encouraging the driver to speed.
56. Construct a probability distribution for the possible number of heads when tossing a coin three times.

Chapter Test

In Exercises 1–4, compute the requested value.

1. $\frac{20!}{18!}$
2. ${}_7C_5$
3. ${}_{12}P_5$
4. The probability of 8 successes in 10 trials when the probability of success on each trial is $\frac{1}{4}$.
5. If someone saw the title of Chapter 11 in this book and asked you "What is probability?" what would you say?
6. Describe a situation where combinations would be used to count possibilities and one where permutations would be used.
7. One company's ID cards consist of five letters followed by two digits. How many cards can be made if repetitions are allowed? If repetitions are not allowed?
8. A physics test consists of 25 true-false questions. How many different possible answer keys can be made?
9. At Blockbuster Video, there are seven DVDs in a box. How many different ways can four DVDs be selected?
10. How many ways can five sopranos and four altos be chosen for the university chorus from seven sopranos and nine altos?
11. Eight students at graduation are in line to take their seats. How many ways can they be seated in a row?
12. A soda machine servicer must restock and collect money from 15 machines, each one at a different location. How many ways can she select four machines to service in one day?
13. How many different ways can three cell phone covers be drawn from a box containing four differently colored cell phone covers, if
    (a) Each cell phone cover is replaced after it has been drawn?
    (b) There is no replacement?
14. If a man can wear a shirt or a sweater and a pair of dress slacks or a pair of jeans, how many different outfits can he wear?
15. When a card is drawn from an ordinary deck, find the probability of getting
    (a) A jack. (b) A 4.

(c) A card less than 6 (an ace is considered above 6).

16. When a card is drawn from a deck, find the probability of getting
    (a) A diamond.
    (b) A 5 or a heart.
    (c) A 5 and a heart.
    (d) A king.
    (e) A red card.
17. At a sporting goods store, 12 people purchased blue rollerblades, 8 purchased green rollerblades, 4 purchased gray rollerblades, and 7 bought black rollerblades. If a customer is selected at random, find the probability he or she purchased
    (a) A pair of blue rollerblades.
    (b) A pair of green or gray rollerblades.
    (c) A pair of green or black or blue rollerblades.
    (d) A pair of rollerblades that was not black.
18. When two dice are rolled, find the probability of getting
    (a) A sum of 6 or 7.
    (b) A sum greater than 3 or greater than 8.
    (c) A sum less than 3 or greater than 8.
    (d) A sum that is divisible by 3.
    (e) A sum of 16.
    (f) A sum less than 11.
19. There are six cards numbered 1, 2, 3, 4, 5, and 6. A person flips a coin. If it lands heads up, he will select a card with an odd number. If it lands tails up, he will select a card with an even number. Draw a tree diagram and find the sample space.
20. Of the physics graduates of a university, 30% received a starting salary of $60,000 or more. If five of the graduates are selected at random, find the probability that all had a starting salary of $60,000 or more.
21. Five cards are drawn from an ordinary deck *without* replacement. Find the probability of getting
    (a) All red cards.
    (b) All diamonds.
    (c) All aces.
22. Four coins are tossed. Find the probability of getting four heads if it is known that two of the four coins landed heads up.
23. A card is drawn from a deck. Find the probability of getting a club if it is known that the card selected was a black card.
24. A die is rolled. Find the probability of getting a four if it is known that the result of the roll was an even number.
25. A coin is tossed and a die is rolled. Find the probability of getting a head on the coin if it is known that the number on the die is even.
26. Nurses at one hospital can be classified according to gender (male, female), income (low, medium, high), and rank (staff nurse, charge nurse, head nurse). Draw a tree diagram and show all possible outcomes.
27. The National Gaming Association can select one of four cities for its Pac Man tournament next year. The cities are Miami, San Francisco, Philadelphia, and Houston. The following year, it can hold the tournament in Chicago or Indianapolis. How many different possibilities are there for the next 2 years? Draw a tree diagram and show all possibilities.
28. Find the odds in favor of an event E when $P(E) = \frac{3}{8}$.
29. Find the odds against an event E when $P(E) = \frac{4}{9}$.
30. Find the probability of an event when the odds against the event are 3:7.
31. There are six cards placed face down in a box. Each card has a number written on it. One is a 4, one is a 5, one is a 2, one is a 10, one is a 3, and one is a 7. A person selects a card. Find the expected value of the draw.
32. A person selects a card from an ordinary deck of cards. If it is a black card, she wins $2. If it is a red card between or including 3 and 7, she wins $10. If it is a red face card, she wins $25, and if it is a black jack, she wins an additional $100. If it is any other card, she wins nothing. Find the expectation of the game.
33. In a soda machine in the student union, there are five Diet Cokes, four Mountain Dews, and two Dr. Peppers. If a person selects three sodas at random, find the probability that the selection will include one Diet Coke, one Mountain Dew, and one Dr. Pepper.
34. At Sally's freshman orientation, there were six computer science majors, four electrical engineering majors, and three architecture majors in her group. If four students are selected at random to receive a free tote bag with the school's logo, find the probability that the selection will include two computer science majors, one electrical engineering major, and one architecture major.
35. The results of a survey revealed that 30% of the people surveyed said that they would buy home electronic equipment at post-holiday sales. If 20 people are selected, find the probability that exactly 8 will purchase home electronic equipment after the holidays.

Projects

1. Make a set of three cards—one with the word "heads" on both sides, one with the word "tails" on both sides, and one with "heads" on one side and "tails" on the other side. With a partner, play the game described in Exercise 23 of Section 11-6 (page 616) 100 times and record how many times your partner wins. (*Note:* Do not change options during the 100 trials.)
   (a) Do you think the game is fair (i.e., does one person win approximately 50% of the time)?
   (b) If you think the game is unfair, explain what the probabilities might be and why.
2. Take a coin and tape a small weight (e.g., part of a paper clip) to one side. Flip the coin 100 times and record the results. Do you think you have changed the probabilities of the results of flipping the coin?
3. This game is called "Diet Fractions." Roll two dice and use the numbers to make a fraction less than or equal to one. Player A wins if the fraction cannot be reduced; otherwise, player B wins.
   (a) Play the game 100 times and record the results.
   (b) Decide if the game is fair or not. Explain why or why not.
   (c) Using the sample space for two dice, compute the probabilities of player A winning and player B winning. Do these agree with the results obtained in part a?

   Source: George W. Bright, John G. Harvey, and Margariete Montaque Wheeler, "Fair Games, Unfair Games." Chapter 8, *Teaching Statistics and Probability. NCTM 1981 Yearbook.* Reston, Virginia: The National Council of Teachers of Mathematics, Inc., 1981, p. 49. Used with permission.
4. Remember looking through cereal boxes for toys when you were a kid? It always seemed like you didn't get the exact one you wanted. Let's say that there was a certain toy you wanted, and five others that you could take or leave, all packed one per box at random. About how may boxes would you expect to have to buy to get the toy you wanted? Of course, you might have gotten it in the first box. Or you might have exhausted mom and dad's savings without ever getting it. These are the extremes.
   (a) You can simulate this experience using a single die, and rolling until a particular number of your choice comes up. Keep track of how many rolls it took, then repeat 99 more times, and find the average number of times it took.
   (b) If there were 10 different choices, you could simulate that by using the ace through 10 of a certain suit. Pick a certain card, shuffle the deck, and start dealing out the cards, keeping track of how many it takes to get the one you picked. Repeat 99 times and find the average.
   (c) Summarize your findings for both experiments.
   (d) Call getting the number or card you wanted a value of 2, and not getting it a value of 1. Then find the probability of each on any given roll or draw, and find the expected value for each experiment. How does it compare to the experimental results?

CHAPTER 12

Statistics

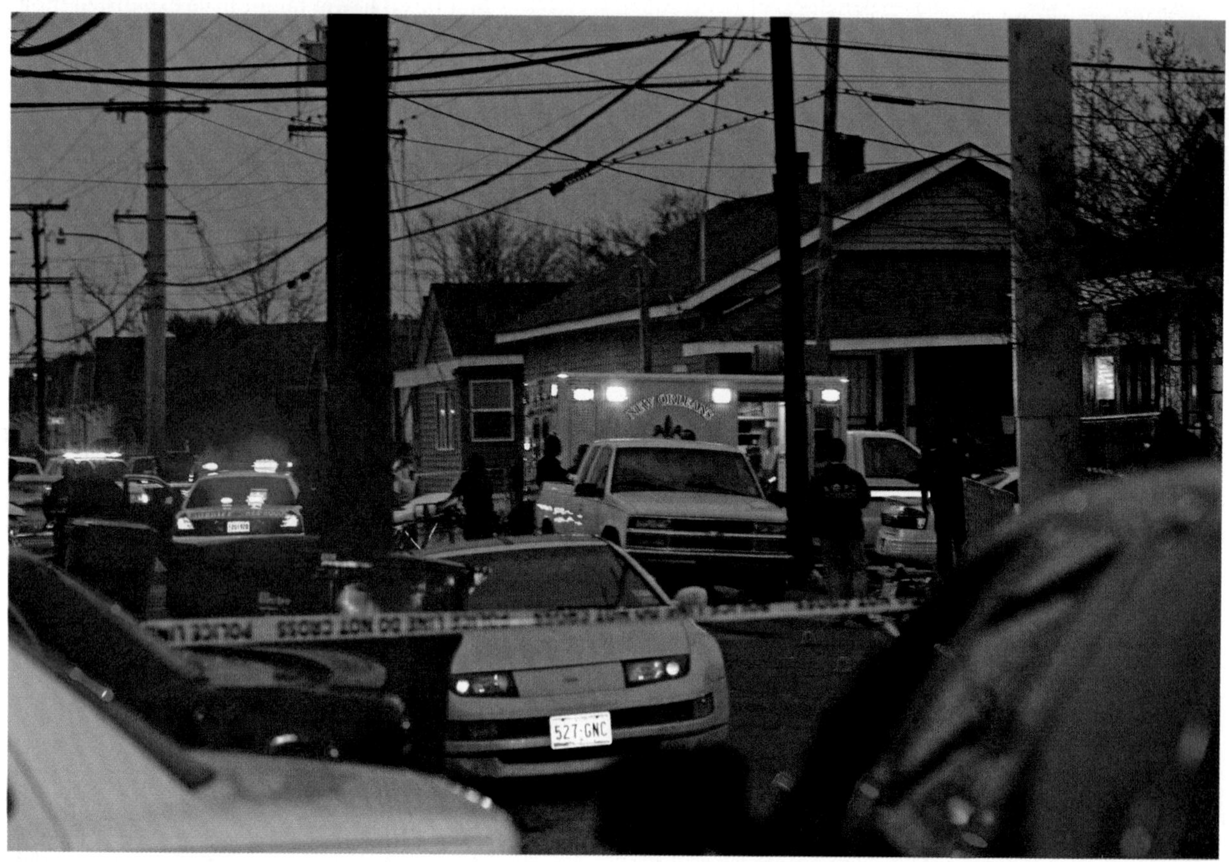

Outline

12-1 Gathering and Organizing Data
12-2 Picturing Data
12-3 Measures of Average
12-4 Measures of Variation
12-5 Measures of Position
12-6 The Normal Distribution
12-7 Applications of the Normal Distribution
12-8 Correlation and Regression Analysis
Summary

MATH IN Sociology

Broadly defined, sociology is the study of human behavior within society. One important branch is criminology. This is not about investigating crimes, but rather studying patterns of criminal behavior and their effect on society. One of the main tools used by sociologists is statistics. This important area of math allows researchers to study patterns of behavior objectively, by analyzing information gathered from a mathematical perspective, not a subjective one.

It all begins with the gathering of data. Data are measurements or observations that are gathered for a study of some sort. Let's look at an example. There is an old adage in police work that violent crime increases with the temperature. A statement like "My cousin Jed is a cop, and he told me that there are more violent crimes during the summer" is not evidence, nor is it data. It's the subjective opinion of one individual. In order to study whether this phenomenon is legitimate, we would need to gather information about the number of violent crimes at different times during the year, as well as temperature information, and study those numbers objectively to see if there appears to be a connection. This is one of the most important topics in this chapter.

Data on its own isn't good for very much unless we develop methods for organizing, studying, and displaying that data. These are the methods that make up the bulk of this chapter. After discussing methods of gathering data, we will learn effective methods for organizing and displaying data so that it can be presented in a meaningful, understandable way. We will then turn our attention to techniques of analyzing data that will help us to unlock the secrets of what sets of data may be trying to tell us.

As for math in sociology, in an attempt to study the phenomena of violent crime and temperature increase, the data below were gathered. The table shows the number of homicides committed in each month of 2008 in Columbus, Ohio, along with the average high temperature for each of those months.

Month	Avg. high temp.	Homicides
1	36	6
2	40	8
3	51	5
4	63	7
5	73	7
6	82	12
7	85	9
8	84	12
9	77	11
10	66	15
11	53	9
12	41	7

After you've finished this chapter, you should be able to answer the following questions.

1. Based on a graph of the data called a scatter plot, does there appear to be a relationship between temperature and homicide rate?
2. Can you conclude, using a measure called the correlation coefficient, with 95% certainty that there is in fact a relationship between the data sets? What about with 99% certainty?
3. Do the results indicate that warmer temperatures cause people to behave in a more violent manner?

For answers, see Math in Sociology Revisited on page 711

Section 12-1 Gathering and Organizing Data

LEARNING OBJECTIVES

- ☐ 1. Define data and statistics.
- ☐ 2. Explain the difference between a population and a sample.
- ☐ 3. Describe four basic methods of sampling.
- ☐ 4. Construct a frequency distribution for a data set.
- ☐ 5. Draw a stem and leaf plot for a data set.

One of the big issues being discussed over the last 10 years or so in higher education is grade inflation—the perception that college students in general are getting much higher grades than they used to. But perception and reality are often two different things, so how could we decide if this is actually taking place? And if so, what are some possible reasons? Are students just getting smarter? Are professors lowering their standards?

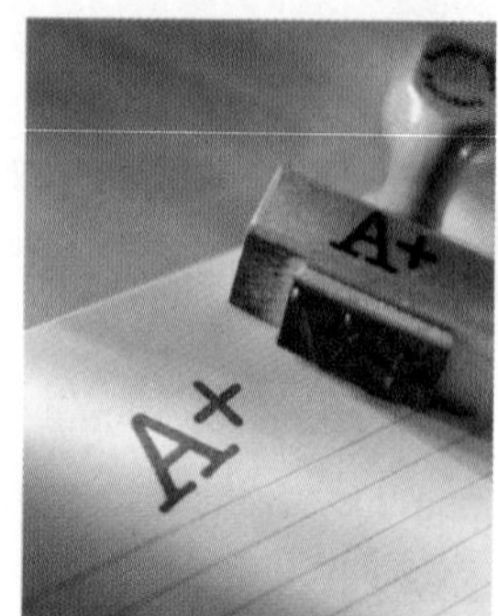

In order to examine questions like this, it would be very valuable to gather some *data*.

Data are measurements or observations that are gathered for an event under study.

Gathering data plays an important role in many different areas. In college sports, recruiters look at past performance to help them decide which high school players will be stars in college. In public health, administrators keep track of the number of residents in a certain area who contract a new strain of the flu. In education, researchers might try to determine if a new method of teaching is an improvement over traditional methods. Media outlets report the results of Nielsen, Harris, and Gallup polls. All of these studies begin with collecting data.

Once data have been collected, in order to get anything of value from it, the data need to be organized, summarized, and presented in a form that allows observers to draw conclusions. This is what the study of statistics is all about:

☑ 1. Define data and statistics.

Statistics is the branch of mathematics that involves collecting, organizing, summarizing, and presenting data and drawing general conclusions from that data.

Populations and Samples

In the case of grade inflation, relevant data might be average grade point averages for students in several different years. Table 12-1 shows some GPA data from the website Gradeinflation.com.

TABLE 12-1 Average GPAs for the Years 1991–1992 to 2006–2007 at Some American Universities

School Year	Average undergraduate GPA
1991–1992	2.93
1996–1997	2.99
2001–2002	3.07
2006–2007	3.11

The numbers certainly appear to indicate that grade inflation is real. But where did the data come from? Does this factor in *all* colleges, or just some?

When statistical studies are performed, we usually begin by identifying the *population* for the study.

A **population** consists of all subjects under study.

In our example, the population is all colleges in the United States. If you were interested in study habits of the students at your school, the population would be all of the students at the school. More often than not, it's not realistic to gather data from every member of a population. In that case, a smaller representative group of the population would be selected. This group is called a *sample*.

☑ 2. Explain the difference between a population and a sample.

A **sample** is a representative subgroup or subset of a population.

The GPA data in Table 12-1 were gathered from a sample of 70 schools. The reason the data are compelling is that the sample is *representative* of the population as a whole. The schools chosen vary in size, cost, and geographic location. Some are public, and some are private. In order to ensure that a sample is representative, researchers use a variety of methods. Four of these sampling methods are explained next.

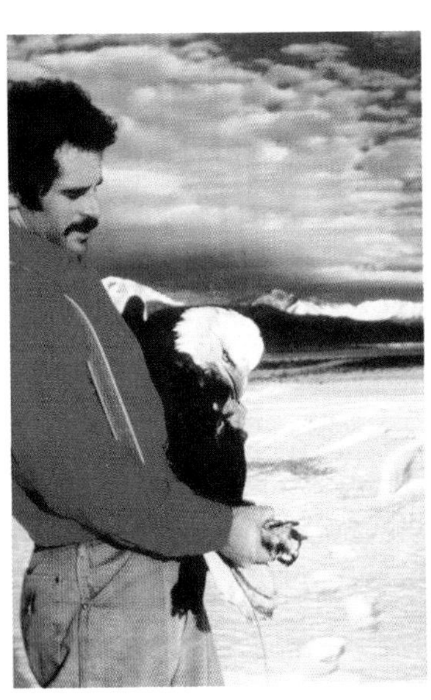

Wildlife biologists capture and tag animals to study how they live in the wild. The population is all the animals of the species in this region. The tagged animals constitute the sample. If the sample is not representative of the species the data will be skewed, and possibly misleading.

Sampling Methods

We will study four basic sampling methods that can be used to obtain a representative sample: random, systematic, stratified, and cluster.

In order to obtain a **random sample,** each subject of the population must have an equal chance of being selected. The best way to obtain a random sample is to use a list of random numbers. Random numbers can be obtained from a table or from a computer or calculator. Subjects in the population are numbered, and then they are selected by using the corresponding random numbers.

Using a random number generator such as a calculator, computer, or table of random numbers is like selecting numbers out of a hat. The difference is that when random numbers are generated by a calculator, computer, or table, there is a better chance that every number has an equally likely chance of being selected. When numbers are placed in a hat and mixed, you can never be sure that they are thoroughly mixed so that each number has an equal chance of being chosen.

A **systematic sample** is taken by numbering each member of the population and then selecting every kth member, where k is a natural number. For example, the researcher from gradeinflation.com might have numbered all of the colleges in the country and chosen every tenth one. When using systematic sampling, it's important that the starting number is selected at random.

When a population is divided into groups where the members of each group have similar characteristics (like large public schools, large private schools, small public schools, and small private schools) and members from each group are chosen at random, the result is called a **stratified sample.** The grade inflation researcher might have decided to choose five schools from each of those groups. Since 20 is a relatively small sample, it's possible that 15 or more came from large public schools. This would jeopardize the study because it may be the case that grade inflation is more or less likely at large public schools than schools in the other categories. So the purpose of a stratified sample is to ensure that all groups will be properly represented.

When an existing group of subjects that represent the population is used for a sample, it is called a **cluster sample.** For example, an inspector may select at random one carton of calculators from a large shipment and examine each one to determine how many are defective. The group in this carton represents a cluster. In this case, the researcher assumes that the calculators in the carton represent the population of all calculators manufactured by the company.

Samples are used in the majority of statistical studies, and if they are selected properly, the results of a study can be generalized to the population.

EXAMPLE 1 Choosing a Sample

A student in an education class is given an assignment to find out how late typical students at his campus stay up to study. He decides to stop by the union before his 9 A.M. class and ask everyone sitting at a table how late they were up studying the night before.

(a) What method of sampling is he using?
(b) Do you think he's likely to get a representative sample?

SOLUTION

(a) Since he is choosing all students in a particular place at a particular time, he has chosen a cluster sample.
(b) The sample is unlikely to be representative. Since he's polling people early in the morning, those that tend to stay up very late studying are less likely to be included in the sample.

▼ Try This One 1

3. Describe four basic methods of sampling.

To study the number of credit hours taken by a typical student, Shawna asks the registrar to provide e-mail addresses for 10 freshmen, 10 sophomores, 10 juniors, and 10 seniors. From each group, she asks for one whose student ID ends in 0, one whose ends in 1, and so forth.

(a) What method of sampling did she use?
(b) Do you think the sample will be representative?

Descriptive and Inferential Statistics

There are two main branches of statistics: descriptive and inferential. Statistical techniques that are used to *describe* data are called **descriptive statistics.** For example, a researcher may wish to determine the average age of the full-time students enrolled in your college and the percentage who own automobiles.

Sidelight MATH IN THE COURTROOM

Mathematicians are often called on to testify in court, and often it's to interpret the statistical likelihood of some event. In one case in Los Angeles, a couple was convicted of robbery based almost entirely on statistics! There were witnesses who saw the robbers leaving the scene, but they could not positively identify the suspects: just certain characteristics, like race, hair color, facial hair, and type of vehicle. A mathematician calculated that there was just one chance in 12 million of a second couple with the exact characteristics as the suspects being in the same area at that time. Despite a lack of any hard evidence, the couple was convicted.

When the conviction was appealed, the appellate judge was somewhat less impressed by the statistical argument than the jury had been. The conviction was overturned based on faulty calculations and the lack of other evidence.

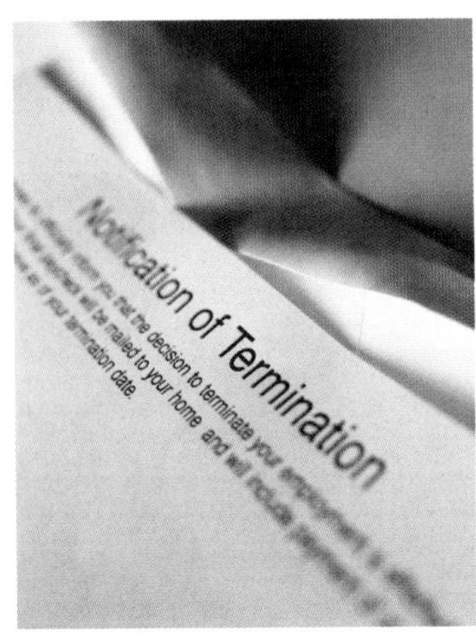

Statistical techniques used to make *inferences* are called **inferential statistics.** For example, every month the Bureau of Labor and Statistics estimates the number of people in the United States who are unemployed. Since it would be impossible to survey every adult resident of the United States, the Bureau selects a sample of adult individuals in the United States to see what percent are unemployed. In this case, the information obtained from a sample is used to estimate a population measure.

Another area of inferential statistics is called *hypothesis testing.* A researcher tries to test a hypothesis to see if there is enough evidence to support it. Here are some research questions that lend themselves to hypothesis testing:

Is one brand of aspirin better than another brand?
Does taking vitamin C prevent colds?
Are children more susceptible to ear infections than adults?

A third aspect of inferential statistics is determining whether or not a relationship exists between two or more variables. This area of statistics is called *correlation and regression;* for example:

Is caffeine related to heart trouble?
Is there a relationship between a person's age and his or her blood pressure?
Is these a connection between hours spent studying and grades on a final exam?

Frequency Distributions

The data collected for a statistical study are called **raw data.** In order to describe situations and draw conclusions, the researcher must organize the data in a meaningful way.

Two methods that we will use are *frequency distributions* and *stem and leaf plots.* There are two types of frequency distributions, the categorical frequency distribution and the grouped frequency distribution.

A categorical frequency distribution is used when the data are categorical rather than numerical. Example 2 shows how to construct a **categorical frequency distribution.**

EXAMPLE 2 Constructing a Frequency Distribution

Twenty-five volunteers for a medical research study were given a blood test to obtain their blood types. The data follow.

A	B	B	AB	O
O	O	B	AB	B
B	B	O	A	O
AB	A	O	B	A
A	O	O	O	AB

Construct a frequency distribution for the data.

SOLUTION

Step 1 Make a table as shown, with all categories represented.

Type	Tally	Frequency
A		
B		
O		
AB		

Step 2 Tally the data using the second column.

Step 3 Count the tallies and place the numbers in the third column. The completed frequency distribution is shown.

Type	Tally	Frequency
A	~~////~~	5
B	~~////~~ //	7
O	~~////~~ ////	9
AB	////	4

▼ Try This One 2

A health-food store recorded the type of vitamin pills 35 customers purchased during a 1-day sale. Construct a categorical frequency distribution for the data.

C	C	C	A	D	E	C
E	E	A	B	D	C	E
C	E	C	C	C	D	A
B	B	C	C	A	A	E
E	E	E	A	B	C	B

Another type of frequency distribution that can be constructed uses numerical data and is called a **grouped frequency distribution.** In a grouped frequency distribution, the numerical data are divided into classes. For example, if you gathered data on the weights of people in your class, there's a decent chance that no two people have the exact same weight. So it would be reasonable to group people into weight ranges, like 100–119 pounds, 120–139 pounds, and so forth. In the 100–119 pound class, we call 100 the lower limit and 119 the upper limit. When deciding on classes, here are some useful guidelines:

1. Try to keep the number of classes between 5 and 15.
2. Make sure the classes do not overlap.
3. Don't leave out any numbers between the lowest and highest, even if nothing falls into a particular class.
4. Make sure the range of numbers included in a class is the same for each one. (For example, don't have the first class go from 100–120 pounds and the second from 121–130 pounds.)

Example 3 demonstrates the steps for building a grouped frequency distribution.

EXAMPLE 3 Constructing a Frequency Distribution

These data represent the record high temperatures for each of the 50 states in degrees Fahrenheit. Construct a grouped frequency distribution for the data.

112	100	127	120	134	105	110	109	112	118
110	118	117	116	118	114	114	105	109	122
107	112	114	115	118	118	122	106	110	117
116	108	110	121	113	119	111	104	111	120
120	113	120	117	105	118	112	114	114	110

Source: *The World Almanac Book of Facts*

SOLUTION

Step 1 Subtract the lowest value from the highest value: 134 − 100 = 34.

Step 2 If we use a range of 5 degrees, that will give us seven classes, since the entire range (34 degrees) divided by 5 is 6.8. This is certainly not the only choice for the number of classes: if we choose a range of 4 degrees, there will be 9 classes.

Step 3 Start with the lowest value and add 5 to get the lower class limits: 100, 105, 110, 115, 120, 125, 130, 135.

Step 4 Set up the classes by subtracting one from each lower class limit except the first lower class limit.

Step 5 Tally the data and record the frequencies as shown.

Class	Tally	Frequency
100–104	//	2
105–109	卌 ///	8
110–114	卌 卌 卌 ///	18
115–119	卌 卌 ///	13
120–124	卌 //	7
125–129	/	1
130–134	/	1

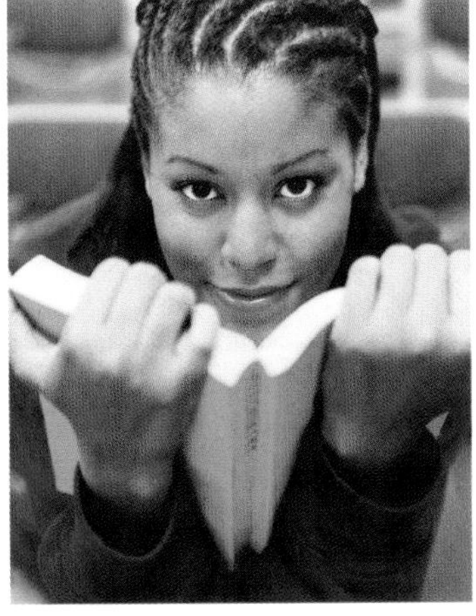

☑ 4. Construct a frequency distribution for a data set.

▼ Try This One 3

In one math class, the data below represent the number of hours each student spends on homework in an average week. Construct a grouped frequency distribution for the data using six classes.

1	2	6	7	12	13	2	6	9	5
18	7	3	15	15	4	17	1	14	5
4	16	4	5	8	6	5	18	5	2
9	11	12	1	9	2	10	11	4	10
9	18	8	8	4	14	7	3	2	6

Stem and Leaf Plots

Another way to organize data is to use a **stem and leaf plot** (sometimes called a stem plot). Each data value or number is separated into two parts. For a two-digit number such as 53, the tens digit, 5, is called the *stem,* and the ones digit, 3, is called its *leaf.* For the number 72, the stem is 7, and the leaf is 2. For a three-digit number, say 138, the first two digits, 13, are used as the stem, and the third digit, 8, is used as the leaf. Example 4 shows how to construct a stem and leaf plot.

EXAMPLE 4 Drawing a Stem and Leaf Plot

The data below show the number of games won by the Chicago Cubs in each of the 21 seasons from 1988–2008, with the exception of 1994, which was a short season because of a player strike. Draw a stem and leaf plot for the data.

97 85 66 79 89 88 67 88 65 67 90 68
76 73 84 78 77 77 93 77

SOLUTION

Notice that the first digit ranges from 6 to 9, so we set up a table with stems 6, 7, 8, 9:

Stems	Leaves
6	
7	
8	
9	

Now we go through the data one value at a time, putting the appropriate leaf next to the matching stem. For 97, we put a 7 under leaves, next to the stem 9 (shown in color below). For 85, we put a 5 under leaves, next to the stem 8, and so on, until all data values have been included.

The stem and leaf plot for number of wins by the Cubs is

Stems	Leaves
6	6 7 5 7 8
7	9 6 3 8 7 7 7
8	5 9 8 8 4
9	7 0 3

This is an efficient way of organizing the data. We can easily see that the most common class of wins is from 70–79, and the least common is 90–99. We can also see that 77 is the number of wins that occurred most often, three times.

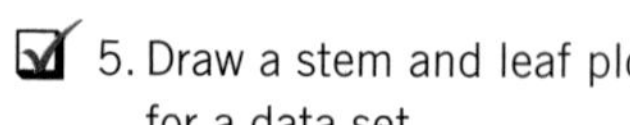
5. Draw a stem and leaf plot for a data set.

Math Note

Remember, when drawing a stem and leaf plot with three-digit numbers, the first two digits are the stem. So for 119, the stem is 11 and the leaf is 9.

▼ Try This One 4

It's no secret that gas prices have fluctuated wildly in the last few years, but from 1980–1999, they were surprisingly stable. According to the Energy Information Administration, the data below represent the average price (in cents) per gallon of regular unleaded gas for those years. Draw a stem and leaf plot of the data.

119	131	122	113	112	86	90	90	100	115
114	113	111	111	115	123	123	123	106	117

Answers to Try This One

1 (a) Stratified sample

(b) The population is likely to be representative since the last digit of a student's ID number is likely to be completely random.

2

Type	Tally	Frequency
A	~~////~~ /	6
B	~~////~~	5
C	~~////~~ ~~////~~ //	12
D	///	3
E	~~////~~ ////	9

3

Class	Tally	Frequency
1–3	~~||||~~ ~~||||~~	10
4–6	~~||||~~ ~~||||~~ ||||	14
7–9	~~||||~~ ~~||||~~	10
10–12	~~||||~~ |	6
13–15	~~||||~~	5
16–18	~~||||~~	5

4

8	6
9	0 0
10	0 6
11	9 3 2 5 4 3 1 1 5 7
12	2 3 3 3
13	1

EXERCISE SET 12-1

Writing Exercises

1. What are *data?*
2. Define *statistics.*
3. Explain the difference between a population and a sample.
4. How is a random sample selected?
5. How is a systematic sample selected?
6. How is a stratified sample selected?
7. How is a cluster sample selected?
8. What are the similarities between a grouped frequency distribution and a stem and leaf plot?

For Exercises 9–14, classify each sample as random, systematic, stratified, or cluster and explain your answer.

9. In a large school district, all teachers from two buildings are interviewed to determine whether they believe the students have less homework to do now than in previous years.
10. Every seventh customer entering a shopping mall is asked to select his or her favorite store.
11. Nursing supervisors are selected using random numbers in order to determine annual salaries.
12. Every hundredth hamburger manufactured is checked to determine its fat content.
13. Mail carriers of a large city are divided into four groups according to gender (male or female) and according to whether they walk or ride on their routes. Then 10 are selected from each group and interviewed to determine whether they have been bitten by a dog in the last year.
14. For the draft lottery that was conducted from 1969 to 1975, the days of the year were written on pieces of paper and put in plastic capsules. The capsules were then mixed in a barrel and drawn out one at a time, with the day recorded.

Real-World Applications

15. At a college financial aid office, students who applied for a scholarship were classified according to their class rank: Fr = freshman, So = sophomore, Jr = junior, Se = senior. Construct a frequency distribution for the data.

Fr	Fr	Fr	Fr	Fr
Jr	Fr	Fr	So	Fr
Fr	So	Jr	So	Fr
So	Fr	Fr	Fr	So
Se	Jr	Jr	So	Fr
Fr	Fr	Fr	Fr	So
Se	Se	Jr	Jr	Se
So	So	So	So	So

16. A questionnaire about how students primarily get news resulted in the following responses from 25 people. Construct a frequency distribution for the data. (I = Internet, N = newspaper, R = radio, T = TV.)

T I I N I I I N I R T N I
I T R I I N I R T N I I

17. Twenty-five fans of reality TV were asked to rate four shows, and the data below reflect which one each rated highest. Construct a frequency distribution for the data. (S = *Survivor,* P = *Project Runway,* B = *Big Brother,* A = *The Amazing Race.*)

S S B P P B S P A B A A S
B A P B A S A P B B S A

18. The new rap artist, Funky Fido, has had some success, although not as much as he'd like. The number of downloads per week of his hit, "Bark Up My Tree," is listed below. Construct a frequency distribution for the data using six classes.

373	254	237	243	308	210	266	253	201	266
239	114	224	373	286	329	236	284	247	273
198	361	416	207	243	326	251	169	360	311
215	189	344	268	363	21	270	165	240	48
150	300	207	314	197	209	210	260	327	

19. The ages of the signers of the Declaration of Independence are shown here. (Age is approximate since only the birth year appeared in the source, and one has been omitted since his birth year is unknown.) Construct a frequency distribution for the data using seven classes.

41	54	47	40	39	35	50	37	49	42	70	32
44	52	39	50	40	30	34	69	39	45	33	52
44	62	60	27	42	34	50	42	52	38	36	45
35	43	48	46	31	27	55	63	46	33	60	62
35	46	45	34	53	50	50					

Source: *The Universal Almanac*

20. The percentage of people killed in car accidents who were not wearing seatbelts in 2006 is shown for 27 states. Construct a frequency distribution using five classes.

56.9	59.1	33.1	43.3	62.3	36.3	74.4	64.0	60.0
56.8	45.5	38.2	71.0	46.6	42.4	39.3	46.0	60.4
52.5	69.2	63.9	60.5	47.5	43.3	62.3	67.1	39.8

Source: *Statemaster.com*

21. The data (in cents) are the cigarette taxes per pack imposed by each state as of April 1, 2009. Construct a frequency distribution. Use 0–39, 40–79, 80–119, etc.

42.5	200	200	115	87	84	200	115	33.9	37
200	57	98	99.5	136	79	60	36	200	200
251	200	150.4	18	17	170	64	80	133	257.5
91	275	35	44	125	103	118	135	246	7
153	62	141	69.5	199	30	202.5	55	177	60

Source: *National Conference at State Legislatures*

22. The acreage (in thousands of acres) of the 39 U.S. National Parks is shown here. Construct a frequency distribution for the data using eight classes.

41	66	233	775	169
36	338	233	236	64
183	61	13	308	77
520	77	27	217	5
650	462	106	52	52
505	94	75	265	402
196	70	132	28	220
760	143	46	539	

Source: *The Universal Almanac*

23. The heights in feet above sea level of the major active volcanoes in Alaska are given here. Construct a frequency distribution for the data using 10 classes.

4,265	3,545	4,025	7,050	11,413
3,490	5,370	4,885	5,030	6,830
4,450	5,775	3,945	7,545	8,450
3,995	10,140	6,050	10,265	6,965
150	8,185	7,295	2,015	5,055
5,315	2,945	6,720	3,465	1,980
2,560	4,450	2,759	9,430	
7,985	7,540	3,540	11,070	
5,710	885	8,960	7,015	

Source: *The Universal Almanac*

24. During the 1998 baseball season, Mark McGwire and Sammy Sosa both broke Roger Maris's home run record of 61. The distances in feet for each home run follow. Construct a frequency distribution for each player using the same eight classes.

McGwire

306	370	370	430
420	340	460	410
440	410	380	360
350	527	380	550
478	420	390	420
425	370	480	390
430	388	423	410
360	410	450	350
450	430	461	430
470	440	400	390
510	430	450	452
420	380	470	398
409	385	369	460
390	510	500	450
470	430	458	380
430	341	385	410
420	380	400	440
377	370		

Sosa

371	350	430	420
430	434	370	420
440	410	420	460
400	430	410	370
370	410	380	340
350	420	410	415
430	380	380	366
500	380	390	400
364	430	450	440
365	420	350	420
400	380	380	400
370	420	360	368
430	433	388	440
414	482	364	370
400	405	433	390
480	480	434	344
410	420		

Source: *USA Today*

25. The data (in millions of dollars) are the values of the 32 National Football League franchises in 2009. Construct a frequency distribution for the data using seven classes.

937	1,178	872	1,064	1,076	941	1,116	1,612
861	1,035	929	1,040	1,016	888	876	1,010
994	839	1,125	1,044	1,170	865	885	1,324
1,061	1,015	1,053	1,538	914	917	1,023	1,062

26. Twenty-nine executives reported the number of telephone calls made during a randomly selected week as shown here. Construct a stem and leaf plot for the data and analyze the results.

22	14	12	9	54	12
16	12	14	49	10	14
8	21	37	28	36	22
9	33	58	31	41	19
3	18	25	28	52	

27. The National Insurance Crime Bureau reported that these data represent the number of registered

vehicles per car stolen for 35 selected cities in the United States. For example, in Miami, one automobile is stolen for every 38 registered vehicles in the city. Construct a stem and leaf plot for the data and analyze the distribution. (The data have been rounded to the nearest whole number.)

38	53	53	56	69	89	94
41	58	68	66	69	89	52
50	70	83	81	80	90	74
50	70	83	59	75	78	73
92	84	87	84	85	84	89

Source: *USA Today*

28. As an experiment in a botany class, 20 plants are placed in a greenhouse, and their growth in centimeters after 20 days is recorded, with the results shown below. Construct a stem and leaf plot for the data.

20	12	39	38
41	43	51	52
59	55	53	59
50	58	35	38
23	32	43	53

29. The data shown represent the percentage of unemployed males for a sample of countries of the world. Using whole numbers as stems and the decimals as leaves, construct a stem and leaf plot.

8.8	1.9	5.6	4.6	1.5
2.2	5.6	3.1	5.9	6.6
9.8	8.7	6.0	5.2	5.6
4.4	9.6	6.6	6.0	0.3
4.6	3.1	4.1	7.7	

Source: *The Time Almanac*

Critical Thinking

For Exercises 30–34, decide whether descriptive or inferential statistics is being used.

30. A recent study showed that eating garlic can lower blood pressure.
31. The average number of students in a class at White Oak University is 22.6.
32. It is predicted that the average number of automobiles each household owns will increase next year.
33. Last year's total attendance at Long Run High School's football games was 8,325.
34. The chance that a person will be robbed in a certain city is 15%.
35. In addition to the four basic sampling methods, other methods are also used. Some of these methods are *sequence sampling, double sampling,* and *multiple sampling*. Look up these methods on the Internet and explain the advantages and disadvantages of each method.
36. For the data in Exercise 25, draw a stem and leaf plot, and compare your result to the frequency distribution. Which do you think provides more useful information, and why? What are the strengths and weaknesses of each?

Section 12-2 Picturing Data

LEARNING OBJECTIVES

- ❑ 1. Draw bar graphs and pie charts.
- ❑ 2. Draw histograms and frequency polygons.
- ❑ 3. Draw time series graphs.

Now we've gathered some data, and we think maybe some of it has an interesting story to tell. How can we most effectively present that data? In Section 12-1, we displayed data in table form, using frequency distributions, and using stem and leaf plots. All are perfectly valid, but… they don't exactly *pop*. The graphic on this page displays data related to identity theft complaints that could have been put in table form:

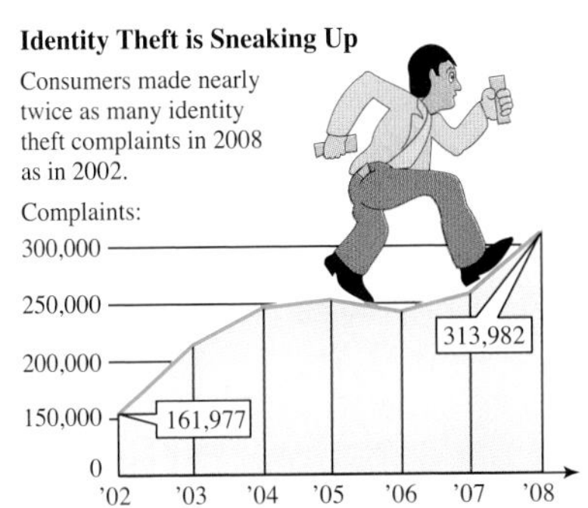

Source: Federal Trade Commission, February 2009

Year	2002	2003	2004	2005	2006	2007	2008
Complaints in thousands	162	215	247	256	246	258	314

The information is the same, but with the graph, just a quick glance is enough to see that identity theft rose sharply for a couple of years, then sort of leveled off, before another sharp rise. Simply put, if you want your data to really catch someone's eye, you can't do much better than a nice graphical representation. In this section, we'll study several methods for accomplishing this goal.

Bar Graphs and Pie Charts

The first type of frequency distribution we studied in Section 12-1 was the categorical frequency distribution. When data are representative of certain categories, rather than numerical, we often use bar graphs or circle graphs (commonly known as pie charts) to illustrate that data. We will start with a bar graph in Example 1.

EXAMPLE 1 Drawing a Bar Graph to Represent Data

The marketing firm Deloitte Retail conducted a survey in 2008 of grocery shoppers. The frequency distribution below represents the responses to the survey question "How often do you bring your own bags when grocery shopping?" Draw a vertical bar graph to represent the data.

Response	Frequency
Always	10
Never	39
Frequently	19
Occasionally	32

SOLUTION

Step 1 Draw and label the axes. We were asked for a vertical bar graph, so the responses go on the horizontal axis, and the frequencies on the vertical.

Step 2 Draw vertical bars with heights that correspond to the frequencies.

The completed graph is shown in Figure 12-1.

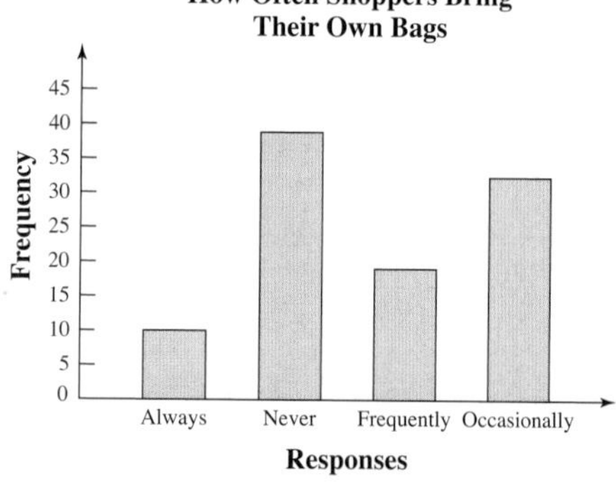

Figure 12-1

Math Note

For clarity, bar graphs are usually drawn with space between the bars.

▼ Try This One 1

The frequency distribution below represents responses to a survey on whether or not the posting of calorie information at fast food restaurants affects what customers order. Draw a vertical bar graph to illustrate the data.

Response	Frequency
Great impact	40
Some impact	42
Not much impact	11
None at all	7

CAUTION

In drawing a vertical bar graph, it's very important that the scale on the vertical axis begins at zero, and has consistent spacing. As you will see in the supplement on misuse of statistics, an improperly labeled axis can lead to a deceptive graph.

Another type of graph that can be drawn for the data that are categorical in nature is the **circle graph** or **pie chart.** A pie chart is a circle that is divided into sections in proportion to the frequencies corresponding to the categories. The purpose of a pie chart is to show the relationship of the parts to the whole by visually comparing the size of the sections. Example 2 shows the procedure for constructing a pie chart.

EXAMPLE 2 Drawing a Pie Chart to Represent Data

Math Note

Spreadsheet programs, like Microsoft Excel, have built-in commands that generate bar graphs and pie charts when you enter frequency distributions in table form.

Draw a pie chart for the frequency distribution from Example 1. The distribution is repeated below.

Response	Frequency
Always	10
Never	39
Frequently	19
Occasionally	32

SOLUTION

Step 1 Find the number of degrees corresponding to each slice using the formula

$$\text{Degrees} = \frac{f}{n} \cdot 360°$$

where f is the frequency for each class and n is the sum of the frequencies. In this case, $n = 100$, so the degree measures are:

Always $\frac{10}{100} \cdot 360° = 36°$

Never $\frac{39}{100} \cdot 360° = 140.4°$

Frequently $\frac{19}{100} \cdot 360° = 68.4°$

Occasionally $\frac{32}{100} \cdot 360° = 115.2°$

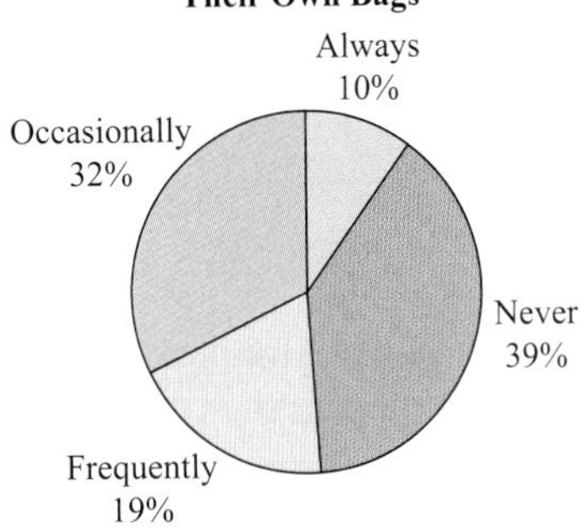

Figure 12-2

Step 2 Using a protractor, graph each section on the circle using the calculated angles, as shown in Figure 12-2. Notice the labeling that makes it clear what each slice represents.

Step 3 Calculate the percent of the circle covered by each slice.

Always: $\frac{10}{100} = 10\%$ Never: $\frac{39}{100} = 39\%$

Frequently: $\frac{19}{100} = 19\%$ Occasionally: $\frac{32}{100} = 32\%$

Label each section with the percent, as shown in Figure 12-2.

▼ Try This One 2

Draw a pie chart that illustrates the frequency distribution from Try This One 1.

☑ 1. Draw bar graphs and pie charts.

Histograms and Frequency Polygons

When data are organized into grouped frequency distributions, two types of graphs are commonly used to represent them: *histograms* and *frequency polygons.*

A **histogram** is similar to a vertical bar graph in that the heights of the bars correspond to frequencies. The difference is that class limits are placed on the horizontal axis, rather than categories. The procedure is illustrated in Example 3.

EXAMPLE 3 Drawing a Histogram

The frequency distribution to the right is for the closing price of General Motors stock in dollars for each of the first 74 trading days of 2009. Draw a histogram for the data.

Class	Frequency	Class	Frequency
1.01–1.50	1	3.01–3.50	12
1.51–2.00	18	3.51–4.00	8
2.01–2.50	14	4.01–4.50	5
2.51–3.00	16		

SOLUTION

Step 1 Write the scale for the frequencies on the vertical axis and the class limits on the horizontal axis.

Step 2 Draw vertical bars with heights that correspond to the frequencies for each class. See Figure 12-3.

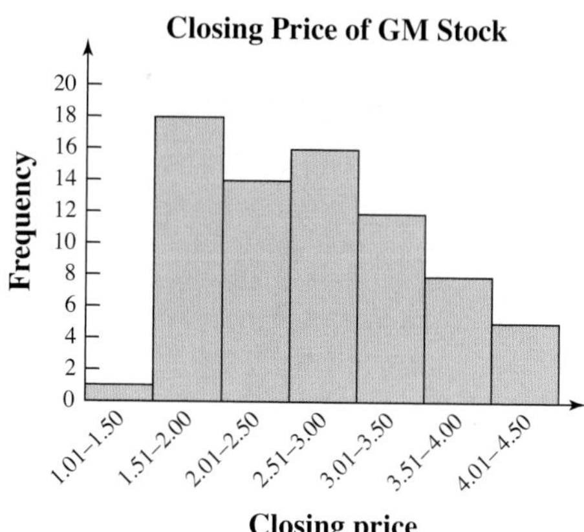

Figure 12-3

Math Note

When drawing a histogram, make sure that the bars touch (unlike a bar graph) and that every class is included, even if the frequency is zero.

▼ Try This One 3

Draw a histogram for the frequency distribution to the right, which represents the number of losses by the team that won the NCAA men's basketball championship for the years from 1939–2009.

Losses	Frequency	Losses	Frequency
0–1	13	6–7	9
2–3	27	8–9	2
4–5	17	10–11	3

A **frequency polygon** is similar to a histogram, but instead of bars, a series of line segments is drawn connecting the midpoints of the classes. The heights of those points match the heights of the bars in a histogram. In Example 4, we'll draw a frequency polygon for the frequency distribution from Example 3 so you can compare the two.

EXAMPLE 4 Drawing a Frequency Polygon

Draw a frequency polygon for the frequency distribution from Example 3. The distribution is repeated here.

Class	Frequency	Class	Frequency
1.01–1.50	1	3.01–3.50	12
1.51–2.00	18	3.51–4.00	8
2.01–2.50	14	4.01–4.50	5
2.51–3.00	16		

Math Note

Once you have the first midpoint, you can get all of the others by just adding the size of each class to the previous. For example, the second midpoint is the first, 1.255, plus the size of the classes, 0.50.

SOLUTION

Step 1 Find the midpoints for each class. This is accomplished by adding the upper and lower limits and dividing by 2. For the first two classes, we get:

$$\frac{1.50 + 1.01}{2} = 1.255$$

$$\frac{2.00 + 1.51}{2} = 1.755$$

The remaining midpoints are 2.255, 2.755, 3.255, 3.755, and 4.255.

Math Note

A frequency polygon should always touch the horizontal axis at both ends to indicate that the frequency is zero for any values not included on the graph.

Step 2 Write the scale for the frequencies on the vertical axis, and label a scale on the horizontal axis so that all midpoints will be included.

Step 3 Plot points at the midpoints with heights matching the frequencies for each class, then connect those points with straight lines.

Step 4 Finish the graph by drawing a line back to the horizontal axis at the beginning and end. The horizontal distance to the axis should equal the distance between the midpoints. In this case, that distance is 0.5, so we extend back to 0.755 and forward to 4.755. See Figure 12-4.

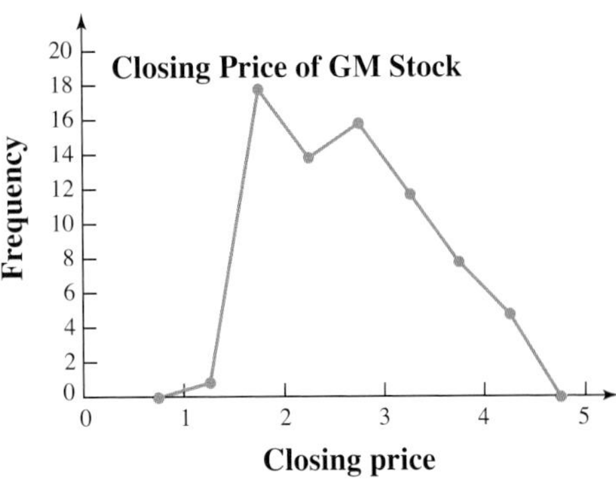

Figure 12-4

▼ Try This One 4

☑ 2. Draw histograms and frequency polygons.

Draw a frequency polygon for the frequency distribution from Try This One 3.

Time Series Graphs

A **time series graph** can be drawn for data collected over a period of time. This type of graph is used primarily to show trends, like prices rising or falling, for the time period. There are three types of trends. Secular trends are viewed over a long period of time, such as yearly. Cyclical trends show oscillating patterns. Seasonal trends show the values of a commodity for shorter periods of the year, such as fall, winter, spring, and summer. Example 5 shows how to draw a time series graph.

EXAMPLE 5 Drawing a Time Series Graph

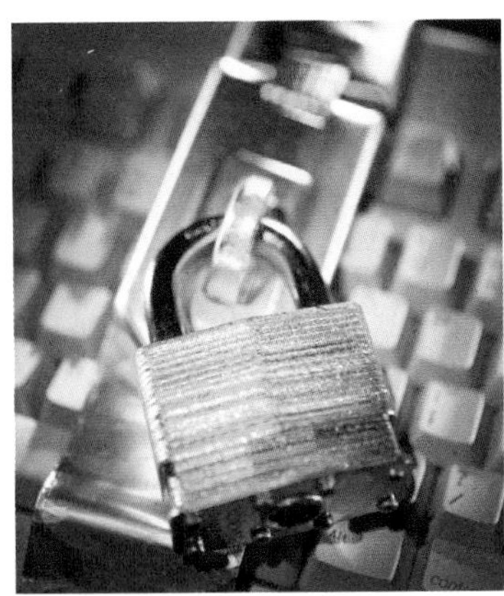

Identity theft has increased as more people are shopping online. The table below shows the number of identify theft complaints made to the FTC in thousands between 2002 and 2008. Draw a time series graph for the data.

Year	2002	2003	2004	2005	2006	2007	2008
Complaints in thousands	162	215	247	256	246	258	314

Source: Federal Trade Commission

SOLUTION

Label the horizontal axis with years, and the vertical axis with the number of complaints in thousands, then plot the points from the table and connect them with line segments. See Figure 12-5.

Identity Theft Complaints by Year

Number (thousands): 0, 50, 100, 150, 200, 250, 300, 350

Year: 2001, 2002, 2003, 2004, 2005, 2006, 2007, 2008, 2009

Figure 12-5

Math Note

Notice how this time series graph matches the one at the beginning of this section.

☑ 3. Draw time series graphs.

Math Note

It's especially important to scale the vertical axis appropriately when drawing a time series graph. See Exercise 27 for some perspective on why.

▼ Try This One 5

The number of bankruptcy filings (in millions) in the United States from 2002 to 2008 is shown in the table. Draw a time series graph for the data.

Year	2002	2003	2004	2005	2006	2007	2008
Filings (millions)	1.58	1.66	1.60	2.08	0.62	0.85	1.12

This section provided just a sampling of the many types of charts and graphs used to picture data. A Google search for the string *picturing data* will help you find a more comprehensive list.

1

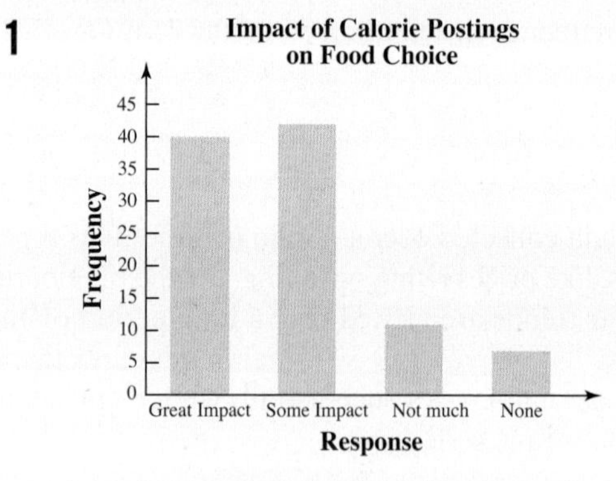

2

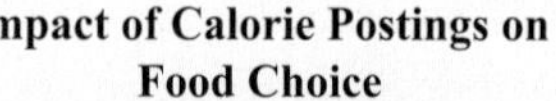

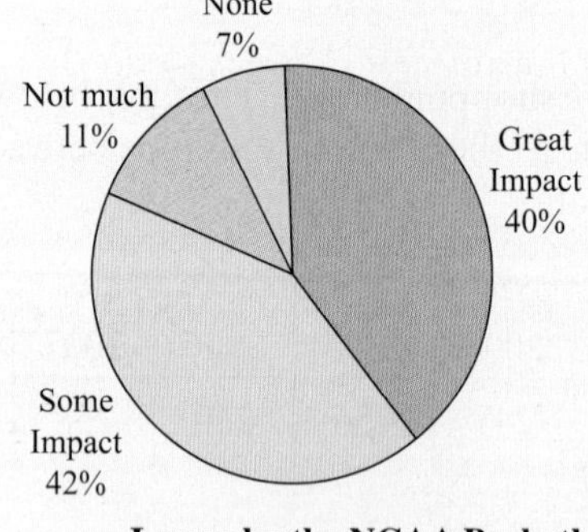

3

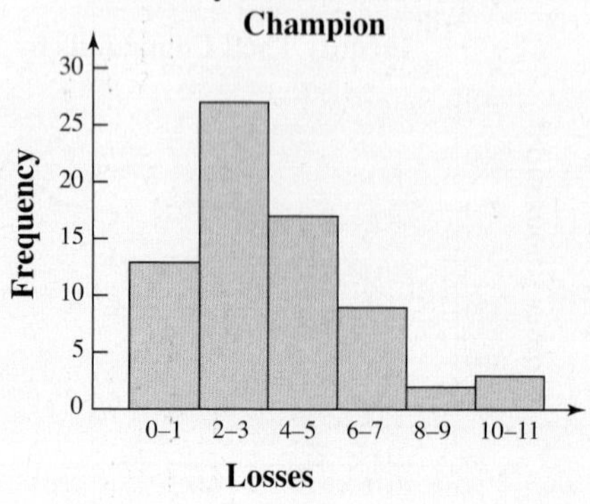

4

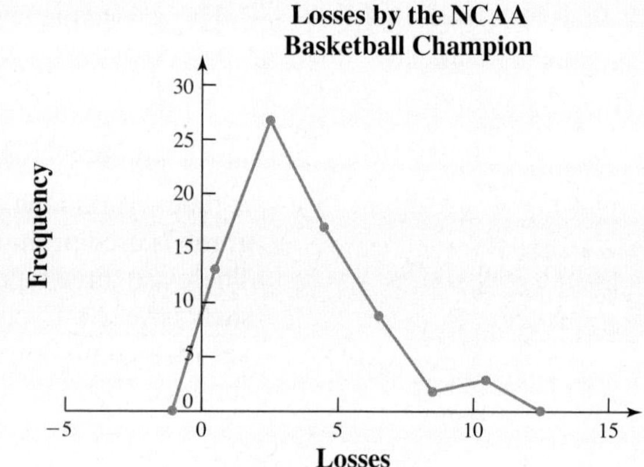

5

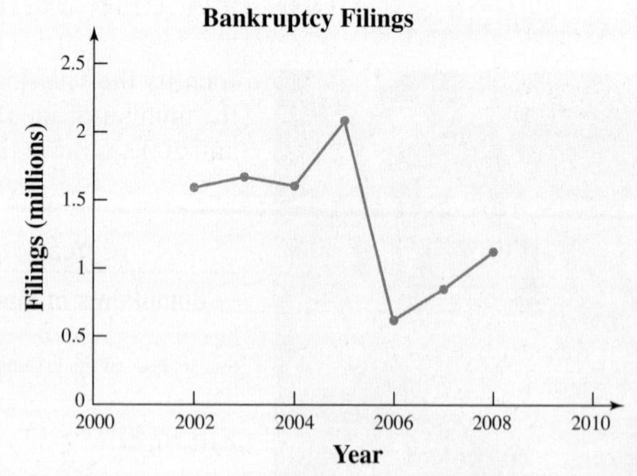

EXERCISE SET 12-2

Writing Exercises

1. Describe how to draw a bar graph and a pie chart given a frequency distribution.
2. Explain why bar graphs and pie charts are typically used for categorical frequency distributions, while histograms are used for grouped frequency distributions.
3. How are histograms and frequency polygons similar? How are they different?
4. Describe the purpose of a time series graph.

Real-World Applications

5. Draw a bar graph for the number of transplants of various types performed in the United States in 2008.

Type	Number
Kidney	13,743
Liver	5,273
Pancreas	370
Heart	1,802
Lung	1,221

Source: *Infoplease.com*

6. The data below show the number of students visiting a university health care center during finals week who suffered from certain conditions. Draw a bar graph to illustrate the data.

Condition	Number
Flu	48
Panic attack	36
Bronchitis	32
Headache	32
Broken bone	19
Cold/sniffles	17
Rash	16
Pneumonia	11
Ear infection	9

7. Draw a bar graph for the number of registered taxicabs in 2008 in the selected cities.

City	Number
New York	13,087
Washington, D.C.	7,000
Chicago	7,000
Los Angeles	2,300
Atlanta	1,600

Source: *Wikipedia*

8. Draw a bar graph for the number of unemployed people in the selected states for February 2009.

State	Number
Texas	772,000
New York	760,000
Pennsylvania	486,000
Florida	886,000
Ohio	567,000

Source: *Bureau of Employment Statistics*

9. The number of students at one campus who had a 4.0 GPA is shown below, organized by class. Draw a pie chart for the data.

Rank	Frequency
Freshman	12
Sophomore	25
Junior	36
Senior	17

10. In an insurance company study of the causes of 1,000 deaths, these data were obtained. Draw a pie chart to represent the data.

Cause of death	Number of deaths
Heart disease	432
Cancer	227
Stroke	93
Accidents	24
Other	224
	1,000

11. In a survey of 100 college students, the numbers shown here indicate the primary reason why the students selected their majors. Draw a pie chart for the data and analyze the results.

Reason	Number
Interest in subject	62
Future earning potential	18
Pressure from parents	12
Good job prospects	8

12. A survey of the students in the school of education at a large university obtained the following data for

students enrolled in specific fields. Draw a pie chart for the data and analyze the results.

Major field	Number
Preschool	893
Elementary	605
Middle	245
Secondary	1,096

13. For 108 randomly selected college applicants, the frequency distribution shown here for entrance exam scores was obtained. Draw a histogram and frequency polygon for the data.

Class	Frequency
90–98	6
99–107	22
108–116	43
117–125	28
126–134	9

14. For 75 employees of a large department store, the distribution shown here for years of service was obtained. Draw a histogram and frequency polygon for the data.

Class	Frequency
1–5	21
6–10	25
11–15	15
16–20	0
21–25	8
26–30	6

15. Thirty cars and trucks were tested by the EPA for fuel efficiency in miles per gallon (mpg). The frequency distribution shown here was obtained. Draw a histogram and frequency polygon for the data.

Class	Frequency
8–12	3
13–17	5
18–22	15
23–27	5
28–32	2

16. In a study of reaction times of dogs to a specific stimulus, an animal trainer obtained these data, given in seconds. Draw a histogram and frequency polygon for the data and analyze the results.

Class	Frequency
2.3–2.9	10
3.0–3.6	12
3.7–4.3	6
4.4–5.0	8
5.1–5.7	4
5.8–6.4	2

17. The data below represent the number of people admitted to graduate programs in mechanical engineering by year. Draw a time series graph for the data.

Year	2004	2005	2006	2007	2008
Number	1,981	2,895	3,027	1,651	1,432

18. The data below represent the number of laptops sold by a local computer store for the years listed. Draw a time series graph for the data.

Year	Number
1999	201
2000	256
2001	314
2002	379
2003	450
2004	576
2005	681
2006	799
2007	873
2008	1,012

19. The data below represent the number of downloads per year of a popular shareware program. Draw a time series graph for the data.

Year	2000	2002	2004	2006	2008
Downloads	12,413	15,160	18,201	20,206	19,143

Critical Thinking

In Exercises 20–25, state which type of graph (bar graph, pie chart, or time series graph) would most appropriately represent the given data.

20. The number of students enrolled at a local college each year for the last 5 years.

21. The budget for the student activities department at your college.

22. The number of students who get to school by automobile, bus, train, or by walking.

23. The record high temperatures of a city for the last 30 years.
24. The areas of the five lakes in the Great Lakes.
25. The amount of each dollar spent for wages, advertising, overhead, and profit by a corporation.
26. Use the Internet or a library to find a reference describing what a pictograph is, then describe its purpose.
27. Redraw the time series graph from Example 5, but make the scale on the vertical axis range from 150 to 350 rather than from 0 to 350. Why does this make the graph deceiving?

Section 12-3 Measures of Average

LEARNING OBJECTIVES

- ☐ 1. Compute the mean of a data set.
- ☐ 2. Compute the median of a data set.
- ☐ 3. Compute the mode of a data set.
- ☐ 4. Compute the midrange of a data set.
- ☐ 5. Compare the four measures of average.

Are you an average college student? What does that even mean? According to a variety of sources found on the Web, the average college student is 20 years old, lives on campus, and works off campus during the school year. She is female, talks to her parents every day, and comes from a family with an annual income between $50,000 and $100,000. She sleeps 6 hours a night, has $2,700 in credit card debt, and will leave college with $21,000 in total debt.

If this doesn't sound exactly like anyone you know, don't feel bad. As we will learn in this section, the word *average* is ambiguous—there are a variety of ways to describe an average. So what one source considers average might not match someone else's thoughts, and with so many different facets to a person, it's possible that nobody exactly meets all of the average criteria.

Our goal in Section 12-3 will be to understand the different measures of average. In casual terms, average means the most typical case, or the center of the distribution. Measures of average are also called *measures of central tendency,* and include the *mean, median, mode,* and *midrange*.

The Mean

The *mean,* also known as the arithmetic average, is found by adding the values of the data and dividing by the total number of values. The Greek letter Σ (sigma) is used to represent the sum of a list of numbers. If we use the letter X to represent data values, then ΣX means to find the sum of all values in a data set. Using this notation:

The **mean** is the sum of the values in a data set divided by the number of values. If $X_1, X_2, X_3, \ldots X_n$ are the data values, we use $\overline{X}$ to stand for the mean, and

$$\overline{X} = \frac{X_1 + X_2 + X_3 + \cdots + X_n}{n} = \frac{\Sigma X}{n}$$

EXAMPLE 1 Finding the Mean of a Data Set

In 2003, there were 12 inmates on death row who were proven innocent and freed. In the 5 years after that, there were 6, 2, 1, 3, and 4. Find the mean number of death row inmates proven innocent for the 6 years from 2003 to 2008.

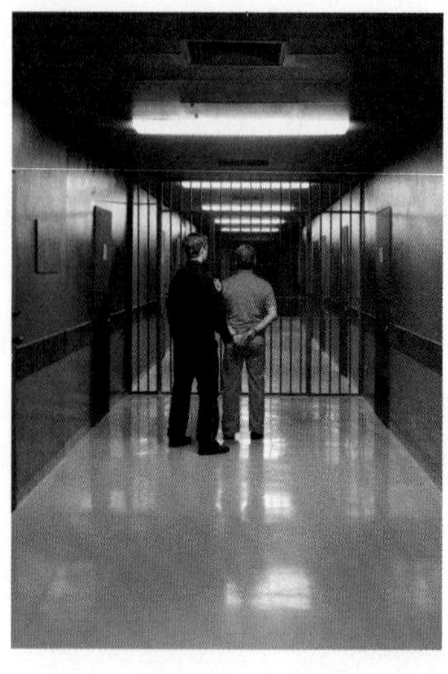

SOLUTION

$$\overline{X} = \frac{\sum X}{n} = \frac{12 + 6 + 2 + 1 + 3 + 4}{6} = \frac{28}{6} \approx 4.7$$

The mean is about 4.7.

▼ Try This One 1

To test a customer's Internet connection, a tech records how long it takes for 10 different websites to load completely. The times in seconds are shown below. Find the mean time.

20.68 16.22 16.76 16.90 20.98 18.84 14.18 40.04 24.92 22.96

The procedure for finding the mean for grouped data uses the midpoints and the frequencies of the classes as shown in Example 2. This procedure will give only an approximate value for the mean, and it is used when the data set is very large or when the original raw data are unavailable but have been grouped by someone else.

Formula for Finding the Mean for Grouped Data:

$$\overline{X} = \frac{\sum(f \cdot X_m)}{n}$$

where f = frequency

X_m = midpoint of each class

$n = \sum f$ or sum of the frequencies

EXAMPLE 2 Finding the Mean for Grouped Data

Find the mean for the price of GM stock for the first 74 days of trading in 2009. The frequency distribution is below.

Class	Frequency	Class	Frequency
1.01–1.50	1	3.01–3.50	12
1.51–2.00	18	3.51–4.00	8
2.01–2.50	14	4.01–4.50	5
2.51–3.00	16		

SOLUTION

We found the midpoint of each class in Example 4 of Section 12-2: the midpoints are 1.255, 1.755, 2.255, 2.755, 3.255, 3.755, and 4.255. Now we multiply each midpoint by the frequency:

Midpoint (X_m)	Frequency (f)	$f \cdot X_m$
1.255	1	1.255
1.755	18	31.59
2.255	14	31.57
2.755	16	44.08
3.255	12	39.06
3.755	8	30.04
4.255	5	21.275
Sums	74	198.87

Now we divide $\Sigma(f \cdot X_m)$ by the sum of the frequencies to get the mean:

$$\overline{X} = \frac{198.87}{74} \approx 2.69$$

The mean closing price was \$2.69.

▼ Try This One 2

1. Compute the mean of a data set.

Find the mean for the number of losses by the NCAA men's basketball champion for the years from 1939–2009. The frequency distribution is below.

Losses	Frequency	Losses	Frequency
0–1	13	6–7	9
2–3	27	8–9	2
4–5	17	10–11	3

The Median

According to payscale.com, the *median* salary for all federal government employees as of early 2009 was \$60,680. This measure of average means that half of all government employees made less than \$60,680, and half made more.

Simply put, the **median** is the halfway point of a data set when it is arranged in order. When a data set is ordered, it is called a **data array.** The median will either be a specific value in the data set, or will fall between two values, as shown in Examples 3 and 4.

Steps in Computing the Median of a Data Set

Step 1 Arrange the data in order, from smallest to largest.

Step 2 Select the middle value. If the number of data values is odd, the median is the value in the exact middle of the list. If the number of data values is even, the median is the mean of the two middle data values.

EXAMPLE 3 Finding the Median of a Data Set

The mean and median age for this group are different. The grandfather's age pulls up the mean, while the four children bring down the median.

The weights of the five starting offensive linemen for the Pittsburgh Steelers in Super Bowl XLIII were 345, 340, 315, 285, and 317. Find the median weight.

SOLUTION

Step 1 Arrange the data in order.

285 315 317 340 345

Step 2 Select the middle value.

The median weight was 317 pounds.

▼ Try This One 3

The high temperatures for seven consecutive days in April, 2009 in Cincinnati were 60°, 49°, 67°, 73°, 76°, 59°, and 59°. Find the median temperature.

EXAMPLE 4 Finding the Median of a Data Set

One of the authors of this book conducts a campus food drive at the end of each semester. The amount of food in pounds gathered over the last eight semesters is shown below. Find the median weight.

1,675 1,209 1,751 1,700 1,532 2,171 2,292 3,211

SOLUTION

Step 1 Arrange the data in order.

1,209 1,532 1,675 1,700 1,751 2,171 2,292 3,211

Step 2 Find the middle value: in this case, it is in between 1,700 and 1,751, so we find the mean of those two values.

$$\frac{1{,}700 + 1{,}751}{2} = 1{,}725.5$$

The median weight is 1,725.5 pounds.

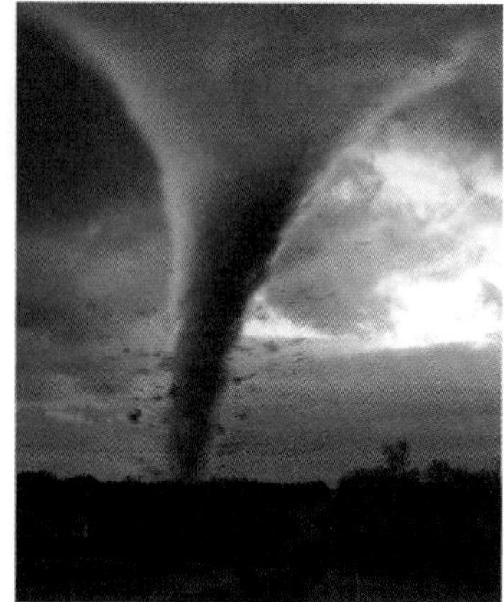

▼ Try This One 4

The data below are the number of fatalities in the United States caused by tornadoes for the years 1999–2008. Find the median number of fatalities.

125 81 67 39 66 54 55 44 41 94

CAUTION When finding the median of a data set, make sure the data values are arranged in order!

☑ 2. Compute the median of a data set.

The Mode

The third measure of average is called the *mode.* The mode is sometimes said to be the most typical case.

The value that occurs most often in a data set is called the **mode.**

A data set can have more than one mode or no mode at all. These situations will be shown in some of the examples that follow.

EXAMPLE 5 Finding the Mode of a Data Set

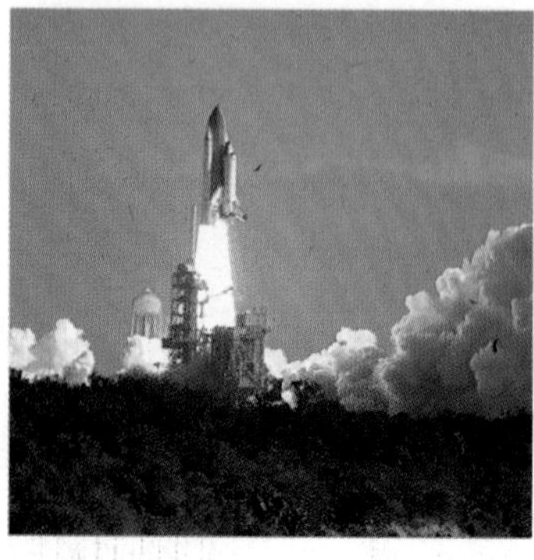

These data represent the duration (in days) of U.S. space shuttle voyages for the years 2002–2008. Find the mode.

10 10 13 10 13 15 13 12 11 12 13 12 15 12 15 13 15

Source: Wikipedia

SOLUTION

If we construct a frequency distribution, it will be easy to find the mode—it's simply the value with the greatest frequency. The frequency distribution for the data is shown to the right, and the mode is 13.

Days	Frequency	Days	Frequency
10	3	13	5
11	1	15	4
12	4		

▼ Try This One 5

The data below represent the number of fatal commercial airline incidents in the United States for each year from 1998–2008. Find the mode.

1 2 3 6 0 2 2 3 2 1 2

EXAMPLE 6 Finding the Mode of a Data Set

Math Note

When there is no mode, don't say that the mode is zero. That would be incorrect because for some data, such as temperature, zero can be an actual value.

Six strains of bacteria were tested by the Centers for Disease Control to see how long they could remain alive outside their normal environment. The time, in minutes, is recorded below. Find the mode.

2 3 5 7 8 10

SOLUTION

Since each value occurs only once, there is no mode.

▼ Try This One 6

The final exam scores for 10 students in a seminar course are listed below. Find the mode.

91 74 66 93 76 85 86 90 71 89

EXAMPLE 7 Finding the Mode of a Data Set

Math Note

When a data set has two modes, it is called *bimodal*.

The number of wins in a 16-game season for the Cincinnati Bengals from 1997–2008 is listed below. Find the mode.

7 3 4 4 6 2 8 8 11 8 7 4

SOLUTION

There are two numbers that occur three times: 4 and 8. No number occurs more than three times, so the data set has two modes, 4 and 8.

▼ Try This One 7

The table below lists the average high temperature in degrees Fahrenheit for each month of the year on the island of Antigua. Find the mode.

Month	Jan	Feb	Mar	Apr	May	Jun	Jul	Aug	Sep	Oct	Nov	Dec
High	81	82	82	83	85	86	87	87	87	86	84	82

The mode is the only measure of central tendency that can be used in finding the most typical case when the data are classified by groups or categories, like those shown in Example 8.

EXAMPLE 8 Finding the Mode for Categorical Data

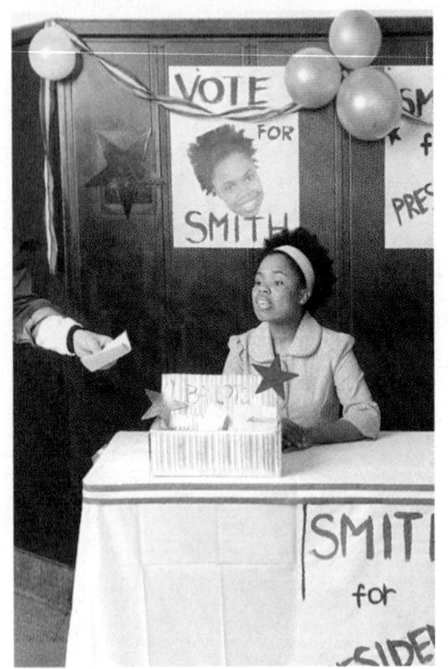

The winner of an election by popular vote is the mode.

A survey of the junior class at Fiesta State University shows the following number of students majoring in each field. Find the mode.

Business	1,425
Liberal arts	878
Computer science	632
Education	471
General studies	95

SOLUTION

The values here are not the numbers, but the categories: business, liberal arts, computer science, education, and general studies. The value that appears most often is business, so that is the mode.

▼ Try This One 8

Five hundred college graduates were asked how much they donate to their alma mater on an annual basis. Find the mode of their responses, summarized below.

\$500 or more	45
Between 0 and \$500	150
Nothing	275
Refused to answer	30

☑ 3. Compute the mode of a data set.

The Midrange

The midrange is a very quick and rough estimate of the middle of a data set. It is found by adding the lowest and highest values and dividing by 2.

Finding the Midrange for a Data Set

$$\text{Midrange} = \frac{\text{Lowest value} + \text{highest value}}{2}$$

The midrange is not terribly reliable since it can be greatly affected by just one extremely high or low data value.

EXAMPLE 9 Finding the Midrange

In Example 1, we saw that the number of death row inmates proven innocent for the years 2003–2008 was 12, 6, 2, 1, 3, and 4. Find the midrange.

SOLUTION

The lowest value is 1 and the highest is 12, so the midrange is

$$\text{Midrange} = \frac{12 + 1}{2} = 6.5.$$

▼ Try This One 9

☑ 4. Compute the midrange of a data set.

The number of earthquakes worldwide measuring at least 6.0 on the Richter scale from 2000–2008 is shown below. Find the midrange.

173 142 143 155 157 151 153 200 178

In Example 1, we found that the mean for the data set in Example 9 is 4.7. The midrange is much higher because the data value 12 is so much higher than the rest. This brings up an interesting question: how do our four measures of average compare for a data set?

EXAMPLE 10 Comparing the Four Measures of Average

The table below lists the number of golfers who finished the Masters tournament with a better score than Tiger Woods between 1997 and 2009.

Year	Number	Year	Number
1997	0	2004	21
1998	7	2005	0
1999	17	2006	2
2000	4	2007	1
2001	0	2008	1
2002	0	2009	5
2003	14		

Find the mean, median, mode, and midrange for the data.

SOLUTION

It will be helpful to arrange the numbers in order:

0 0 0 0 1 1 2 4 5 7 14 17 21

$$\text{Mean} = \frac{\Sigma X}{n} = \frac{0 + 0 + 0 + 0 + 1 + 1 + 2 + 4 + 5 + 7 + 14 + 17 + 21}{13} \approx 5.54$$

$$\text{Median} = 2$$

$$\text{Mode} = 0$$

$$\text{Midrange} = \frac{21 + 0}{2} = 10.5$$

So which is the best indicator of average performance? That's open to interpretation, but all four measures of average are very different.

▼ Try This One 10

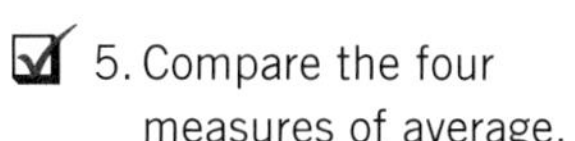

☑ 5. Compare the four measures of average.

Tiger's final scores in relation to par for the 13 Masters tournaments from 1997 to 2009 are listed below. Find the mean, median, mode, and midrange.

−18 −3 +1 −4 −16 −12 +2 +2 −12 −4 +3 −5 −8

To conclude the section, Table 12-2 summarizes some of the strengths and weaknesses of each measure of average.

TABLE 12-2 A Comparison of Measures of Average

Measure	Strengths	Weaknesses
Mean	• Unique—there's exactly one mean for any data set • Factors in all values in the set • Easy to understand	• Can be adversely affected by one or two unusually high or low values • Can be time-consuming to calculate for large data sets
Median	• Divides a data set neatly into two groups • Not affected by one or two extreme values	• Can ignore the effects of large or small values even if they are important to consider
Mode	• Very easy to find • Describes the most typical case • Can be used with categorical data like candidate preference, choice of major, etc.	• May not exist for a data set • May not be unique • Can be very different from mean and median if the most typical case happens to be near the low or high end of the range
Midrange	• Very quick and easy to compute • Provides a simple look at average	• Dramatically affected by extremely high or low values in the data set • Ignores all but two values in the set

Sidelight THE BIRTH OF STATISTICS

The study of statistics has its origin in censuses taken by the Babylonians and Egyptians almost 6,000 years ago. "Birth" is an appropriate word to use in describing the beginning of statistics, since it started with birth records. The infancy of statistics occurred much later, some time around 27 BCE, when the Roman emperor Augustus conducted surveys on births and deaths of citizens, as well as the amount of livestock each owned and the crops each had harvested. Suddenly, a large amount of data was wanted, and methods had to be developed to collect, organize, and summarize that data.

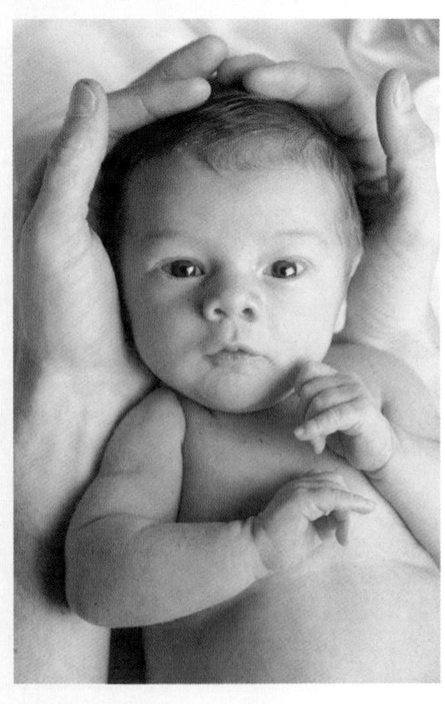

The adolescence of statistics is probably sometime around the 14th century, when people began keeping detailed records of births, deaths, and accidents in order to determine insurance rates. Over the next 600 years or so, statistics grew up, with a wide variety of methods developed to fill needs in biology, physics, astronomy, commerce, and many other areas.

The true adulthood of statistics coincides with the rise of computer technology. Today, with just a few clicks of a mouse, statisticians can process and analyze amounts of data that would have taken the pioneers in the field centuries.

Answers to Try This One

1 21.248 seconds

2 3.63 losses

3 60°

4 60.5 fatalities

5 2 incidents

6 There is no mode.

7 The modes are 82° and 87°.

8 The mode is donating nothing.

9 171 earthquakes

10 Mean: −5.69; median: −4; mode: −12, −4, +2; midrange: −7.5

EXERCISE SET 12-3

Writing Exercises

1. If someone asked you before covering this section what the word "average" meant, what would you have said? What would your response be now?
2. Describe how to find the mean of a data set.
3. Describe how to find the median of a data set.
4. What is meant by the term *mode of a data set?*
5. Describe how to find the midrange of a data set.
6. Which of the measures of average can be used to describe the average for categorical data? Why?

Real-World Applications

For exercises 7–18, find the mean, median, mode, and midrange.

7. These data are the number of on-campus burglaries reported in 2008 for nine western Pennsylvania universities.

   61 11 1 3 2 30 18 3 7

8. The data below show the number of attorneys employed by the ten largest law firms in Pittsburgh.

   87 109 57 221 175 123 170 80 66 80

9. A graduate student in social work surveyed students at 10 colleges in a Midwestern state to find how many are single parents. The results are the data below.

   700 298 638 260 1,380 280 270 1,350 380 570

10. The number of wins for the teams that won college football bowl games at the end of the 2008 season is

    8 7 8 8 7 11 7 7 9 9 9 8 8 10
    8 10 10 8 9 7 8 8 9 8 10 9 7
    8 11 12 10 13 12 13

11. The average number of cigarettes smoked per person in a year in the top ten countries for smoking are shown next.

Greece	4,313	Malta	2,668
Hungary	3,265	Bulgaria	2,574
Kuwait	3,062	Belarus	2,571
Japan	3,023	Belgium	2,428
Spain	2,779	Turkey	2,394

12. The number of existing home sales in millions for the months from August 2007 to February 2009 are shown below.

    5.5 5.1 5.1 5.0 4.9 4.9 5.0 4.9 4.9 5.0
    4.9 5.0 4.9 5.1 4.9 4.5 4.7 4.5 4.7

13. The number of bids for the five most popular products under concert tickets on eBay for a day in April is shown below.

    340 75 123 259 151

14. The number of sales of single-family homes in the largest metropolitan areas of Florida in 2008 was

    321 307 338 200 141 115 518 185 258
    276 31 116 881 72 181 175 490 162
    1,235 369

15. From 1992 to 2004, the number of violent crimes committed per 1,000 students in U.S. schools each year was

    10 12 13 9 9 8 9 7 5 6 6 5

16. The average amount of tuition, fees, room and board for full-time students at public 4-year institutions in 2005–2006 in states starting with "M" was

$12,865 $15,253 $15,199 $14,519 $13,719
$10,040 $12,588 $11,292

17. The annual number of murders in the United States from 1995 to 2007 was as follows:

Year	Murders	Year	Murders
1995	21,610	2002	16,229
1996	19,650	2003	16,528
1997	18,208	2004	16,148
1998	16,914	2005	16,740
1999	15,522	2006	17,030
2000	15,586	2007	16,929
2001	16,037		

18. The 10 states with the highest unemployment rates in February 2009 and their rates of unemployment are given below:

State	Rate
Michigan	12.0
South Carolina	11.0
Oregon	10.8
Rhode Island	10.5
Nevada	10.1
North Carolina	10.7
California	10.5
District of Columbia	9.9
Florida	9.4
Indiana	9.4

For Exercises 19–26, find the mean of the data set.

19. For 50 students in the student union at one campus, the distribution of the students' ages was obtained as shown.

Class	Frequency	Class	Frequency
18–20	20	24–26	8
21–23	18	27–29	4

20. Thirty new automobiles were tested for fuel efficiency by the EPA (in miles per gallon). This frequency distribution was obtained:

Class	Frequency	Class	Frequency
8–12	3	23–27	5
13–17	5	28–32	2
18–22	15		

21. In a study of the time it takes an untrained mouse to run a maze, a researcher recorded these data in seconds.

Class	Frequency
2.1–2.7	5
2.8–3.4	7
3.5–4.1	12
4.2–4.8	14
4.9–5.5	16
5.6–6.2	8

22. Eighty randomly selected lightbulbs were tested to determine their lifetimes (in hours). The frequency distribution was obtained as shown.

Class	Frequency
53–63	6
64–74	12
75–85	25
86–96	18
97–107	14
108–118	5

23. These data represent the net worth (in millions of dollars) of 45 national corporations.

Class	Frequency
10–20	2
21–31	8
32–42	15
43–53	7
54–64	10
65–75	3

24. The cost per load (in cents) of 35 laundry detergents tested by *Consumer Reports* is shown here.

Class	Frequency
13–19	2
20–26	7
27–33	12
34–40	5
41–47	6
48–54	1
55–61	0
62–68	2

25. The frequency distribution shown represents the commission earned (in dollars) by 100 salespeople employed at several branches of a large chain store.

Class	Frequency	Class	Frequency
150–158	5	186–194	20
159–167	16	195–203	15
168–176	20	204–212	3
177–185	21		

26. This frequency distribution represents the data obtained from a sample of 75 copy machine service technicians. The values represent the days between service calls for various copy machines.

Class	Frequency	Class	Frequency
16–18	14	25–27	10
19–21	12	28–30	15
22–24	18	31–33	6

Critical Thinking

27. For each of the characteristics of the average college student listed on page 665, decide which measure of average you think was used, and explain your choices.
28. For these situations, state which measure of average—mean, median, or mode—should be used.
    (a) The most typical case is desired.
    (b) The data are categorical.
    (c) The values are to be divided into two approximately equal groups, one group containing the larger and one containing the smaller values.

For Exercises 29–34, describe which measure of average—mean, median, or mode—was probably used in each situation.

29. Half of the factory workers make more than \$5.37 per hour and half make less than \$5.37 per hour.
30. The average number of children per family in the Plaza Heights is 1.8.
31. Most people prefer red convertibles to any other color.
32. The average person cuts the lawn once a week.
33. The most common fear today is fear of speaking in public.
34. The average age of college professors is 42.3 years.

Section 12-4 Measures of Variation

LEARNING OBJECTIVES

- ☐ 1. Find the range of a data set.
- ☐ 2. Find the variance and standard deviation of a data set.
- ☐ 3. Interpret standard deviation.

Now that we know about measures of average, we'll consider the fact that there's more to the story told by a data set than just the average. For an example, consider the two pictures of dogs on this page.

If we look only at measures of average, particularly the mean, we might be fooled into thinking that the two groups are very similar, when clearly they are not. The difference, of course, is that all of the dogs in the first picture are of similar size, while those in the second picture have many different weights. Because there are some small dogs and one very large one, the mean weight in both groups is probably similar.

In this section we will study *measures of variation,* which will help to describe how the data within a set vary. The three most commonly used measures of variation are *range, variance,* and *standard deviation.*

The two groups of dogs have about the same mean size, but the range of sizes is quite different.

Range

The *range* is the simplest of the three measures of variation that we will study.

The **range** of a data set is the difference between the highest and lowest values in the set.

$$\text{Range} = \text{Highest value} - \text{lowest value}$$

EXAMPLE 1 Finding the Range of a Data Set

The first list below is the weights of the dogs in the first picture, and the second is the weights of the dogs in the second picture. Find the mean and range for each list.

70 73 58 60
30 85 40 125 42 75 60 55

SOLUTION

For the first list,

$$\text{Mean} = \frac{70 + 73 + 58 + 60}{4} = 65.25 \text{ lb}$$

$$\text{Range} = 73 - 58 = 15 \text{ lb}$$

For the second list,

$$\text{Mean} = \frac{30 + 85 + 40 + 125 + 42 + 75 + 60 + 55}{8} = 64 \text{ lb}$$

$$\text{Range} = 125 - 30 = 95 \text{ lb}$$

As we suspected, the means are very close, but the ranges are very different.

▼ Try This One 1

☑ 1. Find the range of a data set.

The monthly average high temperatures from January to December in Aruba are shown below. Find the range.

85 85 86 87 88 89 88 89 89 89 87 85

Variance and Standard Deviation

The range is a limited measure of variation because it ignores all the data except the highest and lowest values. If most of the values are similar, but there's just one unusually high value, the range will make it look like there's a lot more variation than there actually is. For this reason, we will next define *variance* and *standard deviation,* which are much more reliable measures of variation.

Variance and standard deviation are a little tricky to compute. In the box below, we'll explain how to find them. Later in the section, we'll discuss their significance.

Procedure for Finding the Variance and Standard Deviation

Step 1 Find the mean.

Step 2 Subtract the mean from each data value in the data set.

Step 3 Square the differences.

Step 4 Find the sum of the squares.

Step 5 Divide the sum by $n - 1$ to get the variance, where n is the number of data values.

Step 6 Take the square root of the variance to get the standard deviation.

EXAMPLE 2 Finding Variance and Standard Deviation

The heights in inches of the top six scorers for the Cleveland Cavaliers during the 2008–2009 season are listed below. Find the variance and standard deviation.

80 73 87 74 83 74

SOLUTION

Step 1 Find the mean height.

$$\text{Mean} = \frac{\sum X}{n} = \frac{80 + 73 + 87 + 74 + 83 + 74}{6} = 78.5 \text{ inches}$$

Step 2 Subtract the mean from each data value.

$80 - 78.5 = 1.5$ $\quad 87 - 78.5 = 8.5$ $\quad 83 - 78.5 = 4.5$

$73 - 78.5 = -5.5$ $\quad 74 - 78.5 = -4.5$ $\quad 74 - 78.5 = -4.5$

Step 3 Square each result.

$(1.5)^2 = 2.25$ $\quad (8.5)^2 = 72.25$ $\quad (4.5)^2 = 20.25$

$(-5.5)^2 = 30.25$ $\quad (-4.5)^2 = 20.25$ $\quad (-4.5)^2 = 20.25$

Step 4 Find the sum of the squares.

$$2.25 + 30.25 + 72.25 + 20.25 + 20.25 + 20.25 = 165.5$$

Step 5 Divide the sum by $n - 1$ to get the variance, where n is the sample size. In this case, n is 6, so $n - 1 = 5$.

$$\text{Variance} = \frac{165.5}{5} = 33.1$$

Step 6 Take the square root of the variance to get standard deviation.

$$\text{Standard deviation} = \sqrt{33.1} \approx 5.75 \text{ inches}$$

To organize the steps, you might find it helpful to make a table with three columns: The original data, the difference between each data value and the mean, and their squares. Then you just add the entries in the last column and divide by $n - 1$ to get the variance.

Data (X)	X − mean	(X − mean)^2
80	1.5	2.25
73	−5.5	30.25
87	8.5	72.25
74	−4.5	20.25
83	4.5	20.25
74	−4.5	20.25
		165.5

▼ Try This One 2

The heights in inches for the eight nonpitchers in the starting lineup for the New York Mets on opening day 2008 are listed below. Find the variance and standard deviation.

73 74 72 75 73 74 73 71

To understand the significance of standard deviation, we'll look at the process one step at a time.

Step 1 *Compute the mean.* Variation is a measure of how far the data vary from the mean, so it makes sense to begin there.

Step 2 *Subtract the mean from each data value.* In this step, we are literally calculating how far away from the mean each data value is. The problem is that since some are greater than the mean and some less, their sum will always add up to zero. (Try it!) So that doesn't help much.

Step 3 *Square the differences.* This solves the problem of those differences adding to zero—when we square, they're all positive.

Step 4 *Add the squares.* In the next two steps, we're getting an approximate average of the squares of the individual variations from the mean. First we add them, then . . .

Step 5 *Divide the sum by $n - 1$.* It's a bit technical to explain why we divide by $n - 1$ rather than n, but in any case we now have the approximate average of the squares of the individual variations from the mean.

Step 6 *Take the square root of the sum.* This "undoes" the square we did in step 3. It will return the units of our answer to the units of the original data, giving us a good measure of how far the typical data value varies from the mean.

Now we'll formally define these two key measures of variation.

2. Find the variance and standard deviation of a data set.

The **variance** for a data set is an approximate average of the square of the distance between each value and the mean. If X represents individual values, $\overline{X}$ is the mean and n is the number of values:

$$\text{Variance} = \frac{\sum(X - \overline{X})^2}{n - 1}$$

The **standard deviation** is the square root of the variance.

Interpreting Standard Deviation

Since standard deviation measures how far typical values are from the mean, its size tells us how spread out the data are. We'll examine this idea in Example 3.

EXAMPLE 3 Interpreting Standard Deviation

A professor has two sections of Math 115 this semester. The 8:30 A.M. class has a mean score of 74% with a standard deviation of 3.6%. The 2 P.M. class also has a mean score of 74%, but a standard deviation of 9.2%. What can we conclude about the students' averages in these two sections?

SOLUTION

In relative terms, the morning class has a small standard deviation and the afternoon class has a large one. So even though they have the same mean, the classes are quite different. In the morning class, most of the students probably have scores relatively close to the mean, with few very high or very low scores. In the afternoon class, the scores vary more widely, with a lot of high scores and a lot of low scores that average out to a mean of 74%.

▼ Try This One 3

☑ 3. Interpret standard deviation.

For the dogs in the two pictures on page 675, discuss what you think the standard deviations might be in comparison to one another, then check your answer by computing the standard deviation for each group.

Answers to Try This One

1 4°

2 Variance ≈ 1.55 inches; standard deviation ≈ 1.25 inches

3 The second group should have a much larger standard deviation than the first since the weights are more spread out. The actual standard deviation for the first group is 7.37 lb and for the second group it is 30.7 lb.

EXERCISE SET 12-4

Writing Exercises

1. Name three measures of variation.
2. What is the range?
3. Why is the range not usually the best measure of variation?
4. What is the relationship between the variance and standard deviation?
5. Explain the procedure for finding the standard deviation for data.
6. Explain how the variation of two data sets can be compared by using the standard deviations.

For Exercises 7–10, discuss the relative sizes of the standard deviations for the two data sets.

7. Data set 1: 12 15 13 10 16 13 12 13
   Data set 2: 5 26 31 2 10 25 6 33
8. Data set 1: 40-yard dash times for starting running backs in the National Football League
   Data set 2: 40-yard dash times for every student in your math class
9. Data set 1: Average monthly high temperature in Chicago
   Data set 2: Average monthly high temperature in Los Angeles
10. Data set 1: Weights of all the dogs at a golden retriever rescue shelter
    Data set 2: Weights of all the dogs for sale in a pet store

Real-World Applications

For Exercises 11-26, find the range, variance, and standard deviation.

11. These data are the number of junk e-mails Lena received for 9 consecutive days.

    61 1 1 3 2 30 18 3 7

12. The number of hospitals in the five largest hospital systems is shown here.

    340 75 123 259 151

    Source: USA Today

13. Ten used trail bikes are randomly selected from a bike shop, and the odometer reading of each is recorded as follows.

    1,902 103 653 1,901 788 361 216 363 223 656

14. Fifteen students were selected and asked how many hours each studied for the final exam in statistics. Their answers are recorded here.

    8 6 3 0 0 5 9 2 1 3 7 10 0 3 6

15. The weights of nine players from a college football team are recorded as follows.

206 215 305 297 265 282 301 255 261

16. Shown here are the numbers of stories in the 11 tallest buildings in St. Paul, Minnesota.

37 46 32 32 33 25 17 27 21 25 33

Source: The World Almanac and Book of Facts

17. The heights (in inches) of nine male army recruits are shown here.

78 72 68 73 75 69 74 73 72

18. The number of calories in 12 randomly selected microwave dinners is shown here.

560 832 780 650 470 920 1,090 970
495 550 605 735

19. The table below shows the average price of a gallon of regular unleaded gas in U.S. dollars for various cities in June 2007.

City	Price
Chicago	2.69
Paris	6.52
Toronto	3.28
Moscow	2.89
São Paulo	4.89
Seoul	6.06

20. The number of attorneys in 10 law firms in Pittsburgh is 87, 109, 57, 221, 175, 123, 170, 80, 66, and 80.

Source: *Pittsburgh Tribune Review*

21. The stock prices for eight major grocery store chains in early 2009 were: $18.28, $20.32, $9.34, $11.57, $15.21, $48.04, $48.84, and $28.23.
22. The table reflects consumer priorities worldwide for a variety of expenditures:

Global Priority	U.S. Currency (in billions)
Cosmetics in the United States	8
Ice cream in Europe	11
Perfumes in Europe and the United States	12
Pet foods in Europe and the United States	17
Business entertainment in Japan	35
Cigarettes in Europe	50
Alcoholic drinks in Europe	105
Narcotic drugs in the world	400
Military spending in the world	780

Source: www.globalissues.org

23. The number of flu breakouts in 2009 in various regions of the U.S. were

429 690 884 827 914 176 565 479 569

Source: cdc.gov

24. The five teams in baseball's National League eastern division had the following number of wins in 2008: 92, 89, 84, 72, 59.
25. The number of workplace fatalities in the United States from 2000 to 2007 were as follows:

Year	Fatalities	Year	Fatalities
2000	5,920	2004	5,764
2001	5,915	2005	5,734
2002	5,534	2006	5,840
2003	5,575	2007	5,488

Source: *bls.gov*

26. From 1992 to 2004, the number of violent crimes committed per 1,000 students in U.S. schools each year was

10 12 13 9 9 8 9 7 5 6 6 5

Critical Thinking

27. The three data sets have the same mean and range, but is the variation the same? Explain your answer.
    (a) 5 7 9 11 13 15 17
    (b) 5 6 7 11 15 16 17
    (c) 5 5 5 11 17 17 17

28. Using this set—10, 20, 30,40, and 50,
    (a) Find the standard deviation.
    (b) Add 5 to each value and then find the standard deviation.
    (c) Subtract 5 from each value and then find the standard deviation.
    (d) Multiply each value by 5 and then find the standard deviation.
    (e) Divide each value by 5 and then find the standard deviation.
    (f) Generalize the results of (a)–(e).

Section 12-5 Measures of Position

LEARNING OBJECTIVES

- ☐ 1. Compute the percentile rank for a data value.
- ☐ 2. Find a data value corresponding to a given percentile.
- ☐ 3. Use percentile rank to compare values from different data sets.
- ☐ 4. Compute quartiles for a data set.

So you managed to survive 4 years of college and escape with that long-awaited diploma. But maybe the thought of entering the real world starts to look a little less appealing, and you realize that a couple more years of college might suit you. Next step: take the GRE (Graduate Record Examination), which is like the SAT for grad school.

When you get your score back, you see that it's 1120. How do you know if that's a good score? The quickest way is to look at the *percentile rank* on the report, which in this case would probably be 60%. Does this mean that you only got 60% of the questions right? That wouldn't be a very good score! In fact, that's not at all what it means, and you did just fine.

The term percentile is used in statistics to measure the position of a data value in a data set.

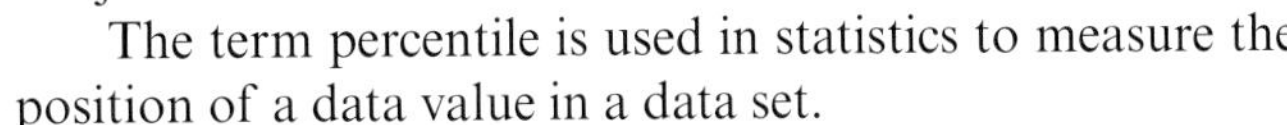

A **percentile,** or percentile rank, of a data value indicates the percent of data values in a set that are below that particular value.

In this case, your percentile rank of 60% means that 60% of all students who took the GRE scored lower than you did. Not bad! Scores on standardized tests are one of the most common uses for percentiles because they help you to put a raw score like 1120 into context.

Percentiles are also used very commonly in health care, especially pediatrics. To monitor a child's development, doctors compare the child's height, weight, and head size at a certain age to measurements of other children at the same age. If a child's percentiles suddenly change radically, the doctor might suspect a problem in his or her development.

Percentiles were originally used to analyze data sets with 100 or more values. But when statistical techniques were developed for smaller data sets, the percentile concept came into use for those sets as well. There are several methods that can be used for computing percentiles, and sometimes the answers vary slightly, especially with small data sets. In this section, we will use a basic method that works well for smaller data sets.

EXAMPLE 1 Finding the Percentile Rank of a Data Value

Math Note

In order to keep our study of percentiles relatively simple, we'll work with data sets with no repeated values. More complicated techniques are used to find percentiles for data sets with repeated values.

Suppose you score 77 on a test in a class of 10 people, with the 10 scores listed below. What was your percentile rank?

93 82 64 75 98 52 77 88 90 71

SOLUTION

Step 1 Arrange the scores in order.

52 64 71 75 77 82 88 90 93 98

Step 2 Find the number of data values below 77. There are 4 values below 77.

Step 3 Divide the number below the score by the total number of data values and change the answer to a percent.

$$\frac{4}{10} = 0.40 = 40\%$$

A test score of 77 is equivalent to the 40th percentile.

1. Compute the percentile rank for a data value.

▼ Try This One 1

The weights in pounds for the 12 members of a college gymnastics team are below. Find the percentile rank of the gymnast who weighs 97 pounds.

101 120 88 72 75 80 98 91 105 97 78 85

In Example 1, we saw how to find the percentile rank for a given data value. In Example 2, we'll examine the opposite question—finding the data value that corresponds to a given percentile.

EXAMPLE 2 Finding a Data Value Corresponding to a Given Percentile

The number of words in each of the last 10 presidential inaugural addresses is listed below. Find the length that corresponds to the 30th percentile.

2,406 2,073 1,571 2,170 1,507 2,283 2,546 2,463 1,087 1,668

SOLUTION

Step 1 We are asked to find the number on the list that has 30% of the numbers below it. There are 10 numbers, and 30% of 10 is 3.

Step 2 Arrange the data in order from smallest to largest, and find the value that has 3 values below it.

1,087 1,507 1,571 1,668 2,073 2,170 2,283 2,406 2,463 2,546

The 30th percentile is the speech that consisted of 1,668 words.

▼ Try This One 2

The average monthly rainfall in inches for St. Petersburg, Florida, is shown in the chart below. Which month is at the 75th percentile?

Jan	Feb	Mar	Apr	May	Jun	Jul	Aug	Sep	Oct	Nov	Dec
2.3	2.8	3.4	1.6	2.6	5.7	7.0	7.8	6.1	2.5	1.9	2.2

2. Find a data value corresponding to a given percentile.

Percentile ranks are particularly useful in comparing data that come from two different sets, as shown in Example 3.

EXAMPLE 3 Using Percentiles to Compare Data from Different Sets

Two students are competing for one remaining spot in a law school class. Miguel ranked 51st in a graduating class of 1,700, while Dustin ranked 27th in a class of 540. Which student's position was higher in his class?

SOLUTION

This is an ideal application of percentile rank. Miguel ranked 51st out of 1,700, so there were 1,700 − 51 = 1,649 students ranked below him. His percentile rank is

$$\frac{1{,}649}{1{,}700} = 0.97 \text{ or } 97\%$$

Dustin had 540 − 27 = 513 students ranked below him, so his percentile rank is

$$\frac{513}{540} = 0.95 \text{ or } 95\%$$

Both are excellent students, but Miguel's ranking is higher.

▼ Try This One 3

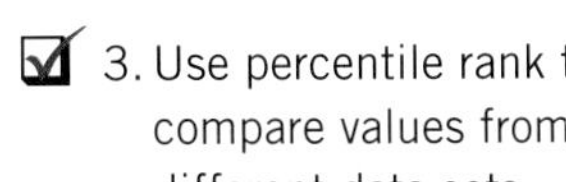
3. Use percentile rank to compare values from different data sets.

In the 2007–2008 school year, West Virginia University finished the season ranked sixth out of 119 Division IA football teams, and ranked 17th out of 346 Division I basketball teams. Which team had the higher percentile rank?

In Section 12-3, we saw that the median of a data set divides the set into equal halves. Another statistical measure we will study is the **quartile,** which divides a data set into quarters. The second quartile is the same as the median, and divides a data set into an upper half and a lower half. The first quartile is the median of the lower half, and the third quartile is the median of the upper half. We use the symbols Q_1, Q_2, and Q_3 for the first, second, and third quartiles respectively.

EXAMPLE 4 Finding Quartiles for a Data Set

Find Q_1, Q_2, and Q_3 for the number of aircraft stolen during a recent 8-year period.

14 11 20 21 42 24 36 35

Source: USA Today

SOLUTION

Step 1 Arrange the data in order.

11 14 20 21 24 35 36 42

Step 2 Find the median. This is Q_2.

11 14 20 21 ↑ 24 35 36 42

$Q_2 = 22.5$

Step 3 Find the median of the data values less than Q_2. This is Q_1.

11 14 ↑ 20 21

$Q_1 = 17$

Step 4 Find the median of the data values above Q_2. This is Q_3.

24 35 ↑ 36 42

$Q_3 = 35.5$

In summary, $Q_1 = 17$, $Q_2 = 22.5$, and $Q_3 = 35.5$.

4. Compute quartiles for a data set.

▼ Try This One 4

Find Q_1, Q_2, and Q_3 for the data shown.

18 32 54 36 27 42 31 15 60 25

In this section, we studied two measures of position, percentile and quartile. In Section 12-6, the concept of position plays a very important role when we study a third measure of position, the *z* score.

Answers to Try This One

1 58th percentile

2 September

3 Basketball

4 $Q_1 = 25$, $Q_2 = 31.5$, $Q_3 = 42$

EXERCISE SET 12-5

Writing Exercises

1. If your score in your math class puts you in the 60th percentile, what exactly does that mean?
2. Does a score in the 90th percentile mean that you got 90% of the questions right on a test? Explain.
3. Explain what quartiles are.
4. What is the connection between the second quartile and the median for a data set? Explain.

Real-World Applications

5. The scores for 20 students on a 50-point math test are 42, 48, 50, 36, 35, 27, 47, 38, 32, 43, 24, 33, 39, 49, 44, 40, 29, 30, 41, and 37.

   (a) Find the percentile rank for a score of 32.
   (b) Find the percentile rank for a score of 44.
   (c) Find the percentile rank for a score of 36.
   (d) Find the percentile rank for a score of 27.
   (e) Find the percentile rank for a score of 49.

6. The heights (in inches) of the 12 students in a seminar course are 73, 68, 64, 63, 71, 70, 65, 67, 72, 66, 60, and 61.

   (a) Find the percentile rank for a height of 67 in.
   (b) Find the percentile rank for a height of 70 in.
   (c) Find the percentile rank for a height of 63 in.
   (d) Find the percentile rank for a height of 68 in.
   (e) Find the percentile rank for a height of 66 in.

7. In a class of 500 students, Carveta's rank was 125. Find her percentile rank.
8. In a class of 400 students, John's rank was 80. Find his percentile rank.
9. In a charity marathon to raise money for AIDS research, Chen finished 43rd out of 200 entrants. Find her percentile rank.
10. In a speed boat race, there were 25 participants. Nate came in 4th place. What is his percentile rank?
11. On an exam, Angela scored in the 20th percentile. If there were 50 students in the class, how many students scored lower than she did?
12. Out of 600 applicants to a graduate program at an Ivy League school, Marissa's GRE score was in the 25th percentile. How many applicants scored lower than she did?
13. Lea's percentile rank on an exam in a class of 600 students is 60. Maurice's class rank is 220. Who is ranked higher?
14. In an English class of 30 students, Audrelia's percentile rank is 20. Maranda's class rank is 20. Whose rank is higher?

15. In an evening statistics class, the ages of 20 students are as follows.

18 24 19 20 33 42 43 27 31 39
21 44 26 32 37 34 23 35 28 25

(a) What is the percentile rank of 33?
(b) What is the rank (from the top) of 33?
(c) What age corresponds to the 20th percentile?

16. Twenty subjects in a psychology class experiment scored the following on an IQ test.

95 107 110 101 122 96 94 104 131 90
111 103 97 100 119 85 108 120 130 88

(a) What is the percentile rank of 96?
(b) What is the rank (from the top) of 96?
(c) What score corresponds to the 40th percentile?

For Exercises 17–22, find the values for Q_1, Q_2, and Q_3.

17. Number of drive-in theaters in nine selected states:

59 20 21 34 52 48 24 29 55

Source: National Association of Theater Owners

18. Average cost in cents per kilowatt hour of producing electricity using nuclear energy for the past 8 years:

1.83 2.18 2.36 2.04 2.10 2.25 2.48 2.57

Source: Nuclear Energy Institute

19. The costs of a 1-hour massage in 10 large cities are:

\$75 \$77 \$78 \$90 \$87 \$93 \$89 \$93 \$95 \$88

20. Number of passengers (in millions) at 10 selected airports in the United States:

32.6 28.9 15.5 16.6 19.4 15.9 31.6 15.4 29.1 15.7

Source: Aviation Statistics

21. The annual number of homicides in Milwaukee for a 10-year period is shown.

122 111 124 121 127 108 107 88 122 103

22. Below are the market caps in billions of dollars of nine of the largest companies that are 30 years old or younger.

Google	\$140.6	Yahoo	\$40.2
Home Depot	\$85.2	EMC	\$31.9
Apple	\$73.0	Starbucks	\$24.5
Dell	\$53.4	Staples	\$19.4
eBay	\$45.8		

Source: *USA Today*

Critical Thinking

23. Is it possible to score 90% on a test and have a percentile rank less than 90th? Is it possible to score 90% and have a percentile rank of exactly 90?

24. What would you have to do in order to rank in the 100th percentile in your college graduating class?

Section 12-6 The Normal Distribution

LEARNING OBJECTIVES

- ☐ 1. Identify characteristics of a data set that is normally distributed.
- ☐ 2. Apply the empirical rule.
- ☐ 3. Compute *z* scores.
- ☐ 4. Use a table and *z* scores to find areas under the standard normal distribution.

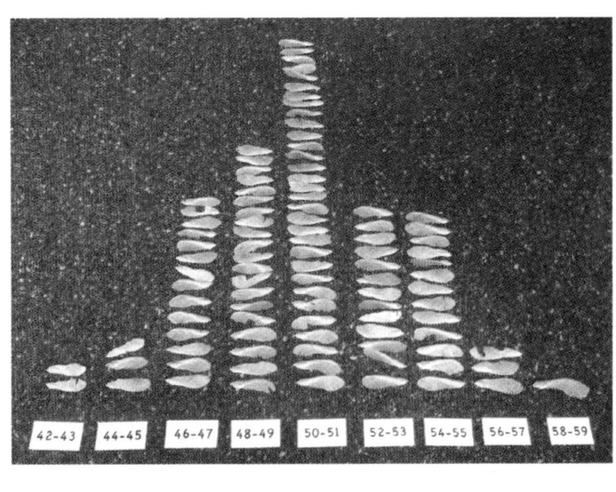

If you grew up anywhere near maple trees, you're probably familiar with the seed pods they drop by the thousands in the spring. Like almost all living things, the individual pods vary in size—on any given tree, there's a typical size, with some pods bigger and some smaller.

On a beautiful spring day, I gathered 100 pods from the maple in my back yard, measured each, then grouped them according to their length. The smallest was 42 mm, the largest 59 mm, and in the photo they're grouped in classes 42–43, 44–45, 46–47, etc. An interesting thing happened—we see that the largest number of pods have lengths somewhere in the middle of the range, and the classes further away from the center have less pods.

A wide variety of quantities in the real world, like sizes of individuals in a population, IQ scores, life spans for batteries of a certain brand, and many others, tend to exhibit this same phenomenon. In fact, it's so common that frequency distributions of this type came to be known as *normal distributions*.

Suppose a researcher selects a random sample of 100 adult women, measures their heights, and constructs a histogram. The researcher would probably get a graph similar to the one shown in Figure 12-6a. If the researcher increases the sample size and decreases the width of the classes, the histograms will look like the ones shown in Figures 12-6b and 12-6c. Finally, if it were possible to measure the exact heights of all adult females in the United States and plot them, the histogram would approach what is called the *normal distribution,* shown in Figure 12-6d. This distribution is also known as a *bell curve* or a *Gaussian distribution,* named for the German mathematician Carl Friedrich Gauss (1777–1855) who derived its equation.

A normal distribution is defined formally as follows.

A **normal distribution** is a continuous, symmetric, bell-shaped distribution.

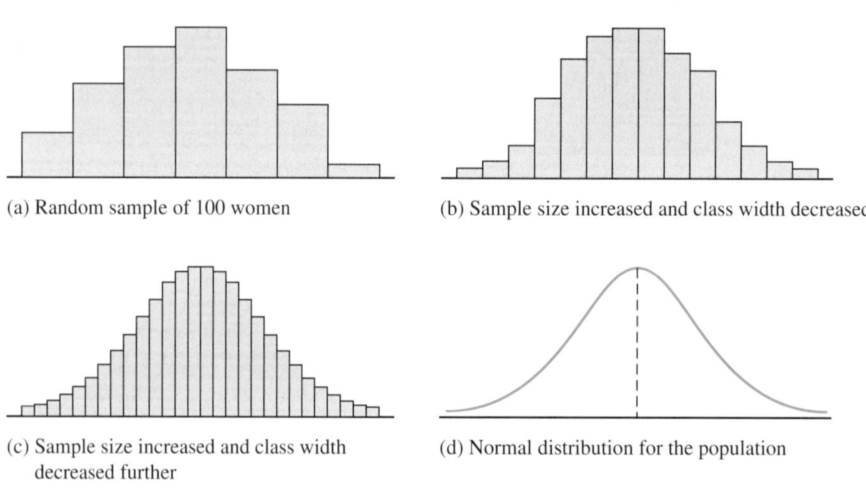

Figure 12-6

Some properties of a normal distribution, including those mentioned in the definition, are explained next.

In a normal distribution, values cluster around a central value and fall off in both directions. Waiting times can be normally distributed.

Some Properties of a Normal Distribution

1. It is bell-shaped.
2. The mean, median, and mode are equal and located at the center of the distribution.
3. It is unimodal (i.e., it has only one mode).
4. It is symmetrical about the mean, which is equivalent to saying that its shape is the same on both sides of a vertical line passing through the center.
5. It is continuous—i.e., there are no gaps or holes.
6. The area under a portion of a normal curve is the percentage (in decimal form) of the data that falls between the data values that begin and end that region. (We will use this property extensively in Section 12-7.)
7. The total area under a normal curve is exactly 1. This makes sense based on property 6, since the area under the whole curve encompasses all data values.

☑ 1. Identify characteristics of a data set that is normally distributed.

The Empirical Rule

When data values are normally distributed, there is a rule that allows us to quickly determine a range in which most of the data values fall. It is known as the *empirical rule.*

The Empirical Rule

When data are normally distributed, approximately 68% of the values are within 1 standard deviation of the mean, approximately 95% are within 2 standard deviations of the mean, and approximately 99.7% are within 3 standard deviations of the mean (see Figure 12-7).

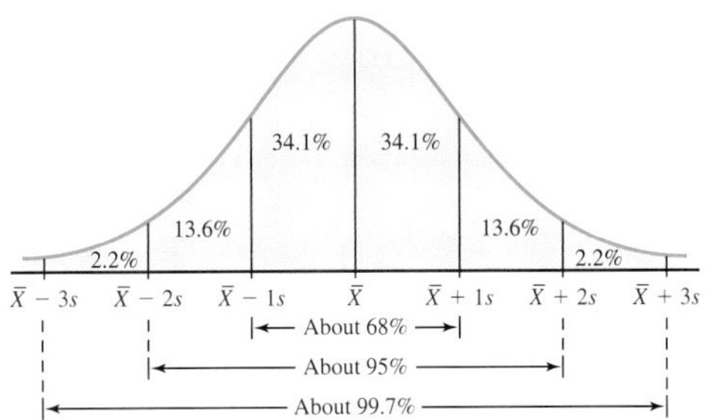

Figure 12-7 $\overline{X}$ = mean, s = standard deviation

EXAMPLE 1 Using the Empirical Rule

According to the website answerbag.com, the mean height for male humans is 5 feet 9.3 inches, with a standard deviation of 2.8 inches. If this is accurate, out of 1,000 randomly selected men, how many would you expect to be between 5 feet 6.5 inches and 6 feet 0.1 inch?

SOLUTION

The given range of heights corresponds to those within 1 standard deviation of the mean, so we would expect about 68% of men to fall in that range. In this case, we expect about 680 men to be between 5 feet 6.5 inches and 6 feet 0.1 inch.

▼ Try This One 1

☑ 2. Apply the empirical rule.

A standard test of intelligence is scaled so that the mean IQ is 100, and the standard deviation is 15. If there are 40,000 people in a stadium, how many would you expect to have an IQ between 70 and 130?

The Standard Normal Distribution

Any set of data that is normally distributed has its own mean and standard deviation, so the exact shape and location of the associated normal curve varies accordingly. This presents a problem: to answer questions related to the number of data values that are likely to fall into a given range, we need a method for finding the area under the curve for each specific case. This is not an easy problem to solve. Instead, statisticians developed a way to standardize the normal curves for different data sets by using the *standard normal distribution.*

The **standard normal distribution** is a normal distribution with mean 0 and standard deviation 1.

The standard normal distribution is shown in Figure 12-8. The values under the curve shown in Figure 12-8 indicate the proportion of area in each section. For example, the area between the mean and 1 standard deviation above or below the mean is about 0.341, or 34.1%.

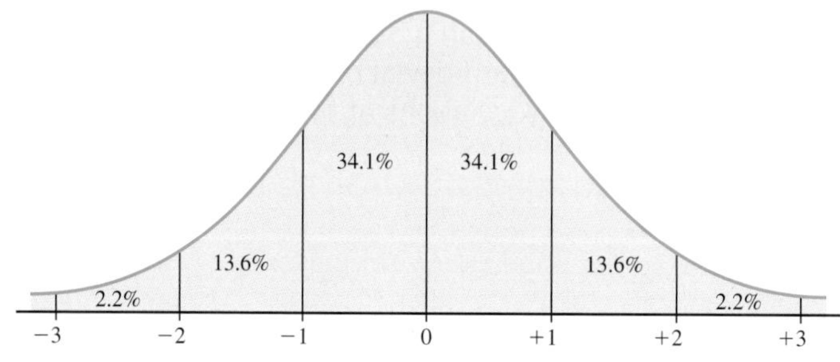

Figure 12-8

Tables have been compiled that help us to find the area under certain portions of a standard normal curve. In order to apply them to any data that are normally distributed, those data have to be altered to make the distribution have mean 0 and standard deviation 1. To accomplish that, we will use *z scores*.

For a given data value from a data set that is normally distributed, we define that value's **z score** to be

$$z = \frac{\text{Data value} - \text{mean}}{\text{Standard deviation}}$$

The z score is a measure of position: it describes how many standard deviations a data value lies above (if positive) or below (if negative) the mean.

EXAMPLE 2 Computing a *z* Score

Math Note

Notice that the units divide out, so z scores have no units.

Based on the information in Example 1, find the z score for a man who is 6 feet 4 inches tall.

SOLUTION

Using the formula for z scores with mean 5 feet 9.3 inches and standard deviation 2.8 inches:

$$z = \frac{76\text{ in.} - 69.3\text{ in.}}{2.8\text{ in.}} \approx 2.4$$

▼ Try This One 2

☑ 3. Compute z scores.

Using the information in Try This One 1, find the z score for a person with an IQ of 91.

The value of z scores is that they will allow us to find areas under a normal curve using only areas under a standard normal curve, which can be read from a table, like Table 12-3.

TABLE 12-3 **Area Under a Normal Distribution Curve Between $z = 0$ and a Positive Value of z**

z	A	z	A	z	A
0.00	0.000	1.10	0.364	2.20	0.486
0.05	0.020	1.15	0.375	2.25	0.488
0.10	0.040	1.20	0.385	2.30	0.489
0.15	0.060	1.25	0.394	2.35	0.491
0.20	0.079	1.30	0.403	2.40	0.492
0.25	0.099	1.35	0.412	2.45	0.493
0.30	0.118	1.40	0.419	2.50	0.494
0.35	0.137	1.45	0.427	2.55	0.495
0.40	0.155	1.50	0.433	2.60	0.495
0.45	0.174	1.55	0.439	2.65	0.496
0.50	0.192	1.60	0.445	2.70	0.497
0.55	0.209	1.65	0.451	2.75	0.497
0.60	0.226	1.70	0.455	2.80	0.497
0.65	0.242	1.75	0.460	2.85	0.498
0.70	0.258	1.80	0.464	2.90	0.498
0.75	0.273	1.85	0.468	2.95	0.498
0.80	0.288	1.90	0.471	3.00	0.499
0.85	0.302	1.95	0.474	3.05	0.499
0.90	0.316	2.00	0.477	3.10	0.499
0.95	0.329	2.05	0.480	3.15	0.499
1.00	0.341	2.10	0.482	3.20	0.499
1.05	0.353	2.15	0.484	3.25*	0.499

*For z scores greater than 3.25, use A = 0.500

Math Note

An area table with more values can be found in Appendix A. If you need the area for a z score between two values in Table 12-3, you can get an approximate area by choosing an area between areas in the table, or use the more extensive table in the appendix.

Finding Areas under the Standard Normal Distribution

Section 12-7 is entirely devoted to solving real-world problems that use areas under a normal distribution. In the remainder of this section, we will focus on simply finding those areas. We will find areas corresponding to z scores using Table 12-3 and the following key facts.

Two Important Facts about the Standard Normal Curve

1. The area under any normal curve is divided into two equal halves at the mean. Each of the halves has area 0.500.
2. The area between $z = 0$ and a positive z score is the same as the area between $z = 0$ and the negative of that z score.

Each of the facts in the colored box is a consequence of the fact that normal distributions are symmetric about the mean. Examples 3, 4, and 5 will illustrate how to find areas under a normal curve.

EXAMPLE 3 Finding the Area between Two z Scores

Math Note

Remember that the area under any portion of a normal distribution has to be non-negative and less than or equal to 1. If you don't get an area in that range, you must have made a mistake.

Find the area under the standard normal distribution

(a) Between $z = 1.55$ and $z = 2.25$.
(b) Between $z = -0.60$ and $z = -1.35$.
(c) Between $z = 1.50$ and $z = -1.75$.

SOLUTION

(a) Draw the picture, label z scores, and shade the requested area. See Figure 12-9.

Using Table 12-3, the area between $z = 0$ and $z = 2.25$ is 0.488, and the area between $z = 0$ and $z = 1.55$ is 0.439. The area we are looking for is the larger area minus the smaller:

$$0.488 - 0.439 = 0.049.$$

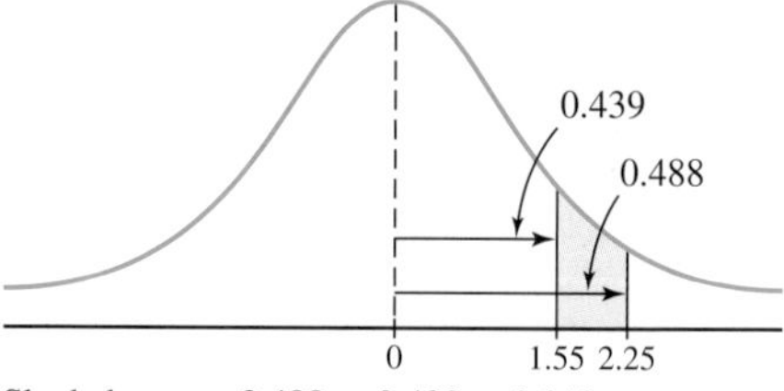

Shaded area = 0.488 − 0.439 = 0.049

Figure 12-9

The area between $z = 1.55$ and $z = 2.25$ is 0.049.

(b) Draw the picture, label z scores, and shade the requested area. See Figure 12-10.

Using key fact 2, we can find the area between $z = 0$ and the two given negative z scores by finding the corresponding positive z scores in the table. The areas are 0.226 for $z = -0.60$, and 0.412 for $z = -1.35$. Again, the area we're looking for is the larger area minus the smaller:

$$0.412 - 0.226 = 0.186$$

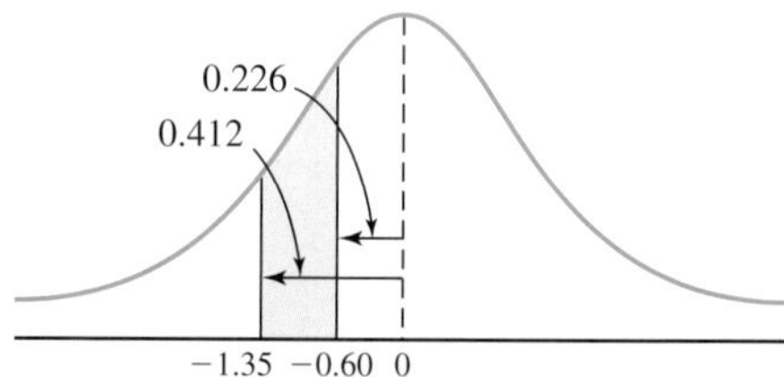

Shaded area = 0.412 − 0.226 = 0.186

Figure 12-10

The area between $z = -0.60$ and $z = -1.35$ is 0.186.

(c) Draw the picture, label the z scores, and shade the requested area. See Figure 12-11.

Since the z values are on opposite sides of the mean, the areas corresponding to the z values are added. The area corresponding to $z = -1.75$ is 0.460, and the area corresponding to $z = 1.50$ is 0.433.

$$0.460 + 0.433 = 0.893$$

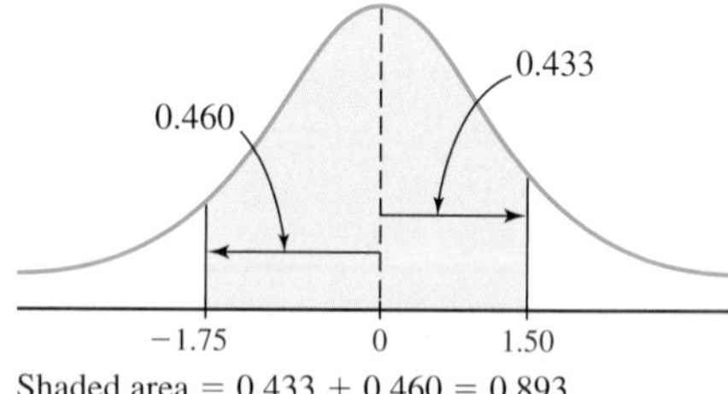

Shaded area = 0.433 + 0.460 = 0.893

Figure 12-11

The area between $z = 1.50$ and $z = -1.75$ is 0.893.

▼ Try This One 3

Find the area under the standard normal distribution

(a) Between $z = 2.05$ and $z = 2.40$.
(b) Between $z = -3.2$ and $z = -2.0$.
(c) Between $z = -0.55$ and $z = 1.6$.

When we want to find the area to the right or left of a single z score, we will use the fact that the area under half of the curve is exactly 0.500, as in Examples 4 and 5.

EXAMPLE 4 Finding the Area to the Right of a z Score

Math Note

Using a sketch and the two important facts on page 689 allows us to find areas without having to learn a variety of rules and formulas for the different possible situations.

Find the area under the standard normal distribution

(a) To the right of $z = 1.70$. (b) To the right of $z = -0.95$.

SOLUTION

(a) Draw the picture, label the z scores, and shade the area. See Figure 12-12.

The area of the entire portion to the right of $z = 0$ is 0.500. According to the table, the area of the portion between $z = 0$ and $z = 1.70$ is 0.455. So the shaded portion is the difference between 0.500 and 0.455:

$$0.500 - 0.455 = 0.045$$

The area to the right of $z = 1.70$ is 0.045.

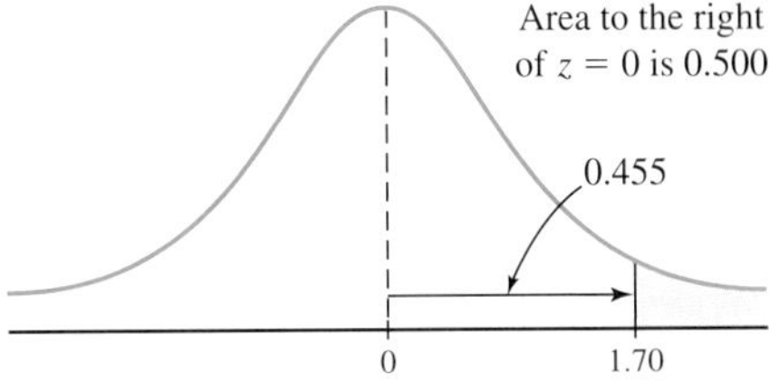

Figure 12-12

(b) Draw the picture, label the z scores, and shade the area. See Figure 12-13.

This time, the area is the sum of 0.500 (the entire right half of the distribution) and the area between $z = 0$ and $= -0.95$, which we find to be 0.329 using Table 12-3.

$$0.500 + 0.329 = 0.829$$

The area to the right of $z = -0.95$ is 0.829.

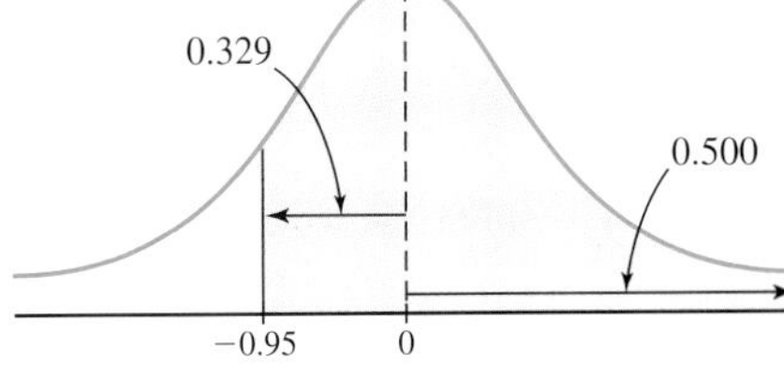

Figure 12-13

▼ Try This One 4

Find the area under the standard normal distribution

(a) To the right of $z = -2.40$. (b) To the right of $z = 0.25$.

EXAMPLE 5 Finding the Area to the Left of a z Score

Find the area under the standard normal distribution

(a) To the left of $z = 2.20$. (b) To the left of $z = -1.95$.

SOLUTION

(a) Draw the picture, label the z scores, and shade the area. See Figure 12-14.

The shaded area is the entire left half (0.500) minus the area between $z = 0$ and $z = -2.2$, which is 0.486 according to Table 12-3.

$$0.500 - 0.486 = 0.014$$

The area to the left of $z = -2.2$ is 0.014.

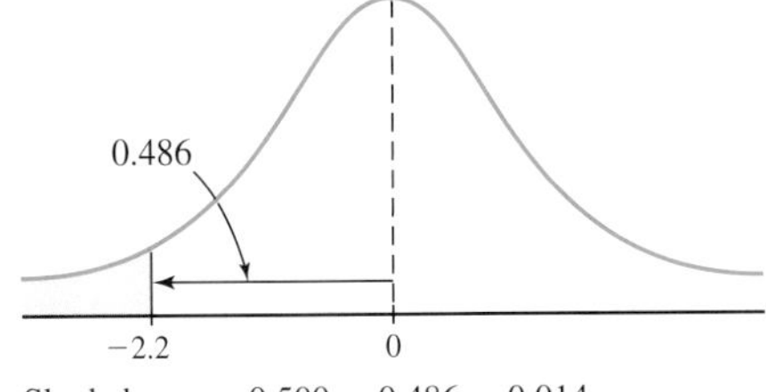

Figure 12-14

(b) Draw the picture, label the z scores, and shade the area. See Figure 12-15.

Table 12-3 gives us an area of 0.474 between $z = 0$ and $z = 1.95$. Adding to the area of the left half (0.500), we get

$$0.500 + 0.474 = 0.974$$

The area to the left of $z = 1.95$ is 0.974.

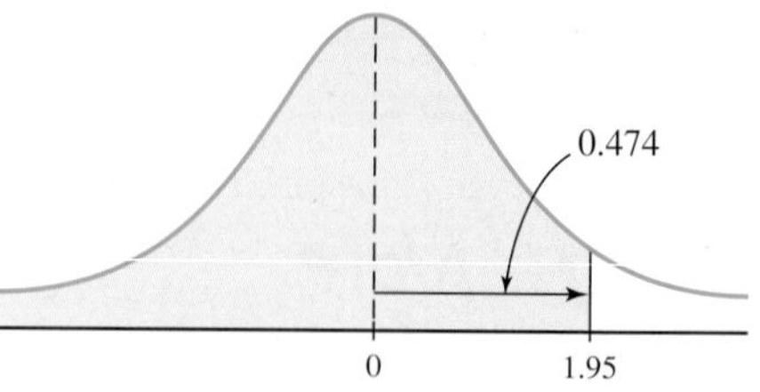

Shaded area = 0.500 + 0.474 = 0.974

Figure 12-15

CAUTION

Don't ignore your graph when writing your final answer! As long as you keep in mind that the total area under the distribution is 1, you can visually check if your answer is reasonable. In Example 5b, just from a quick glance at the graph you can see that most of the distribution is shaded, so your answer should be close to 1.

▼ Try This One 5

☑ 4. Use a table and z scores to find areas under the standard normal distribution.

Find the area under the standard normal distribution

(a) To the left of $z = -1.05$. (b) To the left of $z = 0.1$.

To close the section, we should point out that no data set fits the normal distribution perfectly, since the normal curve is a theoretical distribution that would come from infinitely many data values. But there are many, many quantities that vary from a true normal distribution so slightly that the normal curve can be used to study those quantities very effectively.

Answers to Try This One

1 About 38,000

2 −0.6

3 (a) 0.012 (b) 0.022 (c) 0.654

4 (a) 0.992 (b) 0.401

5 (a) 0.147 (b) 0.540

EXERCISE SET 12-6

Writing Exercises

1. What is the distinguishing characteristic of a quantity that is normally distributed?
2. Write as many properties of a normal curve as you can think of.
3. What does the area under a portion of a normal curve tell you about the data in the associated distribution?
4. Explain what the empirical rule says.
5. What percentage of the area under a normal curve falls to the right of the mean? To the left? Explain.
6. Why does it make sense that the total area under a normal curve is 1?

Computational Exercises

For Exercises 7–10, assume that the data set described is normally distributed with the given mean and standard deviation, and with n total values. Find the approximate number of data values that will fall in the given range.

7. Mean = 12, standard deviation = 1.5, $n = 50$.
   Range: 10.5 to 13.5
8. Mean = 100, standard deviation = 8, $n = 200$.
   Range: 84 to 116
9. Mean = 400, standard deviation = 25, $n = 500$.
   Range: 325 to 475
10. Mean = −14.2, standard deviation = 1.6, $n = 120$.
    Range: −17.4 to −11

For Exercises 11–28, find the area under the standard normal distribution curve

11. Between $z = 0$ and $z = 1.95$.
12. Between $z = 0$ and $z = 0.55$.
13. Between $z = 0$ and $z = -0.5$.
14. Between $z = 0$ and $z = -2.05$.
15. To the right of $z = 1.0$.
16. To the right of $z = 0.25$.
17. To the left of $z = -0.40$.
18. To the left of $z = -1.45$.
19. Between $z = 1.25$ and $z = 1.90$.
20. Between $z = 0.8$ and $z = 1.3$.
21. Between $z = -0.85$ and $z = -0.20$.
22. Between $z = -1.55$ and $z = -1.85$.
23. Between $z = 0.25$ and $z = -1.10$.
24. Between $z = 2.45$ and $z = -1.05$.
25. To the left of $z = 1.20$.
26. To the left of $z = 2.15$.
27. To the right of $z = -1.90$.
28. To the right of $z = -0.20$.

Critical Thinking

29. Find a z value to the right of the mean so that 67.4% of the distribution lies to the left of it.
30. Find a z value to the left of the mean so that 98.6% of the area lies to the right of it.
31. Find two z values, one positive and one negative but having the same absolute value, so that the areas in the two tails (ends) total these values.
    (a) 4%
    (b) 8%
    (c) 1.6%

Section 12-7 Applications of the Normal Distribution

LEARNING OBJECTIVES

- ☐ 1. Use the normal distribution to find percentages.
- ☐ 2. Use the normal distribution to find probabilities.
- ☐ 3. Use the normal distribution to find percentile ranks.

Surely there must be someone out there who doesn't like Oreos, but I haven't met him or her. A standard package of America's favorite cookie contains 510 grams of sweet temptation. Of course, when you buy a package, you expect every one of those 510 grams. But there's variation in just about everything, so some packages will have more than the intended weight, and some will have less. The folks running Nabisco aren't dummies, though—they know that if someone decides to check the weight and finds it to be less than 510 grams, they won't be a very happy customer. So what to do?

This is exactly the sort of situation where weights tend to be normally distributed. The company would likely design its production and packaging process so that the mean is somewhat larger than 510 grams, with a standard deviation that assures that the vast majority of packages will weigh at least 510 grams.

In this section, we will see how the area calculations we practiced in Section 12-6 can be used to solve real-world problems for data that are normally distributed. Our general strategy will be to transform specific data into z scores, then use Table 12-3 to find areas. Finally, we'll interpret the area as it applies to the given data.

The reason this works is indicated by Figure 12-16. Suppose that the weight of Oreos in a package is normally distributed with mean 518 grams and standard deviation 4 grams. The distribution would look like Figure 12-16a. After using the z score formula to standardize, we get the graph in Figure 12-16b; now we can apply our area calculations.

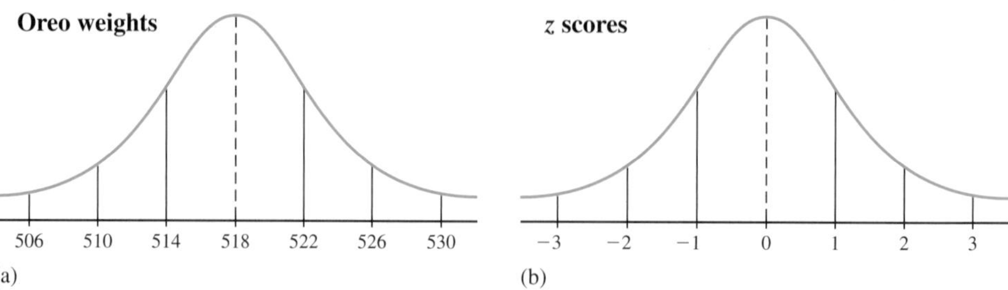

Figure 12-16

EXAMPLE 1 Solving a Problem Using the Normal Distribution

If the weights of Oreos in a package are normally distributed with mean 518 grams and standard deviation 4 grams, find the percentage of packages that will weigh less than 510 grams.

> **Math Note**
>
> It's typical for manufacturers to make sure that the listed weight for products packaged by weight is at least two standard deviations below the mean, ensuring that over 97% of the packages have the listed weight or more.

SOLUTION

Step 1 Draw the figure and represent the area, as shown in Figure 12-17.

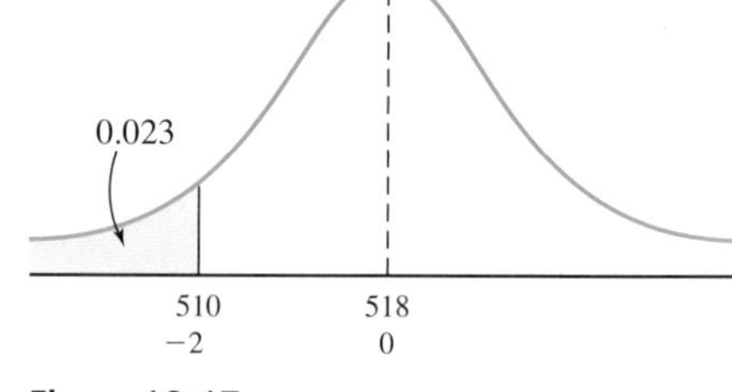

Figure 12-17

Step 2 Find the z score for data value 510.

$$z = \frac{\text{value} - \text{mean}}{\text{standard deviation}}$$

$$= \frac{510 - 518}{4} = -2$$

This tells us that 510 grams is 2 standard deviations below the mean.

Step 3 Find the shaded area using Table 12-3. The shaded area is the area of the left half (0.5) minus the area between $z = 0$ and $z = -2$. The area between $z = 0$ and $z = -2$ is 0.477 (using $z = +2$ from the table).

$$\text{Area} = 0.5 - 0.477 = 0.023$$

Step 4 Interpret the area. In this case, an area of 0.023 tells us that only 2.3% of packages will have weights less than 510 grams.

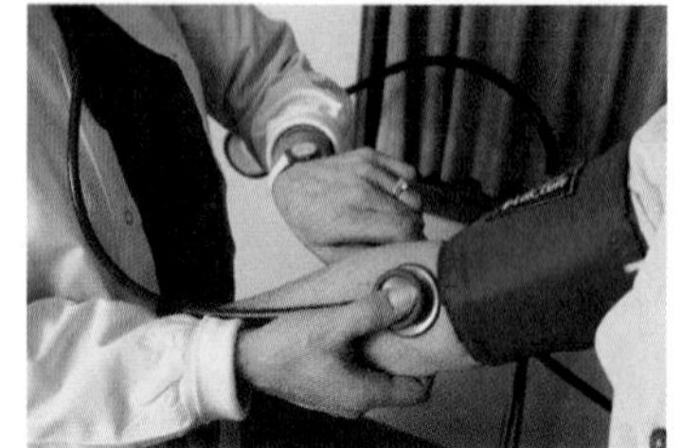

☑ 1. Use the normal distribution to find percentages.

▼ Try This One 1

Based on data compiled by the World Health Organization, the mean systolic blood pressure in the United States is 120, the standard deviation is 16, and the pressures are normally distributed. Find each.

(a) The percent of individuals who have a blood pressure between 120 and 128
(b) The percent of individuals who have a blood pressure above 132
(c) The percent of individuals who have a blood pressure between 112 and 116
(d) The percent of individuals who have a blood pressure between 124 and 144
(e) The percent of individuals who have a blood pressure lower than 104

Our next interpretation for area under a normal distribution is tremendously important because it ties our study of statistics to the main topic of Chapter 11, probability.

Probability and Area under a Normal Distribution

The area under a normal distribution between two data values is the probability that a randomly selected data value is between those two values.

Example 2 illustrates the use of area to find probability.

EXAMPLE 2 Using Area under a Normal Distribution to Find Probabilities

Based on data in the 2009 *Statistical Abstract of the United States,* the average *American* generates 1,679 pounds of garbage per year. Let's assume that the number of pounds generated per person is approximately normally distributed with standard deviation 200 pounds. Find the probability that a randomly selected person generates

(a) Between 1,300 and 2,000 pounds of garbage per year.
(b) More than 2,000 pounds of garbage per year.

SOLUTION

(a) **Step 1** Draw the figure and represent the area. See Figure 12-18 below.

Step 2 Find the z scores for the two given weights.

$$1{,}300 \text{ pounds: } z = \frac{1{,}300 - 1{,}679}{200} \approx -1.9$$

$$2{,}000 \text{ pounds: } z = \frac{2{,}000 - 1{,}679}{200} \approx 1.6$$

Step 3 Find the shaded area. Using Table 12-3, the area between $z = 0$ and $z = 1.6$ is 0.445, and the area between $z = 0$ and $z = -1.9$ is 0.471. So the shaded area is

$$0.471 + 0.445 = 0.916$$

0.471 0.445
1,300 1,679 2,000
−1.9 0 1.6

Figure 12-18

Step 4 Interpret the area. This tells us that if a person is selected at random, the probability that he or she generates between 1,300 and 2,000 pounds of garbage per year is 0.916, or 91.6%.

(b) Most of the work was already done in part a. We can adapt the graph a bit (Figure 12-19).

The shaded area is the area between $z = 0$ and $z = 1.6$ subtracted from the area of the entire right half (0.5).

$$0.5 - 0.445 = 0.055$$

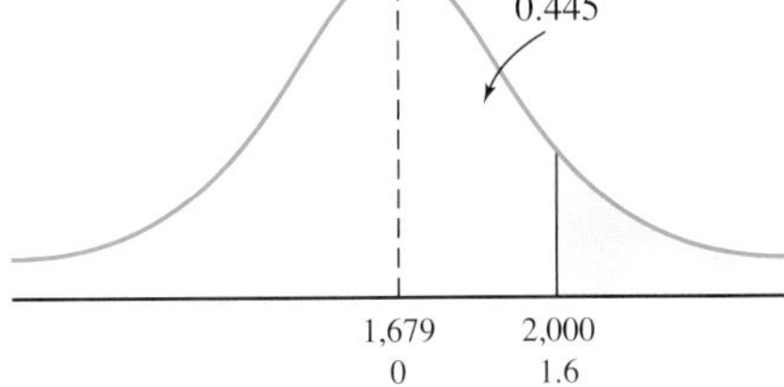

Figure 12-19

The probability that a randomly selected person generates more than 2,000 pounds of garbage per year is 0.055, or 5.5%.

Math Note

The probability that a randomly selected value from a normal distribution is less than some value is the area to the left of that value. The probability of a randomly selected value being greater than some value is the area to the right of that value.

▼ Try This One 2

☑ 2. Use the normal distribution to find probabilities.

The *Statistical Abstract* also indicates that of the 1,679 pounds of garbage generated by the average individual, 913 pounds will end up in a landfill. If these amounts are approximately normally distributed with standard deviation 160 pounds, find the probability that a randomly selected person generates

(a) Less than 600 pounds that end up in a landfill.
(b) Between 600 and 1,000 pounds that end up in a landfill.

Normal distributions can also be used to find percentiles, and answer questions of "how many?" Example 3 illustrates these types of questions.

EXAMPLE 3 Finding Number in a Sample and Percentile Rank

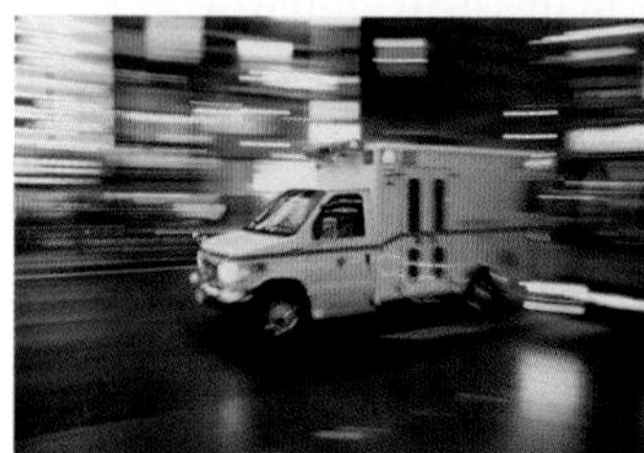

The American Automobile Association reports that the average time it takes to respond to an emergency call is 25 minutes. Assume the response time is approximately normally distributed and the standard deviation is 4.5 minutes.

(a) If 80 calls are randomly selected, approximately how many will have response times less than 15 minutes?
(b) In what percentile is a response time of 30 minutes?

SOLUTION

(a) **Step 1** Draw a figure and represent the area as shown in Figure 12-20.

Step 2 Find the z value for 15.

$$z = \frac{\text{Value} - \text{mean}}{\text{Standard deviation}}$$

$$= \frac{15 - 25}{4.5} \approx -2.2$$

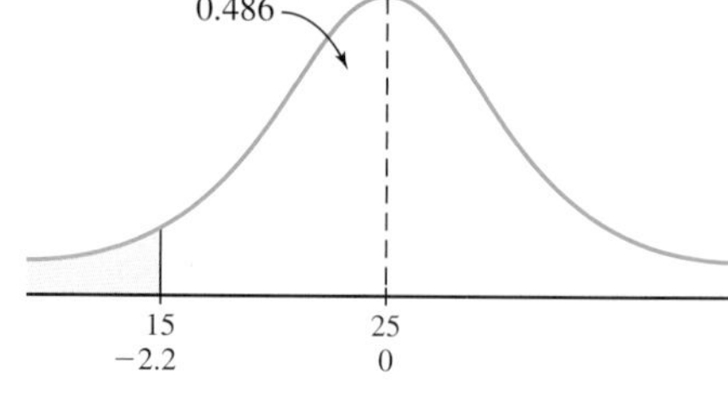

Figure 12-20

Step 3 Find the shaded area. Using Table 12-3, we find that the area between $z = 0$ and $z = -2.2$ is 0.486. The shaded area is the difference of the entire left half (0.5) and 0.486.

$$\text{Area} = 0.5 - 0.486 = 0.014.$$

Step 4 Interpret the area. An area of 0.014 tells us that 1.4% of calls will have a response time of less than 15 minutes. Multiply the percentage by the number of calls in our sample:

$$1.4\% \text{ of } 80 \text{ calls} = (0.014)(80) = 1.12$$

Approximately one call in 80 will have response time less than 15 minutes.

Math Note

Notice that in this case (number of calls), an answer of 1.12 doesn't make sense. We need to round to the nearest whole number.

(b) **Step 1** Draw a figure and represent the area as shown in Figure 12-21. Remember that to find the percentile, we are interested in the percentage of times that are less than 30 minutes.

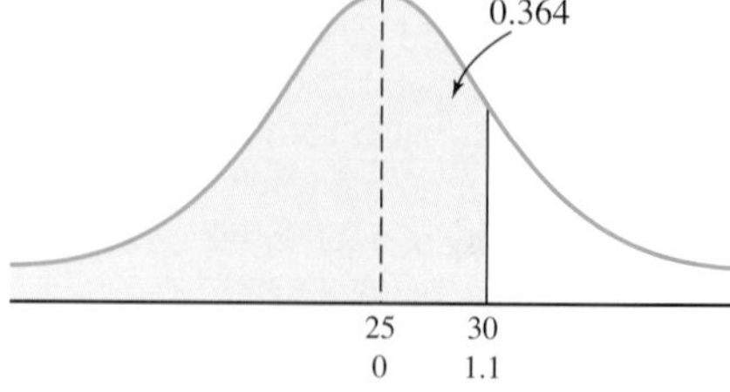

Figure 12-21

Step 2 Find the z score for a response time of 30 minutes.

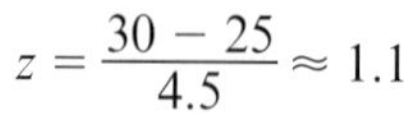

$$z = \frac{30 - 25}{4.5} \approx 1.1$$

Step 3 Find the shaded area. The area to the left of $z = 1.1$ is the area of the left half (0.5) plus the area between $z = 0$ and $z = 1.1$ (0.364 according to Table 12-3).

$$\text{Area} = 0.5 + 0.364 = 0.864$$

Step 4 Interpret the area. An area of 0.864 means that 86.4% of calls will get a response time of less than 30 minutes, so the time of 30 minutes is in the 86th percentile.

▼ Try This One 3

The mean for a reading test given nationwide is 80, and the standard deviation is 8. The variable is normally distributed. If 10,000 students take the test, find each.

☑ 3. Use the normal distribution to find percentile ranks.

(a) The number of students who will score above 90
(b) The number of students who will score between 78 and 88
(c) The percentile of a student who scores 94
(d) The number of students who will score below 76

Answers to Try This One

1 (a) 19.2% (b) 22.7% (c) 9.3% (d) 33.4% (e) 15.9%

2 (a) 0.026 or 2.6% (b) 0.683 or 68.3%

3 (a) 1,060 (b) 4,400 (c) 96th percentile (d) 3,080

EXERCISE SET 12-7

Writing Exercises

1. Explain why the normal distribution can be used to solve many real-life problems.
2. Given a data set, how could you decide if the distribution of the data was approximately normal?

Real-World Applications

3. The average hourly wage of production workers in manufacturing in 2007 was \$17.41. Assume the variable is normally distributed. If the standard deviation of earning is \$3.72, find these probabilities for a randomly selected production worker.

   (a) The production worker earns more than \$18.55.
   (b) The production worker earns less than \$14.00.

   Source: Statistical Abstract of the United States
4. The average cost for two people to go to a movie is \$22. The standard deviation is \$3. Assume the cost is normally distributed. Find the probability that at any given theater, the cost will be more than \$26 for two people to go to a movie.
5. If the mean salary of public school teachers in the United States in 2007 was \$49,294 and the standard deviation was \$9,000, find these probabilities for a randomly selected teacher. Assume the variable is normally distributed.

   (a) The teacher earns more than \$55,000.
   (b) The teacher earns less than \$45,000.
6. Average sales for an online textbook distributor were \$71.12 per customer per purchase. Assume the sales are normally distributed. If the standard deviation of the amount spent on textbooks is \$8.42, find these probabilities for a randomly selected customer of the online textbook distributor.

   (a) He or she spent more than \$60 per purchase.
   (b) He or she spent less than \$80 per purchase.
7. A survey found that people keep their television sets an average of 4.8 years. The standard deviation is 0.89 year. If a person decides to buy a new TV set,

find the probability that he or she has owned the old set for the given amount of time. Assume the variable is normally distributed.

(a) Less than 2.5 years (b) Between 3 and 4 years
(c) More than 4.2 years

8. The average age of CEOs is 56 years. Assume the variable is normally distributed. If the standard deviation is 4 years, find the probability that the age of a randomly selected CEO will be in the given range.

   (a) Between 53 and 59 years old
   (b) Between 58 and 63 years old
   (c) Between 50 and 55 years old

9. The average life of a brand of automobile tires is 30,000 miles, with a standard deviation of 2,000 miles. If a tire is selected and tested, find the probability that it will have the given lifetime. Assume the variable is normally distributed.

   (a) Between 25,000 and 28,000 miles
   (b) Between 27,000 and 32,000 miles
   (c) Between 31,500 and 33,500 miles

10. The average time a person spends in each visit to an online social networking service is 62 minutes. The standard deviation is 12 minutes. If a visitor is selected at random, find the probability that he or she will spend the time shown on the networking service. Assume the times are normally distributed.

    (a) At least 180 minutes (b) At least 50 minutes

11. The average amount of snow per season in Trafford is 44 inches. The standard deviation is 6 inches. Find the probability that next year Trafford will receive the given amount of snowfall. Assume the variable is normally distributed.

    (a) At most 50 inches of snow
    (b) At least 53 inches of snow

12. The average waiting time for a drive-in window at a local bank is 9.2 minutes, with a standard deviation of 2.6 minutes. When a customer arrives at the bank, find the probability that the customer will have to wait the given time. Assume the variable is normally distributed.

    (a) Between 5 and 10 minutes
    (b) Less than 6 minutes or more than 9 minutes

13. The average time it takes college freshmen to complete the Mason Basic Reasoning Test is 24.6 minutes. The standard deviation is 5.8 minutes. Find these probabilities. Assume the variable is normally distributed.

    (a) It will take a student between 15 and 30 minutes to complete the test.
    (b) It will take a student less than 18 minutes or more than 28 minutes to complete the test.

14. A brisk walk at 4 miles per hour burns an average of 300 calories per hour. If the standard deviation of the distribution is 8 calories, find the probability that a person who walks 1 hour at the rate of 4 miles per hour will burn the given number of calories. Assume the variable is normally distributed.

    (a) More than 280 calories
    (b) Less than 293 calories
    (c) Between 285 and 320 calories

15. During September, the average temperature of Laurel Lake is 64.2° and the standard deviation is 3.2°. Assume the variable is normally distributed. For a randomly selected day in September, find the probability that the temperature will be

    (a) Above 62°. (c) Between 65° and 68°.
    (b) Below 67°.

16. If the systolic blood pressure for a certain group of people has a mean of 132 and a standard deviation of 8, find the probability that a randomly selected person in the group will have the given systolic blood pressure. Assume the variable is normally distributed.

    (a) Above 130 (c) Between 131 and 136
    (b) Below 140

17. An IQ test has a mean of 100 and a standard deviation of 15. The test scores are normally distributed. If 2,000 people take the test, find the number of people who will score

    (a) Below 93. (c) Between 80 and 105.
    (b) Above 120. (d) Between 75 and 82.

18. The average size (in square feet) of homes built in the United States in the fourth quarter of 2008 was 2,343. Assume the variable is normally distributed and the standard deviation is 152 square feet. In a sample of 500 recently built homes, find the number of homes that will

    (a) Have between 1,900 and 2,000 square feet.
    (b) Have more than 3,000 square feet.
    (c) Have less than 2,000 square feet.
    (d) Have more than 1,600 square feet.

    Source: National Association of Homebuilders

19. According to the National Association of Realtors, the mean sale price for existing homes in the United States in 2008 was $242,700. Assume that sale prices are normally distributed with a standard deviation of $41,000. If 800 home sales are selected at random, find the number of homes that cost

    (a) More than $300,000.
    (b) Between $200,000 and $300,000.
    (c) Less than $150,000.

20. The average price of Stephen King paperbacks sold at bookstores across the country is $9.52, and the standard deviation is $1.02. Assume the price is normally distributed. If a national bookstore sells 1,000 Stephen King paperbacks during August,

find the number of Stephen King paperbacks that were sold

(a) For less than $8.00.
(b) For more than $10.00.
(c) Between $9.50 and $10.50.
(d) Between $9.80 and $10.05.

21. Refer to Exercise 3. If your buddy Earl gets a job as a manufacturing production worker making $15.20 per hour, what percentile is he in?
22. Refer to Exercise 7. If you buy a TV for your dorm and keep it for your 4 years of school then sell it, what percentile does that put you in?
23. Refer to Exercise 10. Suppose that your grades have been slipping, and you impose a new rule on yourself: no more than 20 minutes at a time on social networking sites. What percentile would that put you in?
24. Refer to Exercise 18. When the one-hit wonder band The Flaming Rogers falls off the charts, their singer is forced to downsize from the 5,000 square foot home he built earlier in the year to a more modest 2,000 square foot model. What was his change in percentile rank?

Critical Thinking

25. If a distribution of raw scores were plotted and then the scores were transformed into z scores, would the shape of the distribution change?
26. An instructor gives a 100-point examination in which the grades are normally distributed. The mean is 60 and the standard deviation is 10. If there are 5% A's and 5% F's, 15% B's and 15% D's, and 60% C's, find the scores that divide the distribution into those categories.
27. A researcher who is in charge of an educational study wants subjects to perform some special skill. Fearing that people who are unusually talented or unusually untalented could distort the results, he decides to use people who scored in the middle 50% on a certain test. If the mean for the population is 100 and the standard deviation is 15, find the two limits (upper and lower) for the scores that would enable a volunteer to participate in the study. Assume the scores are normally distributed.
28. While preparing for their comeback tour, The Flaming Rogers find that the average time it takes their sound tech to set up for a show is 58.6 minutes, with a standard deviation of 4.3 minutes. If the band manager decides to include only the fastest 20% of sound techs on the tour, what should the cutoff time be for concert setup? Assume the times are normally distributed.

Section 12-8 Correlation and Regression Analysis

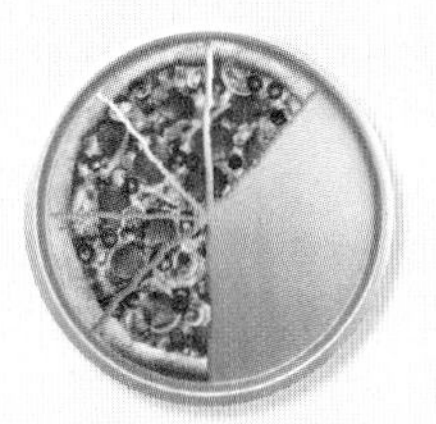

LEARNING OBJECTIVES

- ☐ 1.Construct scatter plots for two data sets.
- ☐ 2. Calculate correlation coefficients.
- ☐ 3. Determine if correlation coefficients are significant.
- ☐ 4. Find a regression line for two data sets.
- ☐ 5. Use regression lines to make predictions.

The Business Insider blog posted an article in November 2008 with the headline "Proof of a Correlation between MySpace Usage and Illiteracy." It certainly seems to imply that MySpace users tend to be illiterate. On what data were these claims based?

The key word in the claim is *correlation*. In statistics, that term describes an attempt to determine if a relationship exists between two sets of data, like literacy rates and MySpace usage.

In order to decide if a relationship exists between two data sets, we have to first gather data in such a way that values from each set can be paired together. In this case, the literacy rate for each state was recorded, as well as data from Google on how likely Internet users from each state were to search for the term "MySpace." We can then plot a graph designating one set of data as the *x variable* or **independent variable** and the other as the *y variable* or **dependent variable.** This graph is called a *scatter plot*.

A **scatter plot** is a graph of the ordered pairs (x, y) consisting of data from two data sets.

After the scatter plot is drawn, we can analyze the graph to see if there is a pattern. If there is a noticeable pattern, such as the points falling in an approximately straight line, then a possible relationship between the two variables may exist.

If a strong relationship appears to exist between two data sets, we might further analyze the relationship by trying to find an equation relating the variables. This is known as *regression analysis,* which we will learn about later in the section. For now, we'll focus on scatter plots.

EXAMPLE 1 Constructing a Scatter Plot

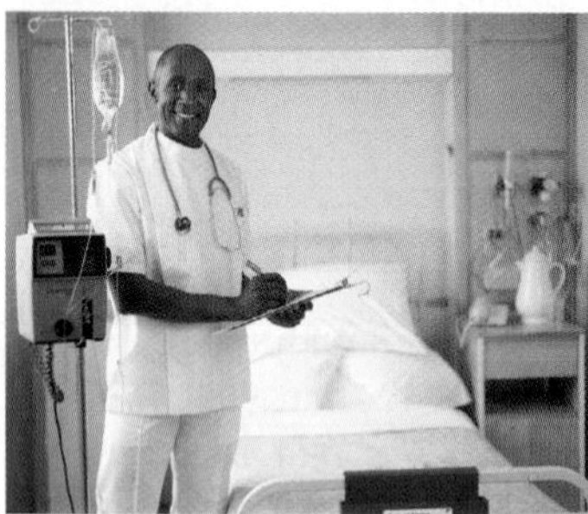

A medical researcher selects a sample of small hospitals in his state and hopes to discover if there is a relationship between the number of beds and the number of personnel employed by the hospital. Construct a scatter plot for the data shown.

No. of beds (x)	28	56	34	42	45	78	84	36	74	95
Personnel (y)	72	195	74	211	145	139	184	131	233	366

SOLUTION

Step 1 Draw and label the x and y axes, as shown.

Step 2 Plot the data pairs. See Figure 12-22.

Figure 12-22

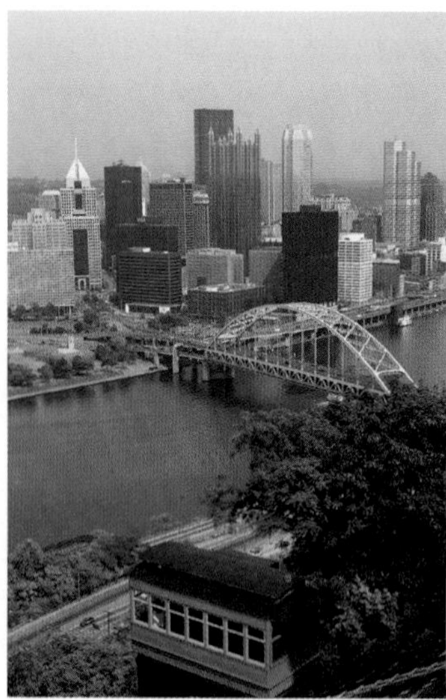

▼ Try This One 1

The data below represent the heights in feet and the number of stories of the tallest buildings in Pittsburgh. Draw a scatter plot for the data and describe the relationship.

Height (x)	485	511	520	535	582	615	616	635	728	841
No. of stories (y)	40	37	41	42	38	45	31	40	54	64

Source: *The World Almanac and Book of Facts*

Analyzing the Scatter Plot

There are several types of relationships that can exist between the x values and the y values in a scatter plot. These relationships can be identified by looking at the pattern of the points on the graphs. The types of patterns and corresponding relationships are:

1. *A positive linear relationship* exists when the points fall approximately in an ascending straight line from left to right, and both the x and y values increase at the same time. See Figure 12-23a on the next page.
2. *A negative linear relationship* exists when the points fall approximately in a descending straight line from left to right. See Figure 12-23b on the next page. The relationship then is as the x values are increasing, the y values are decreasing.

3. *A nonlinear relationship* exists when the points fall in a curved line. See Figure 12-23c. The relationship is described by the nature of the curve.
4. *No relationship* exists when there is no discernible pattern to the points. See Figure 12-23d.

☑ 1. Construct scatter plots for two data sets.

The relationship between the variables in Example 1 might be a positive linear relationship. It looks like as the size of the hospital based on the number of beds increases, the number of personnel also increases.

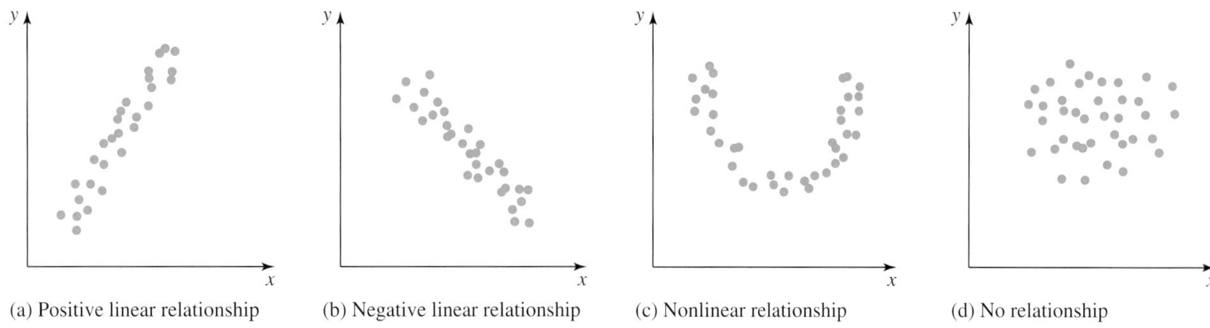

(a) Positive linear relationship (b) Negative linear relationship (c) Nonlinear relationship (d) No relationship

Figure 12-23

The Correlation Coefficient

Deciding whether or not two data sets are related by simply looking at a scatter plot is a pretty subjective process, so it would be nice to have a way to quantify how strongly connected data sets are.

The **correlation coefficient** is a number that describes how close to a linear relationship there is between two data sets. Correlation coefficients range from -1 (perfect negative linear relationship) to $+1$ (perfect positive linear relationship). The closer this number is to one in absolute value, the more likely it is that the data sets are related. A correlation coefficient close to zero indicates that the data are most likely not related at all (see Figure 12-24). We use the letter r to represent the correlation coefficient. It doesn't depend on the units for the two data sets, and it also doesn't depend on which set you choose for the x variable.

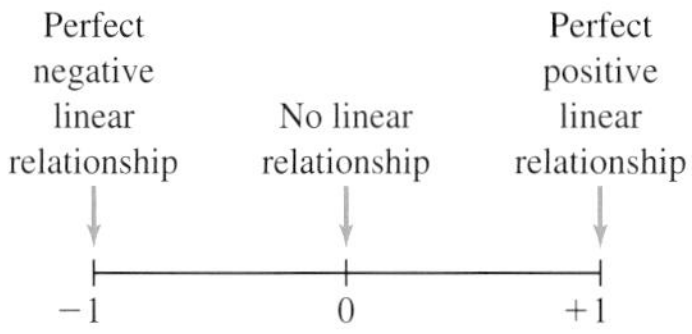

Perfect negative linear relationship | No linear relationship | Perfect positive linear relationship

-1 | 0 | $+1$

Figure 12-24

Calculating the Value of the Correlation Coefficient

In order to find the value of the correlation coefficient, we will use the following formula:

Math Note

The sign of r will indicate the nature of the relationship if one exists. That is, if r is positive, the linear relationship will be positive. If r is negative, the linear relationship will be negative. Looking at the scatter plots in Figure 12-23, for (a) r would be close to $+1$; for (b), r would be close to -1; for (c) and (d), r would be close to zero.

Formula for Finding the Value of r:

$$r = \frac{n(\Sigma xy) - (\Sigma x)(\Sigma y)}{\sqrt{[n(\Sigma x^2) - (\Sigma x)^2][n(\Sigma y^2) - (\Sigma y)^2]}}$$

where

n = the number of data pairs
Σx = the sum of the x values
Σy = the sum of the y values
Σxy = the sum of the products of the x and y values for each pair
Σx^2 = the sum of the squares of the x values
Σy^2 = the sum of the squares of the y values

When calculating the value of r, it is helpful to make a table, as shown in Example 2.

EXAMPLE 2 Calculating a Correlation Coefficient

Find the correlation coefficient for the data in Example 1.

SOLUTION

Step 1 Make a table as shown with the following headings: x, y, xy, x^2, and y^2.

Step 2 Find the values for the product xy, the values for x^2, the values for y^2, and place them in the columns as shown. Then find the sums of the columns.

x	y	xy	x^2	y^2
28	72	2,016	784	5,184
56	195	10,920	3,136	38,025
34	74	2,516	1,156	5,476
42	211	8,862	1,764	44,521
45	145	6,525	2,025	21,025
78	139	10,842	6,084	19,321
84	184	15,456	7,056	33,856
36	131	4,716	1,296	17,161
74	233	17,242	5,476	54,289
95	366	34,770	9,025	133,956
$\Sigma x = 572$	$\Sigma y = 1{,}750$	$\Sigma xy = 113{,}865$	$\Sigma x^2 = 37{,}802$	$\Sigma y^2 = 372{,}814$

Step 3 Substitute into the formula and evaluate (note that there are 10 data pairs):

$$r = \frac{n(\Sigma xy) - (\Sigma x)(\Sigma y)}{\sqrt{[n(\Sigma x^2) - (\Sigma x)^2][n(\Sigma y^2) - (\Sigma y)^2]}}$$

$$= \frac{10(113{,}865) - (572)(1{,}750)}{\sqrt{[10(37{,}802) - (572)^2][10(372{,}814) - (1{,}750)^2]}}$$

$$= \frac{137{,}650}{\sqrt{(50{,}836)(665{,}640)}}$$

$$\approx 0.748$$

The correlation coefficient for the two data sets is 0.748. This appears to confirm what our eyes tell us about the scatter plot in Example 1—it's certainly not a perfect linear relationship, but does tend to indicate that as bed space goes up, the number of personnel does as well.

▼ Try This One 2

Find the correlation coefficient for the data in Try This One 1.

☑ 2. Calculate correlation coefficients.

Next we will tackle the issue of how to interpret the correlation coefficient. In Example 2, the value of r was based on a sample of data. The population is all hospitals in a certain state with fewer than 100 beds, and a sample of such hospitals was chosen. When data from a sample is used, we can't be positive that it represents the entire population. So conclusions drawn, like the conclusion that there's a connection between numbers of beds and personnel, may not be correct.

Sidelight TECHNOLOGY AND THE CORRELATION COEFFICIENT

If there's just one thing you learned from Example 2, it's probably that finding a correlation coefficient is not exactly simple. Even for a reasonably small data set, the calculations involved are a little cumbersome. You can imagine how much worse it would be for larger data sets.

Fortunately, most graphing calculators and spreadsheets have built-in commands to compute correlation coefficients. In Microsoft Excel, enter the two data sets in two columns. Then choose another blank cell to enter the correlation coefficient. Next, choose Function . . . from the Insert menu: in the category dialog box choose Statistical, from the list of functions choose CORREL. You will then be prompted to choose the range of cells that make up each data set, and presto! The correlation coefficient appears in the chosen cell.

Using a standard graphing calculator, the data are entered on the list editing screen (STAT 1). This is shown in Figure 12-25a with the data from Example 1 entered. Next, push STAT right Arrow 8 as shown in Figure 12-25b, then ENTER. The resulting screen shows the correlation coefficient at the bottom, as well as the regression line for the data, which we will learn about shortly (see Figure 12-25c). If your calculator doesn't display *r*, push CATALOG, which is 2nd 0, then scroll down and choose DiagnosticOn.

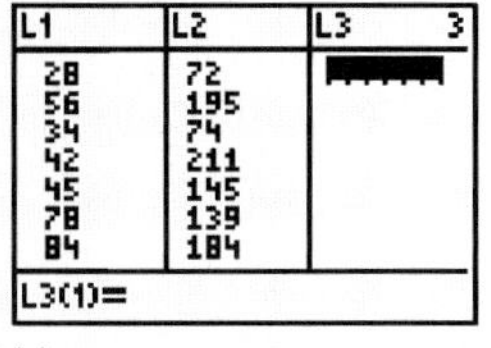

(a)

(b)

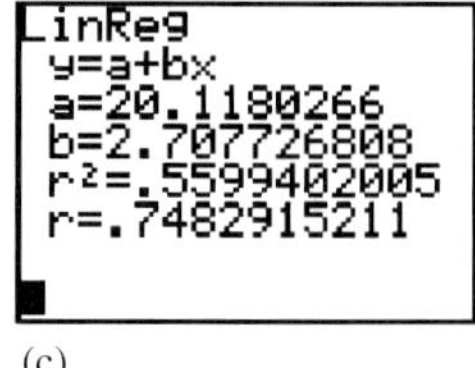

(c)

Figure 12-25

Statisticians have traditionally agreed that when we conclude that two data sets have a relationship, we can be satisfied with that conclusion if there is either a 95% or 99% chance that we're correct. This corresponds to a 5% or 1% chance of being wrong. These percentages are called **significance levels.**

We know that the more data we have, the more likely that data are representative of the population. Not surprisingly, the value of *r* needed to be reasonably sure that two data sets are correlated is higher for small sample sizes, and lower for large sample sizes. In Table 12-4, we see the minimum *r* values needed to have a 5% and 1% chance of being wrong when we conclude that two data sets are related.

TABLE 12-4 Significant Values for the Correlation Coefficient

Sample Size	5%	1%	Sample Size	5%	1%
4	.950	.990	17	.482	.606
5	.878	.959	18	.468	.590
6	.811	.917	19	.456	.575
7	.754	.875	20	.444	.561
8	.707	.834	21	.433	.549
9	.666	.798	22	.423	.537
10	.632	.765	23	.412	.526
11	.602	.735	24	.403	.515
12	.576	.708	25	.396	.505
13	.553	.684	30	.361	.463
14	.532	.661	40	.312	.402
15	.514	.641	60	.254	.330
16	.497	.623	120	.179	.234

Using Significant Values for the Correlation Coefficient

If $|r|$ is greater than or equal to the value given in Table 12-4 for either the 5% or 1% significance level, then we can reasonably conclude that the two data sets are related.

The use of Table 12-4 is illustrated in Example 3.

EXAMPLE 3 Deciding if a Correlation Coefficient is Significant

Determine if the correlation coefficient $r = 0.748$ found in Example 2 is significant at the 5% level.

SOLUTION

Since the sample size is $n = 10$ and the 5% significance level is used, the value of $|r|$ must be greater than or equal to the number found in Table 12-4 to be significant. In this case, $|r| = 0.748$, which is greater than or equal to 0.632 obtained from the table. We can conclude that there is a significant relationship between the data sets.

However, we can't conclude that there is a relationship between the data sets at the 1% significance level since r would need to be greater than or equal to 0.765. In this example, $r = 0.748$, which is less than 0.765.

Researchers could use correlation and regression analysis to find out if there is a correlation between the size of a family's home and the number of children in the family.

▼ Try This One 3

Test the significance of the correlation coefficient obtained from Try This One 2. Use 5%, and then 1%.

CAUTION

Remember, the result of Example 3 doesn't guarantee that the data sets are actually related—there's still a 5% chance that they're not related at all.

Multi-symptom cold and cough relief without drowsiness

INDICATIONS: For the temporary relief of nasal congestion, minor aches, pains, headache, muscular aches, sore throat, and fever associated with the common cold. Temporarily relieves cough occurring with a cold. Helps loosen phlegm (mucus) and thin bronchial secretions to drain bronchial tubes and make coughs more productive.

DIRECTIONS: Adults and children 12 years of age and over, 2 liquid caps every 4 hours, while symptoms persist, not to exceed 8 liquid caps in 24 hours, or as directed by a doctor. Not recommended for children under 12 years of age.

WARNINGS: Do not exceed recommended dosage. If nervousness, dizziness, or sleeplessness occur, discontinue use and consult a doctor. Do not take this product for more than 10 days. A persistent cough may be a sign of a serious condition. If symptoms do not improve or if cough persists for more than 7 days, tends to recur, or is accompanied by rash, persistent headache, fever that lasts for more than 3 days, or if new symptoms occur, consult a doctor. Do not take this product for persistent or chronic cough such as occurs with smoking, asthma, chronic bronchitis, or emphysema, or where cough is accompanied by excessive phlegm (mucus) unless directed by a doctor. If sore throat is severe, persists for more than 2 days, is accompanied or followed by fever, headache, rash, nausea, or vomiting, consult a doctor promptly. Do not take this product if you have heart disease, high blood pressure, thyroid disease, diabetes, or difficulty in urination due to enlargement of the prostate gland unless directed by a doctor. As with any drug, if you are pregnant or nursing

Medication labels help users to become aware of the effects of taking medications.

☑ 3. Determine if correlation coefficients are significant.

How should we choose whether to use the 5% or 1% significance level in a given situation? It depends on the seriousness of the situation and the importance of drawing a correct conclusion. Suppose that researchers think that a new medication helps patients with asthma to breathe easier, but that some patients have experienced side effects. A correlation study would probably be done to decide if there is a relationship between the medication and these particular side effects. If the potential side effects are serious, like heart attacks or strokes, the 1% significance level would be used. If the side effects are mild, like headache or nausea, the 5% significance level would probably be considered sufficient, since the consequences of being wrong are not as dire as in the first case.

Regression

Once we have concluded that there is a significant relationship between the two variables, the next step is to find the *equation* of the **regression line** through the data points.

The regression line is a line that best fits the data. Broadly speaking, we say it is the line that passes through the points in such a way that the overall distance each point is from the line is at a minimum. The regression line is also called the *line of best fit*.

Recall from algebra that the equation of a line in the slope-intercept form is $y = mx + b$, where m is the slope and b is the y intercept. (See Section 7-1.) In statistics, the equation of the regression line is written as $y = a + bn$, where a is the y intercept and b is the slope. This is the equation that will be used here. In order to find the values for a and b, two formulas are used.

Formulas for Finding the Values of *a* and *b* for the Equation of the Regression Line

$$b = \frac{n(\Sigma xy) - (\Sigma x)(\Sigma y)}{n(\Sigma x^2) - (\Sigma x)^2} \quad \text{slope}$$

$$a = \frac{\Sigma y - b(\Sigma x)}{n} \quad y \text{ intercept}$$

EXAMPLE 4 Finding a Regression Line

Math Note

The procedure outlined in the Sidelight on page 703 is a great way to find regression lines using a graphing calculator. Because of potential rounding error, the answer may be a bit different than the one obtained using the procedure in Example 4. This procedure is very sensitive to rounding error, so using at least three decimal places is a good idea.

Find the equation of the regression line for the data in Example 1.

SOLUTION

We already calculated the values needed for each formula in Example 2. Substitute into the first formula to find the value for the slope, b.

$$b = \frac{n(\Sigma xy) - (\Sigma x)\Sigma y}{n(\Sigma x^2) - (\Sigma x)^2} = \frac{10(113{,}865) - (572)(1{,}750)}{10(37{,}802) - (572)^2} = \frac{137{,}650}{50{,}836} \approx 2.7077$$

Substitute into the second formula to find the value for the y intercept, a, when $b = 2.7077$.

$$a = \frac{\Sigma y - b(\Sigma x)}{n} = \frac{(1{,}750) - 2.7077(572)}{10} = \frac{201.20}{10} = 20.120$$

The equation of the regression line is $y = 20.120 + 2.7077x$.

Try This One 4

☑ 4. Find a regression line for two data sets.

Find the equation of the regression line for the data in Try This One 1.

After the equation of the regression line is found, it can be drawn on the scatter plot using one of the methods shown in Chapter 7. For example, you can find the coordinates of two points on the line by selecting two values for x, then substituting each into the regression line equation, and finding the corresponding values for y.

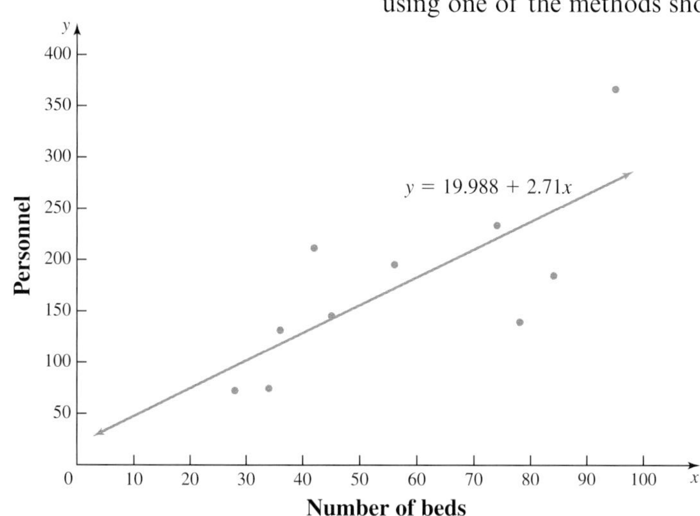

Figure 12-26

In the first case, let $x = 30$; then

$y = 19.988 + 2.71x$
$= 19.988 + 2.71(30)$
$= 101.288$

In the second case, let $x = 70$; then

$y = 19.988 + 2.71x$
$= 19.988 + 2.71(70)$
$= 209.688$

The coordinates of the first point are (30, 101), and the coordinates of the second point are (70, 210) (the y values have been rounded). Next, plot the two points and draw a line through each point. See Figure 12-26.

The Relationship between r and the Regression Line

Two things should be noted about the relationship between the value of r and the regression line. First, the value of r and the value of the slope, b, always have the same sign. Second, the closer the value of r is to $+1$ or -1, the better the points will fit the line. In other words, the stronger the relationship, the better the fit. Figure 12-27 shows the relationship between the correlation coefficient and the regression line.

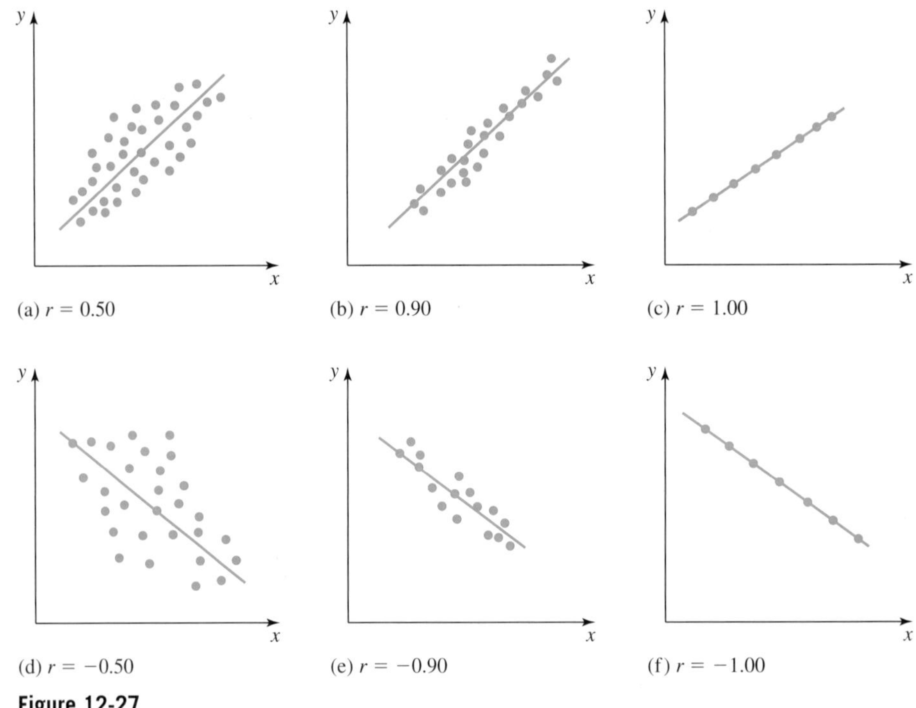

Figure 12-27

Once we have the equation of a regression line for two data sets, we can use it to predict a value for y given a particular value for x, as we will see in Example 5.

EXAMPLE 5 Using a Regression Line to Make a Prediction

Use the equation of the regression line found in Example 4 to predict the approximate number of personnel for a hospital with 65 beds.

SOLUTION

Substitute 65 for x into the equation $y = 20.120 + 2.7077x$ and find the value for y.

$$y = 20.120 + 2.7077(65) = 196.12 \text{ (which can be rounded to 196)}$$

We predict that a hospital with 65 beds will have approximately 196 personnel.

▼ Try This One 5

☑ 5. Use regression lines to make predictions.

Use the equation of the regression line found in Try This One 4 to predict the number of stories for a 670-foot building.

CAUTION

Never use a regression line to predict anything unless you've already checked that the correlation is significant! If it isn't significant, the regression line is meaningless.

Correlation and Causation

As you will see in the Sidelight below, there's more to the story of a relationship between two data sets than simply finding out that they're related. The nature of the relationship then needs to be further explored. The following is a list of possibilities to consider.

Possible Relationships between Data Sets

1. *There is a direct cause-and-effect relationship between two variables.* That is, x causes y. For example, water causes plants to grow, poison causes death, and heat causes ice to melt.

2. *There is a reverse cause-and-effect relationship between the variables.* That is, y causes x. Suppose that we find a correlation between excessive coffee consumption and nervousness. It would be tempting to conclude that drinking too much coffee causes nervousness. But it might actually be the case that nervous people drink a lot of coffee because they think it calms their nerves.
3. *The relationship between the variables may be caused by a third variable.* You could probably find a positive correlation between the amount of ice cream consumed per week and the number of drowning deaths for each week of the year. But this doesn't mean that eating ice cream causes you to drown. Both eating ice cream and swimming are more common during the summer months, and it stands to reason that more people will drown when more people are swimming. In this case, a third variable (seasonal weather) is affecting each of the original two variables.

Sidelight DOES MYSPACE CAUSE ILLITERACY?

At the beginning of this section, we looked at a claim that there's a correlation between using MySpace and illiteracy. How strong is that connection? And what does it mean?

A scatter plot of the data for all 50 states is shown in Figure 12-28. The x variable is the percent of the adult population that lacks basic literacy as measured by the National Assessment of Adult Literacy. The y variable is a scaled score that indicates how often Internet users from each state searched for the term MySpace on Google. The scatter plot looks like there is a positive linear correlation, and the correlation coefficient bears that out: it is almost exactly 0.6, which is well above the 1% significance level for a sample size of 50. So we can definitely conclude that the two data sets are related. (Note that $n = 50$ is not in Table 12-4, but we can see that 0.6 is within the 1% confidence interval for any sample size greater than 17.)

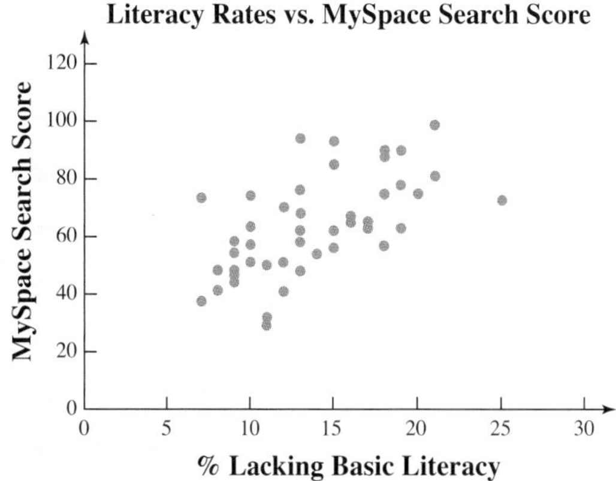

Figure 12.28

Then the most important question should be asked: what does it mean? This is the key point about correlation—there is a big difference between *correlation* and *causation*. When two data sets are related, that tells us nothing about *why* they are related. In this case, it would be inappropriate to conclude that being illiterate causes people to use MySpace, and it would be just plain silly to conclude that using MySpace causes people to be illiterate. There may be some other combination of factors that keep literacy rates low in certain states and make people more likely to search for the term MySpace. Or (and this is my best guess) it might just be one big, amazing coincidence. In any case, it's important to understand that correlation between data sets by itself never tells us anything about *how* or *why* two sets are related.

4. *There may be a complexity of interrelationships among many variables.* For example, a researcher may find a significant relationship between students' high school grades and college grades. But there probably are many other variables involved, such as IQ, hours of study, influence of parents, motivation, age, and instructors.
5. *The relationship might simply be coincidental.* For example, historians have noticed that there is a very strong correlation between the party that wins a presidential election and the result of the Washington Redskins' final game before the election. But common sense dictates that any such relationship has to be coincidental.

Answers to Try This One

1

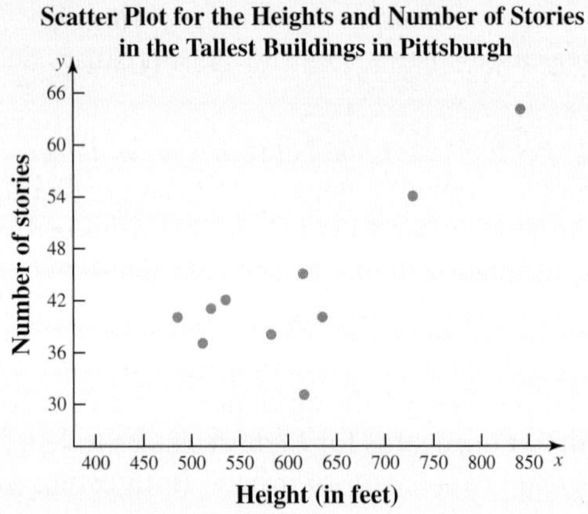
Scatter Plot for the Heights and Number of Stories in the Tallest Buildings in Pittsburgh
y
66
60
54
48
42
36
30
Number of stories
400 450 500 550 600 650 700 750 800 850 x
Height (in feet)

2 $r \approx 0.798$

3 Significant at both levels

4 $y = 1.634 + 0.0685x$

5 47 stories

EXERCISE SET 12-8

Writing Exercises

1. Describe what a scatter plot is and how one is used.
2. Explain what is meant when two variables are positively linearly related. What would the scatter plot look like?
3. Explain what is meant when two variables are negatively linearly related. What would the scatter plot look like?
4. Explain what the value of the correlation coefficient tells you about the relationship between two data sets.
5. What is a regression line for two data sets?
6. Describe how regression lines can be used to make predictions about situations.

Computational Exercises

For the data sets in Exercises 7–12,

(a) Draw a scatter plot.
(b) Find the value for r.
(c) Test the significance of r at the 5% level and at the 1% level.
(d) Find the equation of the regression line and draw the line on the scatter plot, but only if r is significant.
(e) Describe the nature of the relationship if one exists.

7.

x	1	4	6	2	3	5	7
y	8	15	20	10	11	16	25

8.

x	21	25	24	30	36	40
y	12	8	9	5	3	2

9.

x	75	80	85	90
y	10	5	11	4

10.

x	9	12	15	11	10	13
y	50	60	71	55	53	60

11.

x	27	35	48	43	32
y	19	13	8	10	15

12.

x	31	34	37	40	46
y	3	15	2	13	5

Real-World Applications

For the data sets in Exercises 13–18,

(a) Draw a scatter plot. (b) Find the value for r.
(c) Test the significance of r at the 5% level and at the 1% level.
(d) Find the equation of the regression line and draw the line on the scatter plot, but only if r is significant.
(e) Describe the nature of the relationship if one exists.
(f) Make the requested prediction using your regression line.

13. The data represent the heights in feet and the number of stories of the tallest buildings in Cleveland.

Height, x	947	708	658	529	450	446	430	420	419
Stories, y	57	52	46	40	31	28	24	26	32

Source: *The World Almanac and Book of Facts*

Predict the number of stories in a 500-foot building.

14. A researcher hopes to determine whether the number of hours a person jogs per week is related to the person's age.

Age, x	34	22	48	56	62
Hours, y	3.5	7	3.5	3	1

Predict the number of hours for a 35-year-old.

15. A study was conducted to determine if the amount a college student spends per month on recreation is related to the student's income.

Monthly income, x	**Amount, y**	**Monthly income, x**	**Amount, y**
$800	$160	$850	$145
$1,200	$300	$907	$190
$1,000	$260	$1,100	$250
$900	$235		

Predict the amount spent by a student with an income of $925.

16. A researcher hopes to determine if there is a relationship between the number of days an employee missed a year and the person's age.

Age, x	22	30	25	35	65	50	27	53	42	58
Days missed, y	0	4	1	2	14	7	3	8	6	4

Predict the number of days missed by a 56-year-old employee.

17. A statistics instructor plans to see it there's a relationship between the final exam score in Statistics 102 and the final exam scores of the same students who took Statistics 101.

Stat 101, x	87	92	68	72	95	78	83	98
Stat 102, y	83	88	70	74	90	74	83	99

Predict the stat 102 score for a student who got 90 on the stat 101 final.

18. The data shown indicate the number of wins and the number of goals scored for teams in the National Hockey League after the first month of the season.

No. of wins, x	**No. of goals, y**	**No. of wins, x**	**No. of goals, y**
10	23	7	21
9	22	5	16
6	15	9	12
5	15	8	19
4	10	6	16
12	26	6	16
11	26	4	11
8	26		

Source: *USA Today*

Predict the number of goals for a team with 8 wins.

Critical Thinking

19. Explain why when r is significant, you can't conclude that x causes y.

20. Find the value for r, then interchange the values for x and y and find the value of r. Explain the results.

x	1	2	3	4	5
y	3	5	7	9	11

21. Draw a scatter plot and determine if the variables are related. Explain the results. Then find the value for r for the data shown.

x	−3	−2	−1	0	1	2	3
y	9	4	1	0	1	4	9

22. Design your own correlation problem. Think of two quantities that you think might be correlated, then gather some data, and use the techniques of this section to decide if the data sets are correlated.

CHAPTER 12 Summary

Section	Important Terms	Important Ideas
12-1	Data Statistics Population Sample Random sample Systematic sample Stratified sample Cluster sample Descriptive statistics Inferential statistics Raw data Categorical frequency distribution Grouped frequency distribution Stem and leaf plot	**Statistics** is the branch of mathematics that involves the collection, organization, summarization, and presentation of data. In addition, researchers use statistics to make general conclusions from the data. When a study is conducted, the researcher defines a population, which consists of all subjects under study. Since populations are usually large, the researcher will select a representative subgroup of the population, called a sample, to study. There are four basic sampling methods. They are random, systematic, stratified, and cluster. Once the data are collected, they are organized into a frequency distribution. There are two types of frequency distributions. They are categorical and grouped. When the data set consists of a small number of values, a stem and leaf plot can be constructed. This plot shows the nature of the data while retaining the original data values.
12-2	Bar graph Pie chart Histogram Frequency polygon Time series graph	**In order** to represent data pictorially, graphs can be drawn. From a categorical frequency distribution, a bar graph and a pie chart can be drawn. From a grouped frequency distribution, a histogram and a frequency polygon can be drawn to represent the data. To show how data vary over time, a time series graph can be drawn.
12-3	Mean Data array Median Mode Midrange	**In addition** to collecting data, organizing data, and representing data using graphs, various summary statistics can be found to describe the data. There are four measures of average. They are the mean, median, mode, and midrange. The mean is found by adding all of the data values and dividing the sum by the number of data values. The median is the middle point of the data set. The mode is the most frequent data value. The midrange is found by adding the lowest and the highest data values and dividing by two.
12-4	Range Variance Standard deviation	**There are** three commonly used measures of variation. They are the range, variance, and standard deviation. The range is found by subtracting the lowest data value from the highest data value. The standard deviation measures the spread of the data values. When the standard deviation is small, most data values are close to the mean. When the standard deviation is large, the data values are spread out farther away from the mean.
12-5	Percentile Quartile	**The position** of a data value in a data set can be determined by its percentile rank. The percentile rank of a specific data value gives the percent of data values that fall below the specific value. Quartiles divide a distribution into quarters.

Section	Key Terms	Summary
12-6	Normal distribution Empirical rule Standard normal distribution *z* score	**Many variables** have a distribution that is bell-shaped. These variables are said to be approximately normally distributed. Statisticians use the standard normal distribution to describe these variables. The standard normal distribution has a mean of zero and a standard deviation of 1. A variable can be transformed into a standard normal variable by finding its corresponding *z* score. A table showing the area under the standard normal distribution can be used to find areas for various *z* scores.
12-7		**Since many** real-world variables are approximately normally distributed, the standard normal distribution can be used to solve many real-world applications.
12-8	Correlation Independent variable Dependent variable Scatter plot Correlation coefficient Significance levels Regression line	**Statisticians** are also interested in determining whether or not two variables are related. In order to determine this, they draw and analyze a scatter plot. After the scatter plot is drawn, the value of the correlation coefficient is computed. If the correlation coefficient is significant, the equation of the regression line is found. Then a prediction for *y* can be obtained given a specific value of *x*.

MATH IN Sociology REVISITED

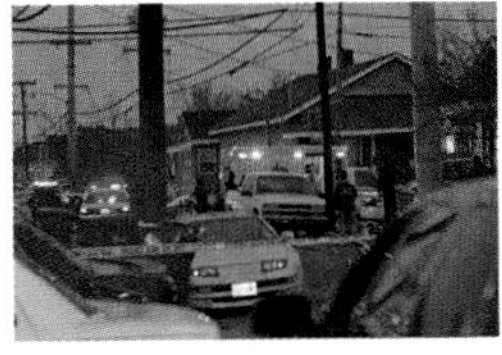

1. The scatter plot is shown below, with temperature as the *x* variable. While it's not at all clear whether there is a real correlation, it looks like there's at least a chance that the two data sets have a positive linear relationship.

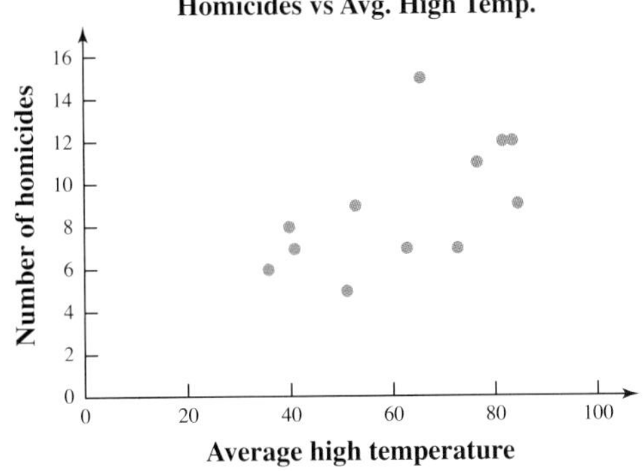

2. The correlation coefficient is about 0.584. According to the significance table, for a sample size of 12 this is significant at the 5% level (just barely), but not at the 1% level. So we can say with 95% confidence that the two data sets are related.
3. Absolutely not. In fact, no matter how strongly the data sets are related, we can't draw any conclusions about whether one causes the other. All we know is that for some reason, they appear to be related. There are a wide variety of theories trying to account for this phenomenon, but none have been proven or even widely accepted.

Review Exercises

Section 12-1

1. A sporting goods store kept a record of sales of five items for one randomly selected hour during a recent sale. Construct a frequency distribution for the data (B = baseballs, G = golf balls, T = tennis balls, S = soccer balls, F = footballs).

F	B	B	B	G	T	F
G	G	F	S	G	T	
F	T	T	T	S	T	
F	S	S	G	S	B	

2. The data set shown below represents the number of hours 25 part-time employees worked at the Sea Side Amusement Park during a randomly selected week in June. Construct a stem and leaf plot for the data and analyze the results.

16	25	18	39	25	17	29	14	37
22	18	12	23	32	35	24	26	
20	19	25	26	38	38	33	29	

3. During June, a local theater company recorded the given number of patrons per day. Construct a grouped frequency distribution for the data. Use six classes.

102	116	113	132	128	117
156	182	183	171	168	179
170	160	163	187	185	158
163	167	168	186	117	108
171	173	161	163	168	182

Section 12-2

4. Construct a bar graph for the number of homicides reported for these cities.

City	Number
New Orleans	179
Washington, D.C.	186
Chicago	509
Baltimore	234
Atlanta	105

Source: FBI

5. The data set shown below represents the time in minutes spent using a computer at Kinko's Copy Center by 25 customers on a randomly selected day. Construct a stem and leaf plot for the data and analyze the results.

16	25	18	39	25	17	29	14	37
22	18	12	23	32	35	24	26	
20	19	25	26	38	38	33	29	

6. Draw a histogram and frequency polygon for the frequency distribution obtained from the data in Exercise 3.

7. The data set shown below indicates how much Janine earned each year from her part-time job at the Otherworld Internet Cafe during the 5 years that she attended college. Draw a time series graph for the data.

Year	Amount
2005	$8,973
2006	$9,388
2007	$11,271
2008	$13,877
2009	$19,203

Section 12-3

8. These data represent the number of deer killed by motor vehicles for eight counties in Southwestern Pennsylvania.

2,343 1,240 1,088 600 497 1,925 1,480 458

Source: *Pittsburgh Post-Gazette*

Find each of these.

(a) Mean
(b) Median
(c) Mode
(d) Midrange

9. Twelve batteries were tested to see how many hours they would last. The frequency distribution is shown here.

Hours	Frequency
1–3	1
4–6	4
7–9	5
10–12	1
13–15	1

Find the mean.

10. In your own words, explain the difference between mean, median, mode, and midrange. Include an explanation of why each is considered a measure of average.

Section 12-4

11. Find the range, variance, and standard deviation for the data in Exercise 8.

12. Which data set do you think would have a greater standard deviation: the ages of everyone currently working out in the gym on your campus, or the ages of everyone currently shopping at the nearest off-campus grocery store? Explain your answer.

Section 12-5

13. The number of previous jobs held by each of six applicants is shown here.

    2 4 5 6 8 9

    (a) Find the percentile for each value.
    (b) What value corresponds to the 30th percentile?

14. The data shown represent the number of days' inventory eight high-tech firms have on hand. Find the values for Q_1, Q_2, and Q_3.

    158 151 91 45 74 118 285 29

Section 12-6

15. Find the area under the standard normal distribution curve.

    (a) Between $z = 0$ and $z = 1.95$
    (b) Between $z = 0$ and $z = 0.40$
    (c) Between $z = 1.30$ and $z = 1.80$
    (d) Between $z = -1.05$ and $z = 2.05$
    (e) Between $z = -0.05$ and $z = 0.55$
    (f) Between $z = 1.10$ and $z = -1.80$
    (g) To the right of $z = 2.00$
    (h) To the right of $z = -1.35$
    (i) To the left of $z = -2.10$
    (j) To the left of $z = 1.70$

16. The weights of players on one college rugby team are normally distributed with mean 210 pounds and standard deviation 12 pounds. Use the empirical rule to estimate the number of players out of 35 that weigh between 186 and 234 pounds.

17. For a data set with 90 values that is normally distributed with mean 200 and standard deviation 25, approximately how many data values will fall in each range?

    (a) The range from 190 to 210.
    (b) The range of values greater than 240.

Section 12-7

18. The average number of years a person takes to complete a graduate degree program is 3. The standard deviation is 4 months or $\frac{1}{3}$ of a year. Assume the data are normally distributed. If an individual enrolls in the program, find the probability that it will take

    (a) More than 4 years to complete the program.
    (b) Less than 3 years to complete the program.
    (c) Between 3.8 and 4.5 years to complete the program.
    (d) Between 2.5 and 3.1 years to complete the program.

19. On the daily run of an express bus, the average number of passengers is 48. The standard deviation is 3. Assume the data are normally distributed. Find the probability that the bus will have

    (a) Between 36 and 40 passengers.
    (b) Fewer than 42 passengers.
    (c) More than 48 passengers.
    (d) Between 43 and 48 passengers.

20. The average weight of an airline passenger's suitcase is 45 pounds. The standard deviation is 2 pounds. Assume the weights are normally distributed. If an airline handles 2,000 suitcases in one day, find the number that will weigh less than 43.5 pounds.

21. The average cost of Cheetah brand running shoes is $83.00 per pair, with a standard deviation of $8.00. If 90 pairs of shoes are sold, how many will cost between $80.00 and $85.00? Assume the cost is normally distributed.

Section 12-8

22. A study is done to see whether there is a relationship between a student's grade point average and the number of hours the students watches television each week. The data are shown here.

Hours, x	10	6	8	15	5	6	12
GPA, y	2.4	4	3.2	1.6	3.7	3.7	3.0

Draw a scatter plot for the data and describe the relationship. Find the value for r and determine whether or not it is significant at the 5% significance level. If yes, find the equation of the regression line, and predict y when $x = 9$.

23. The table below displays the number of homicides committed in Chicago, and the number of wins by the Chicago Bears of the NFL for each year from 2000 to 2008.

Homicides	Wins	Homicides	Wins
628	5	449	11
666	13	452	13
647	4	442	7
598	7	509	9
448	4		

(a) Without doing any calculations, discuss whether or not you think the two data sets should be related.
(b) Find the correlation coefficient for the data sets, and then see if it is significant at the 5% level. Was your prediction from part (a) right or wrong?

Chapter Test

1. A questionnaire about the last 25 peripheral devices purchased at a computer store is shown below. Construct a frequency distribution (L = laser mouse, E = external hard drive, K = wireless keyboard, W = webcam)

L L K E E K L E W K
W W L K W E K W L W
E K K L L

2. Draw a bar graph for the frequency distribution obtained in Exercise 1.
3. Draw a pie chart for the data found in Exercise 2.
4. The data (in millions of dollars) are the values of the 30 Major League baseball franchises. Construct a frequency distribution for the data using eight classes.

1,500	912	320	700	356	277	347	722
401	314	496	390	833	509	450	342
373	486	400	353	319	288	405	445
399	471	426	371	446	406		

Source: *Forbes Magazine*

5. Draw a histogram and frequency polygon using the frequency distribution for the data in Exercise 4.
6. A special aptitude test is given to job applicants. The data shown here represent the scores of 30 applicants. Construct a stem and leaf plot for the data and summarize the results.

204	210	227	218	254
256	238	242	253	227
251	243	233	251	241
237	247	211	222	231
218	212	217	227	209
260	230	228	242	200

7. The given data represent the federal minimum hourly wage in the years shown. Draw a time series graph to represent the data and analyze the results.

Year	Wage	Year	Wage
1960	$1.00	1990	3.80
1965	1.25	1995	4.25
1970	1.60	2000	5.15
1975	2.10	2005	5.15
1980	3.10	2009	7.25
1985	3.35		

8. These temperatures were recorded in Pasadena for a week in April.

87 85 80 78 83 86 90

Find each of these.
(a) Mean
(b) Median
(c) Mode
(d) Midrange
(e) Range
(f) Variance
(g) Standard deviation

9. The distribution of the number of errors 10 students made on a typing test is shown.

Errors	Frequency
0–2	1
3–5	3
6–8	4
9–11	1
12–14	1

Find the mean.

10. The number of credits in business courses eight job applicants had is shown here.

9 12 15 27 33 45 63 72

(a) Find the percentile for each value.
(b) What value corresponds to the 40th percentile?

11. Find the area under the standard normal distribution for each.

(a) Between 0 and 1.50
(b) Between 0 and −1.25
(c) Between 1.56 and 1.96
(d) Between −1.20 and −2.25
(e) Between −0.06 and 0.73
(f) Between 1.10 and −1.80
(g) To the right of $z = 1.75$
(h) To the right of $z = -1.28$
(i) To the left of $z = -2.12$
(j) To the left of $z = 1.36$

12. The mean time it takes for a certain pain reliever to begin to reduce symptoms is 30 minutes, with a standard deviation of 4 minutes. Assuming the time is normally distributed, find the probability that it will take the medication

(a) Between 34 and 35 minutes to begin to work.
(b) More than 35 minutes to begin to work.
(c) Less than 25 minutes to begin to work.
(d) Between 35 and 40 minutes to begin to work.

13. The average height of a certain age group of people is 53 inches. The standard deviation is 4 inches. If the heights are normally distributed, find the probability that a selected individual's height will be

(a) Greater than 59 inches.
(b) Less than 45 inches.
(c) Between 50 and 55 inches.
(d) Between 58 and 62 inches.

14. The average cost of a skateboard at Sk8 Gr8, a local skate shop, is \$55.00, with a standard deviation of \$8.00. The prices are normally distributed. If 200 skateboards are purchased, how many of them cost more than \$60.00?
15. A study is conducted to determine the relationship between a driver's age at the beginning of the study and the number of accidents he or she has over a 1-year period. The data are shown here. Draw a scatter plot for the data and explain the nature of the relationship. Find the value for r and determine whether or not it is significant at the 5% significance level. If r is significant, find the equation of the regression line, and predict y when x is 61.

Age, x	No. of accidents, y
32	1
20	1
55	0
16	3
19	2
17	1
45	0
26	0
32	1
61	0

Projects

1. Survey at least 30 students on your campus to find out how many miles away from the campus they live. Construct a frequency distribution with at least five classes from the data. Draw a histogram and frequency polygon for your data, then compute the mean, median, mode, midrange, and standard deviation for the data. Write a report summarizing the results of your study, including a discussion of whether or not the data appear to be approximately normally distributed.
2. Survey at least 30 students on your campus to find out how many credit hours each is taking, and the estimated number of hours per week that they spend on schoolwork outside of class. Draw a scatter plot for the data. Then find the correlation coefficient and decide if it is significant at the 5% level and at the 1% level. Write a report summarizing your findings. Include a discussion of whether or not you were surprised by the results and why.

CHAPTER 13

Other Mathematical Systems

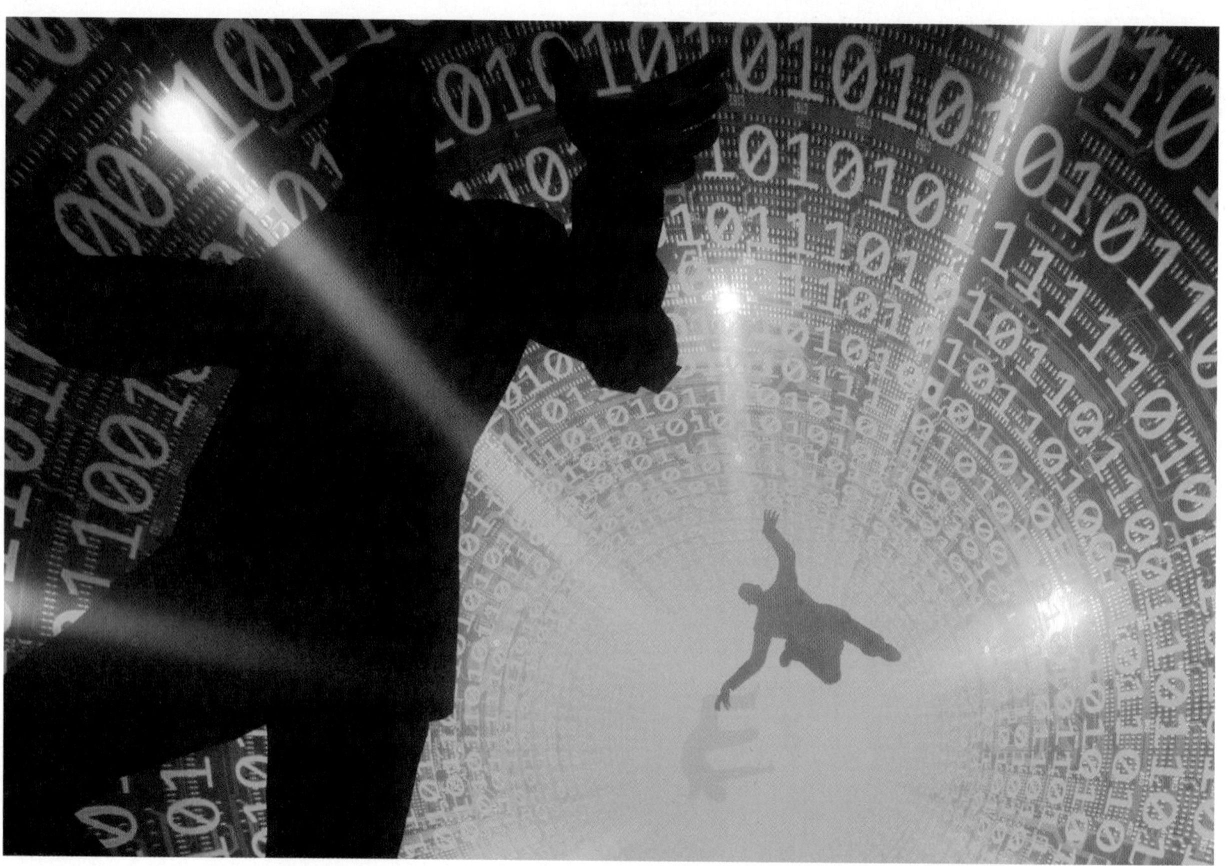

Outline

13-1 Mathematical Systems and Groups
13-2 Clock Arithmetic
13-3 Modular Systems
Summary

MATH IN Encryption

Cryptography is the study of hiding information by using some sort of code. There was a time when codes were of interest mostly to military officers, spies, and grade school kids passing notes they didn't want the teacher to read. But that time passed very quickly with the advent of the computer age. According to some reports, more than half of all Americans made at least one purchase online in 2008. When sensitive financial information and passwords are being transmitted, encryption becomes of supreme importance. Without proper encryption, you could find your identity lost in the endless depths of cyberspace.

In order to be sent via the Internet, information first has to be encoded, often into numeric form. If that information is sensitive, it then needs to be encrypted so that if it's intercepted by a third party, the information remains hidden. There are many methods for encrypting information, and many of them are based on the modular arithmetic we will study in this chapter.

For most of your life, you have worked with one numerical system: the real number system. Within that system, you're familiar with operations like addition and multiplication. In this chapter, we will learn about mathematical systems different than the one you're accustomed to. In most cases, the systems will use a set of numbers you're familiar with, but the operation of addition will be defined differently. This, in turn, will affect the other operations. In some cases, we will look at mathematical systems and operations that don't use numbers at all! An open mind will be key—without one, you will probably have a hard time swallowing computations like $11 + 2 = 1$. But by the time you finish this chapter, computations of that nature will make perfect sense. In particular, after completing Section 13-3 on modular arithmetic, you will be able to answer the questions below.

Most encryption methods based on modular arithmetic are very complicated, by necessity—the simpler the encryption, the easier it is for a computer to break the code. Here's a simple example using some of the same ideas: the first string of numbers below is a verbal message encoded using a modular system. Each letter and space was given a number from 1 (A) to 27 (space) based on position in the alphabet. The numbers were added to 50 and the resulting numbers were replaced with their congruent number modulo 32. To decode the message, find the original numbers and translate back into letters.

The second string of numbers is the license number from a driver's license presented by a college student to a police officer doing spot checks for underage alcohol consumption. In this particular state, the last digit is a check digit: if you drop the last digit of the license number and find the number congruent to the result modulo 29, you should get the last digit. If you don't, the license number is fake. Is this number legitimate?

1. 27 22 23 0 6 27 6 11 13 6 26 23 24 6 13 5 6 27 0 29 5
2. 0411039416

For answers, see Math in Encryption Revisited on page 738

Section 13-1 Mathematical Systems and Groups

LEARNING OBJECTIVES

- ☐ 1. Use an operation table to perform the operation in a mathematical system.
- ☐ 2. Determine which properties of mathematical systems are satisfied by a given system.
- ☐ 3. Decide if a mathematical system is a group.

When you're driving on city streets and come to a four-way intersection, you have four choices: right, left, straight, or U-turn. What you don't realize is that you've just landed in the middle of a *mathematical system*. A **mathematical system** consists of a finite or infinite set of symbols and at least one operation. In this case, the four symbols are R, L, S, and U, representing your four choices at the intersection.

What about the operation? Consider what happens if you make two of those choices consecutively. If you turn right and then left, you end up going in the same direction you started in, so this corresponds to going straight. (For this discussion, we're going to ignore what particular street you're on, and focus only on the direction.) If you make a U-turn, then turn left, you'll end up in the same direction as if you had turned right.

If we use the symbol + to represent combining consecutive turns, we can make a table describing the result of any two consecutive choices. The symbols down the left side of Table 13-1 represent the first choice, and the symbols along the top represent the second choice.

TABLE 13-1 An Operation Table for a Mathematical System

+	R	L	S	U	
R	U	(S)	R	L	*Right + Left = Straight*
L	S	U	L	R	
S	R	L	S	U	
U	L	(R)	U	S	*U-turn + Left = Right*

To find the result of two consecutive turns, look along the row corresponding to the first turn until you reach the column corresponding to the second turn. The symbols in color represent the combinations we already mentioned: R + L = S, and U + L = R. Table 13-1 represents a **finite mathematical system** because there are a finite number of symbols with an operation. The table is called an **operation table.** A system with infinitely many symbols and an operation is called an **infinite mathematical system.** The integers under addition are an example.

EXAMPLE 1 Using an Operation Table

Use Table 13-1 to find the result of each operation and describe what it means physically.

(a) L + S
(b) R + R
(c) U + S
(d) (S + U) + R

SOLUTION

(a) The symbol in row L and column S is L, so L + S = L. If you turn left, then go straight, the direction is the same as just turning left.
(b) The symbol in row R and column R is U, so R + R = U. Turn right twice and your direction is the same as if you had made a U-turn.
(c) U + S = U; a U-turn followed by straight is the same direction as the U-turn itself.
(d) First, we find that S + U = U; then U + R = L. If you follow going straight and making a U-turn with a right turn, it's the same direction as just turning left.

Math Note

If you've spent much time in the Detroit area, you will recognize the string of turns in part d as a "Michigan left." At busy intersections, left turns are not permitted. Instead, those wanting to turn left go straight through the intersection, then make a U-turn and turn right back at the original intersection.

▼ Try This One 1

Use Table 13-1 to find the result of each operation and describe what it means physically.

(a) R + U
(b) U + U
(c) S + S
(d) (U + R) + R

☑ 1. Use an operation table to perform the operation in a mathematical system.

There are five important properties of mathematical systems that we will study: *closure, commutative, associative, identity, and inverse.*

Closure Property

For a system to be closed under an operation, when the operation is performed on any symbol, the result must be another symbol in the system. The system in Table 13-1 satisfies this property, and is called a **closed system** because any combination of two turns results in one of the four basic turns. The integers using operation division provide an example of a system that is not closed: sometimes when you divide two integers, the result (like $\frac{1}{2}$, for example) is not an integer.

Math Note

Notice that Table 13-1 is symmetric about the diagonal from upper left to lower right: that is, the bottom left and upper right triangular portions are mirror images. This is a quick way to check if a system is commutative. If the table displays that symmetry, the system is commutative.

Commutative Property

A system is commutative if the order in which you perform the operation doesn't matter. More formally, a system is commutative if for any two symbols a and b in a system with some operation $*$, $a * b = b * a$. The system in Table 13-1 is commutative because the result of any two consecutive turns is the same regardless of the order you perform them in. An example of a system that is not commutative is the integers under subtraction, because the order in which you subtract affects the result.

Associative Property

A system has the associative property if for any a, b, c in a system with operation $*$, $(a * b) * c = a * (b * c)$. There really isn't a quick way to show that a system is associative: you would have to try every possible combination of three symbols. (For a system with more than two symbols, the number of calculations required can be pretty large.) Typically, we'll check a few examples to try and get an idea of whether or not a system is associative. In Example 1 part d, we saw that (S + U) + R = L. Let's try S + (U + R): first, we find that U + R = L. Then S + L = L, so S + (U + R) and (S + U) + R both equal L. This doesn't prove that the system is associative—just that the property holds in this one case—but it turns out that the system is associative.

Math Note

In Exercise 69, we'll check all of the possibilities to prove that the system in Example 1 is associative.

Identity Property

A system has the identity property if there is a symbol a in the system so that $a * b = b * a = b$ for any other symbol b. In short, if there's a symbol that leaves all the others unchanged when combined using the operation, that symbol is called the **identity element** for the system. For our system of turns, it's easy to see that S is the identity element: combining straight with any other turn results in the same direction as the turn itself.

Inverse Property

When a system has an identity element, the next question to consider is whether every symbol has another symbol that produces the identity element under the operation. That is, for any a in the system, is there always an **inverse element** b so that $a * b$ is the identity element? If there is, the system satisfies the inverse property. We can verify that the system of turns satisfies the inverse property by finding an inverse for each turn. For right, turning left results in straight (the identity element). For left, turning right results in straight. For straight, another straight results in straight, and for U-turns, another U-turn results in straight. We have displayed an inverse for each symbol, so the system satisfies the inverse property.

EXAMPLE 2 Identifying the Properties of a Finite Mathematical System

Which properties does the system defined by the given table exhibit?

&	0	1	2	3
0	2	3	0	1
1	1	2	1	0
2	0	1	2	3
3	1	0	3	2

SOLUTION

Closure property: Since every element in the body of the table is in the set the system is defined on, namely, {0, 1, 2, 3}, the system is closed.

Commutative property: Since the system is not symmetric with respect to the diagonal, it does not have the commutative property. For example, 0 & 1 = 3, but 1 & 0 = 1.

Associative property: Let's try some examples.

(1 & 2) & 3 = 1 & 3 = 0   1 & (2 & 3) = 1 & 3 = 0

(1 & 0) & 3 = 1 & 3 = 0   1 & (0 & 3) = 1 & 1 = 2

We found a counterexample, so the system is not associative.

Identity property: There is an identity element, 2, since 2 & x = x & 2 = x for all $x \in \{0, 1, 2, 3\}$

Inverse Property: 0 and 0 are inverses, 1 and 1 are inverses, 2 and 2 are inverses, and 3 and 3 are inverses. Since every element has an inverse, the system has the inverse property.

Therefore, the system exhibits *closure, identity,* and *inverse* properties.

CAUTION

It takes just one counterexample to prove that a system is *not* associative, but every possible combination must be verified to prove that a system *is* associative.

☑ 2. Determine which properties of mathematical systems are satisfied by a given system.

▼ Try This One 2

Which properties does the following system exhibit?

∘	x	y	z
x	x	x	z
y	x	y	z
z	y	z	x

Sidelight MAGIC SQUARES

Have you ever heard of a magic square? These squares have fascinated and intrigued people ever since the first one was discovered more than 4,000 years ago.

The sum of the numbers in each row in the square below is equal to 15:

$$4 + 9 + 2 = 15$$
$$3 + 5 + 7 = 15$$
$$8 + 1 + 6 = 15$$

The sum of the numbers in each column is also 15:

$$4 + 3 + 8 = 15$$
$$9 + 5 + 1 = 15$$
$$2 + 7 + 6 = 15$$

And the sum of the two diagonals is also 15:

$$4 + 5 + 6 = 15$$
$$2 + 5 + 8 = 15$$

A magic square!

4	9	2
3	5	7
8	1	6

The first magic square appears in the Chinese classic *I-Ching* and is called *lo-shu.* It is said that the Emperor Yu saw the square engraved on the back of a divine tortoise on the bank of the Yellow River in 2200 BCE. The square was the same as the one shown in the above figure except that the numerals were indicated by black and white dots, the even numbers in black and the odd numbers in white.

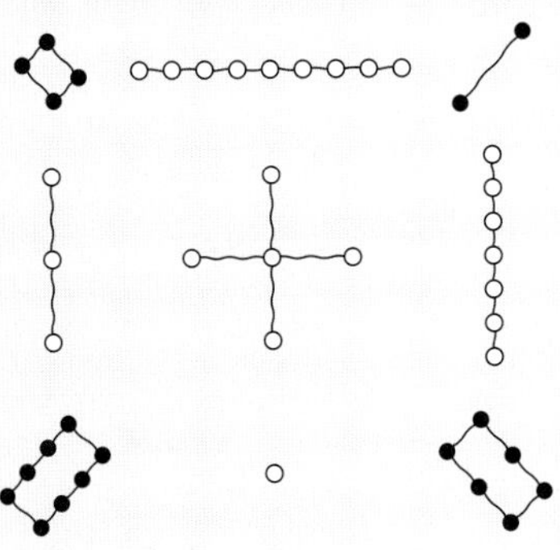

Since some sort of magic was attributed to these squares, they were frequently used to decorate the abodes of gypsies and fortune tellers. They were very popular in India, and they were eventually brought to Europe by the Arabs.

Albrecht Dürer, a 16th-century artist from Germany, used the magic square below in his famous woodcut entitled "Melancholia." This square is indeed magic, for not only are the sums of the rows, columns, and diagonals all equal to 34, but also the sum of the four corners $(16 + 13 + 4 + 1)$ is 34. Furthermore, the sum of the four center cells $(10 + 11 + 6 + 7)$ is also 34, and the sum of the slanting squares $(2 + 8 + 9 + 15$ and $3 + 5 + 12 + 14)$ is 34. Finally, the year in which Dürer made the woodcut appears in the bottom center squares (1514).

16	3	2	13
5	10	11	8
9	6	7	12
4	15	14	1

A great mathematician, Leonhard Euler (1707–1783), constructed the magic square shown next. The sum of the rows and columns is 260. Stopping halfway on each row or column gives a sum of half of 260 or 130. Finally, a knight from a chess game can start on square 1 and proceed in L-shaped moves and come to rest on all squares in numerical order.

1	48	31	50	33	16	63	18
30	51	46	3	62	19	14	35
47	2	49	32	15	34	17	64
52	29	4	45	20	61	36	13
5	44	25	56	9	40	21	60
28	53	8	41	24	57	12	37
43	6	55	26	39	10	59	22
54	27	42	7	58	23	38	11

Magic squares have been a source of fascination for people throughout the ages. Interestingly, they're a lot easier to construct than you would think. An Internet search for *magic square* will identify several sources with methods for building them.

Groups and Abelian Groups

A mathematical system is said to be a **group** if it has closure, associative, identity, and inverse properties. A mathematical system is said to be an **Abelian group** if, in addition to closure, associative, identity, and inverse properties, it also has the commutative property.

EXAMPLE 3 Determining If a Mathematical System Is a Group

Do the natural numbers under the operation of addition form a group? An Abelian group?

SOLUTION

The natural numbers are given by the set $N = \{1, 2, 3, 4, \ldots\}$.

Closure: The natural numbers are closed under addition; that is, the sum of any two natural numbers is a natural number. Therefore the system is *closed.*

Associative: The associative property of addition holds for all real numbers, so it holds for the natural numbers as well. So the system is *associative*.

Identity: There is *no identity* element in the set of natural numbers (the identity element would be 0 under addition, but this element is not in N).

Inverse: Since there is no identity element, there can be *no inverse*. Therefore the inverse property does not hold.

Since the system does not have the four properties required to be a group, it is *not a group*. Since it is not a group, it is also not an Abelian group.

▼ Try This One 3

Do the whole numbers form a group under addition? An Abelian group?

EXAMPLE 4 Determining if a Mathematical System is a Group

Does the set $\{-1, 1\}$ form a group under the operation of multiplication? An Abelian group?

SOLUTION

Closure: The product of any two elements in the set is also an element of the set, therefore the system is *closed*.

Associative: The associative property holds for all elements in the set.

Identity: The identity element is 1.

Inverse: -1 and -1 are inverses as are 1 and 1. Therefore each element has an inverse, so the inverse property holds.

Since the system has *closure, associative, identity,* and *inverse* properties, *it is a group.* Furthermore, since the commutative property holds, *it is also an Abelian group.*

▼ Try This One 4

Does the set $\{-1, 0, 1\}$ form a group under multiplication? An Abelian group?

☑ 3. Decide if a mathematical system is a group.

Answers to Try This One

1 (a) R + U = L; a right turn followed by a U-turn results in the same direction as a left turn.

(b) U + U + S; two consecutive U-turns is the same direction as going straight.

(c) S + S = S; choosing to go straight twice leaves you in the same direction as going straight once.

(d) (U + R) + R = S; following a U-turn with a right turn, then going right again results in the same direction as going straight.

2 Closure, identity. Note that $(z \circ x) \circ x \neq z \circ (x \circ x)$.

3 Not a group; only zero has an inverse

4 No; 0 doesn't have an inverse

EXERCISE SET 13-1

Writing Exercises

1. What is a mathematical system?
2. What is an operation table for a mathematical system?
3. What does it mean for a system to be closed?
4. Explain why the positive integers are closed if the operation is addition, but not subtraction.
5. How can you quickly decide if a mathematical system is commutative by looking at the operation table for the system?
6. How can you tell if a mathematical system is a group?

Computational Exercises

For Exercises 7–21, use the elements C, D, E, and F, and the operation ? as defined by

?	C	D	E	F
C	D	F	C	E
D	F	E	D	C
E	C	D	E	F
F	E	C	F	D

7. $C ? E$
8. $F ? D$
9. $E ? E$
10. $F ? F$
11. $C ? F$
12. $(D ? E) ? D$
13. $E ? (C ? C)$
14. $F ? (D ? C)$
15. $(E ? D) ? E$
16. $C ? (D ? E)$
17. Is the system closed under the operation?
18. Is there an identity for the operation?
19. Is the operation commutative?
20. What are the inverses for each element?
21. Find the value for x when $E ? x = D$.

For Exercises 22–36, use the elements and the operation ∗ as defined by

∗	△	□	○
△	△	□	○
□	□	☆	△
○	○	△	□

22. △ ∗ ○
23. □ ∗ □
24. □ ∗ ○
25. ○ ∗ ○
26. △ ∗ △
27. △ ∗ (□ ∗ ○)
28. □ ∗ (○ ∗ △)
29. (○ ∗ ○) ∗ □
30. (□ ∗ ○) ∗ △
31. (△ ∗ △) ∗ □
32. Is the system closed under ∗?
33. Is ∗ commutative?
34. Is there an identity for the system?
35. Is ∗ associative?
36. What is the inverse of ○?

For Exercises 37–46, determine which properties the given mathematical system exhibits. Identify any systems that are groups or Abelian groups.

37.

ℵ	A	B	C
A	A	A	A
B	B	B	B
C	C	C	C

38.

Ω	1	2	3	4
1	1	2	3	4
2	2	3	4	1
3	3	4	1	2
4	4	1	2	3

39.

•	α	β	γ	δ
α	δ	β	α	γ
β	β	α	β	α
γ	α	β	γ	δ
δ	γ	α	δ	β

40.

@	7	4	3
7	4	3	7
4	3	7	4
3	7	4	3

41.

⊕	q	r	s	t	u
q	q	r	s	t	u
r	s	t	u	r	q
s	r	s	t	u	q
t	u	q	r	s	t
u	t	u	s	r	s

42.

𝒫	1	2
1	1	2
2	2	1

43.

^	a	b	c
a	c	a	b
b	a	b	c
c	b	c	c

44.

#	0	1	2
0	0	0	0
1	0	0	0
2	0	0	0

45.

Π	1	2	3
1	1	2	3
2	2	3	1
3	3	1	2

46.

◊	*P*	*Q*
P	*Q*	*P*
Q	*P*	*Q*

For exercises 47–56, determine whether the given system forms (a) a group and (b) an Abelian group.

47. Integers under addition
48. Integers under multiplication
49. Whole numbers under multiplication
50. Natural numbers under multiplication
51. Rational numbers under multiplication
52. Rational numbers under addition
53. $\{-1, 1\}$ under addition
54. $\{\frac{1}{2}, 1, 2\}$ under multiplication
55. $\{-1, 0, 1\}$ under addition
56. $\{\ldots \frac{1}{256}, \frac{1}{16}, \frac{1}{4}, \frac{1}{2}, 1, 2, 4, 16, 256 \ldots\}$ under multiplication

Real-World Applications

Exercises 57–62 use the mathematical system described next. Many ceiling fans have a four-way switch on a pull chain, with positions off, high speed, medium speed, and low speed. We can build a mathematical system to describe the positions using 0 for off, 1 for high speed, 2 for medium speed, and 3 for low speed. For example, if the fan is off and you pull the chain once, it goes to high speed, so if we use + to represent the operation of combining positions by pulling the chain, $0 + 1 = 1$. If it's on medium speed and you pull the chain three times the result is also high speed (medium → low → off → high), so $2 + 3 = 1$.

57. Fill in the operational table we've started below for the mathematical system described by the ceiling fan pull chain.

*	0	1	2	3
0		1		
1			3	
2	2			1
3		0		

58. Is the mathematical system commutative?
59. Is the mathematical system associative?
60. Is the mathematical system closed?
61. Does the mathematical system satisfy the inverse property?
62. Is the mathematical system a group?

Exercises 63–68 use the mathematical system described next. Mixing paint using the three primary colors red (R), blue (B), and yellow (Y) can be thought of as a mathematical system with the operation being mixing of two primary colors.

M	R	B	Y
R			
B			
Y			

63. Using the symbol M to represent the mixing operator, fill in the operational table for this mathematical system. (Use the symbols P for purple, O for orange, and G for green.)
64. Is the mathematical system commutative?
65. Is the mathematical system associative?
66. Is the mathematical system closed?
67. Does the mathematical system satisfy the inverse property?
68. Is the mathematical system a group?

Critical Thinking

69. Prove that the mathematical system in Example 1 on page 718 is associative by testing the associative property on all possible choices of three symbols. (*Hint:* We already know that the system is commutative, so order doesn't matter.)

A truth table similar to one shown in Chapter 3 for the conjunction is shown in Figure 13-1(a). This can be converted to a mathematical system using T and F as the elements and ∧ as the operation. This is shown in Figure 13-1(b).

p	q	$p \wedge q$
T	T	T
T	F	F
F	T	F
F	F	F

(a)

∧	T	F
T	T	F
F	F	F

(b)

Figure 13-1

70. Construct a table for a mathematical system for $p \vee q$ using ∨ as the operation.
71. Construct a table for a mathematical system for $p \rightarrow q$. What properties are valid for this system?
72. Construct a table for a mathematical system for $p \leftrightarrow q$. What properties are valid for this system?

Section 13-2 Clock Arithmetic

LEARNING OBJECTIVES

- ❑ 1. Perform addition on the 12-hour clock.
- ❑ 2. Find the equivalent of any whole number on the 12-hour clock.
- ❑ 3. Perform multiplication on the 12-hour clock.
- ❑ 4. Perform subtraction on the 12-hour clock.
- ❑ 5. Perform operations in other clock systems.
- ❑ 6. Determine the properties of the 12-hour clock system.

Have you ever stopped to think about how much control the clock has over your life? You could make a pretty strong case that it has more influence over where you go and what you do than anything else. Strange, isn't it?

Because clocks are so familiar, most of us never really think about the unusual math involved in the simple act of telling time. Everywhere else in your life, $11 + 2 = 13$. But when it comes to telling time, you accept as given that a 2-hour final exam beginning at the hour of 11 ends at 1, in which case $11 + 2 = 1$.

In this section, we will study the math involved in working with clocks, and see how it applies to the mathematical systems we studied in Section 13-2. The math used in the 12-hour clock you learned as a child is an excellent example of a mathematical system: the elements are the whole numbers from 1 through 12, and the basic operation is addition. We will also take a look at the operations of multiplication and subtraction as they apply to clocks.

Addition on the 12-Hour Clock

The answers to addition problems on the 12-hour clock can be found by counting clockwise around the face of a clock. If it is 7 o'clock now, in 8 hours, it will be 3 o'clock, as shown in Figure 13-2. We can write this operation as $7 + 8 = 3$.

Figure 13-2

Some other examples of addition on the 12-hour clock are:

$7 + 6 = 1 \qquad 6 + 11 = 5$

$8 + 7 = 3 \qquad 4 + 3 = 7$

EXAMPLE 1 Adding on the 12-Hour Clock

Using the 12-hour clock, find these sums.

(a) $9 + 12$ (b) $6 + 5$ (c) $8 + 8$ (d) $3 + 11$ (e) $5 + 7$

SOLUTION

(a) 9. Start at 9 and count 12 hours clockwise, ending at 9.
(b) 11. Start at 6 and count 5 hours clockwise, ending at 11.
(c) 4. Start at 8 and count 8 hours clockwise, ending at 4.
(d) 2. Start at 3 and count 11 hours clockwise, ending at 2.
(e) 12. Start at 5 and count 7 hours, ending at 12.

▼ Try This One 1

☑ 1. Perform addition on the 12-hour clock.

Using the 12-hour clock, find these sums.

(a) $3 + 12$ (b) $7 + 6$ (c) $11 + 5$ (d) $9 + 3$

We can make the 12-hour clock look just like the mathematical systems we studied in Section 13-1 by making an operation table for addition.

+	1	2	3	4	5	6	7	8	9	10	11	12
1	2	3	4	5	6	7	8	9	10	11	12	1
2	3	4	5	6	7	8	9	10	11	12	1	2
3	4	5	6	7	8	9	10	11	12	1	2	3
4	5	6	7	8	9	10	11	12	1	2	3	4
5	6	7	8	9	10	11	12	1	2	3	4	5
6	7	8	9	10	11	12	1	2	3	4	5	6
7	8	9	10	11	12	1	2	3	4	5	6	7
8	9	10	11	12	1	2	3	4	5	6	7	8
9	10	11	12	1	2	3	4	5	6	7	8	9
10	11	12	1	2	3	4	5	6	7	8	9	10
11	12	1	2	3	4	5	6	7	8	9	10	11
12	1	2	3	4	5	6	7	8	9	10	11	12

Multiplication on the 12-Hour Clock

Multiplication in the real number system is defined in terms of repeated addition. We will use the same idea to define multiplication in our 12-hour clock system. Let's start with an example. In order to multiply 5×3, start at 12 o'clock and count around the clock in 3-hour portions 5 times. This is equivalent to adding $3 + 3 + 3 + 3 + 3$.

$$\begin{aligned} 12 + 3 &= 3 && \textit{1 time} \\ 3 + 3 &= 6 && \textit{2 times} \\ 6 + 3 &= 9 && \textit{3 times} \\ 9 + 3 &= 12 && \textit{4 times} \\ 12 + 3 &= 3 && \textit{5 times} \end{aligned}$$

We conclude that $5 \times 3 = 3$ on the 12-hour clock.

CAUTION

The biggest stumbling block that most students face in this kind of arithmetic is ignoring years of experience that wants you to write $5 \times 3 = 15$. You have to really concentrate!

Converting Whole Numbers to the 12-Hour Clock System

Next, we'll develop an approach that makes multiplication on the 12-hour clock quicker and easier. To convert any whole number to its equivalent on the 12-hour clock, we'll take advantage of the fact that any 12-unit rotation on the clock puts you back where you started. So 24 is equivalent to 12 because 24 represents two full turns, starting and ending at 12. For a number like 43, we can break it up into portions of 12, with some left over:

$$43 = \underbrace{12 + 12 + 12}_{\textit{3 full turns}} + 7$$

This shows that 43 is equivalent to 7 on the 12-hour clock.

A simple way to arrive at this answer is to divide the original number by 12 and find the remainder:

$$\begin{array}{r} 3 \\ 12\overline{)43} \\ -\underline{36} \\ 7 \end{array}$$

Math Note

Since we have no zero on the 12-hour clock, but 12 plays the role of zero since it is our typical starting point, a remainder of zero corresponds to 12 on the 12-hour clock.

☑ 2. Find the equivalent of any whole number on the 12-hour clock.

The remainder is 7, so 43 is equivalent to 7 on the 12-hour clock.

Now we return to the multiplication problem 5×3. We can simply use regular real-number multiplication ($5 \times 3 = 15$) then convert the result to its equivalent number on the 12 hour clock: $15 \div 12 = 1R3$ (1 remainder 3), so we conclude that $5 \times 3 = 3$ on the 12-hour clock. We can think of 1R3 as "one revolution, plus 3 more hours."

EXAMPLE 2 Multiplying on the 12-Hour Clock

Calculator Guide

To convert whole numbers to their equivalent on the 12-hour clock, for example, 40:

Standard Scientific Calculator

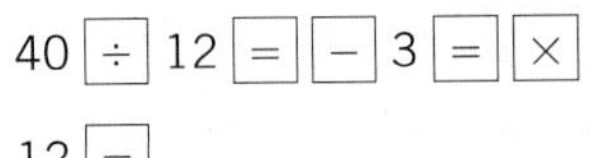

Standard Graphing Calculator

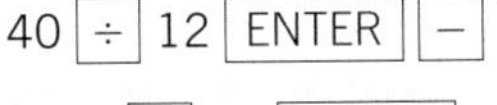

Enter × 12 ENTER.

In each case, subtracting 3 leaves behind only the decimal part of the answer, and multiplying by 12 converts the decimal part back to a remainder.

Perform these multiplications on the 12-hour clock.

(a) 5×8
(b) 12×9
(c) 6×4
(d) 11×5
(e) $10(6 + 9)$

SOLUTION

(a) $5 \times 8 = 40$ and $40 \div 12 = 3$, remainder 4: $5 \times 8 = 4$.
(b) $12 \times 9 = 108$ and $108 \div 12 = 9$, remainder 0. Since 0 corresponds to 12 on the 12-hour clock, the answer is 12: $12 \times 9 = 12$.
(c) $6 \times 4 = 24$ and $24 \div 12 = 2$, remainder 0: $6 \times 4 = 12$.
(d) $11 \times 5 = 55$ and $55 \div 12 = 4$, remainder 7: $11 \times 5 = 7$.
(e) First do the addition inside parentheses: $6 + 9 = 3$. Now multiply: $10 \times 3 = 30$ and $30 \div 12 = 2$, remainder 6. So $10(6 + 9) = 6$.

▼ Try This One 2

Perform these multiplications on the 12-hour clock.

(a) 7×5 (b) 8×9 (c) 10×10 (d) $8(9 + 5)$

☑ 3. Perform multiplication on the 12-hour clock.

A table can be constructed for multiplication in the same manner as the addition table was constructed. The multiplication table is shown here.

×	1	2	3	4	5	6	7	8	9	10	11	12
1	1	2	3	4	5	6	7	8	9	10	11	12
2	2	4	6	8	10	12	2	4	6	8	10	12
3	3	6	9	12	3	6	9	12	3	6	9	12
4	4	8	12	4	8	12	4	8	12	4	8	12
5	5	10	3	8	1	6	11	4	9	2	7	12
6	6	12	6	12	6	12	6	12	6	12	6	12
7	7	2	9	4	11	6	1	8	3	10	5	12
8	8	4	12	8	4	12	8	4	12	8	4	12
9	9	6	3	12	9	6	3	12	9	6	3	12
10	10	8	6	4	2	12	10	8	6	4	2	12
11	11	10	9	8	7	6	5	4	3	2	1	12
12	12	12	12	12	12	12	12	12	12	12	12	12

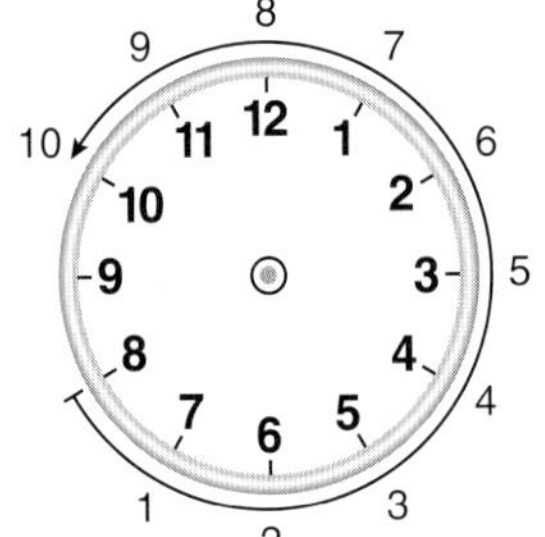

Figure 13-3

Subtraction on the 12-Hour Clock

Subtraction can be performed on the 12-hour clock by counting counterclockwise (backward). For example, $8 - 10$ means that if it is 8 o'clock now, what time was it 10 hours ago? Figure 13-3 shows that if you start at 8 and count counterclockwise, you will end at 10 o'clock; so $8 - 10 = 10$.

EXAMPLE 3 Subtracting on the 12-Hour Clock

> **Math Note**
>
> The simplest way to find the equivalent of a negative number on the 12-hour clock is to add 12 (full revolutions) until you land on a number between 1 and 12. For −14, we would get $-14 + 12 + 12 = 10$, so −14 is equivalent to 10.

Perform these subtraction operations on the 12-hour clock.

(a) $2 - 10$ (b) $12 - 7$ (c) $5 - 9$ (d) $6 - 12$ (e) $4 - 11$

SOLUTION

(a) Starting at 2 on the clock and counting backward 10 numbers, you will get 4.
(b) Starting at 12 and counting 7 numbers backward, you will get 5.
(c) Starting at 5 and counting 9 numbers backward, you will get 8.
(d) Starting at 6 and counting 12 numbers backward, you will get 6.
(e) Starting at 4 and counting 11 numbers backward, you will get 5.

☑ 4. Perform subtraction on the 12-hour clock.

▼ Try This One 3

Perform the following subtractions on the 12-hour clock.

(a) $7 - 2$
(b) $3 - 10$
(c) $9 - 12$

> **Math Note**
>
> There is one key difference between the standard 12-hour clock and clocks with a different number of hours: on all others, we will start at 0 in the 12 o'clock position to match the way it is done on the military clock. We start at 12 on the 12-hour clock so that it matches the 12-hour clock you're so familiar with using.

Clocks with Hours Different from 12

The military does not use a 12-hour clock, but a 24-hour clock. The hours on a military clock start at "zero hundred hours," then to "one hundred hours" and so on up to "23 hundred hours." These are written 0000 hours, 0100 hours, etc., up to 2300 hours. A 24-hour clock is illustrated in Figure 13-4.

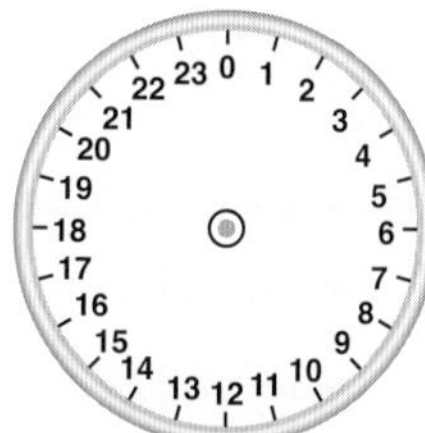

Figure 13-4

Arithmetic on the 24-hour clock can be performed just as on the 12-hour clock. If it is 0500 hours now, then 13 hours from now it will be 5 + 13 or 18 (that is, 1800 hours). If it is 0400 hours now, then 10 hours ago it was $4 - 10 = 18$ (counting counterclockwise 10 units from 4 on the 24-hour clock).

In fact, the same arithmetic that can be performed on a 12-hour clock can be performed on a clock with any number of hours. Suppose, for example, it is 4 o'clock now; what time will it be 5 hours from now on a 6-hour clock? To figure this out we can draw a 6-hour clock, then starting at 4, count 5 hours, and end up at 3 (see Figure13-5); so, on the 6-hour clock, $4 + 5 = 3$. Notice the same shortcut applies that applied to 12-hour clocks; that is, if 4 and 5 are added, the result is 9, and on a 6-hour clock this is 1 revolution plus 3, so $4 + 5 = 3$.

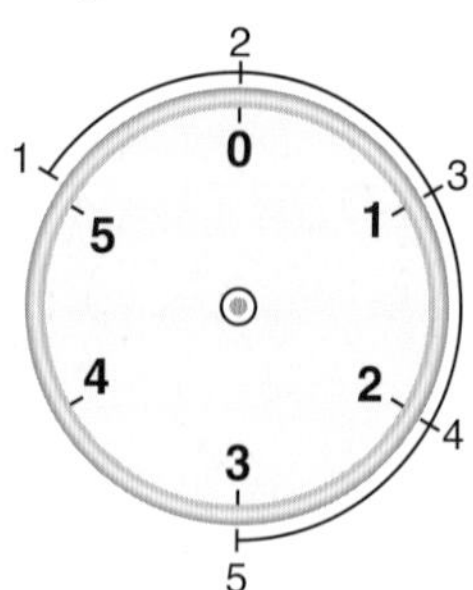

Figure 13-5

EXAMPLE 4 Performing Arithmetic on Other Clocks

Perform the indicated operations:

(a) $7 - 5$ on an 8-hour clock
(b) $5 + 9$ on a 10-hour clock
(c) 2×6 on a 5-hour clock

SOLUTION

(a) $7 - 5 = 2$; on an 8-hour clock this is still 2.
(b) $5 + 9 = 14$; on a 10-hour clock 14 is one revolution plus 4, so $5 + 9 = 4$ on a 10-hour clock.
(c) $2 \times 6 = 12$; on a 5-hour clock 12 is two revolutions plus 2, so $2 \times 6 = 2$ on a 5-hour clock.

▼ Try This One 4

☑ 5. Perform operations in other clock systems.

Perform the indicated operations:

(a) $3 - 5$ on a 7-hour clock
(b) $5 + 7$ on a 4-hour clock
(c) 5×3 on an 11-hour clock

Properties of the 12-Hour Clock System

Next, we'll determine whether the 12-hour clock system satisfies any of the properties of mathematical systems we learned about in Section 13-1.

The closure property for addition and multiplication can be verified by looking at the two operation tables shown previously. The answers for every combination of addition problems and for every combination of multiplication problems are numbers on the clock.

The commutative properties for addition and multiplication can also be verified using the operation tables: both are symmetric about the main diagonal from upper left to lower right, so the system is commutative under both operations.

To verify the associative property, we would need to try every combination of 3 symbols chosen from the 12 in our system. This would require over 200 calculations. (If the system were not commutative, it would be even worse: 1,728 calculations!) In any case, we will just try one sample calculation, for addition: beyond that

Sidelight EVARISTE GALOIS

Evariste Galois (1811–1832) was a brilliant young mathematician who lived a short life with incredible bad luck. When he was 17 years old, he submitted a manuscript on the solvability of algebraic equations to the French Academy of Sciences. Augustin-Louis Cauchy, a famous mathematician of his time, was appointed as a referee to read it. However, Cauchy apparently lost the manuscript. A year later, Galois submitted a revised version to the academy. A new referee was appointed to read it but died before he could read it, and the manuscript was lost a second time.

A year later, Galois submitted the manuscript again for a third time. After a 6-month delay, the new referee, Siméon-Denis Poisson, rejected it, saying it was too vague and recommended that Galois rewrite it in more detail.

Galois at that time was considered a dangerous political radical, and he was provoked into a duel. (Some feel that the challenger was hired by local police to eliminate him.) Galois realized that he would probably die the next morning, so he spent the night trying to revise his manuscript as well as some other papers he had written, but he did not have enough time to finish everything.

The next day, he was shot and killed. Eleven years after his death, Galois's writings were found by Joseph Liouville, who studied them, realized their importance, and published them. At last, Galois was given credit for his work, which made important advances in the study of groups. (We will see shortly that the 12-hour clock system is actually a group.)

you'll have to trust us that the system is in fact associative for both addition and multiplication.

$$(6 + 9) + 10 = 3 + 10 = 1 \qquad 6 + (9 + 10) = 6 + 7 = 1$$

The identity element for addition is 12 since adding 12 hours to any number on the clock brings you back to the same number.

A quick look at the multiplication table shows that 1 is the identity element for multiplication: looking across the row and down the column beginning with 1 shows that all elements are unchanged when multiplied by 1.

Finally, we can use the addition and multiplication tables to check the inverse property. For addition, the identity element is 12, and 12 appears once in every row of the addition table. This shows that every number has exactly one other number that produces 12 when added to it. In other words, every number in the system has an additive inverse. But this is not the case for multiplication. The identity element is 1, but 1 appears only in the rows for 5, 7, and 11. That means the other numbers do not have multiplicative inverses, and the inverse property does not hold for multiplication.

☑ 6. Determine the properties of the 12-hour clock system.

In summary, addition on the 12-hour clock satisfies all of the closure, commutative, associative, identity, and inverse properties, so the 12-hour clock with operation addition forms an Abelian group. Multiplication satisfies the closure, commutative, associative, and identity properties, but not the inverse property, so the 12-hour clock with operation multiplication is not a group.

Answers to Try This One

1 (a) 3 (b) 1 (c) 4 (d) 12

2 (a) 11 (b) 12 (c) 4 (d) 4

3 (a) 5 (b) 5 (c) 9

4 (a) 5 (b) 0 (c) 4

EXERCISE SET 13-2

Writing Exercises

1. Explain why $10 + 5 = 3$ on a 12-hour clock.
2. Describe the procedure for finding 8×10 using remainders on a 12-hour clock.
3. Write out the 12 symbols used in 12-hour clock arithmetic and the 6 symbols used in 6-hour clock arithmetic. Why is zero used for one but not the other?
4. How do we define subtraction in the 12-hour clock system?

Computational Exercises

For Exercises 5–16, find the equivalent number on (a) the 12-hour clock, (b) a 6-hour clock, and (c) an 8-hour clock.

5. 27
6. 92
7. 155
8. 334
9. 18
10. 42
11. 259
12. 3,230
13. −5
14. −10
15. −3
16. −20

For Exercises 17–28, perform the additions on the 12-hour clock.

17. $5 + 9$
18. $10 + 8$
19. $11 + 11$
20. $9 + 7$
21. $12 + 3$
22. $4 + 8$
23. $10 + 20$
24. $9 + 6$
25. $(6 + 5) + 12$
26. $8 + (10 + 9)$
27. $3 + (11 + 8)$
28. $(5 + 7) + 2$

For Exercises 29–40, perform the subtractions on the 12-hour clock.

29. $8 - 6$
30. $12 - 10$
31. $9 - 11$
32. $10 - 12$
33. $0 - 6$
34. $6 - 10$
35. $3 - 12$
36. $0 - 8$
37. $4 - 5$
38. $12 - 8$
39. $3 - 11$
40. $2 - 7$

For Exercises 41–52, perform the multiplications on the 12-hour clock.

41. 3×2
42. 10×10
43. 8×6
44. 9×7
45. 2×5
46. 12×6
47. 3×7
48. 4×5
49. $5 \times (6 \times 9)$
50. $3 \times (2 \times 9)$
51. $(6 \times 4) \times 7$
52. $(8 \times 3) \times 5$

In Exercises 53–64, perform the indicated operation on the indicated clock.

53. $7 + 3$; 8-hour clock
54. $4 + 2$; 5-hour clock
55. $9 + 11$; 10-hour clock
56. $3 + 7$; 4-hour clock
57. $9 - 4$; 10-hour clock
58. $1 - 7$; 6-hour clock
59. $3 - 5$; 5-hour clock
60. $8 - 7$; 4-hour clock
61. 10×5; 3-hour clock
62. 7×4; 13-hour clock
63. 12×11; 6-hour clock
64. 6×6; 15-hour clock

For Exercises 65–74, find the additive inverse for each number on the 12-hour clock.

65. 12
66. 3
67. 5
68. 8
69. 2
70. 9
71. 7
72. 4
73. -5
74. -6

For Exercises 75–80, find the multiplicative inverse if it exists for each number on the 12-hour clock.

75. 4
76. 7
77. 12
78. 9
79. 1
80. 10

For Exercises 81–86, give on example of each using the 12-hour clock.

81. Associative property of addition
82. Commutative property of multiplication
83. Identity property of addition
84. Inverse property of multiplication
85. Distributive property
86. Commutative property of addition

For Exercises 87–96, find the value of y using the 12-hour clock.

87. $5 + y = 3$
88. $9 + y = 2$
89. $y - 5 = 8$
90. $y + 6 = 2$
91. $4 \times (2 + y) = 4$
92. $8 \times 2 = y$
93. $6 \times 9 = y$
94. $9 \times 4 = y$
95. $3 \times (4 + 10) = y$
96. $5 \times (6 - 11) = y$

Real-World Applications

97. A computer simulation takes 3 hours to run. If it is now 11 o'clock A.M., what time will it be when the simulation has run 10 times?

98. A sorority organizes a dance marathon to raise money for charity. It starts at 8 P.M. on Saturday evening. Participants dance for 2 hours, then switch partners. If the goal is for everyone to last through eight 2-hour sessions, at what time will the marathon end?

Time in the military is based on a 24-hour clock. From midnight to noon is designated as 0000 to 1200. From noon until 1 minute before midnight, time is designated as 1200 to 2359, where the first two digits indicate the hour and the last two digits represent the minutes. For example, 1824 means 6:24 P.M. For Exercises 99–105, translate military time into standard time.

99. 0948
100. 0311
101. 0500
102. 1542
103. 1938
104. 2218
105. 2000

For Exercises 106–114, change the standard times into military times.

106. 6:56 A.M.
107. 3:52 A.M.
108. 4:00 A.M.
109. 11:56 A.M.
110. 5:27 P.M.
111. 8:06 P.M.
112. 11:42 P.M.
113. 9:36 P.M.
114. 12:00 A.M.

For Exercises 115–118, find the standard time for each.

115. 0627 + 3 hours and 42 minutes
116. 2342 + 5 hours and 6 minutes
117. 1540 − 1 hour and 4 minutes
118. 1242 − 2 hours and 20 minutes

Critical Thinking

119. Does the distributive property of multiplication over addition hold in the 12-hour clock system? That is, is $a(b + c) = a \times b + a \times c$ for any a, b, and c in the system? Try several examples and draw a conclusion.

120. How could division be defined on the 12-hour clock? (*Hint:* $8 \div 4 =$ ______ can be rewritten as $4 \times$ ______ $= 8$.) Under what conditions will your division be defined within the system?

121. Calculate 4×4, 4×7, and 4×10 on a 12-hour clock. Explain why all of the answers are the same, then use your explanation to find three other numbers b on the 12-hour clock so that $3 \times b = 3 \times 3$.

Section 13-3 Modular Systems

LEARNING OBJECTIVES

- ☐ 1. Find congruent numbers in modular systems.
- ☐ 2. Perform addition, subtraction, and multiplication in modular systems.
- ☐ 3. Solve congruences in modular systems.

For most people, daylight savings time is not that big of a deal—you move all your clocks an hour twice a year, and get an extra hour of daylight for outdoor activities in the evening during the summer. But some people are strongly opposed to the idea for one reason or another, including some logically challenged farmers, who have argued that the extra hour of sun can cause their crops to wilt during the hottest part of the year.

Hopefully, you recognize the humor in that argument: the sun rises and sets at the same time regardless of what number human beings choose to attach to those points in time. In fact, our entire system of keeping time is basically arbitrary. If there were only six hours in a day, you wouldn't have any less time to get stuff done—each "hour" would just be four times as long.

In Section 13-2, we looked very briefly at arithmetic on clocks with numbers of hours other than 12. In this section, these clocks will become our main focus as we define the study of modular arithmetic. A **modular system** is a mathematical system with a specific number of elements in which the arithmetic is analogous to the clock arithmetic from Section 13-2. For example, a modular 5 system, denoted as *mod 5,* has as its five elements 0, 1, 2, 3, 4 and uses the clock shown in Figure 13-6a. A mod 3 system consists of elements 0, 1, 2 and uses the clock shown in Figure 13-6b. A mod 8 system consists of elements 0, 1, 2, 3, 4, 5, 6, 7 and uses the clock shown in

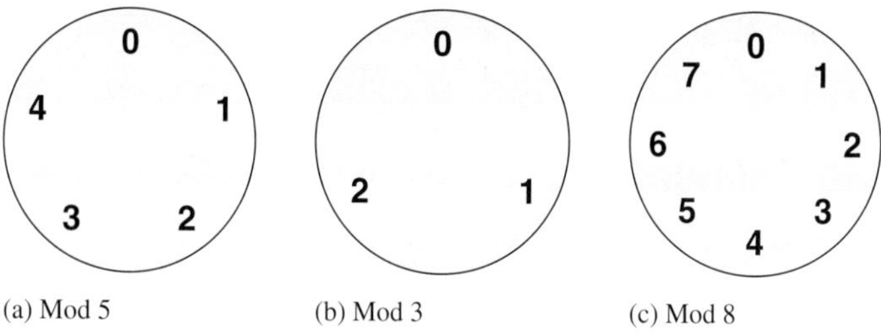

(a) Mod 5 (b) Mod 3 (c) Mod 8

Figure 13-6

Figure 13-6c. In general, a mod m system consists of m whole numbers starting at zero and ending at $m - 1$. The number m is called the **modulus** of the system.

The 12-hour clock system we studied in Section 13-2 is a system with modulus 12; the one difference is that we used the whole numbers from 1 through 12 rather than 0 through 11, but the idea is the same. One of the key things we learned how to do in Section 13-2 is convert any whole number to its equivalent in the 12-hour system. We will now focus on doing so for any modular system using the term *congruence*.

Math Note

In modular systems, 0 is used in place of the modulus. For example, in mod 5, 0 is the starting number, not 5. When we work with mod 12 systems, we will also use 0 rather than 12, making it just a bit different from the 12-hour clock system.

a* is congruent to *b* modulo *m (written $a \equiv b \bmod m$) if a and b have the same remainder when divided by m. Alternately, $a \equiv b \bmod m$ if m divides $b - a$.

To write a whole number in any given modulo system, we will divide that number by the modulus. The remainder tells us the congruent number in the modular system. This is illustrated in Example 1.

EXAMPLE 1 Finding Congruent Numbers in Modular Systems

Find the number congruent to each number in the given modular system.

(a) 19 mod 5 (b) 25 mod 3

Calculator Guide

Use these steps to find the number congruent to a given whole number in a modular system.

Step 1 Divide the number by the modulus.

Step 2 Subtract the part of the result to the left of the decimal point.

Step 3 Multiply by the modulus.

(Occasionally, because of rounding error, the result might not be a whole number. If it isn't, choose the nearest whole number.)

SOLUTION

(a) Divide 19 by 5 and find the remainder, as shown.

$$\begin{array}{r} 3\text{R}4 \\ 5\overline{)19} \\ -\underline{15} \\ 4 \end{array}$$

So 19 = 4 (mod 5). This answer can be verified by starting at 0 on the mod 5 clock and counting around 19 numbers.

(b) Divide 25 by 3 and find the remainder as shown.

$$\begin{array}{r} 8\text{R}1 \\ 3\overline{)25} \\ -\underline{24} \\ 1 \end{array}$$

So 25 = 1 (mod 3).

▼ Try This One 1

Find the number congruent to each number in the given modular system.

(a) 27 mod 5 (b) 13 mod 3 (c) 50 mod 8

To write a negative number in modulo m, you can keep adding m to the negative number until you get a non-negative number. Another way to say this is that you add the smallest multiple of the modulus that yields a nonnegative number. This is illustrated in Example 2.

EXAMPLE 2 Finding a Negative Number in Modulo *m*

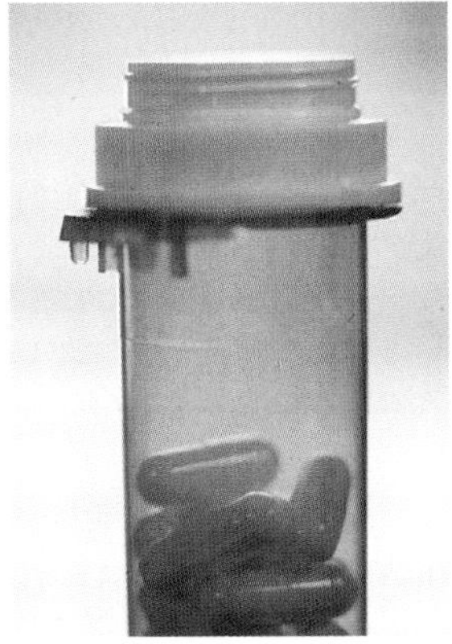

Suppose that you need to take antibiotics three times a day for 10 days. Your first pill was a morning dose. You can't remember if you took your second pill today or not. To find out, count the remaining pills and convert to mod 3: if the answer is 1, you took your pill. (The one remaining pill is your evening dose.) If the answer is 2, you missed a dose. If the answer is 0, you missed two doses.

Find the number congruent to the given number in the given modular system.
(a) −13 mod 7 (b) −1 mod 3 (c) −27 mod 10 (d) 20 mod 5

SOLUTION

(a) Multiples of 7 are 7, 14, 21, etc. The smallest multiple of 7 that can be added to −13 that results in a non-negative number is 14. Therefore, −13 is congruent to $-13 + 14 \equiv 1 \bmod 7$.

(b) $-1 + 3 = 2$, so $-1 \equiv 2 \bmod 3$

(c) $-27 + 30 = 3$, so $-27 \equiv 3 \bmod 10$ *30 is 3 × the modulus*

(d) $-20 + 20 = 0$, so, $-20 \equiv 0 \bmod 5$ *20 is 4 × the modulus*

▼ Try This One 2

Find the number congruent to the given number in the given modular system.

(a) −13 mod 6
(b) −11 mod 4
(c) −42 mod 9
(d) −36 mod 12

Adding, Subtracting, and Multiplying in Modular Systems

The operations of addition, subtraction, and multiplication in modular systems can be performed by adding, subtracting, or multiplying the numbers as usual, then

☑ 1. Find congruent numbers in modular systems.

converting the answers to equivalent numbers in the specified system like we did in Examples 1 and 2.

EXAMPLE 3 Adding and Multiplying in Modular Systems

Find the result of each operation in the given system

(a) 4×4 in mod 5
(b) $6 + 5$ in mod 7
(c) 5×6 in mod 9
(d) $9 + 7$ in mod 8

SOLUTION

(a) $4 \times 4 = 16$ and $16 \div 5 = 3$ remainder 1: $4 \times 4 \equiv 1 \text{ mod } 5$.
(b) $6 + 5 = 11$ and $11 \div 7 = 1$ remainder 4: $6 + 5 \equiv 4 \text{ mod } 7$.
(c) $5 \times 6 = 30$ and $30 \div 9 = 3$ remainder 3: $5 \times 6 \equiv 3 \text{ mod } 9$.
(d) $9 + 7 = 16$ and $16 \div 8 = 2$ remainder 0: $9 + 7 \equiv 0 \text{ mod } 8$.

▼ Try This One 3

Find the result of each operation in the given system

(a) $7 + 11$ in mod 12
(b) 3×7 in mod 8
(c) 5×5 in mod 6
(d) $15 + 12$ in mod 20

EXAMPLE 4 Subtracting in Modular Systems

Perform the following subtractions in the given modular system.
(a) $7 - 10 \text{ mod } 12$ (b) $8 - 3 \text{ mod } 4$ (c) $13 - 21 \text{ mod } 6$

SOLUTION

(a) $7 - 10 = -3$; $-3 + 12 = 9$, so $-3 \equiv 9 \text{ mod } 12$
(b) $8 - 3 = 5$; $5 \equiv 1 \text{ mod } 4$
(c) $13 - 21 = -8$; $-8 + 12 = 4$, so $-8 \equiv 4 \text{ mod } 6$

▼ Try This One 4

Perform the following subtractions in the given modular system.
(a) $13 - 5 \text{ mod } 3$ (b) $6 - 15 \text{ mod } 7$ (c) $8 - 21 \text{ mod } 8$

☑ 2. Perform addition, subtraction, and multiplication in modular systems.

Solving Congruences in Modular Systems

An equation like $x + 2 = 5$ is simple to solve in the real number system because there's only one number that will result in 5 when added to 2, namely 3. But when we use congruence in modular systems rather than equality in the real number system, the picture is more complicated. In mod 6, for example, the congruence $x + 2 \equiv 5 \text{ mod } 6$ has more than one solution. One is $x = 3$, but $x = 9$ is also a solution (because $9 + 2 = 5 \text{ mod } 6$), as is $x = 15$. In fact, we can keep getting new solutions by adding 6 to the previous one. We can use this idea to solve congruences, as we will see in Examples 5 and 6.

EXAMPLE 5 Solving a Congruence in a Modular System

A lamp with a four-way switch (off, low, medium, and high) can be represented by addition in the mod 4 system (0, 1, 2, 3). Turning the knob is like doing addition—two turns moves you up two notches, and so on. What is the inverse of 1 in the system? (That is, if the lamp is on low, how many turns does it take to turn it off?)

Find all natural number solutions to $3x - 5 \equiv 1 \bmod 6$.

SOLUTION

The first whole number that is congruent to 1 mod 6 is 1, so we begin by solving $3x - 5 = 1$.

$$3x - 5 = 1$$
$$3x = 6$$
$$x = 2$$

The next number that is congruent to 1 mod 6 is 7:

$$3x - 5 = 7$$
$$3x = 12$$
$$x = 4$$

The next number that is congruent to 1 mod 6 is 13:

$$3x - 5 = 13$$
$$3x = 18$$
$$x = 6$$

The pattern emerging is that each new solution is found by adding 2 to the previous solution. So the set of solutions is $\{2, 4, 6, 8, 10, \ldots\}$.

▼ Try This One 5

Find all natural number solutions to $2x - 7 \equiv 3 \bmod 8$.

EXAMPLE 6 Solving a Congruence in a Modular System

Find all natural number solutions to $6x \equiv 12 \bmod 3$.

SOLUTION

First, notice that 12 mod 3 is the same as 0 mod 3, so we can rewrite the congruence as $6x \equiv 0 \bmod 3$. Next, note that any multiple of 6 will have remainder 0 when dividing by 3; this means that any natural number is a solution, so the solution is $\{1, 2, 3, 4, 5, \ldots\}$.

▼ Try This One 6

Find all natural number solutions to $4x \equiv 24 \bmod 8$.

☑ 3. Solve congruences in modular systems.

In Section 13-1, we saw that a mathematical system can be defined using any symbols you like as long as you define an operation on those symbols. In Sections 13-2 and 13-3, we learned that it's possible to define new mathematical systems using the numbers you're familiar with by revising the standard operations of addition and multiplication.

Sidelight CHECK DIGITS AND MODULAR ARITHMETIC

One important application of modular arithmetic is the check digit, which is a method that allows computers to recognize whether a code number for some purpose is valid. One example is the nine-digit routing number on the bottom left of all checks. The number's purpose is to identify the financial institution, but only the first eight digits accomplish that. The last is a check digit, with a purpose of making sure that none of the previous eight digits was read incorrectly or transposed. Here's how it works.

If we use the symbol d_1 for the first digit, d_2 for the second, and so on, the nine digits are plugged into the formula

$$7d_1 + 3d_2 + 9d_3 + 7d_4 + 3d_5 + 9d_6 + 7d_7 + 3d_8 + 9d_9$$

If the result is not congruent to 0 mod 10, the computer reading the number knows there was an error. This particular formula allows computers to detect a single-digit error, as well as a switch of two consecutive digits, and most switches of two digits that are one space apart from each other. It can also detect a counterfeit routing number.

A similar formula is used for the check digit on UPC codes (see "Math in Retail Sales" in Chapter 4), and another is used by most states to make sure that drivers' license numbers are not fake. In most states, the last digit of the driver's license number is the remainder when the number without that digit is divided by some modulus, but the modulus varies from state to state, making it hard to counterfeit licenses.

Answers to Try This One

1 (a) 2 (b) 1 (c) 2

2 (a) 5 (b) 1 (c) 3 (d) 0

3 (a) 6 (b) 5 (c) 1 (d) 7

4 (a) 2 (b) 5 (c) 3

5 {5, 9, 13, 17, 21, . . .}

6 {2, 4, 6, 8, 10, . . .}

EXERCISE SET 13-3

Writing Exercises

1. Explain how modular systems apply to clock arithmetic.
2. Describe the process for finding the number congruent to 25 in the mod 7 system.
3. Describe the process for finding the number congruent to −11 in the mod 3 system.
4. Explain how the process you described in Exercise 2 is used to define addition, multiplication, and subtraction in modular systems.

Computational Exercises

For Exercises 5–14, find the values of each number in the given modular system.

5. $32 \equiv$ mod 6
6. $51 \equiv$ mod 4
7. $135 \equiv$ mod 7
8. $48 \equiv$ mod 5
9. $16 \equiv$ mod 9
10. $92 \equiv$ mod 10
11. $326 \equiv$ mod 3
12. $451 \equiv$ mod 5
13. $987 \equiv$ mod 8
14. $1{,}656 \equiv$ mod 11

For Exercises 15–44, perform the following operations in the specified modular system.

15. $4 + 3 \equiv$ mod 5
16. $8 + 6 \equiv$ mod 9
17. $3 + 3 \equiv$ mod 4
18. $5 + 6 \equiv$ mod 7
19. $5 \times 8 \equiv$ mod 9
20. $3 \times 7 \equiv$ mod 8
21. $3 \times 3 \equiv$ mod 4
22. $4 \times 6 \equiv$ mod 7
23. $3 - 8 \equiv$ mod 9
24. $5 - 7 \equiv$ mod 10

25. $2 - 3 \equiv$ mod 4
26. $1 - 9 \equiv$ mod 11
27. $(3 + 5) + 2 \equiv$ mod 7
28. $(4 + 4) + 4 \equiv$ mod 6
29. $2 + (3 + 5) \equiv$ mod 8
30. $2 + (3 + 4) \equiv$ mod 5
31. $4 \times (2 \times 3) \equiv$ mod 6
32. $(2 \times 2) \times 2 \equiv$ mod 3
33. $7 \times (3 \times 5) \equiv$ mod 9
34. $(2 \times 6) + 4 \equiv$ mod 7
35. $6 \times (2 - 5) \equiv$ mod 8
36. $5 \times (8 + 3) \equiv$ mod 9
37. $7 \times (3 - 5) \equiv$ mod 10
38. $4 \times (1 - 7) \equiv$ mod 8
39. $2 - (3 - 5) \equiv$ mod 6
40. $3 - (1 - 4) \equiv$ mod 5
41. $(4 - 7) - 3 \equiv$ mod 9
42. $(2 - 10) - 1 \equiv$ mod 11
43. $8 - (2 - 5) \equiv$ mod 12
44. $(1 - 1) - 1 \equiv$ mod 2

For Exercises 45–54, find all natural number solutions for each congruence.

45. $3 + y \equiv 1$ mod 6
46. $y + 3 \equiv 2$ mod 4
47. $1 - y \equiv 6$ mod 8
48. $3 - y \equiv 5$ mod 9
49. $7 \times y \equiv 6$ mod 8
50. $9 + y \equiv 8$ mod 10
51. $y + 4 \equiv 1$ mod 5
52. $y - (-1) \equiv 7$ mod 8
53. $4 \times y \equiv 6$ mod 7
54. $1 \times y \equiv 6$ mod 7

Real-World Applications

The days of the week can be thought of as a modular system using 0 = Sunday, 1 = Monday, 2 = Tuesday, etc. Using this system, find the answer to each and give it as the day of the week.

55. Sunday + 30 days
56. Monday + 5 days
57. Tuesday + 45 days
58. Friday + 120 days
59. Saturday + 360 days
60. Wednesday + 20 days

For Exercises 61–64, decide if the given nine-digit number is a legitimate bank routing number. (See Sidelight on page 736.)

61. 274070413
62. 741302058
63. 513939274
64. 042000314
65. Jamaal has 34 boxes of soda with 111 cans each, and he wants to break them up into 12-packs to sell. How many will be left over when he makes the 12-packs? Use modular multiplication with modulus 12 to find your answer.
66. Lenore has 44 boxes containing 89 eggs, each delivered fresh from the farm, and she wants to break them up into 6-egg cartons to sell. How many will be left over when she makes the 6-egg cartons? Use modular multiplication with modulus 6 to find your answer.
67. Business days are generally considered to be Monday through Friday of any given week, and can be thought of as the modular system 0 = Monday, 1 = Tuesday, 2 = Wednesday, 3 = Thursday, 4 = Friday. If a transaction will take 32 business days and starts on a Wednesday, on what day will the transaction be completed?
68. It is currently 11 A.M. and the concert you've been waiting to see is just 83 hours from now. At what time does the concert start?
69. Janice has been 21 for exactly 430 hours. If she turned 21 at midnight on April 5, what is today's date, and what time is it?
70. This coming Friday, Macy's will be holding a 100-hour sale that starts at 6 A.M. On what day and time will the sale end?
71. Sven is sailing west when a storm hits and his boat is rotated 1,890 degrees counterclockwise. In which direction does he face after the rotation? (Recall that 360 degrees is one full rotation.)
72. On a game board, there is a spinner that has spaces marked with the numbers 0 through 8 in order. The pointer is on the 3 and Shira spins it so that it moves 60 spaces and lands exactly on a number. What number does the spinner land on?
73. A truck delivers 50 laptops to be set up at a job fair for the employers to use. Each table can accommodate three employers with laptops. How many tables are needed and how many empty spaces will there be?
74. A shipment of 1,345 textbooks of the same size were delivered and a shelf can only hold 60 books. How many books will be on the shelf that is not fully stacked?
75. A machine that dispenses lemonade can hold up to 1,200 fluid ounces. Each cup that the campus café uses can hold 14 fluid ounces of lemonade. How many fluid ounces of lemonade will the last customer get?
76. A fortune teller told Mary that she would be married in 59 months, and she visited the fortune teller in March of 2009. For what month and year should she book a reception hall?

Critical Thinking

77. Look back at the mathematical system of possible turns at an intersection on page 718. Can you turn the system into a modular system by assigning numbers to the turns? Explain why you can or cannot.
78. Consider the congruence $x^2 \equiv 4$ mod 5.
    (a) Check that $x = 2$ and $x = -2$ are solutions.
    (b) Find the next number greater than 4 that is congruent to 4 mod 5 and find two more natural number solutions.
    (c) Make a conjecture about how many other integer solutions there are, and find the next two pairs of solutions.
79. Fighting a nasty upper respiratory infection, you are supposed to take an antibiotic capsule four times a day until the bottle is empty, starting on Wednesday morning. As you might expect, the doctor prescribed a number of pills that is a multiple of 4. The following Monday afternoon, you can't remember if you took your second dose or not. You empty the bottle and find the number of pills is congruent to 3 modulo 4. Did you remember your pill?

CHAPTER 13 Summary

Section	Important Terms	Important Ideas
13-1	Mathematical system Infinite system Finite system Operation table Closed system Identity element Inverse element Group Abelian group	**A mathematical system** consists of a nonempty set of elements and at least one operaton. A finite mathematical system has a specific number of elements, whereas an infinite mathematical system has an unlimited number of elements. A mathematical system is called a group if it is closed under the operation, has an identity element, satisfies the associative property for the operation, and every element has an inverse. If the operation is commutative, then the group is called an Abelian group. It is possible to create mathematical systems using elements other than numbers. Some of these systems are shown in this section.
13-2	Clock arithmetic	**We can define** a finite mathematical system with 12 elements using the numbers on a 12-hour clock, and defining addition to correspond to adding times on the clock. For example, $11 + 3 = 2$ on a 12-hour clock. We can then define multiplication and subtraction on the 12-hour clock. The 12-hour clock system using addition is an Abelian group, but the system with multiplication is not a group. We can also define similar systems using clocks with numbers of hours other than 12.
13-3	Modular system Modulus Congruent modulo m	**A modular system** with modulus m is a system consisting of m numbers in which arithmetic is defined analogously to the clock arithmetic studied in Section 13-2. In any modular system, we can find a number congruent to any other integer with the given modulus. For example, 5 is congruent to 2 modulo 3 because starting at zero and counting 5 units on a 3-hour clock will result in an ending position of 2. Congruence statements containing variables can be solved in a manner similar to solving equations in the real number system.

MATH IN Encryption REVISITED

1. The tricky part is that "undoing" congruence produces many possible answers. For example, the first number in the string (27) is a congruence modulo 32. It could be 27, 59, 81, 113, etc. The key is that the original numbers were all between 1 and 27, so when 50 was added, they ended up between 51 and 77. Of all the numbers congruent to 27 modulo 32, only one of them, 59, is between 51 and 77. Using this idea, the other numbers can be translated to the string

59 54 55 64 70 59 70 75 77 70 58
55 56 70 77 69 70 59 64 61 69

It's then easy to subtract 50 from each number and translate back into letters, giving us the message "Identity theft stinks."

2. Our friend is going to have some explaining to do. When you drop the last digit, the resulting number is 041103941; divide by 29 and subtract off the part to the left of the decimal place. The result is 0.2758621. Multiply by 29 to convert to the remainder: with a stight rounding error, the result is 8, which should be the check digit of the license number. Uh oh!

Review Exercises

Section 13-1

For Exercises 1–11, use the elements A, B, C and operation @ defined by

@	A	B	C
A	C	A	B
B	A	B	C
C	B	C	A

1. $C@B =$
2. $A@C =$
3. $B@B =$
4. $A@(B@C) =$
5. $(C@B)@B =$
6. $B@(C@C) =$
7. Is the system closed under operation @?
8. Is the system commutative?
9. Is the system associative?
10. Is there an identity element? If so what is it?
11. Does the system have the inverse property? If so, name the inverses of *A*, *B*, and *C*.

*For Exercises 12–16, use the elements %, *, &, and # and the operation / defined by*

/	%	*	&	#
%	&	*	%	%
*	%	*	&	#
&	#	&	*	%
#	*	#	#	&

12. Is the system closed under operation /?
13. Is the system commutative?
14. Is the system associative?
15. Is there an identity element? If so what is it?
16. Does every element have an inverse? If so what are the inverses?

For Exercises 17–20, determine if the given system forms (a) a group and (b) an Abelian group.

17. Integers under addition.
18. Rational numbers under multiplication.
19. $\{\frac{1}{2}, 2\}$ under multiplication.
20. $\{-1, 0, 1\}$ under multiplication.

Section 13-2

For Exercises 21–26, perform the indicated operation in the 12-hour clock system.

21. $4 + 11$
22. $4 - 11$
23. 3×8
24. 2×6
25. $5(8 + 10)$
26. $3(2 - 5)$

For Exercises 27–32, perform the indicated operation on the specified clock.

27. $3 - 6$; 5-hour clock
28. $8 + 10$; 11-hour clock
29. 3×7; 15-hour clock
30. $12 + 3$; 10-hour clock
31. $19 - 26$; 24-hour clock
32. $12 + 15$; 16-hour clock
33. At Miami University, final exam week for Monday through Friday classes begins at 8:00 A.M. the Monday after classes end, and lasts for 103 hours. When does the last exam end?
34. The U.S. military uses a 24-hour clock; the period from midnight to noon is 0000 to 1200, and the period from noon to 1 minute before midnight is 1200 to 2359, where the first two digits are the hour and the second two are the minutes. If a military exercise begins at 0720 and lasts for 21 hours, when does it end? If a flight is scheduled to leave Guantanamo Bay, Cuba at 2320 and land in Washington, D.C., at 0240, how long does it last?

Section 13-3

For Exercises 35–54, find the equivalent number for the given modular system.

35. $67 \equiv \text{mod } 5$
36. $41 \equiv \text{mod } 3$
37. $532 \equiv \text{mod } 8$
38. $861 \equiv \text{mod } 6$
39. $22 \equiv \text{mod } 4$
40. $10 \equiv \text{mod } 2$
41. $37 \equiv \text{mod } 10$
42. $999 \equiv \text{mod } 7$
43. $56 \equiv \text{mod } 9$
44. $80 \equiv \text{mod } 5$
45. $173 \equiv \text{mod } 9$
46. $45 \equiv \text{mod } 7$
47. $250 \equiv \text{mod } 10$
48. $64 \equiv \text{mod } 3$
49. $18 \equiv \text{mod } 3$
50. $1{,}235 \equiv \text{mod } 6$
51. $4{,}721 \equiv \text{mod } 8$
52. $856 \equiv \text{mod } 11$
53. $1{,}000 \equiv \text{mod } 12$
54. $25 \equiv \text{mod } 4$

For Exercises 55–74, perform the indicated operation for the given modular system.

55. $5 + 9 \equiv \text{mod } 11$
56. $2 - 10 \equiv \text{mod } 12$
57. $6 \times 6 \equiv \text{mod } 7$
58. $7 + 8 \equiv \text{mod } 9$
59. $3 - 7 \equiv \text{mod } 8$
60. $4 \times 5 \equiv \text{mod } 6$
61. $3 + 2 \equiv \text{mod } 4$
62. $5 - 12 \equiv \text{mod } 13$
63. $6 \times 7 \equiv \text{mod } 10$
64. $10 \times 10 \equiv \text{mod } 12$
65. $3 - 4 \equiv \text{mod } 5$
66. $5 \times 5 \equiv \text{mod } 6$
67. $5 \times (3 + 7) \equiv \text{mod } 8$
68. $2 \times (2 + 9) \equiv \text{mod } 12$
69. $3 - (3 - 5) \equiv \text{mod } 6$
70. $(10 - 6) - 9 \equiv \text{mod } 11$
71. $5 \times (7 - 9) \equiv \text{mod } 12$
72. $8 + 8 + 8 \equiv \text{mod } 10$
73. $4 \times 3 \times 5 \equiv \text{mod } 9$
74. $3 \times (4 + 5) \equiv \text{mod } 7$

For Exercises 75–84, find all natural number solutions.

75. $6 + y \equiv 2 \text{ mod } 8$
76. $y + 7 \equiv 1 \text{ mod } 10$
77. $y + 7 \equiv 1 \text{ mod } 9$
78. $3 - y \equiv 6 \text{ mod } 8$
79. $y - 2 \equiv 5 \text{ mod } 6$
80. $3 \times y \equiv 6 \text{ mod } 8$
81. $y + 2 \equiv 1 \text{ mod } 12$
82. $5 - y \equiv 6 \text{ mod } 9$
83. $3 \times 5 \equiv y \text{ mod } 7$
84. $5 \times (2 + y) \equiv 1 \text{ mod } 12$

85. A bakery has 220 cookies left over from a street fair, and the owners decide to give them away in packs of six to customers until they run out. How many cookies will be in the last package?
86. The owner of an eBay store gets a great deal on 100 cans of tennis balls with three balls in each can, and plans to repackage them for resale, with five balls in each package. Will there be any tennis balls left over? If so, how many?
87. Suppose that one state uses a 10-digit driver's license number, but the 10th digit is a check digit. The number formed by the first nine digits should be congruent to the tenth digit modulo 23. A college student presents a license with the number 0245601183 on it to a bouncer. Is the number legitimate?

Chapter Test

For Exercises 1–8, use the system shown.

$*$	x	y	z
x	x	y	z
y	y	x	z
z	z	z	s

1. $x * z$
2. $(y * x) * z$
3. $z * z$
4. $x * (z * y)$
5. $(z * x) * x$
6. What is the inverse of y?
7. Is the operation $*$ commutative?
8. Is the system closed?

For Exercises 9–24, use this system:

$\cdot$	a	-1	$-a$	1
a	-1	$-a$	1	a
-1	$-a$	1	a	-1
$-a$	1	a	-1	$-a$
1	a	-1	$-a$	1

9. $a \cdot a$
10. $-1 \cdot a$
11. $a \cdot 1$
12. $a \cdot (a \cdot a)$
13. $(-a \cdot a) \cdot (-1)$
14. $(1 \cdot 1) \cdot (-a)$
15. $(-a)^3$
16. Find the value of y when $a \cdot y = 1$.
17. Find the value of y when $a \cdot (y \cdot a) = 1$.
18. Is $(-a \cdot 1) \cdot a = -a \cdot (1 \cdot a)$?
19. Is $a^3 = a^7$?
20. Is the system closed under $\cdot$?
21. Is the system commutative?
22. What is the identity for the system?
23. What is the inverse of a?
24. What is the inverse of -1?

For Exercises 25–27, perform the indicated operation on the 12-hour clock.

25. $6 + 10$
26. 8×5
27. $7(4 - 12)$

For Exercises 28–30, perform the indicated operation on the given clock.

28. $2 + 5$ on a 4-hour clock
29. 8×9 on a 10-hour clock
30. $7(2 - 5)$ on an 8-hour clock

For Exercises 31–34, find the equivalent number for the given modular system.

31. $43 \equiv$ mod 6
32. $518 \equiv$ mod 3
33. $-6 \equiv$ mod 4
34. $-15 \equiv$ mod 5

For Exercises 35–40, perform the indicated operation for the given modular system.

35. $8 + 6 \equiv$ mod 10
36. $3 + 7 \equiv$ mod 9
37. $4 - 6 \equiv$ mod 7
38. $8 - 10 \equiv$ mod 12
39. $5 \times 9 \equiv$ mod 11
40. $4 \times 7 \equiv$ mod 10

For Exercises 41–44, find all natural number solutions for each congruence.

41. $2 + y \equiv 4 \bmod 6$
42. $y + 8 \equiv 6 \bmod 10$
43. $3 \times y \equiv 0 \bmod 12$
44. $3 \times y \equiv 0 \bmod 5$
45. In March 2009, Philips Electric announced a new LED light bulb that lasts for 46,000 hours of use. If that claim is exactly accurate, and you turn one on at noon and leave it on until it burns out, what time of day will that happen?
46. An entire baseball league is signed up for an out-of-town tournament. If there are eight teams with 16 players plus two coaches each, and the league rents buses that will hold 38 people, how many people will be on the bus that isn't full?
47. When a drill sergeant catches Private Pyle sending text messages to his girlfriend during a training exercise, he decides that it would be a good idea for the private to spend 30 straight hours peeling potatoes. If he starts peeling at 2 P.M., at what time can he stop?
48. According to the information in the sidelight on page 736, is 121144612 a legitimate bank routing number?

Projects

1. In the Sidelight on page 736, we learned about a formula that is used to find the check digit for bank routing numbers. Two very common coding systems that also use check digits are UPC numbers on the products we buy and the ISBN number that identifies all books published. (ISBNs can be found on the publisher information page at the beginning of books.) Use the Internet as a resource to find the formulas that are used for the check digits in both UPC and ISBN numbers. Write a brief paper on your findings, then write down the UPC from five products you have in your house/apartment/dorm room and use the formula to verify the check digit. Then do the same for the ISBN on all of the textbooks you currently own or use.
2. A mathematical system can be defined using geometry by rotating a five-pointed star about its center point (as in Figure 13-7). A clockwise rotation of 72° will move one of the points labeled with letters to the next one. We will define an operation on the symbols *A*, *B*, *C*, *D* and *E* using the symbol Ⓡ as follows: to find *B* Ⓡ *C*, start at position *B*, then rotate the number of degrees that correspond to position *C*. In Figure 13-7, we see that this is 144°. The rotation will move *B* two positions, to position *D*, so *B* Ⓡ *C* = *D*. To find *C* Ⓡ *E*, start at position *C* and rotate the star 288°, which is the rotation corresponding to position *E*. This moves *C* four spots to position *B*, so *C* Ⓡ *E* = *B*.

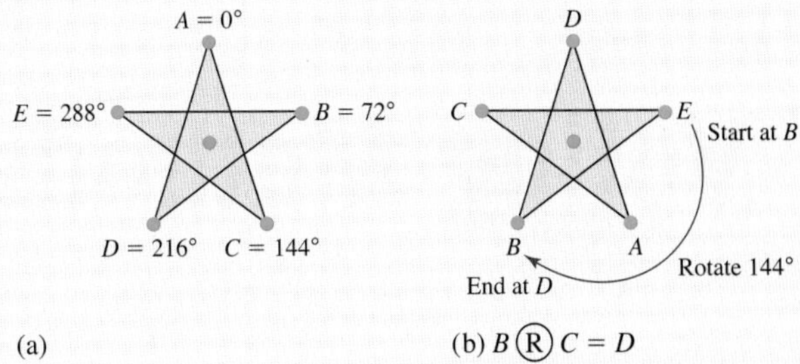

Figure 13-7

Answer these questions:

(a) Perform each operation.
   (i) *B* Ⓡ *B*
   (ii) *C* Ⓡ *A*
   (iii) *D* Ⓡ *E*
   (iv) *E* Ⓡ *C*
   (v) *A* Ⓡ *B*
   (vi) *C* Ⓡ *B*

(b) Construct a table for the operation.

(c) Is the system commutative?

(d) Is there an identity? If so, what is it?

(e) What is the inverse of *B*?

(f) What is the inverse of *C*?

(g) Is the system closed?

(h) The modular system with modulus 5 has the same number of symbols as this system. Are the two systems really the same system with different labeling? Explain.

CHAPTER 14

Voting Methods

Outline

14-1 Preference Tables and the Plurality Method

14-2 The Borda Count Method and the Plurality-with-Elimination Method

14-3 The Pairwise Comparison Method and Approval Voting

14-4 Apportionment

14-5 Apportionment Flaws

Summary

MATH IN College Football

Football and other sports have been used as examples dozens of times throughout this book because sports present excellent examples of a variety of ways that math gets used in our world. There are the obvious ways—keeping score and adding up statistics like yards gained or home runs hit. But there are also many behind-the-scenes examples of the importance of math in sports. Allocation of salaries, ticket prices, devising a schedule that meets the needs of every team in a league, assigning officials to work games . . . these are just four of many such examples.

The mathematics of voting, which we will study in Chapter 14, plays a huge role in college football. Strange, isn't it? Every game has a winner and a loser, so why should voting matter? It turns out that the teams selected to play for the national championship at the end of the season are chosen based on polls voted on by coaches, athletic directors, and members of the media. For championship contenders, the season is really one long election!

Another place where voting is important is in the awarding of the Heisman trophy, which many feel is the most prestigious individual award in amateur sports. It is given each year to the top college football player in America as determined by a voting process that we will study in this chapter.

On the surface, voting seems like a very straightforward and simple process: everyone casts a vote for his or her favorite candidate and whoever gets the most votes wins. But as we will see, there are flaws in any voting system, so a variety of systems have been developed, depending on the needs of the organization holding the election. In this chapter, we will study five voting methods in depth and examine the strengths and weaknesses of each.

We will then turn our attention to the process of apportionment. This is the process that is used to allocate representatives from a legislative body, like a city council or the U.S. House of Representatives, to different regions. We will see that the ideas can also be used for other practical problems of dividing resources.

By the time you finish this chapter, you should be able to answer the following questions about the 2008 Heisman trophy balloting. A total of 902 ballots were returned, and the results for the top five candidates are summarized in the table below. The voting is done using a modified version of the Borda count method, which you will study in Section 14-2. Voters list their top three candidates in order. Any candidate not listed on a ballot gets 0 points; the candidates listed first, second, and third get 3 points, 2 points, and 1 point respectively. The candidate with the most points wins the award.

Player	School	First	Second	Third
Tim Tebow	**Florida**	309	207	234
Sam Bradford	**Oklahoma**	300	315	196
Colt McCoy	**Texas**	266	288	230
Graham Harrell	**Texas Tech**	13	44	86
Michael Crabtree	**Texas Tech**	3	27	53

1. Who won the award?
2. Is the winner the same if the plurality method is used?
3. Discuss whether or not you think any of the fairness criteria in this chapter were violated by the election.

For answers, see Math in College Football Revisited on page 785

Section 14-1 Preference Tables and the Plurality Method

LEARNING OBJECTIVES

- ☐ 1. Interpret the information in a preference table.
- ☐ 2. Determine the winner of an election using the plurality method.
- ☐ 3. Decide if an election violates the head-to-head comparison criterion.

Voting seems like such a simple idea: two candidates both want a position, and whichever one gets the most votes wins. But like most things in the real world, elections rarely turn out to be as simple as they appear. The most obvious complication arises when there are more than two candidates. Should the winner just be the one who gets the most votes, even if less than half of the voters want him or her in office? Maybe voters should rank the candidate in order of preference . . . but then how do we decide on the winner?

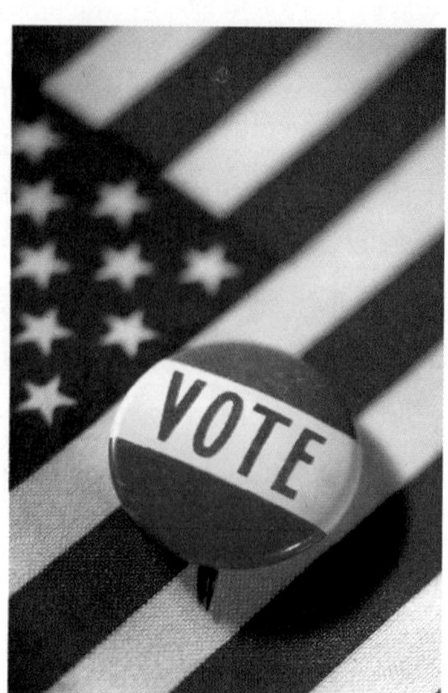

We will begin our study of voting methods by examining a method for summarizing the results when candidates are ranked in order of preference by voters. We will then study the simplest of the methods for determining the winner of an election, and begin a study of the weaknesses inherent in different voting systems.

Preference Tables

Suppose there are three candidates running for club president. We'll call them A, B, and C. Now, instead of voting for the single candidate of your choice, you are asked to rank each candidate in order of preference. This type of ballot is called a *preference ballot.*

In this case, there are six possible ways to rank the candidates, as shown.

First choice	A	A	B	B	C	C
Second choice	B	C	A	C	A	B
Third choice	C	B	C	A	B	A

Now, suppose that the 20 club members voted as follows.

A	B	A	A	A	A	A	B	A	A	B	A	A	A
B	C	B	C	B	C	B	C	B	B	C	B	C	C
C	A	C	B	C	B	C	A	C	C	A	C	B	B

A	C	C	A	A	B
C	B	B	B	B	C
B	A	A	C	C	A

Of the 6 possible rankings, only 4 appear in the 20 ballots. Nine people voted for the candidates in order of preference ABC, five people voted ACB, four people voted BCA, and two people voted CBA.

A **preference table** can be made showing the results.

Number of voters	9	5	4	2
First choice	A	A	B	C
Second choice	B	C	C	B
Third choice	C	B	A	A

The sum of the numbers in the top row indicates the total number of voters. Also note that 9 + 5 or 14 voters selected candidate A as their first choice, 4 selected candidate B as their first choice, and 2 voters selected candidate C as their first choice.

Because no voters cast ballots ranking candidates as BAC or CAB, those possible rankings are not listed as columns in the table.

EXAMPLE 1 Interpreting a Preference Table

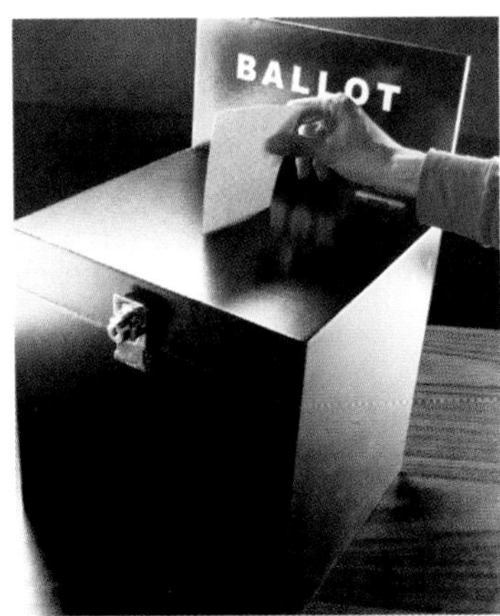

Four candidates, W, X, Y, and Z, are running for student government president. The students were asked to rank all candidates in order of preference. The results of the election are shown in the preference table.

Number of voters	86	42	19	13	40
First choice	X	W	Y	X	Y
Second choice	W	Z	Z	Z	X
Third choice	Y	X	X	W	Z
Fourth choice	Z	Y	W	Y	W

(a) How many students voted?
(b) How many people voted for candidates in the order Y, Z, X, W?
(c) How many students selected candidate Y as their first choice?
(d) How many students selected candidate W as their first choice?

SOLUTION

(a) To find the total number of voters, find the sum of the numbers in the top row.

$$86 + 42 + 19 + 13 + 40 = 200$$

(b) The ranking Y, Z, X, W is in the third column of the table, which is headed by the number 19. This means that 19 voters chose that order.

(c) There were 19 voters who chose Y, Z, X, W (third column) and 40 that chose Y, X, Z, W (fifth column), and those are the only rankings with Y listed first. So $19 + 40 = 59$ voters listed candidate Y first.

(d) Only one ranking order has candidate W first—the one in the second column. There were 42 voters who submitted that order, so 42 people chose candidate W as their first choice.

▼ Try This One 1

☑ 1. Interpret the information in a preference table.

The Student Activities Committee at Camden College is choosing a location for an end-of-year banquet, and they ask all members to list the four possible locations in order of preference. The choices are Airport Restaurant (A), Bob's Bar and Grill (B), The Crab Shack (C), and Dino's (D). The results are shown in the preference table.

Number of voters	19	13	12	9	4	2
First choice	C	B	C	C	A	B
Second choice	B	C	A	B	C	A
Third choice	A	D	B	D	D	D
Fourth choice	D	A	D	A	B	C

(a) How many members voted?
(b) How many members listed The Crab Shack as their first choice?
(c) How many members listed The Crab Shack and Bob's Bar and Grill in their top two?

Math Note

Plurality does not necessarily mean majority; it simply means more votes than any other candidate receives. Majority means more than 50% of the votes cast.

In the remainder of this section, and in Sections 14-2 and 14-3, we will study four common voting methods.

The Plurality Method

The simplest method of determining a winner in an election with three or more candidates is called the *plurality method.*

In an election with three or more candidates that uses the **plurality method** to determine a winner, the candidate with the most first-place votes is the winner.

EXAMPLE 2 Using the Plurality Method

The preference table for a club presidential election consisting of three candidates is shown. Using the plurality method, determine the winner.

Number of votes	4	7	5	4
First choice	B	A	C	B
Second choice	C	C	A	A
Third choice	A	B	B	C

SOLUTION

In this situation, only the first-place votes for each candidate are considered. Candidate A received 7 first-place votes (column 2). Candidate B received 4 + 4 or 8 first-place votes (columns 1 and 4). Candidate C received 5 first-place votes (column 3). Candidate B is the winner since that candidate received the most first-place votes.

▼ Try This One 2

☑ 2. Determine the winner of an election using the plurality method.

An election was held for the chairperson of the Psychology Department. There were three candidates: Professor Jones (J), Professor Kline (K), and Professor Lane (L). The preference table for the ballot is shown.

Number of votes	2	4	1	3
First choice	L	J	K	L
Second choice	J	K	L	K
Third choice	K	L	J	J

Who won the election if the plurality method of voting was used?

In Example 2, the top row consists of the number of voters who ranked the candidates in the order shown in the column. Instead of numbers in the top row, percents can also be used. In order to change the counts to percents, divide the count by the total number of ballots cast and multiply by 100. Using the preference table shown in Example 2, the percent of voters ranking the candidates as BCA is $\frac{4}{20} \times 100\% = 20\%$. The percent of voters ranking the candidates as ACB is $\frac{7}{20} \times 100\% = 35\%$. The percent of voters ranking the candidates as CAB is $\frac{5}{20} \times 100\% = 25\%$. The percent of voters ranking the candidates as BAC is $\frac{4}{20} \times 100\% = 20\%$. So we can write the preference table as

Percent of votes	20%	35%	25%	20%
First choice	B	A	C	B
Second choice	C	C	A	A
Third choice	A	B	B	C

Sidelight DOES HISTORY REPEAT ITSELF?

It seems that the United States' presidents who were elected in a year with "0" at the end and in increments of 20 years have shared some sad coincidences.

1840: William Henry Harrison died in office.

1860: Abraham Lincoln was assassinated.

1880: James A. Garfield was assassinated.

1900: William McKinley died in office.

1920: Warren G. Harding died in office.

1940: Franklin D. Roosevelt died in office.

1960: John F. Kennedy was assassinated.

1980: Ronald Reagan survived an assassination attempt.

George W. Bush, elected in 2000, survived 8 years in
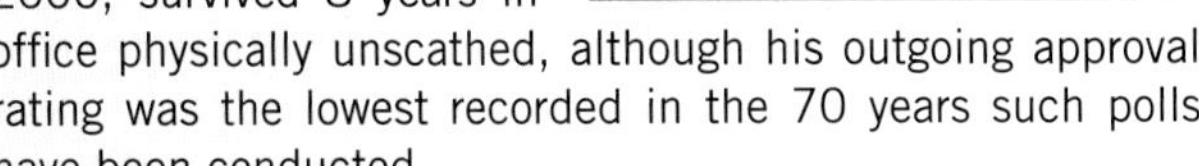
office physically unscathed, although his outgoing approval rating was the lowest recorded in the 70 years such polls have been conducted.

When we have percentages, it's easy to draw a pie chart that represents the percentage of voters that chose each candidate first (see Figure 14-1).

Percentage of First Place Votes

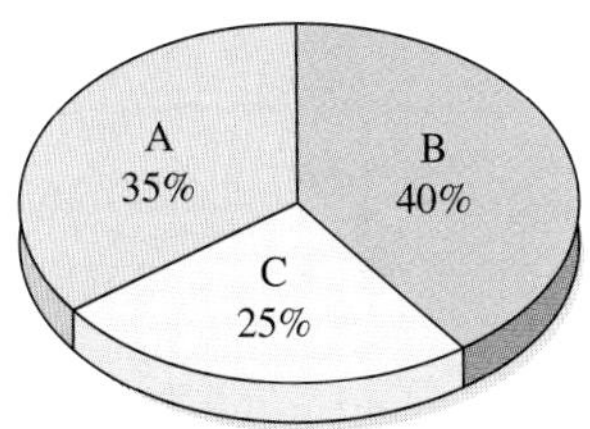

Figure 14-1

The plurality method is a simple way to determine the winner of an election, but it has some flaws. First, some would suggest that a candidate shouldn't win an election if less than half of the voters choose him or her. As we see from Figure 14-1, candidate B wins the election even though less than 44% of the ballots listed him or her first. Second, the possibility of a tie exists, and is greater when there are fewer voters. Third, the method completely ignores information about voters' preferences except for their first-place vote. Fourth, this method can sometimes violate what is called the *head-to-head comparison criterion.*

A *criterion* is a way of measuring or evaluating a situation. In this chapter, we will discuss various criteria for assessing the fairness of voting systems. The first of these is the head-to-head comparison criterion.

The **head-to-head comparison criterion** states that if a particular candidate wins all head-to-head comparisons with all other candidates, then that candidate should win the election.

In Example 3, we will illustrate the head-to-head criterion by seeing if the election in Example 2 violates it.

EXAMPLE 3 The Head-to-Head Comparison Criterion

Does the election in Example 2 violate the head-to-head comparison criterion?

SOLUTION

The idea is to compare all combinations of two candidates at a time to see which is preferred in a head-to-head matchup without the third candidate involved.

The preference table for the club president's election is reprinted here for reference.

Number of votes	4	7	5	4
First choice	B	A	C	B
Second choice	C	C	A	A
Third choice	A	B	B	C

First, compare A with B:

In column 1, B was preferred over A by 4 voters.
In column 2, A was preferred over B by 7 voters.
In column 3, A was preferred over B by 5 voters.
In column 4, B was preferred over A by 4 voters.

So, A was preferred over B by $7 + 5 = 12$ voters; B was preferred over A by $4 + 4 = 8$ voters, meaning in a head-to-head comparison, A wins over B.

Next, compare A with C:

In column 1, C was preferred over A by 4 voters.

In column 2, A was preferred over C by 7 voters.

In column 3, C was preferred over A by 5 voters.

In column 4, A was preferred over C by 4 voters.

So, A was preferred over C by $7 + 4 = 11$ voters; C was preferred over A by $4 + 5 = 9$ voters, meaning in a head-to-head comparison, A wins over C.

Without even comparing B and C, we can see that the head-to-head comparison criterion is violated: candidate A defeats both B and C head-to-head, yet B won the election using the plurality method. (The head-to-head criterion says that any candidate who defeats all opponents should win the election.)

CAUTION

The head-to-head comparison criterion doesn't say that the winner of an election has to defeat every opponent head-to-head. It says that if there is a candidate that *does* defeat all others head-to-head, that candidate should win the election.

▼ Try This One 3

☑ 3. Decide if an election violates the head-to-head comparison criterion.

Does the election in Try This One 2 violate the head-to-head comparison criterion?

The result of Example 3 shows that the plurality method doesn't always satisfy the head-to-head comparison criterion. This is not to say that *every* election conducted by the plurality method violates the head-to-head criterion. We have simply found that *some* do, so we say the method in general doesn't meet the criterion.

The head-to-head criterion is called a *fairness criterion.* It is one of four fairness criteria that we will study in this chapter. Political scientists have come to agree that a truly fair voting system should satisfy these four criteria.

Answers to Try This One

1 (a) 59 (b) 40 (c) 41

2 Professor Lane

3 No. Head-to-head, J wins over K, L wins over J, and K ties L, so there is no candidate who defeats all others head-to-head.

EXERCISE SET 14-1

Writing Exercises

1. Explain how a preference table is used to display voting results.
2. Explain how the winner of an election is determined using the plurality voting method.
3. What is the difference between a plurality and a majority?
4. Explain the head-to-head comparison criterion.

Real-World Applications

5. The preference ballots for the election of a CEO by the board of directors are shown. Make a preference table for the results of the election and answer each question.
   (a) How many people voted?
   (b) How many people voted for the candidates in the order of preference XZY?
   (c) How many people voted for candidate Y as their first choice?
   (d) Using the plurality method, determine the winner of the election.

X	X	Y	Z	X	Y	Z	Z	X	Y	X	Y	X	Y
Y	Z	Z	Y	Y	Z	Y	Y	Y	Z	Z	Z	Y	Z
Z	Y	X	X	Z	X	X	X	Z	X	Y	X	Z	X

X	Y	Z	Z	X	X	Y	Y
Z	Z	Y	Y	Y	Z	Z	Z
Y	X	X	X	Z	Y	X	X

6. The Tube City Talkers Club held its annual speech contest. The preference ballots for the best speaker are shown. The candidates were Cortez (C), Lee (L), and Smith (S). Make a preference table for the results of the election and answer each question.
   (a) How many people voted?
   (b) How many people voted for the candidates in the order of preference CLS?
   (c) How many people voted for Lee for first place?
   (d) Using the plurality method, determine the winner of the election.

C	S	S	L	L	C	S	S	L	L	C	C	S	S
L	C	C	S	S	L	C	C	S	S	L	L	C	C
S	L	L	C	C	S	L	L	C	C	S	S	L	L

C	L	S	L	L
L	S	C	S	S
S	C	L	C	C

7. The preference ballots of the board of directors for the selection of a city in which to hold the next National Mathematics Instructors' Association Conference are shown. The three cities under consideration are Chicago (C), Philadelphia (P), and Miami (M). Make a preference table for the results of the election and answer each question.
   (a) How many people voted?
   (b) How many people voted for the candidates in the order of preference PMC?
   (c) How many people voted for Chicago as their first choice?
   (d) Using the plurality method, determine the winner of the election.

P	C	M	P	P	P	P	C	M	M	P	P	C
M	P	P	M	M	M	M	P	P	P	M	M	P
C	M	C	C	C	C	C	M	C	C	C	C	M

P	M	P	C	M
M	P	M	P	P
C	C	C	M	C

8. A group of club members decide to vote to select the color of their meeting room. The color choices are white (W), light blue (B), and light yellow (Y). Make a preference table for the election and answer each question.
   (a) How many club members voted?
   (b) How many club members voted for white as their first choice?
   (c) How many club members voted for the colors in the order BYW?
   (d) Using the plurality method, determine the winner.

W	B	B	Y	B	Y	W	B	Y	Y	W	B	Y
B	Y	Y	W	Y	W	B	Y	W	W	B	Y	W
Y	W	W	B	W	B	Y	W	B	B	Y	W	B

9. Students at a college were asked to rank three improvements that they would like to see at their college. The choices were build a new gymnasium (G), build a swimming pool (S), or build a baseball/football field (B). The votes are summarized in the preference table.

Number of votes	83	56	42	27
First choice	G	S	S	B
Second choice	S	G	B	S
Third choice	B	B	G	G

   (a) How many students voted?
   (b) What option won if the plurality method was used to determine the winner?

10. A college fraternity plans to provide a free drink stand for people during the Homecoming Parade. They decide to vote on the choice of beverage. The choices are bottled water (W), Gatorade (G), Coke (C), or lemonade (L). The results of the election are shown in the preference table.

Number of votes	8	6	5	3	2
First choice	C	L	W	G	W
Second choice	W	W	G	C	C
Third choice	G	G	C	W	L
Fourth choice	L	C	L	L	G

   (a) How many people voted?
   (b) What drink was selected as the winner if the plurality method was used to determine the winner?

11. Students from a sorority voted to select a floral bouquet for their representative in a homecoming parade. The choices were roses (R), gardenias (G), carnations (C), and daisies (D). The preference table is shown.

Number of votes	3	5	2	6	4
First choice	R	D	C	C	R
Second choice	G	C	R	G	C
Third choice	C	G	G	D	D
Fourth choice	D	R	D	R	G

   (a) How many students voted?
   (b) What flower bouquet won if the plurality method was used to determine the winner?

12. The students in Dr. Lee's math class were asked to vote on the starting time for their final exam. Their choices were 8:00 A.M., 10:00 A.M., 12:00 P.M., or 2:00 P.M. The results of the election are shown in the preference table.

Number of votes	8	12	5	3	2	2
First choice	8	10	12	2	10	8
Second choice	10	8	2	12	12	2
Third choice	12	2	10	8	8	10
Fourth choice	2	12	8	10	2	12

   (a) How many students voted?
   (b) What time was the final exam if the plurality method was used to determine the winner?

13. Using the election results given in Exercise 9, has the head-to-head comparison criterion been violated? Explain your answer.
14. Using the election results given in Exercise 10, has the head-to-head comparison criterion been violated? Explain your answer.
15. Using the election results given in Exercise 11, has the head-to-head criterion been violated? Explain your answer.
16. Using the election results given in Exercise 12, has the head-to-head criterion been violated? Explain your answer.

Critical Thinking

Use the following information for Exercises 17 through 19.

Suppose 100 votes are cast in an election involving three candidates, A, B, and C, and 80 votes are counted so far. The results are

A 36 B 32 C 12

(*Note:* This is not a preference list. Voters are submitting a single name on a ballot.)

17. What is the minimum number of remaining votes candidate A needs to guarantee that he or she wins

the election using the plurality method? Explain your answer.

18. What is the minimum number of remaining votes candidate B needs to guarantee that he or she wins the election using the plurality voting method?
19. Can candidate C win the election using the plurality voting method? Explain your answer.
20. If there are 408 votes cast in an election with four candidates, what is the smallest number of votes a candidate can win with if the plurality method is used?
21. Is it possible to have a tie if the head-to-head comparison method is used? If so, give an example.
22. If an election is held with four candidates and 204 votes are cast, what is the smallest number of votes a candidate can win with if the plurality method is used?
23. Can you think of a circumstance under which an election decided using the plurality method is guaranteed to satisfy the head-to-head criterion?

Section 14-2 The Borda Count Method and the Plurality-with-Elimination Method

LEARNING OBJECTIVES

- ❑ 1. Determine the winner of an election using the Borda count method.
- ❑ 2. Decide if an election violates the majority criterion.
- ❑ 3. Determine the winner of an election using the plurality-with-elimination method.
- ❑ 4. Decide if an election violates the monotonicity criterion.

One of the most controversial examples of voting today comes from the world of college football. Because the weekly polls determine the teams that get to play for a national championship, those polls get an awful lot of attention. But how exactly are the top teams determined? The answer is that coaches, athletic directors, and members of the media send in a weekly ballot listing their top 25 teams in order, and the teams receive points based on where they appear on ballots. The higher they are listed on any given ballot, the more points they get. When the points are added up, the totals determine where teams are ranked.

In this section, the first voting method we will study is the basis for college football polls and a wide variety of other lists and awards.

The Borda Count Method

A second method of voting when there are three or more alternatives is called the Borda count method. This method was developed by a French naval captain and mathematician, Jean-Charles de Borda.

The **Borda count method** of voting requires the voter to rank each candidate from most favorable to least favorable then assigns 1 point to the last-place candidate, 2 points to the next-to-the-last-place candidate, 3 points to the third-from-the-last-place candidate, etc. The points for each candidate are totaled separately, and then the candidate with the most points wins the election.

Example 1 illustrates how to use the Borda count method.

EXAMPLE 1 Using the Borda Count Method

The preference table for a club presidential election consisting of three candidates is shown. Use the Borda count method to determine the winner.

Number of votes	15	8	3	2
First choice	B	A	C	A
Second choice	C	B	A	C
Third choice	A	C	B	B

SOLUTION

Assign 1 point for the third choice, 2 points for the second choice, and 3 points for the first choice. Then multiply the number of votes by the number of the choice for each candidate to get the total points.

For candidate A:

Because A is ranked third in column 1, A gets 1 point from each of the 15 voters at the top of the column. Since A is ranked first in column 2, A gets 3 points from each of the 8 voters. Since A is ranked second in column 3, A gets 2 points from each of the 3 voters. Finally, since A is ranked first in the last column, A gets 3 points from each of the 2 voters at the top. The total points for A are $15 \cdot 1 + 8 \cdot 3 + 3 \cdot 2 + 2 \cdot 3 = 51$.

For candidate B:

Because candidate B is ranked first in column 1, B gets 3 points from each of the 15 voters at the top of the column. Since candidate B is ranked second in column 2, B gets 2 points from each of the 8 voters at the top of the column. Candidate B is ranked third in column 3 and gets 1 point from each of the 3 voters at the top of the column. Candidate B is ranked third in column 4 and gets 1 point from each of the 2 voters at the top of the column. The total points for B are $15 \cdot 3 + 8 \cdot 2 + 3 \cdot 1 + 2 \cdot 1 = 66$.

For candidate C:

Because candidate C was ranked second in column 1, C gets 2 points from each of the 15 voters at the top of the column. Since candidate C is ranked third in column 2, C gets 1 point from each of the 8 voters at the top of the column. Candidate C is ranked first in column 3 and gets 3 points from each of the 3 voters at the top of the column. Candidate C is ranked second in column 4 and gets 2 points from each of the 2 voters at the top of the column. The total points for C are $15 \cdot 2 + 8 \cdot 1 + 3 \cdot 3 + 2 \cdot 2 = 51$.

In this case, candidate B received the most points, so candidate B is the winner.

Math Note

Notice that candidate B in Example 1 would also have won the election if the plurality method had been used.

CAUTION

Make sure that you award the *most* points to the candidate listed first! It's very common to mistakenly award one point for a first place vote, two points for second, and so on.

▼ Try This One 1

☑ 1. Determine the winner of an election using the Borda count method.

There are four candidates for homecoming queen at Johnsonville College: Kia (K), Latoya (L), Michelle (M), and Natalie (N). The preference table for the election is shown next:

Number of votes	232	186	95	306
First choice	M	K	M	L
Second choice	K	L	L	K
Third choice	L	N	K	N
Fourth choice	N	M	N	M

Using the Borda count method, determine who won the election.

The Borda count method, like the plurality method, has its shortcomings. This method sometimes violates the fairness criterion called the *majority criterion*.

Sidelight JEAN-CHARLES DE BORDA (1733–1799)

Jean-Charles de Borda was born in France. As a young man, he served in both the army and the navy. During the Revolutionary War, he commanded the ship *Solitaire* and served with "great distinction."

Besides being a teacher of mathematics, Borda made many contributions to science and architecture. He wrote papers on the mathematics of projectiles, the construction of naval vessels, the physics of hydraulics, and navigation. He founded a school of naval architecture in France.

In 1770, as a member of the Academy of Science, he devised the voting method that bears his name in order to elect officers of the academy.

The **majority criterion** states that if a candidate receives a majority of first-place votes, then that candidate should be the winner of the election.

Example 2 illustrates the majority criterion.

EXAMPLE 2 Checking the Majority Criterion

The staff of an entertainment magazine is voting for their choice for best new show of the 2008 fall season. The choices are *The Mentalist* (M), *Fringe* (F), and *Surviving Suburbia* (S). The results are summarized below. If the winner is chosen using the Borda count method, does the election violate the majority criterion?

Number of votes	11	7	6	4
First choice	M	F	F	M
Second choice	F	S	M	S
Third choice	S	M	S	F

SOLUTION

First, find the winner using the Borda count method:

The Mentalist: $11 \cdot 3 + 7 \cdot 1 + 6 \cdot 2 + 4 \cdot 3 = 64$

Fringe: $11 \cdot 2 + 7 \cdot 3 + 6 \cdot 3 + 4 \cdot 1 = 65$

Surviving Suburbia: $11 \cdot 1 + 7 \cdot 2 + 6 \cdot 1 + 4 \cdot 2 = 39$

Fringe wins using the Borda count method, but of 28 ballots cast, 15 listed *The Mentalist* first. This means that a majority of voters listed *The Mentalist* first, and since it didn't win, the majority criterion is violated.

▼ Try This One 2

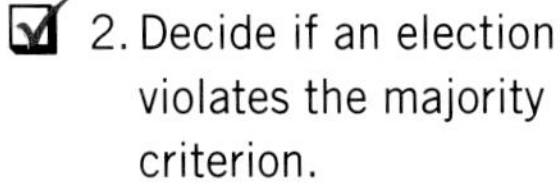

Does the election in Try This One 1 violate the majority criterion?

In addition to its use in college football polls, the Borda count method is used to select the Heisman trophy winner, the most valuable player award in both major baseball leagues, and the country music vocalists of the year, among many others.

The Plurality-with-Elimination Method

The plurality-with-elimination method of voting has been dubbed the "survival of the fittest" method. In this method of voting, if no candidate receives a majority of votes, a series of "rounds" is used. Formally defined,

Math Note

The plurality-with-elimination method is also commonly called *instant runoff voting*, and abbreviated IRV.

In the **plurality-with-elimination method,** the candidate with the majority of first-place votes is declared the winner. If no candidate has a majority of first-place votes, the candidate (or candidates) with the least number of first-place votes is eliminated, then the candidates who were below the eliminated candidate move up on the ballot, and the number of first-place votes is counted again. If a candidate receives the majority of first-place votes, that candidate is declared the winner. If no candidate receives a majority of first-place votes, the one with the least number of first-place votes is eliminated, and the process continues.

Example 3 shows how to use the plurality-with-elimination method.

EXAMPLE 3 Using the Plurality-with-Elimination Method

Use the plurality-with-elimination method to determine the winner of the election shown in the preference table.

Number of votes	6	27	17	9
First choice	A	B	C	D
Second choice	D	A	D	B
Third choice	C	C	B	C
Fourth choice	B	D	A	A

SOLUTION

Round 1: There were 59 votes cast, and no one received a majority of votes (30 or more), so candidate A is eliminated since he has the fewest first-place votes, 6. The other candidates in the first column move up.

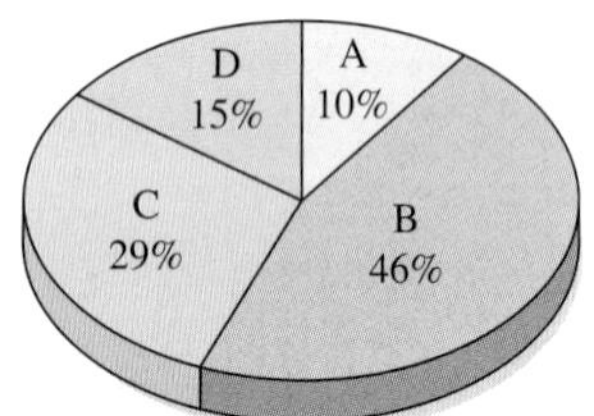

Round 1 results

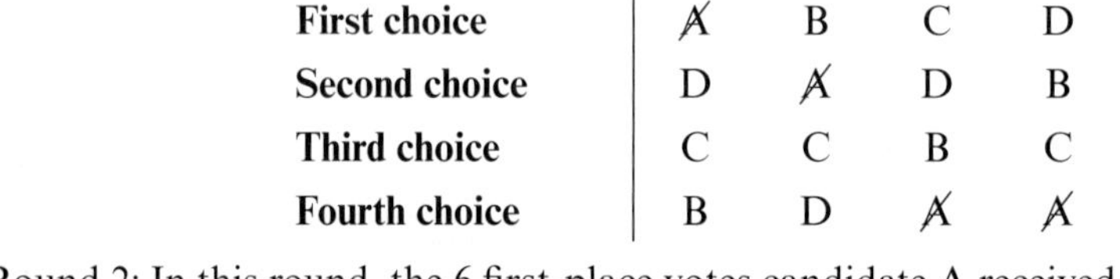

Number of votes	6	27	17	9
First choice	~~A~~	B	C	D
Second choice	D	~~A~~	D	B
Third choice	C	C	B	C
Fourth choice	B	D	~~A~~	~~A~~

Round 2: In this round, the 6 first-place votes candidate A received go to candidate D because she moved up in the first column when candidate A was eliminated. But there is still no candidate with 30 or more votes, so next candidate D is eliminated because she has the fewest first-place votes (15) in this round.

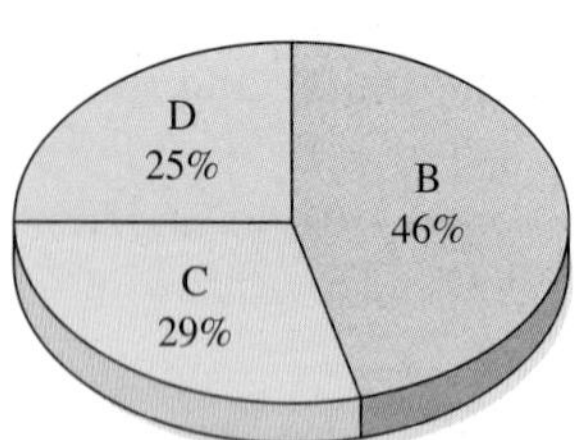

Round 2 results

Number of votes	6	27	17	9
First choice	~~D~~	B	C	~~D~~
Second choice	C	C	~~D~~	B
Third choice	B	~~D~~	B	C

Round 3: The first-place votes candidate D received in column 1 to go to candidate C while the 9 first-place votes candidate D received in the fourth column go to candidate B. With A and D eliminated, the preference table looks like this.

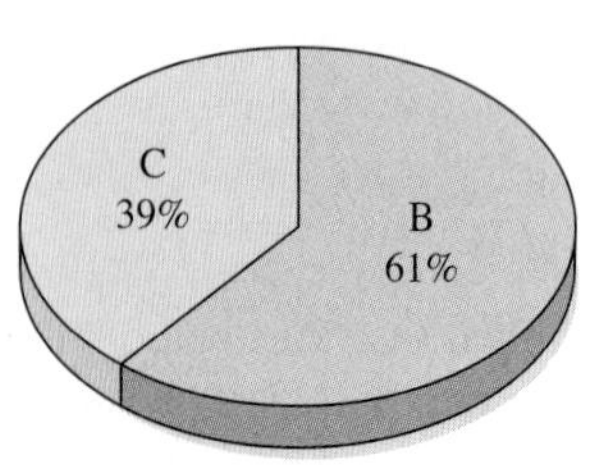

Round 3 results

Number of votes	6	27	17	9
First choice	C	B	C	B
Second choice	B	C	B	C

Now candidate B has 36 (27 + 9) first-place votes and candidate C has 23 (6 + 17); candidate B now has a majority and is declared the winner.

☑ 3. Determine the winner of an election using the plurality-with-elimination method.

▼ Try This One 3

The planning committee for a company's annual picnic votes for an afternoon activity. The choices are a softball game (S), a touch football game (F), a bocce game (B), and a volleyball game (V). The preference table is shown.

Number of votes	3	5	2	1	1
First choice	B	S	V	F	B
Second choice	S	F	S	S	V
Third choice	F	B	B	V	F
Fourth choice	V	V	F	B	S

Determine the winner using the plurality-with-elimination method. Explain your answer.

The plurality-with-elimination method, like the other two methods, has some shortcomings. One shortcoming is that it sometimes fails the fairness criterion known as the *monotonicity criterion.*

The **monotonicity criterion** states that if a candidate wins an election, and a reelection is held in which the only changes in voting favor the original winning candidate, then that candidate should still win the reelection.

Consider this election:

Number of votes	7	13	11	10
First choice	X	Z	Y	X
Second choice	Z	X	Z	Y
Third choice	Y	Y	X	Z

Using the plurality-with-elimination method, candidate Y is eliminated in round 1. With Y eliminated, the preference table looks like this.

Number of votes	7	13	11	10
First choice	X	Z	Z	X
Second choice	Z	X	X	Z

Z wins with 24 first-place votes.

Now suppose the first election was declared invalid for some reason, and on a second election, the voters in column 1 change their ballots in favor of candidate Z and vote ZXY.

The new preference table will be

Number of votes	7	13	11	10
First choice	Z	Z	Y	X
Second choice	X	X	Z	Y
Third choice	Y	Y	X	Z

Here X is eliminated on the first round and the preference table becomes

Number of votes	7	13	11	10
First choice	Z	Z	Y	Y
Second choice	Y	Y	Z	Z

Here Y is the winner with 21 votes compared to 20 votes for Z.

In this case, the plurality-with-elimination method fails the monotonicity criterion. On the second election, even though candidate Z had received seven more first-choice votes, he lost the election! By doing better the second time, the candidate did worse!

EXAMPLE 4 Checking the Monotonicity Criterion

Suppose that all 6 voters from the first column of the election in Example 3 are persuaded to change their ballots to match the 17 voters in the third column. Does this violate the monotonicity criterion?

SOLUTION

First, we should point out that this change favors candidate B, since all 6 ballots move her from fourth to third choice.

Number of voters	27	23	9
First choice	B	C	D
Second choice	A	D	B
Third choice	C	B	C
Fourth choice	B	A	A

No candidate has a majority, so we eliminate the candidate with the fewest first-place votes (A). But this keeps the first-place votes the same, so we still have no majority. Next we eliminate candidate D, who now has the fewest first-place votes. The resulting preference table is

Number of voters	27	23	9
First choice	B	C	B
Second choice	C	B	C

Now B has a majority and wins. Since B was the winner of the original election, the monotonicity criterion is not violated.

▼ Try This One 4

If the one voter who listed softball last in the election in Try This One 3 changes her vote to the order B-S-F-V, does this violate the monotonicity criterion?

☑ 4. Decide if an election violates the monotonicity criterion.

The plurality-with-elimination method is used in some very high profile elections. The International Olympic Committee uses it to select the cities that will host the games, and it is also used by the Academy of Motion Pictures to determine the annual academy award winners.

Answers to Try This One

1 The winner is Kia with 2,548 points.

2 No. None of the candidates received a majority of the first-place votes.

3 The winner is softball.

4 No. Softball still wins.

EXERCISE SET 14-2

Writing Exercises

1. Explain how to determine the winner of an election using the Borda count method.
2. Describe the majority criterion in your own words.
3. Explain how to determine the winner of an election using the plurality-with-elimination method.
4. Describe the monotonicity criterion in your own words.

Real-World Applications

5. A gaming club holds a vote to decide what type of video game they'll play at the next meeting. The choices are sports (S), action (A), or role-playing (R). The preference table for the results is shown here.

Number of votes	10	6	5
First choice	R	S	S
Second choice	A	R	A
Third choice	S	A	R

Using the Borda count method of voting, determine the winner.

6. The McKees' Point Yacht Club Board of Directors wants to decide where to hold their fall business meeting. The choices are the Country Club (C), Frankie's Fine Foods (F), West Oak Golf Club (W), and Rosa's Restaurant (R). The results of the election are shown in the preference table.

Number of votes	8	6	5	2
First choice	R	W	C	F
Second choice	W	R	F	R
Third choice	C	F	R	C
Fourth choice	F	C	W	W

Determine the winner using the Borda count method of voting.

7. A local movie theater asks its patrons which movies they would like to view during next month's "Oldies but Goodies" week. The choices are *Gone with the Wind* (G), *Casablanca* (C), *Anatomy of a Murder* (A), and *Back to the Future* (B). The preference table is shown.

Number of votes	331	317	206	98
First choice	G	A	C	B
Second choice	A	C	B	G
Third choice	C	B	G	A
Fourth choice	B	G	A	C

Use the Borda count voting method to determine the winner.

8. Parents of a kindergarten class are given the option of choosing a location for the annual spring field trip. The choices are the zoo (Z), the local amusement park (A), or the local museum (M). The preference table is shown.

Number of votes	15	9	5	4
First choice	Z	M	A	Z
Second choice	A	Z	M	M
Third choice	M	A	Z	A

Using the Borda count voting method, determine the winner.

9. Students at a college were asked to rank three improvements that they would like to see at their college. Their choices were build a new gymnasium (G), build a baseball/football field (B), or build a swimming pool (S). The votes are summarized in the preference table.

Number of votes	83	56	42	27
First choice	G	S	S	B
Second choice	S	G	B	S
Third choice	B	B	G	G

(a) Using the Borda count method of voting, determine the winner.

(b) Refer to Exercise 9 in Section 14-1. Is the winner the same as the one determined by the plurality method?

10. A college fraternity plans to provide a free drink stand for people during the homecoming parade. They decide to vote on the choice of beverage. The choices are bottled water (W), Gatorade (G), Coke (C), or lemonade (L). The results of the election are shown in the preference table.

Number of votes	8	6	5	3	2
First choice	C	W	W	C	C
Second choice	W	L	G	G	W
Third choice	G	G	C	W	L
Fourth choice	L	C	L	L	G

Using the Borda count method of voting, determine the winner.

11. Does the election in Exercise 5 violate the majority criterion?
12. Does the election in Exercise 6 violate the majority criterion?
13. Does the election in Exercise 7 violate the majority criterion?
14. Does the election in Exercise 8 violate the majority criterion?
15. Does the election in Exercise 9 violate the majority criterion?
16. Does the election in Exercise 10 violate the majority criterion?
17. The English department is voting for a new department chairperson. The three candidates are Professor Greene (G), Professor Williams (W), and Professor Donovan (D). The results of the election are shown in the preference table.

Number of votes	10	8	7	4
First choice	D	W	G	G
Second choice	G	D	W	D
Third choice	W	G	D	W

Using the plurality-with-elimination method of voting, determine the winner.

18. The Association of Self-Employed Working Persons must select a speaker for its next meeting. The choices for a topic are health care (H), investments (I), or advertising (A). The results of the election are shown in the preference table.

Number of votes	6	4	9	2
First choice	H	H	I	I
Second choice	I	A	H	A
Third choice	A	I	A	H

Using the plurality-with-elimination method of voting, determine the winner.

19. Students in a sorority voted to select a floral bouquet for their representative to wear in the homecoming parade. The choices included roses (R), gardenias (G), carnations (C), and daisies (D). The preference table is shown.

Number of votes	3	5	2	6	4
First choice	R	D	C	C	R
Second choice	G	C	R	G	C
Third choice	C	G	G	D	D
Fourth choice	D	R	D	R	G

(a) Using the plurality-with-elimination method, determine the winner.

(b) Is the winner the same as the one determined by the plurality method used in Exercise 11 of Section 14-1?

20. The students in Dr. Lee's math class were asked to vote on the starting time for the final exam. Their choices were 8:00 A.M., 10:00 A.M., 12:00 P.M., or 2:00 P.M. The results of the election are shown in the preference table.

Number of votes	8	12	5	3	2	2
First choice	8	10	12	2	10	8
Second choice	10	8	2	12	12	2
Third choice	12	2	10	8	8	10
Fourth choice	2	12	8	10	2	12

(a) Using the plurality-with-elimination method, determine the winner.

(b) Is the winner the same as the one determined by the plurality method used in Exercise 12 of Section 14-1?

21. Suppose that all 4 voters from the last column of the preference table in Exercise 17 decide to change their vote to the order D-G-W (the order in the first column). Does this violate the monotonicity criterion?
22. Suppose that 2 of the 4 voters from the second column of the preference table in Exercise 18 decide to change their vote to the order H-I-A (the order in the first column). Does this violate the monotonicity criterion?
23. If 2 of the voters from column 1 in Exercise 19 change their vote to column 2, does this violate the monotonicity criterion?
24. If the 3 voters in column 4 in Exercise 20 change their vote to column 1, does this violate the monotonicity criterion?

Critical Thinking

25. Construct a preference table for an election involving three candidates so that candidate A wins the election using the Borda count method, but the majority criterion is violated.
26. Construct a preference table for an election involving three candidates so that candidate B wins using the plurality-with-elimination method, but the monotonicity criterion is violated.
27. Construct a preference table for an election so that one candidate wins the election using the Borda count method and a different candidate wins the same election using the plurality-with-elimination method.
28. If the candidates on a preference ballot are ranked so that the lowest candidate gets 0 points, the next to the lowest candidate gets 1 point and so on, will the winner be the same using the Borda count method as the winner when the candidates are ranked the way they are explained in this section? Explain your answer using an illustration.
29. If the candidates on a preference ballot are ranked by giving the top candidate a score of 1 (first place), the next highest candidate a score of 2 (second place), and so on, can the Borda count method be used to determine a winner? Explain why or why not.

Section 14-3 The Pairwise Comparison Method and Approval Voting

LEARNING OBJECTIVES

- ❑ 1. Determine the winner of an election using the pairwise comparison method.
- ❑ 2. Decide if an election violates the irrelevant alternatives criterion.
- ❑ 3. Describe Arrow's impossibility theorem.
- ❑ 4. Determine the winner of an election using approval voting.

In the modern media age, pretty much everyone with an interest in politics goes to bed on election night knowing who the next president of the United States will be. But this was most definitely not the case on November 7, 2000. At various times throughout the evening, major media outlets projected a win for both candidates, George W. Bush and Al Gore. At issue was the historically close and hotly debated result from Florida. Widespread protests were common for supporters of both candidates, and the election was not finally decided until the Supreme Court ruled that any further recounts were to be terminated, essentially declaring Bush the winner on December 12.

During that crazy 35-day period when the election hung in the balance, a lot of people made claims that the election was unfair. But what exactly does it mean for an election to be unfair, and can any election ever truly be fair? Throughout this chapter, we've been identifying criteria that can be used to determine if an election is fair. In this section, we will examine whether any election can satisfy all these criteria.

But first, we will study a fourth voting method, and a fourth fairness criterion.

The Pairwise Comparison Method

The pairwise comparison method uses a preference table to compare each pair of candidates. For example, if there are four candidates, A, B, C, and D, running in an election, then there would be six comparisons, as shown.

A vs. B	B vs. C
A vs. C	B vs. D
A vs. D	C vs. D

In general, the number of pairwise comparisons when n candidates are running is equal to

$$\frac{n(n-1)}{2} \qquad (1)$$

For example, if there were five candidates running, you would need to make

$$\frac{5(5-1)}{2} = \frac{5 \cdot 4}{2} = \frac{20}{2} = 10 \text{ pairwise comparisons}$$

EXAMPLE 1 Finding the Number of Pairwise Comparisons

Find the number of pairwise comparisons needed if six candidates are running in an election.

SOLUTION

With $n = 6$ in formula (1), we get

$$\frac{6(6-1)}{2} = \frac{6(5)}{2} = 15$$

You would need to make 15 comparisons.

Math Note

Formula (1) can be found using the combination formula in Chapter 11 with $r = 2$:

$${}_nC_2 = \frac{n!}{(n-2)!2!}$$

$${}_nC_2 = \frac{n!}{(n-2)!2!}$$
$$= \frac{n \cdot (n-1)\cancel{(n-2)!}}{\cancel{(n-2)!}2!}$$
$$= \frac{n(n-1)}{2}$$

▼ Try This One 1

Find the number of pairwise comparisons needed if eight candidates are running in an election.

The pairwise comparisons we will make are the same ones we did to check the head-to-head comparison criterion in Section 13-1. For example, to compare candidates A and B using a preference table, add up the number of ballots on which A is preferred over B, and the number of ballots on which B is preferred over A. The candidate with the most ballots in his favor gets one point. In case of a tie, each candidate gets ½ point. After all possible pairwise comparisons have been made, the candidate with the most points wins. Written formally:

The **pairwise comparison method** of voting requires that all candidates be ranked by the voters. Then each candidate is paired with every other candidate in a one-to-one contest. For each one-to-one comparison, the candidate who wins on more ballots gets 1 point. In case of a tie, each candidate gets $\frac{1}{2}$ point. After all possible two-candidate comparisons are made, the points for each candidate are tallied, and the candidate with the most points wins the election.

Example 2 shows how to use the pairwise comparison method.

EXAMPLE 2 Using the Pairwise Comparison Method

Use the pairwise comparison method to find the winner of the elections whose results are shown in the following preference table.

Number of votes	14	13	16	15
First choice	B	A	C	B
Second choice	C	C	A	A
Third choice	A	B	B	C

SOLUTION

We will need to make three pairwise comparisons: A vs. B, A vs. C, and B vs. C.

First, A vs. B:

Number of votes	14	13	16	15
First choice	Ⓑ	Ⓐ	C	Ⓑ
Second choice	C	C	Ⓐ	Ⓐ
Third choice	Ⓐ	Ⓑ	Ⓑ	C

Candidate A is ranked higher than B on 29 ballots (columns 2 and 3), and candidate B is also ranked higher on 29 ballots (columns 1 and 4). This is a tie, so each candidate gets $\frac{1}{2}$ point.

Next, compare A to C:

Number of votes	14	13	16	15
First choice	B	Ⓐ	Ⓒ	B
Second choice	Ⓒ	Ⓒ	Ⓐ	Ⓐ
Third choice	Ⓐ	B	B	Ⓒ

Candidate A is ranked higher than C in columns 2 and 4, so A gets 13 + 15 = 28 votes. Candidate C is ranked higher than A in columns 1 and 3, so C gets 14 + 16 = 30 votes. Since C has more votes, assign 1 point to C.

Finally, compare B to C:

Number of votes	14	13	16	15
First choice	(B)	A	(C)	(B)
Second choice	(C)	(C)	A	A
Third choice	A	(B)	(B)	(C)

Candidate B is ranked higher than C in columns 1 and 4, so B gets 14 + 15 = 29 votes. Candidate C is ranked higher than B in columns 2 and 3, so C gets 13 + 16 = 29 votes. Since this is a tie, assign $\frac{1}{2}$ point to B and $\frac{1}{2}$ point to C.

Now find the totals.

		Total	
Candidate A	$\frac{1}{2}$	$\frac{1}{2}$	*A ties with B.*
Candidate B	$\frac{1}{2} + \frac{1}{2}$	1	*B ties with A and C.*
Candidate C	$1 + \frac{1}{2}$	$1\frac{1}{2}$	*C defeats B and ties with A.*

Candidate C has the most points and is the winner.

Math Note

For the election in Example 1, candidate B wins using the plurality method, candidate C wins using the Borda count method, and candidates B and C tie using the plurality-with-elimination method.

▼ Try This One 2

The members of the Music Appreciation Club vote to decide whether they will attend an opera (O), a symphony (S), or a ballet (B). The results of the election are shown in the preference table.

Number of votes	13	9	6	11
First choice	O	B	S	B
Second choice	S	O	B	S
Third choice	B	S	O	O

Use the pairwise comparison voting method to determine the winning selection.

☑ 1. Determine the winner of an election using the pairwise comparison method.

Mathematicians have been able to prove that the pairwise comparison voting method satisfies the majority criterion, the head-to-head criterion, and the monotonicity criterion. Does that make it the ideal voting method? Sadly, it does not. In fact, it fails the fourth fairness criterion, which is known as the *irrelevant alternatives criterion.*

The **irrelevant alternatives criterion** requires that if a certain candidate X wins an election and one of the other candidates is removed from the ballot and the ballots are recounted, candidate X still wins the election.

EXAMPLE 3 Checking the Irrelevant Alternatives Criterion

Does the election in Example 2 violate the irrelevant alternatives criterion?

SOLUTION

Candidate C won the election, so the two irrelevant alternatives are A and B. If A is eliminated, the preference table is now:

Number of votes	14	13	16	15
First place	B	C	C	B
Second place	C	B	B	C

There are 14 + 15 = 29 voters who preferred B to C, and there are 13 + 16 = 29 voters who preferred C to B; a tie has occurred. This violates the irrelevant alternatives criterion because C should still win no matter which other candidate drops out.

☑ 2. Decide if an election violates the irrelevant alternatives criterion.

▼ Try This One 3

Does the election in Try This One 2 violate the irrelevant alternatives criterion?

Math Note

In all voting methods, the possibility of a tie should be considered before the votes are counted and some way of breaking a tie should be agreed on in advance of the election.

Another problem with the pairwise comparison method is that it is possible for all candidates to tie. Consider the next preference table.

Number of votes	8	5	7
First choice	A	C	B
Second choice	B	A	C
Third choice	C	B	A

Here A wins over B (13 to 7), B wins over C (15 to 5), and C wins over A (12 to 8); this gives us a three-way tie for first place, in which case the election accomplished absolutely nothing!

Table 14-1 summarizes the four voting methods and the marks show which criteria are always satisfied by the voting methods.

TABLE 14-1 Fairness Criteria Satisfied by Various Voting Methods

	Head-to-head criterion	Majority criterion	Monotonicity criterion	Irrelevant alternatives criterion
Plurality		✓	✓	
Borda count			✓	
Plurality-with-elimination		✓		
Pairwise comparison	✓	✓	✓	

Arrow's Impossibility Theorem

Table 14-1 shows that none of the four voting methods is perfectly fair. Each violates one or more of the four fairness criteria: the majority criterion, the head-to-head criterion, the monotonicity criterion, or the irrelevant alternatives criterion.

The obvious question now becomes: What method satisfies all of the fairness criteria? The surprising and somewhat disappointing answer is that there isn't one. In 1951, an economist named Kenneth Arrow was able to prove that there does not exist and never will exist a democratic voting method for three or more alternatives that satisfies all four of the fairness criteria. The result is now known as **Arrow's impossibility theorem.** (The proof is beyond the scope of our study, but don't feel bad—Arrow is a Nobel Prize winner, so he must be awfully smart.)

☑ 3. Describe Arrow's impossibility theorem.

Sidelight KENNETH J. ARROW (1921–)

Kenneth J. Arrow received a B.S. in Social Science from the City College of New York and an M.A. and a Ph.D. from Columbia University. His research interests include the economics of information and organization, collective decision making, and general equilibrium theory.

In 1951, he proved the theorem that states that no voting system will satisfy all of the four fairness criteria. The theorem became known as Arrow's impossibility theorem.

In 1972, he received the Nobel Prize in Economics for his work in social choice theory. In 1986, he won the Von Neuman Theory Prize for his contributions to decision sciences.

He is a professor of economics (emeritus) at Stanford University. He has many professional affiliations including being past president of the Institute of Management Sciences. As of this writing, Dr. Arrow is still alive and kicking at the age of 87.

Math Note

Approval voting is so named because the voter is rating each candidate as either acceptable or unacceptable, indicating whether or not she approves of the candidate.

Approval Voting

In the late 1970s, a new voting method called *approval voting* was introduced.

With **approval voting,** each voter gives one vote to as many candidates on the ballot as he or she finds acceptable. The votes are counted, and the winner is the candidate who receives the most votes.

In this case, voters can select anything from no candidates to all of the candidates. Example 4 shows how approval voting works.

EXAMPLE 4 Using Approval Voting to Determine the Winner of an Election

Five candidates are nominated for the teacher of the year award, and 20 of their colleagues will vote using approval voting. The results are shown in the table. (For example, the first column indicates that 9 voters marked only candidates A and D acceptable.) Which candidate wins?

Number of votes	9	3	2	5	1
Candidate A	/		/		/
Candidate B		/	/	/	
Candidate C		/	/	/	
Candidate D	/	/		/	
Candidate E				/	

SOLUTION

From the table, count the number of votes for each candidate:

Candidate	Votes
A	$9 + 2 + 1 = 12$
B	$3 + 2 + 5 = 10$
C	$3 + 2 + 5 = 10$
D	$9 + 3 + 5 = 17$
E	5

In this election, candidate D received 17 votes and is declared the winner.

4. Determine the winner of an election using approval voting.

▼ Try This One 4

An election was held for an employee of the month award using approval voting. The results are shown in the table. Which candidate won?

Number of votes	20	18	12	4
Candidate F		/		/
Candidate G	/		/	/
Candidate H		/	/	/
Candidate I	/			/
Candidate J			/	/

Several political scientists and analysts independently developed approval voting in the late 1970s. This method is now used to elect the Secretary General of the United Nations, and it is also used to elect the leaders of some academic and professional societies such as the National Academy of Sciences. The advantages of approval voting are that it is simple to use and easy to understand. The ballot is also uncomplicated.

There are some disadvantages to approval voting though. The major one is that there is no ranking or preference of the candidates. Most voters have favorite choices, but there is no way to indicate these preferences on an approval ballot.

Because of this, approval voting is also prone to violating the majority criterion. It's possible that one candidate could be considered the best choice by well more than half of voters, but if they are also left off of the other ballots, a second candidate that is considered marginally acceptable by most of the voters could win. In addition, some societies that tried approval voting dropped it when they found that most voters were only marking one candidate acceptable, in which case the election is essentially decided using the plurality method.

Tie Breaking

Regardless of the voting method selected, the possibility of a tie between two or more candidates always exists. There are many ways to break a tie and a fair tie-breaking method should always be decided upon in advance of an election.

In some cases, the chairperson of a committee does not vote on motions unless there is a tie. In this case, the chairperson would cast the tie-breaking vote. The most obvious method of breaking a tie is, of course, the age-old method of flipping a coin. In other cases, drawing a name from a hat could be used. Using a third-party judge could be considered. For example, if a tie occurs in an election of a department chairperson, the dean could decide the winner. Another possibility might be to consider some other criteria such as seniority, education, or experience of the candidates.

Finally, in case of a tie, another voting method could be used. For example, if a tie occurred using the Borda count method, then perhaps the pairwise comparison method could be used to determine a winner.

Answers to Try This One

1 28

2 B wins with 2 points, compared to 1 for O and 0 for S.

3 No; if either O or S drops out, B still wins.

4 Candidate G wins with 36 votes.

EXERCISE SET 14-3

Writing Exercises

1. Explain how to determine the winner of an election using the pairwise comparison method.
2. Describe the irrelevant alternatives criterion. How is it different from the monotonicity criterion?
3. Describe Arrow's impossibility theorem. How is it connected to the four fairness criteria we have studied?
4. Explain how to determine the winner of an election using approval voting.
5. What are some strengths and weaknesses of approval voting?
6. Which of the five voting methods we studied do you think is easiest to apply? Which do you think is the most fair?

Computational Exercises

7. If there are four candidates in an election, how many pairwise comparisons need to be made in order to determine a winner?
8. If there are seven candidates in an election, how many pairwise comparisons need to be made in order to determine a winner?
9. If there are 10 candidates in an election, how many pairwise comparisons need to be made in order to determine a winner?
10. If there are nine candidates in an election, how many pairwise comparisons need to be made in order to determine a winner?

Real-World Applications

11. A college band was invited to perform at three different shows on the same day. They were Real Town (R), Steel Center (S), and Temple Village (V). Since they could only perform at one show, they voted on which one they would do. The results are shown in the preference table.

Number of votes	26	19	15	6
First choice	R	T	S	R
Second choice	S	S	R	T
Third choice	T	R	T	S

Using the pairwise comparison voting method, determine the winner.

12. The senior class at a small high school votes on the senior class trip. Their selections are Disneyland (D), Epcot Center (E), Sea World (S), and Six Flags Resort (F). The preference table is shown.

Number of voters	19	17	8	6
First choice	D	E	S	D
Second choice	E	S	D	F
Third choice	S	D	F	S
Fourth choice	F	F	E	E

Using the pairwise comparison voting method, determine the winner.

13. The English department is voting for a new department chairperson. The three candidates are Professor Greene (G), Professor Williams (W), and Professor Donovan (D). The results of the election are shown in the preference table.

Number of votes	10	8	7	4
First choice	D	W	G	G
Second choice	G	D	W	D
Third choice	W	G	D	W

(a) Determine the winner using the pairwise comparison voting method.
(b) Compare the winner with the one determined by the plurality-with-elimination method. See Exercise 17 of Section 14-2.

14. A gaming club holds a vote to decide what type of video game they'll play at the next meeting. The choices are sports (S), action (A), or role-playing (R). The preference table for the results is shown here.

Number of votes	10	6	5
First choice	R	S	S
Second choice	A	R	A
Third choice	S	A	R

(a) Determine the winner using the pairwise comparison method.
(b) Compare the winner with the one determined by the Borda count method used in Exercise 5 of Section 14-2.

15. The McKees' Point Yacht Club Board of Directors wants to decide where to hold their fall business meeting. The choices are the Country Club (C), Frankie's Fine Foods (F), West Oak Golf Club (W), and Rosa's Restaurant (R). The results of the election are shown in the preference table.

Number of votes	8	6	5	2
First choice	R	W	C	F
Second choice	W	R	F	R
Third choice	C	F	R	C
Fourth choice	F	C	W	W

(a) Determine the winner using the pairwise comparison method.
(b) Is this the same winner as the one using the Borda count method in Exercise 6 of Section 14-2?

16. The students in Dr. Lee's math class are asked to vote on the starting time for their final exam. Their choices are 8:00 A.M., 10:00 A.M., 12:00 P.M., or 2:00 P.M. The results of the election are shown in the preference table.

Number of votes	8	12	5	3	2	2
First choice	8	10	12	2	10	8
Second choice	10	8	2	12	12	2
Third choice	12	2	10	8	8	10
Fourth choice	2	12	8	10	2	12

(a) Determine the starting time using the pairwise comparison method.
(b) Is this the same starting time as determined by the plurality-with-elimination method used in Exercise 20 of Section 14-2? What about the plurality method in Exercise 12 of Section 14-1?

17. If Professor Donovan was unable to serve as department chairperson in the election shown in Exercise 13 and the votes are recounted, does this election violate the irrelevant alternatives criterion?
18. If the class members decide to eliminate Epcot from consideration and the votes were recounted in the election shown in Exercise 12, is the irrelevant alternatives criterion violated?
19. If the White Oak Golf Club is unavailable and the votes were recounted in the election shown in Exercise 15, is the irrelevant alternatives criterion violated?
20. If a room for Dr. Lee's final exam was not available at 2:00 P.M. and the votes were recounted in the election shown in Exercise 16, is the irrelevant alternatives criterion violated?
21. A sports committee of students needs to choose a team doctor. The result of the voting is shown. The approval method will be used.

Number of votes	15	18	12	10	5
Dr. Michaels	/		/	/	/
Dr. Zhang		/	/	/	/
Dr. Philip	/	/		/	
Dr. Perez	/	/		/	

Which doctor was selected?

22. The students at a college voted on a new mascot for their sports teams. The results of the election are shown. Using the approval voting method, determine the winner.

Number of votes	235	531	436	374
Ravens	/		/	/
Panthers		/	/	/
Killer Bees	/	/	/	/
Termites		/		/

23. A nursing school committee decides to purchase a van for the school. They vote on a color using the approval voting method. The results are shown.

Number of votes	1	2	1	3	2	1
White	/		/	/	/	
Blue	/	/	/			
Green		/	/	/	/	
Silver			/	/		/

Which color was selected?

24. A research committee decides to test market a new flavor for a children's drink. The results of a survey using approval voting at a local mall are shown.

Number of votes	38	32	16	5	3
Strawberry	/		/		/
Lime	/		/	/	/
Grape		/	/		
Orange			/	/	/
Bubble gum	/	/		/	

Which flavor was selected?

25. Due to prison overcrowding, a parole board must release one prisoner on good behavior. After hearing each case, they decide to use the approval voting method. The result is shown here.

Number of votes	1	1	1	1	1	1
Inmate W	/		/		/	/
Inmate X		/	/		/	
Inmate Y	/	/	/			
Inmate Z		/	/	/	/	/

Which inmate was released?

26. The park association committee decided to make one improvement for the local park this spring. The result of an election using approval voting is shown.

Number of votes	2	1	3	4	2
Paint benches	/		/		
Trim bushes			/	/	/
Repair snack bar		/	/	/	
Patch cement walks	/	/	/	/	/

Which repair was made?

Critical Thinking

27. Construct a preference table with three candidates X, Y, and Z so that the same candidate wins the election using all of the plurality, Borda count, plurality-with-elimination, and pairwise comparison methods.
28. Construct a preference table so that one candidate wins using the pairwise comparison method, but that violates the irrelevant alternatives criterion.
29. Explain why Exercise 28 would be impossible if you replace "irrelevant alternatives" with "head-to-head comparison."
30. If you are voting in an election using approval voting and you approve of all of the candidates equally, discuss whether you should give every candidate one vote or every candidate no vote.

Section 14-4 Apportionment

LEARNING OBJECTIVES

- ☐ 1. Compute standard divisors and quotas.
- ☐ 2. Apportion seats using Hamilton's method.
- ☐ 3. Apportion seats using Jefferson's method.
- ☐ 4. Apportion seats using Adams' method.
- ☐ 5. Apportion seats using Webster's method.
- ☐ 6. Apportion seats using the Huntington-Hill method.

You arrive at an 8 A.M. final (ugh!) and find that at least your professor was kind enough to bring donuts. Let's say she brought 4 dozen donuts and there are 24 people in class. How can she fairly distribute them?

The obvious answer is two donuts per person. This is a very simple example of *apportionment.* It's a simple example because everyone gets the same number of donuts and there are none left over. But what if some people are entitled to more donuts for one reason or another? And what if the number of students doesn't evenly divide the number of donuts? Then the situation gets more complicated.

The most common example of apportionment is the U.S. House of Representatives. There are 435 seats in the House, and they are divided among the 50 states. But because some states are more populous than others, they don't all get the same number of seats. For example, Florida has 25 seats, Pennsylvania has 19, and Rhode Island has just 2.

Apportionment is a process of fairly dividing identical indivisible objects among individuals entitled to shares that may be unequal.

For the remainder of this section, we will refer to the objects being divided as "seats." The individuals they are allocated to will be called "states" or "districts."

Standard Divisors and Quotas

Math Note

The term "indivisible" refers to objects that can't be split into fractional parts, like people or seats in Congress.

A good starting point is to find how many people will be represented by each seat. This is known as the *standard divisor.*

The **standard divisor** for an apportionment process is the average number of people per seat:

$$\text{Standard divisor} = \frac{\text{Total population}}{\text{Number of seats}}$$

EXAMPLE 1 Finding the Standard Divisor

A large city has 20 seats to be allocated proportionally among 5 districts according to the population of each district. The populations of the districts are shown in the table. Find the standard divisor.

District	1	2	3	4	5	Total
Population (in thousands)	32	80	41	65	22	240

SOLUTION

The total population is 240,000 and there are 20 seats:

$$\text{Standard divisor} = \frac{\text{Total population}}{\text{Number of seats}} = \frac{240{,}000}{20} = 12{,}000$$

This tells us that there will be on average 1 seat for every 12,000 people.

▼ Try This One 1

District	1	2	3	4	5	6	7	Total
Population	125	89	235	97	102	128	184	960

The table above gives the population in ten thousands of seven districts of a particular county. The county assembly has 30 seats that are to be allocated according to the population of each district. Find the standard divisor.

If fractions of seats were possible, the simplest way to allocate seats would be to divide the population of each district by the standard divisor. This would correspond to the percentage of the total population in each district multiplied by the number of seats:

$$\frac{\text{Population of district}}{\dfrac{\text{Total population}}{\text{Number of seats}}} = \frac{\text{Population of district}}{\text{Total population}} \cdot \text{Number of seats}$$

This value, called the standard quota, will play a key role in our apportionment methods.

The **standard quota** for a district in an apportionment process is the population of the district divided by the standard divisor:

$$\text{Standard quota} = \frac{\text{Population of district}}{\text{Standard divisor}}$$

EXAMPLE 2 Finding Standard Quotas

Find the standard quota for each of the districts in Example 1.

SOLUTION:

District	1	2	3	4	5	Total
Population (in thousands)	32	80	41	65	22	240

Refer to the population chart for the city above. From Example 1, the standard divisor is 12. Therefore the standard quota for each district is as follows:

$$\text{For District 1: Standard quota} = \frac{\text{Population of district}}{\text{Standard divisor}} = \frac{32}{12} \approx 2.67.$$

$$\text{For District 2: Standard quota} = \frac{80}{12} \approx 6.67$$

$$\text{For District 3: Standard quota} = \frac{41}{12} \approx 3.42$$

$$\text{For District 4: Standard quota} = \frac{65}{12} \approx 5.42$$

$$\text{For District 5: Standard quota} = \frac{22}{12} \approx 1.83$$

Math Note

The sum of the standard quotas should equal the number of seats. In this case the sum of the standard quotas is 20.01. The difference from 20 is due to rounding error.

▼ Try This One 2

Find the standard quota for each of the districts in Try This One 1.

As we pointed out, fractional parts of seats don't make sense, so none of the standard quotas in Example 2 can be used as seat numbers. You might guess that numbers of seats would be found using the traditional rules of rounding: round up to the next whole number if the decimal part is 0.5 or more, and round down to the previous whole number if the decimal part is less than 0.5. Let's see how that would work in Example 2:

District	1	2	3	4	5	Total
Seats	3	7	3	5	2	20

This works out perfectly! If it always worked out that way, we would be done. Since the section isn't over, you can guess that it doesn't always work out that way.

Consider the following example. Ten computers (seats) are to be apportioned to three schools (districts) based on the enrollment, which is summarized in the following table.

School	A	B	C	Total
Enrollment	561	1,015	1,324	2,900

The standard divisor is 2,900/10 = 290. The standard quotas are shown below:

School	A	B	C	Total
Standard quota	1.93	3.5	4.57	

According to the traditional rounding rule, School A gets two computers, School B gets four computers, and School C gets five computers. But, 2 + 4 + 5 = 11 and there are only 10 computers available!

Because of this issue, other methods of apportionment will be discussed that use the **upper quota** and **lower quota.** To calculate the upper quota, round the standard quota up to the next whole number (regardless of what comes after the decimal). To calculate the lower quota, round the standard quota down to the previous whole number.

EXAMPLE 3 Finding Upper and Lower Quotas

Find the upper and lower quotas for the districts in Example 2.

SOLUTION

Using the standard quotas from Example 2, we round up to get the upper quota and round down to get the lower quota. The results are shown in the table below.

District	1	2	3	4	5	Total
Population (in thousands)	32	80	41	65	22	240
Standard quota	2.67	6.67	3.42	5.42	1.83	20.01
Lower quota	2	6	3	5	1	17
Upper quota	3	7	4	6	2	22

▼ Try This One 3

Using the information from Try This One 2, find the upper and lower quotas for each district.

☑ 1. Compute standard divisors and quotas.

Notice that if the lower quota is used, only 17 seats are filled but there were 20 to be filled. Which district will get the extra seats? If the upper quota is used, 22 seats are needed so we are short two seats. Who will forfeit a seat? A variety of methods have been developed to answer these questions. In the remainder of the section, we'll study five of them.

Hamilton's Method

This method of apportionment was suggested by Alexander Hamilton and was approved by congress in 1791. His method, however, was subsequently vetoed by President Washington. The method was later adopted by the U.S. Government in 1852 and used until 1911. Hamilton's method was also known as the "method of largest remainders" which is a pretty good description. This method uses the lower quota and assigns surplus seats one by one in descending order of the fractional parts of their standard quota. The steps for using Hamilton's method are summarized below.

Hamilton's Method

1. Calculate the standard divisor.
2. Calculate the standard quota for each district.
3. Calculate the lower quota for each district.
4. Assign each district the number of seats according to its lower quota.
5. Assign any surplus of seats one by one to the districts in descending order of the fractional parts of their standard quotas.

EXAMPLE 4 Using Hamilton's Method

A large city with four districts plans to select 20 council members according to the population of the districts shown in the table. Using Hamilton's method, divide the 20 seats between the four districts.

District	1	2	3	4	Total
Population (in thousands)	150	88	17	65	320

SOLUTION:

1. Calculate the standard divisor: $\frac{320}{20} = 16$.
2. Calculate the standard quota for each district:

   District 1: $\dfrac{150}{16} = 9.375$

   District 2: $\dfrac{88}{16} = 5.5$

   District 3: $\dfrac{17}{16} = 1.0625$

   District 4: $\dfrac{65}{16} = 4.0625$

3. Calculate the lower quota for each district:

   District 1: 9

   District 2: 5

District 3: 1
District 4: 4

4. Initially assign the districts the number of seats equal to their lower quota above.
5. There is a surplus of one seat (since the sum of the lower quotas is 19). Assign the surplus seat to the district with the largest fractional part of the standard quota, which is District 2.

Math Note

Notice that leaving off the thousands has no effect on the quotas.

The distribution of seats is summarized as follows:

District	1	2	3	4	Total
Population (in thousands)	150	88	17	65	320
Seats	9	6	1	4	20

▼ Try This One 4

☑ 2. Apportion seats using Hamilton's method.

An elementary school plans to assign 35 teachers to teach each of the grades K–5. The number of students in each grade is given in the following table. Use Hamilton's method to assign each grade level the appropriate number of teachers.

Grade	K	1	2	3	4	5	Total
Enrollment	123	87	144	72	199	75	700

Jefferson's Method

This method of apportionment is attributed to Thomas Jefferson and rivaled Alexander Hamilton's method. It was used in congress from 1791 until 1842 when it was replaced with Webster's method. Jefferson's method uses a **modified divisor,** which is determined by trial and error and is smaller than the standard divisor. If the number of seats given by the lower quota is less than the number of seats available, we "modify" the standard divisor and recalculate the lower quotas with the modified divisor until the sum equals the number of available seats. Jefferson's method is summarized below.

Jefferson's Method

1. Calculate the standard divisor.
2. Calculate the standard quota for each district.
3. Calculate the lower quota for each district.
4. (a) If the sum of the lower quotas is equal to the number of seats, assign each district the number of seats equal to its lower quota.
   (b) If the sum of the lower quotas is less than the number of seats, choose a modified divisor by trial and error until the sum of the lower quotas equals the number of seats available. Assign each district the number of seats corresponding to its modified lower quota.

The problem of finding a modified divisor is a matter of repeatedly dividing by smaller numbers until the modified lower quotas sum to the correct value. To find the modified divisor we will start by subtracting 0.5 from the standard divisor. If the number of seats is still too few, but close, we might subtract smaller amounts, like 0.2 or 0.1; if it's not close we'll subtract larger amounts. If the number of seats is too many, then too much was subtracted.

EXAMPLE 5 Using Jefferson's Method

A city has 15 seats to divide among three districts according to the population shown in the table. Find the appropriate modified divisor to be used in Jefferson's method.

District	1	2	3	Total
Population (in thousands)	86	191	52	329

SOLUTION

The standard divisor is 21.9. The standard and lower quotas are summarized in the table:

District	1	2	3	Total
Population (in thousands)	86	191	52	329
Standard Quota	3.93	8.72	2.37	
Lower Quota	3	8	2	13

Since the number of seats is too few, try using a modified divisor. Subtracting 0.5 from 21.9, we get 21.4. The modified quotas are summarized below:

District	1	2	3	Total
Population (in thousands)	86	191	52	329
Modified quota	4.02	8.93	2.43	
Modified lower quota	4	8	2	14

This added one seat, but it's still not enough, since we need 15. Try subtracting 0.5 again and use 20.9. The new modified quotas are summarized below:

District	1	2	3	Total
Population (in thousands)	86	191	52	329
Modified quota	4.11	9.14	2.49	
Modified lower quota	4	9	2	15

Since the total of the lower quotas is now the desired number of seats, we can stop here and assign seats according to the last modified lower quota.

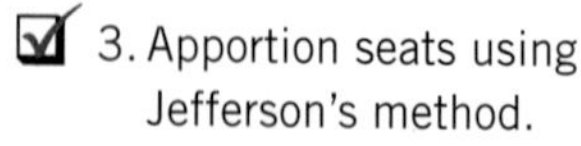
3. Apportion seats using Jefferson's method.

▼ Try This One 5

A county with population given in the table plans to assign 30 representatives according to population. Use Jefferson's method to calculate how many representatives each district will get.

District	1	2	3	4	Total
Population (in thousands)	92	274	79	193	638

Adams' Method

Proposed by John Quincy Adams in 1822, this method was never used in Congress. Adams' method is similar to Jefferson's method, except it uses modified upper quotas rather than lower.

EXAMPLE 6 Using Adams' Method

Assign the 15 seats from example 5 using Adams' Method.

Math Note

Recall that the standard divisor is found by dividing the population by the number of seats, and the standard quota for a district is found by dividing the population of the district by the standard divisor.

SOLUTION

The population information is given in the table.

District	1	2	3	Total
Population (in thousands)	86	191	52	329

The standard divisor is 21.9. The standard and upper quotas are summarized below:

District	1	2	3	Total
Population (in thousands)	86	191	52	329
Standard quota	3.93	8.72	2.37	
Upper quota	4	9	3	16

The upper quotas assign 16 seats, which is one too many. Since we need to have *fewer* seats, we will *raise* the divisor in increments of 0.5 until the modified upper quotas add to 15. All of 22.4, 22.9, and 23.4 leave the upper quotas unchanged, but for 23.9 we get the results below:

District	1	2	3	Total
Population (in thousands)	86	191	52	329
Modified quota	3.60	7.99	2.18	
Modified upper quota	4	8	3	15

Now we can assign the seats according to the modified upper quotas.

▼ Try This One 6

Assign the 30 seats from Try This One 5 using Adams' Method.

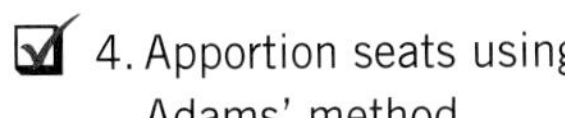
4. Apportion seats using Adams' method.

Webster's Method

Daniel Webster's method of apportionment was used in Congress from 1842 until Hamilton's method was adopted in 1852. Webster's method was readopted in 1911 and was used until 1941, at which time it was replaced by the Huntington-Hill method, which was signed into law by President Roosevelt in 1941. The Huntington-Hill method is still in use in Congress today. Webster's method is a cross between Hamilton's and Adams' methods and is summarized below.

Webster's Method

1. Calculate the standard divisor.
2. Calculate the standard quota.
3. Calculate the lower and upper quotas.
4. Initially assign each district the number of seats corresponding to the upper quota if the fractional part of the standard quota is 0.5 or greater. If the fractional part of the standard quota is less than 0.5, assign the lower quota.
5. Check to see if the sum of the seats is equal to the number of seats available. If so, use the assignments from step 4. If not, use a modified divisor and reassign seats based on the criteria in step 4, then repeat step 5. (If there are too many seats, try a larger divisor. If there are too few seats, try a smaller divisor.)

EXAMPLE 7 Using Webster's Method

Use Webster's method to assign 27 seats according to population for the following city with four districts.

District	1	2	3	4	Total
Population (in thousands)	149	83	92	126	450

SOLUTION

The standard divisor is 450/27 = 16.7. The corresponding standard quotas and initial assignments of seats are summarized in the table.

District	1	2	3	4	Total
Population (in thousands)	149	83	92	126	450
Standard Quota	8.92	4.97	5.51	7.54	
Initial Assignments	9	5	6	8	28

We have one seat too many, so we should try a larger divisor. We're pretty close, so we try a modified divisor of 16.8:

District	1	2	3	4	Total
Population (in thousands)	149	83	92	126	450
Modified quota	8.87	4.94	5.48	7.5	
Modified rounded quota	9	5	5	8	27

Now we can assign the seats according to the modified rounded quota.

☑ 5. Apportion seats using Webster's method.

▼ Try This One 7

A trucking company wants to assign 14 new trucks to four districts with volume of business by truckload indicated in the table. Use Webster's method to apportion the trucks to the districts according to volume of business.

District	1	2	3	4	Total
Volume (by truckload)	230	145	214	198	787

The Huntington-Hill Method

The Huntington-Hill method is the one currently used to apportion Congressional seats. It was developed around 1911 by Joseph Hill, the chief statistician for the Bureau of the Census, and Edward V. Huntington, professor of mechanics and mathematics at Harvard. This method of apportionment uses the *geometric mean.*

The geometric mean of two numbers x and y is given by $\sqrt{xy}$.

For example, the geometric mean of 2 and 8 is $\sqrt{(2)(8)} = \sqrt{16} = 4$. The geometric mean of 15 and 18 is $\sqrt{(15)(18)} = \sqrt{270} \approx 16.43$. The Huntington-Hill method uses the geometric mean to assign seats as follows:

The Huntington-Hill Method

1. Calculate the standard divisor.
2. Calculate the standard, lower, and upper quotas and the geometric mean of the lower and upper quotas for each district.
3. Initially assign the lower quota if the standard quota is less than the geometric mean of the upper and lower quotas; assign the upper quota if the standard quota is greater than the geometric mean of the upper and lower quotas.
4. If the sum of the seats assigned in step 3 is equal to the number of seats available, leave seats assigned according to step 3. If the sum of the seats assigned in step 3 is not equal to the number of seats available, use a modified divisor as in the previous three methods and reassign seats accordingly until the sum equals the total number of seats available.

EXAMPLE 8 Using the Huntington-Hill Method

A club with 400 members is to assign 10 chair positions to represent the members coming from various communities. The distribution of the members is as follows:

Community	A	B	C	Total
Number of Members	153	75	172	400

Use the Huntington-Hill method to assign the 10 seats.

SOLUTION

The standard divisor is 400/10 = 40. The standard quotas and geometric means are given in the table.

Community	A	B	C	Total
Number of members	153	75	172	400
Standard quota	3.825	1.875	4.3	
Geometric mean	3.46	1.41	4.47	

Since community A has a standard quota higher than the geometric mean, assign the upper quota of four seats. Similarly, assign the upper quota of two seats to community B. Finally, since the standard quota of community C is less than the geometric mean, assign the lower quota of four seats.

The sum of seats assigned is 4 + 2 + 4 = 10, so there is no need for a modified divisor and the assignments work with no modification.

▼ Try This One 8

☑ 6. Apportion seats using the Huntington-Hill method.

A university mathematics department hired five new teaching assistants to be assigned to three professors according to the total number of students each professor teaches, which is indicated in the table. Use the Huntington-Hill method to assign the 5 TA's to the professors.

Professor	Hunt	Tingue	Hill	Total
Number of Students	200	250	145	595

Answers to Try This One

1 320,000

2

District	1	2	3	4	5	6	7
Standard quota	3.91	2.78	7.34	3.03	3.19	4	5.75

3

District	1	2	3	4	5	6	7
Lower quota	3	2	7	3	3	4	5
Upper quota	4	3	8	4	4	4	6

4

Grade	K	1	2	3	4	5
Teachers	6	4	7	4	10	4

5 Using modified divisor 19.7:

District	1	2	3	4
Representatives	4	13	4	9

6 Using modified divisor 22.9:

District	1	2	3	4
Representatives	5	12	4	9

7 Using modified divisor 56.7:

District	1	2	3	4
Trucks	4	3	4	3

8

Professor	Hunt	Tingue	Hill
TA's	2	2	1

EXERCISE SET 14-4

Writing Exercises

1. Describe what apportionment is in your own words.
2. Explain how to find the standard divisor for an apportionment process, and describe its significance.
3. Why can we not simply find the standard quota and use traditional rules of rounding to get an apportionment?
4. Describe how to find the upper and lower quotas from a standard quota.
5. Which apportionment methods use a modified divisor? How do those methods differ?
6. If an apportionment using upper or lower quotas results in too few seats, should you raise or lower the standard divisor? Explain your answer.

Real-World Applications

For Exercises 7–9 find:

(a) The standard divisor.
(b) The standard quota for each district.
(c) The upper and lower quotas for each district.
(d) Assignment of seats using Hamilton's method.

7. This year 10 faculty members are to receive promotions at a community college with three campuses. The number of promotions will be apportioned according to the number of full-time faculty members as shown. Determine how many promotions there will be on each of the campuses.

Campus	South	Central	North	Total
Faculty	62	148	110	320

8. A large school district has four high schools with enrollments as shown. Forty school buses need to be allocated to the high schools. Determine how many buses are allocated to each of the high schools.

School	RRHS	CCHS	HHHS	MHS	Total
Enrollment	628	941	1,324	872	3,765

9. A county has three districts; the population of each district in thousands is shown. Nine representatives are to be appointed to the county board. Determine how many representatives each district should receive.

District	1	2	3	Total
Population (in thousands)	242	153	185	580

For Exercises 10–12 find:

(a) The standard divisor.
(b) The standard quota for each district.
(c) The upper and lower quotas for each district.
(d) Assign seats using Jefferson's method and Adams' method.
(e) Write the modified divisor you used for each method.

10. A large chain store needs to assign 12 buyers to its five stores. The number of employees for each store is shown. Determine how many buyers should be assigned to each store.

Store	1	2	3	4	5	Total
Employees	56	32	74	62	86	310

11. A freight company has four terminals. The volume of business in truckloads for each terminal is shown. The owner has purchased 12 new trucks to be allocated to the four terminals. Determine how many trucks each terminal should get based on volume of business.

Terminal	A	B	C	D	Total
Volume (in truck loads)	51	37	65	22	175

12. A large library has six branches; the number of books at each branch is shown. The library has obtained 100 new books. Determine how many books each library should get based on their current volume of books.

Branch	A	B	C	D	E	F	Total
Books	1,236	872	2,035	1,655	3,271	1,053	10,122

For Exercises 13–15, find:

(a) The standard divisor.
(b) The standard quota for each district.
(c) The upper and lower quotas for each district.
(d) Assign seats using Webster's method.
(e) Write the modified divisor you used.

13. An automotive repair company has four stores in a large city. The owner purchased 10 diagnostic computers for the business. The number of cars serviced per store each month is given below. Determine how many computers each store should get based on number of cars serviced.

Store	1	2	3	4	Total
Cars serviced	119	99	186	136	540

14. A hospital has five wards; the average number of patients per ward is shown. Twelve new nurses are hired. Determine how many nurses each ward should receive based on average number of patients.

Ward	A	B	C	D	E	Total
Patients	372	465	558	619	197	2,211

15. Sixteen new police officers are to be assigned to four precincts. The populations of the precincts are shown. Determine how many of the 16 new officers each precinct should receive.

Precinct	1	2	3	4	Total
Population	3,562	8,471	2,146	5,307	19,486

For Exercises 16–18, find:

(a) The standard divisor.
(b) The standard quota for each district.
(c) The upper and lower quotas for each district.
(d) Assign seats using the Huntington-Hill method.
(e) Write the modified divisor you used.

16. A county has three clinics; the weekly number of patients seen by a nurse practitioner in each clinic is shown. The director hires eight new nurse practitioners; assign them to the three clinics based on number of patients seen.

Clinic	A	B	C	Total
Patients	456	616	573	1,645

17. A chain of physical therapy clinics has four offices; the average number of patients treated per month at each office is shown. Six new therapists are hired and apportioned to the offices according to number of patients seen. How many therapists should each office receive?

Office	1	2	3	4	Total
Patients	2,231	996	1,564	1,799	6,590

18. A department store is hiring 10 new sales associates and needs to assign them to four departments according to volume of sales in dollars. The daily sales in dollars are given. Decide how many associates each department should receive.

Department	A	B	C	D	Total
Sales	5,562	4,365	4,012	6,531	20,470

Critical Thinking

19. Suppose that an apportionment method initially assigns the upper quota to each district. Devise a fair plan of your own to eliminate any excess seats that were assigned. Your plan should be different from any in this section.

20. Of the five apportionment methods we studied, which do you think is the most fair? Explain your answer.

Section 14-5 Apportionment Flaws

LEARNING OBJECTIVES

- ☐ 1. Illustrate the Alabama paradox.
- ☐ 2. Illustrate the population paradox.
- ☐ 3. Illustrate the new states paradox.
- ☐ 4. Describe the quota rule.

The U.S. constitution requires the House of Representatives to be reapportioned after each census. Except for a 4-year period after Alaska and Hawaii joined the union, the number of representatives has been fixed at 435 since 1911. But in the 1800s, that number was very fluid as the country experienced rapid population growth. Hamilton's method had been in use to apportion seats since 1852, but after the 1880 census, something funny happened. The chief clerk of the Census Bureau computed apportionment for all house sizes between 275 and 350 (by hand!), and noticed that if there were 299 seats in the House, Alabama was apportioned 8 seats. But if there were 300 seats, Alabama was only apportioned 7. In other words, adding a seat to the house would cause Alabama to *lose* a seat! This is known as a *paradox*—a conclusion that makes no sense even though it was drawn from a sound logical process. This particular result became known as the *Alabama paradox*.

The Alabama Paradox

An increase in the total number of seats to be apportioned causes a district to lose a seat.

Example 1 illustrates the Alabama paradox.

EXAMPLE 1 Illustrating the Alabama Paradox

A county has six districts populated as shown; 15 seats are to be apportioned according to population using Hamilton's method. The county then decides to add a representative and reapportion the seats. Show that the Alabama paradox occurs.

District	A	B	C	D	E	F	Total
Population (in thousands)	39	220	264	163	167	101	954

SOLUTION

By Hamilton's method we get:

District	A	B	C	D	E	F	Total
Population (in thousands)	39	220	264	163	167	101	954
Standard quota	0.61	3.46	4.15	2.56	2.63	1.59	
Lower quota	0	3	4	2	2	1	12
Apportionment	1	3	4	2	3	2	15

Districts A, E, and F get the extra seats because their standard quotas have the three largest decimal parts.

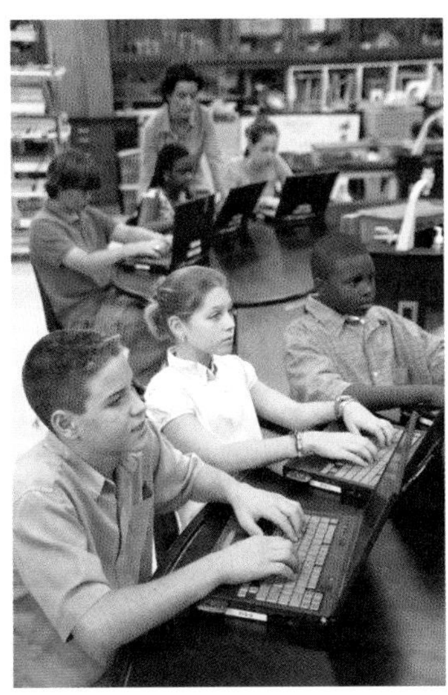

When the 16th seat is added the result is:

District	A	B	C	D	E	F	Total
Population (in thousands)	39	220	264	163	167	101	954
Standard quota	0.65	3.69	4.43	2.73	2.80	1.69	
Lower quota	0	3	4	2	2	1	12
Apportionment	0	4	4	3	3	2	16

Now the four largest decimal parts belong to districts B, D, E, and F. Adding one seat caused A to lose its only representative!

▼ Try This One 1

A large company decided to donate 17 computers to West Hills school district. The director of the computer services decided to allot them to the three middle schools based on number of students, using Hamilton's method. However, when the computers arrived there were actually 18 computers that were then reallocated. Show that in this situation the Alabama Paradox occurred.

☑ 1. Illustrate the Alabama paradox.

School	JDM	DM	AM	Total
Number of students	1,328	1,008	384	2,720

At the time of its discovery, the Alabama paradox was overcome by choosing a number of seats for which the paradox didn't seem to come into play, not exactly a long-term solution. Another flaw in Hamilton's method was discovered in 1900. At that time, Virginia was growing at a rate about 60% faster than Maine was, but Virginia lost a seat, while Maine gained one. This is known as the *population paradox*.

The Population Paradox

An increase in a district's population can cause it to lose a seat to a slower growing district.

EXAMPLE 2 Illustrating the Population Paradox

A county has three districts with populations shown. There are to be 120 representatives elected to represent the three districts. A year later the population increases in two of the districts as indicated, and the seats are reapportioned. Show that if seats are apportioned using Hamilton's method, the population paradox occurs.

District	A	B	C	Total
Original population	29,317	106,350	14,333	150,000
New population	29,604	106,350	14,477	150,431

SOLUTION

For the original population we have the following:

District	A	B	C	Total
Original population	29,317	106,350	14,333	150,000
Standard quota	23.45	85.08	11.47	
Lower quota	23	85	11	119
Apportionment	23	85	12	120

C has the largest decimal part.

For the new population we have a 0.98% increase in population for district A, 0% for district B, and 1% for district C. The new apportionments are as follows:

District	A	B	C	Total
New population	29,604	106,350	14,477	150,431
Standard quota	23.615	84.836	11.548	
Lower quota	23	84	11	118
Apportionment	24	85	11	120

C has the smallest decimal part.

Notice that even though district C had a greater percent increase in population than district A, district C lost a seat and district A gained a seat. This is an example of the population paradox.

▼ Try This One 2

☑ 2. Illustrate the population paradox.

A county with three districts has populations as shown. Six representatives are to be elected and apportioned to the districts according to population. Five years later the population increases as indicated and the six representatives are reapportioned. Show that the population paradox occurs in this situation.

District	A	B	C	Total
Original population (in thousands)	135	253	572	960
New population (in thousands)	173	420	927	1,520

Hamilton's method somehow managed to survive the discovery of both the Alabama and population paradoxes, but another problem was discovered in 1907 when Oklahoma joined the union. At that time there were 386 seats in the house, and it was clear based on its population that Oklahoma should get 5 seats, so the number of seats was increased to 391. The idea was Oklahoma would get the five new seats and the other states would retain the seats they had. However, when the 391 seats were reapportioned mathematically this didn't happen! Maine gained a seat and New York lost a seat. This paradox has become known as the *new states paradox*.

The New States Paradox

Adding a state with its fair share of seats can affect the number of seats due other states.

EXAMPLE 3 Illustrating the New States Paradox

A company wants to distribute 10 new computers to three branch offices with number of employees as shown.

Branch	A	B	C	Total
Employees	21	146	204	371

The company president suddenly remembers that they've recently opened a new branch that has 26 employees. Based on the apportionment of the original 10 computers he decides the new branch should get one computer so he purchases one more for that branch. Show that if the president reapportions the 11 computers to the four branch offices using Hamilton's method the new states paradox occurs.

SOLUTION

The original apportionment would be as follows:

Branch	A	B	C	Total
Employees	21	146	204	371
Standard quota	0.57	3.94	5.50	
Lower quota	0	3	5	8
Apportionment	1	4	5	10

The two "surplus" computers went first to branch B then to branch A by Hamilton's method.

The new apportionment with the addition of branch D is as follows:

Branch	A	B	C	D	Total
Employees	21	146	204	26	397
Standard quota	0.58	4.05	5.65	0.72	
Lower quota	0	4	5	0	9
Apportionment	0	4	6	1	11

The two "surplus" computers now go first to branch D then to branch C by Hamilton's method. This causes branch A to lose a computer. This is the new states paradox.

▼ Try This One 3

☑ 3. Illustrate the new states paradox.

A city has two districts with populations as shown. There are 100 representatives between the two districts.

District	A	B	Total
Population	1,545	8,455	10,000

(a) Apportion the representatives according to population using Hamilton's method.

(b) A new district is annexed with population 625 and correspondingly 6 new representatives are added. Reapportion the 106 seats using Hamilton's method and show that the new states paradox occurs.

Math Note

After this third paradox in Hamilton's method was discovered in 1907, the government, moving with typical speed, finally removed it in favor of Webster's method in 1911.

Just as there are fairness criteria for voting methods, there are fairness criteria for apportionment methods. In a fair method, none of the three paradoxes should occur, and the method should also satisfy the *quota rule*.

The Quota Rule

Every district in an apportionment should be assigned either its upper quota or its lower quota. An apportionment that violates the rule is said to *violate quota*.

The obvious question to close with is "Is there an apportionment method that is completely fair?" The answer is no. We have seen that Hamilton's method can violate all of the paradoxes, but it satisfies the quota rule (since by definition every district gets either its upper or lower quota). It turns out that the other four apportionment methods we studied in Section 14-4 are immune from the paradoxes, but they all use modified divisors, so they occasionally violate quota. (Very occasionally, as it turns out—none of the examples in Section 14-4 violate quota.) In fact, mathematicians

Sidelight SOME INTERESTING FACTS ABOUT VOTING

- You might think that voting is a human creation, but it isn't. Honeybees vote, and they can't even count! When it's time to locate a new hive, scouts return from their search for a good location, and they dance. The bees that dance most vigorously manage to recruit other scouts to their side until one site has the majority of scouts dancing in its favor.
- The first election poll in U.S. history, conducted by the *Harrisburg Pennsylvanian* newspaper, correctly predicted that Andrew Jackson would win the most votes in the upcoming election. There was just one problem: Jackson lost the election in the electoral college, and John Quincy Adams won the presidency. (Project 1 at the end of the chapter examines our system for electing a president.)

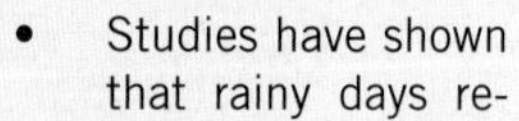

- The lever voting machine was first used in 1892, and at the time it had more moving parts than almost any other device made in America. Ironically, it was marketed as a "plain, simple" way to cast ballots.

- For the 2006 elections, the U.S. Department of Defense paid $830,000 for a Web-based system that allowed those serving in the military oversees to vote easily. Only 63 people used it.
- Studies have shown that rainy days result in lower voter turnout, at about the rate of 0.8% for each inch of rain. Computer simulations indicate that if it had rained in Illinois on election day 1960, Nixon would have defeated Kennedy, and if it had been sunny in Florida on election day 2000, Gore would have defeated Bush.

Source: *Discover Magazine*

Michael L. Balinski and H. Peyton Young proved in 1980 that any apportionment method that doesn't violate the quota rule must produce paradoxes, and any method that never produces paradoxes must violate the quota rule.

☑ 4. Describe the quota rule.

In summary, there is no perfect method for either voting or apportionment: it is always left to the discretion of the people involved to decide on a method that is most fair for their needs.

Answers to Try This One

1 The original apportionment is 8 for JDM, 6 for DM, and 3 for AM. The new apportionment is 9 for JDM, 7 for DM, and 2 for AM. AM lost a computer because of the district apportioning one more.

2 The original apportionment is 1 for A, 2 for B, and 3 for C. The new apportionment is 1 for A, 1 for B, and 4 for C. But B's growth rate was 66% compared to 62% for C, yet B lost a seat to C.

3 The original apportionment is 15 for A, 85 for B. The new apportionment is 16 for A, 84 for B, and 6 for C. B lost a seat because C was added.

EXERCISE SET 14-5

Writing Exercises

1. What is a paradox?
2. Describe the Alabama paradox in your own words.
3. Describe the population paradox in your own words.
4. Describe the new states paradox in your own words.
5. What is the quota rule? Which apportionment methods can violate it?
6. Explain why Hamilton's method cannot violate the quota rule.

Real-World Applications

7. A county library system has three branches. The number of books at each branch is shown. The head librarian purchased 57 new books and apportioned them to the libraries using Hamilton's method. When the book order arrived the librarian noticed there was an extra book added as a "bonus" and she reapportioned the books with the extra book added. Determine if the reapportionment resulted in the Alabama paradox.

Branch	A	B	C	Total
Books	1,556	1,328	365	3,249

8. A school district hired nine elementary teachers to be apportioned to their four elementary schools according to enrollment, which is shown. By mistake the director called 10 of the applicants and told them they were hired! He now needs to reapportion the 10 teachers to the 4 schools. Determine if the Alabama paradox occurs if the teachers are assigned to schools by using Hamilton's method.

School	A	B	C	D	Total
Enrollment	876	586	758	103	2,323

9. Six computers were donated to be distributed to three local churches according to membership. When the computers arrived at the denomination's main office there were actually seven computers. If Hamilton's method was used to decide how to distribute the computers among the churches, determine if the Alabama paradox occurs.

Church	Saint A's	Saint B's	Saint C's	Total
Membership	937	114	622	1,673

10. A city with three districts has an increase in population as indicated in the table. Using Hamilton's method, apportion the 24 seats between the three districts both before and after the population increase. Find the percent increase in population for each district and determine if the population paradox occurs.

District	A	B	C	Total
Original population (in thousands)	988	530	2,242	3,760
New population (in thousands)	1,248	677	2,571	4,496

11. A small business has three branch offices with increase in number of employees indicated. Thirty new desks are to be apportioned according to number of employees at a given branch. Apportion the 30 desks both before and after the increase in number of employees and determine if the population paradox occurs.

Branch	1	2	3	Total
Original number of employees	161	249	490	900
New number of employees	177	275	514	966

12. Eleven legislative seats are to be apportioned to a country's three states according to population using Hamilton's method. Five years later the 11 seats are reapportioned with the indicated increase in population. Determine if the population paradox occurs when the seats are reapportioned.

State	A	B	C	Total
Original population	55	125	190	370
New population	60	150	213	423

13. The table shows the enrollment at two campuses of a community college.

Campus	North	South	Total
Enrollment	1,326	8,163	9,489

(a) Apportion 61 new media podiums to the two campuses according to enrollment using Hamilton's method.
(b) Suppose the college adds a third campus with enrollment of 1,070 and correspondingly added seven podiums. Reapportion the 68 podiums using Hamilton's method.
(c) Determine if the new states paradox occurs.

14. A state with 40 legislative seats has two districts with populations as follows.

District	A	B	Total
Population	153	787	940

(a) Apportion the seats using Hamilton's method.
(b) The state adds a third district with population 139 and adds 6 new seats accordingly. Reapportion the seats using Hamilton's method.
(c) Determine if the new states paradox occurs.

15. A small island nation has 97 legislative seats to be divided among three states with population as follows.

State	A	B	C	Total
Population	2,163	9,504	8,133	19,800

(a) Apportion the seats using Hamilton's method.
(b) The country adds a state with population 2,629 and adds 13 seats accordingly; reapportion the seats using Hamilton's method.
(c) Determine if the new states paradox occurs.

Critical Thinking

16. For the city in Example 4 of Section 14-4, see if you can find an original and new number of seats that would result in the Alabama paradox.

CHAPTER 14 Summary

Section	Important Terms	Important Ideas
14-1	Preference table Plurality method Head-to-head comparison criterion	**This chapter** presented four voting methods people have used to determine the results of an election where there are three or more choices (i.e., candidates, courses of action, etc.). These methods involve ranking the choices in order of preference on a preference ballot. A preference table can be used to summarize the results of the election. Using the plurality method, the candidate with the most first-place votes is the winner. Each method has at least one inherent flaw. That is, it fails to satisfy one of the fairness criteria for a voting method. The head-to-head criterion states that if a candidate wins all head-to-head comparisons with all other candidates, that candidate should win the election.
14-2	Borda count method Majority criterion Plurality-with-elimination method Monotonicity criterion	**In the Borda count method,** candidates are ranked on the ballot by voters. The candidate in last place on the ballot gets 1 point. The candidate in the next to the last place gets 2 points, etc. The points are tallied and the candidate with the most points is the winner. The majority criterion says that if a candidate receives a majority of first-place votes, then that candidate should be the winner. In the plurality-with-elimination method the candidate with a majority of first-place votes is the winner. If no candidate has a majority of first-place votes, the candidate with the least number of votes is eliminated, and all candidates who were below the eliminated candidate move up. The first-place votes are counted again. The process then is repeated until a candidate receives a majority of first-place votes. The monotonicity criterion states that if after an election a reelection is held with the same candidates and the only changes in voting favor the winner of the original election, then he or she should win the reelection.
14-3	Pairwise comparison method Irrelevant alternatives criterion Arrow's impossibility theorem Approval voting	**In the pairwise comparison method,** each candidate is ranked by the voters. Then each candidate is paired with every other candidate in a head-to-head or one-to-one contest. The winner of each contest gets 1 point. In case of a tie, each candidate gets $\frac{1}{2}$ of a point. The candidate with the most points wins the election. The irrelevant alternatives criterion states that if candidate A wins a certain election and if one of the other candidates is removed from the ballot and the ballots are recounted, candidate A should still win the election. The search for a perfect voting method continued until the early 1950s when an economist, Kenneth Arrow, proved a theorem that states that in elections with three or more candidates, a voting method that doesn't violate any of the four fairness criteria is an impossibility. This fact is known as Arrow's impossibility theorem. In the late 1970s several political scientists devised a voting method that is called approval voting. Here each voter votes for as many candidates as he or she finds acceptable. The votes are tabulated and the candidate who receives the most votes is declared the winner.

Section	Terms	Summary
14-4	Apportionment Standard divisor Standard quota Upper and lower quotas Hamilton's method Jefferson's method Adams' method Webster's method The Huntington-Hill method	**Apportionment** is the process of fairly dividing objects among individuals that may be entitled to unequal shares. Hamilton's method uses lower quotas. It initially assigns the lower quota to all districts, then assigns any leftover seats one at a time based on the size of the decimal part of the standard quota. Jefferson's method also begins with lower quotas, but if there are unassigned seats, lower quotas are recalculated using a modified divisor. Adams' method is similar, but uses upper quotas. Webster's method uses quotas rounded using traditional rounding rules, and uses a modified divisor if the number of seats apportioned is incorrect. The Huntington-Hill method is similar to Webster's method, but uses comparison to the geometric mean of the upper and lower quotas rather than traditional rounding rules.
14-5	Alabama paradox Population paradox New states paradox Quota rule	**Hamilton's method** can exhibit three different paradoxes. The Alabama paradox occurs when adding one or more seats to an apportionment causes one district to lose a seat. The population paradox occurs when a district loses a seat to another district that had a smaller percentage growth. The new states paradox occurs when adding a new district with an appropriate number of seats causes an existing district to lose a seat. The quota rule says that for an apportionment process to be fair, every district should be apportioned either its upper or lower quota. Any apportionment method that is immune to paradoxes can violate the quota rule, and any method that never violates the quota rule can exhibit one or more of the paradoxes.

MATH IN College Football REVISITED

1. Using the Borda count method, but awarding only 3 points for first place, 2 for second, and 1 for third, the results were

Sam Bradford, Oklahoma	1,726 points
Colt McCoy, Texas	1,604 points
Tim Tebow, Florida	1,575 points
Graham Harrell, Texas Tech	213 points
Michael Crabtree, Texas Tech	116 points

Sam Bradford won the award, even though he had fewer first-place votes than Tim Tebow.

2. Since Tebow had the most first-place votes, he would have won if the plurality method had been used. (Tebow had the most first-place votes and didn't even finish second!)

3. Because of the large number of ballots, and the many possible combinations, the voting information was not listed in preference table form, so it's difficult to determine if the fairness criteria were violated. The head-to-head comparison criterion may have been violated, since Tebow had the most first-place votes. But the fact that he was listed third on the highest number of ballots means that one of the top two finishers may beat him head-to-head, or maybe even both. The majority criterion was definitely not violated since nobody got a majority of first-place votes. It's hard to guess if the monotonicity criterion was violated, but because of the closeness of the election, it's possible that if some ballots that had Bradford third were changed to ballots that had him second, it could have cost him the election if a large majority of those ballots moved one of McCoy or Tebow to first. It's possible, and maybe even likely, that the irrelevant alternatives criterion was violated. The removal of the candidates not in the top three would have added more votes to those in the top three, and there's most likely some combination that would tip the election in favor of McCoy or Tebow.

Review Exercises

Sections 14-1 to 14-3

Use this information for Exercises 1–4: The preference ballots for an election for the best speaker in a contest are shown. There are three candidates: Peterson (P), Quintana (Q), and Ross (R).

Q	P	R	Q	Q	P	R	P
R	Q	P	R	R	Q	P	Q
P	R	Q	P	P	R	Q	R

R	Q	P	R	Q	R	P
P	R	Q	P	R	P	Q
Q	P	R	Q	P	Q	R

1. Construct a preference table for the results of the election.
2. How many people voted in the election?
3. How many people voted for Ross as the best speaker?
4. How many people voted for the contestants in the order Quintana, Ross, Peterson?

Use this information for Exercises 5–8: A class of students decided to have a pizza party on the last day of math class. They voted to pick a pizza place. The candidates were Pizza Palace (P), Pizza Heaven (H), and Pizza City (C). The preference ballots are shown.

C	P	H	H	H	H	C	P
P	C	P	P	P	P	P	C
H	H	C	C	C	C	H	H

H	H	H	C	C	P	H	C
P	P	P	P	P	C	P	P
C	C	C	H	H	H	C	H

P	H	C	H
C	P	P	P
H	C	H	C

5. Construct a preference table for the results of the election.
6. How many people voted?
7. How many people voted for Pizza City as their first choice?
8. How many people voted for the Pizza Palace, Pizza City, and Pizza Heaven in that order?

Use this information for Exercises 9–17: A large city police department has the option to choose among three styles of bulletproof vests. They are A, B, and C. The preference table is shown.

Number of votes	26	15	10	7
First choice	A	B	C	B
Second choice	B	C	A	A
Third choice	C	A	B	C

9. How many police officers voted?
10. How many votes did the preference CAB receive?
11. Using the plurality method, which style won?
12. Using the Borda count method, which style won?
13. Using the plurality-with-elimination method, which style won?
14. Using the pairwise comparison method, which style won the election?
15. When the plurality method was used, was the head-to-head comparison criterion violated?
16. When the plurality-with-elimination method was used, was the majority criterion violated?
17. If style C was unavailable, and the votes were recounted, was the irrelevant alternatives criterion violated if the pairwise comparison voting method is used?

Use this information for Exercises 18–27: A college theater class must decide which musical to perform as part of their course requirements. Their options are *A Chorus Line* (C), *Guys and Dolls* (G), *Bye Bye Birdie* (B), and *The Music Man* (M). They decide to vote on the choices, and the preference table is shown.

Number of votes	18	17	9	3
First choice	M	B	G	C
Second choice	B	G	C	G
Third choice	G	C	M	B
Fourth choice	C	M	B	M

18. How many votes did the preference GCMB receive?
19. How many students voted in the election?
20. Using the Borda count method which musical was selected?
21. Using the plurality method which musical was selected?
22. Using the pairwise comparison method which musical was selected?
23. Using the plurality-with-elimination method which musical was selected?
24. When the Borda count method was used, was the majority criterion violated?
25. When the plurality method was used, was the head-to-head comparison criterion violated?
26. If *A Chorus Line* was too expensive to produce and had to be removed and the votes were

recounted, was the irrelevant alternatives criterion violated if the pairwise comparison voting method is used?

27. Suppose that the plurality-with-elimination method was used, and the three members who voted CGBM decided to change their votes to MBGC. Would this violate the monotonicity criterion?
28. As a fundraiser for charity, five professors volunteer to spend a weekend in the county jail if students raised $10,000 for the local foodbank. The students get to choose which professor goes behind bars, using approval voting. The results are shown below. Which professor gets locked up?

Number of voters	3	1	1	2	4	1	1
Seubert	/		/		/		/
Glass	/	/	/				
Carothers		/	/		/	/	/
Hern				/	/		
Nguyen		/				/	/

29. The members of the student activities committee vote to decide on a program for parents weekend. The results of the approval voting are shown.

Number of voters	1	5	3	1	1	2
Magician	/			/	/	/
Speaker		/	/		/	
Rock band	/	/	/			
Comedian		/	/	/	/	

What will be on the program?

Section 14-4

30. A community college has purchased 15 laptop computers to be apportioned to three departments with number of faculty as shown. Using Hamilton's method, determine how many computers each department will receive.

Department	Math	English	Science	Total
Number of faculty	25	20	15	60

31. A church has four locations with average weekly attendance as shown. Ten new assistant pastors are to be assigned to each location according to weekly attendance. Using Hamilton's method, determine how many assistant pastors each location should receive.

Location	A	B	C	D	Total
Average weekly attendance	950	265	180	450	1,845

32. Repeat exercise 30 using Jefferson's method.
33. Repeat exercise 31 using Adams' method.
34. Repeat exercise 30 using Webster's method.
35. Repeat exercise 31 using Webster's method.
36. Repeat exercise 30 using the Huntington-Hill method.
37. Repeat exercise 31 using the Huntington-Hill method.

Section 14-5

In Problems 38 and 39, determine if the Alabama paradox occurs if one seat is added. Use Hamilton's method.

38. Originally there were 10 seats; one seat is added to make 11.

District	A	B	C	Total
Population	32	359	433	824

39. Originally there were seven seats; one seat is added to make eight.

District	A	B	C	Total
Population	137	933	1,000	2,070

In Problems 40–41, determine if the population paradox occurs. Use Hamilton's method.

40. There are 24 seats to be assigned.

District	A	B	C	Total
Original population	529	989	2,237	3,755
New population	681	1,249	2,568	4,498

41. As part of a group project in a political science course, students have to choose how to divide up the total points earned by their group. They decide to apportion the points using Hamilton's method based on how many words each wrote for the final report. The group earns 95 points for the project, and no fractional parts of points will be assigned. The number of words written by each student in the first draft and final report are shown below.

Student	Sinead	June	Tabitha	Cameron	Total
Words in draft	3,210	1,975	2,455	4,001	11,641
Words in final report	3,390	2,510	2,500	4,111	12,511

In Problems 42–43, determine if the new states paradox occurs. Use Hamilton's method.

42. There are 15 original seats; one new district with population 1,900 is added and correspondingly 5 seats are added.

District	A	B	Total
Population	981	4,893	5,874

43. There are 55 original seats; one new district is added with population 5,912 and correspondingly 10 seats are added.

District	A	B	C	Total
Population	6,230	16,323	10,101	32,654

Chapter Test

Use this information for Exercises 1–4: Customers in a coffee shop are asked to rank three types of coffee. They are Arabica (A), blended (B), and Colombian (C). The preference ballots are shown.

C	B	B	C	B	A	C	B
B	A	A	B	A	B	B	A
A	C	C	A	C	C	A	C

B	C	B	B
A	B	A	A
C	A	C	C

1. Construct a preference table for the ballots.
2. How many people voted?
3. How many people voted for blended as their first choice?
4. How many people selected CBA as their ranking preference?

Use this information for Exercises 5–14: A small contracting company has been invited to relocate to one of three cities: Pittsburgh (P), Baltimore (B), and Richmond (R). The employees are asked to vote on the city. The preference table is shown.

Number of votes	43	27	18	12
First choice	P	R	B	P
Second choice	R	P	R	B
Third choice	B	B	P	R

5. How many employees voted?
6. How many selected the preference PBR?
7. Using the plurality method which city won?
8. Using the Borda count method which city won?
9. Using the plurality-with-elimination method which city won?
10. Using the pairwise comparison method which city won?
11. When the plurality method was used, was the head-to-head criterion violated?
12. When the Borda count method was used, was the majority criterion violated?
13. When the plurality-with-elimination method was used, was the head-to-head criterion violated?
14. If Baltimore was ruled out and the votes were recounted, was the irrelevant alternatives criterion violated if the Borda count method was used?
15. A website devoted to criticizing TV shows holds a vote among its staff members and loyal readers for the worst reality TV show, using approval voting. A vote indicates support for that show being named worst. The results are shown below. Which show was named worst?

Number of votes	247	193	52	119	220
Celebrity Bocce	/	/	/		
The Canine Bachelor		/	/	/	
Big Brother Nursing Home	/		/	/	/
Dancing with Bears				/	

16. The five players who are named first-team all conference in basketball for the Great Central Conference are eligible for conference player of the year. The winner is chosen using approval voting by the conference coaches and athletic directors. The results are shown below. Who won?

Number of votes	5	3	8	4
Donte Brown	/		/	
Phillip Sparks	/	/		/
Monte Lopez		/	/	
Patrick Miller	/		/	/
Taj Watts			/	/

17. An airline offers nonstop flights from Fort Lauderdale to three different locations. It has recently hired 12 new flight attendants and plans to assign them to the flights according to number of passengers. The number of passengers on each flight is summarized in the table. Apportion the flight attendants using Hamilton's method.

Flight	A	B	C	Total
Number of passengers	156	264	86	506

18. Repeat Problem 17 using Jefferson's method.
19. Repeat Problem 17 using Webster's method.
20. Repeat Problem 17 using the Huntington-Hill method.
21. Repeat Problem 17 using Adams' method.
22. Determine if the Alabama paradox occurs. There were 10 original seats, one seat is added. Use Hamilton's method.

District	A	B	C	D	Total
Population	48	393	454	258	1,153

23. The original and new populations for four districts are shown below. There are 20 seats. Determine if the population paradox occurs. Use Hamilton's method.

District	A	B	C	D	Total
Original population	1,156	2,300	3,005	1,795	8,256
New population	1,215	2,555	3,147	2,101	9,018

24. Determine if the new states paradox occurs. There were two original states as indicated in the table and 40 original seats; an additional state with population 3,400 was added and correspondingly 7 additional seats. Use Hamilton's method.

State	A	B	Total
Population	6,255	13,745	20,000

Projects

1. (a) Using the Internet or a library, write a brief report on the Electoral College system used to elect the president.
   (b) Once the Electoral College has been chosen by the individual states, what voting system does the College use to elect the president?
   (c) What voting system is used by the individual states to allocate their electoral votes?
   (d) Discuss the entire election process as it relates to the four fairness criteria we studied. Which if any of the criteria do you think may be violated by the process?
   (e) As a summary, write an opinion piece on whether you (or your group) think that the Electoral College process should be abandoned, and if so which voting system you think would be preferable.
2. (a) Print up a ballot listing five TV shows that are currently popular among college students and ask everyone in your class to rank the shows from most favorite to least favorite. After collecting the ballots, make a preference table and use all four voting methods from Chapter 14 to determine the most popular show. Did you get the same result for all four methods?
   (b) Now have all of your classmates fill out an approval ballot, marking all of the shows that they like. Tally the results. How does the result compare to the other four voting systems?
   (c) Write a short opinion piece on which method you think provided the most reliable gauge of which show is most popular among your classmates. (If the winner was the same for all five methods, your answer will be pretty short.)
3. Using the Internet as a resource, find the most recent apportionment of seats to the U.S. House of Representatives, then find the population figures for each state. The simplest and quickest method of apportionment is Hamilton's method—use Hamilton's method to apportion the House seats to the 50 states and see if the results match the apportionment you found online.

CHAPTER 15

Graph Theory

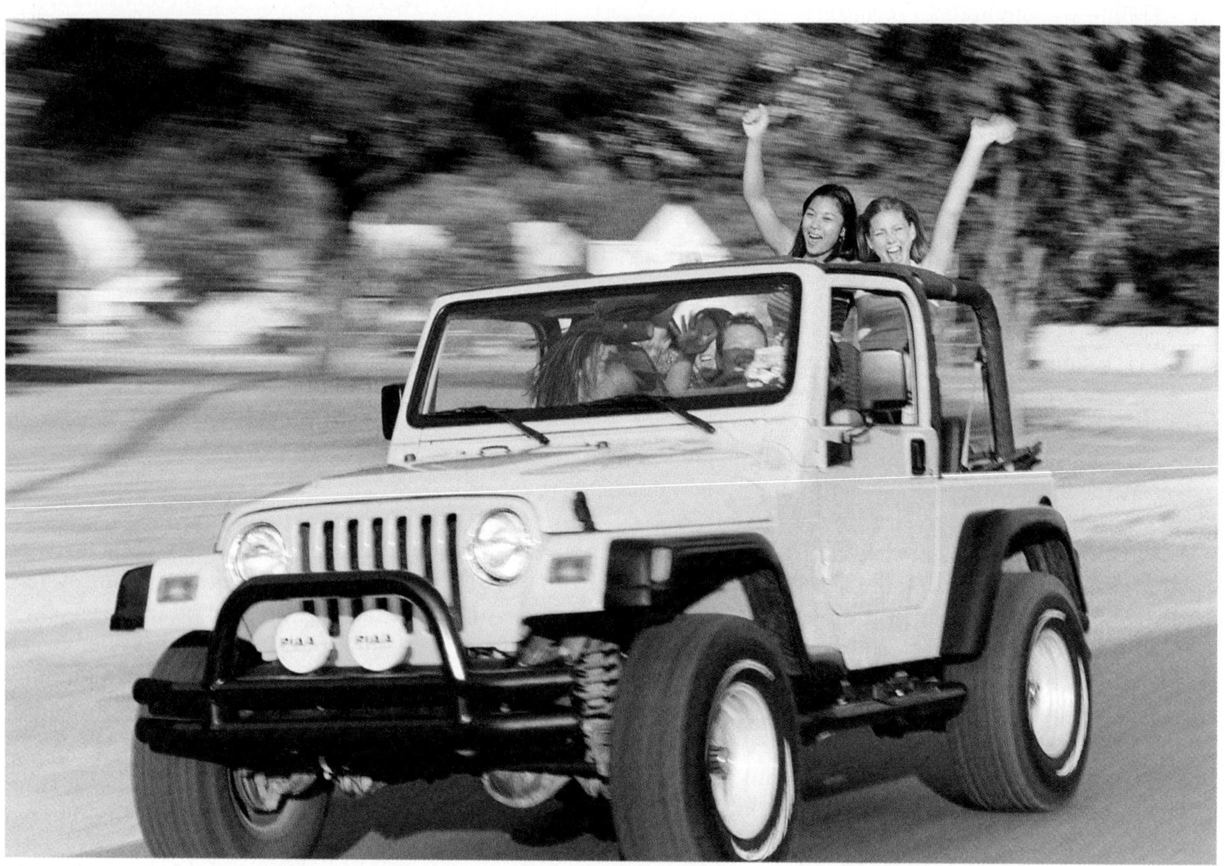

Outline

15-1 Basic Concepts of Graph Theory
15-2 Euler's Theorem
15-3 Hamilton Paths and Circuits
15-4 Trees
Summary

MATH IN Road Trips

The road trip is a great American tradition. There's just something magical about the freedom of heading out to the open road and driving wherever you feel like. It usually takes about an hour for the magic to wear off, though. Then you just want to get to where you're headed as quickly as possible.

The branch of mathematics known as graph theory was created to solve problems involving the most efficient way to travel between different locations. One of the interesting things about graph theory is that we can trace its beginnings to a single problem from the early 1700s. The town of Kaliningrad in modern day Russia was founded as Königsberg, Prussia, in 1255. The Pregel river splits into two branches as it approaches the city, dividing the city into four distinct regions, as shown in the figure.

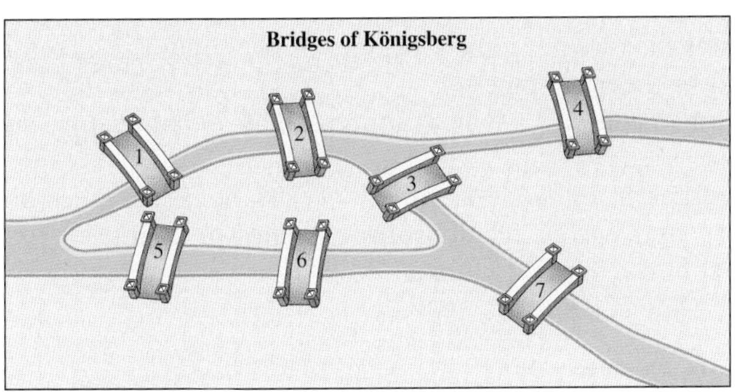

The residents of Königsberg were fond of leisurely walks covering all of the bridges, but none were able to find a route that would enable them to cover all seven exactly once while beginning and ending at the same location.

At some point, someone wrote the great Swiss mathematician Leonhard Euler, who lived not far away in St. Petersburg, to ask if he could prove that such a route either was or was not possible. Although he considered the problem trivial, Euler was intrigued by the fact that it didn't appear to fit into any current branch of math. And so graph theory was born. One of the important results of this chapter bears Euler's name, and it was the one used to solve the Königsberg bridge problem. (The solution is in the sidelight on page 801.)

From that humble beginning, graph theory expanded to cover a wide variety of practical applications, many of which we will study in this chapter. And this is where we return our attention to road trips. Suppose that you're planning to visit friends at three different colleges and want to find the most efficient route to visit all three and then return home. You would probably begin by finding the distance between all of the locations. This information for one road trip is provided in the table below. When you have finished this chapter, you should be able to answer the questions below the table.

	Georgia	Georgia Tech	Auburn	Alabama
Georgia	–	70	185	271
Georgia Tech	70	–	114	200
Auburn	185	114	–	158
Alabama	271	200	158	–

1. Draw a complete weighted graph that represents the distances between these schools.
2. If a student at Georgia Tech plans on visiting friends at the other three schools and then returning home, find the route that would cover the smallest possible number of miles.
3. If the plan was to instead connect the four schools with the shortest possible path without returning to the starting point, what would that path be, and how many miles would it cover?

For answers, see Math in Road Trips Revisited on page 823

Section 15-1 Basic Concepts of Graph Theory

LEARNING OBJECTIVES

- ☐ 1. Define basic graph theory terms.
- ☐ 2. Represent relationships with graphs.
- ☐ 3. Decide if two graphs are equivalent.
- ☐ 4. Recognize features of graphs.

It's time to plan that big spring break trip you've just been *living* for, and the first step is to get to South Padre Island. There's a pretty good chance that you're not going to find a flight directly from the nearest airport to Brownsville, Texas, which is the airport closest to South Padre. Instead, you'll need to make one or more intermediate connections to reach your final destination. (There's a metaphor for life in there somewhere.)

Graph theory is used to study problems like how to arrange connections that most efficiently get you to spring break nirvana. In this section we will study the basic concepts we need to learn about this useful branch of math. The graphs that we will study in this chapter are different from the graphs of equations and functions we've worked with previously. These graphs will be used to show relationships between items like people, cities, streets, etc.

A **graph** consists of a finite set of points called **vertices** (singular: vertex) and line segments called **edges** connecting the points.

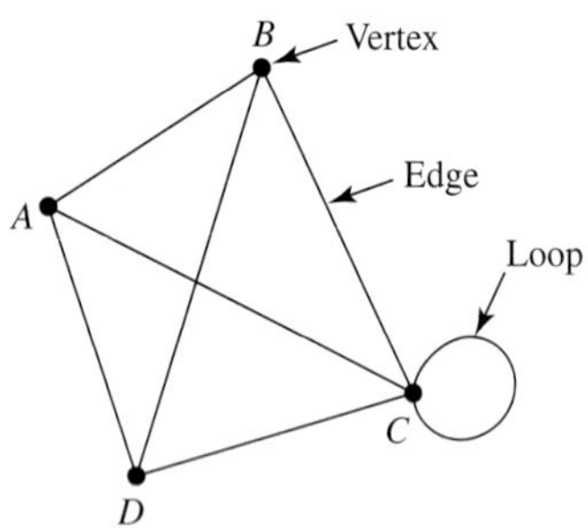

Figure 15-1

☑ 1. Define basic graph theory terms.

Figure 15-1 shows an example of a graph.

We will always draw vertices as black dots. In this case, they are labeled A, B, C, and D. The edges are AB, BC, CD, DA, AC, and BD. Since there is no vertex indicated at the intersection of edges AC and BD, it is assumed that the lines are not connected. You can think of it as two wires that are not joined together or two pipes that are not connected, one pipe passing over the other pipe.

The graph in Figure 15-1 also contains a **loop.** A loop is an edge that begins and ends at the same vertex.

Graphs can be used to represent or model a variety of situations like cities on a map, floor plans for houses, and border relationships of states or countries. Examples 1 through 4 illustrate some of these situations.

Math Note

The physical location of vertices in a graph is not important—just the edges that connect them. You may want to look ahead at Figure 15-14 on page 795 to see another graph for Example 1 that looks different but is equivalent.

EXAMPLE 1 Representing Islands with a Graph

Represent the islands and the bridges shown in Figure 15-2 with a graph.

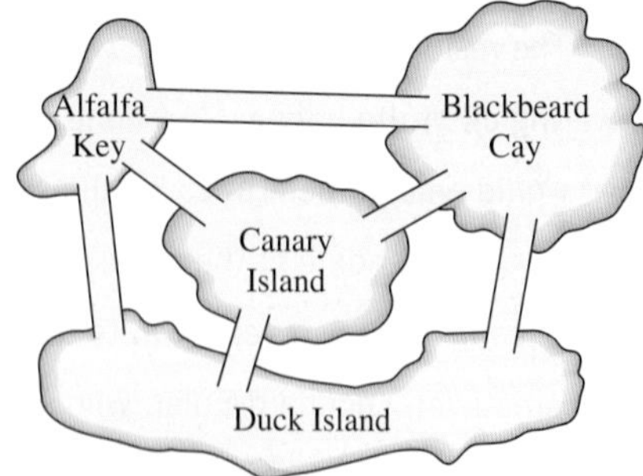

Figure 15-2

SOLUTION

Each island can be represented by a vertex. There are four islands; label them as A, B, C, and D. Next represent each bridge by an edge. For example, edge AD connects island A with island D. Each bridge connects two islands, and should be represented by an edge. The graph is shown in Figure 15-3.

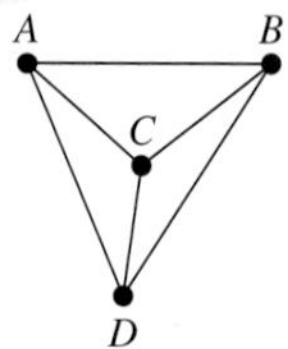

Figure 15-3

▼ Try This One 1

Draw a graph to represent the land and bridges shown in Figure 15-4 on the next page.

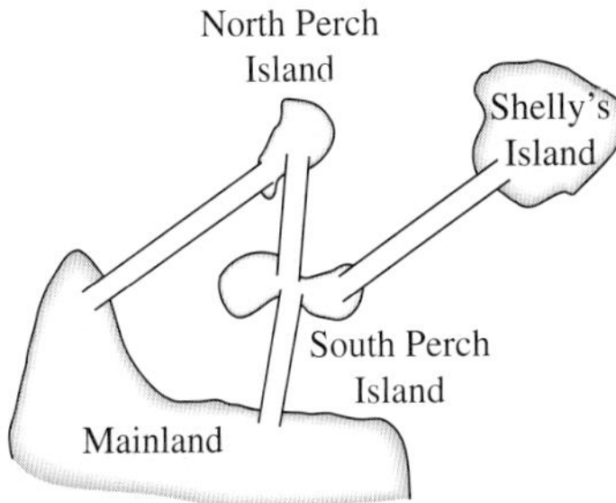

Figure 15-4

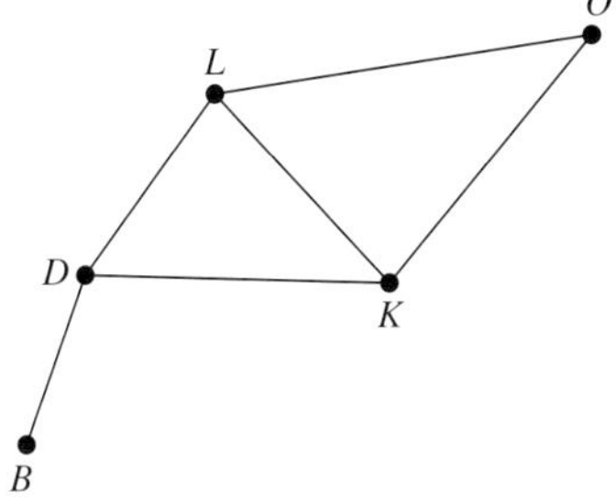

Figure 15-6

EXAMPLE 2 Representing a Floor Plan with a Graph

Represent the floor plan of the house shown in Figure 15-5 by a graph.

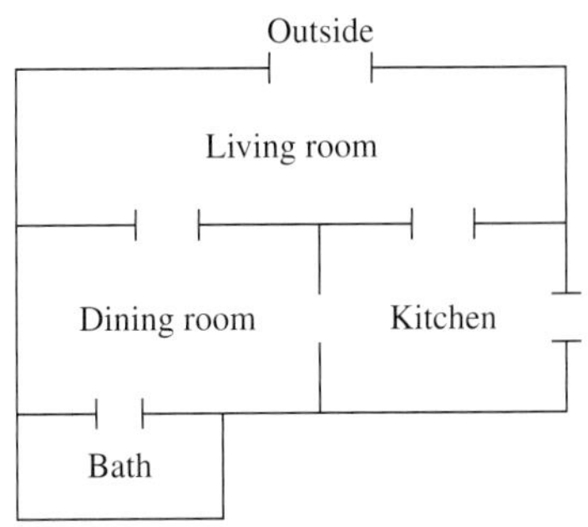

Figure 15-5

SOLUTION

Each room is represented by a vertex labeled with the first letter of the room name. Vertices are connected when there is a doorway from one room to the next. The living room, kitchen, and dining room all have three doors; the outside is accessible only from the living room and kitchen, and the bathroom connects only to the dining room. The graph is shown in Figure 15-6.

▼ Try This One 2

The floor plan shown in Figure 15-7 is for a two-bedroom student apartment in Oxford, Ohio. Draw a graph to represent it.

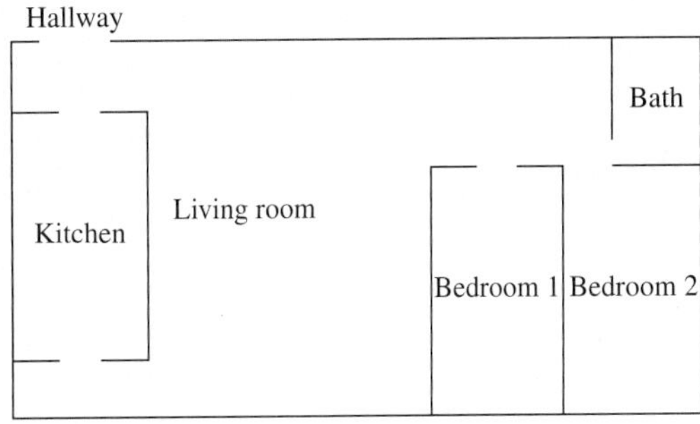

Figure 15-7

Math Note

As we can see in Example 3, a graph describing border relationships is much different from a map! The graph shows only which counties share a border. It doesn't indicate how the counties are situated.

EXAMPLE 3 A Graph Representing Border Relationships

The map shown in Figure 15-8 shows five counties in central Pennsylvania. Draw a graph that shows the counties that share a common border. Let the vertices represent the counties and the edges represent the common borders.

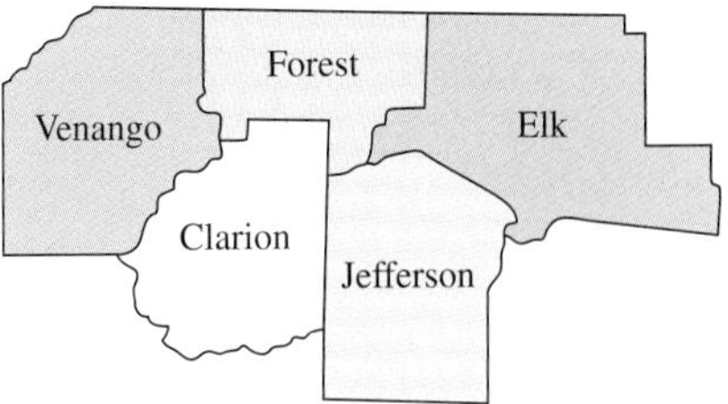

Figure 15-8

SOLUTION

The five counties are represented by five vertices; each is labeled with the first letter of the name of the counties. Notice that Venango County and Clarion County share a common border, so connect V and C with an edge. Venango County and Forest County share a common border, so connect V and F with an edge. Since Venango County does not share a border with Jefferson County or Elk County, no edges connect V with J or E. Next, connect vertices F and E since Forest County and Elk County share a common border, and continue in this manner. The completed graph is shown in Figure 15-9.

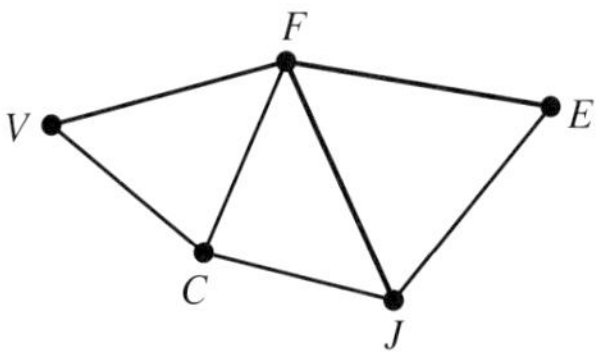

Figure 15-9

Figure 15-10

▼ Try This One 3

Draw a graph similar to the one in Example 3 for the map of Central America shown in Figure 15-10.

EXAMPLE 4 A Graph Representing City Streets

> **Math Note**
>
> In Example 4, we chose to represent intersections with vertices and streets with edges. In Exercise 35, you'll be asked to use vertices for the streets and edges for the intersections.

A plan of streets in a neighborhood is shown in Figure 15-11. Draw a graph for the neighborhood. Use the intersections of the streets as vertices and the streets as the edges.

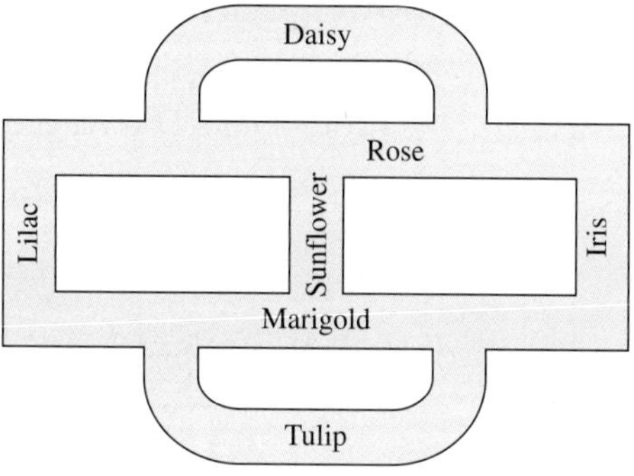

Figure 15-11

SOLUTION

Place a vertex at each point of intersection of any two or more streets. Connect the vertices with edges that follow the streets. See Figure 15-12.

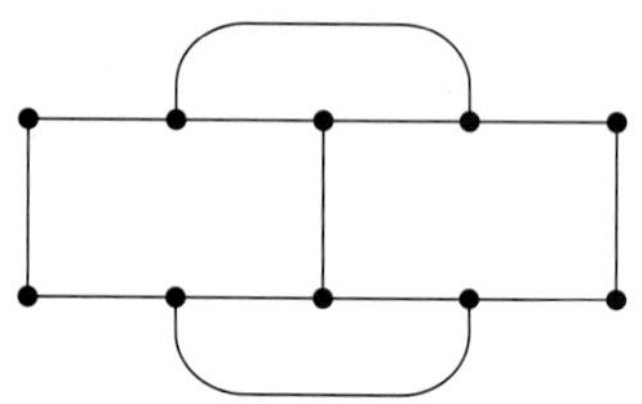

Figure 15-12

☑ 2. Represent relationships with graphs.

▼ Try This One 4

Draw a graph for the neighborhood shown in Figure 15-13. Use vertices for the intersections and edges for the streets.

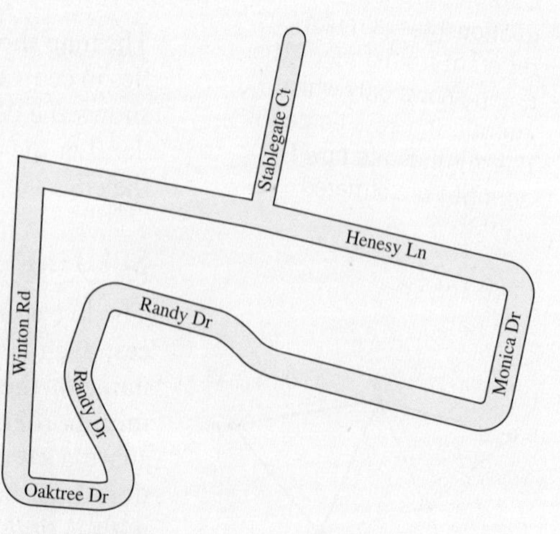

Figure 15-13

Equivalent Graphs

Glance quickly at the two graphs in Figure 15-14. Are they the same? If you answered yes, you should be very proud of yourself. Even though they physically look different, they are really the same graph, because they have the same vertices connected in the same way. The only difference is where the vertices are physically situated.

When two graphs have the same vertices connected in the same way, we call them **equivalent**.

EXAMPLE 5 Recognizing Equivalent Graphs

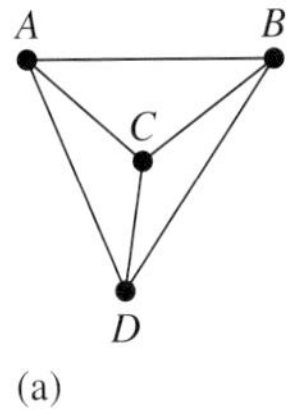

(a)

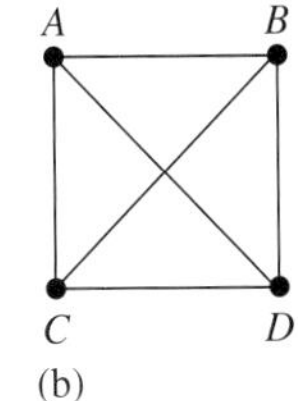

(b)

Figure 15-14

Explain why the two graphs shown in Figure 15-14 are equivalent.

SOLUTION

Each graph contains four vertices labeled A, B, C, and D. In each graph, every vertex is connected to each of the other three vertices. So the graphs have the same vertices and same connections, and they are equivalent.

3. Decide if two graphs are equivalent.

Try This One 5

Are the two graphs in Figure 15-15 equivalent? Explain.

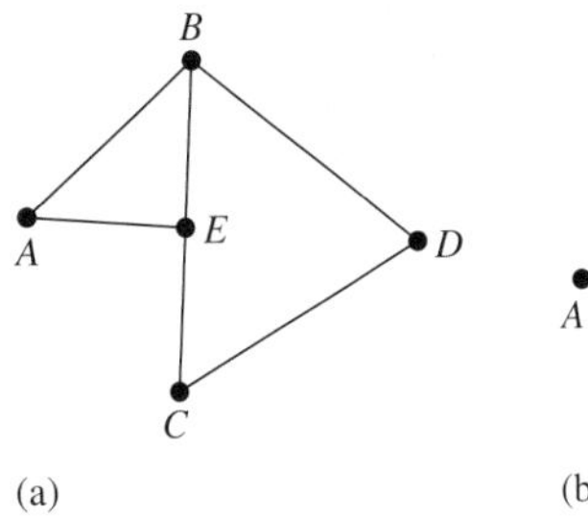

(a)

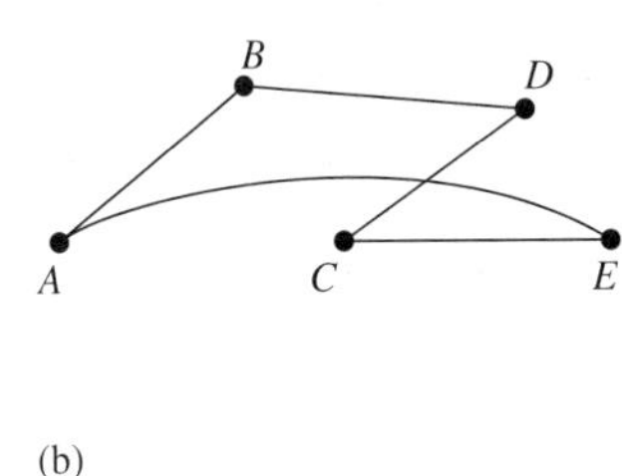

(b)

Figure 15-15

More Graph Theory Terminology

Math Note

Every loop is by default also a circuit, but of course a very short and simple one! Also, every circuit is a path, but not every path is a circuit since paths don't have to begin and end at the same vertex.

The **degree** of a vertex is the number of edges emanating from that vertex. For example, if there are six edges emanating from a vertex, we say it has a degree of 6. An **even vertex** has an even number of edges emanating from it. An **odd vertex** has an odd number of edges emanating from it. Loops are considered to be two edges emanating from the vertex.

Figure 15-16 shows some even and some odd vertices.

Adjacent vertices have at least one edge connecting them. In the graph shown in Figure 15-17 on the next page, vertex A is adjacent to vertices B and E, while vertex E is not adjacent to either vertex B or D since there is no edge connecting them.

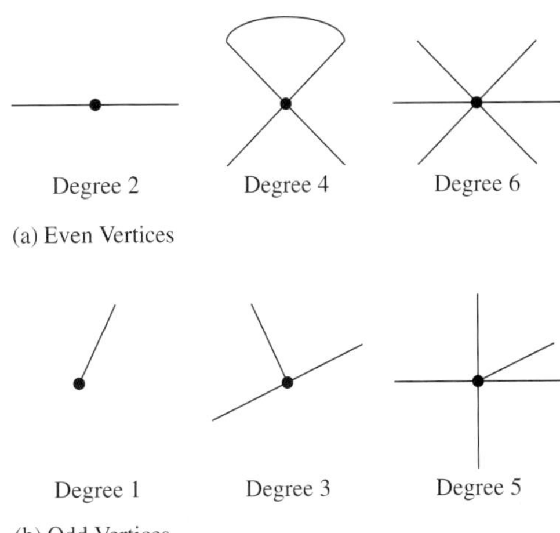

Figure 15-16

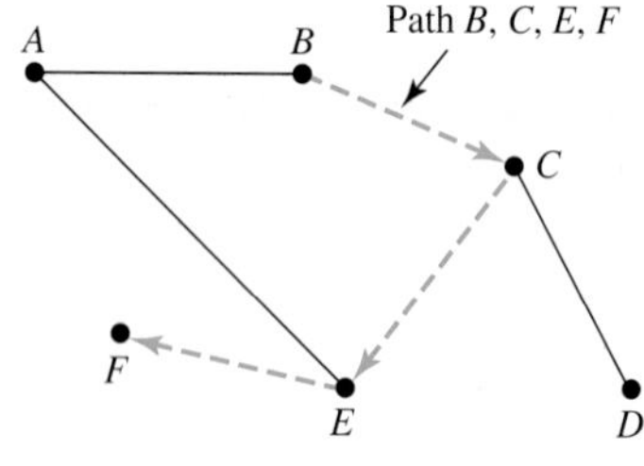

Figure 15-17

A **path** on a graph is a sequence of adjacent vertices and the edges connecting them that uses no edge more than once. When finding a path on a graph, a vertex can be crossed more than once, but an edge can be crossed only once. Figure 15-17 shows a path on a graph. The path starts at vertex *B* and then goes through vertices *C*, *E*, and *F* in that order. The path is then described as *B*, *C*, *E*, *F*. A graph can have many paths, and a path doesn't have to include all of the vertices or edges.

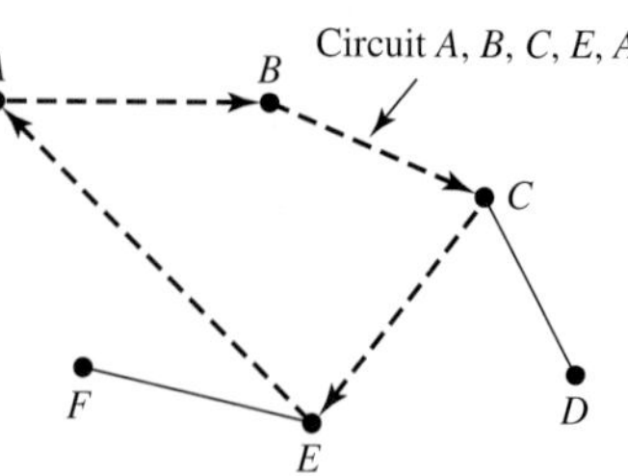

Figure 15-18

A **circuit** is a path that begins and ends at the same vertex. A circuit is shown on the graph in Figure 15-18. It is given by *A*, *B*, *C*, *E*, *A*.

A graph is said to be **connected** if for any two vertices, there is at least one path that connects them. In other words, a graph is connected if you can start at any vertex, follow along the edges and end up at any other vertex without tracing an edge more than once. If a graph is not connected, we say it is **disconnected**. Figure 15-19 shows examples of connected and disconnected graphs.

A **bridge** on a connected graph is an edge that, if removed, makes the graph disconnected. Figure 15-20 shows three graphs that have bridges. Notice that if a bridge is removed, there will be at least one pair of vertices that can no longer be connected with a path.

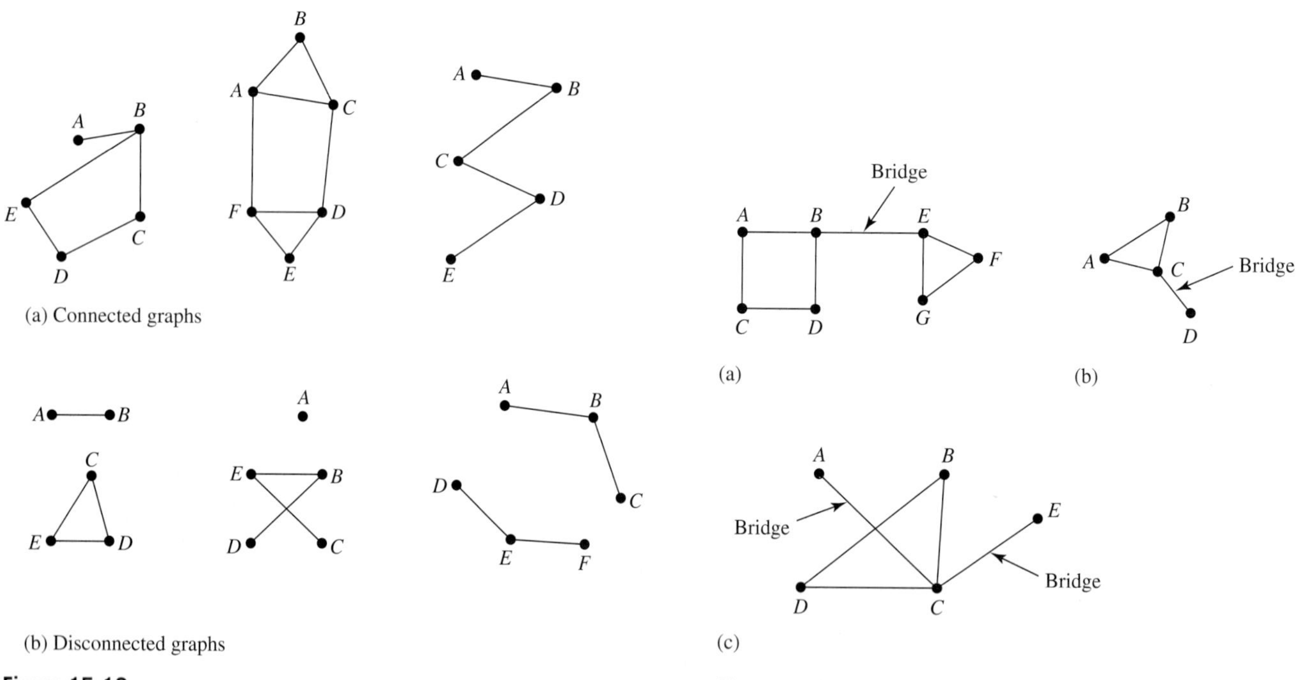

Figure 15-19

Figure 15-20

EXAMPLE 6 Recognizing Features of a Graph

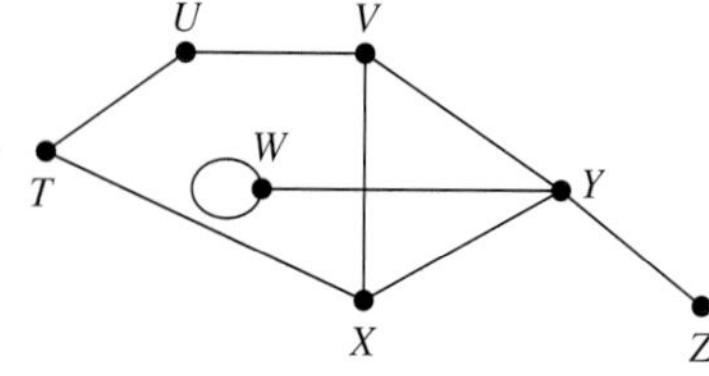

Figure 15-21

Use the graph in Figure 15-21 to answer each question.

(a) List the odd vertices.
(b) List the even vertices.
(c) What vertex has a loop?
(d) Are there any bridges?
(e) Identify at least one circuit.

SOLUTION

(a) Vertices *V*, *W*, and *X* have three edges touching them, so they are odd vertices. Vertex *Z* has just one edge touching it, so it is also odd.
(b) Vertices *T* and *U* have two edges touching them, and vertex *Y* has four edges touching it, so these are even vertices.
(c) There is a loop at vertex *W*.
(d) Removing edge *WY* will isolate vertex *W*, so *WY* is a bridge. Removing edge *YZ* will isolate *Z*, so it is a bridge. These are the only two bridges.
(e) There are dozens of examples of circuits. One example is *T*, *U*, *V*, *X*, *T*.

CAUTION

Don't confuse the use of the word *bridge* in graph theory with bridges in maps and diagrams. In Example 1, the bridges connecting islands are represented by edges on the graph.

▼ Try This One 6

☑ 4. Recognize features of graphs.

Use the graph shown in Figure 15-22 to answer each question.

(a) List the odd vertices.
(b) List the even vertices.
(c) What vertex has a loop?
(d) What edge is a bridge?
(e) Identify a circuit. (There are several correct answers.)

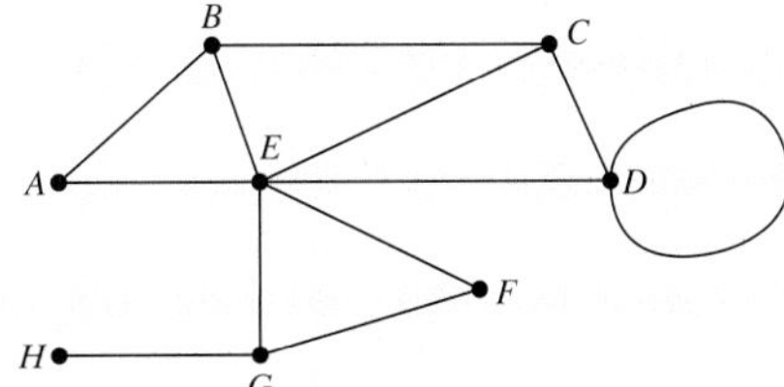

Figure 15-22

Answers to Try This One

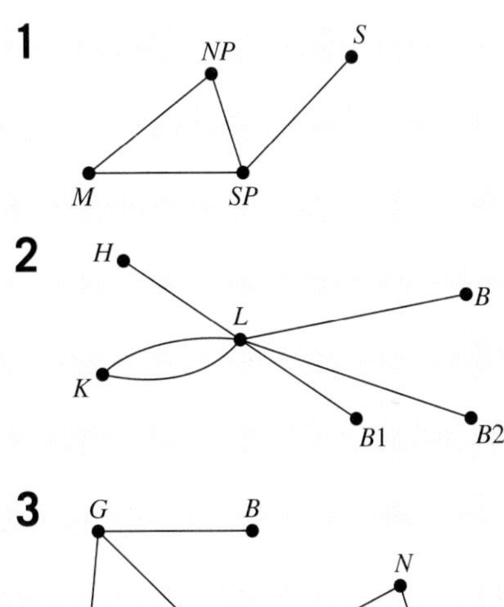

4

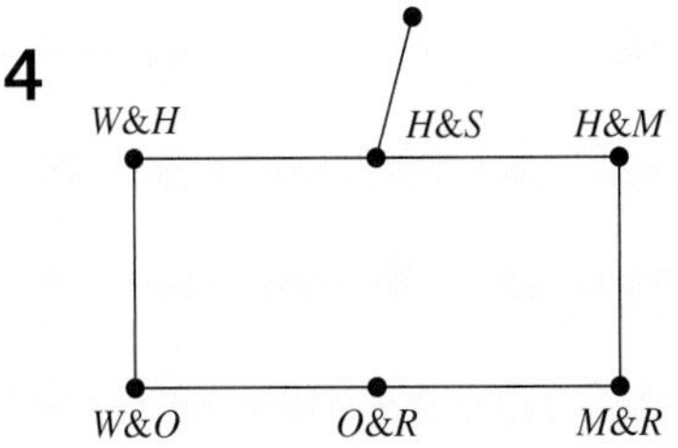

5 No. Vertex *B* connects to *E* in the first graph but not the second.

6 (a) *B*, *C*, *G*, and *H* are odd
(b) *A*, *D*, *E*, and *F* are even
(c) *D*
(d) *GH*
(e) One of many possible answers is *A*, *B*, *C*, *D*, *E*, *A*

EXERCISE SET 15-1

Writing Exercises

1. Describe the difference between the meaning of the word graph in this chapter and the meaning we used earlier in the book.
2. What is the difference between a loop and a circuit?
3. What is the difference between a circuit and a path?
4. Draw two graphs that look physically different but are equivalent, then explain why they are equivalent.
5. How can you tell if a graph is connected?
6. Think about a graph that represents the commercial flight routes for all American cities. Do you think this graph is connected? What does it mean if it's disconnected?
7. What does it mean for two vertices in a graph to be adjacent? What would it mean in the context of Exercise 6?
8. Does every graph have a bridge? Explain.

Computational Exercises

Use the following graph to answer Exercises 9–20.

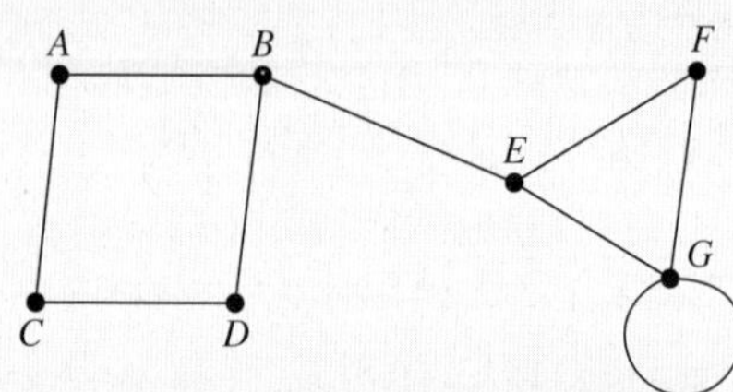

9. List the vertices of the graph.
10. How many edges does the graph have?
11. Name three vertices that are adjacent to vertex *E*.
12. List the even vertices.
13. List the odd vertices.
14. What edge is a bridge?
15. Find a path that contains vertex *E*. (There are several answers.)
16. Find a circuit that includes vertex *D*.
17. Identify a vertex that has a loop.
18. Which edges are not included in the path *A*, *B*, *E*, *G*, *F*?
19. Explain why *A*, *B*, *E*, *D* is a not a path.
20. Explain why *E*, *F*, *G* is not a circuit.

Real-World Applications

For Exercises 21–24, represent each figure using a graph. Use vertices for islands and edges for bridges.

21.

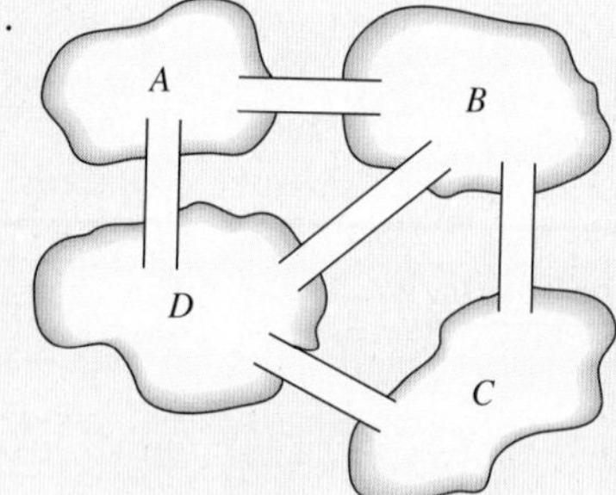

22.

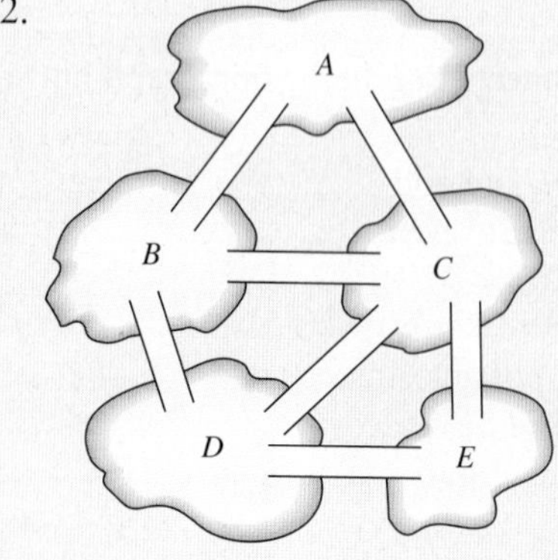

23.

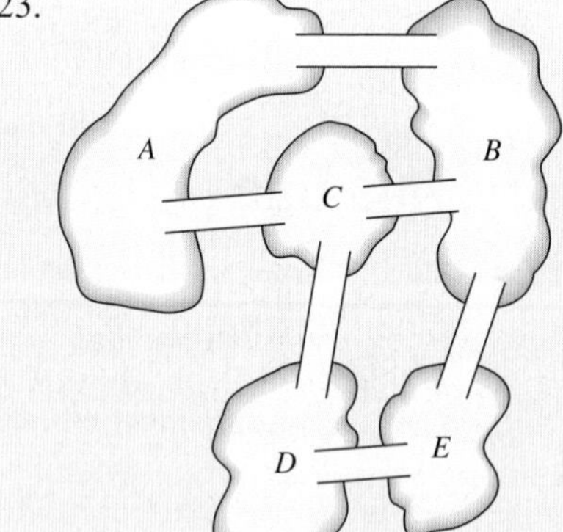

24.

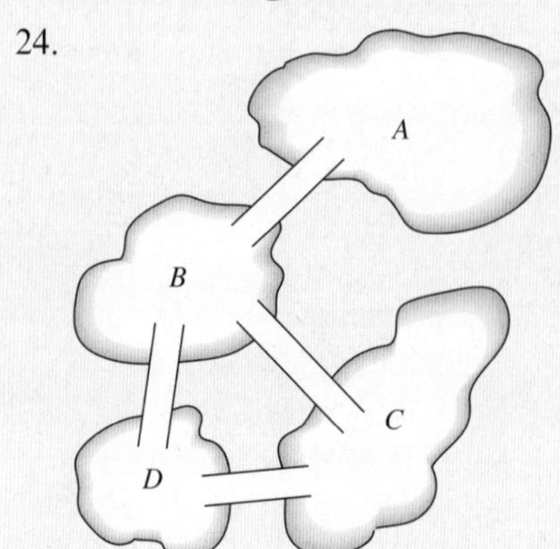

For Exercises 25–28, draw a graph to represent each map. Use vertices to represent the states and edges to represent the common borders.

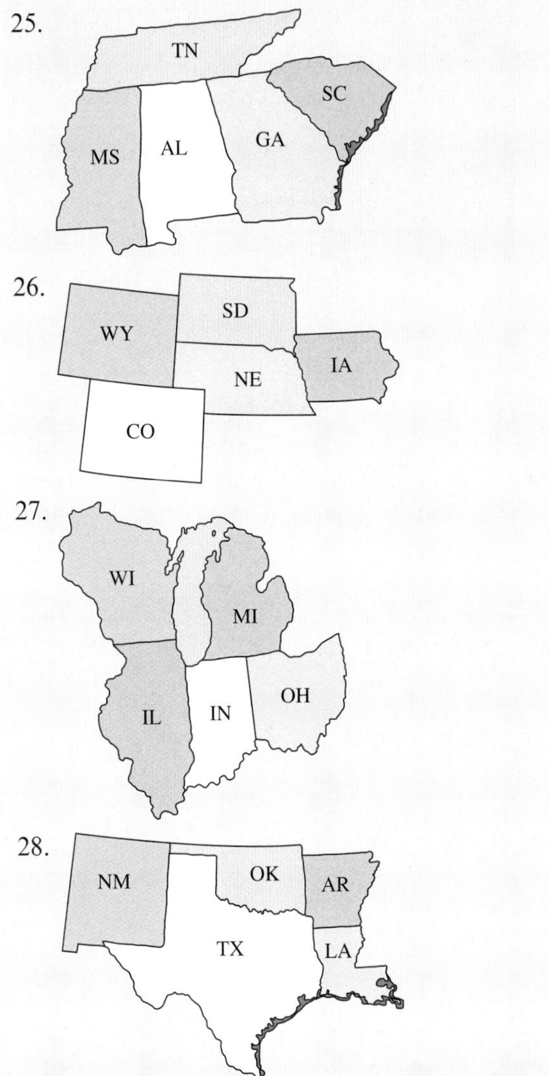

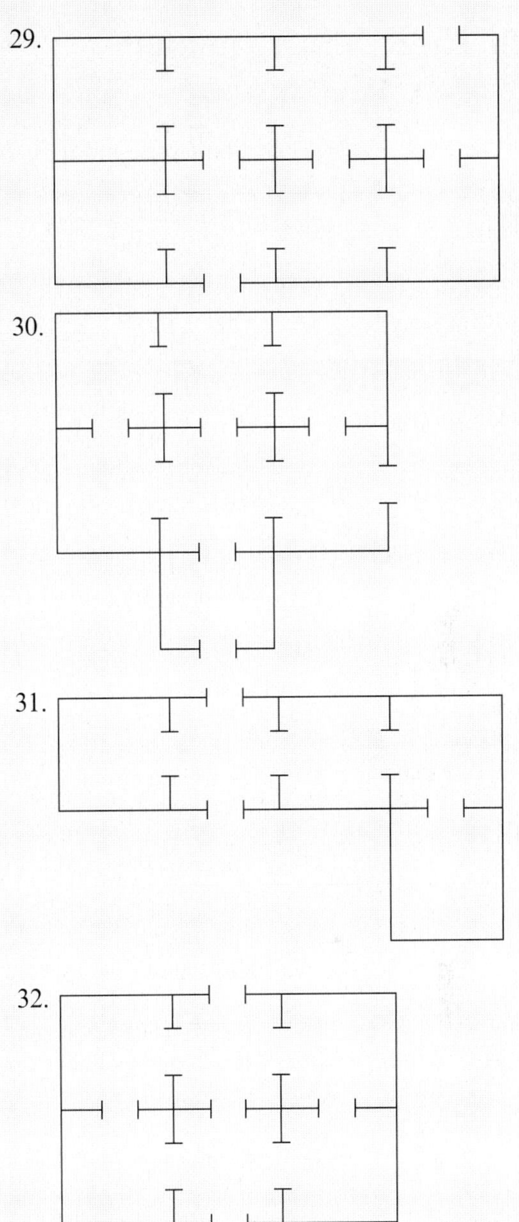

For Exercises 29–32, draw a graph that represents each floor plan. Use vertices to represent the rooms and outside area and edges to represent the connecting doors.

Critical Thinking

33. Draw a circuit with five vertices, a loop, and a bridge.
34. In a dorm, a group of friends consisted of Antoine, Bill, Cathy, Dana, Evita, and Fran. Antoine was friends with Bill and Cathy; Bill was friends with Antoine, Fran, and Dana. Cathy was friends with Evita, Antoine, and Dana. Draw a graph representing the friendships. Who were friends with Evita? Who were friends with Fran?
35. Draw a graph that represents the street map in Example 4, but this time use vertices to represent the streets, and edges to describe which streets intersect. Is your graph different from the one in Example 4? What can you conclude?
36. Can you trace the graph shown in the figure without lifting your pencil or crossing a path twice? If so, show how.

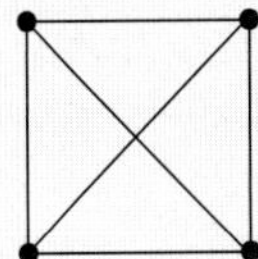

Section 15-2 Euler's Theorem

LEARNING OBJECTIVES

- ☐ 1. Define Euler path and Euler circuit.
- ☐ 2. Use Euler's theorem to decide if an Euler path or Euler circuit exists.
- ☐ 3. Use Fleury's algorithm to find an Euler path or Euler circuit.
- ☐ 4. Solve practical problems using Euler paths or circuits.

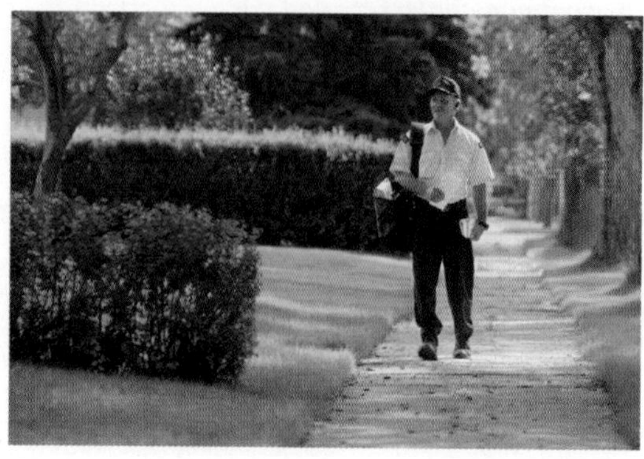

Nobody has ever sat down to do homework and thought "I have *got* to figure out a way to make this take as much time as possible." People are constantly searching for more efficient ways to do things, leaving more time for the fun stuff. Consider the backbone of our postal system, the mail carriers. They have to visit every house and business on their route each day, and obviously it would be helpful to do so with as little backtracking as possible.

If you think about it, this sounds like a graph theory problem. The delivery locations are the vertices, and the routes between them are the edges. An efficient route would visit every vertex without going over any edges more than once. In this section, we will study the work of Leonhard Euler, which addresses problems that involve tracing an entire graph without repeating any edges.

An **Euler path** is a path that passes through each edge exactly once. An **Euler circuit** is a circuit that passes through each edge exactly once.

☑ 1. Define Euler path and Euler circuit.

The difference between a path and an Euler path is that a path can pass through any subset of vertices and edges on a graph, but an Euler path passes through all of the edges exactly once. Figure 15-23 shows an Euler path. As indicated by the arrows, it begins at vertex A and follows the path *A*, *B*, *E*, *D*, *C*, *E*. Figure 15-24 shows an Euler circuit that begins and ends at vertex *B*.

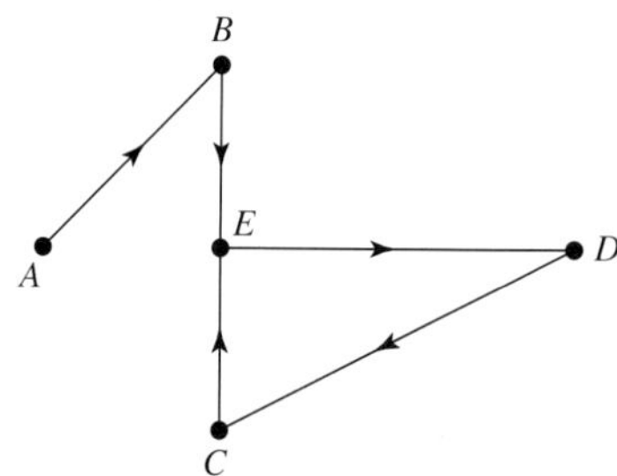

Euler path *A*, *B*, *E*, *D*, *C*, *E*

Figure 15-23

The question now becomes: given a graph, does it have any Euler paths or circuits? Our mail carrier would be interested in this answer—if a graph representing his route has an Euler path, he can cover every street without doubling back at all.

It turns out that some graphs have Euler paths and circuits, and some don't. Euler's theorem allows us to tell if a given graph has an Euler path or Euler circuit without first trying to find one.

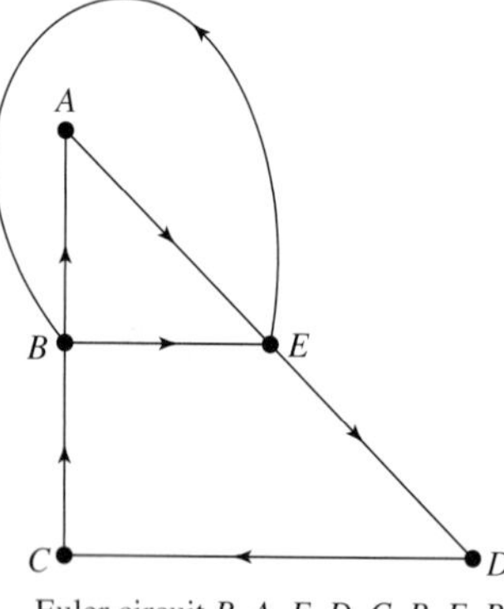

Euler circuit *B*, *A*, *E*, *D*, *C*, *B*, *E*, *B*

Figure 15-24

Math Note

Every Euler circuit is also an Euler path, but there are Euler paths that are not Euler circuits, like the one in Figure 15-23.

Euler's Theorem

For any connected graph:

1. If all vertices are even, the graph has at least one Euler circuit (which is by definition also an Euler path). An Euler circuit can start at any vertex.
2. If exactly two vertices are odd, the graph has no Euler circuits but at least one Euler path. The path must begin at one odd vertex and end at the other odd vertex.
3. If there are more than two odd vertices, the graph has no Euler paths and no Euler circuits.

EXAMPLE 1 Using Euler's Theorem

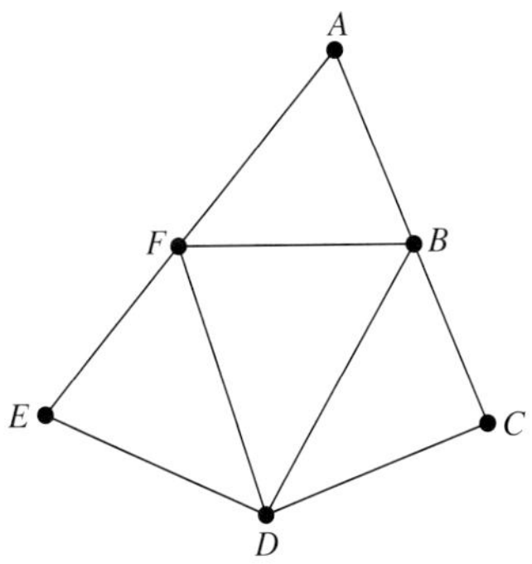

Figure 15-25

Use Euler's theorem to determine if the graph shown in Figure 15-25 has an Euler path or an Euler circuit.

SOLUTION

Vertices *A*, *C*, and *E* have degree 2; vertices *B*, *D*, and *F* have degree 4. Since all vertices have even degree, Euler's theorem guarantees that there is at least one Euler circuit, which is also an Euler path.

▼ Try This One 1

Use Euler's theorem to determine if the graph shown in Figure 15-26 has an Euler path or an Euler circuit.

 2. Use Euler's theorem to decide if an Euler path or Euler circuit exists.

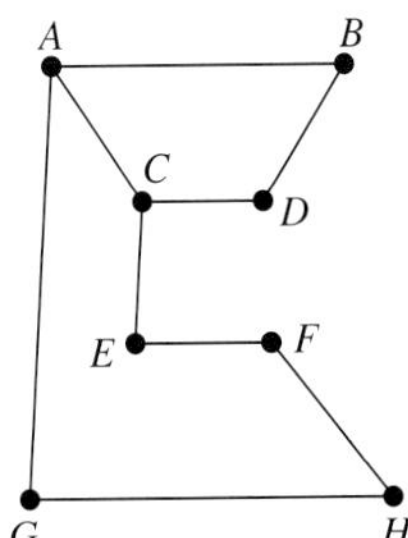

Figure 15-26

Finding Euler Paths and Euler Circuits

Once we know that a graph has an Euler circuit or an Euler path, it seems perfectly reasonable to try to find one. (Think of the mail carrier example—just knowing that there *is* an efficient route to take doesn't help him much!) You can try the old-fashioned trial-and-error method, but this can be very challenging unless there are a small number of vertices and edges. A procedure known as *Fleury's algorithm* has been developed to help save time.

Fleury's Algorithm

To find an Euler path or Euler circuit:

1. If a graph has no odd vertices, start at any vertex. If the graph has two odd vertices, start at either odd vertex.
2. Number the edges as you trace through the graph making sure not to traverse any edge twice.
3. At any vertex where you have a choice of edges, choose one that is not a bridge for the part of the graph that has not yet been numbered.

Math Note

Before using Fleury's algorithm, make sure you've used Euler's Theorem to confirm that the graph actually has an Euler path or Euler circuit.

Sidelight THE BRIDGES OF KÖNIGSBERG

The question in the Chapter 15 introduction about crossing all of the bridges of Königsberg is actually a question about Euler circuits. Residents in search of a path that would cross every bridge exactly once and return them to the location where they began were seeking an Euler circuit for the graph below. On this graph, the land masses are represented by vertices, and the bridges are represented by edges.

Notice that all four vertices are odd. Euler's theorem tells us that the residents were vainly searching for a route that didn't exist—there is no Euler circuit. It's pretty interesting that a problem that spawned an entire branch of math becomes really easy when you know Euler's theorem.

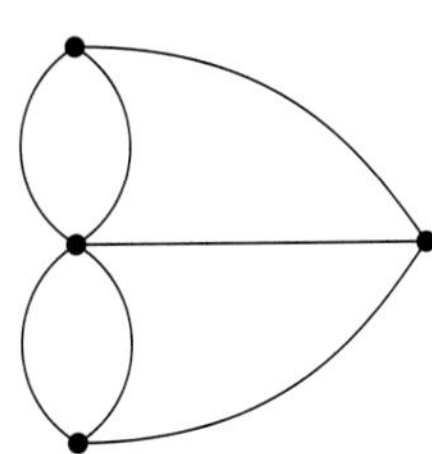

The procedure, particularly point 3, is illustrated by Example 2.

EXAMPLE 2 Finding an Euler Path

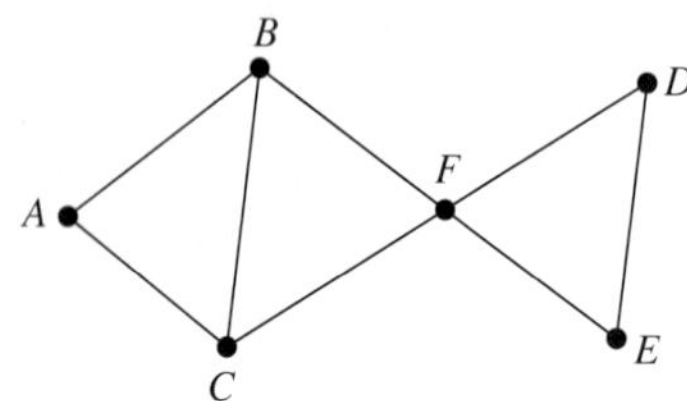

Figure 15-27

Use Fleury's algorithm to find an Euler circuit or path (if one exists) for the graph shown in Figure 15-27.

SOLUTION

The graph has two odd vertices, *B* and *C*, so there is an Euler path but no Euler circuit. Fleury's algorithm tells us to start at one of the odd vertices: we'll choose *B*. We next chose to proceed to *A*, then *C*. At *C*, we can continue to either *B* or *F* since neither *CB* nor *CF* is a bridge. We choose to proceed to *F*. Now we have three choices, but one of them won't work: with the edges we've already numbered out of the graph, edge *BF* is now a bridge (see Figure 15-28). So we continue to *D*, then *E*, back to *F*, then to *B*, and finally to *C*. The completed Euler path is *B*, *A*, *C*, *F*, *D*, *E*, *F*, *B*, *C*, and is shown in Figure 15-29.

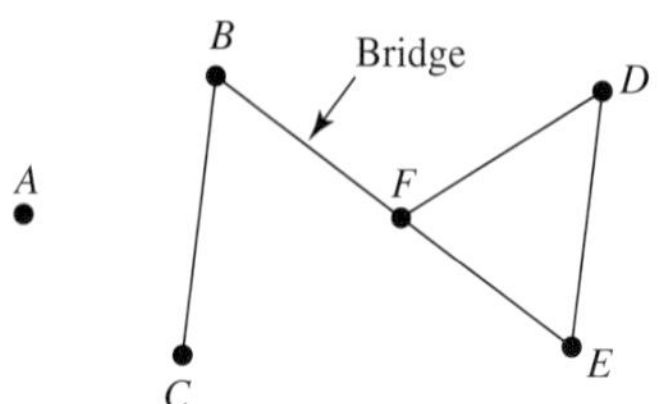

Figure 15-28

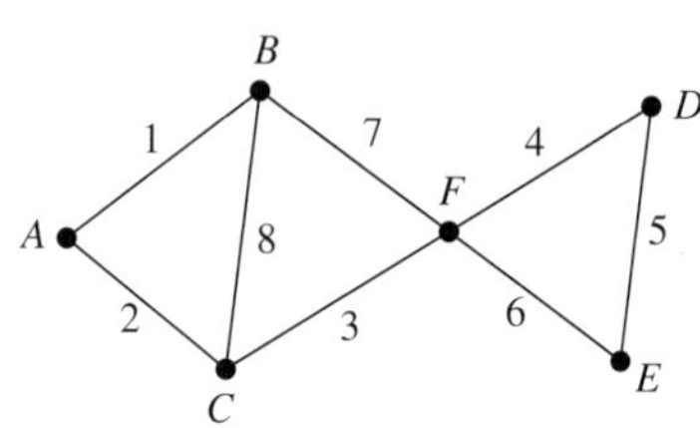

Figure 15-29

▼ Try This One 2

Use Fleury's algorithm to find an Euler circuit or path (if one exists) for the graph shown in Figure 15-30.

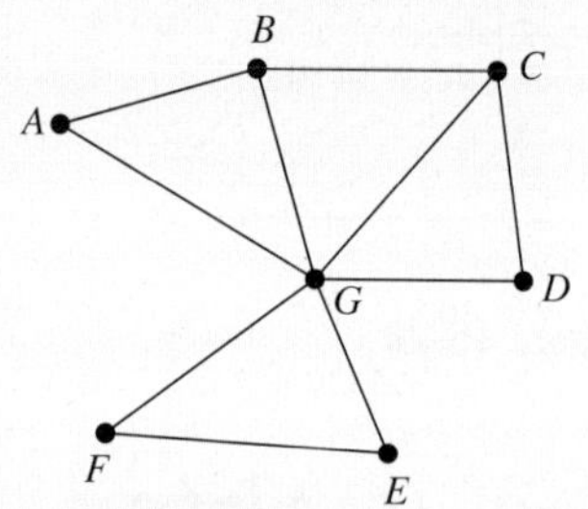

Figure 15-30

☑ 3. Use Fleury's algorithm to find an Euler path or Euler circuit.

The Euler path in Example 2 is not unique; there are several other Euler paths that can be traced on the graph. Also notice that we started at one odd vertex *B* and ended at the other odd vertex *C*.

The remaining example demonstrates a practical application of Euler's theorem.

EXAMPLE 3 An Application of Euler's Theorem

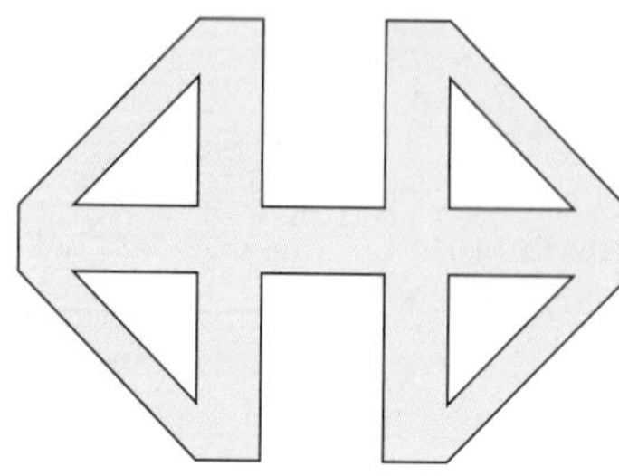

Figure 15-31

A mail carrier has the neighborhood pictured in Figure 15-31 on his route. He wants to cover each street exactly once without retracing any street. Find an Euler path to accomplish this.

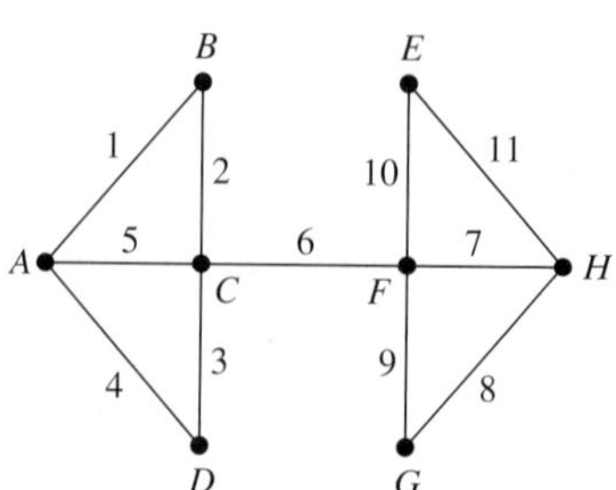

Figure 15-32

SOLUTION

Draw a graph representing the streets as shown in Figure 15-32. There are two odd vertices, *A* and *H*, so there is an Euler path but not an Euler circuit. We should start at one of the odd vertices and finish at the other. If we start at *A*, we can use the path *A*, *B*,

C, D, A, C, F, H, G, F, E, H. Don't forget to number the edges as they are crossed so that none get repeated.

▼ Try This One 3

4. Solve practical problems using Euler paths or circuits.

A mail carrier wants to traverse all streets in the neighborhood shown in Figure 15-33 without retracing any street twice. Find an Euler path that will permit the carrier to do this.

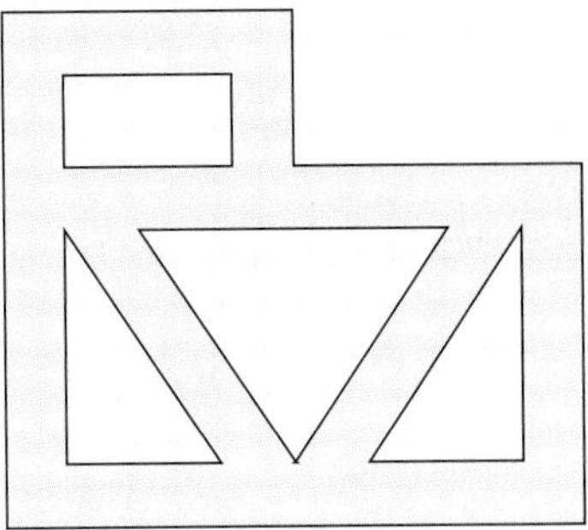

Figure 15-33

Answers to Try This One

1 Vertices *A* and *C* are odd; the rest are even. There is an Euler path, but no Euler circuit.

2 One possible Euler path is *B, A, G, F, E, G, B, C, G, D, C.*

3 With the diagram represented by a graph, one possible Euler path is numbered here.

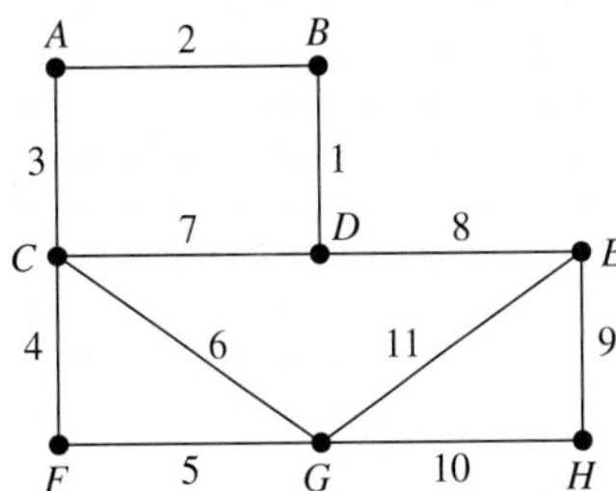

EXERCISE SET 15-2

Writing Exercises

1. What is the difference between a path and an Euler path?
2. Explain the difference between an Euler path and an Euler circuit.
3. Explain how you can tell if a graph has an Euler circuit.
4. What kind of graph has an Euler path but no Euler circuit?
5. What kind of graph has neither an Euler path nor an Euler circuit?
6. What is Fleury's algorithm used for?

Computational Exercises

For Exercises 7–10, decide whether each connected graph has an Euler path, Euler circuit, or neither.

7. The graph has 6 even vertices and no odd vertices.
8. The graph has 2 odd vertices and 6 even vertices.
9. The graph has 2 even vertices and 4 odd vertices.
10. The graph has 2 odd vertices and 3 even vertices.

For Exercises 11–20,

(a) State whether the graph has an Euler path, an Euler circuit, or neither

(b) If the graph has an Euler path or an Euler circuit, find one.

11.

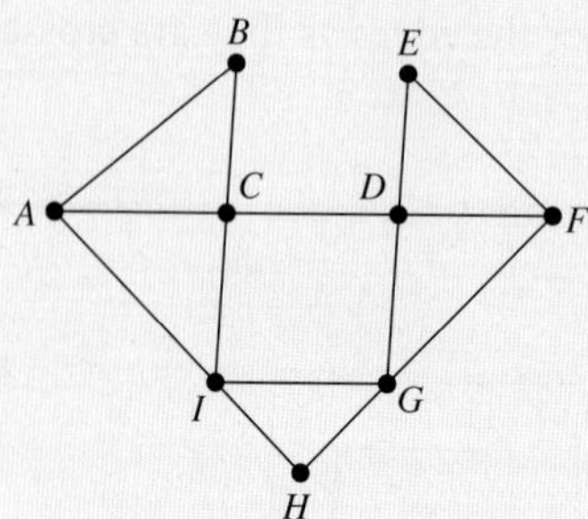

12.

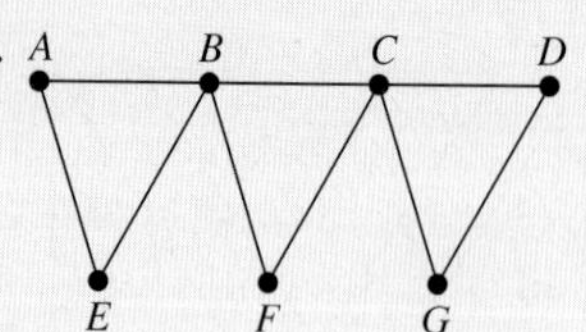

13.

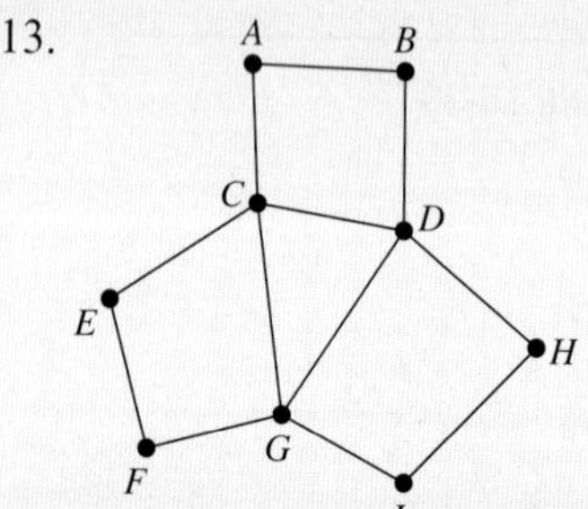

14.

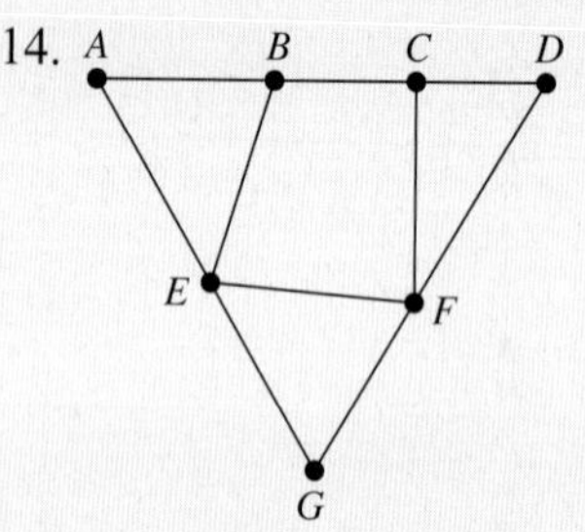

15.

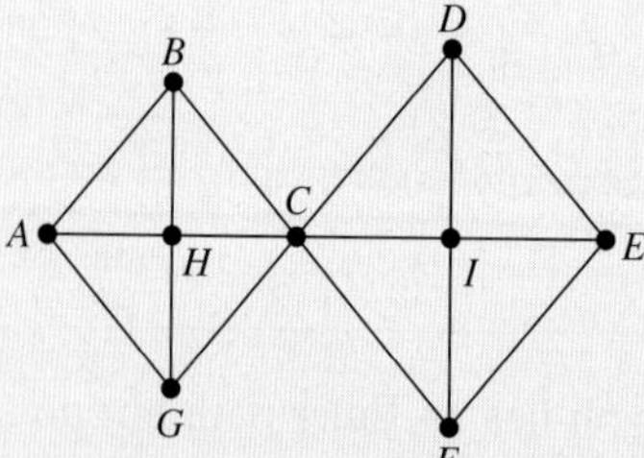

16.

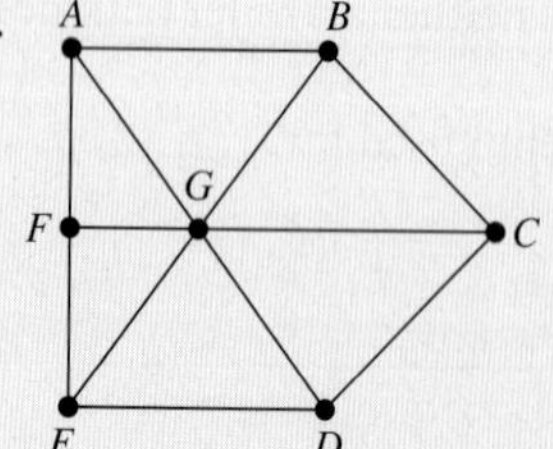

17.

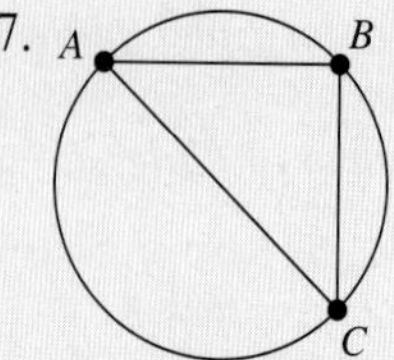

18.

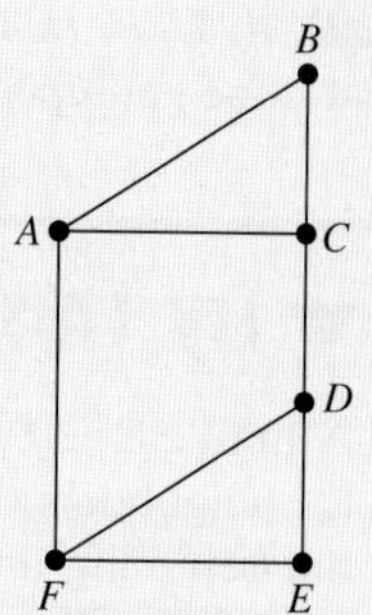

19.

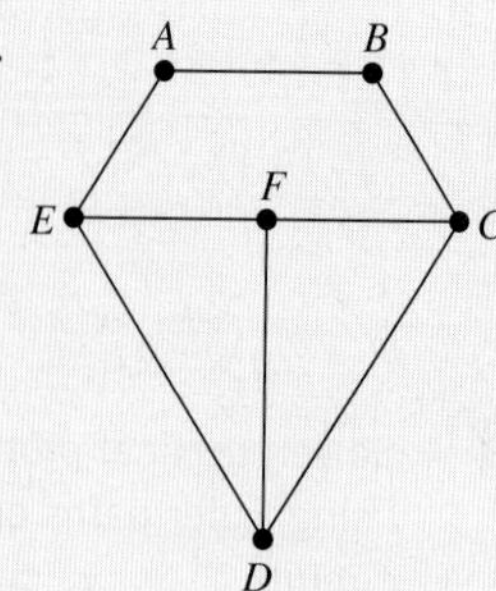

20.

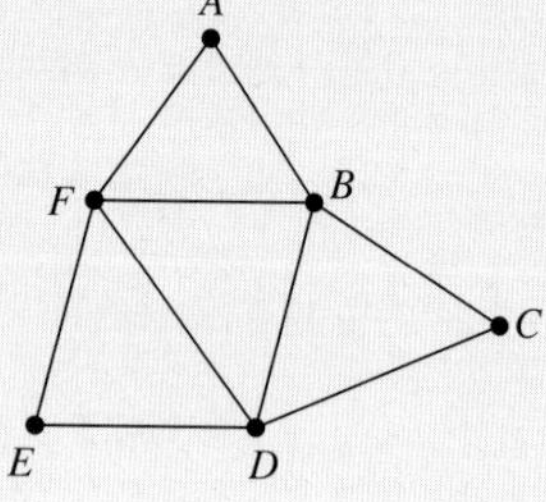

For Exercises 21–26, draw a graph for the figures using vertices for the islands and edges for the bridges. Determine if the graph has an Euler path, an Euler circuit, or neither. If it has an Euler path or Euler circuit, find one.

21.

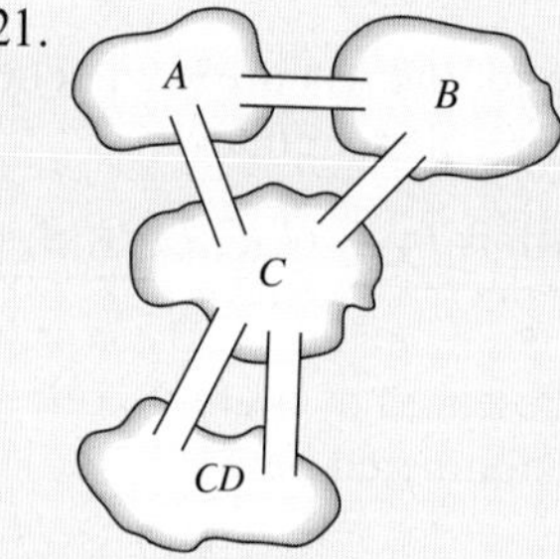

22.

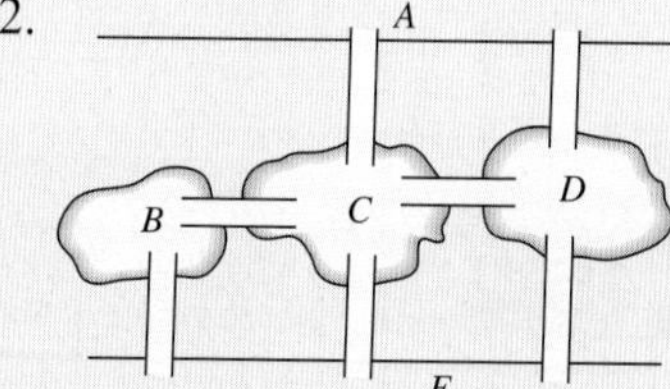

23.

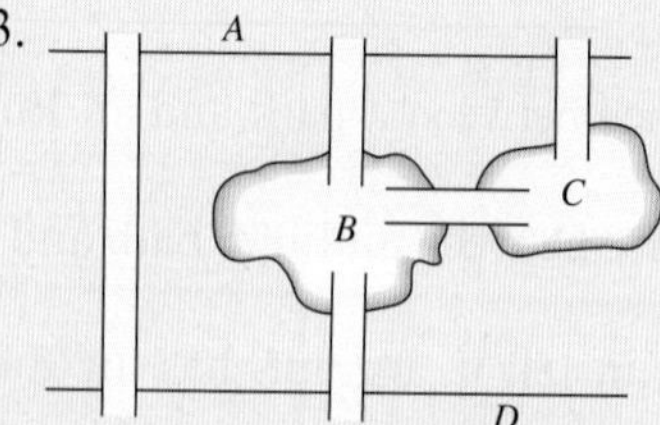

24.

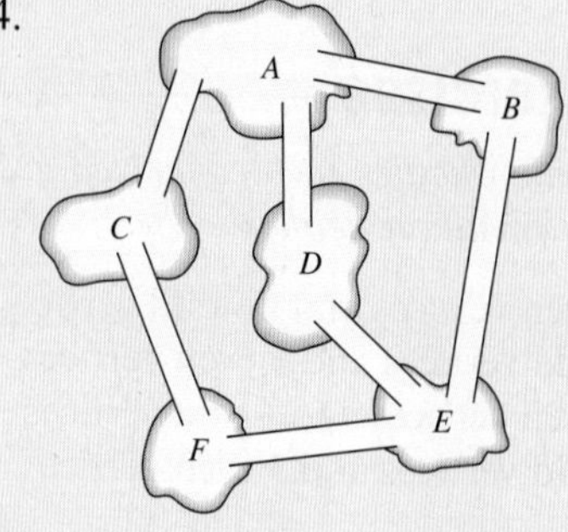

25.

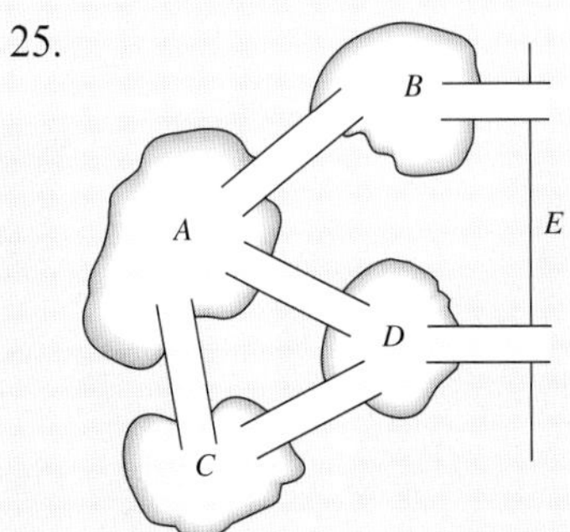

26.

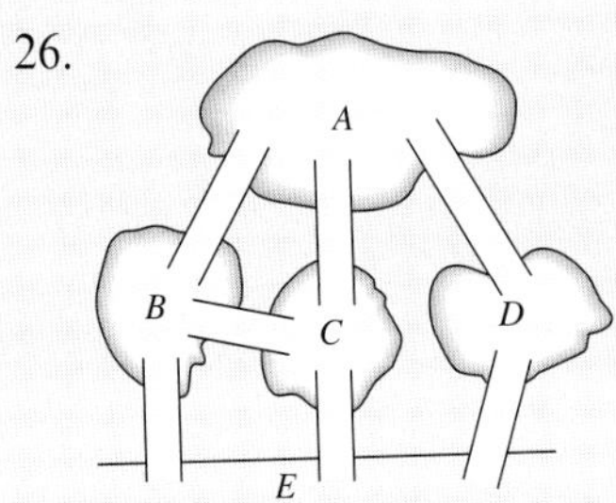

For Exercises 27–30, draw a graph for each floor plan using the rooms and the exterior area as vertices and the door openings as the edges. Determine if the graph has an Euler path, an Euler circuit, or neither. If the graph has an Euler path or an Euler circuit, find one.

27.

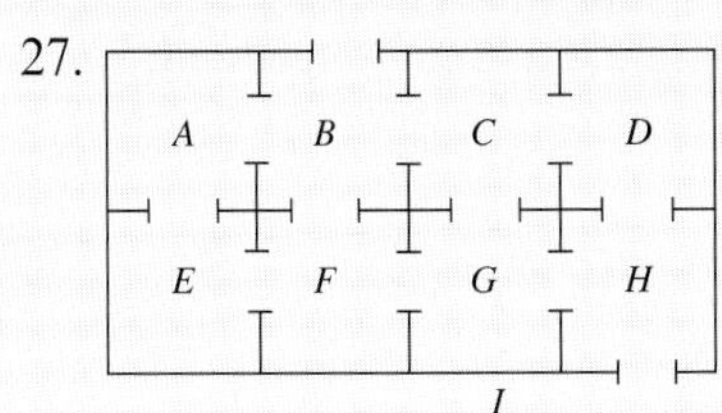

28.

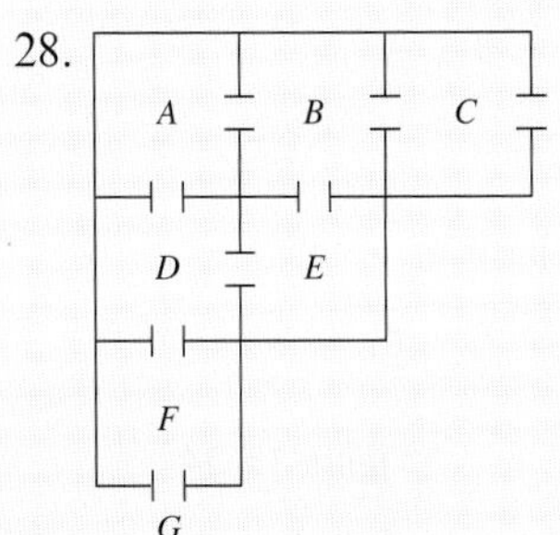

29.

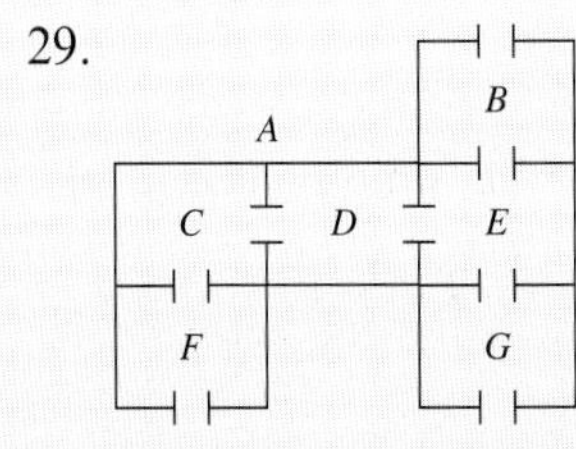

30.

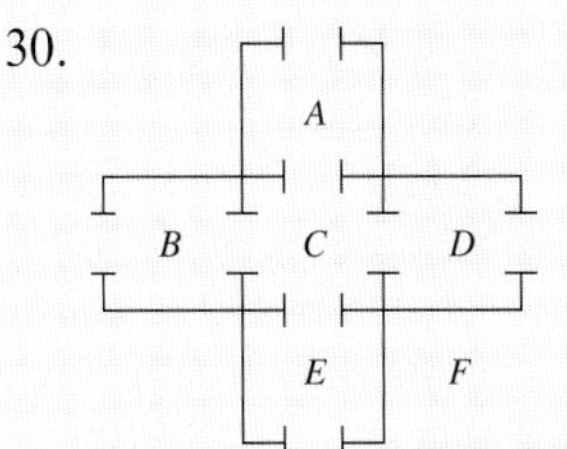

For Exercises 31 and 32, draw a graph for each map using each state as a vertex and each common border as an edge. Determine if the graph has an Euler path or an Euler circuit or neither. If an Euler path or an Euler circuit exists, find one.

31.

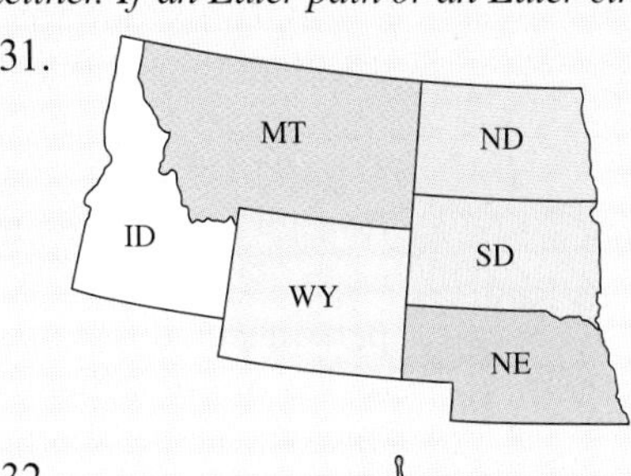

32.

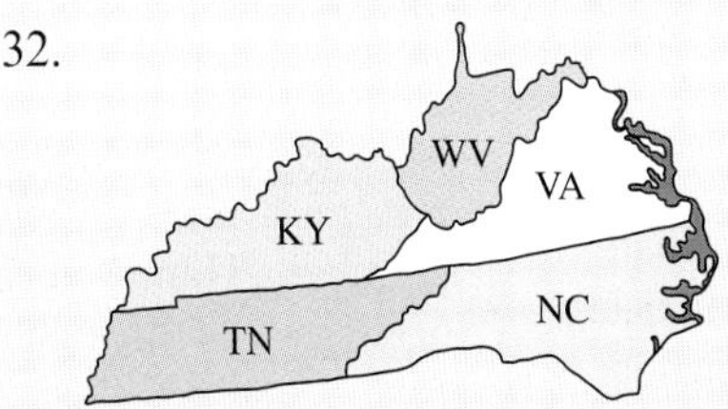

For Exercises 33 and 34, determine if an Euler path or an Euler circuit exists so that a person who plows the roads does not have to pass over any street twice. If an Euler path or an Euler circuit exists, find one. (Hint: Draw a graph and state what each vertex and each edge represents.)

33.

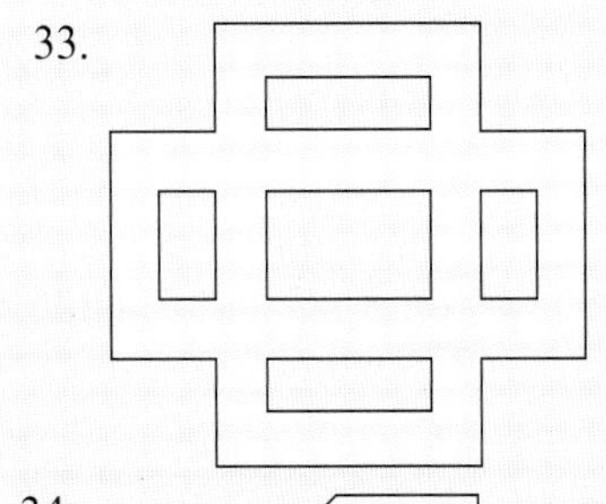

34.

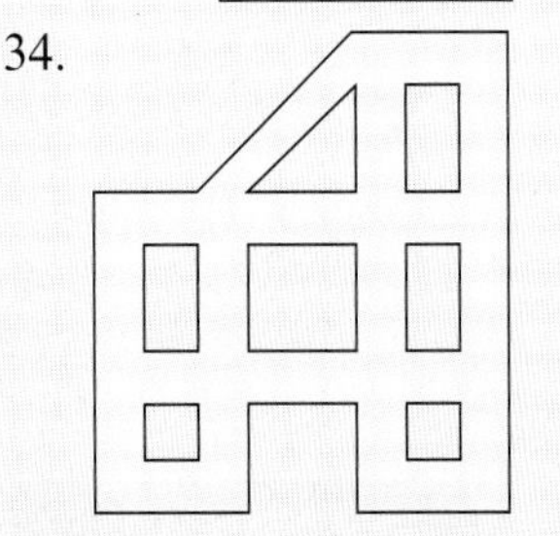

Critical Thinking

35. Explain why the word "connected" is crucial in the statements of Euler's theorem.
36. In Euler's theorem, the case where a connected graph has exactly one odd vertex is omitted. Why?
37. Find as many different Euler paths as you can for the graph in Example 2.

Section 15-3 Hamilton Paths and Circuits

LEARNING OBJECTIVES

- ❑ 1. Find Hamilton paths and Hamilton circuits on graphs.
- ❑ 2. Find the number of Hamilton circuits for a complete graph.
- ❑ 3. Solve a traveling salesperson problem using the brute force method.
- ❑ 4. Find an approximate optimal solution using the nearest neighbor method.
- ❑ 5. Draw a complete weighted graph based on provided information.

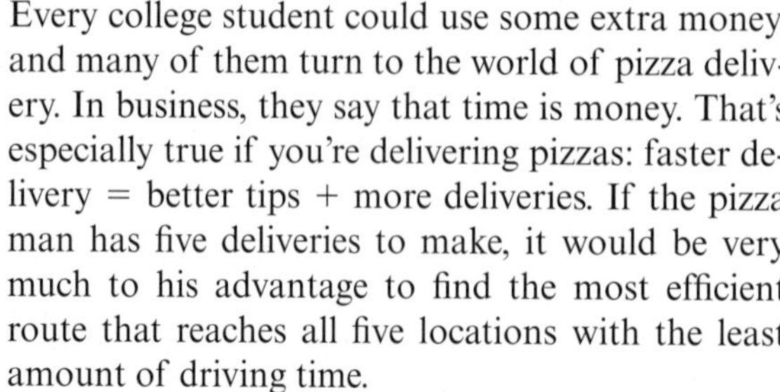

Every college student could use some extra money, and many of them turn to the world of pizza delivery. In business, they say that time is money. That's especially true if you're delivering pizzas: faster delivery = better tips + more deliveries. If the pizza man has five deliveries to make, it would be very much to his advantage to find the most efficient route that reaches all five locations with the least amount of driving time.

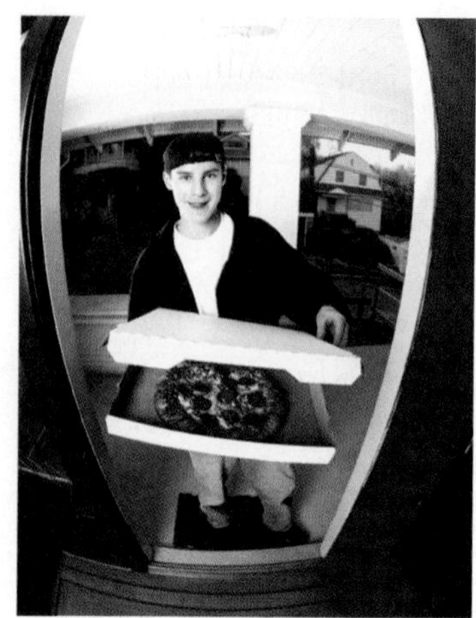

It turns out that this is exactly the type of problem that graph theory was developed to solve (but to be fair, the theory was developed about a hundred years before pizza delivery). In Section 15-2, we were interested in paths that covered every edge. That's not the case here: the delivery person isn't interested in how many roads he or she drives on, or which ones. The deliverer's concern is going to every location exactly once in the most efficient way possible. We will study this problem with the aid of *Hamilton paths.*

A path on a connected graph that passes through every vertex exactly once is called a **Hamilton path**. A Hamilton path that begins and ends at the same vertex, but passes through all other vertices exactly once is called a **Hamilton circuit**.

EXAMPLE 1 Finding a Hamilton Path

Find a Hamilton path that begins at vertex A for the graph shown in Figure 15-34.

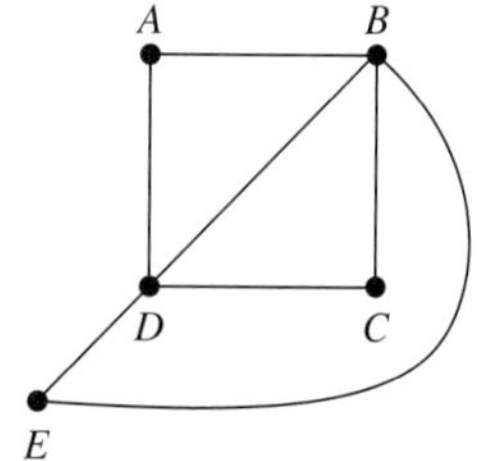

Figure 15-34

SOLUTION

There are many solutions: one of the simplest is alphabetical! The path A, B, C, D, E passes through every vertex exactly once, so it is a Hamilton path.

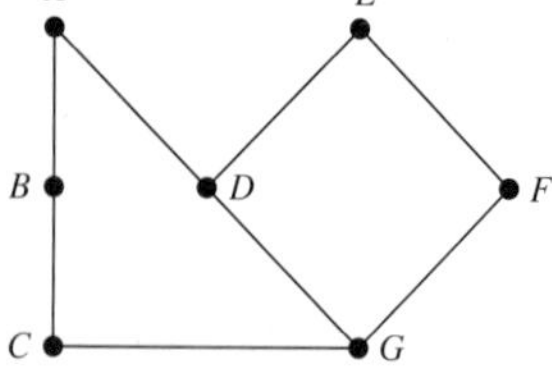

Figure 15-35

▼ Try This One 1

Find a Hamilton path that begins at vertex C for the graph shown in Figure 15-35.

EXAMPLE 2 Finding a Hamilton Circuit

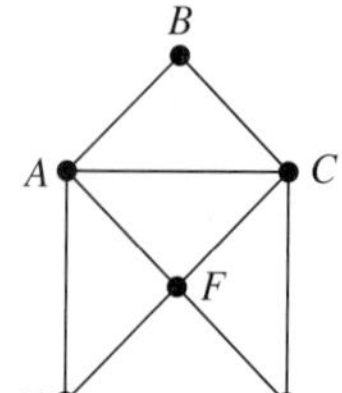

Figure 15-36

Find a Hamilton circuit for the graph shown in Figure 15-36.

SOLUTION

Again, there are many possible answers. If we choose to start and end at vertex A, one choice is A, E, D, F, C, B, A. This path begins and ends at the same vertex, and passes through every other vertex exactly once.

▼ Try This One 2

Find a Hamilton circuit that begins and ends at vertex A for the graph in Try This One 1.

☑ 1. Find Hamilton paths and Hamilton circuits on graphs.

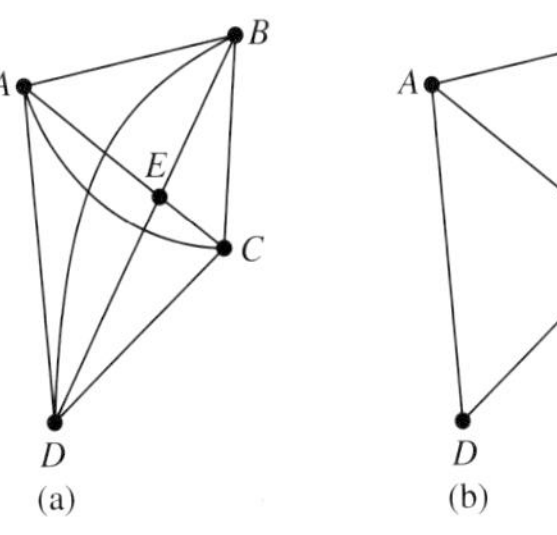

Figure 15-37

Euler's theorem enables us to know exactly when a graph does or does not have an Euler path or circuit. There is no exact analog for Hamilton paths and circuits; that is, there isn't a single theorem that allows us to tell if any graph does or does not have a Hamilton path. There is one condition that guarantees the existence of Hamilton circuits, however.

A **complete graph** is a graph that has an edge connecting every pair of vertices.

Figure 15-37a is a complete graph: every vertex has an edge that connects it to every other vertex. The graph in Figure 15-37b is not complete: there is no edge connecting B and D. A key fact makes complete graphs important in the study of Hamilton circuits.

Math Note

In Exercise 42, you will develop the formula for the number of Hamilton circuits in a complete graph.

Complete Graphs and Hamilton Circuits

Every complete graph with more than two vertices has a Hamilton circuit. Furthermore, the number of Hamilton circuits in a complete graph with n vertices is $(n - 1)!$.

EXAMPLE 3 Finding the Number of Hamilton Circuits in a Graph

How many Hamilton circuits are there in the graph in Figure 15-37a?

Math Note

Two Hamilton circuits are considered to be the same if they pass through the same vertices in the same order, regardless of the vertex where they begin and end.

SOLUTION

First, since the graph is complete, we know it has Hamilton circuits. There are 5 vertices, so the graph has $(5 - 1)!$ Hamilton circuits.

$$(5 - 1)! = 4! = 4 \cdot 3 \cdot 2 \cdot 1 = 24$$

There are 24 Hamilton circuits.

☑ 2. Find the number of Hamilton circuits for a complete graph.

▼ Try This One 3

Find the number of Hamilton circuits for a complete graph with 13 vertices.

Hopefully, the connection between Hamilton circuits and pizza delivery is apparent: a delivery route should start and end at the pizza place, and visit every destination exactly once. A problem of this nature has come to be known as a *traveling salesperson problem.* Folks who travel for business are interested in the most efficient way to visit all of their accounts. The "efficient" part could refer to time, miles, or maybe cost of travel.

The graph in Figure 15-38 shows the distances in miles between four cities that a saleswoman needs to visit in one day.

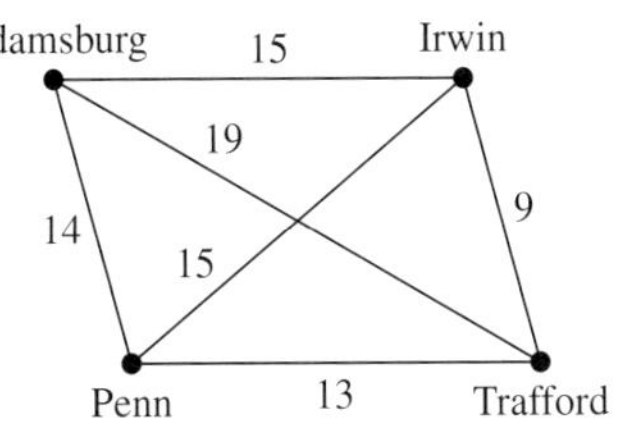

Figure 15-38

Math Note

A graph for a traveling salesperson problem should be complete because the salesperson would want to be able to travel directly between any pair of cities. Of course, this guarantees that there are Hamilton circuits.

This type of graph, with the edges labeled with some quantity of interest, is called a **complete weighted graph**, and the numbers are called **weights.** In this case they are distances, but could also be times, cost of flights, or something else relevant to the situation.

Looking at Figure 15-38, suppose that the saleswoman, who lives in Adamsburg, decides to visit Trafford, then Irwin, then Penn, and then return home. She would travel 19 + 9 + 15 + 14 = 57 miles. This is found by adding the weights of the edges. Is this the shortest route in terms of miles? In this case, the shortest route is called the *optimal solution.*

The **optimal solution** for a traveling salesperson problem is a Hamilton circuit for a complete weighted graph for which the sum of the weights of the edges traversed is the smallest possible number.

The first method we will use to solve a traveling salesperson problem is called the *brute force method.* The brute force involved is mental, not physical:

The Brute Force Method for Solving a Traveling Salesperson Problem

Step 1 Draw a complete weighted graph for the problem.

Step 2 List all possible Hamilton circuits.

Step 3 Find the sum of the weights of the edges for each circuit.

The circuit with the smallest sum is the optimal solution.

EXAMPLE 4 Using the Brute Force Method

For the traveling salesperson problem illustrated by Figure 15-38, find the optimal solution.

SOLUTION

Step 1 Draw the graph. This has been done for us.

Step 2 List all possible Hamilton circuits starting at one particular vertex. In this case, we chose vertex *A*. Note that there are 4 vertices, so there are (4 − 1)!, or 6, different Hamilton circuits. They are listed in the table below.

Step 3 Find the sum of the weights of the edges for each circuit. These are also shown in the table below.

Hamilton Circuit	Sum of Weights
A, *I*, *T*, *P*, *A*	15 + 9 + 13 + 14 = 51 miles
A, *I*, *P*, *T*, *A*	15 + 15 + 13 + 19 = 62 miles
A, *T*, *I*, *P*, *A*	19 + 9 + 15 + 14 = 57 miles
A, *T*, *P*, *I*, *A*	19 + 13 + 15 + 15 = 62 miles
A, *P*, *T*, *I*, *A*	14 + 13 + 9 + 15 = 51 miles
A, *P*, *I*, *T*, *A*	14 + 15 + 9 + 19 = 57 miles

Math Note

It makes perfect sense that there are two optimal solutions to Example 4; the second circuit is simply the first one covered in the opposite order.

There are two optimal solutions: *A*, *I*, *T*, *P*, *A*, and *A*, *P*, *T*, *I*, *A*, both covering 51 miles.

☑ 3. Solve a traveling salesperson problem using the brute force method.

▼ Try This One 4

The driving times in minutes between four cities are shown in the graph in Figure 15-39. Find the optimal solution for a copy machine repair technician starting in city A who has to visit a location in each city, and wants to minimize driving time.

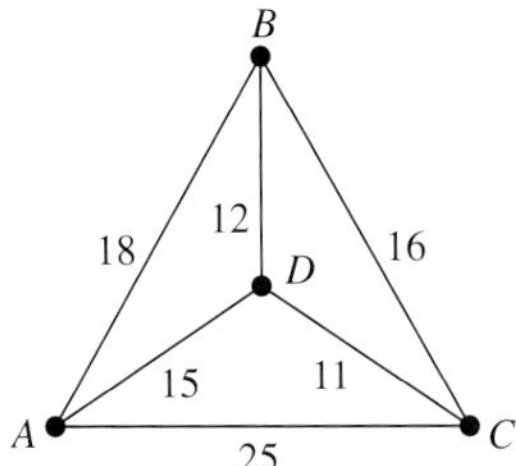

Figure 15-39

The brute force method worked very nicely in Example 4, but that's largely because there are only four vertices. Suppose the salesperson has 10 different companies to visit. Now there would be 9!, or 362,880 different possibilities! Without a computer program that can both find Hamilton circuits and then add their weights, this would obviously be unreasonable. Worse still, if there are 16 locations, there are 15!, or about 1.3 *trillion* different circuits! This would take days even for a very capable computer. And the number of possible circuits grows very rapidly beyond that. Clearly, another method would be nice.

A second method for finding the optimal circuit is called the *nearest neighbor method.* This method does not always give the optimal solution, but it does gives an approximation to the optimal solution. That is, there may be a shorter route than the one given by the nearest neighbor method, but finding it by using the brute force method may be too time-consuming.

The Nearest Neighbor Method for Finding an Approximate Solution to a Traveling Salesperson Problem

Step 1 Draw a complete weighted graph for the problem.

Step 2 Starting at a designated vertex, select the edge with the smallest weight and move to the second vertex.*

Step 3 At the next vertex, select the edge with the smallest weight that does not go to a vertex already used.*

Step 4 Continue until the circuit is completed.

The sum of the weights is an approximation to the optimal solution.

*In case the weights of two edges are the same, choose either one.

EXAMPLE 5 Using the Nearest Neighbor Method

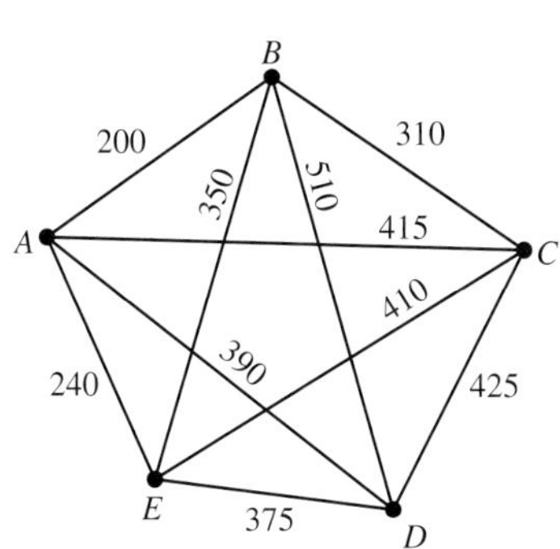

Figure 15-40

A regional manager decides that she needs to visit all of the company locations in her region during the last 2 weeks of May. She lives in city A and needs to visit all of cities B, C, D, and E, then return home. The driving times in minutes between the cities are shown in Figure 15-40. Use the nearest neighbor method to find the route that will give her the least amount of driving time.

SOLUTION

Starting at vertex A, the edge with the shortest driving time takes her to vertex B first. From there, the shortest driving time (without returning to A) is to vertex C. Next, the shortest driving time without returning to B is to vertex E. At that point, the only city left not visited is D, so we go there, then return to A. The circuit is A, B, C, E, D, A. The sum of the weights (driving times) is $200 + 310 + 410 + 375 + 390 = 1{,}685$. The

Sidelight HAMILTON'S PUZZLE

Hamilton circuits are named in honor of the 19th century physicist, astronomer, and mathematician William Rowan Hamilton. Hamilton made a wide variety of important contributions to several areas of math. He didn't spend a lot of time on graph theory, but in introducing a curious puzzle in 1857, he ushered in the study of the circuits that now bear his name.

The puzzle in question was a wooden dodecahedron, which is a three-dimensional figure made of 12 faces that are regular pentagons. There was a peg at each of the 20 vertices, and the point was to attach a string to one peg, then find a route that follows the edges of the pentagons, and has the string visit every peg exactly once.

Hamilton was able to represent the puzzle using the type of graph we've studied in this chapter, and was able to develop a solution using algebraic methods. Unfortunately, his solution could not be generalized to other graphs.

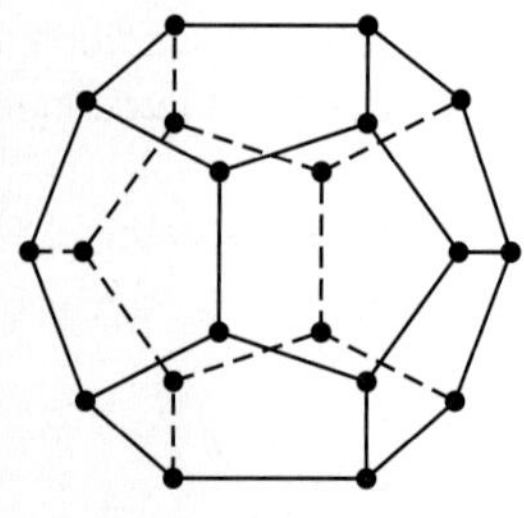

Today, over 150 years later, mathematicians are still searching for a method that can reliably find Hamilton circuits for arbitrary graphs, and the optimal solution to a traveling salesperson problem without using brute force. If you should find the latter, by all means contact delivery services like Federal Express and UPS—you would become a very, very wealthy person very, very quickly.

approximate optimal solution is driving for 1,685 minutes, or 28 hours and 5 minutes, using the circuit A, B, C, E, D, A.

CAUTION

In Example 5, the nearest neighbor method doesn't find the actual optimal solution. There are two paths that have a driving time of 1,550 minutes. Remember that the nearest neighbor method finds an *approximate* optimal solution.

▼ Try This One 5

The owner of a retail electronics franchise has locations in four different cities. He's planning a trip to visit all four, and the one-way plane fares between the cities are shown on the graph in Figure 15-41. Use the nearest neighbor method to find the approximate optimal route in terms of cost to visit the locations starting at city P.

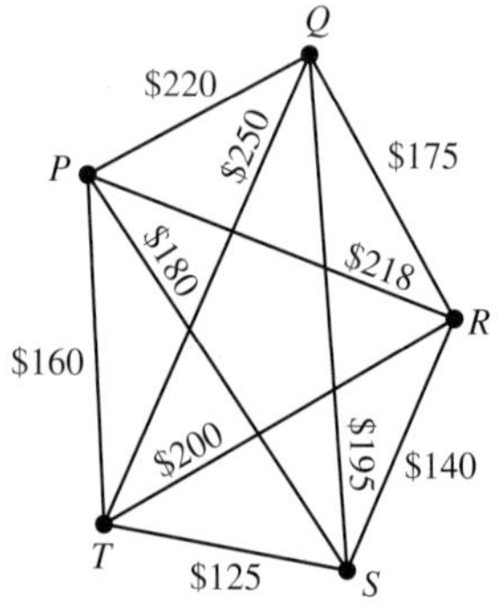

Figure 15-41

☑ 4. Find an approximate optimal solution using the nearest neighbor method.

In the examples we've looked at so far, a complete weighted graph has been provided. If we are instead given information like distances, plane fares, or driving times, we can use them to draw a complete weighted graph.

EXAMPLE 6 Drawing a Complete Weighted Graph

The distances in miles between four cities are shown in the table. Draw a complete weighted graph for the information.

	Corning	Mansfield	Elmira	Towanda
Corning (C)	–	30	21	54
Mansfield (M)	30	–	35	38
Elmira (E)	21	35	–	37
Towanda (T)	54	38	37	–

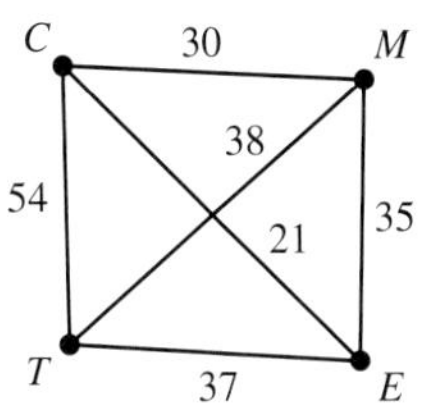

Figure 15-42

SOLUTION

Step 1 Draw and label the four vertices as *C*, *M*, *E*, and *T*.

Step 2 Connect each vertex with every other vertex using line segments (edges).

Step 3 Place the mile numbers (weights) on each segment using the information in the table. See Figure 15-42.

▼ Try This One 6

☑ 5. Draw a complete weighted graph based on provided information.

The distances in miles between four cities are shown in the table. Draw a complete weighted graph for the information.

	Youngwood	Scottdale	Mt. Pleasant	New Stanton
Youngwood	–	10	12	4
Scottdale	10	–	6	8
Mt. Pleasant	12	6	–	11
New Stanton	4	8	11	–

Answers to Try This One

1 One of many possible answers is *C, B, A, D, G, F, E*.

2 The two possibilities are *A, B, C, G, F, E, D, A* and *A, D, E, F, G, C, B, A*.

3 479,001,600

4 Either *A, B, C, D, A* or *A, D, C, B, A*.

5 *P, T, S, R, Q, P*, which costs $820.

6

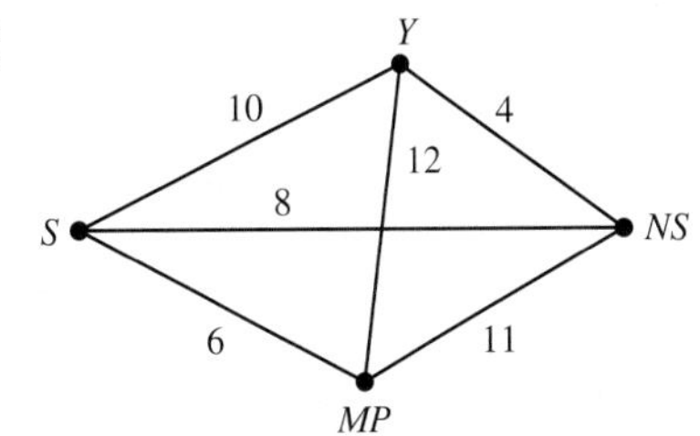

EXERCISE SET 15-3

Writing Exercises

1. What is the difference between a Hamilton path and an Euler path?
2. What is the difference between a Hamilton path and a Hamilton circuit?
3. What does it mean for a graph to be complete?
4. What does it mean for a graph to be weighted?
5. Describe what a typical traveling salesperson problem entails. What is an optimal solution?
6. Describe how to use the brute force method to find an optimal solution.
7. Describe how to use the nearest neighbor method to find an approximate optimal solution.
8. What does it mean to say that the nearest neighbor method finds an approximate optimal solution?

Computational Exercises

For Exercises 9–16, find two different Hamilton paths.

9.

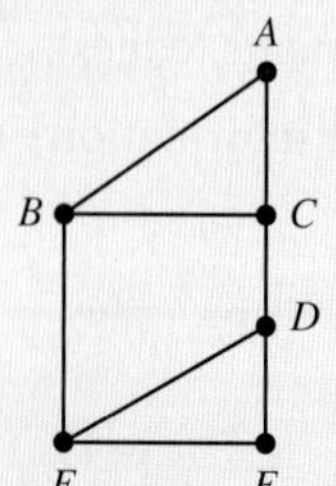

10.

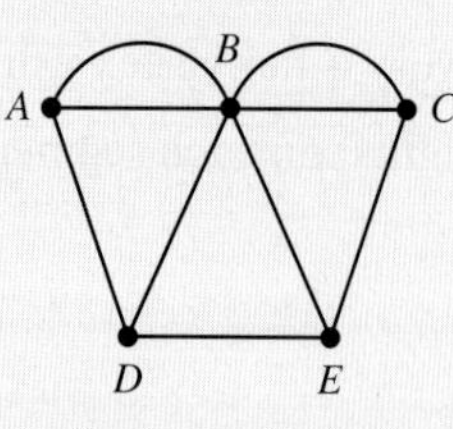

11.

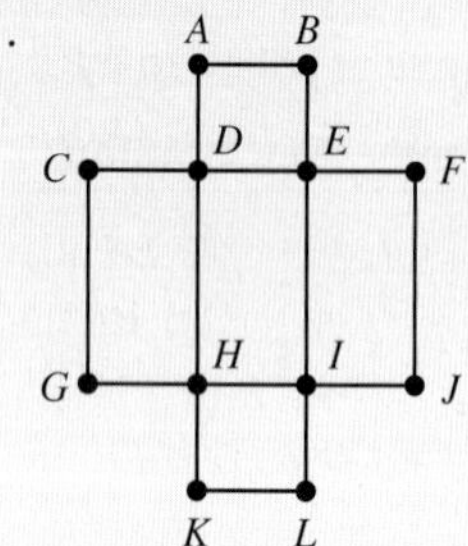

12.

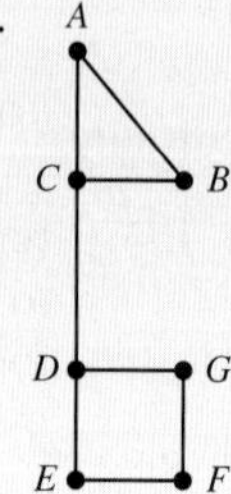

13.

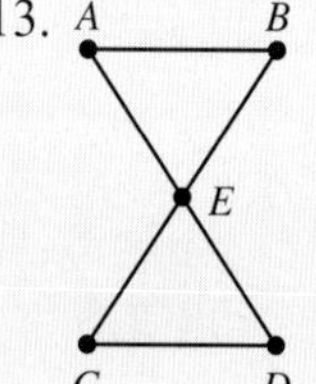

14.

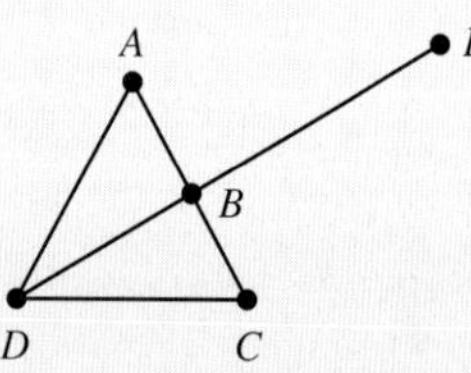

15.

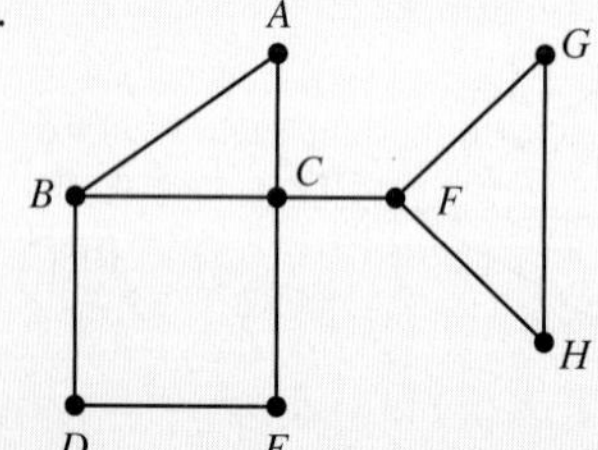

16.

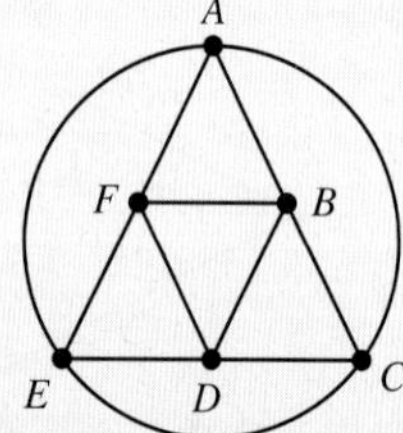

For Exercises 17–22, find two different Hamilton circuits.

17.

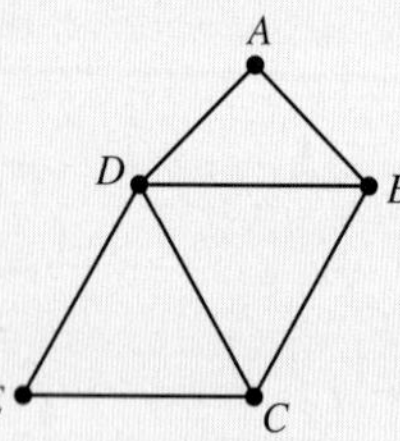

18.

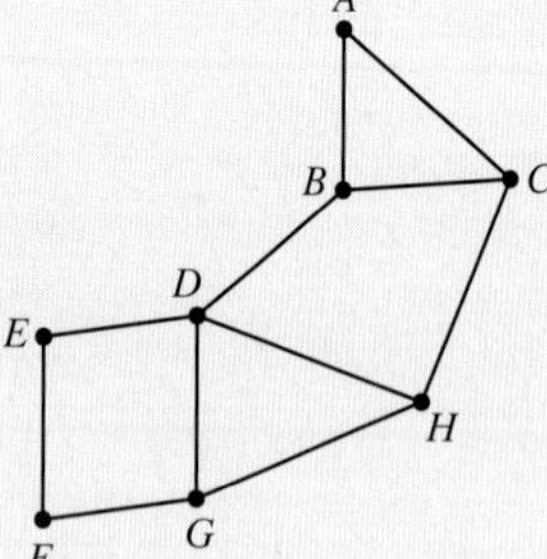

19.

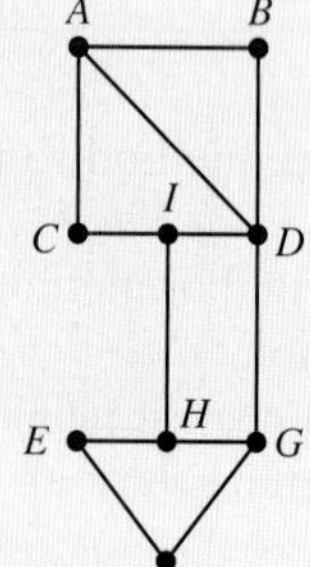

20.

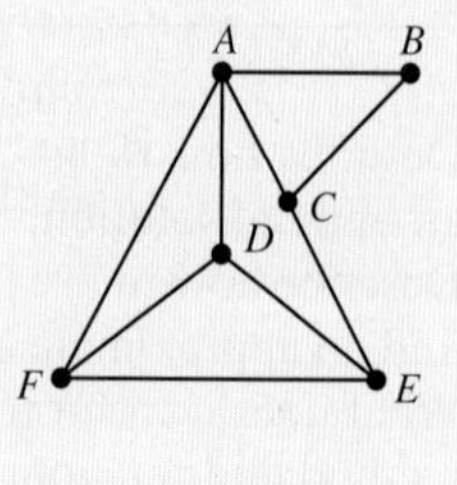

21.

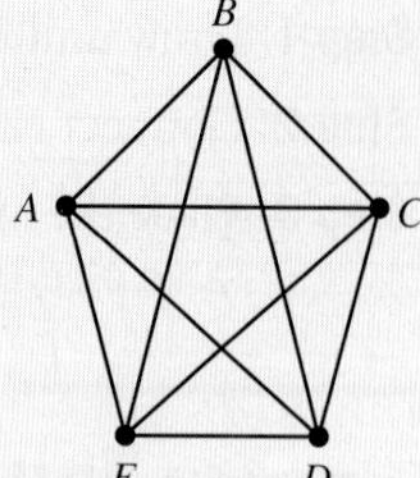

22.

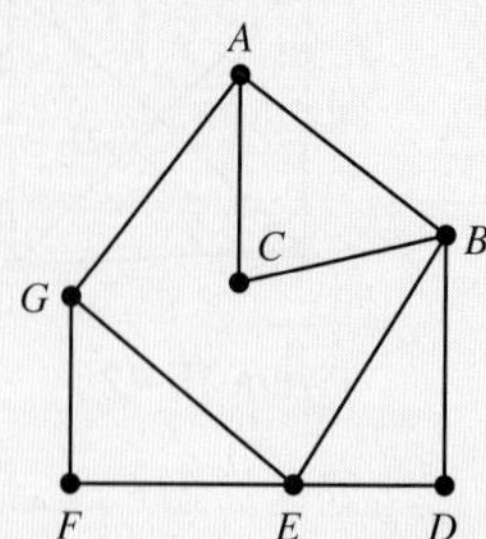

For Exercises 23–26, find the number of Hamilton circuits if a complete graph has the indicated number of vertices.

23. 3
24. 6
25. 9
26. 11

For Exercises 27–28, use the brute force method to find the optimal solution for each weighted graph.

27.

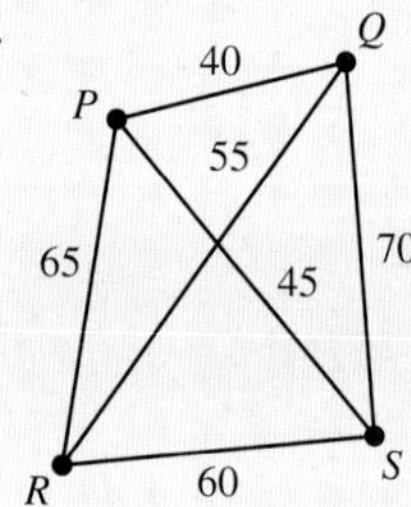

28.

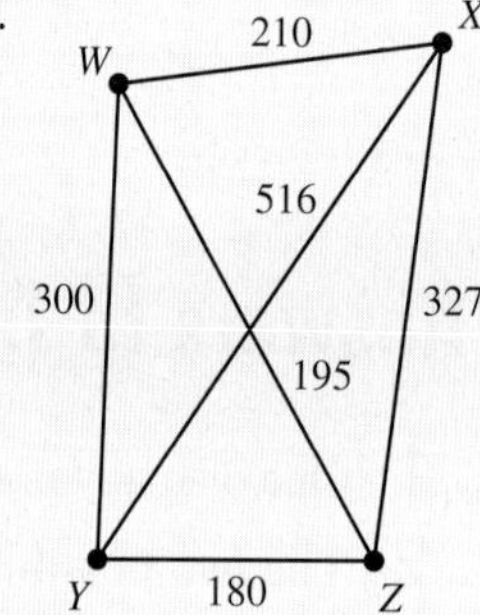

For Exercises 29–30, use the nearest neighbor method to approximate the optimal solution. Start at vertex A in each case.

29.

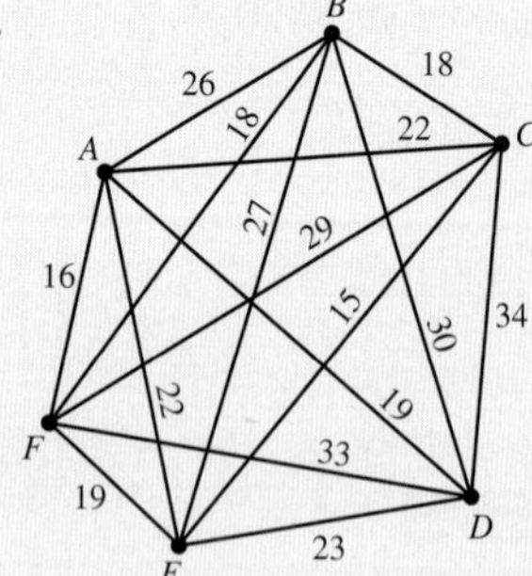

30.

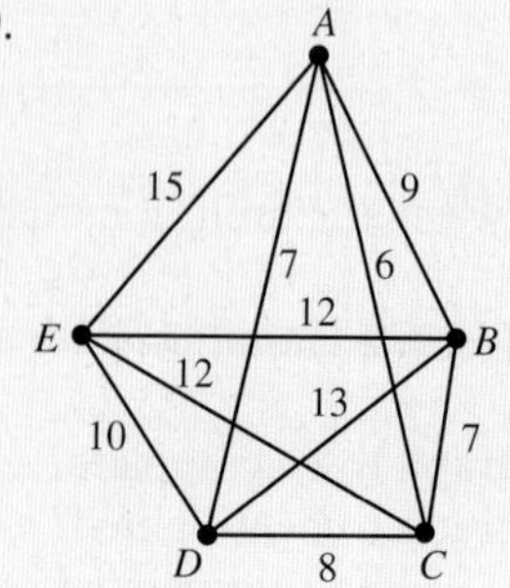

Real-World Applications

For Exercises 31–34, use the information in the table shown.

Distances (in miles) between Cities

	Pittsburgh	Philadelphia	Baltimore	Washington
Pittsburgh	–	305	244	245
Philadelphia	305	–	100	136
Baltimore	244	100	–	38
Washington	245	136	38	–

31. Draw a complete weighted graph for the information in the table.
32. Find the shortest route from Pittsburgh to all other cities and back to Pittsburgh using the brute force method.
33. Use the nearest neighbor method to find an approximation to the shortest route starting and ending at Pittsburgh.
34. What is the minimum distance given by the nearest neighbor method? Is it the same as that found by using the brute force method?

For Exercises 35–38, use the information in the table shown.

Air Fares between Cities

	New York	Cleveland	Chicago	Baltimore
New York	–	\$375	\$450	\$200
Cleveland	\$375	–	\$250	\$300
Chicago	\$450	\$250	–	\$325
Baltimore	\$200	\$300	\$325	–

35. Draw a complete weighted graph for the information in the table.
36. Find the cheapest route from Chicago to all other cities and back to Chicago using the brute force method.
37. Use the nearest neighbor method to find an approximation to the cheapest route from Chicago to all other cities and back to Chicago.
38. What is the least amount a person has to spend on airfare to travel to all cities from Chicago and back to Chicago using the nearest neighbor method? Is this the same amount found by the brute force method?

Critical Thinking

39. Find a road atlas that has a mileage chart. Select five cities and make a table of miles between the cities. Then make a weighted graph and find an approximate optimal circuit.
40. Draw a graph with more than four vertices, then find a Hamilton circuit that is also an Euler circuit.
41. Draw a graph with more than four vertices, then find a Hamilton circuit that is not an Euler circuit.
42. Suppose that you have a complete graph with n vertices, and you choose one vertex to begin and end as many Hamilton circuits as you can find.
    (a) How many different vertices can you choose to visit after the starting vertex? What is it about the graph that ensures you can visit any of the others?
    (b) After you've visited the beginning vertex and one other, how many remaining choices of vertex do you have to visit next?
    (c) Explain how you can continue this process to show that a complete graph with n vertices has $(n - 1)!$ Hamilton circuits beginning at any vertex.

Section 15-4 Trees

LEARNING OBJECTIVES

- ☐ 1. Determine if a graph is a tree.
- ☐ 2. Find a spanning tree for a graph.
- ☐ 3. Find a minimum spanning tree for a weighted graph.
- ☐ 4. Apply minimum spanning trees to real-world problems.

According to the Federal Highway Administration, there are over 4 million miles of road in the United States. Think for a minute about how many different routes you could take to get home from school. Our road system might be incredibly convenient, but it's not exactly what you would call efficient—there are many more routes than we really need to get from one place to another.

In this section, we will look at problems of efficiency: we will try to build graphs that connect vertices with the smallest number of edges between them. The graphs that we will construct are called *trees.*

A **tree** is a graph in which any two vertices are connected by exactly one path.

Math Note

Recall that we used tree diagrams in Chapter 11 to help compute the number of outcomes in probability problems.

Figure 15-43a shows a graph that is a tree, while the graph in Figure 15-43b is not a tree. In the first case, if you pick any pair of vertices, there is exactly one path connecting them. In the second case, given any pair of vertices, there are two different paths connecting them. For example, *A* and *C* are connected by the paths *A*, *B*, *C* and *A*, *D*, *C*.

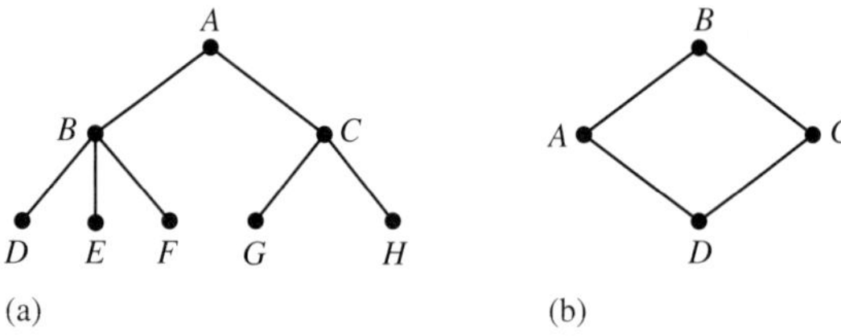

Figure 15-43

There are a number of properties of trees that will help us to recognize when a graph is or is not a tree. They are summarized next.

Properties of Trees

1. A tree has no circuits.
2. Trees are connected graphs.
3. Every edge in a tree is a bridge.
4. A tree with n vertices has exactly $n - 1$ edges.

EXAMPLE 1 Recognizing Trees

Which of the graphs in Figure 15-44 are trees?

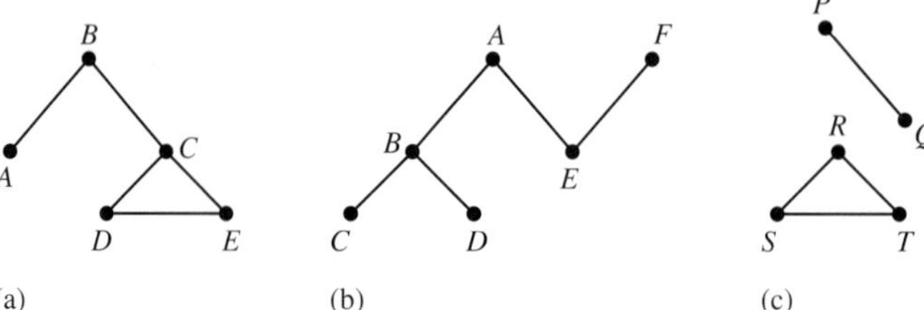

Figure 15-44

SOLUTION

The graph in 15-44a is not a tree. One way to see that is to observe that it contains a circuit: *C*, *E*, *D*, *C*. It also has the same number of edges as vertices, and none of edges *CD*, *DE*, or *EC* is a bridge.

The graph in Figure 15-44b is a tree. There is exactly one path to get from any vertex to any other. The graph is connected, has no circuits, and every edge is a bridge. Finally, there are six vertices and five edges.

The graph in Figure 15-44c is not a tree because it's not connected. It also has five vertices and only three edges.

▼ Try This One 1

☑ 1. Determine if a graph is a tree.

Which of the graphs in Figure 15-45 are trees?

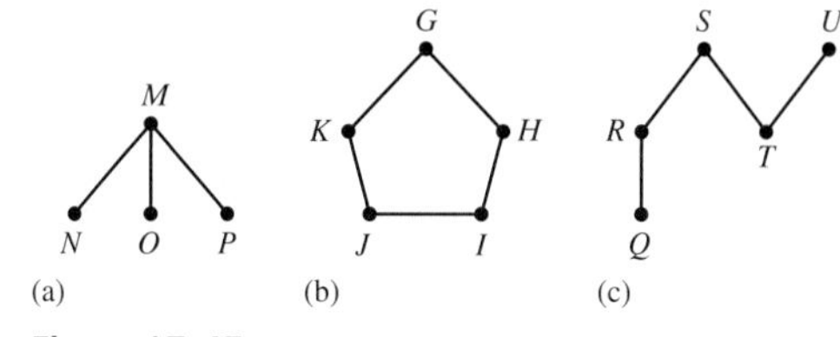

Figure 15-45

Spanning Trees

Suppose that in one county, a legislature intent on not getting re-elected decides that each town is going to be taxed based on how many roads they have. Facing fiscal disaster, the town leaders decide to eliminate all roads except the minimum number necessary to travel between the towns. If the original system of roads is represented by a graph, the problem would entail removing edges until the resulting graph is a tree. Such a graph is called a *spanning tree*.

A **spanning tree** for a graph is a tree that results from the removal of as many edges as possible from the original graph without making it disconnected.

EXAMPLE 2 Finding a Spanning Tree

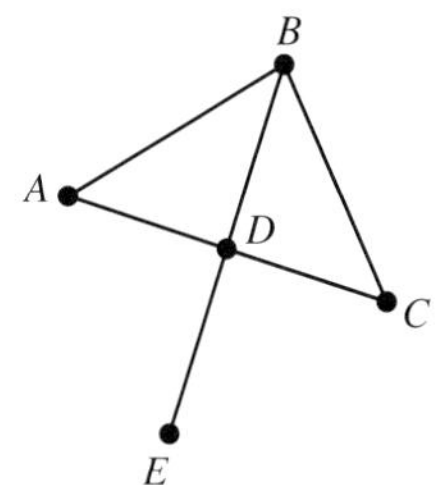

Figure 15-46

Find a spanning tree for the graph shown in Figure 15-46.

SOLUTION

The graph has five vertices, so a spanning tree must have four edges. The original graph has six edges, so we need to remove two edges without making the graph disconnected. One way to accomplish this is to remove edges *AD* and *DC*, resulting in the tree in Figure 15-47.

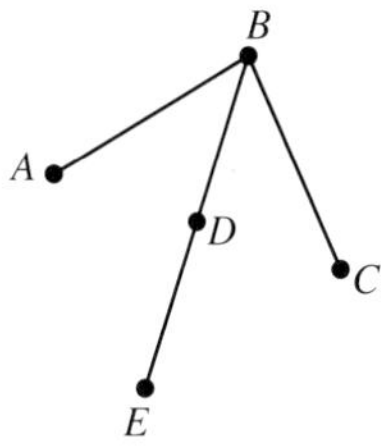

Figure 15-47

▼ Try This One 2

☑ 2. Find a spanning tree for a graph.

Find a spanning tree for the graph shown in Figure 15-48.

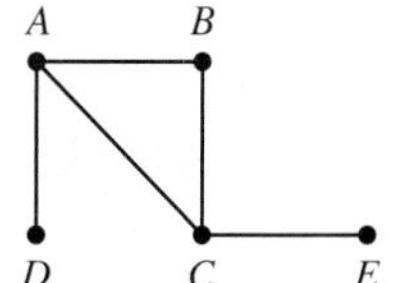

Figure 15-48

Spanning trees can be used to find the most efficient solution to a variety of problems involving cost, distance, and time. In order to do this, we will draw a weighted graph that represents the situation. For example, suppose that a railroad company needs to connect three industrial sites with track. The distances between the sites are shown in Figure 15-49a. There is no need to lay track on all three edges; instead, we will find a spanning tree. Three spanning trees can be drawn, as shown in Figure 15-49b.

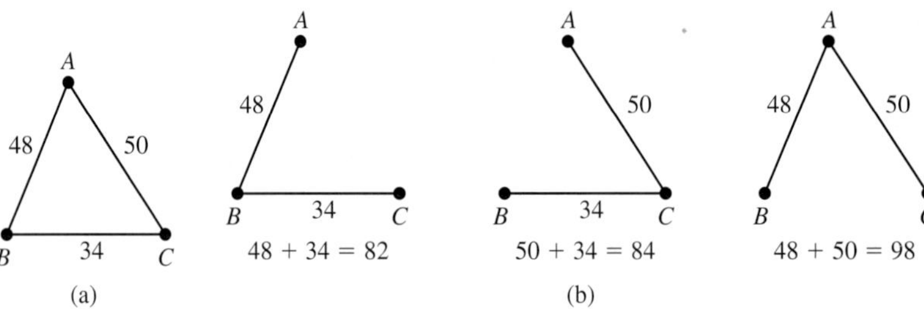

Figure 15-49

Of the three spanning trees shown, the first allows for the minimum amount of track, and is the most efficient solution. We call this a *minimum spanning tree.*

A **minimum spanning tree** for a weighted graph is the spanning tree for that graph that has the smallest possible sum of the weights.

For a very simple weighted graph, the minimum spanning tree can be drawn by observation; however, when the weighted graph is more complicated, a systematic procedure is helpful. A procedure known as Kruskal's algorithm can be used.

Kruskal's Algorithm

To construct a minimum spanning tree for a weighted graph:

Step 1 Choose the edge with the lowest weight and highlight it in color.

Step 2 Choose the edge with the next lowest weight and highlight it in color.

Step 3 Choose the unmarked edge with the next lowest weight that does not form a circuit with the edges already highlighted, and highlight it.

Step 4 Repeat until all vertices have been connected.

Note: If more than one edge could be chosen at any stage, pick one randomly.

Math Note

Don't forget that a minimum spanning tree must have one fewer edge than the number of vertices.

EXAMPLE 3 Finding a Minimum Spanning Tree

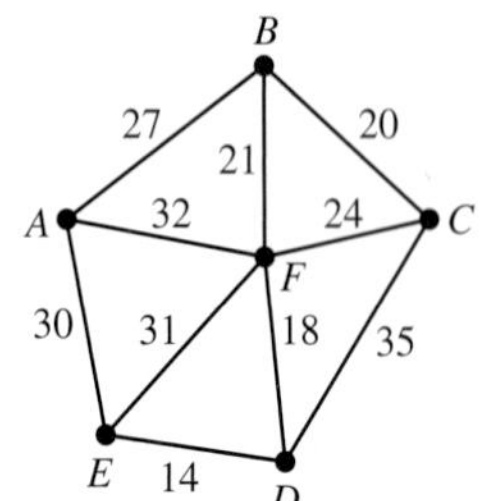

Figure 15-50

Use Kruskal's algorithm to find a minimum spanning tree for the weighted graph shown in Figure 15-50.

SOLUTION

Step 1 Highlight the lowest weighted edge, which is *ED*. See Figure 15-51a on the next page.

Step 2 Highlight the next lowest weighted edge, which is *DF*.

Step 3 The next lowest weight is 20, which is edge *BC*. Highlighting this edge won't make a circuit with the edges already highlighted, so this is a good choice.

Step 4 Edge BF has the next lowest weight, and it also will not form a circuit if highlighted.

Step 5 The next lowest weight is 24, but highlighting edge *FC* would form a circuit. Instead, we go to the next lowest, which is *AB*.

We have now connected all six vertices with five edges, so we have found a minimum spanning tree, shown in Figure 15-51b. The sum of the weights is 14 + 18 + 20 + 21 + 27 = 100.

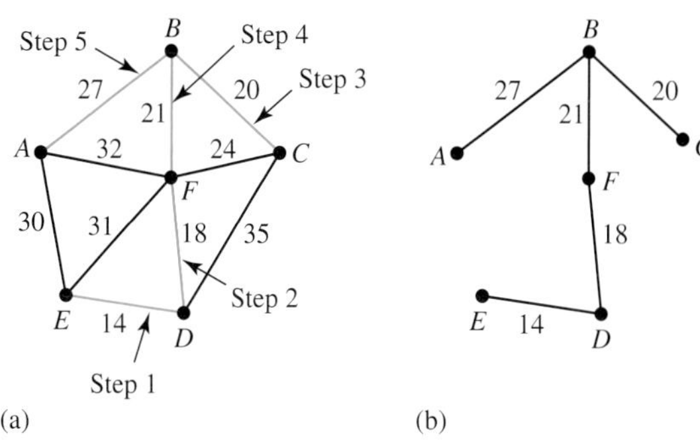

Figure 15-51

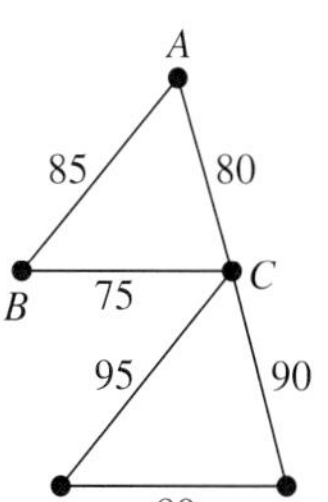

Figure 15-52

▼ Try This One 3

3. Find a minimum spanning tree for a weighted graph.

Find a minimum spanning tree for the graph in Figure 15-52.

The remaining two examples are applications of minimum spanning trees.

EXAMPLE 4 Spanning Trees in Agriculture

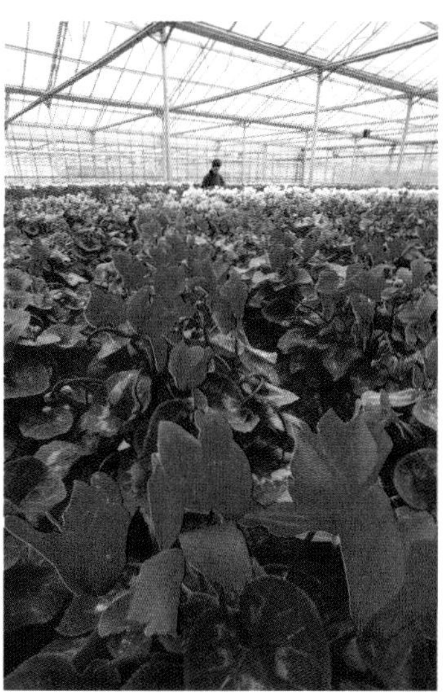

A flower farm has five greenhouses that need an irrigation system. The distances in feet between each of the greenhouses are shown in Figure 15-53. Use a minimum spanning tree to connect the greenhouses and determine the minimum amount of pipe needed.

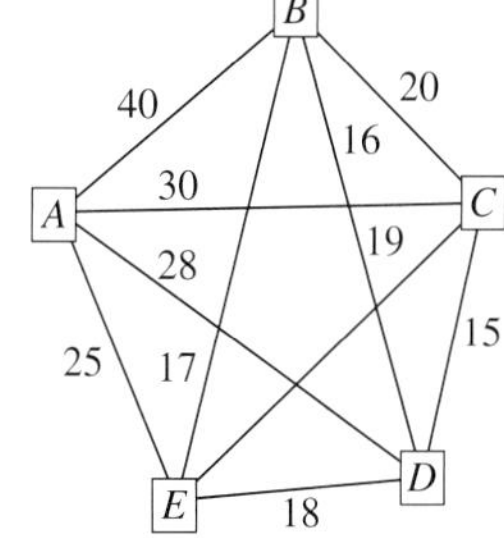

Figure 15-53

SOLUTION

Step 1 Highlight the edge with the lowest weight, *CD* (15 feet). See Figure 15-54a.

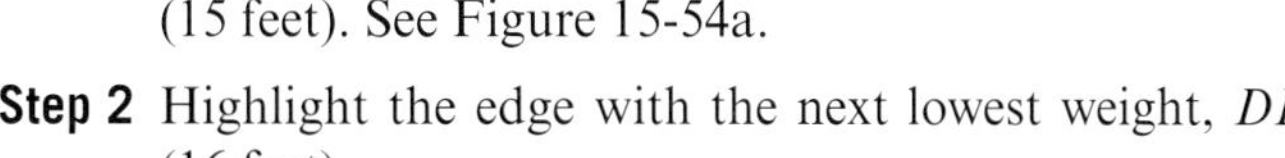

Step 2 Highlight the edge with the next lowest weight, *DB* (16 feet).

Step 3 The next lowest weight is 17 feet (edge *BE*), and choosing this edge will not make a circuit with the edges already highlighted.

Step 4 All of edges ED (18 feet), EC (19 feet), and BC (20 feet) would make a circuit, so we go up to edge *AE* (25 feet) to complete our spanning tree. The tree is shown in Figure 15-54b. The minimum amount of pipe needed is 15 + 16 + 17 + 25 = 73 feet.

Math Note

In Example 4, all of the edges chosen connected to edges that were already highlighted, but it isn't necessary to do so. In Example 3, the highlighted graph was disconnected after step 3.

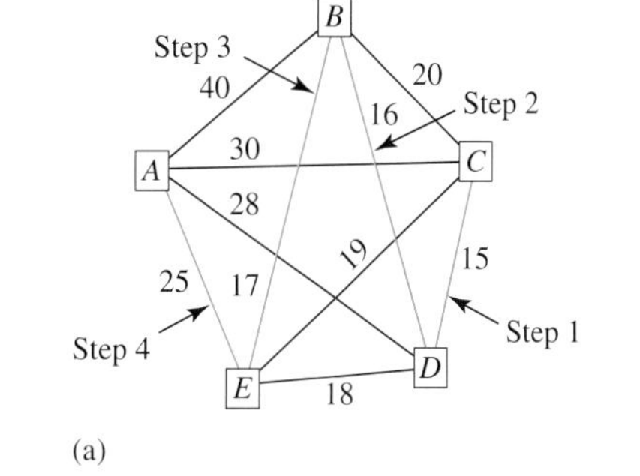

(b)

Figure 15-54

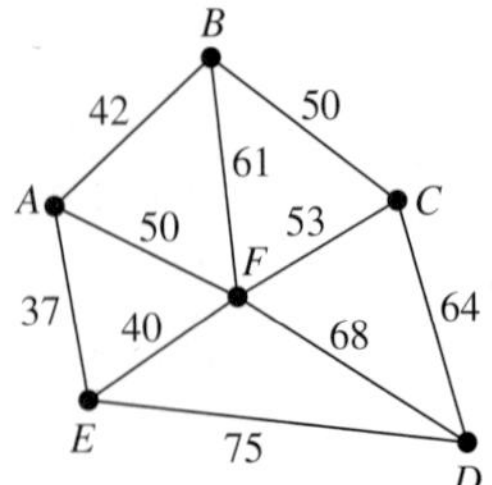

Figure 15-55

▼ Try This One 4

A regional campus is being designed to be environmentally friendly. There will be six buildings, with the lengths in yards between them shown in Figure 15-55. The plan is to construct the minimum length of sidewalks that will connect all of the buildings. Use a minimum spanning tree to connect the buildings and determine the minimum amount of sidewalk needed.

EXAMPLE 5 Spanning Trees in Courier Service

A power company with plants in Pittsburgh, Philadelphia, Baltimore, and Washington, D.C., wants to establish courier service between its plants using the shortest routes. Draw a complete weighted graph and use Kruskal's algorithm to find a minimum spanning tree that would connect each city using the smallest number of miles.

	Pittsburgh	Philadelphia	Baltimore	Washington DC
Pittsburgh	–	305	244	245
Philadelphia	305	–	100	136
Baltimore	244	100	–	38
Washington DC	245	136	38	–

SOLUTION

Draw a complete weighted graph and find the minimum spanning tree as shown in Figure 15-56. The minimum number of miles is 38 + 100 + 244 = 382 miles.

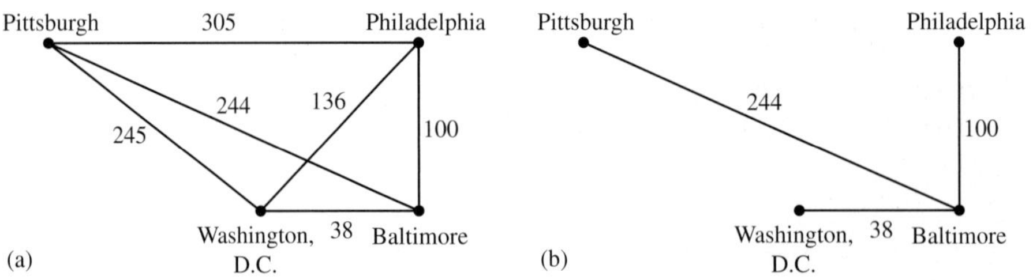

Figure 15-56

▼ Try This One 5

☑ 4. Apply minimum spanning trees to real-world problems.

A bus company plans to run bus service between the cities shown in the table. Draw a complete weighted graph and find a minimum spanning tree to determine the shortest route. Find the mileage of the one-way route.

	Adamsburg	Irwin	Penn	Trafford
Adamsburg	–	15	14	19
Irwin	15	–	15	9
Penn	14	15	–	13
Trafford	19	9	13	–

Answers to Try This One

1 Graphs (a) and (c) are trees.

2 One possibility is the following:

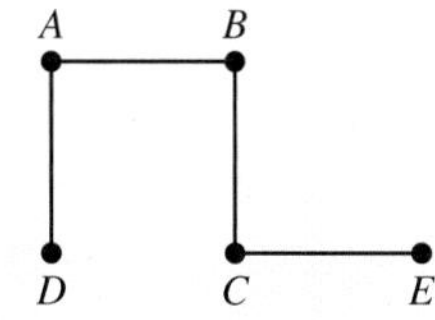

3

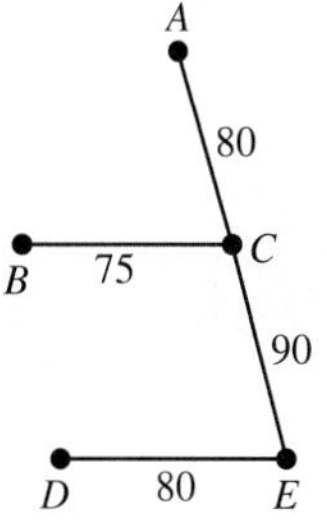

4

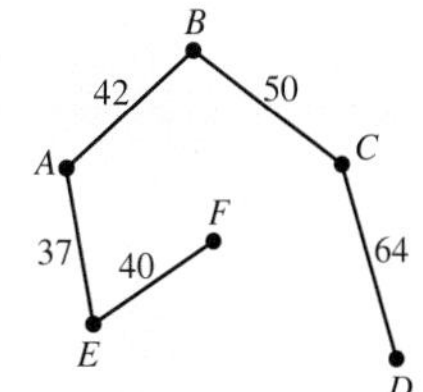

The length of sidewalk needed is 233 yards.

5

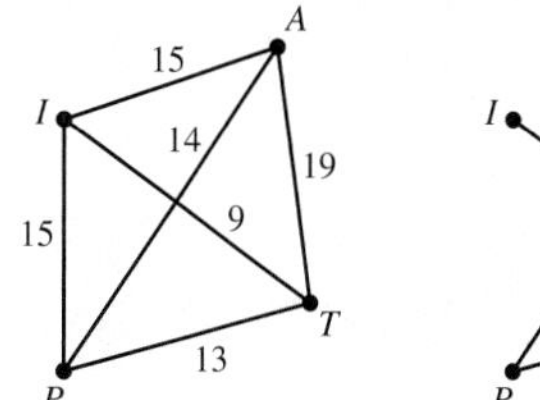

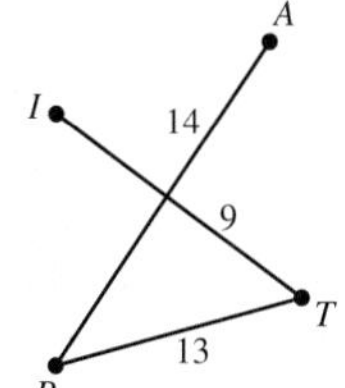

The one-way mileage is 36 miles.

EXERCISE SET 15-4

Writing Exercises

1. Explain the difference between a graph and a tree.
2. Can any tree be a complete graph? Explain.
3. What is a spanning tree for a graph?
4. What is a minimum spanning tree for a weighted graph?
5. Describe Kruskal's algorithm for finding a minimum spanning tree.
6. Describe a real-world problem that can be solved using a minimum spanning tree.

Computational Exercises

For Exercise 7–16, determine whether or not each graph is a tree. If it is not, explain why.

7.

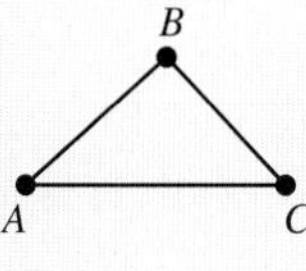

8.

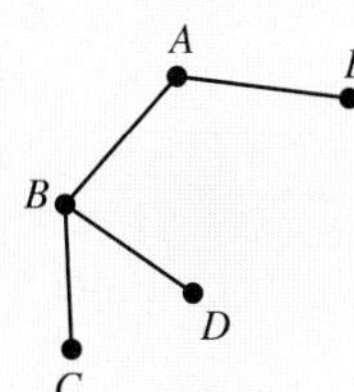

9.

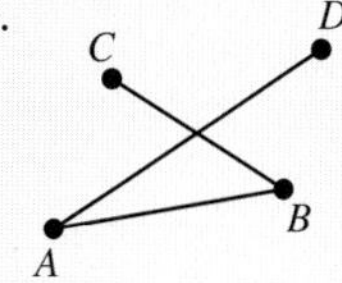

10.

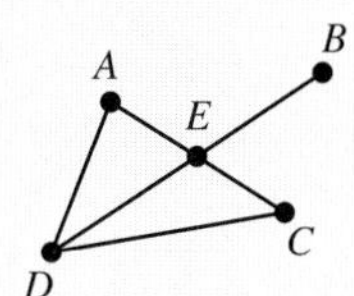

11.

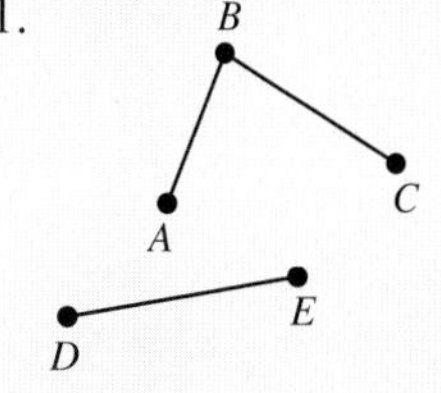

12. A D C E F B G H

13.

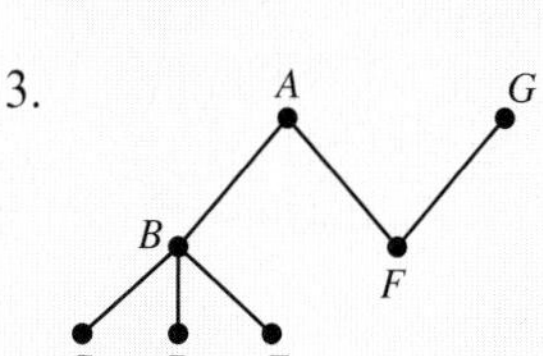

14.

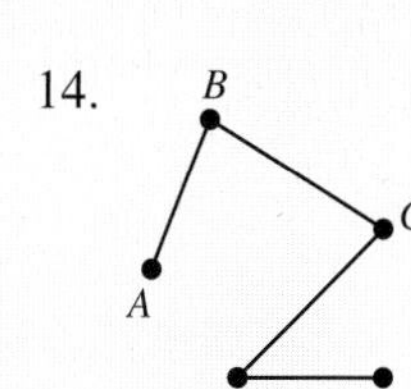

15.

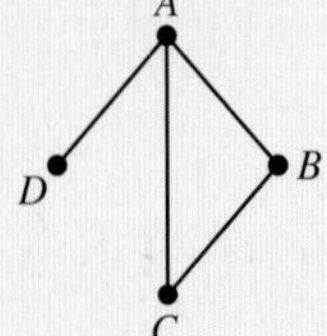

16.

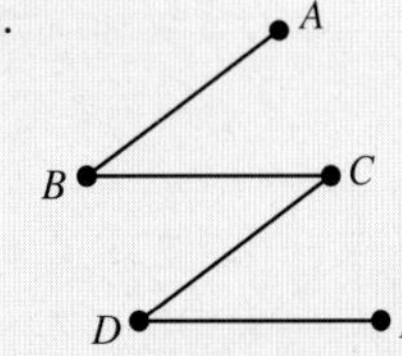

For Exercises 17–22, find a spanning tree for each. (Answers may differ from those given in the answer section.)

17.

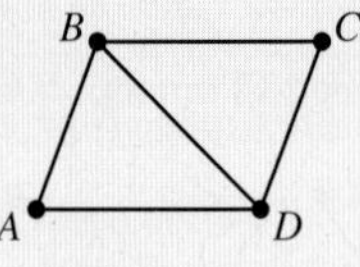

18.

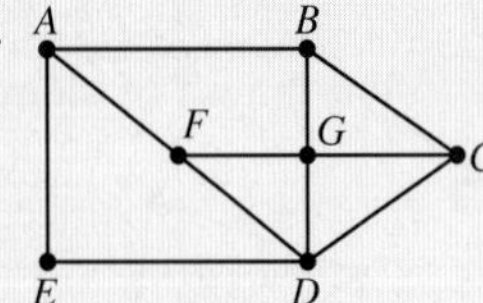

19.

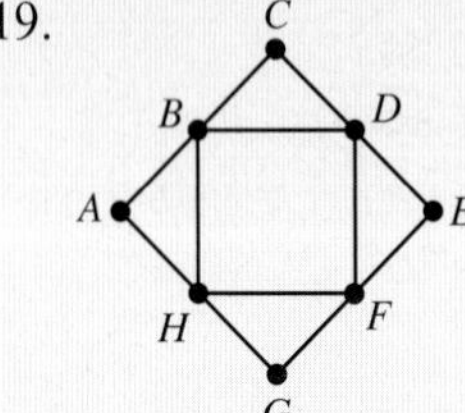

20.

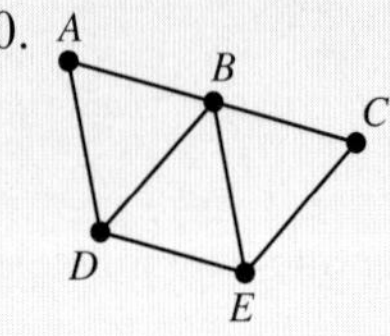

21.

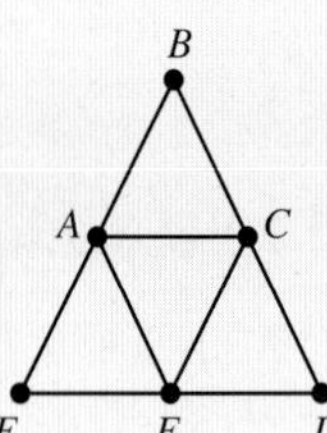

22.

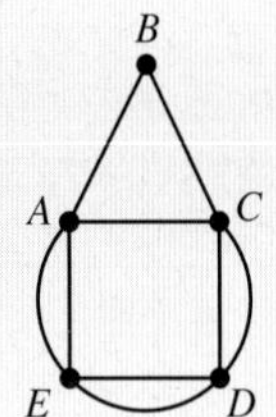

For Exercises 23–30, use Kruskal's algorithm to find a minimum spanning tree for each. Give the total weight for each one.

23.

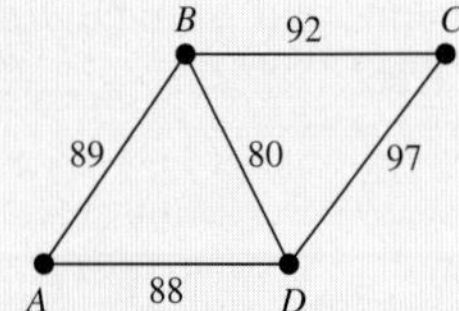

24.

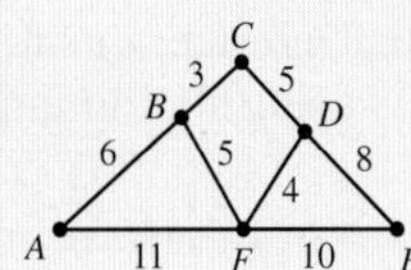

25.

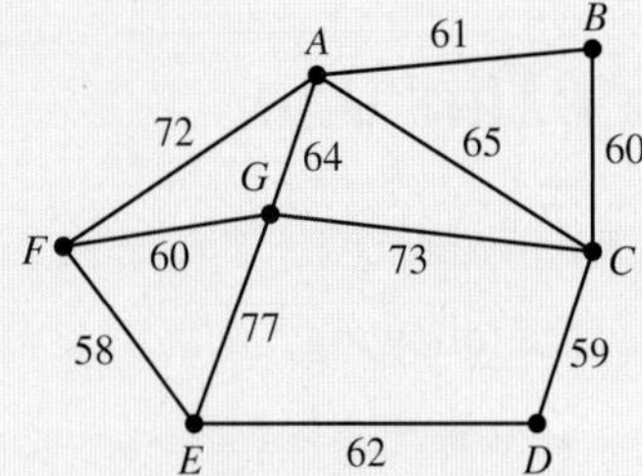

26.

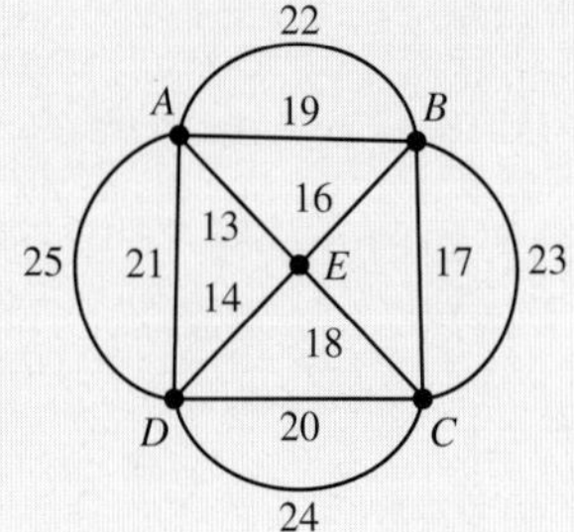

27.

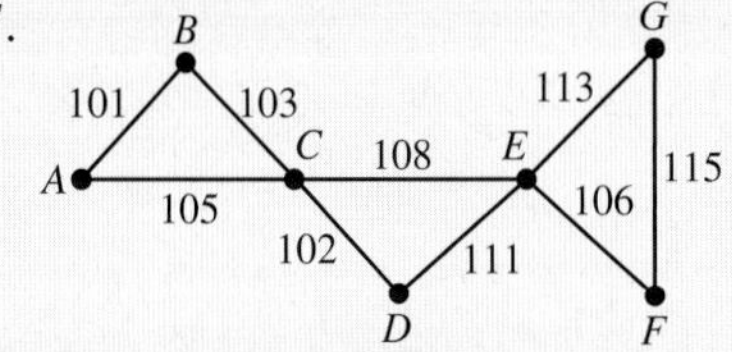

28.

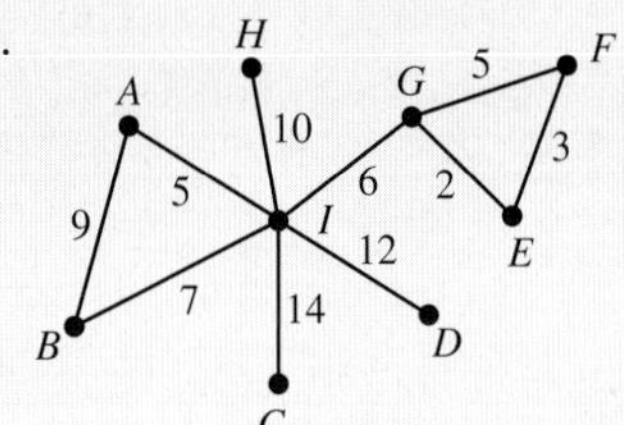

29.

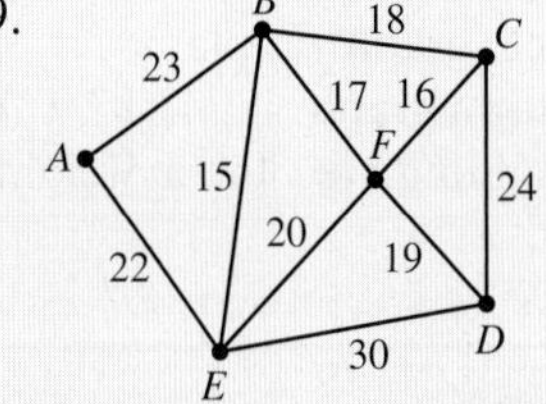

30.

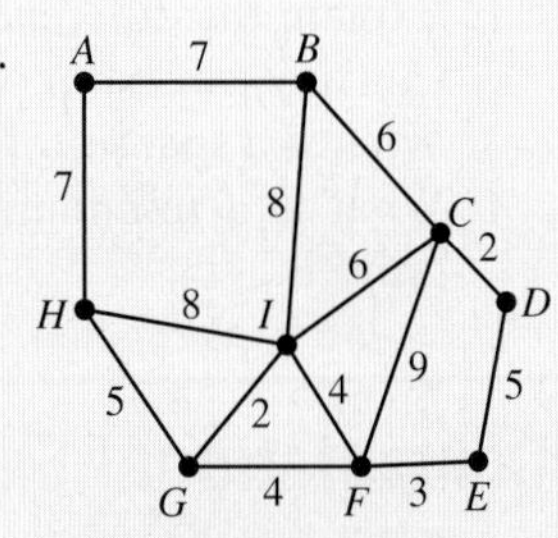

Real-World Applications

31. At a convention, the manager needs to run electricity to each table. The distances in feet are shown. Use a minimum spanning tree to find the shortest amount of wire necessary. How much wire will be needed?

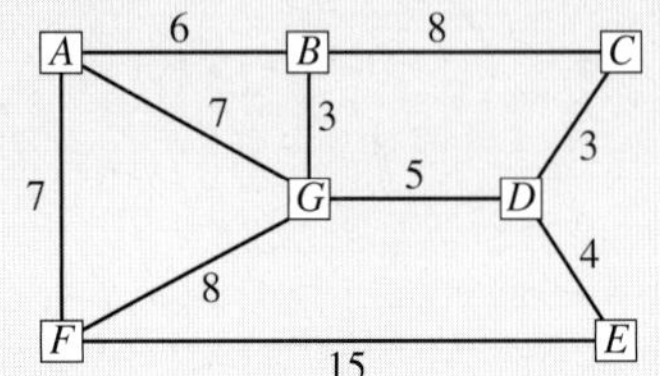

32. Eight buildings in a housing complex are connected by sidewalks. The maintenance manager wants to clear the shortest path connecting all buildings when it snows. Use a minimum spanning tree to find the shortest distance. The distances shown at the top of the next page are given in feet. Find the length of the path.

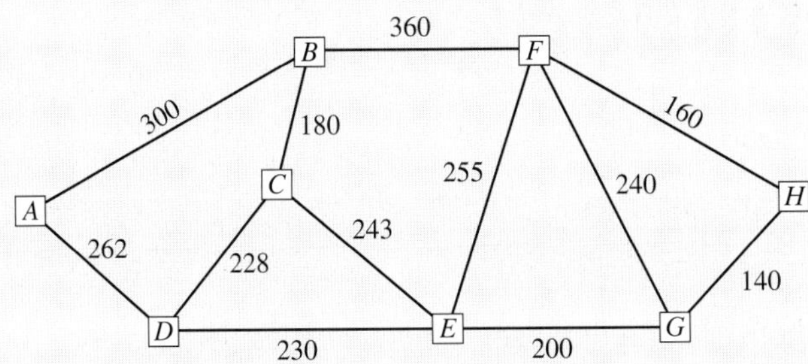

33. A committee decides to build an outdoor fitness facility with five stations in a park. Using a minimum spanning tree, find the shortest distance between all stations. If a path were to be built between the stations, find its length. The distances are given in feet.

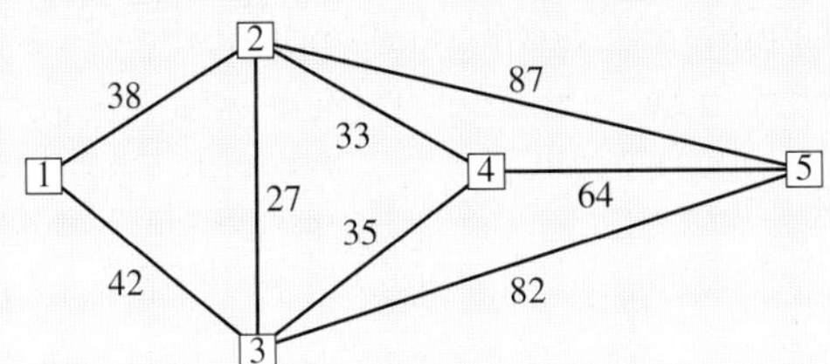

34. A bus company plans to establish a bus route between the cities shown in the table. Draw a complete weighted graph and find the minimum spanning tree for the mileage. What is the number of miles for the route?

	Corning	Mansfield	Elmira	Towanda
Corning (C)	–	30	17	54
Mansfield (M)	30	–	35	38
Elmira (E)	17	35	–	37
Towanda (T)	54	38	37	–

Critical Thinking

35. Using a road map, an atlas, or the Internet, find five cities and determine the distances between them. Then draw a graph connecting the cities. Finally, determine a minimum spanning tree for the graph.

36. Repeat Exercise 35 for five buildings on a campus of a college or university.

37. Repeat Exercise 35 for five historical sites or landmarks in a large city.

CHAPTER 15 Summary

Section	Important Terms	Important Ideas
15-1	Graph Vertex Edge Loop Equivalent graphs Degree of a vertex Odd vertex Even vertex Adjacent vertices Path Circuit Connected graph Disconnected graph Bridge	**A graph** consists of a finite set of points called vertices and line segments called edges connecting the points. A loop is an edge that begins and ends at the same vertex. The degree of a vertex is the number of edges emanating from the vertex. Adjacent vertices have at least one edge connecting them. A path on a graph is a sequence of adjacent vertices and the edges connecting them that uses no edge more than once. A circuit is a path that begins and ends at the same vertex. A graph is said to be connected if there is at least one path that connects any two vertices; otherwise, it is said to be disconnected. A bridge on a connected graph is an edge that, if removed, makes the graph disconnected.
15-2	Euler path Euler circuit Euler's theorem Fleury's algorithm	**Graphs** can be used as models for floor plans, neighborhoods, etc. An Euler path is a path that passes through each edge exactly once. An Euler circuit is a circuit that passes through each edge exactly once. Euler's theorem states that a connected graph that has all even vertices has at least one Euler circuit. If it has exactly two odd vertices, it has at least one Euler path but no Euler circuits. Finally, if a connected graph has more than two odd vertices, it has no Euler circuits or Euler paths. Fleury's algorithm can be used to find an Euler path or Euler circuit.
15-3	Hamilton path Hamilton circuit Complete graph Complete weighted graph Traveling salesperson problem Optimal solution Brute force method Nearest neighbor method Approximate optimal solution	**A path** on a connected graph that passes through each vertex only once is called a Hamilton path. If a Hamilton path begins and ends at the same vertex, it is called a Hamilton circuit. A complete graph is a graph that contains an edge between any two vertices. The number of Hamilton circuits in a complete graph with n vertices is $(n - 1)!$. A complete weighted graph has weights (numbers) on all edges. The optimal solution on a weighted graph is the circuit for which the sum of the weights of the edges is the smallest. A traveling salesperson problem requires you to find a Hamilton circuit for a complete weighted graph such that the sum of the weights of the edges is the smallest possible number. The brute force method to find the optimal circuit is to find all possible Hamilton circuits, add the weights for each circuit, and then select the one with the smallest sum. The nearest neighbor method to find an approximate optimal circuit is to start at a vertex and then select an edge with the smallest weight to the next vertex. Continue by selecting edges with the smallest weights. Do not select any edges that go to a vertex already selected.

15-4	Tree Spanning tree Minimum spanning tree Kruskal's algorithm	**A tree** is a graph in which any pair of vertices is connected by exactly one path. A spanning tree is a tree that has been created by removing edges from a graph. A minimum spanning tree for a weighted graph is the spanning tree (out of all possible spanning trees for the graph) that has the smallest sum of the weights. Kruskal's algorithm is a method that can be used to construct a minimum spanning tree for a weighted graph. Start at the lowest weighted edge and select the next lowest weighted edge that does not form a circuit with previously selected edges. Continue until you have included all vertices.

MATH IN Road Trips REVISITED

1. The graph is shown below, with weights representing the distances. Georgia is *G*, Georgia Tech is *T*, Alabama is *B*, and Auburn is *A*.

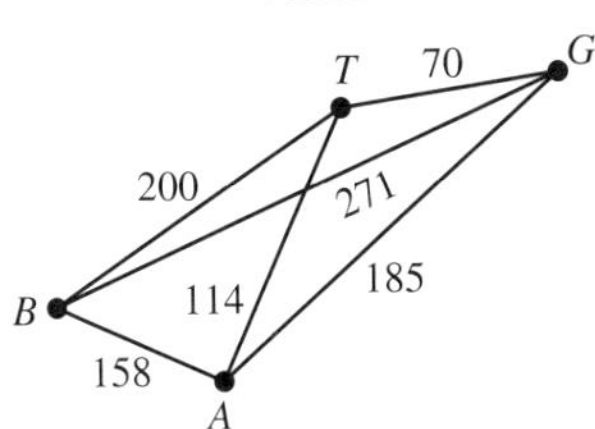

2. This is a question about Hamilton circuits. It turns out that of the six possible routes, four of them are optimal!

Using the brute force method, we find that the four paths listed below all cover 613 miles.

T, G, B, A, T; T, G, A, B, T; T, B, A, G, T; T, A, B, G, T

3. This is a spanning tree problem. The minimum spanning tree is shown below, and covers just 342 miles.

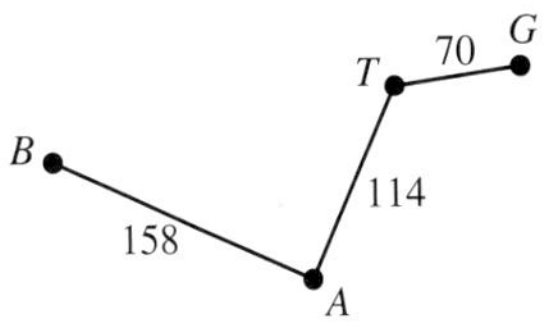

Review Exercises

Section 15-1

Use the graph shown in Figure 15-57 for Exercises 1–8.

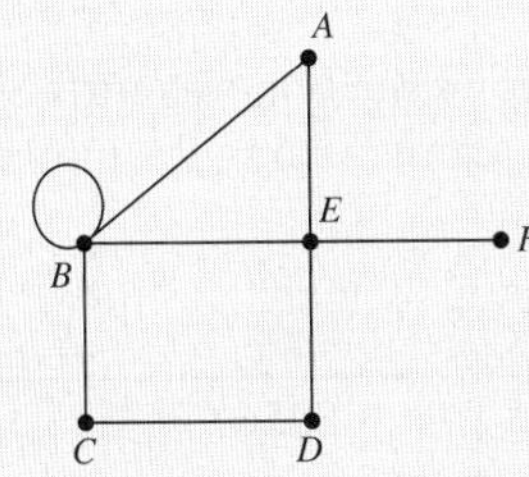

Figure 15-57

1. Identify the vertices.
2. Is vertex *E* an even or odd vertex?
3. Is vertex *A* an even or odd vertex?
4. Which vertex has a loop?
5. Which vertex is adjacent to vertex *F*?
6. Identify an edge that is a bridge.
7. Is the graph connected?
8. How many edges does the graph have?
9. Draw a graph that models the region shown in Figure 15-58. Use vertices for land masses and edges for bridges.

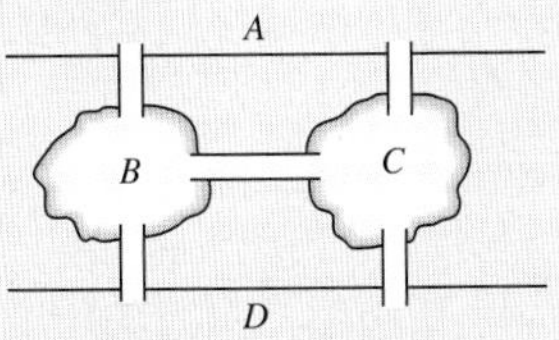

Figure 15-58

10. Draw a graph that models the floor plan shown in Figure 15-59. Use a vertex for each room and the outside area, and an edge for each door.

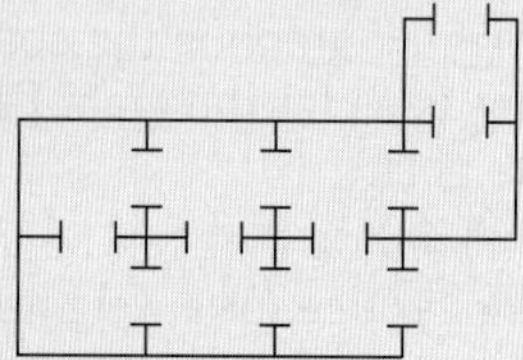

Figure 15-59

11. Draw a graph that models the border relationships among the counties shown in Figure 15-60. Use vertices to represent the counties and edges to represent common borders.

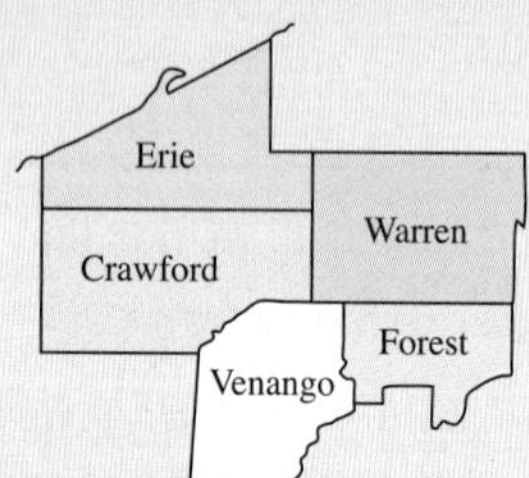

Figure 15-60

Section 15-2

12. For the graph shown in Figure 15-61, find an Euler path.

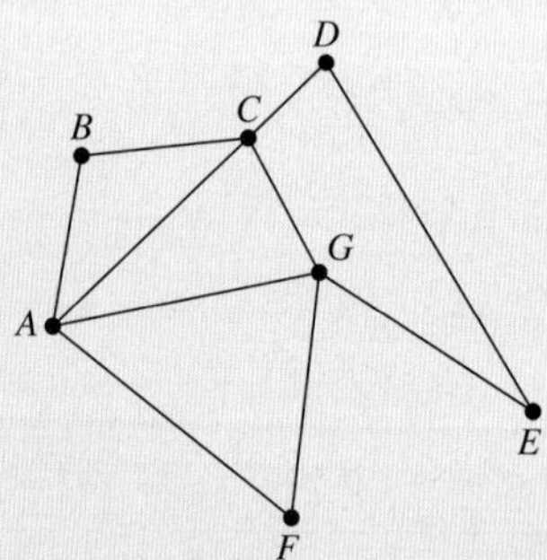

Figure 15-61

13. For the graph shown in Figure 15-62, find an Euler circuit.

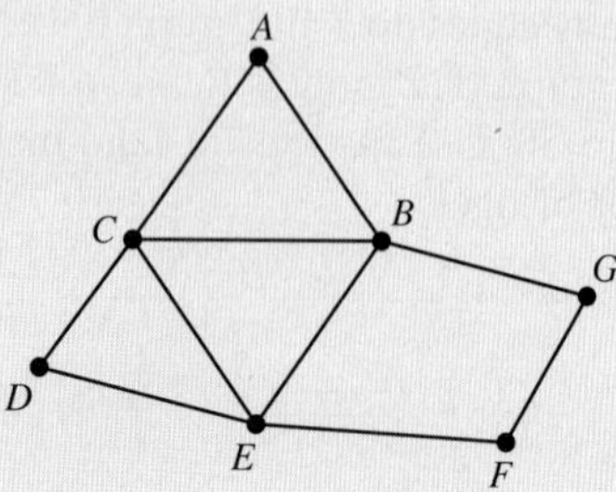

Figure 15-62

Section 15-3

14. For the graph shown in Figure 15-63, find a Hamilton path.

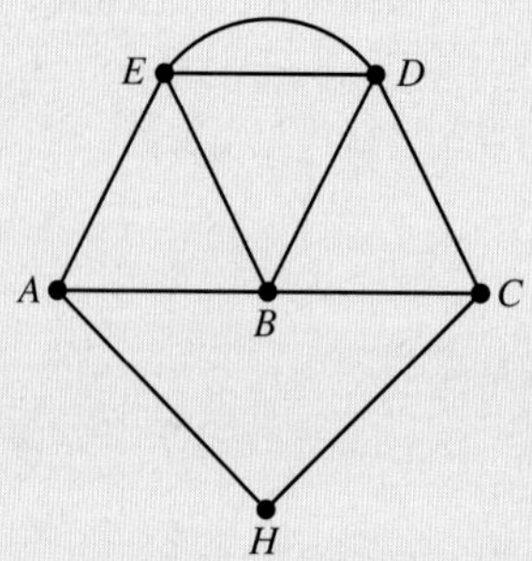

Figure 15-63

15. For the graph shown in Figure 15-64, find a Hamilton circuit.

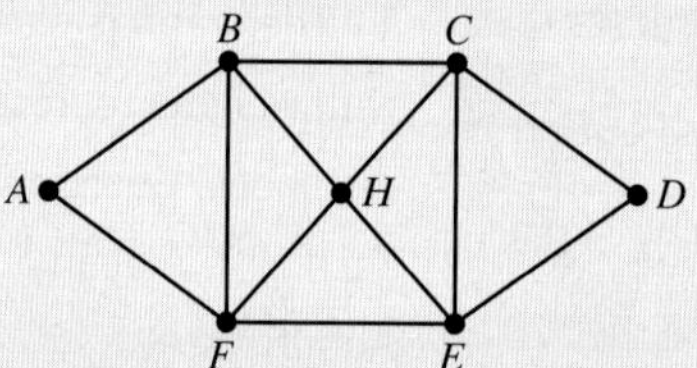

Figure 15-64

16. If a complete graph has 12 vertices, how many distinct Hamilton circuits does it have?
17. For the weighted graph shown in Figure 15-65, find an approximate optimal circuit starting at vertex A using the nearest neighbor method.

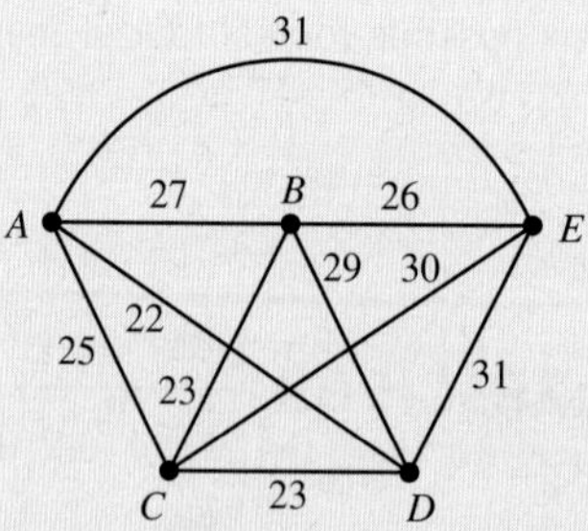

Figure 15-65

18. The table shows the one-way airfares between four cities. Draw a graph using the information in the table.

	C	D	E	F
C	–	$256	$732	$350
D	$256	–	$560	$197
E	$732	$560	–	$230
F	$350	$197	$230	–

19. Find an approximate optimal solution using the nearest neighbor method for the graph in Exercise 18 starting at city *C* and going to all the other cities and returning to city *C*.

Section 15-4

20. Find a spanning tree for the graph shown in Figure 15-66.

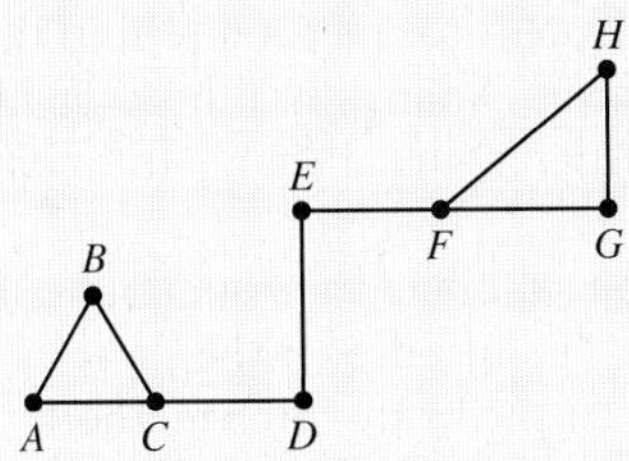

Figure 15-66

21. Use Kruskal's algorithm to find the minimum spanning tree for the weighted graph shown in Figure 15-67.

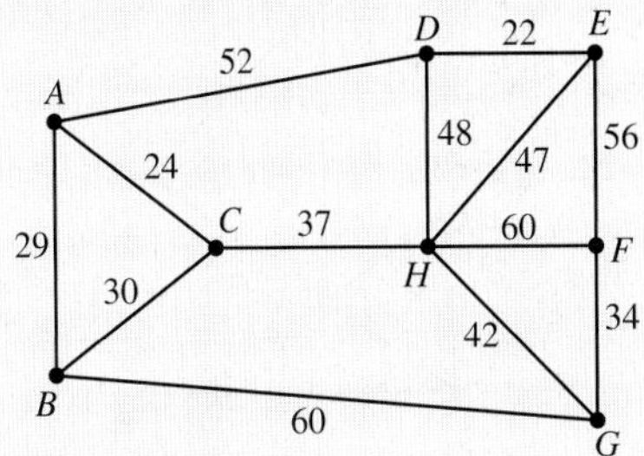

Figure 15-67

Chapter Test

For Exercises 1–6 use Figure 15-68.

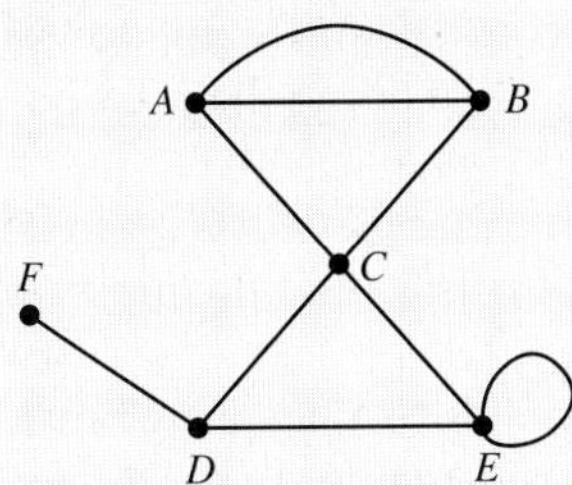

Figure 15-68

1. What is the degree of vertex *C*?
2. Which vertex has a loop?
3. Describe a path that starts at vertex *A* and passes through vertex *E*.
4. Which edge is a bridge?
5. Name four vertices adjacent to vertex *C*.
6. Is vertex *D* even or odd?
7. Draw a disconnected graph.
8. Draw a graph for the states shown in Figure 15-69. Use a vertex for each state and an edge for each common border.

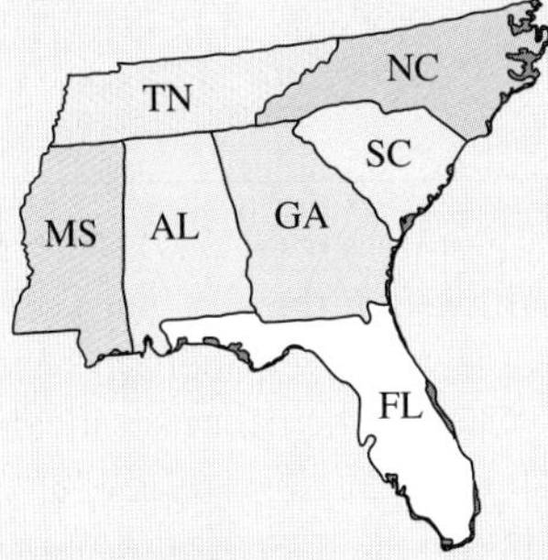

Figure 15-69

9. Draw a graph for the floor plan shown in Figure 15-70. Use a vertex for each room and the outside area, and an edge for each door.

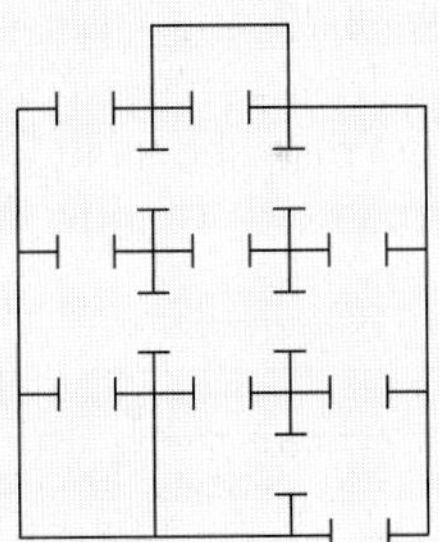

Figure 15-70

10. For the graph shown in Figure 15-71, find an Euler path.

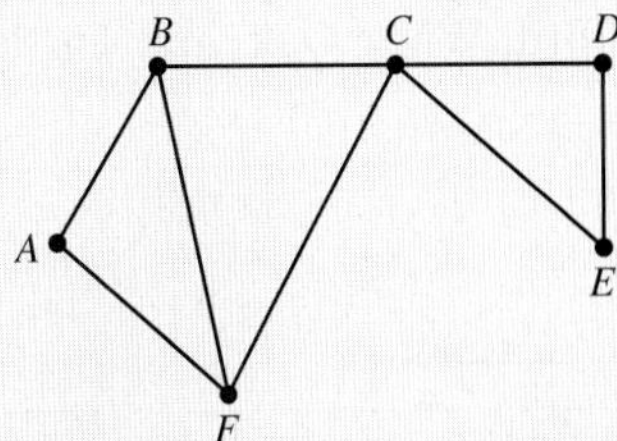

Figure 15-71

11. Find a Hamilton path for the graph shown in Figure 15-72.

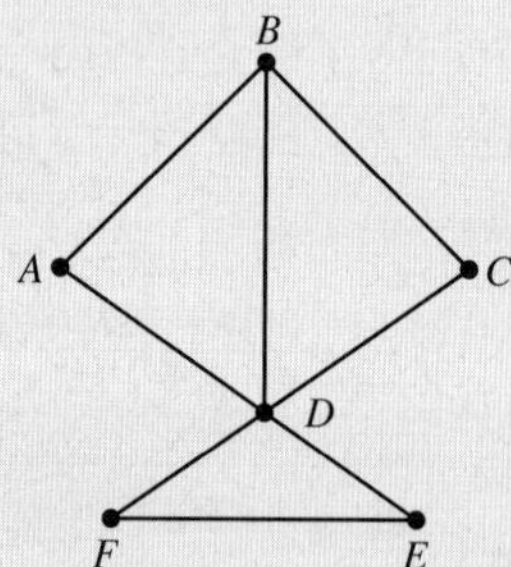

Figure 15-72

12. Make a spanning tree for the graph shown in Figure 15-73.

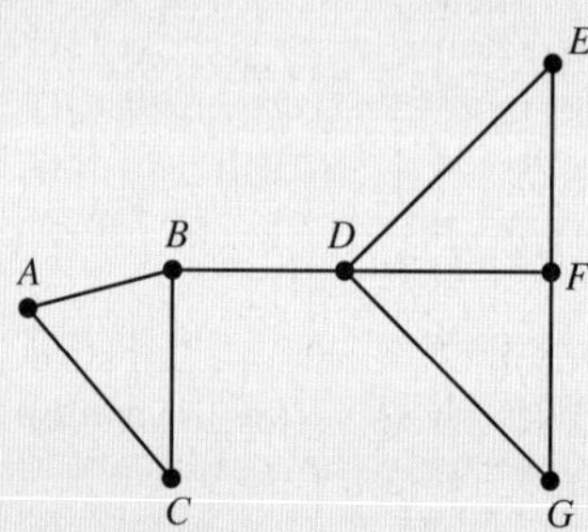

Figure 15-73

13. For the housing plan shown in Figure 15-74, draw a graph and find a way to plow the roads without doing any street twice.

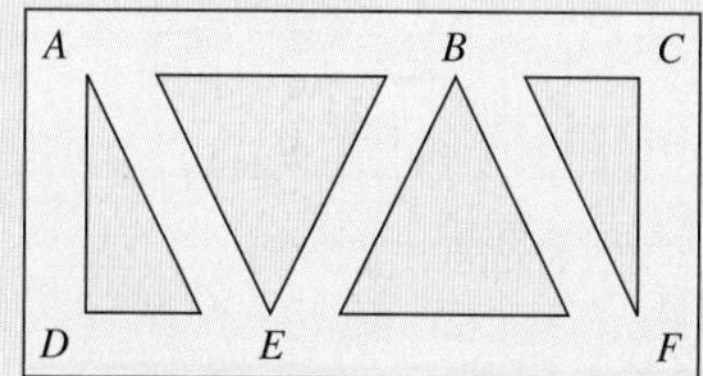

Figure 15-74

14. A salesperson who lives in Adamsburg must visit Trafford, White Oak, and Turtle Creek. The distances between the cities are as follows: Adamsburg to White Oak is 27 miles, Adamsburg to Turtle Creek is 18 miles, Adamsburg to Trafford is 20 miles, Trafford to White Oak is 14 miles, Trafford to Turtle Creek is 19 miles, and White Oak to Turtle Creek is 12 miles. Represent the distances with a complete weighted graph.
15. Use the brute force method to find the shortest distance from Adamsburg through all cities and back to Adamsburg. What is the shortest distance in miles?
16. Use the nearest neighbor method to approximate an optimal solution. Is it the same as the result obtained by the brute force method?

For each problem in Exercises 17–22, decide whether the problem can be solved using Euler paths, Hamilton paths, or trees.

17. A civil engineer needs to drive once over every road in a neighborhood to check for salt damage after an especially snowy winter.
18. A group of friends goes to Epcot in Orlando on a 97 degree day and wants to visit 7 different attractions while doing the least possible amount of walking.
19. A UPS driver has 22 deliveries to make before lunch and wants to find the most time-efficient route.
20. Two roommates plan a spring break road trip. The plan is to visit friends at four different colleges then return home while driving the shortest distance.
21. A police officer on patrol is ordered to drive down every street on the west end of town before returning to the station to clock out.
22. When a new campus building is being planned, the IT team is asked to wire five computer classrooms on the fifth floor to the campus network using the least amount of network cable possible.

Projects

1. When solving distance problems in urban areas, we have to consider the fact that you have to follow city blocks, rather than go straight from one location to another. Figure 15-75 on the next page shows a large city, with seven locations marked. A pizza delivery person based at location *A* has deliveries at the other six locations.

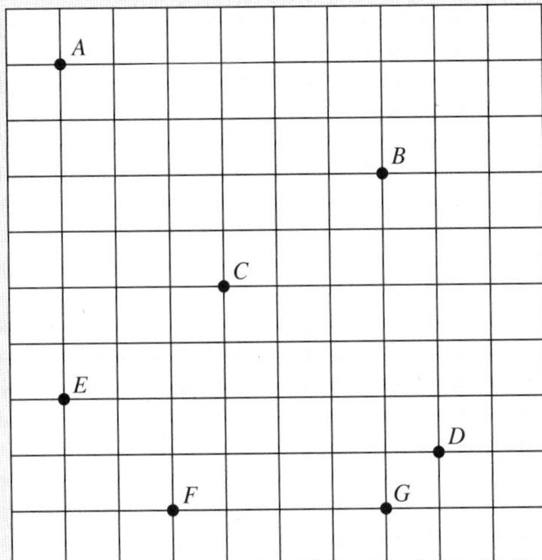

Figure 15-75

(a) Without using any graph theory, draw the path that you think would be shortest to visit all six locations and return to *A*.

(b) Draw a weighted graph for the figure by finding the nearest distance in blocks between each pair of locations.

(c) Using the nearest neighbor method, find an approximate optimal solution to the problem of finding the shortest circuit connecting all seven locations. Is the result shorter or longer than the one you chose in part a?

(d) See if you can find another shorter path using the brute force method. (There are 720 possible paths, so don't try to check them all!)

2. A famous problem in graph theory involves visiting every one of the state capitals in the continental United States (excluding Alaska) with the least possible amount of driving.

   (a) Print a full-page map of the continental United States with all of the capitals marked, then use an atlas or the Internet to mark distances between capitals. You don't have to mark the distance from each capital to every other one–just the ones that are reasonably close to each one.

   (b) Use the nearest neighbor method to find an approximate optimal solution for the graph you've drawn. Since you will not have a complete graph, it's possible that a Hamilton circuit doesn't exist. If you get stuck, you may need to go back and add some more distance measurements.

   (c) Use Kruskal's algorithm to find a minimum spanning tree for your graph. Does the route differ significantly from the circuit you found in part b?

   (d) Using the Internet as a resource, find the actual optimal solution to the problem and see how it compares to the ones you found in parts b and c.

APPENDIX A

Area Under the Standard Normal Distribution

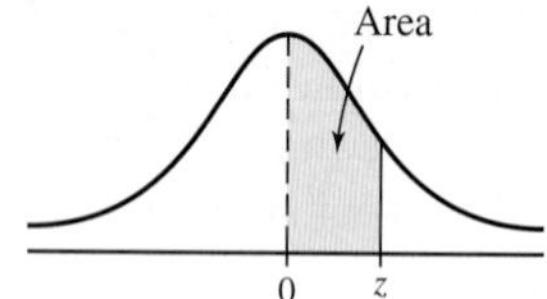

The area in the *A* column is the area under the normal distribution between $z = 0$ and the positive value of z found in the z column.

z	A	z	A	z	A	z	A	z	A	z	A	z	A
.00	.000	.25	.099	.50	.192	.75	.273	1.00	.341	1.25	.394	1.50	.433
.01	.004	.26	.103	.51	.195	.76	.276	1.01	.344	1.26	.396	1.51	.435
.02	.008	.27	.106	.52	.199	.77	.279	1.02	.346	1.27	.398	1.52	.436
.03	.012	.28	.110	.53	.202	.78	.282	1.03	.349	1.28	.400	1.53	.437
.04	.016	.29	.114	.54	.205	.79	.285	1.04	.351	1.29	.402	1.54	.438
.05	.020	.30	.118	.55	.209	.80	.288	1.05	.353	1.30	.403	1.55	.439
.06	.024	.31	.122	.56	.212	.81	.291	1.06	.355	1.31	.405	1.56	.441
.07	.028	.32	.126	.57	.216	.82	.294	1.07	.358	1.32	.407	1.57	.442
.08	.032	.33	.129	.58	.219	.83	.297	1.08	.360	1.33	.408	1.58	.443
.09	.036	.34	.133	.59	.222	.84	.300	1.09	.362	1.34	.410	1.59	.444
.10	.040	.35	.137	.60	.226	.85	.302	1.10	.364	1.35	.412	1.60	.445
.11	.044	.36	.141	.61	.229	.86	.305	1.11	.367	1.36	.413	1.61	.446
.12	.048	.37	.144	.62	.232	.87	.308	1.12	.369	1.37	.415	1.62	.447
.13	.052	.38	.148	.63	.236	.88	.311	1.13	.371	1.38	.416	1.63	.449
.14	.056	.39	.152	.64	.239	.89	.313	1.14	.373	1.39	.418	1.64	.450
.15	.060	.40	.155	.65	.242	.90	.316	1.15	.375	1.40	.419	1.65	.451
.16	.064	.41	.159	.66	.245	.91	.319	1.16	.377	1.41	.421	1.66	.452
.17	.068	.42	.163	.67	.249	.92	.321	1.17	.379	1.42	.422	1.67	.453
.18	.071	.43	.166	.68	.252	.93	.324	1.18	.381	1.43	.424	1.68	.454
.19	.075	.44	.170	.69	.255	.94	.326	1.19	.383	1.44	.425	1.69	.455
.20	.079	.45	.174	.70	.258	.95	.329	1.20	.385	1.45	.427	1.70	.455
.21	.083	.46	.177	.71	.261	.96	.332	1.21	.387	1.46	.428	1.71	.456
.22	.087	.47	.181	.72	.264	.97	.334	1.22	.389	1.47	.429	1.72	.457
.23	.091	.48	.184	.73	.267	.98	.337	1.23	.391	1.48	.431	1.73	.458
.24	.095	.49	.188	.74	.270	.99	.339	1.24	.393	1.49	.432	1.74	.459

Continued

z	A	z	A	z	A	z	A	z	A	z	A	z	A
1.75	.460	1.97	.476	2.19	.486	2.41	.492	2.63	.496	2.85	.498	3.07	.499
1.76	.461	1.98	.476	2.20	.486	2.42	.492	2.64	.496	2.86	.498	3.08	.499
1.77	.462	1.99	.477	2.21	.487	2.43	.493	2.65	.496	2.87	.498	3.09	.499
1.78	.463	2.00	.477	2.22	.487	2.44	.493	2.66	.496	2.88	.498	3.10	.499
1.79	.463	2.01	.478	2.23	.487	2.45	.493	2.67	.496	2.89	.498	3.11	.499
1.80	.464	2.02	.478	2.24	.488	2.46	.493	2.68	.496	2.90	.498	3.12	.499
1.81	.465	2.03	.479	2.25	.488	2.47	.493	2.69	.496	2.91	.498	3.13	.499
1.82	.466	2.04	.479	2.26	.488	2.48	.493	2.70	.497	2.92	.498	3.14	.499
1.83	.466	2.05	.480	2.27	.488	2.49	.494	2.71	.497	2.93	.498	3.15	.499
1.84	.467	2.06	.480	2.28	.489	2.50	.494	2.72	.497	2.94	.498	3.16	.499
1.85	.468	2.07	.481	2.29	.489	2.51	.494	2.73	.497	2.95	.498	3.17	.499
1.86	.469	2.08	.481	2.30	.489	2.52	.494	2.74	.497	2.96	.499	3.18	.499
1.87	.469	2.09	.482	2.31	.490	2.53	.494	2.75	.497	2.97	.499	3.19	.499
1.88	.470	2.10	.482	2.32	.490	2.54	.495	2.76	.497	2.98	.499	3.20	.499
1.89	.471	2.11	.483	2.33	.490	2.55	.495	2.77	.497	2.99	.499	3.21	.499
1.90	.471	2.12	.483	2.34	.490	2.56	.495	2.78	.497	3.00	.499	3.22	.499
1.91	.472	2.13	.483	2.35	.491	2.57	.495	2.79	.497	3.01	.499	3.23	.499
1.92	.473	2.14	.484	2.36	.491	2.58	.495	2.80	.497	3.02	.499	3.24	.499
1.93	.473	2.15	.484	2.37	.491	2.59	.495	2.81	.498	3.03	.499	3.25	.499
1.94	.474	2.16	.485	2.38	.491	2.60	.495	2.82	.498	3.04	.499		*
1.95	.474	2.17	.485	2.39	.492	2.61	.496	2.83	.498	3.05	.499		
1.96	.475	2.18	.485	2.40	.492	2.62	.496	2.84	.498	3.06	.499		

*For z values beyond 3.25 use $A = 0.500$.

SELECTED ANSWERS

CHAPTER 1: PROBLEM SOLVING

Exercise Set 1-1

7. 37 **8.** 3,645 **9.** 10 **10.** 49 **11.** 72 **12.** 17 **13.** 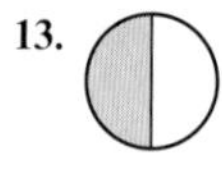
14. **15.** $5 + 13 + 17 = 35$, which is odd.
16. $7(15) + 4 = 109$, which is odd. **17.** $5^2 \div 2 = 12.5$
18. $58(6) = 348$; $3 + 4 + 8 = 15$, which is not divisible by 6.
19. Conjecture: The final answer is -10.
20. Conjecture: The final answer is 7.
21. Conjecture: The final answer is 20.
22. Conjecture: The final answer is the original number minus 5.
23. $12{,}345{,}679 \times 72 = 12{,}345{,}679 \times 9(8) = 888{,}888{,}888$
24. $5^2 + 11 = 6^2$ **25.** $999{,}999 \times 9 = 8{,}999{,}991$
26. $1 + 2 + 3 + 4 + 5 + 6 + 7 + 6 + 5 + 4 + 3 + 2 + 1 = 7^2$
27. $99{,}999 \times 99{,}999 = 9{,}999{,}800{,}001$ **28.** $12{,}345 \times 8 + 5 = 98{,}765$
29. $11{,}111 \times 11{,}111 = 123{,}454{,}321$ **30.** $5 \times 91 = 455$
31. When multiplied by the numbers 1–6 the digits in the answer are a permutation of the original number. But the hypothesis fails when the number is multiplied by 7 and 8.
32. $1 + 3 + 5 + 7 + 9 = 25$, $1 + 3 + 5 + 7 + 9 + 11 = 36$, and $1 + 3 + 5 + 7 + 9 + 11 + 13 = 49$
33. The result will always be 9. **34.** The process always results in 1,089.
35. The next three sums are $\frac{9}{5}$, $\frac{11}{6}$, and $\frac{13}{7}$. **36.** The sum is 56.
37. $5 + 10 + 15 + 20 + 25 = \frac{25(6)}{2}$ **38.** $50{,}505 \times 10 = 505{,}050$
39. $37{,}037 \times 15 = 555{,}555$ **40.** $4 \cdot (1^3 + 2^3 + 3^3 + 4^3) = 4^2(5)^2$
41. Inductive **42.** Deductive **43.** Inductive **44.** Deductive
45. Deductive **46.** Inductive **47.** Inductive **48.** Deductive
49. Inductive **50.** Inductive **51.** Deductive **52.** Inductive
53. Deductive **54.** Inductive **55.** Inductive **56.** Deductive
57. Deductive **58.** Deductive **59.** Inductive **60.** Deductive
61. Inductive **62.** Deductive **63.** Inductive **64.** Inductive
65. It is not possible. **66.** (a) 34, 55, 89 (b) 49, 64, 81 (c) 64, 128, 256 **67.** 1 6 15 20 15 6 1
69. (a) 21, 28, 36 (b) 36, 49, 64 (c) 35, 51, 70 (d) 1, 6, 15, 28
70. $\frac{n(4n - 2)}{2} = n(2n - 1)$

Exercise Set 1-2

7. 2,900 **8.** 732.650 **9.** 3,260,000 **10.** 9,350
11. 63 **12.** 45,000,000 **13.** 200,000 **14.** 900 **15.** 3.67 **16.** 56
17. 327.1 **18.** 83,000,000 **19.** 5,460,000 **20.** 7.12 **21.** 300,000
22. 63.7 **23.** 264.9735 **24.** 2,000,000 **25.** 563.27 **26.** 426.86136
27. \$136 **28.** \$200 **29.** 6 hours **30.** 120 miles **31.** \$72
32. \$36.50 **33.** \$6.50 **34.** \$260 **35.** 9 hours **36.** \$25 per hour
37. \$18,000 per year **38.** \$60 **39.** \$54 **40.** \$48 **41.** \$25
42. \$5,040 **43.** 3,200 acres **44.** 500 acres **45.** 4,400 acres
46. 2,000 acres **47.** 490 people **48.** 770 workers **49.** 65%
50. 81% **51.** 350 billion **52.** 580 billion **53.** 1,940 **54.** 1,955
55. 99 miles **56.** 60 miles **57.** 28% **58.** 19% **59.** 48% **60.** 16%
61. 25% **62.** 850 **63.** \$15 billion **64.** \$7 billion
65. The difference between the cost of milk in 1988 and the cost in 2006 is exaggerated by the fact that the picture changed in all three dimensions, rather than just vertically.
66. The increase appears sharper because the vertical axis covers only values from 10 to 19 billion in the same space the previous graph covered values from 0 to 20 billion.
67. Answers vary **68.** Answers vary

Exercise Set 1-3

5. 8 and 14 **6.** 8 and 15 **7.** 43
8. The daughter is 9 years old and the mother is 36 years old.
9. Mark is 7 years old.
10. Lashanna is 9 years old and Pete is 18 years old.
11. 16 inches and 32 inches **12.** 7 dogs and 5 owners
13. 13 quarters and 18 dimes **14.** 16 quarters and 8 dollar bills
15. 35 girls **16.** Six 2-wheeled mopeds and three 3-wheeled scooters
17. Four \$15 cards and six \$20 cards
18. May earned \$54.38 for working 5 hours
19. 28 feet **20.** 8 posts **21.** 143 inches or about 12 feet
22. 48 inches wide **23.** 50 feet **24.** 100 feet **25.** 38 windows
26. 180 feet **27.** Sam earned \$45 and Pete earned \$15.
28. $\frac{1}{4}$ of the original **29.** 84 pounds
30. 138 cases are needed, and 4 bottles will be left over **31.** \$1,093
32. 20 protein bars **33.** 42,900 calories **34.** 41.2 miles per gallon
35. \$206.35 **36.** 5.9 hours **37.** \$420
38. Mary: \$1,187.50, Jean: \$593.75, Claire: \$296.88, Margie: \$296.87
39. \$49.29 **40.** 10.9 miles per gallon **41.** \$465.50 **42.** \$8
43. 2 hours
44. He cuts the bar at the 1-inch and the 3-inch marks, giving him bars of length 1, 2, and 3 inches. He pays the knight for the first day, then takes the one-inch piece back and gives him the two-inch on the second day (and continues in this fashion for six days).
45. Fill the 3-gallon container; pour it into the 5-gallon container. Fill the 3-gallon container again and pour into the 5-gallon container until it's full. There will then be one gallon left in the 3-gallon container.
46. Three, since the car in front is in front of two cars and the car at the end is behind two cars.
47. \$8,000 **48.** \$2,400

Review Exercises

1. 18, 19, 21 **2.** 10, 15, 12 **3.** q, 1,024, n **4.** 14, D, 12
5. **6.** **7.** $5(7)(11) = 385$, which is odd

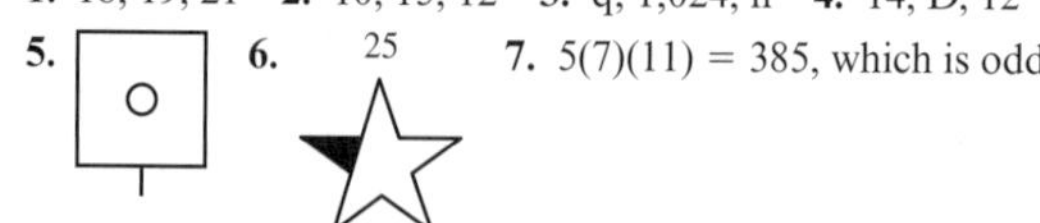

8. $2(5) + 4(5) + 6(5) = 10 + 20 + 30 = 60$, which does not end in a 5.
9. Conjecture: The final answer is 13 more than $\frac{1}{2}$ of the original even number.
10. Conjecture: The final answer equals 3 times the original number.
11. $337 \times 12 = 4{,}044$
$337 \times 15 = 5{,}055$
12. $33{,}333 \times 33{,}333 = 1{,}111{,}088{,}889$
$333{,}333 \times 333{,}333 = 111{,}110{,}888{,}889$
13. 9 **14.** The result is always going to be 153. **15.** Inductive
16. Deductive **17.** Deductive **18.** Inductive **19.** 132,000 **20.** 187
21. 14.6316 **22.** 0.6 **23.** 3,730 **24.** \$1,320 **25.** \$340
26. 300 miles **27.** \$340 **28.** (a) \$6,890 (b) 23 weeks
29. 13 DVDs **30.** 1,172 million pounds **31.** 254 million pounds
32. \$350 **33.** \$520 **34.** 60 miles **35.** 9 **36.** 10 games
37. 110 pounds **38.** Biscotti is \$1.50 and the mocha latte is \$1.90
39. 30 years old **40.** \$48 **41.** \$40

42. A few possible answers are:
$1{,}024 \times 38 = 38{,}912$
$2{,}024 \times 34 = 68{,}816$
$2{,}024 \times 39 = 78{,}936$
43. 20 years old
44. At times during the summer, Alaska has a 3-hour night and a 21-hour day.
45. Debbie is 3 years old. 46. $2 \times 9 + 6 - 7 = 17$
47. Joe has done 2 problems and Tina has done 3.
48.

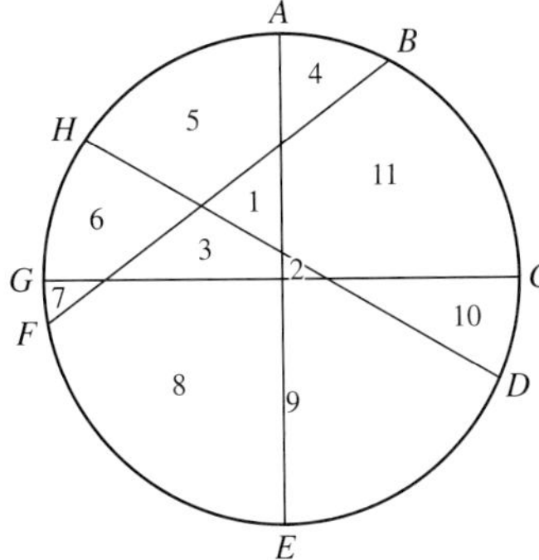

The cuts are *AE*, *BF*, *CG*, and *DH*.
49. 13 triangles 50. Charles is 10 years old. 51. 7.5 and 12.5
52. 4 pounds of the nature mix and 6 pounds of the soy medley
53. Mr. Taylor invested \$800 at 8% and \$200 at 6%.

Chapter 1 Test

* When using estimation, other correct answers are possible.
1. 13, 12, 18 2. 160, 320, 640 3. 8,888,888 4. 44,442,222
5. The final answer is equal to the original number plus 13. 6. 2
7. Eighth 8. E, N
9. Move the last coin on the right on top of another coin.
10. 9 feet (the ship also rises) 11. 84 12. 8 13. 13
14. Add five lines where the dotted lines are shown.

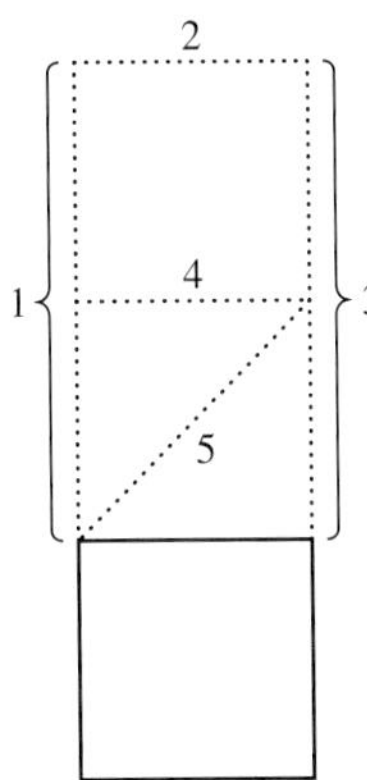

15. First person earns \$36; second person earns \$24.
16. The numbers are 2 and 3. 17. 93% 18. 20,580 feet
19. 65,400 feet 20. 7.2 inches
21. Mark is 17 years old and his mother is 49. 22. 1,700,000
23. 1.38 24. \$146
25. Number of hours per week in 1980 was about 35.5 hours. Number of hours per week in 2005 was about 33.8 hours.
26. 1,823 27. 310 28. 64 miles

CHAPTER 2: SETS

Exercise Set 2-1

7. $S = \{s, t, r, e\}$ 8. $A = \{a, l, b, m\}$
9. $P = \{51, 52, 53, 54, 55, 56, 57, 58, 59\}$
10. $R = \{12, 14, 16, 18, 20, 22, 24, 26, 28, 30, 32, 34, 36, 38\}$
11. $Q = \{1, 3, 5, 7, 9, 11, 13\}$ 12. $M = \{2, 4, 6\}$
13. $G = \{11, 12, 13, \ldots\}$ 14. $B = \{101, 102, 103, \ldots\}$
15. $Y = \{2{,}001, 2{,}002, 2{,}003, \ldots, 2{,}999\}$
16. $Z = \{501, 502, 503, \ldots, 5{,}999\}$
17. W = {Sunday, Monday, Tuesday, Wednesday, Thursday, Friday, Saturday}
18. C = {red, white, blue} 19. D = {hearts, diamonds, spades, clubs}
20. F = {jack, queen, king} 21. The set of even natural numbers.
22. The set of odd natural numbers.
23. The set of the first four multiples of 9.
24. The set of the first four multiples of 5.
25. The set of letters in Mary.
26. The set of letters in Thomas.
27. The set of natural numbers from 100 to 199.
28. The set of natural numbers from 21 to 30.
29. $\{x \mid x$ is a multiple of $10\}$
30. $\{x \mid x$ is a multiple of 5 between 50 and $90\}$
31. $X = \{x \mid x \in N \text{ and } x > 20\}$ 32. $Z = \{x \mid x \in E \text{ and } x < 12\}$
33. $\{x \mid x$ is an odd natural number less than $10\}$
34. $\{x \mid x$ is a multiple of 3 between 15 and $33\}$
35. There are no natural numbers less than zero so $H = \emptyset$
36. {71, 72, 73, 74, 75, 76, 77, 78, 79}
37. {7, 14, 21, 28, 35, 42, 49, 56, 63}
38. {5, 12, 19, 26, 33, 40}
39. {102, 104, 106, 108, 110, 112, 114, 116, 118}
40. {91, 93, 95, 97, 99}
41. Well-defined 42. Well-defined 43. Not well-defined
44. Well-defined 45. Well-defined 46. Not well-defined
47. Not well-defined 48. Well-defined 49. True 50. False
51. True 52. True 53. True 54. False 55. Infinite 56. Finite
57. Finite 58. Finite 59. Infinite 60. Finite 61. Finite
62. Infinite 63. Equal and equivalent 64. Equivalent 65. Neither
66. Neither 67. Equivalent 68. Equal and equivalent 69. Neither
70. Equal and equivalent
71. {10, 20, 30, 40}
↕ ↕ ↕ ↕
{40, 10, 20, 30}
72. {w, x, y, z}
↕ ↕ ↕ ↕
{1, 2, 3, 4}
73. {1, 2, 3, . . . , 25, 26}
↕ ↕ ↕ ↕ ↕
{a, b, c, . . . , y, z}
74. (1, 3, 5, 7, 9)
↕ ↕ ↕ ↕ ↕
(2, 4, 6, 8, 10)
75. $n(A) = 5$ 76. $n(B) = 37$ 77. $n(C) = 7$ 78. $n(D) = 12$
79. $n(E) = 1$ 80. $n(F) = 4$ 81. $n(G) = 0$ 82. $n(H) = 0$ 83. True
84. True 85. True 86. False 87. False 88. False 89. False
90. False
91. (a) {California, New York, Florida}
(b) {Virginia, Massachusetts, Georgia, Maryland}
(c) {California, New York, Florida, Texas, New Jersey}
(d) {Texas, New Jersey, Illinois}
92. (a) {25–34, 35–44, 45–54} People between 25 and 54 have a higher percentage in ISPs, web search portals and data processing companies, than the younger or older age groups.
(b) {16–19, 55–64, 65 and older} Older people and high-school aged people are less likely to be working in these categories.
(c) {10.1, 26.8, 31.8}
(d) {35-44}
(e) {19.6, 7.8}
(f) $\emptyset$
93. (a) {Drunk driving, Injury, Assault}
(b) {Injury, Health problems}
(c) {Injury, Assault, Drunk driving}
(d) {97,000, 1,700, 150,000}
(e) Answers vary
94. (a) {Psychology, Computers, Philosophy}
(b) {Education, Health professions, Engineering, Physical sciences}
(c) {Education, Psychology, Health professions, Engineering}

(d) {Psychology, Health professions, Computers, Physical sciences, Philosophy}
(e) {Education, Psychology, Engineering, Physical sciences, Mathematics, Philosophy}
(f) {Business, Communications, Computers, Philosophy}

95. (a) {Employment fraud, Bank fraud}
(b) {18–29, 30–39, 40–49}
(c) {Utilities/phone fraud, Credit card fraud, Other}
(d) {20, 13, 9}
(e) {Employment fraud, Bank fraud, Utilities/phone fraud}

96. (a) {16–24, 25–34}
(b) {35–44, 45–54, 55+}
(c) {16–24, 25–34}
(d) {28, 15}

97. (a) {2000, 2001, 2002}
(b) {2002, 2003, 2004, 2005}
(c) {2002, 2003, 2004, 2005, 2006}
(d) {2000, 2001, 2002}

98. (a) {2005, 2006, 2007, 2008}
(b) {2004}
(c) {2004, 2005, 2006}
(d) {2007, 2008}

99. Yes **100.** No **101.** Answers vary **102.** Answers vary
103. Answers vary **104.** Answers vary

Exercise Set 2-2

11. $A' = \{2, 3, 17, 19\}$ **12.** $B' = \{3, 5, 7, 11, 13, 17, 19\}$
13. $C' = \{2, 3, 5, 7, 11\}$ **14.** $D' = \{7, 11, 13, 17, 19\}$
15. ∅; {r}; {s}; {t}; {r, s}; {r, t}; {s, t}; {r, s, t}
16. ∅; {2}; {5}; {7}, {2, 5}; {2, 7}; {5, 7};{2, 5, 7}
17. ∅; {1}; {3}; {1, 3} **18.** ∅; {p}; {q}; {p, q} **19.** { } or ∅
20. { } or ∅
21. ∅; {5}; {12}; {13}; {14}; {5, 12}; {5, 13}; {5, 14}; {12, 13}; {12, 14}; {13, 14}, {5, 12, 13}; {5, 12, 14}; {5, 13, 14}; {12, 13, 14}; {5, 12, 13, 14}
22. ∅; {m}; {o}; {r}; {e}; {m, o}; {m, r}; {m, e}; {o, r}; {o, e}; {r, e}; {m, o, r}; {m, o, e}; {m, r, e}; {o, r, e}; {m, o, r, e}
23. ∅; {1}; {10}; {20}; {1, 10}; {1, 20}; {10, 20}
24. ∅; {March}; {April}; {May}; {March, April}; {March, May}; {April, May}
25. ∅ **26.** ∅ **27.** None **28.** None **29.** True **30.** False
31. False **32.** False **33.** False **34.** False **35.** False **36.** True
37. True **38.** False **39.** $2^3 = 8$ **40.** $2^{10} = 1{,}024$ **41.** $2^0 = 1$
42. $2^1 = 2$ **43.** $2^2 = 4$ **44.** $2^5 = 32$ **45.** $U = \{1, 2, 3, 4, 5, 6, 7, 8, 9\}$
46. $A = \{2, 3, 5, 9\}$ **47.** $B = \{5, 6, 7, 8, 9\}$ **48.** $A \cap B = \{5, 9\}$
49. $A \cup B = \{2, 3, 5, 6, 7, 8, 9\}$ **50.** $A' = \{1, 4, 6, 7, 8\}$
51. $B' = \{1, 2, 3, 4\}$ **52.** $(A \cup B)' = \{1, 4\}$
53. $(A \cap B)' = \{1, 2, 3, 4, 6, 7, 8\}$ **54.** $A \cap B' = \{2, 3\}$
55. $A \cup C = \{10, 30, 40, 50, 60, 70, 90\}$ **56.** $A \cap B = \varnothing$
57. $A' = \{20, 40, 60, 80, 100\}$ **58.** $(A \cap B) \cup C = \{30, 40, 50, 60\}$
59. $A' \cap (B \cup C) = \{20, 40, 60, 80, 100\}$ **60.** $(A \cap B) \cap C = \varnothing$
61. $(A \cup B)' \cap C = \varnothing$ **62.** $A \cap B' = A = \{10, 30, 50, 70, 90\}$
63. $(B \cup C) \cap A' = \{20, 40, 60, 80, 100\}$
64. $(A' \cup B') \cup C' = \{10, 20, 30, 40, 50, 60, 70, 80, 90, 100\}$
65. $P \cap Q = \{b, d\}$ **66.** $Q \cup R = \{a, b, c, d, e, f, g\}$
67. $P' = \{a, c, e, h\}$ **68.** $Q' = \{e, f, g, h\}$ **69.** $R' \cap P' = \{a, c, h\}$
70. $P \cup (Q \cap R) = P = \{b, d, f, g\}$ **71.** $(Q \cup P)' \cap R = \{e\}$
72. $P \cap (Q \cap R) = \varnothing$ **73.** $(P \cup Q) \cap (P \cup R) = \{b, d, f, g\}$
74. $Q' \cup R' = \{a, b, c, d, e, f, g, h\}$ **75.** $W \cap Y = \{2, 4, 6\}$
76. $X \cup Z$ = All of U except 4 **77.** $W \cup X$ = the universal set
78. $X \cap Y \cap Z = \{5\}$ **79.** $W \cap X = \varnothing$ **80.** $(Y \cup Z)' = \{7, 9\}$
81. $(X \cup Y) \cap Z = \{2, 5, 6, 11\}$
82. $(Z \cap Y) \cup W = \{2, 4, 5, 6, 8, 10, 12\}$ **83.** $W' \cap X' = \varnothing$
84. $(Z \cup X)' \cap Y = \{4\}$ **85.** $A \cap B = B$
86. $A' \cap C$ = {all even natural numbers that are not multiples of 3} = {2, 4, 8, 10, 14, . . .}
87. $A \cap (B \cup C')$ = {x | x is an odd multiple of 3 or an even multiple of 9} = {3, 9, 15, 18, 21, 27, 33, 36, 39, . . .}
88. $A \cup B = A$ **89.** $A - B = \{20, 110\}$
90. $A - C = \{20, 60\}$ **91.** $B - C = \{60\}$ **92.** $B - A = \{80\}$
93. $C \cap B' = \{110\}$ **94.** $A \cap C' = \{20, 60\}$ **95.** $C - B = \{p\}$
96. $A - C = \{q, s\}$ **97.** $B - C = \{s, u\}$ **98.** $B - A = \{u, v\}$
99. $B \cap C' = \{s, u\}$ **100.** $C \cap A' = \{v\}$
101. $A \times B = \{(9, 1), (9, 2), (9, 3), (12, 1), (12, 2), (12, 3), (18, 1), (18, 2), (18, 3)\}$
102. $B \times A = \{(1, 9), (1, 12), (1, 18), (2, 9), (2, 12), (2, 18), (3, 9), (3, 12), (3, 18)\}$
103. $A \times A = \{(9, 9), (9, 12), (9, 18), (12, 9), (12, 12), (12, 18), (18, 9), (18, 12), (18, 18)\}$
104. $B \times B = \{(1, 1), (1, 2), (1, 3), (2, 1), (2, 2), (2, 3), (3, 1), (3, 2), (3, 3)\}$
105. $A \times B = \{(1, 1), (1, 3), (2, 1), (2, 3), (4, 1), (4, 3), (8, 1), (8, 3)\}$
106. $B \times A = \{(1, 1), (1, 2), (1, 4), (1, 8), (3, 1), (3, 2), (3, 4), (3, 8)\}$
107. $B \times B = \{(1, 1), (1, 3), (3, 1), (3, 3)\}$
108. $A \times A = \{(1, 1), (1, 2), (1, 4), (1, 8), (2, 1), (2, 2), (2, 4), (2, 8), (4, 1), (4, 2), (4, 4), (4, 8), (8, 1), (8, 2), (8, 4), (8, 8)\}$
109. {cell phone, laptop, iPod}, {cell phone, laptop}, {cell phone, iPod}, {laptop, iPod}, {cell phone}, {laptop}, {iPod}, ∅
110. $2^5 = 32$ **111.** $2^7 - 1 = 127$ **112.** $2^6 = 64$ **113.** $2^4 = 16$
114. {treadmill, cycle, stair stepper}; {treadmill, cycle}; {treadmill, stair stepper}; {cycle, stair stepper}; {treadmill}; {cycle}; {stair stepper}
115. Answers vary **116.** Answers vary **117.** Answers vary
118. Answers vary

Exercise Set 2-3

5.

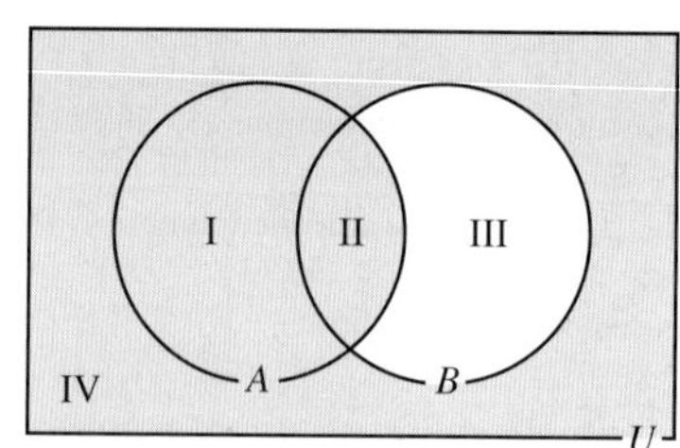

6.

I II III IV A B U

7.

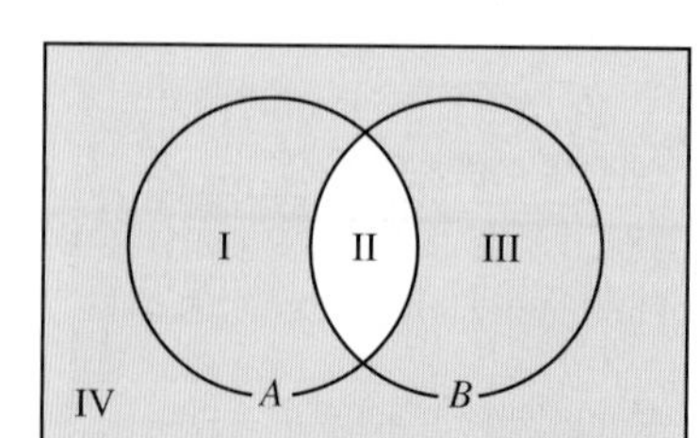

8.

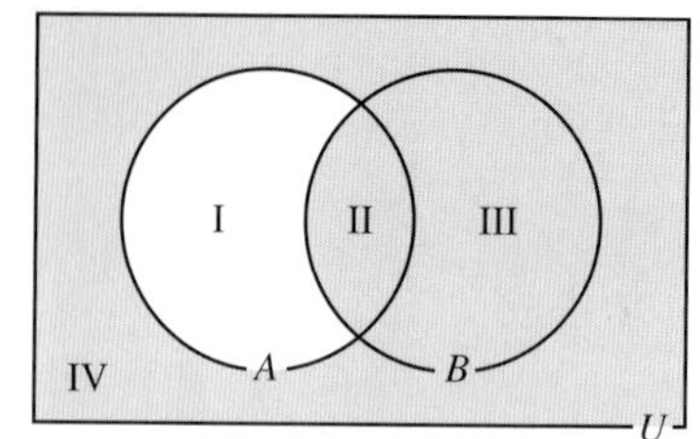

9.

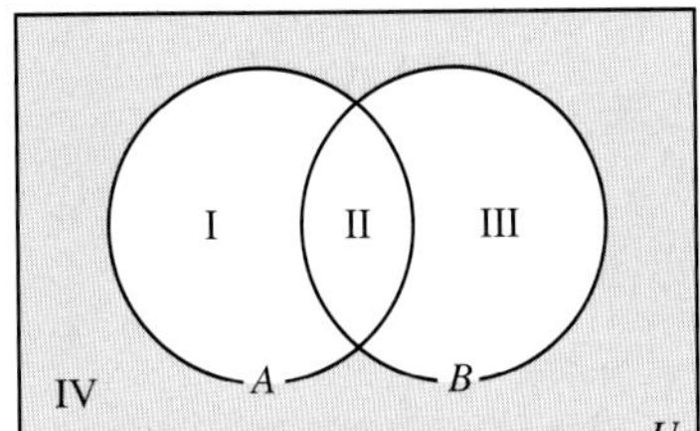

10.

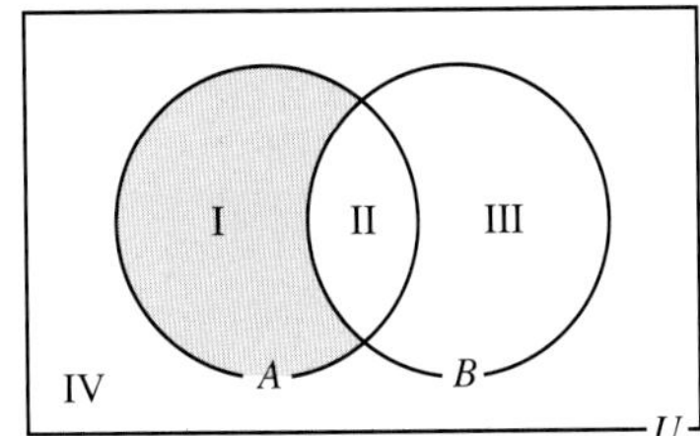

11.

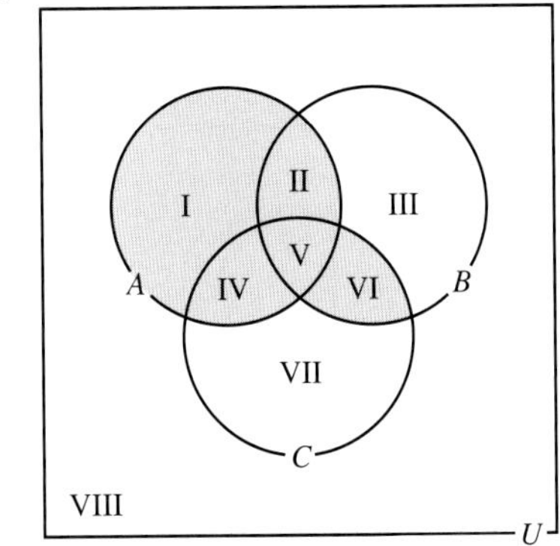

12.

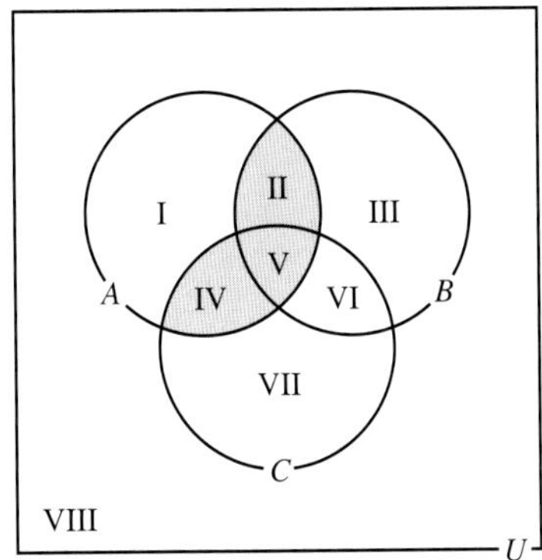

13.

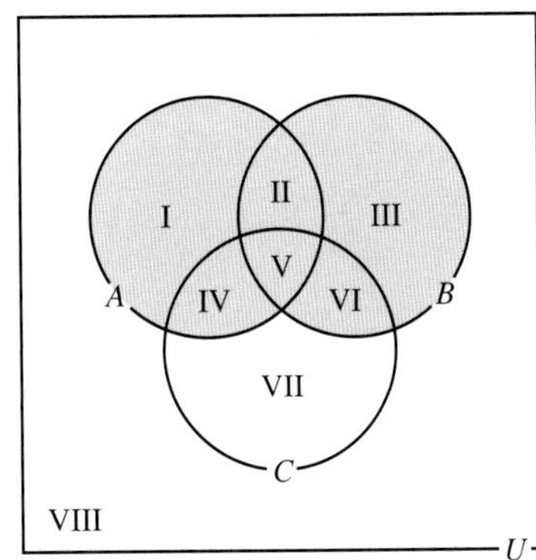

14.

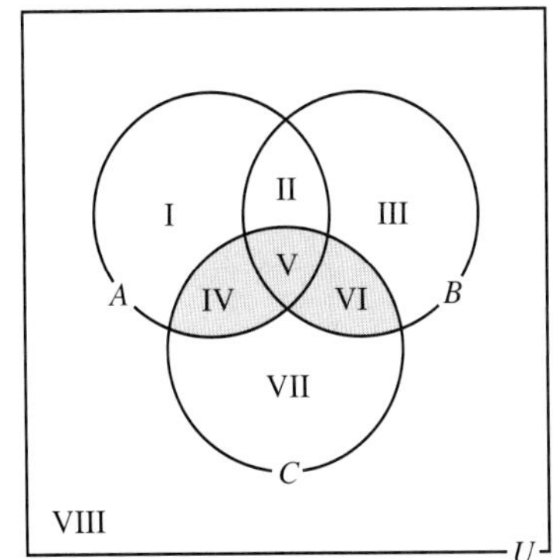

15.

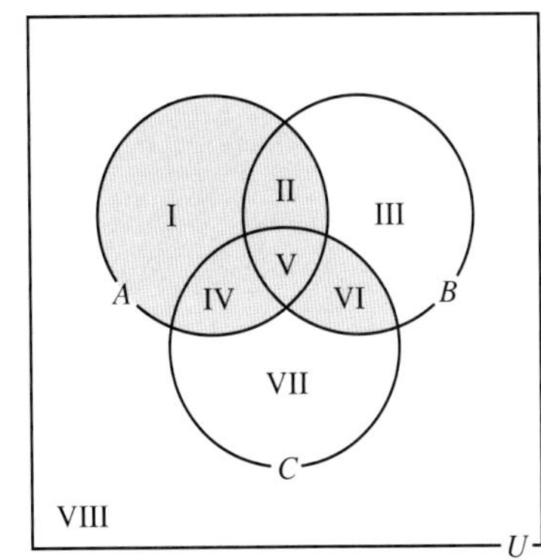

16.

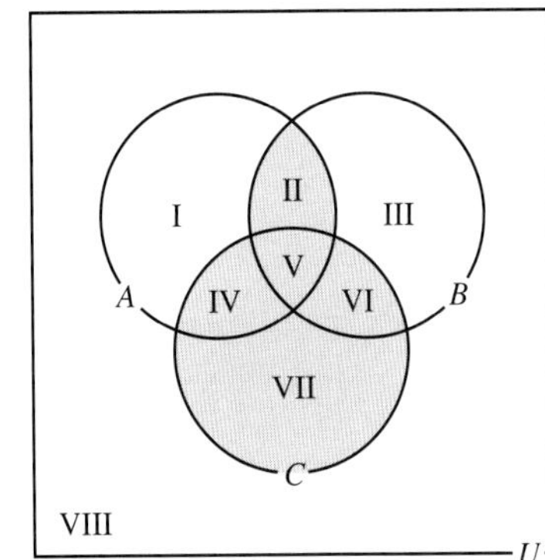

17.

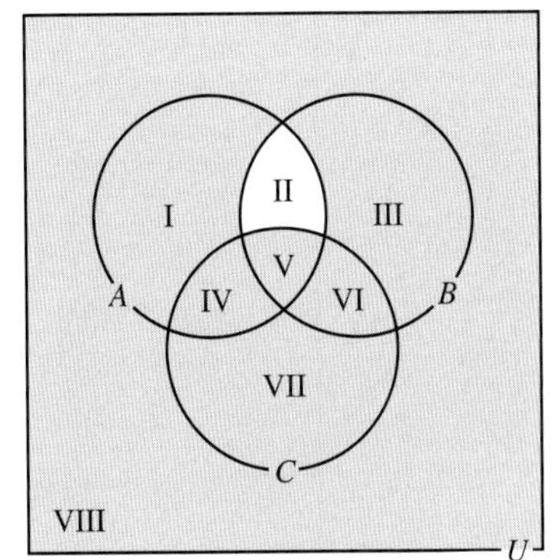

18.

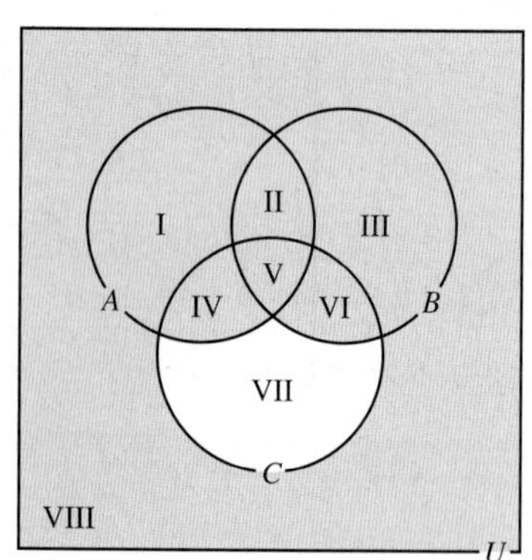

19.

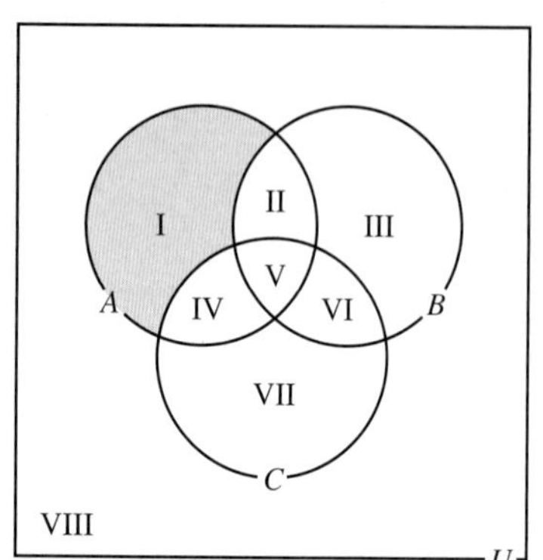

20.

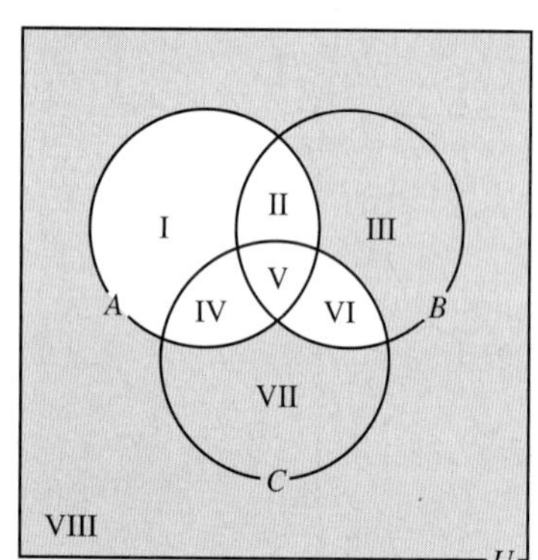

21.

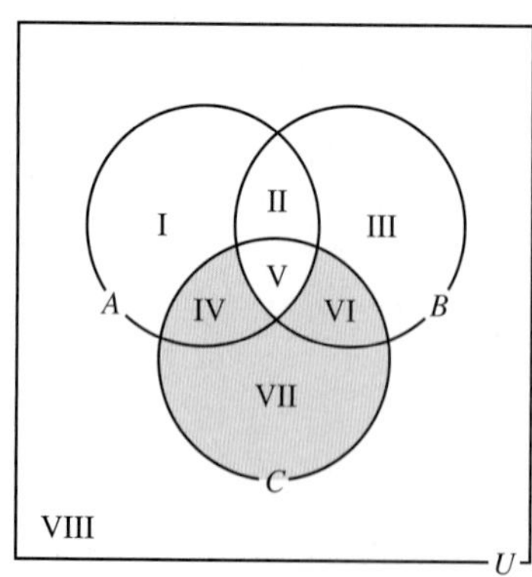

22.

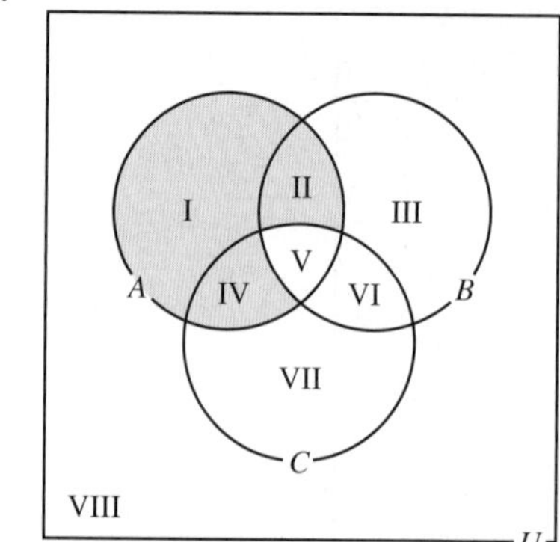

23.

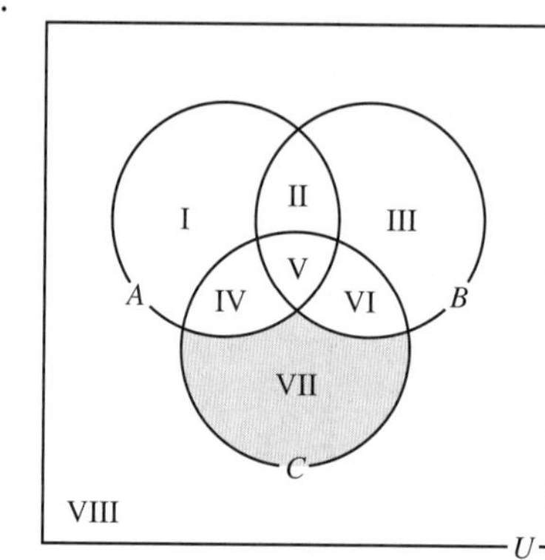

24.

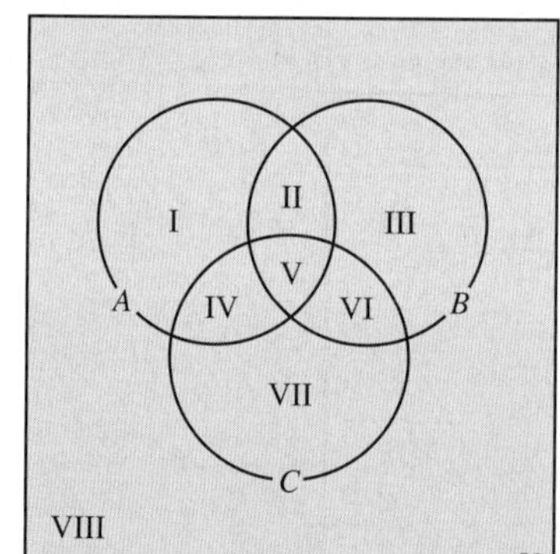

25.

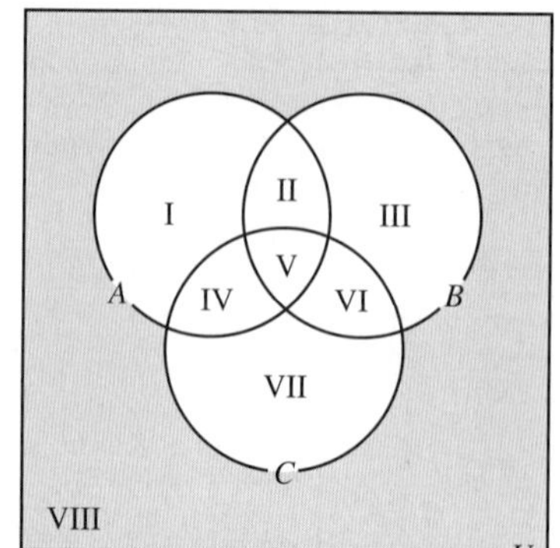

26.

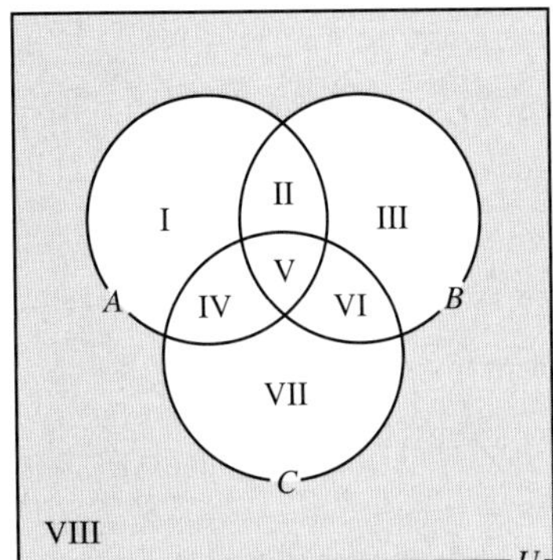

27.

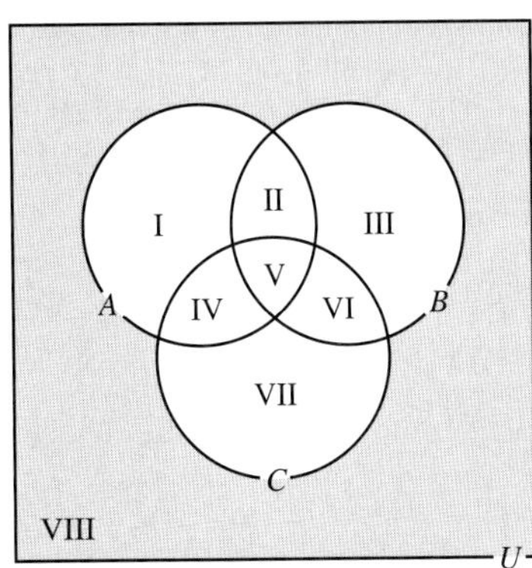

28.

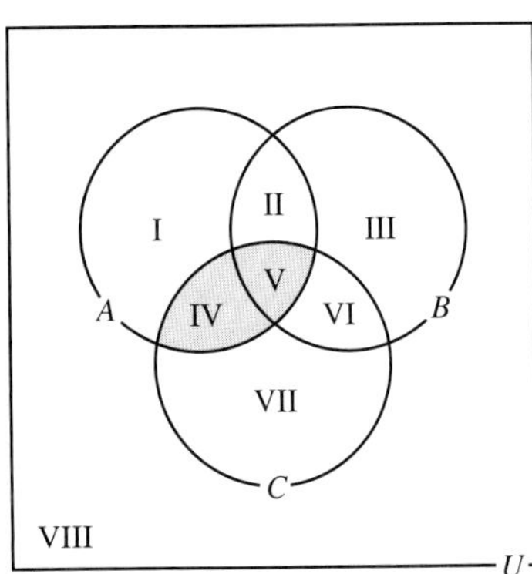

29. equal **30.** not equal **31.** equal **32.** equal **33.** not equal **34.** not equal **35.** not equal **36.** not equal **37.** $n(A) = 6$ **38.** $n(B) = 5$ **39.** $n(A \cap B) = 2$ **40.** $n(A \cup B) = 9$ **41.** $n(A') = 6$ **42.** $n(B') = 7$ **43.** $n(A' \cap B') = 3$ **44.** $n(A' \cup B') = 10$ **45.** $n(A - B) = 4$ **46.** $n(B - A) = 3$ **47.** $n(A) = 6$ **48.** $n(B) = 5$ **49.** $n(A \cap B) = 4$ **50.** $n(A \cup B) = 7$ **51.** $n(A \cap B') = 1$ **52.** $n(A' \cup B) = 10$ **53.** $n(A') = 6$ **54.** $n(B') = 7$ **55.** $n(A - B) = 2$ **56.** $n(B' - A) = 5$

57. People who drive an SUV or a hybrid vehicle.

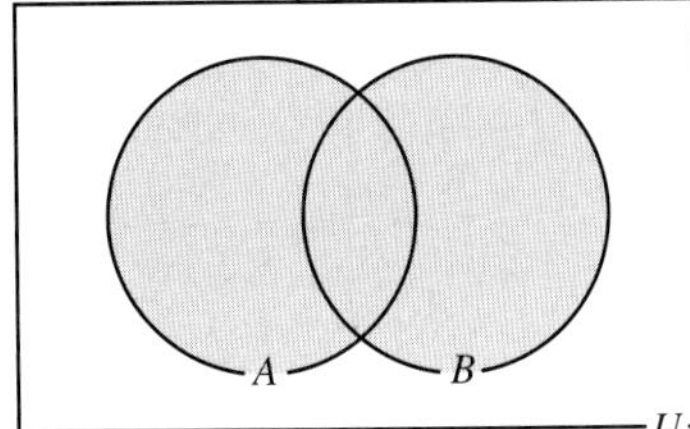

58. People who drive a hybrid SUV.

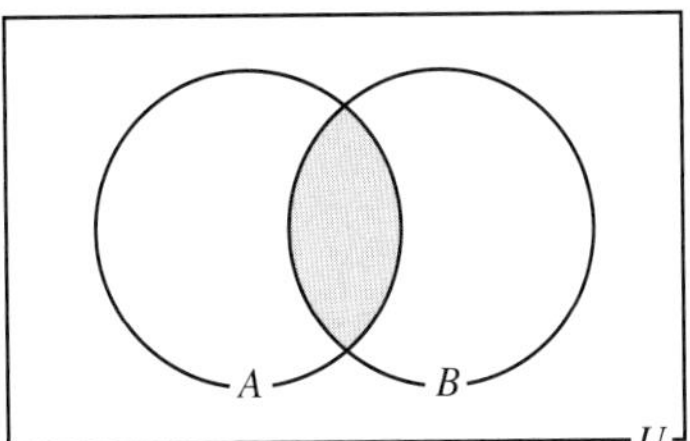

59. People who do not drive an SUV.

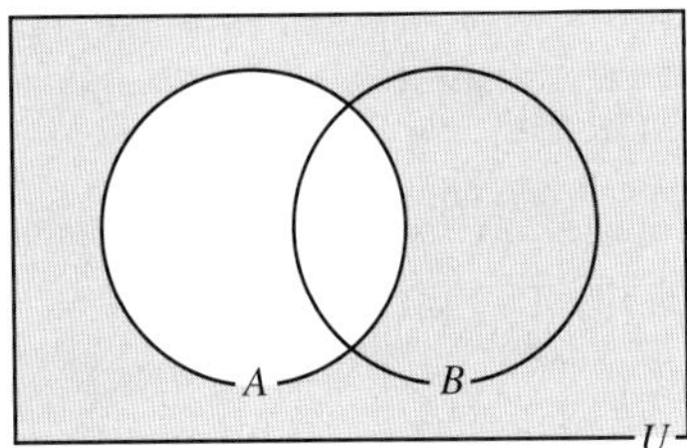

60. People who do not drive a hybrid SUV.

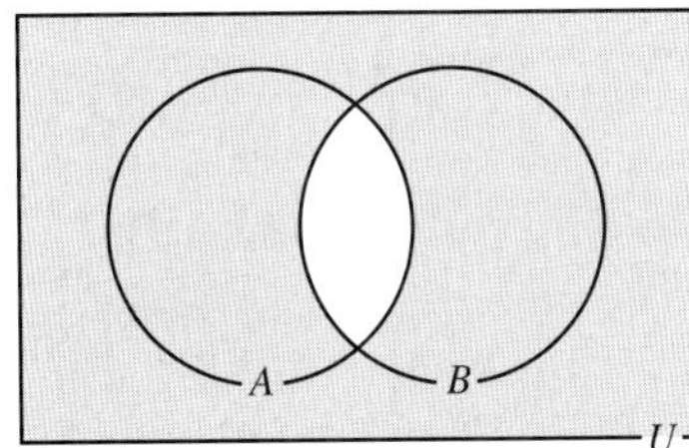

61. Students in online courses and blended or traditional courses.

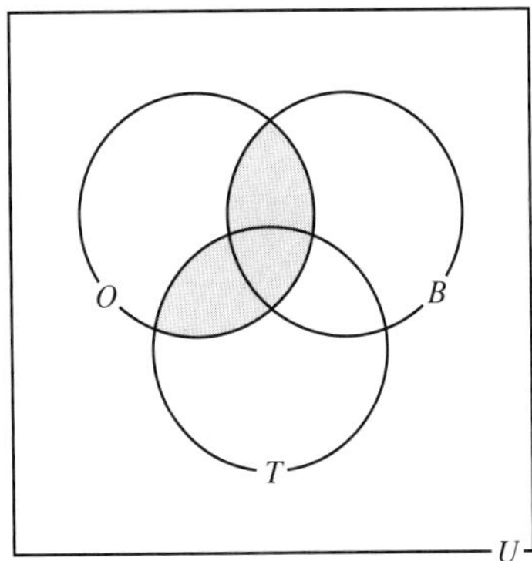

62. Students who are in blended courses or online and traditional courses.

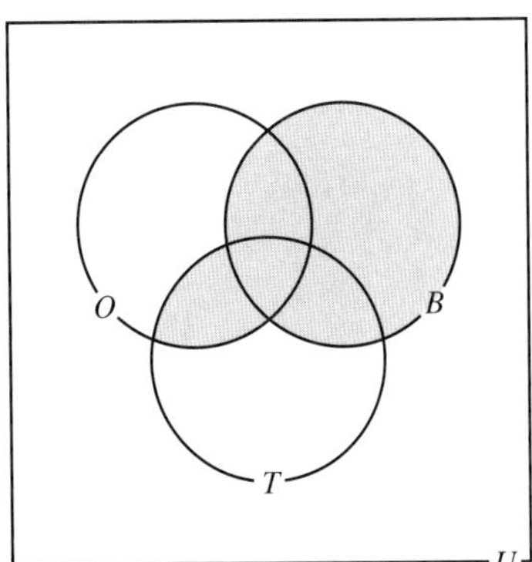

63. Students who are in blended, online, and traditional courses.

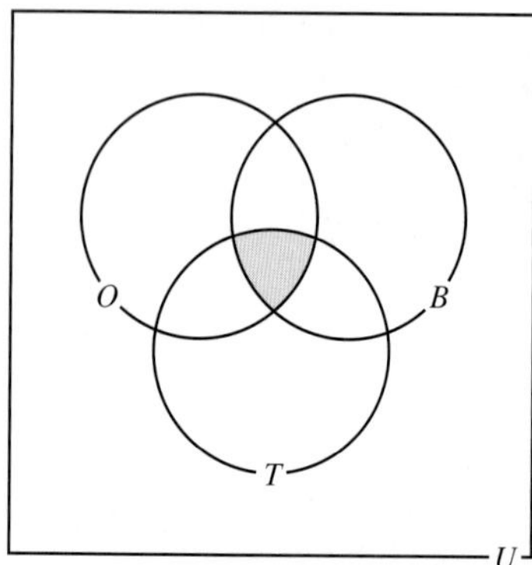

64. Students in blended or online, and traditional or online courses.

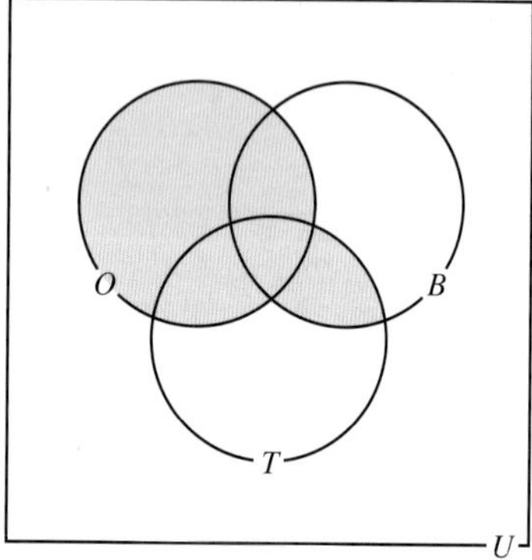

65. Students not voting democrat or voting republican.

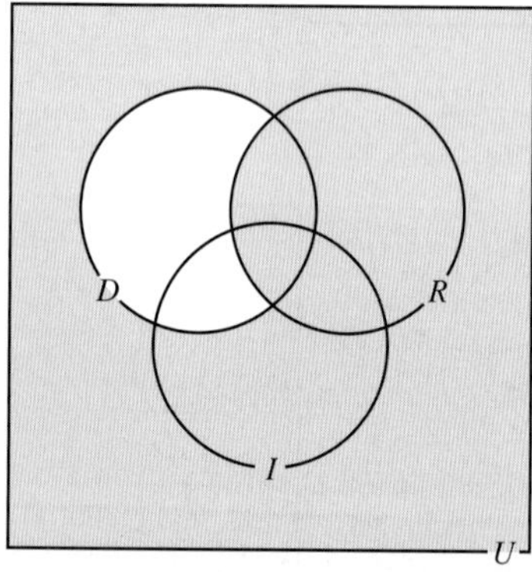

66. Students not voting democrat or independent.

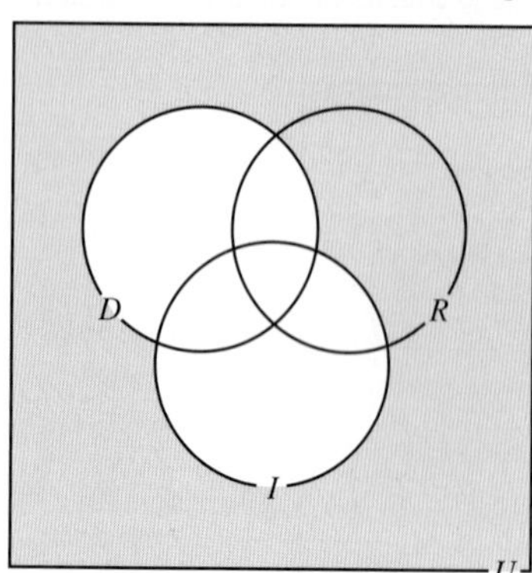

67. Students voting democrat or republican but not independent.

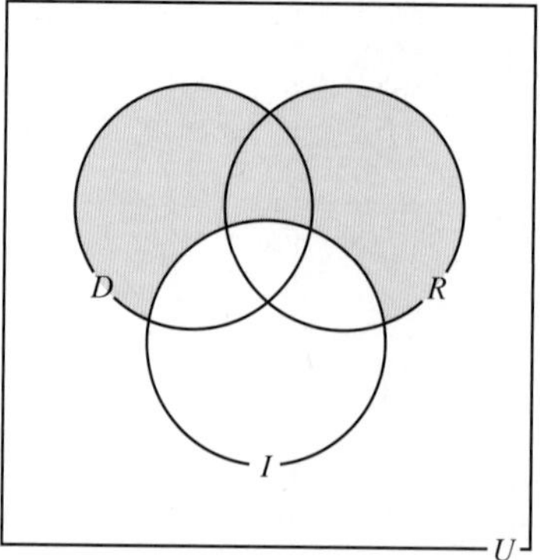

68. Students voting independent but not democrat or republican.

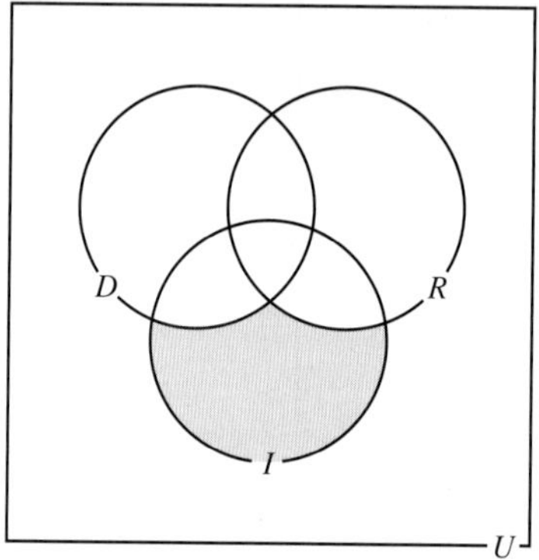

69. People who regularly use Google but not Yahoo!.

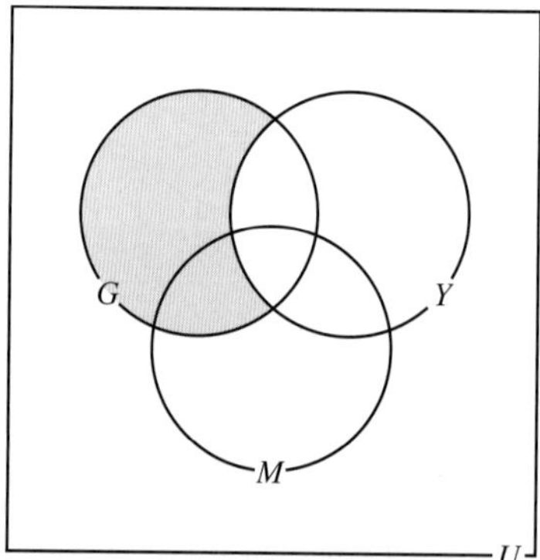

70. People who regularly use Google, but not Yahoo! and MSN Live.

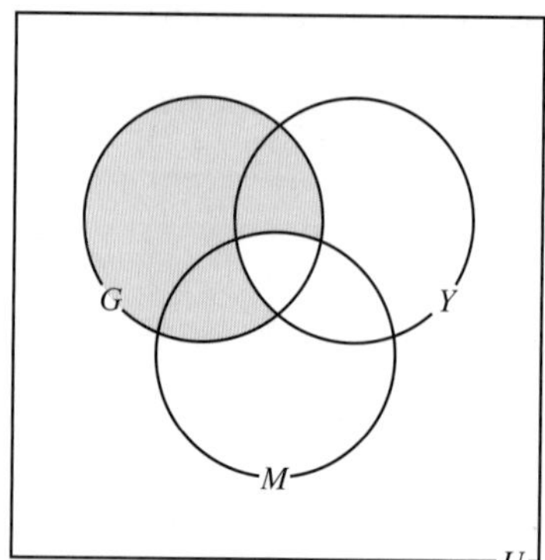

71. People who do not regularly use Google, Yahoo!, or MSN Live.

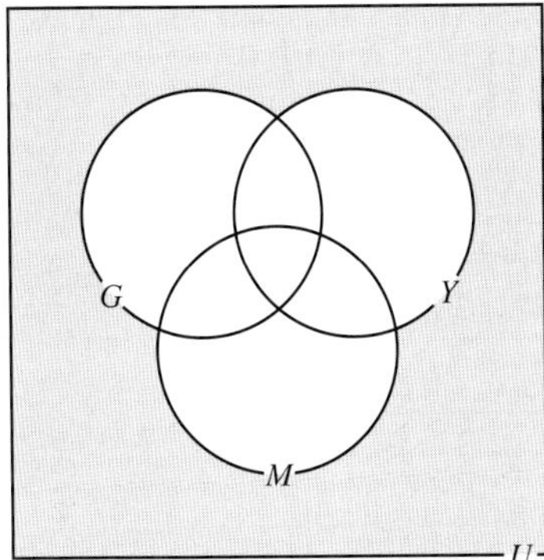

72. People who regularly use Yahoo! and MSN Live or Yahoo! and Google.

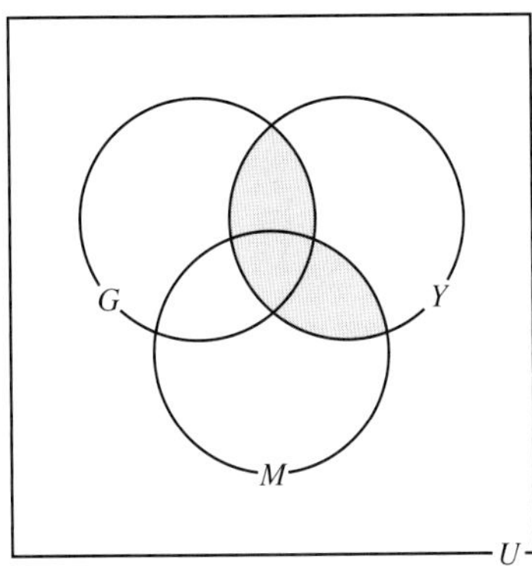

73. Region IV **74.** Region IV **75.** Region VII **76.** Region III **77.** Region V **78.** Region III **79.** No; Answers vary **80.** No; Answers vary

Exercise Set 2-4

1.	(a) 13	(b) 18	(c) 2	
2.	(a) 11	(b) 13	(c) 2	
3.	(a) 8	(b) 5	(c) 0	
4.	(a) 11	(b) 16	(c) 20	
5.	(a) 2	(b) 9	(c) 35	
6.	(a) 14	(b) 11	(c) 8	
7.	(a) 16	(b) 7	(c) 14	(d) 9
8.	(a) 90	(b) 76	(c) 1	
9.	(a) 37	(b) 2	(c) 37	
10.	(a) 36	(b) 3	(c) 45	
11.	(a) 22	(b) 3	(c) 15	
12.	(a) 116	(b) 42	(c) 72	
13.	(a) 32	(b) 37	(c) 58	

14. 10 **15.** 6

16. (a) 62 (b) 69 (c) 71

17. The total of the eight regions is 39 but the researcher surveyed 40 people.

Exercise Set 2-5

5. $7n$ **6.** n^3 **7.** 4^n **8.** n^2 **9.** $-3n$ **10.** $22n$ **11.** $\frac{1}{n+1}$ **12.** $\frac{n}{3}$ **13.** $4n-2$ **14.** $3n-2$ **15.** $\frac{n+1}{n+2}$ **16.** $\frac{1}{n^2}$ **17.** $100n$ **18.** $50n$ **19.** $-3n-1$ **20.** $2n-1$

For 21 through 30 we will show each set is infinite by putting it into a one-to-one correspondence with a proper subset of itself.

21. $\{3, 6, 9, 12, 15, \ldots, 3n, \ldots\}$
$\updownarrow \updownarrow \updownarrow \updownarrow \updownarrow \quad \updownarrow$
$\{6, 12, 18, 24, 30, \ldots, 6n, \ldots\}$

22. $\{10, 15, 20, 25, 30, \ldots, 5n+5, \ldots\}$
$\updownarrow \updownarrow \updownarrow \updownarrow \updownarrow \quad \updownarrow$
$\{15, 25, 35, 45, 55, \ldots, 10n+5, \ldots\}$

23. $\{9, 18, 27, 36, 45, \ldots, 9n, \ldots\}$
$\updownarrow \updownarrow \updownarrow \updownarrow \updownarrow \quad \updownarrow$
$\{18, 36, 54, 72, 90, \ldots, 18n, \ldots\}$

24. $\{4, 10, 16, 22, 28, \ldots, 6n-2, \ldots\}$
$\updownarrow \updownarrow \updownarrow \updownarrow \updownarrow \quad \updownarrow$
$\{10, 22, 34, 46, 58, \ldots, 12n-2, \ldots\}$

25. $\{2, 5, 8, 11, \ldots, 3n-1, \ldots\}$
$\updownarrow \updownarrow \updownarrow \updownarrow \quad \updownarrow$
$\{5, 11, 17, 23, \ldots, 6n-1, \ldots\}$

26. $\{20, 24, 28, \ldots, 16+4n, \ldots\}$
$\updownarrow \updownarrow \updownarrow \quad \updownarrow$
$\{24, 28, 32, \ldots, 20+4n, \ldots\}$

27. $\{10, 100, \ldots, 10^n, \ldots\}$
$\updownarrow \updownarrow \quad \updownarrow$
$\{100, 10{,}000, \ldots, 10^{2n}, \ldots\}$

28. $\{100, 200, 300, 400, \ldots, 100n, \ldots\}$
$\updownarrow \updownarrow \updownarrow \updownarrow \quad \updownarrow$
$\{200, 400, 600, 800, \ldots, 200n, \ldots\}$

29. $\left\{\frac{5}{1}, \frac{5}{2}, \frac{5}{3}, \ldots, \frac{5}{n}, \ldots\right\}$
$\updownarrow \updownarrow \updownarrow \quad \updownarrow$
$\left\{\frac{5}{2}, \frac{5}{3}, \frac{5}{4}, \ldots, \frac{5}{n+1}, \ldots\right\}$

30. $\left\{\frac{1}{2}, \frac{1}{4}, \frac{1}{8}, \frac{1}{16}, \ldots, \frac{1}{2^n}, \ldots\right\}$
$\updownarrow \updownarrow \updownarrow \updownarrow \quad \updownarrow$
$\left\{\frac{1}{4}, \frac{1}{8}, \frac{1}{16}, \frac{1}{32}, \ldots, \frac{1}{2^{n+1}}, \ldots\right\}$

31. $\aleph_0 + 1 = \aleph_0$ and $\aleph_0 + \aleph_0 = \aleph_0$

32. $\{1, 2, 3, 4, 5, \ldots, 2n-1, 2n, \ldots\}$
$\updownarrow \updownarrow \updownarrow \updownarrow \updownarrow \quad \updownarrow \updownarrow$
$\{0, -1, 1, -2, 2, \ldots, n, -n, \ldots\}$

33. The rational numbers can be put into a one-to-one correspondence with the natural numbers. **34.** No

Review Exercises

1. $D = \{52, 54, 56, 58\}$ **2.** $F = \{5, 7, 9, \ldots, 39\}$
3. $L = \{$l, e, t, r$\}$ **4.** $A = \{$a, r, k, n, s$\}$
5. $B = \{501, 502, 503, \ldots\}$ **6.** $C = \{6, 7, 8, 9, 10, 11\}$
7. $M = \varnothing$ **8.** $G = \varnothing$ **9.** $\{x \mid x$ is even and $16 < x < 26\}$
10. $\{x \mid x$ is a multiple of 5 between 0 and 25$\}$
11. $\{x \mid x$ is an odd natural number greater than 100$\}$
12. $\{x \mid x$ is a positive multiple of 8 less than 73$\}$ **13.** Infinite
14. Infinite **15.** Finite **16.** Finite **17.** Finite **18.** Finite
19. False **20.** True **21.** False **22.** False
23. ∅; {r}; {s}; {t}; {r, s}; {r, t}; {s, t}; {r, s, t}
24. ∅; {m}; {n}; {o}; {m, n}; {m, o}; {n, o}; {m, n, o}
25. $2^5 = 32$ subsets; 31 proper subsets
26. $2^6 = 64$ subsets; 63 proper subsets
27. $A \cap B = \{$t, u, v$\}$ **28.** $B \cup C = \{$s, t, u, v, w, x, y, z$\}$
29. $(A \cap B) \cap C = \varnothing$ **30.** $B' = \{$p, q, r, s, w, z$\}$
31. $A - B = \{$p, r$\}$ **32.** $B - A = \{$x, y$\}$
33. $(A \cup B)' \cap C = \{$s, w, z$\}$ **34.** $B' \cap C' = \{$p, q, r$\}$
35. $(B \cup C) \cap A' = \{$s, w, x, y, z$\}$
36. $(A \cup B) \cap C' = \{$p, r, t, u, v, x, y$\}$
37. $(B' \cap C') \cup A' = \{p, q, r, s, w, x, y, z\}$
38. $(A' \cap B) \cup C = \{$s, w, x, y, z$\}$
39. $M \times N = \{$(s, v), (s, w), (s, x), (t, v), (t, w), (t, x), (u, v), (u, w), (u, x)$\}$
40. $N \times M = \{$(v, s), (v, t), (v, u), (w, s), (w, t), (w, u), (x, s), (x, t), (x, u)$\}$
41. $M \times M = \{$(s, s), (s, t), (s, u), (t, s), (t, t), (t, u), (u, s), (u, t), (u, u)$\}$
42. $N \times N = \{$(v, v), (v, w), (v, x), (w, v), (w, w), (w, x), (x, v), (x, w), (x, x)$\}$
43. $A - B$ **44.** $A \cap B$ **45.** $B - A$ **46.** $(A \cup B)'$
47. $(A \cup B) - (A \cap B)$ **48.** B′

49.

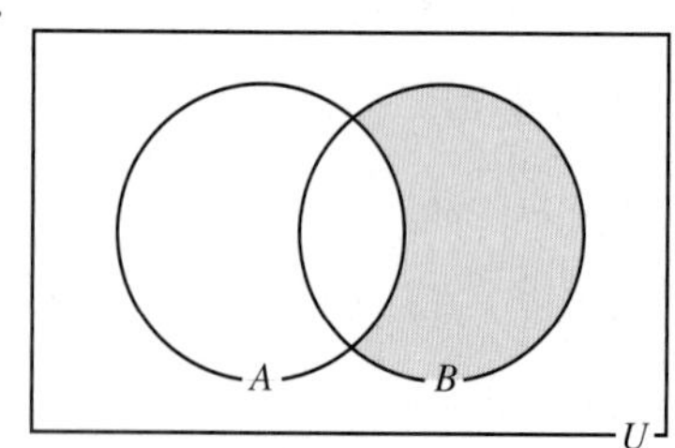

50.

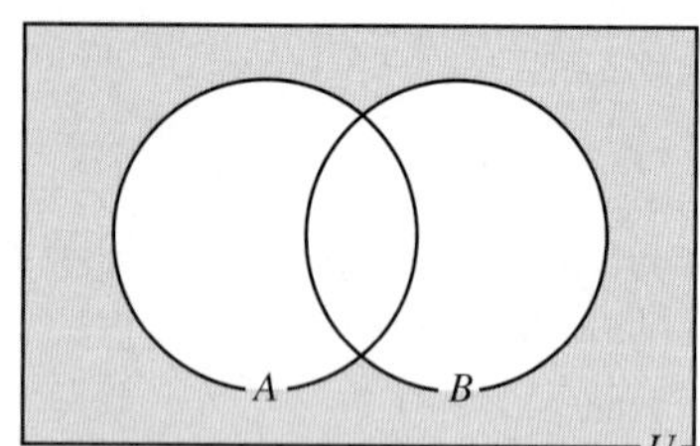

51.

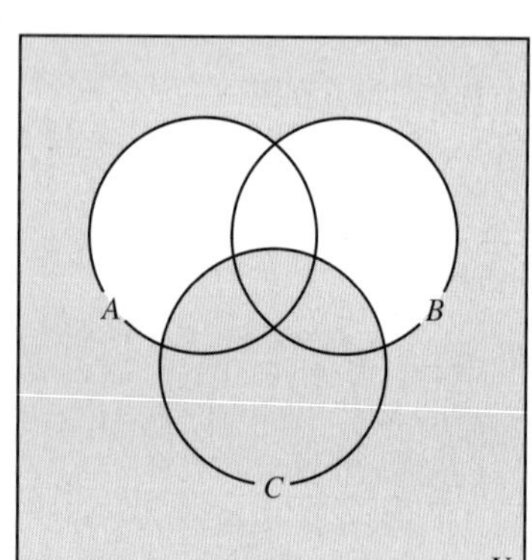

52.

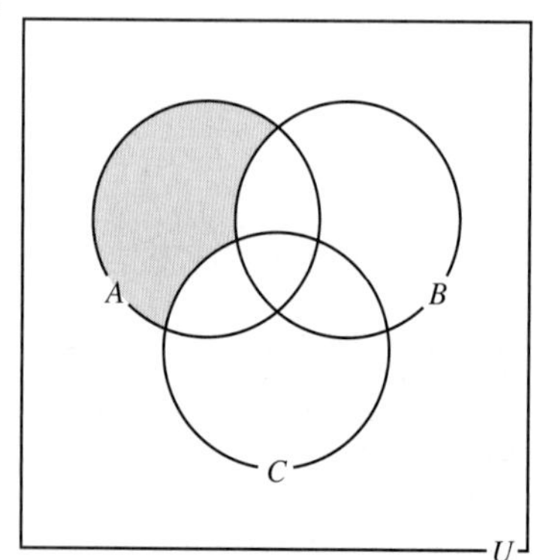

53. 20 **54.** 30 **55.** Region V **56.** Region VII **57.** Region II
58. Region VII **59.** (a) 12 (b) 3 **60.** (a) 15 (b) 6
61. (a) 3 (b) 5 (c) 6 **62.** (a) 9 (b) 5 (c) 7 **63.** $-3 - 2n$
64. $\{12, 24, 36, \ldots, 12n, \ldots\}$
$\updownarrow \ \updownarrow \ \updownarrow \qquad \updownarrow$
$\{24, 48, 72, \ldots, 24n, \ldots\}$

Chapter 2 Test

1. $P = \{92, 94, 96, 98\}$ **2.** $J = \{41, 43, 45, 47, 49\}$
3. $K = \{$e, n, v, l, o, p$\}$ **4.** $W = \{$w, a, s, h, i, n, g, t, o$\}$
5. $X = \{1, 2, 3, 4, \ldots, 79\}$ **6.** $Y = \{17, 18, 19, 20, 21, 22, 23, 24\}$
7. $J = \{$January, June, July$\}$ **8.** $L = \{\ \}$ or $\varnothing$
9. $\{x \mid x \in E$ and $10 < x < 20\}$
10. $\{x \mid x$ is a multiple of 5 between 25 and 50$\}$
11. $\{x \mid x$ is an odd natural number greater than 200$\}$
12. $\{x \mid x = 2^{n+1}$ when n is a natural number less than 7$\}$
13. Infinite **14.** Infinite **15.** Finite **16.** Finite **17.** Finite
18. $\varnothing$; {d}; {e}; {f}; {d, e}; {d, f}; {e, f}; {d, e, f}
19. $\varnothing$; {p}; {q}; {r}; {p, q}; {p, r}; {q, r}; {p, q, r}
20. $2^5 = 32$ subsets; 31 proper subsets
21. $A \cap B = \{$a$\}$ **22.** $B \cup C = \{$a, e, g, h, i, j, k$\}$
23. $B' = \{$b, c, d, e, f, h$\}$ **24.** $(A \cup B)' = \{$c, h$\}$
25. $(A \cap B') \cup C' = \{$a, b, c, d, e, f, g, i, k$\}$
26. $A - B = \{$b, d, e, f$\}$ **27.** $B - C = \{$a, g, i, k$\}$
28. $(A - B) - C = \{$b, d, f$\}$ **29.** $A - C = \{$a, b, d, f$\}$
30.

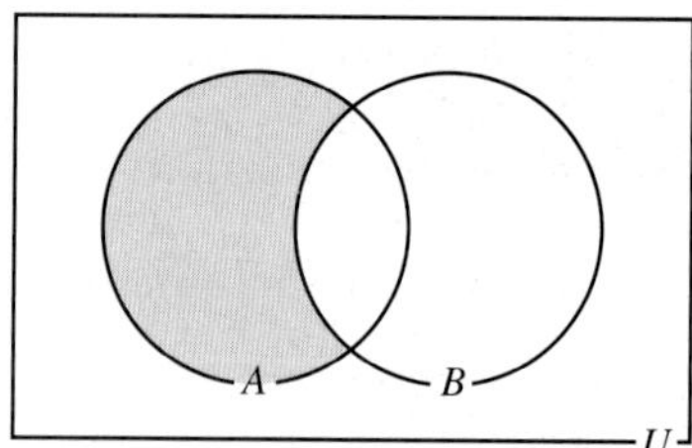

31. $A \times B = \{(@, \pi), (@, \#), (!, \pi), (!, \#), (\alpha, \pi), (\alpha, \#)\}$
32. $B \times A = \{(\pi, @), (\pi, !), (\pi, \alpha), (\#, @), (\#, !), (\#, \alpha)\}$
33. $B \times B = \{(\pi, \pi), (\pi, \#), (\#, \pi), (\#, \#)\}$
34. $A \times A = \{(@, @), (@, !), (@, \alpha), (!, @), (!, !), (!, \alpha), (\alpha, @), (\alpha, !), (\alpha, \alpha)\}$
35.

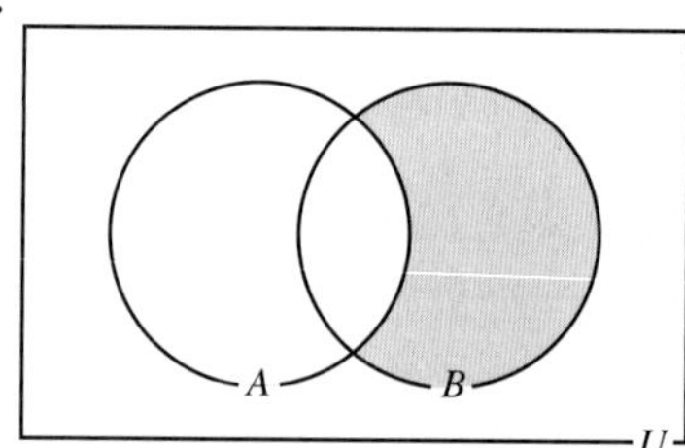

36.

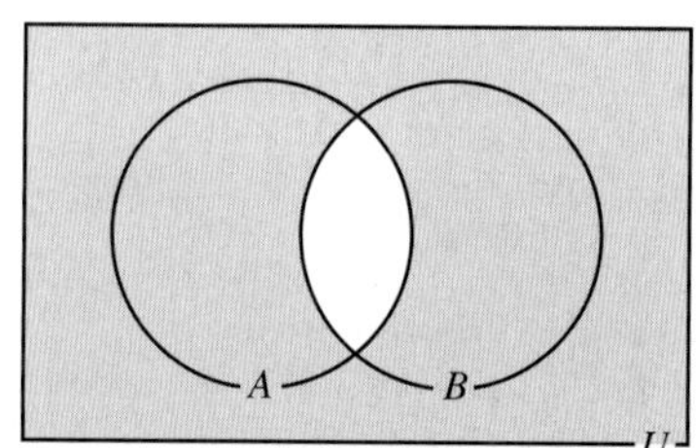

37.

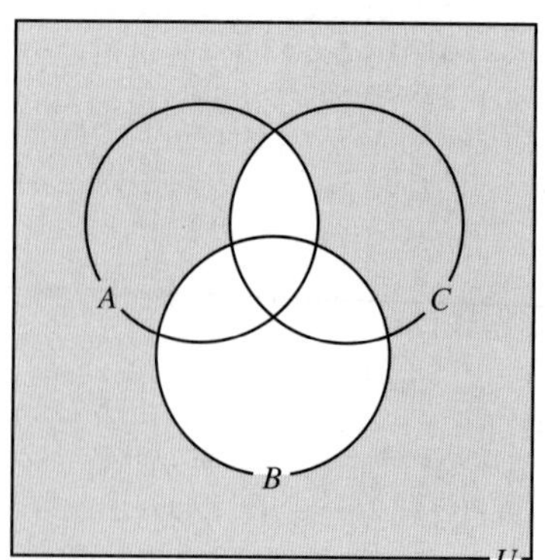

38. $n(A \cup B) = 2{,}300$ **39.** (a) 11 (b) 4 **40.** $15n$
41. $\{1, -1, 2, -2, 3, -3, \ldots, n, -n, \ldots\}$
$\updownarrow \ \updownarrow \updownarrow \ \updownarrow \updownarrow \ \updownarrow \qquad \updownarrow \quad \updownarrow$
$\{1, 2, 3, 4, 5, 6, \ldots, 2n - 1, 2n, \ldots\}$
42. False **43.** True **44.** True **45.** True **46.** True **47.** True
48. False **49.** False **51.** True **52.** False **53.** False **54.** True

CHAPTER 3: LOGIC

Exercise Set 3-1

5. Not a statement **6.** Statement **7.** Statement **8.** Statement **9.** Not a statement **10.** Statement **11.** Not a statement **12.** Statement **13.** Not a statement **14.** Not a statement **15.** Compound statement **16.** Simple statement **17.** Compound statement **18.** Compound statement **19.** Simple statement **20.** Simple statement **21.** Compound statement **22.** Compound statement **23.** Simple statement **24.** Compound statement **25.** Conjunction **26.** Disjunction **27.** Biconditional **28.** Conjunction **29.** Disjunction **30.** Conditional **31.** Biconditional **32.** Conditional **33.** The sky is not blue. **34.** Your computer has a virus. **35.** The dorm room is large. **36.** The class is full. **37.** You will fail this class. **38.** He does not have large biceps. **39.** Universal **40.** Universal **41.** Existential **42.** Existential **43.** Universal **44.** Universal **45.** Existential **46.** Universal **47.** Universal **48.** Universal **49.** Existential **50.** Universal
51. Not all fish swim in water; Some fish do not swim in water.
52. Someone who passes algebra has not studied; Not everyone who passes algebra has studied.
53. No people who live in glass houses throw stones.
54. There is no one in this class that won't pass; Everyone in this class will pass.
55. Not every happy dog wags its tail; Some happy dog does not wag its tail.
56. Some men can join a sorority. **57.** There is no four leaf clover.
58. Some person who participated in the study will not get a hundred dollars; Not every person who participated in the study will get a hundred dollars.
59. Someone with green eyes wears glasses.
60. Not everyone in the class was bored by the professor's lecture; Someone in the class was not bored by the professor's lecture.
61. None of my friends have an iPhone.
62. Someone gets out of here alive. **63.** $p \wedge q$ **64.** $\sim p$ **65.** $\sim q \rightarrow p$ **66.** $\sim(q \vee p)$ **67.** $\sim q$ **68.** $\sim p$ **69.** $q \vee \sim p$ **70.** $\sim q \vee p$ **71.** $q \leftrightarrow p$ **72.** $p \rightarrow q$ **73.** $\sim q$ **74.** $\sim p \wedge \sim q$ **75.** $q \rightarrow p$ **76.** $\sim p$ **77.** $\sim q \vee p$ **78.** $\sim(\sim q \wedge p)$ **79.** $q \leftrightarrow p$ **80.** $\sim(q \vee p)$ or $\sim q \wedge \sim p$ **81.** $\sim q \rightarrow p$ **82.** $\sim p \leftrightarrow \sim q$ **83.** The plane is on time and the sky is clear.
84. The plane is not on time or the sky is clear.
85. If the sky is clear, then the plane is on time.
86. If the sky is clear, then the plane is not on time.
87. The plane is not on time and the sky is not clear.
88. The sky is clear if and only if the plane is on time.
89. The plane is on time or the sky is not clear.
90. The plane is not on time if and only if the sky is not clear.
91. If the sky is clear, then the plane is or is not on time.
92. If the plane is on time, then the sky is clear; or the plane is not on time.
93. Trudy does not live off campus.
94. If Mark lives on campus, then Trudy lives off campus.
95. Mark lives on campus or Trudy does not live off campus.
96. Trudy lives off campus if and only if Mark lives on campus.
97. If Mark does not live on campus, then Trudy does not live off campus.
98. Mark does not live on campus.
99. Mark lives on campus or Trudy lives off campus.
100. Mark does not live on campus or Trudy lives off campus; or Trudy does not live off campus.
101. Trudy lives off campus or Mark lives on campus.
102. If Mark lives on campus or Trudy lives off campus, then it is not the case that Trudy does not live off campus.
103. It cannot be classified as true or false.
104. It cannot be classified as true or false.

Exercise Set 3-2

5. FFFT **6.** TTFT **7.** FFTF **8.** TFTT **9.** FTTF **10.** FFTT **11.** TTFF **12.** TFFF **13.** TFTT **14.** FFTF **15.** TTFF **16.** TFTT **17.** TTFF **18.** TTTTFTTT **19.** TTFFTFFF **20.** TTTTTTTT **21.** TTTTTTTT **22.** TTFTFTFT **23.** TFFFTFTF **24.** TTFTFFFF **25.** TTTFFFTF **26.** TFTFTFFF **27.** FTFTFTTT **28.** TTTTTTTT **29.** FFFTTTTT **30.** FFFTFFTF **31.** FFFFTFFF **32.** TFTTFFTT **33.** TFTFFFFF **34.** FTTFTFTF **35.** False **36.** False **37.** True **38.** True **39.** True **40.** True
41. Let p be "if you take their daily product," q be "you cut your calories intake by 10%, and r be "you lose at least 10 pounds in the next 4 months."
42. $(p \wedge q) \rightarrow r$ **43.** TFTTTTTT **44.** True **45.** True **46.** True
47. Let p be "the attendance for the following season is over 2 million," q be "he will add 20 million dollars to the payroll," and r be "the team will make the playoffs the following year."
48. $p \rightarrow (q \wedge r)$ **49.** TFFFTTTT **50.** False **51.** True **52.** False
53. The statements are equivalent. **54.** Order does matter.
55. False

Exercise Set 3-3

7. Tautology **8.** Neither **9.** Self-contradiction **10.** Neither **11.** Tautology **12.** Self-contradiction **13.** Neither **14.** Neither **15.** Neither **16.** Neither **17.** Equivalent **18.** Negations **19.** Neither **20.** Neither **21.** Neither **22.** Neither **23.** Negations **24.** Equivalent **25.** Neither **26.** Equivalent **27.** $q \rightarrow p$; $\sim p \rightarrow \sim q$; $\sim q \rightarrow \sim p$ **28.** $\sim q \rightarrow \sim p$; $p \rightarrow q$; $q \rightarrow p$ **29.** $p \rightarrow \sim q$; $q \rightarrow \sim p$; $\sim p \rightarrow q$ **30.** $q \rightarrow \sim p$; $p \rightarrow \sim q$; $\sim q \rightarrow p$ **31.** $\sim q \rightarrow p$; $\sim p \rightarrow q$; $q \rightarrow \sim p$ **32.** $p \rightarrow q$; $\sim q \rightarrow \sim p$; $\sim p \rightarrow \sim q$
33. The concert is not long and it is not fun.
34. The soda is not sweet and it is carbonated.
35. It is cold or I am not soaked.
36. I will not walk in the Race for the Cure walkathon or I will not be tired.
37. I will not go to the beach or I will get sunburned.
38. The coffee is not a latte and it is not an espresso.
39. The student is not a girl and the professor is a man.
40. I will not go to college or I will not get a degree.
41. It is not right and it is not wrong.
42. Our school colors are blue and green. **43.** $p \rightarrow q$ **44.** $\sim q \rightarrow \sim p$ **45.** $p \rightarrow q$ **46.** $p \rightarrow q$ **47.** $p \rightarrow q$ **48.** $p \rightarrow q$ **49.** $\sim p \rightarrow \sim q$
50. The statements in exercises 43, 44, 45, 46, 47, and 48 are all equivalent.
51. *Converse*: If he did get a good job, then he graduated with a Bachelor's degree in Management Information Systems.
Inverse: If he did not graduate with a Bachelor's degree in Management Information Systems, then he will not get a good job.
Contrapositive: If he did not get a good job, he did not graduate with a Bachelor's degree in Management Information Systems.
52. *Converse*: If she cannot buy the green Ford Focus, she did not earn \$5,000 this summer as a barista at the coffee house.
Inverse: If she does earn \$5,000 this summer as a barista at the coffee house, she can buy the green Ford Focus.
Contrapositive: If she can buy the green Ford Focus, then she did earn \$5,000 this summer as a barista at the coffee house.
53. *Converse*: If I host a party in my dorm room, then the *American Idol* finale is today.
Inverse: If the *American Idol* finale is not today, then I will not host a party in my dorm room.
Contrapositive: If I do not host a party in my dorm room, then the *American Idol* finale is not today.
54. *Converse*: If I replace the battery, then my cell phone will not charge.
Inverse: If my cell phone will charge, then I will not replace the battery.
Contrapositive: If I do not replace the battery then my cell phone will charge.

55. *Converse*: If I go to Nassau for spring break then I will lose 10 pounds by March 1.
Inverse: If I do not lose 10 pounds by March 1 then I will not go to Nassau for spring break.
Contrapositive: If I do not go to Nassau for spring break then I did not lose 10 pounds by March 1.
56. *Converse*: If the politician goes to jail then he got caught taking kickbacks.
Inverse: If the politician does not get caught taking kickbacks, then he will not go to jail.
Contrapositive: If the politician does not go to jail, then he did not get caught taking kickbacks.
57. $\sim(p \rightarrow q) \equiv p \wedge \sim q$ **58.** $\sim(p \leftrightarrow q) \equiv (p \wedge \sim q) \vee (q \wedge \sim p)$
59. Answers vary **60.** Answers vary **61.** Equivalent
62. Equivalent

Exercise Set 3-4

7. Valid **8.** Valid **9.** Invalid **10.** Valid **11.** Invalid
12. Valid **13.** Valid **14.** Valid **15.** Invalid **16.** Invalid
17. Valid **18.** Valid **19.** Valid **20.** Valid **21.** Invalid
22. Invalid **23.** Valid **24.** Invalid **25.** Invalid **26.** Valid
27. Valid **28.** Invalid **29.** Invalid **30.** Valid **31.** Invalid
32. Valid **33.** Invalid **34.** Valid **35.** Invalid **36.** Invalid
37. Valid **38.** Invalid **39.** Invalid **40.** Valid **41.** Valid
42. Invalid **43.** Invalid **44.** Valid **45.** Valid **46.** Invalid
47. Invalid **48.** Valid **49.** Valid **50.** Invalid **51.** Valid
52. Invalid **53.** Answers vary **54.** Answers vary

Exercise Set 3-5

5.
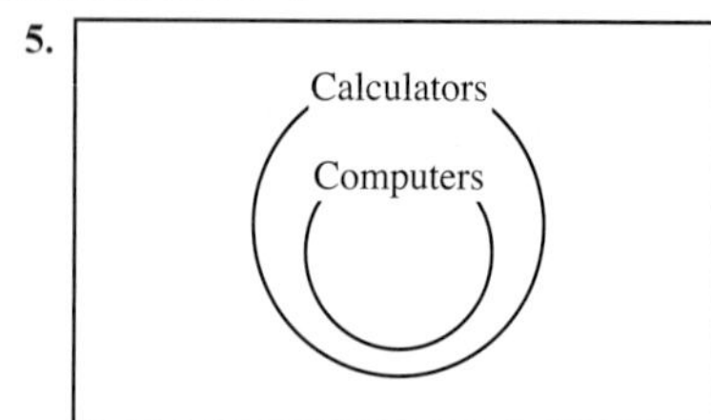

6.
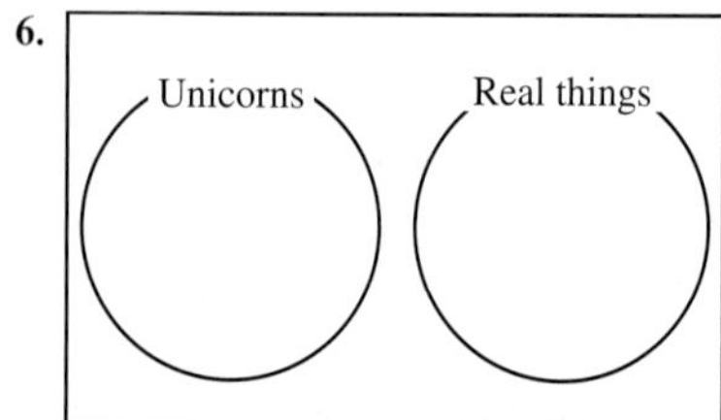

7.
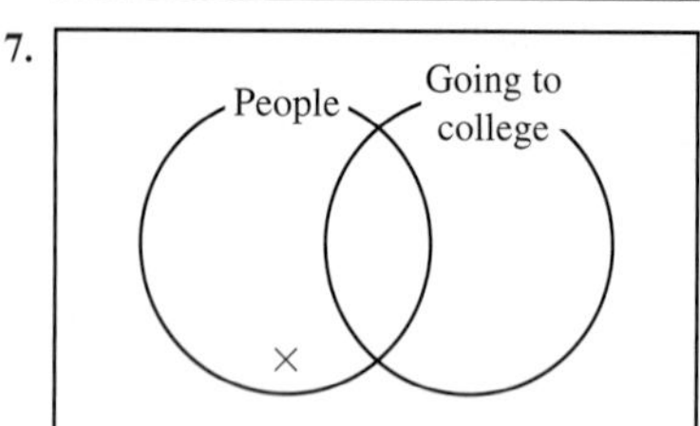

8.
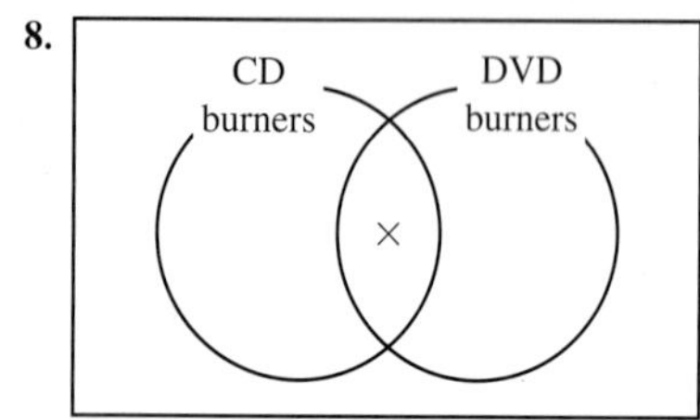

9.
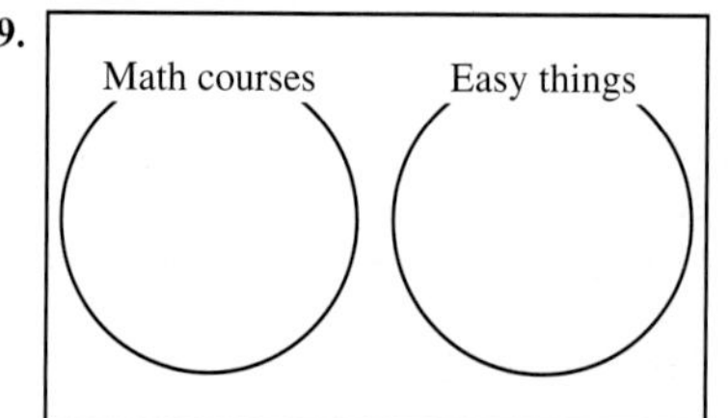

10.
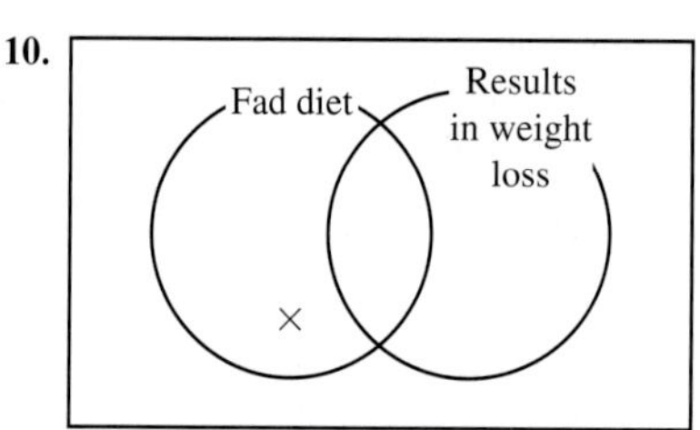

11.
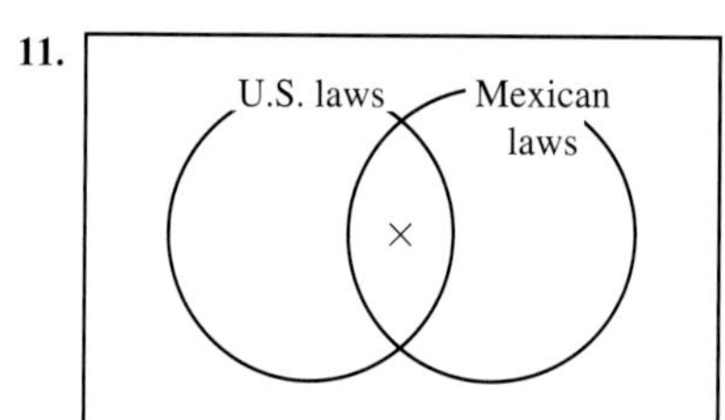

12.
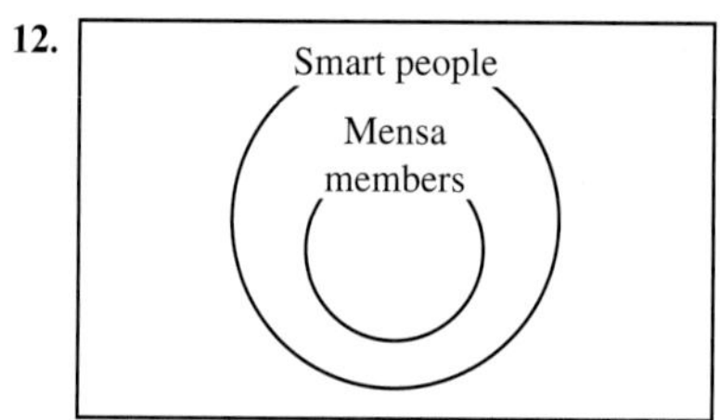

13.
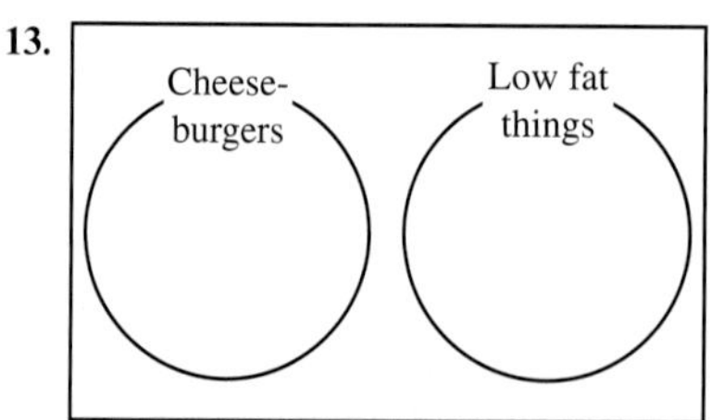

14.
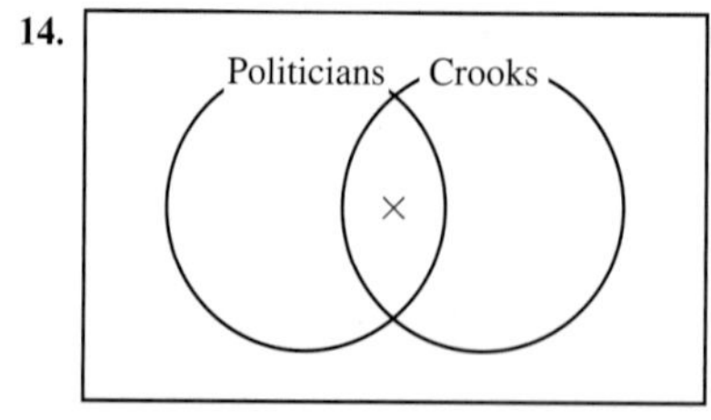

15. Invalid **16.** Invalid **17.** Valid **18.** Invalid **19.** Invalid
20. Invalid **21.** Valid **22.** Valid **23.** Invalid **24.** Invalid
25. Invalid **26.** Valid **27.** Invalid **28.** Invalid **29.** Valid
30. Invalid **31.** Invalid **32.** Invalid **33.** Invalid
34. Invalid **35.** Valid **36.** Valid **37.** Invalid **38.** Valid
39. All *A* is *C*. **40.** No *S* is *P*.
41. No calculators can make breakfast.
42. Some things with brains are prejudiced.

Review Exercises

1. Not a statement **2.** Statement **3.** Not a statement **4.** Statement **5.** Not a Statement **6.** Compound **7.** Simple **8.** Compound **9.** Compound **10.** Simple **11.** It is not scary. **12.** The cell phone is not out of juice. **13.** The popsicle is not green. **14.** Some people who live in glass houses throw stones. **15.** No failing students can learn new study methods. **16.** Not everyone will pass the test on logic; Someone will not pass the test on logic. **17.** All printers have ink. **18.** Some of these links are broken. **19.** None of the contestants will be voted off the island. **20.** Not all SUVs are gas guzzlers; Some SUVs are not gas guzzlers. **21.** Conjunction **22.** Conditional **23.** Disjunction **24.** Biconditional **25.** Conditional **26.** $p \wedge q$ **27.** $q \rightarrow p$ **28.** $q \leftrightarrow p$ **29.** $q \wedge \sim p$ **30.** $\sim p \rightarrow \sim q$ **31.** $\sim(q \wedge p)$ **32.** $\sim(p \rightarrow q)$ **33.** $\sim q \leftrightarrow \sim p$ **34.** $\sim(\sim q)$ **35.** $\sim p \wedge \sim q$ **36.** It is cool or it is not cloudy. **37.** If it is cloudy, then it is cool. **38.** It is cool if and only if it is cloudy. **39.** If it is cool or cloudy, then it is cool. **40.** It is not true that it is not cool or it is cloudy. **41.** FTTT **42.** TTFT **43.** FFFT **44.** TTTT **45.** FFFT **46.** TFTTTTTT **47.** TFTFFFTF **48.** TTFTTTTT **49.** False **50.** True **51.** True **52.** True **53.** Tautology **54.** Neither **55.** Neither **56.** Tautology **57.** Neither **58.** Not equivalent **59.** Not equivalent **60.** Equivalent **61.** The internet connection is not dial-up and it's not DSL. **62.** We will not increase sales and our profit margin will not go down. **63.** The signature is authentic or the check is valid. **64.** It is strenuous or I am not tired. **65.** Let p be the statement "I will be happy" and q be the statement "I get rich." The compound statement is $p \rightarrow q$. **66.** Let p be the statement "having a good career" and q be the statement "having a fulfilling life." The compound statement is $p \rightarrow q$. **67.** *Converse:* If I start riding my bike to work, gas prices will go higher. *Inverse:* If gas prices do not go any higher, I will not start riding my bike to work. *Contrapositive:* If I do not start riding my bike to work, then gas prices will not go any higher. **68.** *Converse:* If my parents kill me, then I didn't pass this class. *Inverse:* If I pass this class, then my parents won't kill me. *Contrapositive:* If my parents don't kill me, then I did pass this class. **69.** *Converse:* If it rains, the festival will move inside the student center. *Inverse:* If the festival did not move inside the student center, then it did not rain. *Contrapositive:* If it does not rain, then the festival will not move inside the student center. **70.** Invalid **71.** Invalid **72.** Invalid **73.** Invalid **74.** Valid **75.** Invalid **76.** Invalid **77.** Invalid **78.** Invalid **79.** Invalid **80.** Valid **81.** Valid **82.** Invalid **83.** Invalid **84.** Invalid **85.** Invalid **86.** Valid

Chapter 3 Test

1. Statement **2.** Statement **3.** Not a statement **4.** Not a statement **5.** The image is not uploading to my online bio. **6.** Not all men have goatees; Some men do not have goatees. **7.** No students ride a bike to school. **8.** Some short people can dunk a basketball. **9.** $p \wedge q$ **10.** $q \rightarrow p$ **11.** $p \leftrightarrow q$ **12.** $p \vee q$ **13.** $\sim(\sim p \wedge q)$ **14.** $\sim q \wedge \sim p$ **15.** It is sunny or not warm. **16.** If it is warm, then it is sunny. **17.** It is sunny if and only if it is warm. **18.** If it is sunny or warm, then it is sunny. **19.** It is not the case that it is not sunny or it is warm. **20.** FTTT **21.** FFTFTFTF **22.** FTTTFTFT **23.** TTFF **24.** TFTTTTTT **25.** Self-contradiction **26.** Neither **27.** Neither **28.** Neither **29.** Tautology **30.** Equivalent **31.** Equivalent **32.** *Converse*: If I am healthy, then I exercise regularly. *Inverse*: If I do not exercise regularly, then I will not be healthy. *Contrapositive*: If I am not healthy, then I do not exercise regularly. **33.** It is cold or it is not snowing. **34.** I am not hungry and I am not thirsty. **35.** Invalid **36.** Invalid **37.** Valid **38.** Invalid **39.** Invalid **40.** Invalid **41.** Valid **42.** Invalid **43.** Invalid **44.** Valid **45.** Valid

CHAPTER 4: NUMERATION SYSTEMS

Exercise Set 4-1

7. 35 **8.** 432 **9.** 20,225 **10.** 200,132 **11.** 30,163 **12.** 1,000,210 **13.** 20,314 **14.** 2,200,000 **15.** 1,112,010 **16.** 2,048

17. |||||||
18. ∩||||||||
19. ∩∩∩|||||||
20. ∩∩∩∩∩||
21. 𓍢∩∩∩∩∩∩||||||||
22. 𓍢𓍢𓍢∩∩∩∩∩∩|||||
23. 𓍢𓍢𓍢𓍢𓍢𓍢𓍢𓍢|
24. 𓍢𓍢𓍢𓍢𓍢𓍢𓍢𓍢𓍢∩∩∩∩∩|||||
25. 𓆼𓍢𓍢∩∩∩∩∩||||||
26. 𓆼𓆼𓆼𓆼𓆼𓆼𓆼𓆼𓍢𓍢∩∩∩∩∩∩|
27. 𓍢∩|||||| **28.** 𓍢𓍢𓍢∩∩∩∩∩∩||||
29. 𓆐𓆐𓆐𓂭𓂭𓂭𓂭𓂭∩∩∩|||||
30. ∩||||||||
31. 𓆼𓆼𓆼𓆼𓆼𓆼𓆼𓆼𓆼𓍢𓍢𓍢𓍢𓍢𓍢𓍢𓍢∩∩∩∩∩∩∩∩∩|||||||||
32. 𓆐𓆐𓆐𓆐𓆐𓆐𓆐𓆐𓆐𓂭𓂭𓂭𓂭𓂭𓂭𓂭𓂭𓂭𓆼𓆼𓆼𓆼𓆼𓆼𓆼𓆼𓍢𓍢𓍢𓍢𓍢𓍢𓍢𓍢𓍢∩∩∩∩∩∩∩∩|||||||||

33. 189 **34.** 3,407 **35.** 52 **36.** 9,834 **37.** 713 **38.** 89

39. 八十九 **40.** 五百六十七 **41.** 二百八十四 **42.** 九千八百五十七 **43.** 二千三百五十六 **44.** 二十一

45. hundreds **46.** thousands **47.** ten thousands **48.** millions **49.** ones **50.** hundred thousands **51.** $8 \times 10^1 + 6$ **52.** $3 \times 10^2 + 2 \times 10^1 + 5$ **53.** $1 \times 10^3 + 8 \times 10^2 + 1 \times 10^1 + 2$ **54.** $3 \times 10^4 + 2 \times 10^3 + 7 \times 10^2 + 1 \times 10^1 + 4$ **55.** $6 \times 10^3 + 2$ **56.** $2 \times 10^4 + 9 \times 10^3 + 3 \times 10^2$ **57.** $1 \times 10^5 + 6 \times 10^4 + 2 \times 10^3 + 8 \times 10^2 + 7 \times 10^1 + 3$ **58.** $2 \times 10^8 + 3 \times 10^5 + 2 \times 10^4 + 1 \times 10^3 + 4 \times 10^2 + 1 \times 10^1 + 6$ **59.** $1 \times 10^7 + 7 \times 10^6 + 5 \times 10^5 + 3 \times 10^4 + 1 \times 10^3 + 8 \times 10^2 + 1$ **60.** $1 \times 10^6 + 3 \times 10^5 + 2 \times 10^4 + 6 \times 10^3 + 4 \times 10^2 + 1 \times 10^1 + 9$ **61.** 12 **62.** 34 **63.** 51 **64.** 35 **65.** 40,871 **66.** 81,613 **67.** 109,284 **68.** 115,871 **69.** 792 **70.** 112,262

71. 𒌋𒌋𒌋𒁹𒁹 **72.** 𒌋𒌋𒁹𒁹𒁹 **73.** 𒁹 𒌋𒁹𒁹𒁹𒁹𒁹𒁹𒁹𒁹
74. 𒁹𒁹 𒌋𒌋𒌋𒁹𒁹𒁹𒁹𒁹𒁹 **75.** 𒁹𒁹𒁹𒁹 𒌋𒌋𒌋𒌋𒌋𒁹𒁹

76. 𒁹𒁹𒁹𒁹𒁹𒁹𒁹𒁹 𒌋𒌋𒌋𒁹𒁹𒁹𒁹

77. 𒌋𒁹𒁹𒁹𒁹𒁹𒁹𒁹 𒁹𒁹𒁹

78. 𒌋𒌋𒁹𒁹𒁹𒁹𒁹𒁹𒁹𒁹𒁹 𒌋𒌋𒌋𒁹𒁹𒁹𒁹𒁹𒁹

79. 𒁹 𒌋𒌋𒁹𒁹𒁹𒁹𒁹𒁹 𒌋𒌋𒌋𒌋𒌋𒁹𒁹𒁹𒁹𒁹𒁹

80. 𒁹𒁹 𒌋𒁹𒁹𒁹𒁹𒁹𒁹 𒌋𒌋𒌋𒌋

81. 17 **82.** 99 **83.** 43 **84.** 221 **85.** 86 **86.** 233 **87.** 418 **88.** 2,917 **89.** 490 **90.** 906 **91.** XXXIX **92.** CXLII **93.** DLXVII **94.** DCCCXCIII **95.** MCCLVIII **96.** MMMDCCXX **97.** MCDLXII **98.** MMCLXX **99.** MMM **100.** MMCCXXII **101.** 1939 **102.** 1942 **103.** 1978 **104.** 1981 **105.** 2005 **106.** 2004 **107.** Answers vary **108.** Answers vary **109.** Because the Babylonian system is an ancient number system which is not in use today. **110.** Answers vary

Exercise Set 4-2

3. 252 **4.** 450 **5.** 391 **6.** 435 **7.** 3,740 **8.** 765 **9.** 448 **10.** 143 **11.** 216 **12.** 245 **13.** 253 **14.** 448 **15.** 368 **16.** 493 **17.** 374 **18.** 714 **19.** 392 **20.** 187 **21.** 270 **22.** 315 **23.** 10,488 **24.** 424,914 **25.** 166,786 **26.** 2,193,541 **27.** 791,028 **28.** 26,235 **29.** 58,625 **30.** 29,797 **31.** 203,912 **32.** 195,224,460 **33.** 4,707 **34.** 5,848 **35.** 1,035 **36.** 3,692 **37.** 5,781 **38.** 22,632 **39.** 36,344 **40.** 87,776 **41.** Answers vary **42.** Answers vary **43.** Answers vary

Exercise Set 4-3

3. 11 **4.** 518 **5.** 33 **6.** 15,289 **7.** 108 **8.** 1,261 **9.** 106 **10.** 51 **11.** 311 **12.** 2,768 **13.** 69 **14.** 25 **15.** 359 **16.** 1,393 **17.** 184 **18.** 782 **19.** 37,406 **20.** 64 **21.** 1,166 **22.** 129 **23.** 11111_{two} **24.** 272_{eight} **25.** 1333_{six} **26.** 100212_{three} **27.** 22_{seven} **28.** 1922_{twelve} **29.** 1017_{nine} **30.** 302101_{four} **31.** 10110_{two} **32.** 4250_{eleven} **33.** 33_{five} **34.** 1201_{four} **35.** 939_{sixteen} **36.** 1100000_{two} **37.** 1042212_{five} **38.** $61A8_{\text{sixteen}}$ **39.** 44_{seven} **40.** 10021_{three} **41.** 100000000_{two} **42.** 13301_{four} **43.** 111010_{two} **44.** 132_{eight} **45.** 81_{twelve} **46.** 11210101_{three} **47.** 310_{four} **48.** 211100_{three} **49.** $1B6_{\text{twelve}}$ **50.** 2335_{six} **51.** 3665_{seven} **52.** 165_{eleven}

53. (a) 3272_{eight} (b) $6BA_{\text{sixteen}}$
54. (a) 135_{eight} (b) $5D_{\text{sixteen}}$
55. (a) 637_{eight} (b) $19F_{\text{sixteen}}$
56. (a) 541_{eight} (b) 161_{sixteen}
57. (a) 2731_{eight} (b) $5D9_{\text{sixteen}}$
58. (a) 3667_{eight} (b) $7B7_{\text{sixteen}}$
59. (a) 3775_{eight} (b) $7FD_{\text{sixteen}}$
60. (a) 4307_{eight} (b) $8C7_{\text{sixteen}}$
61. (a) 6553_{eight} (b) $D6B_{\text{sixteen}}$
62. (a) 36467_{eight} (b) $3D37_{\text{sixteen}}$

63. 1101111000_{two} **64.** 11101100_{two} **65.** $110101000100_{\text{two}}$ **66.** $111110010100001_{\text{two}}$ **67.** 110010000_{two} **68.** 10111100010_{two} **69.** 110001_{two} **70.** 100011010_{two} **71.** $111010001110101_{\text{two}}$ **72.** $100111011001_{\text{two}}$ **73.** $110101000101011_{\text{two}}$ **74.** $1100101011111110001_{\text{two}}$ **75.** 10101011_{two} **76.** 1110100100_{two} **77.** 1101100010_{two} **78.** $1001111001110011_{\text{two}}$ **79.** $101001011101_{\text{two}}$ **80.** $1001010111000001_{\text{two}}$ **81.** $110010101001011_{\text{two}}$ **82.** $101011110011_{\text{two}}$ **83.** 5 lb 7 oz **84.** 7 qt 13 oz **85.** 34 yd 2 ft 8 in. **86.** 23 quarters, 1 dime, 3 pennies **87.** STOP **88.** NOHOMEWORK **89.** CLASSISOVER **90.** RESTFORLUNCH **91.** ITISRAINING **92.** TURNAROUND **93.** 13256 **94.** 91442 **95.** 67138 **96.** 03487 **97.** 44501 **98.** 28905

99. ıı|ı|ı||ııııı||ıı||ıı|ı|ı **100.** ııı||ı|ıı|ııı||ı|ı|ı|ııı|

101. ııı|||ıı|ıı|ıı|ıı|ı|ıı||ı **102.** ıı||ı||ııııı|ı|ııı||ı|ıı|

103. ııı||ııı||ı||ıı|ııı|ıı|ı| **104.** ı|ı|ıı|ıı||ı|ıı||ıııııı||

105. one symbol; same as tally system
106. 3; manufacturer's
107. 6; manufacturer's
108. 2; product
109. 0; product
110. 7; manufacturer's

Exercise Set 4-4

3.

+	0	1	2
0	0	1	2
1	1	2	10
2	2	10	11

4.

+	0	1	2	3
0	0	1	2	3
1	1	2	3	10
2	2	3	10	11
3	3	10	11	12

5.

×	0	1	2	3
0	0	0	0	0
1	0	1	2	3
2	0	2	10	12
3	0	3	12	21

6.

×	0	1	2
0	0	0	0
1	0	1	2
2	0	2	11

7. 32_{five} **8.** 132_{five} **9.** 11212_{four} **10.** $A749_{\text{twelve}}$ **11.** 2020_{six} **12.** 1130_{five} **13.** 6657_{nine} **14.** 3364_{seven} **15.** 31_{five} **16.** 104_{five} **17.** 101_{seven} **18.** $200A_{\text{eleven}}$ **19.** 40040_{nine} **20.** 5605_{seven} **21.** 571_{twelve} **22.** 1122_{three} **23.** 332_{six} **24.** 2136_{seven} **25.** 56067_{nine} **26.** 313041_{five} **27.** $3976A_{\text{twelve}}$ **28.** 131520_{nine} **29.** 482_{nine} remainder 2_{nine} **30.** 212_{three} remainder 1_{three} **31.** 230_{five} remainder 3_{five} **32.** 162_{seven} **33.** 10000_{two} **34.** 227_{eight} **35.** 403_{sixteen} **36.** 100100_{two} **37.** 1001_{two} **38.** 665_{eight} **39.** $4DCA_{\text{sixteen}}$ **40.** 11_{two} **41.** 110010_{two} **42.** 130_{eight} **43.** 2894_{sixteen} **44.** 7066_{eight} **45.** 11_{two} remainder 10_{two} **46.** 57_{eight} remainder 5_{eight} **47.** $B23_{\text{sixteen}}$ remainder 2_{sixteen} **48.** 1111_{two} remainder 1_{two} **49.** base 8 **50.** base 7 **51.** L **52.** U **53.** G **54.** Z **55.** 01000100 01001111 01010010 01001101 **56.** 01010000 01000001 01010010 01010101 01011001 **57.** 01010101 01001110 01001001 01001111 01001110 **58.** 01010001 01010101 01000001 01000100

Review Exercises

1. 1,000,221 **2.** 12,026 **3.** 681 **4.** 1,147 **5.** 419 **6.** 1,882 **7.** 2,604 **8.** 957 **9.** 𓎆𓎆𓎆𓎆𓏺𓏺𓏺𓏺𓏺𓏺𓏺𓏺𓏺 **10.** DCCCXCVI **11.** 𒁹 𒌋𒌋𒁹𒁹𒁹𒁹𒁹𒁹𒁹𒁹 **12.** 𓍢𓎆𓎆𓏺𓏺𓏺𓏺𓏺 **13.** DIII **14.** 八千三百二十五 **15.** 𒁹𒁹 𒌋𒌋𒌋𒌋𒁹𒁹𒁹𒁹𒁹 **16.** 七十四 **17.** 𓆼 𓍢𓍢𓍢𓍢𓍢 𓎆𓎆𓎆𓎆𓎆𓎆𓎆𓎆 𓏺

18. 𓂭 𓆼𓆼𓆼 𓍢𓍢𓍢 𓎆𓎆𓎆𓎆 𓏺𓏺𓏺𓏺𓏺𓏺𓏺𓏺

19. 276 **20.** 104 **21.** 147 **22.** 240 **23.** 330 **24.** 204 **25.** 156
26. 990 **27.** 1,955 **28.** 17,910 **29.** 365,687 **30.** 276,740
31. 2,542 **32.** 3,922 **33.** 13,965 **34.** 70,048 **35.** 119 **36.** 442
37. 17,327 **38.** 45 **39.** 1,246 **40.** 59 **41.** 2,058 **42.** 19,481
43. 215 **44.** 28,208 **45.** 52_{six} **46.** 89_{twelve} **47.** 2663_{nine}
48. 10000_{three} **49.** 101011_{two} **50.** 325_{eight} **51.** 103_{four} **52.** 110011_{two}
53. 1000_{seven} **54.** $62B_{\text{twelve}}$
55. (a) 733_{eight} (b) $1DB_{\text{sixteen}}$ **56.** (a) 2167_{eight} (b) 477_{sixteen}
57. (a) 1547_{eight} (b) 367_{sixteen} **58.** (a) 7075_{eight} (b) $E3D_{\text{sixteen}}$
59. $111011010100_{\text{two}}$ **60.** 110100011_{two} **61.** $1010010110110011_{\text{two}}$
62. $1001111110000111_{\text{two}}$ **63.** 251_{nine} **64.** 1330_{five} **65.** 111011_{two}
66. 11051_{six} **67.** $15AB6_{\text{twelve}}$ **68.** 6813_{nine} **69.** 101100_{two}
70. 201_{three} **71.** 331_{four} **72.** 16541_{seven} **73.** 21331_{nine} **74.** 1331_{five}
75. 1011011_{two} **76.** 3528_{sixteen} **77.** 230_{five} remainder 2_{five}
78. 403_{eight} remainder 1_{eight} **79.** 342_{eight} remainder 6_{eight}
80. $A0_{\text{sixteen}}$ remainder 4_{sixteen}

Chapter 4 Test

1. 2,000,312 **2.** 24,055 **3.** 1,271 **4.** 1966 **5.** 426 **6.** 368
7. 𓎆𓎆𓎆𓎆𓎆𓎆𓎆𓎆𓎆𓎆𓏺𓏺𓏺 **8.** DLXVII
9. 𒌋𒌋𒌋𒌋𒌋𒌋𒁹𒁹𒁹𒁹𒁹𒁹 **10.** 𓍢𓍢𓍢𓍢𓍢𓎆𓎆𓏺
11. DCV **12.** 八百七十三 **13.** 221 **14.** 345 **15.** 267,904 **16.** 26,544
17. 96 **18.** 379 **19.** 17,375 **20.** 54 **21.** 1,231 **22.** 199 **23.** 500
24. 103 **25.** 241 **26.** 28,474 **27.** 133_{five} **28.** 133_{twelve} **29.** 6362_{nine}
30. 2102_{three} **31.** 10001_{two} **32.** 474_{eight} **33.** 1123_{four} **34.** 110000_{two}
35. 1160_{seven} **36.** 621_{twelve}
37. (a) 731_{eight} (b) $1D9_{\text{sixteen}}$ **38.** (a) 3073_{eight} (b) $63B_{\text{sixteen}}$
39. $111011010100_{\text{two}}$ **40.** $10100110110110010010_{\text{two}}$ **41.** 282_{nine}
42. 1104_{five} **43.** 1000111_{two} **44.** 11452_{six} **45.** $14B25_{\text{twelve}}$
46. 6035_{eight} **47.** 10010111_{two} **48.** 1010_{three} **49.** 1030_{four}
50. 10301_{seven} **51.** 1250_{six} **52.** 14223_{five} **53.** 151_{eight} remainder 3_{eight}
54. 121_{three}

CHAPTER 5: THE REAL NUMBER SYSTEM

Exercise Set 5-1

7. 1, 2, 4, 8, 16 **8.** 1, 3, 5, 9, 15, 25, 45, 75, 225
9. 1, 2, 3, 6, 7, 9, 14, 18, 21, 42, 63, 126 **10.** 1, 2, 3, 6, 9, 18, 27, 54
11. 1, 2, 4, 8, 16, 32 **12.** 1, 2, 3, 4, 6, 8, 12, 16, 24, 48 **13.** 1, 3, 9
14. 1, 2, 5, 10 **15.** 1, 2, 3, 4, 6, 8, 12, 16, 24, 32, 48, 96
16. 1, 2, 4, 5, 10, 20, 25, 50, 100 **17.** 1, 17 **18.** 1, 19
19. 1, 2, 4, 8, 16, 32, 64
20. 1, 2, 3, 4, 5, 6, 8, 10, 12, 15, 20, 24, 30, 40, 60, 120
21. 1, 3, 5, 7, 15, 21, 35, 105 **22.** 1, 5, 73, 365 **23.** 1, 2, 7, 14, 49, 98
24. 1, 2, 3, 4, 6, 9, 12, 18, 36 **25.** 1, 71 **26.** 1, 47 **27.** 3, 6, 9, 12, 15
28. 7, 14, 21, 28, 35 **29.** 10, 20, 30, 40, 50 **30.** 12, 24, 36, 48, 60
31. 15, 30, 45, 60, 75 **32.** 20, 40, 60, 80, 100 **33.** 17, 34, 51, 68, 85
34. 19, 38, 57, 76, 95 **35.** 1, 2, 3, 4, 5 **36.** 25, 50, 75, 100, 125
37. 2^4 **38.** 2×3^2 **39.** $2^4 \times 3^4$ **40.** $2^3 \times 5 \times 7^2$ **41.** 17 is prime
42. 19 is prime **43.** 2×5^2 **44.** 2^6 **45.** 2^7 **46.** 13^2 **47.** $2^2 \times 3 \times 5^2$
48. $2^2 \times 5^3$ **49.** $5^2 \times 19$ **50.** 5^4 **51.** 7×2^6 **52.** 7×11
53. 13×19 **54.** $2^3 \times 7$ **55.** $2 \times 3 \times 5^3$ **56.** $3 \times 5^2 \times 11$ **57.** 3
58. 5 **59.** 1 **60.** 1 **61.** 6 **62.** 15 **63.** 21 **64.** 70 **65.** 25
66. 25 **67.** 12 **68.** 5 **69.** 6 **70.** 6 **71.** 12 **72.** 6 **73.** 25
74. 7 **75.** 10 **76.** 24 **77.** 35 **78.** 30 **79.** 126 **80.** 175 **81.** 150
82. 180 **83.** 630 **84.** 390 **85.** 308 **86.** 390 **87.** 60 **88.** 630
89. 36 **90.** 336 **91.** 840 **92.** 132 **93.** 4 students **94.** 8 visits
95. 72 minutes **96.** 90 days
97. 6 members per team; 5 female teams, 6 male teams
98. 4 groups of six pencils and 3 groups of six pictures
99. noon **100.** 60 seconds
101. 4 groups of Republicans, 5 groups of Democrats, 7 groups of Independents
102. 12 boxes **103.** 221 years **104.** 15 years
105. The only even prime number is 2. **106.** Answers vary
107. $3 + 1 = 4$, $3 + 3 = 6$, $3 + 5 = 8$, $5 + 5 = 10$, $5 + 7 = 12$, $7 + 7 = 14$, $3 + 13 = 16$, $5 + 13 = 18$, $3 + 17 = 20$
108. $3 + 3 + 3 = 9$, $3 + 3 + 5 = 11$, $3 + 5 + 5 = 13$, $5 + 5 + 5 = 15$, $3 + 3 + 11 = 17$, $3 + 3 + 13 = 19$, $3 + 5 + 13 = 21$, $5 + 5 + 13 = 23$, $3 + 3 + 19 = 25$
109. Answers vary, 13.
110. (a) less than or equal to (b) greater than or equal to

Exercise Set 5-2

7. 8 **8.** 12 **9.** 10 **10.** 14 **11.** 8 **12.** −27 **13.** −10 **14.** 16
15. 0 **16.** 9 **17.** $<$ **18.** $<$ **19.** $>$ **20.** $>$ **21.** $>$ **22.** $<$
23. $<$ **24.** $>$ **25.** $<$ **26.** $<$ **27.** −1 **28.** −4 **29.** 9 **30.** −12
31. −11 **32.** 5 **33.** −12 **34.** 2 **35.** −13 **36.** −19 **37.** 14
38. 7 **39.** −5 **40.** −6 **41.** 1 **42.** 2 **43.** −5 **44.** −24 **45.** −70
46. −43 **47.** 45 **48.** 42 **49.** −24 **50.** −72 **51.** −36 **52.** −84
53. 42 **54.** 98 **55.** 0 **56.** 0 **57.** 8 **58.** 8 **59.** −5 **60.** −6
61. −4 **62.** −7 **63.** 7 **64.** 5 **65.** 1 **66.** −14 **67.** 0 **68.** 0
69. 0 **70.** 2 **71.** 11 **72.** 28 **73.** 51 **74.** 56 **75.** 36 **76.** 135
77. 6 **78.** −30 **79.** −758 **80.** 24 **81.** −56 **82.** 102 **83.** 59
84. −14 **85.** −492 **86.** 302 **87.** $1,180 **88.** 14,392 feet
89. 1,399 rats **90.** 265 cases
91. Flagler, FL: 24,002, Sumter, FL: 18,797: Paulding, GA: 21,082, Kendall, IL: 13,877, Pinal, AZ: 56,264. **92.** 57 degrees. **93.** 8 inches
94. 25 inches **95.** 7,000 pounds **96.** 200 students
97. No general statements can be made. **98.** Answers vary
99. Answers vary **100.** Answers vary

Exercise Set 5-3

9. $\frac{1}{6}$ **10.** $\frac{1}{3}$ **11.** $\frac{7}{10}$ **12.** $\frac{4}{5}$ **13.** $\frac{5}{6}$ **14.** $\frac{1}{3}$ **15.** $\frac{7}{8}$ **16.** $\frac{4}{7}$ **17.** $\frac{5}{9}$
18. $\frac{19}{48}$ **19.** $\frac{15}{48}$ **20.** $\frac{45}{96}$ **21.** $\frac{43}{49}$ **22.** $\frac{25}{40}$ **23.** $\frac{35}{45}$ **24.** $\frac{9}{30}$ **25.** $\frac{55}{80}$
26. $\frac{12}{28}$ **27.** $\frac{6}{30}$ **28.** $\frac{20}{64}$ **29.** $-\frac{1}{6}$ **30.** $\frac{29}{20}$ or $1\frac{9}{20}$ **31.** $-\frac{37}{24}$ or $-1\frac{13}{24}$
32. $\frac{29}{72}$ **33.** $\frac{7}{24}$ **34.** $\frac{1}{10}$ **35.** $\frac{7}{6}$ or $1\frac{1}{6}$ **36.** $-\frac{35}{92}$ **37.** $\frac{7}{10}$ **38.** $-\frac{315}{256}$
39. $-\frac{1}{16}$ **40.** $\frac{7}{60}$ **41.** $\frac{11}{12}$ **42.** $-\frac{35}{48}$ **43.** $\frac{6}{5}$ or $1\frac{1}{5}$ **44.** $\frac{7}{6}$ or $1\frac{1}{6}$ **45.** $\frac{3}{4}$
46. $\frac{603}{40}$ or $15\frac{3}{40}$ **47.** $\frac{7}{36}$ **48.** $\frac{2}{3}$ **49.** 0.2 **50.** 0.3 **51.** 0.66 . . . or $0.\overline{6}$
52. 1.4 **53.** 2.25 **54.** 1.22 . . . or $1.\overline{2}$ **55.** 0.3055 . . . or $0.30\overline{5}$
56. $1.\overline{714285}$ **57.** 0.75 **58.** 1.875 **59.** $0.\overline{9411764705882352}$
60. $0.708\overline{3}$ **61.** $\frac{4}{25}$ **62.** $\frac{9}{25}$ **63.** $\frac{3}{8}$ **64.** $\frac{37}{40}$ **65.** $\frac{7}{9}$ **66.** $\frac{2}{9}$ **67.** $\frac{6}{11}$
68. $\frac{62}{99}$ **69.** $\frac{34}{75}$ **70.** $\frac{247}{900}$ **71.** 171 miles **72.** 13 miles **73.** 1/7
74. 40 men **75.** 935 people **76.** 110.4 million people
77. 935 people **78.** 190 miles **79.** 12 feet **80.** $\frac{3}{40}$ m or 7.5 cm

81. $\frac{3}{8}$ **82.** $7{,}733.\overline{3}$ tons **83.** $1\frac{1}{4}$ cups of flour and $\frac{1}{3}$ cup of sugar
84. $11\frac{1}{4}$ cups of flour, $4\frac{3}{4}$ cups of both granulated sugar and brown sugar.
85. $\frac{7}{500}$ **86.** $\frac{29}{100}$ **87.** $\frac{13}{100}$ **88.** $\frac{23}{50}$ **89.** $\frac{45}{100}$ **90.** $\frac{17}{10}$
91. Any common denominator will do. **92.** The answer is the same.
93. This question is a paradox, an answer of yes or no leads to a contradiction.
94. No **95.** Add the fractions and divide by two.

Exercise Set 5-4

7. rational **8.** irrational **9.** irrational **10.** rational **11.** irrational
12. rational. **13.** 3 and 4 **14.** 5 and 6 **15.** 9 and 11 **16.** 8 and 9
17. 14 and 15 **18.** 12 and 13 **19.** $2\sqrt{6}$ **20.** $3\sqrt{3}$ **21.** $4\sqrt{5}$ **22.** $5\sqrt{7}$
23. $\sqrt{30}$ **24.** $\sqrt{42}$ **25.** $20\sqrt{5}$ **26.** $8\sqrt{2}$ **27.** $30\sqrt{7}$ **28.** $18\sqrt{2}$
29. $2\sqrt{5}$ **30.** $3\sqrt{10}$ **31.** $3\sqrt{30}$ **32.** $5\sqrt{5}$ **33.** $24\sqrt{3}$ **34.** $60\sqrt{3}$
35. $\sqrt{30}$ **36.** $\sqrt{7}$ **37.** $2\sqrt{2}$ **38.** $\sqrt{5}$ **39.** $12\sqrt{7}$ **40.** $61\sqrt{11}$
41. $-7\sqrt{3}$ **42.** $-6\sqrt{7}$ **43.** $-2\sqrt{3}$ **44.** $-5\sqrt{5}$ **45.** $4\sqrt{5}$ **46.** $7\sqrt{5}$
47. $-6\sqrt{5}$ **48.** $49\sqrt{10}$ **49.** $18\sqrt{2}$ **50.** $9\sqrt{10}$ **51.** $\sqrt{2} + 8\sqrt{3}$
52. $26\sqrt{5} - 20\sqrt{10}$ **53.** $10\sqrt{10} - 4\sqrt{2}$ **54.** $21\sqrt{3} - 18\sqrt{5}$
55. $7\sqrt{15}$ **56.** 3 **57.** $\frac{\sqrt{5}}{5}$ **58.** $\frac{3\sqrt{2}}{4}$ **59.** $\frac{\sqrt{6}}{2}$ **60.** $\sqrt{5}$ **61.** $\frac{\sqrt{21}}{14}$
62. $\frac{1}{3}$ **63.** $\frac{\sqrt{6}}{3}$ **64.** $\sqrt{\frac{14}{4}}$ **65.** $\sqrt{9} + \sqrt{16} = 3 + 4 \neq 5 = \sqrt{25}$
66. $\sqrt{25} + \sqrt{144} = 5 + 12 \neq 13 = \sqrt{169}$
67. $\sqrt{40} + \sqrt{40} \approx 6.32 + 6.32 \neq 8.94 \approx \sqrt{80}$
68. $\sqrt{60} + \sqrt{60} \approx 7.75 + 7.75 \neq 10.95 \approx \sqrt{120}$
69. 3 seconds **70.** 4 seconds **71.** 9.2 seconds **72.** 9.9 seconds
73. 3.5 seconds, 2.5 seconds, the difference is 1 second.
74. 5 seconds, 2.5 seconds, the difference is 2.5 seconds.
75. 20 volts **76.** 60 volts **77.** 141.4 volts **78.** 173.2 volts
79. 4π seconds **80.** $2\pi\sqrt{2}$ seconds **81.** 0.8 foot **82.** 20.2 feet
83. The irrational numbers are not closed under multiplication. The rational numbers are closed under multiplication.
84. $\sqrt{a-b} \neq \sqrt{a} - \sqrt{b}$; a counterexample is $\sqrt{9-4} = \sqrt{5} \neq 3 - 2 = \sqrt{9} - \sqrt{4}$.
85. You cannot compute the square root of a negative number because there is no number, that when squared, is negative.
86. Answers vary
87. Yes, $\sqrt{3+0} = \sqrt{3} + \sqrt{0}$. The conjecture will be true when one of a or b is zero.

Exercise Set 5-5

7. Integer, rational, real **8.** Natural, whole, integer, rational, real
9. Rational, real **10.** Rational, real **11.** Rational, real
12. Rational, real **13.** Irrational, real **14.** Irrational, real
15. Irrational, real **16.** Irrational, real **17.** Rational, real
18. Rational, real **19.** Natural, whole, integer, rational, real
20. Integer, rational, real **21.** Natural, whole, integer, rational, real
22. Natural, whole, integer, rational, real
23. Closure property of addition **24.** Identity property of multiplication
25. Commutative property of addition **26.** Distributive property
27. Commutative property of multiplication
28. Inverse property of addition **29.** Distributive property
30. Closure property of multiplication
31. Inverse property of multiplication
32. Inverse property of addition
33. Commutative property of addition
34. Associative property of addition
35. Identity property of addition
36. Commutative property of addition
37. Commutative property of multiplication
38. Commutative property of multiplication
39. Addition, multiplication **40.** Addition, multiplication
41. Addition, subtraction, multiplication
42. Addition, subtraction, multiplication, division (except by 0)
43. None
44. Addition, subtraction, multiplication, division (except by 0)
45. 27.46, overweight **46.** 26.90, overweight **47.** 23.73, normal
48. 23.67, normal **49.** 17.53, underweight **50.** 21.56, normal
51. 38.51, obese **52.** 36.97, obese **53.** not commutative
54. commutative **55.** commutative **56.** not commutative
57. associative **58.** associative **59.** associative **60.** not associative
61. The results are not the same, subtraction is not associative.
62. The results are not the same, division is not associative.
63. The results are not the same, addition does not distribute over multiplication.
64. The results are the same, multiplication does distribute over subtraction.

Exercise Set 5-6

7. 243 **8.** 1,296 **9.** 1 **10.** 1 **11.** 1 **12.** 1 **13.** $\frac{1}{243}$ **14.** $\frac{1}{1{,}296}$
15. $\frac{1}{64}$ **16.** $\frac{1}{49}$ **17.** $3^6 = 729$ **18.** $5^6 = 15{,}625$ **19.** $4^7 = 16{,}384$
20. $2^{10} = 1{,}024$ **21.** $3^2 = 9$ **22.** $6^2 = 36$ **23.** 2^1 or 2 **24.** $8^2 = 64$
25. $5^6 = 15{,}625$ **26.** $4^8 = 65{,}536$ **27.** $\frac{1}{3^2} = \frac{1}{9}$ **28.** $4^2 = 16$
29. $\frac{1}{5^5} = \frac{1}{3{,}125}$ **30.** $6^0 = 1$ **31.** $\frac{1}{2^2} = \frac{1}{4}$ **32.** $\frac{1}{3^2} = \frac{1}{9}$ **33.** $\frac{1}{4^3} = \frac{1}{64}$
34. $\frac{1}{5^3} = \frac{1}{125}$ **35.** $\frac{1}{7}$ **36.** $\frac{1}{8^2} = \frac{1}{64}$ **37.** 6.25×10^8 **38.** 9.91×10^6
39. 7.3×10^{-3} **40.** 2.61×10^{-1} **41.** 5.28×10^{11} **42.** 2.22×10^6
43. 6.18×10^{-6} **44.** 7.7×10^{-9} **45.** 4.32×10^4 **46.** 5.6×10^4
47. 8.14×10^{-2} **48.** 1.1×10^{-3} **49.** 3.2×10^{13} **50.** 4.35×10^7
51. 59,000 **52.** 6,280,000 **53.** 0.0000375 **54.** 0.0000000009
55. 2,400 **56.** 772,000 **57.** 0.000003 **58.** 0.000000004
59. 1,000 **60.** 22,600 **61.** 8,020,000,000 **62.** 0.0001
63. 7,000,000,000,000 **64.** 133 **65.** $6 \times 10^{10} = 60{,}000{,}000{,}000$
66. $4 \times 10^9 = 4{,}000{,}000{,}000$ **67.** $2.67 \times 10^{-7} = 0.000000267$
68. $6.46 \times 10^{-11} = 0.0000000000646$ **69.** $8.8 \times 10^{-3} = 0.0088$
70. $7.92 \times 10^1 = 79.2$ **71.** $1.5 \times 10^{-9} = 0.0000000015$
72. $9.46 \times 10^{-1} = 0.946$ **73.** $2 \times 10^2 = 200$ **74.** $3 \times 10^4 = 30{,}000$
75. $6 \times 10^0 = 6$ **76.** $8 \times 10^9 = 8{,}000{,}000{,}000$ **77.** $6 \times 10^{-2} = 0.06$
78. $2 \times 10^{12} = 2{,}000{,}000{,}000{,}000$ **79.** 2.58×10^{15} **80.** 1.56×10^{11}
81. 2.4×10^1 **82.** 5×10^{-1} **83.** 1×10^{-12} **84.** 4×10^{-3}
85. 1.116×10^8 miles **86.** 3×10^{22} cells **87.** 56%
88. 3.03×10^{13} grains of sand per ft^2 **89.** 2.4696×10^{13} miles
90. 1.6464×10^{15} miles **91.** 7.96×10^{-4} light years **92.** 8.3 minutes
93. 10 billion **94.** 6.5 million grains **95.** 421 million miles
96. 6.25×10^{21} molecules **97.** $111,000,000 a day **98.** $135
99. 5.7 million tickets **100.** $2 per viewer
101. Each entry is half the previous entry so the last three entries are 1, ½, and ¼.
102. Answers vary **103.** Answers vary **104.** (a) $2^3 \times (10^4)^3$ (b) $a^3 \times (10^n)^3$ (c) Answers vary

Exercise Set 5-7

7. (a) 1, 7, 13, 19, 25 (b) 6 (c) 67 (d) 408
8. (a) 10, 15, 20, 25, 30 (b) 5 (c) 65 (d) 450
9. (a) $-9, -12, -15, -18, -21$ (b) -3 (c) -42 (d) -306
10. (a) $-15, -17, -19, -21, -23$ (b) -2 (c) -37 (d) -312
11. (a) $\frac{1}{4}, \frac{5}{8}, 1, \frac{11}{8}, \frac{7}{4}$ (b) $\frac{3}{8}$ (c) $\frac{35}{8}$ (d) $\frac{111}{4}$
12. (a) $\frac{3}{7}, \frac{4}{7}, \frac{5}{7}, \frac{6}{7}, 1$ (b) $\frac{1}{7}$ (c) 2 (d) $\frac{102}{7}$
13. (a) $4, \frac{11}{3}, \frac{10}{3}, 3, \frac{8}{3}$ (b) $-\frac{1}{3}$ (c) $\frac{1}{3}$ (d) 26
14. (a) $\frac{5}{2}, \frac{9}{4}, 2, \frac{7}{4}, \frac{3}{2}$ (b) $-\frac{1}{4}$ (c) $-\frac{1}{4}$ (d) $\frac{27}{2}$
15. (a) 5, 13, 21, 29, 37 (b) 8 (c) 93 (d) 588
16. (a) 2, 12, 22, 32, 42 (b) 10 (c) 112 (d) 684
17. (a) 50, 48, 46, 44, 42 (b) -2 (c) 28 (d) 468
18. (a) $12, 7, 2, -3, -8$ (b) -5 (c) -43 (d) -186
19. (a) $\frac{1}{8}, \frac{19}{24}, \frac{35}{24}, \frac{17}{8}, \frac{67}{24}$ (b) $\frac{2}{3}$ (c) $\frac{179}{24}$ (d) $\frac{91}{2}$
20. (a) $\frac{1}{2}, \frac{9}{10}, \frac{13}{10}, \frac{17}{10}, \frac{21}{10}$ (b) $\frac{2}{5}$ (c) $\frac{49}{10}$ (d) $\frac{162}{5}$
21. (a) 0.6, 1.6, 2.6, 3.6, 4.6 (b) 1 (c) 11.6 (d) 73.2

22. (a) 0.3, 0.7, 1.1, 1.5, 1.9 (b) 0.4 (c) 4.7 (d) 30
23. (a) 12, 24, 48, 96, 192 (b) 2 (c) 24,576 (d) 49,140
24. (a) 8, 24, 72, 216, 648 (b) 3 (c) 1,417,176 (d) 2,125,760
25. (a) $-5, -\frac{5}{4}, -\frac{5}{16}, -\frac{5}{64}, -\frac{5}{256}$ (b) $\frac{1}{4}$ (c) $-\frac{5}{4,194,304}$ (d) ≈ -6.7
26. (a) $-9, -6, -4, -\frac{8}{3}, -\frac{16}{9}$ (b) $\frac{2}{3}$ (c) $-\frac{2,048}{19,683}$ (d) ≈ -26.79
27. (a) $\frac{1}{6}, -1, 6, -36, 216$ (b) -6 (c) 60,466,176 (d) $-\frac{15,237,476,345}{6}$
28. (a) $\frac{3}{7}, -\frac{9}{7}, \frac{27}{7}, -\frac{81}{7}, \frac{243}{7}$ (b) -3 (c) $\frac{531,441}{7}$ (d) $-56,940$
29. (a) $100, -25, \frac{25}{4}, -\frac{25}{16}, \frac{25}{64}$ (b) $-\frac{1}{4}$ (c) $-\frac{25}{1,048,576}$ (d) ≈ 80
30. (a) $10, -1, \frac{1}{10}, -\frac{1}{100}, \frac{1}{1000}$ (b) $-\frac{1}{10}$ (c) $-\frac{1}{10^{10}}$ (d) $\frac{10^{12}-1}{11 \cdot 10^{10}}$
31. (a) 4, 12, 36, 108, 324 (b) 3 (c) 708,588 (d) 1,062,880
32. (a) 6, 12, 24, 48, 96 (b) 2 (c) 12,288 (d) 24,570
33. (a) $\frac{1}{2}, \frac{1}{4}, \frac{1}{8}, \frac{1}{16}, \frac{1}{32}$ (b) 1/2 (c) ≈ 0.000244 (d) ≈ 0.9998
34. (a) $\frac{2}{3}, \frac{2}{9}, \frac{2}{27}, \frac{2}{81}, \frac{2}{243}$ (b) $\frac{1}{3}$ (c) $\frac{2}{531,441}$ (d) $\frac{531,440}{531,441}$
35. (a) $-3, 15, -75, 375, -1875$ (b) -5 (c) 146,484,375 (d) 122,070,312
36. (a) $-3, 12, -48, 192, -768$ (b) -4 (c) 12,582,912 (d) 10,066,329
37. (a) 1, 3, 9, 27, 81 (b) 3 (c) 177,147 (d) 265,720
38. (a) $8, 2, \frac{1}{2}, \frac{1}{8}, \frac{1}{32}$ (b) $\frac{1}{4}$ (c) $\frac{1}{524,288}$ (d) ≈ 10.67
39. geometric **40.** arithmetic **41.** neither **42.** neither
43. arithmetic **44.** geometric **45.** arithmetic **46.** neither
47. (a) \$500 (b) \$13,000 **48.** (a) \$1,600 (b) \$41,500
49. no **50.** \$60 **51.** \$18,000 **52.** 360 workers **53.** 160 ft
54. 54.4 ft **55.** \$3,766.11 **56.** \$814.45 **57.** 8 questions
58. \$16,971.52 **59.** First job pays \$5,300.65 more in ten years.
60. The first is the better investment.
61. $a_1 = \frac{3}{10}$ and $r = \frac{1}{10}$. The sum is $\frac{1}{3}$. **62.** $\frac{5}{33}$ **63.** Answers vary

Review Exercises

1. 1, 2, 3, 6, 13, 26, 39, 78 **2.** 1, 3, 9, 27, 81 **3.** 1, 3, 5, 9, 15, 45
4. 1, 2, 19, 38 **5.** 1, 2, 4, 5, 7, 10, 14, 20, 28, 35, 70, 140
6. 1, 2, 3, 4, 6, 9, 12, 18, 27, 36, 54, 81, 108, 162, 324
7. 4, 8, 12, 16, 20 **8.** 32, 64, 96,128, 160
9. 9, 18, 27, 36, 45 **10.** 60, 120, 180, 240, 300 **11.** $2^5 \times 3$
12. $2^2 \times 11$ **13.** 2×5^3 **14.** $2^4 \times 3^2 \times 5$ **15.** $2^3 \times 3 \times 5^2$
16. 3×5^2 **17.** 2; 30 **18.** 2; 180 **19.** 5; 280 **20.** 25; 150
21. 20; 1,200 **22.** 9; 216 **23.** 198 months, or 16.5 years **24.** 18
25. -14 **26.** -45 **27.** -4 **28.** 7 **29.** 44 **30.** -6 **31.** 89
32. 2,709 **33.** -157 **34.** \$675 **35.** $\frac{15}{19}$ **36.** $\frac{7}{8}$ **37.** $\frac{4}{5}$ **38.** $\frac{4}{5}$
39. $\frac{23}{24}$ **40.** $\frac{3}{20}$ **41.** $\frac{5}{21}$ **42.** $-\frac{25}{14}$ **43.** $\frac{6}{17}$ **44.** $\frac{51}{80}$ **45.** $\frac{13}{18}$
46. $\frac{21}{16}$ or $1\frac{5}{16}$ **47.** $-\frac{17}{7}$ or $-2\frac{3}{7}$ **48.** 1 **49.** $\frac{41}{40}$ or $1\frac{1}{40}$ **50.** $-\frac{19}{48}$
51. 0.9 **52.** 0.3125 **53.** $0.\overline{857142}$ **54.** $0.\overline{1}$ **55.** $\frac{11}{16}$ **56.** $\frac{11}{50}$
57. $\frac{23}{90}$ **58.** $\frac{5}{11}$ **59.** NFL: 12; MLB: 8; NBA and NHL: 16
60. $4\sqrt{3}$ **61.** $4\sqrt{7}$ **62.** $\frac{7\sqrt{5}}{5}$ **63.** $\frac{\sqrt{5}}{2}$ **64.** $\frac{\sqrt{6}}{4}$ **65.** $\frac{\sqrt{15}}{6}$
66. $-\sqrt{5} + 10\sqrt{3}$ **67.** $22\sqrt{2}$ **68.** $9\sqrt{21}$ **69.** $20\sqrt{3}$ **70.** 2 **71.** $\sqrt{6}$
72. $2\sqrt{3} + \sqrt{30}$ **73.** $14\sqrt{3} - 6\sqrt{7}$ **74.** Rational, real
75. Rational, real **76.** Rational, real **77.** Irrational, real
78. Whole, integer, rational, real
79. Natural, whole, integer, rational, real
80. Inverse property of multiplication
81. Commutative property of addition
82. Closure property of addition **83.** Distributive property
84. 1,024 **85.** 1 **86.** 1 **87.** 1/81 **88.** 1/7,776 **89.** 117,649
90. 625 **91.** 6,561 **92.** 1/4 **93.** 1/7,776 **94.** 3.83×10^3
95. 2.59×10^{10} **96.** 3.27×10^{-6} **97.** 4.8×10^{-4}
98. 580,000,000,000 **99.** 2,330,000,000 **100.** 0.000627
101. 0.0000088 **102.** 9.2×10^{-2} **103.** 2.848×10^{-11}
104. 2×10^{10} **105.** 6×10^{-8}
106. 1.82682×10^{11}; that's about 182.7 billion cans of soda!
107. 1.41×10^4 seconds or about 3.9 hours
108. 8, 18, 28, 38, 48, 58; $a_9 = 88$, $S_9 = 432$
109. $4, 1, -2, -5, -8, -11$; $a_9 = -20$, $S_9 = -72$
110. $-13, -18, -23, -28, -33, -38$; $a_9 = -53$, $S_9 = -297$
111. $-\frac{1}{5}, \frac{3}{10}, \frac{4}{5}, \frac{13}{10}, \frac{9}{5}, \frac{23}{10}$; $a_9 = \frac{19}{5}$, $S_9 = \frac{81}{5}$
112. 7.5, 15, 30, 60, 120, 240; $a_9 = 1,920$, $S_9 = 3,832.5$
113. $-3, -9, -27, -81, -243, -729$; $a_9 = -19,683$, $S_9 = -29,523$
114. $\frac{1}{9}, \frac{1}{36}, \frac{1}{144}, \frac{1}{576}, \frac{1}{2304}, \frac{1}{9,216}$; $a_9 = \frac{1}{589,824}$, $S_9 \approx 0.148$
115. $-\frac{2}{5}, \frac{1}{5}, -\frac{1}{10}, \frac{1}{20}, -\frac{1}{40}, \frac{1}{80}$; $a_9 = -\frac{1}{640}$, $S_9 = -\frac{171}{640}$
116. 49 million people
117. The profit for the sixth year is \$25,525.63, and the total earnings for the six years are \$136,038.26

Chapter 5 Test

1. Integer, rational, real **2.** Rational, real **3.** Rational, real
4. Rational, real **5.** Irrational, real **6.** Irrational, real
7. Whole, integer, rational, real **8.** Rational, real
9. Integer, rational, real **10.** Rational, real **11.** 14; 168
12. 9; 180 **13.** 25; 4,200 **14.** 10; 2,640 **15.** $\frac{3}{7}$ **16.** $\frac{3}{4}$ **17.** $\frac{16}{25}$
18. $\frac{1}{2}$ **19.** $\frac{7}{10}$ **20.** $\frac{49}{64}$ **21.** $4\sqrt{3}$ **22.** $9\sqrt{3}$ **23.** 36 **24.** -5
25. $\frac{15}{16}$ **26.** $-\frac{31}{126}$ **27.** $\frac{27}{8}$ or $3\frac{3}{8}$ **28.** -8 **29.** $2(\sqrt{3} + \sqrt{6})$
30. $2\sqrt{2}$ **31.** 3 **32.** $-2\sqrt{2}$ **33.** $\frac{7}{8}$ **34.** $\frac{16}{25}$ **35.** $\frac{2}{9}$ **36.** $\frac{35}{99}$
37. Commutative property of addition.
38. Closure property of multiplication.
39. Identity property of addition
40. Inverse property of multiplication
41. Associative property of multiplication
42. Distributive property **43.** 4,096 **44.** $\frac{1}{343}$ **45.** 1 **46.** 65,536
47. $\frac{1}{3,125}$ **48.** 5.2×10^7 **49.** 2.36×10^{-3} **50.** 9,770
51. -0.00006 **52.** 1.56×10^4 **53.** 3×10^3
54. 1, 3.5, 6, 8.5, 11, 13.5, 16
$a_{20} = 48.5$, $S_{20} = 495$
55. $\frac{3}{4}, -\frac{1}{8}, \frac{1}{48}, -\frac{1}{288}, \frac{1}{1,728}, -\frac{1}{10,368}, \frac{1}{62,208}$
$a_{15} \approx 9.57 \times 10^{-12}$, $S_{15} \approx 0.643$
56. 22 weeks **57.** \$320; \$620

CHAPTER 6: TOPICS IN ALGEBRA

Exercise Set 6-1

7. $11x$ **8.** $-4x^2$ **9.** $-18y$ **10.** $-5A$ **11.** $9p - q - 17$
12. $7x - 4y - 18$ **13.** $11x^2 - x + 5$ **14.** $-9x^2 + 14x - 4$
15. $30x - 35$ **16.** $27x + 72$ **17.** $-48x + 40$ **18.** $-32m - 56$
19. $x + 27$ **20.** $14x - 19$ **21.** $-26x - 50$ **22.** $-55x - 91$
23. $8x - 23$ **24.** $25b + 36$ **25.** $-x + 13$ **26.** $-4x + 21$
27. $3x - 5$ **28.** $2x - 10$ **29.** $-2x^2 - 10x - 2$ **30.** $-y^2 - 9y - 14$
31. $-4x^2y + 18xy^2 - 9$ **32.** $6ab + 19a^2b - 11b^2$ **33.** $13m - 9$
34. $9k + 6$ **35.** $\frac{7}{2}yz + \frac{3}{4}$ **36.** $-\frac{3}{4}ab + 5$ **37.** $-3.9x + 10$
38. $4x - 5.9$ **39.** $0.75tz$ **40.** $-0.4rs + 2.6 + 0.3tz$ **41.** 83 **42.** -7
43. 16 **44.** -39 **45.** 79 **46.** 1,940 **47.** 466 **48.** 758 **49.** 104
50. 94 **51.** 136 **52.** 62 **53.** 997 **54.** 226 **55.** 205 **56.** 10,973
57. 17 **58.** $\frac{35}{2}$ **59.** $127\frac{7}{12}$ **60.** $149\frac{7}{9}$ **61.** -5.2 **62.** 54.2 **63.** -27.5
64. 4 **65.** 376.8 sq in. **66.** 30 ft **67.** \$6,050 **68.** 6,400 ft
69. 267.95 mm^3 **70.** 615.44 **71.** \$31,876.96 **72.** 306 **73.** 678.24
74. 3.51 **75.** \$100 **76.** \$22.79 per hour **77.** 3 defective items
78. \$2,845,936 **79.** 10°C **80.** 1.47 **81.** 600,000 ergs **82.** 312,375
83. $\frac{8}{9}$ **84.** $\approx$267.95 in. **85.** \$14,281.87 **86.** 1.8×10^{11} joules
87. $\sqrt{13}$ **88.** $\approx$0.0643 **89.** 1,340 feet **90.** \$1,550 **91.** 183 mg
92. about 377 yards using 3.14 as an approximation for π

Exercise Set 6-2

7. not linear **8.** not linear **9.** linear **10.** linear **11.** not linear
12. not linear **13.** linear **14.** linear **15.** yes **16.** yes **17.** no
18. no **19.** yes **20.** yes **21.** $\{26\}$ **22.** $\{36\}$ **23.** $\{59\}$ **24.** $\{45\}$
25. $\{-3\}$ **26.** $\{-6\}$ **27.** $\{3\}$ **28.** $\{7\}$ **29.** $\{-12\}$ **30.** $\{6\}$

31. $\{6\}$ **32.** $\left\{\frac{5}{2}\right\}$ **33.** $\{16\}$ **34.** $\{-8\}$ **35.** $\{20\}$ **36.** $\{-6\}$
37. $\{11\}$ **38.** $\{11\}$ **39.** $\{5\}$ **40.** $\{8\}$ **41.** $\{4\}$ **42.** $\{2\}$ **43.** $\{12\}$
44. $\{7\}$ **45.** $\left\{\frac{20}{3}\right\}$ **46.** $\{9\}$ **47.** $\{3\}$ **48.** $\{2\}$ **49.** $\{5\}$ **50.** $\left\{\frac{7}{3}\right\}$
51. $\{54\}$ **52.** $\left\{\frac{39}{2}\right\}$ **53.** $\{3\}$ **54.** $\{-49\}$ **55.** $\{36\}$ **56.** $\{-8\}$
57. $\{64\}$ **58.** $\{-50\}$ **59.** $\left\{\frac{76}{3}\right\}$ **60.** $\{-48\}$ **61.** $\left\{\frac{180}{7}\right\}$ **62.** $\left\{\frac{72}{17}\right\}$
63. $\left\{\frac{84}{13}\right\}$ **64.** $\left\{\frac{5}{8}\right\}$ **65.** $\left\{\frac{30}{11}\right\}$ **66.** $\left\{\frac{1}{7}\right\}$ **67.** $\left\{\frac{4}{9}\right\}$ **68.** $\left\{\frac{19}{12}\right\}$ **69.** $\left\{-\frac{9}{53}\right\}$
70. $\left\{\frac{5}{2}\right\}$ **71.** $\left\{-\frac{40}{9}\right\}$ **72.** $\left\{\frac{13}{4}\right\}$ **73.** $\left\{\frac{16}{13}\right\}$ **74.** $\left\{\frac{3}{10}\right\}$ **75.** $y = \frac{3x+4}{2}$
76. $x = \frac{5y-2}{3}$ **77.** $x = \frac{7y+16}{5}$ **78.** $y = \frac{3x+8}{5}$ **79.** $y = \frac{9-7x}{2}$
80. $x = \frac{3y-2}{2}$
81. identity; $\{x \mid x \text{ is a real number}\}$ **82.** contradiction; $\varnothing$
83. contradiction; $\varnothing$ **84.** identity; $\{x \mid x \text{ is a real number}\}$
85. contradiction; $\varnothing$ **86.** identity; $\{x \mid x \text{ is a real number}\}$
87. identity; $\{x \mid x \text{ is a real number}\}$ **88.** contradiction; $\varnothing$
89. $L = \frac{Rd^2}{k}$ **90.** $C = ID^2$ **91.** $h = \frac{v}{\pi r^2}$ **92.** $w = \frac{P-2l}{2}$ **93.** $h = \frac{V}{lw}$
94. $m = \frac{E}{c^2}$ **95.** $a = \frac{2d}{t^2}$ **96.** $I = \frac{9R}{E}$ **97.** $H = \frac{AP - S - 2D - 3T}{4}$
98. $h = \frac{2A}{b}$ **99.** $b = 2a - c$ **100.** $r = \frac{mv^2}{F}$ **101.** 5 m/s^2 **102.** 7 m/s^2
103. 4 kg **104.** 6 kg **105.** \$55.50 **106.** \$80
107. about 1,571 minutes **108.** 3,000 minutes **109.** 2,720 kg·m/s
110. 5,304 kg·m/s **111.** about 1.47 m/s **112.** about 5.15 m/s
113. 81.7 **114.** 87.3 **115.** 65 **116.** 95 **117.** 62.5 watts
118. 100 watts **119.** 9 ohms **120.** 1.6 ohms
121. Dividing is the same as multiplying by the reciprocal and subtracting is the same as adding the opposite.
122. You will always get 0 = 0 and not find the solution.
123. This could introduce solutions that weren't solutions before, or eliminate solutions that were solutions before.
124. Answers vary **125.** Answers vary

Exercise Set 6-3

5. $x - 3$ **6.** $x - 17$ **7.** $x + 9$ **8.** $6 + x$ **9.** $11 - x$ **10.** $x + 8$
11. $x - 9$ **12.** $x - 6$ **13.** $7 - x$ **14.** $7x$ **15.** $8 \cdot x$ **16.** $\frac{1}{2}x$
17. $3x + 5$ **18.** $\frac{3x}{6}$ or $3x \div 6$ **19.** $5x + 3$ **20.** $4x$ **21.** $2x$
22. $6x - 4$ **23.** $\frac{x}{14}$ or $x \div 14$ **24.** $\frac{x}{8}$ or $x \div 8$ **25.** 8 **26.** 9 **27.** 16
28. 16 and 26 **29.** 17 and 11 **30.** 3 **31.** 12 **32.** −16 **33.** 28
34. 32
35. There are 27 students in one section and 30 students in the other section.
36. The revenues were \$18 billion for Coca-Cola and \$29 billion for PepsiCo. **37.** \$15,713 **38.** Bill's age is 12 and Pete's age is 36.
39. The bill was \$58.62 in September and \$60.94 in October.
40. \$3,000 **41.** 18 games **42.** 8,754 in 2007 and 8,229 in 2008
43. \$270
44. 1.5 feet (or 18 inches), 2 feet (or 24 inches), and 2.5 feet (or 30 inches). **45.** \$1.25 billion on costumes and \$1.93 billion on candy
46. 60 miles **47.** 8 feet **48.** \$68,333.33 **49.** \$36,000
50. 17 female senators and 78 female members of the House of Representatives. **51.** \$40 **52.** \$3 million **53.** No
54. 4 feet, 4 feet, 7 feet **55.** \$563.20 **56.** \$437.50
57. The losing candidate received 206 votes and the winning candidate received 281 votes. **58.** \$57,142.84
59. wind chill = temperature − 1.5 × wind speed
60. The only solution is not a whole number.

Exercise Set 6-4

5. $\frac{18}{28} = \frac{9}{14}$ **6.** $\frac{5}{12}$ **7.** $\frac{14}{32} = \frac{7}{16}$ **8.** $\frac{40}{75} = \frac{8}{15}$ **9.** $\frac{12}{15} = \frac{4}{5}$ **10.** $\frac{18}{42} = \frac{3}{7}$
11. $\frac{3}{8}$ **12.** $\frac{32}{12} = \frac{8}{3}$ **13.** $\frac{60}{30} = \frac{2}{1}$ **14.** $\frac{12}{20} = \frac{3}{5}$ **15.** $x = \frac{135}{14}$ **16.** $x = 6$
17. $x = 35$ **18.** $x = 40$ **19.** $x = 10$ **20.** $x = 4$ **21.** $x = \frac{31}{3}$
22. $x = \frac{10}{3}$ **23.** $x = \frac{21}{4}$ **24.** $x = -2$ **25.** 15 cones **26.** 15 gallons
27. $\frac{14}{3}$ inches **28.** 21 students **29.** 4 gallons **30.** 119 students
31. 99 **32.** 12,000 miles **33.** 40 iPods **34.** 77 students
35. 8 sections **36.** $17\frac{1}{2}$ feet **37.** 20 professors **38.** \$2,000
39. 4 cans **40.** 72 **41.** \$1,200 **42.** 1,250 tickets **43.** 298 lb
44. 418 lb **45.** 117 lb **46.** 248 lb **47.** 24 lb **48.** 347.2 lb **49.** 3 lb
50. 119 lb **51.** 9 more students **52.** 48 decibels
55. 25 pounds for \$7.49 is a better buy.
56. 24 ounces for \$1.75 is a better buy.
57. 7 ounces for \$1.99 is a better buy.
58. 14 ounces for \$1.50 is a better buy.
59. 34.5 ounces for \$7.49 is a better buy.

Exercise Set 6-5

5. 3 **6.** −2
7. −4 **8.** 0
9. −9 **10.** 1
11. −3, 7 **12.** 4, 10
13. 2, 5 **14.** −3, 0
15. $\{x \mid x < 5\}$ 5
16. $\{x \mid x \le 17\}$ 17
17. $\{x \mid x \le 30\}$ 30
18. $\{x \mid x > 11\}$ 11
19. $\{y \mid y \ge 6\}$ 6
20. $\{y \mid y < 6\}$ 6
21. $\{x \mid x < -35\}$ −35
22. $\{x \mid x \ge -11\}$ −11
23. $\{x \mid x < 27\}$ 27
24. $\{x \mid x \ge 48\}$ 48
25. $\{t \mid t > -3\}$ −3
26. $\{t \mid t \le -4\}$ −4
27. $\{z \mid z \ge 12\}$ 12
28. $\{z \mid z < 9\}$ 9
29. $\{x \mid x \ge -8\}$ −8
30. $\{x \mid x < -10\}$ −10
31. $\{x \mid x \le 21\}$ 21
32. $\left\{x \mid x > -\frac{5}{6}\right\}$ $-\frac{5}{6}$
33. $\left\{y \mid y \le 8\frac{2}{5}\right\}$ $8\frac{2}{5}$
34. $\left\{y \mid y > 2\frac{2}{9}\right\}$ $2\frac{2}{9}$

35. $\left\{x \mid x \le -\frac{6}{5}\right\}$

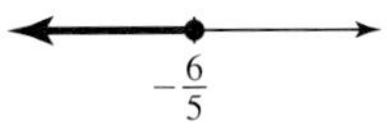

36. $\left\{x \mid x > -\frac{20}{3}\right\}$

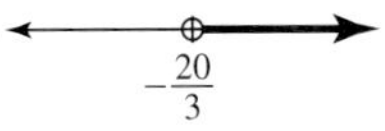

37. $\{x \mid x < 14\}$

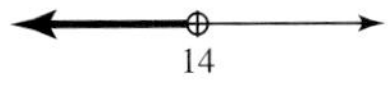

38. $\{n \mid n < 5\}$

39. $\left\{n \mid n \le -10\frac{3}{5}\right\}$

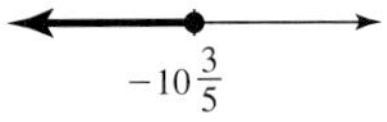

40. $\left\{x \mid x > -5\frac{5}{6}\right\}$

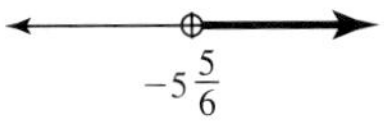

41. $\left\{x \mid x \ge -46\frac{1}{2}\right\}$

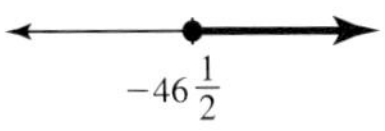

42. $\{x \mid x \le 15\}$

43. $\{x \mid x \ge -20\}$

44. $\left\{x \mid x < \frac{1}{11}\right\}$

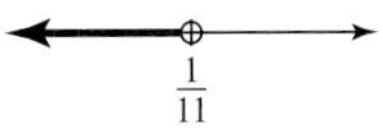

45. $\left\{x \mid x \le -\frac{13}{45}\right\}$

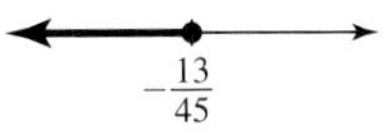

46. $\left\{z \mid z > -\frac{77}{43}\right\}$

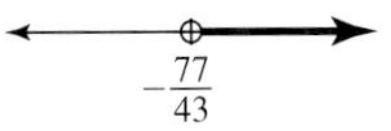

47. $\left\{t \mid t \ge -\frac{46}{65}\right\}$

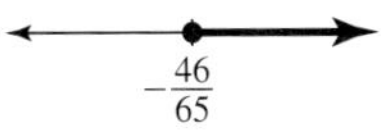

48. $\left\{x \mid x < -\frac{214}{445}\right\}$

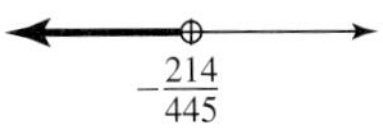

49. $\{x \mid -7 \le x < 8\}$

50. $\{y \mid -16 < y \le -12\}$

51. $\{y \mid -3 < y < 3\}$

52. $\{x \mid 5 \le x \le 9\}$

53. $\left\{z \mid -5 \le z < \frac{5}{4}\right\}$

54. $\left\{x \mid -2 < x \le -\frac{3}{2}\right\}$

55. $\left\{x \mid -1 < x < \frac{3}{2}\right\}$

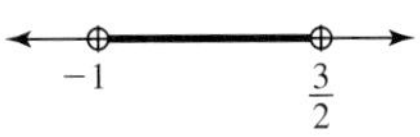

56. $\left\{t \mid -13 \le t \le \frac{11}{2}\right\}$

57. Texas, Florida, Kansas, Colorado
58. Maine, Delaware, Arizona, Maryland, California
59. Missouri, West Virginia
60. Arizona, Maryland, California, Pennsylvania
61. Maryland, California, Pennsylvania, Missouri, West Virginia
62. Colorado, Kansas, Florida **63.** at most $7,364.48
64. at least 93 **65.** 64% or greater **66.** at least $178,494.62
67. at most $18.86 **68.** 93 or better **69.** at least 312.5 miles
70. 23 holiday cards **71.** no more than 944 minutes
72. at least 27.8 hours **73.** at most $76,923.08 **74.** at most $787.50
75. at most 4 feet tall **76.** more than 20 hours **77.** 88% to 100%
78. between 65 and 265 calories **79.** between 20 and 30 hours
80. between $3,076.93 and $4,615.38
81. You can always move the variables to the side of the inequality that will leave the coefficient positive.

Exercise Set 6-6

7. $3x^2 + 2x - 5 = 0; a = 3, b = 2, c = -5$
8. $2x^2 + 7x - 4 = 0; a = 2, b = 7, c = -4$
9. $30x^2 - 10 = 0; a = 30, b = 0, c = -10$
10. $4x^2 - 8 = 0; a = 4, b = 0, c = -8$
11. $2x^2 + 5x = 0; a = 2, b = 5, c = 0$
12. $50x^2 + 100x = 0; a = 50, b = 100, c = 0$
13. $x^2 + 16x + 63$ **14.** $x^2 - 20x + 96$ **15.** $y^2 - 17y + 70$
16. $y^2 + 6y + 8$ **17.** $x^2 - 7x - 120$ **18.** $x^2 + 7x - 30$
19. $14x^2 - 67x + 63$ **20.** $16x^2 - 8x + 1$ **21.** $15z^2 - 19z - 56$
22. $6t^2 + t - 40$ **23.** $\{-2, -3\}$ **24.** $\{-4, -5\}$ **25.** $\{-4, 3\}$
26. $\{-2, 5\}$ **27.** $\{-3, 17\}$ **28.** $\{-4, 5\}$ **29.** $\{-27, 3\}$
30. $\{-4, 16\}$ **31.** $\{3, 5\}$ **32.** $\{2, 10\}$ **33.** $\left\{-3, \frac{7}{2}\right\}$ **34.** $\left\{-6, \frac{3}{5}\right\}$
35. $\left\{-\frac{4}{3}, \frac{3}{2}\right\}$ **36.** $\left\{-4, \frac{3}{4}\right\}$ **37.** $\left\{-\frac{4}{3}, \frac{3}{2}\right\}$ **38.** $\left\{-\frac{5}{2}, \frac{2}{5}\right\}$ **39.** $\left\{-2, \frac{3}{5}\right\}$
40. $\left\{-\frac{3}{5}, 6\right\}$ **41.** $\left\{-\frac{3}{2}, \frac{4}{3}\right\}$ **42.** $\left\{\frac{2}{3}, \frac{3}{2}\right\}$ **43.** $\{-3, 2\}$ **44.** $\{-1, -5\}$
45. $\left\{-\frac{1}{2}, -4\right\}$ **46.** $\left\{-\frac{2}{3}, 3\right\}$ **47.** $\left\{\frac{-1+\sqrt{13}}{6}, \frac{-1-\sqrt{13}}{6}\right\}$ **48.** $\left\{-\frac{1}{4}, 2\right\}$
49. $\left\{-\frac{3}{2}, 4\right\}$ **50.** $\left\{\frac{-5+\sqrt{73}}{2}, \frac{-5-\sqrt{73}}{2}\right\}$ **51.** $\left\{\frac{-5+\sqrt{13}}{6}, \frac{-5-\sqrt{13}}{6}\right\}$
52. $\left\{-1, \frac{3}{5}\right\}$ **53.** $\{-1, 9\}$ **54.** $\left\{-\frac{5}{2}, \frac{7}{3}\right\}$ **55.** $\left\{\frac{-5+\sqrt{37}}{2}, \frac{-5-\sqrt{37}}{2}\right\}$
56. $\left\{\frac{3+\sqrt{5}}{4}, \frac{3-\sqrt{5}}{4}\right\}$ **57.** $\left\{\frac{21+\sqrt{321}}{2}, \frac{21-\sqrt{321}}{2}\right\}$ **58.** $\left\{\frac{1+\sqrt{41}}{8}, \frac{1-\sqrt{41}}{8}\right\}$
59. $\left\{\frac{7+\sqrt{199}}{5}, \frac{7-\sqrt{199}}{5}\right\}$ **60.** $\left\{\frac{3+\sqrt{3{,}309}}{110}, \frac{3-\sqrt{3{,}309}}{110}\right\}$
61. $\left\{\frac{-13+\sqrt{73}}{4}, \frac{-13-\sqrt{73}}{4}\right\}$ **62.** $\left\{\frac{45+\sqrt{1{,}585}}{20}, \frac{45-\sqrt{1{,}585}}{20}\right\}$
63. $\left\{\frac{-8+7\sqrt{6}}{10}, \frac{-8-7\sqrt{6}}{10}\right\}$ **64.** $\left\{\frac{67+\sqrt{8{,}629}}{46}, \frac{67-\sqrt{8{,}629}}{46}\right\}$
65. 16 and 18 or −18 and −16 **66.** −13 and −12 or 12 and 13
67. 9 seconds
68. The base is 2 inches and the height is 8 inches.
69. 4 inches and 10 inches **70.** 10 bus routes and 15 subway stops
71. 9 square feet **72.** 2 feet **73.** 40 gift baskets
74. width 5 inches and length 20 inches
75. width 8 feet and length 24 feet **76.** 5 seconds **77.** 6 feet
78. 1.2 seconds **79.** 19.7 mph **80.** 3 inches **81.** 5 hours
82. 10.2 feet
83. horizontal distance 407 feet and vertical distance 419 feet
84. 30 miles **85.** Since $x = 1$, $x - 1 = 0$. Division by 0 is undefined.
87. Answers vary

Review Exercises

1. $-2x + 5y - 7$ **2.** $2x + 4y + 7$ **3.** $7x - 36$ **4.** $-21x - 30$
5. $13x - 21$ **6.** $3x + 31$ **7.** $x + 1$ **8.** $7x - 5$ **9.** 99 **10.** 5
11. −70 **12.** 37 **13.** 120 **14.** 4,200 **15.** $\{-10\}$ **16.** $\{6\}$
17. $\{20\}$ **18.** $\{-13\}$ **19.** $\{-13\}$ **20.** $\{17\}$ **21.** $\{67\}$ **22.** $\left\{\frac{73}{2}\right\}$
23. $\{-4\}$ **24.** $\{3\}$ **25.** $\{12\}$ **26.** $\{1\}$ **27.** $\left\{\frac{8}{5}\right\}$ **28.** $y = 4 - x$
29. $x = \frac{y+2}{3}$ **30.** $c = P - a - b$ **31.** $r = \frac{1}{pt}$ **32.** $y = \frac{xz}{z+x}$
33. $r = \frac{C}{2\pi}$ **34.** $W = \frac{P-2L}{2}$ **35.** $h = \frac{2A}{b}$ **36.** $8n - 4$ **37.** $6 \cdot 2n$ or $12n$
38. $4n + 3$ **39.** $n + 5$
40. $4\frac{4}{9}$ hours to get home and $3\frac{5}{9}$ hours to get back
41. 6 pounds of mixed soy nuts and 8 pounds of Asian trail mix.
42. 158 tickets at $8 each, 79 tickets at $10 each, and 89 tickets at $12 each were sold. **43.** 270 adults **44.** $30
45. 5 nickels, 20 dimes, 10 quarters **46.** $47 per couple
47. $\frac{82 \text{ miles}}{15 \text{ gallons}}$ **48.** $\frac{16 \text{ ounces}}{\$2.37}$ **49.** $\frac{4 \text{ months}}{24 \text{ months}} = \frac{1}{6}$ **50.** $\frac{18 \text{ minutes}}{120 \text{ minutes}} = \frac{3}{20}$
51. $x = 9$ **52.** $x = 8$ **53.** $x = 72$ **54.** $x = 13.64$ **55.** 750 calories
56. 34 people **57.** wife $10,800; son $7,200 **58.** 25.5 **59.** $4,320
60. 9 pints **61.** 4.44 amps **62.** $450 **63.** $x > 10$ **64.** $x \le -20$
65. $y > 7$ **66.** $x > -21$ **67.** $t \le -\frac{93}{4}$ **68.** $x \ge -\frac{27}{10}$

69. up to 28 months **70.** at most 14 potted plants **71.** $\{-2, 3\}$
72. $\{-13, 2\}$ **73.** $\{-3, 7\}$ **74.** $\left\{\frac{1}{2}, -3\right\}$ **75.** $\left\{-1, \frac{2}{3}\right\}$ **76.** $\left\{\frac{3}{2}, 3\right\}$
77. $\left\{\frac{5+\sqrt{53}}{2}, \frac{5-\sqrt{53}}{2}\right\}$ **78.** $\left\{\frac{7+\sqrt{129}}{10}, \frac{7-\sqrt{129}}{10}\right\}$
79. $\left\{\frac{-7+\sqrt{17}}{8}, \frac{-7+\sqrt{17}}{8}\right\}$ **80.** $\left\{\frac{2+\sqrt{11}}{3}, \frac{2-\sqrt{11}}{3}\right\}$
81. $\left\{\frac{-7+\sqrt{69}}{4}, \frac{-7-\sqrt{69}}{4}\right\}$ **82.** $\left\{\frac{6+\sqrt{11}}{5}, \frac{6-\sqrt{11}}{5}\right\}$
83. -12 and -11 or 11 and 12 **84.** 8 seconds

Chapter 6 Test

1. $5x - 10y + 5$ **2.** $7x - 40$ **3.** 91 **4.** 85,716 **5.** $\left\{\frac{9}{7}\right\}$ **6.** $\left\{-\frac{83}{2}\right\}$
7. $r = \frac{mv^2}{F}$ **8.** $y = \frac{10-3x}{2}$ **9.** $\left\{x \mid x \le -\frac{3}{2}\right\}$ **10.** $\{y \mid y > -6\}$
11. $x = 4$ **12.** $x = 35$ **13.** $2x^2 - 13x - 24$ **14.** $12x^2 - 41x + 35$
15. $\{-3, 17\}$ **16.** $\{-13, 1\}$ **17.** $\left\{-\frac{3}{2}, \frac{4}{3}\right\}$ **18.** $\left\{\frac{1+\sqrt{13}}{6}, \frac{1-\sqrt{13}}{6}\right\}$
19. $\left\{-1, \frac{3}{5}\right\}$
20. \$1,500 was invested at 4% and \$3,500 was invested at 6%
21. \$2,500 was invested at 6% and \$500 was invested at 8%
22. 8 jump drives **23.** 21.9 hours **24.** 200 vibrations per second
25. 12 hours **26.** -26 and -24 or 24 and 26
27. The width is 8 inches and the length is 12 inches.

CHAPTER 7: ADDITIONAL TOPICS IN ALGEBRA

Exercise Set 7-1

7.
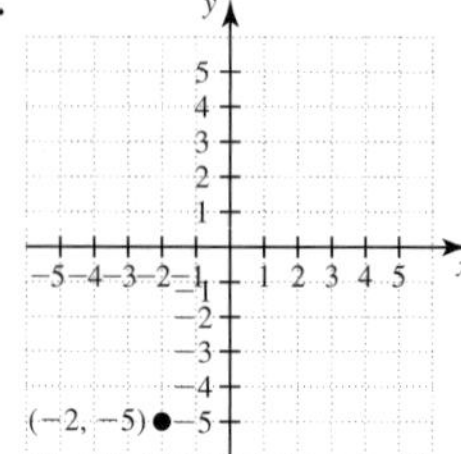

8.
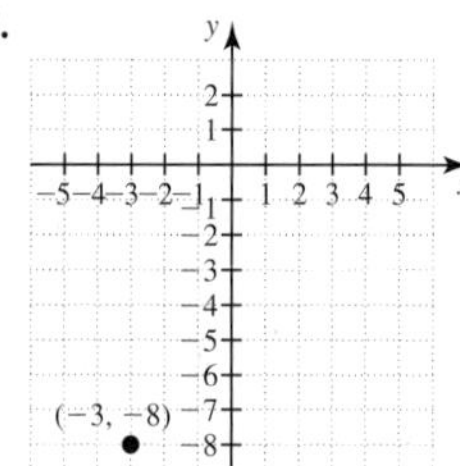

9.
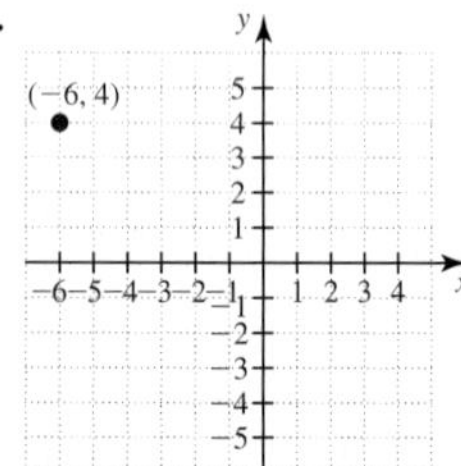

10.
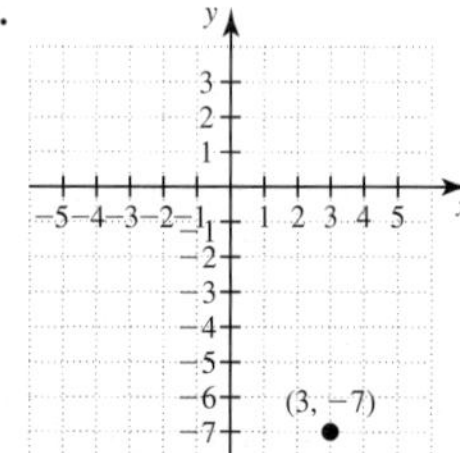

11.
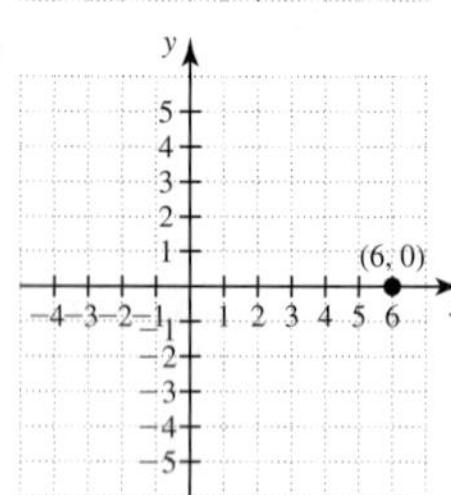

12.
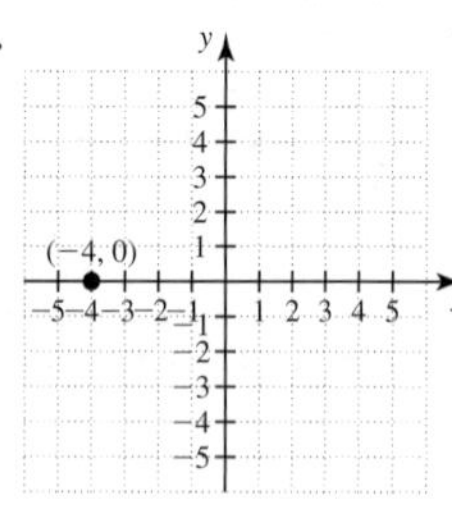

13.
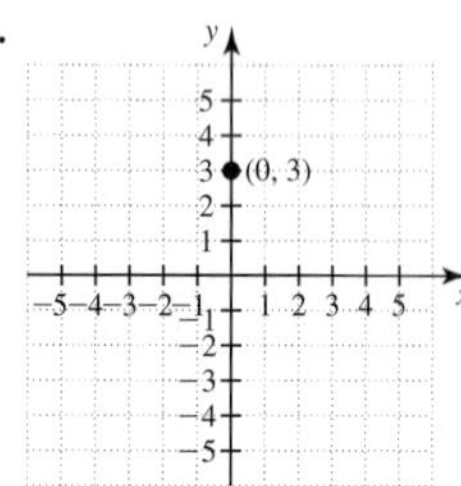

14.
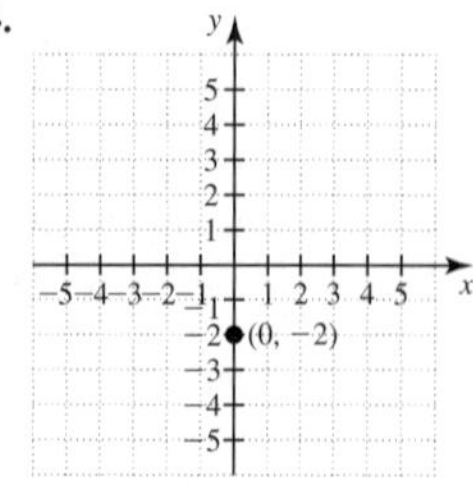

15.
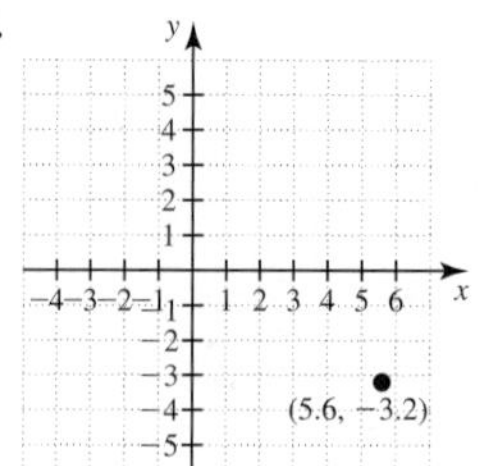

16.
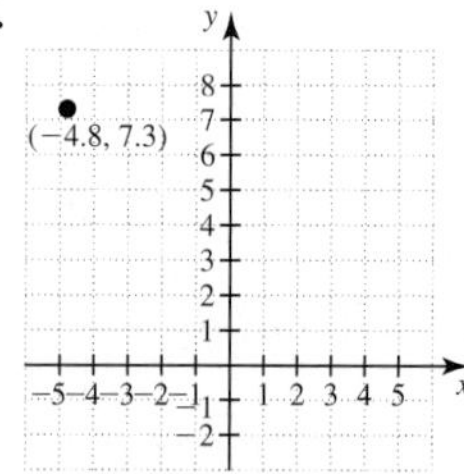

17.
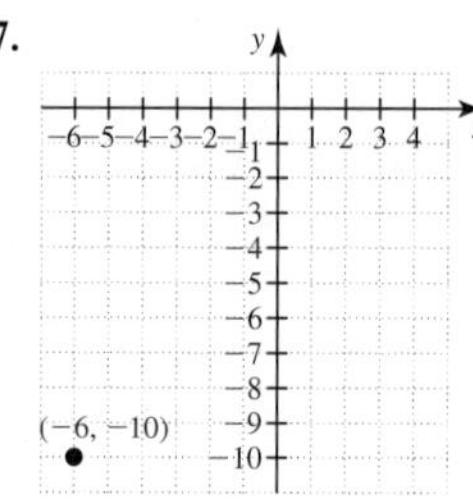

18.
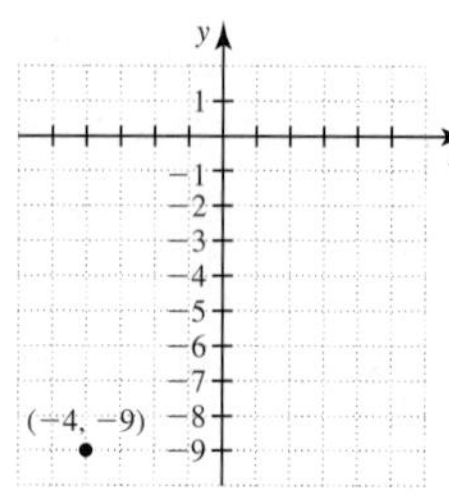

19.
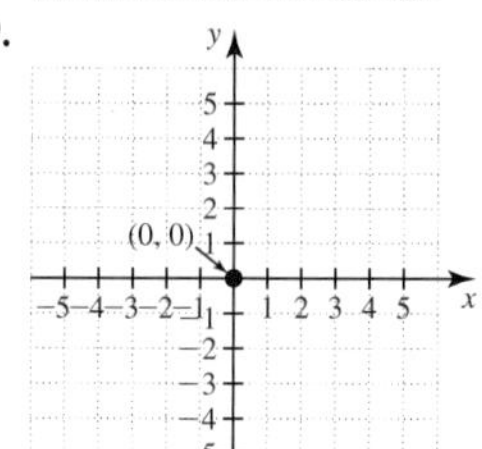

20.
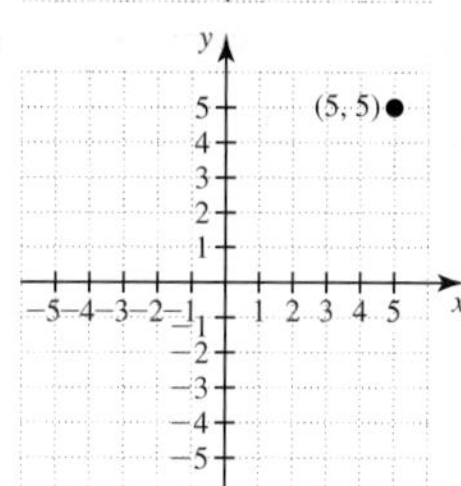

21.
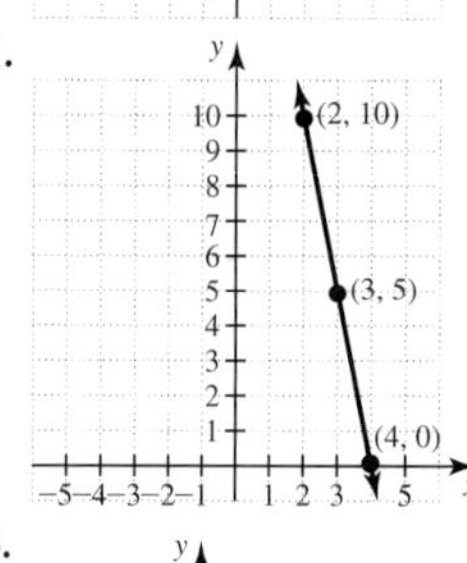

22.
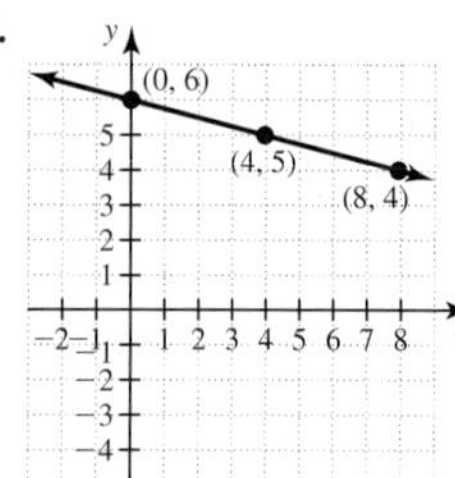

23.
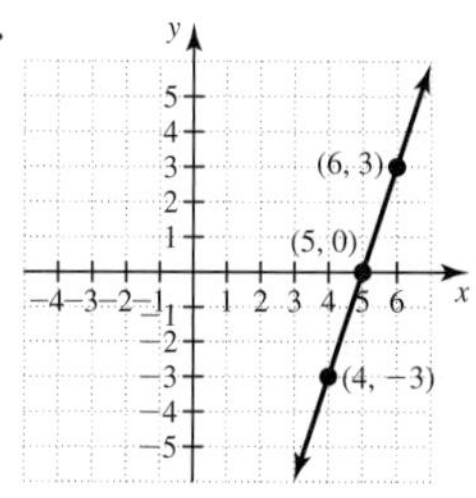

24.
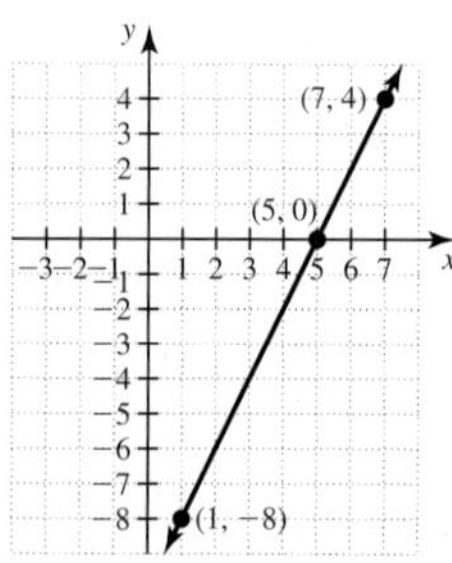

25.
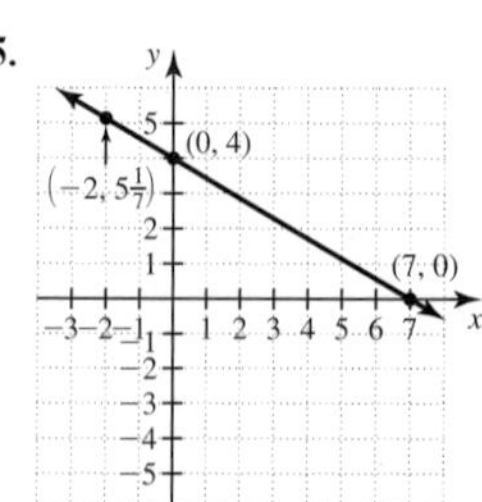

26.
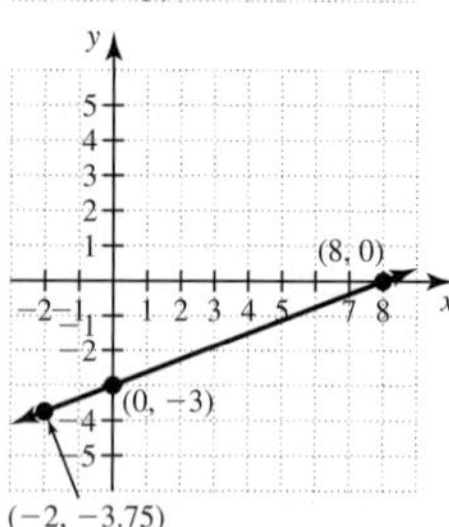

27.
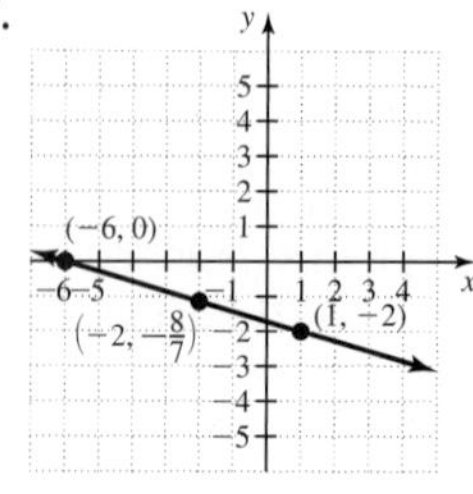

28.
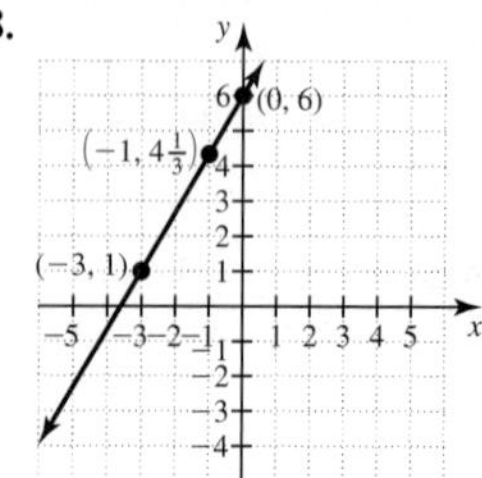

29.

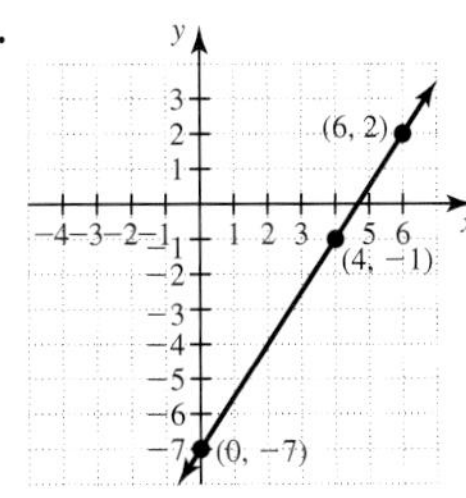

30.

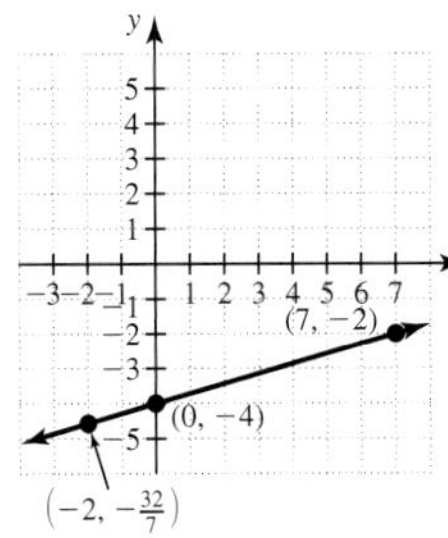

31. 1 **32.** 5 **33.** −0.5 **34.** $-\frac{25}{18}$ **35.** $-\frac{3}{5}$ **36.** $-\frac{43}{16}$ **37.** $\frac{33}{14}$
38. $-\frac{7}{4}$ **39.** x intercept = (8, 0); y intercept = (0, 6)
40. x intercept = (14, 0); y intercept = (0, −4)
41. x intercept = (−6, 0); y intercept = (0, −5)
42. x intercept = (10, 0); y intercept = $(0, \frac{5}{3})$
43. x intercept = (9, 0); y intercept = (0, −18)
44. x intercept = (−4, 0); y intercept = (0, −9)
45. x intercept = (3, 0); y intercept = (0, −7.5)
46. x intercept = (2, 0); y intercept = $\left(0, -\frac{18}{7}\right)$

47. $y = -\frac{7}{5}x + 7$

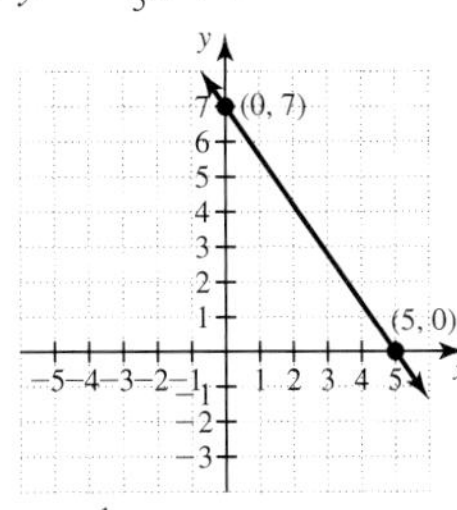

48. $y = \frac{1}{4}x - 4$

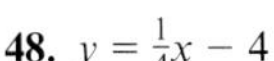
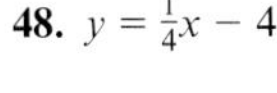

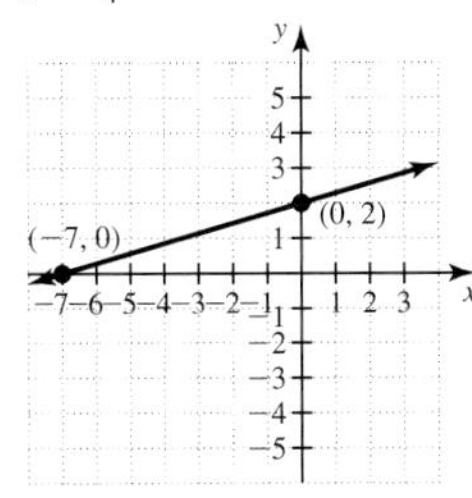

49. $y = \frac{1}{4}x - 4$

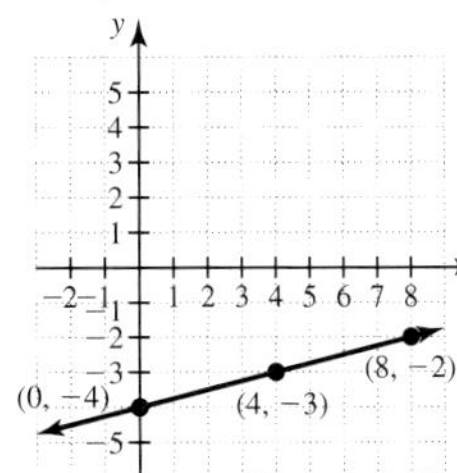

50. $y = \frac{1}{2}x - \frac{15}{8}$

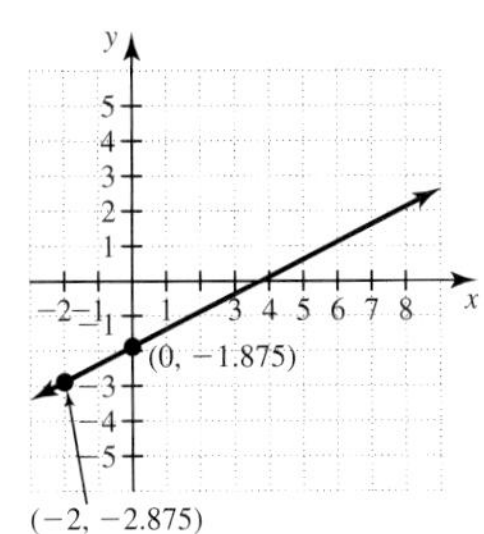

51. $y = \frac{8}{3}x - 8$

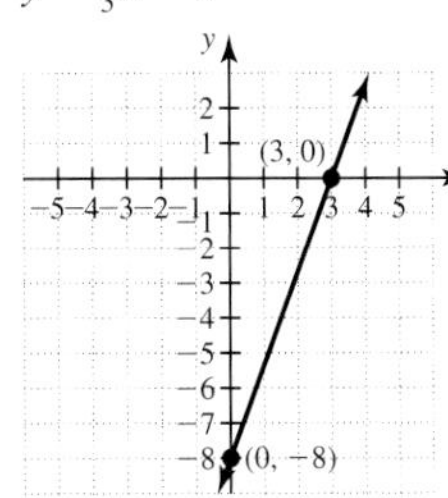

52. $y = \frac{3}{7}x - 2$

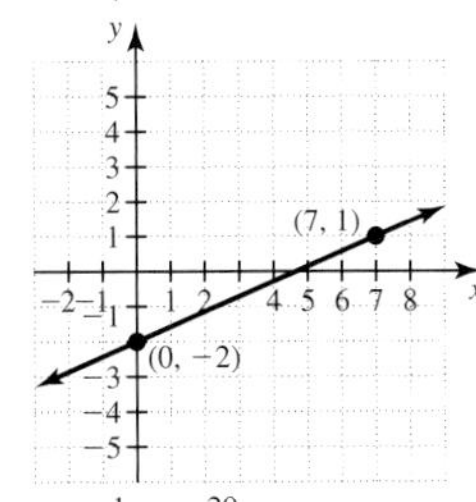

53. $y = 2x - 19$

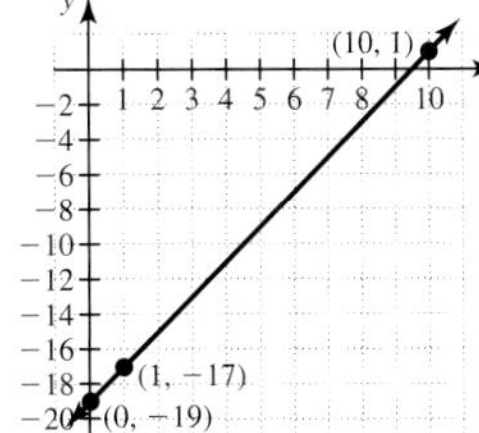

54. $y = \frac{1}{3}x - \frac{20}{9}$

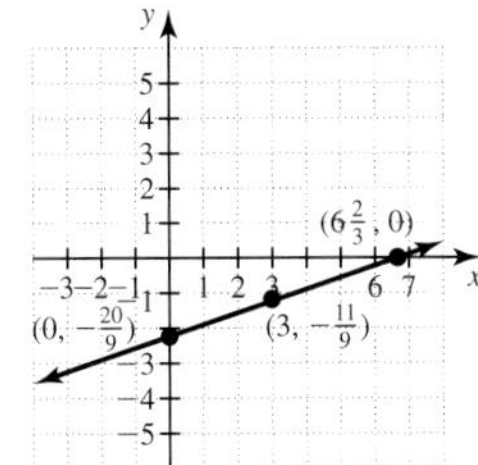

55.

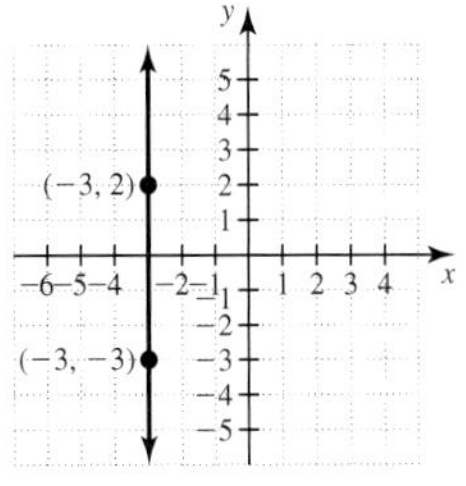

56.

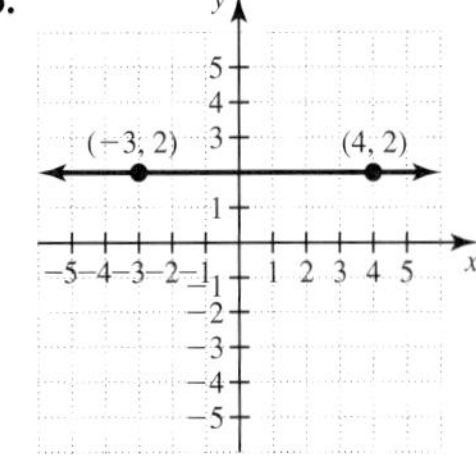

57.

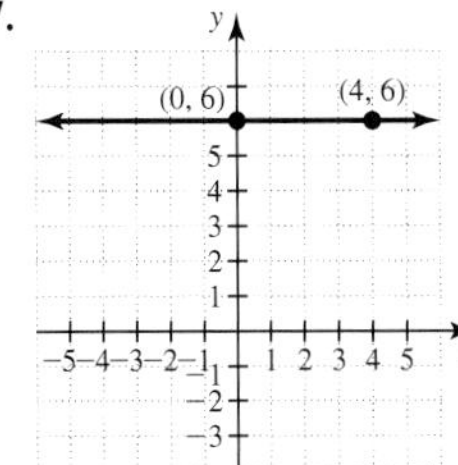

58.

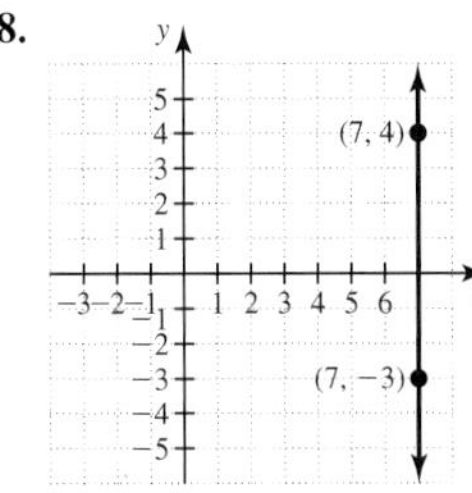

59. (a) \$69.50 (b) \$82.50 (c) \$115 **60.** (a) \$250 (b) \$330 (c) \$410 **61.** (a) \$109.30 (b) \$86.20 (c) \$179.70
62. 240 billion **63.** 31.5 thousand or 31,500 **64.** 2.4%
65. 37.8 million **66.** 13.6% **67.** \$14.64 **68.** 157,003 women
69. (a) \$39.97 (b) \$54.94 (c) \$144.76 **70.** 1,936,000 students
71. $y = 40 + 20x$; 23 hours **72.** $C = 12{,}000 + 1.73x$; \$13,730
73. $w = 160 - 3x$; 10 months **74.** $y = 2{,}000{,}000 + 0.5x$; \$2,125,000.
75. $y = 350 - 5x$; 70 years
76. $y = 1{,}250 + 12x$; 729 days, or almost 2 years
77. The denominator of the fraction for the slope will always be zero.
78. The numerator of the fraction for the slope will always be zero.

Exercise Set 7-2

7. (3, 4) is a solution; (5, 10) is not a solution
8. (4, −1) is not a solution; (−2, 2) is a solution
9. (11, −6) is not a solution; (1/2, 3) is a solution
10. (2, 2/3) is a solution; (3, 3) is not a solution
11. (−10, 25) is a solution; (5, 35) is not a solution
12. (4, 18) not a solution; (−8, 10) is a solution **13.** (7, 0)
14. (3, −1) **15.** (−2, 0) **16.** (−1, 3) **17.** (2, −2)
18 (0, 0) **19.** (6, −2)
20. $\{(x, y) \mid 3x - 5y = -2\}$. The system is dependent.
21. (−3, 3) **22.** (1, 1) **23.** (4, −1) **24.** (5, 3)
25. $\left(\frac{51}{11}, \frac{16}{11}\right)$
26. The system is dependent. The solution set is $\{(x, y) \mid 3x = 5y + 16\}$. **27.** (5, 3) **28.** (−8, 5)
29. $\left(\frac{76}{21}, \frac{3}{7}\right)$ **30.** $\left(\frac{2}{9}, -\frac{80}{9}\right)$ **31.** (2, −1)
32. The system is inconsistent. The solution set is ∅.
33. The system is dependent. The solution set is $\{(x, y) \mid 5x - 2y = 11\}$. **34.** (8, 2) **35.** (4, 2)
36. $\left(-\frac{23}{11}, -\frac{8}{11}\right)$ **37.** consistent; (1, −2) **38.** consistent; $\left(-\frac{9}{17}, -\frac{19}{17}\right)$
39. consistent; (7, −2) **40.** consistent; $\left(\frac{47}{19}, -\frac{21}{19}\right)$
41. consistent; $\left(-\frac{5}{2}, -\frac{11}{2}\right)$ **42.** consistent; (8, 2)
43. consistent; $\left(\frac{41}{19}, -\frac{45}{19}\right)$
44. The system is dependent. The solution set is $\{(x, y) \mid 4x - y = 3\}$.
45. \$220 a week at Job A and \$350 a week at Job B; \$259 into checking
46. 8 pounds of \$3.20 coffee and 12 pounds of \$5.40 coffee
47. 25 adults and 15 children
48. \$150 on Airline 1 and \$175 on Airline 2
49. about 10.2 pounds of chicken and about 11.9 pounds of filet mignon
50. 5 pounds of each type **51.** sandwich \$2.19, fries \$1.15
52. \$1,500 at 9% and \$900 at 6%.
53. 322 students attending, 178 general admission
54. 12 CDs and 6 VCR tapes
55. The tops cost \$4 and the pants cost \$6.
56. The smaller number is 5 and the larger number is 17.
57. \$22 **58.** movie \$20; music video \$4.50

59. 200 carnations and 600 bagels
60. 45 feet of poplar and 155 feet of mahogany
61. paperbacks \$7.22; hard cover books \$15.15
62. 4 hours at 65 mph and 3 hours at 75 mph.
63. Answers vary **64.** (1, 2, 3)
65. Substitution: You have to divide by coefficients to solve and are more likely to end up with fractions.
66. When using graphing, sometimes only an approximate answer can be obtained.

Exercise Set 7-3

5. row echelon form **6.** not in row echelon form
7. not in row echelon form **8.** not in row echelon form
9. row echelon form **10.** not in row echelon form
11. not in row echelon form **12.** row echelon form
13. (2, 1) **14.** (−5, 2) **15.** (3, 7) **16.** (−1, 1)
17. (2, −1, 1) **18.** (3, −1, 2) **19.** (−1, 5, 3)
20. $\left\{\left(\frac{86-11t}{27}, \frac{4t-19}{9}, t\right)\right\}$ for t any real number
21. $\left\{\left(\frac{3t+1}{13}, \frac{11t+21}{13}, t\right)\right\}$ where t is any real number **22.** (1, 3, 2)
23. (1, 1, −1) **24.** $\left(\frac{37}{28}, -\frac{23}{14}, \frac{3}{4}\right)$
25. 10 adult tickets, 22 child tickets, 7 senior citizen tickets
26. 0 nickels, 35 dimes, 2 quarters
27. lawn tickets cost \$40, general admission costs \$80 and ground floor seating costs \$150
28. binders cost \$5.38, graphing calculators \$84.88, and backpacks \$10.41
29. \$500 in Fund A, \$1,200 in Fund B, and \$300 in Fund C
30. 8 college algebra books, 30 liberal arts math books, and 10 intermediate algebra books
31. won 9 games, lost 5 games, tied 2 games
32. \$12 on blackjack, \$20 on Texas hold 'em, \$8 on roulette
33. 100 watched beam, 300 watched floor, 250 watched uneven bars
34. 5 pounds of cashews, 7 pounds of almonds, 3 pounds of pecans
35. Answers vary
36. (a) The solution of the system can be read easily from the last column of the matrix.

Exercise Set 7-4

5.

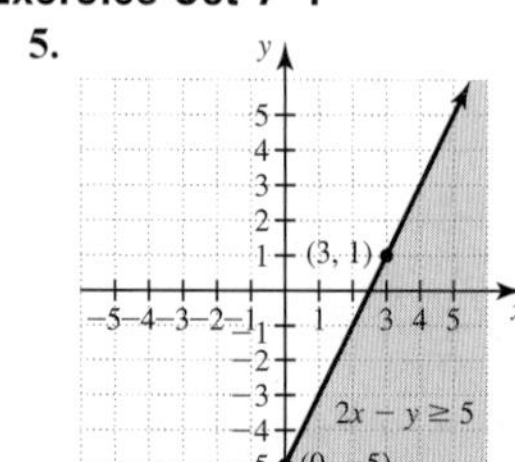

6.

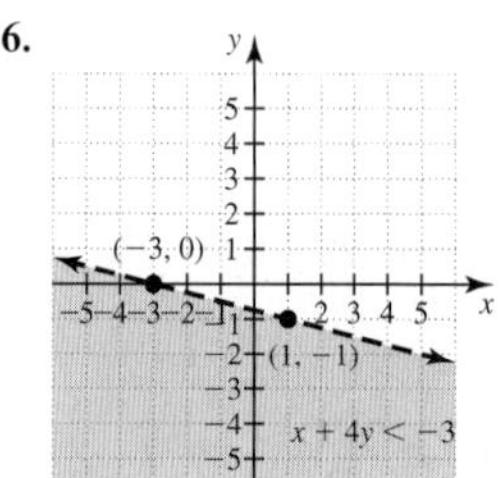

7.

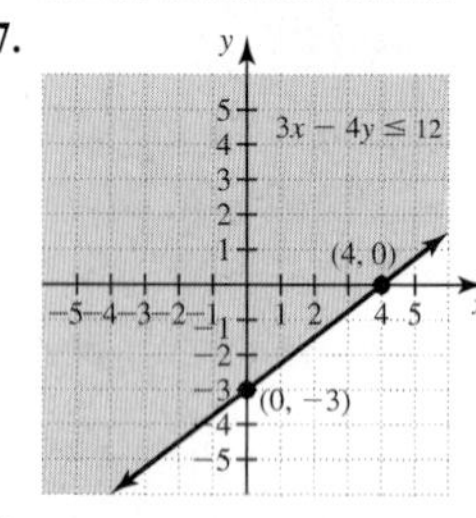

8.

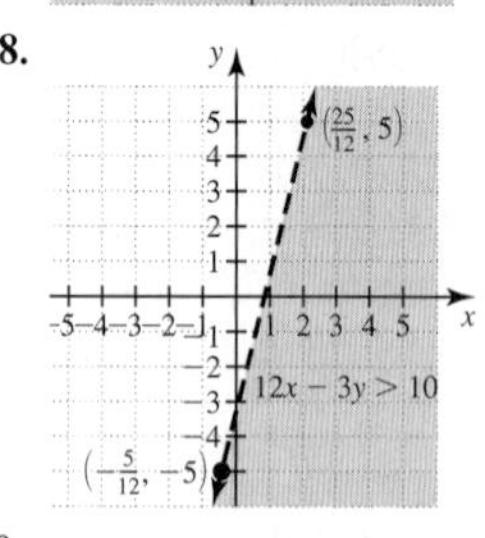

9.

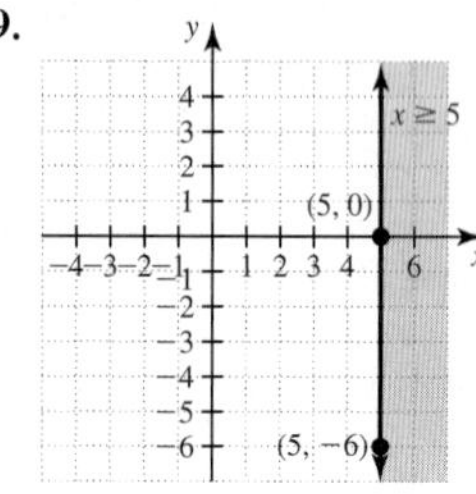

10.

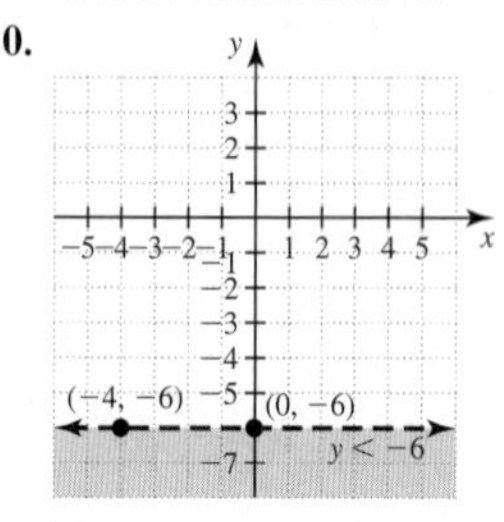

11.

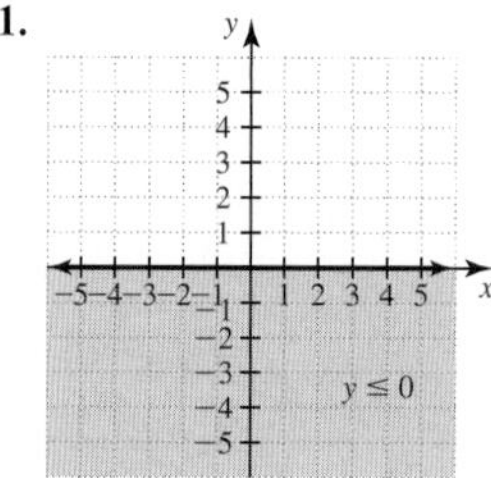

12.

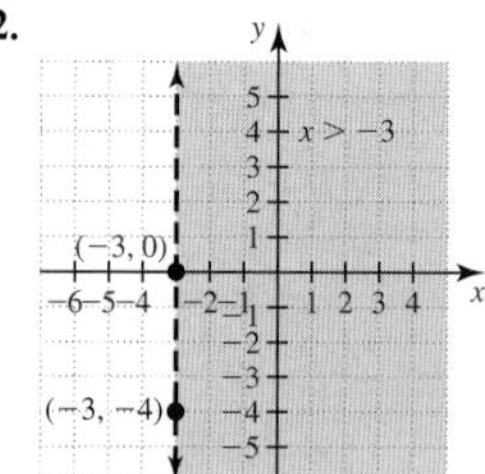

13.

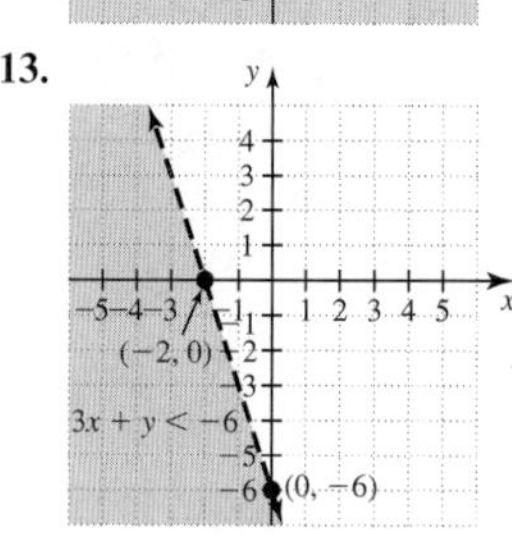

14.

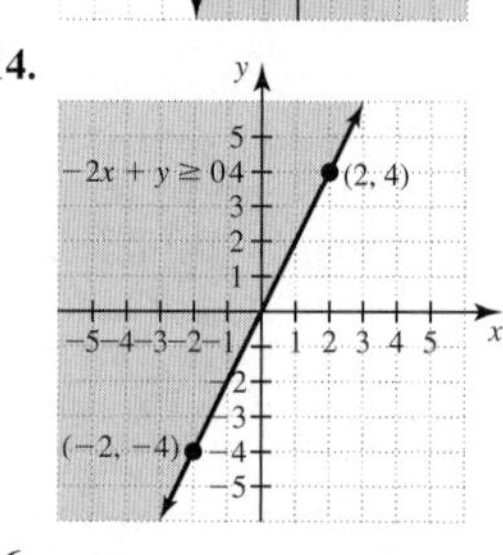

15.

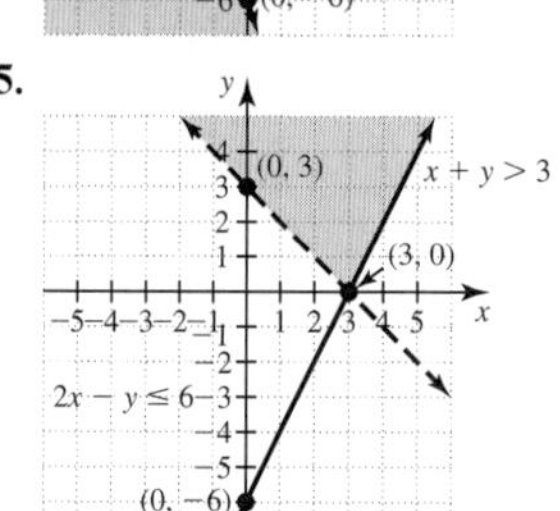

16.

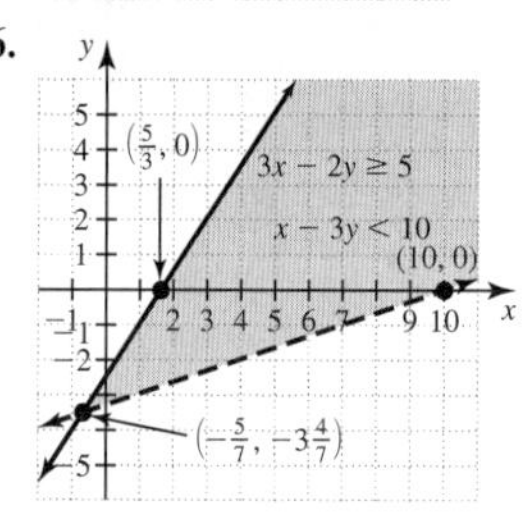

17.

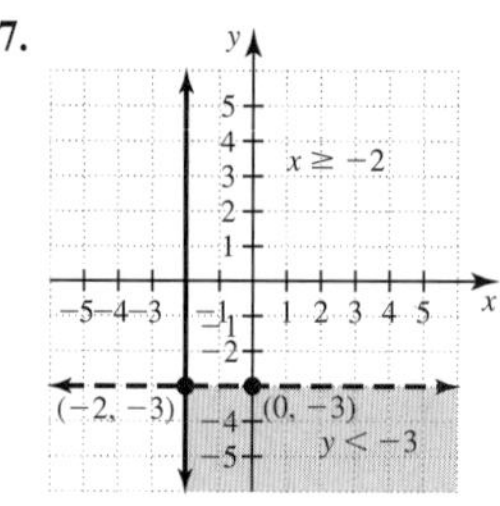

18.

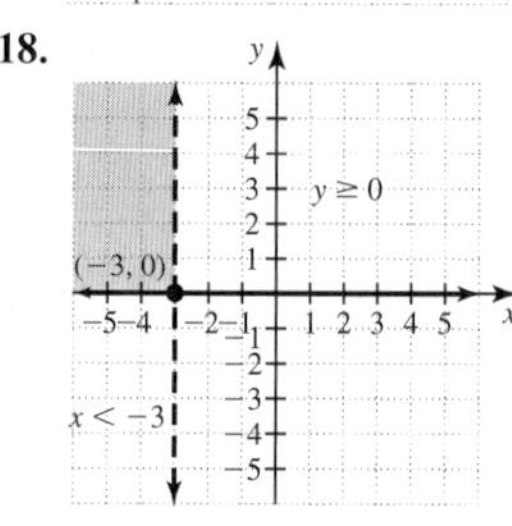

19.

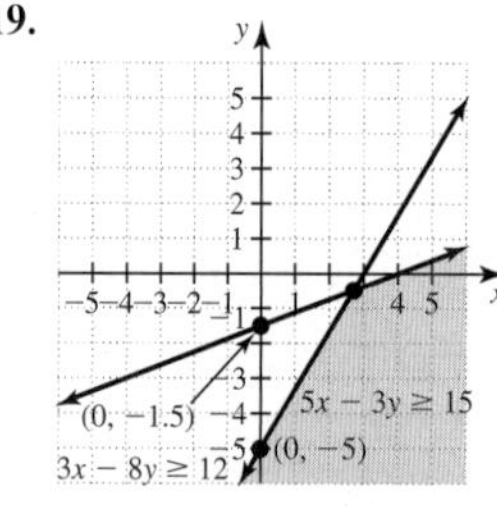

20.

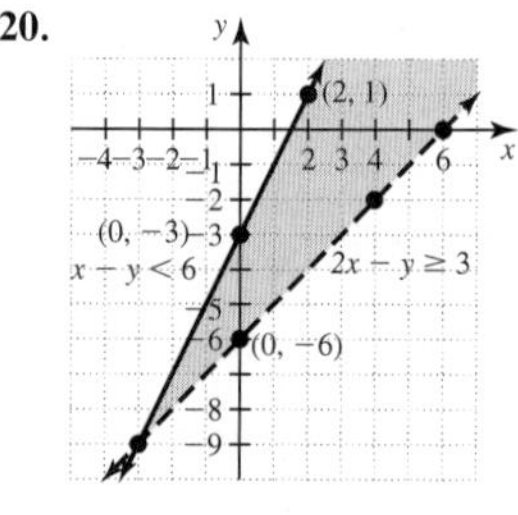

21.

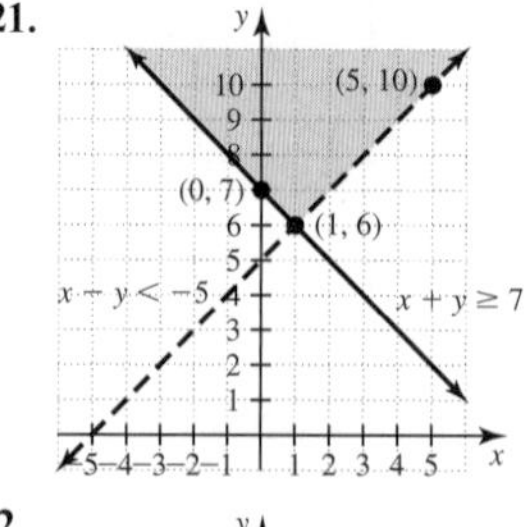

22.

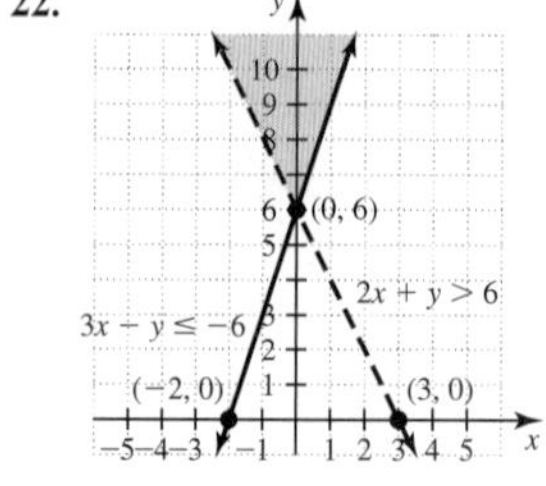

23.

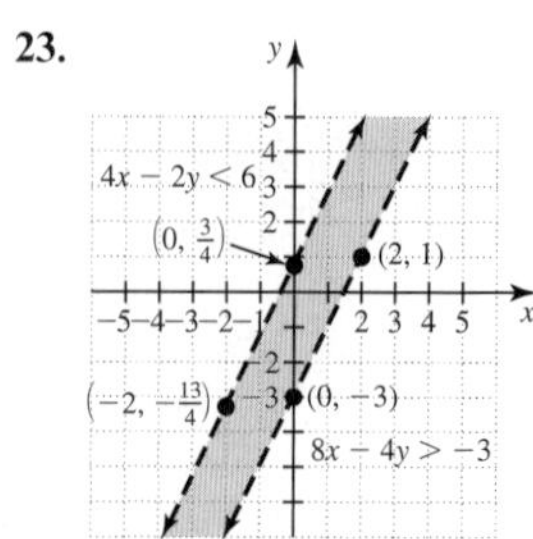

24.

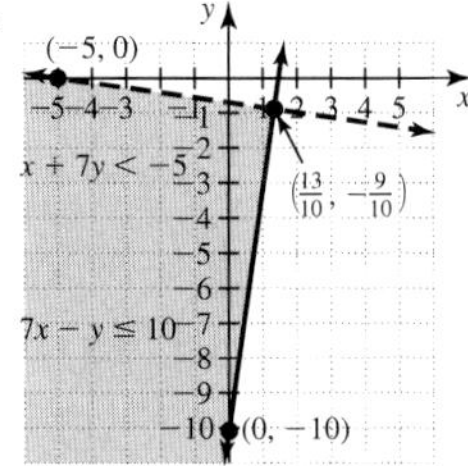

25. x = number of computational problems, y = number of word problems.

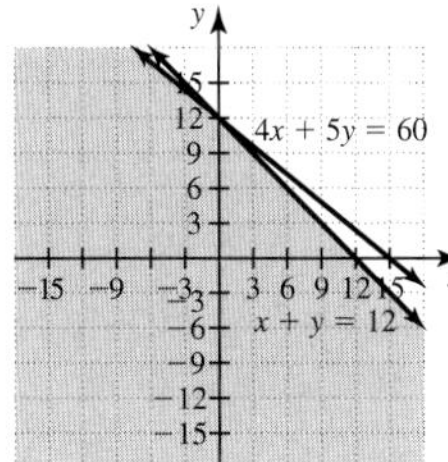

26. w = number of wraps, p = number of pita chips.

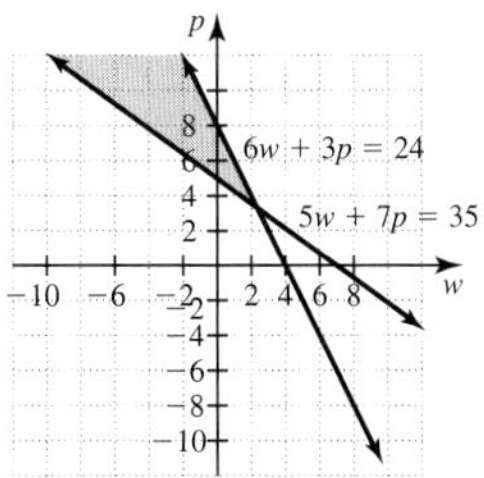

27. L = tubes of body lotion, S = number of sponges.

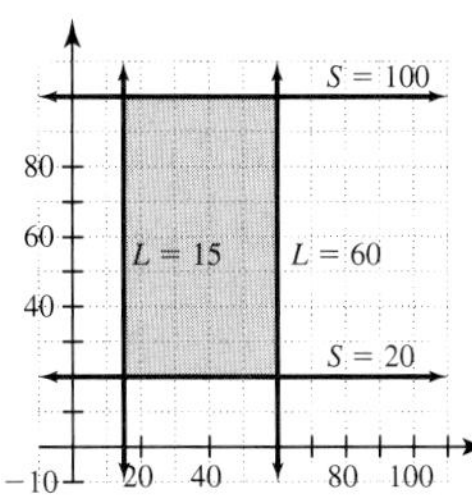

28. A = number of model A units, B = number of model B units.

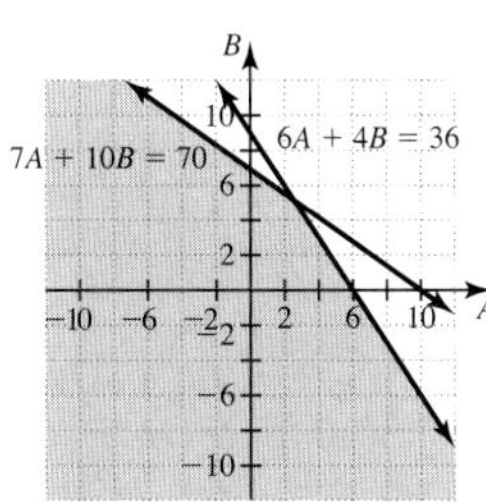

29. g = number of graduate classes, u = number of undergraduate classes.

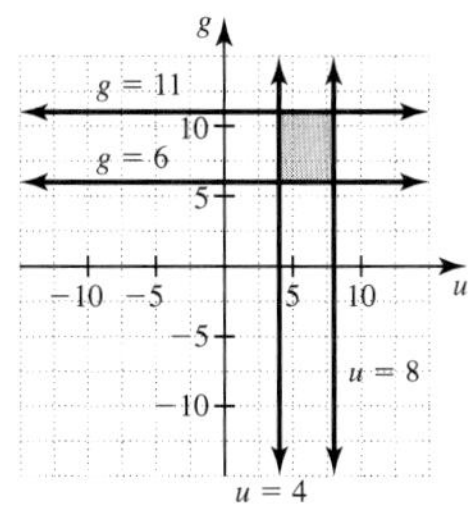

30. S = number of ski lodge models, G = number of golf view models.

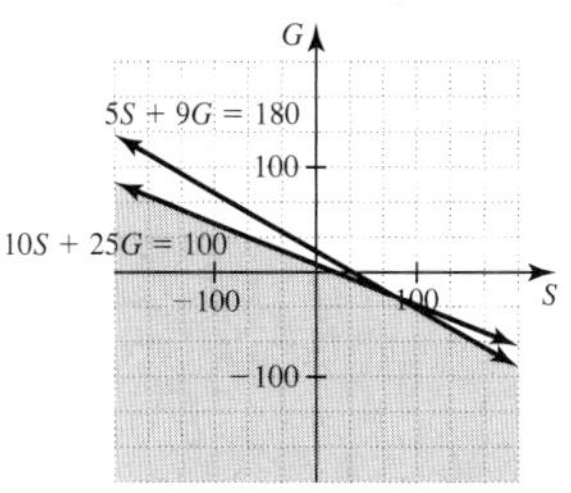

31. x = number of workers at first factory, y = number of workers at second factory.

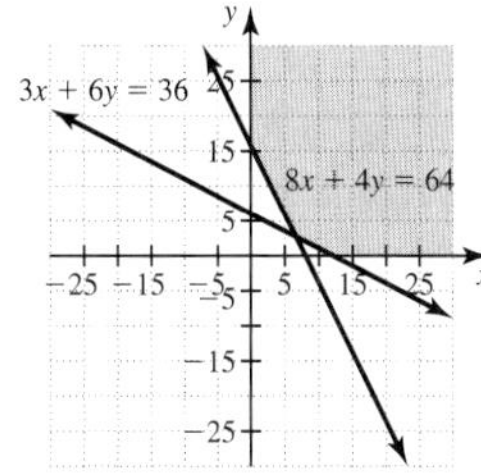

32. C = number of cherry-finish cabinets, W = number of walnut-finish cabinets.

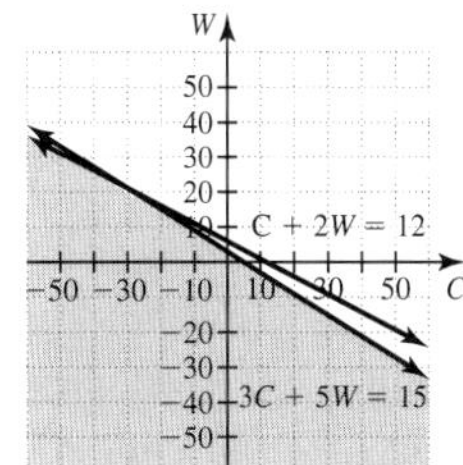

33. C= number of civil suits, M = number of malpractice suits.

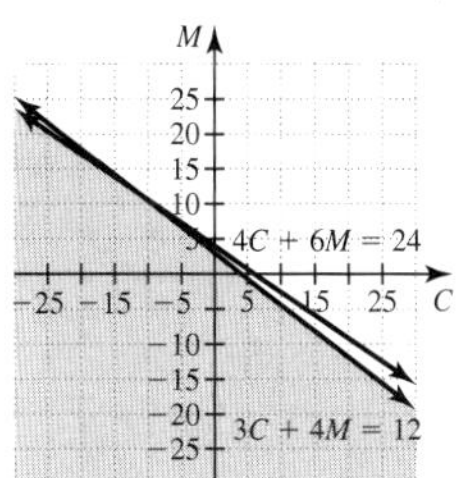

34. C = number of coach passengers, F = number of first-class passengers.

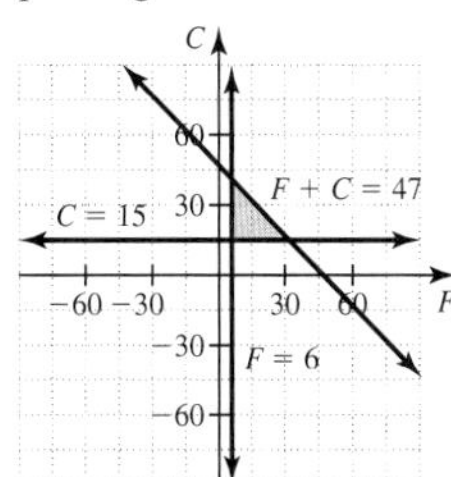

35. When the signs of both inequalities are a combination of $\geq$ or $\leq$.
36. Yes; the half planes may not intersect.
37. The solution is $\varnothing$.

Exercise Set 7-5

5. $x \leq 50$; x = number of calculators
6. $P = 10s + 12.5g$; s = number of scientific calculators sold and g = number of graphing calculators sold
7. $x = 2y$; x = number of pens and y = number of pencils
8. $d \geq l + 20$; d = number of desktop computers and l = number of laptop computers

9. $P = 85d + 130l$; d = number of desktop computers sold and l = number of laptop computers sold
10. $t + h \le 50$; t = number of turkeys and h = number of hams
11. $P = 4t + 3h$; t = number of turkeys sold and h = number of hams sold
12. $t \le h + 10$; t = number of turkeys purchased and h = number of hams purchased
13. $2v + 3c \le 6$; v = number of VCRs assembled and c = number of CD players assembled
14. $P = 32c + 45v$; c = number of CD players sold and v = number of VCRs sold
15. (5, 6) gives a maximum profit of $390
16. (7, 2) gives a maximum profit of $859
17. (7, 0) gives a maximum profit of $105
18. (0, 10) gives a maximum profit of $200
19. (0, 62) gives a maximum profit of $21,080
20. (20, 13) gives a maximum profit of $2,425
21. 20 television sets and 10 VCRs
22. 5 cakes and 25 pies
23. 50 hats and 0 scarves
24. 0 train sets and 10 trucks
25. 18 square feet of tomatoes and 12 square feet of squash
26. There should be 80 student seats reserved
27. 4 cats and 26 dogs
28. 400 Biolage packages and 200 Rusk packages
29. 100 laptops and 200 desktops
30. 500 student and 300 general admission tickets
31. Answers vary
32. Answers vary

Exercise Set 7-6

7. Function **8.** Not a function **9.** Function **10.** Not a function
11. Function **12.** Function **13.** Function **14.** Not a function
15. Function **16.** Not a function **17.** $f(3) = 17; f(-2) = 2$
18. $f(-5) = -5; f(2) = 1$ **19.** $f(10) = 32; f(-10) = -48$
20. $f(2) = 25; f(-4) = 109$ **21.** $f(0) = 0; f(6) = 306$
22. $f(1.5) = -1.75; f(-2.1) = -8.23$
23. $f(-3.6) = 5.56; f(4.5) = 45.25$ **24.** $f(0) = -3; f(-20) = -523$
25. Domain $\{x \mid -\infty \le x < \infty\}$; Range $\{y \mid y \ge 0\}$
26. Domain $\{x \mid x \le 3\}$; Range $\{y \mid y \ge 0\}$
27. Domain $\{x \mid x \ne 2\}$;Range $\{y \mid y \ne 1\}$
28. Domain: $\{x \mid -\infty \le x < \infty\}$; Range: $\{y \mid -\infty \le y < \infty\}$
29. Domain $\{x \mid x \ge 0\}$; Range $\{y \mid y \ge 3\}$
30. Domain $\{x \mid x \ge 0\}$; Range $\{y \mid y \le 0\}$
31. Domain: $\{x \mid -\infty \le x < \infty\}$; Range: $\{y \mid -\infty \le y < \infty\}$
32. Domain $\{x \mid x \ne 0\}$; Range $\{y \mid y \ne 2\}$
33. Domain $\{x \mid x \ge -1\}$; Range $\{y \mid y \le -1\}$
34. Domain $\{x \mid -\infty \le x < \infty\}$;Range $\{y \mid y \ge 0\}$
35. Function **36.** Not a function **37.** Not a function
38. Function
39. $88.08 for a 4-hour shift and $98.16 for an 8-hour shift
40. $10,800 **41.** No, they lose $128; $2,422 **42.** 7
43. $c(x) = 20x$; $100 **44.** $I(x) = 250 + 0.10x$; $750
45. $z(m) = 3m - 10$; $125 **46.** $A(n) = \frac{234.56}{n}$; $39
47. $d(t) = 80 + 65t$; three more hours
48. $p(m) = 500 + 0.40m$; 25 miles
49. $n(r) = 3r$; 135 minutes, or 2 hours and 15 minutes
50. $w(j) = 0.8j$; 96 calories **51.** $B(W) = 2W + 4$; 4 games
52. $A = 3M - 5$; 13 right **53.** $s(p) = 0.8p$; $95.82
54. $p(x) = \frac{15.40}{0.07x}$; $0.11 per unit **55.** $n(x) = \frac{2}{3}x + 4$; 24 cookies
56. $T(s) = 140.28 + 1.59s$; $155.44 **57.** $d(x) = 48 + 7.5x$; $20
58. $w(x) = 0.9x - 5$; 157 pounds
59. $f(3) = 19; f(-3) = 1$;
$f(x + 3) = x^2 + 9x + 19$;
$f(x - 3) = x^2 - 3x + 1$
60. Answers vary

Exercise Set 7-7

9.
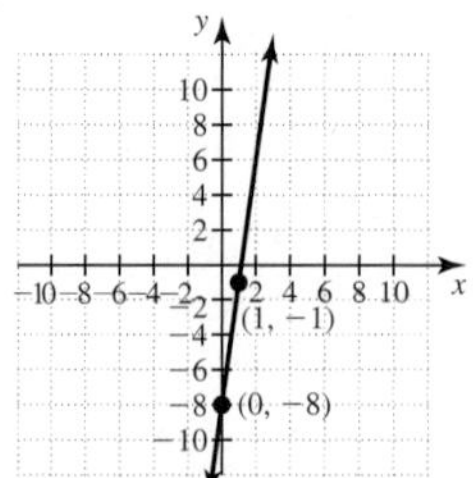

10.
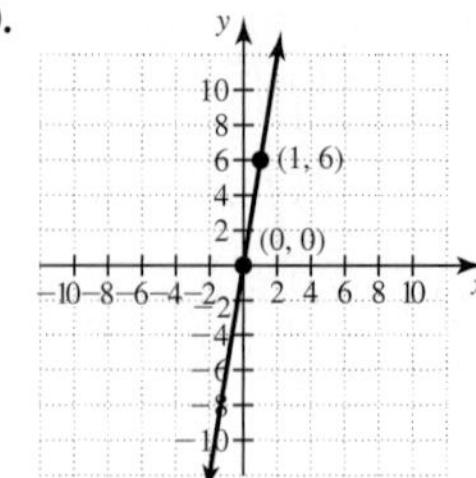

11.
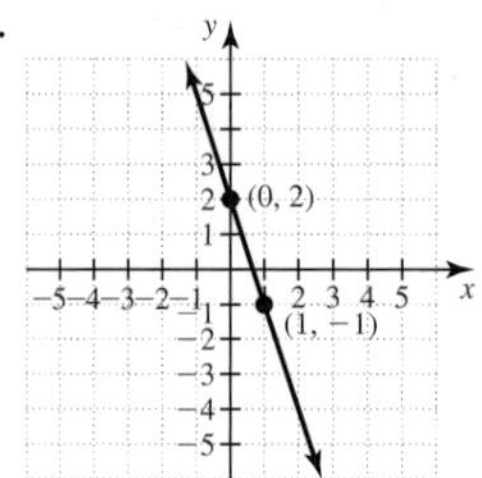

12.
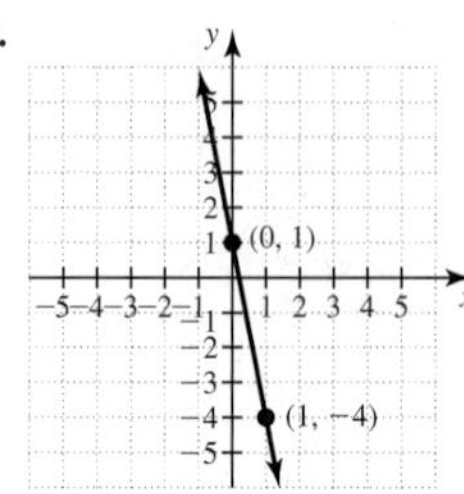

13.
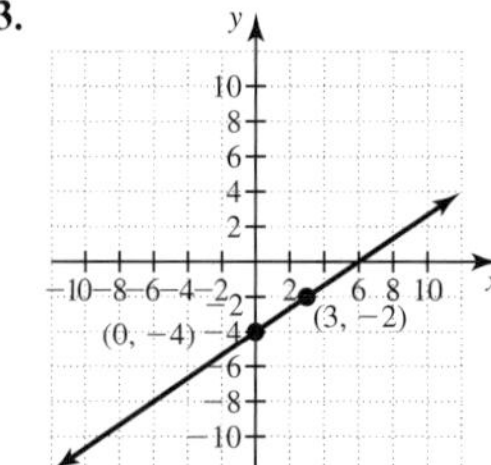

14.
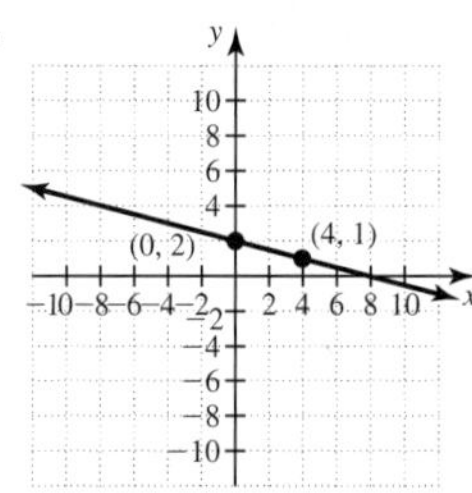

15.
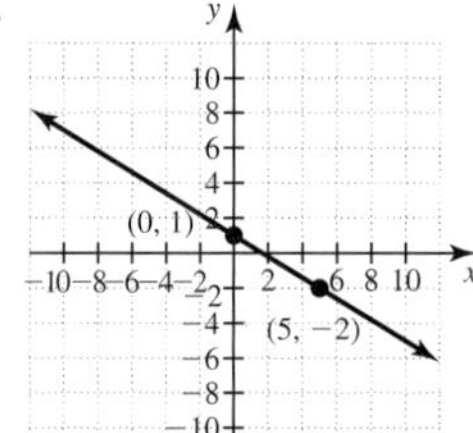

16.
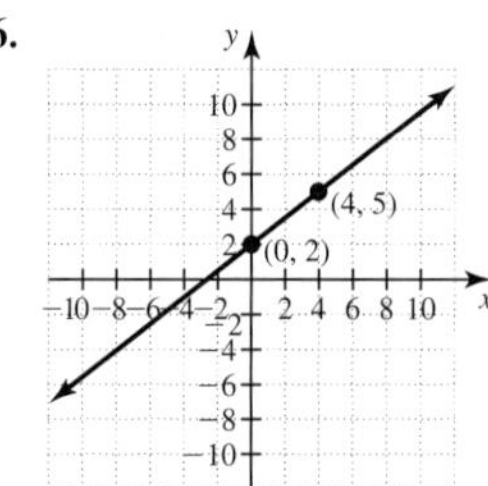

17.
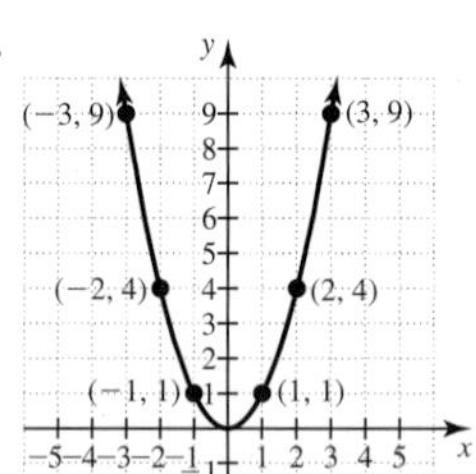

18.
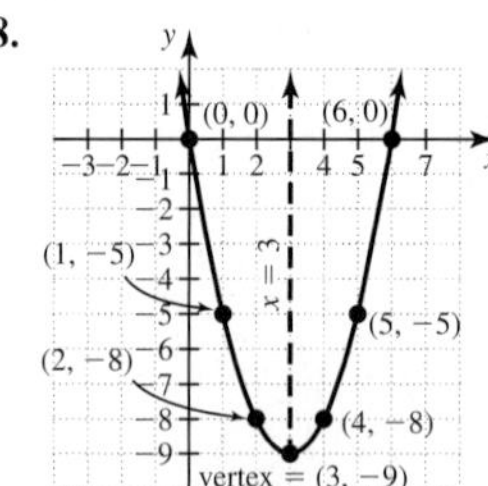

19.
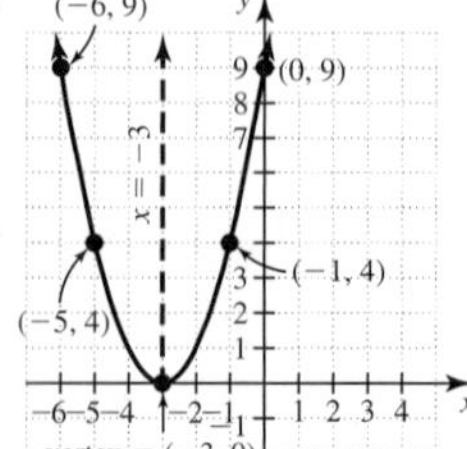

20.
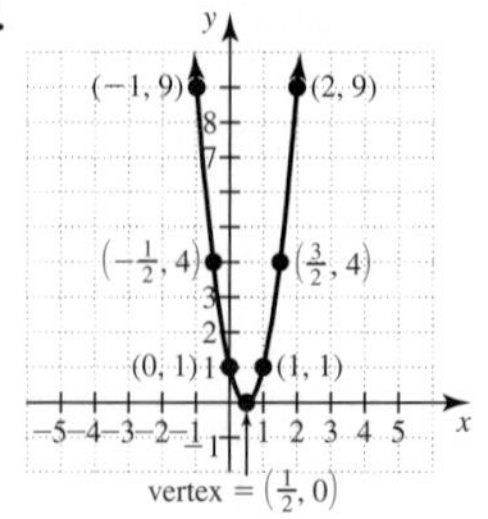

21.

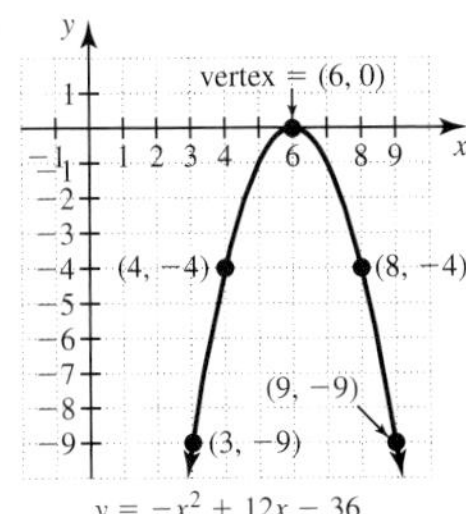

22.

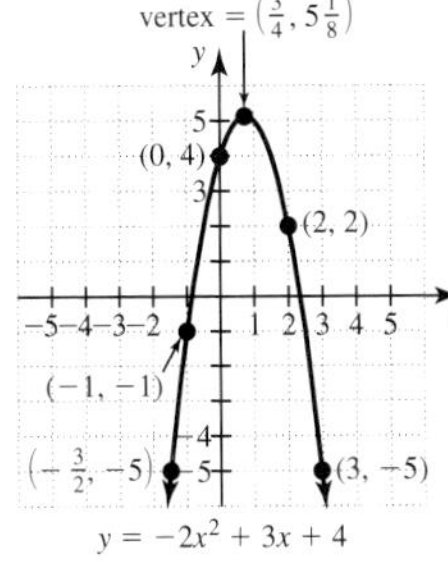

23.

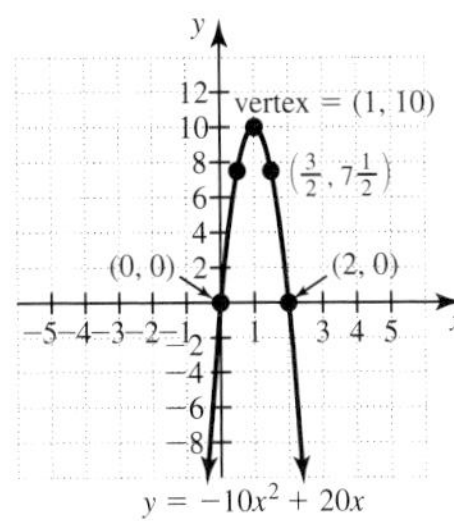

24.

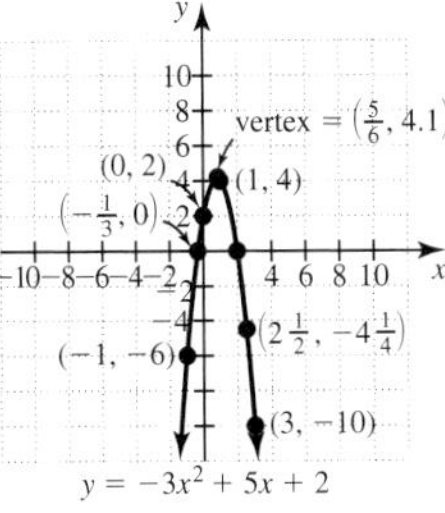

25.

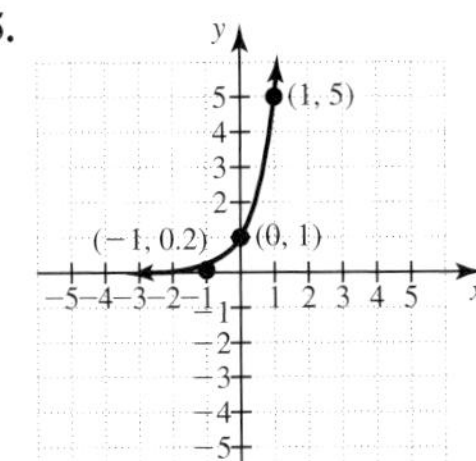

26.

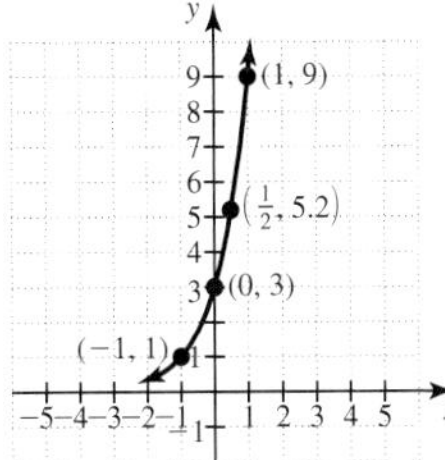

27.

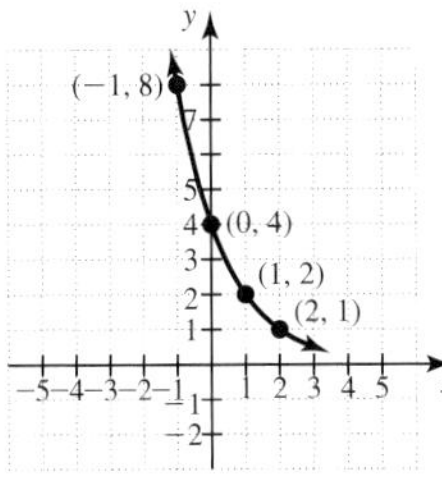

28.

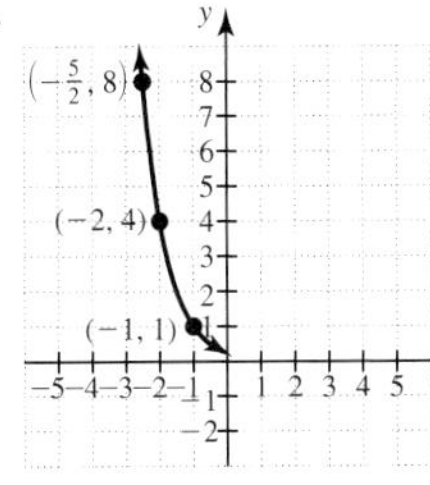

29.

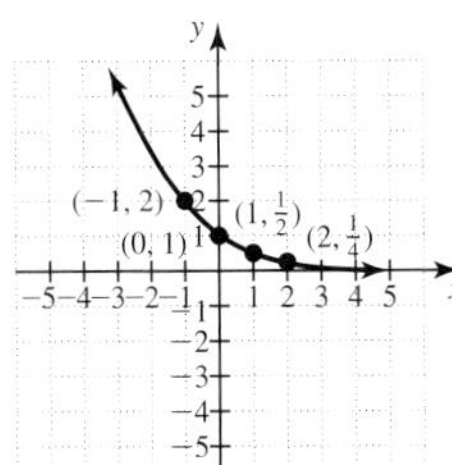

30.

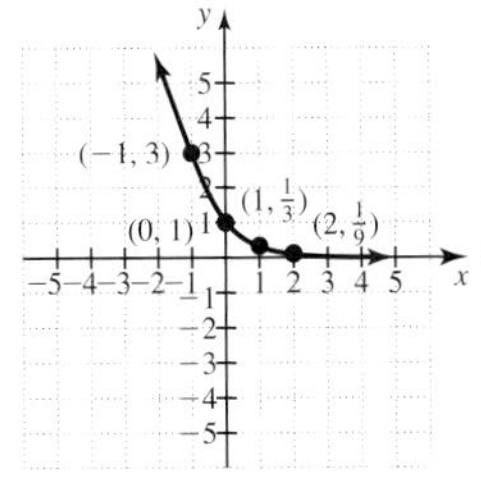

31. a) $C = -20p + 1{,}000$
b)

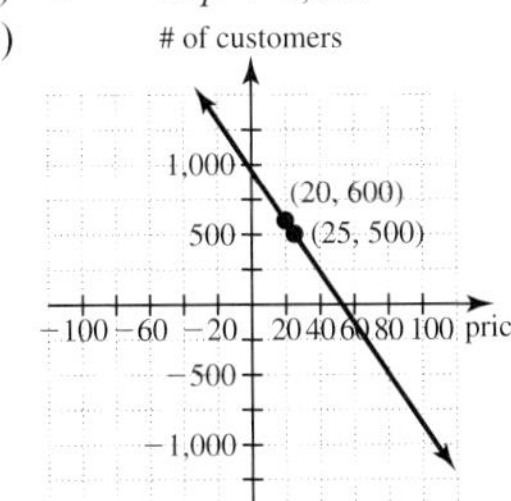

(c) 100 customers
(d) \$35
(e) The p intercept is (50, 0). It is the price of a ticket that would attract 0 customers.
(f) The C intercept is (0, 1,000). It is the number of customers that a price of \$0 would attract.

32. (a) $V(t) = -7{,}000t + 256{,}000$
(b)

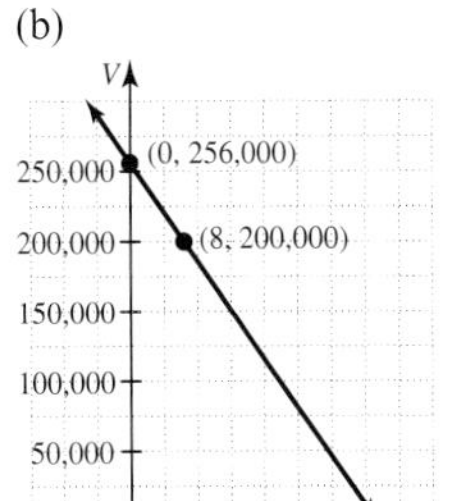

(c) 2020 (d) \$60,000

33. (a) $D(t) = 70t$
(b)

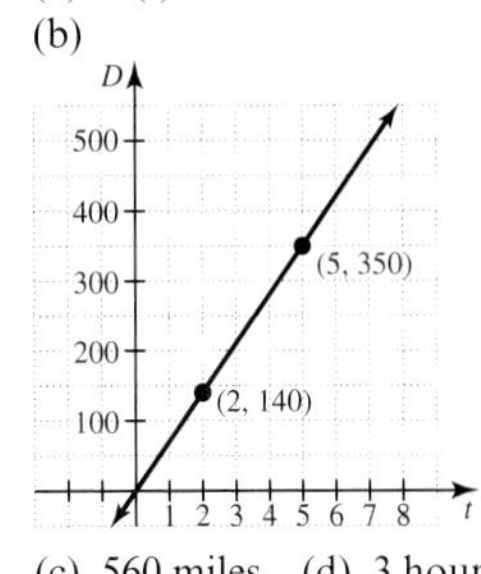

(c) 560 miles (d) 3 hours

34. (a) $S(t) = -4{,}000t + 52{,}000$
(b)

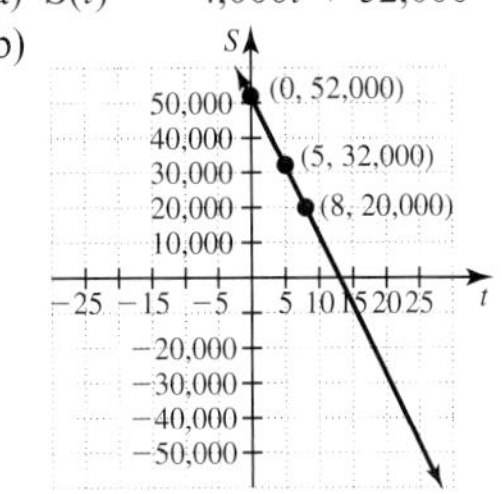

(c) The S intercept is (0, 52,000) and it represents the sales in the year 2000 were \$52,000.
(d) The t intercept is (13, 0) and it represents that in the year 2013 the sales will be \$0.
(e) \$64,000
(f) \$8,000

35. 62.25 feet; about 3.85 seconds **36.** 60 boxes **37.** 144 feet
38. (a) 4 pairs (b) \$146 (c) \$930 (d) 2 or 6 pairs
39. (a) June (b) \$3,600 (c) \$2,700 (d) April and August
40. (a) The most revenue is \$441 and the number of price hikes needed is −3. This means that they need to actually decrease the price by \$1.50 to maximize their revenue. (b) \$392 (c) \$392
41. (a) $A = \frac{5}{2}W^2$ (b) $A = 250$ square inches and $l = 25$ inches (c) width 4 inches; length 10 inches (d) $A = 5w^2$
42. base 8 inches and folded sides of 4 inches each **43.** 4,732,864
44. 293 cells **45.** \$699.21
46.

n	Profit
1	\$1
2	\$1.25
4	\$1.44
12	\$1.61
365	\$1.71

47. about 13 pounds **48.** There is no difference.
49. 0.217 is the rate at which unemployment is increasing per month and 4.40 is the unemployment rate at the beginning of 2008.
50. (9, 0) **51.** The graph is a sideways parabola opening to the right.
52. The graph is a sideways parabola opening to the left.

Review Exercises

1.

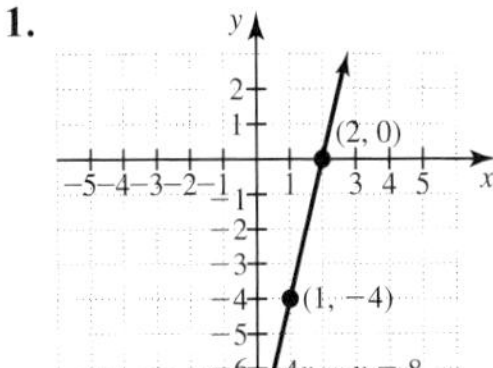

2.

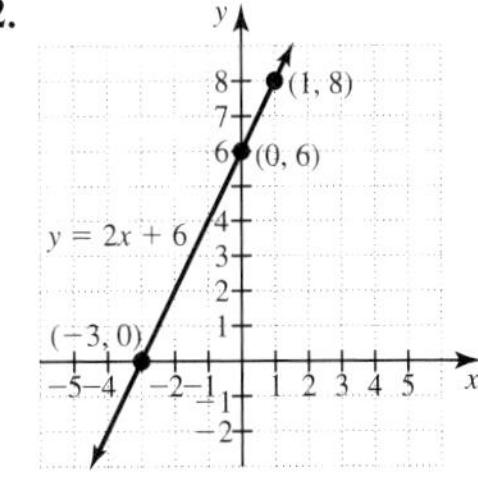

3.

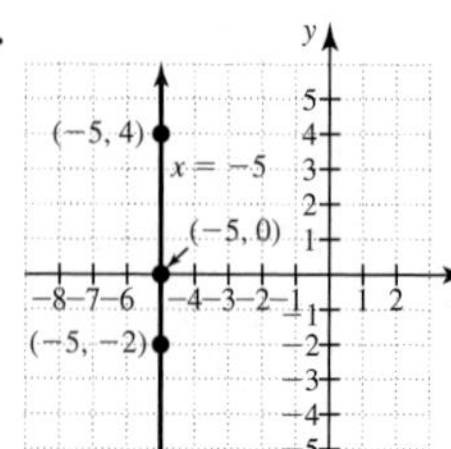

4.

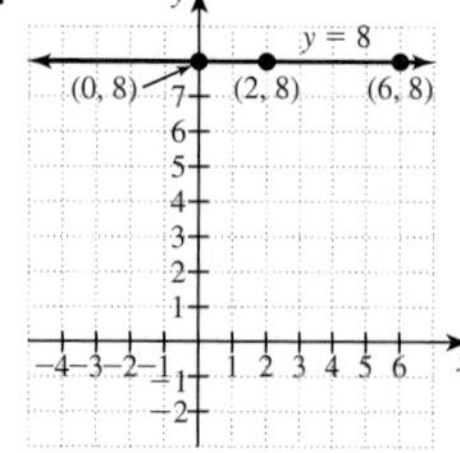

5.

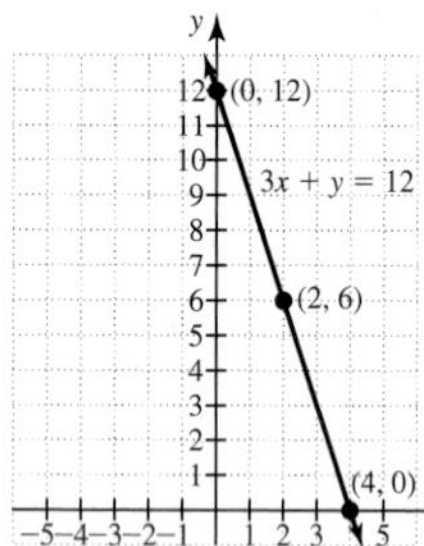

6.

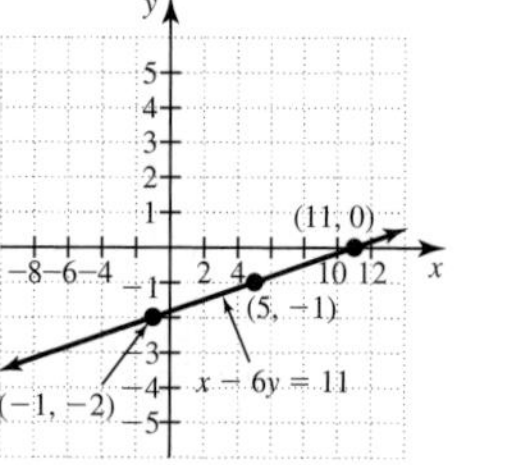

7. $-\frac{1}{4}$ **8.** $\frac{2}{5}$ **9.** 7 **10.** The slope is undefined. **11.** 0 **12.** $\frac{2}{3}$

13. $y = -3x + 12; m = -3$

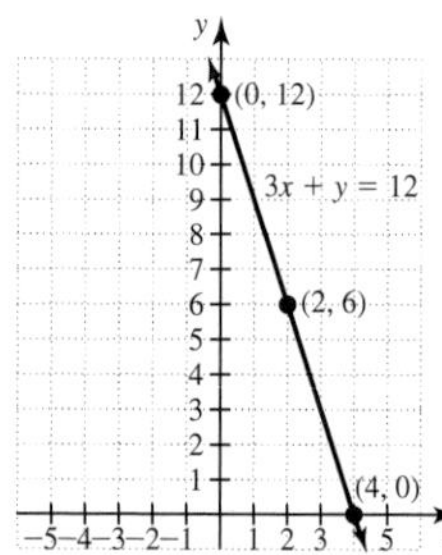

14. $y = \frac{1}{4}x + \frac{15}{8}; m = \frac{1}{4}$

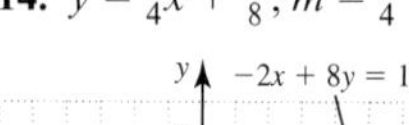

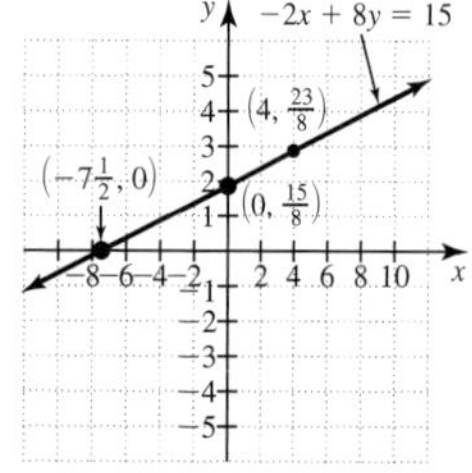

15. $y = \frac{4}{7}x - 4; m = \frac{4}{7}$

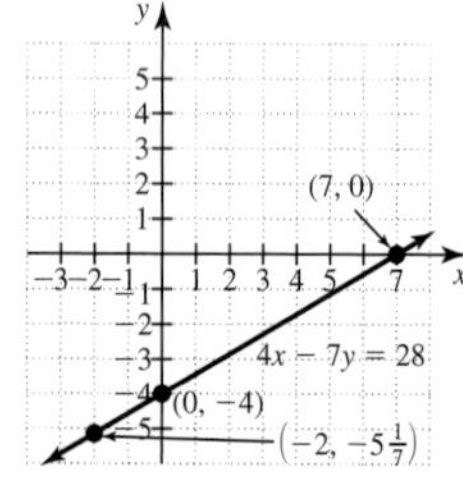

16. $y = \frac{1}{3}x - 3; m = \frac{1}{3}$

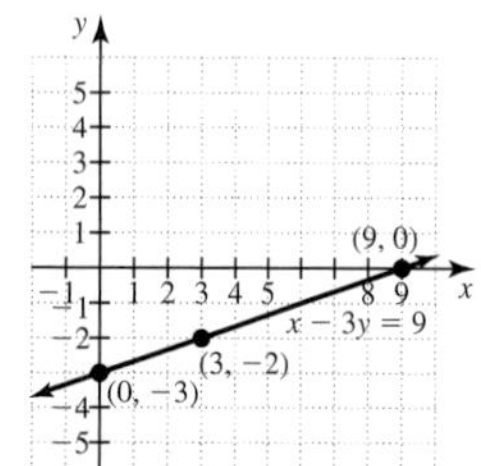

17. Betty collected \$7.64 and Mary collected \$24.36.
18. \$423.81 at 5% and \$847.62 at 8% **19.** (10, 2) **20.** (5, 4)
21. (5, 2) **22.** $\{(x, y) \mid 4x - y = 10\}$. **23.** (10, −1)
24. (3, −1) **25.** (6, 3) **26.** (4, −1) **27.** (4, 0) **28.** $\varnothing$
29. $\left(\frac{3}{2}, -\frac{5}{2}\right)$ **30.** (3, 3)
31. coffee \$2.50 per pound ; tea \$1.75 per pound
32. \$25,000 in store 1 and \$15,000 in store 2

33. $\left[\begin{array}{ccc|c} 2 & -1 & 4 & 10 \\ -3 & -4 & -5 & 8 \\ 1 & 1 & 1 & -8 \end{array}\right]$ **34.** $\left[\begin{array}{ccc|c} 1 & -1 & 1 & -4 \\ 0 & 2 & 1 & 7 \\ 3 & 1 & 0 & 11 \end{array}\right]$ **35.** (−8, 3)

36. $\left(-2, \frac{5}{2}\right)$ **37.** (13, 1, 5) **38.** (−1, −2, −3)

39.

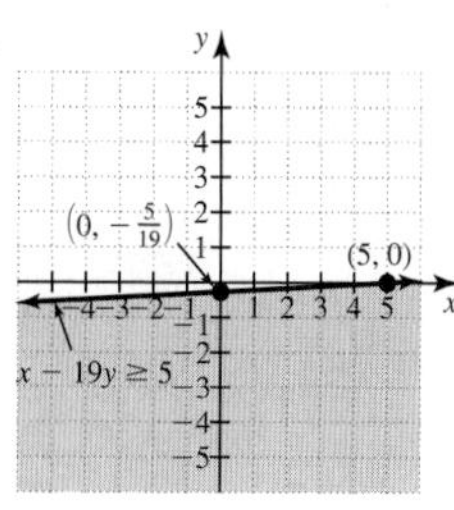

40.

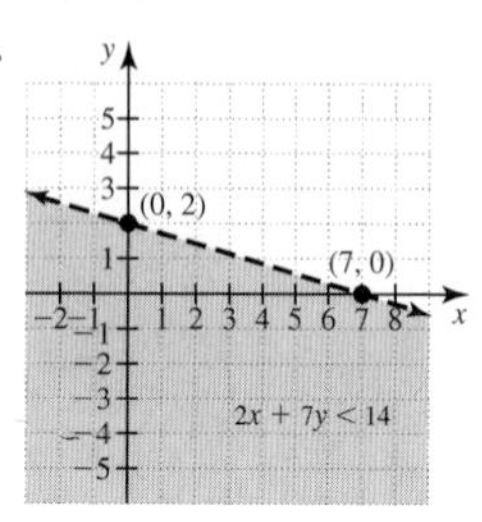

41.

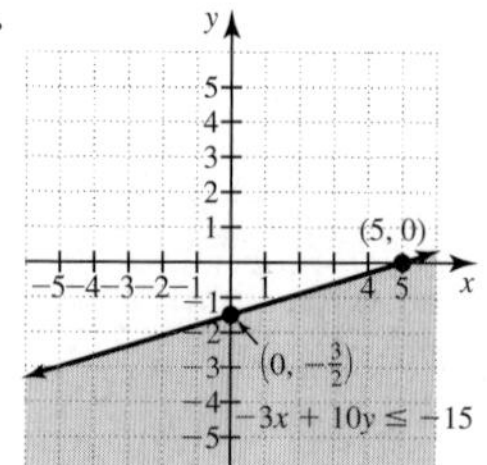

42.

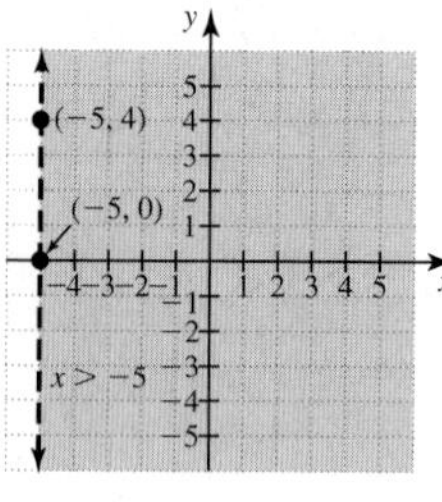

43.

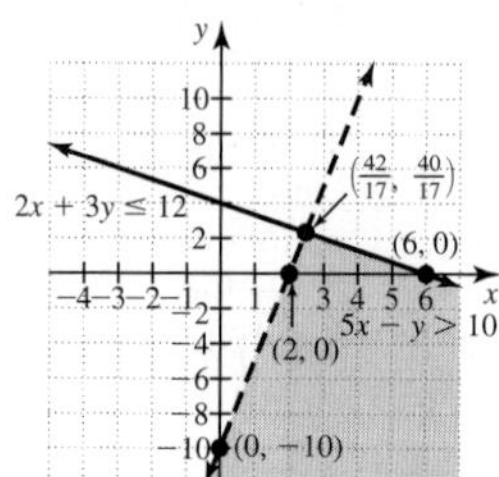

44.

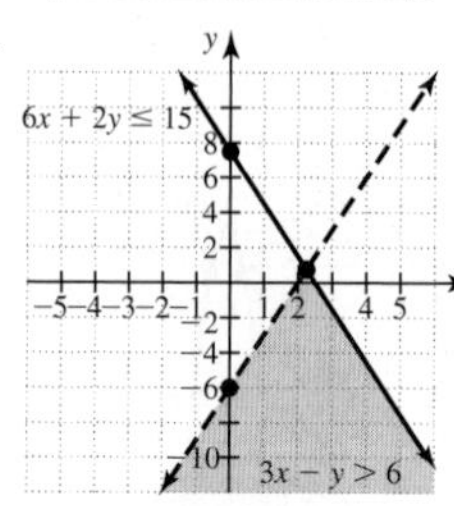

45.

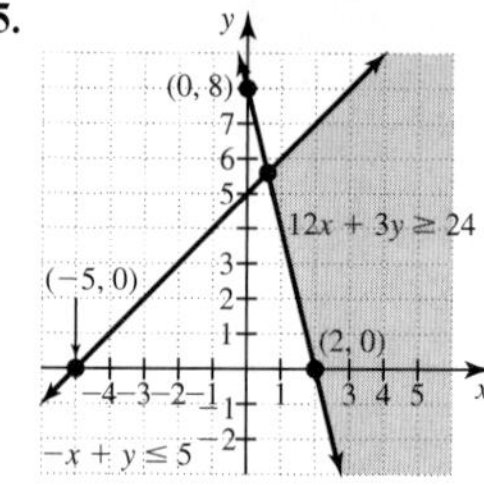

46.

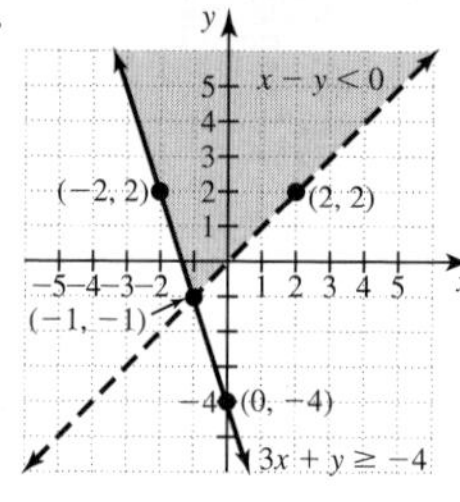

47. 26 of model A and 0 of model B
48. 24 of model I and 12 of model II
49. Domain = {2, 5, 6}; range = {5, −7, −10}; function
50. Domain = {−1, 2}; range = {5, 6, 3}; function
51. Domain = $\{x \mid x \le 3\}$; range = $\{y \mid y \ge 0\}$; function
52. Domain = $\{x \mid -\infty \le x \le \infty\}$; range = $\{y \mid -\infty \le y \le \infty\}$; function
53. Domain = $\{x \mid x \ne 3\}$; range = $\{y \mid y \ne 0\}$; function
54. Domain = $\{x \mid x > 0\}$; Range = $\{x \mid -\infty \le x \le \infty\}$; function
55. $f(5) = 13$ **56.** $f(-3) = 16$ **57.** $f(-10) = 40$ **58.** $f(8) = 104$

59.

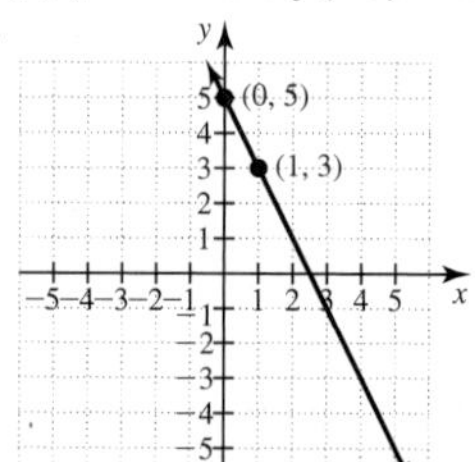

60.

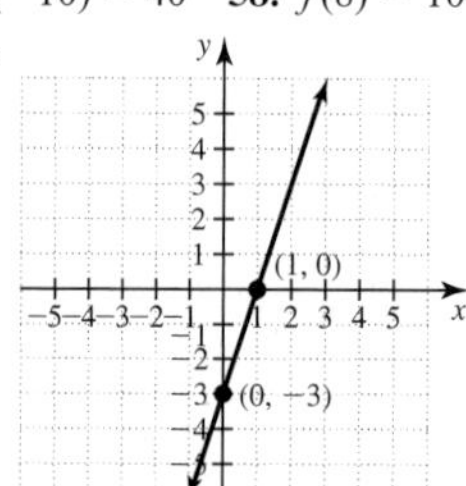

61.

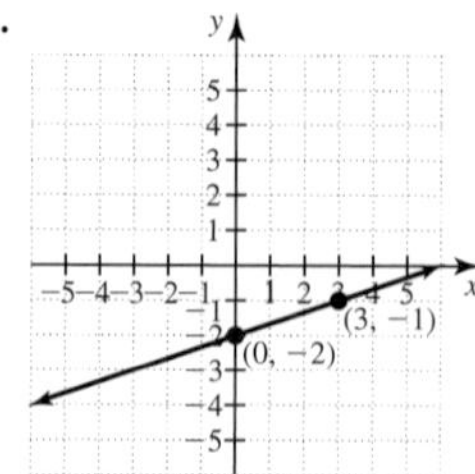

62. $f(x) = -\frac{3}{4}x + 4$

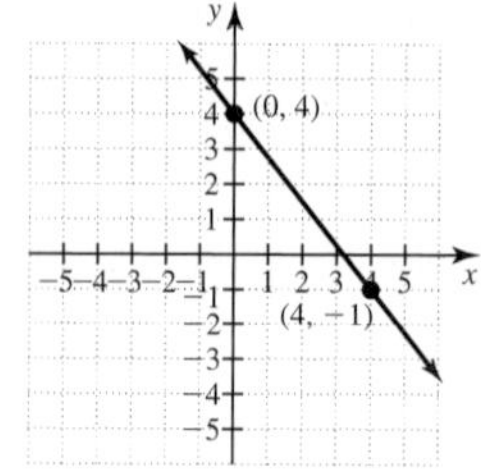

63.

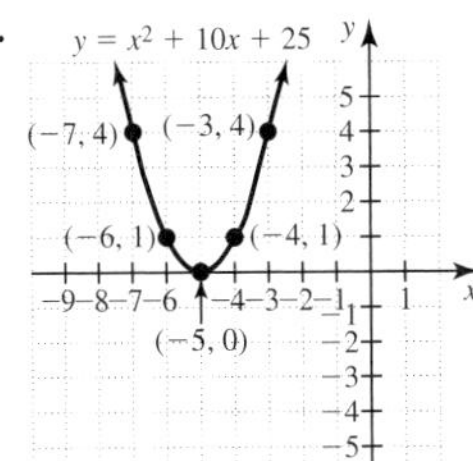

64.

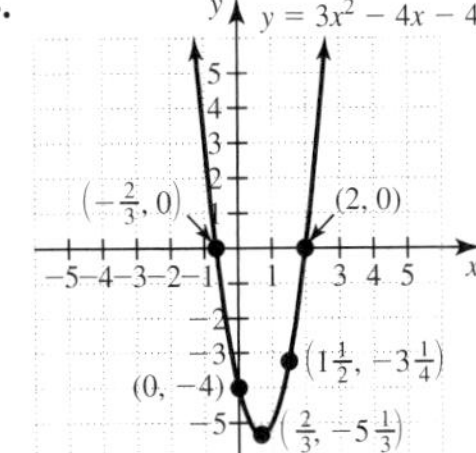

65.

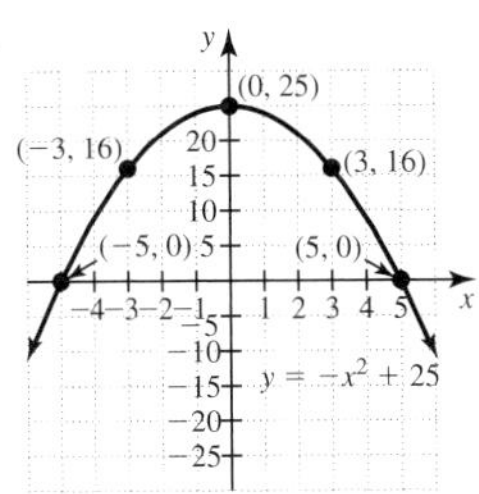

66.

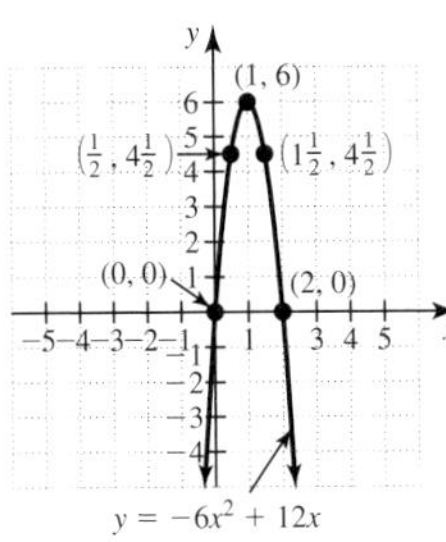

67.

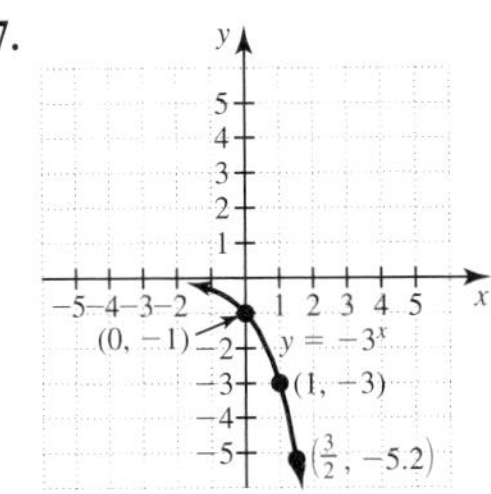

68.

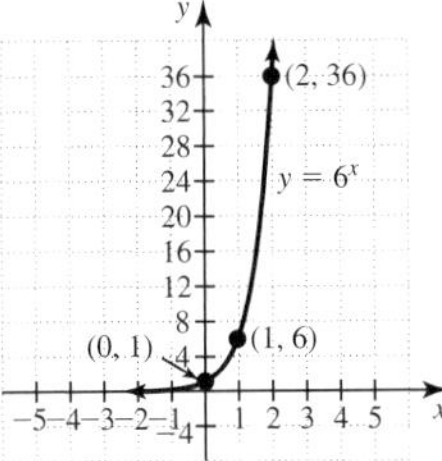

69.

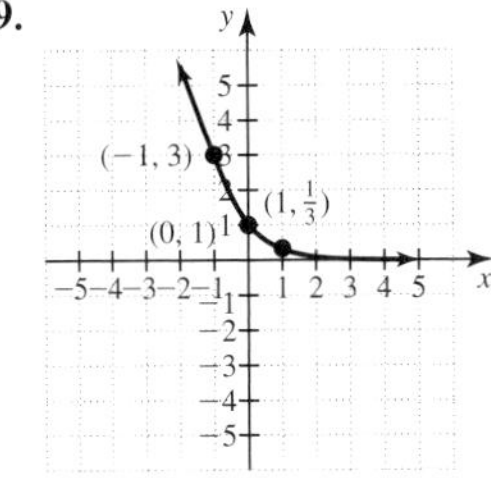

70.

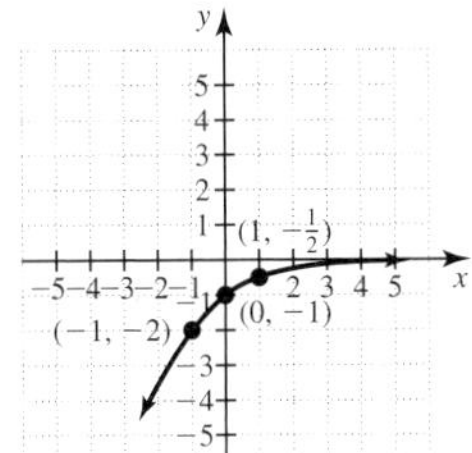

71. 104 feet **72.** 6.25 pounds

Chapter 7 Test

1.

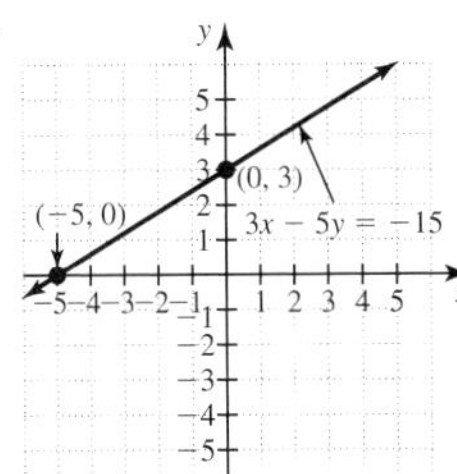

2.

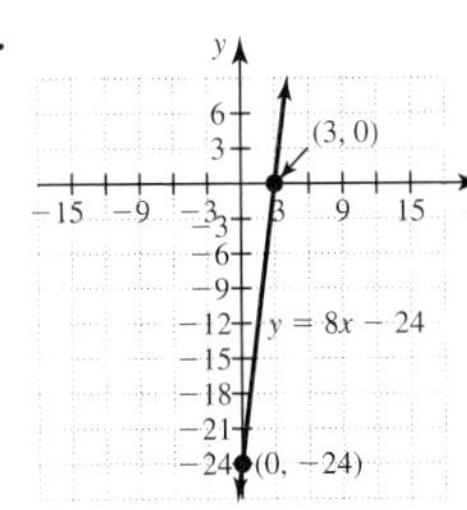

3.

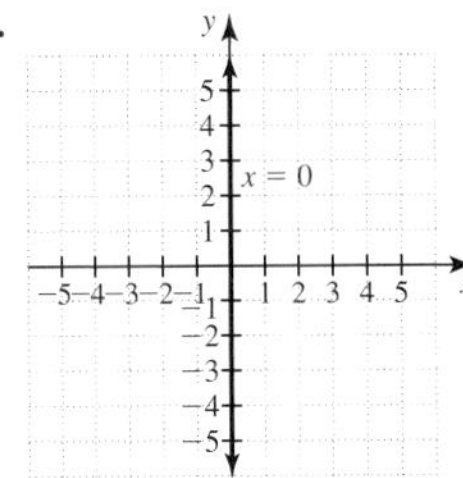

4.

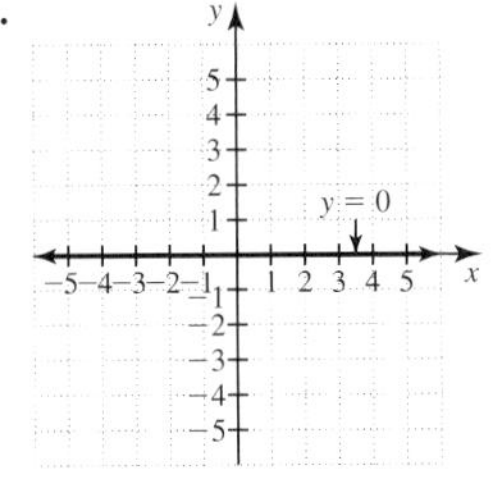

5.

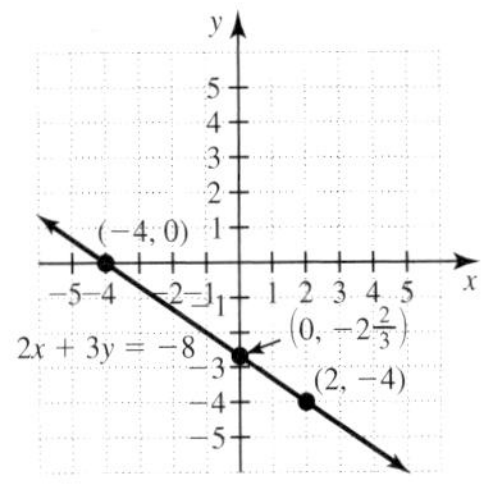

6.

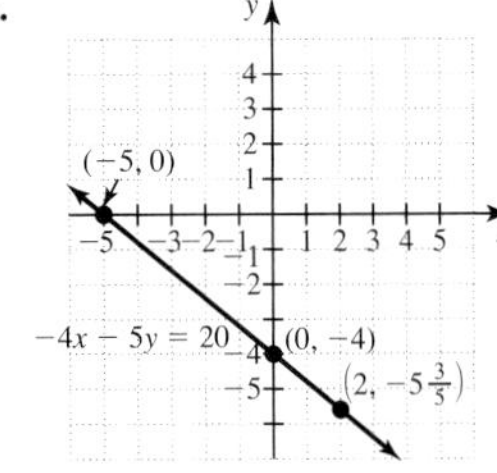

7. $\frac{17}{7}$ **8.** 7
9. $y = -\frac{x}{5} + 4; m = -\frac{1}{5}$

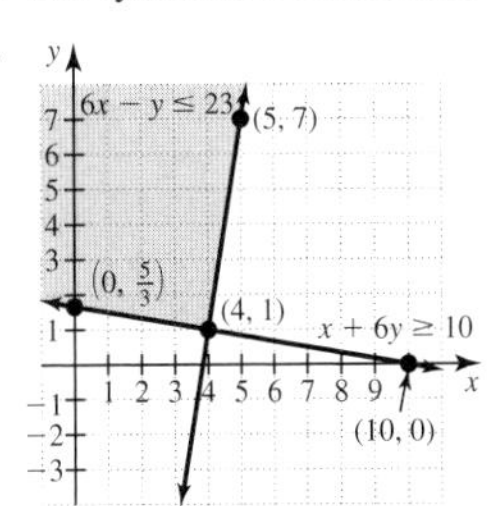

10. $y = \frac{2}{11}x - 2; m = \frac{2}{11}$

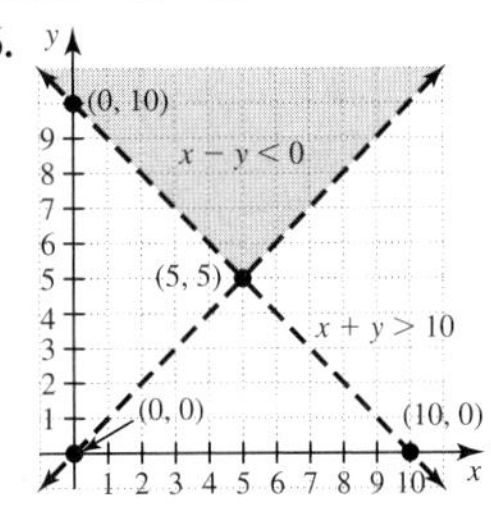

11. (3, −1) **12.** $\left(\frac{23}{5}, -\frac{17}{5}\right)$
13. The system is dependent. The solution set is $\{(x, y) \mid 3x + 4y = 19\}$.
14. The system is inconsistent. The solution set is $\varnothing$.

15.

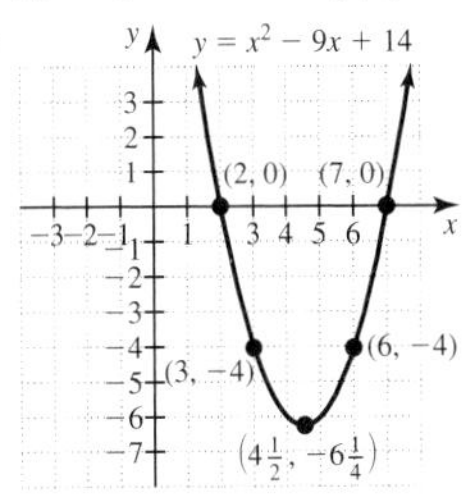

16.

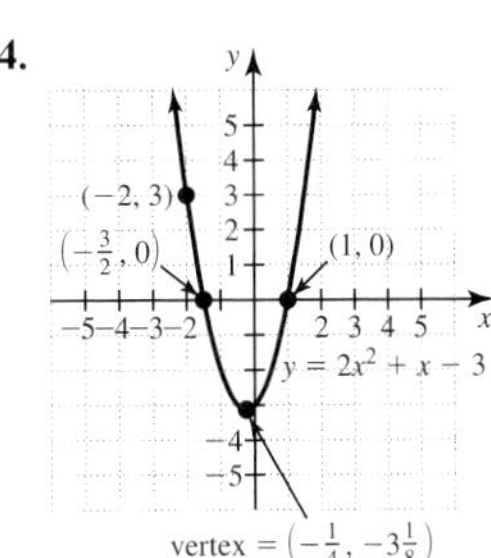

17. (2, −3, 5) **18.** $\left(\frac{70}{13}, \frac{2}{13}, 7\right)$
19. Domain = {−4, −10, 12}; range = {6, 18, 5}; function.
20. Domain = {3, 4, 5}; range = {6, 7, 8, 9}; not a function
21. $f(-15) = 55$ **22.** $f(3) = 31$

23.

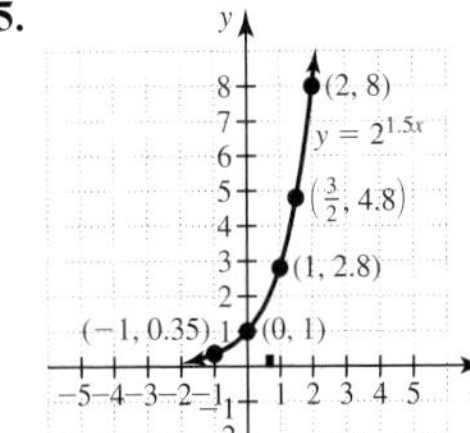

24.

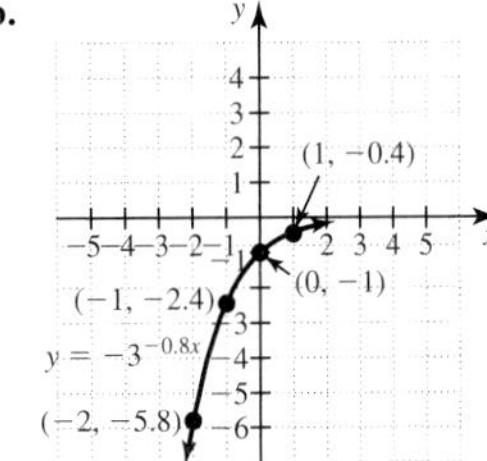

25.

(2, 8)
$y = 2^{1.5x}$
$(\frac{3}{2}, 4.8)$
(1, 2.8)
(−1, 0.35)
(0, 1)

26.

(1, −0.4)
(−1, −2.4)
(0, −1)
$y = -3^{-0.8x}$
(−2, −5.8)

27. 26 and 31 students **28.** 62 *Daily News* and 34 *Tribune*
29. The cooks earn $7.50 per hour and the servers earn $5.50 per hour.
30. 24 snowmen and 0 wreaths
31. base width 12 inches and sides height 6 inches **32.** $2,064.86

CHAPTER 8: CONSUMER MATHEMATICS

Exercise Set 8-1

7. 63% **8.** 87% **9.** 2.5% **10.** 8.72% **11.** 156% **12.** 387.5% **13.** 20% **14.** 62.5% **15.** $66.\overline{6}$% **16.** $16.\overline{6}$% **17.** 125% **18.** 237.5% **19.** 0.18 **20.** 0.23 **21.** 0.06 **22.** 0.02 **23.** 0.625 **24.** 0.756 **25.** 3.2 **26.** 2.75 **27.** $\frac{6}{25}$ **28.** $\frac{9}{25}$ **29.** $\frac{9}{100}$ **30.** $\frac{1}{25}$ **31.** $2\frac{9}{25}$ **32.** $5\frac{1}{5}$ **33.** $\frac{1}{200}$ **34.** $\frac{1}{8}$ **35.** $\frac{1}{6}$ **36.** $\frac{1}{24}$ **37.** \$15; \$314.99 **38.** \$4.20; \$64.15 **39.** \$9.00; \$158.99 **40.** \$0.90; \$20.89 **41.** 60% **42.** 9% **43.** 24% **44.** 10% **45.** \$95.99 **46.** \$126 **47.** \$149.99 **48.** \$300 **49.** \$1,256 **50.** \$1,600 **51.** About 2,446 **52.** \$800 **53.** 8.5% **54.** 74.7% **55.** 60% **56.** 211% **57.** 52.8% **58.** 12.1%

59. No; the total discount is 44%.

60. The stock is worth \$70 the next day. Next week the stock is worth \$91 per share, which means you do not break even.

61. Total discount will be 75%, not 100%

62. Allowing for rounding, neither is lying.

Exercise Set 8-2

7. \$1,440 **8.** \$1,062.50 **9.** 2 years **10.** 3 years **11.** 5% **12.** 7.5% **13.** \$985 **14.** \$62,500 **15.** 3.2% **16.** 6 years **17.** \$128.25 **18.** 2.95% **19.** 6.5 years **20.** \$497.70 **21.** 10% **22.** 9 years **23.** 6% **24.** 0.45 year **25.** 3 years **26.** \$61.43 **27.** \$960 **28.** \$24,450 **29.** \$1,120 **30.** \$1,303.13 **31.** \$490.44 **32.** \$3,729.60 **33.** \$1.65 **34.** \$57 **35.** \$35 **36.** \$35.34 **37.** \$26.47 **38.** \$21 **39.** (a) \$720 (b) \$2,280 (c) 10.5% **40.** (a) \$472.50 (b) \$1,277.50 (c) 6.2% **41.** (a) \$4,400 (b) \$17,600 (c) 6.25% **42.** (a) \$120 (b) \$480 (c) 2.5% **43.** (a) \$421.20 (b) \$358.80 (c) 13% **44.** (a) \$277.50 (b) \$1,572.50 (c) 3.5% **45.** 8.43% **46.** 10% **47.** \$141,509.43 **48.** \$5,000 **49.** 3 years **50.** 0.75 year or 9 months **51.** 4.25 years or 51 months **52.** \$1,350 **53.** \$2,096.25 **54.** \$110.16 **55.** \$7,316 **56.** \$18,480 **57.** \$6 **58.** \$21.38 **59.** (a) \$5,400 (b) \$12,600 (c) 7.1% **60.** (a) \$5,400 (b) \$14,600 (c) 12.3% **61.** The personal loan

62. The rate is 0.49%. The rate seems unreasonably low.

63. The repair loan **64.** The bank loan

65. They are the same regardless of principal.

66. It is worse for the borrower if the term is longer.

Exercise Set 8-3

5. I = \$396.20 and A = \$1,221.20 **6.** I = \$338.26 and A = \$3,588.26 **7.** I = \$14.67 and A = \$89.67 **8.** I = \$640.11 and A = \$2,190.11 **9.** I = \$991.92 and A = \$1,616.92 **10.** I = \$213.36 and A = \$2,788.36 **11.** I = \$688.05 and A = \$2,683.05 **12.** I = \$237.92 and A = \$697.92 **13.** I = \$70.62 and A = \$820.62 **14.** I = \$2,241.32 and A = \$5,741.32 **15.** \$2,604.21 **16.** \$10,644.02 **17.** \$8,167.15 **18.** \$6,596.69 **19.** \$58,541.39 **20.** \$37,422.50 **21.** \$18,072.57 **22.** \$14,642.69 **23.** \$7,710.98 **24.** \$1,006.38 **25.** 6.14% **26.** 10.25% **27.** 6.66% **28.** 9.78%

29. 4.5% compounded semiannually

30. 7.2% compounded semiannually **31.** 3.1% compounded quarterly **32.** 5.74% compounded semiannually **33.** \$12,175.94 **34.** \$82,078.65 **35.** \$335,095.61 **36.** \$14,499.48 **37.** \$20,548.25 **38.** \$8,047.25 **39.** \$27,223.76 **40.** \$855.64 **41.** \$24,664.64 **42.** \$2,761.26 **43.** \$59,556.16 **44.** \$2,546.88 **45.** \$2,095.18 **46.** \$42.13 **47.** \$114,717.32 **48.** \$2,316.64 **49.** \$8,347.30 **50.** \$3,285.86 **51.** \$1,235.34

52. \$20,232.75; \$289.99; \$17,399.40; the annuity costs more for the same yield. But the advantage is you do not have to have a lump sum of money as you would in the first investment.

Exercise Set 8-4

7. \$480

8. Down payment: \$258; Installment Price: \$1,778.48; Monthly payment: \$84.47

9. Down payment: \$56.25; Installment Price: \$383.25; Monthly payment: \$27.25

10. \$867.50 **11.** \$43.17 **12.** \$43.94 **13.** 8.5% **14.** 6.5% **15.** 12% **16.** 11% **17.** 9.5% **18.** 6.5%

19. u = \$709.50; payoff amount: \$10,956.50

20. u = \$131.59; payoff amount: \$4,015.41

21. u = \$14.54; payoff amount: \$482.81

22. u = \$15.77; payoff amount: \$572.23

23. u = \$14.66; payoff amount: \$615.34

24. u = \$604.20; payoff amount: \$9,133.30

25. \$180.13 **26.** \$4.04 **27.** \$8.22 **28.** \$7.56 **29.** \$5.18 **30.** \$33.33 **31.** \$16.65; \$1,124.15 **32.** \$19.80; \$914.01 **33.** \$39.49; \$3,368.60 **34.** \$15.26; \$3,647.63 **35.** \$13.32; \$411.35 **36.** \$9.57; \$490.74 **37.** (a) \$602.14 (b) \$7.23 (c) \$669.15

38. (a) \$1,489.52 (b) \$29.79 (c) \$1,418.88

39. (a) \$370.55 (b) \$5.19 (c) \$350.76

40. (a) \$1,484.23 (b) \$29.68 (c) \$1,382.11

41. (a) \$370.92 (b) \$4.08 (c) \$533.30

42. (a) \$289.52 (b) \$5.50 (c) \$540.53 **43.** \$3,636.36

44. The borrower saves more money when computed equally over 12 months.

Exercise Set 8-5

5. (a) \$21,750 (b) \$123,250 (c) \$871.38 (d) \$138,161

6. (a) \$9,125 (b) \$173,375 (c) \$1,558.64 (d) \$107,180.20

7. (a) \$80,000 (b) \$120,000 (c) \$720.00 (d) \$139,200

8. (a) \$15,000 (b) \$110,000 (c) \$812.90 (d) \$133,870

9. (a) \$32,500 (b) \$292,500 (c) \$1,924.65 (d) \$631,332

10. (a) \$38,500 (b) \$136,500 (c) \$1,018.29 (d) \$107,889.60

11. (a) \$360,000 (b) \$840,000 (c) \$5,779.20 (d) \$547,008

12. (a) \$137,500 (b) \$412,500 (c) \$2,268.75 (d) \$676,500

13. \$1,499.44 **14.** \$1,490.45 **15.** \$1,148.90 **16.** \$1,919.91 **17.** \$4,215.22 **18.** \$1,132.46 **19.** \$3,746.38 **20.** \$3,071.41

21.

Payment number	Interest	Payment on Principal	Balance of Loan
1	\$718.96	\$152.42	\$123,097.58
2	\$718.07	\$153.31	\$122,944.27
3	\$717.17	\$154.21	\$122,790.06

22.

Payment number	Interest	Payment on Principal	Balance of Loan
1	\$1,011.35	\$547.29	\$172,827.71
2	\$1,008.16	\$550.48	\$172,277.23
3	\$1,004.95	\$553.69	\$171,723.54

23.

Payment number	Interest	Payment on Principal	Balance of Loan
1	\$3,850	\$1,929.20	\$838,070.80
2	\$3,841.16	\$1,938.04	\$836,132.76
3	\$3,832.28	\$1,946.92	\$834,185.84

24.

Payment number	Interest	Payment on Principal	Balance of Loan
1	\$2,062.50	\$206.25	\$412,293.75
2	\$2,061.47	\$207.28	\$412,086.47
3	\$2,060.43	\$208.32	\$411,878.15

Exercise Set 8-6

7. \$97.25 per share **8.** \$57.50 per share **9.** \$1.23 per share **10.** \$215.25 **11.** 4,626,000 shares **12.** \$62.06 **13.** \$8.87 per share **14.** \$30,235.63 **15.** \$766.29 **16.** \$61.29 **17.** \$40.08 per share **18.** \$24.75 per share **19.** \$0.04 **20.** \$14.28 **21.** 345,000 shares **22.** \$29.79 **23.** \$1.49 per share **24.** \$38,113.98 **25.** \$156.20 **26.** \$29.40 **27.** \$50.87 per share **28.** \$42.31 per share **29.** \$0.67 **30.** \$456.94 **31.** 9,662,000 shares **32.** \$48.12 **33.** \$2.53 per share **34.** \$45,235.10 **35.** \$820.75 **36.** \$48.22 **37.** 24.91 **38.** 19 **39.** 15.02 **40.** 19.04 **41.** \$0.34 per share **42.** \$3.03 per share **43.** \$1.98 per share **44.** \$1.46 per share **45.** lost \$432.48 **46.** lost \$1,766.25 **47.** lost \$2,022.63 **48.** earned \$3,525.04

49. The second investment would be the better choice.

Review Exercises

1. 0.875; 87.5% **2.** $\frac{27}{50}$; 54% **3.** $\frac{4}{5}$; 0.8 **4.** $0.41\overline{6}$; $41.\overline{6}$% **5.** $\frac{37}{20}$; 1.85 **6.** $\frac{3}{50}$; 6% **7.** 5.75; 575% **8.** $\frac{31}{20}$; 155% **9.** $\frac{91}{200}$; 0.445 **10.** 0.375; 37.5% **11.** 69.12 **12.** 30% **13.** 1,100

14. The tax is \$1.00 and the total cost is \$20.95. **15.** \$60 **16.** \$32,500 **17.** 73.5% **18.** 29.3% **19.** \$2,322 **20.** 4%

21. 2.5 years **22.** \$4.50 **23.** 7% **24.** \$425 **25.** \$1,375 **26.** 12%
27. I = \$1,800 and A = \$7,800 **28.** I = \$16,140 and A = \$29,590
29. 1.4% **30.** \$195.81 **31.** \$25.56 **32.** \$206.60
33. Discount: \$1,080; David received \$4,920
34. Discount: \$4,440; Marla received \$4,810
35. I = \$603.67 and A = \$2,378.67 **36.** I = \$97.19 and A = \$297.19
37. I = \$12.07 and A = \$57.07 **38.** I = \$10,861.67 and A = \$31,861.67
39. 12.55% **40.** \$73,925.08 **41.** \$8,129.93
42. Total installment price: \$865.10
Finance Charge: \$11.10
43. Total installment price: \$788.25
Monthly payment: \$84.47
44. 8% **45.** 10.5% **46.** \$1,139.40
47. u = \$156.44; payoff amount = \$3,873.56
48. u = \$757.80; payoff amount = \$7,017.20 **49.** \$15.24
50. \$9.86; \$786.36 **51.** \$124.13; \$7,424.53
52. (a) \$29,000 (b) \$116,000 (c) \$933.80
(d)

Payment number	Interest	Payment on Principal	Balance of Loan
1	\$821.67	\$112.13	\$115,887.87
2	\$820.87	\$112.93	\$115,774.94

53. \$1,827.94
54. The 52-week high was \$38.28 per share and the 52-week low was \$27.09 per share.
55. \$342 **56.** 5,528,000 shares **57.** \$33.04 **58.** \$1.09 per share
59. She earned \$203.94

Chapter 8 Test

1. 31.25% **2.** 63% **3.** $\frac{7}{25}$ **4.** 0.167 **5.** 80% **6.** 42 **7.** 300
8. \$2.40 **9.** \$2,568 **10.** 25% **11.** \$486 **12.** 5%
13. I = \$8.16 , A = \$443.16. The monthly payment is \$73.86.
14. I = \$17.27, A = \$1,552.27. The monthly payment is \$517.42.
15. I = \$216, A = \$2,016. The monthly payment is \$168.
16. \$33.33
17. Discount: \$5,692.50; Latoya received \$6,957.50
18. I = \$145.79 and A = \$645.79
19. I = \$7,885.08 and A = \$17,635.08
20. \$345 **21.** 8.16% **22.** \$19,043.39
23. Total installment price: \$1,000.45; Monthly payment: \$166.74
24. 6%.
25. The unearned interest is \$438.36 and the payoff amount is \$7,186.64.
26. \$16.15
27. The finance charge is \$20.
The new balance is \$1,030.
28. The finance charge is \$6.63.
The new balance is \$387.13.
29. 11%
30. (a) \$9,000
(b) \$171,000
(c) \$1,026
(d)

Payment number	Interest	Payment on Principal	Balance of Loan
1	\$855	\$171	\$170,829
2	\$854.15	\$171.85	\$170,657.15

31. \$2,834.66
32. The 52-week high is \$36.98 and the 52-week low is \$23.17.
33. \$60 **34.** 1,501,000 **35.** \$27.37 **36.** \$2.29 per share

CHAPTER 9: MEASUREMENT

Exercise Set 9-1

5. 4 yd **6.** 3 ft **7.** $5\frac{2}{3}$ yd **8.** ≈1.96 mi **9.** 18 ft **10.** 8,800 yd
11. $1\frac{3}{4}$ ft **12.** 30 ft **13.** $\frac{1}{2}$ yd **14.** ≈82.67 mi **15.** $24\frac{1}{3}$ yd
16. $60\frac{2}{3}$ yd **17.** 800 cm **18.** 2.5 dm **19.** 120 m **20.** 240 dam
21. 6 hm **22.** 3 dm **23.** 900 dm **24.** 18.426 m **25.** 3.756 m
26. 0.063 km **27.** 0.4053 km **28.** 0.06 hm **29.** 1,200,000 cm
30. 0.5 km **31.** 1,850,000 mm **32.** 0.65 km **33.** 126,200 dm
34. 3,900 cm **35.** 0.8 km **36.** 5,400 cm **37.** ≈196.85 in.
38. ≈12.80 m **39.** 406.4 mm **40.** ≈54.68 yd **41.** ≈7.16 dam
42. ≈2,216.54 in. **43.** ≈4,429.08 ft **44.** ≈143,535 ft **45.** 15.24 mm
46. ≈159 mi **47.** ≈19.68 ft **48.** 137.16 cm **49.** 4,099.56 dm
50. ≈48.12 yd **51.** ≈1.46 mi **52.** ≈1,347.57 km **53.** ≈4.59 yd
54. ≈29.53 in. **55.** ≈8.46 m **56.** ≈49,448,818.90 in. **57.** 15 ft
58. 1,128 in. **59.** 60 yd **60.** 1,620 in. **61.** ≈0.85 mi
62. The mattress will not fit. **63.** $3\frac{1}{2}$ yd
64. 60 cm by 50 cm; 600 mm by 500 mm
65. 2,300 mg **66.** 10 legs **67.** 14 km **68.** ≈6.214 mi
69. ≈103.04 km per hour **70.** 50.8 cm by 60.96 cm **71.** ≈19.16 m
72. 12.7 mm **73.** 38.1 mm by 76.2 mm **74.** $\approx 3.54 \times 10^{12} \frac{\text{ft}}{\text{hr}}$
75. 7.92×10^9 ft **76.** 24 in. **77.** 573 boards **78.** $6{,}666\frac{2}{3}$ mg per serving
79. 20 g **80.** 984 tiles **81.** 488 Canadian dollars **82.** a new car
83. 37.85 francs **84.** \$24.30 **85.** the first runner **86.** dekameters
87. millimeters

Exercise Set 9-2

Note: Answers can vary, depending on rounding and on the conversion factors used.

7. 24 ft^2 **8.** ≈5.6 acres **9.** ≈7.3 mi^2 **10.** ≈14.7 mi^2 **11.** 108 ft^2
12. 4,480 acres **13.** ≈0.07 acre **14.** ≈12.3 acres **15.** ≈505 yd^2
16. 12,960 in.^{2} **17.** ≈116.1 cm^2 **18.** ≈4.84 m^2 **19.** ≈47.85 yd^2
20. ≈186.48 km^2 **21.** ≈0.13 km^2 **22.** ≈277.35 cm^2 **23.** ≈167.4 dm^2
24. ≈14.42 in.^{2} **25.** ≈250.8 dm^2 **26.** ≈167.74 ft^2 **27.** ≈25,451.74 acres
28. ≈12,403.10 in.^{2} **29.** ≈650.32 in.^{2} **30.** ≈0.021 ft^2
31. ≈120.94 dm^2 **32.** ≈29.26 m^2 **33.** ≈8.26 in.^{2} **34.** ≈0.62 km^2
35. ≈83.55 ft^2 **36.** ≈1,196,172.25 yd^2 **37.** ≈2,872.32 fluid oz
38. ≈1,212 gal **39.** ≈53.48 ft^3 **40.** ≈18.44 yd^3 **41.** ≈98.13 gal
42. ≈6.94 ft^3 **43.** ≈164,560 gal **44.** ≈924 in.^{3} **45.** ≈1,616 gal
46. ≈19.62 gal **47.** 4,500 cm^3 **48.** 0.000097 m^3 **49.** 28.5 mL
50. 1,400 dL **51.** 433 cm^3 **52.** 87.25 L **53.** 32,000 cm^3
54. 0.00437 m^3 **55.** 0.00032 m^3 **56.** 1,600 cm^3 **57.** 0.5 L
58. 8,000 L **59.** 920 dL **60.** 4.8 hL **61.** 80 cL **62.** 42,000 mL
63. 67 daL **64.** 0.92 hL **65.** ≈45.28 L **66.** ≈0.5035 gal
67. ≈5.035 gal **68.** ≈584.91 L **69.** ≈17,924.53 mL **70.** ≈1.59 qt
71. ≈22.31 qt **72.** ≈25,471.70 mL **73.** ≈46,375 gal **74.** ≈0.44 kL
75. ≈0.34 kL **76.** ≈22,286.5 gal **77.** ≈31,250 lb **78.** ≈1,472 lb
79. ≈3.2 ft^3 **80.** ≈11.80 ft^3 **81.** 2 gallons **82.** \$16
83. 7.4 cubic yards **84.** 2.5 glasses **85.** 1.56 mi^2 **86.** 24,200 yd^2
87. ≈0.67 ft^3 **88.** 1,250 doses **89.** 576 tiles
90. 5 gallon containers **91.** 520 lb **92.** $7\frac{3}{4}$ square yards
93. 5 bikers **94.** 3.7 planters **95.** ≈$15\frac{1}{2}$ km^2 **96.** 2,021 lb
97. 9 trips **98.** 35 minutes **99.** Answers vary

Exercise Set 9-3

3. 2.0625 lb **4.** 3 lb **5.** 36.8 oz **6.** 49.6 oz **7.** 8,400 lb
8. 7,600 lb **9.** 1.75 T **10.** 1.175 T **11.** ≈1.08 T **12.** 67,200 oz
13. 90 cg **14.** 0.6 cg **15.** 440 hg **16.** 1,634 g **17.** 180 cg **18.** 0.27 g
19. 7.1 dg **20.** 21.5 dag **21.** 50 dg **22.** 320 hg **23.** 325 mg
24. 32.17 g **25.** 4,325,000 g **26.** 0.05 kg **27.** 0.086 g **28.** 2,400 g
29. 0.4 kg **30.** 6,600 g **31.** 56.32 hg **32.** 15 cg **33.** ≈4.29 oz
34. ≈441 g **35.** ≈0.106 lb **36.** ≈59,400 lb **37.** ≈27,272.73 hg
38. ≈0.05 oz **39.** ≈162.36 t **40.** ≈59,090.91 dg **41.** ≈5.23 t
42. ≈17.93 T **43.** ≈59,640 dg **44.** ≈2,909.09 dag **45.** ≈291.07 oz
46. ≈33.64 t **47.** ≈51,240 dg **48.** ≈1,964.29 oz **49.** ≈122,727.27 dg
50. ≈0.03 lb **51.** ≈18,636.36 g **52.** ≈1.5 oz **53.** 57.2°F **54.** 80.6°F
55. 131°F **56.** 212°F **57.** 302°F **58.** 23°F **59.** −0.4°F **60.** −4°F
61. −27.4°F **62.** −58°F **63.** −15°C **64.** ≈−2.78°C **65.** 0°C
66. 70°C **67.** ≈37.78°C **68.** ≈−19.44°C **69.** ≈−23.33°C
70. −30°C **71.** ≈−25.56°C **72.** 100°C
73. Juan's parents are not correct, a 6.25 lb baby is average.
74. Billy's truck will not be able to handle the load.
75. The elevator is more than able to hold their weight.
76. Susan's doctor might suggest she go on a diet.
77. She will not pack a coat. **78.** 1.5 pounds **79.** The cars are too heavy.

80. He can fill 8 jars (almost 9). **81.** 68°F **82.** 101.84°F
83. He has plenty of solution. **84.** 2,240 grams
85. The truck poured too much sand into the foundation.
86. She lost 57.6 oz of fat. **87.** They ate 5.5 lb of steak.
88. They should not let him on!
89. Her heating element will not be hot enough.
90. Yuan did not buy the right cement. **91.** ≈1,545.5 kg
92. The sugar should be melted. **93.** There is not enough to fill 50 bags.
94. 67.5 kg

Review Exercises

1. 12 ft **2.** ≈4.46 mi **3.** 1.75 ft **4.** ≈9.89 yd **5.** ≈130.13 mi
6. 1,000 cm **7.** 3.6 dm **8.** 135,400 dm **9.** 34.5 m **10.** 0.056 hm
11. ≈275.59 in. **12.** ≈38.276 yd **13.** ≈33.46 in. **14.** ≈210.31 dm
15. 1,051.56 dm **16.** ≈31.08 km **17.** 91.44 m **18.** ≈6.21 mi
19. ≈330 km/h; ≈300.67 ft/sec **20.** ≈0.75 h **21.** The Eiffel Tower
22. ≈39.57 ft^2 **23.** ≈0.11 acre **24.** 55,756,800 ft^2 **25.** ≈352.10 yd^2
26. ≈8.82 mi^2 **27.** ≈148.35 cm^2 **28.** ≈213.9 dm^2 **29.** ≈0.50 km^2
30. ≈0.03 ft^2 **31.** ≈3,483,870,968 ft^2 **32.** ≈6,702 oz
33. 72,576,000 in.^{3} **34.** ≈1,212 gal **35.** 3,500 cm^3 **36.** ≈1,299.47 in.^{3}
37. 300,000 mL **38.** 67,300 L **39.** 0.054457 kL **40.** 2.31 L
41. 0.45672 kL **42.** The patient is in danger. **43.** 85 drops
44. $1,044,555.82 **45.** The second is cheaper. **46.** $500
47. 4.0625 lb **48.** 4,800 lb **49.** 17.25 T **50.** 41,600 oz **51.** ≈18 T
52. 70 cg **53.** 457 mg **54.** 5,600 g **55.** 0.04 kg **56.** 343.45 hg
57. ≈4.576 oz **58.** ≈0.10 lb **59.** ≈1,292.61 g **60.** ≈44,318.18 dg
61. ≈0.05 lb **62.** ≈1,605.12 oz **63.** 55.4°F **64.** 73.4°F **65.** 23°F
66. 10.4°F **67.** ≈19.44°C **68.** ≈33.33°C **69.** ≈−29.44°C
70. ≈−22.78°C **71.** ≈7.04 lb **72.** Heinrich has a fever.
73. 66.4°F **74.** 218.4 lb
75. It appears the deal on the Canadian side is better, but only if they're both using the same currency!

Chapter 9 Test

1. ≈3.67 yd **2.** ≈1.64 mi **3.** 12.3 cm **4.** 2,000 dm **5.** ≈13.39 in.
6. 3.6576 m **7.** ≈24.13 ft^2 **8.** ≈5.05 mi^2 **9.** ≈1.55 in.^{2}
10. ≈1.116 m^2 **11.** ≈6.10 in.^{3} **12.** ≈0.25 m^3 **13.** ≈606 gal
14. ≈981.75 in.^{3} **15.** 22.8125 lb **16.** ≈1.08 T **17.** ≈7.66 T
18. 50 cg **19.** ≈8.8 oz **20.** ≈0.06 lb **21.** 55.4°F **22.** ≈29.44°C
23. ≈41.86 km **24.** ≈193.6 lb
25. The first at $1.39 per square foot is cheaper.
26. ≈16,404 ft **27.** ≈ 3.37 hours **28.** ≈226.366 km
29. ≈0.093 yd^2 per min **30.** $32,168.80
31. The agent is wasting her time. **32.** ≈41,818,182 kg **33.** 15.12°F

CHAPTER 10: GEOMETRY

Exercise Set 10-1

7. Ray; $\overrightarrow{AB}$ **8.** Line; $\overleftrightarrow{RS}$ **9.** Line; l **10.** Point; P
11. Segment; $\overline{TU}$ **12.** Half line; $\overset{\circ\!\!\rightarrow}{EF}$ **13.** $\angle RST$; $\angle TSR$; $\angle S$; $\angle 3$
14. $\angle CDE$; $\angle EDC$; $\angle 5$; $\angle D$ **15.** straight **16.** acute **17.** obtuse
18. right **19.** vertical angles **20.** alternate interior angles
21. corresponding angles **22.** vertical angles
23. corresponding angles **24.** alternate exterior angles
25. alternate exterior angles **26.** corresponding angles **27.** 82°
28. 66° **29.** 58° **30.** 34° **31.** 12° **32.** 6° **33.** 24° **34.** 90°
35. 118° **36.** 37° **37.** 60° **38.** 142°
39. $m\angle 1 = 37°$; $m\angle 2 = 143°$; $m\angle 3 = 37°$
40. $m\angle 1 = 152°$; $m\angle 2 = 28°$; $m\angle 3 = 152°$
41. $m\angle 1 = 90°$; $m\angle 2 = 90°$; $m\angle 3 = 90°$
42. $m\angle 1 = 110°$; $m\angle 2 = 70°$; $m\angle 3 = 110°$
43. $m\angle 1 = m\angle 3 = m\angle 5 = m\angle 7 = 15°$
$m\angle 2 = m\angle 4 = m\angle 6 = 165°$
44. $m\angle 1 = m\angle 3 = m\angle 4 = m\angle 7 = 60°$
$m\angle 2 = m\angle 5 = m\angle 6 = 120°$
45. 90° **46.** 180° **47.** 60° **48.** 120° **49.** 104° **50.** 76° **51.** 104°
52. 76° **53.** 104° **54.** 104° **55.** 104° **56.** 76°
57. $m\angle 3 = 55°$; $m\angle 4 = 125°$ **58.** $m\angle 1 = 30°$

Exercise Set 10-2

9. Isosceles; acute **10.** Right **11.** Obtuse **12.** Equilateral; acute
13. Scalene triangle; acute **14.** Equilateral; acute **15.** 70° **16.** 50°
17. 15° **18.** 150° **19.** 40° **20.** 30° **21.** 34 ft
22. 578 in. **23.** 560 cm **24.** 78 mi **25.** 810 yd
26. 270 km **27.** 12 ft **28.** 36 mi **29.** 30 in. **30.** 36 cm **31.** $21\frac{9}{11}$ in.
32. $173\frac{23}{29}$ yd **33.** 17 m **34.** 6 km **35.** 11 ft **36.** $\frac{6}{5}$ yd
37. ≈84.9 ft **38.** ≈45.9 in. **39.** 14 mi **40.** 360 yd **41.** 270 ft
42. 35 m **43.** 10 ft **44.** ≈19.1 ft **45.** ≈212.2 mi **46.** ≈35 ft
47. 180° since the three angles form a straight angle
48. $m\angle 3 = m\angle 5$ since they are alternate interior angles between two parallel lines
49. $m\angle 1 = m\angle 4$ since they are alternate interior angles between two parallel lines
50. $m\angle 2 + m\angle 4 + m\angle 5 = 180°$
51. This proves the result for every triangle.

Exercise Set 10-3

7. Octagon, 1,080° **8.** Quadrilateral, 360° **9.** Triangle, 180°
10. Pentagon, 540° **11.** Hexagon, 720° **12.** Heptagon, 900°
13. Rectangle **14.** Parallelogram **15.** Trapezoid **16.** Rhombus
17. 76 yd **18.** 44 in. **19.** 28 ft **20.** 36 cm **21.** 44 in. **22.** 60 yd
23. 42 ft **24.** 14.3 in. **25.** 42 mi **26.** 42 km **27.** 360 ft **28.** 81 ft
29. 241 ft **30.** $576 **31.** 66 in. **32.** 79 ft **33.** 5.08 times
34. 4.65 times **35.** $392 **36.** $84.60 **37.** ≈1.74 mph **38.** 1.88 mph
40. 72°

Exercise Set 10-4

5. 289 in.^{2} **6.** 264 ft^2 **7.** 450 yd^2 **8.** 7,875 cm^2 **9.** 400 m^2 **10.** 70 ft^2
11. 105 mi^2 **12.** 40.5 in.^{2} **13.** 467.5 in.^{2} **14.** 351.5 km^2 **15.** 72 in.^{2}
16. 60 yd^2 **17.** $C \approx 50.27$ in.; $A \approx 201.06$ in.^{2}
18. $C \approx 62.83$ ft; $A \approx 314.16$ ft^2 **19.** $C \approx 50.27$ m; $A \approx 201.06$ m^2
20. $C \approx 18.85$ cm; $A \approx 28.27$ cm^2
21. $C \approx 131.95$ km; $A \approx 1{,}385.44$ km^2 **22.** $C \approx 28.27$ m; $A \approx 63.62$ m^2
23. 11.11 yd^2 **24.** $577.50 **25.** 40 images **26.** 57,600 ft^2; $1,152,000
27. $1,620 **28.** $206.11 **29.** 7.5 ft^2 **30.** 2 yd^2 **31.** $188.55
32. 98.17 in.^{2} **33.** 11,309.73 mi^2 **34.** 285.66 yd **35.** 16.28 ft^2
36. 50.27 ft^2 **37.** When the radius is 2.
38. The height will be the length of one side of a triangle if it is a right triangle and the base is one of the legs.
39. Answers vary **40.** 30 in.^{2} **41.** 30 in.^{2}

Exercise Set 10-5

7. 125 in.^{3} **8.** 36 m^3 **9.** 210 m^3 **10.** 8 cm^3 **11.** 48 m^3 **12.** 400 in.^{3}
13. 19,704.07 cm^3 **14.** 4,310.27 yd^3 **15.** 1,005.31 ft^3
16. 35,185.84 km^3 **17.** 33,510.32 in.^{3} **18.** 523,598.78 cm^3
19. 461.81 cm^3 **20.** 929.31 ft^3 **21.** 150 in.^{2} **22.** 72 m^2 **23.** 214 m^2
24. 24 cm^2 **25.** 4,046.37 cm^2 **26.** 1,539.38 yd^2 **27.** 5,026.55 in.^{2}
28. 31,415.93 cm^2 **29.** 648 ft^3 **30.** 216 ft^2 **31.** 190,228,656 ft^3
32. 48 ft^3 **33.** 13,684.78 in.^{3} **34.** 314.16 cm^2 **35.** 141.11 in.^{3}
36. 25.13 ft^3 **37.** 55,417,694.41 mi^2 **38.** 2.20×10^{10} km^3 **39.** 2,304 in.^{3}
40. $3,600 **41.** 1,080 in.^{2} **42.** 6,500 cm^2 **43.** 3.4 in. **44.** 5.3 cm
45. two 5-gallon cans **46.** three 1-gallon cans **47.** 1 1-gallon can
48. 1 5-gallon can and 2 1-gallon cans **49.** 154 in.^{3} **50.** 48.62 gallons
51. 96 m^2 **53.** 2,262 ft^2

Exercise Set 10-6

5. $\cos A = \frac{3}{5}$; $\sin A = \frac{4}{5}$; $\tan A = \frac{4}{3}$
6. $\cos A = \frac{48}{73}$; $\sin A = \frac{55}{73}$; $\tan A = \frac{55}{48}$
7. $\cos A = \frac{15}{17}$; $\sin A = \frac{8}{17}$; $\tan A = \frac{8}{15}$
8. $\cos A = \frac{12}{37}$; $\sin A = \frac{35}{37}$; $\tan A = \frac{35}{12}$
9. $\cos A = \frac{\sqrt{51}}{10}$; $\sin A = \frac{7}{10}$; $\tan A = \frac{7}{\sqrt{51}}$
10. $\cos A = \frac{10}{\sqrt{269}}$; $\sin A = \frac{13}{\sqrt{269}}$; $\tan A = \frac{13}{10}$ **11.** 92.71 cm **12.** 150.94 yd
13. 76.88 in. **14.** 20 km **15.** 1.39 mm **16.** 666.71 mi **17.** 122.98 in.
18. 1,257.31 yd **19.** 1,857.41 ft **20.** 10.91 ft **21.** 77.47 mi
22. 1,294.10 ft **23.** ≈ 48° **24.** ≈ 52° **25.** 60° **26.** ≈ 58°
27. ≈ 29° **28.** ≈ 38° **29.** ≈ 45° **30.** ≈ 43° **31.** 23,644.67 ft

32. 390.64 ft **33.** 4,980.76 m **34.** 3° **35.** 1,218.93 ft **36.** 88.11 ft **37.** 68° **38.** 22.28 ft **39.** 352.50 ft **40.** 10.36 ft **41.** 3,932.92 ft **42.** 19° **43.** 1.2° **44.** 35.3° **45.** 6.9° **46.** 7.7° **47.** h = 737.6 ft and d = 673.8 ft **48.** 44.5 ft
49. Use the cosine then use the Pythagorean Theorem.
50. Use the ratio for the sine of A to find c.

Exercise Set 10-7

9. Answers vary **10.** Answers vary **11.** Answers vary
12.

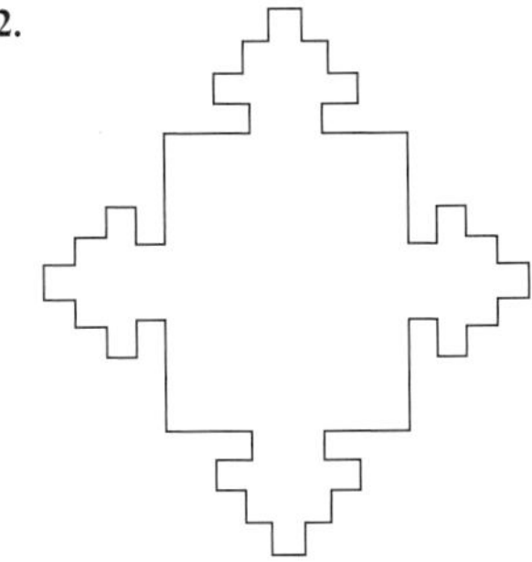

15. A regular polygon will tessellate only if the measure of its angle is a divisor of 360°.

Review Exercises

1. line **2.** ray **3.** line segment **4.** angle **5.** trapezoid **6.** parallelogram **7.** triangle **8.** right triangle **9.** hexagon **10.** cone **11.** obtuse **12.** right angle **13.** vertical angles **14.** corresponding angles **15.** alternate exterior angles **16.** complementary **17.** (a) 63° (b) 2° **18.** (a) 8° (b) 167°
19. $m\measuredangle 2 = 163°$; $m\measuredangle 3 = m\measuredangle 1 = 17°$
20. $m\measuredangle 1 = m\measuredangle 4 = m\measuredangle 5 = m\measuredangle 7 = 138°$; $m\measuredangle 2 = m\measuredangle 3 = m\measuredangle 6 = 42°$
21. 43° **22.** 17 in. **23.** 72 ft **24.** right and scalene **25.** obtuse and scalene **26.** acute and isosceles **27.** acute and equilateral **28.** 13 units **29.** 6 units **30.** a = 6 and b = 14 **31.** a = 15 and b = 18 **32.** 90 ft **33.** 15.8 ft **34.** 17.9 ft **35.** 166.4 ft **36.** 600 ft **37.** decagon; 1,440° **38.** heptagon; 900°. **39.** 16 ft **40.** 26 in. **41.** 28 m **42.** 24 m **43.** 1,376 cm^2 **44.** 60.8 m^2 **45.** 9,160.88 km^2 **46.** 116 ft^2 **47.** C = 51.84 yd; A = 213.82 yd^2 **48.** \$31,200 **49.** \$199.74 **50.** 31.1 minutes **51.** 24,429.02 cm^3 **52.** 25,056 in.^{3} **53.** 84.82 yd^3 **54.** 69,944.42 ft^3 **55.** 5,736 in.^{2} **56.** 9,399.65 ft^2 **57.** 775.71 revolutions **58.** 450 in.^{2} **59.** 38.79 cm^3 **60.** 648 ft^2 **61.** $\sin B = \frac{3}{\sqrt{34}}$; $\cos B = \frac{5}{\sqrt{34}}$; $\tan B = \frac{3}{5}$ **62.** ≈ 20.38 in. **63.** ≈ 2.33 yd **64.** ≈ 47.0° **65.** ≈ 51.8° **66.** 39° **67.** 33.04 ft **68.** 11.1° **69.** 9 minutes **70.** Answers vary
72. circles that pass through both poles

Chapter 10 Test

1. 17° **2.** 31° **3.** $m\measuredangle 2 = 28°$; $m\measuredangle 3 = m\measuredangle 1 = 152°$
4. $m\measuredangle 1 = m\measuredangle 4 = m\measuredangle 5 = 32°$; $m\measuredangle 2 = m\measuredangle 3 = m\measuredangle 6 = m\measuredangle 7 = 148°$
5. 48° **6.** 20 in. **7.** 2.5 ft **8.** 30.4 ft **9.** 1,080° **10.** 16 timbers **11.** 18 in.^{2} **12.** 1,066 km^2 **13.** 432 yd^2 **14.** 5,808.8 ft^2 **15.** 345.96 mi^2 **16.** 81.68 cm **17.** 884.36 in.^{3} **18.** 282.7 cm^3 **19.** 100 ft^3 **20.** 923.63 yd^3 **21.** 150 ft^3 **22.** 33,510.32 m^3 **23.** 552.96 in.^{2} **24.** 472.81 in.^{2} **25.** 346.36 in.^{2}
26. Latitude lines are not lines in elliptical geometry.
27. A brick wall is a tessellation. **28.** 6,280 ft^2 **29.** 24.49 mi **30.** \$41.08 **31.** 0.21 pound **32.** 3,463.32 pounds **33.** 33.3°

CHAPTER 11: PROBABILITY AND COUNTING TECHNIQUES

Exercise Set 11-1

5. 3,628,800 **6.** 120 **7.** 362,880 **8.** 1 **9.** 1 **10.** 24 **11.** 12 **12.** 10,080 **13.** 336 **14.** 72 **15.** 420 **16.** 15,120 **17.** 22,350 **18.** 7,880,400 **19.** 56 **20.** 2,520 **21.** 479,001,600 **22.** 60 **23.** 720 **24.** 1 **25.** 1 **26.** 40,320 **27.** 990 **28.** 30 **29.** 840 **30.** 336 **31.** 151,200 **32.** 5,040 **33.** 5,527,200 **34.** 1,860,480 **35.** 300 **36.** 2,520 **37.** 210 **38.** 120 **39.** 720 **40.** 11,441,304,000 **41.** 57,120 **42.** 24 **43.** 120 **44.** 12,600 **45.** 210 **46.** 60 **47.** 210 **48.** 60 **49.** 2,520 **50.** $\approx 4.359 \times 10^{10}$ **51.** 43,680 **52.** $\approx 7.414 \times 10^{11}$ **53.** $\approx 5.1988 \times 10^{35}$ **54.** 332,640 **55.** 362,880; 40,320 **56.** 30 choices

Exercise Set 11-2

5. 10 **6.** 56 **7.** 35 **8.** 15 **9.** 15 **10.** 1 **11.** 1 **12.** 36 **13.** 66 **14.** 4
15. ${}_8C_5 = 56$; ${}_8P_5 = 6{,}720$ **16.** ${}_5C_3 = 10$; ${}_5P_3 = 60$ **17.** ${}_6C_2 = 15$; ${}_6P_2 = 30$
18. ${}_{10}C_6 = 210$; ${}_{10}P_6 = 151{,}200$ **19.** ${}_9C_9 = 1$; ${}_9P_9 = 362{,}880$ **20.** ${}_{12}C_1 = 12$; ${}_{12}P_1 = 12$
21. Combination **22.** Permutation **23.** Permutation **24.** Combination **25.** Permutation **26.** Combination **27.** Combination **28.** Permutation **29.** 2,598,960 **30.** 20 **31.** 126; 35 **32.** 210 **33.** 120 **34.** 15,504 **35.** 462 **36.** 18,480 **37.** 166,320 **38.** 210; 285 **39.** 14,400 **40.** 560 **41.** 67,200 **42.** 2,970 **43.** 53,130 **44.** 455 **45.** 126 **46.** 45 **47.** 30,045,015 **48.** 24,310

Exercise Set 11-3

9. Yes **10.** Yes **11.** No **12.** Yes **13.** No **14.** Yes **15.** Yes **16.** No **17.** Empirical **18.** Classical **19.** Classical **20.** Empirical
21. (a) $\frac{1}{6}$ (b) $\frac{1}{2}$ (c) $\frac{1}{3}$ (d) 1 (e) 1 (f) $\frac{5}{6}$ (g) $\frac{1}{6}$
22. (a) $\frac{1}{4}$ (b) $\frac{3}{4}$ (c) $\frac{1}{2}$ **23.** (a) $\frac{1}{7}$ (b) $\frac{3}{7}$ (c) $\frac{3}{7}$ (d) 1 (e) 0
24. (a) $\frac{1}{2}$ (b) $\frac{3}{10}$ (c) $\frac{7}{10}$ (d) $\frac{7}{10}$ (e) $\frac{1}{2}$ **25.** $\frac{4}{9}$ **26.** $\frac{7}{50}$ **27.** $\frac{9}{16}$
28. 0.33 **29.** 0.91 **30.** 0.78 **31.** 0.84 **32.** (a) $\frac{9}{16}$ (b) $\frac{7}{16}$
33. $\frac{3}{13}$ **34.** $\frac{1}{32}$ **35.** $\frac{5}{11}$ **36.** 0.32 **37.** $\frac{17}{25}$ **38.** $\frac{1}{3}$ **39.** $\frac{13}{35}$ **40.** $\frac{3}{7}$
41. 0 **42.** (a) 0.14812 (b) 0.06414 (c) 0.09732
43. (a) ≈0.26 (b) ≈0.96 **44.** (a) 0.51 (b) 0.485 (c) 0.44
45. (a) 0.42 (b) 90% **46.** (a) 0.24; 0.09 (b) 0.17 (c) 0.91
47. It is likely that the penny is unbalanced.
48. The statement is incorrect.
49. (a) $\frac{6}{25}$ (b) $\frac{2}{5}$ (c) $\frac{9}{25}$ (d) $\frac{12}{25}$ (e) $\frac{1}{5}$ **50.** $\frac{1}{36}$

Exercise Set 11-4

3. (a) $\frac{3}{8}$ (b) $\frac{1}{8}$ (c) $\frac{7}{8}$ **4.** (a) $\frac{1}{4}$ (b) $\frac{1}{4}$ (c) $\frac{1}{3}$
5. (a) $\frac{5}{12}$ (b) $\frac{1}{4}$ (c) $\frac{1}{6}$ **6.** (a) $\frac{1}{4}$ (b) $\frac{1}{4}$ (c) $\frac{3}{8}$
7. (a) $\frac{1}{3}$ (b) $\frac{1}{3}$ (c) $\frac{1}{9}$ (d) $\frac{4}{9}$ (e) $\frac{1}{3}$
8. (a) $\frac{1}{4}$ (b) $\frac{1}{16}$ (c) $\frac{11}{16}$ (d) $\frac{11}{16}$ (e) $\frac{15}{16}$
9. (a) $\frac{1}{4}$ (b) $\frac{3}{4}$ (c) $\frac{1}{4}$ (d) $\frac{1}{8}$
10. (a) $\frac{1}{4}$ (b) $\frac{3}{4}$ (c) $\frac{7}{16}$ (d) $\frac{1}{16}$ (e) $\frac{1}{8}$
11. (a) $\frac{3}{5}$ (b) $\frac{1}{2}$ (c) $\frac{4}{5}$ **12.** (a) $\frac{1}{2}$ (b) $\frac{1}{9}$ (c) $\frac{1}{18}$ (d) $\frac{1}{3}$ (e) $\frac{2}{9}$
13. (a) $\frac{1}{2}$ (b) $\frac{1}{6}$ (c) $\frac{1}{4}$ **14.** (a) $\frac{1}{2}$ (b) $\frac{3}{5}$ (c) $\frac{1}{5}$ (d) $\frac{2}{5}$ (e) $\frac{3}{5}$
15. (a) $\frac{1}{13}$ (b) $\frac{1}{4}$ (c) $\frac{1}{52}$ (d) $\frac{2}{13}$ (e) $\frac{4}{13}$ (f) $\frac{4}{13}$ (g) $\frac{1}{2}$ (h) $\frac{1}{26}$ (i) $\frac{7}{13}$ (j) $\frac{1}{26}$
16. (a) $\frac{1}{52}$ (b) $\frac{1}{2}$ (c) $\frac{1}{13}$ (d) $\frac{1}{26}$ (e) $\frac{7}{13}$ (f) $\frac{1}{52}$ (g) $\frac{2}{13}$ (h) $\frac{3}{4}$ (i) $\frac{3}{52}$ (j) $\frac{1}{26}$
17. (a) $\frac{1}{9}$ (b) $\frac{2}{9}$ (c) $\frac{1}{6}$ (d) $\frac{5}{18}$ (e) $\frac{11}{36}$ (f) $\frac{1}{2}$ (g) $\frac{3}{4}$ (h) 1
18. (a) $\frac{5}{36}$ (b) $\frac{5}{12}$ (c) $\frac{11}{36}$ (d) $\frac{7}{12}$ (e) $\frac{1}{36}$ (f) 0 (g) $\frac{1}{18}$
19. The outcomes are not equally likely because there is a $\frac{1}{4}$ probability of getting WW; each of the other six outcomes has a probability of $\frac{1}{8}$.
20. The probability of getting the first wrong is $\frac{1}{2}$ but the probability of the second wrong is $\frac{1}{4}$ etc. So each outcome is not equally likely.
21. 216 **22.** 10 **23.** $\frac{5}{108}$

Exercise Set 11-5

3. (a) 0.0004 (b) 0.15 (c) 0.383 (d) 0.45
4. (a) 0.114 (b) 0.0286 (c) 0.343 (d) 0.514

5. (a) 0.033 (b) 0.0083 (c) $\frac{3}{10}$ (d) $\frac{3}{40}$ (e) $\frac{3}{20}$
6. (a) 0.108 (b) 0.011 (c) 0.377 (d) 0.126 (e) 0.377
7. (a) 0 (b) 0.51 (c) 0.018 **8.** 0.00082
9. (a) 0.07 (b) 0.42 (c) 0.015 (d) 0.16 (e) 0.336
10. (a) 0.123 (b) 0.029 (c) 0.022 (d) 0.176 (e) 0.308
11. 0.01 **12.** 0.363 **13.** 0.42 **14.** 1.5×10^{-6} **15.** 0.476 **16.** 0.115
17. 0.00144 **18.** 0.00198 **19.** 0.0226 **20.** 0.00024 **21.** 0.0000015
22. 0.0000139 **23.** 0.0000628

Exercise Set 11-6

7. (a) 1:11 (b) 1:35 (c) 5:1 (d) 17:1 (e) 1:5
8. (a) 1:5 (b) 1:5 (c) 1:1 (d) 1:1 (e) 1:1
9. (a) 1:12 (b) 3:10 (c) 3:1 (d) 1:12 (e) 1:1
10. (a) 1:7 (b) 1:7 (c) 5:3 (d) 5:3 (e) 7:1
11. (a) $\frac{7}{11}$ (b) $\frac{5}{7}$ (c) $\frac{3}{4}$ (d) $\frac{4}{5}$ **12.** (a) $\frac{3}{7}$ (b) $\frac{7}{8}$ (c) $\frac{5}{9}$ (d) $\frac{6}{11}$
13. $\frac{5}{14}$ **14.** odds in favor 1:5 odds against 5:1 **15.** −$3.00
16. $7.25 **17.** $0.83 **18.** −$1.06 **19.** −$1.00 **20.** –$2.00
21. –$0.50; −$0.52 **22.** −$32

Exercise Set 11-7

2. No **3.** No **4.** Yes **5.** No **6.** No **7.** Yes **8.** Yes **9.** $\frac{1}{6}$
10. $\frac{9}{13}$ **11.** $\frac{11}{19}$ **12.** $\frac{17}{22}$ **13.** (a) $\frac{17}{20}$ (b) $\frac{11}{20}$ (c) $\frac{3}{5}$
14. (a) $\frac{9}{19}$ (b) $\frac{7}{19}$ (c) $\frac{11}{19}$ **15.** (a) $\frac{8}{17}$ (b) $\frac{6}{17}$ (c) $\frac{9}{17}$ (d) $\frac{12}{17}$
16. (a) $\frac{4}{13}$ (b) $\frac{1}{2}$ (c) $\frac{7}{13}$ **17.** (a) $\frac{6}{7}$ (b) $\frac{4}{7}$ (c) 1
18. (a) $\frac{33}{40}$ (b) $\frac{13}{20}$ **19.** (a) $\frac{67}{118}$ (b) $\frac{81}{118}$ (c) $\frac{44}{59}$
20. (a) $\frac{1}{28}$ (b) $\frac{113}{168}$ (c) $\frac{125}{168}$ **21.** (a) $\frac{38}{45}$ (b) $\frac{22}{45}$ (c) $\frac{2}{3}$
22. (a) $\frac{1}{5}$ (b) $\frac{2}{15}$ (c) $\frac{2}{3}$ **23.** (a) $\frac{14}{31}$ (b) $\frac{23}{31}$ (c) $\frac{19}{31}$
24. (a) $\frac{467}{1,392}$ (b) $\frac{47}{58}$ (c) $\frac{833}{1,392}$ **25.** (a) $\frac{467}{1,392}$ (b) $\frac{25}{32}$ (c) $\frac{1,955}{2,784}$
26. (a) $\frac{4}{9}$ (b) $\frac{1}{3}$ (c) $\frac{5}{12}$ **27.** $\frac{7}{10}$ **28.** (a) $\frac{1}{36}$ (b) $\frac{1}{36}$ **29.** 0.06
30. 0.10 **31.** 0.30 **32.** (b) 36 (c) 7 (d) $\frac{7}{36}$ (e) $\frac{1}{9}$ (f) $\frac{1}{12}$ (g) $\frac{7}{36}$

Exercise Set 11-8

5. Independent **6.** Dependent **7.** Dependent **8.** Dependent
9. Independent **10.** Dependent **11.** Dependent **12.** Independent
13. 0.0058 **14.** 0.3136 **15.** 0.4746 **16.** 0.0016 **17.** 0.0625
18. $\frac{1}{144}$ **19.** $\frac{1}{12}$ **20.** 0.00058 **21.** $\frac{1}{8}$ **22.** $\frac{1}{133,225}$ **23.** $\frac{1}{15}$ **24.** $\frac{4}{15}$
25. 0.1311 **26.** 0.2306 **27.** (a) 0.00018 (b) 0.0129 (c) 0.1176
28. 0.1786 **29.** 0.0179 **30.** 0.188 **31.** $\frac{1}{6}$ **32.** $\frac{1}{13}$ **33.** $\frac{2}{11}$ **34.** $\frac{1}{4}$
35. $\frac{1}{4}$ **36.** $\frac{7}{11}$ **37.** $\frac{2}{3}$ **38.** $\frac{1}{3}$ **39.** 1 **40.** $\frac{2}{3}$ **41.** 0.61 **42.** 0.16
43. $\frac{1}{2}$ **44.** $\frac{28}{59}$ **45.** $\frac{25}{216}$ **46.** $\frac{31}{32}$ **47.** $\frac{1}{133,225}$

Exercise Set 11-9

5. 0.288 **6.** 0.2048 **7.** 0.0616 **8.** 0 **9.** 0.0000304 **10.** 0.00162
11. 0.0994 **12.** 0.1570 **13.** 0.00615 **14.** 0.3115 **15.** 0.00463
16. (a) 0.2963 (b) 0.4444 (c) 0.0370 **17.** 0.0750 **18.** 0.1951
19. 0.1546 **20.** 0.1851 **21.** 0.2461 **22.** 0.0879 **23.** 0.2340
24. 0.3241 **25.** 0.1906 **26.** 0.3811 **27.** 0.1585 **28.** 0.2013
29. 0.1593 **30.** 0.0434 **31.** 0.1541 **32.** 0.2801 **33.** 0.6870
34. 0.00707 **35.** Answers vary

Review Exercises

1. 2,184 **2.** 665,280 **3.** 175,760,000; 88,583,040 **4.** 120 **5.** 120
6. 40 **7.** 84 **8.** ${}_{10}P_4 = 5,040$; ${}_{10}C_4 = 210$
9. In a permutation order matters, in a combination order does not matter. **10.** 60; no **11.** 78 **12.** 792 **13.** 729 **14.** 15
15. b, c, and e **16.** (a) $\frac{1}{6}$ (b) $\frac{1}{6}$ (c) $\frac{2}{3}$
17. (a) $\frac{1}{4}$ (b) $\frac{1}{52}$ (c) $\frac{4}{13}$ (d) $\frac{1}{13}$ (e) $\frac{1}{2}$ **18.** $\frac{16}{45}$
19. (a) 0 (b) $\frac{1}{2}$ (c) $\frac{1}{3}$ **20.** $\frac{17}{30}$ **21.** (a) $\frac{9}{35}$ (b) $\frac{23}{35}$ (c) $\frac{19}{35}$ (d) $\frac{19}{35}$
22. (a) 0.1 (b) $\frac{11}{30}$ (c) $\frac{13}{15}$ (d) $\frac{13}{15}$
23. (a) $\frac{1}{4}$ (b) $\frac{1}{6}$ (c) $\frac{1}{4}$ (d) $\frac{1}{4}$ (e) 0 (f) 1 **24.** $\frac{2}{11}$
25. S = {1H, 1T, 2H, 2T, 3H, 3T, 4H, 4T, 5H, 5T, 6H, 6T, 7H, 7T, 8H, 8T}
26. (a) S = {En,P,W; En,P,Ec; En,S,W; En,S,Ec; M,P,W; M,P,Ec; M,S,W; M,S,Ec; C,P,W; C,P,Ec; C,S,W; C,S,Ec}
(b) S = {En,P,W; En,P,Ec; En,S,W; En,S,Ec; M,P,W; M,P,Ec; M,S,W; M,S,Ec; C,P; C,S}
27. $\frac{2}{25}$ **28.** $\frac{1}{2}$ **29.** $\frac{12}{55}$ **30.** $\frac{1}{8}$ **31.** $\frac{33}{182}$ **32.** 1:3 **33.** 1:5 **34.** $\frac{3}{5}$
35. 10.5 **36.** 18.2¢ **37.** $7.23 **38.** Not mutually exclusive
39. Mutually exclusive **40.** Mutually exclusive
41. (a) $\frac{17}{50}$ (b) $\frac{3}{25}$ **42.** $\frac{5}{13}$ **43.** Dependent **44.** Independent
45. Dependent **46.** Independent **47.** $\frac{1}{7}$ **48.** 0.289 **49.** 0.016
50. (a) 0.118 (b) 0.013 (c) 1.81×10^{-4} **51.** (a) $\frac{1}{26}$ (b) $\frac{1}{4}$ (c) $\frac{1}{8}$
52. (a) $\frac{249}{340}$ (b) $\frac{279}{290}$ (c) $\frac{285}{452}$ **53.** 0.0165 **54.** 0.2429 **55.** 0.1684
56. Let x represent the number of heads.

x	$P(x)$
0	$\frac{1}{8}$
1	$\frac{3}{8}$
2	$\frac{3}{8}$
3	$\frac{1}{8}$

Chapter 11 Test

1. 380 **2.** 21 **3.** 95,040 **4.** 0.00039
5. Probability is a number that represents the likelihood of an event.
6. Answers vary **7.** 1,188,137,600; 710,424,000 **8.** 33,554,432
9. 35 **10.** 2,646 **11.** 40,320 **12.** 1,365 **13.** (a) 64 (b) 24
14. 4 **15.** (a) $\frac{1}{13}$ (b) $\frac{1}{13}$ (c) $\frac{4}{13}$
16. (a) $\frac{1}{4}$ (b) $\frac{4}{13}$ (c) $\frac{1}{52}$ (d) $\frac{1}{13}$ (e) $\frac{1}{2}$
17. (a) $\frac{12}{31}$ (b) $\frac{12}{31}$ (c) $\frac{27}{31}$ (d) $\frac{24}{31}$
18. (a) $\frac{11}{36}$ (b) $\frac{11}{12}$ (c) $\frac{11}{36}$ (d) $\frac{1}{3}$ (e) 0 (f) $\frac{11}{12}$
19. S = {H1, H3, H5, T2, T4, T6} **20.** 0.00243
21. (a) 0.025 (b) 0.000495 (c) 0 **22.** $\frac{1}{4}$ **23.** $\frac{1}{2}$ **24.** $\frac{1}{3}$ **25.** $\frac{1}{2}$
26. S = {Ma,L,S; Ma,L,C; Ma,L,He; Ma,Me,S; Ma,Me,C; Ma,Me,He; Ma,Hi,S; Ma,Hi,C; Ma,Hi,He; F,L,S; F,L,C; F,L,He; F,Me,S; F,Me,C; F,Me,He; F,Hi,S; F,Hi,C; F,Hi,He}
27. S = {M,CH; M,IN; SF,CH; SF,IN; P,CH; P,IN; H,CH; H,IN}
28. 3:5 **29.** 5:4 **30.** $\frac{7}{10}$ **31.** $5\frac{1}{6}$ **32.** $9.65 **33.** $\frac{8}{33}$ **34.** $\frac{36}{143}$
35. 0.1144

CHAPTER 12: STATISTICS

Exercise Set 12-1

9. Cluster **10.** Systematic **11.** Random **12.** Systematic
13. Stratified **14.** Random

15.

Rank	Frequency
Fr	18
So	12
Jr	6
Se	4

16.

Source	Frequency
I	13
N	5
R	3
T	4

17.

Show	Frequency
S	6
P	5
B	7
A	7

18.

Class	Frequency
21–86	2
87–152	2
153–218	12
219–284	18
285–350	9
351–416	6

19.

Class	Frequency
27–33	7
34–40	14
41–47	14
48–54	12
55–61	3
62–68	3
69–75	2

20.

Class	Frequency
33.1–41.3	5
41.4–49.6	7
49.7–57.9	3
58–66.2	8
66.3–74.5	4

21.

Class	Frequency
0–39	8
40–79	10
80–119	10
120–159	7
160–199	3
200–239	8
240–279	4

22.

Class	Frequency
5–101	17
102–198	6
199–295	6
296–392	2
393–489	2
490–586	3
587–683	1
684–780	2

23.

Class	Frequency
150–1,276	2
1,277–2,403	2
2,404–3,530	5
3,531–4,657	8
4,658–5,784	7
5,785–6,911	3
6,912–8,038	7
8,039–9,165	3
9,166–10,292	3
10,293–11,419	2

24.

Class	McGwire	Sosa
306–336	1	0
337–367	6	10
368–398	19	16
399–429	15	21
430–460	18	15
461–491	6	3
492–522	3	1
523–553	2	0

25.

Class	Frequency
839–949	12
950–1,060	9
1,061–1,171	7
1,172–1,282	1
1,283–1,393	1
1,394–1,504	0
1,505–1,615	2

26.

Stems	Leaves
0	3 8 9 9
1	0 2 2 2 4 4 4 6 8 9
2	1 2 2 5 8 8
3	1 3 6 7
4	1 9
5	2 4 8

27.

Stems	Leaves
3	8
4	1
5	0 0 2 3 3 6 8 9
6	6 8 9 9
7	0 0 3 4 5 8
8	0 1 3 3 4 4 4 5 7 9 9 9
9	0 2 4

28.

Stems	Leaves
1	2
2	0 3
3	2 5 8 8 9
4	1 3 3
5	0 1 2 3 3 5 8 9 9

29.

Stems	Leaves
0	3
1	5 9
2	2
3	1 1
4	1 4 6 6
5	2 6 6 6 9
6	0 0 6 6
7	7
8	7 8
9	6 8

30. Inferential **31.** Descriptive **32.** Inferential **33.** Descriptive **34.** Inferential **35.** Answers vary **36.** Answers vary

Exercise Set 12-2

5.

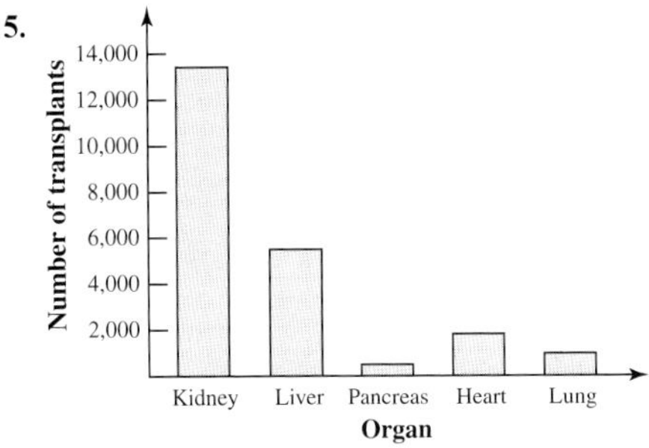

6.

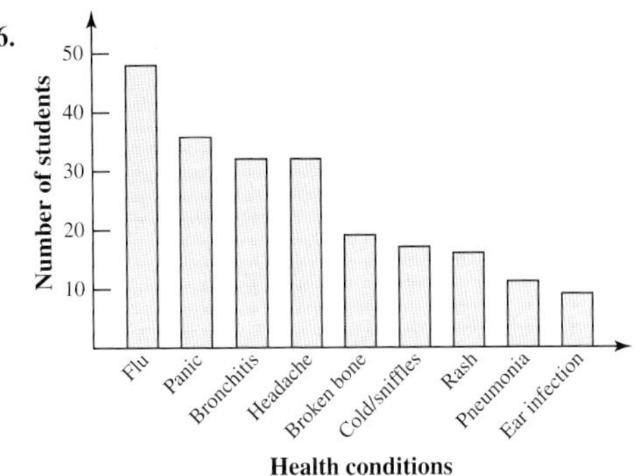

7.

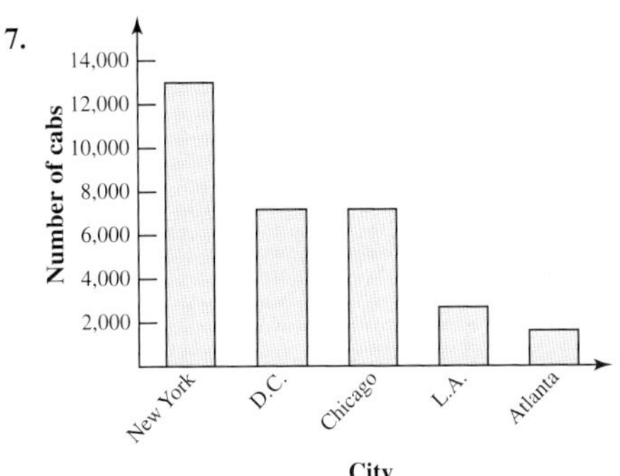

8.

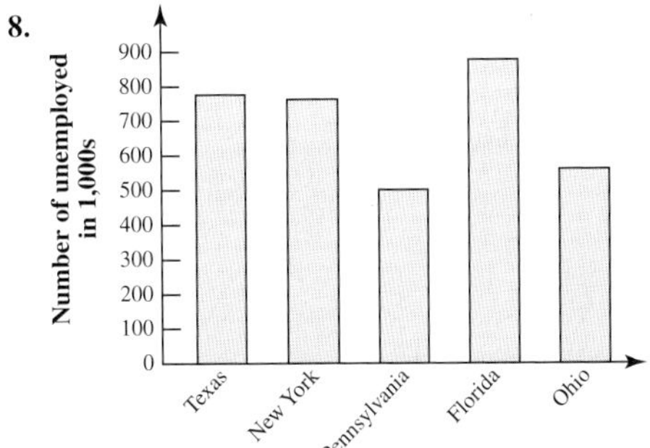

9. **10.**

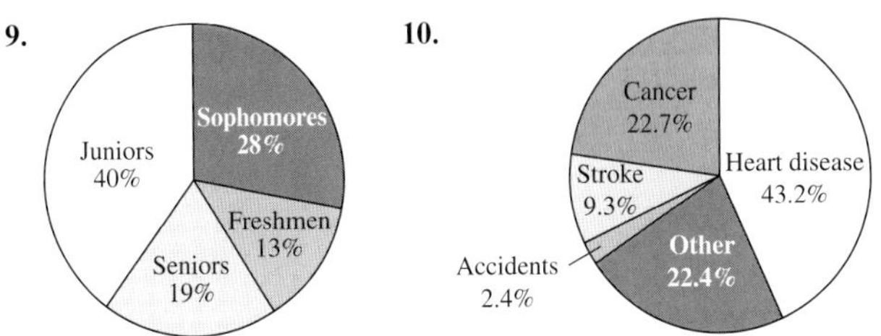

11.

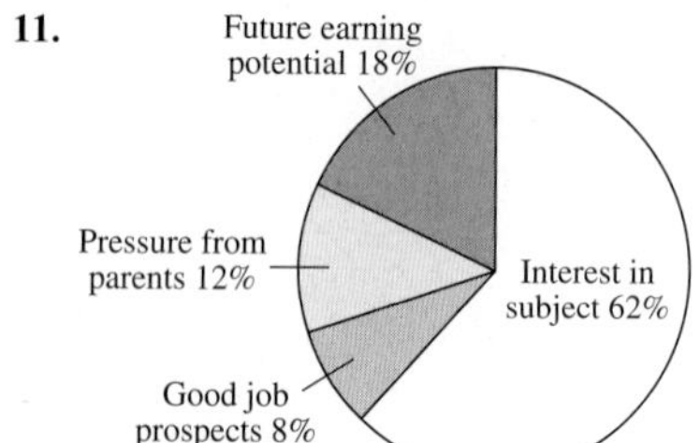

12.

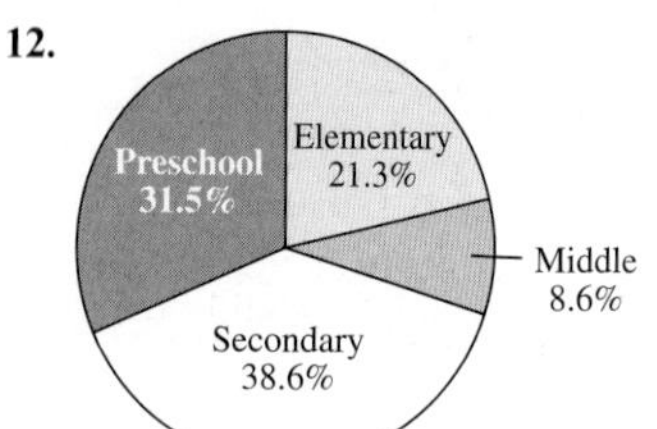

13.

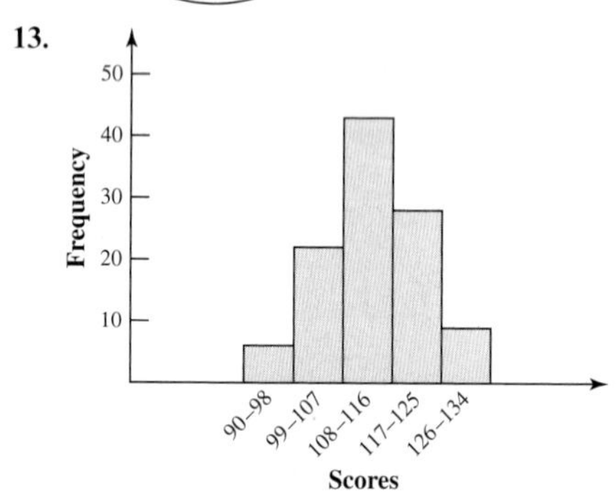

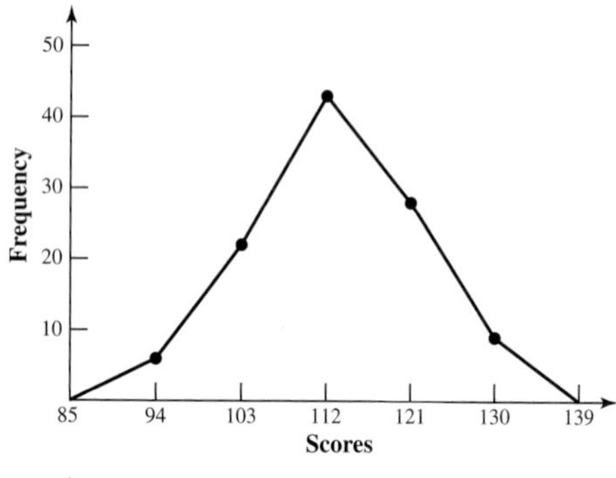

14.

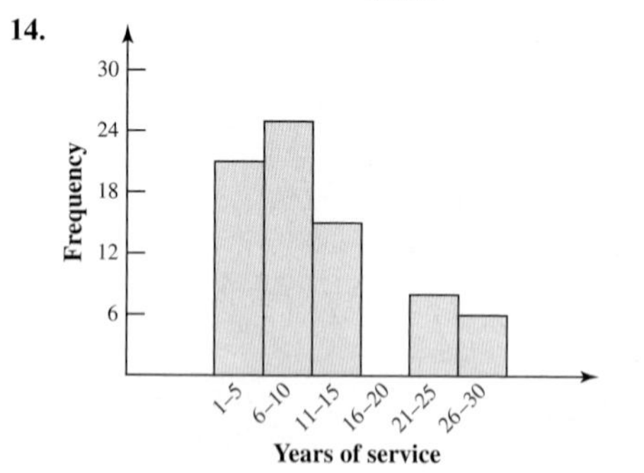

15.

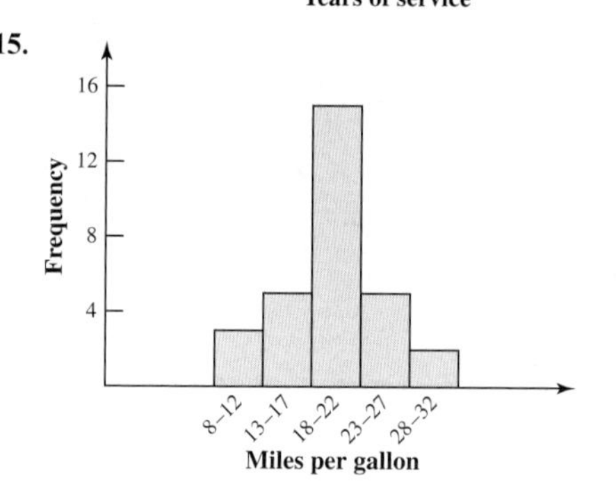

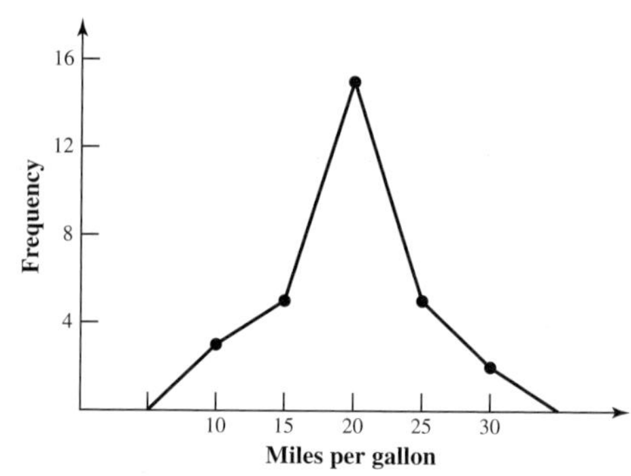

16.

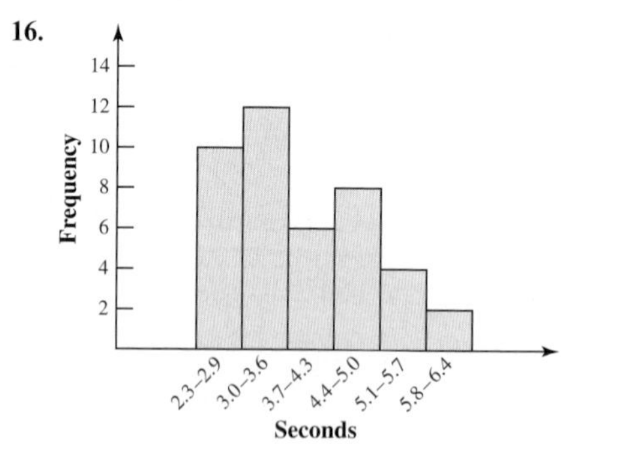

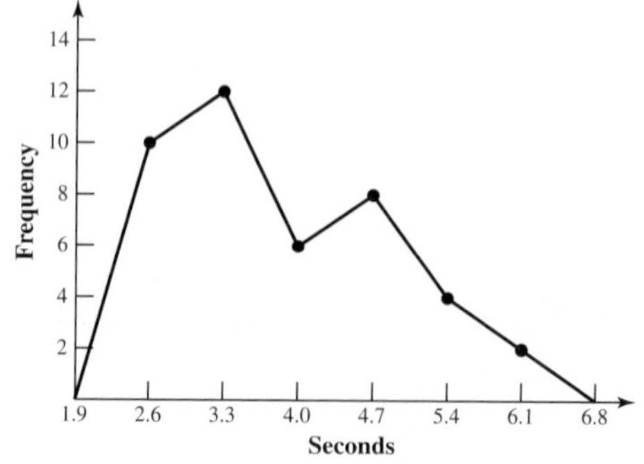

17.

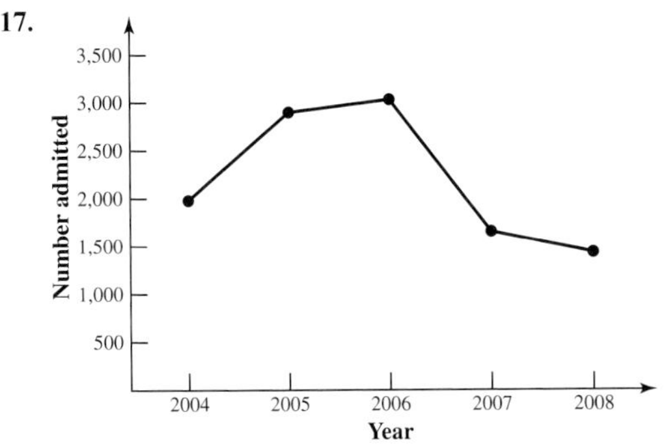

18.

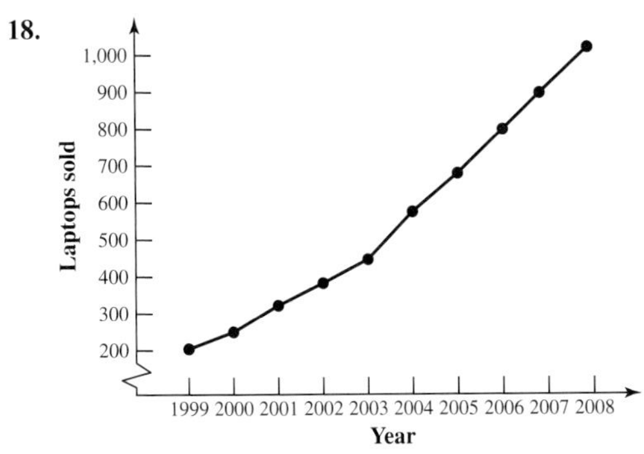

19.

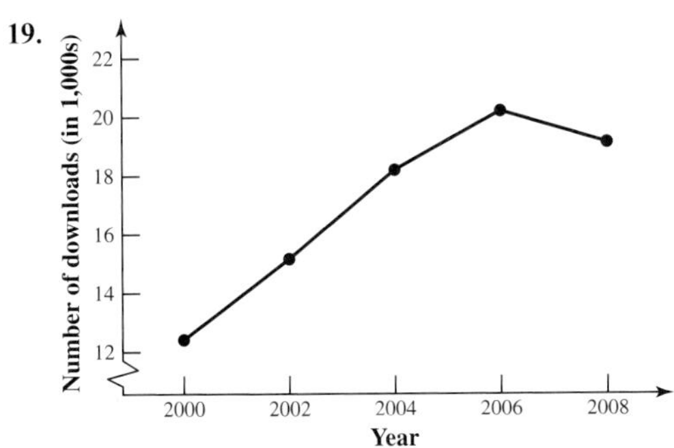

20. Time series graph **21.** Pie chart **22.** Bar graph
23. Time series graph **24.** Pie chart **25.** Pie chart
26. Answers vary **27.** The scale makes the increase look greater.

Exercise Set 12-3

7. mean ≈15.11, median = 7, mode = 3, midrange = 31
8. mean = 116.8, median = 98, mode = 80, midrange = 139
9. mean = 612.6, median = 475, no mode, midrange = 820
10. mean = 9, median = 8.5, mode = 8, midrange = 10
11. mean = 2,907.7, median = 2,723.5, no mode, midrange = 3,353.5
12. mean ≈ 4.92, median = 4.9, mode = 4.9, midrange = 5
13. mean = 189.6, median = 151, no mode, midrange = 219.5
14. mean = 318.55, median = 229, no mode, midrange = 633
15. mean = 8.25, median = 8.5, mode = 9, midrange = 9
16. mean ≈ 13,184.4, median =$13,292, no mode, midrange = 12,646.5
17. mean ≈ 17,163.9, median = 16,740, no mode, midrange = 18,566
18. mean ≈ 10.4, median = 10.5, modes = 9.4 and 10.5, midrange = 10.7
19. mean = 21.76 **20.** mean ≈ 19.67 mpg **21.** mean ≈ 4.4 seconds
22. mean ≈ 85.09 hours **23.** mean ≈ 42.87 or $42.87 million
24. mean = 33.8¢ **25.** mean = $ 180.28 **26.** mean = 23.72 days
27. Answers vary **28.** (a) Mode (b) Mode (c) Median
29. Median **30.** Mean **31.** Mode **32.** Mode **33.** Mode
34. Mean

Exercise Set 12-4

7. $s_1 < s_2$ **8.** $s_1 < s_2$ **9.** $s_1 > s_2$ **10.** $s_1 < s_2$
11. R = 60, variance = 406.75, s ≈ 20.17
12. R = 265, variance = 11,623.8, s ≈ 107.81
13. R = 1,799, variance = 438,113.6, s ≈ 661.90
14. R = 10, variance ≈ 11.31, s ≈ 3.36
15. R = 99, variance ≈ 1,288.19, s ≈ 35.89
16. R = 29, variance ≈ 63.96, s ≈ 8.00
17. R = 10, variance = 9, s = 3
18. R = 620, variance ≈ 40,135.90, s ≈ 200.34
19. R = $3.83, variance ≈ 2.79, s ≈ 1.67
20. R = 164, variance ≈ 2,991.96, s ≈ 54.7
21. R = $39.50, variance ≈ 242.59, s ≈ 15.58
22. R = 772, variance = 70,044.2778, s ≈ 264.66
23. R = 738, variance ≈ 57,844.95, s ≈ 240.5
24. R = 33, variance = 185.7, s ≈ 13.6
25. R = 432, variance ≈ 29,141.36, s ≈ 170.7
26. R = 8, variance = 6.75, s ≈ 2.6
27. The variation is not the same
28. (a) s ≈ 15.8 (b) s ≈ 15.8 (c) s ≈ 15.8 (d) s ≈ 79.1 (e) s ≈ 3.2 (f) Answers vary

Exercise Set 12-5

5. (a) 20th percentile (b) 75th percentile (c) 35th percentile (d) 5th percentile (e) 90th percentile
6. (a) 50th percentile (b) 67th percentile (c) 17th percentile (d) 58th percentile (e) 42nd percentile **7.** 75th percentile
8. 80th percentile **9.** 79th percentile **10.** 84th percentile
11. 10 **12.** 150 **13.** Maurice is ranked higher.
14. Maranda's rank is higher.
15. (a) 60th percentile (b) 8 (c) 23 years
16. (a) 25th percentile (b) 15 (c) 101
17. Q_1 = 22.5, Q_2 = 34, Q_3 = 53.5
18. Q_1 = 2.07, Q_2 = 2.215, Q_3 = 2.42
19. Q_1 = 78, Q_2 = 88.5, Q_3 = 93 **20.** Q_1 = 15.7, Q_2 = 18, Q_3 = 29.1
21. Q_1 = 107, Q_2 = 116, Q_3 = 122
22. Q_1 = 28.2, Q_2 = 45.8, Q_3 = 79.1 **23.** yes; yes **24.** not possible

Exercise Set 12-6

7. 34 **8.** 190 **9.** 499 **10.** 114 **11.** 0.474 **12.** 0.209 **13.** 0.192
14. 0.480 **15.** 0.159 **16.** 0.401 **17.** 0.345 **18.** 0.073 **19.** 0.077
20. 0.115 **21.** 0.223 **22.** 0.029 **23.** 0.463 **24.** 0.846 **25.** 0.885
26. 0.984 **27.** 0.971 **28.** 0.579 **29.** $z = +0.45$ **30.** $z = -2.2$
31. (a) $z = \pm 2.05$ (b) $z = \pm 1.75$ (c) $z = \pm 2.40$

Exercise Set 12-7

3. (a) 0.378 (b) 0.179 **4.** 0.092 **5.** (a) 0.264 (b) 0.316
6. (a) 0.907 (b) 0.853 **7.** (a) 0.005 (b) 0.162 (c) 0.749
8. (a) 0.546 (b) 0.268 (c) 0.334
9. (a) 0.153 (b) 0.774 (c) 0.187 **10.** (a) 0 (b) 0.841
11. (a) 0.841 (b) 0.067 **12.** (a) 0.569 (b) 0.641
13. (a) 0.776 (b) 0.405 **14.** (a) 0.994 (b) 0.189 (c) 0.964
15. (a) 0.758 (b) 0.816 (c) 0.286
16. (a) 0.599 (b) 0.841 (c) 0.244
17. (a) 638 (b) 184 (c) 1,074 (d) 136
18. (a) 5 (b) 0 (c) 6 (d) 500
19. (a) 65 (b) 616 (c) 10 **20.** (a) 68 (b) 319 (c) 340 (d) 93
21. 28th percentile **22.** 18th percentile **23.** 0th percentile
24. The change in percentile rank is 98. **25.** No
26. A's above 76, Bs between 76 and 69, Cs between 68 and 52, Ds between 51 and 44, Fs 43 and below. **27.** 90, 110
28. 55 minutes

Exercise Set 12-8

7. (a)

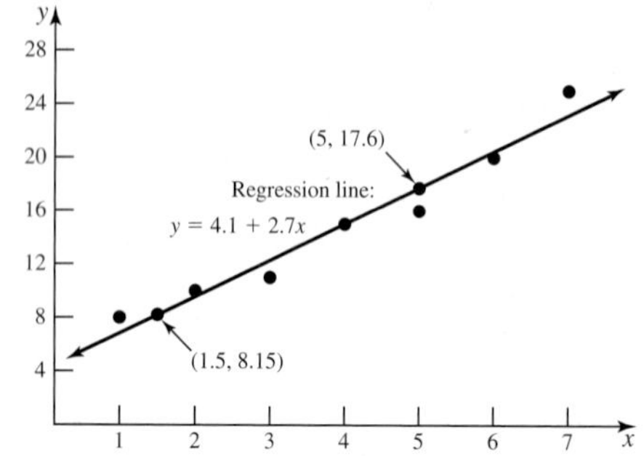

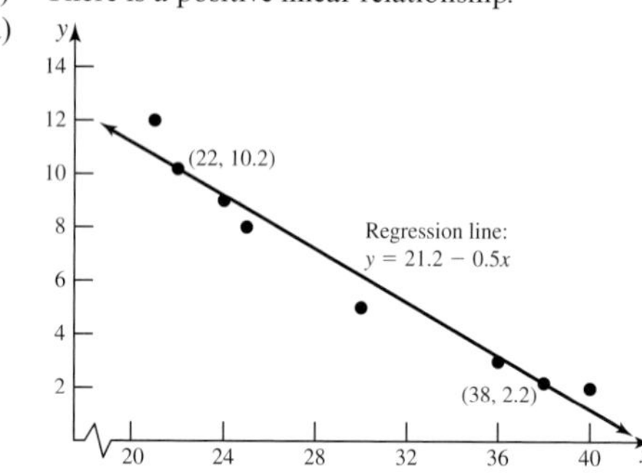

(b) $r \approx 0.977$
(c) r is significant at the 5% and the 1% level
(d) $y = 4.1 + 2.7x$
(e) There is a positive linear relationship.

8. (a)

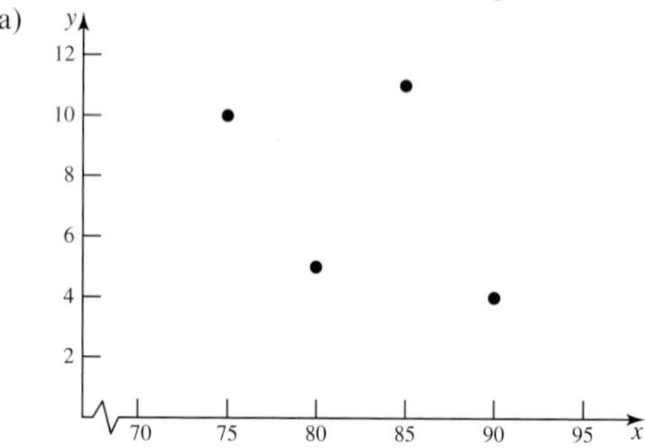

(b) $r \approx -0.970$
(c) r is significant at the 5% and the 1% level.
(d) $y = 21.2 - 0.5x$
(e) There is a negative linear relationship.

9. (a)

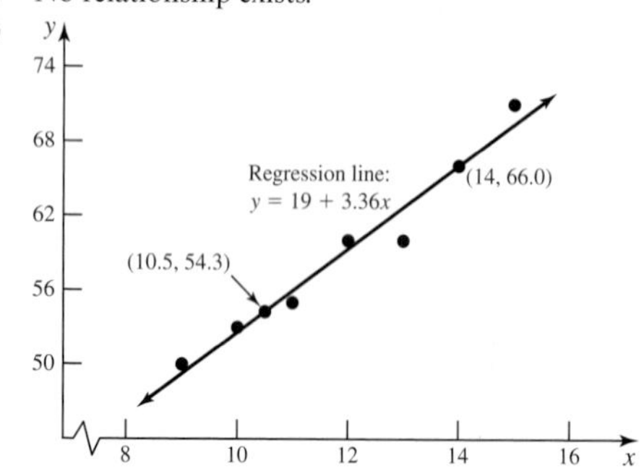

(b) $r \approx -0.441$
(c) r is not significant at 5% nor at 1% level.
(d) no regression line
(e) No relationship exists.

10. (a)

(b) $r \approx 0.978$
(c) r is significant at the 5% and 1% level.
(d) $y = 19 + 3.36x$
(e) There is a positive linear relationship.

11. (a)

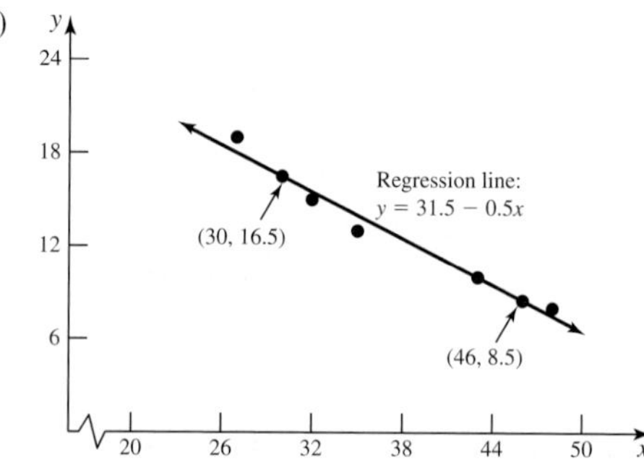

(b) $r \approx -0.983$
(c) r is significant at the 5% and 1% level.
(d) $y = 31.5 - 0.5x$
(e) There is a negative linear relationship.

12. (a)

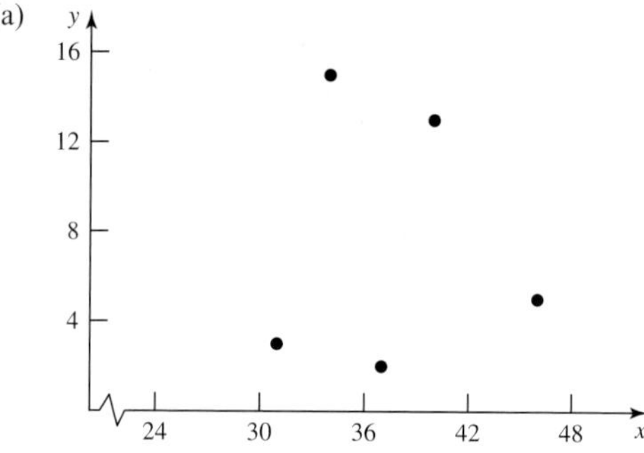

(b) $r \approx -0.013$
(c) r is not significant at the 5% nor the 1% level.
(d) no regression line
(e) No relationship exists.

13. (a)

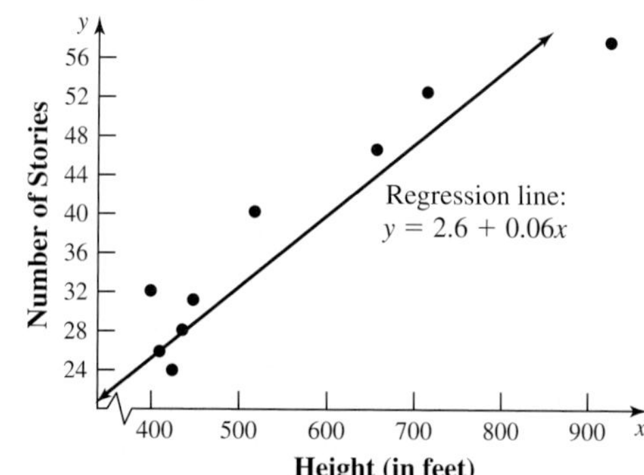

(b) $r \approx 0.942$
(c) r is significant at the 5% and 1% level.
(d) $y = 2.76 + 0.06x$
(e) There is a positive linear relationship. (f) 33 Stories

14. (a)

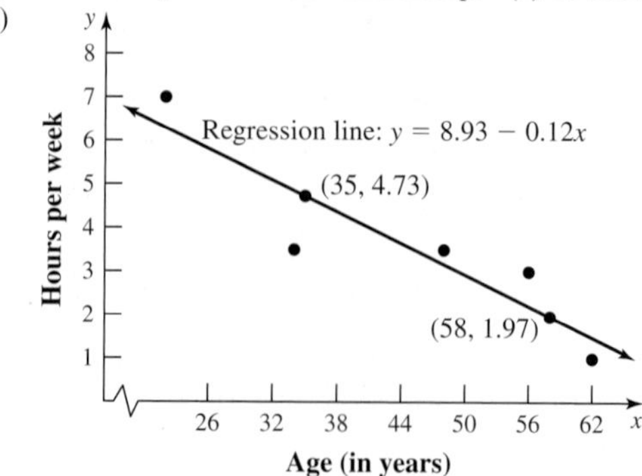

(b) $r \approx -0.907$
(c) r is significant at the 5% level, but not at the 1% level.
(d) $y = 8.93 - 0.12x$
(e) There is a negative linear relationship. (f) 4.7 hr.

15. (a)

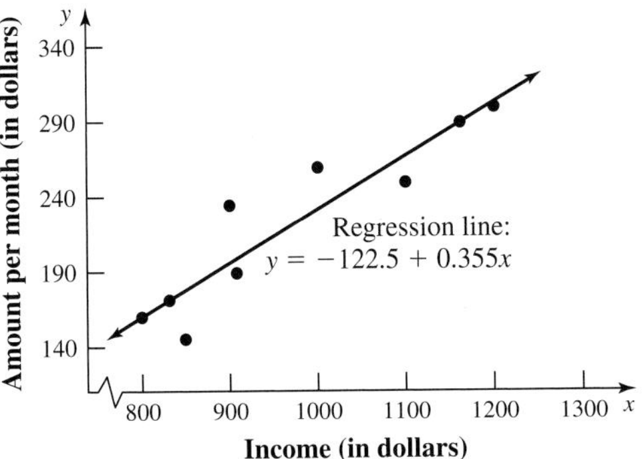

(b) $r \approx 0.896$
(c) r is significant at the 5% and 1% level.
(d) $y = -122.5 + 0.355x$
(e) There is a positive linear relationship. (f) $205.88

16. (a)

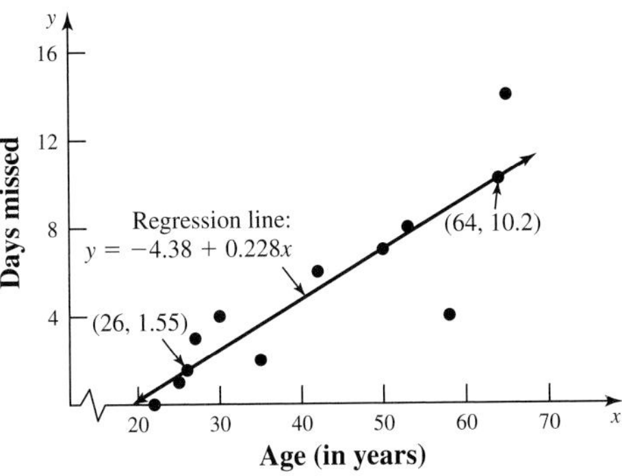

(b) $r \approx 0.842$
(c) r is significant at the 5% and 1% level.
(d) $y = -4.38 + 0.228x$
(e) There is a positive linear relationship. (f) 8.4 days

17. (a)

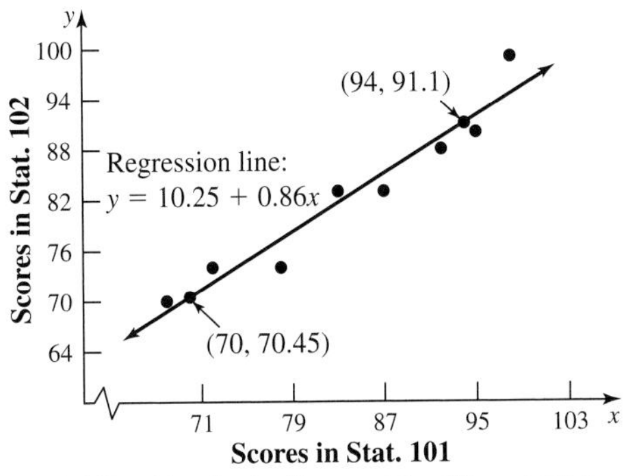

(b) $r \approx 0.963$ (c) r is significant at the 5% and 1% level.
(d) $y = 10.25 + 0.86x$
(e) There is a positive linear relationship. (f) 88

18. (a)

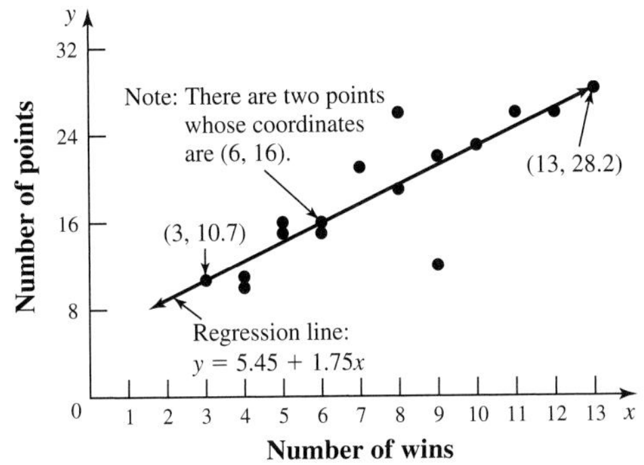

(b) $r \approx 0.796$ (c) r is significant at the 5% and 1% level.
(d) $y = 5.45 + 1.75x$
(e) There is a positive linear relationship. (f) 19 goals

19. Answers vary **20.** $r = 1$ in both cases; the points lie on a line.
21. Non-linear relationship; $r = 0$ **22.** Answers vary

Review Exercises

1.

Item	Frequency
B	4
F	5
G	5
S	5
T	6

2.

Stems	Leaves
1	2 4 6 7 8 8 9
2	0 2 3 4 5 5 5 6 6 9 9
3	2 3 5 7 8 8 9

3.

Rank	Frequency
102–116	4
117–131	3
132–146	1
147–161	4
162–176	11
177–191	7

4.

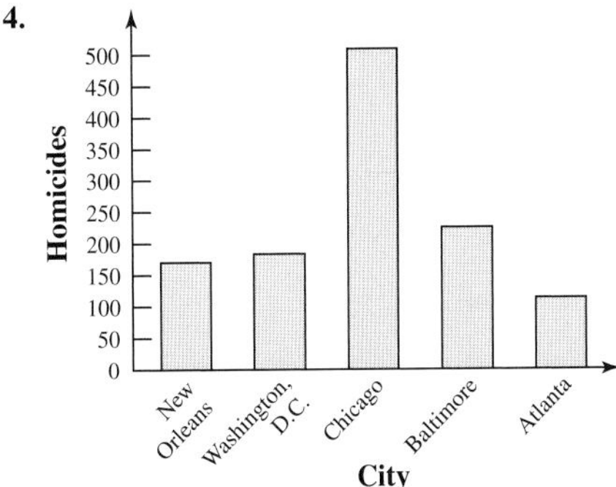

5.

Stems	Leaves
1	2 4 6 7 8 8 9
2	0 2 3 4 5 5 5 6 6 9 9
3	2 3 5 7 8 8 9

6.

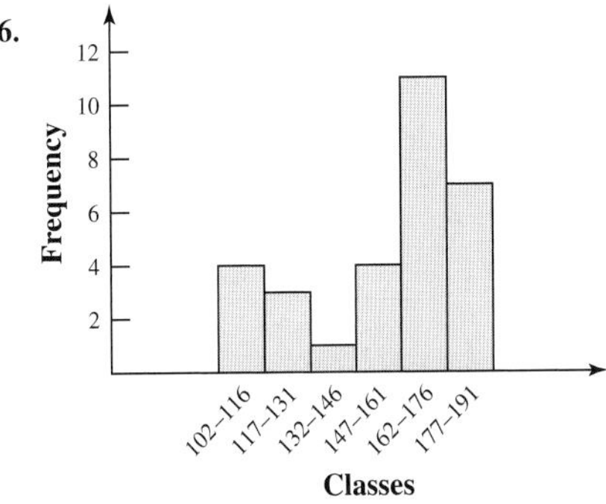

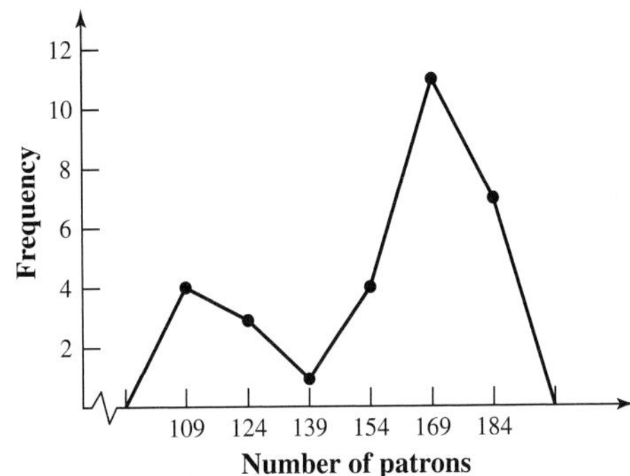

7.

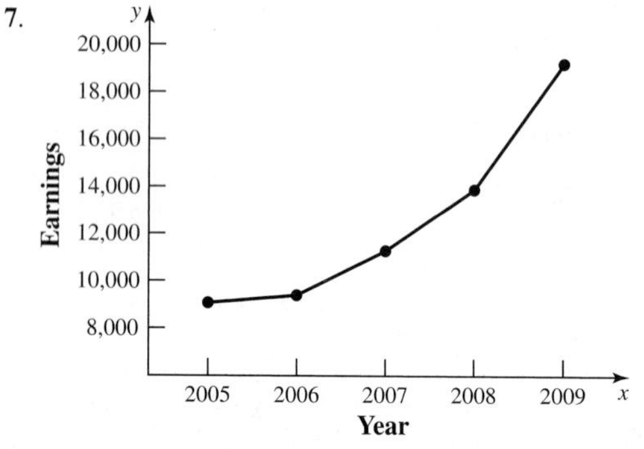

8. (a) mean ≈ 1,203.9 (c) no mode
(b) median = 1,164 (d) midrange = 1,400.5

9. mean = 7.25 10. Answers vary

11. $R = 1{,}885$, variance ≈ 475,610.1, $s \approx 689.6$

12. Answers vary

13. (a)

Value	Percentile
2	0th
4	17th
5	33rd
6	50th
8	67th
9	83rd

(b) 5

14. $Q_1 = 59.5$, $Q_2 = 104.5$, $Q_3 = 154.5$

15. (a) 0.474 (f) 0.828
(b) 0.155 (g) 0.023
(c) 0.061 (h) 0.912
(d) 0.833 (i) 0.018
(e) 0.229 (j) 0.955

16. 33 17. (a) 28 (b) 5 18. (a) 0.001 (b) 0.5 (c) 0.008 (d) 0.551 19. (a) 0.004 (b) 0.023 (c) 0.5 (d) 0.324

20. 454 21. 22

22.

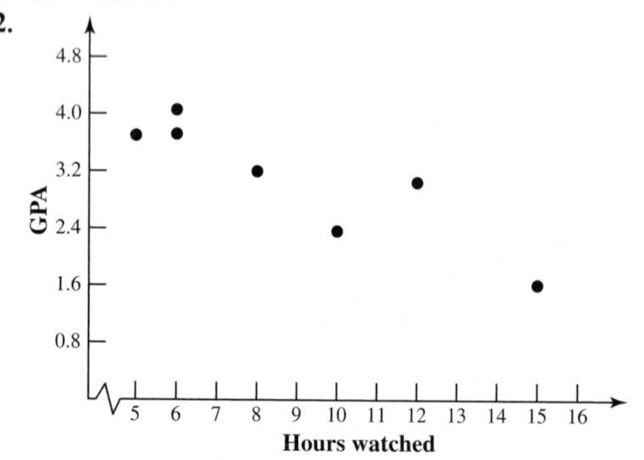

$r \approx -0.914$, r is significant at the 5% level.
$y = 4.95 - 0.21x$
$y = 3.06$ when $x = 9$

23. (a) Answers vary
(b) $r \approx -0.144$; answers vary

Chapter 12 Test

1.

Source	Frequency
W	6
L	7
K	7
E	5

2.

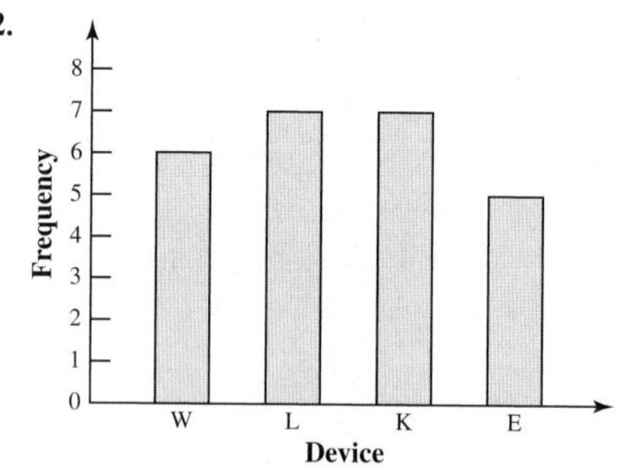

3.

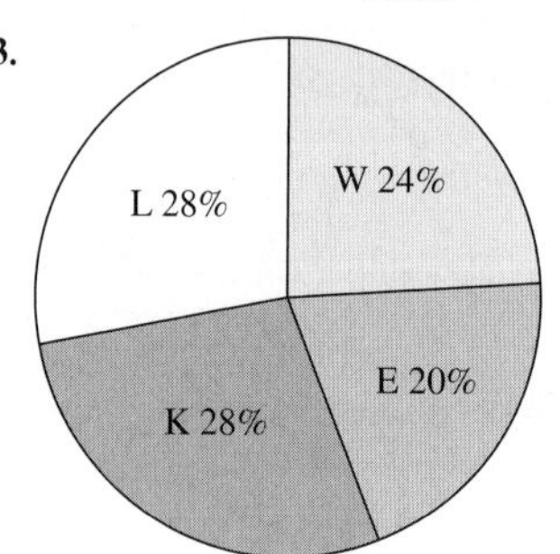

4.

Class	Frequency
277–429	18
430–582	7
583–735	2
736–888	1
889–1,041	1
1,042–1,194	0
1,195–1,347	0
1,348–1,500	1

5.

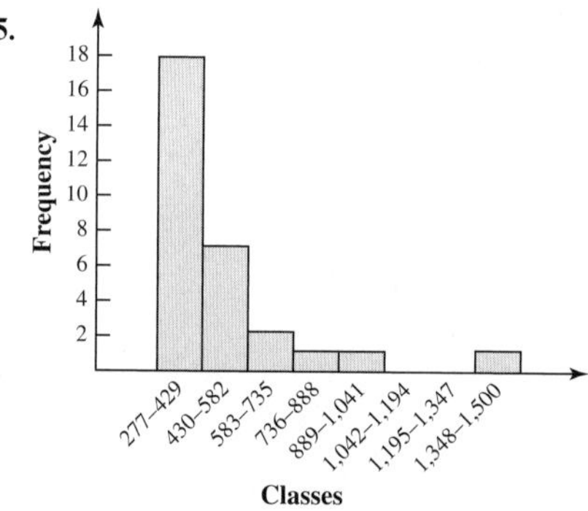

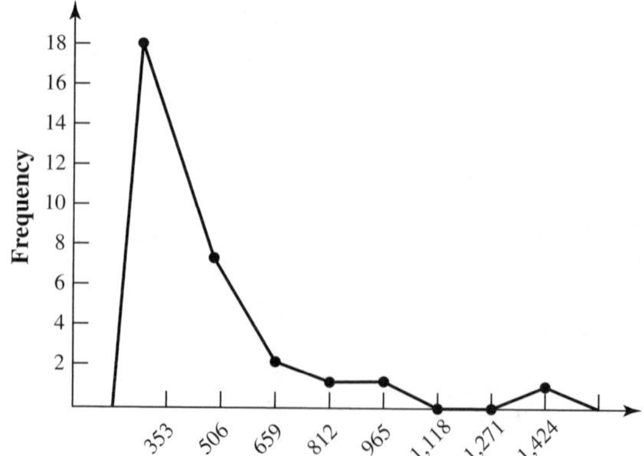

6.

Stems	Leaves
20	0 4 9
21	0 1 2 7 8 8
22	2 7 7 7 8
23	0 1 3 7 8
24	1 2 2 3 7
25	1 1 3 4 6
26	0

7.

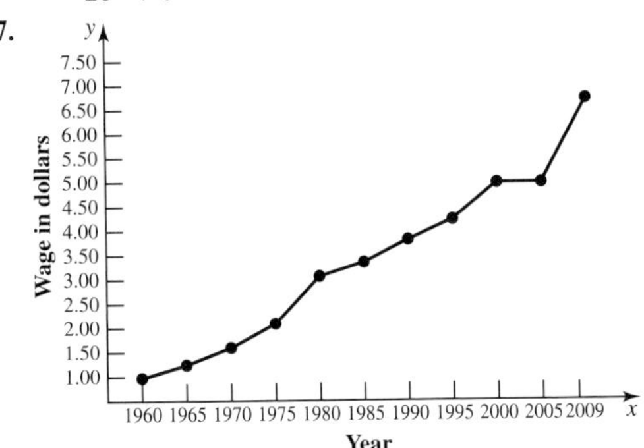

8. (a) mean ≈ 84.1 (b) median = 85 (c) no mode (d) midrange = 84 (e) range = 12 (f) variance ≈ 17.1 (g) $s \approx 4.14$

9. mean = 6.4

10. (a)

Value	Percentile
9	0th
12	13th
15	25th
27	38th
33	50th
45	63rd
63	75th
72	88th

(b) 27

11. (a) 0.433 (b) 0.394 (c) 0.035 (d) 0.103 (e) 0.293 (f) 0.828 (g) 0.040 (h) 0.903 (i) 0.016 (j) 0.912

12. (a) 0.053 (b) 0.106 (c) 0.106 (d) 0.1

13. (a) 0.067 (b) 0.023 (c) 0.465 (d) 0.094

14. 53

15.

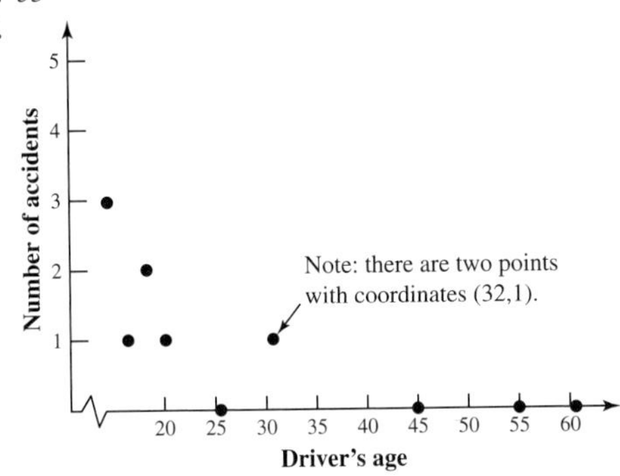

$r \approx -0.714$, r is significant at the 5% level.
$y = 2.31 - 0.04x$
$y \approx 0$ when $x = 61$

CHAPTER 13: OTHER MATHEMATICAL SYSTEMS

Exercise Set 13-1

7. *C* **8.** *C* **9.** *E* **10.** *D* **11.** *E* **12.** *E* **13.** *D* **14.** *D* **15.** *D* **16.** *F* **17.** Yes **18.** Yes: *E* **19.** Yes

20. *C* is the inverse of *F*
D is the inverse of *D*
E is the inverse of *E*
F is the inverse of *C*

21. *D* **22.** ○ **23.** ☆ **24.** △ **25.** □ **26.** △ **27.** △ **28.** △ **29.** ☆ **30.** △ **31.** □ **32.** No **33.** Yes **34.** Yes: △ **35.** No

36. □ is the inverse of ○ **37.** closure, associative

38. closure, commutative, identity, inverse, associative; Abelian group

39. closure, commutative, identity

40. closure, commutative, identity, inverse, associative; Abelian group

41. closure

42. closure, commutative, identity, inverse, associative; Abelian group

43. closure, commutative, identity, inverse

44. closure, commutative, associative

45. closure, commutative, identity, inverse, associative; Abelian group

46. closure, commutative, identity, inverse, associative; Abelian group

47. Abelian group **48.** not a group **49.** not a group **50.** not a group **51.** not a group **52.** Abelian group **53.** not a group **54.** not a group **55.** not a group **56.** Abelian group

57.

*	0	1	2	3
0	0	1	2	3
1	1	2	3	0
2	2	3	0	1
3	3	0	1	2

58. yes **59.** yes **60.** yes

61. yes, 0 is its own inverse, 2 is its own inverse, and 1 and 3 are inverses. **62.** Abelian group

63.

M	R	B	Y
R	R	P	O
B	P	B	G
Y	O	G	Y

64. yes **65.** no **66.** no **67.** no **68.** no

70.

∨	T	F
T	T	T
F	T	F

71.

→	T	F
T	T	F
F	T	T

Closure

72.

↔	T	F
T	T	F
F	F	T

Closure, commutative, associative, identity, inverse

Exercise Set 13-2

5. (a) 3 (b) 3 (c) 3 **6.** (a) 8 (b) 2 (c) 4 **7.** (a) 11 (b) 5 (c) 3 **8.** (a) 10 (b) 4 (c) 6 **9.** (a) 6 (b) 0 (c) 2 **10.** (a) 6 (b) 0 (c) 2 **11.** (a) 7 (b) 1 (c) 3 **12.** (a) 2 (b) 2 (c) 6 **13.** (a) 7 (b) 1 (c) 3 **14.** (a) 2 (b) 2 (c) 6 **15.** (a) 9 (b) 3 (c) 5 **16.** (a) 4 (b) 4 (c) 4 **17.** 2 **18.** 6 **19.** 10 **20.** 4 **21.** 3 **22.** 12 **23.** 6 **24.** 3 **25.** 11 **26.** 3 **27.** 10 **28.** 2 **29.** 2 **30.** 2 **31.** 10 **32.** 10 **33.** 6 **34.** 8 **35.** 3 **36.** 4 **37.** 11 **38.** 4 **39.** 4 **40.** 7 **41.** 6 **42.** 4 **43.** 12 **44.** 3 **45.** 10 **46.** 12 **47.** 9 **48.** 8 **49.** 6 **50.** 6 **51.** 12 **52.** 12 **53.** 2 **54.** 1 **55.** 0 **56.** 2 **57.** 5 **58.** 0 **59.** 3 **60.** 1 **61.** 2 **62.** 2 **63.** 0 **64.** 6 **65.** 12 **66.** 9 **67.** 7 **68.** 4 **69.** 10 **70.** 3 **71.** 5 **72.** 8 **73.** 5 **74.** 6 **75.** None **76.** 7 **77.** None **78.** None **79.** 1 **80.** None **81.** Answers vary **82.** Answers vary **83.** Answers vary **84.** Answers vary **85.** Answers vary **86.** Answers vary **87.** $y = 10$ **88.** $y = 5$ **89.** $y = 1$ **90.** $y = 8$ **91.** $y = 11$, 2, 5 and 8 **92.** $y = 4$ **93.** $y = 6$ **94.** $y = 12$ **95.** $y = 6$ **96.** $y = 11$ **97.** 5:00 P.M. **98.** 12 P.M. on Sunday **99.** 9:48 A.M. **100.** 3:11 A.M. **101.** 5:00 A.M. **102.** 3:42 P.M. **103.** 7:38 P.M. **104.** 10:18 P.M.

105. 8:00 P.M. **106.** 0656 **107.** 0352 **108.** 0400 **109.** 1156 **110.** 1727 **111.** 2006 **112.** 2342 **113.** 2136 **114.** 0000 **115.** 10:09 A.M. **116.** 4:48 A.M. **117.** 2:36 P.M. **118.** 10:22 A.M.
119. The property holds.
120. Define division as $a \div b = c$ if $c \times b = a$ on a 12-hour clock. The value of c would have to be a non-negative integer.

Exercise Set 13-3

5. 2 **6.** 3 **7.** 2 **8.** 3 **9.** 7 **10.** 2 **11.** 2 **12.** 1 **13.** 3 **14.** 6 **15.** 2 **16.** 5 **17.** 2 **18.** 4 **19.** 4 **20.** 5 **21.** 1 **22.** 3 **23.** 4 **24.** 8 **25.** 3 **26.** 3 **27.** 3 **28.** 0 **29.** 2 **30.** 4 **31.** 0 **32.** 2 **33.** 6 **34.** 2 **35.** 6 **36.** 1 **37.** 6 **38.** 0 **39.** 4 **40.** 1 **41.** 3 **42.** 2 **43.** 11 **44.** 1 **45.** {4, 10, 16, 22, 28, . . . }
46. {3, 7, 11, 15, 19, . . . } **47.** {3, 11, 19, 27, 35, . . . }
48. {7, 16, 25, 34, 43, . . . } **49.** {2, 10, 18, 26, 34, . . . }
50. {9, 19, 29, 39, 49, . . . } **51.** {2, 7, 12, 17, 22, . . . }
52. {6, 14, 22, 30, 38, . . . } **53.** {5, 12, 19, 26, 33, . . . }
54. {6, 13, 20, 27, 34, . . . } **55.** Tuesday **56.** Saturday **57.** Friday **58.** Saturday **59.** Tuesday **60.** Tuesday **61.** Valid **62.** Invalid **63.** Invalid **64.** Valid **65.** 6 cans **66.** 4 eggs **67.** Friday **68.** 10 P.M. **69.** It's April 22nd, 10 P.M. **70.** Tuesday at 10 A.M. **71.** south **72.** 1 **73.** 17 tables; one empty space **74.** 25 books **75.** 10 oz **76.** February of 2014 **77.** Answers vary
78. (b) 3, −3 (c) infinitely many; 7, −7
79. You did not take your second dose.

Review Exercises

1. C **2.** B **3.** B **4.** B **5.** C **6.** A **7.** yes **8.** yes **9.** yes **10.** yes; B **11.** yes; A and C are inverses, B is its own inverse. **12.** yes **13.** no **14.** no **15.** no **16.** no **17.** Abelian group **18.** not a group **19.** not a group **20.** not a group **21.** 3 **22.** 5 **23.** 12 **24.** 12 **25.** 6 **26.** 3 **27.** 2 **28.** 7 **29.** 6 **30.** 5 **31.** 17 **32.** 11
33. Friday at 3 P.M.
34. 4:20 A.M. the following day; 3 hours and 20 minutes **35.** 2 **36.** 2 **37.** 4 **38.** 3 **39.** 2 **40.** 0 **41.** 7 **42.** 5 **43.** 2 **44.** 0 **45.** 2 **46.** 3 **47.** 0 **48.** 1 **49.** 0 **50.** 5 **51.** 1 **52.** 9 **53.** 4 **54.** 1 **55.** 3 **56.** 4 **57.** 1 **58.** 6 **59.** 4 **60.** 2 **61.** 1 **62.** 6 **63.** 2 **64.** 4 **65.** 4 **66.** 1 **67.** 2 **68.** 10 **69.** 5 **70.** 6 **71.** 2 **72.** 4 **73.** 6 **74.** 6 **75.** {4, 12, 20, 28, 36, . . . } **76.** {4, 14, 24, 34, 44, . . . }
77. {3, 12, 21, 30, 39, . . . } **78.** {5, 13, 21, 29, 37, . . . }
79. {1, 7, 13, 19, 25, . . . } **80.** {2, 10, 18, 26, 34, . . . }
81. {11, 23, 35, 47, 59, . . . } **82.** {8, 17, 26, 35, 44, . . . }
83. $y = 1$ **84.** {3, 15, 27, 39, 51, . . . } **85.** 4 cookies
86. 0 tennis balls **87.** no

Chapter 13 Test

1. z **2.** z **3.** s **4.** z **5.** z **6.** y is the inverse of y **7.** Yes **8.** No **9.** −1 **10.** $-a$ **11.** a **12.** $-a$ **13.** −1 **14.** $-a$ **15.** $-a$ **16.** $y = -a$ **17.** $y = -1$ **18.** Yes **19.** Yes **20.** Yes **21.** Yes **22.** 1 **23.** $-a$ is the inverse of a **24.** −1 is the inverse of −1 **25.** 4 **26.** 4 **27.** 4 **28.** 3 **29.** 2 **30.** 3 **31.** 1 **32.** 2 **33.** 2 **34.** 0 **35.** 4 **36.** 1 **37.** 5 **38.** 10 **39.** 1 **40.** 8
41. {2, 8, 14, 20, 26, . . . } **42.** {8, 18, 28, 38, 48, . . . }
43. {0, 4, 8, 12, 16, . . . } **44.** {0, 5, 10, 15, 20, . . . } **45.** 4 A.M.
46. 30 people **47.** 8 P.M. the following day **48.** yes

CHAPTER 14: VOTING METHODS

Exercise Set 14-1

5. (a) 22 (b) 4 (c) 8 (d) X
6. (a) 19 (b) 5 (c) 7 (d) Lee and Smith
7. (a) 18 (b) 9 (c) 4 (d) Philadelphia
8. (a) 13 (b) 3 (c) 5 (d) Blue and Yellow
9. (a) 208 (b) Swimming pool **10.** (a) 24 (b) Coke **11.** (a) 20 (b) Carnations **12.** (a) 32 (b) 10 A.M. **13.** No **14.** Yes **15.** No **16.** No **17.** 9 **18.** 13 **19.** No **20.** 103 **21.** Yes **22.** 52 **23.** Answers vary

Exercise Set 14-2

5. role-playing (R) **6.** Rosa's Restaurant (R)
7. *Anatomy of a Murder* (A) **8.** zoo (Z)
9. (a) swimming pool (S) (b) Yes **10.** water (W) **11.** Yes **12.** No **13.** No **14.** No **15.** No **16.** Yes
17. Professor Donovan (D) **18.** investments (I)
19. (a) carnations (C) (b) Yes **20.** (a) 10 A.M. (b) Yes **21.** Yes **22.** No **23.** No **24.** No **25.** Answers vary **26.** Answers vary **27.** Answers vary **28.** Yes **29.** Yes

Exercise Set 14-3

7. 6 **8.** 21 **9.** 45 **10.** 36 **11.** Steel Center (S) **12.** Disneyland (D)
13. (a) There is a three-way tie. (b) The results are different.
14. (a) Sports (S) (b) The results are different.
15. (a) Rosa's Restaurant (R) (b) Yes **16.** (a) 10:00 A.M. (b) Yes **17.** No **18.** Yes **19.** No **20.** No **21.** Dr. Zhang **22.** Killer Bees **23.** green **24.** bubble gum **25.** inmate Z **26.** patch cement walks **27.** Answers vary **28.** Answers vary **29.** Using pairwise comparison puts the candidates head-to-head against each other.
30. Your vote wouldn't matter.

Exercise Set 14-4

7. a) 32
b) South: 1.9375
Central: 4.625
North: 3.4375
c)

Campus	South	Central	North
Lower	1	4	3
Upper	2	5	4

d)

Campus	South	Central	North
Promotions	2	5	3

8. a) 94.125
b) RRHS: 6.672
CCHS: 9.997
HHHS: 14.066
MHS: 9.264
c)

School	RRHS	CCHS	HHHS	MHS
Lower	6	9	14	9
Upper	7	10	15	10

d)

School	RRHS	CCHS	HHHS	MHS
Buses	7	10	14	9

9. a) 64,444
b) District 1: 3.755
District 2: 2.374
District 3: 2.871
c)

District	1	2	3
Lower	3	2	2
Upper	4	3	3

d)

District	1	2	3
Representatives	4	2	3

10. a) 25.8
b) Store 1: 2.171
Store 2: 1.240
Store 3: 2.868
Store 4: 2.403
Store 5: 3.333
c)

Store	1	2	3	4	5
Lower quota	2	1	2	2	3
Upper quota	3	2	3	3	4

d) Jefferson's Method:

Store	1	2	3	4	5
Buyers	2	1	3	2	4

Adams' Method:

Store	1	2	3	4	5
Buyers	2	2	3	2	3

e) 21.5; 31.1

11. a) 14.6

b) Terminal A: 3.493
Terminal B: 2.534
Terminal C: 4.542
Terminal D: 1.507

c)

Terminal	A	B	C	D
Lower	3	2	4	1
Upper	4	3	5	2

d) Jefferson's Method:

Terminal	A	B	C	D
Trucks	4	2	5	1

Adams' Method:

Terminal	A	B	C	D
Trucks	3	3	4	2

e) 12.7; 17.1

12. a) 101

b) Branch A: 12.238
Branch B: 8.634
Branch C: 20.149
Branch D: 16.386
Branch E: 32.386
Branch F: 10.426

c)

Branch	A	B	C	D	E	F
Lower quota	1	8	2	1	3	10
Upper quota	1	9	2	1	3	11

d) Jefferson's Method:

Branch	A	B	C	D	E	F
Books	12	8	2	1	3	10

Adams' Method:

Branch	A	B	C	D	E	F
Books	12	9	20	16	32	11

e) 97; 104

13. a) 54

b) Store 1: 2.204
Store 2: 1.833
Store 3: 3.444
Store 4: 2.519

c)

Store	1	2	3	4
Lower	2	1	3	2
Upper	3	2	4	3

d)

Store	1	2	3	4
Computers	2	2	3	3

e) None

14. a) 184.25

b) Ward A: 2.019
Ward B: 2.524
Ward C: 3.028
Ward D: 3.360
Ward E: 1.069

c)

Ward	A	B	C	D	E
Lower	2	2	3	3	1
Upper	3	3	4	4	2

d)

Ward	A	B	C	D	E
Nurses	2	3	3	3	1

e) None

15. a) 1,217.875

b) Precinct 1: 2.925
Precinct 2: 6.956
Precinct 3: 1.762
Precinct 4: 4.358

c)

Precinct	1	2	3	4
Lower	2	6	1	4
Upper	3	7	2	5

d)

Precinct	1	2	3	4
Officers	3	7	2	4

e) None

16. a) 205.625

b) Clinic A: 2.218
Clinic B: 2.996
Clinic C: 2.787

c)

Clinic	A	B	C
Lower	2	2	2
Upper	3	3	3

d)

Clinic	A	B	C
Nurse practitioners	2	3	3

e) None

17. a) 1,098

b) Office 1: 2.032
Office 2: 0.907
Office 3: 1.424
Office 4: 1.638

c)

Office	1	2	3	4
Lower	2	0	1	1
Upper	3	1	2	2

d)

Office	1	2	3	4
Therapists	2	1	1	2

e) 1,106

18. a) 2,047

b) Dept A: 2.717
Dept B: 2.132
Dept C: 1.960
Dept D: 3.191

c)

Department	A	B	C	D
Lower	2	2	1	3
Upper	3	3	2	4

d)

Department	A	B	C	D
Associates	3	2	2	3

e) None

19. Answers vary

20. Answers vary

Exercise Set 14-5

7. Alabama paradox occurred **8.** Alabama paradox occurred

9. Alabama paradox occurred **10.** population paradox occurred

11. population paradox did not occur

12. population paradox occurred

13. a)

Campus	North	South
Podiums	9	52

b)

Campus	North	South	Campus 3
Podiums	8	53	7

c) new states paradox occurred

14. a)

District	A	B
Seats	7	33

b)

District	A	B	C
Seats	6	34	6

c) new states paradox occurred

15. a)

State	A	B	C
Seats	11	46	40

b)

State	A	B	C	D
Seats	10	47	40	13

c) the new states paradox occurred

16. Answers vary

Review Exercises

1.

Number of votes	5	5	5
First choice	Q	P	R
Second choice	R	Q	P
Third choice	P	R	Q

2. 15 **3.** 5 **4.** 5

5.

Number of votes	6	4	10
First choice	C	P	H
Second choice	P	C	P
Third choice	H	H	C

6. 20 **7.** 6 **8.** 4 **9.** 58 **10.** 10 **11.** Style A **12.** Style B
13. Style A **14.** Style A **15.** No **16.** No **17.** No **18.** 9
19. 47 **20.** *Bye Bye Birdie* (B) **21.** *Music Man* (M)
22. tie between *Guys and Dolls* (G) and *Bye Bye Birdie* (B)
23. *Music Man* (M) **24.** No **25.** Yes **26.** No **27.** No
28. Seubert **29.** comedian

30.

Department	Math	English	Science
Laptops	6	5	4

31.

Location	A	B	C	D
Asst. Pastors	5	1	1	3

32.

Department	Math	English	Science
Laptops	6	5	4

33.

Location	A	B	C	D
Asst. Pastors	5	2	1	2

34.

Department	Math	English	Science
Laptops	6	5	4

35.

Location	A	B	C	D
Asst. Pastors	5	1	1	3

36.

Department	Math	English	Science
Laptops	6	5	4

37.

Location	A	B	C	D
Asst. Pastors	5	2	1	2

38. Alabama paradox occurred **39.** Alabama paradox occurred
40. population paradox occurred
41. population paradox did not occur
42. new states paradox occurred
43. new states paradox did not occur

Chapter 14 Test

1.

Number of votes	1	7	4
First choice	A	B	C
Second choice	B	A	B
Third choice	C	C	A

2. 12 **3.** 7 **4.** 4 **5.** 100 **6.** 12 **7.** Pittsburgh (P)
8. Pittsburgh (P) **9.** Pittsburgh (P) **10.** Pittsburgh (P)
11. No **12.** No **13.** No **14.** No
15. *Big Brother Nursing Home* **16.** Patrick Miller

17.

Flight	A	B	C
Flight Attendants	4	6	2

18.

Flight	A	B	C
Flight Attendants	4	6	2

19.

Flight	A	B	C
Flight Attendants	4	6	2

20.

Flight	A	B	C
Flight Attendants	4	6	2

21.

Flight	A	B	C
Flight Attendants	4	6	2

22. Alabama paradox occurred
23. population paradox did not occur
24. new states paradox occurred

CHAPTER 15: GRAPH THEORY

Exercise Set 15-1

9. *A, B, C, D, E, F, G* **10.** 9 **11.** *B, G, F* **12.** *A, C, D, F, G*
13. *B* and *E* **14.** *BE* **15.** Answers vary; one such path is *A, B, E, F*
16. *A, B, D, C, A* **17.** *G* **18.** *AC, CD, DB, EF* and the loop at *G*
19. There is no edge connecting *E* and *D*.
20. It doesn't begin and end at the same vertex.

21. **22.** **23.**

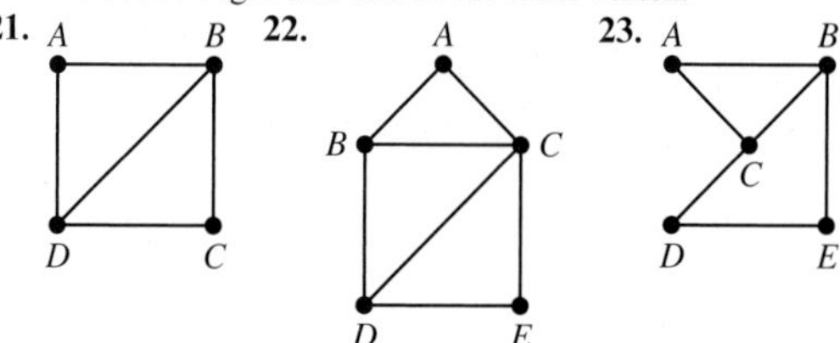

24. **25.**

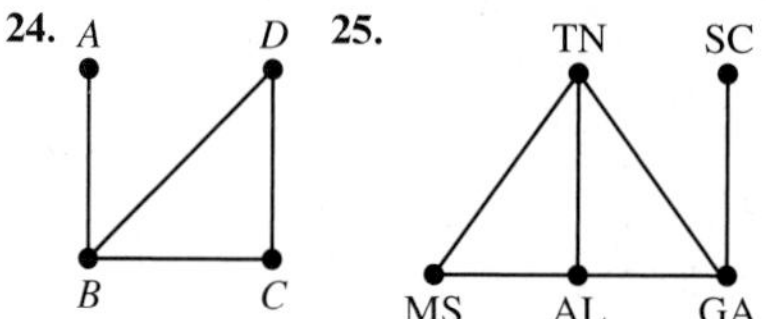

26. **27.**

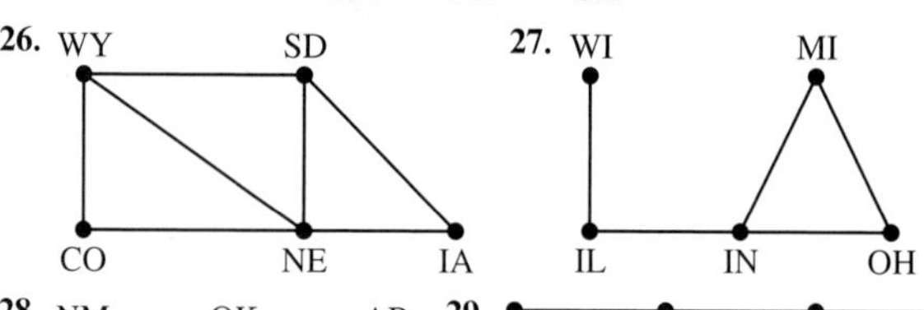

28. **29.**

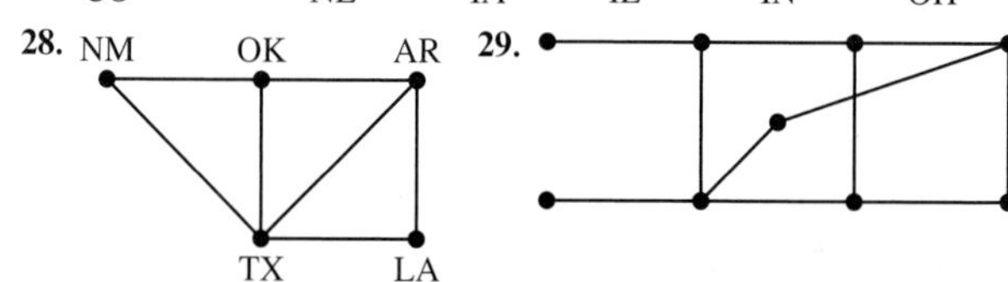

30. **31.**

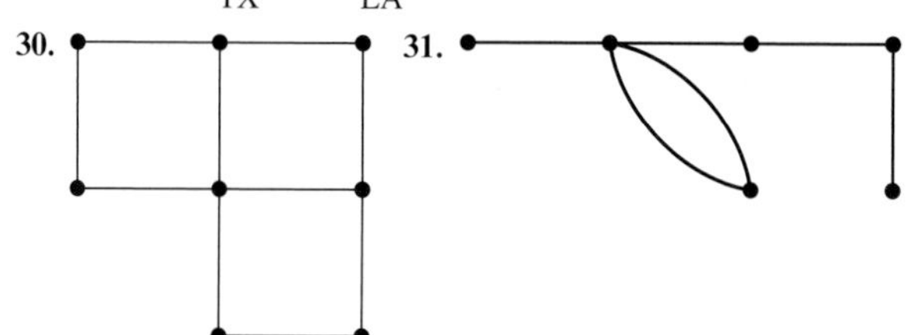

32. **33.** Answers vary

34.

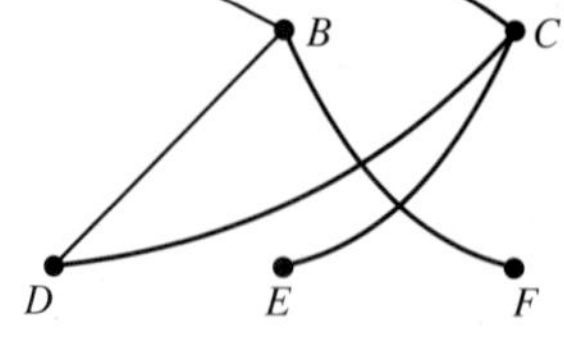

; Cathy; Bill

35.

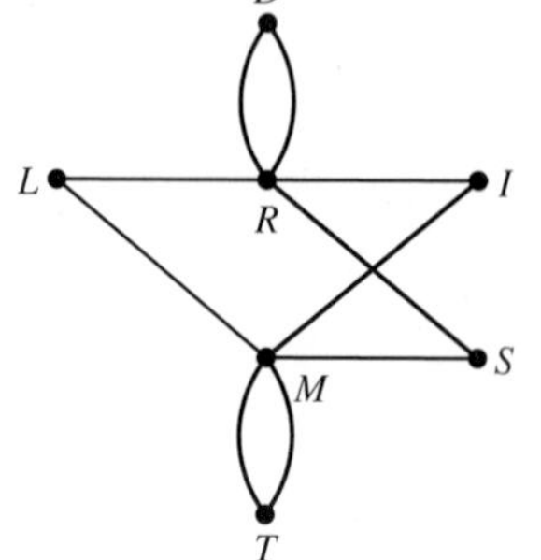

36. It is not possible.

Exercise Set 15-2

7. Euler circuit **8.** Euler path **9.** Neither **10.** Euler path

11. (a) Euler path (b) *A, B, C, A, I, C, D, G, I, H, G, F, E, D, F*

12. (a) Euler circuit (b) *A, E, B, F, C, G, D, C, B, A*

13. (a) Euler circuit (b) *A, B, D, H, I, G, D, C, G, F, E, C, A*

14. (a) Euler path (b) *B, A, E, G, F, E, B, C, F, D, C*

15. (a) Neither **16.** (a) Neither

17. (a) Euler circuit (b) *A, B, A, C, B, C, A*

18. (a) Neither **19.** (a) Neither

20. (a) Euler circuit (b) *A, B, C, D, B, F, D, E, F, A*

21. Euler circuit: *A, B, C, D, C, A*

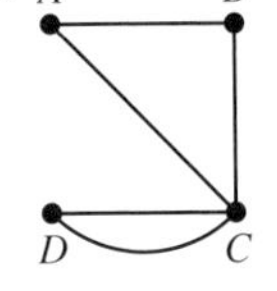

22. Euler path: *E, B, C, E, D, C, A, D*

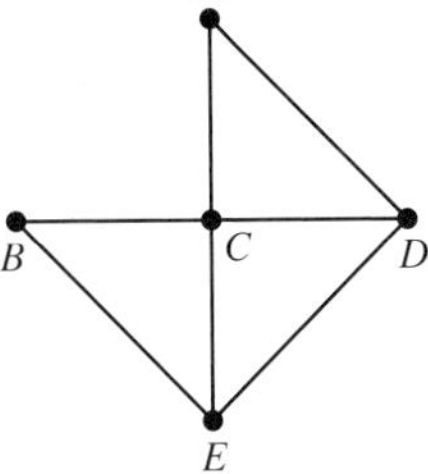

23. Euler path: *A, B, C, A, D, B*

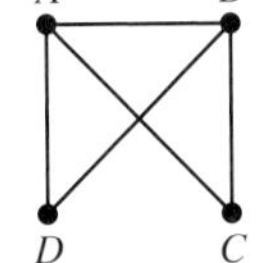

24. Euler path: *A, B, E, D, A, C, F, E*

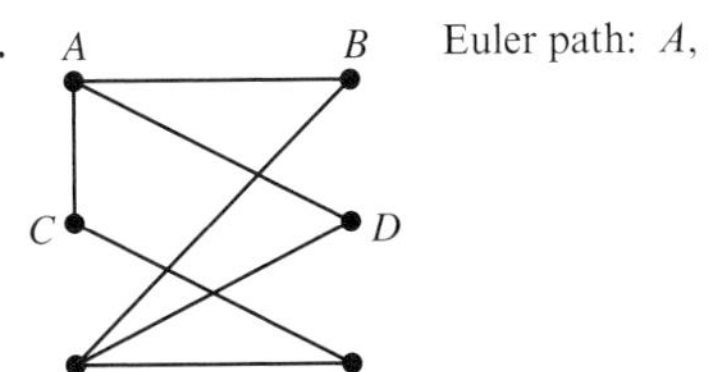

25. Euler path: *A, B, E, D, A, C, D*

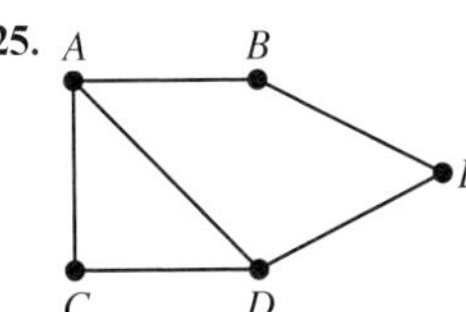

26. Neither

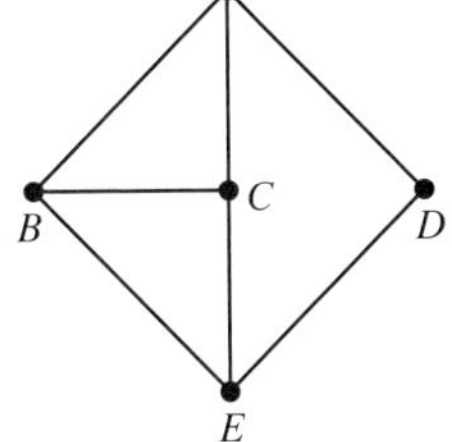

27. Neither

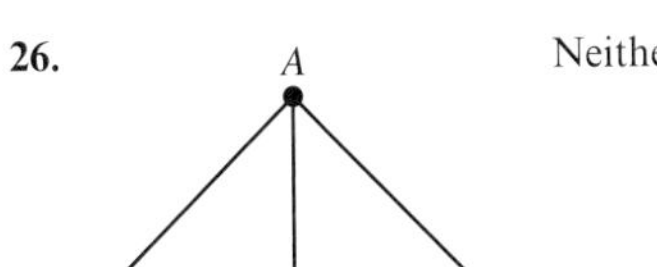

28. Euler path: *B, A, D, E, B, C, G, F, D*

29. Euler path: *A, B, E, D, C, F, A, G, E*

30. Euler circuit: *A, F, D, C, E, F, B, C, A*

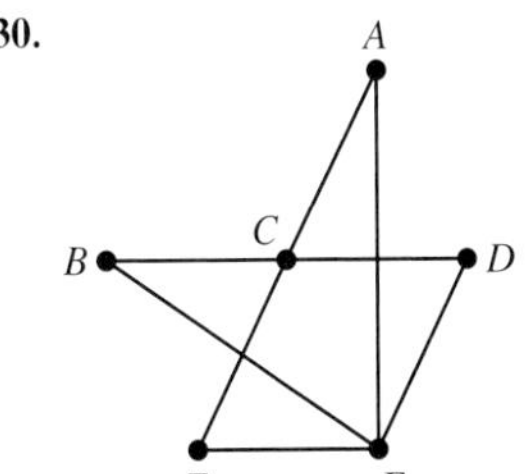

31. Euler circuit: WY, ID, MT, ND, SD, MT, WY, NE, SD, WY

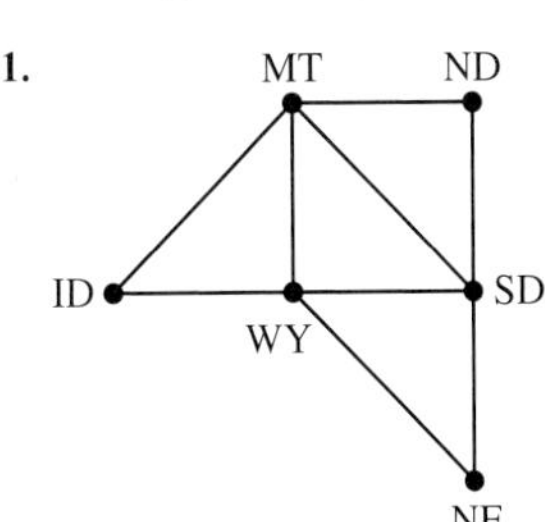

32. Euler path: KY, TN, VA, KY, WV, VA, NC, TN

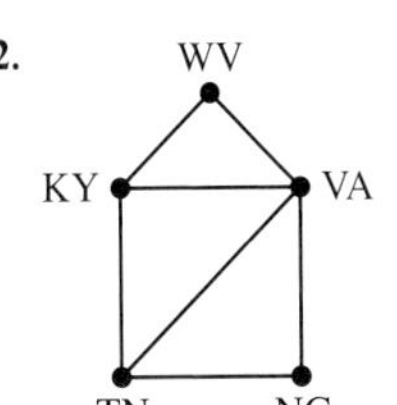

33. Euler circuit: *A, B, E, F, J, I, L, K, H, I, E, D, H, G, C, D, A*

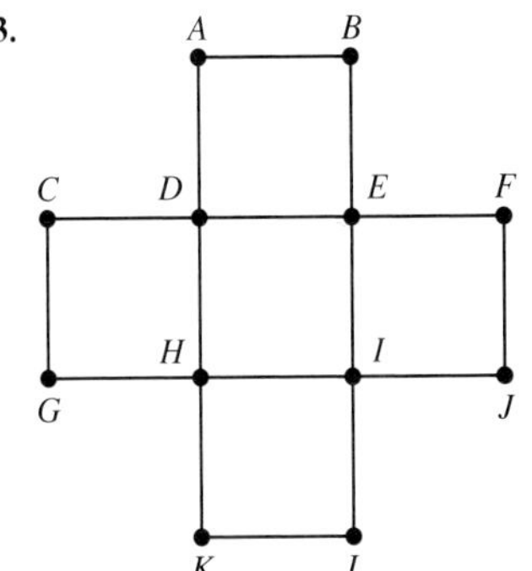

34.

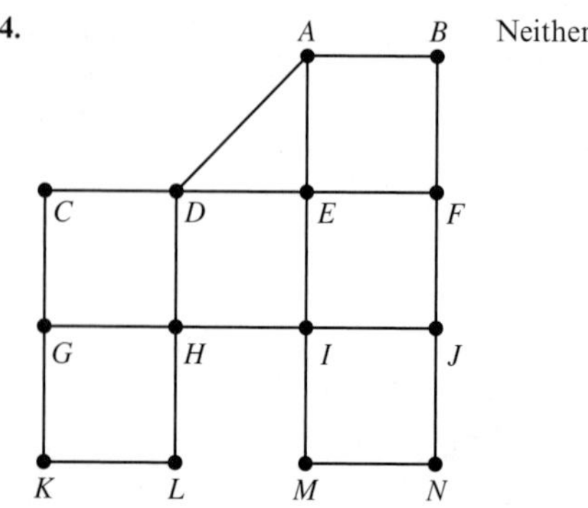

Neither

35. There would be no way to traverse every edge if the graph was not connected.
36. It is not possible to have exactly one odd vertex in a connected graph.
37. Answers vary

Exercise Set 15-3

9. Answers vary, two possibilities are: *A, B, C, D, E, F* and *A, C, D, E, F, B.*
10. Answers vary, two possibilities are: *A, B, C, E, D* and *A, D, E, C, B.*
11. Answers vary, two possibilities are: *A, B, E, F, J, I, L, K, H, G, C, D* and *A, B, E, D, C, G, H, K, L, I, J, F.*
12. Answers vary, two possibilities are: *B, A, C, D, E, F, G* and *A, B, C, D, G, F, E.*
13. Answers vary, two possibilities are: *A, B, E, C, D* and *A, B, E, D, C.*
14. Answers vary, two possibilities are: *E, B, A, D, C* and *A, D, C, B, E.*
15. Answers vary, two possibilities are: *A, B, D, E, C, F, G, H* and *A, B, D, E, C, F, H, G.*
16. Answers vary, two possibilities are: *A, B, C, D, E, F* and *A, F, B, C, D, E.*
17. Answers vary, two possibilities are: *A, B, C, E, D, A* and *C, B, A, D, E, C.*
18. Answers vary, two possibilities are: *A, B, D, E, F, G, H, C, A* and *B, A, C, H, G, F, E, D, B.*
19. Answers vary, two possibilities are: *A, B, D, G, F, E, H, I, C, A* and *D, B, A, C, I, H, E, F, G, D.*
20. Answers vary, two possibilities are: *A, B, C, E, D, F, A* and *B, C, E, F, D, A, B.*
21. Answers vary, two possibilities are: *A, B, C, D, E, A* and *A, D, B, E, C, A.*
22. Answers vary, two possibilities are: *A, C, B, D, E, F, G, A* and *G, F, E, D, B, C, A, G.*
23. 2 **24.** 120 **25.** 40,320 **26.** 3,628,800
27. *P, Q, R, S, P* and *P, S, R, Q, P;* 200
28. *W, X, Z, Y, W* and *W, Y, Z, X, W;* 1,017
29. *A, F, B, C, E, D, A;* 109 **30.** *A, C, B, E, D, A;* 42
31. Let *T* = Pitt, *L* = Phil, *B* = Balt and *W* = Wash.

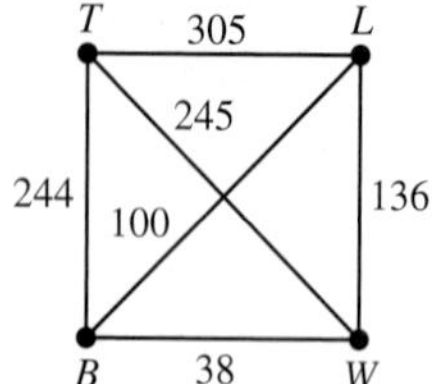

32. *T, L, B, W, T* and *T, W, B, L, T;* 688 miles
33. *T, B, W, L, T;* 723 miles **34.** 723 miles; no
35. Let *N* = New York, *D* = Cleveland, *O* = Chicago and *B* = Baltimore.

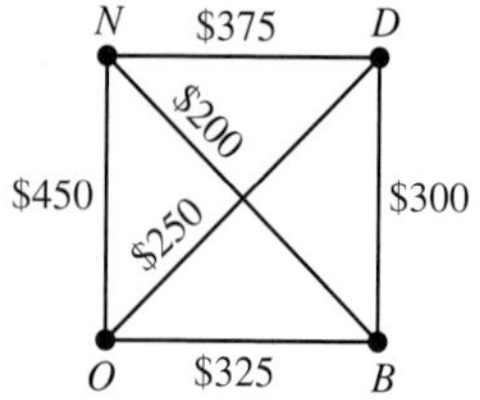

36. *O, B, N, D, O* and *O, D, N, B, O;* $1,150
37. *O, D, B, N, O;* $1,200 **38.** $1,200; no **39.** Answers vary
40. Answers vary **41.** Answers vary
42. (a) $n - 1$; the graph is complete
(b) $n - 2$

Exercise Set 15-4

7. The graph is not a tree because it contains a circuit.
8. The graph is a tree. **9.** The graph is a tree.
10. The graph is not a tree because it contains a circuit.
11. The graph is not a tree because it is disconnected.
12. The graph is a tree. **13.** The graph is a tree. **14.** The graph is a tree.
15. The graph is not a tree because it contains a circuit.
16. The graph is a tree.

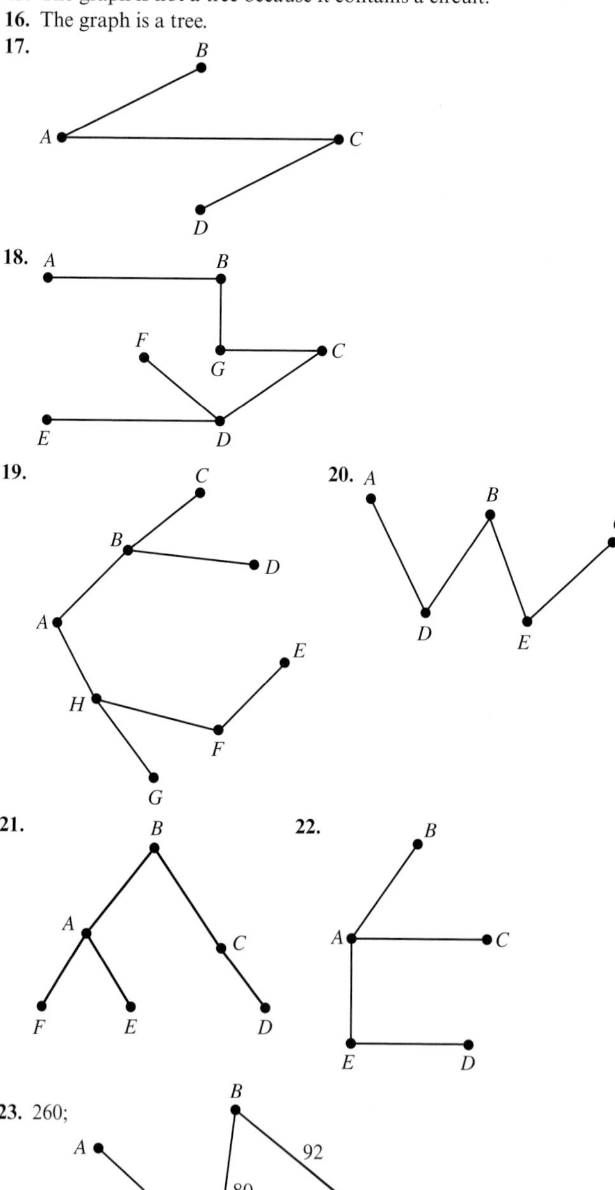

24. 26;

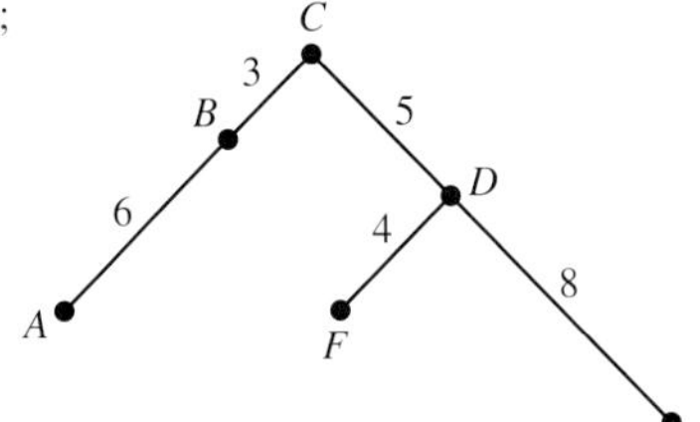

Note: Edge *BF* could replace edge *CD*.

25. 360;

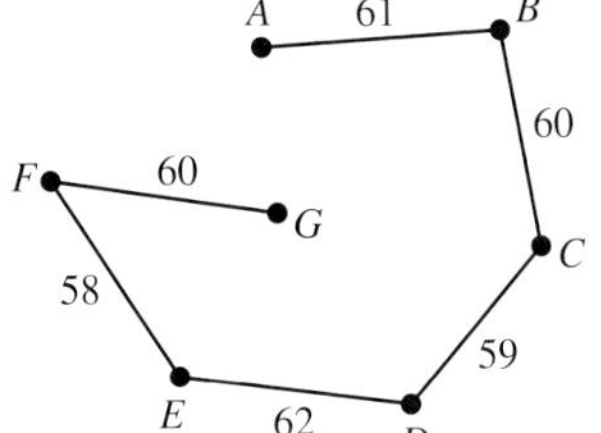

26. 60;

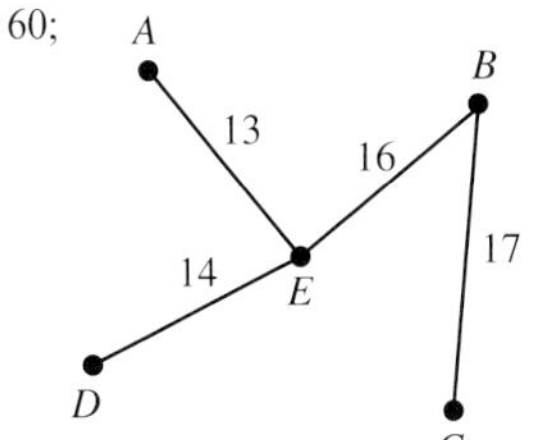

27. 633;

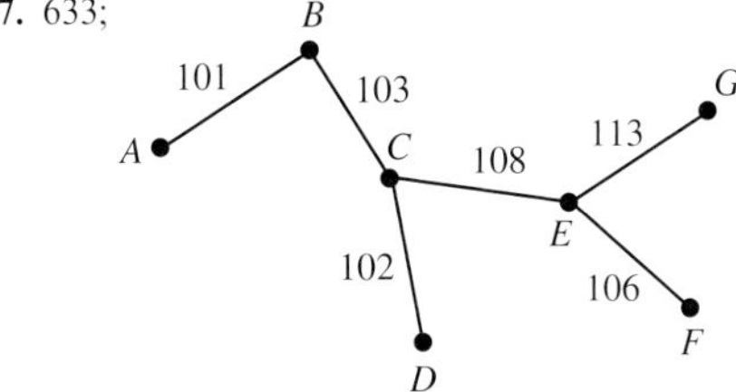

28. 59;

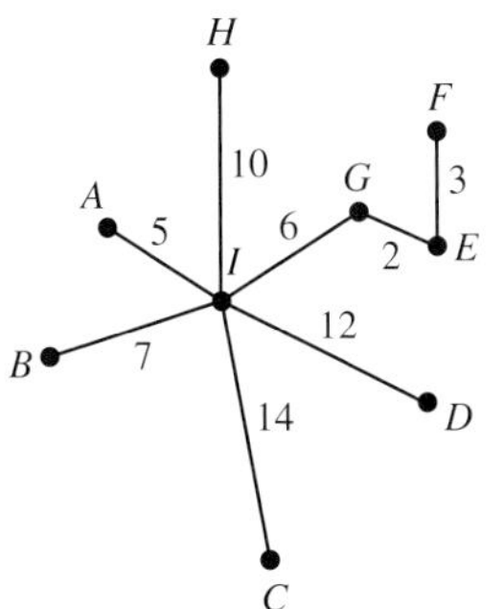

29. 89;

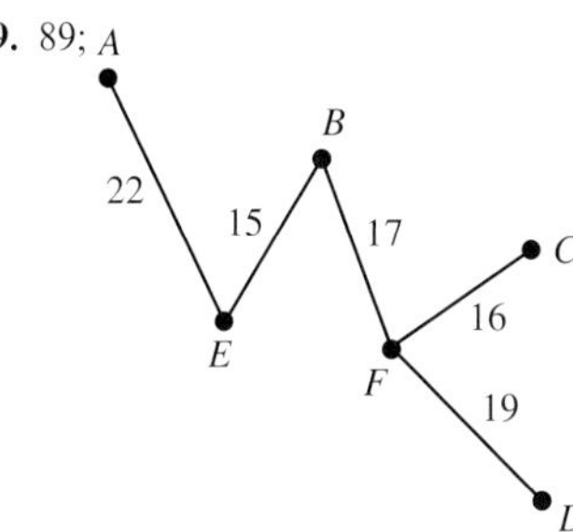

30. 34;

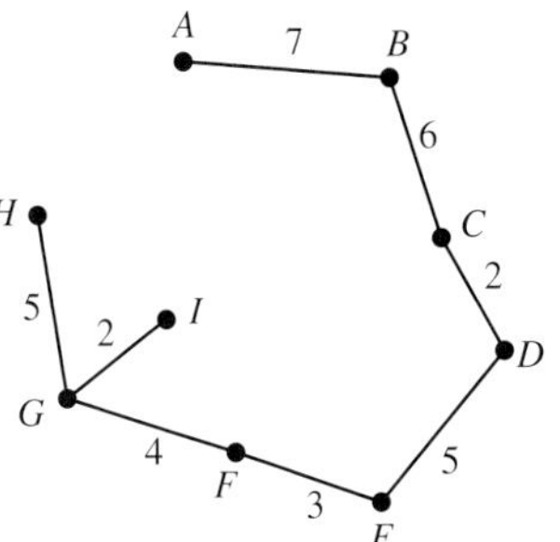

Notes: Edge *AH* could replace edge *AB*, and edge *FI* could replace edge *FG*.

31. 28 feet;

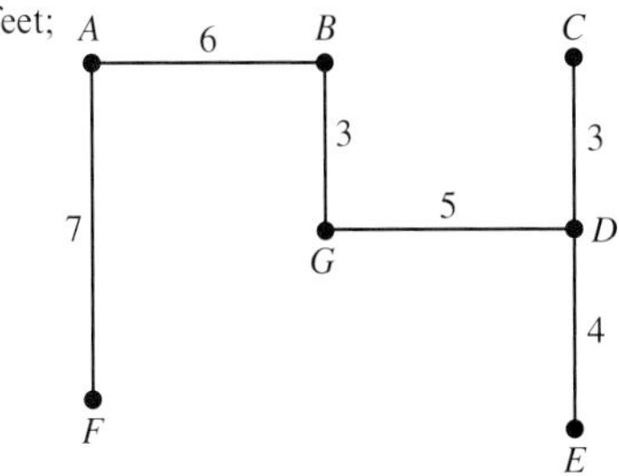

32. 1,400 feet;

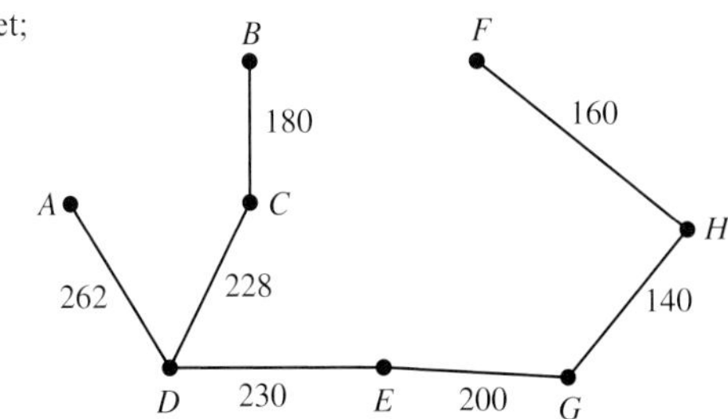

33. 162 feet;

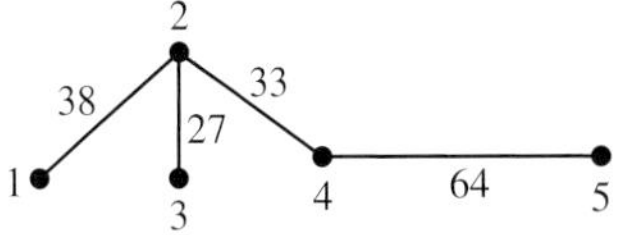

34.

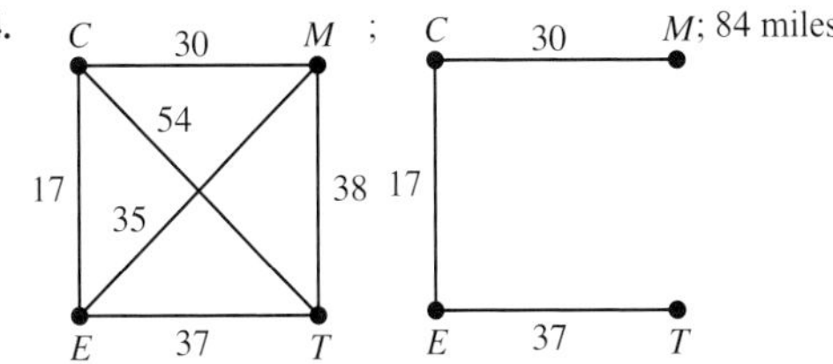

35. Answers vary **36.** Answers vary **37.** Answers vary

Review Exercises

1. *A*, *B*, *C*, *D*, *E*, *F* **2.** Even **3.** Even **4.** *B* **5.** *E* **6.** *EF*
7. Yes **8.** 8 **9.**

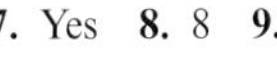

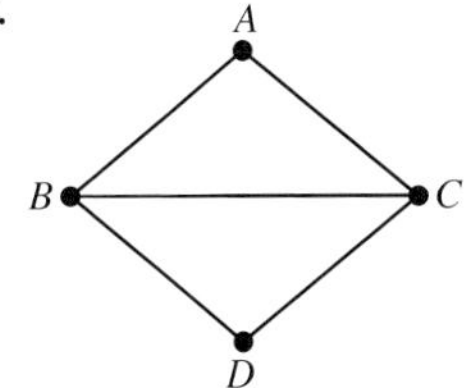

10.

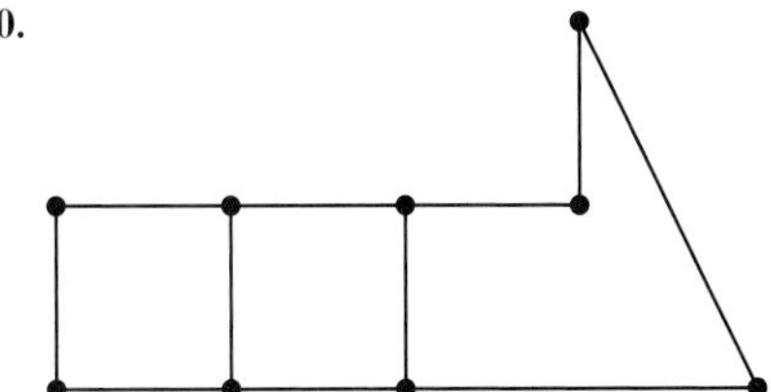

11.

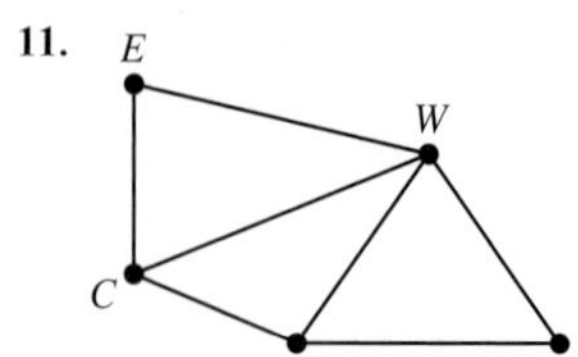

12. Answers vary; one Euler path is $D, E, G, F, A, G, C, A, B, C, D$.
13. Answers vary; one Euler circuit is $D, C, A, B, C, E, B, G, F, E, D$.
14. Answers vary; one Hamilton path is A, E, B, D, C, H.
15. Answers vary; one Hamilton circuit is A, B, C, D, E, H, F, A.
16. 39,916,800 **17.** A, D, C, B, E, A for a total of 125
18.

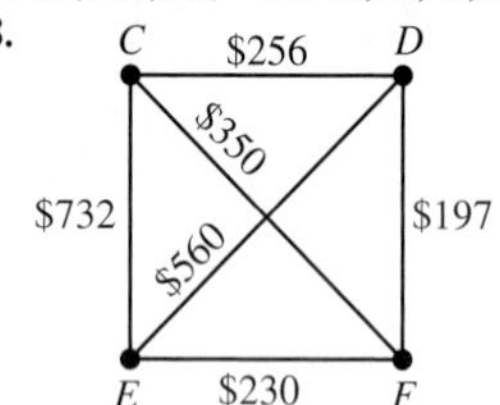

19. C, D, F, E, C; $1,415
20.

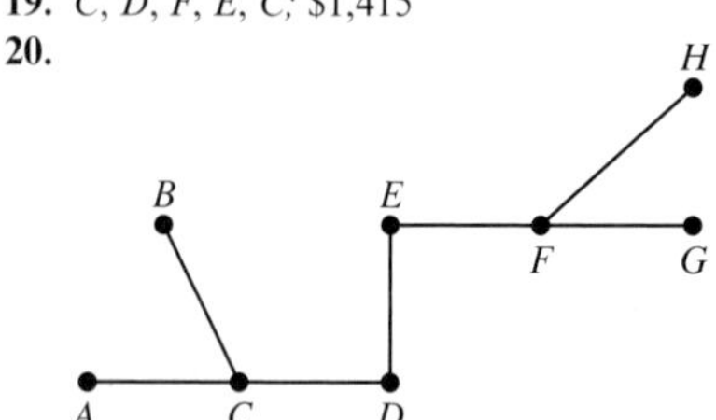

21.

D 22 E
A 24 37 47 F
29 C H 34
B 42
G

Chapter 15 Test

1. 4 **2.** E **3.** Answers vary; one possible path is A, B, C, E, D, F.
4. FD **5.** A, B, D, E **6.** Odd **7.** Answers vary
8.

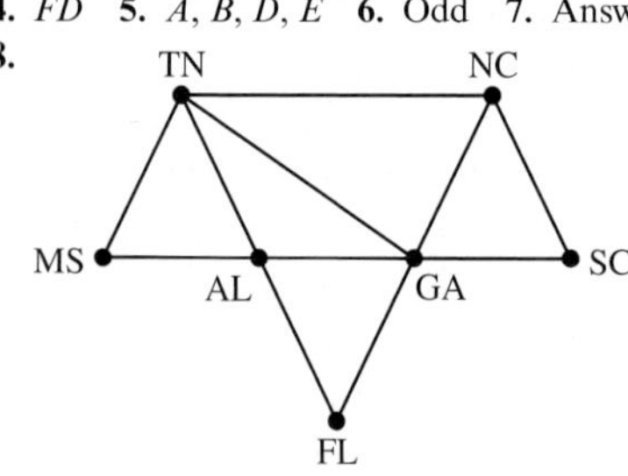

9.

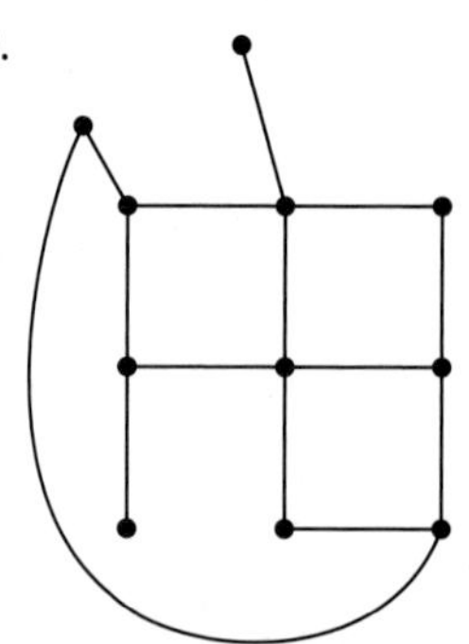

10. One Euler path is $B, A, F, B, C, D, E, C, F$.
11. Answers vary; one Hamilton path is A, B, C, D, E, F.
12.

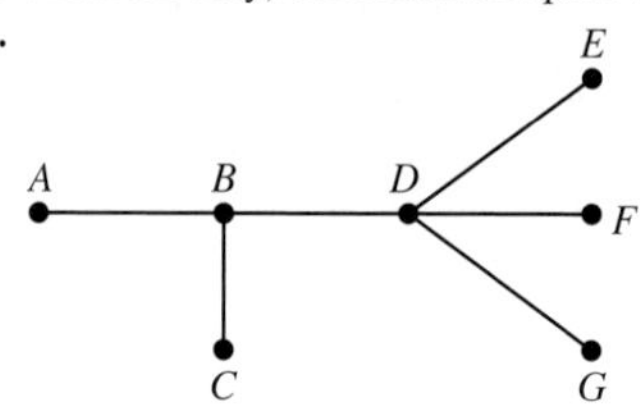

13. A B C ; $A, D, E, A, B, E, F, B, C, F$

D E F

14. Let A = Adamsburg, T = Trafford, W = White Oak and C = Turtle Creek.

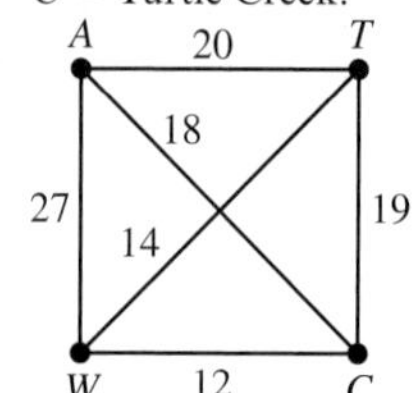

15. 64 miles **16.** 64 miles; yes **17.** Euler path **18.** Hamilton path
19. Hamilton path **20.** Hamilton circuit **21.** Euler path
22. minimum spanning tree

PHOTO CREDITS

Design Elements

(Pizza for Learning Objectives): MBPHOTO/iStock photo; (Silver pan under pizza): Laurent Renault/iStock Photo; (Calculator): iStock Photo.

Chapter 1

Opener: © Robert Voets/CBS Photo Archive via Getty Images; p. 4: © The McGraw-Hill Companies, Inc./C.P. Hammond, photographer; p. 5: © Photodisc/Getty RF; p. 8: © Mark Burnett/Stock Boston/PictureQuest; p. 16: © Photodisc/Alamy RF; p. 19: © JupiterImages RF; p. 28(top): © ETH-Bibliothek Zurich, Image Archive; p. 28(bottom): © Alinari Archives/Corbis; p. 30: © Thinkstock Images/JupiterImages RF; p. 31(top): © De Agostini/Getty Images; p. 31(bottom): © Bruce Laurance/Getty Images; p. 35: © Robert Voets/CBS Photo Archive via Getty Images.

Chapter 2

Opener: © Lester Cohen/WireImage/Getty Images; p. 42: © PhotoLink/Getty Images RF; p. 43: © Phil Walter/Getty Images; p. 48: © Royalty-Free/Corbis; p. 55: © The McGraw-Hill Companies, Inc./Barry Barker, photographer; p. 61: © Momentum Creative Group/Alamy RF; p. 71: © Digital Vision/Getty Images RF; p. 74: © Digital Vision Ltd./SuperStock RF; p. 76: © Digital Vision RF; p. 77: © BananaStock Ltd. RF; p. 82: © Royalty-Free/Corbis; p. 84: © IMS Communications Ltd./Capstone Design/FlatEarth Images RF; p. 87: © Lester Cohen/WireImage/Getty Images.

Chapter 3

Opener 3a: © Muskopf Photography, LLC/Alamy; Opener 3b: © Gino's Premium Images/Alamy; p. 92: © The McGraw-Hill Companies, Inc./Jill Braaten, photographer RF; p. 94: National Library of Medicine, #81192; p. 97: © Life File/Getty Images RF; p. 98: © Cat Sobecki; p. 101: © The McGraw-Hill Companies, Inc./John Flournoy, photographer; p. 104: © Comstock Images/Alamy RF; p. 105: © Photodisc/Getty Images RF; p. 106: © Creatas/PictureQuest RF; p. 107: © Getty Images RF; p. 111: © Robert Landau/Corbis; p. 115: © The McGraw-Hill Companies, Inc./Jill Braaten, photographer RF; p. 116, 119: © Cat Sobecki; p. 121: © Getty Images RF; p. 124: © Christina Lane; p. 126: © Royalty-Free/Corbis; p. 128: © StockTrek/Getty Images RF; p. 129: © Liquidlibrary/PictureQuest RF; p. 134: Library of Congress, Prints and Photographs Division [LC-USZ62-13016]; p. 135: © Royalty-Free/Corbis; p. 141(left): © Muskopf Photography, LLC/Alamy; p. 141(right): © Gino's Premium Images/Alamy.

Chapter 4

Opener: © Spencer Platt/Getty Images; p. 148: © The Studio Dog/Getty Images RF; p. 157: © Scala/Art Resource, NY. Iraq Museum, Baghdad, Iraq; p. 159(top): © Think-Stock/JupiterImages RF; p. 159(bottom), 163: © Digital Vision/Getty Images RF; p. 167: © SSPL/The Image Works; p. 169: © Hulton Archive/Getty Images; p. 172: National Radio Astronomy Observatory; p. 176: © Kevin Winter/Getty Images; p. 178: © Getty Images RF; p. 180: © The McGraw-Hill Companies, Inc./Jill Braaten, photographer; p. 190: Courtesy, Naval Historical Center; p. 193: © Spencer Platt/Getty Images.

Chapter 5

Opener: © Brand X Pictures/PunchStock RF; p. 198: © Photodisc/Getty Images RF; p. 206: © Granger Collection; p. 210: © Steve Bronstein/Getty Images; p. 211: © Christina Lane; p. 216: © Creatas/PunchStock RF; p. 218: © C Squared Studios/Getty Images RF; p. 221(top): © Jamie Squire/Getty Images; p. 221(bottom): © David Young-Wolff/PhotoEdit; p. 228(top): © Cat Sobecki; p. 228(bottom): © Royalty-Free/Corbis; p. 236: © Hulton Archive/Getty Images; p. 244: © Stockdisc/Getty Images RF; p. 247: © Getty Images RF; p. 251: © Nick Laham/Getty Images; p. 258(top): © Digital Vision/Getty Images RF; p. 258(bottom), 261: © Getty Images RF; p. 263: © Thinkstock Images/JupiterImages RF; p. 264: © Getty Images RF; p. 272: © Brand X Pictures/PunchStock RF.

Chapter 6

Opener: © Scott Olson/Getty Images; p. 278: © Getty Images RF; p. 282: © The Trustees of the British Museum; p. 283: © Blend Images/Getty Images RF; p. 296(top): © Digital Vision/Getty Images RF; p. 296(bottom): © The McGraw-Hill Companies, Inc./Ken Cavanagh, photographer; p. 304: © Royalty-Free/Corbis; p. 305(top): © Getty Images RF; p. 305(bottom): © Royalty-Free/Corbis; p. 311: © 1999 Copyright IMS Communications Ltd./Capstone Design. All Rights Reserved. RF;

p. 312: © PhotoLink/Getty Images RF; p. 313(top): USDA; p. 313(bottom), 314: © Life File/Getty Images RF; p. 315: © Photodisc/PunchStock RF; p. 319: © Stockbyte/Getty Images RF; p. 324: © Getty Images RF; p. 325: © The McGraw-Hill Companies, Inc./Christopher Kerrigan, photographer RF; p. 326: © Getty Images RF; p. 328: © Brand X Pictures/JupiterImages RF; p. 341: © Scott Olson/Getty Images.

Chapter 7

Page 349: © Lowell Georgia/Corbis; p. 350: © C Squared Studios/Getty Images RF; p. 353, 356(top): © Getty Images RF; p. 356(bottom): © liquidlibrary/Dynamic Graphics/JupiterImages RF; p. 363: © PhotoLink/ Getty Images RF; p. 370: © Getty Images RF; p. 374: © StockTrek/Getty Images RF; p. 380: © Digital Vision/ Getty Images RF; p. 389: © Royalty-Free/Corbis; p. 390: © Susan Van Etten/PhotoEdit; p. 391: © PhotoLink/Getty Images RF; p. 394: © Royalty-Free/Corbis; p. 401: © Getty Images RF; p. 402: © Joel Gordon; p. 404: © Comstock Images/Corbis; p. 405: © Comstock Images/Alamy RF; p. 408(top,bottom): © Getty Images RF; p. 409: © James L. Amos/Photo Researchers, Inc.

Chapter 8

Opener: © Royalty-Free/Corbis; p. 420: © PhotoLink/ Getty Images RF; p. 423: © Photodisc/Getty Images RF; p. 424: © Royalty-Free/Corbis; p. 425: © Getty Images RF; p. 429: © Stockbyte/PunchStock RF; p. 430–431: © PhotoLink/Getty Images RF; p. 432: © Royalty-Free/ Corbis; p. 435: © Beathan/Corbis RF; p. 439: © Jeff Greenberg/PhotoEdit; p. 446: © SW Productions/Getty Images RF; p. 447(top): © Cat Sobecki; p. 447(bottom): © Getty Images RF; p. 453(top): © Royalty-Free/Corbis; p. 453(bottom): © Getty Images RF; p. 458: © Brand X Pictures RF; p. 460: © PhotoLink/Getty Images RF; p. 461: © Getty Images RF; p. 462: © Photodisc/Getty Images RF; p. 465: © Comstock Images RF; p. 467: © Royalty-Free/Corbis; p. 470: © The McGraw-Hill Companies, Inc./Jill Braaten, photographer; p. 475: © Royalty-Free/Corbis.

Chapter 9

Opener: © Panoramic Images/Getty Images; p. 482: © Stockbyte/PunchStock RF; p. 483: © Pixtal/SuperStock RF; p. 484: © PhotoLink/Getty Images RF; p. 485, 486: © Brand X Pictures RF; p. 487: © Cat Sobecki; p. 490, 491: © Royalty-Free/Corbis; p. 492: © Getty Images RF; p. 494: © Erica S. Leeds RF; p. 499(top): © AFP/Getty Images; p. 499(bottom): © Creatas/PictureQuest RF; p. 500–501: © Royalty-Free/Corbis; p. 502(top): © Creatas/PunchStock RF; p. 502(bottom): © The McGraw-Hill Companies, Inc./Ken Cavanagh, photographer; p. 506: © Panoramic Images/Getty Images.

Chapter 10

Opener 10a–d, p. 512(top): © Cat Sobecki; p. 512(bottom): © Getty Images RF; p. 516: © Hulton Archive/Getty Images; p. 521: © Life File/Getty Images RF; p. 524: © Royalty-Free/Corbis; p. 525: © Brand X Pictures RF; p. 527: © Brand X Pictures/PunchStock RF; p. 531: © Digital Vision/Getty Images RF; p. 532(left): © Ingram Publishing/Alamy RF; p. 532(middle): © Image Club RF; p. 532(right): © Stockbyte/Getty Images RF; p. 534: © Digital Vision RF; p. 536: © Comstock Images/PunchStock RF; p. 539(top, bottom): © Getty Images RF; p. 541: NASA; p. 544: © Comstock Images/JupiterImages/ Alamy RF; p. 548: © Cat Sobecki; p. 549(left): © The McGraw-Hill Companies, Inc./Ken Cavanagh, photographer; p. 549(right): © Getty Images RF; p. 552: © Brand X Pictures RF; p. 556: © Purestock/Alamy RF; p. 557: © Royalty-Free/Corbis; p. 559: © Photodisc/Getty Images RF; p. 563: M.C. Escher's "Symmetry Drawing E110" © 2009 The M.C. Escher Company-Holland. All rights reserved. www.mcescher.com; p. 566 (1–4): © Cat Sobecki.

Chapter 11

Opener 11a: © Royalty-Free/Corbis; Opener 11b: © Mark Steinmetz/Amanita Pictures RF; Opener 11c: © Photodisc/ Getty Images RF; Opener 11d: © Stockbyte/PunchStock RF; Opener 11e: © Royalty-Free/Corbis; p. 574: © Corbis/agefotostock RF; p. 575(top): © C Squared Studios/Getty Images RF; p. 577: © MedioImages RF; p. 580: © PhotoLink/Getty Images RF; p. 582, p. 584: © Christina Lane; p. 587: © Royalty-Free/Corbis; p. 588: © Photodisc/Getty Images RF; p. 589: © Getty Images RF; p. 590: © Stockbyte/Alamy RF; p. 591, 592: © Getty Images RF; p. 593: © The McGraw-Hill Companies, Inc./ C.P. Hammond, photographer; p. 597: © MedioImages RF; p. 599: © Bettmann/Corbis; p. 600: © Royalty-Free/ Corbis; p. 601: © The McGraw-Hill Companies, Inc./C.P. Hammond, photographer; p. 605(top): © Stockbyte/ PunchStock RF; p. 605(bottom): © BananaStock/JupiterImages RF; p. 606: © F Schussler/PhotoLink/Getty Images RF; p. 609: © Rob Carr/AP Photo; p. 613: © PhotoLink/Getty Images RF; p. 614: © STAN HONDA/ AFP/Getty Images; p. 616: Courtesy of the Office of the Historian, U.S. House of Representatives; p. 620: © Getty Images RF; p. 623: © The McGraw-Hill Companies, Inc./ Barry Barker, photographer; p. 624: © moodboard/ agefotostock RF; p. 627: © Digital Vision/SuperStock RF; p. 628: © Royalty-Free/Corbis; p. 632: © Veer RF; p. 635: © BananaStock/agefotostock RF; p. 640a: © Royalty-Free/ Corbis; p. 640b: © Amanita Pictures RF; p. 640c: © PhotoDisc/Getty Images RF; p. 640d: © Stockbyte/ PunchStock RF; p. 640e: © Royalty-Free/Corbis.

Chapter 12

Opener: © Julie Dermansky/Photo Researchers, Inc.; p. 648: © Royalty-FreeCorbis; p. 649: U.S. Fish & Wildlife

Service/Mike Lockhart; p. 650: © Brand X Pictures RF; p. 651: © Getty Images RF; p. 653: © Digital Vision/Getty Images RF; p. 658: © Getty Images RF; p. 660: © Digital Vision/Getty Images RF; p. 661, 665, 666: © Getty Images RF; p. 667: © Purestock/PunchStock RF; p. 668(top): © Getty Images RF; p. 668(bottom): NASA; p. 670: © Image Source/PunchStock RF; p. 671: © WireImage/Getty Images; p. 672: © Frare/Brand X Pictures/Corbis RF; p. 675(top): © Tim Davis/Corbis; p. 675(bottom): © Chris Collins/Corbis; p. 681(top): © Comstock Images/JupiterImages RF; p. 681(bottom): © Getty Images RF; p. 682(top): United States Navy; p. 682(bottom): © Comstock Images/JupiterImages RF; p. 685: © Cat Sobecki; p. 686: © Getty Images RF; p. 693: © The McGraw-Hill Companies, Inc./John Flournoy, photographer; p. 694: © PhotoLink/Getty Images RF; p. 695: © PhotoDisc/Getty Images RF; p. 696: © Life File/Getty Images RF; p. 699: © The McGraw-Hill Companies, Inc./John Flournoy, photographer; p. 700(top): © Digital/SuperStock RF; p. 700(bottom): © Royalty-Free/Corbis; p. 704(top): © Getty Images RF; p. 704 (bottom): © The McGraw-Hill Companies, Inc./C.P. Hammond, photographer; p. 707: © The McGraw-Hill Companies, Inc./John Flournoy, photographer; p. 711: © Julie Dermansky/Photo Researchers, Inc.; p. S1: © Tracy Kahn/Corbis RF; p. S2 (top): © Stockbyte/Getty Images RF; p. S2 (center): © Jim Bourg/Reuters; p. S2 (bottom): © The McGraw-Hill Companies, Inc./C.P. Hammond, photographer; p. S4: © SuperStock RF.

Chapter 13

Opener: © Chad Baker/Getty Images RF; p. 725: © Royalty Free/Corbis; p. 729: http://commons.wikimedia.org/wiki/File:Evariste_galois.jpg#filelinks; p. 732: © Royalty Free/Corbis; p. 733: © Getty Images RF; p. 735: © Photodisc/Getty Images RF; p. 738: © Chad Baker/Getty Images RF.

Chapter 14

Opener: © Nick Laham/Getty Images; p. 744: © Comstock Images/PunchStock RF; p. 745: © Digital Vision Ltd./SuperStock RF; p. 747: © The McGraw-Hill Companies, Inc./Jill Braaten, photographer; p. 751: © Doug Benc/Getty Images; p. 755: © The McGraw-Hill Companies, Inc./Lars A. Niki, photographer; p. 759: © The McGraw-Hill Companies, Inc./John Flournoy, photographer; p. 761: © Royalty-Free/Corbis; p. 763: © Kenneth J. Arrow; p. 764: © Visions of America/Corbis RF; p. 767(top): © Getty Images RF; p. 767(bottom): © Royalty-Free/Corbis; p. 770: © BananaStock Ltd. RF; p. 774: © Comstock Images/Alamy RF; p. 778: The Library of Congress, Prints and Photographs Division [LC-USZC4-422]; p. 779: © Creatas/PunchStock RF; p. 782: © IT Stock/Alamy RF; p. 785: © Nick Laham/Getty Images.

Chapter 15

Opener: © Digital Vision RF; p. 792: © Stockdisc/Digital Vision RF; p. 800: © Design Pics Inc./Alamy RF; p. 806: © Getty Images RF; p. 807: © Digital Vision RF; p. 814: © Comstock Images/Alamy RF; p. 817: © PhotoLink/Getty Images RF; p. 823: © Digital Vision RF.

INDEX

A

absolute value, 211
absolute zero, 502
acres, 491–492
actuarial method, 450
acute angle, 514
acute triangle, 522
Adams' method, 772–773
addition
  associative property of, 247
  in base 16, 186
  in base five, 183–185
  in base two, 185
  on clock, 725–726
  closure property of, 246
  commutative property of, 246
  in Egyptian system, 151
  of fractions, 226–227
  of integers, 212–213
  inverse property of, 248
  in modular systems, 734
  of radicals, 239–241
addition property of equality, 290
addition property of inequality, 321
addition rule 1, 617–618
addition rule 2, 618–620
addition/subtraction method, 365–369
additive inverse, 247
adjacent vertices, 795
Ahmes, 282
Aiken, Howard, 190
Alabama paradox, 778–779
algebra
  fundamentals of, 278–285
  history of, 282
algebraic expression, 278
alternate exterior angles, 518
alternate interior angles, 518
amicable numbers, 214
amortization schedule, 462–464
amplitude, S7.1
analytic geometry, 349
Anelian groups, 721–722
angle(s)
  acute, 514
  alternate exterior, 518
  alternate interior, 518
  complementary, 514, 515
  corresponding, 518
  definition of, 513
  of depression, 556
  of elevation, 556
  exterior, 517
  formed by transversal, 518–519
  interior, 517
  measurement of, 514
  naming of, 513–514
  obtuse, 514
  pairs of, 514–519
  right, 514
  straight, 514
  sum of, in triangle, 522
  supplementary, 514, 515–516
  of triangle, finding, 523
  trigonometric ratios and, 555
  vertical, 516, 517
annual percentage rate, 448–451
annuities, 441–443
apportionment, 767–775
apportionment flaws, 778–781
approval voting, 763–764
Arabic numeration system, 154–156
arbitrary, 10
Archimedes, 31
area
  calculus and, 539
  of circle, 541
  conversions of, 490–493
  from English to metric units, 492–493
  of parallelogram, 538
  of polygons, 534–539
  of rectangle, 537
  of square, 537
  surface, 544–549
  of trapezoid, 539
  of triangle, 538
  units of, 490
arguments
  common valid forms of, 128–129
  fallacies, 123–124, 129–130
  invalid, 123–124
  logical, 123–131
  valid, 123–124, 125, 136–137
Aristotle, 94
arithmetic sequences, 261–264
Arrow, Kenneth, 762, 763
Arrow's impossibility theorem, 762–763
associative property of addition, 247
associative property of multiplication, 247
augmented matrix, 374
average
  ambiguous, S12.2
  measures of, 665–672
average daily balance method, 452, 454
axis of symmetry, 402

B

Babylonian numeration system, 156–158
banker's rule, 433–434
bar graph, 19–20, 658–659
base, of exponent, 155, 252
base five system, 171–174, 183–185, 186–187, 187–189, 189–190
base number systems, 170–180, 183–190
base 16 system, 186
base ten system, 172–174, 176–177
base two system, 185, 187–189
biconditional, 93, 96
biconditional statement, 104
binary system, 174, 179, 180
binomial distribution, 632–636
binomial probability formula, 633–636
binomials
  definition of, 329
  multiplication of, 329–330
bonds, 465–472
Boole, George, 94
Borda, Jean Charles de, 751, 753
Borda count method, 751–753
bridge, on graph, 796
brute force method, 808–809

C

calculation, problem solving by, 30–32
calculus, area and, 539
"cancel," 224
Cantor, Georg, 82
capacity, 493–494, 496
cardinality of set, 47
cardinality of union, 71
cardinal number of set, 47
cardinal number of union, 71
Cartesian plane, 348
Cartesian product, 62
Cataldi, Pietro, 202
categorical frequency distribution, 651
Cauchy, Augustin-Louis, 729
causation, 707
Celsius scale, 502–503
center
  of circle, 540
  of sphere, 547
centigrade, 502
Chinese numeration system, 153–154
circle graph, 20–21, 658–659
circles, 540–542
circuit
  Euler, 800
  on graph, 800
  Hamilton, 806–811
circuit, on graph, 796
circular reasoning, 130
circumference, of circle, 540
classical probability, 588–591
clock arithmetic, 725–730
closed figure, 521
closed system, 719
closed under multiplication, 244
closure property, 719
closure property of addition, 246
closure property of multiplication, 246
cluster sample, 648
coefficient, 239
  numerical, 278
combination rule, 583
combinations, 582–585, 604–608
common difference, 261
common ratio, 264
commutative property, 719
commutative property of addition, 246
commutative property of multiplication, 246
complement, of set, 54
complementary angles, 514, 515
complete graph, 807
complete weighted graph, 808, 810–811
composite numbers, 198–201
compound interest, 437–444
compound statement, 93, 104–105, 119
conditional, 93, 96
conditional probability, 623–630
conditional statement, 102–103, 119, 121
cone, volume of, 547
congruences, 734–735
congruent numbers, 732–733
conjecture, 4
conjunction, 93, 96, 100–101
connected graph, 796
connective, 93, 96, 108–109
consistent system, 361
constant of proportionality, 314
contradictions, 297–298
contrapositive, 119–120
converse, 119–120
coordinates, 349
coordinate system, 348–357
correlation, 699–708
correlation coefficient, 701–704
correspondence, one-to-one, 49
corresponding angles, 518
cosine, 552, 554
countable set, 84
counterexample, 6–7
cube, surface area of, 548
cube roots, 244
cubic feet, 494
cubit, 483
cylinder
  surface area of, 548
  volume of, 545, 546

D

data
  array, 667
  definition of, 648
  picturing, 657–662
  raw, 651
  in stem and leaf plot, 653–654
decimals
  converting from percents to, 421–422
  converting scientific notation to, 255
  converting to percents, 420–421
  fractions and, 228–231
  repeating, 230
deductive reasoning, 8–12
degrees
  of angles, 514
  of vertices, 795
De Morgan, Augustus, 68, 94
De Morgan's laws, 68–69, 118
denominator, 221
dependent event, 624
dependent system, 361, 362, 369
dependent variable, 395, 699
depression angle, 556
Descartes, René, 282, 349
descriptive method, 44
descriptive statistics, 650–651
detached statistics, S12.2
diagram, for problem solving, 27
diameter
  of circle, 540
  of sphere, 547
difference, of sets, 61–62
digits (numerical), 155
digit (unit of measurement), 483
dimensional analysis, 482, 487
Diophantus, 282
direct variation, 314
disconnected graph, 796
discounted loan, 434–435
disjoint, 59
disjunction, 93, 96, 101–103
distributive property, 279
  of multiplication, 248
dividend, stock, 466
divisibility tests, 199
divisible, 199
division
  in base five, 189–190
  of fractions, 224–226
  of integers, 216
  by primes, 207
  in scientific notation, 257–258
division method, of factorization, 202
division property of equality, 291
division property of inequality, 321
divisors, 199
domain, of function, 396–397
duration, S7.1
Dürer, Albrecht, 721

E

edges, of graphs, 792
effective rate, 439–441
Egyptian algorithm, 163–164

Egyptian numeration system, 149–152
element, of set, 42
elevation angle, 556
elimination
  Gaussian, 374–380
  solving systems of equations with, 365–369
ellipsis, 45–46
empirical probability, 591–593
empirical rule, 687
empty set, 47
endpoints, 512
English system of measurement, 482
entries, in matrix, 374
equal set, 48, 69–70
equations
  definition of, 288
  equivalent, 290
  with fractions, 294–295
  linear, 288–298, 301–306
  quadratic, 328–338
  solution of, 288
  systems of linear, 359–371
equilateral triangle, 521
equivalent equations, 290
equivalent fractions, 223
equivalent graphs, 795
equivalent set, 48
Escher, M. C., 563
estimation, 16–23
Euclid, 94, 516
Euler, Leonhard, 94, 202, 721
Euler circles, 134–138
Euler circuit, 800
Euler path, 800, 802
Euler's theorem, 800–803
evaluation, of algebraic expressions, 281–283
event, definition of, 588
even vertex, 795
expanded notation, 155–156
expectation, 609–614
expected value, 612–613
exponent, 155, 251–258
exponential functions, 405–410
expression
  algebraic, 278
  exponential, 155, 252–255
exterior angle, 517

F

factorial notation, 576
factoring
  solving quadratic equations with, 334–335
factoring, of trinomials, 331–334
factors
  finding, 198–199
  greatest common, 203–204, 207
  prime, 201–202
Fahrenheit scale, 502–503
fallacies, 123–124, 129–130
feet
  converting square inches to, 491
  converting to miles from, 483
  converting yards to, 482–483
  cubic, 494
Fermat, Pierre de, 206, 214
Fermat numbers, 206
Fibonacci, Leonardo, 282
Fibonacci sequence, 264, 313
finance amount, 446
finance charge, 446
finite mathematical system, 718
finite set, 46
five base system, 171–174, 183–185, 186–187, 187–189, 189–190
fixed installment loans, 446–447
FOIL method, 330
fractal geometry, 561–562
fractions. *see also* rational numbers
  addition of, 226–227
  converting from percents to, 422
  converting percents to, 421
  decimals and, 228–231
  equivalent, 223
  improper, 222
  multiplication of, 224–226
  music and, 228
  proper, 221
  reducing, 223
  solving equations with, 294–295
  subtraction of, 226–227
  unit, 482
frequency, S7.3
frequency distributions, 651–653
frequency polygons, 659–661
function(s)
  definition of, 395
  domain of, 396–397
  evaluating, 396
  exponential, 405–410
  identifying based on equations, 395–396
  linear, 401–402
  notation, 395–396
  objective, 389
  quadratic, 402–405
  range of, 396–397
  relations and, 394
  vertical line test for, 397–398
fundamental counting principle, 574–575
fundamental theorem of arithmetic, 201
future value, 429, 430, 441–443

G

gallons, 494, 496
Galois, Evariste, 729
Gauss, Carl Friedrich, 378
Gaussian elimination, 374–380
general term, 83, 84
geometric mean, 774
geometric sequence, 264–267
geometry, analytic, 349
Goldbach, Christian, 209
Golden Ratio, 313
googol, 18
gram, 485
graph(s)
  bar, 19–20, 658–659
  complete, 807
  complete weighted, 808, 810–811
  connected, 796
  definition of, 792
  disconnected, 796
  equivalent, 795
  of exponential functions, 406–407
  of inequalities, 320–321
  interpretation of, 16–23
  line, 21–22
  of linear equation in two variables, 351
  of linear functions, 401–402
  of linear inequality in two variables, 383–384
  misleading, S12.3–S12.4
  path on, 796
  pie chart, 20–21, 658–659
  of quadratic function, 403–404
  slope-intercept form for, 354
  systems of linear equations on, 360–363
  theory, basic concepts, 792–797
  time series, 21–22, 661–662
greatest common factors, 203–204, 207
grouped frequency distribution, 652
grouping systems, 149–152
groups, 721–722

H

half lines, 512
Halley, Sir Edmond, 541
Halley's comet, 541
Hamilton, Alexander, 770

Hamilton circuit, 806–811
Hamilton path, 806–811
Hamilton's method, 770–771
Hamilton's puzzle, 810
harmonics, S7.3–S7.5
head-to-head comparison criterion, 747–748
hexadecimal system, 175, 179, 180
hierarchy of connectives, 108–109
Hindu-Arabic numeration system, 154–156
histograms, 659–661
home ownership, 457–464
Hopper, Grace Murray, 190
horizontal lines, 355
Huntington-Hill method, 774–775
hyperbolic geometry, 560–561
hypotenuse, 523

I

identities, 297–298
identity element, 719
identity property, 719
implication, 102
implied connections, S12.2
improper fraction, 222
inconsistent system, 368–369
independence, of events, 623
independent system, 361, 362
independent variable, 395, 699
inductive reasoning, 4–8
inequalities
  linear, 319–326, 382–387
  systems of linear, 384–387
inferential statistics, 650–651
infinite mathematical system, 718
infinite set, 46, 82–85
infinity, 84
installment buying, 446–448
installment price, 446
integer(s)
  addition of, 212–213
  definition of, 210
  division of, 216
  multiplication of, 215–216
  subtraction of, 213–214
intercepts, 351–352
interest
  compound, 437–444
  definition of, 429
  effective rate of, 439–441
  rate, 429, 432
  simple, 429–435
  total, 458–461
  unearned, 449
interior angles, 517
interpretation, of graphs, 16–23
intersection, of sets, 58–59
invalid arguments, 123–124
inverse, 119–120
inverse element, 720
inverse property, 720
inverse property of addition, 248
inverse property of multiplication, 248
inverse variation, 315
irrational numbers, 235–242
irrelevant alternatives criterion, 761–762
isosceles triangle, 521
iteration, 561

J

Jefferson's method, 771–772

K

Kelvin scale, 502
Kruskal's algorithm, 816

L

lattice method, 165–167
law of contraposition, 128
law of detachment, 128
law of disjunctive syllogism, 128
law of syllogism, 128
least common denominator, 227
least common multiples, 205–206
legs, of right triangle, 523
length
  converting between English and metric, 488
  English and metric equivalents for, 487
  in English system, 482
  measurement, 482–488
like radicals, 239
like terms, 279, 280–281
Lincoln, Abraham, 134
line(s)
  definition of, 512
  half, 512
  horizontal, 355
  parallel, 517
  slope of, 353
  vertical, 355
linear equation in one variable, 289
linear equation in two variables, 350–351
linear equations, 288–298, 301–306
  systems of, 359–371
linear function, 401–402
linear inequalities, 319–326, 382–387
  system of, 384–387
linear programming, 389–392
linear units, 490
line graph, 21–22
line segment, 512
liter, 485, 496
loans
  discounted, 434–435
  down payment on, 433
  fixed installment, 446–447
  principal of, 431–432
  term of, 430, 433
  unpaid balance of, 452
logic
  history of, 94
  symbolic, 92
logical arguments, 123–131
logically equivalent statement, 117
loop, on graph, 792
lower quota, 769

M

magic squares, 721
majority criterion, 753
mathematical system, 718–722
matrices, 374–380
maturity date, of bond, 471
maximize, 389
Mayans, 149
McGrevey, John D., 591
mean, 665–667, 672
  geometric, 774
measurement
  of angles, 514
  English system of, 482
  history of, 483
  of length, 482–488
  metric system of, 484–488
  of temperature, 502–503
  of weight, 485, 499–501
median, 667–668, 672
member, of set, 42
meters, 485
metric conversions, 486–488
metric prefixes, 485
metric system, 484–488
midrange, 670–671, 672
miles
  converting acres to square, 491
  converting feet to, 483
minimize, 389
minimum spanning tree, 816
mixed number, 222
Mobius strip, 561

mode, 668–670, 672
modified divisor, 771
modular systems, 732–736
modulus, 732
monotonicity criterion, 755
monthly interest compounding, 438
monthly payments, 458–461
mortgage crisis, 461
mortgages, 458
multiples, least common, 205–206
multiplication
  associative property of, 247
  of binomials, 329–330
  on clock, 726, 727
  closed under, 244
  closure property of, 246
  commutative property of, 246
  distributive property of, 248
  of fractions, 224–226
  of integers, 215–216
  inverse property of, 248
  in modular systems, 734
  in scientific notation, 256–257
  of square roots, 238–239
multiplication, in bases five and two, 187–189
multiplication property of equality, 291
multiplication property of inequality, 321
multiplication rule 1, 624–625
multiplication rule 2, 627–628
multiplicative grouping system, 152–153
multiplicative inverse, 225
music, 228
mutual exclusivity, 616–620
mutual fund, 466, 471

N

Napier, John, 167, 169
Napier's bones, 167–169
natural numbers, 198–207
nearest neighbor method, 809–810
negation, 95, 100, 118, 119–120
new states paradox, 780–781
non-Euclidean geometry, 559–563
normal distribution, 685–697
notation
  expanded, 155–156
  factorial, 576
  function, 395–396
  scientific, 253–258
  set, 43–44
  set-builder, 44
  subset, 57
  symbolic, 96–97
number(s)
  amicable, 214
  cardinal of set, 47
  cardinal of union, 71
  composite, 198–201
  congruent, 732–733
  definition of, 148
  in Egyptian notation, 150–151
  Fermat, 206
  irrational, 235–242
  mixed, 222
  natural, 198–207
  opposite of, 210
  perfect, 202
  prime, 198–201
  rational, 221–223
  real, 244–249
  rounding, 16–17
  whole, 210
number line, 320–321
numeral
  types of, 159
numeral, definition of, 148
numeration systems, 148–161
  Arabic, 154–156
  Babylonian, 156–158
  Chinese, 153–154
  Egyptian, 149–152
  Hindu-Arabic, 154–156
  Roman, 158–161
numerator, 221
numerical coefficient, 278

O

objective function, 389
obtuse angle, 514
obtuse triangle, 522
octal system, 174, 179
odds, 609–614
one-to-one correspondence, 49
operation table, 718–722
opposite, of number, 210
optimal solution, 808
order of operations, 216–217
outcome, definition of, 587

P

Pacioli, Luca, 282
Paganini, Nicolo, 214
pairs, of angles, 514–519
pairwise comparison method, 759–762
palm (unit of measurement), 483
parabola, 402
parallel line, 517
parallelogram, area of, 538
parallel postulate, 560
path
  Euler, 800, 802
  on graph, 796, 800, 802
  Hamilton, 806–811
Pentagon (government building), 534
P/E ratio, 468–471
percent conversions, 420–422
percent decrease, 426
percentile, 681–684
percent increase, 426
percents, 420–427
perfect numbers, 202
perfect square, 236
perimeter, 531–534
  of polygon, 534
permutations, 577–580, 582, 604–608
phrases, operations and, 302
pi, 237, 540
picturing data, 657–662
pie chart, 20–21, 658–659
pitch, S7.1
place values, 155
plan, in problem solving, 26
plane, 512
plotting points, 349
plurality method, 744–748
plurality-with-elimination method, 753–756
points, 512
Polya, George, 26, 28
polygonal region, 389
polygons, 531–534
  area of, 534–539
  frequency, 659–661
polyhedron, 544
population, 648–649
population paradox, 779–780
portion, finding percentage from, 423
positional system, 154
position measures, 681–684
postulate, 560
power rule, 253
preference tables, 744–748
prime factorization, 201–202
prime numbers, 198–201
principal, of loan, 431–432
probability
  classical, 588–591
  combinations and, 604–608
  conditional, 623–630
  distribution, 636
  empirical, 591–594
  event in, 588

expectation and, 609–614
experiment, 587
odds and, 609–614
permutations and, 604–608
sample spaces and, 587–588, 597–598
sets and, 591, 620
problem solving, 26–32
by calculation, 30–32
by diagram, 27
by trial and error, 29
product, 215
product rule
for exponents, 253
for square roots, 236
programming, linear, 389–392
proper fraction, 221
proper subset, 56
proportionality, constant of, 314
proportions, 310–311
protractor, 514
pyramids, 524, 546–547
Pythagoras, 202, 235
Pythagorean theorem, 523–525
Pythagorean triple, 524

Q

quadrants, 348
quadratic equations, 328–338
quadratic formula, 336–337
quadratic functions, 402–405
quantified statement, 95
quarterly interest compounding, 438
quartile, 683
quota rule, 781
quotas
lower, 769
standard, 768–770
upper, 769
quotient, 216
quotient rule
for exponents, 253
for square roots, 239

R

radicals, simplification of, 236–242
radicand, 239
radius
of circle, 540
of sphere, 547
random sample, 648
range
of function, 396–397
as variation measure, 675–676
rate of change, 356
rate of interest, 429, 432
ratio
common, 264
trigonometric, 552–555
with units, 309
writing, 309
rationalizing the denominator, 241
rational numbers, 221–223
raw data, 651
real numbers, 244–249
reasoning
circular, 130
deductive, 8–12
inductive, 4–8
mathematical, 4–12
reciprocal, 225
rectangle, area of, 537
rectangular coordinate system, 348–357
rectangular solid
surface area of, 548
volume of, 544–545
reducing fractions, 223
regression, 407, 704–708
regression analysis, 699–708
regression line, 704–708
relation, 394
repeating decimal, 230
right angle, 514
right triangle, 522, 523
right triangle trigonometry, 552–557
Roman numeration system, 158–161
roster method, 42, 43
rounding numbers, 16–17
row echelon form, 375–376
rule of 72, 443
rule of 78, 451
Russian peasant method, 164–165

S

samples, 648–649, S12.1–S12.2
sample spaces, 587–588, 597–598
sampling methods, 649–650
scalene triangle, 521
scatter plot, 699–701
scientific notation, 253–258
self-contradiction, 115–116
semiannual interest compounding, 438
sequence
arithmetic, 261–264
geometric, 264–267
set(s)
cardinality of, 47
complement of, 54
countable, 84
definition of, 42
descriptive method for, 44
element of, 42
empty, 47
equal, 48, 69–70
equivalent, 48
finite, 46
infinite, 46, 82–85
intersection of, 58–59
notation, 43–44
in one-to-one correspondence, 49
probability and, 591, 620
of solutions, 288
subset, 55–56
subtraction, 61–62
uncountable, 84
union of, 58, 59
universal, 54
set-builder notation, 44
set operation, 58, 60–61, 65–72
shareholder, 466
sides, of angle, 513
significance levels, 703–704
similar triangles, 525–527
simple interest, 429–435
simple statement, 93
simplification
of algebraic expressions, 279–281
of radicals, 236–242
sine, 552
slope, 352–354, 356
slope-intercept form, 354
solution, definition of, 288
solution set, 288, 319
solving
definition of, 288
linear equations, 288–298
sound, S7.1–S7.8
sound wave characteristics, S7.2
spanning trees, 815–818
speed units, 484
sphere, 547–548
square, area of, 537
square inches, 491
square miles, 491
square root, 235
square units, 490
square yards, 492
standard deviation, 676–679
standard divisors, 767–768
standard form of quadratic equation, 328–329
standard normal distribution, 687–692, A1–A2
standard quota, 768–770
statement
biconditional, 104
compound, 93, 104–105, 119

conditional, 102–103, 119, 121
definition of, 92
logically equivalent, 117
negation of, 95
quantified, 95
recognizing, 93
self-contradictory, 115–116
simple, 93
tautological, 115–116
translation of, 97–98
types of, 115–122
statistics
definition of, 648
descriptive, 650–651
detached, S12.2
history of, 672
inferential, 650–651
misuse of, S12.1–S12.5
population in, 648–649
samples in, 648–649
sampling methods in, 649–650
stem and leaf plot, 653–654
stock broker, 466
stock exchange, 466
stocks, 465–472
straight angle, 514
stratified sample, 648
subset, 55–56
substitution, solving linear systems of equations with, 363–365
subtraction
in base five, 186–187
on clock, 727–728
in Egyptian system, 152
of fractions, 226–227
of integers, 213–214
in modular systems, 734
of radicals, 239–241
set, 61–62
subtraction property of equality, 290
subtraction property of inequality, 321
supplementary angle, 514, 515–516
surface area, 544–549
survey questions, faulty, S12.4–S12.5
suspect samples, S12.1–S12.2
symbolic logic, 92
symbolic notation, 96–97
systematic sample, 648
systems of linear equations, 359–371
consistent, 361
dependent, 361, 362, 369
inconsistent, 368–369
independent, 361, 362
with matrices, 378–380
systems of linear inequalities, 384–387

T

tables, 600–602
tally system, 148–149
tangent, 552, 553–554
tautologies, 115–116
temperature measurement, 502–503
ten base system, 172–174
term(s)
general, 83, 84
like, 279, 280–281
unlike, 280
term of loan, 430, 433
tessellation, 562, 563
theorem, 523
tie breaking, 764
tiling, 562
timbre, S7.1
time, of loan, 430, 433
time series graph, 21–22, 661–662
total interest, 458–461
transformational geometry, 559–563
transversal, 517, 518–519
trapezoid, area of, 539
tree diagrams, 597–600
tree method, 201–202
trees, 814–818
trial and error, 29
triangles, 521–527
area of, 538
equilateral, 521
finding angle in, 523
finding side with cosine, 554
finding side with tangent, 553–554
isosceles, 521
obtuse, 522
right, 522, 523
scalene, 521
similar, 525–527
sum of angles in, 522
types of, 521–522
trigonometric ratios, 552–555
trigonometry, 552–557
trinomials
definition of, 331
factoring, 331–334
truth tables, 100–113, 124–125
twin primes, 201
two base system, 185, 187–189

U

uncountable set, 84
understanding problem, 26
unearned interest, 449
union, of sets, 58, 59
unit fraction, 482
universal set, 54
unlike terms, 280
unpaid balance method, 452
upper quota, 769

V

valid arguments, 123–124, 125, 128–129, 136–137
variable
definition of, 44, 278
dependent, 395, 699
independent, 395, 699
variance, 676–678
variation, 314–317
variation measures, 675–679
Venn, John, 41, 69, 94
Venn diagram, 41, 54, 65–72
vertex (vertices)
adjacent, 795
of angle, 513
even, 795
in graphs, 792
odd, 795
of triangle, 521
vertical angle, 516, 517
vertical lines, 355
vertical line test, 397–398
Vieta, François, 282
volume, 493–494, 544–549

W

Webster's method, 773–774
weight, 485
weight conversions, 499–500
weight measurement, 499–501
weights, on graph, 808
whole numbers, 210
Widman, Johann, 282
Woods, Tiger, 148
word problems, 301–306, 324–327

X

x intercept, 351

Y

yards
converting to acres, 492
converting to feet, 482–483
yearly interest compounding, 438
y intercept, 351

Z

z score, 688–692